T0201958

Handbook on Data Centers

Samee U. Khan • Albert Y. Zomaya
Editors

Handbook on Data Centers

Editors
Samee U. Khan
Department of Electrical
and Computer Engineering
North Dakota State University
Fargo
North Dakota
USA

Albert Y. Zomaya
School of Information Technologies
The University of Sydney
Sydney
New South Wales
Australia

ISBN 978-1-4939-2091-4
ISBN 978-1-4939-2092-1 (eBook)
DOI 10.1007/978-1-4939-2092-1

Library of Congress Control Number: 2014959415

Springer New York Heidelberg Dordrecht London

Printed on acid-free paper

Springer is part of Springer Science+Business Media (www.springer.com)

Preface

Data centers play an important role in modern IT infrastructures. A data center is a home to computational power, storage, and applications necessary to support an enterprise business. Data centers process billions of Internet transactions every day. Nowadays, data centers are the beating hearts of the companies they serve. Large data centers with thousands of servers have been deployed by renowned ICT organizations, like IBM, Microsoft, Amazon, Yahoo, E-Bay, and Google, to provide cloud computing services. The phenomenal increase in the size and number of data centers and resultant increase in operational cost has stimulated the research in various domains of data centers, such as energy efficiency, resource management, networking, and security. This detailed *Handbook on Data Centers* covers in a succinct and orderly manner all aspects pertaining to data centers technologies.

Given the fast growing expansion of data centers, it is not surprising that a variety of methods are now available to researchers and practitioners striving to improve data center performance. This handbook aims to organize all major concepts, theories, methodologies, trends, challenges and applications of data centers into a coherent and unified repository. Moreover, the handbook provides researchers, scholars, students and professionals with a comprehensive, yet concise source of reference to data centers design, energy efficiency, resource management, and scalability issues. The handbook consists of nine parts, where each part consists of several chapters. The following topics spanning the data center technologies are covered in detail: (a) energy efficiency, (b) networking, (c) cloud computing, (d) hardware, (e) modeling and simulation, (f) security, (g) data services, (h) monitoring, and (i) resource management.Each part describes the state of the art methods, as well as the extensions and novel methodologies developed recently for data centers.

The first part describes energy efficiency methods for data centers and addresses the designing of energy efficient scheduling mechanisms for high performance computing environments. The authors have discussed energy-aware algorithms for task graph scheduling, replica placement, check-pointing, numerical processing, and to estimate energy consumption for the given workload. Also, this part addresses the problem of minimizing energy consumption with schedule length constraint on multicore processors. The following chapters covers the energy efficiency aspects of data centers presented in the handbook: *(a) Energy-Efficient and High-Performance*

Processing of Large-Scale Parallel Applications in Data Centers, (b) Energy-aware algorithms for task graph scheduling, replica placement and checkpoint strategies, (c) Energy efficiency in HPC Data Centers: Latest Advances to Build the Path to Exascale, (d) Techniques to achieve energy proportionality in data centers: a survey, (e) A Power-Aware Autonomic Approach for Performance Management of Scientific Applications in a Data Center Environment, (f) CoolEmAll: Models and Tools for Planning and Operating Energy Efficient Data Centres, (g) Smart Data center, (h) Power and Thermal Efficient Numerical Processing, and *(k) Providing Green Services in HPC Data Centers: A Methodology based on Energy Estimation.*

The second part of the handbook provides a study of various communication and networking methodologies for data centers. The network virtualization concepts, the architecture of optical networks for data centers, and network scalability issues are discussed in detail. Moreover, an emphasis is drawn over packet classification in multicore platforms. The routing techniques for data center networks are discussed in detail along with study on TCP congestion control in data center networks. The following chapters provide a detailed overview in terms of networking technologies for data centers are included in this part: *(a) Network Virtualization in Data Centers: A Data Plane Perspective, (b) Optical data center networks: Architecture, performance, and energy efficiency, (c) Scalable Network Communication using Unreliable RDMA, (d) Packet Classification on Multi-core Platforms, (e) Optical Interconnects for Data Center Networks, (f) TCP Congestion Control in Data Center Networks,* and *(g) Routing Techniques in Data Center Networks.*

The third part of the handbook discusses the role of data centers in cloud computing and highlights various challenges faced in ensuring the data integrity, reliability, and privacy in cloud computing environments. The role of trusted third parties in performing monitoring and auditing of service level agreements is illustrated, as well as integrity of big data in cloud computing is discussed. The data intensive applications in cloud are discussed along with storage challenges. This part includes the following chapters: *(a) Auditing for Data Integrity and Reliability in Cloud Storage, (b) I/O and File Systems for Data-Intensive Applications, (c) Cloud resource pricing under tenant rationality, (d) Online Resource Management for Carbon-Neutral Cloud Computing, (e) A Big Picture of Integrity Verification of Big Data in Cloud Computing, (f) An Out-of-Core Task-based Middleware for Data-Intensive Scientific Computing, (g) Building Scalable Software for Data Centers: An Approach to Distributed Computing at Enterprise Level, and (h) Cloud Storage over Multiple Data Centers.*

The fourth part of the handbook shed a light on data centers emerging hardware technologies. The issues pertaining to the efficient data storage on redundant array of independent disks (RAID) technologies, the data synchronization challenges on many-cores, and hardware approaches to transactional memory on chip multiprocessors are detailed. The following chapters are included in this part: *(a) Realizing Accelerated Cost-Effective Distributed RAID, (b) Efficient Hardware-Supported Synchronization Mechanisms for Many-cores,* and *(c) Hardware Approaches to Transactional Memory in Chip Multiprocessors*

The fifth part of the handbook discusses modeling and simulation techniques for data centers and include following chapters: *(a) Data Center Modeling and Simulation Using OMNeT++, (b) Power-Thermal Modeling and Control of Energy-Efficient Servers and Datacenters, (c) Thermal modeling and management of storage systems in data centers,* and *(d) Modeling and Simulation of Data Center Networks.* This portion highlights various techniques to model power consumption, thermal response, storage systems, and communication in data center networks. The authors utilized discrete-time simulator OMNet++ to model the data center traffic.

The sixth part of the handbook provides a discussion on various security and privacy techniques for data centers. A model is proposed to detect and mitigate the covert channels in data centers. The privacy issues regarding data center outsourcing are discussed and a survey of various privacy attacks and their counter measures is presented. The following chapters are included in this portion: *(a) C2Hunter: Detection and Mitigation of Covert Channels in Data Centers, (b) Selective and Private Access to Outsourced Data Centers, and (c) Privacy in Data Centers: A Survey of Attacks and Countermeasures.*

The seventh part details the data services and their management in data center. The quality of service requirements for processing stream data for city applications is discussed. The data management and querying of big data is surveyed in detail. The authors also discussed the various methods of constructing on the fly data centers in wireless ad hoc network environments. The following chapters are included in this part: *(a) Quality-of-Service in Data Center Stream Processing for Smart City Applications, (b) Opportunistic Databank: A context aware on-the-fly data center for mobile networks, (c) Data Management:State-of-the-Practice at Open-Science Data Centers, and (d) Data Summarization Techniques for Big Data—A Survey.*

The eighth part of handbook illustrates various hardware and software-based monitoring solutions for data centers. The use of wireless sensor networks technology to sense the thermal activities and efficient circulation of cooling air is explored. Traffic monitoring is another vital topic discussed with a survey on network intrusion detection systems for data centers. The following chapters are covered in this portion: *(a) Central Management of Datacenters, (b) Monitoring of Data Centers using Wireless Sensor Networks, (c) Network Intrusion Detection Systems in Data Centers,* and *(d) Software Monitoring in Data Centers.*

Lastly, the ninth part of book chapter elaborates the resource management aspects of data centers and present a detailed case study on usage patterns in multi-tenant data centers. A discussion is also performed on scheduling of distributed transactional memory and the various tradeoffs and techniques in distributed memory management are elaborated in detail. Moreover, an emphasis is drawn on the resource scheduling in data-centric systems. This part includes the following chapters: *(a) Usage Patterns in Multi-tenant Data Centers: a Large-Case Field Study, (b) On Scheduling in Distributed Transactional Memory: Techniques and Tradeoffs, (c) Dependability-Oriented Resource Management Schemes for Cloud Computing Data Centers,* and *(d) Resource Scheduling in Data-Centric Systems.*

Contents

Part I
Energy Efficiency

Energy-Efficient and High-Performance Processing of Large-Scale Parallel Applications in Data Centers

Keqin Li

1 Introduction

1.1 Motivation

Next generation supercomputers require drastically better energy efficiency to allow these systems to scale to exaflop computing levels. Virtually all major processor vendors and companies such as AMD, Intel, and IBM are developing high-performance and highly energy-efficient multicore processors and dedicating their current and future development and manufacturing to multicore products. It is conceivable that future multicore architectures can hold dozens or even hundreds of cores on a single die [3]. For instance, Adapteva's Epiphany scalable manycore architecture consists of hundreds and thousands of RISC microprocessors, all sharing a single flat and unobstructed memory hierarchy, which allows cores to communicate with each other very efficiently with low core-to-core communication overhead. The number of cores in this new type of massively parallel multicore architecture can be up to 4096 [1]. The Epiphany manycore architecture has been designed to maximize floating point computing power with the lowest possible energy consumption, aiming to deliver 100 and more gigaflops of performance at under 2 watts of power [4].

Multicore processors provide an ultimate solution to power management and performance optimization in current and future high-performance computing. A multicore processor contains multiple independent processors, called cores, integrated onto a single circuit die (known an a chip multiprocessor or CMP). An m-core processor achieves the same performance of a single-core processor whose clock frequency is m times faster, but consumes only $1/m^{\phi-1}$ ($\phi \geq 3$) of the energy of the single-core

The author can be reached at phone: (845) 257-3534, fax: (845) 257-3996.

K. Li (✉)
Department of Computer Science, State University of New York,
New Paltz, NY 12561, USA
e-mail: lik@newpaltz.edu

S. U. Khan, A. Y. Zomaya (eds.), *Handbook on Data Centers*,
DOI 10.1007/978-1-4939-2092-1_1

processor. The performance gain from a multicore processor is mainly from parallelism, i.e., multiple cores' working together to achieve the performance of a single faster and more energy-consuming processor. A multicore processor implements multiprocessing in a single physical package. It can implement parallel architectures such as superscalar, multithreading, VLIW, vector processing, SIMD, and MIMD. Intercore communications are supported by message passing or shared memory. The degree of parallelism can increase together with the number m of cores. When m is large, a multicore processor is also called a manycore or a massively multicore processor.

Modern information technology is developed into the era of cloud computing, which has received considerable attention in recent years and is widely accepted as a promising and ultimate way of managing and improving the utilization of data and computing resources and delivering various computing and communication services. However, enterprise data centers will spend several times as much on energy costs as on hardware and server management and administrative costs. Furthermore, many data centers are realizing that even if they are willing to pay for more power consumption, capacity constraints on the electricity grid mean that additional power is unavailable. Energy efficiency is one of the most important issues for large-scale computing systems in current and future data centers. Cloud computing can be an inherently energy-efficient technology, due to centralized energy management of computations on large-scale computing systems, instead of distributed and individualized applications without efficient energy consumption control [10]. Moreover, such potential for significant energy savings can be fully explored with balanced consideration of system performance and energy consumption.

As in all computing systems, increasing the utilization of a multicore processor becomes a critical issue, as the number of cores increases and as multicore processors are more and more widely employed in data centers. One effective way of increasing the utilization is to take the approach of multitasking, i.e., allowing multiple tasks to be executed simultaneously in a multicore processor. Such sharing of computing resources not only improves system utilization, but also improves system performance, because more users' requests can be processed in the same among of time. Such performance enhancement is very important in optimizing the quality of service in a data center for cloud computing, where multicore processors are employed as servers. Partitioning and sharing of a large multicore processor among multiple tasks is particularly important for large-scale scientific computations and business applications, where each computation or application consists of a large number of parallel tasks, and each parallel task requires several cores simultaneously for its execution.

When a multicore processor in a data center for cloud computing is shared by a large number of parallel tasks of a large-scale parallel application simultaneously, we are facing the problem of allocating the cores to the tasks and schedule the tasks, such that the system performance is optimized or the energy consumption is minimized. Furthermore, such core allocation and task scheduling should be conducted with energy constraints or performance constraints. Such optimization problems need to be formulated and efficient algorithms need to be developed and their performance need to be analyzed and evaluated. The motivation of the present chapter is to investigate energy-efficient and high-performance processing of large-scale parallel applications on multicore processors in data centers. In particular, we study low-power scheduling of precedence constrained parallel tasks on multicore processors. Our approach is to define combinatorial optimization problems, develop heuristic algorithms, analyze their performance, and validate our analytical results by simulations.

1.2 Our Contributions

In this chapter, we address scheduling precedence constrained parallel tasks on multicore processors with dynamically variable voltage and speed as combinatorial optimization problems. In particular, we define the problem of minimizing schedule length with energy consumption constraint and the problem of minimizing energy consumption with schedule length constraint on multicore processors. Our scheduling problems are defined in such a way that the energy-delay product is optimized by fixing one factor and minimizing the other. The first problem emphasizes energy efficiency, while the second problem emphasizes high performance.

We notice that energy-efficient and high-performance scheduling of parallel tasks with precedence constraints has not been investigated before as combinatorial optimization problems. Furthermore, all existing studies are on scheduling sequential tasks which require one processor to execute, or independent tasks which have no precedence constraint. Our study in this chapter makes some initial attempt to energy-efficient and high-performance scheduling of parallel tasks with precedence constraints on multicore processors with dynamic voltage and speed.

Our scheduling problems contain four nontrivial subproblems, namely, precedence constraining, system partitioning, task scheduling, and power supplying. Each subproblem should be solved efficiently, so that heuristic algorithms with overall good performance can be developed. These subproblems and our strategies to solve them are described as follows.

- *Precedence Constraining*—Precedence constraints make design and analysis of heuristic algorithms more difficult. We propose to use level-by-level scheduling algorithms to deal with precedence constraints. Since tasks in the same level are independent of each other, they can be scheduled by any of the efficient algorithms previously developed for scheduling independent tasks. Such decomposition of scheduling precedence constrained tasks into scheduling levels of independent tasks makes analysis of level-by-level scheduling algorithms much easier and clearer than analysis of other algorithms.
- *System Partitioning*—Since each parallel task requests for multiple cores for its execution, a multicore processor should be partitioned into clusters of cores to be assigned to the tasks. We use the harmonic system partitioning and core allocation scheme, which divides a multicore processor into clusters of equal sizes and schedules tasks of similar sizes together to increase core utilization.

- *Task Scheduling*—Parallel tasks are scheduled together with system partitioning and precedence constraining, and it is NP-hard even scheduling independent sequential tasks without system partitioning and precedence constraint. Our approach is to divide a list (i.e., a level) of tasks into sublists, such that each sublist contains tasks of similar sizes which are scheduled on clusters of equal sizes. Scheduling such parallel tasks on clusters is no more difficult than scheduling sequential tasks and can be performed by list scheduling algorithms.
- *Power Supplying*—Tasks should be supplied with appropriate powers and execution speeds, such that the schedule length is minimized by consuming given amount of energy or the energy consumed is minimized without missing a given deadline. We adopt a four-level energy/time/power allocation scheme for a given schedule, namely, optimal energy/time allocation among levels of tasks (Theorems 6 and 10), optimal energy/time allocation among sublists of tasks in the same level (Theorems 5 and 9), optimal energy allocation among groups of tasks in the same sublist (Theorems 4 and 8), and optimal power supplies to tasks in the same group (Theorems 3 and 7).

The above decomposition of our optimization problems into four subproblems makes design and analysis of heuristic algorithms tractable. A unique feature of our work is to compare the performance of our algorithms with optimal solutions analytically and validate our results experimentally, not to compare the performance of heuristic algorithms among themselves only experimentally. Such an approach is consistent with traditional scheduling theory.

The remainder of the chapter is organized as follows. In Sect. 2, we review related research in the literature. In Sect. 3, we present background information, including the power and task models, definitions of our problems, and lower bounds for optimal solutions. In Sect. 4, we describe our methods to deal with precedence constraints, system partitioning, and task scheduling. In Sect. 5, we develop our optimal four-level energy/time/power allocation scheme for minimizing schedule length and minimizing energy consumption, analyze the performance of our heuristic algorithms, and derive accurate performance bounds. In Sect. 6, we demonstrate simulation data, which validate our analytical results. In Sect. 7, we summarize the chapter and give further research directions.

2 Related Work

Increased energy consumption causes severe economic, ecological, and technical problems. Power conservation is critical in many computation and communication environments and has attracted extensive research activities. Reducing processor energy consumption has been an important and pressing research issue in recent years. There has been increasing interest and importance in developing high-performance and energy-efficient computing systems [15–17]. There exists an explosive body of literature on power-aware computing and communication. The reader is referred to [5, 9, 45, 46] for comprehensive surveys.

Software techniques for power reduction are supported by a mechanism called *dynamic voltage scaling* [2]. Dynamic power management at the operating system level refers to supply voltage and clock frequency adjustment schemes implemented while tasks are running. These energy conservation techniques explore the opportunities for tuning the energy-delay tradeoff [44]. In a pioneering paper [47], the authors first proposed the approach to energy saving by using fine grain control of CPU speed by an operating system scheduler. In a subsequent work [49], the authors analyzed offline and online algorithms for scheduling tasks with arrival times and deadlines on a uniprocessor computer with minimum energy consumption. These research have been extended in [7, 12, 25, 33–35, 50] and inspired substantial further investigation, much of which focus on real-time applications. In [6, 20, 21, 24, 27, 36–40, 42, 43, 48, 52–55] and many other related work, the authors addressed the problem of scheduling independent or precedence constrained tasks on uniprocessor or multiprocessor computers where the actual execution time of a task may be less than the estimated worst-case execution time. The main issue is energy reduction by slack time reclamation.

There are two considerations in dealing with the energy-delay tradeoff. On the one hand, in high-performance computing systems, power-aware design techniques and algorithms attempt to maximize performance under certain energy consumption constraints. On the other hand, low-power and energy-efficient design techniques and algorithms aim to minimize energy consumption while still meeting certain performance goals. In [8], the author studied the problems of minimizing the expected execution time given a hard energy budget and minimizing the expected energy expenditure given a hard execution deadline for a single task with randomized execution requirement. In [11], the author considered scheduling jobs with equal requirements on multiprocessors. In [14], the authors studied the relationship among parallelization, performance, and energy consumption, and the problem of minimizing energy-delay product. In [18], the authors addressed joint minimization of carbon emission and maximization of profit. In [23, 26], the authors attempted joint minimization of energy consumption and task execution time. In [41], the authors investigated the problem of system value maximization subject to both time and energy constraints. In [56], the authors considered task scheduling on clusters with significant communication costs.

In [28–32], we addressed energy and time constrained power allocation and task scheduling on multiprocessors with dynamically variable voltage and frequency and speed and power as combinatorial optimization problems. In [28, 31], we studied the problems of scheduling independent sequential tasks. In [29, 32], we studied the problems of scheduling independent parallel tasks. In [30], we studied the problems of scheduling precedence constrained sequential tasks. In this chapter, we study the problems of scheduling precedence constrained parallel tasks.

3 Preliminaries

In this section, we present background information, including the power and task models, definitions of our problems, and lower bounds for optimal solutions.

3.1 Power and Task Models

Power dissipation and circuit delay in digital CMOS circuits can be accurately modeled by simple equations, even for complex microprocessor circuits. CMOS circuits have dynamic, static, and short-circuit power dissipation; however, the dominant component in a well designed circuit is dynamic power consumption p (i.e., the switching component of power), which is approximately $p = aCV^2 f$, where a is an activity factor, C is the loading capacitance, V is the supply voltage, and f is the clock frequency [13]. In the ideal case, the supply voltage and the clock frequency are related in such a way that $V \propto f^{\phi}$ for some constant $\phi > 0$ [51]. The processor execution speed s is usually linearly proportional to the clock frequency, namely, $s \propto f$. For ease of discussion, we will assume that $V = bf^{\phi}$ and $s = cf$, where b and c are some constants. Hence, we know that power consumption is $p = aCV^2 f = ab^2Cf^{2\phi+1} = (ab^2C/c^{2\phi+1})s^{2\phi+1} = \xi s^{\alpha}$, where $\xi = ab^2C/c^{2\phi+1}$ and $\alpha = 2\phi + 1$. For instance, by setting $b = 1.16$, $aC = 7.0$, $c = 1.0$, $\phi = 0.5$, $\alpha = 2\phi + 1 = 2.0$, and $\xi = ab^2C/c^{\alpha} = 9.4192$, the value of p calculated by the equation $p = aCV^2 f = \xi s^{\alpha}$ is reasonably close to that in [22] for the Intel Pentium M processor.

Assume that we are given a parallel computation or application with a set of n precedence constrained parallel tasks. The precedence constraints can be specified as a partial order $\prec$ over the set of tasks $\{1, 2, ..., n\}$, or a task graph $G = (V, E)$, where $V = \{1, 2, ..., n\}$ is the set of tasks and E is a set of arcs representing the precedence constraints. The relationship $i \prec j$, or an arc (i, j) from i to j, means that task i must be executed before task j, i.e., task j cannot be executed until task i is completed. A parallel task i, where $1 \le i \le n$, is specified by π_i and r_i explained below. The integer π_i is the number of cores requested by task i, i.e., the *size* of task i. It is possible that in executing task i, the π_i cores may have different execution requirements (i.e., the numbers of core cycles or the numbers of instructions executed on the cores) due to imbalanced load distribution. Let r_i represent the maximum execution requirement on the π_i cores executing task i. The product $w_i = \pi_i r_i$ is called the *work* of task i.

We are also given a multicore processor with m homogeneous and identical cores. To execute a task i, any π_i of the m cores of the multicore processor can be allocated to task i. Several tasks can be executed simultaneously on the multicore processor, with the restriction that the total number of active cores (i.e., cores allocated to tasks being executed) at any moment cannot exceed m.

In a more general setting, we can consider scheduling u parallel applications represented by task graphs $G_1, G_2, ..., G_u$ respectively, on v multicore processors $P_1, P_2, ..., P_v$ in a data center with $m_1, m_2, ..., m_v$ cores respectively (see Fig. 1). Notice that multiple task graphs can be viewed as a single task graph with disconnected components. Therefore, our task model can accommodate multiple parallel applications. However, scheduling on multiple multicore processors is significantly different from scheduling on a single multicore processor. In this chapter, we focus on scheduling parallel applications on a single multicore processor, and leave

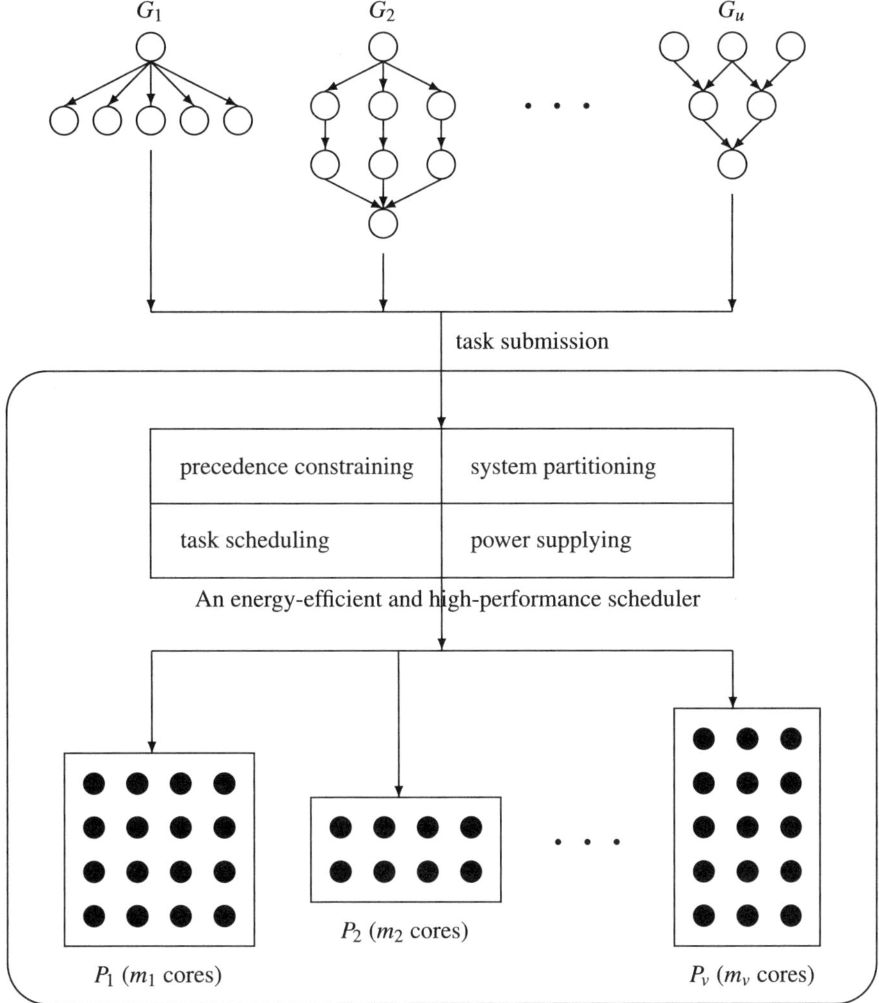

Fig. 1 Processing of parallel applications in a data center

the study of scheduling parallel applications on multiple multicore processors as a further research topic.

We use p_i to represent the power supplied to task i and s_i to represent the speed to execute task i. It is noticed that the constant ξ in $p_i = \xi s_i^\alpha$ only linearly scales the value of p_i. For ease of discussion, we will assume that p_i is simply s_i^α, where $s_i = p_i^{1/\alpha}$ is the execution speed of task i. The execution time of task i is $t_i = r_i/s_i = r_i/p_i^{1/\alpha}$. Note that all the π_i cores allocated to task i have the same speed s_i for duration t_i, although some of the π_i cores may be idle for some time. The energy consumed to execute task i is $e_i = \pi_i p_i t_i = \pi_i r_i p_i^{1-1/\alpha} = \pi_i r_i s_i^{\alpha-1} = w_i s_i^{\alpha-1}$, where $w_i = \pi_i r_i$ is the amount of work to be performed for task i.

3.2 Problems

Our combinatorial optimization problems solved in this chapter are formally defined as follows.

Given n parallel tasks with precedence constraints $\prec$, task sizes $\pi_1, \pi_2, ..., \pi_n$, and task execution requirements $r_1, r_2, ..., r_n$, the problem of *minimizing schedule length with energy consumption constraint* E on an m-core processor is to find the power supplies $p_1, p_2, ..., p_n$ (equivalently, the task execution speeds $s_1, s_2, ..., s_n$) and a nonpreemptive schedule of the n tasks on the m-core processor, such that the schedule length is minimized and that the total energy consumed does not exceed E. This problem aims at achieving energy-efficient processing of large-scale parallel applications with the best possible performance.

Given n parallel tasks with precedence constraints $\prec$, task sizes $\pi_1, \pi_2, ..., \pi_n$, and task execution requirements $r_1, r_2, ..., r_n$, the problem of *minimizing energy consumption with schedule length constraint* T on an m-core processor is to find the power supplies $p_1, p_2, ..., p_n$ (equivalently, the task execution speeds $s_1, s_2, ..., s_n$) and a nonpreemptive schedule of the n tasks on the m-core processor, such that the total energy consumption is minimized and that the schedule length does not exceed T. This problem aims at achieving high-performance processing of large-scale parallel applications with the lowest possible energy consumption.

The above two problems are NP-hard even when the tasks are independent (i.e., $\prec = \emptyset$) and sequential (i.e., $\pi_i = 1$ for all $1 \le i \le n$) [28]. Thus, we will seek fast heuristic algorithms with near-optimal performance.

3.3 Lower Bounds

Let $W = w_1 + w_2 + \cdots + w_n = \pi_1 r_1 + \pi_2 r_2 + \cdots + \pi_n r_n$ denote the total amount of work to be performed for the n parallel tasks. We define T^* to be the length of an optimal schedule, and E^* to be the minimum amount of energy consumed by an optimal schedule.

The following theorem gives a lower bound for the optimal schedule length T^* for the problem of minimizing schedule length with energy consumption constraint.

Theorem 1 *For the problem of minimizing schedule length with energy consumption constraint in scheduling parallel tasks, we have the following lower bound,*

$$T^* \ge \left(\frac{m}{E} \left(\frac{W}{m} \right)^{\alpha} \right)^{1/(\alpha-1)}$$

for the optimal schedule length.

Table 1 Summary of our methods to solve the subproblems

Subproblem	Method
Precedence constraining	Level-by-level scheduling algorithms
System partitioning	Harmonic system partitioning and core allocation scheme
Task scheduling	List scheduling algorithms
Power supplying	Four-level energy/time/power allocation scheme

The following theorem gives a lower bound for the minimum energy consumption E^* for the problem of minimizing energy consumption with schedule length constraint.

Theorem 2 *For the problem of minimizing energy consumption with schedule length constraint in scheduling parallel tasks, we have the following lower bound,*

$$E^* \geq m \left(\frac{W}{m} \right)^{\alpha} \frac{1}{T^{\alpha-1}}$$

for the minimum energy consumption.

The above lower bound theorems were proved for independent parallel tasks [29], and therefore, are also applicable to precedence constrained parallel tasks. The significance of these lower bounds is that they can be used to evaluate the performance of heuristic algorithms when their solutions are compared with optimal solutions (see Sects. 5.1.4 and 5.2.4).

4 Heuristic Algorithms

In this section, we describe our methods to deal with precedence constraints, system partitioning, and task scheduling, i.e., our methods to solve the first three subproblems. Table 1 gives a summary of our strategies to solve the subproblems.

4.1 Precedence Constraining

Recall that a set of n parallel tasks with precedence constraints can be represented by a partial order $\prec$ on the tasks, i.e., for two tasks i and j, if $i \prec j$, then task j cannot start its execution until task i finishes. It is clear that the n tasks and the partial order $\prec$ can be represented by a directed task graph, in which, there are n vertices for the n tasks and (i, j) is an arc if and only if $i \prec j$. We call j a successor of i and i a predecessor of j. Furthermore, such a task graph must be a *directed acyclic graph* (dag). An arc (i, j) is redundant if there exists k such that (i, k) and (k, j) are also arcs in the task graph. We assume that there is no redundant arc in the task graph.

A dag can be decomposed into levels, with v being the number of levels. Tasks with no predecessors (called initial tasks) constitute level 1. Generally, a task i is in level l if the number of nodes on the longest path from some initial task to task i is l, where $1 \le l \le v$. Note that all tasks in the same level are independent of each other, and hence, they can be scheduled by any of the algorithms (e.g., those from [29, 32]) for scheduling independent parallel tasks. Algorithm LL-H_c-A, where A is a list scheduling algorithm, standing for *level-by-level* scheduling with algorithm H_c-A, schedules the n tasks level by level in the order level 1, level 2, ..., level v. Tasks in level $l+1$ cannot start their execution until all tasks in level l are completed. For each level l, where $1 \le l \le v$, we use algorithm H_c-A developed in [29] to generate its schedule (see Fig. 2).

The details of algorithm H_c-A is given in the next two subsections.

4.2 System Partitioning

Our algorithms for scheduling independent parallel tasks are called H_c-A, where "H_c" stands for the *harmonic* system partitioning scheme with parameter c to be presented below, and A is a list scheduling algorithm to be presented in the next subsection.

To schedule a list of independent parallel tasks in level l, algorithm H_c-A divides the list into c sublists $(l,1),(l,2),...,(l,c)$ according to task sizes (i.e., numbers of cores requested by tasks), where $c \ge 1$ is a positive integer constant. For $1 \le j \le c-1$, we define sublist (l,j) to be the sublist of tasks with

$$\frac{m}{j+1} < \pi_i \le \frac{m}{j},$$

i.e., sublist (l,j) contains all tasks whose sizes are in the interval $I_j = (m/(j+1), m/j)$. We define sublist (l,c) to be the sublist of tasks with $0 < \pi_i \le m/c$, i.e., sublist (l,c) contains all tasks whose sizes are in the interval $I_c = (0, m/c)$. The partition of $(0,m)$ into intervals $I_1, I_2, ..., I_j, ..., I_c$ is called the *harmonic system partitioning scheme* whose idea is to schedule tasks of similar sizes together. The similarity is defined by the intervals $I_1, I_2, ..., I_j, ..., I_c$. For tasks in sublist (l,j), core utilization is higher than $j/(j+1)$, where $1 \le j \le c-1$. As j increases, the similarity among tasks in sublist (l,j) increases and core utilization also increases. Hence, the harmonic system partitioning scheme is very good at handling small tasks.

Algorithm H_c-A produces schedules of the sublists sequentially and separately (see Fig. 2). To schedule tasks in sublist (l,j), where $1 \le j \le c$, the m cores are partitioned into j *clusters* and each cluster contains m/j cores. Each cluster of cores is treated as one unit to be allocated to one task in sublist (l,j). This is basically the harmonic system partitioning and core allocation scheme. The justification of the scheme is from the observation that there can be at most j parallel tasks from sublist

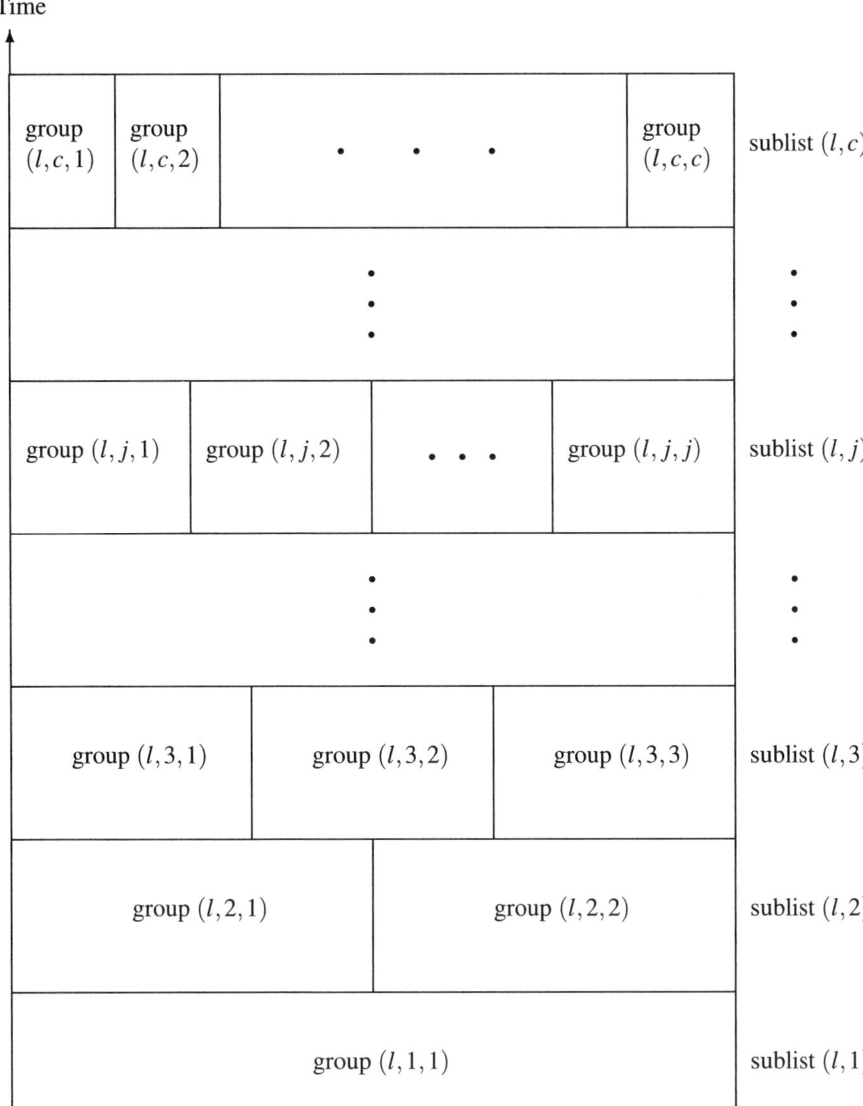

Fig. 2 Scheduling of level l

(l, j) to be executed simultaneously. Therefore, scheduling parallel tasks in sublist (l, j) on the j clusters, where each task i has core requirement π_i and execution requirement r_i, is equivalent to scheduling a list of sequential tasks on j processors where each task i has execution requirement r_i. It is clear that scheduling of a list of sequential tasks on j processors (i.e., scheduling of a sublist (l, j) of parallel tasks on j clusters) can be accomplished by using algorithm A, where A is a list scheduling algorithm to be elaborated in the next subsection.

4.3 *Task Scheduling*

When a multicore processor with m cores is partitioned into $j \geq 1$ clusters, scheduling tasks in sublist (l, j) is essentially dividing sublist (l, j) into j groups $(l, j, 1), (l, j, 2), ..., (l, j, j)$ of tasks, such that each group of tasks are executed on one cluster (see Fig. 2). Such a partition of sublist (l, j) into j groups is essentially a schedule of the tasks in sublist (l, j) on m cores with j clusters. Once a partition (i.e., a schedule) is determined, we can use the methods in the next section to find optimal energy/time allocation and power supplies.

We propose to use the list scheduling algorithm and its variations to solve the task scheduling problem. Tasks in sublist (l, j) are scheduled on j clusters by using the classic *list scheduling* algorithm [19] and by ignoring the issue of power supplies and execution speeds. In other words, the task execution times are simply the task execution requirements r_1, r_2, ..., r_n, and tasks are assigned to the j clusters (i.e., groups) by using the list scheduling algorithm, which works as follows to schedule a list of tasks $1, 2, 3 \cdots$.

- List Scheduling (LS): Initially, task k is scheduled on cluster (or group) k, where $1 \leq k \leq j$, and tasks $1, 2, \cdots, j$ are removed from the list. Upon the completion of a task k, the first unscheduled task in the list, i.e., task $j + 1$, is removed from the list and scheduled to be executed on cluster k. This process repeats until all tasks in the list are finished.

Algorithm LS has many variations, depending on the strategy used in the initial ordering of the tasks. We mention several of them here.

- Largest Requirement First (LRF): This algorithm is the same as the LS algorithm, except that the tasks are arranged such that $r_1 \geq r_2 \geq \cdots \geq r_n$.
- Smallest Requirement First (SRF): This algorithm is the same as the LS algorithm, except that the tasks are arranged such that $r_1 \leq r_2 \leq \cdots \leq r_n$.
- Largest Size First (LSF): This algorithm is the same as the LS algorithm, except that the tasks are arranged such that $\pi_1 \geq \pi_2 \geq \cdots \geq \pi_n$.
- Smallest Size First (SSF): This algorithm is the same as the LS algorithm, except that the tasks are arranged such that $\pi_1 \leq \pi_2 \leq \cdots \leq \pi_n$.
- Largest Task First (LTF): This algorithm is the same as the LS algorithm, except that the tasks are arranged such that $\pi_1^{1/\alpha} r_1 \geq \pi_2^{1/\alpha} r_2 \geq \cdots \geq \pi_n^{1/\alpha} r_n$.
- Smallest Task First (STF): This algorithm is the same as the LS algorithm, except that the tasks are arranged such that $\pi_1^{1/\alpha} r_1 \leq \pi_2^{1/\alpha} r_2 \leq \cdots \leq \pi_n^{1/\alpha} r_n$.

We call algorithm LS and its variations simply as list scheduling algorithms.

Table 2 Overview of the optimal energy/time/power allocation scheme

Level	Method	Theorems
1	Optimal power supplies to tasks in the same group	3 and 7
2	Optimal energy allocation among groups of tasks in the same sublist	4 and 8
3	Optimal energy/time allocation among sublists of tasks in the same level	5 and 9
4	Optimal energy/time allocation among levels of tasks	6 and 10

5 Optimal Energy/Time/Power Allocation

In this section, we develop our optimal four-level energy/time/power allocation scheme for minimizing schedule length and minimizing energy consumption, i.e., our method to solve the last subproblem. We also analyze the performance of our heuristic algorithms and derive accurate performance bounds.

Once the n precedence constrained parallel tasks are decomposed into v levels, 1, 2, ..., v, and tasks in each level l are divided into c sublists $(l, 1), (l, 2), ..., (l, c)$, and tasks in each sublist (l, j) are further partitioned into j groups $(l, j, 1), (l, j, 2), ..., (l, j, j)$, power supplies to the tasks which minimize the schedule length within energy consumption constraint or the energy consumption within schedule length constraint can be determined. We adopt a four-level energy/time/power allocation scheme for a given schedule, namely,

- Level 1—optimal power supplies to tasks in the same group (l, j, k) (Theorems 3 and 7);
- Level 2—optimal energy allocation among groups $(l, j, 1), (l, j, 2), ..., (l, j, j)$ of tasks in the same sublist (l, j) (Theorems 4 and 8);
- Level 3—optimal energy/time allocation among sublists $(l, 1), (l, 2), ..., (l, c)$ of tasks in the same level l (Theorems 5 and 9);
- Level 4—optimal energy/time allocation among levels 1, 2, ..., l of tasks of a parallel application (Theorems 6 and 10).

Table 2 gives an overview of our energy/time/power allocation scheme. We will give the details of the above optimal four-level energy/time/power allocation scheme for the two optimization problems separately.

5.1 Minimizing Schedule Length

5.1.1 Level 1

We first consider optimal power supplies to tasks in the same group. Notice that tasks in the same group are executed sequentially. In fact, we consider a more general case, i.e., n parallel tasks with sizes $\pi_1, \pi_2, ..., \pi_n$ and execution requirements $r_1, r_2, ..., r_n$

to be executed sequentially one by one. Let us define

$$M = \pi_1^{1/\alpha} r_1 + \pi_2^{1/\alpha} r_2 + \cdots + \pi_n^{1/\alpha} r_n.$$

The following result [29] gives the optimal power supplies when the n parallel tasks are scheduled sequentially.

Theorem 3 *When n parallel tasks are scheduled sequentially, the schedule length is minimized when task i is supplied with power $p_i = (E/M)^{\alpha/(\alpha-1)}/\pi_i$, where $1 \le i \le n$. The optimal schedule length is $T = M^{\alpha/(\alpha-1)}/E^{1/(\alpha-1)}$.*

5.1.2 Level 2

Now, we consider optimal energy allocation among groups of tasks in the same sublist. Again, we discuss group level energy allocation in a more general case, i.e., scheduling n parallel tasks on m cores, where $\pi_i \le m/j$ for all $1 \le i \le n$ with $j \ge 1$. In this case, the m cores can be partitioned into j clusters, such that each cluster contains m/j cores. Each cluster of cores are treated as one unit to be allocated to one task. Assume that the set of n tasks is partitioned into j groups, such that all the tasks in group k are executed on cluster k, where $1 \le k \le j$. Let M_k denote the total $\pi_i^{1/\alpha} r_i$ of the tasks in group k. For a given partition of the n tasks into j groups, we are seeking an optimal energy allocation and power supplies that minimize the schedule length. Let E_k be the energy consumed by all the tasks in group k. The following result [29] characterizes the optimal energy allocation and power supplies.

Theorem 4 *For a given partition M_1, M_2, ..., M_j of n parallel tasks into j groups on a multicore processor partitioned into j clusters, the schedule length is minimized when task i in group k is supplied with power $p_i = (E_k/M_k)^{\alpha/(\alpha-1)}/\pi_i$, where*

$$E_k = \left(\frac{M_k^\alpha}{M_1^\alpha + M_2^\alpha + \cdots + M_j^\alpha} \right) E,$$

for all $1 \le k \le j$. The optimal schedule length is

$$T = \left(\frac{M_1^\alpha + M_2^\alpha + \cdots + M_j^\alpha}{E} \right)^{1/(\alpha-1)},$$

for the above energy allocation and power supplies.

5.1.3 Level 3

To use algorithm H$_c$-A to solve the problem of minimizing schedule length with energy consumption constraint E, we need to allocate the available energy E to the c sublists. We use E_1, E_2, ..., E_c to represent an energy allocation to the c sublists, where sublist j consumes energy E_j, and $E_1 + E_2 + \cdots + E_c = E$. By using any of the list scheduling algorithms to schedule tasks in sublist j, we get a partition of the tasks in sublist j into j groups. Let R_j be the total execution requirement of tasks in sublist j, and $R_{j,k}$ be the total execution requirement of tasks in group k, and $M_{j,k}$ be the total $\pi_i^{1/\alpha} r_i$ of tasks in group k, where $1 \le k \le j$. Theorem 5 [29] provides optimal energy allocation to the c sublists for minimizing schedule length with energy consumption constraint in scheduling parallel tasks by using scheduling algorithms H$_c$-A, where A is a list scheduling algorithm.

Theorem 5 *For a given partition $M_{j,1}, M_{j,2}, ..., M_{j,j}$ of the tasks in sublist j into j groups produced by a list scheduling algorithm A, where $1 \le j \le c$, and an energy allocation $E_1, E_2, ..., E_c$ to the c sublists, the length of the schedule produced by algorithm H_c-A is*

$$T = \sum_{j=1}^{c} \left(\frac{M_{j,1}^{\alpha} + M_{j,2}^{\alpha} + \cdots + M_{j,j}^{\alpha}}{E_j} \right)^{1/(\alpha-1)}.$$

The energy allocation $E_1, E_2, ..., E_c$ which minimizes T is

$$E_j = \left(\frac{N_j^{1/\alpha}}{N_1^{1/\alpha} + N_2^{1/\alpha} + \cdots + N_c^{1/\alpha}} \right) E,$$

where $N_j = M_{j,1}^{\alpha} + M_{j,2}^{\alpha} + \cdots + M_{j,j}^{\alpha}$, for all $1 \le j \le c$, and the minimized schedule length is

$$T = \frac{(N_1^{1/\alpha} + N_2^{1/\alpha} + \cdots + N_c^{1/\alpha})^{\alpha/(\alpha-1)}}{E^{1/(\alpha-1)}},$$

by using the above energy allocation.

5.1.4 Level 4

To use a level-by-level scheduling algorithm to solve the problem of minimizing schedule length with energy consumption constraint E, we need to allocate the available energy E to the v levels. We use $E_1, E_2, ..., E_v$ to represent an energy allocation to the v levels, where level l consumes energy E_l, and $E_1 + E_2 + \cdots + E_v = E$.

Let $R_{l,j,k}$ be the total execution requirement of tasks in group (l, j, k), i.e., group k of sublist (l, j) of level l, and $R_{l,j}$ be the total execution requirement of tasks in sublist (l, j) of level l, and R_j be the total execution requirement of tasks in sublist (l, j) of all levels, and $M_{l,j,k}$ be the total $\pi_i^{1/\alpha} r_i$ of tasks in group (l, j, k), where $1 \le l \le v$ and $1 \le j \le c$ and $1 \le k \le j$.

By Theorem 5, for a given partition $M_{l,j,1}, M_{l,j,2}, ..., M_{l,j,j}$ of the tasks in sublist (l, j) of level l into j groups produced by a list scheduling algorithm A, where

$1 \le l \le v$ and $1 \le j \le c$, and an energy allocation $E_{l,1}, E_{l,2}, ..., E_{l,c}$ to the c sublists of level l, where

$$E_{l,j} = \left(\frac{N_{l,j}^{1/\alpha}}{N_{l,1}^{1/\alpha} + N_{l,2}^{1/\alpha} + \cdots + N_{l,c}^{1/\alpha}} \right) E_l,$$

with $N_{l,j} = M_{l,j,1}^{\alpha} + M_{l,j,2}^{\alpha} + \cdots + M_{l,j,j}^{\alpha}$, for all $1 \le l \le v$ and $1 \le j \le c$, the scheduling algorithm H$_c$-A produces schedule length

$$T_l = \frac{(N_{l,1}^{1/\alpha} + N_{l,2}^{1/\alpha} + \cdots + N_{l,c}^{1/\alpha})^{\alpha/(\alpha-1)}}{E_l^{1/(\alpha-1)}},$$

for tasks in level l, where $1 \le l \le v$. Since the level-by-level scheduling algorithm produces schedule length $T = T_1 + T_2 + \cdots + T_v$, we have

$$T = \sum_{l=1}^{v} \frac{(N_{l,1}^{1/\alpha} + N_{l,2}^{1/\alpha} + \cdots + N_{l,c}^{1/\alpha})^{\alpha/(\alpha-1)}}{E_l^{1/(\alpha-1)}}.$$

Let $S_l = (N_{l,1}^{1/\alpha} + N_{l,2}^{1/\alpha} + \cdots + N_{l,c}^{1/\alpha})^{\alpha}$, for all $1 \le l \le v$. By the definition of S_l, we obtain

$$T = \left(\frac{S_1}{E_1}\right)^{1/(\alpha-1)} + \left(\frac{S_2}{E_2}\right)^{1/(\alpha-1)} + \cdots + \left(\frac{S_v}{E_v}\right)^{1/(\alpha-1)}.$$

To minimize T with the constraint $F(E_1, E_2, ..., E_v) = E_1 + E_2 + \cdots + E_v = E$, we use the Lagrange multiplier system

$$\nabla T(E_1, E_2, ..., E_v) = \lambda \nabla F(E_1, E_2, ..., E_v),$$

where λ is the Lagrange multiplier. Since $\partial T/\partial E_l = \lambda \partial F/\partial E_l$, that is,

$$S_l^{1/(\alpha-1)} \left(-\frac{1}{\alpha-1} \right) \frac{1}{E_l^{1/(\alpha-1)+1}} = \lambda,$$

$1 \le l \le v$, we get

$$E_l = S_l^{1/\alpha} \left(\frac{1}{\lambda(1-\alpha)} \right)^{(\alpha-1)/\alpha},$$

which implies that

$$E = (S_1^{1/\alpha} + S_2^{1/\alpha} + \cdots + S_v^{1/\alpha}) \left(\frac{1}{\lambda(1-\alpha)} \right)^{(\alpha-1)/\alpha},$$

and

$$E_l = \left(\frac{S_l^{1/\alpha}}{S_1^{1/\alpha} + S_2^{1/\alpha} + \cdots + S_v^{1/\alpha}} \right) E,$$

for all $1 \le l \le v$. By using the above energy allocation, we have

$$
\begin{aligned}
T &= \sum_{l=1}^{v} \left(\frac{S_l}{E_l}\right)^{1/(\alpha-1)} \\
&= \sum_{l=1}^{v} \frac{S_l^{1/(\alpha-1)}}{\left(\left(\dfrac{S_l^{1/\alpha}}{S_1^{1/\alpha} + S_2^{1/\alpha} + \cdots + S_v^{1/\alpha}}\right) E\right)^{1/(\alpha-1)}} \\
&= \sum_{l=1}^{v} \frac{S_l^{1/\alpha}(S_1^{1/\alpha} + S_2^{1/\alpha} + \cdots + S_v^{1/\alpha})^{1/(\alpha-1)}}{E^{1/(\alpha-1)}} \\
&= \frac{(S_1^{1/\alpha} + S_2^{1/\alpha} + \cdots + S_v^{1/\alpha})^{\alpha/(\alpha-1)}}{E^{1/(\alpha-1)}}.
\end{aligned}
$$

For any list scheduling algorithm A, we have $R_{l,j,k} \le R_{l,j}/j + r^*$, for all $1 \le l \le v$ and $1 \le j \le c$ and $1 \le k \le j$, where $r^* = \max(r_1, r_2, ..., r_n)$ is the maximum task execution requirement. Since $\pi_i \le m/j$ for every task i in group (l, j, k) of sublist (l, j) of level l, we get

$$M_{l,j,k} \le \left(\frac{m}{j}\right)^{1/\alpha} R_{l,j,k} \le \left(\frac{m}{j}\right)^{1/\alpha} \left(\frac{R_{l,j}}{j} + r^*\right).$$

Therefore,

$$N_{l,j} \le m \left(\frac{R_{l,j}}{j} + r^*\right)^{\alpha},$$

and

$$N_{l,j}^{1/\alpha} \le m^{1/\alpha} \left(\frac{R_{l,j}}{j} + r^*\right),$$

and

$$N_{l,1}^{1/\alpha} + N_{l,2}^{1/\alpha} + \cdots + N_{l,c}^{1/\alpha} \le m^{1/\alpha} \left(\left(\sum_{j=1}^{c} \frac{R_{l,j}}{j}\right) + cr^*\right).$$

Consequently,

$$S_l \le m \left(\left(\sum_{j=1}^{c} \frac{R_{l,j}}{j}\right) + cr^*\right)^{\alpha},$$

and

$$S_l^{1/\alpha} \le m^{1/\alpha} \left(\left(\sum_{j=1}^{c} \frac{R_{l,j}}{j}\right) + cr^*\right),$$

and

$$S_1^{1/\alpha} + S_2^{1/\alpha} + \cdots + S_v^{1/\alpha} \le m^{1/\alpha} \left(\left(\sum_{j=1}^{c} \frac{R_j}{j}\right) + cvr^*\right),$$

which implies that

$$T \leq m^{1/(\alpha-1)} \left(\left(\sum_{j=1}^{c} \frac{R_j}{j} \right) + cvr^* \right)^{\alpha/(\alpha-1)} \frac{1}{E^{1/(\alpha-1)}}.$$

We define the *performance ratio* as $\beta = T/T^*$ for heuristic algorithms that solve the problem of minimizing schedule length with energy consumption constraint on a multicore processor. By Theorem 1, we get

$$\beta = \frac{T}{T^*} \leq \left(\left(\left(\sum_{j=1}^{c} \frac{R_j}{j} \right) + cvr^* \right) \Big/ \left(\frac{W}{m} \right) \right)^{\alpha/(\alpha-1)}.$$

Theorem 6 provides optimal energy allocation to the v levels for minimizing schedule length with energy consumption constraint in scheduling precedence constrained parallel tasks by using level-by-level scheduling algorithms LL-H$_c$-A, where A is a list scheduling algorithm.

Theorem 6 *For a given partition $M_{l,j,1}$, $M_{l,j,2}$, ..., $M_{l,j,j}$ of the tasks in sublist (l, j) of level l into j groups produced by a list scheduling algorithm A, where $1 \leq l \leq v$ and $1 \leq j \leq c$, and an energy allocation E_1, E_2, ..., E_v to the v levels, the level-by-level scheduling algorithm LL-H$_c$-A produces schedule length*

$$T = \sum_{l=1}^{v} \frac{(N_{l,1}^{1/\alpha} + N_{l,2}^{1/\alpha} + \cdots + N_{l,c}^{1/\alpha})^{\alpha/(\alpha-1)}}{E_l^{1/(\alpha-1)}},$$

where $N_{l,j} = M_{l,j,1}^{\alpha} + M_{l,j,2}^{\alpha} + \cdots + M_{l,j,j}^{\alpha}$, for all $1 \leq l \leq v$ and $1 \leq j \leq c$. The energy allocation E_1, E_2, ..., E_v which minimizes T is

$$E_l = \left(\frac{S_l^{1/\alpha}}{S_1^{1/\alpha} + S_2^{1/\alpha} + \cdots + S_v^{1/\alpha}} \right) E,$$

where $S_l = (N_{l,1}^{1/\alpha} + N_{l,2}^{1/\alpha} + \cdots + N_{l,c}^{1/\alpha})^{\alpha}$, for all $1 \leq l \leq v$, and the minimized schedule length is

$$T = \frac{(S_1^{1/\alpha} + S_2^{1/\alpha} + \cdots + S_v^{1/\alpha})^{\alpha/(\alpha-1)}}{E^{1/(\alpha-1)}},$$

by using the above energy allocation. The performance ratio is

$$\beta \leq \left(\left(\left(\sum_{j=1}^{c} \frac{R_j}{j} \right) + cvr^* \right) \Big/ \left(\frac{W}{m} \right) \right)^{\alpha/(\alpha-1)},$$

where $r^* = \max(r_1, r_2, ..., r_n)$ is the maximum task execution requirement.

Theorems 4 and 5 and 6 give the power supply to the task i in group (l, j, k) as

$$\frac{1}{\pi_i}\left(\frac{E_{l,j,k}}{M_{l,j,k}}\right)^{\alpha/(\alpha-1)} = \frac{1}{\pi_i}\left(\left(\frac{M_{l,j,k}^\alpha}{M_{l,j,1}^\alpha + M_{l,j,2}^\alpha + \cdots + M_{l,j,j}^\alpha}\right)\right.$$

$$\left.\left(\frac{N_{l,j}^{1/\alpha}}{N_{l,1}^{1/\alpha} + N_{l,2}^{1/\alpha} + \cdots + N_{l,c}^{1/\alpha}}\right)\left(\frac{S_l^{1/\alpha}}{S_1^{1/\alpha} + S_2^{1/\alpha} + \cdots + S_v^{1/\alpha}}\right)\frac{E}{M_{l,j,k}}\right)^{\alpha/(\alpha-1)},$$

for all $1 \le l \le v$ and $1 \le j \le c$ and $1 \le k \le j$.

We notice that the performance bound given in Theorem 6 is loose and pessimistic mainly due to the overestimation of the π_i's in sublist (l, j) to m/j. One possible remedy is to use the value of $(m/(j+1) + m/j)/2$ as an approximation to π_i. Also, as the number of tasks gets large, the term cvr^* may be removed. This gives rise to the following performance bound for β:

$$\left(\left(\sum_{j=1}^{c} \frac{R_j}{j}\left(\frac{2j+1}{2j+2}\right)^{1/\alpha}\right) \Big/ \left(\frac{W}{m}\right)\right)^{\alpha/(\alpha-1)}. \tag{1}$$

Our simulation shows that the modified bound in (1) is more accurate than the performance bound given in Theorem 6.

5.2 *Minimizing Energy Consumption*

5.2.1 Level 1

The following result [29] gives the optimal power supplies when n parallel tasks are scheduled sequentially.

Theorem 7 *When n parallel tasks are scheduled sequentially, the total energy consumption is minimized when task i is supplied with power* $p_i = (M/T)^\alpha/\pi_i$, *where* $1 \le i \le n$. *The minimum energy consumption is* $E = M^\alpha/T^{\alpha-1}$.

5.2.2 Level 2

The following result [29] gives the optimal energy allocation and power supplies that minimize energy consumption for a given partition of n tasks into j groups on a multicore processor.

Theorem 8 *For a given partition* M_1, M_2, ..., M_j *of n parallel tasks into j groups on a multicore processor partitioned into j clusters, the total energy consumption is minimized when task i in group k is executed with power* $p_i = (M_k/T)^\alpha/\pi_i$, *where*

$1 \le k \le j$. The minimum energy consumption is

$$E = \frac{M_1^\alpha + M_2^\alpha + \cdots + M_j^\alpha}{T^{\alpha-1}},$$

for the above energy allocation and power supplies.

5.2.3 Level 3

To use algorithm H$_c$-A to solve the problem of minimizing energy consumption with schedule length constraint T, we need to allocate the time T to the c sublists. We use T_1, T_2, ..., T_c to represent a time allocation to the c sublists, where tasks in sublist sublist j are executed within deadline T_j, and $T_1 + T_2 + \cdots + T_c = T$. Theorem 9 [29] provides optimal time allocation to the c sublists for minimizing energy consumption with schedule length constraint in scheduling parallel tasks by using scheduling algorithms H$_c$-A, where A is a list scheduling algorithm.

Theorem 9 *For a given partition $M_{j,1}$, $M_{j,2}$, ..., $M_{j,j}$ of the tasks in sublist j into j groups produced by a list scheduling algorithm A, where $1 \le j \le c$, and a time allocation T_1, T_2, ..., T_c to the c sublists, the amount of energy consumed by algorithm H_c-A is*

$$E = \sum_{j=1}^{c} \left(\frac{M_{j,1}^\alpha + M_{j,2}^\alpha + \cdots + M_{j,j}^\alpha}{T_j^{\alpha-1}} \right).$$

The time allocation T_1, T_2, ..., T_c which minimizes E is

$$T_j = \left(\frac{N_j^{1/\alpha}}{N_1^{1/\alpha} + N_2^{1/\alpha} + \cdots + N_c^{1/\alpha}} \right) T,$$

where $N_j = M_{j,1}^\alpha + M_{j,2}^\alpha + \cdots + M_{j,j}^\alpha$, for all $1 \le j \le c$, and the minimized energy consumption is

$$E = \frac{(N_1^{1/\alpha} + N_2^{1/\alpha} + \cdots + N_c^{1/\alpha})^\alpha}{T^{\alpha-1}},$$

by using the above time allocation.

5.2.4 Level 4

To use a level-by-level scheduling algorithm to solve the problem of minimizing energy consumption with schedule length constraint T, we need to allocate the time T to the v levels. We use T_1, T_2, ..., T_v to represent a time allocation to the v levels, where tasks in level l are executed within deadline T_l, and $T_1 + T_2 + \cdots + T_v = T$.

By Theorem 9, for a given partition $M_{l,j,1}$, $M_{l,j,2}$, ..., $M_{l,j,j}$ of the tasks in sublist (l, j) of level l into j groups produced by a list scheduling algorithm A, where

$1 \le l \le v$ and $1 \le j \le c$, and a time allocation $T_{l,1}, T_{l,2}, ..., T_{l,c}$ to the c sublists of level l, where

$$T_{l,j} = \left(\frac{N_{l,j}^{1/\alpha}}{N_{l,1}^{1/\alpha} + N_{l,2}^{1/\alpha} + \cdots + N_{l,c}^{1/\alpha}} \right) T_l,$$

with $N_{l,j} = M_{l,j,1}^{\alpha} + M_{l,j,2}^{\alpha} + \cdots + M_{l,j,j}^{\alpha}$, for all $1 \le l \le v$ and $1 \le j \le c$, the scheduling algorithm H_c-A consumes energy

$$E_l = \frac{(N_{l,1}^{1/\alpha} + N_{l,2}^{1/\alpha} + \cdots + N_{l,c}^{1/\alpha})^{\alpha}}{T_l^{\alpha-1}},$$

for tasks in level l, where $1 \le l \le v$. Since the level-by-level scheduling algorithm consumes energy $E = E_1 + E_2 + \cdots + E_v$, we have

$$E = \sum_{l=1}^{v} \frac{(N_{l,1}^{1/\alpha} + N_{l,2}^{1/\alpha} + \cdots + N_{l,c}^{1/\alpha})^{\alpha}}{T_l^{\alpha-1}}.$$

By the definition of S_l, we obtain

$$E = \frac{S_1}{T_1^{\alpha-1}} + \frac{S_2}{T_2^{\alpha-1}} + \cdots + \frac{S_v}{T_v^{\alpha-1}}.$$

To minimize E with the constraint $F(T_1, T_2, ..., T_v) = T_1 + T_2 + \cdots + T_v = T$, we use the Lagrange multiplier system

$$\nabla E(T_1, T_2, ..., T_v) = \lambda \nabla F(T_1, T_2, ..., T_v),$$

where λ is the Lagrange multiplier. Since $\partial E / \partial T_l = \lambda \partial F / \partial T_l$, that is,

$$S_l \left(\frac{1-\alpha}{T_l^{\alpha}} \right) = \lambda,$$

$1 \le l \le v$, we get

$$T_l = S_l^{1/\alpha} \left(\frac{1-\alpha}{\lambda} \right)^{1/\alpha},$$

which implies that

$$T = (S_1^{1/\alpha} + S_2^{1/\alpha} + \cdots + S_v^{1/\alpha}) \left(\frac{1-\alpha}{\lambda} \right)^{1/\alpha},$$

and

$$T_l = \left(\frac{S_l^{1/\alpha}}{S_1^{1/\alpha} + S_2^{1/\alpha} + \cdots + S_v^{1/\alpha}} \right) T,$$

for all $1 \le l \le v$. By using the above time allocation, we have

$$
\begin{aligned}
E &= \sum_{l=1}^{v} \frac{S_l}{T_l^{\alpha-1}} \\
&= \sum_{l=1}^{v} \frac{S_l}{\left(\left(\dfrac{S_l^{1/\alpha}}{S_1^{1/\alpha} + S_2^{1/\alpha} + \cdots + S_v^{1/\alpha}}\right) T\right)^{\alpha-1}} \\
&= \sum_{l=1}^{v} \frac{S_l^{1/\alpha}(S_1^{1/\alpha} + S_2^{1/\alpha} + \cdots + S_v^{1/\alpha})^{\alpha-1}}{T^{\alpha-1}} \\
&= \frac{(S_1^{1/\alpha} + S_2^{1/\alpha} + \cdots + S_v^{1/\alpha})^{\alpha}}{T^{\alpha-1}}.
\end{aligned}
$$

Similar to the derivation in Sect. 5.1.4, we have

$$
S_1^{1/\alpha} + S_2^{1/\alpha} + \cdots + S_v^{1/\alpha} \le m^{1/\alpha} \left(\left(\sum_{j=1}^{c} \frac{R_j}{j} \right) + cvr^* \right),
$$

which implies that

$$
E \le m \left(\left(\sum_{j=1}^{c} \frac{R_j}{j} \right) + cvr^* \right)^{\alpha} \frac{1}{T^{\alpha-1}}.
$$

We define the *performance ratio* as $\beta = E/E^*$ for heuristic algorithms that solve the problem of minimizing energy consumption with schedule length constraint on a multicore processor. By Theorem 2, we get

$$
\beta = \frac{E}{E^*} \le \left(\left(\left(\sum_{j=1}^{c} \frac{R_j}{j} \right) + cvr^* \right) \Big/ \left(\frac{W}{m} \right) \right)^{\alpha}.
$$

Theorem 10 provides optimal time allocation to the v levels for minimizing energy consumption with schedule length constraint in scheduling precedence constrained parallel tasks by using level-by-level scheduling algorithms LL-H$_c$-A, where A is a list scheduling algorithm.

Theorem 10 *For a given partition $M_{l,j,1}$, $M_{l,j,2}$, ..., $M_{l,j,j}$ of the tasks in sublist (l, j) of level l into j groups produced by a list scheduling algorithm A, where $1 \le l \le v$ and $1 \le j \le c$, and a time allocation T_1, T_2, ..., T_v to the v levels, the level-by-level scheduling algorithm LL-H$_c$-A consumes energy*

$$E = \sum_{l=1}^{v} \frac{(N_{l,1}^{1/\alpha} + N_{l,2}^{1/\alpha} + \cdots + N_{l,c}^{1/\alpha})^{\alpha}}{T_l^{\alpha-1}},$$

where $N_{l,j} = M_{l,j,1}^{\alpha} + M_{l,j,2}^{\alpha} + \cdots + M_{l,j,j}^{\alpha}$, *for all* $1 \le l \le v$ *and* $1 \le j \le c$. *The time allocation* $T_1, T_2, ..., T_v$ *which minimizes* E *is*

$$T_l = \left(\frac{S_l^{1/\alpha}}{S_1^{1/\alpha} + S_2^{1/\alpha} + \cdots + S_v^{1/\alpha}} \right) T,$$

where $S_l = (N_{l,1}^{1/\alpha} + N_{l,2}^{1/\alpha} + \cdots + N_{l,c}^{1/\alpha})^{\alpha}$, *for all* $1 \le l \le v$, *and the minimized energy consumption is*

$$E = \frac{(S_1^{1/\alpha} + S_2^{1/\alpha} + \cdots + S_v^{1/\alpha})^{\alpha}}{T^{\alpha-1}},$$

by using the above time allocation. The performance ratio is

$$\beta \le \left(\left(\left(\sum_{j=1}^{c} \frac{R_j}{j} \right) + cvr^* \right) \Big/ \left(\frac{W}{m} \right) \right)^{\alpha},$$

where $r^* = \max(r_1, r_2, ..., r_n)$ *is the maximum task execution requirement.*

Theorems 8 and 9 and 10 give the power supply to the task i in group (l, j, k) as

$$\frac{1}{\pi_i} \left(\frac{M_{l,j,k}}{T_{l,j}} \right)^{\alpha} = \frac{1}{\pi_i} \left(\left(\frac{N_{l,1}^{1/\alpha} + N_{l,2}^{1/\alpha} + \cdots + N_{l,c}^{1/\alpha}}{N_{l,j}^{1/\alpha}} \right) \right.$$

$$\left. \left(\frac{S_1^{1/\alpha} + S_2^{1/\alpha} + \cdots + S_v^{1/\alpha}}{S_l^{1/\alpha}} \right) \frac{M_{l,j,k}}{T} \right)^{\alpha},$$

for all $1 \le l \le v$ and $1 \le j \le c$ and $1 \le k \le j$.

Again, we adjust the performance bound given in Theorem 10 to

$$\left(\left(\sum_{j=1}^{c} \frac{R_j}{j} \left(\frac{2j+1}{2j+2} \right)^{1/\alpha} \right) \Big/ \left(\frac{W}{m} \right) \right)^{\alpha}. \tag{2}$$

Our simulation shows that the modified bound in (2) is more accurate than the performance bound given in Theorem 10.

6 Simulation Data

To validate our analytical results, extensive simulations have been conducted. In this section, we demonstrate some numerical and experimental data for several example task graphs. The following task graphs are considered in our experiments.

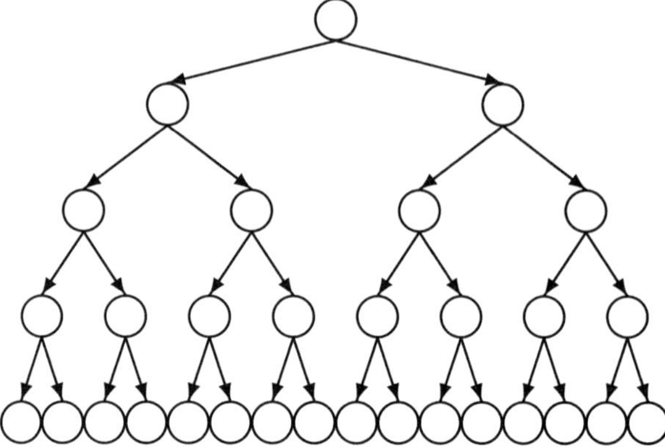

Fig. 3 CT(b, h): a complete binary tree with $b = 2$ and $h = 4$

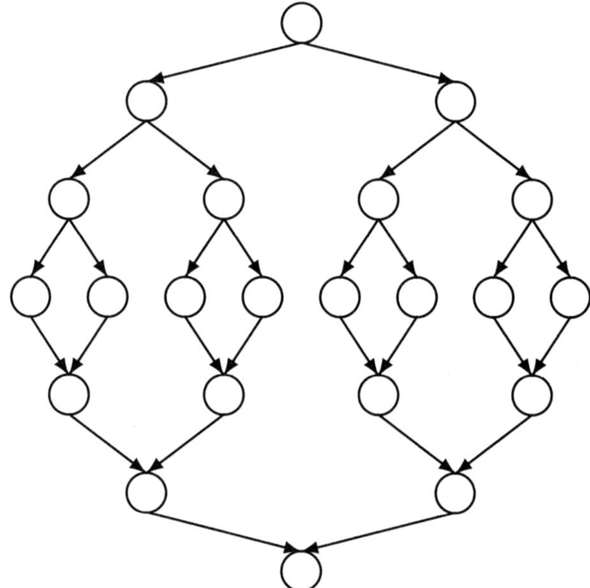

Fig. 4 PA(b, h): a partitioning algorithm with $b = 2$ and $h = 3$

- *Tree-Structured Computations.* Many computations are tree-structured, including backtracking search, branch-and-bound computations, game-tree evaluation, functional and logical programming, and various numeric computations. For simplicity, we consider CT(b, h), i.e., complete b-ary trees of height h (see Fig. 3 where $b = 2$ and $h = 4$). It is easy to see that there are $v = h + 1$ levels numbered as 0, 1, 2, ..., h, and $n_l = b^l$ for $0 \leq l \leq h$, and $n = (b^{h+1} - 1)/(b - 1)$.
- *Partitioning Algorithms.* A partitioning algorithm PA(b, h) represents a divide-and-conquer computation with branching factor b and height (i.e., depth of recursion) h (see Fig. 4 where $b = 2$ and $h = 3$). The dag of PA(b, h) has

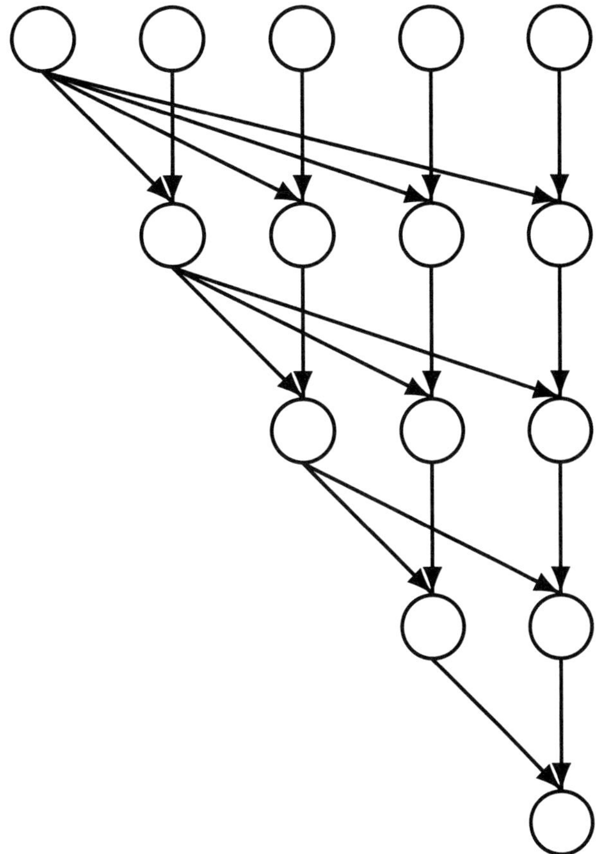

Fig. 5 LA(v): a linear algebra task graph with $v = 5$

$v = 2h + 1$ levels numbered as 0, 1, 2, ..., $2h$. A partitioning algorithm proceeds in three stages. In levels $0, 1, ..., h - 1$, each task is divided into b subtasks. Then, in level h, subproblems of small sizes are solved directly. Finally, in levels $h + 1, h + 2, ..., 2h$, solutions to subproblems are combined to form the solution to the original problem. Clearly, $n_l = n_{2h-l} = b^l$, for all $0 \le l \le h - 1$, $n_h = b^h$, and $n = (b^{h+1} + b^h - 2)/(b - 1)$.

- *Linear Algebra Task Graphs.* A linear algebra task graph LA(v) with v levels (see Fig. 5 where $v = 5$) has $n_l = v - l + 1$ for $l = 1, 2, ..., v$, and $n = v(v + 1)/2$.
- *Diamond Dags.* A diamond dag DD(d) (see Fig. 6 where $d = 4$) contains $v = 2d - 1$ levels numbered as $1, 2, ..., 2d - 1$. It is clear that $n_l = n_{2d-l} = l$, for all $1 \le l \le d - 1$, $n_d = d$, and $n = d^2$.

Since each task graph has at least one parameter, we are actually dealing with classes of task graphs.

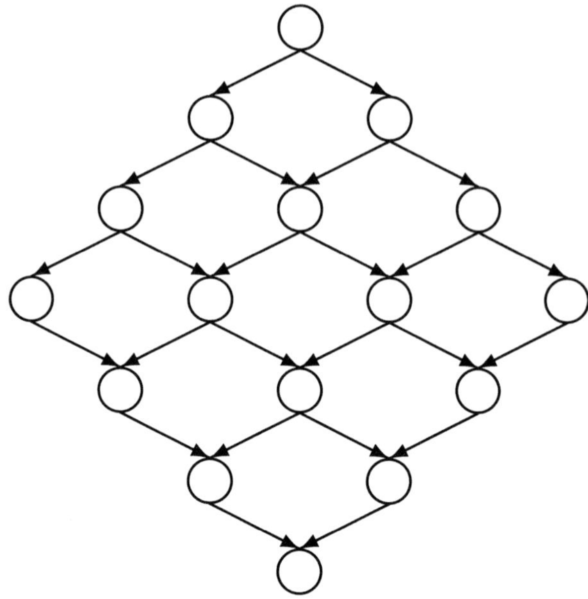

Fig. 6 DD(d): a diamond dag with $d = 4$

We define the *normalized schedule length* (NSL) as

$$\mathrm{NSL} = \frac{T}{\left(\dfrac{m}{E}\left(\dfrac{W}{m}\right)^{\alpha}\right)^{1/(\alpha-1)}}.$$

When T is the schedule length produced by a heuristic algorithm LL-H$_c$-A according to Theorem 6, the normalized schedule length is

$$\mathrm{NSL} = \left(\frac{(S_1^{1/\alpha} + S_2^{1/\alpha} + \cdots + S_v^{1/\alpha})^{\alpha}}{m\left(\dfrac{W}{m}\right)^{\alpha}}\right)^{1/(\alpha-1)}.$$

NSL is an upper bound for the performance ratio $\beta = T/T^*$ for the problem of minimizing schedule length with energy consumption constraint on a multicore processor. When the π_i's and the r_i's are random variables, T, T^*, β, and NSL all become random variables. It is clear that for the problem of minimizing schedule length with energy consumption constraint, we have $\bar{\beta} \le \overline{\mathrm{NSL}}$, i.e., the expected performance ratio is no larger than the expected normalized schedule length. (We use $\bar{x}$ to represent the expectation of a random variable x.)

We define the *normalized energy consumption* (NEC) as

$$\mathrm{NEC} = \frac{E}{m\left(\dfrac{W}{m}\right)^{\alpha}\dfrac{1}{T^{\alpha-1}}}.$$

When E is the energy consumed by a heuristic algorithm LL-H$_c$-A according to Theorem 10, the normalized energy consumption is

$$\mathrm{NEC} = \frac{(S_1^{1/\alpha} + S_2^{1/\alpha} + \cdots + S_v^{1/\alpha})^{\alpha}}{m\left(\dfrac{W}{m}\right)^{\alpha}}.$$

NEC is an upper bound for the performance ratio $\beta = E/E^*$ for the problem of minimizing energy consumption with schedule length constraint on a multicore processor. For the problem of minimizing energy consumption with schedule length constraint, we have $\bar{\beta} \leq \overline{\mathrm{NEC}}$.

Notice that for a given task graph, the expected normalized schedule length $\overline{\mathrm{NSL}}$ and the expected normalized energy consumption $\overline{\mathrm{NEC}}$ are determined by A, c, m, α, and the probability distributions of the π_i's and the r_i's. In our simulations, the algorithm A is chosen as LS; the parameter c is set as 20; the number of cores is set as $m = 128$; and the parameter α is set as 3. The particular choices of these values do not affect our general observations and conclusions. For convenience, the r_i's are treated as independent and identically distributed (i.i.d.) continuous random variables uniformly distributed in [0, 1). The π_i's are i.i.d. discrete random variables. We consider three types of probability distributions of task sizes with about the same expected task size $\bar{\pi}$. Let a_b be the probability that $\pi_i = b$, where $b \geq 1$.

- Uniform distributions in the range [1..u], i.e., $a_b = 1/u$ for all $1 \leq b \leq u$, where u is chosen such that $(u + 1)/2 = \bar{\pi}$, i.e., $u = 2\bar{\pi} - 1$.
- Binomial distributions in the range [1..m], i.e.,

$$a_b = \frac{\binom{m}{b} p^b (1-p)^{m-b}}{1-(1-p)^m},$$

for all $1 \leq b \leq m$, where p is chosen such that $mp = \bar{\pi}$, i.e., $p = \bar{\pi}/m$. However, the actual expectation of task sizes is

$$\frac{\bar{\pi}}{1-(1-p)^m} = \frac{\bar{\pi}}{1-(1-\bar{\pi}/m)^m},$$

which is slightly greater than $\bar{\pi}$, especially when $\bar{\pi}$ is small.
- Geometric distributions in the range [1..m], i.e.,

$$a_b = \frac{q(1-q)^{b-1}}{1-(1-q)^m},$$

for all $1 \leq b \leq m$, where q is chosen such that $1/q = \bar{\pi}$, i.e., $q = 1/\bar{\pi}$. However, the actual expectation of task sizes is

Table 3 Simulation data for expected NSL and NEC on CT(2,12)

	Uniform		Binomial		Geometric	
$\bar{\pi}$	Simulation	Analysis	Simulation	Analysis	Simulation	Analysis
10	1.1772602	1.1850145	1.1127903	1.0635657	1.2695944	1.3183482
20	1.1609754	1.1485746	1.1046696	1.0817685	1.2527372	1.2739448
30	1.2032217	1.2026955	1.1401395	1.1407631	1.2827662	1.3051035
40	1.3783493	1.4501456	1.2111586	1.2364135	1.2959831	1.3174113
50	1.3977418	1.4592250	1.2498124	1.2784298	1.2998132	1.3175610
60	1.3278814	1.3437082	1.2799084	1.3180794	1.3030358	1.3200509
99 % confidence interval ±0.365 %)						
10	1.3816853	1.4002241	1.2386909	1.1314678	1.6180743	1.7403012
20	1.3471473	1.3204301	1.2223807	1.1720051	1.5698000	1.6194065
30	1.4504859	1.4461415	1.2989038	1.2983591	1.6412385	1.6968020
40	1.9023971	2.1084568	1.4683900	1.5308593	1.6805737	1.7387274
50	1.9592480	2.1352965	1.5604366	1.6323378	1.6883269	1.7364845
60	1.7623788	1.8044903	1.6405732	1.7409541	1.6957874	1.7386959
(99 % confidence interval ±0.687 %)						

$$\frac{1/q - (1/q + m)(1-q)^m}{1-(1-q)^m} = \frac{\bar{\pi} - (\bar{\pi} + m)(1 - 1/\bar{\pi})^m}{1-(1-1/\bar{\pi})^m},$$

which is less than $\bar{\pi}$, especially when $\bar{\pi}$ is large.

In Tables 3, 4, 5 and 6, we show and compare the analytical results with simulation data. For each task graph in { CT(2,12), PA(2,12), LA(2000), DD(2000) }, and each $\bar{\pi}$ in the range 10, 20, .., 60, and each probability distribution of task sizes, we generate *rep* sets of tasks, produce their schedules by using algorithm LL-H_c-LS, calculate their NSL (or NEC) and the bound (1) (or bound (2)), report the average of NSL (or NEC) which is the experimental value of $\overline{\text{NSL}}$ (or $\overline{\text{NEC}}$), and report the average of bound (1) (or bound (2)) which is the numerical value of analytical results. The number *rep* is large enough to ensure high quality experimental data. The 99 % confidence interval of all the data in the same table is also given.

We have the following observations from our simulations.

- $\overline{\text{NSL}}$ is less than 1.41 and $\overline{\text{NEC}}$ is less than 1.98. Therefore, our algorithms produce solutions reasonably close to optimum. In fact, $\overline{\text{NSL}}$ and $\overline{\text{NEC}}$ reported here are very close to those for independent parallel tasks reported in [29].
- The performance of algorithm LL-H_c-A for A other than LS is very close (within ±1 %) to the performance of algorithm LL-H_c-LS. Since these data do not provide further insight, they are not shown here.
- The performance bound (1) is very close to $\overline{\text{NSL}}$ and the performance bound (2) is very close to $\overline{\text{NEC}}$.

Table 4 Simulation data for expected NSL and NEC on PA(2,12)

	Uniform		Binomial		Geometric	
$\bar{\pi}$	Simulation	Analysis	Simulation	Analysis	Simulation	Analysis
10	1.1940250	1.1841913	1.1287074	1.0635894	1.2918262	1.3185661
20	1.1710935	1.1489358	1.1120907	1.0822820	1.2628233	1.2735483
30	1.2121712	1.2032254	1.1414699	1.1396784	1.2893692	1.3044971
40	1.3838241	1.4505296	1.2130609	1.2377678	1.3006607	1.3152063
50	1.4034276	1.4608829	1.2497254	1.2777187	1.3052527	1.3182187
60	1.3319146	1.3448578	1.2799201	1.3177687	1.3067475	1.3179615
(99 % confidence interval ±0.284 %)						
10	1.4280855	1.4053089	1.2756771	1.1309478	1.6643757	1.7374005
20	1.3687912	1.3196764	1.2362757	1.1716339	1.5959196	1.6185853
30	1.4680717	1.4464946	1.3037462	1.3007006	1.6629560	1.7012833
40	1.9143602	2.1021764	1.4697836	1.5294041	1.6933298	1.7328875
50	1.9717267	2.1383667	1.5614395	1.6318344	1.7026727	1.7361106
60	1.7748939	1.8095803	1.6402284	1.7397315	1.7084739	1.7376521
(99 % confidence interval ±0.565 %)						

Table 5 Simulation data for expected NSL and NEC on LA(2000)

	Uniform		Binomial		Geometric	
$\bar{\pi}$	Simulation	Analysis	Simulation	Analysis	Simulation	Analysis
10	1.1392509	1.1841096	1.0771624	1.0638363	1.2300726	1.3179978
20	1.1430859	1.1491148	1.0989144	1.0823187	1.2321125	1.2722681
30	1.1954796	1.2028781	1.1372623	1.1399934	1.2686012	1.3032303
40	1.3729227	1.4497884	1.2109722	1.2375699	1.2858406	1.3161030
50	1.3964647	1.4610101	1.2488649	1.2779096	1.2930727	1.3191233
60	1.3272967	1.3445859	1.2802743	1.3187192	1.2959390	1.3182489
(99 % confidence interval ±0.085 %)						
10	1.2974381	1.4020482	1.1602487	1.1313969	1.5137571	1.7379887
20	1.3062497	1.3200333	1.2076518	1.1715685	1.5175999	1.6178453
30	1.4292225	1.4470430	1.2933014	1.2994524	1.6099920	1.6995260
40	1.8847470	2.1014650	1.4664142	1.5315937	1.6530311	1.7317472
50	1.9501571	2.1348479	1.5596494	1.6330611	1.6715971	1.7392715
60	1.7624447	1.8088376	1.6389275	1.7388263	1.6797186	1.7382355
(99 % confidence interval ±0.204 %)						

Table 6 Simulation data for expected NSL and NEC on DD(2000)

	Uniform		Binomial		Geometric	
$\bar{\pi}$	Simulation	Analysis	Simulation	Analysis	Simulation	Analysis
10	1.1393071	1.1842982	1.0770276	1.0636933	1.2303693	1.3183983
20	1.1429980	1.1490295	1.0989960	1.0822466	1.2316570	1.2714949
30	1.1955924	1.2030593	1.1372779	1.1400176	1.2690205	1.3039776
40	1.3726198	1.4493161	1.2109189	1.2375156	1.2859527	1.3162776
50	1.3962951	1.4607530	1.2487413	1.2777190	1.2932855	1.3193741
60	1.3274819	1.3447974	1.2803877	1.3189128	1.2962310	1.3186892
(99 % confidence interval ± 0.054 %)						
10	1.2978774	1.4023671	1.1597583	1.1313744	1.5144683	1.7391638
20	1.3063526	1.3202184	1.2076968	1.1715103	1.5179540	1.6182936
30	1.4292362	1.4470899	1.2934523	1.2996875	1.6099667	1.6996302
40	1.8840943	2.1007925	1.4659063	1.5308111	1.6536717	1.7325694
50	1.9501477	2.1345382	1.5596254	1.6330039	1.6719013	1.7398729
60	1.7625789	1.8090184	1.6405736	1.7412621	1.6799813	1.7386383
(99 % confidence interval ± 0.155 %)						

7 Summary and Future Research

We have emphasized the significance of investigating energy-efficient and high-performance processing of large-scale parallel applications on multicore processors in data centers. We addressed scheduling precedence constrained parallel tasks on multicore processors with dynamically variable voltage and speed as combinatorial optimization problems. We pointed out that our scheduling problems contain four nontrivial subproblems, namely, precedence constraining, system partitioning, task scheduling, and power supplying. We described our methods to deal with precedence constraints, system partitioning, and task scheduling, and developed our optimal four-level energy/time/power allocation scheme for minimizing schedule length and minimizing energy consumption. We also analyzed the performance of our heuristic algorithms, and derived accurate performance bounds. We demonstrated simulation data, which validate our analytical results.

Further research can be directed toward employing more effective and efficient algorithms to deal with independent tasks in the same level. Notice that the approach in this chapter (i.e., algorithm LL-H_c-A) belongs to the class of post-power-determination algorithms. Such an algorithm first generates a schedule, and then determines power supplies [31, 32]. The classes of pre-power-determination and hybrid algorithms are worth of investigation [30]. Our study in this chapter can also be extended to multiple multicore/manycore processors in data centers and discrete speed levels.

References

1. http://en.wikipedia.org/wiki/Adapteva
2. http://en.wikipedia.org/wiki/Dynamic_voltage_scaling
3. http://www.intel.com/multicore/
4. http://www.multicoreinfo.com/2011/10/adapteva-2/
5. S. Albers, "Energy-efficient algorithms," *Communications of the ACM*, vol. 53, no. 5, pp. 86–96, 2010.
6. H. Aydin, R. Melhem, D. Mossé, and P. Mejía-Alvarez, "Power-aware scheduling for periodic real-time tasks," *IEEE Transactions on Computers*, vol. 53, no. 5, pp. 584–600, 2004.
7. N. Bansal, T. Kimbrel, and K. Pruhs, "Dynamic speed scaling to manage energy and temperature," *Proceedings of the 45th IEEE Symposium on Foundation of Computer Science*, pp. 520–529, 2004.
8. J. A. Barnett, "Dynamic task-level voltage scheduling optimizations," *IEEE Transactions on Computers*, vol. 54, no. 5, pp. 508-520, 2005.
9. L. Benini, A. Bogliolo, and G. De Micheli, "A survey of design techniques for system-level dynamic power management," *IEEE Transactions on Very Large Scale Integration (VLSI) Systems*, vol. 8, no. 3, pp. 299–316, 2000.
10. A. Berl, E. Gelenbe, M. Di Girolamo, G. Giuliani, H. De Meer, M. Q. Dang, and K. Pentikousis, "Energy-efficient cloud computing," *The Computer Journal*, vol. 53, no. 7, pp. 1045–1051, 2010.
11. D. P. Bunde, "Power-aware scheduling for makespan and flow," *Proceedings of the 18th ACM Symposium on Parallelism in Algorithms and Architectures*, pp. 190–196, 2006.
12. H.-L. Chan, W.-T. Chan, T.-W. Lam, L.-K. Lee, K.-S. Mak, and P. W. H. Wong, "Energy efficient online deadline scheduling," *Proceedings of the 18th ACM-SIAM Symposium on Discrete Algorithms*, pp. 795–804, 2007.
13. A. P. Chandrakasan, S. Sheng, and R. W. Brodersen, "Low-power CMOS digital design," *IEEE Journal on Solid-State Circuits*, vol. 27, no. 4, pp. 473–484, 1992.
14. S. Cho and R. G. Melhem, "On the interplay of parallelization, program performance, and energy consumption," *IEEE Transactions on Parallel and Distributed Systems*, vol. 21, no. 3, pp. 342–353, 2010.
15. D. Donofrio, L. Oliker, J. Shalf, M. F. Wehner, C. Rowen, J. Krueger, S. Kamil, and M. Mohiyuddin, "Energy-efficient computing for extreme-scale science," *Computer*, vol. 42, no. 11, pp. 62–71, 2009.
16. W.-c. Feng and K. W. Cameron, "The green500 list: encouraging sustainable supercomputing," *Computer*, vol. 40, no. 12, pp. 50–55, 2007.
17. V. W. Freeh, D. K. Lowenthal, F. Pan, N. Kappiah, R. Springer, B. L. Rountree, and M. E. Femal, "Analyzing the energy-time trade-off in high-performance computing applications," *IEEE Transactions on Parallel and Distributed Systems*, vol. 18, no. 6, pp. 835–848, 2007.
18. S. K. Garg, C. S. Yeo, A. Anandasivam, and R. Buyya, "Environment-conscious scheduling of HPC applications on distributed cloud-oriented data centers," *Journal of Parallel Distributed Computing*, vol. 71, no. 6, pp. 732–749, 2011.
19. R. L. Graham, "Bounds on multiprocessing timing anomalies," *SIAM J. Appl. Math.*, vol. 2, pp. 416-429, 1969.
20. I. Hong, D. Kirovski, G. Qu, M. Potkonjak, and M. B. Srivastava, "Power optimization of variable-voltage core-based systems," *IEEE Transactions on Computer-Aided Design of Integrated Circuits and Systems*, vol. 18, no. 12, pp. 1702–1714, 1999.
21. C. Im, S. Ha, and H. Kim, "Dynamic voltage scheduling with buffers in low-power multimedia applications," *ACM Transactions on Embedded Computing Systems*, vol. 3, no. 4, pp. 686–705, 2004.
22. Intel, *Enhanced Intel SpeedStep Technology for the Intel Pentium M Processor – White Paper*, March 2004.

23. S. U. Khan and I. Ahmad, "A cooperative game theoretical technique for joint optimization of energy consumption and response time in computational grids," *IEEE Transactions on Parallel and Distributed Systems*, vol. 20, no. 3, pp. 346–360, 2009.
24. C. M. Krishna and Y.-H. Lee, "Voltage-clock-scaling adaptive scheduling techniques for low power in hard real-time systems," *IEEE Transactions on Computers*, vol. 52, no. 12, pp. 1586–1593, 2003.
25. W.-C. Kwon and T. Kim, "Optimal voltage allocation techniques for dynamically variable voltage processors," *ACM Transactions on Embedded Computing Systems*, vol. 4, no. 1, pp. 211–230, 2005.
26. Y. C. Lee and A. Y. Zomaya, "Energy conscious scheduling for distributed computing systems under different operating conditions," *IEEE Transactions on Parallel and Distributed Systems*, vol. 22, no. 8, pp. 1374–1381, 2011.
27. Y.-H. Lee and C. M. Krishna, "Voltage-clock scaling for low energy consumption in fixed-priority real-time systems," *Real-Time Systems*, vol. 24, no. 3, pp. 303–317, 2003.
28. K. Li, "Performance analysis of power-aware task scheduling algorithms on multiprocessor computers with dynamic voltage and speed," *IEEE Transactions on Parallel and Distributed Systems*, vol. 19, no. 11, pp. 1484–1497, 2008.
29. K. Li, "Energy efficient scheduling of parallel tasks on multiprocessor computers," *Journal of Supercomputing*, vol. 60, no. 2, pp. 223–247, 2012.
30. K. Li, "Scheduling precedence constrained tasks with reduced processor energy on multiprocessor computers," *IEEE Transactions on Computers*, vol. 61, no. 12, pp. 1668–1681, 2012.
31. K. Li, "Power allocation and task scheduling on multiprocessor computers with energy and time constraints," *Energy-Efficient Distributed Computing Systems*, A. Y. Zomaya and Y. C. Lee, eds., Chapter 1, pp. 1-37, John Wiley & Sons, 2012.
32. K. Li, "Algorithms and analysis of energy-efficient scheduling of parallel tasks," *Handbook of Energy-Aware and Green Computing*, Vol. 1 (Chapter 15), I. Ahmad and S. Ranka, eds., pp. 331-360, CRC Press/Taylor & Francis Group, 2012.
33. M. Li, B. J. Liu, and F. F. Yao, "Min-energy voltage allocation for tree-structured tasks," *Journal of Combinatorial Optimization*, vol. 11, pp. 305–319, 2006.
34. M. Li, A. C. Yao, and F. F. Yao, "Discrete and continuous min-energy schedules for variable voltage processors," *Proceedings of the National Academy of Sciences USA*, vol. 103, no. 11, pp. 3983–3987, 2006.
35. M. Li and F. F. Yao, "An efficient algorithm for computing optimal discrete voltage schedules," *SIAM Journal on Computing*, vol. 35, no. 3, pp. 658–671, 2006.
36. J. R. Lorch and A. J. Smith, "PACE: a new approach to dynamic voltage scaling," *IEEE Transactions on Computers*, vol. 53, no. 7, pp. 856–869, 2004.
37. R. N. Mahapatra and W. Zhao, "An energy-efficient slack distribution technique for multimode distributed real-time embedded systems," *IEEE Transactions on Parallel and Distributed Systems*, vol. 16, no. 7, pp. 650–662, 2005.
38. B. C. Mochocki, X. S. Hu, and G. Quan, "A unified approach to variable voltage scheduling for nonideal DVS processors," *IEEE Transactions on Computer-Aided Design of Integrated Circuits and Systems*, vol. 23, no. 9, pp. 1370–1377, 2004.
39. G. Quan and X. S. Hu, "Energy efficient DVS schedule for fixed-priority real-time systems," *ACM Transactions on Embedded Computing Systems*, vol. 6, no. 4, Article no. 29, 2007.
40. N. B. Rizvandi, J. Taheri, and A. Y. Zomaya, "Some observations on optimal frequency selection in DVFS-based energy consumption minimization," *Journal of Parallel Distributed Computing*, vol. 71, no. 8, pp. 1154–1164, 2011.
41. C. Rusu, R. Melhem, D. Mossé, "Maximizing the system value while satisfying time and energy constraints," *Proceedings of the 23rd IEEE Real-Time Systems Symposium*, pp. 256-265, 2002.
42. D. Shin and J. Kim, "Power-aware scheduling of conditional task graphs in real-time multiprocessor systems," *Proceedings of the International Symposium on Low Power Electronics and Design*, pp. 408–413, 2003.

43. D. Shin, J. Kim, and S. Lee, "Intra-task voltage scheduling for low-energy hard real-time applications," *IEEE Design & Test of Computers*, vol. 18, no. 2, pp. 20–30, 2001.
44. M. R. Stan and K. Skadron, "Guest editors' introduction: power-aware computing," *IEEE Computer*, vol. 36, no. 12, pp. 35–38, 2003.
45. O. S. Unsal and I. Koren, "System-level power-aware design techniques in real-time systems," *Proceedings of the IEEE*, vol. 91, no. 7, pp. 1055–1069, 2003.
46. V. Venkatachalam and M. Franz, "Power reduction techniques for microprocessor systems," *ACM Computing Surveys*, vol. 37, no. 3, pp. 195–237, 2005.
47. M. Weiser, B. Welch, A. Demers, and S. Shenker, "Scheduling for reduced CPU energy," *Proceedings of the 1st USENIX Symposium on Operating Systems Design and Implementation*, pp. 13–23, 1994.
48. P. Yang, C. Wong, P. Marchal, F. Catthoor, D. Desmet, D. Verkest, and R. Lauwereins, "Energy-aware runtime scheduling for embedded-multiprocessor SOCs," *IEEE Design & Test of Computers*, vol. 18, no. 5, pp. 46–58, 2001.
49. F. Yao, A. Demers, and S. Shenker, "A scheduling model for reduced CPU energy," *Proceedings of the 36th IEEE Symposium on Foundations of Computer Science*, pp. 374–382, 1995.
50. H.-S. Yun and J. Kim, "On energy-optimal voltage scheduling for fixed-priority hard real-time systems," *ACM Transactions on Embedded Computing Systems*, vol. 2, no. 3, pp. 393–430, 2003.
51. B. Zhai, D. Blaauw, D. Sylvester, and K. Flautner, "Theoretical and practical limits of dynamic voltage scaling," *Proceedings of the 41st Design Automation Conference*, pp. 868-873, 2004.
52. X. Zhong and C.-Z. Xu, "Energy-aware modeling and scheduling for dynamic voltage scaling with statistical real-time guarantee," *IEEE Transactions on Computers*, vol. 56, no. 3, pp. 358–372, 2007.
53. D. Zhu, R. Melhem, and B. R. Childers, "Scheduling with dynamic voltage/speed adjustment using slack reclamation in multiprocessor real-time systems," *IEEE Transactions on Parallel and Distributed Systems*, vol. 14, no. 7, pp. 686–700, 2003.
54. D. Zhu, D. Mossé, and R. Melhem, "Power-aware scheduling for AND/OR graphs in real-time systems," *IEEE Transactions on Parallel and Distributed Systems*, vol. 15, no. 9, pp. 849–864, 2004.
55. J. Zhuo and C. Chakrabarti, "Energy-efficient dynamic task scheduling algorithms for DVS systems," *ACM Transactions on Embedded Computing Systems*, vol. 7, no. 2, Article no. 17, 2008.
56. Z. Zong, A. Manzanares, X. Ruan, and X. Qin, "EAD and PEBD: two energy-aware duplication scheduling algorithms for parallel tasks on homogeneous clusters," *IEEE Transactions on Computers*, vol. 60, no. 3, pp. 360–374, 2011.

Energy-Aware Algorithms for Task Graph Scheduling, Replica Placement and Checkpoint Strategies

Guillaume Aupy, Anne Benoit, Paul Renaud-Goud and Yves Robert

1 Introduction

The *energy consumption* of computational platforms has recently become a critical problem, both for economic and environmental reasons [35]. To reduce energy consumption, processors can run at different speeds. Faster speeds allow for a faster execution, but they also lead to a much higher (superlinear) power consumption. Energy-aware scheduling aims at minimizing the energy consumed during the execution of the target application, both for computations and for communications. The price to pay for a lower energy consumption usually is a much larger execution time, so the energy-aware approach makes better sense when coupled with some prescribed performance bound. In other words, we have a bi-criteria optimization problem, with one objective being energy minimization, and the other being performance-related.

In this chapter, we discuss several problems related to data centers, for which energy consumption is a crucial matter. Indeed, statistics showed that in 2012, some data centers consume more electricity than 250,000 european houses. If the *cloud* was a country, it would be ranked as the fifth world-wide rank in terms of demands in electricity, and the need is expected to be multiplied by three before 2020. We

G. Aupy (✉) · A. Benoit · Y. Robert
LIP, Ecole Normale Supérieure de Lyon, Lyon, France
e-mail: Guillaume.Aupy@ens-lyon.fr

A. Benoit · Y. Robert
Institut Universitaire de France, Paris, France
e-mail: Anne.Benoit@ens-lyon.fr

P. Renaud-Goud
Chalmers University of technology, Gothenburg, Sweden
e-mail: goud@chalmers.se

Y. Robert
University Tennessee Knoxville, Knoxville, USA
e-mail: Yves.Robert@ens-lyon.fr

S. U. Khan, A. Y. Zomaya (eds.), *Handbook on Data Centers*,
DOI 10.1007/978-1-4939-2092-1_2

focus mainly on the energy consumption of processors, although a lot of electricity is now devoted to cooling the machines, and also for network communications.

Energy models are introduced in Sect. 2. Depending on the different research areas, several different energy models are considered, but they all share the same core assumption: there is a static energy consumption, which is independent on the speed at which a processor is running, and a dynamic energy consumption, which increases superlinearly with the speed. The most common models for speeds are either to use continuous speeds in a given interval, or to consider a set of discrete speeds (the latter being more realistic for actual processors). We discuss further variants of the discrete model: in the VDD-hopping model, the speed of a task can be changed during execution, hence allowing to simulate the continuous case; the incremental model is similar to the discrete model with the additional assumption that the different speeds are spaced regularly. Finally, we propose a literature survey on energy models, and we provide an example to compare models.

The first case study is about task graph scheduling (see Sect. 3). We consider a task graph to be executed on a set of processors. We assume that the mapping is given, say by an ordered list of tasks to execute on each processor, and we aim at optimizing the energy consumption while enforcing a prescribed bound on the execution time. While it is not possible to change the allocation of a task, it is possible to change its speed. Rather than using a local approach such as backfilling, we consider the problem as a whole and study the impact of several speed variation models on its complexity. For continuous speeds, we give a closed-form formula for trees and series-parallel graphs, and we cast the problem into a geometric programming problem for general directed acyclic graphs. We show that the classical dynamic voltage and frequency scaling (DVFS) model with discrete speeds leads to an NP-complete problem, even if the speeds are regularly distributed (an important particular case in practice, which we analyze as the incremental model). On the contrary, the VDD-hopping model leads to a polynomial solution. Finally, we provide an approximation algorithm for the incremental model, which we extend for the general DVFS model.

Then in Sect. 4, we discuss a variant of the replica placement problem aiming at an efficient power management. We study optimal strategies to place replicas in tree networks, with the double objective to minimize the total cost of the servers, and/or to optimize power consumption. The client requests are known beforehand, and some servers are assumed to pre-exist in the tree. Without power consumption constraints, the total cost is an arbitrary function of the number of existing servers that are reused, and of the number of new servers. Whenever creating and operating a new server has higher cost than reusing an existing one (which is a very natural assumption), cost optimal strategies have to trade-off between reusing resources and load-balancing requests on new servers. We provide an optimal dynamic programming algorithm that returns the optimal cost, thereby extending known results from Wu, Lin and Liu [33, 43] without pre-existing servers. With power consumption constraints, we assume that servers operate under a set of M different speeds depending upon the number of requests that they have to process. In practice, M is a small number, typically 2 or 3, depending upon the number of allowed voltages [24, 23]. Power consumption includes a static part, proportional to the total number of servers, and a dynamic part, proportional to a constant exponent of the server speed, which depends upon the model for power. The cost function becomes a more complicated function that takes into account reuse and creation as before, but also upgrading or downgrading an existing server from one speed to another. We show that with an arbitrary number of speeds, the power minimization problem is NP-complete, even without cost constraint, and without static power. Still, we provide an optimal dynamic programming algorithm that returns the minimal power, given a threshold value on the total cost; it has exponential complexity in the number of speeds M, and its practical usefulness is limited to small values of M. However, experiments conducted with this algorithm show that it can process large trees in reasonable time, despite its worst-case complexity.

The last case study investigates checkpointing strategies (see Sect. 5). Nowadays, high performance computing is facing a major challenge with the increasing frequency of failures [18]. There is a need to use fault tolerance or resilience mechanisms to ensure the efficient progress and correct termination of the applications in the presence of failures. A well-established method to deal with failures is *checkpointing*: a checkpoint is taken at the end of the execution of each chunk of work. During the checkpoint, we check for the accuracy of the result; if the result is not correct, due to a transient failure (such as a memory error or software error), the chunk is re-executed. This model with transient failures is one of the most used in the literature, see for instance [17, 48]. In this section, we aim at minimizing the energy consumption when executing a divisible workload under a bound on the total execution time, while resilience is provided through checkpointing. We discuss several variants of this multi-criteria problem. Given the workload, we need to decide how many chunks to use, what are the sizes of these chunks, and at which speed each chunk is executed (under the continuous model). Furthermore, since a failure may occur during the execution of a chunk, we also need to decide at which speed a chunk should be re-executed in the event of a failure. The goal is to minimize the expectation of the total energy consumption, while enforcing a deadline on the execution time, that should be met either in expectation (soft deadline), or in the worst case (hard deadline). For each problem instance, we propose either an exact solution, or a function that can be optimized numerically.

Finally, we provide concluding remarks in Sect. 6.

2 Energy Models

As already mentioned, to help reduce energy dissipation, processors can run at different speeds. Their power consumption is the sum of a static part (the cost for a processor to be turned on, and the leakage power) and a dynamic part, which is a strictly convex function of the processor speed, so that the execution of a given amount of work costs more power if a processor runs in a higher speed [23]. More precisely, a processor running at speed s dissipates s^3 watts [4, 12, 15, 25, 38] per time-unit, hence consumes $s^3 \times d$ joules when operated during d units of time. Faster speeds allow for a faster execution, but they also lead to a much higher (superlinear) power consumption.

In this section, we survey different models for dynamic energy consumption, taken from the literature. These models are categorized as follows:

CONTINUOUS model. Processors can have arbitrary speeds, and can vary them continuously within the interval $[s_{min}, s_{max}]$. This model is unrealistic (any possible value of the speed, say $\sqrt{e^{\pi}}$, cannot be obtained) but it is theoretically appealing [5]. In the CONTINUOUS model, a processor can change its speed at any time during execution.

DISCRETE model. Processors have a discrete number of predefined speeds, which correspond to different voltages and frequencies that the processor can be subjected to [36]. These speeds are denoted as $s_1, ..., s_m$. Switching speeds is not allowed during the execution of a given task, but two different tasks scheduled on a same processor can be executed at different speeds.

VDD-HOPPING model. This model is similar to the DISCRETE one, with a set of different speeds $s_1, ..., s_m$, except that switching speeds during the execution of a given task is allowed: any rational speed can be simulated, by simply switching, at the appropriate time during the execution of a task, between two consecutive speeds [34]. In the VDD-HOPPING model, the energy consumed during the execution of one task is the sum, on each time interval with constant speed s, of the energy consumed during this interval at speed s.

INCREMENTAL model. In this variant of the DISCRETE model, there is a value δ that corresponds to the minimum permissible speed increment, induced by the minimum voltage increment that can be achieved when controlling the processor CPU. Hence, possible speed values are obtained as $s = s_{min} + i \times \delta$, where i is an integer such that $0 \leq i \leq \frac{s_{max} - s_{min}}{\delta}$. Admissible speeds lie in the interval $[s_{min}, s_{max}]$. This model aims at capturing a realistic version of the DISCRETE model, where the different speeds are spread regularly between $s_1 = s_{min}$ and $s_m = s_{max}$, instead of being arbitrarily chosen. It is intended as the modern counterpart of a potentiometer knob.

After the literature survey in Sect. 2.1, we provide a simple example in Sect. 2.2, in order to illustrate the different models.

2.1 *Literature Survey*

Reducing the energy consumption of computational platforms is an important research topic, and many techniques at the process, circuit design, and micro-architectural levels have been proposed [22, 30, 32]. The dynamic voltage and frequency scaling (DVFS) technique has been extensively studied, since it may lead to efficient energy/performance trade-offs [5, 14, 20, 26, 29, 42, 45]. Current microprocessors (for instance, from AMD [1] and Intel [24]) allow the speed to be set dynamically. Indeed, by lowering supply voltage, hence processor clock frequency, it is possible to achieve important reductions in power consumption, without necessarily increasing the execution time. We first discuss different optimization problems that arise in this context, then we review energy models.

2.1.1 DVFS and Optimization Problems

When dealing with energy consumption, the most usual optimization function consists of minimizing the energy consumption, while ensuring a deadline on the execution time (i.e., a real-time constraint), as discussed in the following papers.

In [36], Okuma et al. demonstrate that voltage scaling is far more effective than the shutdown approach, which simply stops the power supply when the system is inactive. Their target processor employs just a few discretely variable voltages. De Langen and Juurlink [31] discuss leakage-aware scheduling heuristics that investigate both DVS and processor shutdown, since static power consumption due to leakage current is expected to increase significantly. Chen et al. [13] consider parallel sparse applications, and they show that when scheduling applications modeled by a directed acyclic graph with a well-identified critical path, it is possible to lower the voltage during non-critical execution of tasks, with no impact on the execution time. Similarly, Wang et al. [42] study the slack time for non-critical jobs, they extend their execution time and thus reduce the energy consumption without increasing the total execution time. Kim et al. [29] provide power-aware scheduling algorithms for bag-of-tasks applications with deadline constraints, based on dynamic voltage scaling. Their goal is to minimize power consumption as well as to meet the deadlines specified by application users.

For real-time embedded systems, slack reclamation techniques are used. Lee and Sakurai [32] show how to exploit slack time arising from workload variation, thanks to a software feedback control of supply voltage. Prathipati [37] discusses techniques to take advantage of run-time variations in the execution time of tasks; the goal is to determine the minimum voltage under which each task can be executed, while guaranteeing the deadlines of each task. Then, experiments are conducted on the Intel StrongArm SA-1100 processor, which has eleven different frequencies, and the Intel PXA250 XScale embedded processor with four frequencies. In [44], the goal of Xu et al. is to schedule a set of independent tasks, given a worst case execution cycle (WCEC) for each task, and a global deadline, while accounting for time and energy penalties when the processor frequency is changing. The frequency of the processor can be lowered when some slack is obtained dynamically, typically when a task runs faster than its WCEC. Yang and Lin [45] discuss algorithms with preemption, using DVS techniques; substantial energy can be saved using these algorithms, which succeed to claim the static and dynamic slack time, with little overhead.

Since an increasing number of systems are powered by batteries, maximizing battery life also is an important optimization problem. Battery-efficient systems can be obtained with similar techniques of dynamic voltage and frequency scaling, as described by Lahiri et al. in [30]. Another optimization criterion is the energy-delay product, since it accounts for a trade-off between performance and energy consumption, as for instance discussed by Gonzalez and Horowitz in [21].

2.1.2 Energy Models

Several energy models are considered in the literature, and they can all be categorized in one of the four models investigated in this paper, i.e., CONTINUOUS, DISCRETE, VDD-HOPPING or INCREMENTAL.

The CONTINUOUS model is used mainly for theoretical studies. For instance, Yao et al. [46], followed by Bansal et al. [5], aim at scheduling a collection of tasks (with release time, deadline and amount of work), and the solution is the time at which each task is scheduled, but also, the speed at which the task is executed. In these papers, the speed can take any value, hence following the CONTINUOUS model.

We believe that the most widely used model is the DISCRETE one. Indeed, processors have currently only a few discrete number of possible frequencies [1, 24, 36, 37]. Therefore, most of the papers discussed above follow this model. Some studies exploit the continuous model to determine the smallest frequency required to run a task, and then choose the closest upper discrete value, as for instance [37] and [47].

Recently, a new local dynamic voltage scaling architecture has been developed, based on the VDD-HOPPING model [6, 7, 34]. It was shown in [32] that significant power can be saved by using two distinct voltages, and architectures using this principle have been developed (see for instance [28]). Compared to traditional power converters, a new design with no needs for large passives or costly technological options has been validated in a STMicroelectronics CMOS 65-nm low-power technology [34].

The INCREMENTAL model was introduced in [2]. The main rationale is that future technologies may well have an increased number of possible frequencies, and these will follow a regular pattern. For instance, note that the SA-1100 processor, considered in [37], has eleven frequencies that are equidistant, i.e., they follow the INCREMENTAL model. Lee and Sakurai [32] exploit discrete levels of clock frequency as f, $f/2$, $f/3$, ..., where f is the master (i.e., the higher) system clock frequency. This model is closer to the DISCRETE model, although it exhibits a regular pattern similarly to the INCREMENTAL model.

2.2 *Example*

Energy-aware scheduling aims at minimizing the energy consumed during the execution of the target application. Obviously, it makes better sense only if it is coupled with some performance bound to achieve. For instance, whenever static energy can

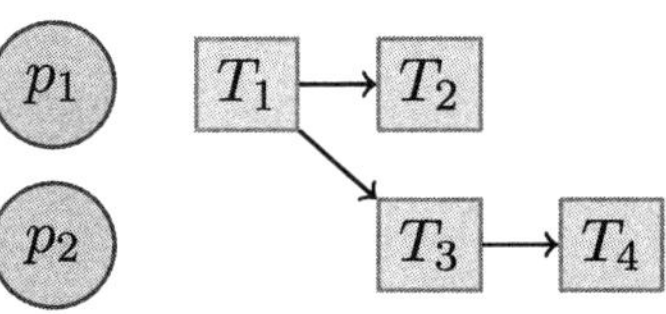

Fig. 1 Execution graph for the example

be neglected, the optimal solution always is to run each processor at the slowest possible speed. In the following, we do neglect static energy and discuss how to minimize dynamic energy consumption when executing a small task graph onto processors.

Consider an application with four tasks of costs $w_1 = 3$, $w_2 = 2$, $w_3 = 1$ and $w_4 = 2$, and three precedence constraints, as shown in Fig. 1. We assume that T_1 and T_2 are allocated, in this order, onto processor P_1, while T_3 and T_4 are allocated, in this order, on processor P_2. The deadline on the execution time is $D = 1.5$.

We set the minimum and maximum speeds to $s_{min} = 0$ and $s_{max} = 6$ for the CONTINUOUS model. For the DISCRETE and VDD-HOPPING models, we use the set of speeds $s_1^{(d)} = 2$, $s_2^{(d)} = 5$ and $s_3^{(d)} = 6$. Finally, for the INCREMENTAL model, we set $\delta = 2$, $s_{min} = 2$ and $s_{max} = 6$, so that possible speeds are $s_1^{(i)} = 2$, $s_2^{(i)} = 4$ and $s_3^{(i)} = 6$. We aim at finding the optimal execution speed s_i for each task T_i ($1 \leq i \leq 4$), i.e., the values of s_i that minimize the energy consumption.

With the CONTINUOUS model, the optimal speeds are non rational values, and we obtain:

$$s_1 = \frac{2}{3}(3 + 35^{1/3}) \simeq 4.18; s_2 = s_1 \times \frac{2}{35^{1/3}} \simeq 2.56; s_3 = s_4 = s_1 \times \frac{3}{35^{1/3}} \simeq 3.83.$$

Note that all speeds are in the interval $[s_{min}, s_{max}]$. These values are obtained thanks to the formulas derived in Sect. 3.2 below. The energy consumption is then $E_{opt}^{(c)} = \Sigma_{i=1}^{4} w_i \times s_i^2 = 3.s_1^2 + 2.s_2^2 + 3.s_3^2 \simeq 109.6$. The execution time is $\frac{w_1}{s_1} + \max\left(\frac{w_2}{s_2}, \frac{w_3+w_4}{s_3}\right)$, and with this solution, it is equal to the deadline D (actually, both processors reach the deadline, otherwise we could slow down the execution of one task).

For the DISCRETE model, if we execute all tasks at speed $s_2^{(d)} = 5$, we obtain an energy $E = 8 \times 5^2 = 200$. A better solution is obtained with $s_1 = s_3^{(d)} = 6$, $s_2 = s_3 = s_1^{(d)} = 2$ and $s_4 = s_2^{(d)} = 5$, which turns out to be optimal: $E_{opt}^{(d)} = 3 \times 36 + (2 + 1) \times 4 + 2 \times 25 = 170$. Note that $E_{opt}^{(d)} > E_{opt}^{(c)}$, i.e., the optimal energy consumption with the DISCRETE model is much higher than the one achieved with the CONTINUOUS model. Indeed, in this case, even though the first processor executes during $3/6 + 2/2 = D$ time units, the second processor remains idle since $3/6 + 1/2 + 2/5 = 1.4 < D$. The problem turns out to be NP-hard (see Sect. 3.3.2), and the solution was found by performing an exhaustive search.

With the VDD-HOPPING model, we set $s_1 = s_2^{(d)} = 5$; for the other tasks, we run part of the time at speed $s_2^{(d)} = 5$, and part of the time at speed $s_1^{(d)} = 2$ in order to use the idle time and lower the energy consumption. T_2 is executed at speed $s_1^{(d)}$ during time $\frac{5}{6}$ and at speed $s_2^{(d)}$ during time $\frac{2}{30}$ (i.e., the first processor executes during time

$3/5+5/6+2/30 = 1.5 = D$, and all the work for T_2 is done: $2 \times 5/6+5 \times 2/30 = 2 = w_2$). T_3 is executed at speed $s_2^{(d)}$ (during time $1/5$), and finally T_4 is executed at speed $s_1^{(d)}$ during time 0.5 and at speed $s_2^{(d)}$ during time $1/5$ (i.e., the second processor executes during time $3/5 + 1/5 + 0.5 + 1/5 = 1.5 = D$, and all the work for T_4 is done: $2 \times 0.5+5 \times 1/5 = 2 = w_4$). This set of speeds turns out to be optimal (i.e., it is the optimal solution of the linear program introduced in Sect. 3.3.1), with an energy consumption $E_{opt}^{(v)} = (3/5 + 2/30 + 1/5 + 1/5) \times 5^3 + (5/6 + 0.5) \times 2^3 = 144$. As expected, $E_{opt}^{(c)} \leq E_{opt}^{(v)} \leq E_{opt}^{(d)}$, i.e., the VDD-HOPPING solution stands between the optimal CONTINUOUS solution, and the more constrained DISCRETE solution.

For the INCREMENTAL model, the reasoning is similar to the DISCRETE case, and the optimal solution is obtained by an exhaustive search: all tasks should be executed at speed $s_2^{(i)} = 4$, with an energy consumption $E_{opt}^{(i)} = 8 \times 4^2 = 128 > E_{opt}^{(c)}$. It turns out to be better than DISCRETE and VDD-HOPPING, since it has different discrete values of energy that are more appropriate for this example.

3 Minimizing the Energy of a Schedule

In this section, we investigate energy-aware scheduling strategies for executing a task graph on a set of processors. The main originality is that we assume that the mapping of the task graph is given, say by an ordered list of tasks to execute on each processor. There are many situations in which this problem is important, such as optimizing for legacy applications, or accounting for affinities between tasks and resources, or even when tasks are pre-allocated [39], for example for security reasons. In such situations, assume that a list-schedule has been computed for the task graph, and that its execution time should not exceed a deadline D. We do not have the freedom to change the assignment of a given task, but we can change its speed to reduce energy consumption, provided that the deadline D is not exceeded after the speed change. Rather than using a local approach such as backfilling [37, 42], which only reclaims gaps in the schedule, we consider the problem as a whole, and we assess the impact of several speed variation models on its complexity. We give the main complexity results without proofs (refer to [2] for details).

3.1 Optimization Problem

Consider an application task graph $\mathcal{G} = (V, \mathcal{E})$, with $n = |V|$ tasks denoted as $V = \{T_1, T_2, \ldots, T_n\}$, and where the set $\mathcal{E}$ denotes the precedence edges between tasks. Task T_i has a cost w_i for $1 \leq i \leq n$. We assume that the tasks in $\mathcal{G}$ have been allocated onto a parallel platform made up of identical processors. We define the *execution graph* generated by this allocation as the graph $G = (V, E)$, with the following augmented set of edges:

- $\mathcal{E} \subseteq E$: if an edge exists in the precedence graph, it also exists in the execution graph;
- if T_1 and T_2 are executed successively, in this order, on the same processor, then $(T_1, T_2) \in E$.

The goal is to the minimize the energy consumed during the execution while enforcing a deadline D on the execution time. We formalize the optimization problem in the simpler case where each task is executed at constant speed. This strategy is optimal for the CONTINUOUS model (by a convexity argument) and for the DISCRETE and INCREMENTAL models (by definition). For the VDD-HOPPING model, we reformulate the problem in Sect. 3.3.1. Let d_i be the duration of the execution of task T_i, t_i its completion time, and s_i the speed at which it is executed. We obtain the following formulation of the MINENERGY(G, D) problem, given an execution graph $G=(V, E)$ and a deadline D; the s_i values are variables, whose values are constrained by the energy model:

$$\begin{array}{lll} \text{Minimize} & & \sum_{i=1}^{n} s_i^3 \times d_i \\ \text{subject to} & \text{(i)} & w_i = s_i \times d_i \text{ for each task } T_i \in V \\ & \text{(ii)} & t_i + d_j \leq t_j \text{ for each edge } (T_i, T_j) \in E \\ & \text{(iii)} & t_i \leq D \text{ for each task } T_i \in V \end{array} \qquad (1)$$

Constraint (i) states that the whole task can be executed in time d_i using speed s_i. Constraint (ii) accounts for all dependencies, and constraint (iii) ensures that the execution time does not exceed the deadline D. The energy consumed throughout the execution is the objective function. It is the sum, for each task, of the energy consumed by this task, as we detail in the next section. Note that $d_i = w_i/s_i$, and therefore the objective function can also be expressed as $\Sigma_{i=1}^{n} s_i^2 \times w_i$.

3.2 The CONTINUOUS Model

With the CONTINUOUS model, processor speeds can take any value between s_{min} and s_{max}. We assume for simplicity that $s_{min} = 0$, i.e., there is no minimum speed. First we prove that, with this model, the processors do not change their speed during the execution of a task:

Lemma 1 (constant speed per task) *With the* CONTINUOUS *model, each task is executed at constant speed, i.e., a processor does not change its speed during the execution of a task.*

We derive in Sect. 3.2.1 the optimal speed values for special execution graph structures, expressed as closed form algebraic formulas, and we show that these values may be irrational (as already illustrated in the example in Sect. 2.2). Finally, we formulate the problem for general DAGs as a convex optimization program in Sect. 3.2.2.

3.2.1 Special Execution Graphs

Consider the problem of minimizing the energy of n independent tasks (i.e., each task is mapped onto a distinct processor, and there are no precedence constraints in the execution graph), while enforcing a deadline D.

Proposition 1 (independent tasks) *When G is composed of independent tasks $\{T_1, \dots, T_n\}$, the optimal solution to* MINENERGY*(G,D) is obtained when each task T_i ($1 \le i \le n$) is computed at speed $s_i = \frac{w_i}{D}$. If there is a task T_i such that $s_i > s_{max}$, then the problem has no solution.*

Consider now the problem with a linear chain of tasks. This case corresponds for instance to n independent tasks $\{T_1, \dots, T_n\}$ executed onto a single processor. The execution graph is then a linear chain (order of execution of the tasks), with $T_i \to T_{i+1}$, for $1 \le i < n$.

Proposition 2 (linear chain) *When G is a linear chain of tasks, the optimal solution to* MINENERGY*(G,D) is obtained when each task is executed at speed $s = \frac{W}{D}$, with $W = \Sigma_{i=1}^n w_i$.*
If $s > s_{max}$, then there is no solution.

Corollary 1 *A linear chain with n tasks is equivalent to a single task of cost $W = \Sigma_{i=1}^n w_i$.*

Indeed, in the optimal solution, the n tasks are executed at the same speed, and they can be replaced by a single task of cost W, which is executed at the same speed and consumes the same amount of energy.

Finally, consider fork and join graphs. Let $V = \{T_1, \dots, T_n\}$. We consider either a fork graph $G = (V \cup \{T_0\}, E)$, with $E = \{(T_0, T_i), T_i \in V\}$, or a join graph $G = (V \cup \{T_0\}, E)$, with $E = \{(T_i, T_0), T_i \in V\}$. T_0 is either the source of the fork or the sink of the join.

Theorem 1 (fork and join graphs) *When G is a fork (resp. join) execution graph with $n + 1$ tasks $T_0, T_1, \dots, T_n$, the optimal solution to* MINENERGY*(G,D) is the following:*

- *the execution speed of the source (resp. sink) T_0 is $s_0 = \dfrac{\left(\Sigma_{i=1}^n w_i^3\right)^{\frac{1}{3}} + w_0}{D}$;*
- *for the other tasks T_i, $1 \le i \le n$, we have $s_i = s_0 \times \dfrac{w_i}{\left(\Sigma_{i=1}^n w_i^3\right)^{\frac{1}{3}}}$ if $s_0 \le s_{max}$.*

Otherwise, T_0 should be executed at speed $s_0 = s_{max}$, and the other speeds are $s_i = \frac{w_i}{D'}$, with $D' = D - \frac{w_0}{s_{max}}$, if they do not exceed s_{max} (Proposition 1 for independent tasks). Otherwise there is no solution.

If no speed exceeds s_{max}, the corresponding energy consumption is

$$\mathbf{minE}(G, D) = \frac{\left(\left(\sum_{i=1}^n w_i^3\right)^{\frac{1}{3}} + w_0\right)^3}{D^2}.$$

Corollary 2 (equivalent tasks for speed) *Consider a fork or join graph with tasks T_i, $0 \le i \le n$, and a deadline D, and assume that the speeds in the optimal solution* *to* MINENERGY*(G,D) do not exceed s_{max}. Then, these speeds are the same as in the optimal solution for n+1 independent tasks $T'_0, T'_1, \ldots, T'_n$, where $w'_0 = \left(\Sigma_{i=1}^n w_i^3\right)^{\frac{1}{3}} + w_0$, and, for $1 \leq i \leq n$, $w'_i = w'_0 \cdot \frac{w_i}{\left(\Sigma_{i=1}^n w_i^3\right)^{\frac{1}{3}}}$.*

Corollary 3 (equivalent tasks for energy) *Consider a fork or join graph G and a deadline D, and assume that the speeds in the optimal solution to* MINENERGY*(G,D) do not exceed s_{max}. We say that the graph G is equivalent to the graph $G^{(eq)}$, consisting of a single task $T_0^{(eq)}$ of weight $w_0^{(eq)} = \left(\Sigma_{i=1}^n w_i^3\right)^{\frac{1}{3}} + w_0$, because the minimum energy consumption of both graphs are identical:* **minE**(G, D)=**minE**$(G^{(eq)}, D)$.

3.2.2 General DAGs

For arbitrary execution graphs, we can rewrite the MINENERGY(G, D) problem as follows:

$$\begin{array}{lll} \text{Minimize} & & \sum_{i=1}^{n} u_i^{-2} \times w_i \\ \text{subject to} & \text{(i)} & t_i + w_j \times u_j \leq t_j \text{ for each edge } (T_i, T_j) \in E \\ & \text{(ii)} & t_i \leq D \text{ for each task } T_i \in V \\ & \text{(iii)} & u_i \geq \frac{1}{s_{max}} \text{ for each task } T_i \in V \end{array} \quad (2)$$

Here, $u_i = 1/s_i$ is the inverse of the speed to execute task T_i. We now have a convex optimization problem to solve, with linear constraints in the non-negative variables u_i and t_i. In fact, the objective function is a posynomial, so we have a geometric programming problem (see [10, Sect. 4.5]) for which efficient numerical schemes exist. However, as illustrated on simple fork graphs, the optimal speeds are not expected to be rational numbers but instead arbitrarily complex expressions (we have the cubic root of the sum of cubes for forks, and nested expressions of this form for trees). From a computational complexity point of view, we do not know how to encode such numbers in polynomial size of the input (the rational task weights and the execution deadline). Still, we can always solve the problem numerically and get fixed-size numbers that are good approximations of the optimal values.

3.3 *Discrete Models*

In this section, we present complexity results on the three energy models with a finite number of possible speeds. The only polynomial instance is for the VDD-HOPPING model, for which we write a linear program in Sect. 3.3.1. Then, we give NP-completeness and approximation results in Sect. 3.3.2, for the DISCRETE and INCREMENTAL models.

3.3.1 The VDD-HOPPING Model

Theorem 2 *With the* VDD-HOPPING *model,* MINENERGY*(G,D) can be solved in polynomial time.*

Proof Let G be the execution graph of an application with n tasks, and D a deadline. Let $s_1, \ldots, s_m$ be the set of possible processor speeds. We use the following rational variables: for $1 \leq i \leq n$ and $1 \leq j \leq m$, b_i is the starting time of the execution of task T_i, and $\alpha_{(i,j)}$ is the time spent at speed s_j for executing task T_i. There are $n + n \times m = n(m + 1)$ such variables. Note that the total execution time of task T_i is $\Sigma_{j=1}^{m} \alpha_{(i,j)}$. The constraints are:

- $\forall 1 \leq i \leq n,\ b_i \geq 0$: starting times of all tasks are non-negative numbers;
- $\forall 1 \leq i \leq n,\ b_i + \Sigma_{j=1}^{m} \alpha_{(i,j)} \leq D$: the deadline is not exceeded by any task;
- $\forall 1 \leq i, i' \leq n$ such that $T_i \rightarrow T_{i'},\ t_i + \Sigma_{j=1}^{m} \alpha_{(i,j)} \leq t_{i'}$: a task cannot start before its predecessor has completed its execution;
- $\forall 1 \leq i \leq n,\ \Sigma_{j=1}^{m} \alpha_{(i,j)} \times s_j \geq w_i$: task T_i is completely executed.

The objective function is then $\min \left(\Sigma_{i=1}^{n} \Sigma_{j=1}^{m} \alpha_{(i,j)} s_j^3 \right)$.

The size of this linear program is clearly polynomial in the size of the instance, all $n(m + 1)$ variables are rational, and therefore it can be solved in polynomial time [40]. □

3.3.2 NP-Completeness and Approximation Results

Theorem 3 *With the* INCREMENTAL *model (and hence the DISCRETE model),* MINENERGY*(G,D) is NP-complete.*

Next we explain, for the INCREMENTAL and DISCRETE models, how the solution to the NP-hard problem can be approximated. Note that, given an execution graph and a deadline, the optimal energy consumption with the CONTINUOUS model is always lower than that with the other models, which are more constrained.

Theorem 4 *With the* INCREMENTAL *model, for any integer $K > 0$, the* MINENERGY*(G,D) problem can be approximated within a factor $(1+\frac{\delta}{s_{min}})^2(1+\frac{1}{K})^2$, in a time polynomial in the size of the instance and in K.*

Proposition 3

- *For any integer $\delta > 0$, any instance of* MINENERGY*(G,D) with the* CONTINUOUS *model can be approximated within a factor $(1+\frac{\delta}{s_{min}})^2$ in the* INCREMENTAL *model with speed increment δ.*
- *For any integer $K > 0$, any instance of* MINENERGY*(G,D) with the* DISCRETE *model can be approximated within a factor $(1 + \frac{\alpha}{s_1})^2(1 + \frac{1}{K})^2$, with $\alpha = \max_{1 \leq i < m}\{s_{i+1} - s_i\}$, in a time polynomial in the size of the instance and in K.*

3.4 Final Remarks

In this section, we have assessed the tractability of a classical scheduling problem, with task preallocation, under various energy models. We have given several results related to CONTINUOUS speeds. However, while these are of conceptual importance, they cannot be achieved with physical devices, and we have analyzed several models enforcing a bounded number of achievable speeds. In the classical DISCRETE model that arises from DVFS techniques, admissible speeds can be irregularly distributed, which motivates the VDD-HOPPING approach that mixes two consecutive speeds While computing optimal speeds is NP-hard with discrete speeds, it has polynomial complexity when mixing speeds. Intuitively, the VDD-HOPPING approach allows for smoothing out the discrete nature of the speeds. An alternate (and simpler in practice) solution to VDD-HOPPING is the INCREMENTAL model, where one sticks with unique speeds during task execution as in the DISCRETE model, but where consecutive speeds are regularly spaced. Such a model can be made arbitrarily efficient, according to our approximation results. Altogether, these results have laid the theoretical foundations for a comparative study of energy models.

4 Replica Placement

In this section, we revisit the well-known replica placement problem in tree networks [8, 16, 43], with two new objectives: reusing pre-existing replicas, and enforcing an efficient power management. In a nutshell, the replica placement problem is the following: we are given a tree-shaped network where clients are periodically issuing requests to be satisfied by servers. The clients are known (both their position in the tree and their number of requests), while the number and location of the servers are to be determined. A client is a leaf node of the tree, and its requests can be served by one internal node. Note that the distribution tree (clients and nodes) is fixed in the approach. This key assumption is quite natural for a broad spectrum of applications, such as electronic, ISP, or VOD service delivery (see [16, 27, 33] and additional references in [43]). The root server has the original copy of the database but cannot serve all clients directly, so a distribution tree is deployed to provide a hierarchical and distributed access to replicas of the original data.

In the original problem, there is no replica before execution; when a node is equipped with a replica, it can process a number of requests, up to its capacity limit. Nodes equipped with a replica, also called servers, serve all the clients located in their subtree (so that the root, if equipped with a replica, can serve any client). The rule of the game is to assign replicas to nodes so that the total number of replicas is minimized. This problem is well understood: it can be solved in time $O(N^2)$ (dynamic programming algorithm of [16]), or even in time $O(N \log N)$ (optimized greedy algorithm of [43]), where N is the number of nodes.

We study in this section a more realistic model of the replica placement problem, for a dynamic setting and accounting for the energy consumption. The first contribution is to tackle the replica placement problem when the tree is equipped with pre-existing replicas before execution. This extension is a first step towards dealing with dynamic replica management: if the number and location of client requests evolve over time, the number and location of replicas must evolve accordingly, and one must decide how to perform a configuration change (at what cost?) and when (how frequently reconfigurations should occur?).

Another contribution of this section is to extend replica placement algorithms to cope with power consumption constraints. Minimizing the total power consumed by the servers has recently become a very important objective, both for economic and environmental reasons [35]. To help reduce power dissipation, processors equipped with Dynamic Voltage and Frequency Scaling technique are used, and we assume that they follow the DISCRETE model. An important result of this section is that minimizing power consumption is an NP-complete problem, independently of the incurred cost (in terms of new and pre-existing servers) of the solution. In fact, this result holds true even without pre-existing replicas, and without static power: balancing server speeds across the tree already is a hard combinatorial problem.

The cost of the best power-efficient solution may indeed be prohibitive, which calls for a bi-criteria approach: minimizing power consumption while enforcing a threshold cost that cannot be exceeded. We investigate the case where there is only a fixed number of speeds and show that there are polynomial-time algorithms capable of optimizing power for a bounded cost, even with pre-existing replicas, with static power and with a complex cost function. This result has a great practical significance, because state-of-the-art processors can only be operated with a restricted number of voltage levels, hence with a few speeds [23, 24].

Finally, we run simulations to show the practical utility of our algorithms, despite their high worst-case complexity. We illustrate the impact of taking pre-existing servers into account, and how power can be saved thanks to the optimal bi-criteria algorithm.

The rest of the section is organized as follows. Section 4.1 is devoted to a detailed presentation of the target optimization problems, and provides a summary of new complexity results. The next two sections are devoted to the proofs of these results: Section 4.2 deals with computing the optimal cost of a solution, with pre-existing replicas in the tree, while Sect. 4.3 addresses all power-oriented problems. We report the simulation results in Sect. 4.4. Finally, we state some concluding remarks in Sect. 4.5.

4.1 Framework

This section is devoted to a precise statement of the problem. We start with the general problem without power consumption constraints, and next we recall the DISCRETE model of power consumption. Then we state the objective functions (with or without power), and the associated optimization problems. Finally we give a summary of all complexity results that we provide in the section.

4.1.1 Replica Servers

We consider a distribution tree whose nodes are partitioned into a set of clients $\mathcal{C}$, and a set of N nodes, $\mathcal{N}$. The clients are leaf nodes of the tree, while $\mathcal{N}$ is the set of internal nodes. Each client $i \in \mathcal{C}$ (leaf of the tree) is sending r_i requests per time unit to a database object. Internal nodes equipped with a replica (also called *servers*) will process all requests from clients in their subtree. An internal node $j \in \mathcal{N}$ may have already been provided with a replica, and we let $\mathcal{E} \subseteq \mathcal{N}$ be the set of pre-existing servers. Servers in $\mathcal{E}$ will be either reused or deleted in the solution. Note that it would be easy to allow *client-server* nodes which play both the rule of a client and of a node (possibly a server), by dividing such a node into two distinct nodes in the tree.

Without power consumption constraints, the problem is to find a *solution*, i.e., a set of servers capable of handling all requests, that minimizes some cost function. We formally define a valid solution before detailing its cost. We start with some notations. Let r be the root of the tree. If $j \in \mathcal{N}$, then $\mathsf{children}_j \subseteq \mathcal{N} \cup \mathcal{C}$ is the set of children of node j, and $\mathsf{subtree}_j \subseteq \mathcal{N} \cup \mathcal{C}$ is the subtree rooted in j, excluding j. A solution is a set $\mathcal{R} \subseteq \mathcal{N}$ of servers. Each client i is assigned a single server $\mathsf{server}_i \in \mathcal{R}$ that is responsible for processing all its r_i requests, and this server is restricted to be the first ancestor of i (i.e., the first node in the unique path that leads from i up to the root r) equipped with a server (hence the name *closest* for the request service policy). Such a server must exist in $\mathcal{R}$ for each client. In addition, all servers are identical and have a limited capacity, i.e., they can process a maximum number W of requests. Let req_j be the number of requests processed by $j \in \mathcal{R}$. The capacity constraint writes

$$\forall j \in \mathcal{R}, \;\; \mathsf{req}_j = \sum_{i \in \mathcal{C} \mid j = \mathsf{server}_i} r_i \leq W. \tag{3}$$

Now for the cost function, because all servers are identical, the cost of operating a server can be normalized to 1. When introducing a new server, there is an additional cost create, so that running a new server costs $1 + \mathsf{create}$ while reusing a server in $\mathcal{E}$ only costs 1. There is also a deletion cost delete associated to deleting each server in $\mathcal{E}$ that is not reused in the solution. Let $E = |\mathcal{E}|$ be the number of pre-existing servers. Let $R = |\mathcal{R}|$ be the total number of servers in the solution, and $e = |\mathcal{R} \cap \mathcal{E}|$ be the number of reused servers. Altogether, the cost is

$$\mathsf{cost}(\mathcal{R}) = R + (R - e) \times \mathsf{create} + (E - e) \times \mathsf{delete}. \tag{4}$$

This cost function is quite general. Because of the create and delete costs, priority is always given to reusing pre-existing servers. If $\mathsf{create} + 2 \times \mathsf{delete} < 1$, priority is given to minimizing the total number of servers R: indeed, if this condition holds, it is always advantageous to replace two pre-existing servers by a new one (if capacities permit).

4.1.2 With Power Consumption

With power consumption constraints, we assume that servers may operate under a set $\mathcal{M} = \{W_1, \ldots, W_M\}$ of different speeds, depending upon the number of requests that they have to process per time unit. Here speeds are indexed according to increasing values, and $W_M = W$, the maximal capacity. If a server $j \in \mathcal{R}$ processes req_j requests, with $W_{i-1} < \mathsf{req}_j \leq W_i$, then it is operated at speed W_i, and we let $\mathsf{speed}(j) = i$. The power consumption of a server $j \in \mathcal{R}$ obeys the DISCRETE model

$$\mathcal{P}(j) = \mathcal{P}^{(\mathsf{static})} + W^3_{\mathsf{speed}(j)}.$$

The total power consumption $\mathcal{P}(\mathcal{R})$ of the solution is the sum of the power consumption of all server nodes:

$$\mathcal{P}(\mathcal{R}) = \sum_{j \in \mathcal{R}} \mathcal{P}(j) = R \times \mathcal{P}^{(\mathsf{static})} + \sum_{j \in \mathcal{R}} W^3_{\mathsf{speed}(j)}. \tag{5}$$

With different power speeds, it is natural to refine the cost function, and to include a cost for changing the speed of a pre-existing server (upgrading it to a higher speed, or downgrading it to a lower speed). In the most detailed model, we would introduce create_i, the cost for creating a new server operated at speed W_i, $\mathsf{changed}_{i,i'}$, the cost for changing the speed of a pre-existing server from W_i to $W_{i'}$, and delete_i, the cost for deleting a pre-existing server operated at speed W_i.

Note that it is reasonable to let $\mathsf{changed}_{i,i} = 0$ (no change); values of $\mathsf{changed}_{i,i'}$ with $i < i'$ correspond to upgrade costs, while values with $i' < i$ correspond to downgrade costs. In accordance with these new cost parameters, given a solution $\mathcal{R}$, we count the number of servers as follows:

- n_i, the number of new servers operated at speed W_i;
- $e_{i,i'}$, the number of reused pre-existing servers whose operation speeds have changed from W_i to $W_{i'}$; and
- k_i, the number of pre-existing server operated at speed W_i that have not been reused.

The cost of the solution $\mathcal{R}$ with a total of $R = \Sigma_{i=1}^M n_i + \Sigma_{i=1}^M \Sigma_{i'=1}^M e_{i,i'}$ servers becomes:

$$\begin{aligned}\mathsf{cost}(\mathcal{R}) = R + \sum_{i=1}^{M} \mathsf{create}_i \times n_i + \sum_{i=1}^{M} \mathsf{delete}_i \times k_i \\ + \sum_{i=1}^{M} \sum_{i'=1}^{M} \mathsf{changed}_{i,i'} \times e_{i,i'}.\end{aligned} \tag{6}$$

Of course, this complicated cost function can be simplified to make the model more tractable; for instance all creation costs create_i can be set identical, all deletion costs delete_i can be set identical, all upgrade and downgrade values $\mathsf{changed}_{i,i'}$ can be set identical, and the latter can even be neglected.

4.1.3 Objective Functions

Without power consumption constraints, the objective is to minimize the cost, as defined by Eq. (4). We distinguish two optimization problems, either with pre-existing replicas in the tree or without:

- MINCOST-NOPRE, the classical cost optimization problem [16] without pre-existing replicas. Indeed, in that case, Eq. (4) reduces to finding a solution with the minimal number of servers.
- MINCOST-WITHPRE, the cost optimization problem with pre-existing replicas.

With power consumption constraints, the first optimization problem is MINPOWER, which stands for minimizing power consumption, independently of the incurred cost. But the cost of the best power-efficient solution may indeed be prohibitive, which calls for a bi-criteria approach: MINPOWER-BOUNDEDCOST is the problem to minimize power consumption while enforcing a threshold cost that cannot be exceeded. This bi-criteria problem can be declined in two versions, without pre-existing replicas (MINPOWER-BOUNDEDCOST-NOPRE) and with pre-existing replicas (MINPOWER-BOUNDEDCOST-WITHPRE).

4.1.4 Summary of Results

In this section, we prove the following complexity results for a tree with N nodes:

Theorem 5 MINCOST-WITHPRE can be solved in polynomial time with a dynamic programming algorithm whose worst case complexity is $O(N^5)$.

Theorem 6 MINPOWER is NP-complete.

Theorem 7 With a constant number M of speeds, both versions of MINPOWER-BOUNDEDCOST can be solved in polynomial time with a dynamic programming algorithm. The complexity of this algorithm is $O(N^{2M+1})$ for MINPOWER-BOUNDEDCOST-NOPRE and $O(N^{2M^2+2M+1})$ for MINPOWER-BOUNDEDCOST-WITHPRE.

Note that MINPOWER remains NP-complete without pre-existing replicas, and without static power: the proof of Theorem 6 (see Sect. 4.3.2) shows that balancing server speeds across the tree already is a hard combinatorial problem. On the contrary, with a fixed number of speeds, there are polynomial-time algorithms capable of optimizing power for a bounded cost, even with pre-existing replicas, with static power and with a complex cost function. These algorithms can be viewed as pseudo-polynomial solutions to the MINPOWER-BOUNDEDCOST problems.

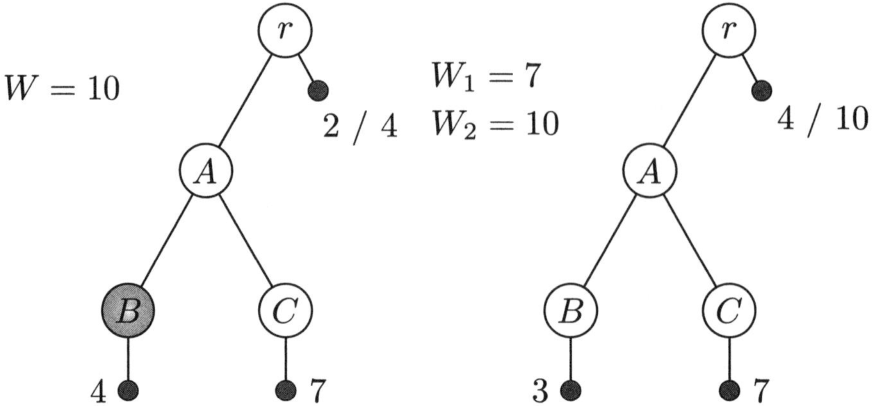

Reusing pre-existing replicas. Minimizing power consumption.
a **b**

Fig. 2 Examples

4.2 Complexity Results: Update Strategies

In this section, we focus on the MINCOST-WITHPRE problem: we need to update the set of replicas in a tree, given a set of pre-existing servers, so as to minimize the cost function.

In Sect. 4.2.1, we show on an illustrative example that the strategies need to trade-off between reusing resources and load-balancing requests on new servers: the greedy algorithm proposed in [43] for the MINCOST-NOPRE problem is no longer optimal. We provide in Sect. 4.2.2 a dynamic programming algorithm that returns the optimal solution in polynomial time, and we prove its correctness.

4.2.1 Running Example

We consider the example of Fig. 2a. There is one pre-existing replica in the tree at node B, and we need to decide whether to reuse it or not. For taking decisions locally at node A, the trade-off is the following:

- either we keep server B, and there are 7 requests going up in the tree from node A;
- either we remove server B and place a new server at node C, hence having only 4 requests going up in the tree from node A;
- either we keep the replica at node B and add one at node A or C, thereby having no traversing request any more.

The choice cannot be made locally, since it depends upon the remainder of the tree: if the root r has two client requests, then it was better to keep the pre-existing server B.

However, if it has four requests, two new servers are needed to satisfy all requests, and one can then remove server B which becomes useless (i.e., keep one server at node C and one server at node r).

From this example, it seems very difficult to design a greedy strategy to minimize the solution cost, while accounting for pre-existing replicas. We propose in the next section a dynamic programming algorithm that solves the MINCOST-WITHPRE problem.

4.2.2 Dynamic Programming Algorithm

Let W be the total number of requests that a server can handle, and r_i the number of requests issued by client $i \in \mathcal{C}$.

At each node $j \in \mathcal{N}$, we fill a table of maximum size $(E + 1) \times (N - E + 1)$ which indicates, for exactly $0 \leq e \leq E$ existing servers and $0 \leq n \leq N - E$ new servers in the subtree rooted in j (excluding j), the solution which leads to the minimum number of requests that have not been processed in the subtree. This solution for (e, n) values at node j is characterized by the minimum number of requests that is obtained, $minr^j_{(e,n)}$, and by the number of requests processed at each node $j' \in \mathsf{subtree}_j$, $req^j_{(e,n)}(j')$. Note that each entry of the table has a maximum size $O(N)$ (in particular, this size is reached at the root of the tree). The req variables ensure that it is possible to reconstruct the solution once the traversal of the tree is complete.

First, tables are initialized to default values (no solution). We set $minr^j_{(e,n)} = W + 1$ to indicate that there is no solution, because in any valid solution, we have $minr^j_{(e,n)} \leq W$. The main algorithm then fills the tables while performing a bottom-up traversal of the tree, and the solution can be found within the table of the root node. Initially, we fill the table for nodes j which have only client nodes: $minr^j_{(0,0)} = \Sigma_{i \in \mathsf{children}_j \cap \mathcal{C}} r_i$, and $minr^j_{(k,l)} = W + 1$ for $k > 0$ or $l > 0$. There are no nodes in the subtree of j, thus no req variables to set. The variable $client(j)$ keeps track of the number of requests directly issued by a client at node j. Also, recall that the decision whether to place a replica at node j or not is not accounted for in the table of j, but when processing the parent of node j.

Then, for a node $j \in \mathcal{N}$, we perform the same initialization, before processing children nodes one by one. To process child i of node j, first, we copy the current table of node j into a temporary one, with values $tminr$ and $treq$. Note that the table is initially almost empty, but this copy is required since we process children one after the other, and when we merge the kth children node of j, the table of j already contains information from the merge with the previous $k - 1$ children nodes. Then, for $0 \leq e \leq E$ and $0 \leq n \leq N - E$, we need to compute the new $minr^j_{(e,n)}$, and to update the $req^j_{(e,n)}$ values. We try all combinations with e' existing replicas and n' new replicas in the temporary table (i.e., information about children already processed), $e - e'$ existing replicas and $n - n'$ new replicas in the subtree of child i. We furthermore try solutions with a replica placed at node i, and we account for it in the value of e if $i \in \mathcal{E}$ (i.e., for a given value e', we place only $e - e' - 1$ replica in the subtree of i, plus one on i); otherwise we account for it in the value of n. Each time we find a solution which is better than the one previously in the table (in terms of $minr$), we copy the values of req from the temporary table and the table of i, in order to retain all the information about the current best solution. The key of the algorithm resides in the fact that during this *merging* process, the optimal solution will always be one which lets the minimum of requests pass through the subtree (see Lemma 2).

The solution to the replica placement problem with pre-existing servers MINCOST-WITHPRE is computed by scanning all solutions in order to return a valid one of minimum cost. To prove that the algorithm returns an optimal solution, we show first that the solutions that are discarded while filling the tables, never lead to a better solution than the one that is finally returned:

Lemma 2 Consider a subtree rooted at node $j \in \mathcal{N}$. If an optimal solution uses e pre-existing servers and places n new servers in this subtree, then there exists an optimal solution of same cost, for which the placement of these servers minimizes the number of requests traversing j.

Proof Let $\mathcal{R}_{opt}$ be the set of replicas in the optimal solution with (e, n) servers (i.e., e pre-existing and n new in $\mathsf{subtree}_j$). We denote by $rmin$ the minimum number of requests that must traverse j in a solution using (e, n) servers, and by $\mathcal{R}_{loc}$ the corresponding (local) placement of replicas in $\mathsf{subtree}_j$.

If $\mathcal{R}_{opt}$ is such that more than $rmin$ requests are traversing node j, we can build a new global solution which is similar to $\mathcal{R}_{opt}$, except for the subtree rooted in j for which we use the placement of $\mathcal{R}_{loc}$. The cost of the new solution is identical to the cost of $\mathcal{R}_{opt}$, therefore it is an optimal solution. It is still a valid solution, since $\mathcal{R}_{loc}$ is a valid solution and there are less requests than before to handle in the remaining of the tree (only $rmin$ requests traversing node j).

This proves that there exists an optimal solution which minimizes the number of requests traversing each node, given a number of pre-existing and new servers. □

The algorithm computes all local optimal solutions for all values (e, n). During the merge procedure, we try all possible numbers of pre-existing and new servers in each subtree, and we minimize the number of traversing requests, thus finding an optimal local solution. Thanks to Lemma 2, we know that there is a global optimal solution which builds upon these local optimal solutions.

We can show that the execution time of this algorithm is in $O(N \times (N - E + 1)^2 \times (E+1)^2)$, where N is the total number of nodes, and E is the number of pre-existing nodes. This corresponds to the N calls to the merging procedure. The algorithm is therefore of polynomial complexity, at most $O(N^5)$ for a tree with N nodes. This concludes the proof of Theorem 5. For a formalization of the algorithm and the details about its execution time, please refer to [9].

4.3 Complexity Results with Power

In this section, we tackle the MINPOWER and MINPOWER-BOUNDEDCOST problems. First in Sect. 4.3.1, we use an example to show why minimizing the number of requests traversing the root of a subtree is no longer optimal, and we illustrate the difficulty to take local decisions even when restricting to the simpler mono-criterion MINPOWER problem. Then in Sect. 4.3.2, we prove the NP-completeness of the latter problem with an arbitrary number of speeds (Theorem 6). However, we propose a pseudo-polynomial algorithm to solve the problem in Sect. 4.3.3. This algorithm turns out to be polynomial when the number of speeds is constant, hence usable in a realistic setting with two or three speeds (Theorem 7).

4.3.1 Running Example

Consider the example of Fig. 2b. There are two speeds, $W_1 = 7$ and $W_2 = 10$, and we focus on the power minimization problem. We assume that the power consumption of a node running at speed W_i is $400 + W_i^3$, for $i = 1, 2$ (400 is the static power). We consider the subtree rooted in A. Several decisions can be taken locally:

- place a server at node A, running at speed W_2, hence minimizing the number of traversing requests. Another solution without traversing requests is to have two servers, one at node B and one at node C, both running at speed W_1, but this would lead to a higher power consumption, since $800 + 2 \times 7^3 > 400 + 10^3$;
- place a server running at speed W_1 at node C, thus having 3 requests going through node A.

The choice cannot be made greedily, since it depends upon the rest of the tree: if the root r has four client requests, then it is better to let some requests through (one server at node C), since it optimizes power consumption. However, if it has ten requests, it is necessary to have no request going through A, otherwise node r is not able to process all its requests.

From this example, it seems very hard to design a greedy strategy to minimize the power consumption. Similarly, if we would like to reuse the algorithm of Sect. 4.2 to solve the MINPOWER-BOUNDEDCOST-WITHPRE bi-criteria problem, we would need to account for speeds. Indeed, the best solution of subtree A with one server is no longer always the one that minimizes the number of requests (in this case, placing one server on node A), since it can be better for power consumption to let three requests traverse node A and balance the load upper in the tree.

We prove in the next section the NP-completeness of the problem, when the number of speeds is arbitrary. However, we can adapt the dynamic programming algorithm, which becomes exponential in the number of speeds, but hence remains polynomial for a constant number of speeds (see Sect. 4.3.3).

4.3.2 NP-Completeness of MINPOWER

In this section, we prove Theorem 6, i.e., the NP-completeness of the MINPOWER problem, even with no static power, when there is an arbitrary number of speeds.

Proof of Theorem 6 We consider the associated decision problem: given a total power consumption $\mathcal{P}$, is there a solution that does not consume more than $\mathcal{P}$?

First, the problem is clearly in NP: given a solution, i.e., a set of servers, and the speed of each server, it is easy to check in polynomial time that no capacity constraint is exceeded, and that the power consumption meets the bound.

To establish the completeness, we use a reduction from 2-Partition [19]. We consider an instance $\mathcal{I}_1$ of 2-Partition: given n strictly positive integers $a_1, a_2, \ldots, a_n$, does there exist a subset I of $\{1, \ldots, n\}$ such that $\Sigma_{i \in I} a_i = \Sigma_{i \notin I} a_i$? Let $S = \Sigma_{i=1}^{n} a_i$; we assume that S is even (otherwise there is no solution).

We build an instance $\mathcal{I}_2$ of our problem where each server has $n + 2$ speeds. We assume that the a_is are sorted in increasing order, i.e., $a_1 \leq \cdots \leq a_n$. The speeds are then, in increasing order:

- $W_1 = K$;
- $\forall 1 \leq i \leq n,\ W_{i+1} = K + a_i \times X$;
- $W_{n+2} = K + S \times X$;

where the values of K and X will be determined later.

We furthermore set that there is no static power, and the power consumption for a server running at capacity W_i is therefore $\mathcal{P}_i = W_i^3$. The idea is to have K large and X small, so that we have an upper bound on the power consumed by a server running at capacity W_{i+1}, for $1 \leq i \leq n$:

$$W_{i+1}^3 = (K + a_i \times X)^3 \leq K^3 + a_i + \frac{1}{n}. \tag{7}$$

To ensure that Eq. (7) is satisfied, we set

$$X = \frac{1}{3 \times K^2},$$

and then we have $(K + a_i \times X)^3 = K^3(1 + \frac{a_i}{3K^3})^3$, with $K > S$ and therefore $\frac{a_i}{3K^3} < 1$. We set $x_i = \frac{a_i}{3K^3}$, and we want to ensure that:

$$(1 + x_i)^3 \leq 1 + 3 \times x_i + \frac{1}{n \times K^3}. \tag{8}$$

To do so, we study the function

$$f(x) = (1 + x)^3 - (1 + 3 \times x) - 5x^2,$$

and we show that $f(x) \leq 0$ for $x \leq \frac{1}{2}$ (thanks to the term in $-5x^2$). We have $f(0) = 0$, and $f'(x) = 3(1 + x)^2 - 3 - 10x$. We have $f'(0) = 0$, and $f''(x) = 6(1 + x) - 10$.

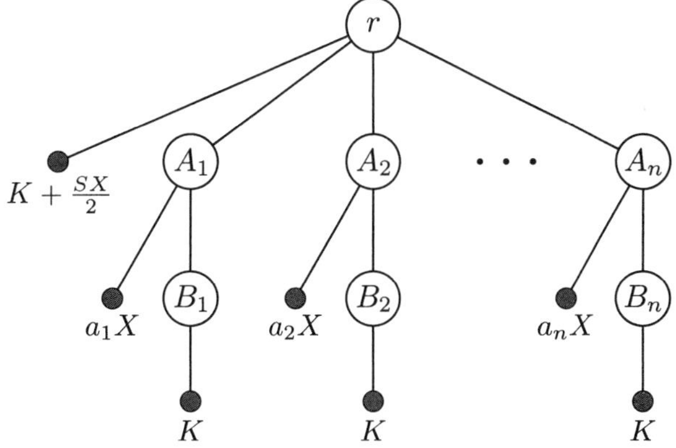

Fig. 3 Illustration of the NP-completeness proof

For $x \leq \frac{1}{2}$, $f''(x) < 0$. We deduce that $f'(x)$ is non increasing for $x \leq \frac{1}{2}$, and since $f'(0) = 0$, $f'(x)$ is negative for $x \leq \frac{1}{2}$.

Finally, $f(x)$ is non increasing for $x \leq \frac{1}{2}$, and since $f(0) = 0$, we have $(1+x)^3 < (1 + 3 \times x) + 5x^2$ for $x \leq \frac{1}{2}$.

Equation (8) is therefore satisfied if $5x_i^2 \leq \frac{1}{n \times K^3}$, i.e., $K^3 \geq \frac{5a_i^2 \times n}{3^2}$. This condition is satisfied for

$$K = n \times S^2,$$

and we then have $x_i < \frac{1}{2}$, which ensures that the previous reasoning was correct. Finally, with these values of K and X, Eq. (7) is satisfied.

Then, the distribution tree is the following: the root node r has one client with $K + \frac{S}{2} \times X$ requests, and n children $A_1, \ldots, A_n$. Each node A_i has a client with $a_i \times X$ requests, and a children node B_i which has K requests. Figure 3 illustrates the instance of the reduction.

Finally, we ask if we can find a placement of replicas with a maximum power consumption of:

$$\mathcal{P}_{max} = (K + S \times X)^3 + n \times K^3 + \frac{S}{2} + \frac{n-1}{n}.$$

Clearly, the size of $\mathcal{I}_2$ is polynomial in the size of $\mathcal{I}_1$, since K and X are of polynomial size. We now show that $\mathcal{I}_1$ has a solution if and only if $\mathcal{I}_2$ does.

Let us assume first that $\mathcal{I}_1$ has a solution, I. The solution for $\mathcal{I}_2$ is then as follows: there is one server at the root, running at capacity W_{n+2}. Then, for $i \in I$, we place a server at node A_i running at capacity W_{1+i}, while for $i \notin I$, we place a server at node B_i running at capacity W_1. It is easy to check that all capacity constraints are satisfied for nodes A_i and B_i. At the root of the tree, there are $K + \frac{S}{2} \times X + \Sigma_{i \notin I} a_i \times X$, which sums up to $K + S \times X$. The total power consumption is then $\mathcal{P} = (K + S \times X)^3 + \Sigma_{i \in I}(K + a_i \times X)^3 + \Sigma_{i \notin I} K^3$. Thanks to Eq. (7), $\mathcal{P} \leq (K + S \times X)^3 + \Sigma_{i \in I} \left(K^3 + a_i + \frac{1}{n}\right) + \Sigma_{i \notin I} K^3$, and finally, $\mathcal{P} \leq (K + S \times X)^3 + n \times K^3 + \Sigma_{i \in I} a_i + \frac{n-1}{n}$. Since I is a solution to 2-Partition, we have $\mathcal{P} \leq \mathcal{P}_{max}$. Finally, $\mathcal{I}_2$ has a solution.

Suppose now that $\mathcal{I}_2$ has a solution. There is a server at the root node r, which runs at speed W_{n+2}, since this is the only way to handle its $K + \frac{S}{2} \times X$ requests. This server has a power consumption of $(K + S \times X)^3$. Then, there cannot be more than n other servers. Indeed, if there were $n + 1$ servers, running at the smallest speed W_1, their power consumption would be $(n + 1)K^3$, which is strictly greater than $n \times K^3 + \frac{S}{2} + 1$. Therefore, the power consumption would exceed $\mathcal{P}_{max}$. So, there are at most n extra servers.

Consider that there exists $i \in \{1, \ldots, n\}$ such that there is no server, neither on A_i nor on B_i. Then, the number of requests at node r is at least $2K$; however, $2K > W_{n+2}$, so the server cannot handle all these requests. Therefore, for each $i \in \{1, \ldots, n\}$, there is exactly one server either on A_i or on B_i. We define the set I as the indices for which there is a server at node A_i in the solution. Now we show that I is a solution to $\mathcal{I}_1$, the original instance of 2-Partition.

First, if we sum up the requests at the root node, we have:

$$K + \frac{S}{2} \times X + \sum_{i \notin I} a_i \times X \leq K + S \times X.$$

Therefore, $\Sigma_{i \notin I} a_i \leq \frac{S}{2}$.

Now, if we consider the power consumption of the solution, we have:

$$(K + S \times X)^3 + \sum_{i \in I} (K + a_i \times X)^3 + \sum_{i \notin I} K^3 \leq \mathcal{P}_{max}.$$

Let us assume that $\Sigma_{i \in I} a_i > \frac{S}{2}$. Since the a_i are integers, we have $\Sigma_{i \in I} a_i \geq \frac{S}{2} + 1$. It is easy to see that $(K + a_i \times X)^3 > K^3 + a_i$. Finally, $\Sigma_{i \in I}(K + a_i \times X)^3 + \Sigma_{i \notin I} K^3 \geq n \times K^3 + \Sigma_{i \in I} a_i \geq n \times K^3 + \frac{S}{2} + 1$. This implies that the total power consumption is greater than $\mathcal{P}_{max}$, which leads to a contradiction, and therefore $\Sigma_{i \in I} a_i \leq \frac{S}{2}$.

We conclude that $\Sigma_{i \notin I} a_i = \Sigma_{i \in I} a_i = \frac{S}{2}$, and so the solution I is a 2-Partition for instance $\mathcal{I}_1$. This concludes the proof. □

4.3.3 A Pseudo-polynomial Algorithm for MINPOWER-BOUNDEDCOST

In this section, we sketch how to adapt the algorithm of Sect. 4.2 to account for power consumption. As illustrated in the example of Sect. 4.3.1, the current algorithm may lead to a non-optimal solution for the power consumption if used only with the higher speed for servers. Therefore, we refine it and compute, in each subtree, the optimal solution with, for $1 \leq j, j' \leq M$,

- exactly n_j new servers running at speed W_j;
- exactly $e_{j,j'}$ pre-existing servers whose operation speeds have changed from W_j to $W_{j'}$.

Recall that we previously had only two parameters, N the number of new servers, and E the number of pre-existing servers, thus leading to a total of $(N - E + 1)^2 \times (E + 1)^2$

iterations for the merging. Now, the number of iterations is $(N - E + 1)^{2M} \times (E + 1)^{2M^2}$, since we have $2 \times M$ loops of maximum size $N - E + 1$ over the n_j and n'_j, and $2 \times M^2$ loops of maximum size $E + 1$ over the $e_{j,j'}$ and $e'_{j,j'}$.

The new algorithm is similar, except that during the merge procedure, we must consider the type of the current node that we are processing (existing or not), and furthermore set it to all possible speeds: we therefore add a loop of size M. The principle is similar, except that we need to have larger tables at each node, and to iterate over all parameters. The complexity of the N calls to this procedure is now in $O(N \times M \times (N - E + 1)^{2M} \times (E + 1)^{2M^2})$.

Of course, we need also to update the initialization and main procedures to account for the increasing number of parameters. For the algorithm, first we compute all costs, accounting for the cost of changing speeds, and then we scan all solutions, and return one whose cost is not greater than the threshold, and which minimizes the power consumption. The most time-consuming part of the algorithm is still the call to the merging procedure, hence a complexity in $O(N \times M \times (N - E + 1)^{2M} \times (E + 1)^{2M^2})$.

With a constant number of capacities, this algorithm is polynomial, which proves Theorem 7. For instance, with $M = 2$, the worst case complexity is $O(N^{13})$. Without pre-existing servers, this complexity is reduced to $O(N^5)$.

4.4 *Simulations*

In this section, we compare our algorithms with the algorithms of [43], which do not account for pre-existing servers and for power consumption. First in Sect. 4.4.1, we focus on the impact of pre-existing servers. Then we consider the power consumption minimization criterion in Sect. 4.4.2. Note that experiments have been run sequentially on an Intel Xeon 5250 processor.

4.4.1 Impact of Pre-existing Servers

In this set of experiments, we randomly build a set of distribution trees with $N = 100$ internal nodes of maximum capacity $W = 10$. Each internal node has between 6 and 9 children, and clients are distributed randomly throughout the tree: each internal node has a client with a probability 0.5, and this client has between 1 and 6 requests.

In the first experiment, we draw 200 random trees without any existing replica in them. Then we randomly add $0 \leq E \leq 100$ pre-existing servers in each tree. Finally, we execute both the greedy algorithm (GR) of [43], and the algorithm of Sect. 4.2 (DP) on each tree, and since both algorithms return a solution with the minimum number of replicas, the cost of the solution is directly related to the number of pre-existing replicas that are reused. Figure 4a shows the average number of pre-existing servers that are reused in each solution over the 200 trees, for each value of the number E of pre-existing servers. When the tree has a very small ($E \approx 0$) or very large ($E \approx N$) number of pre-existing replicas, both algorithms return the same

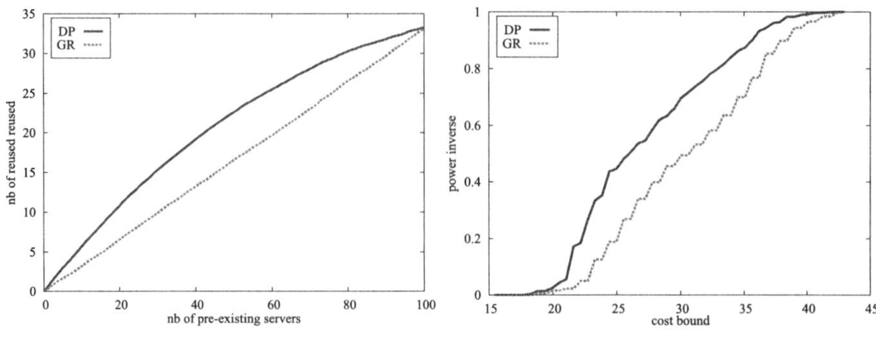

Experiment 1: increasing number of pre-existing servers.
a

Experiment 3: Power minimization.
b

Fig. 4 Experiments 1 and 3

solution. Still, DP achieves an average reuse of 4.13 more servers than GR, and it can reuse up to 15 more servers.

In a second experiment, we study the behavior of the algorithms in a *dynamic* setting, with 20 update steps. At each step, starting from the current solution, we update the number of requests per client and recompute an optimal solution with both algorithms, starting from the servers that were placed at the previous step. Initially, there are no pre-existing servers, and at each step, both algorithms obtain a different solution. However, they always reach the same total number of servers since they have the same requests; but after the first step, they may have a different set of pre-existing servers. Similarly to Experiment 1, the simulation is conducted on 200 distinct trees, and results are averaged over all trees. In Fig. 5 (left), at each step, we compare the number of existing replicas in the solutions found by the two algorithms, and hence the cost of the solutions. We plot the cumulative number of servers that have been reused so far (hence accounting for all previous steps). As expected, the DP algorithm makes a better reuse of pre-existing replicas. Figure 5 (right) compares, at each step, the number of pre-existing servers reused by DP and by GR. We count the average number of steps (over 20) at which each value is reached. It occasionally happens that the greedy algorithm performs a better reuse, because it is not starting from the same set of pre-existing servers, but overall this experiment confirms the better reuse of the dynamic programming algorithm, even when the algorithms are applied on successive steps.

Note however that taking pre-existing replicas into account has an impact on the execution time of the algorithm: in these experiments, GR runs in less than one second per tree, while DP takes around forty seconds per tree. Also, we point out that the shape of the trees does not seem to modify the results: we present in [9] similar results with trees where each node has between 2 and 4 children.

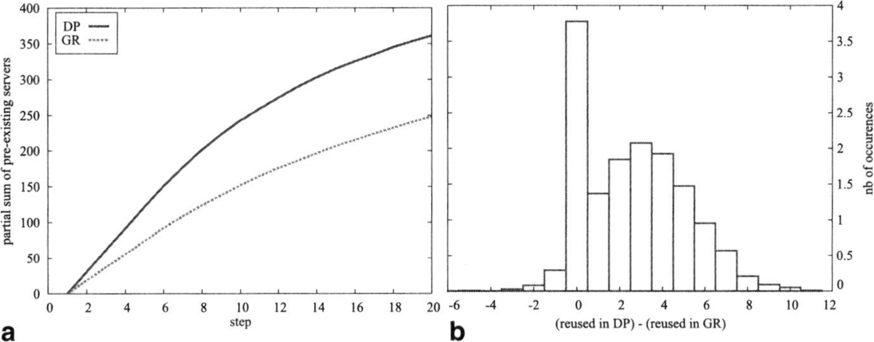

Fig. 5 Experiment 2: consecutive executions of the algorithms

4.4.2 With Power Consumption

To study the practical applicability of the bi-criteria algorithm (DP) for the MinPower-BoundedCost problem (see Sect. 4.3.3), we have implemented it with two speeds $W_1 = 5$ and $W_2 = 10$, and compared it with the algorithm in [43]; this algorithm does not account for power minimization, but minimizes the value of the maximal capacity W when given a cost bound. More precisely, in the experiment we try all values $5 \leq W \leq 10$, and compute the corresponding cost and power consumption. To be fair, when a server has 5 requests or less, we operate it under the first speed W_1. Given a bound on the cost, we keep the solution that minimizes the power consumption. We call GR this version of the algorithm in [43] modified for power as explained above.

We randomly build 100 trees with 50 nodes each, and we select 5 nodes as pre-existing servers. Clients have between 1 and 5 requests, so that a solution with replicas in the first speed can always be found. The cost function is such that, for any $i, i' \in \{1, 2\}$, $\mathsf{create}_i = 0.1$, $\mathsf{delete}_i = 0.01$ and $\mathsf{changed}_{i,i'} = 0.001$. The power consumed by a server in speed i is $\mathcal{P}_i = \frac{1}{10} W_1^3 + W_i^3$. In Fig. 4b, we plot the inverse of the power of a solution, given a bound on the cost (the higher the better). If the algorithm fails to find a solution for a tree, the value is 0, and we average the inverse of the power over the 100 trees, for both algorithms. For intermediate cost values, our algorithm is much better than the version of [43] in terms of power consumption: GR consumes in average more than 30 % more power than DP, when the cost bound is between 29 and 34.

Here again, it takes more time to obtain the optimal solution with DP than to run the greedy algorithm several times: GR runs in around 1 s per tree, while DP takes around 5 min per tree. Also, we have performed some more experiments with slightly different parameters, but got no significant differences, as is shown in [9], in particular with no pre-existing replicas at all.

4.4.3 Running Time of the Algorithms

Recall that the theoretical complexity of GR is of order $O(N \log N)$ (without power and without pre-existing servers), while DP is of order $O(N^5)$, both for the version with power (two speeds) but without pre-existing servers, and for the version without power but with pre-existing servers. In practice, the execution times of GR are always very small (a few milliseconds). For DP, we have plotted its execution time as a function of N (see [9]). Run time measurements show that the experimental values have a shape in N^5, which confirms the theoretical complexity. Moreover, our DP algorithms run in less than N^5 microseconds for reasonable values of N, which allows the use of these algorithms in practical situations.

Indeed, without power, we are able to process trees with 500 nodes and 125 pre-existing servers in 30 min; with power and no pre-existing server, we can process trees with 300 nodes in 1 h. The algorithm with power and pre-existing servers is the most time-consuming: it takes around 1 h to process a tree with 70 nodes and 10 pre-existing servers.

4.5 *Concluding Remarks*

In this section, we have addressed the problem of updating the placement of replicas in a tree network. We have provided an optimal dynamic programming algorithm whose cost is at most $O(N^5)$, where N is the number of nodes in the tree. This complexity may seem high for very large problem sizes, but our implementation of the algorithm is capable of managing trees with up to 500 nodes in half an hour, which is reasonable for a large spectrum of applications (e.g., such as database updates during the night).

The optimal placement update algorithm is a first step towards dealing with dynamic replica management. When client requests evolve over time, the placement of the replicas must be updated at regular intervals, and the overall cost is a trade-off between two extreme strategies: (i) "lazy" updates, where there is an update only when the current placement is no longer valid; the update cost is minimized, but changes in request volume and location since the last placement may well lead to poor resource usage; and (ii) systematic updates, where there is an update every time step; this leads to an optimized resource usage but encompasses a high update cost. Clearly, the rates and amplitudes of the variations of the number of requests issued by each client in the tree are very important to decide for a good update interval. Still, establishing the cost of an update is a key result to guide such a decision. When unfrequent updates are called for, or when resources have a high cost, the best solution is likely to use our optimal but expensive algorithm. On the contrary, with frequent updates or low-cost servers, we may prefer to resort to faster (but sub-optimal) update heuristics.

Our main contribution is to have provided the theoretical foundations for a single step reconfiguration, whose complexity is important to guide the design of lower-cost heuristics. Also, we have done a first attempt to take power consumption into account, in addition to usual performance-related objectives. Power consumption has become a very important concern, both for economic and environmental reasons, and it is important to account for it when designing replica placement strategies. Even though the optimal algorithms have a high worst-case complexity, we have successfully implemented all of them, including the most time-consuming scheme capable of optimizing power while enforcing a bounded cost that includes pre-existing servers. We were able to process trees with a reasonable number of nodes.

As future work, we plan to design polynomial-time heuristics with a lower complexity than the optimal solution. The idea would be to perform some local optimizations to better load-balance the number of requests per replica, with the goal of minimizing the power consumption. These heuristics should be tuned for dedicated applications, and should (hopefully!) build upon the fundamental results (complexity and algorithms) that we have provided in this section. Finally, it would be interesting to add more parameters in the model, such as the cost of routing, or the introduction of quality of service constraints.

5 Checkpointing Strategies

In this section, we give a motivating example of the use of the CONTINUOUS energy model introduced in Sect. 2. We aim at minimizing the energy consumption when executing a divisible workload under a bound on the total execution time, while resilience is provided through checkpointing. We discuss several variants of this multi-criteria problem. Given the workload W, we need to decide how many chunks to use, what are the sizes of these chunks, and at which speed each chunk is executed. Furthermore, since a failure may occur during the execution of a chunk, we also need to decide at which speed a chunk should be re-executed in the event of a failure. Using more chunks leads to a higher checkpoint cost, but smaller chunks imply less computation loss (and less re-execution) when a failure occurs. We assume that a chunk can fail only once, i.e., we re-execute each chunk at most once. Indeed, the probability that a fault would strike during both the first execution and the re-execution is negligible. The accuracy of this assumption is discussed in [3].

Due to the probabilistic nature of failure hits, it is natural to study the expectation $\mathbb{E}(E)$ of the energy consumption, because it represents the average cost over many executions. As for the bound D on execution time (the deadline), there are two relevant scenarios: either we enforce that this bound is a *soft deadline* to be met in expectation, or we enforce that this bound is a *hard deadline* to be met in the worst case. The former scenario corresponds to flexible environment where task deadlines can be viewed as average response times [11], while the latter scenario corresponds to real-time environments where task deadlines are always strictly enforced [41]. In both scenarios, we have to determine the number of chunks, their sizes, and the speed at which to execute (and possibly re-execute) every chunk. The different models are then compared through an extensive set of experiments.

5.1 Framework

First we formalize this important multi-objective problem. The general problem consists of finding n, the number of chunks, as well as the speeds for the execution and the re-execution of each chunk, both for soft and hard deadlines. We identify and discuss two important sub-cases that help tackling the most general problem instance: (i) a single chunk (the task is atomic); and (ii) re-execution speed is always identical to the first execution speed. The main notations are as follows: W is the total amount of work; s is the processor speed for first execution; σ is the processor speed for re-execution; T_C is the checkpointing time; and E_C is the energy spent for checkpointing.

5.1.1 Model

Consider first the case of a single chunk (or atomic task) of size W, denoted as SINGLECHUNK. We execute this chunk on a processor that can run at several speeds. We assume continuous speeds, i.e., the speed of execution can take an arbitrary positive real value. The execution is subject to failure, and resilience is provided through the use of checkpointing. The overhead induced by checkpointing is twofold: execution time T_C, and energy consumption E_C.

We assume that failures strike with uniform distribution, hence the probability that a failure occurs during an execution is linearly proportional to the length of this execution. Consider the first execution of a task of size W executed at speed s: the execution time is $T_{\text{exec}} = W/s + T_C$, hence the failure probability is $P_{\text{fail}} = \lambda T_{\text{exec}} = \lambda(W/s + T_C)$, where λ is the instantaneous failure rate. If there is indeed a failure, we re-execute the task at speed σ (which may or may not differ from s); the re-execution time is then $T_{\text{reexec}} = W/\sigma + T_C$ so that the expected execution time is

$$\begin{aligned}\mathbb{E}(T) &= T_{\text{exec}} + P_{\text{fail}} T_{\text{reexec}} \\ &= (W/s + T_C) + \lambda(W/s + T_C)(W/\sigma + T_C). \end{aligned} \tag{9}$$

Similarly, the worst-case execution time is

$$\begin{aligned}T_{wc} &= T_{\text{exec}} + T_{\text{reexec}} \\ &= (W/s + T_C) + (W/\sigma + T_C). \end{aligned} \tag{10}$$

Remember that we assume success after re-execution, so we do not account for second and more re-executions. Along the same line, we could spare the checkpoint after re-executing the last task in a series of tasks, but this unduly complicates the analysis. In [3], we show that this model with only a single re-execution is accurate up to second order terms when compared to the model with an arbitrary number of failures that follows an Exponential distribution of parameter λ.

What is the expected energy consumed during execution? The energy consumed during the first execution at speed s is $Ws^2 + E_C$, where E_C is the energy consumed during a checkpoint. The energy consumed during the second execution at speed σ is $W\sigma^2 + E_C$, and this execution takes place with probability $P_{\text{fail}} = \lambda T_{\text{exec}} = \lambda(W/s + T_C)$, as before. Hence the expectation of the energy consumed is

$$\mathbb{E}(E) = (Ws^2 + E_C) + \lambda\,(W/s + T_C)\left(W\sigma^2 + E_C\right). \tag{11}$$

With multiple chunks (MULTIPLECHUNKS model), the execution times (worst case or expected) are the sum of the execution times for each chunk, and the expected energy is the sum of the expected energy for each chunk (by linearity of expectations).

We point out that the failure model is coherent with respect to chunking. Indeed, assume that a divisible task of weight W is split into two chunks of weights w_1 and w_2 (where $w_1 + w_2 = W$). Then the probability of failure for the first chunk is $P^1_{\text{fail}} = \lambda(w_1/s + T_C)$ and that for the second chunk is $P^2_{\text{fail}} = \lambda(w_2/s + T_C)$. The probability of failure $P_{\text{fail}} = \lambda(W/s + T_C)$ with a single chunk differs from the probability of failure with two chunks only because of the extra checkpoint that is taken; if $T_C = 0$, they coincide exactly. If $T_C > 0$, there is an additional risk to use two chunks, because the execution lasts longer by a duration T_C. Of course this is the price to pay for a shorter re-execution time in case of failure: Equation 9 shows that the expected re-execution time is $P_{\text{fail}} T_{\text{reexec}}$, which is quadratic in W. There is a trade-off between having many small chunks (many T_Cs to pay, but small re-execution cost) and a few larger chunks (fewer T_Cs, but increased re-execution cost).

5.1.2 Optimization Problems

The optimization problem is stated as follows: given a deadline D and a divisible task whose total computational load is W, the problem is to partition the task into n chunks of size w_i, where $\Sigma_{i=1}^n w_i = W$, and choose for each chunk an execution speed s_i and a re-execution speed σ_i in order to minimize the expected energy consumption:

$$\mathbb{E}(E) = \sum_{i=1}^{n} (w_i s_i^2 + E_C) + \lambda\left(\frac{w_i}{s_i} + T_C\right)\left(w_i \sigma_i^2 + E_C\right),$$

subject to the constraint that the deadline is met either in expectation or in the worst case:

EXPECTED-DEADLINE $\quad \mathbb{E}(T) = \sum_{i=1}^n \left(\frac{w_i}{s_i} + T_C + \lambda\left(\frac{w_i}{s_i} + T_C\right)\left(\frac{w_i}{\sigma_i} + T_C\right)\right) \leq D$

HARD-DEADLINE $\quad T_{wc} = \sum_{i=1}^n \left(\frac{w_i}{s_i} + T_C + \frac{w_i}{\sigma_i} + T_C\right) \leq D$

The unknowns are the number of chunks n, the sizes of these chunks w_i, the speeds for the first execution s_i and the speeds for the second execution σ_i. We consider two variants of the problem, depending upon re-execution speeds:

- SINGLESPEED: in this simpler variant, the re-execution speed is always the same as the speed chosen for the first execution. We then have to determine a single speed for each chunk: $\sigma_i = s_i$ for all i.
- MULTIPLESPEEDS: in this more general variant, the re-execution speed is freely chosen, and there are two different speeds to determine for each chunk.

We also consider the variant with a single chunk (SINGLECHUNK), i.e., the task is atomic and we only need to decide for its execution speed (in the SINGLESPEED model), or for its execution and re-execution speeds (in the MULTIPLESPEEDS model). We start the study in Sect. 5.2 with this simpler problem.

5.2 *With a Single Chunk*

In this section, we consider the SINGLECHUNK model: given a non-divisible workload W and a deadline D, find the values of s and σ that minimize

$$\mathbb{E}(E) = (Ws^2 + E_C) + \lambda \left(\frac{W}{s} + T_C \right) \left(W\sigma^2 + E_C \right),$$

subject to

$$\mathbb{E}(T) = \left(\frac{W}{s} + T_C \right) + \lambda \left(\frac{W}{s} + T_C \right) \left(\frac{W}{\sigma} + T_C \right) \leq D$$

in the EXPECTED-DEADLINE model, and subject to

$$\frac{W}{s} + T_C + \frac{W}{\sigma} + T_C \leq D$$

in the HARD-DEADLINE model. We first deal with the SINGLESPEED model, where we enforce $\sigma = s$, before moving on to the MULTIPLESPEEDS model.

Note that the formal proofs of this section can be found in [3].

5.2.1 SINGLESPEED Model

In this section, we express $\mathbb{E}(E)$ as functions of the speed s. That is, $\mathbb{E}(E)(s) = (Ws^2 + E_C)(1 + \lambda(W/s + T_C))$. The following result is valid for both EXPECTED-DEADLINE and HARD-DEADLINE models.

Lemma 3 $\mathbb{E}(E)$ *is convex on* $\mathbb{R}_+^\star$. *It admits a unique minimum* $s^\star$ *which can be computed numerically.*

EXPECTED-DEADLINE: In the SINGLESPEED EXPECTED-DEADLINE model, we denote $\mathbb{E}(T)(s) = (W/s + T_C)(1 + \lambda(W/s + T_C))$ the constraint on the execution time.

Lemma 4 *For any D, if $T_C + \lambda {T_C}^2 \geq D$, then there is no solution. Otherwise, the constraint on the execution time can be rewritten as $s \in (s_0, +\infty($, where*

$$s_0 = W\frac{1 + 2\lambda T_C + \sqrt{4\lambda D + 1}}{2(D - T_C(1 + \lambda T_C))}. \tag{12}$$

Proposition 4 *In the* SINGLESPEED *model, it is possible to numerically compute the optimal solution for* SINGLECHUNK *as follows:*

1. *If $T_C + \lambda {T_C}^2 \geq D$, then there is no solution;*
2. *Else, the optimal speed is* $\max(s_0, s^\star)$.

HARD-DEADLINE In the HARD-DEADLINE model, the bound on the execution time can be written as $2\left(\frac{W}{s} + T_C\right) \leq D$.

Lemma 5 *In the* SINGLESPEED HARD-DEADLINE *model, for any D, if $2T_C \geq D$, then there is no solution. Otherwise, the constraint on the execution time can be rewritten as $s \in \left[\frac{W}{\frac{D}{2} - T_C}; +\infty\right($.*

Proposition 5 *Let $s^\star$ be the solution indicated in Lemma 3. In the* SINGLESPEED HARD-DEADLINE *model if $2T_C \geq D$, then there is no solution. Otherwise, the minimum is reached when $s = \max\left(s^\star, \frac{W}{\frac{D}{2} - T_C}\right)$.*

5.2.2 MULTIPLESPEEDS Model

In this section, we consider the general MULTIPLESPEEDS model. We use the following notations:

$$\mathbb{E}(E)(s, \sigma) = (Ws^2 + E_C) + \lambda(W/s + T_C)(W\sigma^2 + E_C).$$

EXPECTED-DEADLINE: The execution time in the MULTIPLESPEEDS EXPECTED-DEADLINE model can be written as

$$\mathbb{E}(T)(s, \sigma) = (W/s + T_C) + \lambda(W/s + T_C)(W/\sigma + T_C).$$

We start by giving a useful property, namely that the deadline is always tight in the MULTIPLESPEEDS EXPECTED-DEADLINE model:

Lemma 6 *In the* MULTIPLESPEEDS EXPECTED-DEADLINE *model, in order to minimize the energy consumption, the deadline should be tight.*

This lemma allows us to express σ as a function of s:

$$\sigma = \frac{\lambda W}{\frac{D}{\frac{W}{s}+T_C} - (1 + \lambda T_C)}.$$

Also we reduce the bi-criteria problem to the minimization problem of the single-variable function:

$$s \mapsto Ws^2 + E_C + \lambda \left(\frac{W}{s} + T_C \right) \left(W \left(\frac{\lambda W}{\frac{D}{\frac{W}{s}+T_C} - (1 + \lambda T_C)} \right)^2 + E_C \right), \quad (13)$$

which can be solved numerically.

HARD-DEADLINE In this model we have similar results as with EXPECTED-DEADLINE. The constraint on the execution time writes: $\frac{W}{s} + T_C + \frac{W}{\sigma} + T_C \leq D$.

Lemma 7 *In the* MULTIPLESPEEDS EXPECTED-DEADLINE *model, in order to minimize the energy consumption, the deadline should be tight.*

This lemma allows us to express σ as a function of s:

$$\sigma = \frac{W}{(D - 2T_C)s - W} s$$

Finally, we reduce the bi-criteria problem to the minimization problem of the single-variable function:

$$s \mapsto Ws^2 + E_C + \lambda \left(\frac{W}{s} + T_C \right) \left(W \left(\frac{W}{(D - 2T_C)s - W} s \right)^2 + E_C \right), \quad (14)$$

which can be solved numerically.

5.3 Several Chunks

In this section, we deal with the general problem of a divisible task of size W that can be split into an arbitrary number of chunks. We divide the task into n chunks of size w_i such that $\Sigma_{i=1}^n w_i = W$. Each chunk is executed once at speed s_i, and re-executed (if necessary) at speed σ_i. The problem is to find the values of n, w_i, s_i and σ_i that minimize

$$\mathbb{E}(E)=\sum_i \left(w_i s_i^2 + E_C\right) + \lambda \sum_i \left(\frac{w_i}{s_i} + T_C\right)\left(w_i \sigma_i^2 + E_C\right),$$

subject to

$$\sum_i \left(\frac{w_i}{s_i} + T_C\right) + \lambda \sum_i \left(\frac{w_i}{s_i} + T_C\right)\left(\frac{w_i}{\sigma_i} + T_C\right) \leq D$$

in the EXPECTED-DEADLINE model, and subject to

$$\sum_i \left(\frac{w_i}{s_i} + T_C\right) + \sum_i \left(\frac{w_i}{\sigma_i} + T_C\right) \leq D$$

in the HARD-DEADLINE model. We first deal with the SINGLESPEED model, where we enforce $\sigma_i = s_i$, before dealing with the MULTIPLESPEEDS model.

Note that the formal proofs of this section can be found in [3].

5.3.1 Single Speed Model

EXPECTED-DEADLINE In this section, we deal with the SINGLESPEED EXPECTED-DEADLINE model and consider that for all i, $\sigma_i = s_i$. Then:

$$\mathbb{E}(T)(\cup_i (w_i, s_i, s_i)) = \sum_i \left(\frac{w_i}{s_i} + T_C\right) + \lambda \sum_i \left(\frac{w_i}{s_i} + T_C\right)^2$$

$$\mathbb{E}(E)(\cup_i (w_i, s_i, s_i)) = \sum_i \left(w_i s_i^2 + E_C\right)\left(1 + \lambda \left(\frac{w_i}{s_i} + T_C\right)\right)$$

Theorem 8 *In the optimal solution to the problem with the* SINGLESPEED EXPECTED-DEADLINE *model, all n chunks are of equal size W/n and executed at the same speed s.*

Thanks to this result, we know that the problem with n chunks can be rewritten as follows: find s such that

$$n\left(\frac{W}{ns} + T_C\right) + n\lambda \left(\frac{W}{ns} + T_C\right)^2 = \frac{W}{s} + nT_C + \frac{\lambda}{n}\left(\frac{W}{s} + nT_C\right)^2 \leq D$$

in order to minimize

$$n\left(\frac{W}{n}s^2 + E_C\right) + n\lambda\left(\frac{W}{ns} + T_C\right)\left(\frac{W}{n}s^2 + E_C\right) = \left(Ws^2 + nE_C\right)\left(1 + \frac{\lambda}{n}\left(\frac{W}{s} + nT_C\right)\right).$$

One can see that this reduces to the SINGLECHUNK problem with the SINGLESPEED model (Sect. 5.2.1) up to the following parameter changes:

- $\lambda \leftarrow \frac{\lambda}{n}$ • $T_C \leftarrow nT_C$ • $E_C \leftarrow nE_C$

If the number of chunks n is given, we can express the minimum speed such that there is a solution with n chunks:

$$s_0(n) = W \frac{1 + 2\lambda T_C + \sqrt{4\frac{\lambda D}{n} + 1}}{2(D - nT_C(1 + \lambda T_C))}. \quad (15)$$

We can verify that when $D \leq nT_C(1 + \lambda n)$, there is no solution, hence obtaining an upper bound on n. Therefore, the two variables problem (with unknowns n and s) can be solved numerically.

HARD-DEADLINE: In the HARD-DEADLINE model, all results still hold, they are even easier to prove since we do not need to introduce a second speed.

Theorem 9 *In the optimal solution to the problem with the* SINGLESPEED HARD-DEADLINE *model, all n chunks are of equal size W/n and executed at the same speed s.*

5.3.2 Multiple Speeds Model

EXPECTED-DEADLINE In this section, we still deal with the problem of a divisible task of size W that we can split into an arbitrary number of chunks, but using the more general MULTIPLESPEEDS model. We start by proving that all re-execution speeds are equal:

Lemma 8 *In the* MULTIPLESPEEDS *model, all re-execution speeds are equal in the optimal solution: $\exists\sigma, \forall i, \sigma_i = \sigma$, and the deadline is tight.*

We can now redefine

$$\mathbb{E}(T)(\cup_i (w_i, s_i, \sigma_i)) = T(\cup_i (w_i, s_i), \sigma)$$
$$\mathbb{E}(E)(\cup_i (w_i, s_i, \sigma_i)) = E(\cup_i (w_i, s_i), \sigma)$$

Theorem 10 *In the* MULTIPLESPEEDS *model, all chunks have the same size $w_i = \frac{W}{n}$, and are executed at the same speed s, in the optimal solution.*

Thanks to this result, we know that the n chunks problem can be rewritten as follows: find s such that

- $\frac{W}{s} + nT_C + \frac{\lambda}{n}\left(\frac{W}{s} + nT_C\right)\left(\frac{W}{\sigma} + nT_C\right) = D$
- in order to minimize $Ws^2 + nE_C + \frac{\lambda}{n}\left(\frac{W}{s} + nT_C\right)\left(W\sigma^2 + nE_C\right)$

One can see that this reduces to the SINGLECHUNK MULTIPLESPEEDS EXPECTED-DEADLINE task problem where:

- $\lambda \leftarrow \frac{\lambda}{n}$ • $T_C \leftarrow nT_C$ • $E_C \leftarrow nE_C$

and allows us to write the problem to solve as a two-parameter function:

$$(n,s)\mapsto Ws^2+nE_C+\frac{\lambda}{n}\left(\frac{W}{s}+nT_C\right)\left(W\left(\frac{\frac{\lambda}{n}W}{\frac{D}{\frac{W}{s}+nT_C}-(1+\lambda T_C)}\right)^2+nE_C\right), \tag{16}$$

which can be minimized numerically.

Hard-Deadline In this section, the constraint on the execution time can be written as:

$$\sum_i\left(\frac{w_i}{s_i}+T_C+\frac{w_i}{\sigma_i}+T_C\right)\leq D.$$

Lemma 9 *In the* MULTIPLESPEEDS HARD-DEADLINE *model with divisible chunk, the deadline should be tight.*

Lemma 10 *In the optimal solution, for all* i, j, $\lambda\left(\frac{w_i}{s_i}+T_C\right)\sigma_i^3=\lambda\left(\frac{w_j}{s_j}+T_C\right)\sigma_j^3$.

Lemma 11 *If we enforce the condition that the execution speeds of the chunks are all equal, and that the re-execution speeds of the chunks are all equal, then all chunks should have same size in the optimal solution.*

We have not been able to prove a stronger result than Lemma 11. However we conjecture the following result:

Conjecture 1 *In the optimal solution of* MULTIPLESPEEDS HARD-DEADLINE, *the re-execution speeds are identical, the deadline is tight. The re-execution speed is equal to* $\sigma=\frac{W}{(D-2nT_C)s-W}s$. *Furthermore the chunks should have the same size* $\frac{W}{n}$ *and should be executed at the same speed* s.

This conjecture reduces the problem to the SINGLECHUNK MULTIPLESPEEDS problem where

- $\lambda\leftarrow\frac{\lambda}{n}$
- $T_C\leftarrow nT_C$
- $E_C\leftarrow nE_C$

and allows us to write the problem to solve as a two-parameter function:

$$(n,s)\mapsto Ws^2+nE_C+\frac{\lambda}{n}\left(\frac{W}{s}+nT_C\right)\left(W\left(\frac{W}{(D-2nT_C)s-W}s\right)^2+nE_C\right) \tag{17}$$

which can be solved numerically.

5.4 Simulations

5.4.1 Simulation Settings

We performed a large set of simulations in order to illustrate the differences between all the models studied in this paper, and to show to which extent each additional

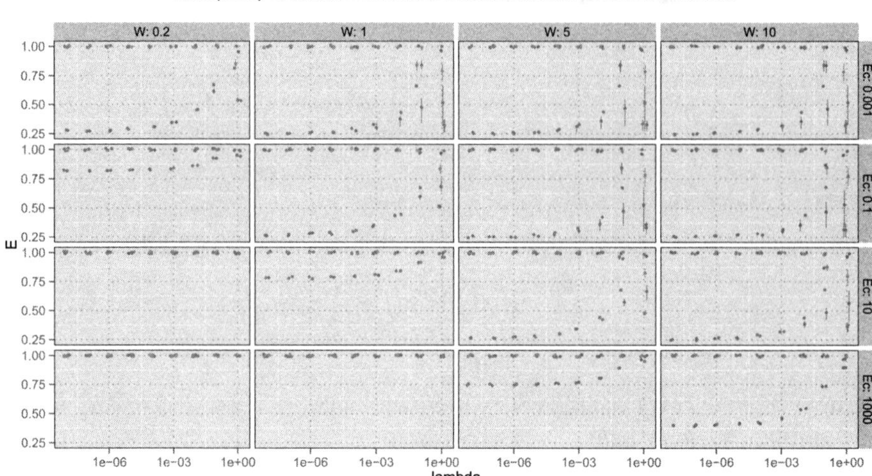

Fig. 6 Comparison with single chunk single speed

degree of freedom improves the results, i.e., allowing for multiple speeds instead of a single speed, or for multiple smaller chunks instead of a single large chunk. All these simulations are conducted under both constraint types, expected and hard deadlines.

We envision reasonable settings by varying parameters within the following ranges:

- $\frac{W}{D} \in [0.2, 10]$;
- $\frac{T_C}{D} \in [10^{-4}, 10^{-2}]$;
- $E_C \in [10^{-3}, 10^{3}]$;
- $\lambda \in [10^{-8}, 1]$.

In addition, we set the deadline to 1. Note that since we study $\frac{W}{D}$ and $\frac{T_C}{D}$ instead of W and T_C, we do not need to study how the variation of the deadline impacts the simulation, this is already taken into account.

We use the Maple software to solve numerically the different minimization problems. Results are showed from two perspectives: on the one hand (Fig. 6), for a given constraint (HARD-DEADLINE or EXPECTED-DEADLINE), we normalize all variants according to SINGLESPEED SINGLECHUNK, under the considered constraint. For instance, on the plots, the energy consumed by MULTIPLECHUNKS MULTIPLESPEEDS (denoted as MCMS) for HARD-DEADLINE is divided by the energy consumed by SINGLECHUNK SINGLESPEED (denoted as SCSS) for HARD-DEADLINE, while the energy of MULTIPLECHUNKS SINGLESPEED (denoted as MCSS) for EXPECTED-DEADLINE is normalized by the energy of SINGLECHUNK SINGLESPEED for EXPECTED-DEADLINE.

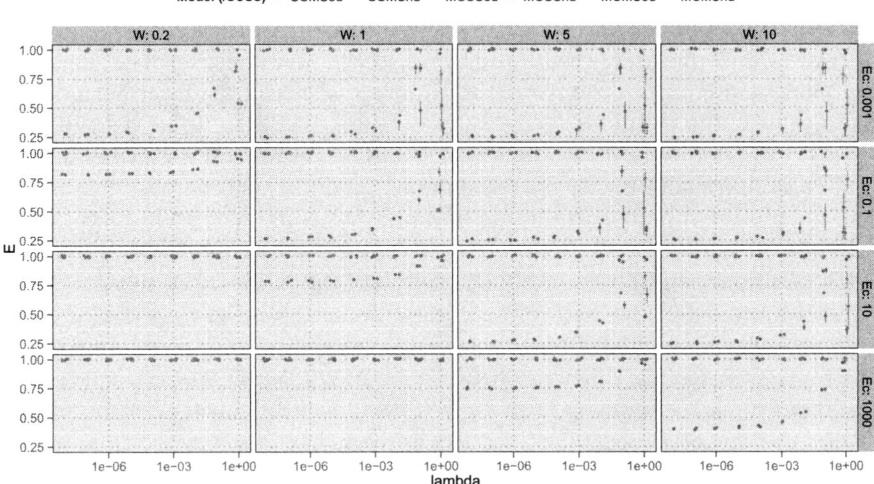

Fig. 7 Comparison hard deadline versus expected deadline

On the other hand (Fig. 7), we study the impact of the constraint hardness on the energy consumption. For each solution form (SINGLESPEED or MULTIPLESPEEDS, and SINGLECHUNK or MULTIPLECHUNKS), we plot the ratio energy consumed for EXPECTED-DEADLINE over energy consumed for HARD-DEADLINE.

Note that for each figure, we plot for each function different values that depend on the different values of T_C/D (hence the vertical intervals for points where T_C/D has an impact). In addition, the lower the value of T_C/D, the lower the energy consumption.

5.4.2 Comparison with Single Speed

At first, we observe that the results are identical for any value of W/D, up to a translation of E_C (see $(W/D = 0.2, E_C = 10^{-3})$ vs. $(W/D = 5, E_C = 1000)$ or $(W/D = 1, E_C = 10^{-3})$ vs. $(W/D = 5, E_C = 0.1)$ on Fig. 6, for instance).

Then the next observation is that for EXPECTED-DEADLINE, with a small λ ($< 10^{-2}$), MULTIPLECHUNKS or MULTIPLESPEEDS models do not improve the energy ratio. This is due to the fact that, in both expressions for energy and for execution time, the re-execution term is negligible relative to the execution one, since it has a weighting factor λ. However, when λ increases, if the energy of a checkpoint is small relative to the total work (which is the general case), we can see a huge improvement (between 25 and 75 % energy saving) with MULTIPLECHUNKS.

On the contrary, as expected, for small λs, re-executing at a different speed has a huge impact for HARD-DEADLINE, where we can gain up to 75 % energy when the failure rate is low. We can indeed run at around half speed during the first execution (leading to the $1/2^2 = 25$ % saving), and at a high speed for the second one, because the very low failure probability avoids the explosion of the expected energy consumption. For both MULTIPLECHUNKS and SINGLECHUNK, this saving ratio increases with λ (the energy consumed by the second execution cannot be neglected any more, and both executions need to be more balanced), the latter being more sensitive to λ. But the former is the only configuration where T_C has a significant impact: its performance decreases with T_C; still it remains still it remains strictly better than SINGLECHUNK MULTIPLESPEEDS.

5.4.3 Comparison Between EXPECTED-DEADLINE and Hard-Deadline

As before, the value of W/D does not change the energy ratios up to translations of E_C. As expected, the difference between the EXPECTED-DEADLINE and HARD-DEADLINE models is very important for the SINGLESPEED variant: when the energy of the re-execution is negligible (because of the failure rate parameter), it would be better to spend as little time as possible doing the re-execution in order to have a speed as slow as possible for the first execution, however we are limited in the SINGLESPEED HARD-DEADLINE model by the fact that the re-execution time is fully taken into account (its speed is the same as the first execution, and there is no parameter λ to render it negligible).

Furthermore, when λ is minimum, MULTIPLESPEEDS consumes the same energy for EXPECTED-DEADLINE and for HARD-DEADLINE. Indeed, as expected, the λ in the energy function makes it possible for the re-execution speed to be maximal: it has little impact on the energy, and it is optimal for the execution time; this way we can focus on slowing down the first execution of each chunk. For HARD-DEADLINE, we already run the first execution at half speed, thus we cannot save more energy, even considering EXPECTED-DEADLINE instead. When λ increases, speeds of HARD-DEADLINE cannot be lowered but the expected execution time decreases, making room for a downgrade of the speeds in the EXPECTED-DEADLINE problems.

5.5 Concluding Remarks

In this section, we have studied the energy consumption of a divisible computational workload on volatile platforms under the CONTINUOUS speed model. In particular, we have studied the expected energy consumption under different deadline constraints: a soft deadline (a deadline for the expected execution time), and a hard deadline (a deadline for the worst case execution time).

As stated in Sect. 2, the CONTINUOUS speed model is theoretically appealing, and allowed us to show mathematically, for all cases but one, that when using the MULTIPLECHUNKS model, then (i) every chunk should be equally sized; (ii) every execution speed should be equal; and (iii) every re-execution speed should also be equal. This problem remains open in the MULTIPLESPEEDS HARD-DEADLINE variant.

Through a set of extensive simulations we have shown the following: (i) when the fault parameter λ is small, for EXPECTED-DEADLINE constraints, the SINGLECHUNK SINGLESPEED model leads to almost optimal energy consumption. This is not true for the HARD-DEADLINE model, which accounts equally for execution and re-execution, thereby leading to higher energy consumption. Therefore, for the HARD-DEADLINE model and for small values of λ, the model of choice should be the SINGLECHUNK MULTIPLESPEEDS model, and that is not intuitive. When the fault parameter rate λ increases, using a single chunk is no longer energy-efficient, and one should focus on the MULTIPLECHUNKS MULTIPLESPEEDS model for both deadline types.

An interesting direction for future work is to extend this study to the case of an application workflow: instead of dealing with a single divisible task, we would deal with a DAG of tasks, that could be either divisible (checkpoints can take place anytime) or atomic (checkpoints can only take place at the end of the execution of some tasks). Again, we can envision both soft or hard constraints on the execution time, and we can keep the same model with a single re-execution per chunk/task, at the same speed or possibly at a different speed. Deriving complexity results and heuristics to solve this difficult problem is likely to be very challenging, but could have a dramatic impact to reduce the energy consumption of many scientific applications.

6 Conclusion

In this chapter, we have discussed several energy-aware algorithms aiming at decreasing the energy consumption in data centers. We have started with a description of various energy models, ranking from the most theoretical model of continuous speeds to the more realistic discrete model. Indeed, processor speeds can be changed thanks to the DVFS technique (Dynamic Voltage and Frequency Scaling), hence decreasing energy consumption when running at a lower speed. Of course, performance should not be sacrificed for energy, and a bound on the performance should always be enforced.

We have first illustrated these models on a task graph scheduling problem where we can reclaim the energy of a schedule by running some non-critical tasks at a lower speed. Depending upon the model, the complexity of the problem varies: while several optimality results can be obtained with continuous speeds, the problem with discrete speeds is NP-hard. Through this study, we have laid the theoretical foundations for a comparative study of energy models.

We have then targeted a problem typical of data centers, namely the replica placement problem. The root server has the original copy of the database but cannot serve all clients directly, so a distribution tree is deployed to provide the clients with a hierarchical and distributed access to replicas of the original data. The problem is to decide where to place replicas, and where to serve each client. We have provided an optimal dynamic programming algorithm that works in a dynamic setting: we assume that client requests can evolve over time, and hence some replicas are already placed in the network. It is more efficient to re-use some of these replicas if possible.

We have also added a criterion of power consumption to the problem, and proved the NP-completeness of this problem with a discrete energy model. In addition, some practical solutions have been proposed.

Finally, a rising concern in data centers, apart from energy consumption, is higher failure rate. We have therefore discussed checkpointing strategies, in the case of a divisible workload. Two deadline constraints have been studied: a hard deadline scenario corresponding to real-time environments where task deadlines are always strictly enforced, and a soft deadline scenario corresponding to a more flexible environment, where an average response time must be enforced. We have conducted this study under the continuous model, which enabled us to derive theoretical results: we proved that every chunk should be equally sized, and that every speed should be equal. In case of failure, we re-execute a chunk, and all re-execution speeds should also be equal.

Through these three case studies, we have demonstrated the importance and the complexity of proposing energy-aware solutions to problems that occur in data centers. We have provided a first step towards energy-efficient data centers by first discussing energy models, and then designing energy-efficient algorithms for some typical problems. Several research directions have been opened.

Acknowledgments This work was supported in part by the ANR *RESCUE* project.

References

1. AMD processors. http://www.amd.com.
2. G. Aupy, A. Benoit, F. Dufossé, and Y. Robert. Reclaiming the energy of a schedule: models and algorithms. *Concurrency and Computation: Practice and Experience*, 2012.
3. G. Aupy, A. Benoit, R. Melhem, P. Renaud-Goud, and Y. Robert. Energy-aware checkpointing of divisible tasks with soft or hard deadlines. In *Proceedings of the International Green Computing Conference (IGCC)*, Arlington, USA, June 2013.
4. H. Aydin and Q. Yang. Energy-aware partitioning for multiprocessor real-time systems. In *Proceedings of the International Parallel and Distributed Processing Symposium (IPDPS)*, pages 113–121. IEEE CS Press, 2003.
5. N. Bansal, T. Kimbrel, and K. Pruhs. Speed scaling to manage energy and temperature. *Journal of the ACM*, 54(1):1–39, 2007.
6. E. Beigne, F. Clermidy, J. Durupt, H. Lhermet, S. Miermont, Y. Thonnart, T. Xuan, A. Valentian, D. Varreau, and P. Vivet. An asynchronous power aware and adaptive NoC based circuit. In *Proceedings of the 2008 IEEE Symposium on VLSI Circuits*, pages 190–191, June 2008.
7. E. Beigne, F. Clermidy, S. Miermont, Y. Thonnart, A. Valentian, and P. Vivet. A Localized Power Control mixing hopping and Super Cut-Off techniques within a GALS NoC. In *Proceedings of the IEEE International Conference on Integrated Circuit Design and Technology and Tutorial (ICICDT)*, pages 37–42, June 2008.
8. A. Benoit, V. Rehn-Sonigo, and Y. Robert. Replica placement and access policies in tree networks. *IEEE Trans. Parallel and Distributed Systems*, 19(12):1614–1627, 2008.
9. A. Benoit, P. Renaud-Goud, and Y. Robert. Power-aware replica placement and update strategies in tree networks. In *Proceedings of the International Parallel and Distributed Processing Symposium (IPDPS)*, Anchorage, USA, May 2011.
10. S. Boyd and L. Vandenberghe. *Convex Optimization*. Cambridge University Press, 2004.

11. G. Buttazzo, G. Lipari, L. Abeni, and M. Caccamo. *Soft Real-Time Systems: Predictability vs. Efficiency*. Springer series in Computer Science, 2005.
12. A. P. Chandrakasan and A. Sinha. JouleTrack: A Web Based Tool for Software Energy Profiling. In *Design Automation Conference*, pages 220–225, Los Alamitos, CA, USA, 2001. IEEE Computer Society Press.
13. G. Chen, K. Malkowski, M. Kandemir, and P. Raghavan. Reducing power with performance constraints for parallel sparse applications. In *Proceedings of the International Parallel and Distributed Processing Symposium (IPDPS)*, page 8 pp., Apr. 2005.
14. J.-J. Chen and C.-F. Kuo. Energy-Efficient Scheduling for Real-Time Systems on Dynamic Voltage Scaling (DVS) Platforms. In *Proceedings of the International Workshop on Real-Time Computing Systems and Applications*, pages 28–38, Los Alamitos, CA, USA, 2007. IEEE Computer Society.
15. J.-J. Chen and T.-W. Kuo. Multiprocessor energy-efficient scheduling for real-time tasks. In *Proceedings of International Conference on Parallel Processing (ICPP)*, pages 13–20. IEEE CS Press, 2005.
16. I. Cidon, S. Kutten, and R. Soffer. Optimal allocation of electronic content. *Computer Networks*, 40:205–218, 2002.
17. V. Degalahal, L. Li, V. Narayanan, M. Kandemir, and M. J. Irwin. Soft errors issues in low-power caches. *IEEE Trans. Very Large Scale Integr. Syst.*, 13:1157–1166, October 2005.
18. J. Dongarra, P. Beckman, P. Aerts, F. Cappello, T. Lippert, S. Matsuoka, P. Messina, T. Moore, R. Stevens, A. Trefethen, and M. Valero. The international exascale software project: a call to cooperative action by the global high-performance community. *Int. J. High Perform. Comput. Appl.*, 23(4):309–322, 2009.
19. M. R. Garey and D. S. Johnson. *Computers and Intractability; A Guide to the Theory of NP-Completeness*. W. H. Freeman & Co., New York, NY, USA, 1990.
20. R. Ge, X. Feng, and K. W. Cameron. Performance-constrained distributed DVS scheduling for scientific applications on power-aware clusters. In *Proceedings of the ACM/IEEE conference on SuperComputing (SC)*, page 34. IEEE Computer Society, 2005.
21. R. Gonzalez and M. Horowitz. Energy dissipation in general purpose microprocessors. *IEEE Journal of Solid-State Circuits*, 31(9):1277–1284, Sept. 1996.
22. P. Grosse, Y. Durand, and P. Feautrier. Methods for power optimization in SOC-based data flow systems. *ACM Trans. Des. Autom. Electron. Syst.*, 14:38:1–38:20, June 2009.
23. Y. Hotta, M. Sato, H. Kimura, S. Matsuoka, T. Boku, and D. Takahashi. Profile-based optimization of power performance by using dynamic voltage scaling on a pc cluster. In *Proceedings of the International Parallel and Distributed Processing Symposium (IPDPS)*, page 340, Los Alamitos, CA, USA, 2006. IEEE Computer Society Press.
24. Intel XScale technology. http://www.intel.com/design/intelxscale.
25. T. Ishihara and H. Yasuura. Voltage scheduling problem for dynamically variable voltage processors. In *Proceedings of International Symposium on Low Power Electronics and Design (ISLPED)*, pages 197–202. ACM Press, 1998.
26. R. Jejurikar, C. Pereira, and R. Gupta. Leakage aware dynamic voltage scaling for real-time embedded systems. In *Proceedings of the 41st annual Design Automation Conference (DAC)*, pages 275–280, New York, NY, USA, 2004. ACM.
27. K. Kalpakis, K. Dasgupta, and O. Wolfson. Optimal placement of replicas in trees with read, write, and storage costs. *IEEE Trans. Parallel and Distributed Systems*, 12(6):628–637, 2001.
28. H. Kawaguchi, G. Zhang, S. Lee, and T. Sakurai. An LSI for VDD-Hopping and MPEG4 System Based on the Chip. In *Proceedings of the International Symposium on Circuits and Systems (ISCAS)*, May 2001.
29. K. H. Kim, R. Buyya, and J. Kim. Power-Aware Scheduling of Bag-of-Tasks Applications with Deadline Constraints on DVS-enabled Clusters. In *Proceedings of the IEEE International Symposium on Cluster Computing and the Grid (CCGRID)*, pages 541–548, May 2007.

30. K. Lahiri, A. Raghunathan, S. Dey, and D. Panigrahi.: a new frontier in low power design. In *Proceedings of the 7th Asia and South Pacific Design Automation Conference and the 15th International Conference on VLSI Design (ASP-DAC)*, pages 261–267, 2002.
31. P. Langen and B. Juurlink. Leakage-aware multiprocessor scheduling. *J. Signal Process. Syst.*, 57(1):73–88, 2009.
32. S. Lee and T. Sakurai. Run-time voltage hopping for low-power real-time systems. In *Proceedings of DAC'2000, the 37th Conference on Design Automation*, pages 806–809, 2000.
33. P. Liu, Y.-F. Lin, and J.-J. Wu. Optimal placement of replicas in data grid environments with locality assurance. In *International Conference on Parallel and Distributed Systems (ICPADS)*. IEEE Computer Society Press, 2006.
34. S. Miermont, P. Vivet, and M. Renaudin. A Power Supply Selector for Energy- and Area-Efficient Local Dynamic Voltage Scaling. In *Integrated Circuit and System Design. Power and Timing Modeling, Optimization and Simulation*, volume 4644 of *Lecture Notes in Computer Science*, pages 556–565. Springer Berlin / Heidelberg, 2007.
35. M. P. Mills. The internet begins with coal. *Environment and Climate News*, page., 1999.
36. T. Okuma, H. Yasuura, and T. Ishihara. Software energy reduction techniques for variable-voltage processors. *IEEE Design Test of Computers*, 18(2):31–41, Mar. 2001.
37. R. B. Prathipati. Energy efficient scheduling techniques for real-time embedded systems. Master's thesis, Texas A&M University, May 2004.
38. K. Pruhs, R. van Stee, and P. Uthaisombut. Speed scaling of tasks with precedence constraints. *Theory of Computing Systems*, 43:67–80, 2008.
39. V. J. Rayward-Smith, F. W. Burton, and G. J. Janacek. Scheduling parallel programs assuming preallocation. In P. Chrétienne, E. G. Coffman Jr., J. K. Lenstra, and Z. Liu , editors, *Scheduling Theory and its Applications*. John Wiley and Sons, 1995.
40. A. Schrijver. *Combinatorial Optimization: Polyhedra and Efficiency*, volume 24 of *Algorithms and Combinatorics*. Springer-Verlag, 2003.
41. J. A. Stankovic, K. Ramamritham, and M. Spuri. *Deadline Scheduling for Real-Time Systems: EDF and Related Algorithms*. Kluwer Academic Publishers, Norwell, MA, USA, 1998.
42. L. Wang, G. von Laszewski, J. Dayal, and F. Wang. Towards Energy Aware Scheduling for Precedence Constrained Parallel Tasks in a Cluster with DVFS. In *Proceedings of the IEEE/ACM International Conference on Cluster, Cloud and Grid Computing (CCGRID)*, pages 368–377, May 2010.
43. J.-J. Wu, Y.-F. Lin, and P. Liu. Optimal replica placement in hierarchical Data Grids with locality assurance. *Journal of Parallel and Distributed Computing (JPDC)*, 68(12):1517–1538, 2008.
44. R. Xu, D. Mossé, and R. Melhem. Minimizing expected energy consumption in real-time systems through dynamic voltage scaling. *ACM Trans. Comput. Syst.*, 25(4):9, 2007.
45. L. Yang and L. Man. On-Line and Off-Line DVS for Fixed Priority with Preemption Threshold Scheduling. In *Proceedings of the International Conference on Embedded Software and Systems (ICESS)*, pages 273–280, May 2009.
46. F. Yao, A. Demers, and S. Shenker. A scheduling model for reduced CPU energy. In *Proceedings of the 36th Annual Symposium on Foundations of Computer Science (FOCS)*, page 374, Washington, DC, USA, 1995. IEEE Computer Society.
47. Y. Zhang, X. S. Hu, and D. Z. Chen. Task scheduling and voltage selection for energy minimization. In *Proceedings of the 39th annual Design Automation Conference (DAC)*, pages 183–188, New York, NY, USA, 2002. ACM.
48. D. Zhu, R. Melhem, and D. Mossé. The effects of energy management on reliability in real-time embedded systems. In *Proceedings of the IEEE/ACM International Conference on Computer-Aided Design (ICCAD)*, pages 35–40, 2004.

Energy Efficiency in HPC Data Centers: Latest Advances to Build the Path to Exascale

Sébastien Varrette, Pascal Bouvry, Mateusz Jarus and Ariel Oleksiak

1 Introduction

Nowadays, moderating energy consumption and building eco-friendly computing infrastructure is a major goal in large data centers. Moreover, data center energy usage has risen dramatically over the past decade and will continue to grow in-step with the High Performance Computing (HPC) intensive workloads which are at the heart of our modern life. The recent advances in the technology has driven the data center into a new phase of expansion featuring solutions with higher density. To this end, much has been done to increase server efficiency and IT space utilization. In this chapter, we will provide a state-of-the-art overview as regards energy-efficiency in High Performance Computing (HPC) facilities while describing the open challenges the research community has to face in the coming years to enable the building and usage of an Exascale platform by 2020.

S. Varrette (✉) · P. Bouvry
Computer Science and Communication (CSC) Research Unit, University of Luxembourg, 6, rue Richard Coudenhove-Kalergi, L-1359 Luxembourg, Luxembourg
e-mail: sebastien.varrette@uni.lu

P. Bouvry
e-mail: pascal.bouvry@uni.lu

M. Jarus · A. Oleksiak
Poznań Supercomputing and Networking Center, Noskowskiego 10, Poznań, Poland
e-mail: jarus@man.poznan.pl

A. Oleksiak
e-mail: ariel@man.poznan.pl

S. U. Khan, A. Y. Zomaya (eds.), *Handbook on Data Centers*,
DOI 10.1007/978-1-4939-2092-1_3

Table 1 Main HPC performance metrics

Type	Metric
Computing capacity/speed	**F**loating **p**oint **o**perations per seconds (Flops)
	1 *GFlops* = 10^9 Flops
	1 *TFlops* = 10^{12} Flops
	1 *PFlops* = 10^{15} Flops
Storage capacity	Multiples of *bytes* = 8 *bits*
	1 *GB* = 10^9 bytes 1 *GiB* = 1024^3 bytes
	1 *TB* = 10^{12} bytes 1 *TiB* = 1024^4 bytes
	1 *PB* = 10^{15} bytes 1 *PiB* = 1024^5 bytes
Transfer rate on a medium	*Mb/s* or *MB/s*
I/O performance	Sequential vs Random *R/W speed, IOPS*
Computing energy performance	Flops/watt

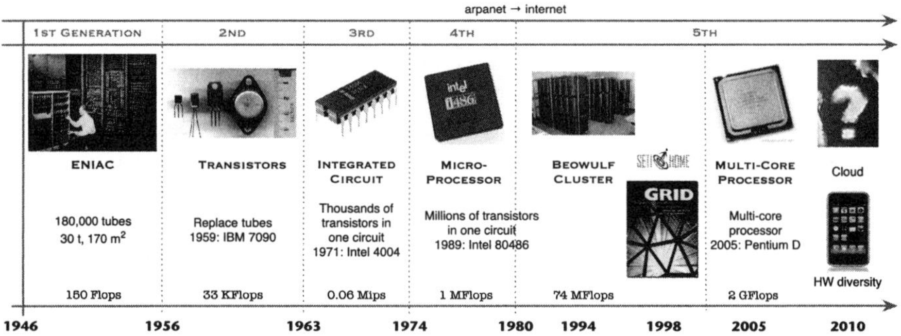

Fig. 1 Evolution of computing systems

2 Computing Systems Architectures

2.1 Architecture of the Current HPC Facilities

Since the advent of computer sciences, applications have been intrinsically restricted by the computing power available on execution. It led to a race to build more and more efficient supercomputers, opening the area of High Performance Computing (HPC). The main performance metrics used in this context are summarized in the Table 1.

Computing systems used to evolve according to successive generations summarized in the Fig. 1. As of today, computing systems are **multi-core** *i.e.* they embedded several computing units which operate in parallel. Also, the Cloud Computing (CC) paradigm opens new perspective as regards computing facilities, when mobile processors are called to as the leading technology in the processor market for the coming

Table 2 Top500 milestones (Gordon Bell Prize)

Date	Computing capacity	HPC system	
1988	1 GFlops	Cray Y-MP	8 processors
1998	1 TFlops	Cray T3E	1024 processors
2008	1 PFlops	Cray XT5	1.5×10^5 processors
2018?	1 EFlops	n/a	

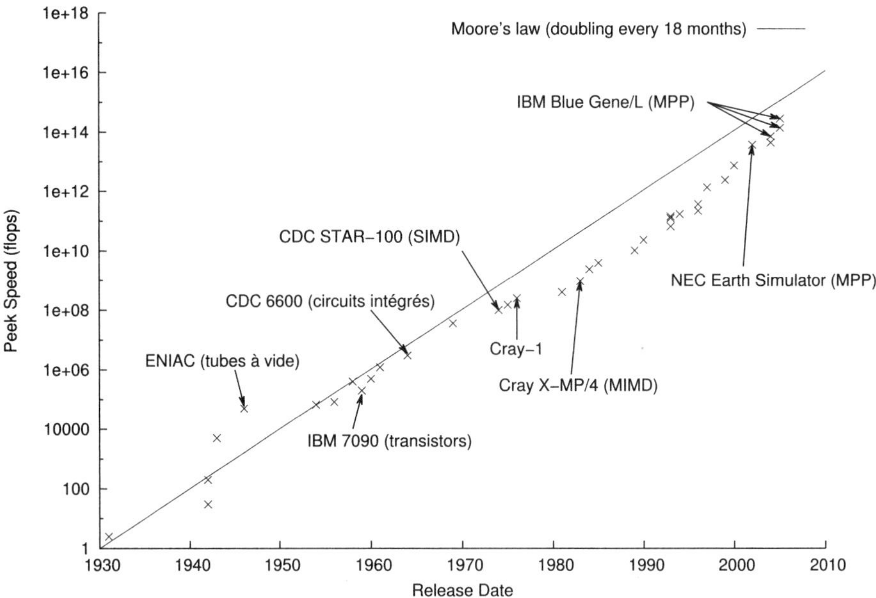

Fig. 2 Moore's law illustrated on supercomputer

years—the reasons for this market change will become self-explanatory by the end of this chapter.

Figure 2 illustrates a famous empirical law due to an engineer at Fairchild Semiconductor named Gordon Moore in the 60s, stating that the density of transistors in a micro-processor are doubling every 18 months. For a couple or years, this law has been reformulated as follows: the number of cores embedded in an micro-processor are doubling every 18 months. Nonetheless, this law is validated since the 1990s by the Top500 [14] project which rank the world's 500 most powerful computers using the High-Performance Linpack (HPL) benchmark. We will have the opportunity to describe more precisely this benchmark, together with its "green" derivative project named *Green 500*, in the section 2.3. Nevertheless, it raises the set milestones of relevance for this chapter summarized in the Table 2.

Different hardware architectures used to permit these milestones to be reached. If we focus on today's systems relative to the seminal classification of Flynn [25], we are now relying on two architectures:

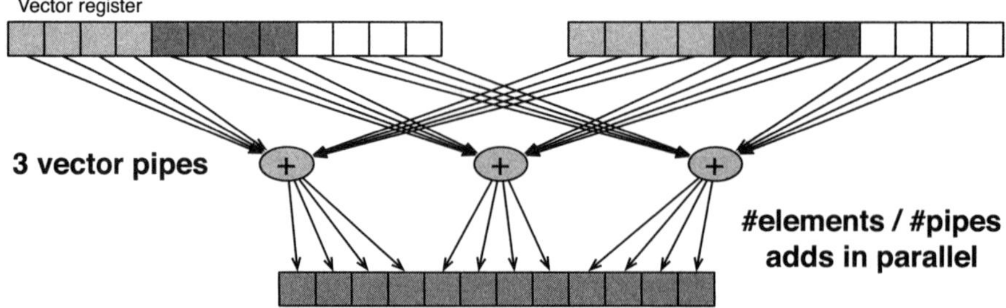

Fig. 3 MMX extension on x86 architectures

1. Single Instruction Multiple Data (SIMD) *i.e.* vector machines
2. Multiple Instruction Multiple Data (MIMD) with shared or distributed memory.

Vector Machines—General-Purpose Graphics Processing Unit (GPGPU) This architecture is tuned to performe element-wise operations on entire vectors in a single instruction. For instance, Fig. 3 illustrates a parallel reduction operated using the MMX extension on x86 architectures or the AltiVec on PowerPC.

These operations are now typically implemented in General-Purpose Graphics Processing Unit (GPGPU) cards (such as nVidia Tesla or ATI Radeon cards) which offer impressive computing performances for a relatively low power consumption. For instance, in 2010, the ATI Radeon HD 5970 used to feature 3200 stream procs running at a frequency of 725 MHz (thus achieving a theoretical computing performance of 4.64 TFlops), for a maximum power consumption of 294 W. For some application, it may be worth to adapt to the capabilities of these devices to ensure large power savings—we will have the opportunity to come back to this later in this chapter.

MIMD with Shared Memory: Symmetric Multi-Processor (SMPs) Under this category falls Symmetric Multi-Processor (SMPs) where all processors access the same memory and I/O. Thus it also applies to multi-core machines. Up to now, most HPC systems are built on such general purpose multi-core processors that use the x86 and Power instruction sets (both to ensure backward productivity and enhance programmers productivity). They are mainly provided by three vendors: Intel (around 71 % of the systems listed in the latest Top500 list[1]), AMD (12 %) and IBM (11 %). While initially designed to target the workstation and laptop market, these processors admittedly offer very good single-thread performance (typically eight operations per cycle @ 2 GHz *i.e.* 16 GFlops), yet at the price of a relative low energy efficiency. For instance, the Table 3 details the Thermal Design Power (TDP) of the top four processors technologies present in the latest Top500 list.

Other systems, often referred to as Massively Parallel Processors (MPPs), feature a virtual shared-memory with *global address space* over physically distributed

[1] Top500 List of November 2012—see http://top500.org.

Table 3 TDP of the top 4 processors technologies present in the Top500 List (Nov. 2012)

Processor technology	Top500 count	Model example	Max. TDP	
Intel Nehalem	225 (45 %)	Xeon X5650 6C 2.66 GHz	85 W	14.1 W/core
Intel Sandybridge	134 (26.8 %)	Xeon E5-2680 8C 2.7 GHz	130 W	16.25 W/core
AMD x86_64	61 (12.2 %)	Optcron 6200 16C "Intcrlagos"	115 W	7.2 W/corc
IBM PowerPC	53 (10.6 %)	Power BQC 16C 1.6 GHz	65 W	4.1 W/core

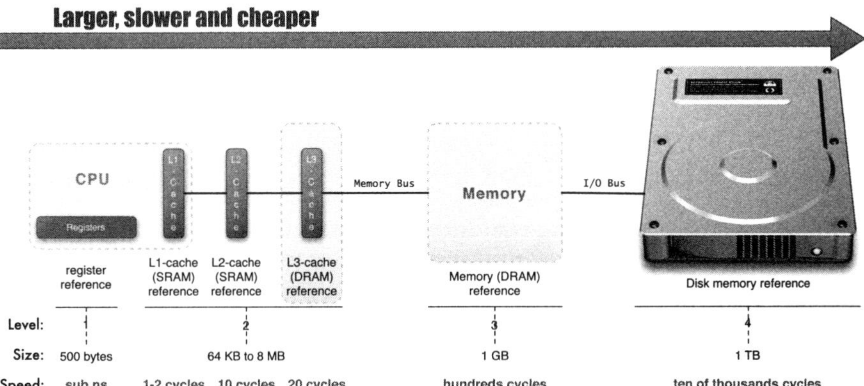

Fig. 4 Overview of the memory hierachy in a computing system

memory. This type of architecture, generally quite expansive, corresponds to 16 % of the computers present in the latest Top500 list.

MIMD with Distributed Memory: Clusters and Grids A low-cost alternative to MPPs feature large-scale distributed systems such as clusters and grids. It corresponds to 83.4 % of the systems listed in the latest Top500 list.

2.2 *Overview of the Main HPC Components*

At the heart of HPC systems rely the computing components *i.e.* CPUs (see Table 3) or GPGPUs. Another important component is the local memory which exists at a different level as illustrated in the Fig. 4.

The interconnect backbone ensures high-performing communications between the resources of an HPC data center. The two key criteria at this level are the *latency* (*i.e.* the time to send a minimal (0 byte) message from A to B) and the *bandwidth* (the maximum amount of data communicated per unit of time). There exist several technologies, the main one being listed in the Table 4. The most represented interconnect family in the latest Top500 list corresponds to the Infiniband technology (around 41 %).

Table 4 Overview of the main interconnect technologies.

Technology	Effective	Bandwidth	Latency (μs)
Gigabit ethernet	1 Gb/s	125 MB/s	40–300
Myrinet (Myri-10G)	9.6 Gb/s	1.2 GB/s	2.3
10 Gigabit Ethernet	10 Gb/s	1.25 GB/s	4–5
Infiniband QDR	40 Gb/s	5 GB/s	1.29–2.6
SGI NUMAlink	60 Gb/s	7.5 GB/s	1

Of course, a software stack is mandatory to operate and exploit efficiently an HPC platform. Nearly every reasonable infrastructure features a Linux Operating System (OS) (95.2 % of the systems ranked in the latest Top500 list). As regards the type of architecture deployed in HPC data centers, most of them corresponds to clusters (83.4 % of the Top500 list). A computing clusters are generally organized in the configuration illustrated on Fig. 5, thus features

- an `access` server used as an SSH interface for the user to the cluster that grants the access to the cluster internals;
- a user `frontend` (eventually merged with the access node), used to reserve nodes on the cluster etc.
- an `adminfront`, often virtualised (typically over the Xen hypervisor [18]) which host the different services required to manage the cluster (either to deploy the computing nodes or to manage various configuration aspects on the cluster such as the user authentication (generally via an LDAP directory) or the Resource and Job Management System (RJMS) (MOAB, OAR, SLURM etc.).
- a shared storage area, typically over a network File system such as NFS, GPFS or Lustre, used for data sharing (homedirs etc.) among the computing nodes and the user frontend
- the computing nodes and the fast interconnect equipment (typically based on the technologies listed in the Table 4).

Data Center Cooling Technologies One of the key factors to operate an HPC platform and obtain the expected performance is the cooling. There are many aspects that must be considered before deciding on a cooling approach to use in a given data center. Energy usage, installation specifics such as the location of the data center itself (does it stand in a cold area etc.), the density of the data center on a per rack and kilowatt per square meter level and other user-specific requirements will all impact this decision. We now review the main cooling technologies that exist.

Historically supercomputers used to be only air-cooled. As the density of computing equipment increases, cooling became a significant challenge. Although more performance is expected to be delivered by next computing systems, their power consumption (and thus also power drawn by cooling) should ideally at least stay at the same level. What is more, high density enclosures often create the potential for hot spots—places with significantly higher temperatures not able to be cooled by traditional chillers.

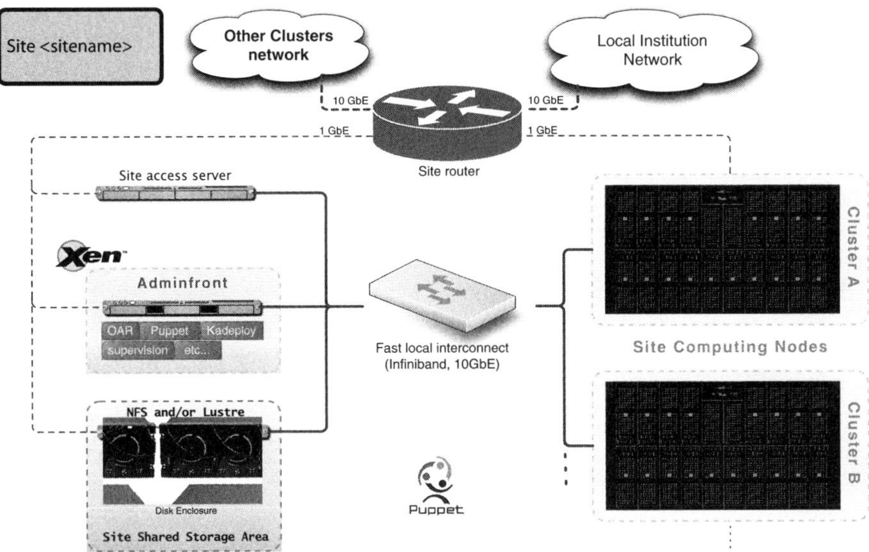

Fig. 5 General organization of a computing clusters

The power consumed by single racks can vary dramatically, with an average around 1.7 kW up to 20 kW in high density servers. Server systems require specific amount of cool air at the intake, depending on their power consumption, and exhausts the same amount of heated air at the outake. In case the room is not capable of providing this quantity of air, the server will draw in its own exhaust air. This situation results in overheating the equipment. Proper cooling design is therefore inevitable for an uninterraptable server operation.

There are three basic cooling architectures: room-, row- and rack-oriented. In the first one Computer Room Air Conditioners (CRAC) units are associated with the room. Cool air supplied by the conditioners may be unrestricted or partially restricted by ducts or vents. The full rated capacity of the CRAC unit cannot be utilized in most cases. It results from the fact that the air supply uniformity is poor due to specific room designs, such as its shape or obstructions. In row-oriented architectures CRAC units are associated with a row. Its performance is higher, as the airflow paths are shorter. As a result, required CRAC fan power supply is smaller, which also decreases the cost of energy. In the last architecture, rack-oriented, CRAC units are associated with the rack. It allows the cooling to be accurately adjusted to the needs of servers. The drawback is, on the hand, that it requires a large number of air conditioning devices.

These three types of cooling architectures may be described in more detail by taking a closer look at the specific air distribution systems. All of them consist of a supply system, which distributes the cool air from the CRAC unit to the computing architectures, and a return system that takes the exhaust air from the loads back to the CRAC. Each of these types has capabilities and benefits, which cause them to be preferred for various applications. Both the supply and return systems may be designed in three ways:

- flooded
- locally ducted
- fully ducted

The first one is the most basic in its design, as no ductwork is used to move the air. This scenario presents the highest risk of mixing hot and cold air in the room. The separation of the exhaust and intake air is crucial as it significantly increases the efficiency and capacity of the cooling system, therefore fully flooded options should be only used in environments drawing less than 40 kW of power, up to 3 kW per rack. The advantage of this solution is a low cost of deployment.

In locally ducted designs some ductwork is used to partially supply cold air or receive hot air from computing systems. It becomes necessary in cases where power density increases. Locally ducted designs are best suited for computing systems up to 5 kW per rack.

When specialized equipment provides for direct input air ducting, fully ducted supply systems may be used. They are also typically used in rooms with raised floor environments, where cold air may be supplied to servers under floor. Fully ducted supplies are appropriate for racks drawing up to 15 kW of power.

These architectures may be mixed, resulting in e.g. locally ducted supplies and fully ducted returns or flooded supplies and locally ducted returns. Altogether they give nine different cooling systems, all of which are presented in Fig. 6. More information about them can be found in [36].

Of course despite choosing the right cooling architecture, equally important are other factors, such as quantity and location of vents, size of the ductwork or location of CRAC units. All of these aspects must be carefully considered during the server room design phase.

Despite traditional air-cooled servers, liquid cooling is also recently gaining popularity. In January 2012 IBM and Leibniz Supercomputing Centre announced the start of hot-water cooled supercomputer, SuperMUC [12]. The computing performance of this supercomputer is 3 petaflops. The water that cools the computing resources is conducted away from the machines, carrying heat with it to an exchange in which it's used in heating the human-occupied areas of the building.

Similar solutions are also used in other research centers. A prototype supercomputer at the Tokyo Institute of Technology, named Tsubame-KFC, is submerged in a tank of mineral oil. The heat absorbed by oil is transferred into water loops via heat exchangers. The warm water releases its heat into the air via a cooling tower. Tsubame-KFC was ranked 1st on the world's Green500 List of computer systems as of November 2013 [13].

Different liquids are also used to accomplish the task of cooling systems. Iceotope [6] submerges computers in liquid fluoroplastic, called Novec 1230. Just like mineral oil, it does not conduct electricity. Therefore there is no risk of short-circuiting the equipment or damaging it. Many research experiments proved this type of cooling to be very efficient. An Iceoptope-based server deployed at Poznan Supercomputing

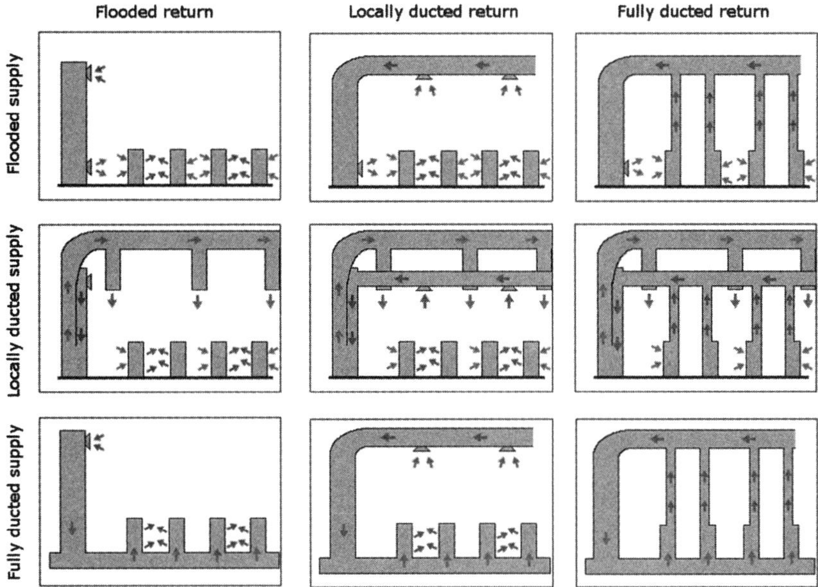

Fig. 6 Different types of cooling systems

and Networking Center (PSNC) proved that this type of cooling is 20 times more efficient than traditional air-based chillers [28].

2.3 HPC Performance and Energy Efficiency Evaluation

The way that performance is measured is critical, as it can determine if a server or other equipment will meet a consumer's needs or be eligible for utility rebates or required as part of federal or state procurement requirements. Developing performance metrics for even the simplest types of equipment can prove difficult and controversial. For this reason, in 1988 the System Performance Evaluation Cooperative (now named Standard Performance Evaluation Corporation, SPEC) was founded. It is a non-profit organization that develops benchmarks for computers and is continuously working on energy-performance protocols—for small to medium-sized servers. Currently it is one of the more successful performance standardization bodies. It develops suites of benchmarks intended to measure computer efficiency, aimed to test "real-life" situations. They are publicly available for a fee covering development and administrative costs. Thanks to these standardized benchmarks it is possible to compare the performance of different machines and rank them.

The High-Performance Linpack (HPL) Benchmark and the Green500 Challenge There are a few initiatives aimed at providing the list of computers in terms of performance or energy efficiency. One of the most popular is the TOP500 project [14] mentioned previously. Since 2003 it ranks twice a year the 500 most powerful computer systems in the world. Current fastest supercomputer, Tianhe-2, achieved a score which is 567,000 times better than the one obtained by the fastest supercomputer in 1993.

The rank in itself relies on results obtained by the running the High-Performance Linpack (HPL) reference benchmark [35]. HPL is a software package that solves a (random) dense linear system in double precision (64 bits) arithmetic on distributed-memory computers. This suit was chosen thanks to it widespread use. The result does not reflect the overall performance of a given system, but rather reflects the performance of a dedicated system for solving a dense system of linear equations. The list was often misinterpreted, often regarded as a general rank that is valid for all applications. To fully examine the performance of the system, an approach consisting of different benchmarks, testing different parts of a supercomputer, is required.

One of the main advantages of Linpack is its scalability. It's been used since the beginning of the list, making it possible to benchmark systems that cover a performance range of 12 orders of magnitude. Moreover, it also measures the reliability for new HPC systems. Some systems were not able to run the Linpack benchmark because they were not stable enough.

Many of these supercomputers consume vast amounts of electrical power and produce so much heat that large cooling facilities must be constructed to ensure proper performance.

In parallel and for decades, the notion of HPC performance has been synonymous with speed (as measured in Flops). In order to raise awareness of other performance metrics than the pure computing speed (as measured in Flops for instance by the HPL suit), the Green500 project [37] was launched in 2005 to evaluate the "Performance per Watt" (PpW) and energy efficiency for improved reliability. More precisely, The PpW metric is defined as follows:

$$\mathrm{PpW} = \frac{R_{\max} \quad (\text{in } \mathbf{MFlops})}{\mathrm{Power}(R_{\max}) \quad (\text{in } \mathbf{W})}$$

This metric is particularly interesting because it is somehow independent of the actual number of physical nodes.

Current most energy-effective (Green500 list from June) supercomputer is Eurora, with 3208.83 MFLOPS/W. On the TOP500 list it takes 467th place. By extrapolating its performance to exascale results, it would result in 312 MW machine. The electricity bills for such a system would be more than US$ 300 million per year. Since current requirements for such a system require the power draw not higher than 20–30 MW, it clearly shows the scale of the challenge that is going to be faced before building an exascale system.

Since the launch of the Green500 list, the energy-efficiency of the highest-ranked machines has improved by only about 11 %. Their performance has increased at a higher rate. It was due to the fact that for decades there was an emphasis on speed as the most important metric. The Green500 seeks to raise awareness in energy efficiency of supercomputers and treat them equally with speed.

The [Green] Graph500 Benchmark Suit Emerging large-data problems have different performance characteristics and archietctural requirements than the floating point performance oriented problems. Supercomputers are typically optimized for the 3D simulation of physics. For this reason in 2010 the Graph 500 list was created to provide information on the suitability of supercomputing systems for data intensive applications. Three key classes of graph kernels with multiple possible implementations were proposed: Search, Optimization and Edge Oriented [33]. Paralelly the Green Graph 500 was created, to complement the Graph 500 list with an energy metric for data intensive computing.

The emergence of such "green" lists clearly shows much greater importance attached to the power consumed by computing systems. There have been also many metrics devised that try to quantitatively describe the energy-effectiveness of whole data centers. One of the most widely used is Power Usage Effectiveness (PUE) [16]. It is the recommended metric for characterizing and reporting overall data centre infrastructure efficiency. Its value indicates the relation between the fraction of power used only for components of the IT (servers, racks, ...) and the complete power consumption of a data centre:

The Challenge (HPCC) Recently, the HPL benchmark has been integrated in a more general benchmark suite, named HPC Challenge (HPCC), which quickly became the industry standard suite used to stress the performance of multiple aspects of an HPC system, from the pure computing power to the disk/RAM usage or the network interface efficiency. More precisely, HPCC basically consists of seven tests:

1. `HPL` (the High-Performance Linpack benchmark), which measures the floating point rate of execution for solving a linear system of equations.
2. `DGEMM` - measures the floating point rate of execution of double precision real matrix-matrix multiplication.
3. `STREAM` - a simple synthetic benchmark program that measures sustainable memory bandwidth (in gigabytes per second) and the corresponding computation rate for simple vector kernel.
4. `PTRANS` (parallel matrix transpose)—exercises the communications where pairs of processors communicate with each other simultaneously. It is a useful test of the total communications capacity of the network.
5. `RandomAccess`—measures the rate of integer random updates of memory (GUPS).
6. `FFT`—measures the floating point rate of execution of double precision complex one-dimensional Discrete Fourier Transform (DFT).
7. Communication bandwidth and latency—a set of tests to measure latency and bandwidth of a number of simultaneous communication patterns.

I/O Performance Evaluation: IOZone and IOR IOZone [11] is a complete cross-platform suite that generates and measures a variety of file operations. Iozone is useful for performing a broad filesystem analysis of a given computing platform, covering tests for file I/O performances for many operations (Read, write, re-read, re-write, read backwards/strided, mmap etc.)

IOR [10] is another I/O benchmark which is of interest when evaluating HPC Data center components that generally feature shared storage based on distributed and parallel FS! (FS!) such as Lustre or GPFS. In this context, IOR permits to benchmark parallel file systems using POSIX, MPIIO, or HDF5 interfaces.

Measuring Data Center Energy Efficiency: The PUE Metric PUE = (Total data centre energy consumption or power / IT energy consumption or power)

A PUE of 1.0 is the best possible value and only theoretically reachable, since in that case no power can be spend for cooling and facility. Today, the average data centre has a PUE of 1.5–2.0, heavily optimized data centres can reach a PUE of 1.1. When calculating PUE, IT energy consumption should, at a minimum, be measured at the output of the uninterruptible power supply (UPS). However, the industry should progressively improve measurement capabilities over time so that measurement of IT energy consumption directly at the IT load (e.g., servers, storage, network, etc.) becomes the common practice.

There are also many others metrics used in data centers. GEC is a metric that quantifies the portion of a facility's energy that comes from green sources. GEC is computed as the green energy consumed by the data centre (kilowatt-hour) divided by total energy consumed by the data centre (kilowatt-hour). For the purposes of GEC, Green energy is defined as any form of renewable energy for which the data centre owns the rights to the green energy certificate or renewable energy certificate, as defined by a local/regional authority. Total energy consumed at the data centre is the total source energy, calculated identically to the numerator of PUE. ERF is a metric that identifies the portion of energy that is exported for reuse outside of the data centre. ERF is computed as reuse energy divided by total energy consumed by the data centre. Reuse energy is measured as it exits the data centre control volume. Total energy consumed by the data centre is the total source energy, calculated identically to the numerator of PUE. CUE is a metric that enables an assessment of the total GHG emissions of a data centre, relative to its IT energy consumption. CUE is computed as the total carbon dioxide emission equivalents (CO2eq) from the energy consumption of the facility divided by the total IT energy consumption, for data centres with electricity as the only energy source this is mathematically equivalent to multiplying the PUE by the data centre's carbon emission factor (CEF). Many others metrics include: Data Centre infrastructure Efficiency (DCiE), Fixed to Variable Energy Ratio metric (FVER), Water Usage Effectiveness (WUE), etc.

3 Energy-Efficiency in HPC Data-Center: Overview & Challenges

3.1 The Exascale Challenge

There are a lot of challenges that need to be faced by the HPC community in the area of low-power computing devices. One of them is to build an exascale HPC system by

2020. The currently fastest supercomputer based on the top500 list, Tianhe-2 [14], consumes almost 18 MW of power. Currently the most efficient system needs one to 2 MW per petaflop/s. By multiplying it by 1000, to get the exascale, the required power becomes unaffordable. The most optimistic current predictions for exascale computers in 2020 envision a power consumption of 20 to 30 MW. It is, however, about 1 million times less power efficient than human brain, which consumes 20–40 W. This comparison shows many science challenges that the computing community needs to take on.

Current measures within a typical blade server estimate that 32.5 % of its supplied power are distributed to the processor. Thus, some simple arithmetic calculations permit to estimate the average consumption per core in such an EFlops system: around 6.4 MW would be dedicated to the computing elements. Their number can be quantified by dividing the target computing capacity (1 EFlops) by one of the current computing cores (16 GFlops), thus leading to approximately 62.5×10^6 cores within an Exascale system. Consequently, such a platform requires a maximal power consumption of 0.1 W per core.

To achieve this goal, alternative low-power processor architectures are required. There are two main directions currently explored: (1) relying on General-Purpose Graphics Processing Unit (GPGPU) accelerators or (2) using the low-power processors. They are often combined, such as in the European Mont Blanc project, which aims at building an Exascale HPC system. It plans to use ARM CPUs combined with Nvidia GPUs to achieve high processing speed at low power consumption. The project aims at decreasing the power consumption at least 15- to 20-fold compared to current fastest supercomputers.

3.2 Hardware Approaches Using Low-Power processors

Instead of high power and high performance CPUs that have been long used in datacenters, low-energy processors are more often used in some selected areas. It is not surprising, taking into account the growth of power consumption in recent years. All of the main CPU producers released their own low-power architectures to the market. Some of them targetted mainly the mobile and embedded devices. However, server-targeted processors are becoming more and more popular. Although the size of this market still remains unclear, the demand for such products will definitely grow. Taking into account only the tablets and smartphones shows the popularity of embedded processors. At the end of 2011 there were 71 million tablets running ARM processors, which is currently the most popular producer of embedded processors. This trend is also visible in case of data centers—those that combine many low-power chips instead of a few high-performance processors provide more computing power for less money and use less electricity.

The growing popularity of low-power processors is recently particularly well visible. In June 2013 AMD announced the new Seattle ARM processor for launch in the second half of 2014, built specifically for servers [2]. Currently the ARM CPUs

Fig. 7 CoolEmAll RECS platform—top view of one example rack featuring different processors

mainly targeted the smartphones and other smaller electronic devices market. This release marks a step in moving away from power-hungry chips. The competition in the field of server machines, dominated so far by Intel and AMD processors, is becoming increasingly fierce, especially in case of low-power CPUs. Other producers also try to enter this market. Applied Micro Circuits, Hewlett-Packard and Dell are expected to make low-power ARM servers.

This trend is not a surprise. The average processor requires about 80–100 W of power in idle state. On the other hand, a multi-core ARM System on Chip (which integrates also other components, not just the processor) draws only about 4 W of power. Moreover, such a system produces much less heat, requiring much less energy to keep the servers cool. It also means that more chips can be installed in the same server box.

PSNC conducts research on such low-power processors in the CoolEmAll project [40]. The CoolEmAll Resource Efficient Computing System (RECS) is a platform is capable to condense capacity of several hundred of servers in high density rack [39]. It consumes only 35 kW, reducing the operating costs by 75 %. Every RECS unit has up to 18 energy efficient computing-nodes. The density of this approach is 4–10 times higher than blade servers. Figure 7 presents an example of RECS unit with three different processor types.

Currently there are three RECS units featuring different models of processors being monitored and benchmarked. All of them contain 18 nodes based on Intel Core i7-3615QE [8], Intel Atom N2600 [7] or AMD G-T40N [1] processors. Several benchmarks were used to measure their performance and energy-efficiency [29]. The experiments performed using a few benchmarks from Phoronix Test Suite, OSU Micro-Benchmarks, HPL, CoreMark, Fhourstones, Whetstone and Linpack confirm high power-efficiency of these processors, yet at the same time they time they prove to be good candidates for HPC environments.

Traditional Intel and AMD CPUs are often equipped with additional mechanisms that aim at reducing the active and static power consumption or providing additional performance when required. One of them is the Dynamic Voltage and Frequency Scaling (DVFS), where the clock frequency of the processor is decreased to allow a corresponding reduction in the supply voltage. As a result, the power consumption is reduced. However, some tests suggest that the effectiveness of DVFS on server-class platforms is questionable. The experiments conducted on AMD CPUs in [32] show that while DVFS is effective on the older platforms, it actually increases energy usage in some cases on the most recent platform.

Another mechanism are the low-power sleep modes. When the CPU is idle, it can be commanded to enter a low-power mode, thus saving energy. Each processor has several power modes and they are called "C-states" or "C-modes". In each of them the clock signal and power is cut from idle units inside the CPU. The more units are stopped, the more energy is saved, but also more time is required for the CPU to wake up.

Dynamic Power Switching allows power management to supply minimum power to those domains of the CPU that are not loaded during any given period of time. Processing in all of the domains must be constantly monitored to change its state to a lower power mode when necessary.

Despite all those mechanisms, there are several areas where microservers with low-power processors are the best option. Servers based on low-energy CPUs can be widely used in data centers running massive web applications, providing hosting, static workloads, cloud gaming and performing other similar functions. System, where a modest amount of computing power is required with lots of disk storage attached to the node is called cold storage. The CPU often runs in a very low power state. OVH and 1&1, global web-hosting services companies, have tested Intel Atom C2000 SoCs and plan to deploy them in their entry-level dedicated hosting services in the last quarter of 2013 [9]. This CPU features up to eight cores, a range of 6–20 W TDP, integrated ethernet and support for up to 64 GB of memory. Such processors are also ideal for entry networking platforms that address the specialized needs for securing and routing Internet traffic. They are ideally suited for routers and security appliances.

3.3 Energy Efficiency of Virtualization Frameworks over HPC Workloads

Virtualization is emerging as the prominent approach to mutualize the energy consumed by a single server running multiple Virtual Machines (VMs) instances. However, little understanding has been obtained about the potential overhead in energy consumption and the throughput reduction for virtualized servers and/or computing resources, nor if it simply suits an environment as high-demanding as a HPC platform. Actually, this question is connected with the rise of Cloud Computing (CC) increasingly advertised as THE solution to most IT problems. Several voices (most probably commercial ones) emit the wish that CC platforms could also serve HPC needs and eventually replace in-house HPC platforms.

The central component of any virtualization framework, and thus Cloud middleware, remain the *hypervisor* or Virtual Machine Manager. Subsequently, a VM running under a given hypervisor is called a guest machine. There exist two types of hypervisors, either *native* or *hosted*, yet only the first class (also named bare-metal) presents an interest for the HPC context. This category of hypervisor runs directly on the host's hardware to control the hardware and to manage guest operating systems. Among the many potential approaches of this type available today, the virtualization technology of choice for most open platforms over the past 7 years has been the Xen hypervisor [18]. More recently, the Kernel-based Virtual Machine (KVM) [30] and VMWare ESXi [17] have also known a widespread deployment within the HPC community, unlike the remaining frameworks available (such as Microsoft's Hyper-V or OpenVZ).

In this context, a couple of recent studies demonstrate that the overhead induced by the Cloud hypervisors cannot be neglected for a pure HPC workload. For instance, in [27, 38], the authors propose a performance evaluation and modeling over HPC benchmarks for the three most widespread hypervisors at the heart of most if not all CC middleware: Xen, KVM and VMWare ESXi. At this level, the Grid'5000 platform [5] was used to perform the experimental study. Grid'5000 is a scientific instrument for the study of large scale parallel and distributed systems. It aims at providing a highly reconfigurable, controllable and monitorable experimental platform to support experiment-driven research in all areas of computer science related to parallel, large-scale or distributed computing and networking, including the CC environments [19].

The study proposed in [38] is relevant of the computing and power performance of the different hypervisors when compared to a "bare-metal" configuration (also referred to as the `baseline` environment corresponding to classical HPC computing nodes) running in native mode *i.e.* without any hypervisor.

The scalability of each virtualization middleware is then evaluated under two perspectives:

1. for a fixed number of physical hosts that run an increasing number of VMs (from 1 to 12)—see Fig. 8. It perfectly illustrates the obvious limitation raised by a multi-VM environment as the performance is bounded by the maximal capacity of the physical host running the hypervisors. Also, we can see that for a computing application as demanding as HPL, the VMWare ESXi hypervisor performs generally better even if this statement is balanced by the fact that the VMWare environment appeared particularly unstable (it was impossible to complete successfully runs for more than four VMs) when Xen and KVM frameworks both offer unmatched scalability;
2. for a fixed number of VM (between 1 and 12), increasing the number of physical hosts (between 1 and 8)—see Fig. 9. It highlights a rather good scalability of the hypervisors when physical nodes are added.

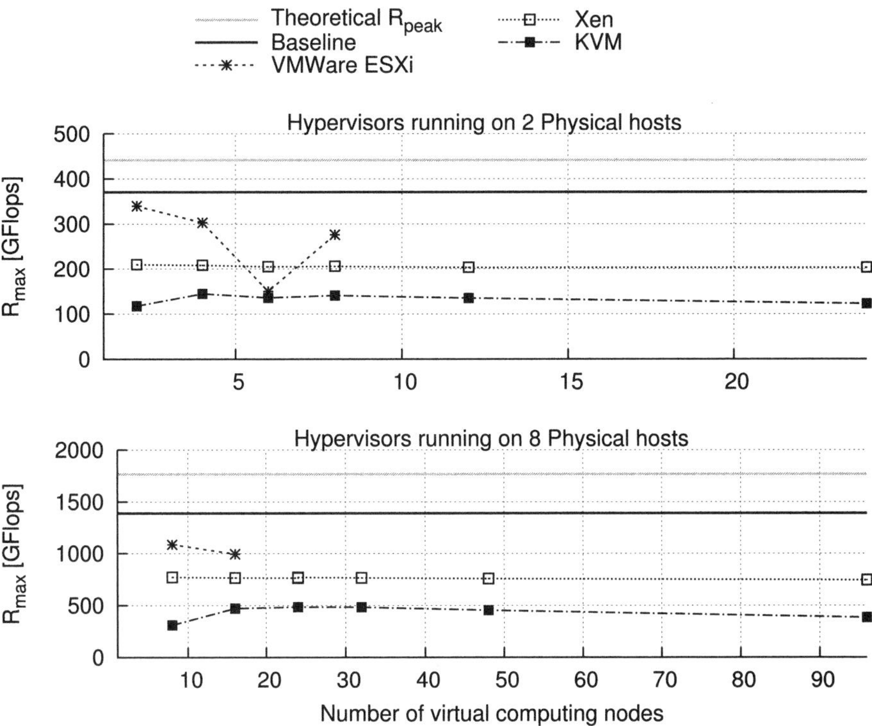

Fig. 8 HPL Performances for fixed numbers of physical nodes with increasing number of VMs per physical host. Baseline execution uses the number of actual physical nodes

Of importance for this book chapter is the analysis using the virtual resources as a basis for the comparisons. It means hypervisor executions on N nodes with V VMs per nodes are compared to baseline executions on $N \times V$ physical nodes. As this approach might appear unfair as the hardware capabilities are not the same, this illustrates the point-of-view of the user that may not know the underlying hardware his application is running on in a virtualized environment. In particular, the best obtained results are displayed in the Fig. 10 which demonstrates the fast degradation in the computing efficiency when the number for computing nodes is artificially increased through virtualization.

To highlight the relative performance of each computing node, the *iso-efficiency* $ISO_{\text{effic.}}(n)$ metric was defined for a given number of computing nodes n. This measure is based on the following definitions:

- $\text{Perf}_{\text{base}}n$: HPL performance of the baseline environment involving n computing nodes;
- $\text{Perf}_{\text{base}}1$: Normalized performance of a bare-metal single node. For this study, as we only started our measures with two hosts, we approximate this value by $\text{Perf}_{\text{base}}1 = \frac{\text{Perf}_{\text{base}}2}{2}$

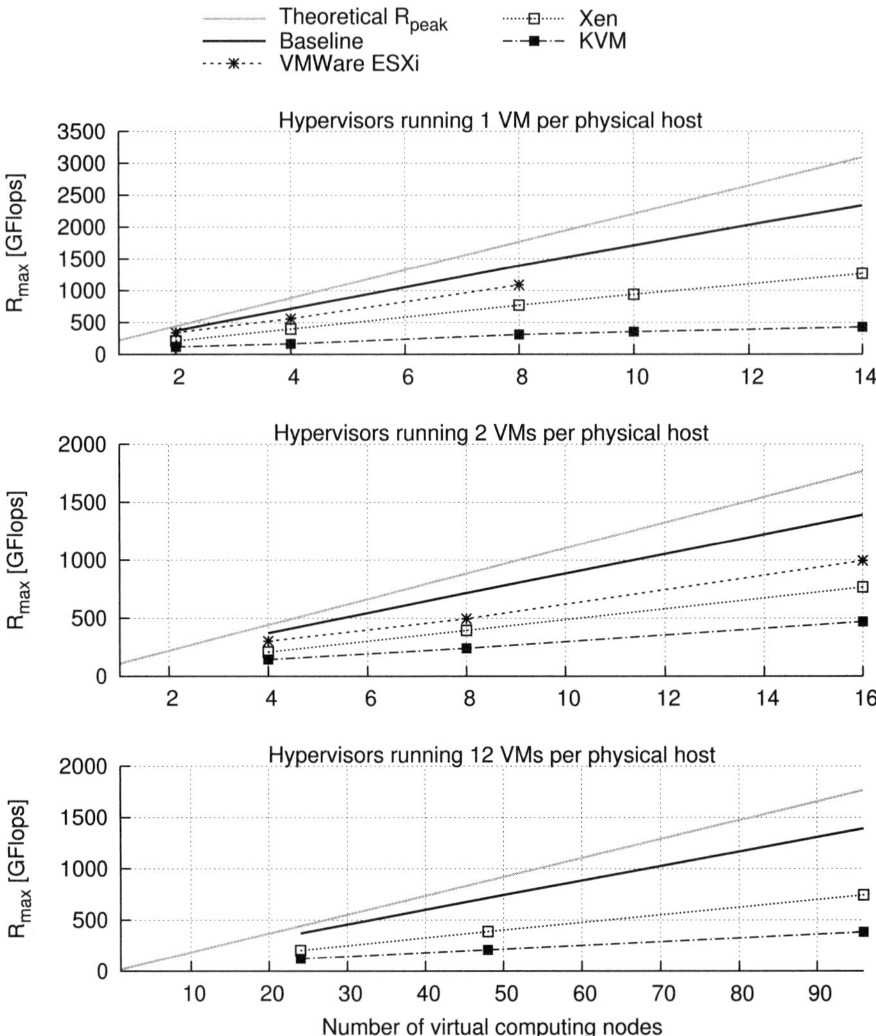

Fig. 9 HPL Performances for fixed numbers of VMs with increasing number of physical nodes. Baseline execution uses the number of actual physical nodes

- $\mathrm{Perf}_{\mathrm{hyp}}n$: Maximal HPL Performance of the virtualized environment based on the hypervisor hyp that feature a total of n computing nodes.

Then for a given hypervisor hyp:

$$ISO_{\mathrm{eff}}^{\mathrm{hyp}}(n) = \frac{\mathrm{Perf}_{\mathrm{hyp}}n}{n \times \mathrm{Perf}_{\mathrm{base}}1}$$

This definition should not be confused with the classical iso-efficiency metric used in parallel programs, the objective being here to simply normalize the hypervisor

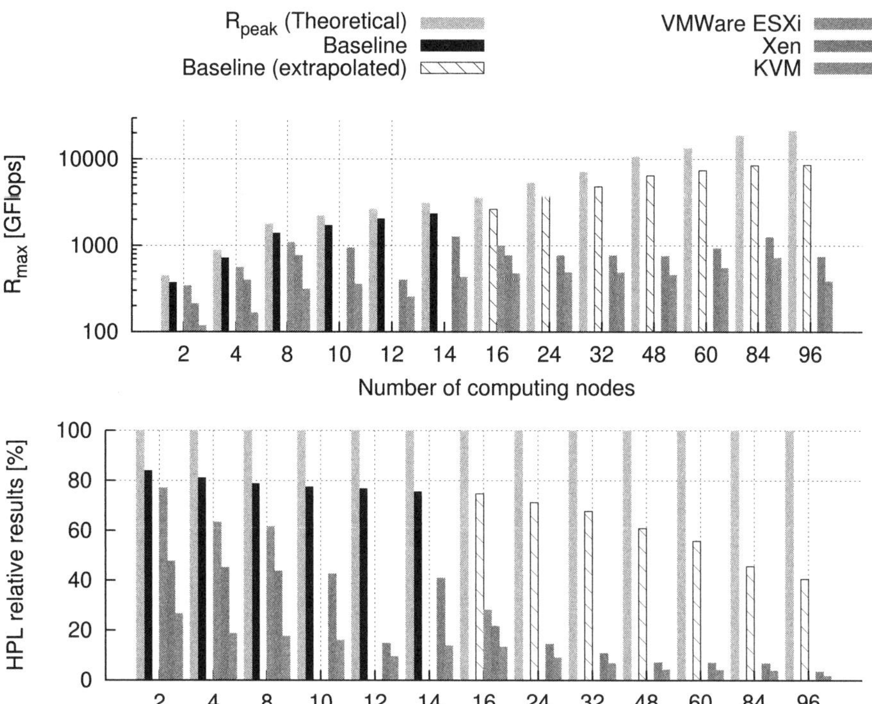

Fig. 10 HPL Efficiency of the considered hypervisors when compared to the baseline environment.

performance with regards to the best performance that can be obtained on a baseline environment *i.e.* $\text{Perf}_{\text{base}}1$. Figure 11 expounds the evolution of $ISO_{\text{eff}}^{\text{hyp}}(n)$ with n. Again, this measure confirms that HPC workloads do not suit virtualized environments from a pure computing capacity point of view. The virtualized environment shows more available processors to the application. However, this computing resources have reduced performance compared to actual physical processors because they are shared for different VMs. This is perfectly highlighted by the HPL benchmark whose performance are mainly bounded by the performance of the processors.

All these performance evaluations confirm what other studies suggested in the past, *i.e.* that the overhead induced by virtualized environments do not suit HPC workloads. Of interest for this book chapter is the energy-efficiency of the virtualized environments when running HPC workloads. For instance, Fig. 12 illustrates the total power profile of a run involving each considered environment in a large scale execution.

Figure 13 details the evolution of the PpW metric (as defined in the Green500 benchmark—see Sect. 2.3) over the baseline environment for an increasing number of computing nodes. We have compared these values with the cases where we have the corresponding PpW measure in the hypervisor environments. This figure outlines many interesting aspects. First of all, with a PpW measure comprised between 700

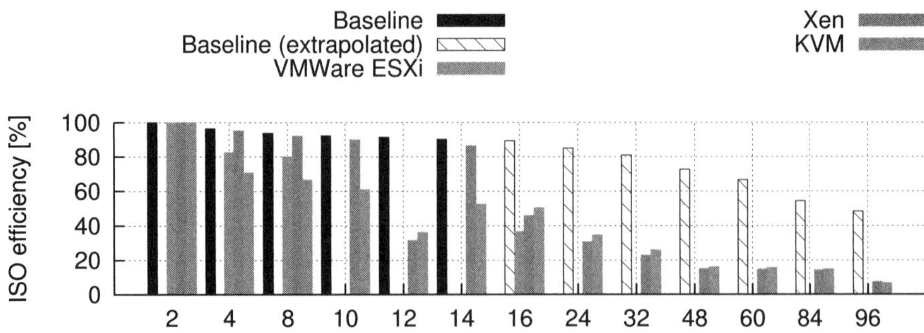

Fig. 11 Iso-efficiency evaluation for an increasing number of computing nodes

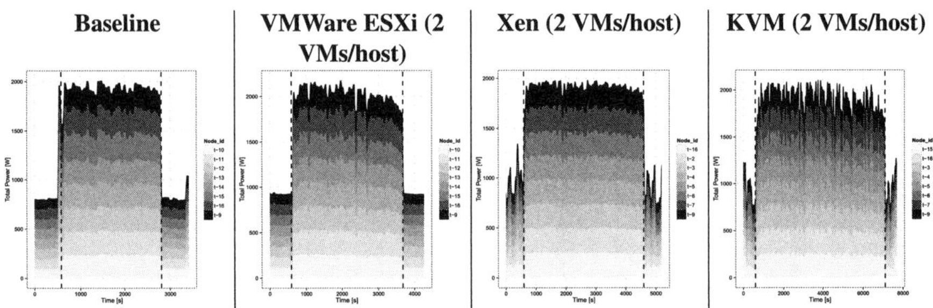

Fig. 12 Stacked traces of the power draw of hosts for selected runs with 8 physical hosts

and 800 MFlops/W, the baseline platform would be ranked between the 93 and 112 position of the Green500 list. While surprising at first glance, this result is easily explained by the usage of cutting-edge processors (Sandy-bridge) and the limited number of involved resources—the linear decrease is evident in the figure. The second conclusion that can be raised from this figure is that **virtualized environments do not even ensure a more energy-efficient usage during an HPC workload.**

3.4 *Energy Efficiency in Resource and Job Management Systems (RJMSs)*

The performance of an HPC system and in particular its energy efficiency, is obviously determined by the unitary performance of the subsystems that compose it, but also by the efficiency of their interactions and management by the middleware. In these kind of systems, a central component called the Resource and Job Management System (RJMS) is in charge of managing the users' tasks (jobs) on the system's computing resources. The RJMS has a strategic position in the whole HPC software stack as it has a constant knowledge of both workload and resources.

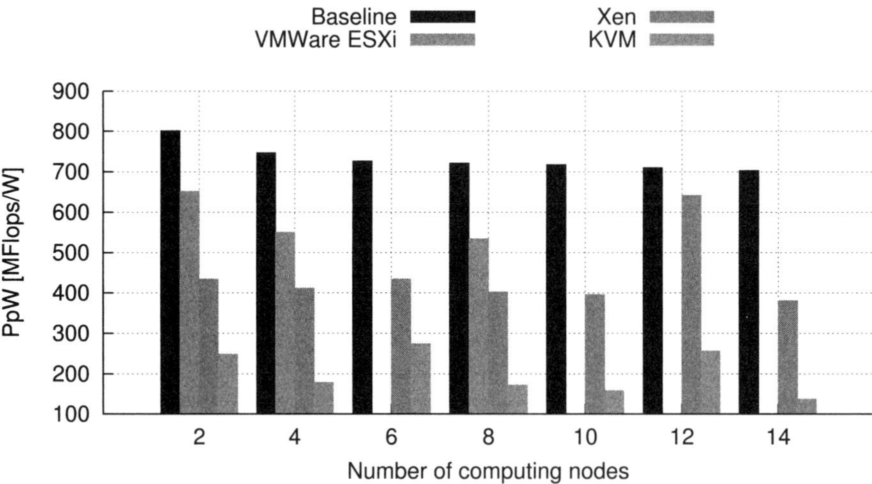

Fig. 13 Green500 PpW metric for the HPL runs over the virtualized frameworks and comparison to the baseline environments

RJMS middleware has to deal with several problematics to be efficient. First, it has to be able to equitably distribute the computing resources to user applications. This is one of the prerequisites for user satisfaction. Then, it has to keep a fairly high level of utilization of the platform resources and avoid utilization "holes" as much as possible. In order to provide the best service to the users, RJMS configuration and scheduling policies have to reflect the user needs and their input workload. Also by its central position, the RJMS is key to operate energy-efficient decisions, assuming that the user workload is better modeled and characterized.

Actually, the understanding of users workloads has motivated the study of production platform through the collection of traces from such systems and to the proposal of a standard: The StandardWorkload Format (SWF) [21]. It is an initiative to make workload data on parallel machines freely available and presented in a common format. This work, along with workload data collection, are presented in the Parallel Workload Archive (PWA)[2]. The idea is to collect and to redistribute several traces from real production systems built from the logs of computing clusters. With SWF, one can work with several logs with the same tools and the format enables to be abstracted from the complexity of mastering different ad-hoc logs from batch schedulers. These contributions enabled the study of numerous workload traces. This also led to the construction of several models, based on the statistical analysis of the different workloads from the collection.

In many of the workloads provided in the PWA, it was observed that the system resources utilization showed unused periods and that the service rate is higher than the arrival rate [22]. If the low utilization comes from a bad management of the

[2] See http://www.cs.huji.ac.il/labs/parallel/workload.

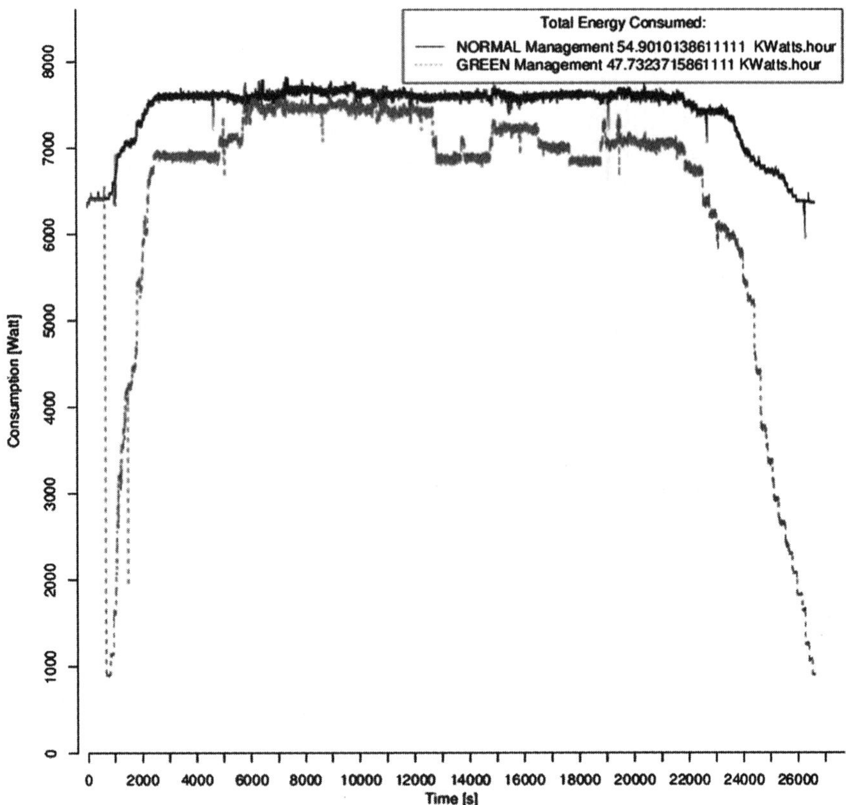

Fig. 14 Energy consumption for NAS BT benchmark upon a 32 nodes (biCPU) cluster with OAR. (Courtesy of Yiannis Georgiou; Source: [26])

resources, this will impact the users (and administrators) satisfaction. A platform not fully used while many jobs are still waiting to be granted for computing resources can be a symptom of this problem. If the low utilization is caused by an intrinsically low workload by the users, several techniques to take benefit of this underutilization can be used, such as energy saving that idles unused resources and save electrical power [23, 34]. The objective here is to define hibernation strategies that grant computing node to be powered off when idle during specific intervals. This mode is more and more present in recent RJMS such as OAR [20] (the RJMS used in Grid5000 or the UL HPC platform for instance), the idea being to wake up "*sleeping*" nodes when there is a job that need the "Powered OFF" resources.

The benefit from such energy saving techniques has already been proved by the GREEN-NET framework[23] works, as depicted in the Fig. 14. At this level, the intelligent placement of tasks and the shutdown of unused resources enabled an interesting energy gain, even with a high utilization of the platform.

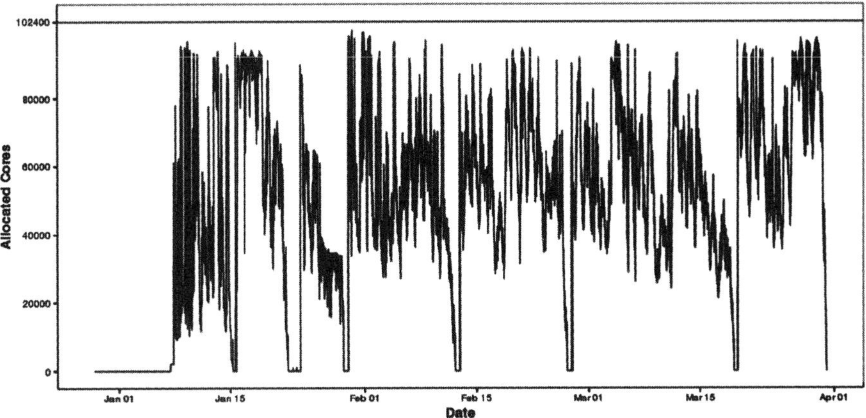

Fig. 15 System Utilization for TACC Stampede Cluster. (Source: [24])

Research on this topic aims at building an energy efficient scheduling through an accurate workload prediction model. Indeed, energy demand in cluster environment is directly proportional to the size of the cluster and the typical usage of the machines varies with time. This behavior is well visible in Fig. 15, representing TACC Stampede[3] cluster System Utilization in its first production months. During daytime, the load is likely to be more than during night. Similarly, the load drastically decreases over the weekend. Of course workloads can change upon different cluster configurations and energy saving can occur if this pattern can be captured. Hence a need for a prediction model arises and the research community currently focuses its effort onto the characterization of this model. The objective at this level remains to scan for current and future workload and tries to correlate it with the past load history to design an accurate energy saving policy.

Nevertheless, having an accurate workload model is not enough to increase the acceptance level of the users when an energy-saving mode is activated. In practice, the hibernation strategy described above generally degrades the average slowdown of the platform. This quantity corresponds for a given job to the ratio between the time spent by it in the system (wait time + run time) over the effective computation time (run time). A slowdown of 1 means that the job didn't wait, a slowdown of 2 means that the job waited as much as it ran. Thus the slowdown is affected by the delay required to wake up (*i.e.* boot) the computing resource. Improving the slowdown of a platform when an energy-saving mode is enabled assumes the continuous optimization of two key parameters associated to every computing resources of the platform: `idle_time` and `sleep_time` as illustrated in the Fig. 16. The first parameter affects, for a given resource, the time to wait once the last job has finished before powering off the node. The second depicts how long the machine should remain in a

[3] http://www.tacc.utexas.edu/resources/hpc/stampede.

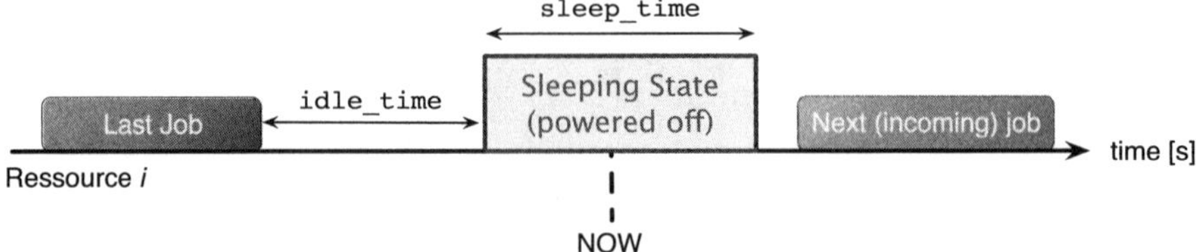

Fig. 16 Illustration of the two parameters to be continiously optimized when an hibernation strategy is in place

sleeping state *i.e.* powered off. The key challenge is to avoid the next job scheduled on that resource to be penalized by a delay to powering up such that in the ideal case, the arrival of a new job is statistically anticipated to power up the machine even before the submission so as to make the resource available at the time of submission.

4 Conclusion: Open Challenges

In this chapter we presented an overview of the latest advances in energy-efficient computing systems. Current HPC facilities were described, together with the methods of evaluating their performance and energy-efficiency. The research community is about to face many challenges on its way to build faster and more power-efficient systems. We presented some examples of technological and economical barriers that have to be overcome before building an exascale system.

The continued demand for new data center capacity and computing power requires particularly careful consideration of computational infrastructure's power consumption. The ambitions to create an exascale system by 2020 need to face many challenges in the design of supercomputers. However, even if this goal will not be reached, the attempts to build such a system would certainly push forward the frontiers of knowledge and contribute to the development of new, useful technology, just like CERN's Large Hadron Collider would have been successful even without discovering the Higgs boson. A few projects worldwide have already been focused directly on exascale systems, such as the Mont-Blanc project [4] (which aims at designing new computer architecture that will deliver exascale performance), FastForward [3] (which supports the development of technology that reduces economic and manufacturing barriers to constructing exaflop-sustained systems) or X-Stack program [15] (which targets the advances in the system software stack that are needed for transitioning to Exascale computing platforms). The outcomes of these projects have the potential to impact many areas of computing systems, from low-power embedded processors to cloud computing.

The exascale resources will bring competitive advantage to the country that will first reach them. The exascale digital design and prototyping will enable rapid delivery of new products. It will be the results of minimizing expensive or dangerous testing. Exascale computing will be an important part of national security. On the other hand, the concern of growing energy demand is becoming more and more visible nowadays. U.S. government austerity restrictions exert pressure to reduce investments in exascale computing. The industry will continue to move forward on its own on exascale projects, however progress will be much slower than with government assistance. The U.S. government is currently involved in an extensive consolidation program that promotes the use of Green IT, reduces the cost of data center hardware, software and operations and shifts IT investments to more efficient computing platforms and technologies. [31] Such initiatives clearly show the importance of research on energy-efficient computing systems.

Acknowledgments The research presented in this paper is partially funded by a grant from Polish National Science Center under award number 2013/08/A/ST6/00296.
The experiments presented in this paper were carried out using the HPC facility of the University of Luxembourg and Poznan Supercomputing and Networking Center.

References

1. AMD G-T40N. http://www.amd.com/us/products/embedded/processors/Pages/g-series.aspx.
2. AMD Unveils Server Strategy and Roadmap. http:/www.amd.com/us/press-releases/Pages/amd-unveils-2013june18.aspx.
3. DOE Extreme-Scale Technology Acceleration FastForward. https://asc.llnl.gov/fastforward/.
4. European Mont-Blanc Project. http://www.montblanc-project.eu/.
5. Grid'5000. [online] http://grid5000.fr.
6. Iceotope Servers. http://www.iceotope.com/.
7. Intel Atom Processor N2600. http://ark.intel.com/products/58916/intel-atom-processor-n2600-(1m-cache-1_6-ghz).
8. Intel Core i7-3615QE. http://ark.intel.com/products/65709/Intel-Core-i7-3615QE-Processor-(6M-Cache-up-to-3_30-GHz).
9. Intel unveils new technologies for efficient cloud datacenters. http://newsroom.intel.com/community/intel_newsroom/blog/2013/09/04/intel-unveils-new-technologies-for-efficient-cloud-datacenters.
10. IOR HPC benchmark. [online] http://sourceforge.net/projects/ior-sio/.
11. Iozone filesystem benchmark. [online] http://www.iozone.org/.
12. SuperMUC - First Commercial IBM Hot-Water Cooled Supercomputer. http://www-03.ibm.com/press/us/en/pressrelease/38065.wss.
13. The Green500 List - November 2013. http://green500.org/lists/green201311.
14. Top500. [online] http://www.top500.org.
15. X-Stack Software. http://www.xstack.org/.
16. PUE (tm): A comprehensive examination of the metric. White paper, The Green Grid, 2012.
17. Q. Ali, V. Kiriansky, J. Simons, and P. Zaroo. Performance evaluation of HPC benchmarks on VMware's ESXi server. In *Proceedings of the 2011 international conference on Parallel Processing*, Euro-Par'11, pages 213–222, Berlin, Heidelberg, 2012. Springer-Verlag.
18. P. Barham, B. Dragovic, K. Fraser, S. Hand, T. Harris, A. Ho, R. Neugebauer, I. Pratt, and A. Warfield. Xen and the art of virtualization. In *Proceedings of the nineteenth ACM symposium on Operating systems principles*, SOSP '03, pages 164–177, New York, NY, USA, 2003. ACM.
19. R. Bolze, F. Cappello, E. Caron, M. Daydé, F. Desprez, E. Jeannot, Y. Jégou, S. Lanteri, J. Leduc, N. Melab, G. Mornet, R. Namyst, P. Primet, B. Quetier, O. Richard, E.-G. Talbi, and I. Touche. Grid'5000: A large scale and highly reconfigurable experimental grid testbed. *Int. J. High Perform. Comput. Appl.*, 20(4):481–494, Nov. 2006.

20. N. Capit and al. A batch scheduler with high level components. In *Cluster computing and Grid 2005 (CCGrid05)*, 2005.
21. S. J. Chapin, W. Cirne, D. G. Feitelson, J. P. Jones, S. T. Leutenegger, U. Schwiegelshohn, W. Smith, and D. Talby. Benchmarks and standards for the evaluation of parallel job schedulers. In D. G. Feitelson and L. Rudolph, editors, *JSSPP*, pages 67–90. 1999. Lect. Notes Comput. Sci. vol. 1659.
22. D. T. D. G. Feitelson and D. Krakov. Experience with the parallel workloads archive. Technical report, School of Computer Science and Engineering, The Hebrew University of Jerusalem, 2012.
23. G. Da-Costa, J.-P. Gelas, Y. Georgiou, L. Lefèvre, A.-C. Orgerie, J.-M. Pierson, O. Richard, and K. Sharma. The green-net framework: Energy efficiency in large scale distributed systems. In *HPPAC 2009*, 2009.
24. J. Emeras. *Workload Traces Analysis and Replay in Large Scale Distributed Systems*. PhD thesis, LIG, Grenoble - France, To be defended October 1st 2013. currently available at: https://forge.imag.fr/docman/view.php/359/754/thesis_emeras_28aug13.pdf.
25. M. Flynn. Some computer organizations and their effectiveness. *IEEE Transactions on Computers*, C(21):948–960, 1972.
26. Y. Georgiou. *Contributions for Resource and Job Management in High Performance Computing*. PhD thesis, LIG, Grenoble - France, Sep 2010.
27. M. Guzek, S. Varrette, V. Plugaru, J. E. Sanchez, and P. Bouvry. A Holistic Model of the Performance and the Energy-Efficiency of Hypervisors in an HPC Environment. In *Proc. of the Intl. Conf. on Energy Efficiency in Large Scale Distributed Systems (EE-LSDS'13)*, volume 8046 of *LNCS*, Vienna, Austria, Apr 2013. Springer Verlag.
28. R. Januszewski, N. Meyer, and J. Nowicka. Evaluation of the impact of direct warm-water cooling of the HPC servers on the data center ecosystem. In To appear *in International Supercomputing Conference 2014, Leipzig, Germany*, 2014.
29. M. Jarus, S. Varette, A. Oleksiak, and P. Bouvry. Performance Evaluation and Energy Efficiency of High-Density HPC Platforms Based on Intel, AMD and ARM Processors. In *Energy Efficiency in Large Scale Distributed Systems*, Lecture Notes in Computer Science, pages 182–200. Springer Berlin Heidelberg, 2013.
30. A. Kivity and al. kvm: the Linux virtual machine monitor. In *Ottawa Linux Symposium*, pages 225–230, July 2007.
31. V. Kundra. Federal data center consolidation initiative. Memorandum for chief information officers, Office of Management and Budget of the USA, 2010.
32. E. Le Seur and G. Heiser. Dynamic voltage and frequency scaling: the laws of diminishing returns. In *HotPower'10 Proceedings of the 2010 international conference on Power aware computing and systems*, California, USA, 2010. USENIX Association Berkeley.
33. R. C. Murphy, K. B. Wheeler, B. W. Barrett, and J. A. Ang. Introducing the graph 500. In *Cray User Group*, 2010.
34. A.-C. Orgerie, L. Lefèvre, and J.-P. Gelas. Save watts in your grid: Green strategies for energy-aware framework in large scale distributed systems. In *14th IEEE International Conference on Parallel and Distributed Systems (ICPADS)*, Melbourne, Australia, Dec. 2008.
35. A. Petitet, C. Whaley, J. Dongarra, A. Cleary, and P. Luszczek. HPL - A Portable Implementation of the High-Performance Linpack Benchmark for Distributed-Memory Computers.
36. N. Rasmussen. Air Distribution Architecture Options for Mission Critical Facilities Whitepaper #55. Technical report, American Power Conversion, 2003.
37. S. Sharma, C.-H. Hsu, and W. chun Feng. Making a case for a Green500 list. In *Parallel and Distributed Processing Symposium, 2006. IPDPS 2006. 20th International*, pages 8 pp.–, 2006.
38. S. Varrette, M. Guzek, V. Plugaru, X. Besseron, and P. Bouvry. HPC Performance and Energy-Efficiency of Xen, KVM and VMware Hypervisors. In *Proc. of the 25th Symposium on Computer Architecture and High Performance Computing (SBAC-PAD 2013)*, Porto de Galinhas, Brazil, Oct. 2013. IEEE Computer Society.

39. M. vor dem Berge, J. Buchholz, L. Cupertino, G. Da Costa, A. Donoghue, G. Gallizo, M. Jarus, L. Lopez, A. Oleksiak, E. Pages, W. Piatek, J.-M. Pierson, T. Piontek, D. Rathgeb, J. Salom, L. Siso, E. Volk, W. U., and T. Zilio. CoolEmAll: Models and Tools for Planning and Operating Energy Efficient Data Centres. *To appear in: Samee Khan, Albert Zomaya (eds.) Handbook on Data Centers*.
40. M. vor dem Berge, G. Da Costa, A. Kopecki, A. Oleksiak, J.-M. Pierson, T. Piontek, E. Volk, and S. Wesner. Modeling and Simulation of Data Center Energy-Efficiency in CoolEmAll. Energy Efficient Data Centers. *Lecture Notes in Computer Science*, 7396:25–36, 2012.

Techniques to Achieve Energy Proportionality in Data Centers: A Survey

Madhurima Pore, Zahra Abbasi, Sandeep K. S. Gupta and Georgios Varsamopoulos

1 Introduction

A data center is a set of physical and possibly virtual machines along with other components such as storage, network, cooling, power supplies and management software, that function together to serve data and information to facilitate information services to a business or organization. It consists of computing and data dissemination as the main functions, however there are several other physical elements such as cooling management and power budgeting that interact with the computing elements, thusly making a data center to exhibit both a cyber and a physical behavior (see Fig. 1).

The "greening" of data centers has been a focus of both the industrial and academic research. This is due to the large amount of energy consumed to power the large number of computers and other equipment in data centers. This enormous energy consumption imposes a huge cost to data center providers. Specifically, with the growing demand on data and internet services, data center providers are faced with

This work has been funded in parts by NSF grant CSR #1218505 and CRI #0855277

M. Pore (✉) · Z. Abbasi · S. K. S. Gupta · G. Varsamopoulos
Department of Computer Science and Engineering, School of Computing and Informatics, Ira A. Fulton School of Engineering, Arizona State University, Tempe, AZ 85281, USA
e-mail: madhurima.pore@asu.edu

Z. Abbasi
e-mail: zahra.abbasi@asu.edu

S. K. S. Gupta
e-mail: sandeep.gupta@asu.edu

G. Varsamopoulos
e-mail: georgios.varsamopoulos@asu.edu

S. U. Khan, A. Y. Zomaya (eds.), *Handbook on Data Centers*,
DOI 10.1007/978-1-4939-2092-1_4

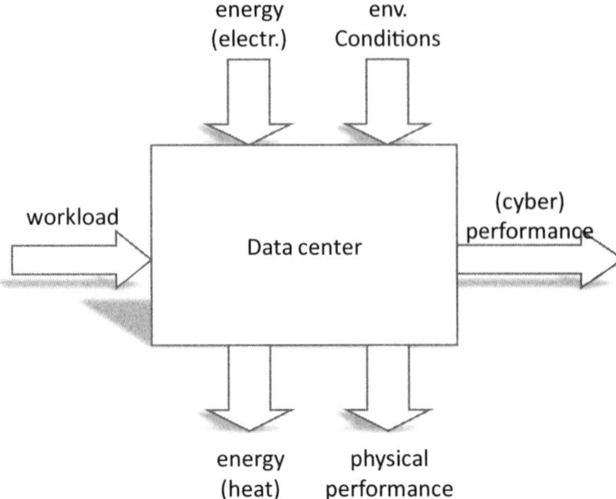

Fig. 1 A data center has both a cyber (computing) performance aspect as well as a physical (energy) performance. Examples of computing performance are throughput and response delay, whereas examples of physical performance are energy efficiency, PUE (power usage effectiveness) and carbon footprint

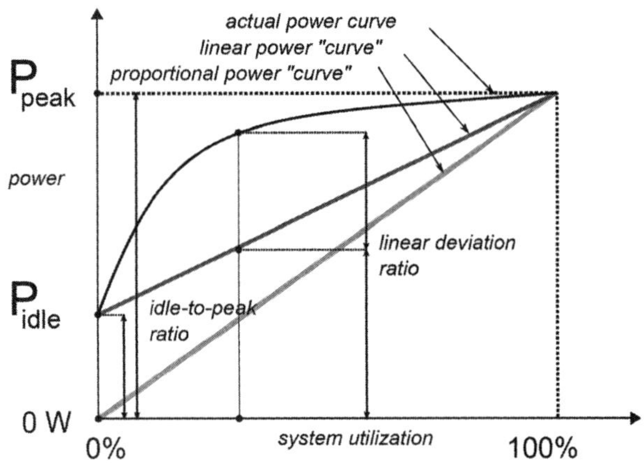

Fig. 2 Power profile of ideal energy-proportional system

huge computing and cooling energy costs. The challenge is to curb these costs while accommodating future expansions. In response, computer vendors are developing power management techniques for computing, storage and networking equipments. These techniques aim to provide energy-proportional computing at server and data center level (Fig. 2).

Power management techniques at the component and server level are designed to (a) switch server components off during idle times, (b) adapt the rate of operation of active server components to the offered load, and (c) exploit low power states of the server and its components during low utilization levels.

Further, several power-managing and workload-shaping techniques are proposed at the data center level or at the cluster level to (i) assign workload in a more energy efficient way, (ii) move workload off of the under-utilized servers and then switching those servers to low power modes, i.e., consolidate workload and apply server provisioning, (iii) select *active servers* such that less cooling energy is incurred, (iv) consolidate heterogeneous applications in fewer servers through virtualization, and (v) manage workload to leverage electricity price across data centers. The efficiency of power and workload management techniques at different data center levels comes from the following:

- Current servers have non-zero idle power and a non-linear power-utilization, i.e. they are not ***ideally energy-proportional***, despite the technological trend toward more energy-proportional servers. Ideally energy-proportional servers consume zero power at idle and have a linear increase of power with respect to utilization.
- Data center workload is not uniform over time and exhibits significant variation, with the peak workload being two to three times greater than the average [1, 2].

While much has been published in the area of data center, server, and component level power management, there is no work, to our knowledge, that evaluates these approaches with respect to each other and analyzes the gaps in the research. In this chapter we focus on different *software management aspects* that use the existing hardware capability to build energy-proportional servers and data centers. We give an overview of the existing solutions, present their associated challenges and provide taxonomy of the current research. Further, we compare the data center, server, and component level power management techniques with respect to their implementation, reliability, scalability, performance degradation and energy proportionality. The rest of the chapter is organized as follows: We introduce energy proportionality and review the need for energy proportionality in Sect. 2. Component level power management and energy proportionality are discussed in Sect. 3, followed by power management techniques at server level in Sect. 4. The chapter goes to further depth in *server provisioning* schemes (Sect. 5.1), *VM management* (Sect. 5.2) and other data center level power management techniques (Sect. 5.3). We compare the data center level and server level power management techniques in Sect. 8. The chapter concludes in Sect. 9.

2 Energy Proportionality

The concept of energy proportionality emphasizes the increase in power consumption of a system in proportion to the work done by the system. The notion of energy-proportional computing was first coined by Hölzle and Barroso [1] which is inspired by the near-energy-proportional behavior of the human body: during sleep, the human body consumes approximately 81 W; while actively working, such as sprinting or running, it consumes more than 1000 W [3]. An energy-proportional system will have a low idle power and wide dynamic range as well as linear power increase with respect to utilization (see Fig. 4a for an ideally energy-proportional server). If the energy efficiency of a system is the maximum work done per Joule of energy, then "Energy proportionality" implies *constant energy efficiency* independent of the utilization level (see Fig. 4a). Thus, energy proportionality or power proportionality (used interchangeably hereafter) is important to achieve overall good energy efficiency in

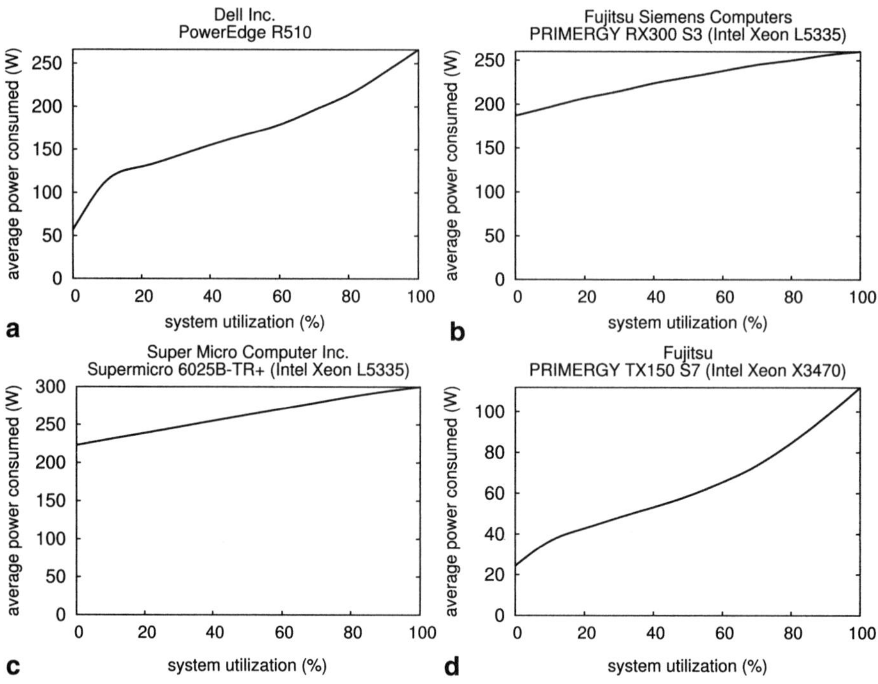

Fig. 3 Power profile of various server platforms from SpecPower [7]

data centers, since data centers have periods of low utilization. A study by Google shows that most of the servers are operated in the lower utilization spectrum at about 30 % [1]. The main reason for servers being so under utilized is that most of the data centers over-provision their computing resources to match workload peaks which are two to three times the average workload rate [1, 2]. Current servers are far from being ideally energy-proportional: they have non-zero idle power and their power-utilization curve is nonlinear (see Fig. 3). Therefore, research efforts led to design of power management techniques both at the server and the data center level to increase energy proportionality of servers and data centers, respectively. The following subsections introduces energy proportionality metrics for server level and data center level.

2.1 *Energy Proportionality at the Server Level*

A server's energy proportionality can be described by its power curve with respect to its utilization. The power-utilization curve depends on the hardware characteristics of a server as well as the type of workload that is offered to the server [4–6]. Therefore,

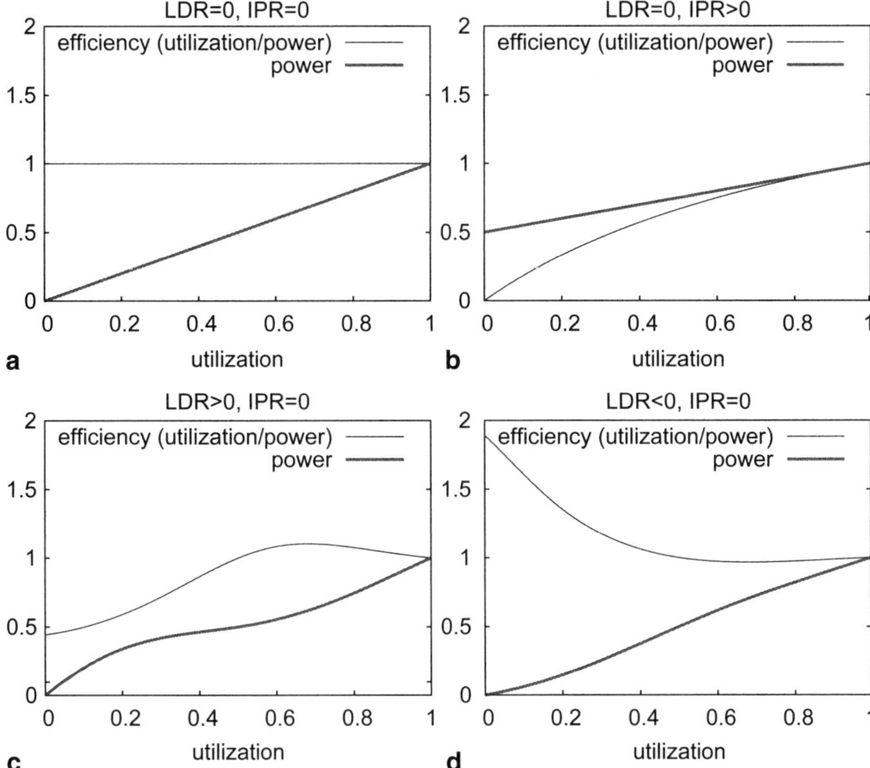

Fig. 4 Server power profiles showing different cases of energy proportionality [4]

the energy proportionality metrics of a server can be defined according to the power-utilization curve and the type of workload.

Varsamopoulos and Gupta [8] suggested two metrics for measuring energy proportionality according to the servers' power-utilization curve: The *idle-to-peak power ratio* (IPR), which measures how close the idle power of a server is to zero, and the *linear deviation ratio* (LDR) which measures the deviation of the *actual power curve* from a hypothetical *linear power curve* as shown in Fig. 2. These metrics can be used to compare the energy proportionality of servers, moreover they help to determine the utilization level where the system has the most energy efficiency.

Varsamopoulos et al. [8] performed an extensive study of power profiles of current servers and observed the trend of reduction in the idle power of the server but the power curve is observed to become more nonlinear. For example, some of the recent servers have (IPR, LDR) of (0.3, 0.4). Figure 4 shows examples of how energy efficiency at different utilization levels changes over various IPR and LDR.

Figure 4a shows the power-utilization curve of an ideal energy-proportional server. It can be seen that the energy efficiency is constant for all the workload utilization range. Figure 4b shows the power utilization curve of a server with positive IPR value which indicates non-zero idle power. Similarly, Fig. 4c shows the power utilization curve of a server with positive LDR which indicates that the system has higher energy efficiency at higher utilization while Fig. 4d shows the power profile of a server with negative LDR which indicates that the server will be more energy efficient at lower utilization levels.

The power curves shown in Fig. 3 are based on utilizing the entire server using CPU-Memory based transactional workload evaluated by SpecPower [7]. However, there are also some studies that show the dependency of the energy proportionality of a server to the type of workload offered. Chun et al. [9] observed that different platforms have different energy proportionality characteristics and exhibit different power performance with respect to the type of workload (e.g. I/O intensive, CPU intensive and transactions). In case of database analytic workloads on a modern server, the CPU power used by different operators can vary widely by upto 60 % for the same CPU utilization, and that the CPU power is not linear with respect to the utilization [6]. Feng et al. [5] observed that the power consumption varies for different components according to the workload type such as memory-bound, CPU-bound, network or disk-bound hence they motivate the need for power management techniques at the server level to incorporate the type of workload and consider platform architecture for improving the energy-proportionality of data centers. Given the wide variety of heterogeneous applications with different workload types, the task of determining energy proportionality metrics is a tedious task.

2.2 Energy Proportionality at Data Center Level

The power consumption of a data center consists of the power consumed by the computing equipment, cooling systems and other sources of power consumption such as lighting and power losses [10–14]. Power Usage Effectiveness, PUE, is a widely used metric to measure the energy efficiency of non computing equipment in data centers [15]. It is calculated as the total power consumed by a data center facility over the power consumed by the computing equipment only:

$$\mathrm{PUE} = \frac{\text{Total Power into the data center}}{\text{IT equipment power}}. \tag{1}$$

A large PUE is a strong indication of large cooling power consumption, since the cooling system is the biggest consumer of the non-computing power in a data center (followed by power conversion and other losses). Power proportionality of a data center depends on both the power proportionality of servers and the magnitude of PUE. Ideally, data centers should have a PUE of one. If data centers have PUE of greater than one, even if the servers are ideally energy-proportional, the data center is not ideally energy-proportional. To show this, we did a simulation study where a data center with PUE of 1.3 is fully utilized. In this set up, the thermal profile of the simulated data center in [11] is used (see Sect. 5.1 on how data centers are thermally modeled and profiled). Further a linear power-utilization curve for servers

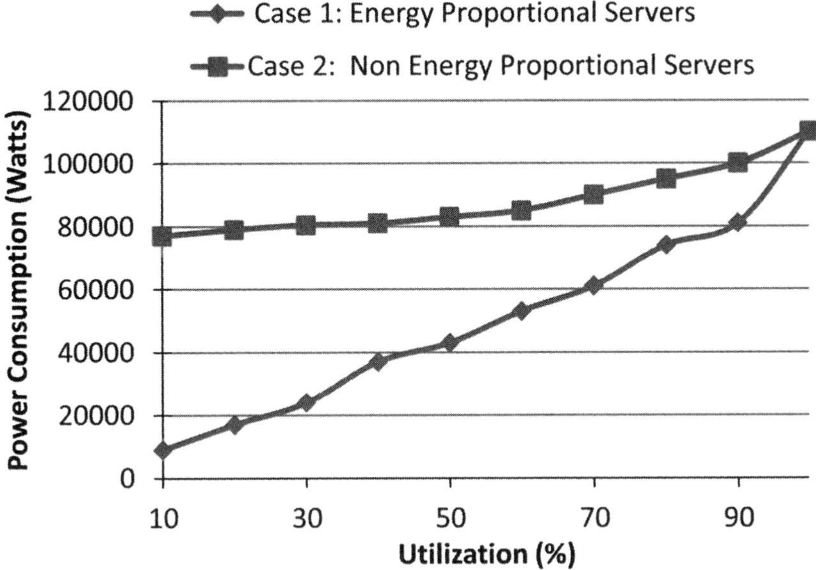

Fig. 5 Power utilization curve of a data center with PUE = 1.3 for two cases of servers with peak power of 300 W: case 1, IPR = 0, and case 2, IPR = 0.6

are assumed and their peak power is set to 300 W. We performed two experiments, where in the first case the servers' have IPR of zero (i.e., ideally energy-proportional), and in the second case the servers have IPR of 0.6. Figure 5 shows that the power utilization curve of the data center with ideally energy proportional servers is not linear because of non-ideal PUE. Figure 6, further shows that at low utilizations, power proportional servers significantly reduces the total power consumption of the data center. Prominent data center owners such as Yahoo, Facebook and Google have reported PUEs of around 1.1, while data center statistics show an overall national PUE of 1.9 in 2011 [16].

2.3 Overview on Power Proportionality Techniques at Different Data Center Levels

Given existing non-power proportional servers and large PUE of data centers, current research efforts are towards proposing power management techniques at different levels, ranging from component to data centers, thermal-aware job placement and workload shaping to achieve energy proportionality at data centers, summarized below (also see Table 1):

- **Component level**: Power management techniques at this level leverage the dynamic power state transitioning capabilities available at different components of a

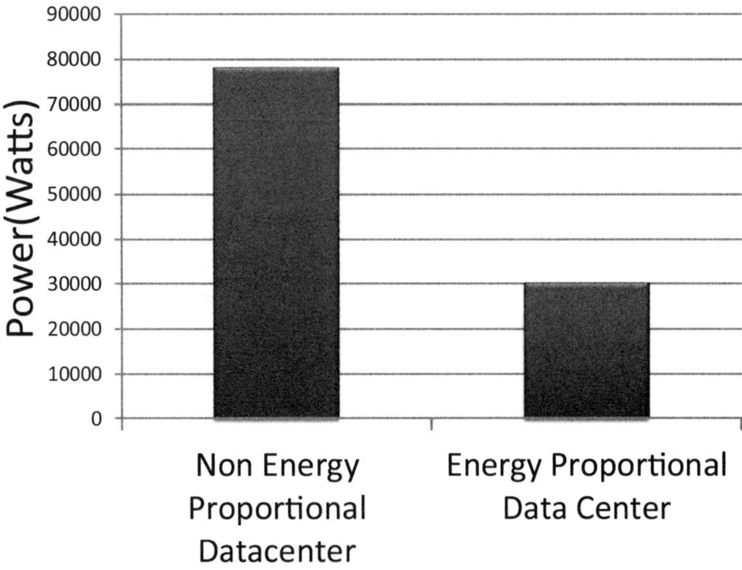

Fig. 6 Power consumption of two data centers at 30 % workload with PUE = 1.3 for two cases of servers with peak power of 300 W: case 1, IPR = 0, and case 2, IPR = 0.6

Table 1 A taxonomy of research in energy proportionality at different levels in data centers

Level	Power management technique	Articles
Component-CPU	Active power transition i.e., DVFS	[2, 17–20]
Component-Memory	Architecture modification i.e. rank subsetting, DVFS	[21–23]
Component-Network	Sleep mode transitioning, architectural modifications	[24–28]
Component-Cooling	Fan speed control	[29]
Server	Inactive power mode transitioning	[30]
Data center/cluster	Server provisioning	[12, 31–37]
	VM management	[38–41]
	Workload shaping	[12, 34, 42, 43]
	Hybrid data center	[9, 36, 44, 45]
Global level	Workload and server management across data centers	[46–52]

server (such as CPU and memory), in order to improve the energy proportionality of the server.

- **Server level**: Modern servers have different power states including hibernate, active mode and inactive mode, with fast power state transitioning capabilities enabled with Wake-On-LAN technique. Existing server level power management techniques leverage this capability to control the power state of the servers depending on the availability of the workload at a given time interval.

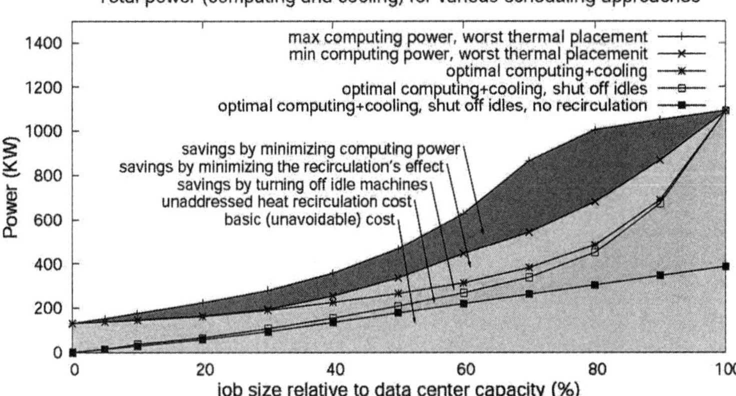

Fig. 7 Power profile of an example data center as a result of using different management schemes. The data center consist of mixed equipment of different power and workload-capacity ratings and cyber physical interactions

- **Data center/cluster level**: In the warehouse scale of computing, data center applications require large number of servers to run, with the total number of servers required is decided based on the peak workload. Hardware and software resources in such a massive scale work continuously and efficiently to maintain Internet service performance. Design and deployment of their power management schemes at data center/cluster level (server power provisioning) leverage the variation of the workload, physical layout of the data centers, and heterogeneity of servers to dynamically decide on the active server set, in order to remove unnecessary idle power and reduce the cooling power. Workload shaping and Virtual Machine (VM) management are also performed in the data center level, where the former concerns power efficient workload distribution across active servers, and the later concerns, power efficient VM placement/ migration across servers of a cluster.
- **Global level (across data centers)**: Large scale Internet applications tends to be replicated across several geo-distributed data centers primarily to provide timely, reliable, and scalable services to their global users. Power and cost management techniques at this level concerns designing workload distribution policies to leverage the spatio-temporal variation of data centers power and energy cost without compromising the desired quality of service.

The above solutions often complement each other to make power proportional data centers. Power efficiency of each of these techniques is constrained by the hardware power management capability available at each level, workload characteristics and the physical layout of the data center. Figure 7 gives an example of how a mix of such management techniques can help to improve the power proportionality of data center. In the following sections, we give a detailed description of the above power management solutions.

Table 2 Breakdown of power consumption in servers. (Data source: Fan et al. [54])

Component	Peak power (W)
CPU	80
Memory	36
Disks	12
Peripheral slots	50
Motherboard	25
Fan	10
Power supply unit	38
Total	251

3 Energy Proportionality at Component Level

Many of the current server components support only two power states, i.e., on-off, yet the technology has started allowing component-level energy proportionality by providing multiple levels of operating frequency (clock) of the device, duty cycling, or reducing the operating voltage [53]. Traditionally, CPU has been the largest, yet not dominant, contributor to the power consumption (Table 2). However currently energy consumption of other components such as the disk drives and network and cooling equipment is becoming significant (see Table 2).

The components in the server can be used in different power and performance states. The specifications of power states are given by the Advanced Configuration and Power Interface (ACPI) which provides a standard interface for hardware vendors and developers to make use of the available power and performance management at global level (for entire server system) or component level [55]. ACPI defines a scales of P which are performance states, P0, P1 etc., and C which are sleep states, C0, C1 etc.

3.1 Energy Proportionality at the CPU

1) Dynamic Voltage Frequency Scaling (DVFS): It is a mechanism that adjusts the clock frequency of the CPU. Reducing the operating frequency of CPU, the power consumption incurred by the CPU can be reduced. However, the lowest operating frequency is limited by the stable voltage requirements of the circuit. The power consumption by DVFS at a frequency f is given by:

$$P(f) = CN_{\text{sw}}V_{\text{dd}}^2 f,$$

where C, is the capacitance of the circuit, a significant percentage of which is wire-related, N_{sw} is the average number of circuit switches per clock cycle and V_{dd} is the supply voltage to the CPU [56]. As the maximum frequency is linearly dependent on the supply voltage, DVFS has a cubic effect on the power savings. The estimated power using DVFS is given by:

$$P(f) = P_{\max}\left(1 - \left(\frac{f}{f_{\max}}\right)^3\right). \tag{2}$$

DVFS reduces the instantaneous power consumption, yet increases the execution time of the application or the response time of the requests. It is worth noting that, the overall energy efficiency of DVFS scheme is because power reduction is cubic effect of frequency at the cost of increase in the execution time which is inversely proportional to the frequency. In the case of delay sensitive applications (e.g., internet services) applications it is essential to maintain the response time within a certain limits. Therefore, the challenging task is to design an online DVFS control scheme that scales frequency proportional to the input workload without violating the response time. The design of such a control system requires to take into account different factors such as (i) adaption time of the system i.e. the time required by the server to adjust the new frequency setting, (ii) workload characteristics such as workload arrival rate, and (iii) application's response time constraints. Existing solutions adapt techniques from control theory and queuing theory (e.g., M/G/1) to explore the energy delay trade-off of DVFS schemes.

Similarly, in the case of delay tolerant applications (e.g., compution intensive scientific applications), DVFS increases the overall execution time of the task. As a result, even if the instantaneous power savings may be obtained there exist a penalty in the form of increase in the execution time. The energy delay trade-off can be explored to maximize the energy savings without exceeding the deadline.

2) Core Power Gating Power Gating involves removing the power supply to a CPU core. This is achieved by inserting a sleep transistor in path of power supply and core. The gated block is entirely cut off from the power supply which results in no power consumption in the gated block. In other words, it eliminates the idle power of the core entirely. Many of the server machines do not provide a knob to the user to use the power gating. However, if this knob is provided, it is one more step towards reducing the CPU power by consolidating the workload during low utilization on fewer cores. The main considerations to design a power management scheme for power gating is the delay to get the core back to active state. Power management based on workload prediction can benefit from power gating to adjust the available CPU resources according to the varying intensity of workload.

Examples from the Literature

A wide research has been done in the past few years in order to design online control schemes for CPU power management (see Table 3). Several algorithms such as model predictive control (MPC) [65], control loop, preemptive or reactive control [10], or proportional-integrative-derivative (PID) control [66] have been used to obtain power and energy savings using DVFS. Study by Andrew et al. [57, 58] suggest that, if the mean value of workload is known, then better energy efficiency is achieved, e.g., using algorithms such as shortest remaining time first [57]. In this

Table 3 Taxonomy of researches on CPU power management

Scheme	Types	Articles
DVFS control	Static, dynamic scheduling	[57, 58]
Heterogenous workload	Memory, CPU	[59, 60]
Multiprocessor DVFS		[58, 61]
Thermal-awareness for single and multicore		[62, 63]
Power gating		[64]

(static) scheduling scheme, a single value of frequency is set in DVFS. However, for bursty workload, dynamic speed scaling solutions are proven to be better. Analytical study by Cho et al. [67, 68] obtained the performance for multicore when DVFS is used for energy minimization of parallel applications. Although DVFS dominantly affects the performance of CPU-bound applications, its effect on memory-bound applications is shown by Dhiman and Rosing [60]. An online control algorithm is suggested for heterogeneous applications (i.e. a mixture of CPU-bound and memory-bound) by Ge et al. [59]. Apart from managing performance versus energy savings trade-off, DVFS is also used for thermal management of servers [63, 69].

Few works also use core power gating. In particular, a group from HP [64], demonstrated the efficacy of power gating for multi core architectures comparing energy savings and performance degradations for several database applications used in their data centers.

3.2 Energy Proportionality at the Memory

Main memory now contributes to a significant portion of a server's power consumption; in contemporary systems, it accounts for almost 40 % of the power consumption. Recent technology progress has allowed a few major companies to develop memory with various power states. Any memory power management should ensure the performance of memory if DRAM low power states are used.

The main memory (DRAM) exhibits *static* and *dynamic* energy consumption. Dynamic energy is consumed in the decoding of address and fetching the data from the memory. The static energy is consumed during the active period amortized over number of data transfers. If E_{rw} is energy per read or write, BW_{rw} is the read or write bandwidth, D are the total DRAM channels, E_{AP} is energy to activate and pre-charge, f_{AP} their frequency then the energy consumption for each DRAM channel is given by [22]:

$$E_{DRAM} = StaticEnergy + E_{rw}BW_{rw} + DE_{AP}f_{AP}. \tag{3}$$

The time required to retrieve data from main memory affects the performance of the memory based applications. For fetching data from main memory, the probability of hit and miss of the previous level of memory i.e. caches can be considered. If the hit probability is p_{hit}, miss probability is given by $p_{\text{miss}} = 1 - p_{\text{hit}}$. The time required to fetch a data from main memory with one level of cache is given by:

$$t_{\text{DRAM}} = p_{\text{hit}} t_{\text{access}} + p_{\text{miss}} t_{\text{miss}}. \tag{4}$$

1) Memory Architecture Modifications One of the promising energy saving technique for improving energy consumption is using memory rank subset where memory is divided into smaller chunks. A small hardware modification enables the access to the memory rank subsets. Dividing the memory into ranks and using smaller subset of memory instead of whole memory results into saving of activation and pre-charge energy associated with the rank subsets that are not accessed. However, immediate effect of this technique is that the data path for each access becomes longer. The design of memory scheme includes different factors such as load balancing across memory ranks, number of memory ranks that affect the effective bandwidth as well as the application characteristics. Several other methods of power saving include managing the refresh rates of memory, use of memory buffer, etc.

2) Memory Low Power Modes The memory now has more power states e.g., RDRAM (Rambus DRAM) provides four different power states: active, standby, nap and shutdown. Power management schemes for the memory use these states to reduce the energy consumption.

- *Static power management:* In this scheme, the memory is assigned a low power state. When the memory access occurs, the chip has to resume to the active power state.
- *Dynamic Power management:* In this scheme, the time interval of a low power state is varied according to the access pattern. The threshold time interval after which the memory is in low power state is a critical design aspect of the power management. This threshold is determined such that energy savings are improved but delay are within the time constraints of the application.

Examples from the literature

Control strategies using proactive and reactive techniques limit the delay within constraints of the applications to increase the power savings [70–72]. In the Rank subset scheme proposed by Ahn et al. [73], frequency scaling of access has been used to gain energy savings. Further, Deng et al. [21] proposed reducing the voltage across memory channels in addition to actively changing the frequency of memory access depending on the workload, i.e., performing both DVFS and DFS for the memory. Using such a scheme they target the idle memory power consumption as well as dynamically scale the power consumption with changing rate of memory accesses. More energy savings are obtained with techniques that co-ordinate the memory accesses, reduce memory conflicts and exploit more data locality (e.g., using address mapping schemes or transaction order schemes).

3.3 Energy Proportionality at the Disk

Most of the storage disk have capability to transition to on-off power states. When the disks are not in use, they are either in idle state, standby state or off state . Consider that d_n total number of memory fetches in the storage disk, P_{active} is proportional to d_n by a constant factor d, P_{standby} is the power dissipated during disk I/O in the low power state, t_{active} is time spent in active state and t_{standby} is time spent in the low power state, then the energy consumption is given by:

$$E_{\text{disk}} = d P_{\text{active}} t_{\text{active}} + P_{\text{standby}} t_{\text{standby}}. \tag{5}$$

The time required to fetch data from disk is important in designing a disk power management scheme. If t_{seek} is seek time, t_{RL} is the rotational latency and t_{tt} is the transfer time from disk to higher level cache, then the time required to fetch a data from disk is given by:

$$t_{disk} = d_n(t_{seek} + t_{RL} + t_{tt}). \tag{6}$$

1) Disk Spinning Down The most common technique of power management in the disks is spinning down (i.e., switching the power off) of the disk when not in use. However, the time to restore the disk to active state is of the order of few seconds and there are sudden variations in the data center workload, such a scheme may degrade the performance of delay sensitive applications severely. In other words, the reactive schemes such as simple spin down with timeout may not be effective for data center with high workload variations. Proactive power management schemes for disk can be designed for specific workload type, application constraints, workload arrival rates. Such schemes can use prediction based techniques to schedule the disk spinning down in the idle period in the workloads thereby reducing the performance degradation of the applications due to power state transitions.

2) Managing Data Storage and Replication Data center applications usually have a large data set stored in multiple storage disks. In applications that involve popular data (e.g., search engines) some of the data is more frequently accessed than the rest of the data. The popular data can be identified and stored on fewer disks and replicated for performance, where as the remaining data is stored on remaining disks. The disk with popular data are always in active state while more power management schemes are applied to remaining disks. Other power management scheme consist of using hybrid disk types such as combination of Solid State Drives (SSD), Flash Storage Devices and Dynamic Random Access Memory (DRAM) to manage the data storage based on the combination of their power and performance characteristics and their costs. Popular data is migrated to more energy efficient disks. However, moving the data frequently may exceed the savings obtained by a spinning down of disks.

Examples from the Literature In analytical study of disk based storage devices, the performance effect of disk spin down technique is evaluated by modeling the probability of transitions of disk into different power states using a queuing model [74]. Popular power management schemes proposed for disk power proportionality in the research are:

- *Consolidation:* The data is moved to a fewer number of storages [75] and the use of scale out (data not shared) architecture is discussed by Tsirogiannis et al. [76].
- *Migration:* Data is stored in more power efficient devices, e.g., SSD [76, 77], hybrid disks such as NAND flash storage and DRAM [78] or according to popularity of data [79].
- *Aggregation:* Read or writes are postponed in order to increase the idle times between the operations to create more opportunity for energy savings [25].
- *Disk spin down:* Spinning down the disk during idle periods is used independently or in combination of other management policies [75].
- *Compression:* Compression of data is used in some cases of workload [25].

Many works in this area suggest that combination of different techniques that can give better energy-proportional behavior for disk storage devices for different workload types and application requirements [75, 80].

3.4 Energy Proportionality at the Networking Interface

The average utilization in the data center is very low and idle networks components such as ports, line cards, switches, are one of the significant consumers of energy in the low utilization periods [25]. Popular power management technique for network devices is to switch off the network components during idle periods [28]. To model the switching of network components, a linear power model is assumed. When the network components are not in use they are either in idle state or standby state. If P_{active} is the active state power and P_{standby} is the power in the standby state, t_{active} is time spent in the active state and t_{standby} is the idle period, E_s is the energy required for switching between the power states and n_s is the number of switching that occurred, then the energy consumption is given by:

$$E_{\text{net}} = P_{\text{active}} t_{\text{active}} + P_{\text{standby}} t_{\text{standby}} + n_s E_s. \tag{7}$$

The time required to transfer the data through a network component is given by:

$$t_{\text{net}} = t_{transfer} + t_{switching}. \tag{8}$$

1) Switching off the Network Component The most popular technique to save network power switch off network devices during the idle times. Some of the techniques use a reactive scheme that switches-off the network component for a certain time after observing that there is no workload for few seconds. Some techniques involve proactive schemes where network interfaces are continuously monitored to learn the inter-arrival time between packets in a window based scheme. Based on the history of workload, the network devices are switched off until the predicted arrival of workload.

2) Managing the Workload These scheme involve managing the data path of the workload by routing the network data through a certain section of network. Hence, the network components in the remaining sections can be switched off. If the application deadlines are not stringent the data can be aggregated, stored in buffers for some time and sent. This will allow the network components to be switched off during idle period. The main focus of such algorithms is to create more opportunity of idle period in network components with minimal performance degradation. Unpredictable workload is a major hurdle in the design of network power management schemes. Some of the schemes require modifications to the existing network architecture. In such cases, the deployment of such a scheme may not be feasible.

Examples from the Literature Various techniques have been explained in the literature to save energy in the network devices but network components used have a limited capability to support multiple low power states.

- *Sleep:* In this scheme, the network components such as switches, routers are put to sleep or switched off in the idle period in between of workload arrivals [24, 25].
- *Aggregation:* Modifying the network topology Nedevschi et al. [24] consolidated the network flow on fewest possible routes such that the data is sent on minimum active set of network devices.
- *Rate adaptation:* Use of rate adaptation technique is demonstrated by adjusting the workload rate such that traffic is serviced within the required time constraints [25].
- *Traffic shaping:* In ElasticTree scheme [27], the traffic is split into bursts, such that traffic to same destinations is buffered before it is routed. This scheme increased the idle periods between the traffic bursts used to transition the network devices into low power states.

Also Nedevschi et al. [24] performed a study comparing energy savings from different power management schemes such as sleep and link rate adaptation using supported hardware. Their schemes show energy savings obtained are comparable to the optimal scheme. Above mentioned techniques give significant energy savings in data center networks where devices are mostly utilized at around 20 %.

Different techniques to reduce the power in various components of servers are summarized in Table 4.

4 Power Management Techniques at Server Level

The low power states available in current servers may not contribute to significant power savings. Hence servers are put into sleep state or power down to achieve more power savings. However the time required to reinstate to active working state is

Table 4 A Taxonomy of researches for energy proportionality at different levels in data centers

Level	Power management technique	Articles
Server	Inactive power mode transitioning	[30]
Component-CPU	Static and dynamic DVFS	[10, 65, 66]
Component-Memory	Power state management	[70–72]
	Rank sub-setting (modifying architecture)	[73]
	Frequency scaling	[21]
Component-Storage Disk	Consolidation	[75]
	Migration	Devices [77, 78], data popularity [79]
	Aggregation	[80]
	Disk spin down	[80]
Component-Network	Sleep	[25]
	Aggregation	[24]
	Rate adaptation	[24]
	Traffic shaping	[24]
	Low power states	[25]
Component-Cooling	Fan speed control	[29]

critical in the design of power management scheme. To obtain energy proportionality inactive power transition scheme is used where idle servers are put in the low power state, and they wake up on arrival of workload. On arrival of workload, if the system is active, then workload is processed immediately else it processed after transition delay.

For workload $x(t)$, arriving at time t, the system state is:

$$\begin{aligned} \text{State} &= S_{\text{active}} && x(t) \neq 0, \\ &= S_{\text{idle}} && x(t) = 0, \end{aligned}$$

where, S_{active} and S_{idle} are the active and idle server states, respectively.

The server is modeled as M/G/1 queue. The workload can be modeled using various parameters such as the rate of workload and transition time to active server state and service rate. It can be evaluated for energy savings using performance constraints for servicing that workload. The main issue is that currently such highly efficient server machines with much lower transition rates as comparable to inter arrival periods of workloads are not available. The model can then be used to optimize power consumption, energy consumption, performance or energy delay trade-off.

Server Power Model Many works that model data center, commonly assume linear power models for servers [4, 81]. The power at idle state is constant, and it linearly increases with utilization. If u_t is CPU utilization of server at time t, α is the proportional constant and P_{idle} is the idle power of the server when no workload is running, server power is given by:

$$P_{Server} = \alpha u_t + P_{\text{idle}}. \tag{9}$$

However more servers are becoming non linear as shown by study of several SpecPower results by Varsamopoulos et al. [4]. Also, a group of researchers from Oak Ridge National laboratories have investigated CPU intensive applications and suggest that the nonlinearity of power is due to several factors such as number of cores, sleep time due to power management techniques [81].

Server Performance Model Many of the works show analytical study of server performance by modeling it as M/G/1 queue. Considering the transition overheads of sleep time, transition time, wake up, service time within the constraint of response time, this study shows the performance of these models for different transition time using the Poisson arrival [82].

Examples from the Literature The low power active working states are not available for the current server systems hence the power management algorithms try to put the server into lowest power consumption mode during the idle periods in workload. Initial work for servers showed promising energy saving benefits for switching of the servers when idle [83]. Many research work have shown different server provisioning algorithms that use consolidation of workload on fewer servers while switching off the remaining servers (See Sect. 5.1). Very few works focus on server power transition [29, 30]. To obtain energy savings, the server is transitioned to a lower power state between the bursts of workloads as well as during the idle period between the arrival of workloads. The decision to transition a server state into a sleep state e.g. PowerNap or use one of the active low power state e.g. DVFS depends on the trade-off between the time required for reinstate the server, the rate of workload arrival and the SLA constraints [82]. Meisner et al. [30] investigate the observed idle times for different types of data centers workloads. Detailed study of live traces of data center showed that in the current data center workload, the idle periods (mean 100ms) particularly for web and IT workloads can be exploited to obtain power savings if fast transitions to low power state are available for a server while sleep states are not very useful in data base analytic workload [84]. Using live data traces and analysis, Gandhi et al. [85] show that power savings can be obtained for different characteristics of traces. On a multicore platform, the challenge is to co-ordinate the sleep time of the cores. Few of the research use control theory based approach to predict the arrival of workload and control the sleep time (or time in the low power state) [86, 87] such that the response time constraints are satisfied.

5 Data Center/Cluster Level Power Management

Warehouse scale of computers is a huge data center of servers equipped with supporting resources such as network, cooling infrastructure for providing continuous services. Energy consumption of such data center is a huge problem and the data center operators require to employ different power management techniques to address this. This section describes different schemes for increasing the energy efficiency of the data center, their trade-offs in maintaining the quality of service requirements and their design challenges in detail.

5.1 Server Provisioning in Internet Data Centers (IDCs)

The idea of server provisioning is to adjust the number of active servers in a server farm to the offered workload and suspend the rest. This is based on the assumption that all servers of the server farm are capable of executing the incoming workload albeit at different speed. The effectiveness of this method is based on (i) the difference in traffic intensity between *workload peaks* and low periods; peak traffic is about two to three times as intense [1, 31, 88]; and (ii) current computing systems are not energy-proportional. Server provisioning has some challenges. Especially suspending the servers incurs energy costs, as servers consume energy to be turned back on. Also, there is a switching delay, that may violate the availability of service, because of suspending the servers which are still in service. Frequent on-off transitions increase the wear and tear of the server components. Further, server provisioning is a proactive approach, where a prediction of the incoming workload is used to determine the number of active servers. However, accurate prediction of workload under stochastic and dynamic nature of the data center workload is highly unlikely. Therefore, server provisioning suffers from over-provisioning and under-provisioning problems. This section presents the challenges regarding its modeling, the recent developments, and the proposed algorithms.

1) Modeling Issues in Server Provisioning To optimize the number of active servers, one needs to provide performance and SLA model, power consumption model as well as the switching cost model.

Power Consumption Model of a Server The power model of a server in a data center specifies how the server consumes power with respect to its utilization (i.e., average utilization for all components). The power model is a key in the server provisioning problem, as correct modeling of the system power behavior at varying utilization levels provides a key to select appropriate energy-efficient servers. The modeling of the utilization level is a challenge in itself. Some research efforts suggest that CPU utilization is a good estimator for power usage [14, 31, 89]. In this way, power consumption of a server can be calculated through CPU utilization which is an indication of total power consumption of a typical server [31, 89]; the CPU is not the dominant contributor in power consumption of a server though [1, 54] (see Table 2). There are also other studies that assume a general utilization of the server which is either derived by a queuing model of the server [34] or empirical data such as the ones that are reported by SpecPower [4, 81]. The power-utilization relationship as already mentioned in Eq. 9 is mostly modeled as a linear function [14, 31–33, 89] where the power consumption increases linearly with respect to the utilization, i.e., $p = ap_{\text{util}} + p_{\text{idle}}$, where p_{idle} is the idle power consumption,

$0 \leqslant a \leqslant 1$, and $p_{\text{idle}} + p_{\text{util}}$ denotes the peak power consumption of a server. The linear power consumption model simplifies the solution for the server provisioning problem, however recent literature [4, 81] show that power-utilization model of a server exhibits nonlinearity. As a side effect, the authors show that the peak energy efficiency may not be at the maximum utilization. This affects the server provisioning problem, since an absolute minimum number of active servers may not be the most energy-efficient set.

Performance Modeling In Internet data centers, performance is usually expressed in throughput, response time and turn-around time. In this context, the SLA statistically imposes an upper bound on the response time: Prob[*response_time* > response_threshold$_{\text{SLA}}$] < probability_threshold$_{\text{SLA}}$.

To model the response time as a function of number of active servers, the literature suggests using a queuing model of servers such as $GI/G/n$ [32, 34, 90] and $M/M/n$ [37] as well as linear models [12, 31, 33]. The $GI/G/n$ is n-server queuing system serving requests with generalized arrival and service time distribution. There exist some approximation models [91] for $GI/G/n$ which can capture the average response time as a function of coefficient of variation and average of arrival and service time as well as number of servers. The $M/M/n$ is based on exponential distribution assumption for inter arrival time of requests as well as their service time. Although the Internet workload does not exhibit Poisson distribution [92], the $M/M/n$ is frequently used in the literature due to its simplicity [30, 37]. Finally, the linear model is based on the strong correlation between CPU utilization, throughput and service time of requests which is observed by empirical studies [12, 33]. The model suggests that, ***by posing a bound to the utilization (e.g., CPU utilization)***, i.e., preventing overloading the server so that its utilization does not exceed a threshold value, one automatically poses a bound on the average response time, and that the utilization linearly increases with the workload as long as the server is not overloaded. In the cost-optimization model, the revenue generated by each service is specified via a pricing scheme or service-level agreement (SLA) that relates the achieved response time to a dollar value that the client is willing to pay. The pricing model is usually a step-wise function which specifies the provider's revenue for various classes of response time.

Switching Cost Switching of servers increases the wear and tear of servers [35], incurs extra power consumption [32], and imposes migration cost for turning off extra servers that are still in service [29]. However, there is no general model that can accurately capture these costs. Kusic et al. [32] profile the extra power consumption caused by switching a server on, and incorporate it into the total power consumption cost. Guenter et al. [35] model reliability costs of server components due to on-off cycles as follows: First, they divide the total cost (procurement and replacement) per component by its MTTF (Mean Time to Failure), available via datasheets or empirical analysis. Second, they calculate the sum of per-cycle costs across all components. Lin et al. [37] assume that switching cost is incurred when servers are switched on, and model this cost as a constant parameter and derive theoretical results. However, this is not true in all cases, since switching servers off sometimes incurs a migration cost.

2) Algorithmic Issues of Server Provisioning Proposed approaches for server provisioning range from analytical-oriented work that focus on developing algorithms with provable guarantees to systems-oriented work that focus purely on implementation. While much has been published in the systems area, there is very little work in the analytical area. Examples of systems work is as follows: Chen et al. [31] proposed *dynamic server provisioning* for long-lived TCP-based services. Their study used data traces of Windows Live Messenger and built a forecasting model to periodically estimate the number of required servers, at a period of about 30 min. Chase et al. [33], propose a server provisioning scheme for hosting data centers where they provide different levels of service for different customers. They use an economic approach, where services "bid" for resources based on their SLA utility function. The objective is to maximize the profit according to a cost-benefit utility. Krioukov et al. [36] try to achieve a energy proportional cluster by using the set of heterogeneous machines such as low power platforms e.g., Atom and high performance servers. They design a power-aware cluster, namely NapSAC, which assigns the average smooth part of workload to high performance servers and workload tails to the Atom servers. Guenter et al. [35] proposed Marlowe, an automated server provisioning system that aims to meet workload demand while minimizing energy consumption and reliability in data centers. They predict workload in a large window of prediction and optimize the number of active servers over the length of window.

Lin et al. [37] perform an analytical study on the server provisioning problem for a homogeneous data centers. Due to the switching cost in the server provisioning problem, solution at every time depends on the solutions at other times. For that, an online algorithm is suggested that has competitive bound of 3 compared to the offline optimal solution. The authors further extend their work on heterogeneous data centers, and prove that the competitive ratio in this case depends on the prediction window size over which the online server provisioning is performed [93].

3) Server Provisioning Decision Interval One of the design challenges for server provisioning is the *period* (i.e. granularity) of the decision making. A low-bounding factor on this period are: (i) the delay of server state transition between off and on, typically considered to be between 30 s and 2 min, and (ii) switching cost, because short time interval increases the frequency of switching and thusly the switching cost. An upper-bound factor is the inefficiency in active server set selection caused by the coarse granularity in decision making, leading to energy wastage. Krioukov et al. [36] and Guenter et al. [35] propose using low power modes which draw idle power (2–5 W), and incur a delay of 30 s–60 s to boot up the machine (e.g., hibernation mode). Other studies such as [12, 31, 32] assume decision intervals of half an hour or more.

4) Workload Prediction The efficiency of server provisioning problem comes from variation of workload which consists of the number of periods of low workload. Intensity variation in web traffic has been witnessed in several research efforts [94–96]. The variation originates from the various sizes of files communicated, which

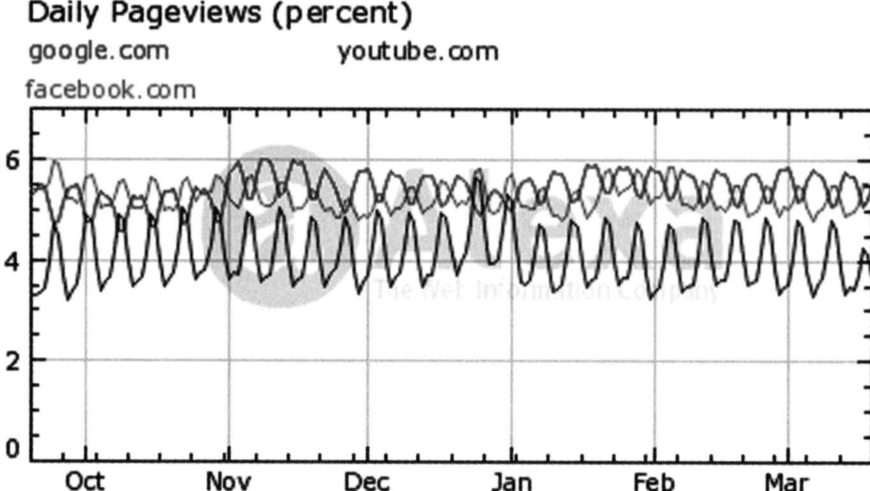

Fig. 8 Demonstration of the variation and cyclic behavior of web traffic for three popular web sites (source: www.alexa.com)

forms a fine-scale variation (fluctuation in time scale of a few seconds), and user behavior, which forms a coarse-scale (daily or weekly) cyclic variation (see example in Fig. 8).

Due to the switching cost, and to maintain performance goals, server provisioning is either managed proactively [12, 31–33, 35], or as a hybrid of proactive and reactive approaches [34, 36, 97]. In proactive approaches, the workload (request arrival rate) is predicted over one period of decision interval, and number of active servers are determined accordingly. The number of active servers is such that they can afford the peak workload during the server provisioning decision interval. Therefore, the peak workload arrival rate during the decision interval should be predicted. For that, workload is usually predicted in two stages. First average workload arrival rate is predicted, and then peak arrival rate is estimated from the average workload arrival rate. There are two approaches that are frequently used to predict average workload arrival rate: offline [31, 97], and online [12, 32, 35, 36]. In the offline approaches, time series prediction models and offline regression models are used to build a prediction model from historical data. For example Chen et al. [31] used data traces of Windows Live Messenger, and built a forecasting model (i.e., seasonal regression model) to periodically predict the average workload arrival rate. In the online model, the prediction model is adaptively changed by observing new data over time. Examples of such models are Kalman Filtering [12, 32], exponential weighted average [36], and online least square [35]. A simple example of Kalman Filtering approach is building an AR (autoregressive) model, where the state variables in the Kalman Filtering models the coefficients of the AR model, and are adaptively updated by observing every new data.

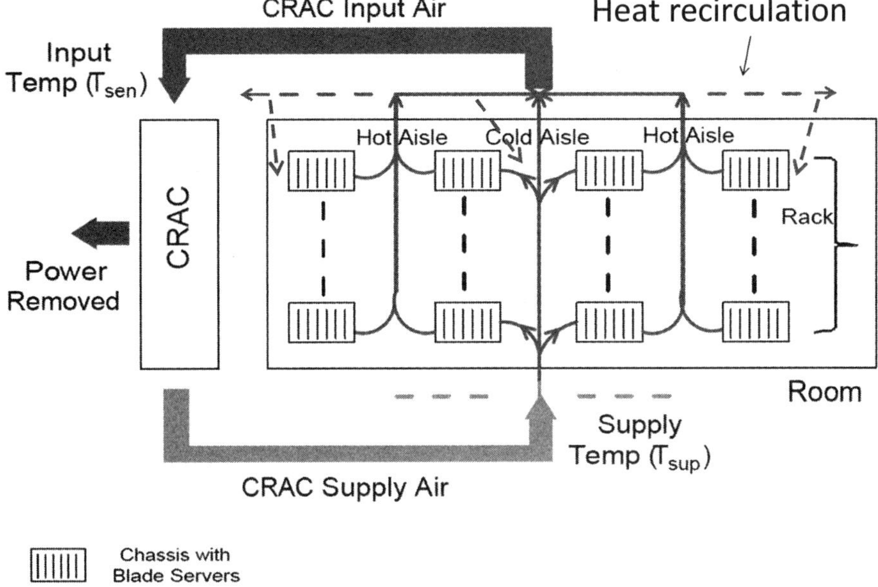

Fig. 9 Cold-hot aisle server arrangement and heat recirculation within data center room

The peak workload is either a predicted average workload estimated using the standard deviation of workload [12, 31, 32] or setting a fixed slack of the average workload [34], or is obtained from offline regression models. The insight behind using standard deviation is Chebyshev inequality which statistically bounds the deviation from the average. Also setting a fixed slack to the average workload arrival rate is due to empirical data that suggest peak arrival rate is usually a certain value more than average [34].

The workload prediction scheme cannot perform prediction perfectly, especially when the workload in a time period deviates from its behavior in the past. Further, periods of load spikes or flash crowds as well as sudden periods of low workload are unpredictable phenomena. Therefore, reactive provisioning is used to swiftly handle to such unforeseen events [34, 36, 97]. In this model, workload is closely monitored and whenever a spike is observed, more servers are activated in response. However, the server activation delay, may degrade the performance of reactive approaches. For that, Bodik et al. [97] propose to maintain a pool of standby servers to swiftly react to surge in workload. Such a scheme, called NapSAC [36], used Atom servers to handle sudden spikes in the workload. The authors argue that Atom servers have a very less off-on transition delay compared to high performance servers, and can be used to swiftly react to the surge in workload.

5) Thermal-Aware Server Provisioning Thermal-aware workload and power management in data centers deals with the minimization of heat *interference* among servers. The heat interference amongst the servers comes from the physical layout of contemporary air-cooled data centers, usually it being the hot-aisle, cold-aisle arrangement. In that layout, the air flow makes *recirculation* of the hot air from the air outlet of the computing servers into their inlet air (see Fig. 9). The heat recirculation forces data center operators to operate their *computer room air conditioners* (CRACs) to supply cold air, denoted as T_{sup}, at a much lower temperature than the servers' redline temperature, T_{red}, the manufacturer-specified highest temperature for the safe operation. The amount of heat recirculation depends on the physical layout/airflow of data center room and is not uniform. In other words, servers do not equally contribute in the heat recirculation nor do they equally receive it. While computer server provisioning that is power-aware only deals with minimization of computing power, thermal-aware server provisioning deals with minimizing the total sum of computing and cooling power. Thermal-aware server provisioning problem can be stated as follows: given a data center with its physical layout, a set of servers, a time varying workload, determine how many and which physical servers are required during each decision interval to service the workload and minimize the total power (cooling and computing). Similarly, thermal-aware workload distribution aims to distribute workload among servers such that thermal hot spots and cooling power are minimized.

There are research works that focus on thermal-aware workload and server management. Notable conclusion from the literature are: (i) due to non-uniform temperature distribution and to power heterogeneity of servers, in some data centers [99, 100], ***active server set selection*** affects the total power [4, 12], and (ii) in data centers with high PUE, consolidating workload to fewer servers may incur higher cooling power due to hot spots created by concentrating the data center load on fewer active servers [34]. Therefore, the cooling power increase caused by server provisioning may outweigh the computing power decrease, especially when modern low idle power servers are used.

Example of Thermal-Aware Server Provisioning from Literature Moore et al. [100, 101], and Bash and Forman [13] show that thermal-aware workload placement can save energy in data centers. Mukherjee et al. [11] and Tang et al. [99] model the heat that is recirculated among the servers; using this model, they propose spatio-temporal thermal-aware job scheduling algorithms for HPC batch job data centers.

Thermal-aware workload and server management for Internet Data Centers (IDCs) is also investigated in some papers. Sharma et al. [43], introduced thermal load balancing and show that dynamic thermal management based upon asymmetric workload placement can promote uniform temperature distribution and reduces the cooling energy. Parolini et al. [102] provide analytical formulation to manage workload distribution among servers. Faraz et al. [34] proposed *PowerTrade-d*, a dynamic thermal aware server provisioning which trades off the idle power and cooling power for each other. They argued, reducing the active server set size may not always reduce the total power, as it may increase the cooling power. PowerTrade-d manages the trade-off through a dynamic refinement process such that whenever a change in the size of active server set is required, extra servers are activated or deactivated one by one to ensure the desired balance between cooling power and idle power. Finally,

Abbasi et al. [12] designed TASP and TAWD, thermal-aware server provisioning and workload distribution. TASP predicts the peak workload in an interval of about 1 h and then solves an optimization problem to choose the optimal active server set that services the predicted workload and minimizes the total power in the data center.

Challenges and Trade-Offs in Holistic, Thermal-Aware Server Provisioning From the above, it is logical to conclude that a holistic approach constituting of server provisioning and workload management (i.e. workload distribution) would yield synergistic benefits on saving energy. Such a holistic approach should take into account the nature of the workload, the energy proportionality of systems, the heterogeneity of systems, the impact of server provisioning on the cooling power and the non-uniform temperature distribution within a data center. There are some intrinsic challenges that relate to trade-offs in a data center:

- **The trade-off between QoS and energy efficiency:** As web traffic intensity varies over time [94, 96], the active server set should be determined according to the peak traffic to be observed during the operation of that active set to ensure QoS [31]. This practice results in over-provisioning during all non-peak periods of web traffic as the servers are under-utilized over that active server set's duration.
- **The trade-off between cooling and computing power:** According to a recent study, there is a trade-off between idle power and cooling power that can manifest when performing server provisioning to decrease total idle power [34]. This trade-off is due to the consolidation of the workload to one "side" of the data center, thus causing un-evenness of heat distribution, which may make it hard for the cooling units to address.
- **The trade-off between finding an efficient active server set and the time to find it:** The problem of server and workload management is exacerbated when considering the impact of active server selection on the cooling energy. Such consideration makes the server selection problem nonlinear because of the nonlinear relationship between computing and cooling power and the nonlinear nature of cooling system's coefficient of performance (CoP), i.e., the ratio of the power extracted over the power required by the cooling unit [100]. These nonlinear effects increase the solution time for the online problem of server and workload management.

Abbasi et al. [103] propose to address the first trade-off by enforcing a two-tier resource management architecture called TACOMA (see Fig. 10); the first tier performs server provisioning by deciding on the active server set for each period, whereas the second tier works at a finer time granularity and distributes the workload among active servers to further minimize the "over-provisioning" effect during that period. They show that the trade-off between computing power and cooling power does not always manifest, and provide necessary and sufficient conditions to the occurrence of the trade-off, based on the structure of the *heat recirculation matrix*, an abstract model of the heat recirculated in the data center; they also argue that high IPR and low PUE disfavor the occurrence of the trade-off. Lastly, they propose heuristics

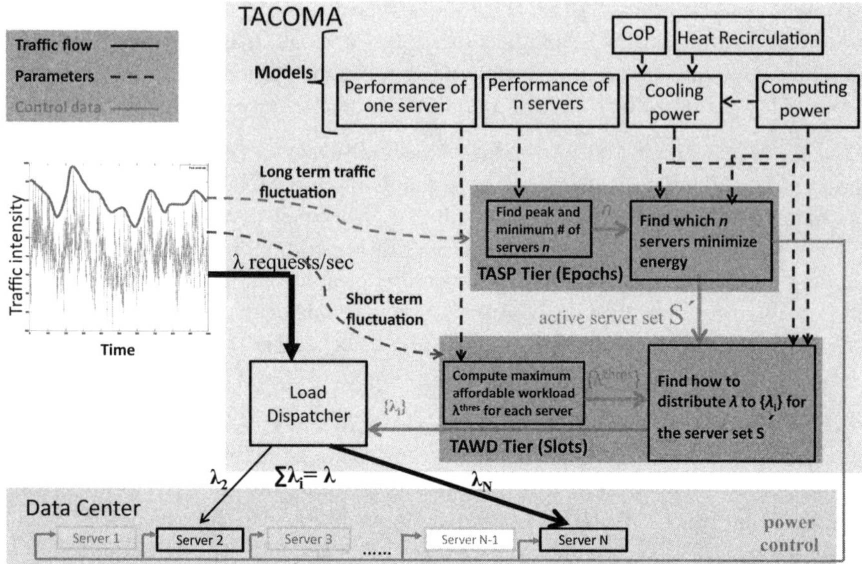

Fig. 10 A two-tier architecture for thermal aware server and workload management of Internet data centers. The first tier works on long time intervals, determines the minimum number of active servers and chooses the active server set to minimize the total energy. The second tier works on shorter time intervals and decides on the workload distribution across the active servers to minimize their energy consumption. Legend: λ is the rate of incoming workload; λ_i is the workload share for server i; λ^{thres} is the maximum sustainable workload rate for any server (assuming identical servers)

to the thermal-aware server provisioning and workload distribution problems that address the trade-off between computation time and solution efficacy.

7) Taxonomy of Server Provisioning Research Summary of research in server provisioning is shown in Table 5 indicates that the energy efficiency of server provisioning under various applications is shown by lot of studies. Efficient solutions to the problem are proposed in many articles, however, most of the solutions are either heuristic or greedy, whose efficiency is not analytically demonstrated. There are studies that analyze the problem theoretically [37], but they make simplified assumptions that rendered the studies impractical. Moreover, many studies show the switching cost overhead [32, 35, 37], but the accurate modeling of the switching cost has not yet been studied (i.e., cost of switching a server from on to off and vice versa). Finally, server provisioning and server level active power management are not orthogonal. However a comparison of their energy efficiency and their applicability in various applications has not been sufficiently addressed in the research.

Table 5 Taxonomy of research on server provisioning

Modeling issue	Approach	Articles
Power consumption	Linear (non zero idle power)	[12, 31–34, 36]
	Nonlinear	[4]
	Linear proportional (zero idle power)	[37]
	Full power consumption	[35]
Performance modeling	Linear model	[4, 12, 31–33, 35, 36]
	$M/M/n$	[37, 98]
	$GI/G/n$	[32, 34, 90]
Algorithmic issues	System	[12, 31, 33, 35]
	Analytic	[37, 98]
Workload prediction	Adaptive linear (Kalman Filter, least square,moving average)	[12, 32, 36, 35]
	Offline regression	[31]
Thermal-awareness	Using heat recirculation matrix	[4, 12, 34]

5.2 *Virtual Machine Management*

Cloud Computing enables the user to access the services, data center hardware over Internet at low cost as the user pays only for the used resources. Many Internet applications now run as service in the cloud in the form of their virtual machine applications. Virtualization allows consolidation of heterogeneous applications in few active servers, it thusly helps to minimize the power consumption within a cloud infrastructure. However, to deploy efficient virtualization schemes, it is necessary to decide on the assignment of a VM to a physical machine and on live VM migration during overloading conditions. In practice, these problems are challenging due to the uncertainty in the workload and resource requirements of applications, and the combinatorial nature of problems where the decision vector (where and when a VM should be assigned or migrated to) is discrete variables.

Many modeling issues of VM management in cloud computing such as performance model and workload prediction can be managed similar to server provisioning, thus we don't go into their details in this section.

VM Management Problem Statement Consider a virtualized data center with servers that host a subset of applications by providing a *virtual machine* (VM) for every application hosted on it. An application may have multiple tiers and there may be multiple instances of each tier running across different VMs in the data center (see Fig.11). The VM management decides on the following: how many VMs are required for each tier of an application, how workload should be distributed among multiple instances of the application for each tier, how resource should be shared among VMs

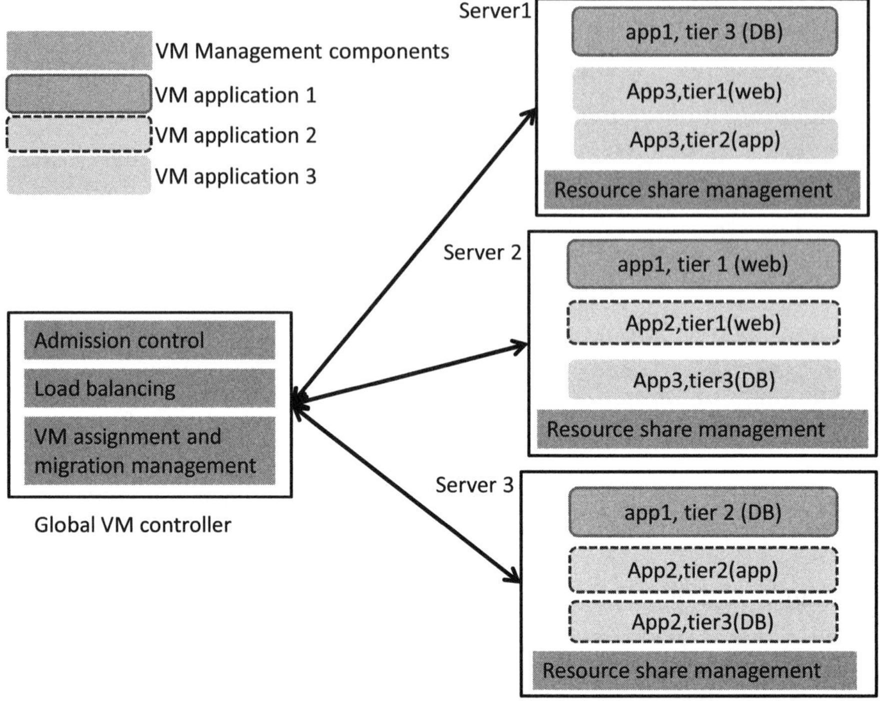

Fig. 11 A pictorial example of VM management components

that are collocated in a VM, when and how VM migration should be performed, how resources should be shared among collocated VMs.

1) VM Assignment and Migration Consider a discrete time system, where at each time the assignment of VMs to a set of (possibly heterogeneous) servers is re-computed. If we assume that VM migration cost is negligible, and the resource requirements of VMs are known at the beginning of each time interval, then VM assignment in each time interval is a bin-packing problem which is a well-known NP-hard problem [38]. In practice, VM assignment is even harder due to the VM migration overhead and the uncertainty associated with the resource requirements of the VMs. The overhead of VM migration incurs a cost of VM assignment which not time-independent. Therefore, the optimal solution for the VM assignment problem in which migration cost is non-negligible, can be only found offline where the information of resource requirement of VMs are known over all times in advance.

Since VM assignment is very similar to job (preemptive) scheduling problem, the theoretical study in that area is also applicable to VM assignment such as the one proposed by [104].

There has been much work to address the problem of VM assignment and migration in the research related to cloud computing that may or may not provide theoretical guarantee. Overview of this work is provided in the following section.

Examples from the Literature Dhiman et al. [40] identify the difference in the power consumption for CPU-intensive and IO-intensive applications through experiments. They argue that since applications have different resource requirement at different times, VM allocation in physical server can be performed such that their peak power does not simultaneously happen. They also claim that collocating a hybrid of CPU-intensive and I/O-intensive application in a physical server incurs less power consumption compared to any other combination. They design an open control loop to manage the VMs and support their claims through experiments.

The problem of allocation virtual machines to physical machines is formulated as constraint programming in [38]. The objective is to have *fully utilized physical machines* such that the wastage of resources in active physical machines can be minimized. The problem of having fully utilized machines is that the VMs need to be dynamically managed, as the resource requirement of transaction based applications (e.g. web based application) and the resource requirement of batch jobs are different. In case of batch jobs, when a job ends, its resource requirement may not match the resource requirement for the new job, hence, VMs should be reallocated to the physical machines to have fully utilized machines. Hermenier et al., develop a dynamic constraint programming based VM manager namely *Entropy*. Entropy manages VMs such that whenever an unallocated VM is available, it is assigned to a physical machine that can meet its resource requirement. Every new assignment may cause some live virtual machine migration in order to satisfy the objective function.

A recent experimental-based study [39] suggests that VM migration not only degrades the performance of the migrated VM but also the performance of other VMs that are collocated with the migrated VM in the source and destination machine. They also suggest a VM migration management scheme that is aware of this behavior.

2) VM Dynamic Resource Allocation The cloud provides autonomous management of available physical resources make the services in the form of VM applications available. There are two approaches for autonomic resource allocation to VMs: (i) static resource allocation [38], and (ii) dynamic resource allocation [87, 105, 106]. The static resource allocation model is based on either of following assumptions: (i) the resource requirement of VMs is known in advance and that the VMs' resource requirement does not change significantly during their life. This assumption is mainly applicable for batch jobs, where an estimation of resource requirement of the job can be either provided by users or obtained from its runtime history, and (ii) VM assignment is performed according to peak resource requirement of the application. The dynamic resource allocation is mainly studied for web based application where their traffic behavior is highly fluctuating and varies over time. The idea of resource management here is to dynamically change CPU share, memory share and other resource among VMs such that the applications' performance goal is guaranteed, and other system objectives such as power capping goals are satisfied. Since different applications have different peak time traffic, in practice dynamic resource allocation is very beneficial. The dynamic resource allocation can be managed in two tiers. The first tier deals with workload distribution over multiple instances of an application, and the second tier deals with resource assignment among VMs that are collocated in a single physical machine.

The proposed schemes in literatures for dynamic resource allocation range from optimizing a utility model that captures both SLA revenue cost as well as energy cost [41, 90, 105], machine learning technique to learn resource requirement of applications [107], and control theory approaches [87].

Examples from the Literature Padala et al. [105], used an adaptive control scheme to decide on VM resource share of multi-tier web based applications. They assume web based applications can be developed in multiple tiers where each tier can be assigned to a VM. The problem is how to manage the resource allocation of multiple VMs of applications to meet their performance goal. They design *Autocontrol*, a two level controller schemes. *AutoControl* is a combination of an online model estimator and a multi-input, multi-output (MIMO) resource controller. The model at the top level, operating across applications, estimates and captures the complex relationship between application performance and resource allocation. While the MIMO controller acting in physical machines allocates the right amount of resources to achieve applications' SLOs (Service Level Objectives).

Urgaonkar et al. [90] use an optimal control technique to decide resource allocation and power management for time-varying workloads and heterogeneous applications. They make use of the queuing information available in the system to make online control decisions by using Lyapunov Optimization technique. The algorithm decides on the number of physical servers, online admission control, routing, and resource allocation algorithm for VMs by maximizing a joint utility of the average application throughput and energy costs of the data center.

Power capping in a virtualized cluster is studied by Wang et al. [87] to design a closed control loop by using Model Predictive Control and PI controller to manage both the power consumption and performance goals of applications in a coordinated way. The controller at the high level loop is a cluster-level power control loop, which uses MPC method to manage the frequency of CPU to meet the power budget by power capping. The controller at the second level is a performance control loop for each virtual machine, which uses PID method to control response time of applications by controlling CPU share of each VM.

Ardagna et al. [41] study VM assignment and resource allocation in a coordinated way. They consider a multi-tier virtualized system with the goal of maximizing the SLAs revenue while minimizing energy cost. They developed resource management to actuate the allocation of virtual machines to servers, load balancing, capacity allocation, server power tuning and dynamic voltage scaling. The authors show that the resource allocation is NP-hard mixed integer non-linear programming problem, and propose a local search procedure to solve it. They evaluated their scheme through experimental and simulation based study.

Urgoankar et al. [90], propose a dynamic provisioning technique for multi-tier Internet applications that employs a queuing model to determine how much resource

Table 6 A taxonomy of research in VM management

VM Management aspect	Articles
VM assignment	[38, 39]
VM migration	[38, 39]
Adaptive resource management	[41, 87, 90, 105]

to allocate to each tier of application and a hybrid of predictive and reactive methods that determine when to provision these resources.

Taxonomy of Researches for VM Management VM assignment and migration management has been widely addressed in the literatures which suggest incorporating the type of jobs and as well VM migration overhead in the VM assignment problem (see Table 6). Also many works address the need for dynamic resource share management using adaptive control theory based approaches shown to be efficient solutions (see Table 6). However, there are still challenges in the VM management. A recent research shows that VM migration overhead degrades not only the performance of the migrated VM but also the performance of the VMs collocated in the source and destination physical machine [39]. However, many studies based on constant cost per each migration [104]. Moreover, VM assignment and resource management are shown to be NP-hard, however most of the proposed solutions are greedy or heuristic whose approximation ratio compared to optimal solutions are not derived. Finally virtualization technology is introduced to facilitate dynamic power management and reduce power consumption, however the applicability of the virtualization under various applications (e.eg, real-time applications) are not well studied. This is important, since due to the VM overhead, the delay requirement of some application in a underutilized VM may not be respected.

5.3 Other Data Center Level Power Management Techniques

1) Workload Shaping The idea of workload shaping is to distribute workload in such a way that the efficiency of power management technique increases. Some examples of workload shaping that are proposed in literatures include the following:

- Isolation of smooth part of workload from the tail in order to prevent performance degradation of large part of workload in the presence of surge and to decrease number of active servers for power minimization purpose [42]. Long term server provisioning is usually used along with over provisioning, where the number of active servers is determined according to the peak workload during the decision time interval. Liu et al. [42] proposed a workload shaping algorithm to alleviate this problem. The idea is to relax the performance requirement for a small fraction of the workload which form the peak part of the workload. The resource provisioning is carried out according to the well-behaved portions of the request stream; the portions of the workload comprising the tail are identified and isolated so that their effects are localized. To this end, authors design an algorithm as follows: each input request is placed in one of two reserved queues, where the response time of requests in the first queue is guaranteed and the response time of requests in the second queue is not guaranteed. The algorithm accepts workload in the first queue as long as the queue length does not exceed the max queue length, where the server can provide the queued requests with guaranteed response time. All other requests are placed in the second queue which can be served through another reserved server or during low traffic time. The main idea behind this approach is that peak workload rate is usually many times greater than average and form a small fraction of the workload over a long time, thus by sacrificing the performance of a small fraction of workload without compromising the quality of service of the rest of workload, significant power can be saved.

- Workload allocation in thermal oriented way, such that the temperature within data center room is uniformly distributed and thusly the energy efficiency of cooling system is increased [43, 100].
- Skew the workload toward energy efficient servers to tackle over provisioning problem in server provisioning. The authors propose two tier architecture, where the first tier determines number of active servers according to peak workload in the long decision time interval called "epoch". Under server provisioning, servers are still underutilized on average due to short term workload fluctuation. Hence, authors design a workload shaping algorithm to predict workload at fine time slots and skew it toward thermal and power efficient servers, thus increasing energy efficiency of systems through increasing the per-server utilization.
- Queue the energy consuming task together and then send them at once for achieving energy-proportional networking [24, 53]. The example of aggregation is implemented for network data by Nedevschi et al. [24], where the authors propose to queue tasks such that the idle time of network equipment is maximized. Then they proposed to use inactive power mode and link rate adaptation to reduce the power in the network, thus achieving energy-proportional networking.

2) Heterogeneity in Workload Collocation In virtualized data center consolidation of workload is done in order to increase the utilization of servers. In this scenario, different tiers of an application e.g. data base tier, front end, management tier are collocated and result in different workload type such as CPU, I/O, network, Memory workloads being served by underlying physical hardware which may also be vary in terms of cores, memory configuration, etc. Research in this area is targeted to analyze the performance degradation due to collocation in order to avoid SLA violation while some attempt energy aware collocation of heterogeneous workload. The contention in shared resources of the underlying platform may result in the performance degradation of applications, termed as "Interference". The variations in performance can be observed not only due to impact of co-runner application (i.e. application that runs simultaneously) but also due to the heterogeneity of the underlying hardware. This heterogeneity in data center applications running on heterogeneous hardware can be leveraged for more energy savings by collocating the applications to maximize the use of server resources without causing the performance degradation of applications.

Examples from the Literature

- Interference in Collocated Applications:
  The interference effect of collocated applications on their performance has been studied in some recent works [108, 109]. Fedorova et al. [110] attribute the reason for contention to misses in the last level cache, as well as other shared resources, such as front-side bus, prefetching resources, and memory controller. Mars et al. [108, 109, 111] estimate the interference effect and hence the performance degradation of the collocated applications by profiling each application for the pressure they exert on the memory subsystem as well applications sensitivity to memory pressure. Chiang and Huang [112] characterized the interference of I/O in data intensive application in virtual environment by statistical modeling methods to estimate the performance degradation while other group empirically characterized the interference effects of communication over computation, correlating computation size, communication packet size, frequency of communication with sending and receiving using empirical methods [113]. In pScimapper [114], a workload collocation scheme for scientific applications, interference amongst applications is measured offline using correlation analysis and further used to address the VM management as a hierarchical clustering problem. Network intensive I/O intensive workload results in excess of context switching [115], while database, memory or file I/O workload interference is shown by increase in last level cache accesses [116]. Using such statistical parameters, interference aware management scheme is developed for live Google data center applications [117].
- Energy aware Collocation of VMs: With increasing use of VM consolidation to address the energy concerns in data centers, many researchers have considered energy impact of collocation of applications. Most common approach is to increase collocation by correctly estimating and avoiding performance degradation, thereby consolidating more VMs on single hardware. Few of these research efforts emphasize on estimating energy footprints with different collocation policies [40, 118, 119]. Scheme proposed by vGreen [40] is designed as an open control loop to manage the application assignments (VM) to the physical servers such that a collocation of different workload types is always preferred over the other policies. Some research efforts propose resource allocation while considering the energy impact. Verma et al. use the dynamic cache footprint and and working set size to manage power in virtual environment [120] while Buyya et al. [121] focus on CPU usage while managing VMs. Merkel et al. [122], address the problem of energy efficient management of collocated heterogeneous application by dividing the applications into tasks, based on their resource requirements. The idea behind sorted co-scheduling is to group the cores into pairs of two and to execute tasks with complementary resource demands on each of them.
- Collocation of application with heterogeneous hardware: Data center inherently possess hardware heterogeneity due to the servers, other data center hardware that are replaced over time result in mix of hardware in the data center. Few of

Table 7 A Taxonomy of research in collocation of workload in VM environments

Category	Collocation	Interference	Performance	Energy	Heterogeneous workload
Interference in collocated application [108, 109, 113, 122]	✓	✓	✓		
Workload type aware Collocation management for VM [40, 118]		✓	✓	✓	✓
Collocation of application with heterogeneous hardware [123, 36]		✓	✓	✓	✓

the research groups have recently suggested the use of low power machines for servicing data center workload [36]. Mars et al. [123] evaluate a application's sensitivity on the basis of collocated applications as well as underlying heterogeneous hardware. They develop a collocation scheme such that the most sensitive applications are least affected. In order to account for the heterogeneity of underlying hardware, the authors estimated the co-runner application's interference when run on the same core versus cross core [109, 123].

Table 7 shows different research work in the area for performance and energy management of collocation of heterogeneous workloads. In summary, the management policy to minimize the interference of application collocation has to consider (i) sensitivity to the co-runner application's (ii) sensitivity of application to the underlying micro architecture (hardware) (iii) pressure created by the application on the shared subsystem.

3) Hybrid Data Center The notion of hybrid data center is proposed to increase energy efficiency of data centers. The power performance of servers in terms of idle power magnitude, power-utilization curve and on-off transition delay are different. Further, power-utilization curve of servers for different types of workload are not identical. Therefore, some of the research propose to match the workload to suitable type of platform [9, 36]. The difference in power performance of platforms with respect to type of workload is observed by some researchers. Experiments have shown that, the CPU power used by different operations (i.e. in case of different database applications) can vary widely, by up to 60 % for the same CPU utilization, and that the CPU power is not linear with utilization [6]. In a similar experimental study, authors observed that different platforms have different energy proportionality characteristics and exhibit different power performance with respect to the type of workload (e.g. memory intensive, CPU intensive and transactions) [9]. Hence, the authors proposed to design a hybrid data-center that mix low power platforms with high performance ones. Finally, Krioukov et al. [36] designed a power-proportional cluster by considering the power-performance of systems consisting of a power-aware cluster and a set of heterogeneous machines including high performance servers and

Atoms. They proposed to use Atom platforms to handle spikes in the workload leveraging its lower wake up time and use high performance servers to service the smooth part of workload. Use of alternative hardware such as embedded processors is suggested in [44, 45]. The assignment of appropriate workload on specific type of platform gives more energy proportionality.

6 Energy Cost Minimization Through Workload Distribution Across Data Centers

Cloud computing is an emerging paradigm based on virtualization which facilitates a dynamic, demand-driven allocation of computation load across physical servers and data centers [124]. Leveraging the cloud infrastructure, recent researchers have shown a growing interest in optimizing a geo-distributed data centers' operational cost and carbon footprint through exploring temporal and spatial diversities among participating data centers [125].

Current large-scale Internet services tend to be replicated over several data centers around the world. These geo-distributed data centers are primarily deployed by large-scale Internet service providers such as Google to serve users across the world efficiently and reliably. Accordingly, the workload distribution policy across data centers is conventionally performed to minimize the delay experienced by the users. However, recent literature propose energy aware global workload management schemes which not only meet the quality of service requirements of users (e.g., delay) in different locations, but also reduce the electricity cost (dollar per Joule) and reduce the carbon footprint (CO_2 emission per Joule) of the cloud. The idea is to shift the workload toward data centers that offer green power or low electricity cost at a given time, and adjust the number of active servers in proportion to the input workload.

Data centers in a cloud are usually diverse in terms of their energy efficiency (e.g, MIPS/joule), electricity cost and carbon emission factors. Servers in different data centers have different computation capabilities and power consumption characteristics. Therefore, depending upon the types of physical servers in data centers, computing power performance of data centers can be different. Also, *Power Usage Efficiency (PUE)* (see Eq. 1), may vary for different data centers. For example, according to US Department of Energy [126] data centers usually have an average PUE of 1.7, whereas Google's modern data centers have PUE of 1.18 [127].

Further, data centers get their primary power from the grid. Various parameters such as the availability of fuel type, the market, the environment, and the time of day affect the electricity cost and the carbon emission of utilities. Furthermore, many data centers utilize on-site renewable energy sources, with solar and wind energy being the most popular ones. Despite progress in growing renewable-energy-powered data centers, utilizing the available on-site renewable energy sources is challenging without large-scale Energy Storage Devices (ESDs). This is due to the intermittent nature of the renewable energy sources as well as fluctuation in the power demand.

In response, energy aware global workload management schemes which are developed on "follow-the-moon" philosophy [128], leveraging the spatial and temporal variation of the aforementioned factors, help to increase the utilization of the available renewable energy sources without the need to use large-scale energy storage devices. The energy aware global workload management scheme deals with workload distribution of each front-end to data centers such that a set of energy related objectives (e.g., energy cost and carbon footprint) is optimized without violating the delay requirement of the users (see Fig. 13). Many modeling issues of the global workload management such as performance model and workload prediction can be managed similar to server provisioning, thus we don't go into their details in this section. However, some more modeling, algorithmic, and implementation challenges come into play when designing global workload management compared to server/workload management for an individual data centers, as such: (i) the performance/cost model needs to be aware of communication overhead between users and the different data centers, (ii) the algorithm needs to be aware of migration overhead of stateful applications (stateful applications stores the state of users/applications, e.g., game applications) across data centers, and (iii) the global workload management can be implemented in either central way, or distributed way across data centers and front-ends, each associated with pros and cons including network overhead, scalability, and confidentiality regarding the data centers' information exchange. Some of these problems and challenges are addresses in the literature as described below.

Example from Literatures The result of the current literature highlights that workload management across data centers can significantly reduce the electricity bill [49, 125, 129, 130–132], and can potentially be a significant aid in reducing the carbon footprint of data centers without requiring large-scale energy storage devices [125, 130, 133–136].

Qureshi et al. [49] identified the temporal and spatial fluctuation of electricity price (see Fig. 12). They used heuristics to quantify the potential economic gain of considering electricity price in the location of computation. Through simulation of realistic historical electricity price and real workload, they report that judicious location of computation load may save millions of dollars on the total operation costs of data centers. Another scheme for workload scheduling across data centers has been developed by Ley et al. [47], where the problem is modeled as a linear programming problem. This problem is further studied by Abbasi et al. [132], where authors prove the NP-hardness of the problem and evaluate the proposed greedy solutions for energy cost management of stateful and stateless Internet applications. Dynamically hosting applications in data centers should be aware of the network delay and bandwidth overhead during migration (e.g., user state data). This overhead depends on the type of Web applications, which can be either stateless or stateful. In stateless applications, for example, search engines, the state of online users is not recorded; whereas stateful applications, for example, multiplayer online games, keep track of the state of users. Therefore, stateful applications tend to induce higher migration cost. Abbasi et al. studied the problem modeling and energy cost benefit of global workload management for both stateful and stateless applications [132].

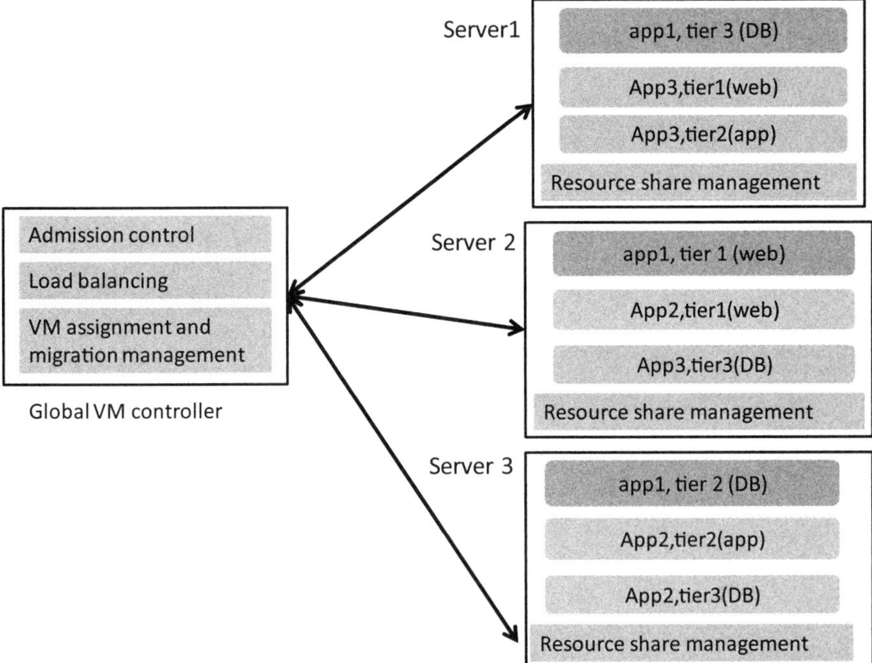

Fig. 12 Hourly electricity price data for three major location of google IDCs on May 2nd, 2009 [46]

In an analytical and experimental study on geographical load balancing Liu et al. [134] derive two distributed algorithms for achieving optimal geographical load balancing. The algorithms allow front ends and data centers separately decide on workload distribution, number of active servers as well as voltage scaling of CPUs. The authors also show that if electricity is dynamically priced in proportion to the instantaneous fraction of the total energy that is brown, then geographical load balancing provides significant reductions in brown energy use.

Buchbinder et al. [52] propose online algorithms for migrating batch jobs between data centers to handle the fundamental trade-off between energy and bandwidth costs. They argue migration overhead for stateful jobs may be significant. They provide competitive-analysis, to establish worst case performance bounds for the proposed online algorithm. Authors also propose a practical, easy-to-implement version of the online algorithm, and evaluate it through simulations on real electricity pricing and job workload data.

The related works also highlight that global workload management can help to efficiently utilize renewable energy. Liu et al. [134] propose a convex-optimization framework to study the economic and environmental benefits of renewable energy when using geographical load balancing. Using a trace-based simulation study Lui et al. investigate how workload management across data centers can reduce the required size of energy storage devices to maximally utilize renewables. Finally, Akoush et al. [130] propose to maximize the use of renewable energy by workload migration.

The variation of electricity price has been also leveraged for cost efficient energy buffering in data centers. The idea is to store energy in UPS batteries during valleys -periods of lower demand, which can be drained during peaks periods of higher demand [50, 51]. Particularly, Urgaonkar et al. develop an on-line control algorithm using Lyapunov optimization to exploit UPS devices to reduce cost in data centers [51]. Govindan et al. perform a comprehensive study on the feasibility of utilizing UPS to store low-cost energy, and design a Markovian based solution to schedule batteries [50]. Palasamudram et al. perform a trace-based simulation using Akamai CDN workload traces to investigate the energy cost saving that can be achieved by using batteries to shave the peak power draw from the grid [137]. Govindan et al., propose to leverage existing UPSes to temporarily augment the utility supply during emergencies (i.e., peak power) [138]. Finally, Kontorinis et al. [139] presents an energy buffering management policy for distributed per-server UPSes to smoothen power draw from grid. Wang et al. investigate how data centers can leverage the existing huge set of heterogeneous ESDs [140]. The authors in this work study the physical characteristics of different types of ESDs, and their cost-benefit for utilizing them in data centers. The authors also develop an offline optimization framework to decide on how heterogeneous set of ESDs can be placed in different levels of data centers power hierarchy (i.e., data center, rack, and server levels) in order to minimize the data center operational energy cost. The management scheme is mainly developed for the data center design, not for dynamic workload management, since it is based on the assumption that both short and long term variation of power demand for a long time horizon is given in advance. Finally, Abbasi et al. propose two-tier online workload and energy buffering management scheme which is aware of the long-term and short term variations of the workload workload, the available renewable power, the electricity pricing as well as the existing set of heterogeneous energy storage devices. The authors, propose an analytical study of multi-tier workload and energy buffering management technique that frames each tier as an optimization problem and solves them in an online and proactive way using Receding Horizon Control (RHC). This study shows that multi-tier energy buffering management increases the utilization of the renewables by upto two times compared to one-tier management (Fig.13)

Finally, there are some recent works which propose joint optimization of energy cost and carbon footprint across data centers [136, 141–146]. This is challenging, primarily because usually there is no coordination between energy cost and carbon footprint of utilities across different locations. Le et al., devised a heuristic online global workload management to dynamically solve green and brown energy mix of data centers in a cloud in order to minimize the electricity cost while operating under carbon cap-and-trade policy. Similarly, Gao et al. [142] and Doyle [146] utilize multi-object optimization and Vornoi partitions, respectively, to determine how to balance the workload across data centers based on the cloud operator's priorities on minimizing the network delay, the electricity cost, and the carbon footprint. Ren et al., and Mahmud et al., focused on designing a Lyapunov based online electricity cost

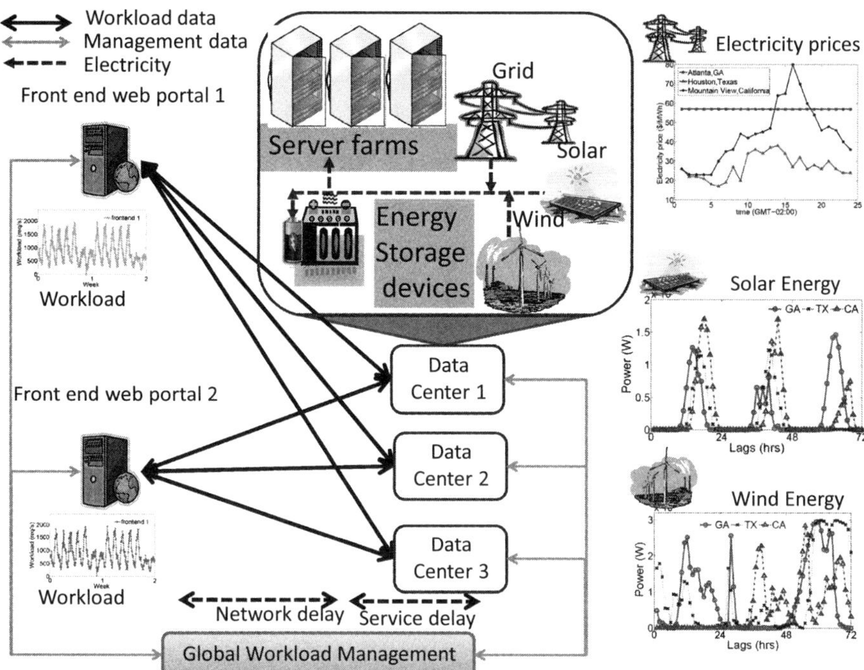

Fig. 13 Pictorial view of workload distribution considering the spatio-temporal variations across data centers

aware workload management to achieve carbon neutrality for a single data center [143, 144]. The authors prove the optimality of their solutions through analytical and experimental studies. Lyapunov based optimization is further used by Zhou et al. and Abbasi et al. to design cost aware and carbon aware global workload management [136, 145, 146]. In particular, Abbasi et al. propose an online global workload management to minimize the electricity cost without violation of carbon footprint target of the cloud specified by the cloud operator. The authors prove that their online solution achieves a near optimal operational cost (electricity cost) compared to the optimal algorithm with future information, while bounding the potential violation of carbon footprint target, depending on the Lyapunov control parameter, namely V. The authors also give a heuristic for finding the Lyapunov control parameter (i.e., V).

Taxonomy of Researches for Workload Distribution Across Data Center There are significant amount of work which deals with identifying and leveraging the opportunities in cloud and data centers by developing cost and carbon efficient workload placement algorithms. A taxonomy of the aforementioned researches according to their covered problems is shown in Table 8. Research in designing energy efficient techniques in data centers is still in the preliminary stages. In particular, most of the existing solutions are evaluated using only trace based simulation studies. Further,

Table 8 A Taxonomy of research in power aware workload management across data centers

Application type	Management type	Articles
Stateless	Workload management	[47, 49]
	Cost aware workload and server management	[46, 134]
	Cost aware workload and energy buffering management	[50, 51]
	Cost and carbon aware workload and server management	[136, 141–146]
	Cost and renewable energy aware workload and server management	[130, 133–135]
Statefull	Cost aware server management	[48, 52, 132, 147]

the cost aware and carbon aware global workload management favors an offline solution, due to the time coupling to manage energy storage and carbon capping. Yet the online algorithms are often designed to address each of the aforementioned coupling factors separately, disregarding their management implications on each other and the practical considerations.

7 Data Center Simulation Tools

To assess the energy efficiency of the existing techniques, researchers either use experimental studies on an actual data center, prototypes, or simulation environment. Testing the solutions in an actual data center is impractical because it requires exclusive use of the data center while testing, something that cannot be easily granted with production data centers. Further, prototype evaluation studies often provide partial evaluation of the solutions depending on their scales and design. Simulation tools, however, are easy to deploy and can provide a holistic evaluation environment depending on the capability of the simulation tool and its accuracy. In this context, the simulation tool needs to holistically account for data centers' cyber (e.g., servers' compution capabilities) and physical factors (e.g., data center physical layout), as both of which affect the performance and the energy consumption.

A tool that partially simulates data center processes, similar to prototype evaluation, might lead to unrealistic results. For example, a simulation tool which only considers the cyber factors cannot capture the physical phenomenon such as server shutdown, throttling, or physical damage in the case of thermal failures.

A holistic simulation tool is also important to account for the interrelation of different power management techniques, when they are deployed together in a data center/cloud. In other words, a combination of the energy management techniques, ranging from component level power management to data center thermal aware power management schemes can be utilized depending on data center circumstances in order to improve the energy proportionality. However, when combined, one needs to account for their interaction on each other due to the cyber and the physical energy and performance factors, e.g., response time which is affected by both DVFS and server provisioning and energy which is affected by both computing and cooling power. Therefore, their evaluation must be holistically considered to effectively assess their efficiency (in improving energy proportionality without compromising the quality of service) for their combined deployment.

Despite importance, the data center simulation tools have not been sufficiently studied in the literature. In particular, most of the existing tools provide partial (either cyber or physical) simulation of a data center. In the following we give an overview on these tools.

Overview on the Existing Data Center Simulation Tools Prior work exists in the area of data center simulation, with some focus on the end goal of investigation for energy savings. For the purpose of simulating cyber-physical interdependencies, however, these tools tend to be piecemeal.

On the cyber side, Lim et al. [148] develop MDCSim to evaluate the data center power efficiency through simulating the cyber aspects of data centers. The tool uses steady-state queuing models to simulate a data center servicing multi-tier web applications. BigHouse [149] further improves the MDCSim models though introducing stochastic queuing system which provides a stochastic discrete time simulation of generalized queuing models to simulate the performance of the data center applications in more detail than that of MDCSim [148]. CloudSim, is capable of simulating the per-server power and performance metrics, where the main abstraction unit is a VM [150]. The above tools ignore simulating the physical aspect of data centers and the potential interaction between cyber and physical aspects. They rely on the steady-state models to estimate the power consumption and performance of servers, lacking simulating the transient processes that can have long-time effects on data center (e.g., redlining).

On the physical side, computational fluid dynamics (CFD) simulators can be used to test the efficiency of a physical design [151–153] and can be used for thermal-map model (temperature distribution within data center room) learning in an offline setting [101]. However, each of these tools requires domain specific expertise, and are very time consuming. There exists some previous work attempting to alleviate this problem. Weatherman by Moore et al. [101] avoids this problem by inducing a physical model of the data center in a "learning process". This model can thereafter be used in lieu of computationally expensive CFD simulation.

GDCSim, is a recently proposed simulator which simulate both the data center cyber and physical processes and their interactions [154, 155]. Similar, to Weatherman, GDCSim utilizes a light-weight heat recirculation model to simulate the temperature distribution in a data center. It also unifies the operationally relevant computation subsystems (e.g., servers, and workload) of a data center required to simulate and characterize the overall efficiency (i.e., performance and energy efficiency) of the data center system. GDCSim is developed as a part of a research project named BlueTool whose goal is to provide an open platform for the analysis and development of state-of-the-art techniques. GDCSim, in its current state, does not account for a combination of power management techniques and their interactions. Further, it accounts for both batch and transactional jobs, however it relies on steady-state models to estimate the metrics related to cyber and physical aspect of data centers.

Lastly, Banerjee et al. [156] propose a cyber-physical hybrid simulator to simulate both interactive and batch workload, while considering the transient models of cyber factors (transient workload variation) and physical factors of data centers (transient temperature variation). The authors also provide an error analysis to specify the confidence intervals of the simulation results. The proposed hybrid simulator does not provide a capability to evaluate a combination deployment of power management techniques at different levels.

8 Performance of Server and Data Center Level Power Management Techniques

As mentioned earlier, the aforementioned techniques are often complementary and can be implemented together in a system. In practice, however, not all of the technique are feasible to deploy depending on the application requirements, and data center infrastructure. One need to decide the feasible and the most power efficient technique for a given data center and a given application based on pros and cons associated with each technique, as given below:

SLA violation/Response Time
All power management and workload shaping techniques trade performance for power saving. The amount of response time overhead however varies over the application and the power management technique. For example, the response time overhead of server level power management is incurred because of transition delay from inactive mode to the active mode. This method is reactive approach and there is no need for workload prediction. In other words, if transition delay is very less than an average response time, then its response time overhead is negligible. In case for DVFS, the response time overhead incurs due to transition between different states, as well as decreasing the frequency. If the scale is inappropriately adjusted, the response time can be severely affected. However, since scaling is managed at a very short time interval, this is unlikely to happen. Data center level power management techniques are more prone to violate the response time. The reason is the unpredictable variations in the workload.

Energy Proportionality
In an ideal case both server provisioning in data center level and server level power management schemes should provide ideal energy-proportional servers and data center, however in reality it is unlikely to happen. For data centers which have non-uniform temperature distribution within the data center room, ideal energy-proportional servers cannot make an energy-proportional data center [4]. Further, the server provisioning is usually with under provisioning and over provisioning problem due to uncertainty associated with workload. The performance of low power mode schemes such as PowerNap, are also degraded by the nature of workload and PDUs. For the same condition of workload, where both low power mode schemes and server provisioning are applicable, and live migration is no longer required (stateless and

short transaction), server provisioning is more preferred due to its higher reliability, higher efficiency and also because of thermal-awareness of the policies implemented in that data center.

Since active power modes such as DVFS does not remove idle power, the energy savings obtained is less than server level power management schemes for the condition where they are both applicable.

Reliability

Frequent on-off cycles increase the wear-and-tear of server components (e.g., CPU disks, fans), incur costs for their procurement and replacement, and affects the performance of current running services [35]. Therefore the reliability of power management techniques can be evaluated according to their impact on the frequency of on-off cycles. That frequency depends on the decision time interval. Since decision time interval of server provisioning are much higher than server level power management schemes, they are less prone to affect the reliability of servers compared to low power mode schemes. In low utilization periods in the workload (i.e. order of milliseconds) server level power management schemes can result in frequent on-off cycles.

Thermal-Awareness in Workload Scheduling and Power Management Techniques

The incorporation of thermal-awareness and its efficiency over various power and workload management is different. Thermal-awareness can be either applied to the server provisioning or workload shaping, however its efficiency is increased when it is applied to the server provisioning compared to workload shaping [12, 34].

Applicability

The power management technique applied depends on data center hardware, type of workload and the applications. In case for server level power management techniques, the applicability depends on the workload statistics. For example the inter arrival times between the requests is a factor that can decide the applicability of server level power management scheme. If the average inter-arrival time of requests is less than power mode transition time, the performance of the server level power management is almost zero. For example Meisner et al. [84] did extensive data analysis of Google data centers and observed that server level power management technique is inappropriate for the Google data intensive workload classes because periods of full-system idleness are scarce in such workloads.

The applicability of active power mode also depends on the latency-power tradeoff. The scaling should be in such a way that the energy consumption over time does not increase.

The applicability of server provisioning is based on fluctuation of workload between low and peak periods. It also depends on the type of applications. For example according to Meisner et al. [84] server provisioning is inapplicable to Google data intensive services because the number of servers provisioned in a cluster is fixed. In this application cluster sizing is determined primarily based on data set size instead of incoming workload arrival rates. For a cluster to process a transaction data set for even a single query with acceptable latency, the data set must be partitioned over thousands of servers that act in parallel.

Ease of Implementation
Implementation of server provisioning techniques requires extra hardware and software to shutdown servers remotely and to perform live migration whenever it is required. Therefore, it may not be applicable for some data centers whose manager kit does not support it. Most of processors support DVFS, therefore DVFS is very easy to implement. In case for server level power management, current modern system use Wake-On-Lan capability of current network interfaces to transition from idle to active state. However, there are challenges for implementing server level power management. The efficiency of PDU is higher for larger power consumption but degrades for lower loads. Hence by using server level power management techniques such as PowerNap [30], the overall power consumption over a cluster may not change significantly. The efficiency of PDUs' can be improved by techniques proposed in PowerNap by using extra hardware to combine several PDUs.

Scalability
Scalability of techniques can be analyzed from two perspectives: software and hardware implementation, and the performance of schemes in terms of computation overhead. The computation overhead of server level power management techniques are usually negligible, since they are applied for an individual server. Server level power managements may need to combine PDUs [30] to increase energy efficiency (PDUs are more energy efficient at high power consumption level) which may not be scalable for large data centers. In case for server provisioning, the scalability is tightly correlated with the server provisioning algorithm. The computation overhead of server provisioning algorithms is usually a function of number of servers, and it may not be scalable depending on the algorithm. The scalability of the server provisioning is also limited for the extra hardware and software that is required for shutting down servers and live migration.

9 Conclusions

To conclude, we highlight the importance of workload and power management techniques to achieve energy proportionality and cost efficiency at data centers. This is due to characteristics of computing systems which has non zero energy consumption when they are idle, which reduce the energy efficiency of data centers in the periods of low utilization. Further, current data centers have PUE of greater than one which motivates the use of thermal-aware workload and power management techniques to improve cooling energy efficiency. The research community proposes workload and power management techniques to increase the energy efficiency of data centers ranging from component level to data center level. While much has been published in this area, the research is still active to expand the dynamic power management for different systems components such as memory, and network, and improve solutions for VM management and server provisioning to have theoretical performance bound and capture the accurate behavior of the system (i.e., VM migration overhead). Green data center schemes are designed to manage the workload across data centers where some aspect of problem such as reducing the energy cost by energy buffering during periods of low energy cost in the workload management are addressed. However, to achieve the ultimate goal of green and sustainable data centers, sophisticated power and workload management which is aware of the cooling energy, applications performance goals, the energy cost and the uncertainty associated with availability of green energy sources is required. The other remarkable conclusion from the above literature survey is that to achieve the energy efficient data centers, power and workload management at different levels is required. Further, the applicability and efficiency of power management techniques depends on the workload offered to the data center.

References

1. L. A. Barroso and U. Hölzle, "The case for energy-proportional computing," *Computer*, vol. 40, no. 12, pp. 33–37, Dec. 2007.
2. P. Ranganathan, P. Leech, D. Irwin, and J. Chase, "Ensemble-level power management for dense blade servers," *SIGARCH Comput. Archit. News*, vol. 34, pp. 66–77, May 2006. [Online]. Available: http://doi.acm.org/10.1145/1150019.1136492.
3. T. Starner, "Human-powered wearable computing," *IBM Systems Journal*, vol. 35, no. 3.4, pp. 618–629, 1996.
4. G. Varsamopoulos, Z. Abbasi, and S. K. S. Gupta, "Trends and effects of energy proportionality on server provisioning in data centers," in *International Conference on High performance Computing (HiPC2010)*, Goa, India, Dec. 2010.
5. X. Feng, R. Ge, and K. W. Cameron, "Power and energy profiling of scientific applications on distributed systems," in *Proceedings. 19th IEEE International Parallel and Distributed Processing Symposium, 2005*. IEEE, 2005, p. 34.
6. D. Tsirogiannis, S. Harizopoulos, and M. A. Shah, "Analyzing the energy efficiency of a database server," in *Proceedings of the 2010 international conference on Management of data SIGMOD'10*. ACM, 2010, pp. 231–242.
7. "Standard performance evaluation corporation specweb 2009."
8. G. Varsamopoulos and S. K. Gupta, "Energy proportionality and the future: Metrics and directions," in *Proceedings of the 2010 39th International Conference on Parallel Processing Workshops*, ser. ICPPW'10. Washington, DC, USA: IEEE Computer Society, 2010, pp. 461–467. [Online]. Available: http://dx.doi.org/10.1109/ICPPW.2010.68.
9. B.-G. Chun, G. Iannaccone, G. Iannaccone, R. Katz, G. Lee, and L. Niccolini, "An energy case for hybrid datacenters," *ACM SIGOPS Operating Systems Review*, vol. 44, no. 1, pp. 76–80, 2010.
10. R. Raghavendra, P. Ranganathan, V. Talwar, Z. Wang, and X. Zhu, "No "power" struggles: coordinated multi-level power management for the data center," *SIGARCH Comput. Archit. News*, vol. 36, pp. 48–59, March 2008. [Online]. Available: http://doi.acm.org/10.1145/1353534.1346289.
11. T. Mukherjee, A. Banerjee, G. Varsamopoulos, S. K. S. Gupta, and S. Rungta, "Spatio-temporal thermal-aware job scheduling to minimize energy consumption in virtualized heterogeneous data centers," *Computer Networks*, June 2009. [Online]. Available: http://dx.doi.org/10.1016/j.comnet.2009.06.008.

12. Z. Abbasi, G. Varsamopoulos, and S. K. S. Gupta, "Thermal aware server provisioning and workload distribution for internet data centers," in *ACM International Symposium on High Performance Distributed Computing (HPDC10)*, Chicago, IL, June 2010.
13. C. Bash and G. Forman, "Cool job allocation: Measuring the power savings of placing jobs at cooling-efficient locations in the data center," HP Laboratories Palo Alto, Tech. Rep. HPL-2007-62, August 2007.
14. X. Fan, W.-D. Weber, and L. A. Barroso, "Power provisioning for a warehouse-sized computer," in *Proceedings of the 34th annual international symposium on Computer architecture*. ACM, 2007, pp. 13–23.
15. J. Racino, "PUE." [Online]. Available: http://www.thegreengrid.org.
16. "Environmental protection agency, energy star program, report to congress on server and data energy efficiency," 2007. [Online]. Available: http://www.energystar.gov/ia/partners/prod_development/downloads/EPA_Datacenter_Report_Congress_Final1.pdf.
17. A. Gandhi, M. Harchol-Balter, R. Das, and C. Lefurgy, "Optimal power allocation in server farms," in *Proceedings of the eleventh international joint conference on Measurement and modeling of computer systems*, ser. SIGMETRICS '09. New York, NY, USA: ACM, 2009, pp. 157–168. [Online]. Available: http://doi.acm.org/10.1145/1555349.1555368.
18. R. Ge, X. Feng, W. Chun Feng, and K. Cameron, "CPU MISER: A performance-directed, run-time system for power-aware clusters," in *International Conference on Parallel Processing, 2007. ICPP 2007*, Sept. 2007, p. 18.
19. D. Meisner, B. T. Gold, and T. F. Wenisch, "The powernap server architecture," *ACM Trans. Comput. Syst.*, vol. 29, pp. 3:1–3:24, February 2011. [Online]. Available: http://doi.acm.org/10.1145/1925109.1925112.
20. J. E. Moreira and J. P. Karidis, "The case for full-throttle computing: An alternative datacenter design strategy," *Micro, IEEE*, vol. 30, no. 4, pp. 25–28, July–Aug. 2010.
21. Q. Deng, D. Meisner, L. Ramos, T. F. Wenisch, and R. Bianchini, "Memscale: active low-power modes for main memory," in *Proceedings of the sixteenth international conference on Architectural support for programming languages and operating systems*, ser. ASPLOS'11. New York, NY, USA: ACM, 2011, pp. 225–238. [Online]. Available: http://doi.acm.org/10.1145/1950365.1950392.
22. J. H. Ahn, N. P. Jouppi, C. Kozyrakis, J. Leverich, and R. S. Schreiber, "Future scaling of processor-memory interfaces," in *Proceedings of the Conference on High Performance Computing Networking, Storage and Analysis*, ser. SC'09. New York, NY, USA: ACM, 2009, pp. 42:1–42:12. [Online]. Available: http://doi.acm.org/10.1145/1654059.1654102.
23. H. Zheng, J. Lin, Z. Zhang, E. Gorbatov, H. David, and Z. Zhu, "Mini-rank: Adaptive dram architecture for improving memory power efficiency," in *Proceedings of the 41st annual IEEE/ACM International Symposium on Microarchitecture*, ser. MICRO 41. Washington, DC, USA: IEEE Computer Society, 2008, pp. 210–221. [Online]. Available: http://dx.doi.org/10.1109/MICRO.2008.4771792.
24. S. Nedevschi, L. Popa, G. Iannaccone, S. Ratnasamy, and D. Wetherall, "Reducing network energy consumption via sleeping and rate-adaptation," in *Proceedings of the 5th USENIX Symposium on Networked Systems Design and Implementation*, ser. NSDI'08. Berkeley, CA, USA: USENIX Association, 2008, pp. 323–336. [Online]. Available: http://dl.acm.org/citation.cfm?id=1387589.1387612.
25. M. Gupta and S. Singh, "Using low-power modes for energy conservation in ethernet lans," in *INFOCOM 2007. 26th IEEE International Conference on Computer Communications. IEEE*, may 2007, pp. 2451–2455.
26. G. Ananthanarayanan and R. H. Katz, "Greening the switch," in *Proceedings of the 2008 conference on Power aware computing and systems*, ser. HotPower'08. Berkeley, CA, USA: USENIX Association, 2008, pp. 7–7. [Online]. Available: http://portal.acm.org/citation.cfm?id=1855610.1855617.
27. B. Heller, S. Seetharaman, P. Mahadevan, Y. Yiakoumis, P. Sharma, S. Banerjee, and N. McKeown, "Elastictree: saving energy in data center networks," in *Proceedings*

of the 7th USENIX conference on Networked systems design and implementation, ser. NSDI'10. Berkeley, CA, USA: USENIX Association, 2010, pp. 17–17. [Online]. Available: http://portal.acm.org/citation.cfm?id=1855711.1855728.

28. M. Gupta and S. Singh, "Greening of the internet," in *Proceedings of the 2003 conference on Applications, technologies, architectures, and protocols for computer communications*, ser. SIGCOMM'03. New York, NY, USA: ACM, 2003, pp. 19–26. [Online]. Available: http://doi.acm.org/10.1145/863955.863959.
29. N. Tolia, Z. Wang, M. Marwah, C. Bash, P. Ranganathan, and X. Zhu, "Delivering energy proportionality with non energy-proportional systems: optimizing the ensemble," in *Proceedings of the 2008 conference on Power aware computing and systems*. USENIX Association, 2008, pp. 2–2.
30. D. Meisner, B. T. Gold, and T. F. Wenisch, "Powernap: eliminating server idle power," *SIGPLAN Notices*, vol. 44, pp. 205–216, March 2009. [Online]. Available: http://doi.acm.org/10.1145/1508284.1508269.
31. G. Chen, W. He, J. Liu, S. Nath, L. Rigas, L. Xiao, and F. Zhao, "Energy-aware server provisioning and load dispatching for connection-intensive internet services," in *NSDI'08: Proceedings of the 5th USENIX Symposium on Networked Systems Design and Implementation*. Berkeley, CA, USA: USENIX Association, 2008, pp. 337–350.
32. D. Kusic, J. O. Kephart, J. E. Hanson, N. Kandasamy, and G. Jiang, "Power and performance management of virtualized computing environments via lookahead control," *Cluster Computing*, vol. 12, pp. 1–15, 2009.
33. J. Chase, D. Anderson, P. Thakar, A. Vahdat, and R. Doyle, "Managing energy and server resources in hosting centers," in *SOSP'01: Proceedings of the eighteenth ACM symposium on Operating systems principles*. New York, NY, USA: ACM, 2001, pp. 103–116.
34. A. Faraz and T. Vijaykumar, "Joint optimization of idle and cooling power in data centers while maintaining response time," *ACM SIGARCH Computer Architecture News*, vol. 38, no. 1, pp. 243–256, 2010.
35. B. Guenter, N. Jain, and C. Williams, "Managing cost, performance, and reliability tradeoffs for energy-aware server provisioning," in *Proc. IEEE INFOCOM, Shanghai, China*. IEEE, 2011, pp. 702–710.
36. A. Krioukov, P. Mohan, S. Alspaugh, L. Keys, D. Culler, and R. Katz, "Napsac: design and implementation of a power-proportional web cluster," in *Proceedings of the first ACM SIGCOMM workshop on Green networking*. ACM, 2010, pp. 15–22.
37. M. Lin, A. Wierman, L. L. H. Andrew, and E. Thereska, "Dynamic right-sizing for power-proportional data centers," in *Proc. IEEE INFOCOM, Shanghai, China*, 2011, pp. 10–15.
38. F. Hermenier, X. Lorca, J. M. Menaud, G. Muller, and J. Lawall, "Entropy: a consolidation manager for clusters," in *ACM SIGPLAN/SIGOPS International Conference on Virtual Execution Environment*, Washington, DC, USA, March 2009, pp. 41–50.
39. S.-H. Lim, J.-S. Huh, Y. Kim, and C. R. Das, "Migration, assignment, and scheduling of jobs in virtualized environment," in *HotCloud, June*, 2011.
40. G. Dhiman, G. Marchetti, and T. Rosing, "vgreen: A system for energy-efficient management of virtual machines," *ACM Trans. Des. Autom. Electron. Syst.*, vol. 16, pp. 6:1–6:27, November 2010. [Online]. Available: http://doi.acm.org/10.1145/1870109.1870115.
41. D. Ardagna, B. Panicucci, M. Trubian, and L. Zhang, "Energy-aware autonomic resource allocation in multi-tier virtualized environments," *IEEE Transactions on Services Computing*, vol. 99, no. PrePrints, 2010.
42. L. Lu, P. J. Varman, and K. Doshi, "Decomposing workload bursts for efficient storage resource management," *IEEE Transactions on Parallel and Distributed Systems*, pp. 860–873, 2010.
43. R. K. Sharma, C. E. Bash, C. D. Patel, R. J. Friedrich, and J. S. Chase, "Balance of power: Dynamic thermal management for internet data centers," *IEEE Internet Computing*, pp. 42–49, 2005.
44. V. Vasudevan, D. Andersen, M. Kaminsky, L. Tan, J. Franklin, and I. Moraru, "Energy-efficient cluster computing with fawn: workloads and implications," in *Proceedings of the 1st*

International Conference on Energy-Efficient Computing and Networking, ser. e-Energy'10. New York, NY, USA: ACM, 2010, pp. 195–204. [Online]. Available: http://doi.acm.org/10.1145/1791314.1791347.
45. K. Lim, P. Ranganathan, J. Chang, C. Patel, T. Mudge, and S. Reinhardt, "Understanding and designing new server architectures for emerging warehouse-computing environments," *SIGARCH Comput. Archit. News*, vol. 36, pp. 315–326, June 2008. [Online]. Available: http://doi.acm.org/10.1145/1394608.1382148.
46. L. Rao, X. Liu, L. Xie, and W. Liu, "Minimizing electricity cost: optimization of distributed internet data centers in a multi-electricity-market environment," in *INFOCOM, 2010 Proceedings*. IEEE, 2010, pp. 1–9.
47. K. Ley, R. Bianchiniy, M. Martonosiz, and T. D. Nguyeny, "Cost and energy aware load distribution across data centers," in *SOSP Workshop on Power Aware Computing and Systems(HotPower'09)*, 2009.
48. Z. Abbasi, T. Mukherjee, G. Varsamopoulos, and S. K. S. Gupta, "Dynamic hosting management of web based applications over clouds," in *International Conference on High performance Computing (HiPC2011)*, India, Dec. 2011.
49. A. Qureshi, R. Weber, H. Balakrishnan, J. Guttag, and B. Maggs, "Cutting the electric bill for internet-scale systems," in *Proceedings of the ACM SIGCOMM 2009 conference on Data communication*. ACM, 2009, pp. 123–134.
50. S. Govindan, A. Sivasubramaniam, and B. Urgaonkar, "Benefits and limitations of tapping into stored energy for datacenters," in *Proc. The 38th International Symposium on Computer Architecture (ISCA)*, San Jose, CA, USA, June 2011.
51. R. Urgaonkar, B. Urgaonkar, M. J. Neely, and A. Sivasubramaniam, "Optimal power cost management using stored energy in data centers," *Arxiv preprint arXiv:1103.3099*, 2011.
52. N. Buchbinder, N. Jain, and I. Menache, "Online job-migration for reducing the electricity bill in the cloud," *NETWORKING 2011*, pp. 172–185, 2011.
53. P. Ranganathan, "Recipe for efficiency: Principles of power-aware computing," in *Commun. ACM*, vol. 53, no. 4. New York, NY, USA: ACM, Apr. 2010, pp. 60–67. [Online]. Available: http://doi.acm.org/10.1145/1721654.1721673.
54. X. Fan, W.-D. Weber, and L. A. Barroso, "Power provisioning for a warehouse-sized compute r," *SIGARCH Comput. Archit. News*, vol. 35, pp. 13–23, June 2007. [Online]. Available: http://doi.acm.org/10.1145/1273440.1250665.
55. D. Wei, "ACPI advanced configuration and power interface," March 2013. [Online]. Available: http://www.acpi.info/.
56. S. Zhuravlev, J. C. Saez, S. Blagodurov, A. Fedorova, and M. Prieto, "Survey of energy-cognizant scheduling techniques," *IEEE Transactions on Parallel and Distributed Systems*, vol. 99, no. PrePrints, 2012.
57. L. L. Andrew, M. Lin, and A. Wierman, "Optimality, fairness, and robustness in speed scaling designs," in *Proceedings of the ACM SIGMETRICS international conference on Measurement and modeling of computer systems*, ser. SIGMETRICS'10. New York, NY, USA: ACM, 2010, pp. 37-48. [Online]. Available: http://doi.acm.org/10.1145/1811039.1811044.
58. A. Wierman, L. Andrew, and A. Tang, "Power-aware speed scaling in processor sharing systems," in *INFOCOM 2009, IEEE*, April 2009, pp. 2007–2015.
59. R. Ge, X. Feng, W. chun Feng, and K. Cameron, "Cpu miser: A performance-directed, run-time system for power-aware clusters," in *Parallel Processing, 2007. ICPP 2007. International Conference on*, Sept. 2007, p. 18.
60. G. Dhiman and T. S. Rosing, "Dynamic voltage frequency scaling for multi-tasking systems using online learning," in *Proceedings of the 2007 international symposium on Low power electronics and design*, ser. ISLPED'07. New York, NY, USA: ACM, 2007, pp. 207–212. [Online]. Available: http://doi.acm.org/10.1145/1283780.1283825.
61. M. Ghasemazar, E. Pakbaznia, and M. Pedram, "Minimizing energy consumption of a chip multiprocessor through simultaneous core consolidation and DVFS," in *Proceedings of 2010*

IEEE International Symposium on Circuits and Systems (ISCAS), 30 2010-June 2 2010, pp. 49–52.
62. P. Bailis, V. Reddi, S. Gandhi, D. Brooks, and M. Seltzer, "Dimetrodon: Processor-level preventive thermal management via idle cycle injection," in *Design Automation Conference (DAC), 2011 48th ACM/EDAC/IEEE*, June 2011, pp. 89–94.
63. K. Kang, J. Kim, S. Yoo, and C.-M. Kyung, "Temperature-aware integrated DVFS and power gating for executing tasks with runtime distribution," *IEEE Transactions on Computer-Aided Design of Integrated Circuits and Systems*, vol. 29, no. 9, pp. 1381–1394, Sept. 2010.
64. J. Leverich, M. Monchiero, V. Talwar, P. Ranganathan, and C. Kozyrakis, "Power management of datacenter workloads using per-core power gating," *IEEE Comput. Archit. Lett.*, vol. 8, no. 2, pp. 48–51, July 2009. [Online]. Available: http://dx.doi.org/10.1109/L-CA.2009.46.
65. X. Wang and Y. Wang, "Coordinating power control and performance management for virtualized server clusters," *IEEE Transactions on Parallel and Distributed Systems*, vol. 22, pp. 245–259, 2011.
66. R. Ayoub, U. Ogras, E. Gorbatov, Y. Jin, T. Kam, P. Diefenbaugh, and T. Rosing, "OS-level power minimization under tight performance constraints in general purpose systems," in *International Symposium on Low Power Electronics and Design (ISLPED) 2011*, Aug. 2011, pp. 321–326.
67. S. Cho and R. G. Melhem, "Corollaries to Amdahl's law for energy," *Computer Architecture Letters*, vol. 7, no. 1, pp. 25–28, 2008.
68. S. Herbert and D. Marculescu, "Analysis of dynamic voltage/frequency scaling in chip-multiprocessors," in *ACM/IEEE International Symposium on Low Power Electronics and Design (ISLPED), 2007*, Aug. 2007, pp. 38–43.
69. S. Zhuravlev, J. C. Saez, S. Blagodurov, A. Fedorova, and M. Prieto, "Survey of energy-cognizant scheduling techniques," *IEEE Transactions on Parallel and Distributed Systems*, vol. 99, no. PrePrints, 2012.
70. X. Fan, C. Ellis, and A. Lebeck, "Memory controller policies for dram power management," in *Proceedings of the 2001 international symposium on Low power electronics and design*, ser. ISLPED'01. New York, NY, USA: ACM, 2001, pp. 129–134. [Online]. Available: http://doi.acm.org/10.1145/383082.383118.
71. H. Huang, P. Pillai, and K. G. Shin, "Design and implementation of power-aware virtual memory," in *Proceedings of the annual conference on USENIX Annual Technical Conference*. Berkeley, CA, USA: USENIX Association, 2003, pp. 5–5. [Online]. Available: http://portal.acm.org/citation.cfm?id=1247340.1247345.
72. X. Li, Z. Li, F. David, P. Zhou, Y. Zhou, S. Adve, and S. Kumar, "Performance directed energy management for main memory and disks," *SIGARCH Comput. Archit. News*, vol. 32, pp. 271–283, October 2004. [Online]. Available: http://doi.acm.org/10.1145/1037947.1024425.
73. J. H. Ahn, N. P. Jouppi, C. Kozyrakis, J. Leverich, and R. S. Schreiber, "Future scaling of processor-memory interfaces," in *Proceedings of the Conference on High Performance Computing Networking, Storage and Analysis*, ser. SC'09. New York, NY, USA: ACM, 2009, pp. 42:1–42:12. [Online]. Available: http://doi.acm.org/10.1145/1654059.1654102.
74. Q. Zou, "An analytical performance and power model based on the transition probability for hard disks," in *3rd International Conference on Awareness Science and Technology (iCAST), 2011*, Sept. 2011, pp. 111–116.
75. A. Verma, R. Koller, L. Useche, and R. Rangaswami, "Srcmap: energy proportional storage using dynamic consolidation," in *Proceedings of the 8th USENIX conference on File and storage technologies*, ser. FAST'10. Berkeley, CA, USA: USENIX Association, 2010, pp. 20–20. [Online]. Available: http://dl.acm.org/citation.cfm?id=1855511.1855531.
76. D. Tsirogiannis, S. Harizopoulos, and M. A. Shah, "Analyzing the energy efficiency of a database server," in *Proceedings of the 2010 international conference on Management of data*, ser. SIGMOD'10. New York, NY, USA: ACM, 2010, pp. 231–242. [Online]. Available: http://doi.acm.org/10.1145/1807167.1807194.

77. T. Härder, V. Hudlet, Y. Ou, and D. Schall, "Energy efficiency is not enough, energy proportionality is needed!" in *Proceedings of the 16th international conference on Database systems for advanced applications*, ser. DASFAA'11. Berlin, Heidelberg: Springer-Verlag, 2011, pp. 226–239. [Online]. Available: http://dl.acm.org/citation.cfm?id=1996686.1996716.
78. Y. Deng, "What is the future of disk drives, death or rebirth?" *ACM Comput. Surv.*, vol. 43, no. 3, pp. 23:1–23:27, Apr. 2011. [Online]. Available: http://doi.acm.org.ezproxy1.lib.asu.edu/10.1145/1922649.1922660.
79. H. Amur, J. Cipar, V. Gupta, G. R. Ganger, M. A. Kozuch, and K. Schwan, "Robust and flexible power-proportional storage," in *Proceedings of the 1st ACM symposium on Cloud computing*, ser. SoCC'10. New York, NY, USA: ACM, 2010, pp. 217–228. [Online]. Available: http://doi.acm.org.ezproxy1.lib.asu.edu/10.1145/1807128.1807164.
80. J. Guerra, W. Belluomini, J. Glider, K. Gupta, and H. Pucha, "Energy proportionality for storage: impact and feasibility," *SIGOPS Oper. Syst. Rev.*, vol. 44, pp. 35–39, March 2010. [Online]. Available: http://doi.acm.org.ezproxy1.lib.asu.edu/10.1145/1740390.1740399.
81. C. H. Hsu and S. W. Poole, "Power signature analysis of the specpower_ssj2008 benchmark," in *IEEE International Symposium on Performance Analysis of Systems and Software (ISPASS), 2011*. IEEE, 2011, pp. 227–236.
82. S. Wang, J. Liu, J.-J. Chen, and X. Liu, "Powersleep: A smart power-saving scheme with sleep for servers under response time constraint," *IEEE Journal on Emerging and Selected Topics in Circuits and Systems*, vol. 1, no. 3, pp. 289–298, Sept. 2011.
83. E. Elnozahy, M. Kistler, and R. Rajamony, "Energy-efficient server clusters," in *Power-Aware Computer Systems*, ser. Lecture Notes in Computer Science, B. Falsafi and T. Vijaykumar, Eds. Springer Berlin / Heidelberg, 2003, vol. 2325, pp. 179–197.
84. D. Meisner, C. M. Sadler, L. A. Barroso, W.-D. Weber, and T. F. Wenisch, "Power management of online data-intensive services," in *Proceeding of the 38th annual international symposium on Computer architecture*, ser. ISCA'11. New York, NY, USA: ACM, 2011, pp. 319–330. [Online]. Available: http://doi.acm.org/10.1145/2000064.2000103.
85. A. Gandhi, M. Harchol-Balter, and M. A. Kozuch, "The case for sleep states in servers," in *Proceedings of the 4th Workshop on Power-Aware Computing and Systems*, ser. HotPower'11. New York, NY, USA: ACM, 2011, pp. 2:1–2:5. [Online]. Available: http://doi.acm.org/10.1145/2039252.2039254.
86. Y. Wang, X. Wang, M. Chen, and X. Zhu, "Power-efficient response time guarantees for virtualized enterprise servers," in *Real-Time Systems Symposium, 2008*, 30 2008-Dec. 3 2008, pp. 303–312.
87. X. Wang and Y. Wang, "Coordinating Power Control and Performance Management for Virtualized Server Clusters," IEEE Transactions on Parallel and Distributed Systems, pp. 245–259, 2010.
88. P. Ranganathan, P. Leech, D. Irwin, and J. Chase, "Ensemble-level power management for dense blade servers," in *Computer Architecture, 2006. ISCA'06. 33rd International Symposium on*, 0-0 2006, pp. 66–77.
89. T. Mukherjee, G. Varsamopoulos, S. Sandeep K. Gupta, and S. Rungta, "Measurement-based power profiling of data center equipment," in *IEEE International Conference on Cluster Computing.*, Austin, Texas, USA, Sept. 2007, pp. 476–477.
90. B. Urgaonkar, P. Shenoy, A. Chandra, P. Goyal, and T. Wood, "Agile dynamic provisioning of multi-tier internet applications," *ACM Trans. Auton. Adapt. Syst.*, vol. 3, pp. 1:1–1:39, March 2008. [Online]. Available: http://doi.acm.org/10.1145/1342171.1342172.
91. A. O. Allen, *Probability, statistics and queuing theory with computer science applications.* Academic Press Inc., 1990.
92. S. Saroiu, K. P. Gummadi, R. J. Dunn, S. D. Gribble, and H. M. Levy, "An analysis of Internet content delivery systems," *ACM SIGOPS Operating Systems Review*, pp. 315–327, 2002.
93. M. Lin, Z. Liu, A. Wierman, and L. L. H. Andrew, "Online algorithms for geographical load balancing," in *Proc. of International Green Computing Conference (IGCC11)*. IEEE, June 2012.

94. P. Bohrer, E. N. Elnozahy, T. Keller, M. Kistler, C. Lefurgy, C. McDowell, and R. Rajamony, "The case for power management in web servers," pp. 261–289, 2002.
95. P. Barford and M. Crovella, "Generating representative web workloads for network and server performance evaluation," in *Proceedings of the 1998 ACM SIGMETRICS joint international conference on Measurement and modeling of computer systems*, ser. SIGMETRICS'98/PERFORMANCE'98. New York, NY, USA: ACM, 1998, pp. 151–160. [Online]. Available: http://doi.acm.org/10.1145/277851.277897.
96. Y. Chen, A. Das, W. Qin, A. Sivasubramaniam, Q. Wang, and N. Gautam, "Managing server energy and operational costs in hosting centers," *SIGMETRICS Performance Evaluation Review*, vol. 33, no. 1, pp. 303–314, 2005.
97. P. Bodik, R. Griffith, C. Sutton, A. Fox, M. I. Jordan, and D. A. Patterson, "Automatic exploration of datacenter performance regimes," in *Proceedings of the 1st workshop on Automated control for datacenters and clouds*, ser. ACDC'09. New York, NY, USA: ACM, 2009, pp. 1–6. [Online]. Available: http://doi.acm.org/10.1145/1555271.1555273.
98. I. Cunha, I. Viana, J. Palotti, J. Almeida, and V. Almeida, "Analyzing security and energy tradeoffs in autonomic capacity management," in *Network Operations and Management Symposium, 2008. NOMS 2008. IEEE*. IEEE, 2008, pp. 302–309.
99. Q. Tang, S. K. S. Gupta, and G. Varsamopoulos, "Energy-efficient thermal-aware task scheduling for homogeneous high-performance computing data centers: A cyber-physical approach," *IEEE Trans. Parallel Distrib. Syst.*, vol. 19, no. 11, pp. 1458–1472, 2008.
100. J. Moore, J. Chase, P. Ranganathan, and R. Sharma, "Making scheduling "cool": temperature-aware workload placement in data centers," in *ATEC'05: Proceedings of the annual conference on USENIX Annual Technical Conference*. Berkeley, CA, USA: USENIX Association, 2005, pp. 5–5.
101. J. Moore, J. Chase, and P. Ranganathan, "Weatherman: Automated, online, and predictive thermal mapping and management for data centers," in *IEEE International Conference on Autonomic Computing (ICAC)*, June 2006, pp. 155–164.
102. L. Parolini, N. Toliaz, B. Sinopoliy, and B. H. Kroghy, "A cyber-physical systems approach to energy management in data centers," in *ACM ICCPS'10*, Stockholm, Sweden, April 2010.
103. Z. Abbasi, G. Varsamopoulos, and S. K. S. Gupta, "TACOMA: Server and workload management in internet data centers considering cooling-computing power trade-off and energy proportionality," *ACM Trans. Archit. Code Optim.*, vol. 9, no. 2, pp. 11:1–11:37, June 2012.
104. P. Sanders, N. Sivadasan, and M. Skutella, "Online scheduling with bounded migration," *Math. Oper. Res.*, vol. 34, no. 2, pp. 481–498, May 2009. [Online]. Available: http://dx.doi.org/10.1287/moor.1090.0381.
105. P. Padala, K.-Y. Hou, K. G. Shin, X. Zhu, M. Uysal, Z. Wang, S. Singhal, and A. Merchant, "Automated control of multiple virtualized resources," in *Proceedings of the 4th ACM European conference on Computer systems*. ACM, 2009, pp. 13–26.
106. R. Nathuji and K. Schwan, "Virtualpower: coordinated power management in virtualized enterprise systems," *SIGOPS Oper. Syst. Rev.*, vol. 41, pp. 265–278, Oct. 2007. [Online]. Available: http://doi.acm.org/10.1145/1323293.1294287.
107. T. Gerald, K. J. Nicholas, D. Rajarshi, and N. B. Mohamed, "A hybrid reinforcement learning approach to autonomic resource allocation," in *IEEE International Conference on Autonomic Computing*. IEEE, 2006, pp. 65–73.
108. J. Mars, L. Tang, R. Hundt, K. Skadron, and M. Soffa, "Bubble-up: Increasing utilization in modern warehouse scale computers via sensible co-locations," in *Proceedings of the 44th Annual IEEE/ACM International Symposium on Microarchitecture*, ser. MICRO-44. New York, NY, USA: ACM, 2011, pp. 248–259. [Online]. Available: http://doi.acm.org/10.1145/2155620.2155650.
109. J. Mars, L. Tang, and M. L. Soffa, "Directly characterizing cross core interference through contention synthesis," in *Proceedings of the 6th International Conference on High Performance and Embedded Architectures and Compilers*, ser. HiPEAC'11. New York, NY, USA: ACM, 2011, pp. 167–176. [Online]. Available: http://doi.acm.org/10.1145/1944862.1944887.

110. A. Fedorova, S. Blagodurov, and S. Zhuravlev, "Managing contention for shared resources on multicore processors," *Commun. ACM*, vol. 53, no. 2, pp. 49–57, Feb. 2010. [Online]. Available: http://doi.acm.org/10.1145/1646353.1646371.
111. L. Tang, J. Mars, and M. L. Soffa, "Contentiousness vs. sensitivity: improving contention aware runtime systems on multicore architectures," in *Proceedings of the 1st International Workshop on Adaptive Self-Tuning Computing Systems for the Exaflop Era*, ser. EXADAPT'11. New York, NY, USA: ACM, 2011, pp. 12–21. [Online]. Available: http://doi.acm.org/10.1145/2000417.2000419.
112. R. C. Chiang and H. H. Huang, "Tracon: interference-aware scheduling for data-intensive applications in virtualized environments," in *Proceedings of 2011 International Conference for High Performance Computing, Networking, Storage and Analysis*, ser. SC'11. New York, NY, USA: ACM, 2011, pp. 47:1–47:12. [Online]. Available: http://doi.acm.org/10.1145/2063384.2063447.
113. B. Kreaseck, L. Carter, H. Casanova, and J. Ferrante, "On the interference of communication on computation in Java," *International Parallel and Distributed Processing Symposium*, vol. 15, p. 246, 2004.
114. Q. Zhu, J. Zhu, and G. Agrawal, "Power-aware consolidation of scientific workflows in virtualized environments," in *Proceedings of the 2010 ACM/IEEE International Conference for High Performance Computing, Networking, Storage and Analysis*, ser. SC'10. Washington, DC, USA: IEEE Computer Society, 2010, pp. 1–12. [Online]. Available: http://dx.doi.org/10.1109/SC.2010.43.
115. X. Pu, L. Liu, Y. Mei, S. Sivathanu, Y. Koh, C. Pu, Y. Cao, and L. Liu, "Who is your neighbor: Net i/o performance interference in virtualized clouds," vol. PP, no. 99, 2012, pp. 1–1.
116. I. Paul., S. Yalamanchili., and L. K. J. John, "Performance impact of virtual machine placement in a datacenter," in *Performance Computing and Communications Conference (IPCCC), 2012 IEEE 31st International*, 2012, pp. 424–431.
117. M. Kambadur, T. Moseley, R. Hank, and M. A. Kim, "Measuring interference between live datacenter applications," in *Proceedings of the International Conference on High Performance Computing, Networking, Storage and Analysis*, ser. SC'12. Los Alamitos, CA, USA: IEEE Computer Society Press, 2012, pp. 51:1–51:12. [Online]. Available: http://dl.acm.org/citation.cfm?id=2388996.2389066.
118. F. Hermenier, X. Lorca, J.-M. Menaud, G. Muller, and J. Lawall, "Entropy: a consolidation manager for clusters," in *Proceedings of the 2009 ACM SIGPLAN/SIGOPS international conference on Virtual execution environments*, ser. VEE'09. New York, NY, USA: ACM, 2009, pp. 41–50. [Online]. Available: http://doi.acm.org/10.1145/1508293.1508300.
119. M. Pore, Z. Abbasi, S. Gupta, and G. Varsamopoulos, "Energy aware colocation of workload in data centers," in *19th International Conference on High Performance Computing (HiPC), 2012*, 2012, pp. 1–6.
120. A. Verma, P. Ahuja, and A. Neogi, "pmapper: power and migration cost aware application placement in virtualized systems," in *Proceedings of the 9th ACM/IFIP/USENIX International Conference on Middleware*, ser. Middleware'08. New York, NY, USA: Springer-Verlag New York, Inc., 2008, pp. 243–264. [Online]. Available: http://dl.acm.org/citation.cfm?id=1496950.1496966.
121. R. Buyya, A. Beloglazov, and J. H. Abawajy, "Energy-efficient management of data center resources for cloud computing: A vision, architectural elements, and open challenges," *CoRR*, vol. abs/1006. 0308, 2010.
122. A. Merkel, J. Stoess, and F. Bellosa, "Resource-conscious scheduling for energy efficiency on multicore processors," in *Proceedings of the 5th European conference on Computer systems*, ser. EuroSys'10. New York, NY, USA: ACM, 2010, pp. 153–166. [Online]. Available: http://doi.acm.org/10.1145/1755913.1755930.
123. J. Mars, L. Tang, and R. Hundt, "Heterogeneity in homogeneous warehouse-scale computers: A performance opportunity," *Computer Architecture Letters*, vol. 10, no. 2, pp. 29–32, July–Dec. 2011.

124. M. Armbrust, A. Fox, R. Griffith, A. D. Joseph, R. H. Katz, A. Konwinski, G. Lee, D. A. Patterson, A. Rabkin, and M. Zaharia, "Above the clouds: A berkeley view of cloud computing," in *Technical Report No. UCB/EECS-2009-28,* University of California at Berkley, USA, February 2009.
125. K. Le, O. Bilgir, R. Bianchini, M. Martonosi, and T. D. Nguyen, "Managing the cost, energy consumption, and carbon footprint of internet services," in *Proceedings of the ACM SIGMETRICS international conference on Measurement and modeling of computer systems*, ser. SIGMETRICS'10. New York, NY, USA: ACM, 2010, pp. 357–358.
126. "Quick start guide to increase data center energy efficiency," U.S. Department of Energy, Tech. Rep., September 2010.
127. "Google data center more efficient that the industry average," http://gigaom.com/2008/10/01/google-data-centers-more-efficient-than-the-industry-average/.
128. J. Caruso, "Follow the moon, and save millions researchers highlight possibilities of data center energy savings." March 2014. [Online]. Available: http://www.networkworld.com/newsletters/lans/2009/081709lan2.html.
129. L. Rao, X. Liu, L. Xie, and W. Liu, "Minimizing electricity cost: Optimization of distributed internet data centers in a multi-electricity-market environment," in *INFOCOM, 2010 Proceedings IEEE*, March 2010, pp. 1–9.
130. S. Akoush, R. Sohan, A. Rice, A. W. Moore, and A. Hopper, "Free lunch: exploiting renewable energy for computing," in *Proceedings of HotOS*, 2011.
131. M. Etinski, M. Martonosi, K. Le, R. Bianchini, and T. D. Nguyen, "Optimizing the use of request distribution and stored energy for cost reduction in multi-site internet services," in *Sustainable Internet and ICT for Sustainability (SustainIT), 2012*. IEEE, 2012, pp. 1–10.
132. Z. Abbasi, T. Mukherjee, G. Varsamopoulos, and S. K. S. Gupta, "Dahm: A green and dynamic web application hosting manager across geographically distributed data centers," *J. Emerg. Technol. Comput. Syst.*, vol. 8, no. 4, pp. 34:1–34:22, Nov. 2012.
133. Y. Zhang, Y. Wang, and X. Wang, "Greenware: greening cloud-scale data centers to maximize the use of renewable energy," *Middleware 2011*, pp. 143–164, 2011.
134. Z. Liu, M. Lin, A. Wierman, S. H. Low, and L. L. H. Andrew, "Geographical load balancing with renewables," *ACM SIGMETRICS Performance Evaluation Review*, vol. 39, no. 3, pp. 62–66, 2011.
135. C. Stewart and K. Shen, "Some joules are more precious than others: Managing renewable energy in the datacenter," in *Proceedings of the Workshop on Power Aware Computing and Systems*, 2009.
136. Z. Abbasi, M. Pore, and S. K. Gupta, "Online server and workload management for joint optimization of electricity cost and carbon footprint across data centers," in *28th IEEE International Parallel & Distributed Processing Symposium (IPDPS)*, 2014.
137. D. S. Palasamudram, R. K. Sitaramanz, B. Urgaonkar, and R. Urgaonkar, "Using batteries to reduce the power costs of internet-scale distributed networks," in *Proceedings of 2012 ACM Symposium on Cloud Computing*. ACM, Oct. 2012, Palasamudram 2012 using.
138. S. Govindan, D. Wang, Anand, Sivasubramaniam, and B. Urgaonkar, "Leveraging stored energy for handling power emergencies in aggressively provisioned datacenters," in *Proceedings of the seventeenth international conference on Architectural Support for Programming Languages and Operating Systems (ASPLOS)*. ACM, 2012, pp. 75–86.
139. V. Kontorinis, L. E. Zhang, B. Aksanli, J. Sampson, H. Homayoun, E. Pettis, D. M. Tullsen, and T. S. Rosing, "Managing distributed ups energy for effective power capping in data centers," in *Proceedings of the 39th International Symposium on Computer Architecture*, ser. ISCA'12. Piscataway, NJ, USA: IEEE Press, 2012, pp. 488–499. [Online]. Available: http://dl.acm.org/citation.cfm?id=2337159.2337216.
140. D. Wang, C. Ren, A. Sivasubramaniam, B. Urgaonkar, and H. Fathy, "Energy storage in datacenters: what, where, and how much?" in *Proceedings of the 12th ACM SIGMETRICS/PERFORMANCE joint international conference on Measurement and Modeling of*

Computer Systems, ser. SIGMETRICS'12. New York, NY, USA: ACM, 2012, pp. 187–198, wang2012-energy-storage.
141. K. Le, R. Bianchini, T. D. Nguyen, O. Bilgir, and M. Martonosi, "Capping the brown energy consumption of internet services at low cost," in *Green Computing Conference, 2010 International*. IEEE, 2010, pp. 3–14.
142. P. X. Gao, A. R. Curtis, B. Wong, and S. Keshav, "It's not easy being green," *ACM SIGCOMM Computer Communication Review*, vol. 42, no. 4, pp. 211–222, 2012.
143. S. Ren and Y. He, "Coca: Online distributed resource management for cost minimization and carbon neutrality in data centers," in *Proceedings of the International Conference on High Performance Computing, Networking, Storage and Analysis*, ser. SC'13. New York, NY, USA: ACM, 2013, pp. 39:1–39:12. [Online]. Available: http://doi.acm.org/10.1145/2503210.2503248.
144. A. H. Mahmud and S. Ren, "Online capacity provisioning for carbon-neutral data center with demand-responsive electricity prices," *ACM SIGMETRICS Performance Evaluation Review*, vol. 41, no. 2, pp. 26–37, 2013.
145. Z. Zhou, F. Liu, Y. Xu, R. Zou, H. Xu, J. C. S. Lui, and H. Jin, "Carbon-aware load balancing for geo-distributed cloud services," in *Proceedings of the 2013 IEEE 21st International Symposium on Modelling, Analysis & Simulation of Computer and Telecommunication Systems*, ser. MASCOTS'13. Washington, DC, USA: IEEE Computer Society, 2013, pp. 232–241. [Online]. Available: http://dx.doi.org/10.1109/MASCOTS.2013.31.
146. J. Doyle, R. Shorten, and D. O'Mahony, "Stratus: Load balancing the cloud for carbon emissions control," *Cloud Computing, IEEE Transactions on*, vol. 1, no. 1, pp. 1–1, Jan 2013.
147. D. Xu and X. Liu, "Geographic trough filling for internet datacenters," in *IEEE Proceedings INFOCOM*. IEEE, 2012, pp. 2881–2885.
148. S.-H. Lim, B. Sharma, G. Nam, E. K. Kim, and C. Das, "Mdcsim: A multi-tier data center simulation, platform," in *Cluster Computing and Workshops, 2009. CLUSTER'09. IEEE International Conference on*, Aug. 2009, pp. 1–9.
149. D. Meisner, J. Wu, and T. Wenisch, "Bighouse: A simulation infrastructure for data center systems," in *Performance Analysis of Systems and Software (ISPASS), 2012 IEEE International Symposium on*, April 2012, pp. 35–45.
150. R. N. Calheiros, R. Ranjan, A. Beloglazov, C. A. De Rose, and R. Buyya, "Cloudsim: a toolkit for modeling and simulation of cloud computing environments and evaluation of resource provisioning algorithms," *Software: Practice and Experience*, vol. 41, no. 1, pp. 23–50, 2011.
151. C. Patel, C. E. Bash, C. Belady, L. Stahl, and D. Sullivan, "Computational fluid dynamics modeling of high compute density data centers to assure system inlet air specifications," in *ASME International Electronic Packaging Technical Conference and Exhibition (IPACK'01)*, 2001, patel_ipack2001.
152. L. Marshall and P. Bems, "Using cfd for data center design and analysis," Applied Math Modeling, Tech. Rep., Jan. 2011, White Paper.
153. U. Singh, "Cfd-based operational thermal efficiency improvement of a production data center," in *Proceedings of the First USENIX conference on Sustainable information technology*, ser. SustainIT'10. Berkeley, CA, USA: USENIX Association, 2010, pp. 6–6, sI2010USENIX.
154. S. K. S. Gupta, R. R. Gilbert, A. Banerjee, Z. Abbasi, T. Mukherjee, and G. Varsamopoulos, "GDCSim - an integrated tool chain for analyzing green data center physical design and resource management techniques," in *International Green Computing Conference (IGCC)*, Orlando, FL, 2011, pp. 1–8.
155. S. K. S. Gupta, A. Banerjee, Z. Abbasi, G. Varsamopoulos, M. Jonas, J. Ferguson, R. R. Gilbert, and T. Mukherjee, "Gdcsim: A simulator for green data center design and analysis," *ACM Trans. Model. Comput. Simul.*, vol. 24, no. 1, pp. 3:1–3:27, Jan. 2014. [Online]. Available: http://doi.acm.org/10.1145/2553083.
156. A. Banerjee, J. Banerjee, G. Varsamopoulos, Z. Abbasi, and S. K. Gupta, "Hybrid simulator for cyber-physical energy systems," in *2013 Workshop on Modeling and Simulation of Cyber-Physical Energy Systems (MSCPES)*. IEEE, 2013, pp. 1–6.

A Power-Aware Autonomic Approach for Performance Management of Scientific Applications in a Data Center Environment

Rajat Mehrotra, Ioana Banicescu, Srishti Srivastava and Sherif Abdelwahed

1 Introduction

The amount of electricity used by servers and data centers has become an important issue in recent years because the demands for high performance computing (HPC) services have become widespread. The recent advancements in the HPC system size and processing power at computing nodes led to a significant increase in power consumption. A study commissioned by the U.S. Environmental Protection Agency estimated that the worldwide power consumed by servers increased by a factor of two between 2000 and 2006 worldwide [1]. In addition to this, an increase in power consumption results in increased temperature, which in turn translates into increased heat dissipation issues, resulting in increased system failure rate that leads to a downtime penalty of extremely high values for the service providers [2]. According to Arrhenius' equation when applied to HPC hardware, a ten degree increase in system's temperature doubles the system failure rate [3]. Therefore, an increase in system temperature resulting from an increase in power consumption can lead to a degradation in the execution performance of scientific applications running on HPC systems. Such HPC systems with enormous heat dissipation need aggressive

R. Mehrotra (✉) · S. Abdelwahed
Department of Electrical and Computer Engineering, NSF Center for Cloud and Autonomic Computing, Mississippi State University, MS, USA
e-mail: rajat.meh@gmail.com

S. Abdelwahed
e-mail: sherif@ece.msstate.edu

I. Banicescu · S. Srivastava
Department of Computer Science and Engineering, NSF Center for Cloud and Autonomic Computing, Mississippi State University, MS, USA

I. Banicescu
e-mail: ioana@cse.msstate.edu

S. Srivastava
e-mail: srishti@hpc.msstate.edu

S. U. Khan, A. Y. Zomaya (eds.), *Handbook on Data Centers*,
DOI 10.1007/978-1-4939-2092-1_5

cooling systems. However, the additional deployment of aggressive cooling systems contribute even more to the total power consumption of the infrastructure, resulting in an increased operational cost. Another solution to the heat dissipation problem is to increase the physical space between different computing nodes. However, this approach results in a high infrastructure cost, due to the use of larger space for fewer computing nodes. Both of these solutions ignore the performance versus operational cost and performance versus space cost metrics of the system, which have become significant to the service providers in the current scenario.

Solutions to the increased concerns of outrageous amounts of power consumption in HPC environments have led to the development of *Green Supercomputing* or *Energy Efficient Computing* technologies. Green Destiny, the first major instantiation of the supercomputing in the Small Spaces Project at Los Alamos National Laboratory was the first supercomputing system built with energy efficiency as its guiding principle [4]. The main approach is to use low power components with performance that could be optimized for supercomputing. Green Supercomputing provides two major approaches for minimizing power consumption in HPC environments while maintaining the desired performance: a low-power and a power-aware approach.

In a *low-power* approach, the HPC system consists of many low power nodes which can be configured for a supercomputing environment. *Green Destiny* was the first low power supercomputer developed at Los Alamos National Laboratory [4]. It consists of 240 computing nodes working at the rate of 240 gigaflops on a linux-based cluster. It fits into a six square feet surface, consumes only 3.2 KW of power, and extra cooling or space are not required. The design of this supercomputer led to a breakthrough in the technology of HPC environments, and shifted the focus towards efficiency, availability, and reliability of computing systems, in addition to their speed.

In a *power-aware* approach, the HPC system is considered to be *aware* of the power requirements of the system on which the scientific applications are executing. In these systems, the power can be dynamically adjusted based on the demands of the application, while maintaining the performance at desired levels [5–7]. In most HPC systems, the *power-aware* functionality is achieved through dynamic voltage and frequency scaling (DVFS) approach in which the power consumption by a processor directly depends upon the frequency and the square of the CPU supplied voltage. The frequency or the voltage across the processor can be decreased when CPU is mostly idle (not doing any or much useful work). This approach translates into considerable power savings.

In addition, the primary objective for designing high performance computing (HPC) systems is performance-to solve scientific problems in minimum time with efficient or maximum utilization of the allocated computing resources. An efficient utilization of the allocated computing resources is achieved by following two objectives: (i) decreasing the computational and interprocess communication overheads, and (ii) ensuring that computing resources are effectively utilized in doing useful work at their peak performance all the time.

In general, scientific applications are large, highly irregular, computationally intensive, and data parallel. One of the major challenges in achieving performance objectives when running scientific applications in heterogeneous HPC environments is often their stochastic behavior [8]. This stochastic behavior results in a high load imbalance, which severely degrades the performance of the overall HPC system, and becomes a potential threat for missing a pre-specified execution deadline. Moreover, the variations in the availability of the non-dedicated computing nodes, due to unpredictable changes in the usage patterns of other applications running on the same computing node in a space shared manner, represents an additional source of overhead for meeting the deadline in HPC systems. Therefore, to address these challenges, a runtime monitoring and corrective management module is required that can reallocate or reconfigure the computational resources for achieving an optimal execution performance of scientific applications, such that the corrective employed strategy incurs minimal overhead.

Parallel loops are the dominant source of parallelism in scientific computing applications. For minimizing the computation time of an application, the loop iterations need to be executed efficiently over the network of available computing nodes. In recent years, various dynamic loop scheduling techniques based on probabilistic analyses have been developed and applied to effectively allocate these iterations to different computing nodes and improve the performance of scientific applications via dynamic load balancing. Details regarding these techniques can be found in the related literature on dynamic loop scheduling [8]. However, these loop scheduling algorithms do not consider power-awareness, reallocation of computing resources, and application deadline.

In this chapter, a *control theoretic, model-based, and power-aware* approach for parallel loop execution is presented, where system performance and power requirements can be adjusted dynamically, while maintaining the predefined quality of service (QoS) goals in the presence of perturbations such as load imbalance due to variations in the availability of computational resources. In this approach, a target turnaround time for the application is identified or supplied, and it is responsible for guiding an appropriate adjustment of the voltage and frequency. The adjustment is performed with the help of a control theoretic approach at the computational resource to ensure that the application finishes execution by the pre-specified deadline (within an acceptable or chosen tolerance value). The control theoretic approach also ensures a load-balanced execution of the scientific application because all the processing nodes simultaneously finish execution within an acceptable tolerance value. The general framework of this work considers deadline driven applications executing over a set of non-dedicated computational resources. This approach is autonomic, performance directed, dynamically controlled, and independent of (does not interfere with) the execution of the application. A few initial results were presented in [9].

The rest of the chapter is organized as follows. Preliminary information regarding power consumption issues and power management techniques in HPC systems are presented in Sect. 2. In the same section, the earlier research efforts on parallel loop scheduling with and without power-aware knowledge are discussed. In addition to this, key elements of a control system are described with their prior applications. Details of the proposed power-aware autonomic approach are presented in Sect. 3. The basic architecture and key components of the proposed approach are also discussed in this section. A case study for managing response time and power consumption in a parallel loop execution environment by using the proposed approach is discussed in Sect. 5. The significance of applying this approach is demonstrated by using three different experiments, where the performance of the hosted applications with different priorities are adjusted according to the response time and power consumption in varying operational settings. The benefits of the proposed approach and major contributions are highlighted in Sect. 5. An extension of the proposed approach that combines dynamic loop scheduling techniques with the proposed approach, is discussed in Sect. 6. Finally, conclusions and future work are presented in Sect. 5.

2 Background

In this section, past and ongoing research efforts are discussed, such as approaches based on power-aware computing, loop scheduling, and control theory for HPC systems.

A. Power Consumption Management in HPC Systems
Since the beginning of the HPC systems in 1960, processing power and performance of the HPC systems were the only concern among designers of these systems. The power consumption cost and carbon footprint of these systems were never in consideration while designing the size and computational power of these systems. As a result, over the years, the power consumption and the cooling cost of these HPC systems kept increasing exponentially. These additional costs further increased the setting up and operating cost of the infrastructure. In addition, the CO_2 emission of the HPC systems also increased its share in the total global emission. Therefore, researchers in academia and industry were inspired to design HPC systems that consume less energy while keeping the processing capabilities within similar range. The basic idea for designing an energy efficient HPC system is to increase the energy efficiency by utilizing the maximum amount of the available energy for doing useful work (computation) and a minimum amount of it for overheads and infrastructure. The primary reasons of power consumption and power consumption management in a computing system are described in the following subsection.

1) Power Consumption in Computing Systems Modern electronic components are in general built using the Complementary Metal Oxide Semiconductor (CMOS) technology. Advances in CMOS technology during the last decade has allowed chip vendors to increase clock rates by stacking more transistors on the die, and that led to an increase in power consumption. At the transistor level, the power consumption can be attributed to three factors: *switching (dynamic) power consumption*, *leakage (static) current power consumption*, and *short circuit power consumption*. Dynamic power consumption is the amount of power consumed while charging the capacitor present in CMOS Field Effect Transistor (FET). Static current power consumption is due to the leakage current flowing through the transistor while it is in the "off" state. A short circuit power consumption is present in CMOS due to a short circuit current on a short circuit path between supply rails and ground. Currently, power loss due to leakage current is 40 % of the total power budget. Lowering the voltage across the chip increases the leakage current. Increasing the leakage current in transistors results increasing power consumption of the microprocessor [10]. In addition, high operating temperature of a microprocessor increases the power consumption significantly due to leakage current. In the past, dynamic power loss has been the main component of the total power loss. Moreover, lately the percentage of static power loss is also increasing as feature sizes in CMOS have been decreasing. These factors are applicable to all the electronic systems of the computer, including the CPU, the memory and even the hard drive. In the hard drive, there are some other mechanical factors that lead to increased power consumption. Discussion on these factors is beyond the scope of this chapter.

2) Power Management in Computing Systems The power consumption issues of the computing system infrastructure have been handled by researchers in academia and industry in the following ways by using: efficient thermal management, energy-aware HPC design, microprocessor power management, and application level power management.

Efficient thermal management techniques are used in the HPC infrastructure for managing the overall power consumption including the power consumed inside the cooling infrastructure. The increased demand for higher computational power in HPC systems has resulted in increased power consumption and heat dissipation. This increased power consumption and heat generation requires an effective and aggressive cooling infrastructure to lower the temperature of the components of the HPC systems. This aggressive cooling infrastructure contributes to the infrastructure cost, as well as to the operating cost due to its own power consumption. According to an earlier study, the power consumption due to the cooling infrastructure accounts for two thirds of the actual HPC infrastructure power consumption [11]. As a results, the total cost of ownership of HPC infrastructure reaches extremely high values and becomes a primary concern for the infrastructure providers. Recently, a few efficient cooling infrastructures are designed by the researchers. These efficient cooling infrastructures suggest efficient heat management in the HPC systems, either through airflow designs [12, 13], or through redistributing the workload in processing [14–16]. A proactive approach for efficient thermal management uses the workload behaviour to minimize the energy consumption in a cooling infrastructure [17].

In energy-aware HPC design, these systems are designed in a space and power efficient way that reduces the total cost of ownership for the HPC systems. The first low power supercomputer "*Green Destiny*" was developed at Los Alamos National Laboratory in 2001 [4]. This project was instrumental in ushering a new research area of "system-on-chips", which was further taken up by IBM by developing an entire family of supercomputers named "Blue Gene" [18]. These supercomputers are developed using low power and reliable embedded chips connected through specialized networks. These supercomputers are designed using racks which can be scaled up or down according to the requirements. The "Blue Gene" family of supercomputers delivered speed, scalability, and power efficiency in a single design. Another approach of designing energy efficient clusters was developed, that used warm water to cool down processors instead of using a traditional air cooling system [19]. In this approach, warm water of temperature $140°F$ is used to transfer heat from the processing nodes. After heat transfer, once the final temperature of the water reaches about $185°F$, it is further utilized in the building for other purposes. This approach shows that 80 % of the recovered heat (from the supercomputers) can be utilized for heating up the office building, which in turn reduces the operating cost of the infrastructure.

The primary focus of the academia and industry with respect to power management in computing systems has been to target the power consumption of the microprocessors. Various methods, such as dynamic power switching, standby leakage management, and dynamic voltage and frequency scaling have been proposed to control the power consumption in microprocessors through system and application level techniques.

The dynamic power switching (DPS) approach tries to maximize the system idle time that in turn forces a processor to make transition to idle or low power mode for reducing the power consumption [10]. The only concern is to keep track of the wakeup latency for the processor. It tries to finish the assigned tasks as quickly as possible, such that the rest of the time can be considered as idle time for the processor. This method reduces only the leakage current power consumption, while at the same time it increases the dynamic power consumption due to excessive mode switching of the processor.

The standby leakage management (SLM) method is close to the strategy used in DPS by keeping the system in low power mode [10]. However, this strategy comes into the effect only when there is no application executing in the system and the system just needs to take care of its responsiveness towards user related wake up events (e.g. GUI interaction, key press, or mouse click).

In contrast to the DPS, in the dynamic voltage and frequency scaling (DVFS) method, the voltage across the chip and the clock frequency of the transistor are varied (increased or decreased), to lower the power consumption and maintain the processing speed at the same time [10]. This method is helpful in preventing the computer systems from an overheating that can result in a system crash. However, the voltage applied should be kept at the level suggested by the manufacturer, to keep the system stable for safe operation. DVFS reduces the processor idle time by lowering the voltage or frequency, by allowing the assigned tasks to continue to be processed within a permissible amount of time, with minimum possible power consumption. This approach reduces the dynamic power loss.

B. Power-Aware Approaches with DVFS

Power-Aware computing has recently gained attention in the HPC research communities with the purpose of lowering the power consumption, for increasing the overall system availability and reliability. The proposed power-aware approaches attempt to model the power consumption pattern of the scientific application, or that of the entire HPC system, based on the application and/or system performance, to minimize the power consumption of the HPC system with minimal or no impact on the application and system performance.

Recently, application developers have started developing applications using energy efficient algorithms that either require less energy or can take advantage of the energy efficient features of the system hardware. An effort to minimize the power consumption of a HPC system through identifying the different execution phases (memory access, I/O access, and system idle) and their performance requirements while executing scientific applications is highlighted in [5]. In this approach, a minimum frequency for each execution phase is determined and applied to the system for achieving the desired system performance. Another approach presents a speedup model to minimize the power consumption while maintaining the similar application performance through identifying the relationship between the parallel overhead and the power requirement of an application via their impact on the execution time [6]. Through this model, the parallel overhead and the power-aware performance can be predicted over several system configurations. A DVFS based approach detects the level of CPU-Boundness of the scientific application at runtime via dynamic regression and adjusts the CPU frequency accordingly [20]. Another approach that utilizes the multiple power-performance state and shows that a power scalable system can save significant amount of energy with negligible time penalty while maintaining the system QoS parameters [21]. A detailed survey of the energy efficient HPC systems is presented in [22].

C. Loop Scheduling

In the past years, extensive work has been performed in academia and industry to improve the performance of scientific applications through achieving load balancing via scheduling of parallel loops prevalent in these applications. There are two primary methods for loop scheduling: *static loop scheduling* and *dynamic loop scheduling (DLS)*.

In case of static loop scheduling, the iterations of a parallel loop are assigned to multiple processing elements (computing nodes) in chunks of fixed sizes equal to the ratio between the number of loop iterations and the number of processing elements. These chunks of loop iterations are executed uninterrupted until completion on the designated processing elements. The chunks contain iterations of variable execution times, thus causing load imbalance among processors. In dynamic loop scheduling, the parallel loop iterations are assigned at runtime one by one in a group of iterations or chunks. Each processing element executes a chunk of loop iterations until all of them are finished, after which, according to a scheduling policy, it receives a new assigned chunk of loop iterations. The simplest of these schemes is self-scheduling [23] where each processing element executes only one iteration of the loop at a time, until all the iterations are completed. This scheme achieves near-optimal load balancing among the processing elements at the cost of a high scheduling overhead. The scheduling techniques, static and self scheduling, discussed above are extreme examples. Other scheduling methods have been developed, that attempt to minimize the cumulative contribution of uneven processor finishing times leading to load imbalance, and that of the scheduling overhead. Such techniques schedule iterations in chunks of sizes greater than one, where a size is the number of iterations in the chunk. Both fixed-size and variable-size chunking methods have been proposed. These approaches, presented in [24–27], consider the profile of the integrations, availability of processing elements, chunk size, or locality of the data elements, when assigning the iterations to processors at runtime.

In the past, the loop scheduling methods addressed load imbalance and did not take into account the fact that the application performance may vary both due to algorithmic characteristic of the application and system related issues (interference by other applications running on the same system, thus taking away a part of the computational power of the resources). However, recent techniques are based on probabilistic analysis and take into account these factors when assigning loop iterations to processors at runtime [24–28].

There are also several adaptive approaches offering better performances, as described in [8, 29–31]. These approaches consider the processor speed and performance while distributing the loop iterations, resulting into better results. In [28], the ratio of job distribution is determined based on the relative processor speed. However, it does not take into account the fact that the processor speed may vary due to the algorithmic characteristics of the application, or due to system related issues (such as, data access latency, operating system interference, and others). A similar kind of approach is described in [29–31], where the distribution of loop iterations depends upon the processor performance at runtime. Each processor is assigned a weight based on its performance at the last sample period and receives a chunk of appropriate size, such that all the processors finish at the same time with high probability. These weights are dynamically computed each time a processor is allocated a chunk of loop iterations. The above mentioned DLS methods utilize a variable size chunking scheme, where the chunks are scheduled in decreasing size, or the technique assigns decreasing size chunks of loop iterations to processors in batches. Further, a more recent DLS method with a higher degree of generality has been developed to allocate variable size (increasing or decreasing) chunks of loop iterations [32]. The advanced DLS methods employ a probabilistic and statistical model for the dynamic allocation of chunks of loop iterations to each processor, such that the average finishing time for completion of all chunks occurs within optimal time with high probability.

Most loop scheduling methods on message-passing distributed systems are implemented using a *foreman-worker* strategy. A straight-forward foreman-worker parallel programming strategy is illustrated in Fig. 1. The foreman is responsible for computing start and size, keeping track of the remaining iterations, directing the movement of data if necessary, and detecting loop completion. These administrative computations do not justify a dedicated processor, hence the foreman is also a worker. An MPI implementation of this strategy utilizes a partitioned work queue of iterations and interleaves computations with communications. A disadvantage of the foreman-worker parallel programming strategy, especially in message- passing systems, is its limited practical scalability. When the number of processors is increasing, the

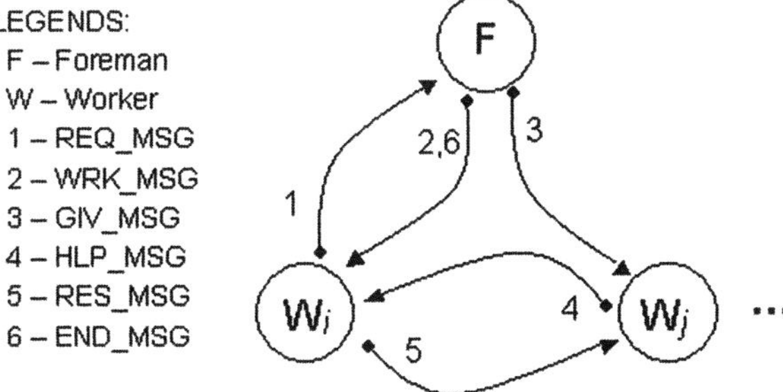

Fig. 1 Basic foreman worker strategy [33]

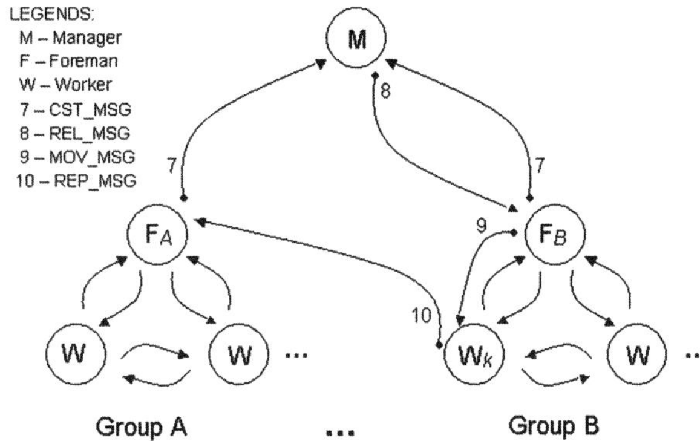

Fig. 2 Foreman worker strategy: two level scheduling [33]

foreman becomes a communication bottleneck, resulting in increased idle times for some workers. To address the bottleneck limitations, a useful modification of the strategy is to utilize multiple foremen as illustrated in Fig. 2. Initially, the processors are divided into a few groups, where each group is assigned a portion of the iteration space. Each group executes the foreman-worker âŁœfirst levelâŁž strategy similar to the one shown in Fig. 1. Load balancing is achieved within each group for the work owned by the group, and a foreman communicates only with the processors in its group. However, if the computational requirements in different regions of the iteration space vary, or if the processors are effectively heterogeneous, then some groups may finish earlier than others. Thus, load balancing is also necessary among the groups. Fig. 2 illustrates the required coordination in this "second level" strategy for load balancing among groups. A more elaborated description of the interactions and coordination among processors using the foreman-worker setting is given in [33].

D. DVFS Based Loop Scheduling

DVFS based loop scheduling techniques consider multiple power modes that processing elements can have for addressing load imbalance in the execution of the scientific application. The power consumption of the executing system is reduced by lowering down the speed or shutting down the idle processing element. These approaches utilize the notion of DVFS by using one of the following three techniques: shut down based, DVFS with static scheduling, and DVFS with dynamic scheduling.

In the shut down based techniques, a fixed size chunks of loop iterations are assigned to each processor within a group of processors at the beginning, after which, the processors start executing their assignment. As soon as a processor finishes the execution of its assigned iterations, the system lowers the processor frequency to its minimum frequency (standby). This scheme does not ensure load balancing (simultaneous completion of the execution of iterations by processors). However, it offers minimal power consumption by lowering the frequency of the idle processors to their minimum frequency, which leaves them in the a low power mode.

In the DVFS with static scheduling, fixed size chunks of loop iterations are assigned to each processor within a group of the processors before they start executing their assignments. Each processor is assigned the optimal frequency of system operation, where the optimal frequency is calculated with respect to the execution time taken by the slowest processor (at the beginning). The processors then proceed with the execution of their fixed size assigned chunks [34]. This approach needs prior information regarding the execution time of loop iterations at different processors to select the slowest processor. In another approach, at the beginning of the execution, processors are assigned the optimal frequency considering the total execution of the slowest processor, which is the limiting factor [34]. Each of the processors is re-assigned a new fixed chunk of loop iterations every time it finishes the previous chunk of iterations.

In the DVFS with dynamic scheduling scheme, to minimize the power consumption, the chunks of loop iterations are assigned to a group of processors at runtime, and the processor that finishes before the slowest processor is kept at a minimum frequency. This approach is an extension of the shut down based techniques by including dynamic loop scheduling.

E. Elements of Control Theory

Control theory concepts offer a powerful tool to enable resource management, and analyze the impact of uncertain changes, as well as system disturbance issues. Recently, control theoretic approaches have successfully been applied to selected resource management problems including task scheduling [35], bandwidth allocation, QoS adaptation in web servers [36, 37], multi-tier websites [38–40], load balancing in e-mail and file servers [38], and processor power management [41]. Control theory provides a taxonomy for designing an automated, self-managed and effective resource management or partition scheme by a continuous monitoring of the system states, of the changes in the environmental input, and of the system response to these changes. This scheme ensures that the system is always operating in a region of safe operational states, while maintaining the QoS demands of the service provider.

A typical control system consists of the components shown in Fig. 3 [42]. The *System Set Point* is the desired state of the system considered as a target to achieve during its operation. The *Control Error* indicates the difference between the desired system set point and the measured output during system operation. The *Control Inputs*

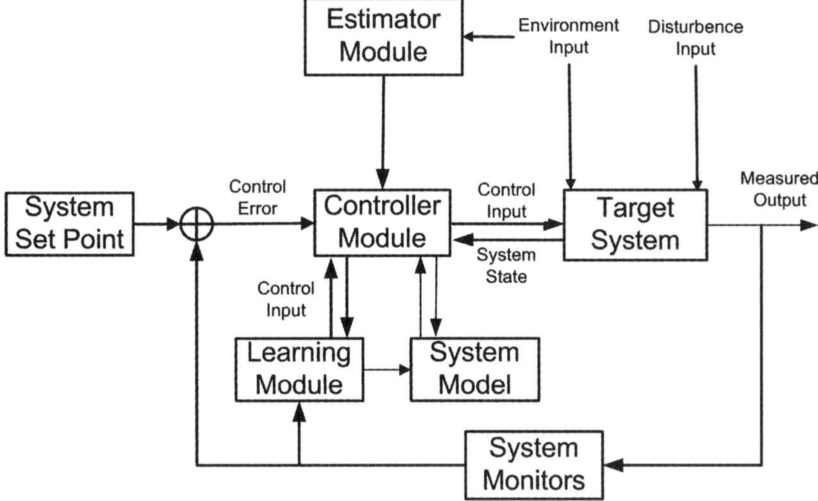

Fig. 3 A general structure of a control system

are the set of system parameters which are dynamically applied to the system for dynamically changing the performance level. The *Controller Module* monitors the measured output and provides the optimal combination of different control inputs to achieve the desired set point. The *Estimator Module* provides estimates of the unknown parameters for the system based on the previous history using statistical methods. The *Disturbance input* can be considered as the environment input that affects the system performance. The *Target system* is the system in consideration, while the *System Model* is the mathematical model of the system, which defines the relationship between its input and output variables. The *Learning Module* collects the output through the monitor and extracts information based on statistical methods. Typically, the *System State* defines the relationship between the control or the input variables, and the performance parameters of the system.

3 An Online Look-Ahead Control-based Management Approach

The design of a generic control structure has been presented in [43] for managing the performance of the application instance hosted on a computing system. In this control structure, the controller optimizes the multi-dimensional QoS objective function that is a sum of the cost of utilizing computational resources (operating cost) and QoS violation penalties (loss). This optimization is performed by choosing an appropriate control action after continuous monitoring of the system performance with respect to the disturbance (environment input) in the system. The chosen value of the control action maximizes the system profitability by minimizing the system operating cost, while at the same time maximizing the profitability.

The Parallel Loop Execution Problem A recent work on scheduling large number of parallel loop iterations in multiprocessor environment utilized DVFS for minimizing the parallel loop execution time while addressing load imbalance in the execution environment [34]. In this approach, N loop iterations are equally distributed to each of the P processors initially. Each of these processors are assigned various values of CPU core frequencies (f_i, where $i \in [1...P]$) according to their relative speed of execution which enables these processors to simultaneously reach the goal of finishing the execution of loop iterations with minimum load imbalance and minimum total execution time. However, this approach ignores the fluctuations in the availability of computational resources (such as CPU), which can be caused by the utilization of a same resource by other applications (such as I/O or OS), when executing on the same computing node. In addition, this perturbation in the amount of available CPU at different computing nodes may result in severe load imbalance, which in turn may lead to deadline violation. Therefore, the loop execution management approach should also consider the variation in available computational resources and perform control actions accordingly.

The performance management approach presented in this chapter employs an effective technique for continuous monitoring of the loop execution environment at a prespecified rate (sample time T), and for re-adjusting the CPU core frequency. This approach facilitates achieving the desired execution deadline T_d with minimum power consumption. This approach also ensures that all of the processing nodes finish the execution at the same time for minimizing the load imbalance. This control action of changing the CPU core frequency value ensures that sufficient amount of CPU cycles are available to the loop execution environment to finish the assigned loop iterations within the deadline T_d, while minimizing the multi-dimensional utility function that includes the execution time as QoS parameter and the power consumption as operating cost at the computing node.

The proposed management structure is shown in Fig. 4 and contains two levels to effectively manage the loop execution environment. At the top level, the incoming tasks (loop iterations) with the execution deadline (T_d) are assigned to the processors executing at the bottom level. In case of static scheduling, each of these processors (processing elements PE) is assigned an equal number of loop iterations. The bottom level monitors the performance of the individual processors with respect to the number of loop iterations executed during a time sample (T), and the remaining number of loop iterations from the total number of iterations assigned at the beginning of the computation. The bottom level optimizes the performance of the loop execution environment, and that of the HPC system through choosing an appropriate value of the CPU core frequency that minimizes the power consumption, while at the same time meeting the requirement of the execution deadline. The bottom level consists of two layers with different functionalities: *an application layer* and *a control layer*. The application layer contains the loop execution environment and uses independent processors to execute an assigned chunk of loop iterations. This layer also contains

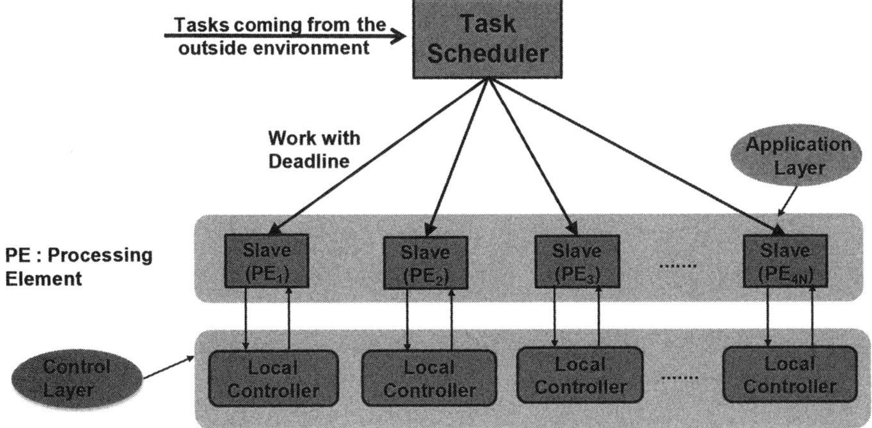

Fig. 4 The proposed two-level approach for performance management

the monitoring elements that record the performance logs of the execution environment regarding the finishing times of each chunk of iterations in each sample and the remaining iterations. The control layer contains the developed controllers that execute the optimization functions (with inputs from the application layer's monitoring element) to calculate the appropriate CPU core frequency and assigns the appropriate value to the processing element. The control layer receives the deadline of the loop execution from the top level as performance specification. Each local controller interacts with the processing element at the same node for exchanging performance measurements and assigning optimal values for the CPU core frequency.

The proposed control framework consists of the following key components:

The System Model This component describes the dynamics of an active state processing element. The state update dynamics can be described through the following state space equation:

$$x(k+1) = \phi(x(k), u(k), \omega(k)) \tag{1}$$

where $x(k)$ is the system state at time k, $u(k)$ is the set of user controlled system control inputs (CPU core frequency at time k), and $\omega(k)$ is the environment input (the percentage of the CPU available to the loop execution environment) at time k. The number of loop iterations that finished execution at time k are shown as $l(k)$, and the number of loop iterations remaining unexecuted at time k are shown as $L(k)$. Here, $x(k) \subset R^n$ and $u(k) \subset R^m$, where R^n and R^m represent the set of system states and control inputs, respectively. System state $x(k)$ at time k can be defined as the set of loop iterations executed at current time $l(k)$, with the remaining number of loop iterations $L(k)$,

$$x(k) = [l(k)\ L(k)] \tag{2}$$

where,

$$l(k) = \frac{\omega(k)}{100} \frac{\alpha(k)}{\hat{W}_f} * T \tag{3}$$

and

$$L(k+1) = L(k) - l(k) \tag{4}$$

$\alpha(k)$ is a scaling factor defined as $\frac{u(k)}{u^m}$, where $u(k) \in U$ is the frequency at time k (U is the finite set of all possible frequencies that the system can take), u^m is the maximum supported frequency of the processor. $\hat{W}_f$ is the predicted average service time (work factor in units of time) required to execute a single loop iteration at maximum frequency u^m, and T is the sampling time of the system.

The Power Consumption The power consumed by a processing element is directly proportional to the supplied core frequency and the square of the applied voltage across the processor. Experiments have shown that this relationship is close to linear [44]. In addition, in a production environment, only the total power consumption of a computing system due to all of the devices attached inside the system (e.g. CPU, memory, hard disk, CD-Rom, CPU cooling fan etc.) can be measured through an external wattmeter. In the case of executing CPU intensive applications, the obtained measurements show a nonlinear relationship with the CPU core frequency and CPU utilization [42]. Therefore, a look-up table with near neighbor interpolation is a best fit power consumption model for a physical computing node. In a previous work [42], a power consumption model was developed using multiple CPU core frequencies, CPU utilization, and corresponding power consumption values. A loop execution environment utilizes the maximum available CPU that always results in 100 % CPU utilization. Therefore, the power consumption model used in the proposed approach is only dependent upon the CPU core frequency for keeping CPU utilization constant at 100 %. We denote $E(k)$ as the power consumed by the processor at current frequency $u(k)$.

Estimating the CPU Availability In the proposed management approach, an estimation of the available percentage of CPU (CPU cycles) for the loop execution process is necessary for computing the system state estimation according to Eq. (3). An autoregressive integrated moving average (ARIMA) filter is used for estimating the percentage of CPU availability for the loop execution process, according to Eq. (5). In this approach, the estimation of the CPU availability for the loop execution process is estimated indirectly, by estimating the CPU utilized by other applications executing on the computing node. If these applications utilize $\sigma(k)$ percentage of the CPU at time k, the available CPU of the loop execution process is equal to $100 - \sigma(k)$, and that is considered as $\omega(k)$. The estimation of CPU utilization for the loop execution process is $\sigma(k+1)$, which is estimated by the ARIMA filter as:

$$\sigma(k+1) = \beta\, \sigma(k) + (1-\beta)\, \sigma_{avg}(k-1, r), \tag{5}$$

where β is the weight on the available CPU utilization in the previous sampling time. A high value of β pushes the estimate towards current CPU utilization by other applications. A low value of β shows a bias towards the average CPU utilization in a past history window by the other applications. $\sigma_{avg}(k-1, r)$ represents the average value of CPU utilization between the time samples $(k-1)$ and $(k-1-r)$. Instead of using static β, an adaptive estimator can be used for better estimation by accommodating the error in estimation at previous sample [43].

$$\delta(k) = \gamma\,\delta + (1-\gamma)\;|\sigma(k-1) - \sigma(k)|, \tag{6}$$

where $\delta(k)$ denotes the error between the observed and the estimated CPU availability at time t, δ denotes the mean error over a certain history period and γ is determined by the experiments. The weight $\beta(k)$ for the ARIMA filter can be calculated as:

$$\beta(k) = 1 - \frac{\delta(k)}{\delta_{max}}, \tag{7}$$

where δ_{max} denotes the maximum error observed over a certain historical period.

Model Based Control Algorithm and Performance Specification of the Loop Execution Environment At each time sample k, the controller on each of the processors P calculates the optimal value of the CPU core frequency f for the next time interval from k to $(k+1)$ that minimizes the cost function $J(k+1)$, as per Eq. (8). This cost function $J(k+1)$ combines the QoS violation penalty (x_s) and the operating cost $E(k)$ with different relative priorities. In this approach, x_s is considered as the expected number of loop iterations that are executed by the processor P during the time interval $k+1$ to achieve the deadline T_d from all of the assigned iterations $(\frac{N}{P})$. Therefore, the controller updates the x_s after every time sample according to the remaining unexecuted number of loop iterations at each processor.

In this approach, $x_s = [l^*, L^*]$, where l^* is the optimal number of loop iterations desired for execution in the given time interval k, and L^* is the optimal number of loop iterations remaining for execution at time sample k, in order to finish the execution by the deadline T_d. The cost function $J(k+1)$ used at control is formulated as follows:

$$J(x(k), u(k)) = \|x(k) - x_s\|_Q + \|E(u(k)\|_R \tag{8}$$

where Q and R are user specified relative weights for the drift from the optimal number of executed loop iterations x_s, and operating cost (power consumption), respectively. The optimization problem from the controller can be described as minimizing the total cost of operating the system J, using a look-ahead prediction horizon with $k = 1, 2, 3, ...H$ steps. Finally, the value of a chosen control input $u(k_0)$ at time sample k_0 is:

$$u(k_0) = \arg\min_{u(k)\in U} \left\{ \sum_{k=k_0+1}^{k=k_0+H} J(x(k), u(k)) \right\} \tag{9}$$

After calculating the value of the control input $u(k_0)$, this value will be assigned to the CPU core for the next time interval $k_0 + 1$.

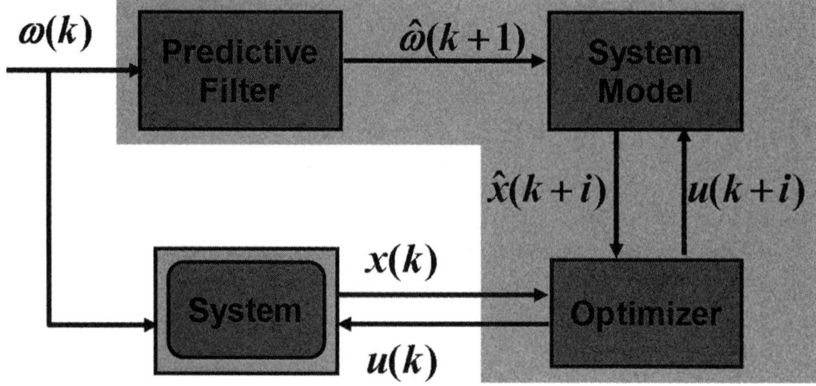

Fig. 5 The main components of the proposed control structure

A schematic representation of the controller and its various components is shown in Fig. 5. In this figure, *Predictive Filter* represents the ARIMA estimator, which estimates the available percentage of CPU to the process computing loop iterations. *System Model* represents the mathematical representation of the loop execution process in terms of loop iterations $l(k)$ executed at time instant k, with the remaining number of loop iterations $L(k)$. *Optimizer* represents the optimization library, which contains various tree search techniques (greedy, pruning, heuristics, and A^*) to compute the optimal value of control inputs [45]. *System* represents the actual process, which is computing the loop iterations.

4 Case Study: Performance Management of a Parallel Loop Execution Environment

The proposed approach is simulated in MATLAB R2010 for managing loop execution of a parallel application by applying static loop scheduling at the top level and DVFS at the lower level for maintaing the prespecified execution deadlines. Simulations details are described in the following subsections.

A. Simulation Setup

A cluster of four computing nodes is simulated on a 3.0 GHz machine with 3 GB of RAM for demonstrating the performance of the proposed approach. These computing nodes support five discrete values of CPU frequencies (1.0,1.2,1.4,1.7,2.0) in GHz. The sample time (T) for observation is considered as 30 seconds, and the execution time (work factor) for each individual loop iteration is fixed at $2x10^{-4}$ seconds. In addition, the look ahead horizon (H) for the simulation is kept constant at two steps for balancing the performance and computing overhead. The total number of loop iterations executed at four computing nodes are equal to 10^8. During these simulations, synthetic time series data are utilized that represent the CPU resources

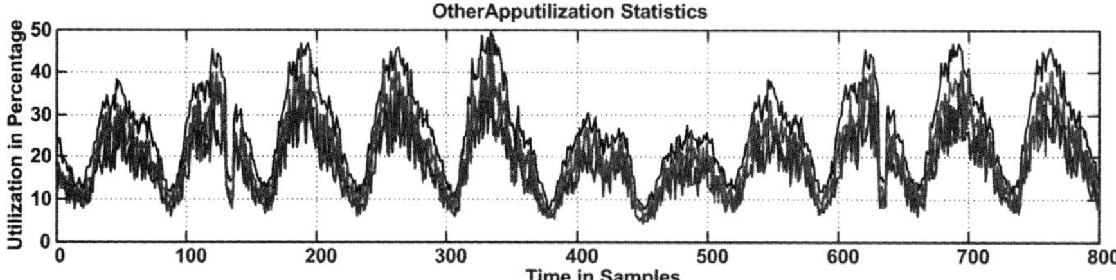

Fig. 6 Percentage of CPU utilized by OS Applications (other than the loop execution process) at four computing nodes. Each of the computing nodes has different CPU utilization characteristic represend by different colors (*red*, *green*, *blue*, and *black*)

consumed by the other OS applications executing at each computing nodes. This synthetic time series were generated using random functions in MATLAB for each computing nodes, and shown as percentages of the total CPU availability in Fig. 6. The deadline for the execution is varied between 200 to 800 samples (where 1 sample is equal to T seconds) depending upon the experiment settings. Experiments with this proposed framework are performed with various amount of perturbations at different processors, and the results demonstrate whether the QoS objectives (deadline and power consumption) have or not been achieved.

Performance of the proposed approach is exhibited using three different experiments with specific purposes as follows: *Experiment-1* demonstrates the impact of the perturbations related to CPU availability on the parallel loop execution deadlines; *Experiment-2* demonstrates the impact of the proposed approach when the loop execution time suffers due to perturbations in CPU availability, and its impact on deadline violations; and *Experiment-3* demonstrates the impact of choosing various priorities of deadline violation and power consumption inside the controller on the QoS objectives (deadline and power consumption);

Various experiments, their settings, and their observations are described in the following subsections.

B. Experiment-1

During this experiment, 10^8 loop iterations are assigned to a cluster of four computing nodes with a deadline of 500 samples, where each sample corresponds to 30 seconds. All the CPUs are assigned their average frequencies (1.4 GHz.) from the set of supported discrete frequency values. This experiment is performed in the following two steps, to show the impact of the perturbations in CPU availability on the loop execution deadline.

Simulation Without Perturbations This experiment is performed when other applications executing on the processing nodes consume negligible amount of computational resources, and almost the entire CPU is available for the loop execution environment. The results of this experiment are presented in Fig. 7 under the tag "No Disturbance". The results from each of the processors are plotted separately indicating the number of loop iterations executed at each time sample, and the number of time samples spent before finishing the execution of total numbers of loop

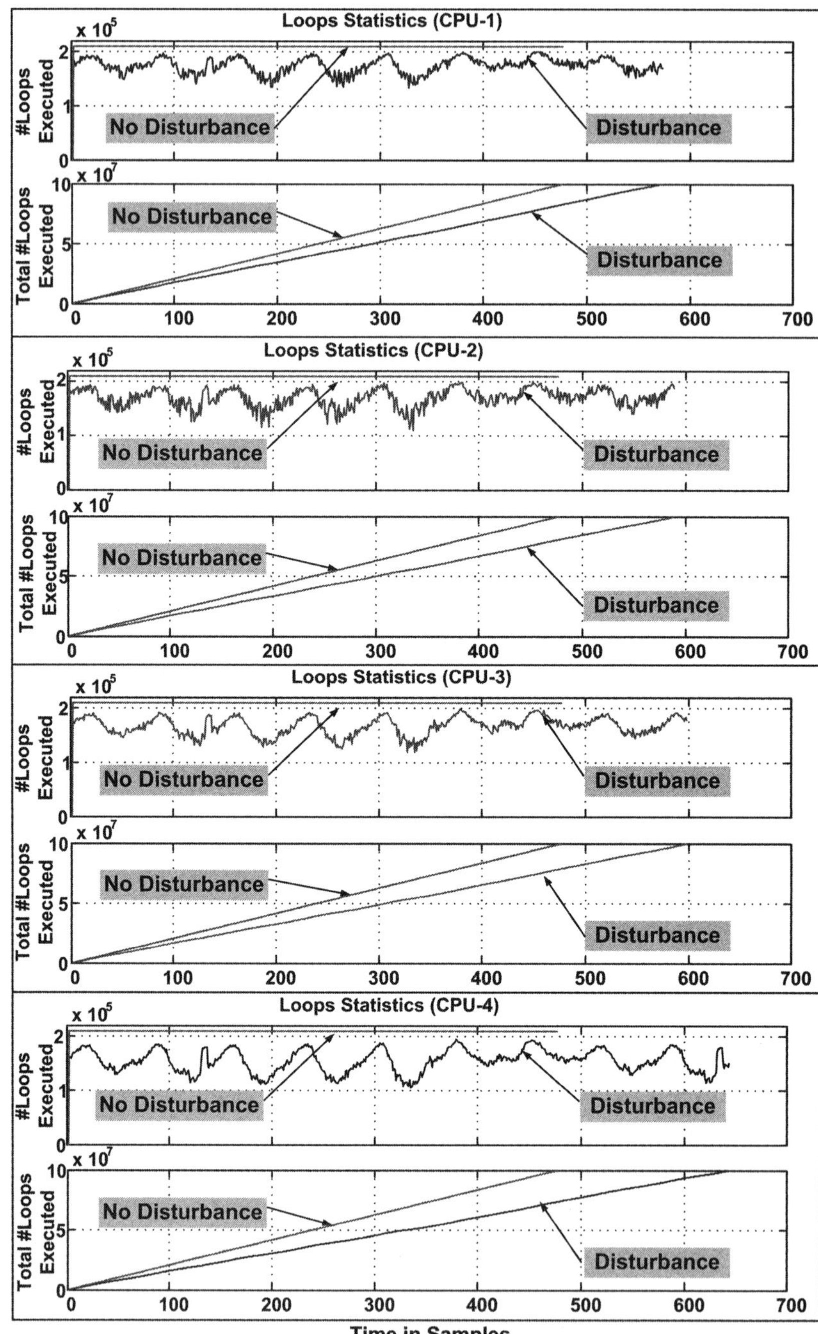

Fig. 7 Experiment-1 performed with and without perturbation in CPU availability with deadline = 500 samples

iterations (10^8). These results indicate that the execution deadline (of 500 samples) of the loop execution is achieved. In addition, each processors finish the execution almost simultaneously indicating near perfect load balancing.

Simulation with Perturbations This experiment is performed in similar settings as the previous one, with the addition of CPU perturbations at each processor due to a significant computational resource (CPU cycles) utilized by other local OS applications executing on the same node. Each of the processing nodes has different amount of perturbations that result in different amount of CPU availability (shown in Fig. 6) for the execution environment. The results of this experiment are shown in Fig. 7 under the tag "With Disturbance". The results for each of the processors are plotted separately in sub-figures. According to Fig. 7, in the presence of perturbations in CPU availability, processors fail severely to achieve the execution deadline (of 500 samples). The processors with low perturbation values for CPU availability miss the deadline less severely compared to the processors having high perturbation values. Also, severe load imbalance is observed as all of the four processors finish their execution at different time samples.

To address these issues, the easiest solution is to assign the highest operating CPU core frequency of execution to the processors, which will ensure the availability of maximum computational resources (CPU cycles) to the loop execution environment. However, this highest operating CPU core frequency results in extremely high value for the power consumption and reliability risk at the processor node. Therefore, a monitoring and reconfiguring approach is required, that can reconfigure the computational resources as and when required to meet the execution deadlines.

C. Experiment-2

This experiment is performed with similar simulation settings as the previous one, in the presence of the model-based online controller as described in the Sect. 3. During this experiment, the controller is deployed at each computing node and monitors the loop iterations executed at each processor with respect to the execution deadline and available CPU resources. This controller re-assigns the optimal value of CPU core frequency that results in achieving the deadline of execution with minimum power consumption. The results of this experiment for each processor are shown in Fig. 8. These plots show the optimal value of the CPU core frequency assigned by the controller and the loop iterations completed by each processor at different time samples. All the processors finish execution within 500 time samples. As shown in Fig. 8, the loop execution deadline is easily achieved even in the presence of perturbations by re-assigning the computational resources according to the control algorithm developed in the proposed approach. Moreover, the power consumption is also lowered in this case, because the processors are not executing at their highest frequencies all the time, compared to the cases in which processors execute at the highest frequency for maximum availability of computational resources, even in the presence of perturbations. Furthermore, all the processors finish at the same time sample (of 500) indicating the near perfect load balancing.

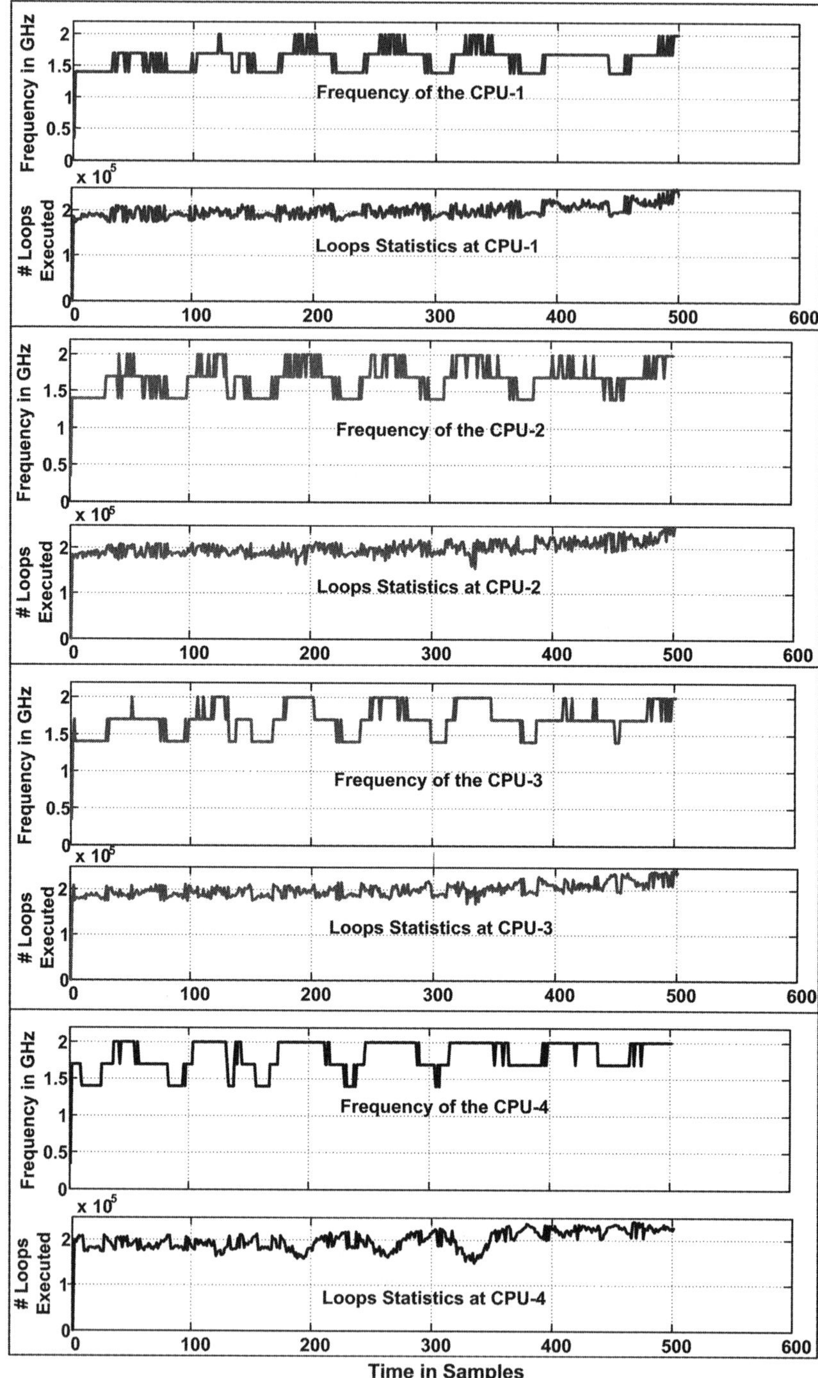

Fig. 8 Experiment-2 performed with perturbation in CPU availability with deadline = 500 samples

In addition to achieving the execution deadline, Fig. 8 shows that the controllers do not change the CPU core frequency of the computing node too frequently even if the CPU availability is varying continuously (as shown in Fig. 6). This demonstrates that the developed control approach is not extremely sensitive to the changes in the CPU availability. The controllers suggest a change in CPU core frequency (increase or decrease) only when the controller predicts a possible deadline violation by increasing the frequency, or a significant power saving by decreasing the frequency. Furthermore, the frequency of changes in CPU core frequency can be further minimized by adding frequency switching cost in the overall cost function J in Eq. (8).

D. Experiment-3

This experiment is performed to exhibit the impact of choosing different relative priorities for the loop execution deadline and power consumption as described by Eq. (8). Three different sets of simulations are performed during this experiment with different relative priorities of deadline versus power consumption as $(1:1, 2:1, 4:1)$ to demonstrate their impact on achieving the deadline of 800 samples. The results of these experiments at one of the processors (CPU-4) are plotted in Fig. 9 with different labels corresponding to the chosen priority ratio. According to this figure, the controller keeps the processor at the minimum frequency (1.0 GHz.) to keep the power consumption low in case of $1:1$ priority due to the high relative priority assigned to the power consumption compared to the priorities $2:1$ and $4:1$. However, this conservative selection of the CPU core frequency leads to an execution of a lower number of loop iterations in the case of $1:1$ and $2:1$ priorities. This slow rate of loop execution results in violation of deadlines (of 800 samples) in the case of $1:1$ and $2:1$ priorities. However, in the case of the $4:1$ relative priority, the controller selects higher values for the CPU core frequency at different time samples to execute a higher number of loop iterations, which finally leads to meeting the loop execution deadlines. In a real environment (HPC settings), these relative priorities can be chosen according to the violation penalties (in dollar amount) and power consumption cost (in dollar per KWh) to compute the total cost of hosting the loop execution environment in dollar values.

E. Other Experiments

A few more experiments are performed with different values of deadlines and amount of perturbations in computational resources. These simulations use tougher execution deadlines, 400 samples or even only 200 samples. In both of these approaches, the controller either assigns higher values for CPU core frequencies compared to the ones in the Experiment 2, or the highest value of frequency for the extreme (may be unrealistic) deadline until all the loop iterations are completed.

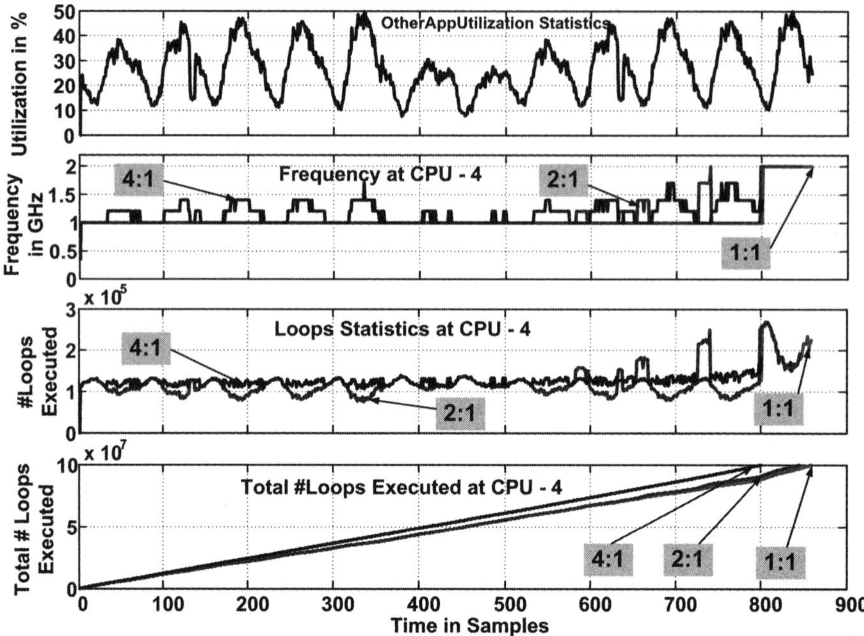

Fig. 9 Experiment-3 shows the impact of relative weights to the deadline and the power consumption with perturbation in CPU availability and deadline = 800 samples

5 Benefits of the Proposed Approach

The proposed two-level approach is using a model-based control theoretic framework for performance management of a parallel loop execution environment with fixed execution deadline and power consumption, and in the presence of computational resource related perturbations. The proposed management approach is using a model-based predictive controller for performance management of the parallel loop execution environment with respect to its execution deadline and power consumption budget, by considering perturbations in the availability of computational resources.

This approach does not interfere with the execution of the scientific application. The controller executes its routines separate from the application, and monitors only the application performance dynamically while the application is executing. The proposed approach uses performance logs related to the completion of loop iterations and to the availability of computational resources in the execution environment. It tunes the performance of the executing application for achieving loop execution deadline while minimizing the power consumption. Thus, it can be used for executing scientific applications of high complexity on a cluster of heterogeneous computing nodes. Moreover, there is no need for code profiling and code modification in the executing application, except for creating the log environment for the application performance statistics. According to the simulations presented in the previous sections, the proposed approach allocates the optimal value of the CPU core frequency to the processors in order to minimize the power consumption and achieve the desired deadline of loop execution, with consideration given to the chosen relative priorities towards achieving their target. Moreover, the execution using this approach leads to an adequate load balancing among the computing nodes. This approach can be applied to any cluster of computing nodes that supports DVFS techniques. For the HPC clusters, this approach is a trade-off between response time and power consumption, and can be considered as one of the solutions for achieving multi-dimensional objectives.

6 Combining DLS Techniques with the Proposed Approach

The proposed approach (Fig. 4) is two-level, *hierarchical, model-based,* and *power-aware*. As an extension to the approach proposed in this paper, another approach can be used, where the present approach is combined with one of the state-of-the-art DLS techniques for performance optimization of scientific applications running in heterogeneous environments. In this extended approach, DLS techniques are applied at the higher level, while the lower level uses a power-aware approach as described in Sect. 3. In this extended approach, each level works independently towards achieving its own individual objective, while at the same time the combined approach achieves the overall performance objective of the application. In other words, the top level ensures that the coefficient of variation of processors finishing times will be very low via an effective load-balancing, while the bottom level ensures that the desired application response time at each processor is achieved with minimum power consumption. The value of sampling time T_h at the top level is much higher than that at the bottom level T.

This difference in sampling times ensures that at the higher level, changes in the allocation size of the iterations can be made after monitoring the performance at the bottom level for a few sampling times. This extended approach is expected to offer superior performance compared to the one obtained when using either of the above mentioned techniques in isolation. Descriptions of the strategies used at the top and at the bottom levels are given in the next paragraphs.

The strategy at the top level, is an implementation of an optimal task distribution with algorithmic guarantee for a scientific application running on a HPC system by using DLS techniques, which are based on probabilistic analyses. These DLS algorithms ensure that the application will complete its execution at the cluster nodes within the optimal time with high probability while using the available computational resources. While executing the scientific application, the top level is primarily responsible for achieving an optimal execution time, as well as achieving load-balancing with the given set of computational resources. At the top level, an extension of a dynamic load balancing tool is implemented. In the current approach, the dynamic load balancing tool is used for being integrated within the scientific application for the purpose of scheduling, and also for maintaining interactions with the bottom level with the purpose to compute the appropriate size of the chunks of iterations to be distributed, as well as to meet the recommended deadlines. The top level estimates the work factor ($W_f = \frac{Response\ Time}{No.\ of\ Iterations}$) of iterations by continuously monitoring the execution of the processor with respect to the number of iterations submitted for execution and the corresponding response time. The work factor information is passed to the bottom level together with the size of the chunk of iterations, with the purpose to calculate the recommended response time to the bottom level.

At the bottom level, an effective resource allocation strategy is implemented. This allocation strategy is based on a control theoretic approach by processing the assigned task within the given constraints of an optimal response time and minimum power consumption. This level optimizes the application execution as well as the HPC system performance through balancing the need between using the minimum power consumption and a chosen average response time. The bottom level receives performance specification for the scientific application in terms of a recommended deadline (T_d) from the top level, and attempts to achieve this objective with minimum power consumption.

7 Conclusion

In this chapter, a model-based control theoretic performance management approach is presented for executing a parallel scientific application hosted in a data center environment that is using loop scheduling. This approach is well suited for the scientific applications of high complexity with prespecified deadline. According to the simulation results presented in this chapter, the proposed approach leads to a minimized power consumption of the deployment, while achieving the target execution deadline. In addition, the proposed approach provides a tuning option for the system administrator for choosing the best or the appropriate trade-off between the QoS specifications (response time) and the operating cost (power consumption) of the deployment. A possible extension of the proposed approach is also discussed in this chapter, where this approach can be combined with various dynamic loop scheduling methods for increasing the performance of the overall system in terms of minimizing the total loop or application execution time.

Acknowledgment The authors would like to thank the National Science Foundation (NSF) for its support of this work through the grant NSF IIP-1034897.

References

1. Report to congress on server and data center energy efficiency public law 109-431. Technical report, U.S. Environmental Protection Agency ENERGY STAR Program, August 2 2007.
2. A simple way to estimate the cost of downtime. In *Proceedings of the 16th USENIX conference on System administration (LISA '02)*, pages 185–188, Berkeley, CA, USA, 2002. USENIX Association.
3. Wu chun Feng, Xizhou Feng, and Rong Ge. Green supercomputing comes of age. *IT Professional*, 10(1):17–23, 2008.
4. W. Feng. Green destiny + mpiblast = bioinfomagic. In *10th International Conference on Parallel Computing (PARCO)*, pages 653–660, 2003.
5. Rong Ge, Xizhou Feng, Wu-chun Feng, and Kirk W. Cameron. Cpu miser: A performance-directed, run-time system for power-aware clusters. In *Proceedings of the 2007 International Conference on Parallel Processing (ICPP '07)*, page 18, Washington, DC, USA, 2007. IEEE Computer Society.
6. R. Ge and K.W. Cameron. Power-aware speedup. In *Proceedings of the IEEE International on Parallel and Distributed Processing Symposium (IPDPS).*, pages 1–10, March 2007.
7. Chung-hsing Hsu and Wu-chun Feng. A power-aware run-time system for high-performance computing. In *Proceedings of the ACM/IEEE conference on Supercomputing (SC '05)*, page 1, Washington, DC, USA, 2005. IEEE Computer Society.
8. Ioana Banicescu and Ricolindo L. Carino. Addressing the stochastic nature of scientific computations via dynamic loop scheduling. Electronic Transactions on Numerical Analysis 21:66-80, 2005.
9. Rajat Mehrotra, Ioana Banicescu, and Srishti Srivastava. A utility based power-aware autonomic approach for running scientific applications. In *Proceedings of IEEE 26th International Parallel and Distributed Processing Symposium (IPDPS)*, pages 1457–1466, 2012.
10. David A. Patterson and John L. Hennessy. *Computer Organization and Design, The Hardware/Software Interface, 4th Edition*. Morgan Kaufmann, 2008.
11. Yongpeng Liu and Hong Zhu. A survey of the research on power management techniques for high-performance systems. *Software: Practice and Experience*, 40(11):943–964, October 2010.
12. M. Nakao, H. Hayama, and M. Nishioka. Which cooling air supply system is better for a high heat density room: underfloor or overhead? In *Proceedings of Telecommunications Energy Conference, (INTELEC '91)*, pages 393–400, 1991.
13. H. Hayama and M. Nakao. Air flow systems for telecommunications equipment rooms. In *Proceedings of Telecommunications Energy Conference (INTELEC '89)*, pages 8.3/1–8.3/7 vol.1, 1989.
14. Taliver Heath, Ana Paula Centeno, Pradeep George, Luiz Ramos, Yogesh Jaluria, and Ricardo Bianchini. Mercury and freon: temperature emulation and management for server systems. In *Proceedings of the 12th international conference on Architectural support for programming languages and operating systems*, ASPLOS XII, pages 106–116, New York, NY, USA, 2006. ACM.
15. Justin Moore, Jeff Chase, Parthasarathy Ranganathan, and Ratnesh Sharma. Making scheduling "cool": temperature-aware workload placement in data centers. In *Proceedings of the annual conference on USENIX Annual Technical Conference*, ATEC '05, pages 5–5, Berkeley, CA, USA, 2005. USENIX Association.
16. Tridib Mukherjee, Ayan Banerjee, Georgios Varsamopoulos, Sandeep K. S. Gupta, and Sanjay Rungta. Spatio-temporal thermal-aware job scheduling to minimize energy consumption in virtualized heterogeneous data centers. *Computer Networks*, 53(17):2888–2904, December 2009.
17. Eun Kyung Lee, Indraneel Kulkarni, Dario Pompili, and Manish Parashar. Proactive thermal management in green datacenters. *Journal of Supercomput.*, 60(2):165–195, May 2012.
18. Blue gene. http://www-03.ibm.com/ibm/history/ibm100/us/en/icons/bluegene/ [May 2013].

19. Severin Zimmermann, Ingmar Meijer, Manish K. Tiwari, Stephan Paredes, Bruno Michel, and Dimos Poulikakos. Aquasar: A hot water cooled data center with direct energy reuse. *Energy*, 43(1):237–245, 2012. 2nd International Meeting on Cleaner Combustion (CM0901-Detailed Chemical Models for Cleaner Combustion).
20. Chung-Hsing Hsu and Wu-Chun Feng. Effective dynamic voltage scaling through cpu-boundedness detection. In *In Workshop on Power Aware Computing Systems*, pages 135–149, 2004.
21. Vincent W. Freeh, David K. Lowenthal, Feng Pan, Nandini Kappiah, Rob Springer, Barry L. Rountree, and Mark E. Femal. Analyzing the energy-time trade-off in high-performance computing applications. *IEEE Trans. Parallel Distrib. Syst.*, 18:835–848, June 2007.
22. Michael Knobloch. Chapter 1 - energy-aware high performance computing—a survey. In Ali Hurson, editor, *Green and Sustainable Computing: Part II*, volume 88 of *Advances in Computers*, pages 1–78. Elsevier, 2013.
23. B. J. Smith. Architecture and applications of the hep multiprocessor computer system. In *SPIE - Real-Time Signal Processing IV*, pages 241–248, 1981.
24. Clyde P. Kruskal and Alan Weiss. Allocating independent subtasks on parallel processors. *IEEE Trans. Softw. Eng.*, 11(10):1001–1016, 1985.
25. T. H. Tzen and L. M. Ni. Trapezoid self-scheduling: A practical scheduling scheme for parallel compilers. *IEEE Trans. Parallel Distrib. Syst.*, 4(1):87–98, 1993.
26. Susan Flynn Hummel, Edith Schonberg, and Lawrence E. Flynn. Factoring: a method for scheduling parallel loops. *Communication of ACM*, 35(8):90–101, 1992.
27. Ioana Banicescu and Susan Flynn Hummel. Balancing processor loads and exploiting data locality in n-body simulations. In *Proceedings of the 1995 ACM/IEEE Conference on Supercomputing, Supercomputing '95 (on CDROM)*, pages 43–55, New York, NY, USA, 1995. ACM.
28. Susan Flynn Hummel, Jeanette Schmidt, R. N. Uma, and Joel Wein. Load-sharing in heterogeneous systems via weighted factoring. In *Proceedings of the eighth annual ACM symposium on Parallel algorithms and architectures (SPAA '96)*, pages 318–328, New York, NY, USA, 1996. ACM.
29. Ioana Banicescu and Vijay Velusamy. Performance of scheduling scientific applications with adaptive weighted factoring. In *Proceedings of the 15th International Parallel & Distributed Processing Symposium (IPDPS '01)*, page 84, Washington, DC, USA, 2001. IEEE Computer Society.
30. Ricolindo L. Carino Cariño and Ioana Banicescu. Dynamic load balancing with adaptive factoring methods in scientific applications. *The Journal of Supercomputing*, 44(1):41–63, 2008.
31. Ioana Banicescu, Vijay Velusamy, and Johnny Devaprasad. On the scalability of dynamic scheduling scientific applications with adaptive weighted factoring. *Cluster Computing*, 6(3):215–226, 2003.
32. Ioana Banicescu and Vijay Velusamy. Load balancing highly irregular computations with the adaptive factoring. In *16th International Parallel and Distributed Processing Symposium (IPDPS 2002), 15-19 April 2002, Fort Lauderdale, FL, USA, CD-ROM/Abstracts Proceedings*. IEEE Computer Society, 2002.
33. Ricolindo Cariño, Ioana Banicescu, Thomas Rauber, and Gudula Rünger. Dynamic loop scheduling with processor groups. In *Proceedings of the ISCA Parallel and distributed Computing Symposium (PDCS)*, pages 78–84, 2004.
34. Yong Dong, Juan Chen, Xuejun Yang, Lin Deng, and Xuemeng Zhang. Energy-oriented openmp parallel loop scheduling. In *Proceedings of the 2008 IEEE International Symposium on Parallel and Distributed Processing with Applications*, pages 162–169, Washington, DC, USA, 2008. IEEE Computer Society.
35. Anton Cervin, Johan Eker, Bo Bernhardsson, and Karl-Erik Arzen. Feedback–feedforward scheduling of control tasks. *Real-Time Systems*, 23(1/2):25–53, 2002.

36. T.F. Abdelzaher, K.G. Shin, and N. Bhatti. Performance guarantees for web server end-systems: a control-theoretical approach. *IEEE Transactions on Parallel and Distributed Systems*, 13(1):80–96, Jan 2002.
37. R. Mehrotra, A. Dubey, S. Abdelwahed, and W. Monceaux. Large scale monitoring and online analysis in a distributed virtualized environment. In *8th IEEE International Conference and Workshops on Engineering of Autonomic and Autonomous Systems (EASe), 2011*, pages 1–9, 2011.
38. Chenyang Lu, Guillermo A. Alvarez, and John Wilkes. Aqueduct: Online data migration with performance guarantees. In *FAST '02: Proceedings of the 1st USENIX Conference on File and Storage Technologies*, page 21, Berkeley, CA, USA, 2002. USENIX Association.
39. R. Mehrotra, A. Dubey, S. Abdelwahed, and A. Tantawi. Integrated monitoring and control for performance management of distributed enterprise systems. In *2010 IEEE International Symposium on Modeling, Analysis Simulation of Computer and Telecommunication Systems (MASCOTS)*, pages 424–426, 2010.
40. Rajat Mehrotra, Abhishek Dubey, Sherif Abdelwahed, and Asser Tantawi. *A Power-aware Modeling and Autonomic Management Framework for Distributed Computing Systems*. CRC Press, 2011.
41. Dara Kusic, Nagarajan Kandasamy, and Guofei Jiang. Approximation modeling for the online performance management of distributed computing systems. In *ICAC '07: Proceedings of the Fourth International Conference on Autonomic Computing*, page 23, Washington, DC, USA, 2007. IEEE Computer Society.
42. Rajat Mehrotra, Abhishek Dubey, Sherif Abdelwahed, and Asser Tantawi. Model identification for performance management of distributed enterprise systems. (ISIS-10-104), 2010.
43. S. Abdelwahed, Nagarajan Kandasamy, and Sandeep Neema. Online control for self-management in computing systems. In *Proceedings of Real-Time and Embedded Technology and Applications Symposium,(RTAS) 2004.*, pages 368–375, 2004.
44. Abhishek Dubey, Rajat Mehrotra, Sherif Abdelwahed, and Asser Tantawi. Performance modeling of distributed multi-tier enterprise systems. *SIGMETRICS Performance Evaluation Review*, 37(2):9–11, 2009.
45. S. Abdelwahed, Jia Bai, Rong Su, and Nagarajan Kandasamy. On the application of predictive control techniques for adaptive performance management of computing systems. *IEEE Transactions on Network and Service Management*, 6(4):212–225, 2009.

CoolEmAll: Models and Tools for Planning and Operating Energy Efficient Data Centres

Micha vor dem Berge, Jochen Buchholz, Leandro Cupertino, Georges Da Costa, Andrew Donoghue, Georgina Gallizo, Mateusz Jarus, Lara Lopez, Ariel Oleksiak, Enric Pages, Wojciech Piątek, Jean-Marc Pierson, Tomasz Piontek, Daniel Rathgeb, Jaume Salom, Laura Sisó, Eugen Volk, Uwe Wössner and Thomas Zilio

1 Introduction

The need to improve how efficiently data centre operate is increasing due to the continued high demand for new data centre capacity combined with other factors such as the increased competition for energy resources. The financial crisis may have

M. vor dem Berge (✉)
christmann informationstechnik + medien GmbH & Co. KG,
Ilseder Huette 10c, 31241 Ilsede, Germany
e-mail: micha.vordemberge@christmann.info

J. Buchholz · G. Gallizo · D. Rathgeb · E. Volk · U. Wössner
High Performance Computing Center Stuttgart (HLRS), University of Stuttgart,
Nobelstr. 19, 70569 Stuttgart, Germany
e-mail: buchholz@hlrs.de

G. Gallizo
e-mail: ggallizo@gmail.com

D. Rathgeb
e-mail: rathgeb@hlrs.de

E. Volk
e-mail: volk@hlrs.de

U. Wössner
e-mail: woessner@hlrs.de

L. Cupertino · G. Da Costa · J.-M. Pierson · T. Zilio
Institute for Research in Informatics of Toulouse (IRIT),
Université Paul Sabatier, 118 Route de Narbonne,
31062 Toulouse Cedex 9, France
e-mail: fontoura@irit.fr

G. Da Cost
e-mail: dacosta@irit.fr

S. U. Khan, A. Y. Zomaya (eds.), *Handbook on Data Centers*,
DOI 10.1007/978-1-4939-2092-1_6

dampened data centre demand temporarily, but current projections indicate strong growth ahead. By 2020, it is estimated that annual investment in the construction of new data centres will rise to $ 50 bn in the US, and $ 220 bn worldwide [23].

According to a survey by the Uptime Institute in 2011, approximately 80 % of data centre owners and operators have built or renovated data centre space in the past 5 years [56]. Furthermore, 36 % of survey respondents reported that they would run out of capacity by 2012, and a significant proportion of those indicated that they are planning to build a new facility in the near future. Despite this overall trend toward expansion, large organizations are also increasingly aware of the need to improve the productivity of existing facilities and consolidate excess or inefficient capacity. The US government, for example, driven by the need to tackle its considerable deficit, is currently involved in an extensive consolidation program, which will see more than 800 of its 2000-plus data centres shut by 2014 [41].

J.-M. Pierson
e-mail: pierson@irit.fr

T. Zilio
e-mail: zilio@irit.fr

A. Donoghue
Paxton House (5th floor)30, Artillery Lane, E1 7LS, London, UK
e-mail: andrew.donoghue@451research.com

M. Jarus · A. Oleksiak · W. Piatek · T. Piontek
Poznan Supercomputing and Networking Center (PSNC),
Applications Department, ul. Noskowskiego 10, 61-704, Poznan, Poland
e-mail: jarus@man.poznan.pl

A. Oleksiak
e-mail: ariel@man.poznan.pl

W. Piatek
e-mail: piatek@man.poznan.pl

T. Piontek
e-mail: piontek@man.poznan.pl

L. Lopez · E. Pages
Atos Spain, S.A. (ATOS), Albarracín, 25, 28037 Madrid, Spain
e-mail: lara.lopez@atos.net

E. Pages
e-mail: enric.pages@atos.net

J. Salom · L. Sisó
Catalonia Institute for Energy Research (IREC),
Jardins de les dones de negre, 1, 2 floor (08930) Sant Adrià de Besòs,
Barcelona, Spain
e-mail: jsalom@irec.cat

L. Sisó
e-mail: lsiso@irec.cat

Some recent high-profile cloud-related data centre builds by the likes of Apple, Google and Facebook have increased awareness of both the financial and environmental costs of data centres. They have also raised awareness of the efficiencies that can be realized by smarter data centre design and operation. Environmental campaigners have begun to target carbon-intensive data centres with campaign tactics traditionally reserved for heavy industry. Environmentalists argue that data centre owners, and particularly cloud providers, could be more transparent about their energy use and efforts to improve efficiency. Greenpeace, for example, is concerned about whether providers are putting electricity costs above carbon intensity when constructing new facilities in locations with cheap, coal-based electricity generation. Data centre operators have responded to this scrutiny with marketing initiatives, but also by developing some genuinely innovative sustainability initiatives, such as Facebook's Open Compute strategy. Cloud providers also maintain that concentrating IT in large centralized facilities has enormous benefits in terms of how energy is consumed and managed. The emergence of more professionally run "information factories" could ultimately mean fewer inefficient private facilities. Cloud service suppliers also point to their innovations around data centre energy efficiency. Meanwhile, government regulators have also begun to take more notice of data centre energy use. For example, the UK's CRC Energy Efficiency Scheme represented the first real "carbon tax" to include data centres. Other governments are expected to follow suit.

All of these factors are gradually shifting the traditional uptime-and-availability-focused view of data centre owners and operators to encompass energy efficiency and sustainability issues. This rise in energy awareness has helped drive the development of an increasing range of eco-efficient data centre strategies and technologies by traditional data centre suppliers, specialist start-ups and consultants. Implemented in the correct way, these tools and approaches can help achieve financial savings, "reputational savings" associated with environmental issues, and operational flexibility associated with freed-up energy.

Development and selection of right energy-savings strategies is driven by identification and optimization of energy-waste causing processes and components. In particular, in many current data centers the actual IT equipment uses only half of the total energy while the remaining part is required for cooling and air movement. This results in poor cooling-efficiency and energy-efficiency, leading to significant CO2 emissions. For this purpose issues related to cooling, heat transfer, IT infrastructure configuration, management, location as well as workload management strategies are more and more carefully studied during planning and operation of data centres. The goal of the CoolEmAll project is to enable designers and operators of next generation data center to reduce its energy impact by combining optimization of IT, cooling and workload management in a holistic and comprehensive way, taking aspects into account that were considered traditionally separately. In the next section we describe objectives and outcomes of the CoolEmAll project.

1.1 The CoolEmAll Project

CoolEmAll is a European Commission funded project which addresses the complex problem of how to make data centres more energy, and resource, efficient. CoolEmAll is developing a range of tools to enable data centre d esigners, operators, suppliers and researchers to plan and operate facilities more efficiently. The participants in the project include a range of scientific and commercial organisations with expertise in data centres, high performance computing, energy efficient server design, and energy efficient metrics.

The defining characteristic of the CoolEmAll project is that it will bridge this traditional gap between IT and facilities approaches to efficiency. The main outcomes of CoolEmAll will be based on a holistic rethinking of data centre efficiency that is crucially based on the interaction of all the factors involved rather just one set of technologies. The expected results of the project include a data centre monitoring, simulation and visualisation software, namely SVD toolkit, designs of energy efficient IT hardware, contribution to existing (and help define new) energy efficiency metrics.

Some commercial suppliers (most notably Data centre Infrastructure Management suppliers) and consultants have recently begun to take a more all-encompassing approach to the problem by straddling both IT and facilities equipment. However, few suppliers or researchers up to now have attempted to include the crucial role of workloads and applications. That is beginning to change, and it is likely that projects such as CoolEmAll can advance the state of the art in this area.

As noted [60], the objective of the CoolEmAll project is to enable designers and operators of a data centre to reduce its energy impact by combining the optimization of IT, cooling and workload management. For this purpose CoolEmAll project investigates in a holistic approach how cooling, heat transfer, IT infrastructure, and application-workloads influence overall cooling- and energy-efficiency of data-centres, taking aspects into account that traditionally have been considered separately.

In order to achieve this objective CoolEmAll provides two main outcomes:

- Design of diverse types of data centre efficiency building blocks (DEBBs) reflecting configuration of IT equipment and data centre facilities on different granularity levels, well defined by hardware specifications, physical dimensions, position of components, and energy efficiency metrics, assessing energy- & cooling-efficiency of building blocks.
- Development of the Simulation, Visualization and Decision support toolkit (SVD Toolkit) enabling the analysis and user driven optimization of data centres and IT infrastructures built of these building blocks.

Both building blocks and the toolkit take into account four aspects that have a major impact on the actual energy consumption: characteristics of building blocks under variable loads, cooling models, properties of applications and workloads, and workload and resource management policies. To simplify selection of right building

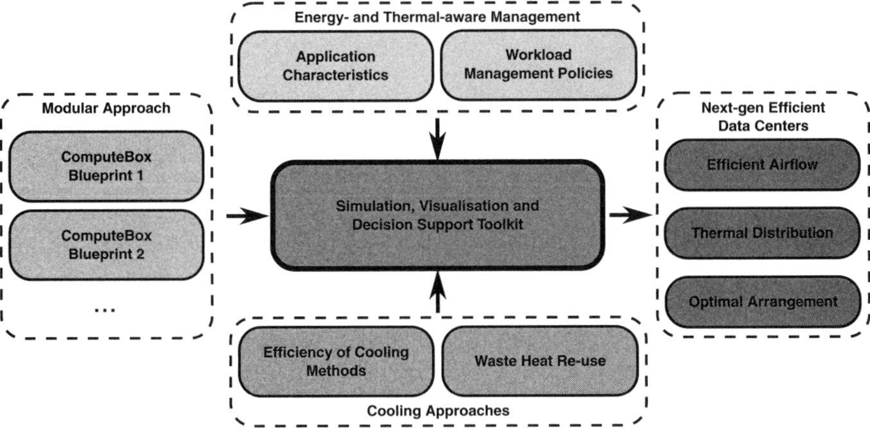

Fig. 1 The CoolEmAll concept

blocks used to design data centres adjusted to particular needs, data centre efficiency building blocks are precisely defined by a set of metrics expressing relations between the energy efficiency and essential factors listed above. In addition to common static approaches, the CoolEmAll approach also enables studies and assessment of dynamic states of data centres based on changing workloads, management policies, cooling methods, environmental conditions and ambient temperature. This enables assessment and optimization of data centre energy/cooling efficiency also for low and variable loads rather than just for peak loads as it is usually done today. The main concept of the project is presented in Fig. 1.

As pointed out, the CoolEmAll approach is realized by the SVD Toolkit, which allows assessment of energy- & cooling efficiency and facilitates optimization of DEBBs (reflecting various configurations of a data centre and its components on various scale level) by means of coupled workload and thermal-airflow simulation. Both, SVD Toolkit and models described in scope of DEBBs are verified within the validation scenarios using testing environment based on Resource Efficient Computing & Storage (RECS) servers, allowing fine grained monitoring and control of resources.

This chapter is based on CoolEmAll's project deliverables [26, 27, 47, 48, 50, 62] and is organized as follows. Section 2 introduces the architectural concept of the main software output of the CoolEmAll project: the Simulation, Visualisation and Decision support (SVD) Toolkit. Section 3 presents the data centre infrastructure description through the Data centre Efficiency Building Blocks (DEBB). Section 4 shows the metrics used to define energy efficiency in a data centre. Section 5 describes the hardware and methodology used to validate the different aspects of the project. In Sect. 6, a report on the impact of the projects' outputs on the market is done. Finally, Sect. 7 summarizes the status of the project as well as future perspectives.

1.2 Related Work

Issues related to cooling, heat transfer, IT infrastructure configuration, IT-management, arrangement of IT-infrastructure as well as workload management are gaining more and more interest and importance, as reflected in many ongoing works both in industry and research, described in [17]. There are already software tools available on the market capable to simulate and analyze thermal processes in data centers. Examples of such software are simulation codes along with more than 600 models of servers from Future Facilities [32] with its DC6sigma products [31], CA tools [21], or the TileFlow [36] application. In most cases these simulation tools are complex and expensive solutions that allow modeling and simulation of heat transfer processes in data centers. To simplify the analysis process Romonet [52] introduced a simulator, which concentrates only on costs analysis using simplified computational and cost models, disclaiming analysis of heat transfer processes using Computational Fluid Dynamics (CFD) simulations. Common problem in case of commercial data center modeling tools is that they use closed limited databases of data center hardware. Although some of providers as Future Facilities have impressive databases, extensions of these databases and use of models across various tools is limited. To cope with this issue Schneider have introduced the GENOME Project that aims at collecting "genes" which are used to build data centers. They contain details of data center components and are publicly available on the Schneider website [30]. Nevertheless, the components are described by static parameters such as "nameplate" power values rather than details that enable simulating and assessing their energy efficiency in various conditions. Another initiative aiming at collection of designs of data centers is the Open Compute Project [29]. Started by Facebook which published its data center design details, consists of multiple members describing data centers' designs. However, Open Compute Project blueprints are designed for description of good practices rather than to be applied to simulations.

In addition to industrial solutions significant research effort was performed in the area of energy efficiency modeling and optimization. For example, models of servers' power usage were presented in [51] whereas application of these models to energy-aware scheduling in [44]. Additionally, authors in [33, 61] proposed methodologies of modeling and estimation of power by specific application classes. There were also attempts to use thermodynamic information in scheduling as in [45]. Nevertheless, the above works are focused on research aspects and optimization rather than providing models to simulate real data centers. In [49], the authors propose a power management solution that coordinates different individual approaches. the solution is validated using simulations based on 180 server traces from nine different real-world enterprises. Second, using a unified architecture as the base, they perform a quantitative sensitivity analysis on the impact of different architectures, implementations, workloads, and system design choices. Shah [53] explores the possibility of globally staggering compute workloads to take advantage of local climatic conditions as a means to reduce cooling energy costs, by performing an in-depth analysis of the environmental and economic burden of managing the thermal infrastructure of a globally connected data centre network. SimWare [63] is a data warehouse simulator which compute its energy efficiency by: (a) decoupling the fan power from the computer power by using a fan power model; (b) taking into account the air travel time from the CRAC to the nodes; (c) considering the relationship between nodes by the use of a heat distribution matrix. CoolEmAll's SVD Toolkit approach differs from SimWare once it uses computation fluid dynamics instead of a static heat distribution matrix. This makes the simulation more precise and susceptible to flow changes due to changes of fan speeds.

2 Simulation, Visualisation and Decision Support Toolkit

The Simulation, Visualization and Decision Support Toolkit (SVD Toolkit) is a platform for interactive analysis and user driven optimization of energy- & cooling efficiency in data centres. It can be used to optimize design of new data centres or to improve operation of existing ones. The simulation platform follows a holistic approach and integrates models of applications, workloads, workload scheduling policies, hardware characteristics and cooling, capable to perform simulation on various scale level. The results of this platform include (i) estimations of energy consumption of workloads obtained from the workload simulation, (ii) thermal air-flows distribution obtained from the computational fluid dynamics (CFD) simulations, and (iii) energy- and cooling efficiency metrics assessing various configurations of data centres on various granularity levels. The SVD Toolkit provides also means for advanced visualization and interactive steering of simulation parameters, facilitating user-driven optimization process.

The overall optimization cycle consists of several steps. Before the simulation process can be started, Data centre Efficiency Building Blocks (DEBBs) are loaded from the repository or generated using DEBB configuration GUI. As noted, DEBBs reflect configuration of IT equipment or data centre facilities on different granularity levels, containing models necessary for simulation. In the next step, particular application profiles are loaded from the repository, or generated using application profiler. Profiles resemble the requirements that particular applications usually have. With these application profiles, synthetic workloads are generated and used by workload simulator to determine the power dissipated by individual hardware components, based on power characteristics described in DEBBs. Dissipated power is turned into heat and is used as input for a Computational Fluid Dynamics (CFD) Solver that calculates heat-flow distribution described by temperature- and airflow distribution map. The results of the workload simulation and CFD simulation are stored in a central database, ready to be processed by other components. In the final step, these results are assessed by efficiency metrics calculated by metrics-calculator and visualized by CoolEmAll web GUI. After getting efficiency assessment results, a user might decide to change configuration or particular parameters (i.e. position of the racks in the room) and repeat the simulation process again to evaluate new settings. In this way a user can adjust various parameters (i.e. server-room layout) at each iteration,

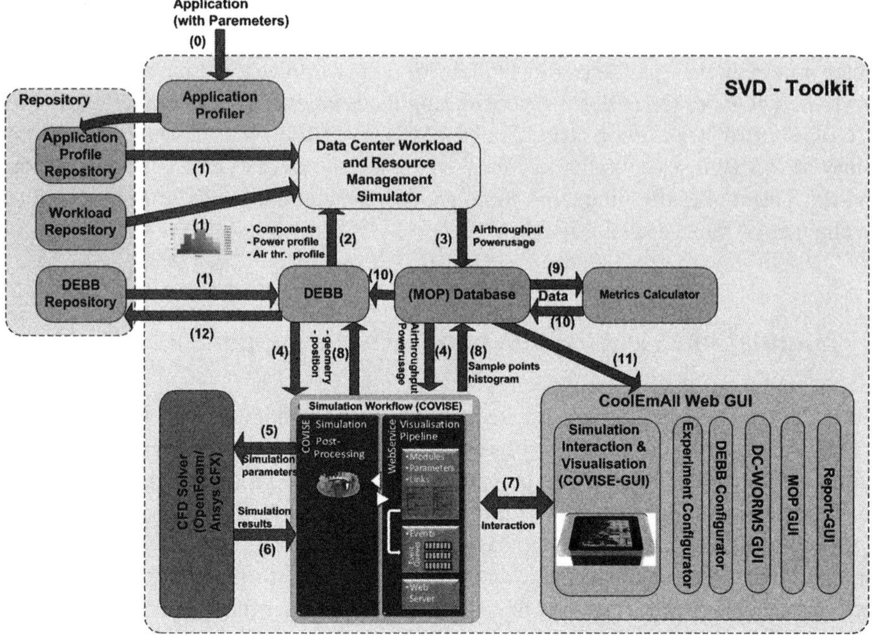

Fig. 2 SVD Toolkit architecture overview [62]

to optimize the design and/or operation of a data centre, driven by visualization and evaluation of results for each cycle-iteration, defined as a trial.

In this section we describe architecture of the SVD Toolkit, presenting its core components and interaction between them.

2.1 Architecture

The SVD Toolkit is a set of loosely coupled tools which, either combined with each other or standalone, provide the user with a set of features to perform simulation and visualisation tasks. This section introduces each component of the SVD Toolkit. Figure 2 summarizes the SVD Toolkit architecture, showing the interactions between its major building blocks along with their data flow. Each step of the data flow is indicated by the numbers in bracket.

The *Application Profiler* is capable of analysing (step 0) applications running on a reference hardware and generating application profiles. Such profiles describe the impact of different application phases (tasks) on the resources that execute it. An application profile consists of a sequence of application phases, each described by usage level of resources, such as CPU, Memory, Disk and Network. Application profiles are stored in the Repository, and are referenced in workloads, used by the "Data centre Workload and Resource Management Simulator" to calculate power-usage (heat dissipation) of workloads being virtually executed (simulated) on hardware resources represented by DEBBs, stored also in the Repository.

The *Repository* allows storing, editing and accessing of files remotely, while ensuring consistency of several files belonging to the same version, representing the configuration for a single iteration of the optimization process. The repository is realized by Apache Subversion and contains application profiles, workloads, and DEBBs, each of them in a distinct repository.

The *Data centre Workload and Resource Management Simulator (DCworms)* is a simulation tool for experimental studies of resource management and scheduling policies in distributed computing systems, providing assessment of the power consumption of a workload (step 1). Within the workload simulation, a workload is "virtually executed" on scheduled "virtual nodes", specified by a power-profile described in a *DEBB* (step 2). The power profile contains the power usage of the "virtual node" for all possible usage-levels of CPU, Memory, Network and Disk. To calculate the actual power-usage of the workload, DCworms maps the usage level in the application profile of the workload, to the usage level in the power-profile of the "virtual node" described in DEBB. In addition to power assessment, the Data centre simulator allows also to calculate air throughput on inlets/outlets of the physical resource (server, rack) represented in *DEBBs*. Results of the workload simulation (power usage and air throughput) are written into the *Database* (step 3), being ready to be processed by the Computational Fluid Dynamics (CFD) Solver.

The *Database* is responsible for accessing, managing and updating following data: (a) workload simulation results (air throughput and power dissipation) updated by DCworms (step 3) and retrieved by Simulation Workflow/COVISE for the CFD simulation (step 4); (b) CFD simulation results (provided by COVISE in step 8) containing temperature history for particular sample points; and (c) energy- and heat-aware metrics provided by the Metric Calculator assessing simulation results. All the results of the simulations (workload and CFD) as well as the assessment of the simulation results are retrieved and visualized by the CoolEmAll Web GUI (step 11).

The *Simulation Workflow COVISE* (Collaborative Visualisation and Simulation Environment) [14], is an extensible distributed software environment capable to integrate simulations, post-processing and visualization functionalities in a seamless manner. The CFD Solver performing CFD simulation is directly integrated into the COVISE workflow, including all necessary pre- and post-processing tasks. COVISE offers a networked SOAP based API and is accessible by all components that can make use of Web Service based components. In CoolEmAll, COVISE firstly retrieves simulation relevant data (step 4) from the DEBB repository (containing geometry data and position of objects) and from the Database (containing results from DCworms, i.e., power usage and air throughput), passes over these data to the CFD Solver (step 5), receives results from the CFD Solver (step 6), post processes and visualizes simulation results allowing at the same time modification of certain parameters (step 7) such as the arrangement of objects. Results of the simulation are written back into the Database (step 8), while modified geometrical parameters and arrangement of objects are used to update DEBBs (step 8), to be stored in the DEBB repository (step 12). Using COVISE, users can analyse their datasets intuitively and interactively in a fully immersible environment through state of the art visualization techniques, including volume and fast sphere rendering.

The *Computational Fluid Dynamics (CFD) Solver* is directly integrated into the COVISE workflow and enables to simulate and analyse complex heat flow and dissipation processes, and their consequences on flow guiding structures, such as compute-building blocks (DEBBs) in data centres. For this purpose a heat flow model defined by partial differential equations is defined. CFD solvers are using this model to calculate and simulate the interaction of liquids and gases with surfaces defined by boundary conditions of DEBB's geometry and other parameters (step 5). The results of a simulation, a heat-flow distribution map, are passed over to Simulation Workflow/COVISE (step 6) and can be visualized using COVISE GUI. In addition, the temperature and air-flow on inlet/outlets of the building blocks are extracted from the heat-flow distribution map and stored in the Database (step 8).

The *Metric Calculator* is responsible for the assessment of the simulation results. Based on metrics identified and defined in Sect. 4, it assesses and calculates energy- and heat-efficiency of building blocks (DEBBs) under corresponding boundary conditions. The calculation itself is based on data/metrics that are retrieved from the Database (step 9). Results of the calculation are written back (step 10) into the database to be retrieved and visualized by CoolEmAll Web GUI (step 11).

The *CoolEmAll Web GUI* provides a web based user interface allowing to interact with the SVD toolkit and visualize its results. It comprises several GUIs integrated into common web based GUI environment. It consists of the following GUIs: Experiment configuration GUI (for configuration of the simulation process/experiment/trial), DEBB configuration GUI (for configuration of DEBB), DCworms GUI (for configuration of DCworms), MOP GUI (for visualization and comparison of results provided by simulation and real measurements from the execution of real experiments), COVISE GUI (for configuration of CFD simulation, adjusting of CFD related parameters and visualization of heat-flow processes), Report GUI (presenting efficiency-metrics for assessment of the entire experiment).

In next sections we describe more detailed the core components of the SVD Toolkit.

2.2 Application Profiler

The focus of CoolEmAll simulations is on power-, energy- and thermal-impact of decisions on the system. In order to have realistic simulations, a precise evaluation of resource consumption is necessary. The Application Profiler is used to create profiles of applications that can be read by DCworms for simulation purpose. It uses data obtained during runtime and stored in MOP Database by the monitoring infrastructure as inputs to the power model described in Sect. 4. These data contains information regarding dissipated power, disk IO, memory and CPU usage, NIC sent/received bytes and CPU frequency. Using these data, it creates a description of applications based on their phases.

The profile of an application is a set of technical attributes that characterize and delivers details of a given application. The application profile purpose is enable finding of the "best" match; it means to get the best performance of the application on the different computing building blocks. Performance is taken here from the point of view of the CoolEmAll project, i.e. taking into account speed, but also power, energy and heat metrics.

Generally application profiling is used for applications that require long executions and a high amount of resources to run. Hence, it is true that application profiling is mostly applied to HPC applications or cloud services that run on data centres. Profiling techniques at development and testing time will allow the developer to optimize and identify different bottlenecks during the software execution.

Having a better understanding of a running application is a key feature for both application developers and hosting platform administrators. While the former have access to the source codes of their application, the latter have usually no a-priori clue on the actual behaviour of an application. Having such information allows for a better and more transparent evaluation of the resource usage per application when several customers share the same physical infrastructure. Platform providers (and the underlying management middleware) can better consolidate applications on a smaller number of actual nodes. Platform provider can provide token-free license where the observation of the system permits to determine the usage of a commercial application without bothering users with the token management.

Classifying applications using a limited number of parameters allows for a fast response on their characterization, suitable for real-time usage. The impact of the monitoring infrastructure is an important characteristic in order not to disturb the production applications.

2.3 Data Center Workload and Resource Management Simulator

Data Center Workload and Resource Management Simulator (DCworms) is a simulation tool based on GSSIM framework [18]. GSSIM has been proposed to provide an automated tool for experimental studies of various resource management and scheduling policies in distributed computing systems. DCworms extends its basic functionality and adds supplementary features providing complex energy-aware simulation environment. In the following sections we will introduce the functionality of the simulator in terms of modelling and simulation of energy efficient data centres.

2.3.1 Architecture

DCworms is a Java-based, event-driven simulation tool. In general, input data for the simulator consist of a description of workload and resources. These characteristics

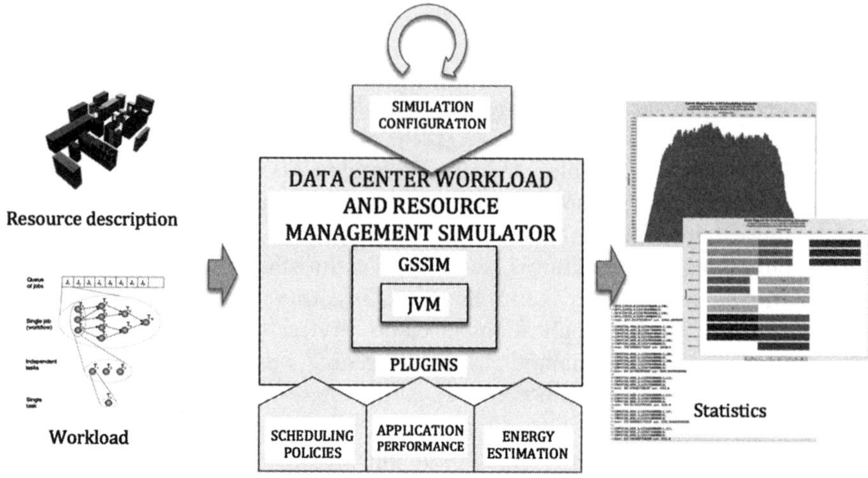

Fig. 3 DCworms architecture [62]

can be defined by user, read from real traces or generated using the generator module facilitating the process of synthetic workload creation. However, the key elements of the presented architecture are plugins. They allow a researcher to configure and adapt the simulation framework to his/her experiment scenario starting from modelling job performance, through energy estimations up to implementation of resource management and scheduling policies. Politics and models provided by the plugins affects the simulated environment and are applied after each change of its state. Plugins can be implemented independently and plugged into a specific experiment. Results of experiments are collected, aggregated, and visualized using the statistics tool. Due to a modular and plug-able architecture DCworms enables adapting it to specific resource management problems and users' requirements. Figure 3 presents the overall architecture of the workload simulator.

2.3.2 Workload Modelling

Experiments performed in DCworms require a description of applications that will be scheduled during the simulation. As a basic description, DCworms uses files in the Standard Workload Format (SWF) [1] or its extension Grid Workload Format (GWF) [37]. In addition to the SWF file, some more detailed description of a job and task can be provided in an additional XML file. This form of description provides the scheduler with more detailed information about application profile, task requirements, user preferences. In addition, DCworms enables reading traces from real resource management systems like SLURM [54] and Torque [57]. Further, the simulator is complemented with an advanced workload generator tool that allows creating synthetic workloads.

DCworms provides user flexibility in defining the application model. Considered workloads may have various shapes and levels of complexity that range from multiple independent jobs, through large-scale parallel applications, up to whole workflows containing time dependencies between jobs. Moreover, DCworms is able to handle rigid and mouldable jobs, as well as pre-emptive jobs. Each job may consist of one or more tasks. Thus, if preceding constraints are defined, a job may be a whole workflow. To model the particular application profile in more detail, DCworms follows the DNA approach proposed in [22]. Accordingly, each task can be presented as a sequence of phases, which shows the impact of this task on the resources that run it. Phases are then periods of time where the system is stable (load, network, memory) given a certain threshold. This form of representation allows users to define a wide range of workloads: HPC (long jobs, cpu-intensive, hard to migrate) or virtualization (short requests) that are typical for cloud data centres environments.

2.3.3 Resource Description

The second part of input data that must be delivered to simulation is a description of the simulated resources. It contains information concerning available resources and scheduling entities with their characteristics. Additionally, DCworms is able to handle DEBB description file format by transforming it to the native format supported by the simulator.

The resource description provides structure and parameters of available resources. Flexible resource definition allows modelling various computing entities consisting of compute nodes, processors and cores. In addition, new resources and computing entities can easily be added. Moreover, detailed location of the given resources can be provided in order to group them and form physical structures such as racks and containers. Each of the components may be described by different parameters specifying available memory, storage capabilities, processor speed etc. In this manner, power distribution system and cooling devices can be defined as well. Due to an extensible description, users are able to define a number of experiment-specific and hypothetical characteristics. With every component, a specific profile can be associated that determines, among others, power, thermal and air throughput properties. The energy estimation plugin can be bundled with each resource.

Scheduling entities allow providing data related to the queuing system characteristics. Thus, information about available queues, resources assigned to them and their parameters like priority can be defined. Moreover, allocation strategy for each scheduling level can be introduced in form of the reference to an appropriate plugin.

In this way, DCworms allows simulating a wide scope of physical and logical architectural patterns that may span from a single computing resource up to whole data centres or geographically distributed grids and clouds. In particular, it supports simulating complex distributed architectures containing models of the whole data centres, containers, racks, nodes, etc. Granularity of such topologies may also differ from coarse-grained to very fine-grained modelling single cores, memory hierarchies and other hardware details.

2.3.4 Simulation of Energy Efficiency

DCworms allows researchers to take into account energy efficiency and thermodynamic issues in distributed computing experiments. That can be achieved by the means of appropriate models and profiles. In general, the main goal of the models is to emulate the behaviour of the real computing resources, while profiles support models by providing required data. Introducing particular models into the simulation environment is possible through selection or implementation of dedicated energy plugins that contain methods to calculate power usage of resources, their temperature and air throughput values. Presence of detailed resource usage information, current resource energy and thermal state description and a functional management interface enables an implementation of energy and thermal aware scheduling algorithms. Energy efficient metrics become in this context an additional criterion in the resource management process. Scheduling plugins are provided with dedicated interfaces, which allow them to collect detailed information about computing resource components and to affect their behaviour.

Power Management Concept

DCworms provides a functionality to define the energy efficiency of resources, dependency of energy consumption on resource load and specific applications, and to manage power modes of resources. Furthermore, it extends the energy management concept presented in GSSIM [40] by proposing a much more granular approach with the possibility of plugging energy consumption models and profiles into each resource level.

Power profiles allow introducing information about power usage of resources. Depending on the accuracy of a model, users may provide additional information about power states which are supported by the resources, amounts of energy consumed in these states, as well as general power profiles that provide means to calculate the total energy consumed by the resource during runtime. The above parameter categories may be defined for each element of a computing resource system. It is possible to define any number of resource specific states, e.g. the P-states in which processor can operate.

Power consumption models emulate the behaviour of the real computing resources and the way they consume energy. Due to a rich functionality and flexible environment description, DCworms can be used to verify a number of theoretical assumptions and develop new energy consumption models. The energy estimation plugin can then be used to calculate energy consumption based on information about the resources power profile, resource utilization, and the application profile including energy consumption and heat production metrics. Users can easily switch between the given models and incorporate new, visionary scenarios.

Air Throughput Management Concept
The presence of an air throughput concept addresses the issue of resource air-cooling facilities provisioning. Using the air throughput profiles and models allows anticipating the air flow level on output of the fans and other air-cooling devices, being the effect of their management policy.

The air throughput profile, analogously to the power profile, allows specifying known and available operating states of air-cooling devices. Each air throughput state definition can consists of an air flow value and a corresponding power draw. It can represent, for instance, a fan working state. Possibility of introducing additional parameters makes the air throughput description extensible for new characteristics.

Similar to energy consumption models, the user is provided with a dedicated interface that allows him to describe the resulting air throughput of the computing system components like cabinets or server fans. Accordingly, air flow estimations can be based on detailed information about the involved resources and cooling devices, including their working states and temperature level.

Thermal Management Concept
The primary motivation behind the incorporation of thermal aspects in DCworms is to exceed the commonly adopted energy use-cases and apply more sophisticated scenarios. By the means of dedicated profiles and interfaces, it is possible to perform experimental studies involving thermal-aware workload placement.

Thermal profile expresses the thermal specification of resources. It may consist of the definition of the thermal design power (TDP), thermal resistance and thermal states that describe how the temperature depends on dissipated heat. For the purposes of more complex experiments, introducing of new, user-defined characteristics is supported. The aforementioned values may be provided for all computing system components distinguishing them, for instance, according to their material parameters and models.

Thermal profile, complemented with the implementation of temperature prediction model can be used to introduce temperature sensors emulation. In this way, users have means to approximately estimate the temperature of the simulated objects. The proposed approach assumes some simplifications that ignore heating and cooling processes that can be addressed in detail by the CFD simulations.

2.3.5 Application Performance Modelling

In general, DCworms models applications by describing their computational, communicational and power characteristics. Additionally, it also provides means to include complex and specific application performance models during simulations. These models can be plugged into the simulation environment through a dedicated API. Implementation of this plugin allows researchers to introduce specific ways of calculating task execution time. The number of parameters including: task length (number of CPU instructions); task requirements; detailed description of allocated resources (processor type and parameters, available memory); input data size; and

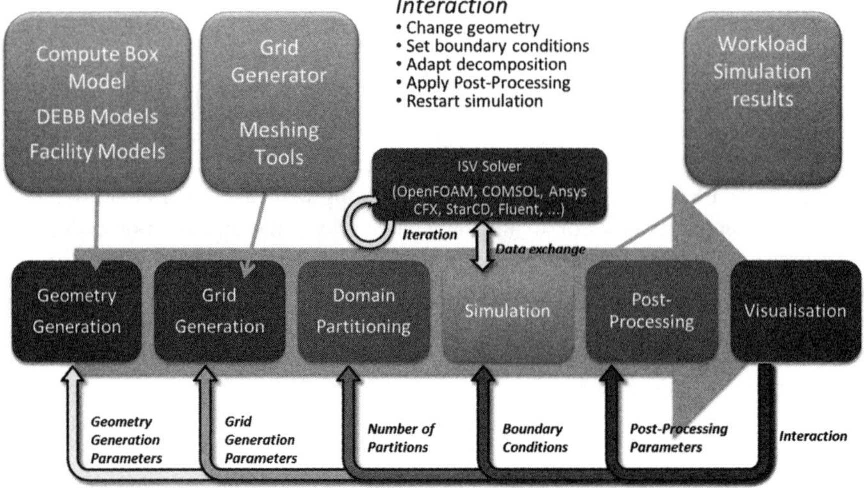

Fig. 4 Simulation workflow steering CFD simulation [62]

network parameters can be applied to specify the execution time of a task. Using these parameters, user can for instance take into account the architectures of the underlying systems, such as multi-core processors, or virtualization overheads, and their impact on the final performance of applications.

2.4 Interactive Computational Fluid Dynamics Simulation

One of the main capabilities of the SVD toolkit is the interactive simulation of airflow and temperature distribution within server rooms, racks and servers to identify hot-spots. To this end, the heat-flow simulation requires the execution of several steps, as presented in Fig. 4, involving: geometry and grid generation, boundary condition definition, domain partitioning, simulation, post-processing, data rendering and storage. Most of the components are modelled as COVISE Modules which allows defining and executing the whole workflow. Data is passed from one module to the other automatically and data conversion is carried out as needed, even if the different processing steps run on physically distributed machines, such as Pre- and Post-Processing servers, Supercomputers, Visualisation workstations or clusters.

The geometry generation step takes as input the DEBB with the geometry models of the concrete resources involved in the simulated environment. It parses the DEBBs models, extracts all relevant info and merges the individual geometry files of the individual components into one overall dataset for the next step. Geometry definition is read from the models specified in DEBBs in two ways: (a) simple box type modelling where components can be composed from an arbitrary number of boxes where the size of each box cannot be smaller than the discretization of the computational mesh; (b) each component is modelled as STL files containing any number of patches. A name must be assigned to each of these patches which allows referencing them in order to assign boundary conditions.

Two types of grid generation modules are developed. In the first one, box type meshes are automatically meshed by a custom grid generator which creates an unstructured grid, consisting of hexahedron cells only. In the second one, meshes for STL geometry is created using SnappyHexMesh from OpenFOAM. This also creates an unstructured grid consisting of mostly hexahedron cells but between the surface mesh and different mesh resolutions, it adds Polyhedron cells.

The definition of boundary conditions are not modelled as a separate module but are carried out directly by the grid generation module. Global boundary conditions such as initial temperature and velocity field, as well as the individual parameters such as air vent velocities or power consumption of a certain node or rack are read from the database and assigned to the relevant surface patches. This allows simulating current conditions by accessing life measurements as well as simulated conditions as defined by the workload simulation which will also store the relevant parameters in the database.

In order to carry out parallel simulations on todays distributed memory HPC systems, the computational mesh needs to be split into smaller parts, one for each compute node. Depending on the solver, this process is either an integral part of the solver itself or it is modelled as a separate COVISE module. When using ANSYS CFX as the solver, domain decomposition is carried out by the solver itself, CFX does not support reading pre partitioned meshes. The first release of the SVD Toolkit uses the open source simulation framework OpenFOAM to do the CFD simulation. Therefore, the parallel version of the OpenFOAM domain decomposition will be integrated as a module within COVISE.

For a initial prototype, the CFD simulation used integrated ANSYS CFX as solver within the SVD toolkit. In the final Version of the SVD Toolkit the Open Source Package OpenFOAM will be used as a solver, integrated into COVISE as mentioned before. This approach makes sure that the architecture can easily be extended by new types of solvers in the future and allows accelerating the development by reusing many existing components.

Existing COVISE post processing modules can be reused to interactively extract data, compute particle traces and visualize 3D flow phenomena. An additional post processing module will be developed which extract measurement values from the simulation results. Those values will be stored in the database for further calculation of assessment metrics and visualization in the CoolEmAll WEB-GUI.

Direct 3D Rendering with COVISE Virtual Reality environment renderer allows interactive in depth analysis of the 3D Air Flow for engineers to optimize the data centre during the planning phase or for modifications. During the operational phase the 3D Rendering will be omitted and only key values will be extracted and presented in the CoolEmAll WEB-GUI.

Interaction by the end user are possible, after obtaining the results from the initial simulation, consisting in the following options: change geometry, e.g. re-arrange racks within server room, modify boundary conditions, e.g. change an inlet air temperature, adapt decomposition, apply post-processing and restart simulation. In this way various configuration of server-room can be evaluated interactively, to find the best solution.

2.5 *Visualization*

As noted, the CoolEmAll Web GUI provides a web based user interface which allows interacting with the SVD toolkit and visualizing its results. It comprises several GUIs integrated into common web based GUI environment. Experiment configuration GUI supports configuration of the experiment/trial specifying simulation parameters. The DEBB configuration GUI allows defining, selecting and configuring DEBBs on various granularity levels. DCworms GUI allows selecting applications, workloads and scheduling policies, and presenting the results of the workload scheduling in a GUI. COVISE GUI presents entire simulation results of the CFD-simulation, visualizing air flow across all building blocks (DEBBs) and enabling interaction with the simulation, allowing to interactively change the simulation parameters that affects position (arrangement) of objects. MOP GUI visualizes DEBB surfaces along with data stored in the Database, containing workload simulation results, CFD simulation results limited to temperature and airflow history for particular sample points on object surfaces, and heat- and energy-efficiency metrics assessing experiment result. As both, real data obtained from measurements (e.g. temperatures, system load) and simulated data from CFD and Workload Simulation will be stored in the database, both of them will be displayed in the MOP GUI to compare results and validate models. Finally, the Report page provides assessment metrics that evaluate energy- and cooling efficiency of the simulation results, applying metrics calculated by the metric calculator to assess configurations.

To facilitate insight into heat-flow processes and conditions within the compute room or rack it is necessary to post-process and visualize the data generated by the CFD-simulation in an appropriate manner. As noted, COVISE already includes extensive visualization capabilities that can be used to an advantage. All necessary post-processing steps can be included into the simulation workflow for immediate analysis. The simulation progress will be constantly monitored even while the simulation is running. The in-situ approach allows early assessments of the simulation results and shortens the evaluation cycles. COVISE visualization results can be explored on the desktop or in Virtual Reality, as well as on mobile devices. It is also possible to use Augmented Reality techniques to compare the simulation results with actual experimental data.

3 Data centre Efficiency Building Blocks

As noted previously, one of the main outcomes of the CoolEmAll project are Data centre Efficiency Building Blocks (DEBBs), designed to model data-centre building blocks on different granularity levels and containing models necessary for workload, heat and airflow simulation. These granularities reach from a single node up to a complete data centre and will help users to model and simulate a virtual data centre for e.g. planning or reviewing processes.

Most data centres are based on server racks, aligned as rows. Each rack contains a variety of servers, storage-systems and peripheral equipment. These racks are in most cases build up step by step and there is some effort to invest to integrate all components. For building up big data centres, some vendors and data centre providers have done some efforts for new concepts of a higher integration level for the computing and storage infrastructure, mostly based on container-modules. Using DEBB concept and DEBB configuration tool, various configurations of the data centres and their components can be modelled on various scale level, to assess their energy and thermal behaviour in simulations, under various boundary conditions. In the following subsections we describe the DEBB concept and models used to realize simulations.

3.1 *DEBB Concept and Structure*

A DEBB is an abstract description of a piece of hardware and other components, reflecting a data-centre configuration/assembly on different granularity levels. A DEBB contains hardware- and thermodynamic models used by SVD toolkit [16] to simulate workload, heat- and airflow, enabling (energy-efficiency) assessment and optimization of different configurations of data centres built of these building blocks (DEBBs).

A DEBB is organized hierarchically and can be described on the following granularity levels:

1. Node Unit reflects the finest granularity of building blocks to be modelled i.e., a single blade CPU module, a so-called “pizza box”, or a RECS CPU module. A Node Unit consists of the following components: Main-Boards (including network), potentially additional cards put in slot of the MB, CPUs (including possible operating frequencies), memory modules, cooling elements (including fan), and, optionally, storage elements.
2. Node Group reflects an assembled unit of building blocks of level 1 (node units), e.g. a complete blade centre or a complete RECS unit (currently consisting of 18 node-units), main pane to which all node units are connected, and might contain a power supply unit (PSU).
3. ComputeBox1 (CB1) reflects a typical rack within a data centre, consisting of the building blocks of level 2 (Node Groups), secondary components (such as

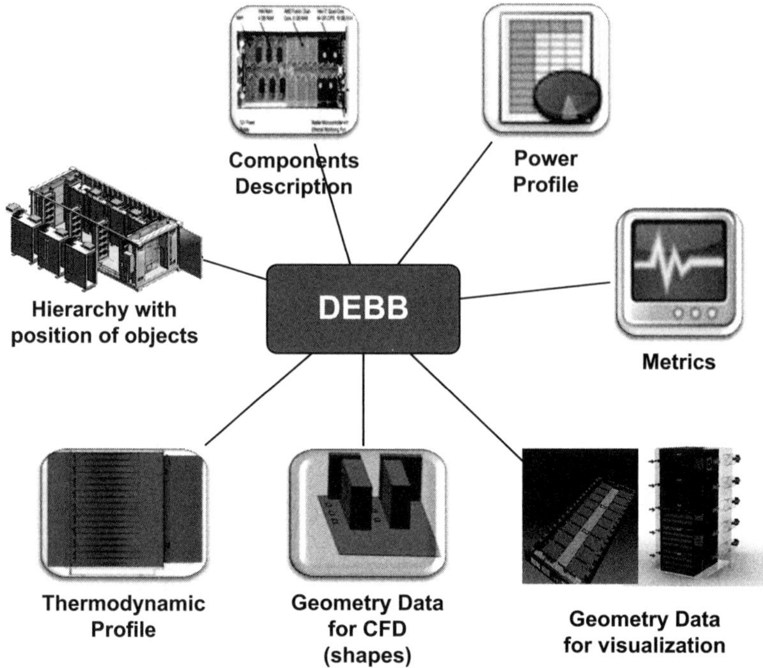

Fig. 5 DEBB concept [62]

interconnect), power distribution unit (PDU), power supply units (PSUs), and integrated cooling devices. CB1 contains in principle the same information as a node group but only on a higher level.

4. ComputeBox2 (CB2) building blocks are assembled of units of level 3, e.g. reflecting a container or even complete compute rooms, filled with racks, Uninterruptable Power System (UPS), cooling devices, etc.. Additionally to the same content as CB1 it contains ICT infrastructure interconnecting multiple ComputeBox1 and facility cooling devices. Since cooling devices may be part of rooms (CB2) and racks (CB1) and in rare cases even within node groups or nodes and represent a main part of the energy consumption depending on the IT load, they are treated separately and then handled like any other.

A DEBB on each level (starting from level 2) can be described by the formal DEBB specification. The formal structure of the DEBB, presented in Fig. 5 and fully described in [35], consists of the specification:

- *DEBB hierarchy* with a reference to and positions of its objects or their aggregation (lower level DEBBs) within the scene (high level DEBB builds up by the hierarchy). The DEBB hierarchy is described in PLMXML format, allowing references to description of models or profiles in different formats, listed below.
- *Geometrical data* describing object-shapes, necessary for CFD simulation, are expressed in STL format, and is referenced from the object description in PLMXML

file. The combination of these two formats: PLMXML for description of the DEBB hierarchy with position of its objects (lower level DEBBs) and STL for description of object-shapes, enables to model any scene definition (needed for CFD simulation) on different granularity levels, such as a server-room consisting of cooling components, racks, power-units, and other devices.

- *DEBB Components* describing a single object as it is delivered like server, cooling device, empty rack containing only rack specific equipment like sensors or power distribution etc., empty room etc. This description is defined in a way that a manufacturer can provide it for any type of hardware, based on naming conventions used in CIM format. The component file describes in details the type and technical attributes references to 3D models if added, and any type of profiles useful to describe the behaviour and power consumption of the component. The PLMXML file, describing DEBB hierarchy, contains for each object a corresponding reference to the component and therefore to its technical description. This allows a workload simulator to identify the node type being selected for the workload execution and correlate it with its power-usage profile.
- *Power-usage profiles* necessary for the workload simulation is embedded into DEBB Component. A power-profile describes for each state/load level (i.e. 20 % rise per level)) of a particular component its corresponding power-usage, enabling calculating and simulating power consumption and heat load for different utilization levels during the simulation of the workload execution. This allows assessing power-usage of workload being executed on particular component types, such as node-types.
- *Thermodynamic profile*, stating air-throughput of fans for different levels and cooling capacity of cooling devices is defined in scope of DEBB Component definition. Thermodynamic profile is used by workload simulator to calculate air flow-initial boundary conditions necessary for airflow and heat-distribution simulation. The entire XSD schema for specification of thermodynamic-profile is described in scope of Component Description schema, in D3.2.
- *Geometrical data for visualization* of DEBB and their shapes in MOP-GUI is described in VRML format. It is referenced from PLMXML in the same manner as geometric shapes (STL format) objects for CFD simulation.
- *Metrics* assessing energy-efficiency of building blocks (DEBB) are embedded into PLMXML file (specifying DEBB hierarchy) with user-defined values.

As mentioned, a DEBB contains models used by SVD toolkit to simulate (i) workload- and (ii) airflow, enabling assessment and optimization of different configurations of data centres built of building blocks (lower level DEBBs). Thereby, a simulation of a DEBB on level n (i.e. ComputeBox2 level), requires DEBBs of level n-1 (i.e. ComputeBox1). For example, in order to enable a simulation of power, heat and airflow behaviour within the computing room (the ComputeBox2 level), corresponding objects must be defined or referenced on the ComputeBox1 level within the DEBB-hierarchy. These objects can be racks or cooling devices (e.g. compute room air conditional, air handler, chillers, heat-exchanger, fans, etc.). As the focus of CoolEmAll is to simulate energy and thermal behaviour of DEBBs in order to assess their efficiency and optimise design, it is modelled as the smallest unit in the thermodynamic modelling process. Therefore, the complete Node Unit is the smallest feature that will be present in a simulation. The thermodynamic processes within a Node Group are modelled using Node-Unit models (node-unit DEBB), enabling to simulate accurate heat distribution within the Node-Group. The ComputeBox1 simulations will require-besides the arrangement of the Node Groups-the velocity field and temperature at the Node Group outlets over time as inbound boundary condition and will provide the room temperature over time at the outlet of the Node Group as outgoing boundary condition. Similarly, the simulation of a computing room (ComputeBox2) or container will require the velocity field and temperature on inlets and outlets of racks (ComputeBox1), reducing simulation models to the required level [17].

3.2 Hardware Models for Workload Simulation

In this section we present hardware models (hardware profiles) used to estimate power consumption of computing resources, air flow values of cooling equipment and air temperature. Section 3.2.1 presents the hardware modelling in DCworms workload simulator [42], while Sect. 3.2.2 presents profiles created from data collected on the CoolEmAll RECS platform.

Hardware profiles are measures that define the changes in energy consumption, air temperature or other parameters of hardware under different conditions. Profiles consist of the energy efficiency (power profile models), air temperature (thermodynamic profile) and air throughput (air throughput profiles). The energy efficiency was carefully examined in the tests, giving exact values of hardware power consumption under changing conditions. Air temperature is precisely simulated by CFD simulations. However, simple models can also be created based on the values from temperature sensors located in the front and back of servers. The air throughput could not be precisely measured due to the limitations of the sensors. They only provide information about the presence or absence of air flow, with no exact values. When more accurate sensors will be purchased in the future, additional experiments will be conducted to create air throughput profiles.

3.2.1 Hardware Modelling in DCworms Workload Simulator

Being part of hardware description, power and thermodynamic profile are provided as an input to the workload simulator. Therefore, they can be used to estimate the power draw of computing system components and to calculate the air flow values resulting from the cooling equipment. Moreover, they can become an additional criterion in the workload management process.

The main aim of incorporating power profiles to the workload simulator is to provide a definition of energy efficiency of resources, dependency of energy consumption on resource load and specific applications, and to manage power modes of resources.

As mentioned in 2.3.4, power profiles allow introducing information about power usage of resources. Depending on the accuracy of a model and available data (both derived from experiments or manufacturers specification), information about supported power states and corresponding power consumption can be provided. That allows optimizing the simulation model with respect to the real-world data. For particular computing resources, it is possible to define resource specific states, for example so called P-states, in which processor can operate. Additionally, power draw can be associated with particular load levels of the system or cooling devices working states.

Based on the delivered power profiles, workload simulator will estimate the power consumption of particular resources and, thus the whole data centre. Hence, calculating and simulating power usage will follow changes of resource power states and real-time resource usage. To ensure the appropriate accuracy, DCworms takes differences in the amount of energy required for executing various types of applications at diverse computing resources. It considers all defined system elements (processors, memory, disk, cooling devices, lighting, etc.), which are significant in total energy consumption. Moreover, it also assumes that each of these components can be utilized in a different way during the experiment and thus have different impact on total energy consumption.

Using the air throughput profiles and models allows anticipating the air flow within the data centre, being the result of air-cooling equipment management. Since the air throughput profile can be bounded with fans and other air-cooling facilities, the air flow value related to the particular server outlets can be estimated during the workload simulation. The air flow level can be determined with respect to the current resource state but also to its temperature. Moreover, mutual impact of several air flows may be considered.

Beside the detailed simulation of power consumption and calculations of air flow, profiles specification can be used during the workload and resource management process. To this end, DCworms provides access to the profiles data which allows acquiring detailed information concerning current system performance, power consumption and air throughput conditions. Moreover, it is possible to perform various operations on the given resources, including dynamically changing the frequency level of a single processor, turning off unused resources and managing cooling devices working states.

The outcome of the workload simulation phase is a distribution of power usage and air throughput for the hardware components specified within the DEBB. These statistics may be analyzed directly and/or provided as an input to the CFD simulation phase. The former case allows studying how the above metrics change over time, while the latter harness CFD simulations to identify, for example, temperature differences between the computing modules, called hot spots.

Table 1 Technical specification of CoolEmAll RECS 2.0 processors

RECS	Count	Processor	Clock rate (GHz)	RAM memory (GB)	Number of cores
1	18	AMD G-T40N	1.0	4	2
2	14	Intel Atom N2600	1.6	2	2 (4 logical)
2	4	Intel Atom D510	1.6	4	2 (4 logical)
3	4	Intel i7-2715QE	2.1	16	4 (8 logical)
3	14	Intel i7-3615QE	2.3	16	4 (8 logical)

3.2.2 Hardware Power Profiles

The power profiles were created on the CoolEmAll RECS 2.0 testbed used in the project. It consists of three RECS, each featuring different types of processors. The technical specifications of these processors are presented in Table 1.

To create the power profiles of presented processors, the power usage of the nodes were monitored while executing different applications. Separate tests were performed for different CPU frequency values (ranging from the lowest to the highest available) and processor load (25, 50, 75 and 100 % in case of Intel i7 and 25, 50 and 100 % in case of the rest of CPUs). The following applications were selected:

- C-ray: a simple ray-tracing benchmark, usually involving only a small amount of data. This software measures floating-point CPU performance. The test is configured with a significantly big scene, requiring about 60s of computation but the resulting image is written to `/dev/null` to avoid disk overhead [58].
- Hmmer: a sequence analysis software. Its general usage is to identify homologous protein or nucleotide sequences. This type of problem was chosen because it requires a relatively big input size (hundreds of MB) and requires specific types of operations related to sequences [28].
- Pybench: a benchmark to measure the performance of Python implementations. In the past it has been used to track down performance bottlenecks or to demonstrate the impact of optimization and new features in Python. In contrast to the other benchmarks, it was run on one core only to test the power profile of servers running single-threaded applications.

The power profiles model the CPU power consumption for a given processor load and P-State (with different clock rates). Using these profiles it is possible to estimate average values of power consumption or power usage while running a given application.

3.2.3 Electrical Model of the Power Supply Unit 2.0

To calculate the actual power usage of the rack (ComputeBox1), DCworms requires in addition to hardware power profiles also models of the power supply units. The Power Supply Unit 2.0 which is used to supply the needed 12 V DC

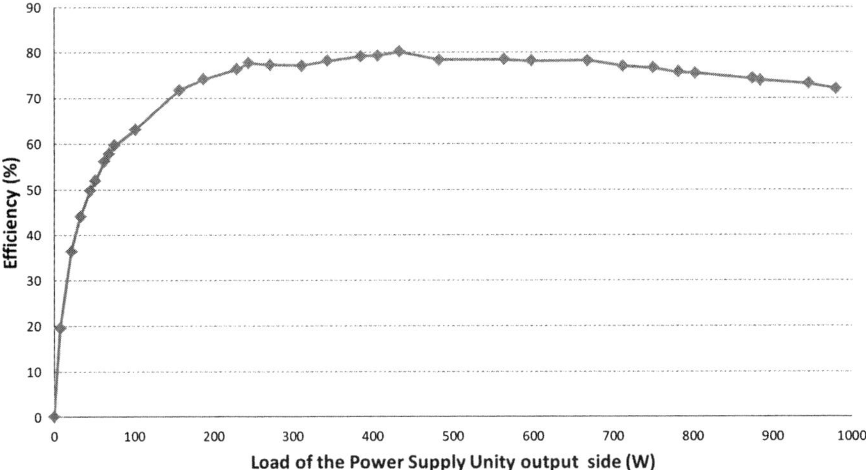

Fig. 6 Efficiency of the power supply unit (Box power unit 2.0)

for the RECS 2.0 can mainly be described via the energy efficiency and a schematic sketch. Internally, the Power Supply Unit 2.0 is based on six single Power Units by the German manufacturer Block Transformatoren, the PVSE 230/12-15 (BLOCK Transformatoren-Elektronik GmbH). To model the electrical characteristics, a series of measurements has been done. These measurements compare the input power to the output power which can be used to determine the load dependent efficiency. The difference between the input and output power is power dissipation which is converted to heat emitted to the air. The result is shown in Fig. 6.

3.3 Hardware Models for Thermodynamic Profiles and Cooling Equipment

The node DEBB will be modelled in CoolEmAll as the smallest unit in the thermodynamic modelling process. As such, the complete node unit is the smallest feature that will be present in a simulation. The thermodynamic processes within a node group are only coarsely modelled as they are merely interesting for providing boundary conditions for the ComputeBox1 and ComputeBox2 simulations. The ComputeBox1 simulations requires—besides the arrangement of the node groups—the velocity field and temperature at the Node Group outlets over time as inbound boundary condition and will provide the room temperature over time at the outlet of the Node Group as outgoing boundary condition. Additionally, the heat generation of the PSU, switches, and other components have to be specified for a complete model of the ComputeBox1. For ComputeBox2, the same boundary conditions have to be defined as for the ComputeBox1 along with all other heat supplicants present in the room. Using the DEBB concept, the thermodynamic modelling can be tackled in a hierarchical way, reducing the complexity of the resulting overall model.

The objective of cooling models is to assess how the power consumption of cooling equipment varies with the variation of power usage related with IT workload. The elements that have power consumption are the following: fans at RECS level; CRAH (Compute room air-handling unit composed by fans and cooling coil); compressor of chiller; heat rejection equipment (dry cooler or cooling tower); auxiliary elements (pumping and automatic valves).

However, not only the IT workload affects the model but also several surrounding factors influence the approach of the model. A first step is to identify these factors in order to discriminate the ones relevant for power consumption calculation and the ones suitable to be neglected. These are the following:

- Thermal load (Qload): it is generated not only by IT equipment but also by other elements (thermal load of fans, lights, electricity equipment, control and security equipment) in the control volume.
- Cooling system type: cooling medium can be air, water or refrigerant, mainly. Different operation temperatures will be required depending on the type of cooling systems. Besides that, different elements will compose the cooling system, with different efficiency of each one and more or less losses depending of the different number of heat exchanger processes. Also, free-cooling option will be different depending on the type of cooling configuration (i.e. direct air free-cooling, indirect free cooling with dry cooler or indirect free cooling with cooling tower).
  - Heat-rejection type: the type of heat rejection affects directly the temperature of operation of condenser of the chiller and the chiller efficiency, as it is explained in the next point (ambient temperatures item). The heat rejection can be done by air, with dry-cooler, or by water, with a wet cooling tower. Wet cooling towers are necessary when temperature of condenser is lower and the climate is hot (particularly on summer). However, the lower temperature level of operation of condenser improves the efficiency of chiller. Besides that, the power consumption is lower in wet cooling towers than in dry-coolers. The main disadvantage of wet cooling towers is the high amount of water consumption.
- Efficiency of the chiller: Energy Efficiency Ratio (EER): energy efficiency ratio is a function of four parameters, at least.
  - Compressor chiller type and nominal capacity: different kind of compressor is appropriate for different levels of nominal cooling capacity. The EER is directly related with the compressor type. There are also variations between manufacturers.
  - Partial load: EER will be directly related with partial cooling load; the cooling load variation will be directly related with IT load and therefore IT power consumption.
  - Supply temperature of chiller: this parameter influences directly the EER. The higher supply temperature of the chiller, the better EER.
  - Ambient temperatures: outside temperature of ambient air will affect the temperature of operation of condenser of the chiller, as well as the suitable type of

heat rejection element (dry cooler, wet cooling tower) and it directly influences the EER. For instance, if it is considered a heat rejection temperature of 35 °C, in soft climates as in Central Europe, dry coolers can operate in appropriate conditions in summer, but in Mediterranean climates, are necessary wet cooling towers to guarantee an energy efficient system. It is due to cooling tower uses wet-bulb temperature of air instead of dry-bulb temperature, which is a lower value. The lower ambient temperature, the lower condenser temperature, the better EER.

- Fan operation: air circulation is directly related with chiller consumption or chiller EER. To show an example, increasing the air flow, fan power consumption will increase but delta-T of air could be lower, then supply temperature of chiller will increase and the chiller would consume less energy.
- Dehumidification requirement: the control of humidity has also a relevant influence on the cooling load. Moreover, the need of dehumidification causes a decreasing of supply temperature of chiller so that a reduction of EER.

The definition of each of the models will be referred to a control volume. The control volume is the reference space that determines the boundary conditions to calculate a thermodynamic energy balance and the processes of heat transfer. This limits the internal thermal loads that will affect the heat dissipation and will permit to determine the boundary conditions regarding flows of energy going inside and outside of the control volume.

3.4 Hardware Models for CFD Simulation

As noted previously, CFD simulation is simulating airflow and temperature distribution within server rooms, racks and servers. Such a CFD simulation requires the execution of several steps, detailed in the following bullets:

- Geometry Generation, taking as input the defined DEBB with geometry models of the concrete resources involved in the simulated environment.
- Grid Generation, with the support of meshing tools, resulting in the corresponding computational mesh.
- Definition of boundary conditions which are queried from the Database and assigned to the relevant surface patches. Additionally, boundary conditions are set based on results obtained by simulation of workload execution and specific application profiles. These results include power usage and outlet air throughput of IT equipment.
- Domain Partitioning, taking as input the previously generated mesh, together with the boundary conditions.
- Simulation, performed by the CFD Solver.
- Post-processing is performed with the resulting data from the simulation.
- Rendering is performed for the visualization of the simulation results. Some of this is, in fact, already part of the visualization usage phase (see below).

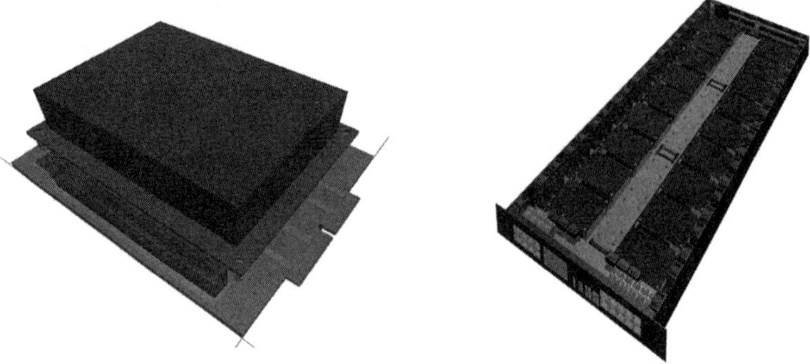

Fig. 7 Geometry data of a node and a node-group (RECS server) [59]

As pointed out, the generation of the geometry data is extracted from the DEBB (main PLMXLM file), containing references to geometric objects specified in STL files. The geometric objects are composed of faces. There are four significant faces for CFD, hat are handled in simulation in different way:

- inlet (source of airflow)
- outlet (exhausting airflow)
- heatsink (source of heat)
- wall (surface reflecting the airflow)

For specification of the boundary patch, such as "inlet", the name attribute of the corresponding ProductInstance-Element (describing reference to particular object) within the PLMXML file contains the keyword, specifying the face-type (for inlet this keyword is "inlet"). In case of absence of face-type, "wall" face-type is presumed. To distinguish between different types boundary conditions (such as airflow, temperature) for particular face-types, these are explicitly defined as user values (within PLMXML file) with title attribute "airflow-volume-sensor" for airflow, and "temperature-sensor" for temperature. In order to setup simulation with right parameters (boundary conditions) belonging to corresponding geometry-object, such as airspeed at "inlet" of a rack, these parameters are queried from the database using full object-path to particular geometrical object. Full object-path is built as a concatenation of all object-names in the hierarchy of PLMXML file.

An example of the geometry model of a node and a node-group is presented in Fig. 7. The node-board consists of connection (serving as a wall surface) and the heat sink (heat sink surface). Nodes with different CPUs and heat-sink have different geometry models (STL files). The node-group shows geometry of the RECS server equipped with 18 nodes.

3.5 Assessment of DEBBs

As mentioned before, a (DEBB) concept describes energy efficiency of data-centre building block on different levels of system granularity. A possible characterization of DEBBs in the terms of energy efficiency can be done according to Green Performance Indicators (GPIs), as defined in GAMES project [38, 39]. GPIs are a measurement of the index of greenness of an IT system indicating the energy consumption, energy efficiency, energy saving potential and all energy related factors on different systems levels within IT service centre, including application and execution environment. In order to assess the global greenness of an application and IT infrastructure, GPIs were classified into four clusters: IT Resource Usage GPI, Application Lifecycle KPIs, Energy Impact GPIs and Organizational GPIs. Such a classification enables assessment of the energy efficiency of an IT centre from the business (organisational) level down to the technical level. GPIs allow also for considering, among others, the trade-off between performance and energy consumption at facility, application and compute node (IT infrastructure) level. As a DEBB is an abstraction of data centre building blocks on various levels of granularity, GPIs can be used for characterization of DEBBs on node level, node-group level, rack level and container level or entire IT centre. Section 4 describes metrics used in the project for the assessment of the cooling- and energy-efficiency of DEBBs within various boundary conditions evaluated within the experiments.

4 Energy Efficiency Metrics

The present tendencies on assessing the Energy Efficiency of data centres are based on global metrics, mainly focused on power consumption as, for instance, the Power Usage Effectiveness (PUE) metric [13]. It is usual to consider the peak or average loads in a static analysis, focusing interest in power consumption instead of energy consumption. This strategy faces many limits as it does not allow predictions of energy performance to improve the energy efficiency. With this strategy cooling demand is only a part of all the components that influences the metrics related with power. As a consequence, cooling impact cannot be properly identified. Only assessing the power consumption do not permit to detect the origin of high heat transfer issues.

CoolEmAll tries to address an energy and thermal based assessment of data centres. In that sense the main goal of the project is to enable designers and operators of data centre to reduce the energy footprint by combining optimization of IT, cooling and workload management. To facilitate that, CoolEmAll provides some particular metrics that enable minimization of data centre energy consumption also for low and variable loads rather than just for peak loads as it is usually done nowadays.

In this chapter, it has been included a summary of the state of the art about energy-efficiency metrics for data centres and the description of the metrics that have been selected to be included in the tool.

4.1 State of the Art

Day after day the energy footprint is becoming one relevant issue in the management department of data centres. The awareness of data centre energy consumption and the increase in energy prices have driven research about metrics suitable to define and quantify the energy efficiency of data centres. A first step towards this direction was taken by the Uptime Institute in 2003 with the introduction of the Triton Coefficient of Effectiveness, defined as the total utility power required for operating a data centre divided by the critical load in the computer room [55]. Up to this date, the assessment was just based on raw performance, defined by simple metrics such as operation per second [25] or request per second [20]. Since then, the urgent need of defining energy efficiency evaluation methodologies pushed the stakeholders in the field to set own metrics, leading to the advent of several figures. Among them, the PUE (Power Usage Effectiveness) [43] has been widely adopted since 2007, with the support of The Green Grid institution [2]. Actually the wide adoption of PUE has resulted in a corresponding common understanding or a shared calculation methodology for PUE [13].

The Task Forces are platforms that permit the cooperation between industry players and government agencies to establish common metrics. The Data Centre Metrics Coordination Task Force (U.S. Regional Task Force) [10] and the Global Harmonization of Data Centre Efficiency Metrics Task Force (Global Task Force) [7, 12] are the main of them. They endorse metrics for measuring infrastructure energy efficiency in data centres and define the methodology and criteria for an appropriate calculation. These Task Forces agreed on using PUE as a standard metric. Besides that, the Global Task Force promotes other metrics focused on renewable energy systems and re-use of energy to reduce carbon emissions as ERE (Energy Reuse Effectiveness) [5], CUE (Carbon Usage Effectiveness) [4], OGE (On-site Energy Generation Efficiency) and ECI (Energy Carbon Intensity) [9]. Also the water consumption is in the focus of those organizations and WUE (Water Usage Effectiveness) is the recommended metric [11].

One important issue to evaluate is the energy necessary to keep the system available but without any useful work produced. In that sense, FVER (Fixed to Variable Energy Ratio) [46], introduced by the British Computer Society, correlates the energy directly related to useful work to the energy susceptible to be eliminated [46].

KPI (Key Performance Indicator) [19] is a combination of up to four indicators: KPI_{EC} (Energy Consumption), KPI_{REN} (Renewable Energy), KPI_{REUSE} (Energy Reuse) and KPI_{TE} (Task Efficiency) and reflects energy impact at global level. Other holistic metric is DPPE (Data centre Performance per Energy) developed by the Green IT Promotion Council (GIPC) [6, 8]. DPPE aims to integrate several energy efficiency parameters in one. It includes the assessment of facility efficiency using PUE, the CO_2 emissions associated to energy purchased, the efficiency features of IT equipment and the IT equipment utilization.

Regarding heat-aware on data centre some metrics already exist on literature background and they are also used in some tools. One of them is the RCI (Rack

Cooling Index). It was proposed by Herrlin [34] as an indicator of how effectively the racks are cooled within industrial thermal standards or guidelines. This metric is useful for CFD results analysis due to the difficulty of getting conclusions from the great amount of data produced by a CFD simulation.

Some other metrics points to configuration of IT server. A relevant one reflects the degree of hardware used during a period of time, named DH-UR (Deployed Hardware Utilization Ration). Other metric is SWAP (Space, Watts and Performance) which links space occupation, rated power usage and rated performance of a single server [38].

Several metrics at the level of IT components appear in the literature. In that sense, it can be found power-based metrics or resource usage metrics. Regarding power-based metrics, the Node Power Usage [39] assesses the ratio of power used regarding the maximum capacity. In [38] indicator frequency/Watt assesses how efficiently is used a certain CPU (Central Processing Unit) or indicator bandwidth/Watt shows how much data is moved per unit of time and power. In case of resource usage metrics several metrics are stated on [3] as percentage over the maximum capacity of storage usage, memory usage, network bandwidth used and activity of the CPU. Other interesting metric about resource usage is the total amount of CPU used regarding the CPU allocated, named CPU Usage on [38].

CoolEmAll has collected the present status of metrics based on the state-of-the-art literature. Many significant metrics were selected for implementation in the CoolEmAll SVD (Simulation, Visualisation and Decision) toolkit. Additionally, some new metrics are proposed in order to introduce heat-aware and holistic analysis on a transient period of time.

4.2 Selected Metrics for CoolEmAll

The selected metrics for Coolemall approach focus on heat-aware of data centre in order to optimize its performance directly from the former site where energy consumption is originated, the node. In that sense, four levels of granularity (node, node-group, rack level and room level) will be permitted on the analysis in order to optimize the performance taking into account different scenarios. Metrics will be defined in these four levels:

- *Node unit* is the smallest element of a data centre to be modelled. This unit reflects a single computing node, e.g. a single blade CPU module.
- *Node-group* reflects an assembled unit of node units, of level 1, e.g. a complete blade centre or a rack unit consisting of 18 server nodes.
- *Rack level* reflects the well-known element within an IT service centre, including blocks of node-groups, power supply units and integrated cooling devices.
- *Room of a data centre* is considered as joined units of racks, placed in a container or even complete compute rooms, with the corresponding CRAC/CRAH (Compute

Room Air Conditioner or Air-Handling Unit), chiller, power distribution units, lighting and other auxiliary facilities.

Metrics of CoolEmAll will provide the design concept with the highest energy efficiency and the lowest green-house emissions from the node level to the room level. Most of them are already on use on other applications but some ones are new metrics developed on CoolEmAll. The new ones are the following listed:

- Imbalance of Temperature of CPU, node-group and racks.
- Rack Cooling Index adapted to a node-group.

The next subsections describe the selected metrics. These are classified in three main groups depending on the focus of the assessment:

- *Resource usage metrics* refers to the ratio of use of a certain resource (CPU, memory, bandwidth, storage capacity, etc) respect the total amount of that resource, concerning a component (node) or a set of components (node-group, rack).
- *Energy based metrics* are defined as the consumption of power along a period of time.
- *Heat-aware metrics* take into account temperature as main variable to define the behavior of data centre.

4.2.1 Resource Usage Metrics

CPU Usage: this metric provides the percentage of amount of time that the allocated CPU spends for processing the instructions of the applications [38].

$$CPU\_Usage = \frac{Amount\_CPU_{used}}{Amount\_CPU_{allocated}}. \tag{1}$$

Power Usage: referred to the ratio of power used regarding the maximum capacity, or power rated given by the manufacturer.

$$PowerUsage = \frac{P_{node}}{P_{node,rated}}, \tag{2}$$

where P means power consumption in a given time step expressed in Watt (W).

Deployed Hardware Utilization Ratio (DH_UR): in a node-group it reflects the degree of hardware used during execution time period. Knowing the quantity of nodes running live applications from the total number of deployed nodes is an indicator of the energy consumption of computing equipment required for the different workloads. For instance if 50 % of the nodes are sufficient to handle average load and are running all the time, then the rest could be shut-down to save energy [38]. Efficient deployments should reach DH-UR as close to the unit as possible.

$$DH\_UR = \frac{N_{useful\_node}}{N}, \tag{3}$$

where N_{useful_node} is the quantity of nodes in the node-group that are running applications producing useful work and N is the total number of nodes in the node-group.

4.2.2 Energy Based Metrics

Productivity of node: this depends on the type of service provided and can be calculated as the ratio of the measurable produced work and the energy consumed by the node during the time execution.

$$Productivity\_node = \frac{W_{node}}{E_{node}}, \tag{4}$$

where W means units of useful work and E means energy in Watts per hour (Wh). The useful work produced by the node depends on the services provided. For instance, on HPC environments it is measured in FLOPS and in the Cloud it is usually measured in number of service invocations and in general-purpose services it can be measured in number of transactions. This metric can be applied at all levels of granularity.

Power Usage Effectiveness (PUE) this metrics consist on dividing power used by the data centre between power used by the IT equipment. It can be defined at instantaneous or aggregated level on a period of time. The level of accuracy of the metric is related with the point of measurement of IT power, that can be the UPS (Uninterruptible Power Supply Unit), the PDU (Power Distribution Unit) or the IT itself.

$$PUE_1 = E_{DC}/E_{UPS}, \tag{5}$$

$$PUE_2 = E_{DC}/E_{PDU}, \tag{6}$$

$$PUE_3 = E_{DC}/E_{IT}, \tag{7}$$

where E means energy.

Data centre infrastructure efficiency (DCiE): corresponds to the inverse of PUE and it is usually referred to instantaneous power consumption

$$DCiE = P_{IT}/P_{DC}, \tag{8}$$

where P means instantaneous power consumption.

Fixed to variable energy ratio (FVER): is calculated comparing the fixed and the variable energy consumption.

$$FVER = 1 + (Fixed\ Energy/Variable\ Energy) \tag{9}$$

Developed by the BCS (British Computer Society) [46], it provides information about how much energy is directly related to the useful work produced in a data centre and how much could be eliminated. It is a fact that even a data centre is not producing any useful work it can consume around 80 % of the peak energy consumption showing a flat behaviour during all data centre operation time. This evidences that a lot of energy can be saved during the idle periods. As it is affirmed in the cited paper, despite all efforts trying to measure the useful work in a data centre, no method of measuring provides values which can be compared between different data centres due to the subjectivity of the methodology, being the principle obstacle for a wide adoption.

4.2.3 Heat-Aware Metrics

Node cooling index high (HI) and low (LO): these metrics aim to provide early detection of cooling requirements from measuring the temperature rise at node level over a certain threshold.

$$NCI_{node,LO} = T_{CPU}/T_{CPU,min-all}, \tag{10}$$

$$NCI_{node,HI} = T_{CPU}/T_{CPU,max-all}, \tag{11}$$

where, T_{CPU} is the temperature of the CPU; $T_{CPU,min-all}$ is the CPU minimum allowable temperature according manufacturer specifications; $T_{CPU,max-all}$ is the CPU maximum allowable temperature according manufacturer specifications.

The metric named *Imbalance of temperature of CPU* provides a quick overview of a possible unbalanced distribution of workload at node-group level. This fact usually causes an overheating of a certain node, loosing capacity of other nodes not used, with bad resulting consequences. To avoid this situation, the managers of data centres keep low operation temperatures on the cooling system with the corresponding increase of energy consumption. An early and quick foreseen of this situation can help the designers and operators to manage more efficiently their data centres.

Imbalance of temperature of CPU: this metric is calculated as the difference between the maximum and the minimum values of CPU temperature divided by the average of all nodes in a same time step. Values close to zero of this metric indicate a good temperature balance.

$$Im_{NG,temp} = \frac{T_{CPU,max} - T_{CPU,min}}{T_{CPU,avg}} \times 100, \tag{12}$$

$$T_{CPU,avg} = \frac{1}{N} \sum_{i}^{N} T_{CPU,i}, \tag{13}$$

where $T_{CPU,max}$ is the maximum temperature reached by the CPU in the node-group in a certain time step; $T_{CPU,min}$ is the minimum temperature in the same time step; $T_{CPU,avg}$ is the average temperature between the nodes in the same time step; $T_{CPU,i}$ is the temperature of each i CPU in the node-group in this time step and N is the number of nodes in the group. This metric is assessed continuously on time of execution of a certain test. The same metric can be extended to rack level or data centre level. In this case, the imbalance considers the average temperature of the CPU of each node-group or rack respectively.

Rack cooling index high (HI) and low (LO): provide information about the distribution of temperatures in a group of nodes to detect if they are operated in an acceptable range according the limits recommended and allowed by certain standard. This metric was originally proposed by Herrlin as an indicator of how effectively the racks are cooled [34]. Combination of this metric with CFD will provide a general overview about where are the points of a node-group where temperature is higher than expected. The early detection will permit to implement strategies for minimizing energy consumption. For instance, in case of a data centre operator this approach is useful to manage the workload foreseen which will be the energy performance.

$$RCI_{HI} = 1 - \frac{\sum \left(T_{rack,x} - T_{max-rec}\right)_{T_{rack,x} > T_{max-rec}}}{(T_{max-all} - T_{max-rec}) * n}, \tag{14}$$

$$RCI_{LO} = 1 - \frac{\sum \left(T_{min-rec} - T_{rack,x}\right)_{T_{rack,x} < T_{min-rec}}}{(T_{min-rec} - T_{min-all}) * n}, \tag{15}$$

where, $T_{rack,x}$ is the temperature at rack air intake (average of node-group), in a certain time step, max means maximum, min means minimum, rec means recommended, all means allowed and n is the total number of node-groups. The same metric can be adapted to node-group level. Then $T_{rack,x}$ is substituted by $T_{NG,x}$ that is the node air intake temperature in a certain time step.

4.3 Application Power Model

Application profiling is generally based on a few parameters of the application and as said before the main concept behind an application profiling is the optimization of the execution of the application and the identification of problems at runtime.

The main issue when modelling the power of application is to validate it since it is impossible to directly measure its power with a watt meter. Therefore, one needs to make some assumptions to correlate the power drained by the machine with the one dissipated by the applications that it executes. In our case, we assume that the power of each application running on a machine is independent and that they can be aggregated in order to sum the total power consumption of the entire machine as follows:

$$P_{mac} = \sum_{pid \in RP} P_{pid}, \tag{16}$$

where P_{mac} and P_{pid} are, respectively, the machine and process power, and RP is the set of all processes currently running on the machine. Based on this conjecture one can validate its model by using a power meter and comparing the measured power with the sum of all the processes' estimation. Some authors prefer to consider the idle power to be independent of the process, we believe that if the machine is turned on, it must be running at least one process, even if it is a kernel process and this process needs to account the idle power, otherwise this machine should be shut down.

Good models need accurate power indicators, i.e. variables. For this purpose, we implemented an open source library, namely *libec* [24], with several power indicators which uses system information and hardware performance counters to generate high level variables. At first the power estimation is done based on a CPU proportional model given by:

$$P_{pid} = \frac{P_{min}}{RP} + (P_{max} - P_{min})\, CPU\_Usage_{pid}, \tag{17}$$

where RP is the number of running processes in a given time, P_{min} and P_{max} are the minimum and maximum power dissipated by the machine, CPU_Usage_{pid} is the CPU time usage of process pid and P_{pid} is the estimated power of the same process. Although the CPU is responsible for most of the power dissipated on a machine, more accurate model will be available soon, considering not only the CPU, but also memory, network and disk usage.

5 Validation of the CoolEmAll Approach

In this section we present description of the validation approach, provide detailed definition and evaluation of CoolEmAll trials for SVD toolkit, specifying their settings and parameters, and, present results of these trials.

5.1 Validation Approach

The general purpose of validation scenarios is to validate and verify the CoolEmAll approach, demonstrating its capability to optimize ComputeBoxes under various boundary conditions, i.e. increase their energy and cooling efficiency. The scenarios execute and evaluate experiments on physical testbed and simulated environment using Module Operation Platform (MOP) and SVD Toolkit, respectively. Experiments are defined by boundary conditions (fixed parameters) and variable parameters that can be changed independently at the beginning of the experiment, and can be adapted after each execution-step. The selection of these parameters depends on particular purpose and focus of the optimization. The general optimization process, described according to Deming optimisation life-cycle, follows *plan*, *do*, *check* and *act* phases. The purpose of the *plan* phase is to establish the objectives of the experiment necessary to deliver results in accordance with the expected output. The *do* phase implements the plan, and execute the experiment according to settings and parameters of the plan phase. During the *check* phase, the actual results of the

Table 2 Executing experiment in simulated environment (Optimization loop by simulation)

Phase	Description	SVD Components
Plan (prepare experiment)	DEBB (HW type, profile, geometry, position)	Var
	(Re)Arrangement	Var
	Workload	Var
	Policies	Var
	Cooling	Var
Do (simulate experiment)	Simulate workload	DCworms
	Simulate CFD	COVISE / CFD Solver
Check (visualize & assess experiment results)	Monitoring experiment	MOP Database
	Assessing results of experiment	Metric calculator
	Vizualize results	CoolEmAll Web-GUI
Act	Decide on changes	Done by human

experiment are visualized and analysed. The analysis involves comparison of the experiment results against expected results. During the *act* phase, deviation between actual and planned experiment results is assessed, and corrective actions to improve them are determined.

The adapted optimization process within the CoolEmAll can be described as follow:

- Plan: during the plan phase, user decides on focus of optimisation and on parameters that are fixed or variable (marked as var) involving: Selection of DEBBs, (Re) arrangement of components, Selection of Workload, Selection of resource management and scheduling policies, Selection of cooling techniques
- Do: during the do phase, experiments with preselected settings are executed in simulation environment using DCworms and/or CFD component.
- Check: during the execution of experiments in simulated environment, after each simulation run, simulation results are written into the MOP database. During the check phase, collected simulation results are assessed and visualized
- Act: during the act-phase, user decides whether the experiment results are acceptable or not. If not, he/she determines parameter changes, to be applied in next optimisation-cycle iteration.

During the execution of experiments in simulated environment, several components of SVD-toolkit are involved. The overview on usage of SVD-toolkit components during the execution of experiments is presented in Table 2.

The above mentioned approach is detailed and elaborated in main use-cases and trials. The three main CoolEmAll Use-Cases aiming at validating SVD Toolkit and CoolEmAll approach are:

- Capacity management based on coupled simulations of dynamic workloads and heat transfer with the goal to select the optimal configuration of hardware and

management software for given application types factoring in performance and energy-efficiency constraints.
- Optimisation of rack arrangement in a server room using open data centre building blocks with the goal to find an optimal arrangement of racks and aisles containment to prevent hot and cold air mixing and minimise risk of hot spots.
- Analysis of free cooling efficiency for various inlet temperatures with the goal to find a maximum inlet temperature in which data centre can operate for given workloads.

These scenarios are detailed in the following subsections.

5.1.1 Capacity Management

The goal of the capacity management is to ensure that an IT capacity meets current and future business requirements in a cost-effective manner. In case of data centres it must include an analysis of both performance and energy efficiency for specific workloads. Using SVD Toolkit, data centre planners and operators who plan to extend or exchange IT equipment can analyse several options (at a CPU, server or rack level) to check if the required performance is delivered without exceeding predefined thermal envelopes and power usage limits. The goal thereby is to select the optimal configuration of hardware and management software for given application types factoring in performance and energy and cooling efficiency constraints.

Capacity management is a common process for IT managers. CoolEmAll will enable an unprecedented level of analysis of this process through the integration of IT equipment and infrastructure simulations. The project software will enable users to simulate the execution of specific workloads on a range of IT hardware configurations. This will then be seamlessly combined with simulations of the resulting thermal processes. Users will also be able to model dynamic processes such as changes of workload in time or impact of management policies.

SVD Toolkit provides advanced support to capacity management by combining simulations of hardware, applications, and heat transfer. It introduces new higher levels of customization of workloads, application profiles, management policies, and hardware models. In this way data centre operators can analyse energy efficiency of IT equipment for HPC, service-based, and virtualized workloads. SVD Toolkit also allows users to observe impact of management policies, for instance effect of consolidation on energy consumption but also temperature distribution and heat transfer. More details of capacity management steps and examples of analysis are presented in Sect. 5.3.1.

5.1.2 Optimisation of Rack Arrangement in a Compute Room Using Open Data Centre Building Blocks

One of the most common problems in data centres is ensuring that the heat produced by servers is dissipated as efficiently as possible. In most cases, servers are located in racks arranged in rows within a server room. The efficiency of the cooling systems within the facility is heavily dependent on a range of factors including the arrangement of the racks, their heat density, placement of aisle containment, and how hot air is ducted from cabinets. Using SVD Toolkit, optimisation of existing or new facilities can be supported by simulating how various arrangements of the equipment in a computing room affect heat transfer. The goal thereby is to find an optimal arrangement of racks and aisles containment to prevent hot and cold air mixing and minimize risk of hot spots.

Although this is a quite common use case in today's data centres and DC planning software, the CoolEmAll approach provides a number of advantages. First of all, a server room model can be built upon predefined open data centre building blocks. These building blocks will be based on a specification designed by CoolEmAll and freely available from the project website. These building blocks will include on one hand low-power processor servers and, on the other hand, larger data centre modules such as shipping containers. This approach should facilitate and speed up the process of server room design and rearrangement. CoolEmAll's approach will also enable users to perform simulations in an interactive manner to see results quickly after introducing changes. Finally, an advanced 3D visualisation environment will allow users to modify a server room and watch results in an intuitive and detailed way.

5.1.3 Analysis of Free Cooling Efficiency for Various Inlet Temperatures

Data centre designers and operators are increasingly investigating the benefits of so-called free cooling (the use of outside air rather than mechanical cooling systems) to dissipate heat produced by IT equipment. This approach is closely coupled to another trend – raising the operating temperature (inlet temperature) of the facility (according to recent ASHRAE recommendations concerning server inlet temperatures). Raising operating temperatures sufficiently could allow a data centre planner to design a facility that does not require expensive mechanical chillers which has significant capital and operating costs implications. However, the use of such approaches has traditionally been held back by concerns over impacts on uptime and availability (if IT equipment gets too hot it can fail). Thanks to the predefined building blocks developed in the project, SVD Toolkit users will be able to analyse the impact of input temperature (and humidity) easily, allowing to find a maximum inlet temperature in which a data centre can operate safely for given workloads.

5.2 Testbed

Energy efficient operation of servers requires infrastructure allowing monitoring, controlling and managing cluster servers energy efficiently, adapting to fluctuating resource-demands, applications and environmental conditions. In this section we describe RECS servers with integrated monitoring and controlling capabilities, and

Table 3 Physical interfaces of the RECS cluster system

Connector/Button	Placement
USB	On each baseboard and two at the front panel of the server enclosure (for Compute Node 9)
2x SATA	On each baseboard
VGA	On each baseboard and one at the front panel of the server enclosure (for Compute Node 9)
18x Gigabit Ethernet	Front panel of the server enclosure
Fast Ethernet for monitoring	Front panel of the server enclosure
Power connector 12V	Back side of the server enclosure and on each baseboard
Power & reset button	On each baseboard
Control buttons for monitoring	Front panel of the server enclosure
LCD display for monitoring	Front panel of the server enclosure

provide an architecture allowing monitoring and managing rack consisting of several RECS and other components.

As described in the CoolEmAll deliverable 3.1 [15], the cluster server RECS consists of 18 single CPU modules, each of them can be treated as an individual PC. The mainboards are COM Express based CPU modules, each mounted on a standardized baseboard which makes it possible to use every available COM Express mainboard that has the "basic" size. In CoolEmAll we will evaluate which CPU module will be the best for each particular use-case. Each baseboard is connected to a central backplane. This backplane has two functions, first it forwards each Gigabit Ethernet Network of the CPU modules to the front panel of the server, and second it connects the baseboards' microcontrollers to the central master-microcontroller. For debugging purposes it has been quite useful in the past to have direct access to single mainboards, therefore every baseboard has several connectors as listed in Table 3.

All components within the cluster server share a common Power Supply Unit (PSU) that provides 12 V with a typical efficiency of more than 92 %. The several potentials needed for the mainboard chipset, CPUs and other components are provided by both, the baseboards and the mainboard potential transformers in the cluster server itself.

The novel monitoring approach of the RECS Cluster System is to reduce network load, avoid the dependency of polling every single compute node at operation system layer and build up a basis on which new monitoring- and controlling-concepts can be developed. Therefore the status of each compute node of the RECS Cluster Server is connected to an additional independent microcontroller in order to manage the measured data. The main advantage of the RECS Cluster System is to avoid the potential overheads caused by measuring and transferring data, which would consume lots of computing capabilities; in particular in a large-scale environment this approach can play a significant role. On the other hand, the microcontrollers also consume additional energy. Comparing with the potential saved energy, it is expected that the additional energy consumption could be neglected. This microcontroller-based monitoring architecture is accessible to the user by a dedicated network port and has to be read out only once to get all information about the installed computing nodes. If a user monitors e.g. 10 metrics on all 18 nodes, he would have to perform 180 pulls which can now be reduced to only one. This example shows the immense capabilities of a dedicated, aggregating monitoring architecture.

The monitoring architecture is realized by a master-slave microcontroller architecture which collects data from connected sensors and reads out the information every mainboard provides via SMBus or I C. Each baseboard is equipped with a thermal and current sensor. All sensor data are read out by one microcontroller per baseboard which acts as a slave and thus waits to be pulled by the master microcontroller. The master microcontroller and thus the monitoring- and controlling-architecture, are accessible to the user by a dedicated network port and additionally by a LCD display at the front of the server enclosure.

Additionally to the monitoring approach, the described infrastructure can be used to control every single compute node. Right now it is possible to virtually press the power- and reset-button of each mainboard. It is even possible to have a mixed setup of energy consumption where some nodes are under full load, others are completely switched off and some nodes are waiting in a low-energy state for computing tasks.

5.3 *Analysis and Optimization of Data Centre Efficiency*

5.3.1 Capacity Management

SVD Toolkit enables data centre operators and IT equipment suppliers to perform capacity management on various levels data centre architecture taking into account all major aspects that impact on energy efficiency. Users can study efficiency of specific IT equipment configuration with respect to various aspects such as characteristics of the workload, types of applications to be executed, management policies, and parameters of cooling devices.

The usual steps that are performed to do capacity management using SVD Toolkit are as follows:

1. **Selection of a DEBB.** Users select existing data centre building blocks from a repository to use them for simulations or they can use the DEBB Configurator tool to create to blocks (e.g. a rack filled with a given model of servers).
2. **Selection of workload.** Users select a workload to simulated in an infrastructure. They can adjust its level (e.g. % of utilization) or details (e.g. number of tasks per time unit, size of tasks, etc.). Workload can be modelled using probabilistic distributions and can reflect various situations in a data centre such as variability of load.
3. **Selection of application models.** In addition, users can select models of specific applications associated to tasks of the workload. They can select models of cloud

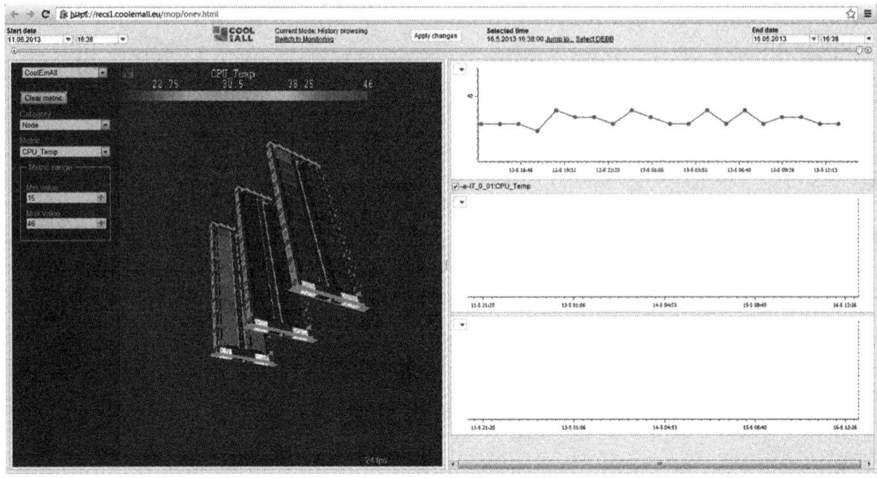

Fig. 8 Visualization of node temperatures in a single RECS system

or High Performance Computing applications depending on a purpose of the data centre. Application profiles describes details such as detailed characteristics of application (e.g. CPU- or IO-intensive) and its impact on power usage.

4. **Selection of management policy.** Users can select a management policy to model how applications in their data centre are distributed. For example, they can investigate various workload consolidation schemas that may have significant impact on heat distribution in a data centre.
5. **Running workload simulation.** Users run simulation of selected workload on IT hardware. As a result they obtain execution times, resource utilization, power usage in time, and further details needed to further analysis and heat transfer simulations.
6. **Visualization of results.** Users can watch obtained results using 3D model of simulated data centre building blocks and charts of metrics values in time. Example of visualization for a single racks with three RECS systems is illustrated in Fig. 8. Users can analyze values of resource utilization, power usage, and energy-efficiency metrics for selected parts of hierarchy. If needed users can come back to previous steps to change for instance hardware configuration or workload and see updated results (after repeating workload simulations). Additionally, two different configurations can be compared using SVD Toolkit visualization component.
7. **Calculation of airflows and temperature distribution for selected time points.** Users can select interesting point in time (e.g. peak load, consolidation of workload) and start a heat transfer simulation for this point in time.
8. **Analysis of results.** Users can watch results in the same way as in step 6. but including data about airflows and temperature distribution. Results can be also compared for various DEBBs and workloads using statistical data and graphical charts. If validation of results is needed users can compare simulation results with real measurements.

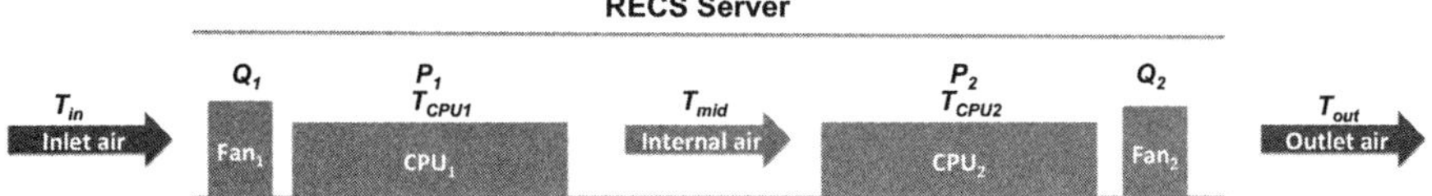

Fig. 9 Airflow through a couple of nodes in a RECS system

9. **Calculation of a report with values of metrics.** In addition to detailed analysis using graphical tools in previous steps, users obtain a report with values of important metrics such as total energy consumption, maximum power usage, mean/max temperatures, temperature/heat imbalance, nodes/rack cooling index, performance per Watt/Joule, pPUE, and others (see Sect. 4 for detailed list of selected metrics).

Of course, users can switch between these steps more times, e.g. to analyze several diverse workloads, application types, management algorithms, or hardware architectures.

In this Section we present an example of capacity management performed for a rack consisting of RECS systems. Architecture of these systems is not typical as presented in 5.2. Computing nodes are located with a high density and have separated fans. Moreover, airflow goes through couples of nodes located closer to the air inlet (called inlet nodes) and air outlet (outlet nodes). The airflow going through a couple of RECS nodes is presented in Fig. 9 whereas thermodynamic formula describing this flow is given in (18) (for more details see [17]).

$$T_{out} = T_{in} + \delta_1 \frac{P_1}{\rho \cdot Q_1 \cdot C} + \delta_2 \frac{P_2}{\rho \cdot Q_2 \cdot C}, \tag{18}$$

Proposed architecture provides some advantages (such as better energy- and resource-efficiency), however causes also additional dependencies and heat effects. Therefore, analysis of performance, energy consumption and related heat transfer processes is important even on the level of particular chassis of a rack. Figure 10 illustrates one of possible extreme loads of RECS. In this case, inlet nodes are idle (but switched on) while outlet nodes loaded. In other cases nodes can be also switched off and obviously load can be distributed in a different way. All these facts affects total energy usage and temperature distribution as switching nodes on and off causes also switching on and off fans. As presented in [17] for various configurations we observed differences in outlet temperatures depending on states and level of loads.

For instance, between the state presented in the figure and its opposite configuration (inlet nodes loaded, outlet nodes idle) are negligible. However, for the latter state are much higher (2–2.5 C) than for state in which idle nodes are switched off. Similarly, for switched off inlet nodes are significantly higher then for switched off outlet nodes (0.6–2.6 C). Interesting case is the difference between inlet nodes idle and switched off. For the highest load outlet temperatures are higher in the latter case (by around 0.5 C) than in former case while for lower loads opposite occurs. For

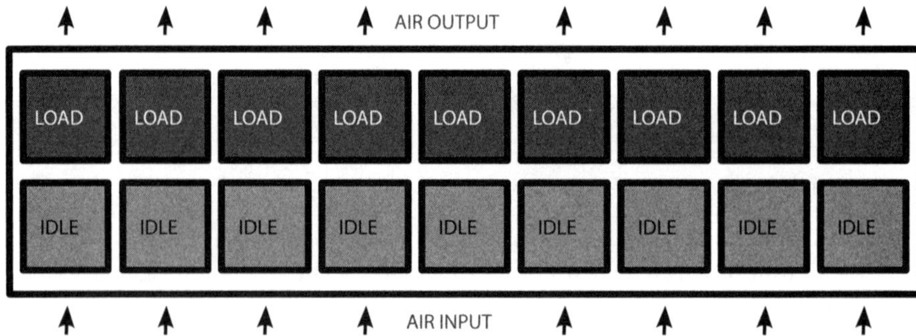

Fig. 10 Example of RECS load: inlet nodes idle, outlet nodes loaded

loads 0.75, 0.5, 0.25 and 0.125, outlet temperature in case with switched off is lower than in state 1 by 0.3, 1.0, 1.1 and 1.5 C, respectively. This uncommon behavior can be explained by a support in removing hot air by a second fan of idle node in a state with idle nodes on. If load of the outlet node decreases gain from additional fan is reduced compared to heat dissipated by the idle node. Additionally, we noticed usual increase of temperatures for nodes under significant load close to measurement points (0.1−0.7 C). As it also happened for inlet temperatures it suggests that this change is caused by heat dissipated in other ways than passed by flowing air.

Impact of Workloads and Management Policies
The first phase to perform the capacity management using the CoolEmAll approach is modelling and simulation of a workload execution in given IT hardware configuration. Based on the models obtained for the considered set of resources and applications we evaluated a set of resource management strategies in terms of energy consumption needed to execute four workloads varying in load intensity (10 %, 30 %, 50 %, 70 %). The differences in the load were obtained by applying various intervals (3000, 1200, 720 and 520 s, respectively) related to submission times of two successive tasks. In all cases the number of tasks was equal to 1000. Moreover, we differentiated the applications in terms of number of cores allocated by them and their type. Further details of the applied workloads can be found in [42].

For these workloads we defined two resource management policies that take into account differences in applications and hardware profiles by trying to find the most energy efficient assignment. The first policy-Energy Usage Optimization-assumes that there is no possibility to switch off unused nodes, thus for the whole time needed to execute workload nodes consume at least power for idle state. Taking into account that the system is running al the time, first we try to assign tasks to the nodes in the manner that results in the lowest increase of energy consumption for the given type of node and class of application. In other words, we investigate (for the given workload a resource configuration) whether the energy gain from executing the particular application on the given type of node can compensate possible energy losses from prolongation of the overall workload execution time. To evaluate this approach, tasks have to be assigned to the nodes of type for which the difference

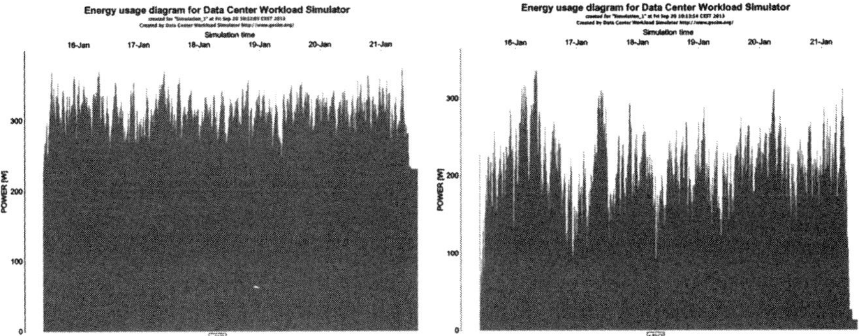

Fig. 11 Power usage in time using the EN-OPT (*left*) and EN-OPT-NODE-ON-OFF policy (*right*)

between energy consumption for the node running the application and in the idle state is minimal. As mentioned, we assign tasks to nodes minimizing the value of expression:

$$(P - P_{idle}) * exec\_time, \tag{19}$$

where P denotes observed power of the node running the particular application and $exec\_time$ refers to the measured application running time.

The second policy-Energy Usage Optimization with switching off unused nodes-makes the assignment of task to the node, we still take into consideration application and hardware profiles, but in that case we assume that the system supports possibility of switching off unused nodes. In this case the minimal energy consumption is achieved by assigning the task to the node for which the product of power consumption and time of execution is minimal. In other words we minimized the following expression:

$$P * exec\_time, \tag{20}$$

All tasks were assigned to nodes with the condition that they can be assigned only to nodes of the type on which the application was able to run (in other words-we had the corresponding value of power consumption and execution time). Differences between power usage in time for these both policies are presented in Fig. 11. It can be easily seen that that peaks are on the same level whereas in periods with lower load the Energy Usage Optimization with switching off unused nodes provides significantly lower power usage.

The last considered by use case is a modification of the random strategy. We assume that tasks do not have deadlines and the only criterion that is taken into consideration is the total energy consumption. In this experiment we configured the simulated infrastructure for the lowest possible frequencies of CPUs. The experiment was intended to check if the benefit of running the workload on less power-consuming frequency of CPU is not levelled by the prolonged time of execution of the workload. The values of the evaluated criteria are as follows: workload completion

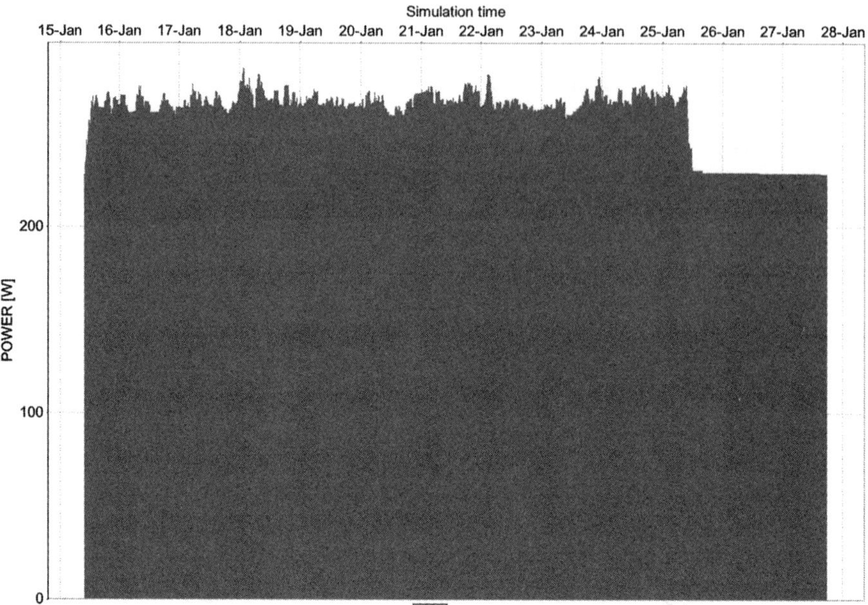

Fig. 12 Power usage in time using the DFS policy

time: 1 065 356 s and total energy usage: 77.109 kWh. As we can see, for the given load of the system (70 %), the cost of running the workload that requires almost twice more time, cannot be compensated by the lower power draw. Moreover, it can be observed that the execution times on the slowest nodes (Atom D510) visibly exceed the corresponding values on other servers. Compared to the two previous policies power usage is less variable with lower peak usage however without periods with significantly lower power usage as can be seen in Fig. 12.

As we were looking for the trade-off between total completion time and energy usage, we were searching for the workload load level that can benefit from the lower system performance in terms of energy-efficiency. For the frequency downgrading policy, we noticed the improvement on the energy usage criterion only for the workload resulting in 10 % system load. For this threshold we observed that slowdown in task execution does not affect the subsequent tasks in the system and thus the total completion time of the whole workload. More details can be found in [42].

Analysis of Heat Transfer

To understand heat-flow distribution within a RECS, a CFD simulation was done. Figure 13 presents the temperature distribution and air-flow inside RECS geometry equipped with Intel i7 nodes, and the corresponding initial and boundary conditions.

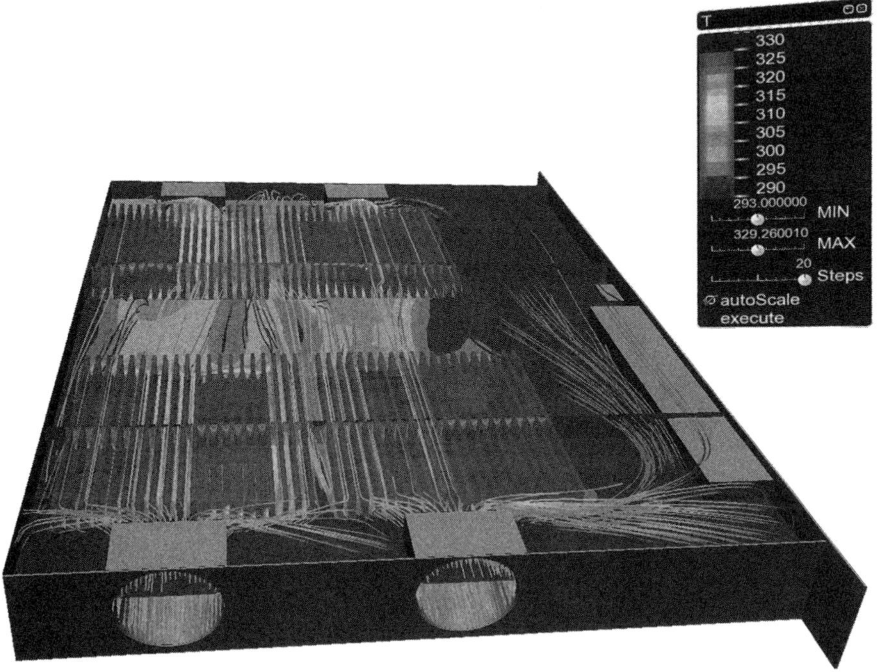

Fig. 13 Heat- and air-flow distribution within one quarter of the RECS of i7 nodes, load at 50 W (each node), room-temperature 22.5 °C

During the tests of the SVD toolkit several simulations were done. For speedup purposes and symmetry reasons there was one quarter of a standard RECS simulated, consisting of 4 nodes of i7, running at 50 W load each node. It is obvious that the temperature is not evenly distributed inside the geometry. The channels of the heat sink which are blocked are facing a much higher temperature than the parts of the heat sinks where the flow can easily pass through. Also it is easily visible how the flow heats while passing over the heat sink. This is done by observing the temperature at different places of the heat sinks. The locations downstream have a considerably higher temperature. This shows how the heat is transferred from the heat sink to the flow. If the flow is observed parallel to the temperature distribution it is easily visible how flow and heat transfer a coupled. In regions where there is hardly any flow visible the temperature is high because there is almost no heat transferred. This fact leads to higher temperatures at places of the heat sinks where the actual flow channels are blocked. However, comparing the temperature of the heat sink (229 K) with the maximum permissible operating range of CPU (378 K)-we can see that it is still within the permissible range.

Next it is a good idea to have a comparing look at the RECS cooled with air at higher input temperature (27.5 °C), presented in Fig. 14.

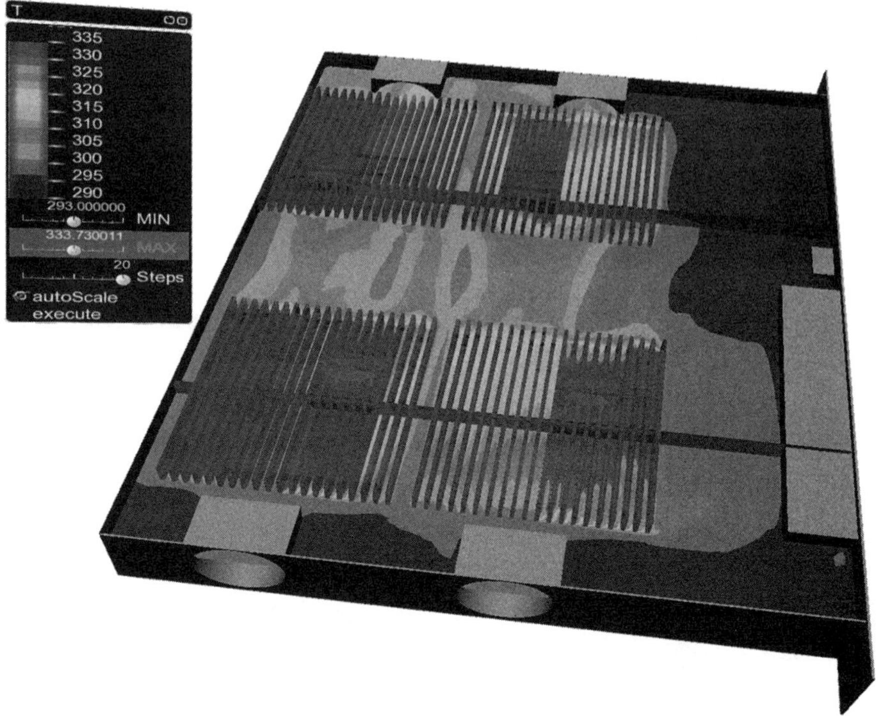

Fig. 14 Heat- and air-flow distribution within one quarter of the RECS of i7 nodes, load at 50 W (each node), room-temperature 27.5 °C

Again there are no surprised compared to the temperature distribution at lower temperature levels. The temperature downstream again is higher than upstream as the flow heats up while passing over the heat sinks. This was expected by earlier simulation. But again, the maximum temperature is about the value higher the inlet temperature is increased. The deviation is caused by keeping the wall temperature constant.

As a conclusion, it can be said, that the RECS at higher temperatures behaves exactly as it behaves at lower temperatures. This means that an increase in inlet or cooling fluid temperature leads directly to an increase in component temperature. So it is possible to increase the inlet temperature of the cooling fluid for energy efficiency reasons as long as the maximum component temperature for safe and stable operation is not reached.

5.3.2 Analysing Cooling Efficiency in Compute-room

In this section we describe scenarios showing benefits of SVD Toolkit to optimize cooing and energy-efficiency in data-centres, and present recently achieved results.

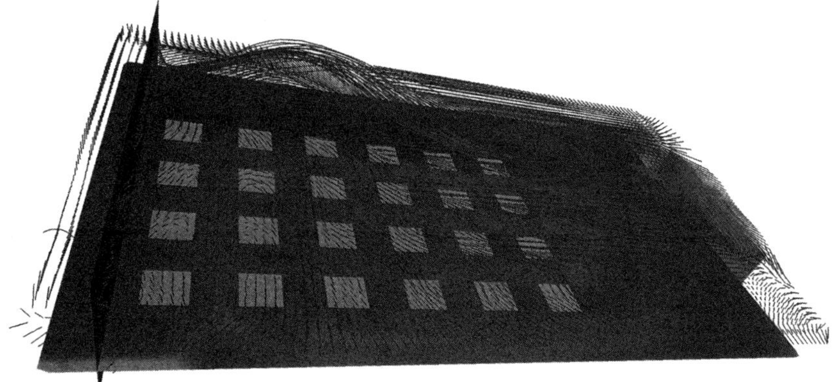

Fig. 15 Heat and airflow simulation within the compute-room with uneven distribution of heat

Evaluation results presented in this section refers to optimization of rack arrangement in a compute room use-case, described in 5.1.2, assessing cooling efficiency in compute room, described in [60]. Basic case evaluated during the development of the SVD Toolkit was a generic compute room: 12 m wide, 20 m long and 3 m high. Inside there are 24 racks located, placed in one half of the room to achieve an uneven distribution. Air is considered as a cooling fluid and enters the rooms via the tops of the racks. The outlet for the cooling fluid is located in a side wall of the room on the opposite site compared with the racks [60].

Figure 15 presents overview of the flow in the room. As one can see the air enters trough the grey coloured squared resembling the top of the racks and exits through the grey opening to the right, with velocity distribution represented by streamlines and velocity vectors. Colour resembles the values, blue for low values and red for high values. Heat distribution is represented by the colour of the cutting plane. The colours are chosen the same way as for flow. In the upper half of the picture one can see an accumulation of heat although there are almost no racks located. This proves the use of a tool like the SVD Toolkit viable, as nobody had expected heat accumulating in that place.

Figure 16 presents heat and airflow distribution for another compute-room configuration. Colours correspond to temperature, according to colour-chart depicted. We can observe that there are several hotspots in the compute-room due to insufficient airflow. The upper left and the most right rack in the figure have quite high temperatures, indicating deficits on airflow circulation leading to heat accumulation.

In this fashion it is possible to model and assess all types of compute rooms. In particular it is possible to simulate compute rooms with hot and cold aisles, allowing to evaluate different cooling methods.

Fig. 16 Heat and airflow simulation within the compute-room

6 Business Impact

The defining characteristic of the CoolEmAll project is that it will bridge this traditional gap between IT and facilities approaches to efficiency. The main outcomes of CoolEmAll project are: Simulation, Visualization and Decision support (SVD) Toolkit; Data centre Efficiency Building Blocks (DEBBs); Enhanced data centre efficiency metrics; and Module Operation Platform (Not primary outcome but will allow for visualisation of monitoring data). Some commercial suppliers (most notably Data Center Infrastructure Management (DCIM) suppliers) and consultants have recently begun to take a more all-encompassing approach to the problem by straddling both IT and facilities equipment. However, few suppliers or researchers up to now have attempted to include the crucial role of workloads and applications. That is beginning to change, and it is likely that projects such as CoolEmAll can advance the state of the art in this area.

The consortium describes the SVD toolkit as a data centre modelling, simulation and decision supporting tool. Using a combination proprietary code plus elements of existing tools (COVISE and OpenFOAM), the SVD Toolkit will allow data centre planners to model the energy efficiency implications of physical placement of servers within the facility or different approaches to cooling. Some of these functions are found in existing DCIM tools; however the SVD Toolkit will add an applications and workload simulation functions not currently found in existing DCIM tools. How each of the planned components of the SVD Toolkit compares with existing tools (based on information from the market assessment deliverable) is outlined below.

Computational fluid dynamics (CFD) creates a detailed mathematical model of airflows, temperatures and other environmental variables within a space. It is likely that the bulk of the SVD Toolkit will be developed from elements of the open source CFD application OpenFOAM and a simulation and visualization tool developed by project partner HLRS—The Collaborative Visualization and Simulation Environment (COVISE). However, a wide range of such tools exists and will be evaluated through the course of the project. (The consortium maintains that there currently is no one offering that covers all the functions the project plans to develop.)

However, most CFD modelling applications are not real-time tools (although some suppliers are trying to develop these) and are used for prediction rather than monitoring purposes. Many suppliers in this are still questioning how to use them with dynamic, virtualized environments in which IT heat output changes with varying workload. CoolEmAll will look to tackle this issue of real-time monitoring and how thermal characteristics change with workload. Suppliers of commercial CFD tools include Applied Math Modelling (CoolSim), Future Facilities (6SigmaDC), Innovative Research Inc (TileFlow), and Data Research (CoolitDC).

A range of tools already exists that can be used to help data centre owners and operators capture energy use information (for use in calculating data centre PUE for example). These include DC-Pro from the US EPA, the Data Centre Efficiency calculator from APC (Schneider Electric) or the BCS Data centre Strategy Group/Carbon Trust Data centre Simulator. These tools are limited in scope, however, and do not addresses the application or workload layer contribution in the granular way it is intended the CoolEmAll SVD Toolkit will.

The emerging, and important, data centre infrastructure management sector (DCIM) has some broad similarities with the holistic approach that underpins the CoolEmAll project. DCIM is difficult to define precisely. It is multifunctional, has many components, attempts to address various technical and business issues, and may consist of numerous subsystems that appear to duplicate or overlap with other systems. However, CoolEmAll is more focused on data centre planning and simulation than the operational focus that most DCIM systems take. DCIM suppliers include nlyte Software, iTRACS, Schneider, Emerson Network Power, CA Technologies and Modius.

The consortium also plans to include thermal and energy-aware resource management functions in the SVD Toolkit—moving virtual workloads between servers or even between data centres for energy-efficiency reasons. Tools such as VMware vSphere, or Ovirt or Platform VM Orchestrator, enable VMs to be managed according to set policies and some of these suppliers have integrated their technology with existing DCIM tools. However, the level of integration between traditional DCIM functions and virtual machine management is still very immature, and CoolEmAll hopes to advance the state of the art in this area.

The SVD Toolkit may also overlap with technology separate but closely aligned to DCIM, which is sometimes called DCPM. DCPM tools, such as Romonet's Prognose and Lumina Decision Systems' Analytical Data Center Capacity Planning Tool, can be used for detailed data centre planning. Prognose, for example, allows a user to create a detailed model of a facility and then run a mathematical simulation to predict energy and cost performance. By changing the model and rerunning the simulation, users can experiment with different data centre designs or operational strategies.

A data centre designer might use the tool to compare the predicted performance of different cooling technologies, and to see how the answer might change in different climate zones. A data centre operator with an active site might use the model to estimate cost and energy savings from a change in temperature set point or an efficiency retrofit. The BCS and Carbon Trust developed an open source DCPM tool, called the Data Centre Simulatoriii. (Romonet's Prognose tool was originally based on the Data centre Simulator, but has been significantly developed since.) It is possible that some DCPM functions could be added to the SVD toolkit by integrating with elements of the open source Data centre Simulator.

Other data centre monitoring and management tools that should be considered when developing the SVD Toolkit include those developed by IT suppliers—most notable Intel (Data Centre Manager) but also server suppliers including HP and Dell. These tools include functions for monitoring and measuring power and temperature, but also some degree of control including power capping.

The other main outcome of the project will be a set of hardware designs. These designs/blueprints are defined by the project as Data centre Efficiency Building Blocks (DEBBs). The DEBB is effectively an abstraction for computing and storage hardware and describes energy efficiency of data-centre building block on different granularity-levels.

Where relevant, the consortium will look to collaborate with existing standards bodies and other organisations in the development of existing and new metrics. This will help ensure that the technology that results from the project is useful to the wider data centre industry, both in terms of research but also in commercial adoption by data centre technology suppliers, and data centre owners and operators.

The project will look to engage with organisations developing new metrics such as Green Grid, as well as efforts to harmonize the development of metrics internationally through groups such as the Global Harmonisation of Metrics Task Force (EC, METI/GIPC, US EPA/DOE and Green Grid). Relevant developments in metrics include the likelihood that Power Usage Effectiveness will be accepted as an ISO Standard in 2012/2013. This may have implications for how PUE is measured and reported which should be integrated into the metrics work within the project.

7 Summary

In this chapter we presented an overview of CoolEmAll models and tools that can be used for optimization of data centres' energy-efficiency and reduction of their carbon footprint. The CoolEmAll's Simulation, Visualization and Decision Support Toolkit (SVD Toolkit) enables careful planning of both data centre hardware and software configuration. To this end, the project took a holistic approach by integrating effects of cooling, IT equipment, and workloads into the analysis. We presented some examples of the SVD Toolkit application to scenarios such as capacity management in a data centre as well as optimization of cooling efficiency in a server room.

The future work includes developing a larger spectrum of hardware models, workload profiles, and power consumption models to enable comparison of various alternatives of data centre equipment. We are also working on energy- and thermal-aware management policies, a final set of benchmarks, and further evaluation of energy-efficiency metrics. The final prototype of CoolEmAll models and tools will also contain integrated web-based user graphical interfaces allowing users to perform the modeling and simulation flow remotely. In the last phase of the project we also plan more tight collaboration with end users and suppliers.

Acknowledgment The results presented in this chapter were funded by the European Commission under contract 288701 through the project CoolEmAll.

References

1. Parallel Workload Archive. (2006). URL http://www.cs.huji.ac.il/labs/parallel/workload/.
2. Green grid data center power efficiency metrics: PUE and DCIE. Tech. rep., The Green Grid (2008).
3. Productivity indicator. Tech. rep., The Green Grid (2008).
4. Carbon usage effectiveness (cue): A green grid data center sustainability metric. Tech. rep., The Green Grid (2010).
5. ERE: A metric for measuring the benefit of reuse energy from a data center. White paper, The Green Grid (2010).
6. Enhancing the energy efficiency and use of green energy in data centers. Tech. rep., Green IT Promotion Council (2011).
7. Harmonizing global metrics for data center energy efficiency. Global taskforce reaches agreement on measurement protocols for PUE continues discussion of additional energy efficiency metrics. Tech. rep., Global Metrics Harmonization Task Force (2011).
8. New data center energy efficiency evaluation index. dppe (datacenter performance per energy). measurement guidelines (ver 2.05). Tech. rep., Green IT Promotion Council (2011).
9. On-site energy generation efficiency (oge) and energy carbon intensity (eci). Tech. rep., Green IT Promotion Council (2011).
10. Recommendations for measuring and reporting overall data center efficiency. version 2 – measuring PUE for data centers. Tech. rep., Data Center Efficiency Task Force (2011).
11. Water usage effectiveness. Tech. rep., The Green Grid (2011).
12. Global taskforce reaches agreement on measurement protocols for GEC, ERF, and CUE – continues discussion of additional energy efficiency metrics. Tech. rep., Global Metrics Harmonization Task Force (2012).
13. PUE (tm): A comprehensive examination of the metric. confidential report. White paper, The Green Grid (2012).
14. Aumueller, M., Schulze-Doebold, J., Lang, R., Rainer, D., Werner, A., Woessner, U., Wol, P.: COVISE User's Guide (2013).
15. vor dem Berge, M.: First definition of the flexible rack-level compute box with integrated cooling. Tech report, CoolEmAll (2012).

16. vor dem Berge, M., Christmann, W., Volk, E., Wesner, S., Oleksiak, A., Piontek, T., Costa, G.D., Pierson, J.M.: CoolEmAll - Models and tools for optimization of data center energy-efficiency. In: Sustainable Internet and ICT for Sustainability (SustainIT), pp. 1–5 (2012).
17. vor dem Berge, M., Da Costa, G., Jarus, M., Oleksiak, A., Piatek, W., Volk, E.: Modeling Data Center Building Blocks for Energy-efficiency and Thermal Solutions. Springer (2013).
18. Bąk, S., Krystek, M., Kurowski, K., Oleksiak, A., Piątek, W., Wąglarz, J.: GSSIM – A tool for distributed computing experiments. Scientific Programming **19**(4), 231–251 (2011). DOI 10.3233/SPR-2011-0332.
19. Bolla, R.: STF439 - global KPIs for energy efficiency of deployed broadband. In: ETSI Workshop on Energy Efficiency (2012).
20. Bosque, A., Ibañez, P., Viñals, V., Stenström, P., Llabería, J.M.: Characterization of Apache web server with Specweb2005. In: Proceedings of the 2007 workshop on MEmory performance: DEaling with Applications, systems and architecture, MEDEA '07, pp. 65–72. ACM, New York, NY, USA (2007). DOI 10.1145/1327171.1327179. URL http://doi.acm.org/10.1145/1327171.1327179.
21. CA: Web-page of the ca company (2014). URL www.ca.com.
22. Chetsa, G.L.T., Lefevre, L., Pierson, J.M., Stolf, P., Da Costa, G.: DNA-inspired scheme for building the energy profile of HPC systems. In: International Workshop on Energy-Efficient Data Centres, Springer (2012).
23. Christian, L., Belady, P.: Projecting annual new data center construction market size. Tech. rep., Microsoft Global Foundation Services (2011).
24. Cupertino, L.F., Costa, G., Sayah, A., Pierson, J.M.: Energy consumption library. In: J.M. Pierson, G. Da Costa, L. Dittmann (eds.) Energy Efficiency in Large Scale Distributed Systems, Lecture Notes in Computer Science, pp. 51–57. Springer Berlin Heidelberg (2013). DOI 10.1007/978-3-642-40517-4_4. URL http://dx.doi.org/10.1007/978-3-642-40517-4_4.
25. Dongarra, J.J., Meuer, H.W., Strohmaier, E., et al.: Top500 supercomputer sites. Supercomputer **67**, 89–111 (1997).
26. Donoghue, A.: Market assessment report. Tech report, CoolEmAll (2012).
27. Donoghue, A.: Preliminary exploitation plan. Tech report, CoolEmAll (2012).
28. Eddy, S.R., Wheeler, T.J.: Hmmer user's guide: Biological sequence analysis using profile hidden markov models (2013). URL http://www.hmmer.org/.
29. Electric, S.: Web-page of the data center genome project (2014). URL http://datacentergenome.com.
30. Facebook: Web-page of the open compute project (2014). URL http://www.opencompute.org.
31. Facilities, F.: Dc6sigma products of future facilities (2014). URL http://www.futurefacilities.com/.
32. Facilities, F.: Web-site of future facilities company (2014). URL http://www.futurefacilities.com/.
33. Georges, D.C., Helmut, H., Karin, H., Jean-Marc, P.: Modeling the Energy Consumption of Distributed Applications. CRC Press (2012).
34. Herrlin, M.: Rack cooling effectiveness in data centers and telecom central offices: The rack cooling index (RCI). In: ASHRAE Transactions [0001-2505], pp. 725 –731 (2005).
35. Hoyer, M., vor dem Berge, M., Volk, E., Gallizo, G., Buchholz, J., Fornós, R., L. Siso, W.P.: First definition of the modular compute box with integrated cooling. Tech report, CoolEmAll (2012).
36. Innovative Research, I.: Tileflow product of innovative research inc. (2014). URL http://inres.com/products/tileflow/overview.html.
37. Iosup, A., Li, H., Dumitrescu, C., Wolters, L., Epema, D.: The Grid Workload Format (2006).
38. Jiang, T., Kipp, A., Cappiello, C., Fugini, M., Gangadharan, G., Ferreira, A.M., Pernici, B., Plebani, P., Salomie, I., Cioara, T., Anghel, I., Christmann, W., Henis, E., Kat, R., Lazzaro, M., Ciuca, A., Hatiegan, D.: Layered green performance indicators definitions. Project deliverable, GAMES project (2010).

39. Kipp, A., Jiang, T., Fugini, M., Salomie, I.: Layered green performance indicators. Future Gener. Comput. Syst. **28**(2), 478–489 (2012). DOI 10.1016/j.future.2011.05.005. URL http://dx.doi.org/10.1016/j.future.2011.05.005.
40. Krystek, M., Kurowski, K., Oleksiak, A., Piatek, W.: Energy-aware simulations with GSSIM. In: Energy Efficiency in Large Scale Distributed Systems (EE-LSDS), pp. 55–58 (2010).
41. Kundra, V.: Federal data center consolidation initiative. Memorandum for chief information officers, Office of Management and Budget of the USA, Washington, DC (2010).
42. Kurowski, K., Oleksiak, A., Piatek, W., Piontek, T., Przybyszewski, A., Weglarz, J.: DCWoRMS - a tool for simulation of energy efficiency in distributed computing infrastructures. Simulation Modelling Practice and Theory (2013). DOI 10.1016/j.simpat.2013.08.007. URL http://dx.doi.org/10.1016/j.simpat.2013.08.007.
43. Malone, C., Belady, C.: Metrics to characterize data center & IT equipment energy use. In: Proceedings of the Digital Power Forum (2006).
44. Mammela, O., Majanen, M., Basmadjian, R., Meer, H.D., Giesler, A., Homberg, W.: Energy-aware job scheduler for high-performance computing. Computer Science - Research and Development **27**(4), 265–275 (2012).
45. Mukherjee, T., Banerjee, A., Varsamopoulos, G., Gupta, S.K.S.: Model-driven coordinated management of data centers. Comput. Networks (2010).
46. Newcombe, L., Limbuwala, Z., Latham, P., Smith, V.: Data center fixed to variable energy ratio metric dc-fver. Tech. rep., BCS The Chartered Institute for IT (2012).
47. Prieto, J.L., Costa, G.D.: Energy and heat-aware classification of application. Tech report, CoolEmAll (2013).
48. Prieto, J.L., Gallizo, G., Oleksiak, A.: Validation scenarios, methodology and metrics. Tech report, CoolEmAll (2012).
49. Raghavendra, R., Ranganathan, P., Talwar, V., Wang, Z., Zhu, X.: No "power" struggles: coordinated multi-level power management for the data center. In: Proceedings of the 13th international conference on Architectural support for programming languages and operating systems, ASPLOS XIII, pp. 48–59. ACM, New York, NY, USA (2008). DOI 10.1145/1346281.1346289. URL http://doi.acm.org/10.1145/1346281.1346289.
50. Rathgeb, D., Volk, E.: First release of the simulation and visualisation toolkit. Tech report, CoolEmAll (2013).
51. Robert, B., Ali, N., Florian, N., de Meer, H., Giuliani, G.: A methodology to predict the power consumption of servers in data centers. Proceedings of the 2nd international conference on energy-efficient computing and networking (2011).
52. Romonet: Romonet products overview (2014). URL http://www.romonet.com/overview.
53. Shah, A., Krishnan, N.: Optimization of global data center thermal management workload for minimal environmental and economic burden. Components and Packaging Technologies, IEEE Transactions on **31**(1), 39–45 (2008). DOI 10.1109/TCAPT.2007.906721.
54. Slurm: Slurm workload manager (2013).
55. Stanley, J.R., Brill, K.G., Koomey, J.: Four metrics define data center "greenness". Tech. rep., Uptime Institute (2007).
56. Stansberry, M.: Data center industry survey results 2011. Tech. rep., Uptime Institute (2011).
57. Torque: Torque resource manager (2013).
58. Tsiombikas, J.: C-Ray simple raytracing tests (2008).
59. Volk, E., Piątek, W., Jarus, M., Costa, G.D., Sisó, L., vor dem Berge, M.: First definition of the hardware and software models. Tech report, CoolEmAll (2012).
60. Volk, E., Rathgeb, D., Oleksiak, A.: Coolemall – optimising cooling efficiency in data centres. Computer Science - Research and Development (2013). DOI 10.1007/s00450-013-0246-4.
61. Witkowski, M., Oleksiak, A., Piontek, T., Weglarz, J.: Practical power consumption estimation for real life hpc applications. Future Generation Computer Systems (2012).
62. Woessner, U., Volk, E., Gallizo, G.: Design of the CoolEmAll simulation and visualisation environment. Tech report, CoolEmAll (2012).
63. Yeo, S., Lee, H.H.: SimWare: A Holistic Warehouse-Scale Computer Simulator. Computer **45**(9), 48–55 (2012). DOI 10.1109/MC.2012.251.

Smart Data Center

Muhammad Usman Shahid Khan and Samee U. Khan

1 Introduction

All the internet services available these days are dependant and running in data centers. Companies like Google, Facebook, and Microsoft hosts millions of servers in their data centers to provide services to their users [19]. The enormous size of data centers leads to huge energy consumption. According to a news article, Google drew 260 MW of power in 2011 [6] that cost millions of dollars.

Recently, the researchers have focused on reducing the data center energy cost. The researchers have focused on migration of the workload from one geographical location to another to use the time and location dependent electricity prices [2] [21]. Similarly, researchers have also focused on the use of Uninterrupted Power Supply (UPS) in data centers to shave off the peak power demands [24]. UPS has also been used to safe the data center from the unexpected power outages. The power outages also cost millions of dollars to data centers. Amazon was hit a severe power outage in 2012 that cost Amazon millions of dollars [17].

The modern smart grid provides the needed electricity to the data centers. Smart grids provide different pricing schemes for electricity based on different time scales [10, 18]. Due to huge electricity demands, the data centers acquires electricity from grids using long term contracts in day ahead market. The long term contracts cost lower than the real time market price of electricity [18]. In this paper, we propose the idea to buy electricity from more than one smart grid. The local power grid will act as the main power source for the data center. However, data center will also be powered by the remote grid with the surplus power. The data center can purchase the available surplus power from remote grids at lower prices than local grid long term

M. U. S. Khan (✉) · S. U. Khan
Electrical and Computer Engineering Department,
North Dakota State University, 58102 Fargo, ND, USA
e-mail: ushahid.khan@ndsu.edu

S. U. Khan
e-mail: samee.khan@ndsu.edu

S. U. Khan, A. Y. Zomaya (eds.), *Handbook on Data Centers*,
DOI 10.1007/978-1-4939-2092-1_7

market and real time market prices. The sale of surplus energy is advantageous to the remote grids as the surplus energy is mostly wasted [14]. The amount of available surplus power could vary over time. UPS available in the data center for backup can be used to store surplus power from the remote grid or when price from the local grid is low. When the surplus power from the remote grid is insufficient or the price of electricity at local grid is high, the stored power in the UPS batteries can be used.

In this work, we have targeted the key problem in the data center that how to minimize the long term running cost of the data center? Several sub problems are investigated to answer the key problem. How much power should be purchased from the local grid in long term and real time price rates? How to efficiently use the available surplus power from the remote grid? How to best use the UPS for power saving and for backup while saving the life of battery for longer time? To optimally utilize the data center with multiple sources while minimizing the operational cost is really challenging task. There are numerous uncertainties both in power demand and supply side. The power demands of the data center are time varying and job dependent. Each job can consume different amounts of power as they may utilize different number of machines. Similarly on the supply side, availability of surplus energy is an uncertain and long term and real time prices from local grids can change with time.

Previous works on reducing the power consumption and cost of electricity for data center, assume the prior knowledge of the power demand to predict the future power demands [25]. The previous works do not consider the scenario of providing the power to the data center from multiple power grids. In contrast to the previous works, we aim to design efficient strategy to reduce the long term operational cost of the data center while having the constraints of dynamic power demand with no previous knowledge and uncertain availability of surplus power from remote grid.

We develop an algorithm titled "Smart Data center" to make a data center smarter using two stage Lyapunov optimization techniques. Smart Data center computes the amount of power to be purchased from the local grid in a long term contract. The amount of electricity to be purchased from the local grid on real time market rate and amount of the electricity to be stored and retrieved from the UPS are also computed by the Smart Data center algorithm. We analyze the performance of the Smart Data center algorithm through rigorous theoretical analysis in this work.

2 System Model

We assume a discrete time model for the working of a data center. The notations and their meanings in the model are presented in the Table 1. Time for the model is divided into k slots each of length T. The length T depends on the intervals provided by the grids in long term contracts. Each time slot is further divided into fine grained slots of length L. We also assumed that power demand of the data center $d(t)$ and available surplus power of remote grid $r(t)$ are random variables. The operations of the data center in a system model include following key decisions.

Table 1 Notations and their meanings

Notation	Meaning
t	Coarse grain time slot
τ	Fine grain time slot
$d(t)$	Power demand
$r(t)$	Surplus power at remote grid
$Pmax$	Maximum purchasing power capacity
$E_{full}(t)$	Electric power units purchased in long term for time t
$E_{lt}(t)$	Electric power units purchased in long term for time τ
$r(t)$	Units of surplus power purchased from remote grid
$p_{lt}(t)$	Unit price of electricity in long term market
$p_{rt}(t)$	Unit price of electricity in real time market
$p_s(t)$	Unit price of surplus power from remote grid
P_{grid}	Maximum capacity of local grid
$E_u(\tau)$	Level of power in the UPS
$E_u\ max$, $E_u\ min$	Maximum and minimum capacity of UPS
$D(\tau)$	Amount of power discharge from battery at time τ
$R(\tau)$	Amount of power (Re)charged in battery at time τ
η	Efficiency of the UPS

2.1 Long Term Power Purchase

The data center takes notes of the power demand *d(t)* and available surplus power at the remote grid *r(t)* at the start of each coarse grained time slot *t*. The data center is provided with a maximum threshold limit *Pmax* as a maximum purchasing power capacity. Based on the observations, the data center takes the decision that how much electric power units E_{full} *(t)* should be purchased from the local grid at price $p_{lt}(t)$ within the purchasing capacity at the start of coarse grain time slot. After the purchase, the data center divides the electric power units equally to be used in all the fine grain time slots.

$$E_{lt}(t) = \frac{E_{full}(t)}{L}. \tag{1}$$

For example, suppose the data center decides to purchase 720 *KW* when the length of the coarse grained time slot is one day and fine grained time slot is 1 h. In the above mentioned case, the data center will distribute the 500 *KW* equally, i.e., 500/24 = 30 *KW* for each fine grained time slot.

2.2 Real Time Power Purchase

We have assumed that the cost of the surplus power at the remote grid is lower than the local grid long term and real time power purchase. Whenever there is a surplus power available on the remote grid in time slot t, the data center tries to use it as much as possible. In case when surplus power is more than the power demand, the excess power is used to charge the UPS. At each fine grained time slot τ, the UPS will not be needed to charge or discharge if the sum of long term power purchase from local grid and surplus power from the remote grid is less than the total power demand from the data center.

$$E_{lt}(t) + r(\tau) \geq d(\tau). \tag{2}$$

Otherwise, if the power demand is more than the sum (left hand side of the Equation) than the data center has to make the decision to discharge the power from the batteries $D(\tau)$ of the UPS. If the UPS power is not enough for the remaining power demand, more electric power units $E_{rt}(\tau)$ are purchased from the local grid at real time price rate $p_{rt}(\tau)$ To balance out the equation, any surplus purchased power is used to charge the batteries of the UPS C(τ). We have an overall equation of the data center as

$$\begin{aligned} E_{lt}\ (t) + \ E_{rt}(\tau) + D(\tau) + r(\tau) - C(\tau) = d(\tau), \\ 0 \leq E_{lt}\ (t) + E_{rt}(\tau) \leq Pgrid. \end{aligned} \tag{3}$$

3 Constraints

There are a number of constraints that must be satisfied by the data center.

3.1 Purchasing Accuracy and Cost

The price of surplus electricity from the remote grid is lower than the electricity prices in the long term contract and real time market price rates from local grids.

$$p_{rt}(\tau)\ >\ p_{lt}\ (t) > p_s(\tau). \tag{4}$$

However, availability of surplus electricity from the remote grid is dynamic in nature. Similarly, the data center can purchase electricity from real time market but that is the most expensive. Therefore, the data center has to make a decision of purchase of electricity with accuracy to keep the overall cost of the electricity purchased to be minimized.

3.2 Data Center Availability

Let $E_u(\tau)$ be the level of the power in the UPS batteries at time τ. Power in the batteries of the UPS is affected by the efficiency of USP (dis)charging. We assumed that efficiency for discharging and charging $\eta \in [0,1]$ is same. The dynamics of the UPS power level can be expressed by the following equation

$$E_u(\tau + 1) = E_u(\tau) + \eta R(\tau) - \frac{D(\tau)}{\eta}. \tag{5}$$

To guarantee the availability of the data center in case of power outages, minimum level of power must be maintained in the batteries of the UPS. If the maximum power storage capacity of the UPS is $E_u\, max$ than we have

$$E_u\, min \ < \ E_u(\tau) < E_u\, max. \tag{6}$$

3.3 UPS Lifetime

At given time t, the amount of power that can be stored or retrieved from the batteries of the UPS is limited by their maximum amounts

$$0 \ \leq D(t) \leq D\, max, 0 \ \leq R(t) \leq R\, max. \tag{7}$$

The lifetime of the UPS is constrained within the number of cycles of UPS charging and discharging [24]. The operating cost of the UPS also depends upon UPS charging and discharging cycles. We assume that cost of UPS C_r is same in both cases of charging and discharging. If the purchase cost of a new UPS is $C_{purchase}$ that can sustain M_{cycles} than we have

$$C_r = \frac{C_{purchase}}{M_{cycles}}. \tag{8}$$

If the life of UPS is defined as Life, than the maximum number of times the batteries of the UPS are allowed to charge and discharge over a longer period of time [0, t − 1] and $t \in kT$, will be

$$N_{max} = \frac{M_{cycles} * kT}{Life}. \tag{9}$$

The variable kT is the total time for modeling, i.e., k coarse grain slots of length T. Therefore, N_{max} satisfies the following equation

$$0 \leq \sum_{\tau=0}^{t-1} \partial(\tau) \leq N_{max}. \tag{10}$$

In the above equation, $\partial(\tau)$ denotes the usability of the batteries of UPS in time τ. The variable $\partial(\tau)$ will be 1 if the discharge or recharge occurs otherwise the variable takes the zero value. The operational cost of the UPS can now be calculated as the product of usability of the batteries of UPS and cost of UPS in time slot t.

$$Cost\ of\ UPS_{opertional} = \partial(t) * Cr. \tag{11}$$

4 Cost Minimization

The operational cost of the data center at a fine grained time slot τ is the sum of the costs for purchasing electricity from the local grid, remote grid, and the operational cost of the UPS.

$$\begin{aligned} &Cost\ of\ data\ center_{opertiaonal}(\tau) \\ &= E_{lt}(t)p_{lt}(t) + E_{rt}(\tau)p_{rt}(\tau) + r(t)p_s(t) + \partial(t)Cr. \end{aligned} \tag{12}$$

In this work, we aimed at designing the algorithm that can make decisions by solving the following minimization problem

$$\min Cost\ of\ data\ center_{avg} \cong \lim_{t\to\infty} \begin{array}{l} \frac{1}{t}\sum_{\tau=0}^{t-1} Cost\ of\ data\ center_{opertiaonal}(\tau), \\ \forall t\text{: } Constraints\ (3)\ (6)\ (7)\ (8) \end{array} \tag{13}$$

5 Algorithm Design

We design our algorithm using the Lyapunov optimization technique to achieve the near optimal solution. The algorithm does not use the prior knowledge of power demand. To guarantee the availability of data center, the algorithm has to track the status of power level in the batteries of the UPS. Tracking the status of power level in the batteries is necessary as we want to ensure that each time the power is discharged or charged from the batteries of the UPS, there should be enough power remain in the battery that can be used during blackouts as backup. To track the battery power of the UPS, we use the supporting variable *X(t)* defined as follows:

$$X(t) = E_u(t) - \frac{VP_{max}}{T} - E_u \min - \frac{D max}{\eta}. \tag{14}$$

In the above equation, V is a control variable that ensures that whenever batteries of UPS is charged or discharged, the power in the batteries should lie in the minimum and maximum level. With increment in the time slot t, the variable $X(t)$ changes as

$$X(t+1) = X(t) + \eta R(\mathrm{t}) - \frac{D(\tau)}{\eta}. \tag{15}$$

We consider the constraint of availability of power level in the batteries of the UPS as a queue problem and transform the constraint into queue stability problem, similar to the work presented in [23]. We define the Lyapunov function to represent the scalar metric of queue congestion as

$$L(t) \cong \frac{1}{2} X^2(t). \tag{16}$$

We use the Lyapunov drift to stabilize the system that pushes the Lyapunov function towards lower congestion state. The Lyapunov drift over time period T is defined as

$$\Delta LD_T \cong L(t+T) - L(t)|X(t). \tag{17}$$

We obtained the drift penalty term by following the Lyapunov drift penalty framework [5]. In every time frame of length T, the Smart Data center algorithm makes a decision to minimize the upper bound on the drift plus penalty. The upper bound can be obtained by adding the operational cost to the drift plus penalty as:

$$\Delta LD_T(t) + V * \sum_{\tau=0}^{t+T-1} \mathit{Cost\ of\ data\ center}_{opertiaonal}(\tau) \mid X(t). \tag{18}$$

The data center chooses the control parameter V to adjust the tradeoff between the level of power in the UPS for backup and minimizing the operational cost of the data center. For optimal cost minimization, V has to be set high and for more power back up, the value of V needs to be small.

5.1 Drift Plus Penalty Upper Bound

A key question is to find out the upper bound for the value of V. The upper bound of drift plus penalty helps in finding the maximum operational cost of the data center that can be saved under the constraint of keeping the power in the batteries of the UPS for backup. To find out the upper bound we assume that Lyapunov function $L(0) > \infty$, $t = KT$, $\tau \in [t, t+T-1]$, and $V > 0$. We take the squares of the Eq. (15) on both sides.

$$X(\tau+1) = X(\tau) + \eta R(\mathrm{t}) - \frac{D(\tau)}{\eta},$$

$$X^2(\tau+1) = X^2(\tau) + 2 * X(\tau) * \left[\eta R(\mathrm{t}) - \frac{D(\tau)}{\eta}\right] + \left[\eta R(\mathrm{t}) - \frac{D(\tau)}{\eta}\right]^2,$$

$$\frac{\left[X^2(\tau+1) - X^2(\tau)\right]}{2} = X(\tau) * \left[\eta R(\mathrm{t}) - \frac{D(\tau)}{\eta}\right] + \frac{\left[\eta R(\mathrm{t}) - \frac{D(\tau)}{\eta}\right]^2}{2}.$$

As $R(t) \in [0, R_{max}]$ and $D(t) \in [0, D_{max}]$, the above equation is transformed into the following equation

$$\frac{\left[X^2(\tau+1) - X^2(\tau)\right]}{2} \leq X(\tau) * \left[\eta R(\mathrm{t}) - \frac{D(\tau)}{\eta}\right] + \frac{1}{2} max\left[\eta^2 R^2(\mathrm{t}), \frac{D^2(\tau)}{\eta^2}\right]. \tag{19}$$

We get the 1-time slot conditional Laypunov drift by taking the expectation over power demand, available surplus power and its price in the remote grid, and the price of the electricity in long term contract and real time market in the local grid on the auxiliary variable $X(t)$ as

$$\Delta LD_1(t) \leq X(\tau) * \left[\eta R(\mathrm{t}) - \frac{D(\tau)}{\eta}\right] + \frac{1}{2} max\left[\eta^2 R^2(\mathrm{t}), \frac{D^2(\tau)}{\eta^2}\right]. \tag{20}$$

By taking the sum of all inequalities over $\tau \in [t, t+1, \ldots\ldots t+T-1]$, we obtain the T-time slot Laypunov drift

$$\Delta LD_T(t) \leq X(\tau) * \left[\eta R(\mathrm{t}) - \frac{D(\tau)}{\eta}\right] + T * \left\{\frac{1}{2} max\left[\eta^2 R^2(\mathrm{t}), \frac{D^2(\tau)}{\eta 2}\right]\right\}. \tag{21}$$

Finally we add the operational cost on both sides of the equation and get the upper bound on the T-time slot Lypunov drift plus penalty.

$$\begin{aligned} &\Delta LD_T(t) + V * \sum_{\tau=0}^{t+T-1} Cost\ of\ data\ center_{opertiaonal}(\tau) \mid X(t) \\ &\leq X(\tau) * \left[\eta R(\mathrm{t}) - \frac{D(\tau)}{\eta}\right] + T \\ &* \left\{\frac{1}{2} max\left[\eta^2 R^2(\mathrm{t}), \frac{D^2(\tau)}{\eta^2}\right]\right\} + V \\ &* \sum_{\tau=0}^{t+T-1} Cost\ of\ data\ center_{opertiaonal}(\tau) \mid X(t). \end{aligned} \tag{22}$$

The Smart data center algorithm follows the drift plus penalty principle and tries to minimize the right hand side of the Equation.

5.2 Relaxed Optimization

In order to minimize the right hand side of the Eq. (22), the data center needs to know the queue backlog *X(t)* over time $\tau \in [t, t+T-1]$. The amount of available surplus power in the remote grid, the power level in the batteries of the UPS, and the power demand affects the queue *X(t)*. Moreover, the dynamic nature of electricity prices, available surplus power, and power demand are major constraints for taking the decision. The researchers have used forecasting techniques to predict the variable nature of the parameters. However, one day head forecasting techniques causes daily mean errors of approximately 8.7 % [16]. Therefore, in order to remove the need of forecasting techniques we used the near-future queue blog statistics. We used the current values of the queue, i.e., $X(\tau) = X(t)$ for the time period $t < \tau \leq t+T-1$ for the backlog statistics. However, the use of near future queue backlog result in slightly "loosening" of the upper bound on the drift plus penalty term. We have proved this loosening of the upper bound in Corollary 1.

Corollary 1 (Loosening Drift plus penalty bound) Suppose the control parameter V is positive and for some nonnegative integer K, the time slot t is equal to KT. By changing the time period from τ to t in the queue X, the drift plus penalty satisfies:

$$
\begin{aligned}
&\Delta LD_T(t) + V\mathbb{E}\{ \sum_{\tau=t}^{t+T-1} \mathit{Cost\ of\ data\ center}_{opertiaonal}(\tau)\,|X(t)\} \\
&\leq \left\{ \frac{1}{2} max\left[\eta^2 R^2(\mathrm{t}),\ \frac{D^2(\tau)}{\eta^2} \right] \right\} \\
&+ \frac{T(T-1)\left[\eta^2 R^2(\mathrm{t}),\ \frac{D^2(\tau)}{\eta^2} \right]}{2} \\
&+ \mathbb{E}\left\{ \sum_{\tau=t}^{t+T-1} X(t) * \left[\eta R(\tau) - \frac{D(\tau)}{\eta} \right] |X(t) \right\}
\end{aligned} \tag{23}
$$

Proof According to the Eq. (15), for any $\tau \in [t, t+T-1]$, we get

$$
X(t) - \frac{(\tau - t)Dmax}{\eta} \leq X(\tau),
$$

$$
\text{and } X(\tau) \leq (\tau - t)\eta Rmax.
$$

Therefore, recalling each term in Eq. (22), we have

$$
\begin{aligned}
&\sum_{\tau=t}^{t+T-1} X(\tau)\left[R(\tau)\ \eta - \mathrm{D}(\tau)/\eta\right] \\
&\leq \sum_{\tau=t}^{t+T-1} \left[X(t) + (\tau - t)\,\eta Rmax\right] R(\tau)\eta
\end{aligned}
$$

$$- \sum_{\tau=t}^{t+T-1} [X(t) - (\tau - t)\, Dmax/\eta]\, D(\tau)/\eta$$

$$=> \sum_{\tau=t}^{t+T-1} X(\tau)\,[R(\tau)\,\eta\ - D(\tau)\,/\eta]$$

$$+ \sum_{\tau=t}^{t+T-1} \left[(\tau - t)\eta^2 Rmax\, R(\tau) - Dmax\, D(\tau)/\eta^2\right]$$

$$\leq \sum_{\tau=t}^{t+T-1} X(\tau)[R(\tau)\ \eta - D(\tau)/\eta]$$

$$+ \frac{T(T-1)\left[\eta^2 R^2\,(\mathrm{t}),\ \frac{D^2(\tau)}{\eta^2}\right]}{2}$$

By substituting the above inequalities into Eq. (22), the corollary is proved.

5.3 Two Timescale Smart Data Center Algorithm

We see that the upper bound that can be achieved using Eq. (23) is larger than the one in Eq. (22). The Smart Data center algorithm aims to make the decision to minimize the right hand side of the Eq. (23). Depending on the available surplus power at the remote grid *r(t)*, the algorithm has to make the decision to purchase $E_{full}(t)$ at the start of the each coarse grained timeslot *t*. Moreover, at the beginning of each fine grain time slot τ, the Smart Data center algorithm has to make the decision for $E_{rt}(\tau), D(\tau), and R(\tau)$. Consequently, the problem can be separated into two timescales as two subproblems. In the coarse grain time slot, the algorithm has to make the decision to ensure that current energy demand is fullfiled and batteries of the UPS should be charged with enough power for the future use. The decisions for UPS charging and discharging along with purchase of electricity on real time rate from the local grid are made by algorithm at the start of each fine grain timeslot. The queue statistics are updated at the end of each time slot.

Algorithm 1 The Smart Data center Algorithm

1. *Long term planning*: The data center decides the optimal power purchase $E_{full}(t)$ at the start of each coarse-grained time slot $t = kT$ where k is nonnegative integer. The long term ahead power purchase is to minimize the following problem

$$\min \mathbb{E}\left\{\sum_{\tau=t}^{t+T-1} V[E_{lt}(t)p_{lt}\,(t) + E_{rt}(\tau)p_{rt}\,(\tau) + r(\tau)p_s\,(\tau)]\,|X(t)\right\}$$

$$+ \mathbb{E}\left\{\sum_{\tau=t}^{t+T-1} X(\tau)[R(\tau)\,\eta - D(\tau)/\eta]|X\,(\mathrm{t})\right\}$$

$$s.t.\ (3)$$

2. *Real time power balancing*: The data center divides the power purchased in long term equally $E_{lt}(t) = \frac{E_{full}(t)}{L}$ among all the fine grained time slots τ ε [t, t + T − 1]. The data center decides real time purchase of power $E_{rt}(\tau)$ from the local grid, charging $R(\tau)$ and discharging $D(\tau)$ of batteries of the UPS to minimize the following problem

$$\min V\, E_{rt}(\tau)p_{rt}(\tau) + r(\tau)p_s(\tau) + X(t)\,[R(\tau)\eta - \mathrm{D}(\tau)/\eta]$$
$$s.t.\ (3)(6)(7)(8)$$

3. *Queue update*: Update the queues using Eqs. (5) and (15).

6 Performance Analysis

In this section, we analyze the performance bound of the Smart Data center algorithm.

Theorem (Performance Bound) The time-averaged cost $\eta R\text{max}$ achieved by the Smart Data center algorithm based on accurate knowledge of $X(\tau)$ in the future coarse-grained interval satisfies the following bound with any nonnegative value of decision parameter V:

1. The time-average cost $Cost\ of\ data\ center_{avg}$ achieved by the algorithm satisfies the following bound:

$$\begin{aligned} Cost\ of\ data\ center_{avg} &\cong \lim_{t\to\infty} 1/t \sum_{\tau=0}^{t+T-1} \mathbb{E}\left[Cost\ of\ data\ center_{opertiaonal}(\tau)\right] \\ &\le \emptyset^{opt} + \left[\left\{\frac{1}{2}max\left[\eta^2 R^2\,(\mathrm{t}),\ \frac{D^2(\tau)}{\eta^2}\right]\right\}\right. \\ &\left.+\frac{T(T-1)\left[\eta^2 R^2\,(\mathrm{t}),\ \frac{D^2(\tau)}{\eta_2}\right]}{2}\right]/V \end{aligned}$$

Where, $\emptyset^{opt}$ is an optimal solution

Proof: Let $t = kT$ for nonnegative k and $\tau \in [t, t+T-1]$. We first look at the optimal solution. In optimal solution, all the future statistics including power demand, surplus energy from the grid and energy prices are kown to the data center in advance. Due to knowledge of future, the data center can manage to reduce the real time purchase to zero. We can say the optimal solution is

$$\begin{aligned} \emptyset^{opt} &\cong \min\{E_{lt}\,(t)\,p_{lt}\,(t) + r(\tau)p_s\,(\tau) + \partial\,(t) * Cr\} \\ &s.t.\ E_{lt}\,(t) + D(\tau) + r(\tau) - C(\tau) = d(\tau), \\ &0 \le E_{lt}\,(t) \le Pgrid, \\ &\forall t\text{: } constraints\ (6)\ (7)\ (10). \end{aligned}$$

By using the optimal solution in right hand side of the Eq. (23), we get

$$\Delta LD_T(t) + V\mathbb{E}\left\{\sum_{\tau=t}^{t+T-1} Cost\ of\ data\ center_{opertiaonal}(\tau)|X(t)\right\}$$
$$\leq \left\{\frac{1}{2}max\left[\eta^2 R^2(\mathrm{t}),\ \frac{D^2(\tau)}{\eta^2}\right]\right\}$$
$$+\frac{T(T-1)\left[\eta^2 R^2(\mathrm{t}),\ \frac{D^2(\tau)}{\eta^2}\right]}{2} + V\O^{opt}$$

Taking the expectation of the both sides and rearranging terms we get

$$\mathbb{E}\{L(t+T) - L(t)\} + VT\mathbb{E}\left\{\sum_{\tau=t}^{t+T-1} Cost\ of\ data\ center_{opertiaonal}(\tau)|X(t)\right\}$$
$$\leq \left\{\left\{\frac{1}{2}max\left[\eta^2 R^2(\mathrm{t}),\ \frac{D^2(\tau)}{\eta^2}\right]\right\}\right.$$
$$\left.+\frac{T(T-1)\left[\eta^2 R^2(\mathrm{t}),\ \frac{D^2(\tau)}{\eta^2}\right]}{2}+\right\} T + VT\O^{opt}.$$

By taking the sum over $t = kT$, $k = 0,1,2,\ldots, k - 1$ and dividing both sides by *VKT*, we get

$$\frac{1}{kT}\mathbb{E}\left\{\sum_{\tau=0}^{kT-1} Cost\ of\ data\ center_{opertiaonal}(\tau)\right\}$$
$$\leq \frac{\left\{\left\{\frac{1}{2}max\left[\eta^2 R^2(\mathrm{t}),\ \frac{D^2(\tau)}{\eta^2}\right]\right\} + \frac{T(T-1)\left[\eta^2 R^2(\mathrm{t}),\ \frac{D^2(\tau)}{\eta^2}\right]}{2} +\right\}}{V} + \O^{opt}.$$

As the variable k approaches to infinity, $k \to \infty$, the theorem is proved.

2. The UPS battery level $E_u(t)$ is bounded in the range $[E_u min,\ E_u max]$. There is always power remained in the batteries for backup in case of black out.

   Proof: We first prove that

$$-\frac{VPmax}{T} - \frac{Dmax}{\eta} \leq X(t) \leq E_u\mathrm{max} - \frac{VPmax}{T} - E_u\mathrm{min} - \frac{Dmax}{\eta}$$

We prove this by induction. For t = 0 we have

$$X(0) = E_u(0) - \frac{VPmax}{T} - E_u\mathrm{min} - \frac{Dmax}{\eta}$$

and $E_u\min \leq E_u(0) \leq E_u max$. So we get

$$-\frac{VPmax}{T} - \frac{Dmax}{\eta} \leq X(0) \leq E_u\max - \frac{VPmax}{T} - E_u\min - \frac{Dmax}{\eta}$$

Now we consider $0 \leq X(t) \leq E_u\max - \frac{VPmax}{T} - E_u\min - \frac{Dmax}{\eta}$, therefore, there is no battery recharging, i.e., $R(t) = 0$. The maximum amount of power that can be discharged each time is $\frac{Dmax}{\eta}$.

Now we have

$$-\frac{VPmax}{T} - \frac{Dmax}{\eta} < -\frac{Dmax}{\eta} < X(t+1) \leq X(t)$$
$$\leq E_u\max - \frac{VPmax}{T} - E_u\min - \frac{Dmax}{\eta}.$$

For the case when $-\frac{VPmax}{T} < X(t) \leq 0, D(t) = 0$. The amount of power that can be charged and discharged at maximum each time are ηRmax and $\frac{Dmax}{\eta}$, respectively. We get

$$-\frac{VPmax}{T} - \frac{Dmax}{\eta} < X(t+1) \leq X(t) + \eta Rmax$$
$$\leq E_u\max - \frac{VPmax}{T} - E_u\min - \frac{Dmax}{\eta}.$$

Finally consider the case, when $-\frac{VPmax}{T} - \frac{Dmax}{\eta} \leq X(t) \leq -\frac{VPmax}{T}$ again D(t) = 0 as $X(t) \leq -\frac{VPmax}{T}$. We get

$$-\frac{VPmax}{T} - \frac{Dmax}{\eta} < X(t) \leq X(t+1) \leq E_u\max - \frac{VPmax}{T} - E_u\min - \frac{Dmax}{\eta}.$$

Using Equ. 14, we have

$$-\frac{VPmax}{T} - \frac{Dmax}{\eta} \leq X(t) = E_u(t) - \frac{VP_{max}}{T}$$
$$-E_u\min - \frac{Dmax}{\eta} \leq E_u\max - \frac{VPmax}{T} - E_u\min - \frac{Dmax}{\eta}.$$

From all the cases, we can conclude that

$$E_u\, min < E_u(\tau) < E_u\, max.$$

3. All decisions are feasible.

   The smart data center algorithm makes decision to satisfy all the constraints. Therefore, the Smart Data center algorithm is feasible.

7 Related Work

The past decade has witnessed the enormous growth in the online applications and services. The online applications and services are hosted in data centers. With the increase demand of online services, the cost of power consumption in the data centers is increasing significantly. There is extensive existing research on the power management of data centers [1, 13, 22]. Most of the works focus on the reducing the power consumption in the data center using different schemes like voltage scaling, frequency scaling, and dynamic shutdown. However, the earlier works have not focused on reducing the overall cost of the power used in the data center.

Recently, the researchers have started to focus on reducing the cost of power utilized in the data center. *Ref.* [2, 20, 21], focused on migration of workload between different data centers to utilize the low electricity prices in different geographical locations. However, the emphasis is not on reducing the cost of a single data center.

For reducing the cost of a single data center, the researchers have emphasis on the power storage in the data centers. In [7–9, 23] and [24], the researchers have shown the importance of using UPS in the data center for reducing the overall cost of electricity in a single data center. However, the aforementioned works have not considered the multiple price markets to power up the data center.

In [10–12, 15] the authors have worked on energy procurement from long term, intermediate, real time markets. However, the approaches in the aforementioned schemes depends upon the forecasting techniques, such as dynamic programming and Markov decisions to know the power demand in advance. Similar to our work, Deng et al. [3, 4] have used two timescale Lyapunov optimization technique to reduce the cost of a single data center. They have utilized the long term and real time price market of a smart grid along with the On-Site wind or solar green energy. However, they have ignored the cost of On-Site wind or solar energy. We have considered the scenario of providing the power to the data center from multiple power grids, local grid for long term and real time market, whereas remote grid for low cost surplus energy.

8 Conclusions

In this work, we have targeted the key problem that how to minimize the cost of power consumption in the data center? We proposed the new idea to power up the data center from more than one Smart grid. We exploited the long term and real time price market from the local grid and low cost surplus power from the remote grid. We developed the algorithm titled "Smart Data center" that decide how much power to be purchased from the long term and real time market. We also utilized the Uninterrupted Power Supply (UPS) as back up in the data center. The performance of the "Smart Data center" algorithm is analyzed using theoretical analysis. The performance analysis of the algorithm using real world traces are left for future work.

References

1. K. Bilal, S. U. R. Malik, O. Khalid, A. Hameed, E. Alvarez, V. Wijaysekara, R. Irfan, S. Shrestha, D. Dwivedy, M. Ali, U. S. Khan, A. Abbas, N. Jalil, and S. U. Khan, "A Taxonomy and Survey on Green Data Center Networks," *Future Generation Computer Systems*, 2013.
2. N. Buchbinder, N. Jain, and I. Menache. "Online job-migration for reducing the electricity bill in the cloud." In *NETWORKING,* Springer, *2011*, pp. 172–185.
3. W. Deng, F. Liu, H. Jin, and C. Wu. "SmartDPSS: Cost-Minimizing Multi-source Power Supply for Datacenters with Arbitrary Demand." In *Proceedings of the 13th International Conference on Distributed Computing Systems (ICDCS-13)*. 2013.
4. W. Deng, F. Liu, H. Jin, C. Wu, and X. Liu, "MultiGreen: cost-minimizing multi-source datacenter power supply with online control," In *Proceedings of the fourth international conference on Future energy systems*, pp. 149–160. ACM, 2013.
5. L. Georgiadis, M. J. Neely, and L. Tassiulas, *Resource allocation and cross-layer control in wireless networks*. Now Publishers Inc, 2006.
6. J.Glanz, "Google details, and defends, its use of electricity," The New York Times, 2011, http://www.nytimes.com/2011/09/09/technology/google-details-and-defends-its-use-of-electricity.html, accessed August 2013.
7. S. Govindan, A. Sivasubramaniam, and B. Urgaonkar, "Benefits and limitations of tapping into stored energy for datacenters," In *38th Annual International Symposium on Computer Architecture (ISCA), 2011*, pp. 341–351. IEEE, 2011.
8. S. Govindan, D. Wang, A. Sivasubramaniam, and B. Urgaonkar, "Leveraging stored energy for handling power emergencies in aggressively provisioned datacenters," In *ACM SIGARCH Computer Architecture News*, ACM, Vol. 40, No. 1, 2012, pp. 75–86.
9. Y. Guo, Z. Ding, Y. Fang, and D. Wu, "Cutting down electricity cost in internet data centers by using energy storage," In *Global Telecommunications Conference (GLOBECOM 2011), 2011 IEEE*, pp. 1–5. IEEE, 2011.
10. M. He, S. Murugesan, and J. Zhang, "Multiple timescale dispatch and scheduling for stochastic reliability in smart grids with wind generation integration," In *INFOCOM, 2011 Proceedings IEEE*, 2011, pp. 461–465.
11. L. Huang, J. Walrand, and K. Ramchandran, "Optimal power procurement and demand response with quality-of-usage guarantees," In *Power and Energy Society General Meeting, IEEE*, 2012, pp. 1–8.
12. L. Jiang and S. Low, "Multi-period optimal procurement and demand responses in the presence of uncrtain supply," In *Proceedings of IEEE Conference on Decision and Control (CDC)*. 2011.
13. D. Kliazovich, P. Bouvry, and S. U. Khan, "GreenCloud: A Packet-level Simulator of Energy-aware Cloud Computing Data Centers," *Journal of Supercomputing*, Vol. 62, No. 3, pp. 1263–1283, 2012.
14. J. A. P. Lopes, F. J. Soares, P. M. Almeida, and M. Moreira da Silva, "Smart charging strategies for electric vehicles: Enhancing grid performance and maximizing the use of variable renewable energy resources," In *EVS24 Intenational Battery, Hybrid and Fuell Cell Electric Vehicle Symposium, Stavanger, Norveška*. 2009.
15. J. Nair, S. Adlakha, and A. Wierman, *Energy procurement strategies in the presence of intermittent sources*. Caltech Technical Report, 2012.
16. F. Nogales and J. Conttreas, "Forecasting Next Day Electricity Prices by Time series Models", *IEEE Transaction on power systems*, Vol. 17, No, 2, May 2002.
17. F. Paraiso, P. Merle, and L. Seinturier, "Managing elasticity across multiple cloud providers," In *Proceedings of the 2013 international workshop on Multi-cloud applications and federated clouds*, pp. 53–60. ACM, 2013.
18. A. Qureshi, "Power-demand routing in massive geo-distributed systems," PhD diss., Massachusetts Institute of Technology, 2010.

19. A. Qureshi, R. Weber, H. Balakrishnan, J. Guttag, and B. Maggs. "Cutting the electricity bill for Internet-scale systems." *ACM SIGCOMM Computer Communication Review,* Vol. 39, No. 4, 2009, pp. 123–134.
20. L. Rao, X. Liu, L. Xie, and W. Liu, "Minimizing electricity cost: optimization of distributed internet data centers in a multi-electricity-market environment," In *INFOCOM, 2010 Proceedings IEEE*, pp. 1–9. IEEE, 2010.
21. L. Rao, Lei, X. Liu, M. D. Ilic, and Jie Liu, "Distributed coordination of internet data centers under multiregional electricity markets." *Proceedings of the IEEE, Vol.* 100, No. 1,2012 pp. 269–282.
22. R. Raghavendra, P. Ranganathan, V. Talwar, Z. Wang, and X. Zhu, "No 'power' struggles: Coordinated multi-level power management for the data center," *ACM SIGARCH Computer Architecture News*, Vol. 36, Mar. 2008, pp. 48–59.
23. R. Urgaonkar, B. Urgaonkar, M. l. J. Neely, and A. Sivasubramaniam, "Optimal power cost management using stored energy in data centers," In *Proceedings of the ACM SIGMETRICS joint international conference on Measurement and modeling of computer systems*, pp. 221–232. ACM, 2011.
24. D. Wang, C. Ren, A. Sivasubramaniam, B. Urgaonkar, and H. Fathy, "Energy storage in datacenters: what, where, and how much?," In *ACM SIGMETRICS Performance Evaluation Review*, Vol. 40, No. 1, ACM, 2012, pp. 187–198.
25. X. Lu, Xin, Z. Y. Dong, and X. Li. "Electricity market price spike forecast with data mining techniques." *Electric power systems research*, Vol. 73, No. 1, 2005, pp. 19–29.

Power and Thermal Efficient Numerical Processing

Wei Liu and Alberto Nannarelli

1 Introduction

Numerical processing is at the core of applications in many areas ranging from scientific and engineering calculations to financial computing. These applications are usually executed on large servers or supercomputers to exploit their high speed, high level of parallelism and high bandwidth to memory.

As of 2013, the performance of the world's top supercomputers are measured at petaflops ($\sim 10^{15}$ floating-point operations per second, or FLOPS). To reach the next level of computing, which is *exa*-scale (exa is $E = 10^{18}$), the performance has to increase by 30 times. In particular, the U.S. Department of Energy has asked the industry to reach that goal while staying within a 20 MW power envelope. This directly leads to a performance efficiency requirement of 50 GFLOPS/W, which is about 20 times higher than where we are today. This gap cannot be fulfilled by semiconductor process evolution alone.

High power dissipation might result in excessive heating. Because silicon is not a good heat conductor, "*hotspots*" might form on the die in areas with high power density.

One of the consequence of increased die temperature is an increased leakage power that contributes to the rise of temperature in the hotspot. This is clearly a positive-feedback loop which might compromise (burn-down) the device.

Moreover, high temperatures have a negative impact on reliability. Beside the extreme effect of burn-down, with device scaling positive/negative biased temperature instability (P/NBTI) is becoming one of the major reliability concerns that can limit the device's lifetime. The NBTI effect primarily affects PMOS transistors (PBTI

W. Liu (✉)
Oticon A/S, Smørum, Denmark
e-mail: wli@oticon.dk

A. Nannarelli
DTU Compute, Technical University of Denmark, Kongens Lyngby, Denmark
e-mail: alna@dtu.dk

S. U. Khan, A. Y. Zomaya (eds.), *Handbook on Data Centers*,
DOI 10.1007/978-1-4939-2092-1_8

affects NMOS transistors) and can lead to a significant shift in the threshold voltage over time. The delay increase induced by P/NBTI aging can severely degrade performance and, in the worst case, result in system failure [1, 2].

Another not negligible aspect of running data centers is the cost of electricity. Electrical power is not only used to perform the computation, but also to cool down the machines [3]. At today's power efficiency, the cost of running an exascale supercomputer would be more than $ 2 billion per year [4].

Therefore, to meet the power efficiency challenge, we need to think at all levels of abstraction.

Floating-point units (FPUs) are a good case study for power and thermal aware design. FPUs are found in a wide variety of processors ranging from server and desktop microprocessors, to graphic processing units (GPUs), to digital signal processors (DSPs), mobile Internet devices and embedded systems.

Floating-point operations are much more complex than their integer and fixed-point counterparts. Consequently, FPUs usually occupy a significant amount of silicon area and can consume a large fraction of power and energy in a chip. For scientific and graphics intensive applications, the high power consumption in the FPU can make it the hotspot on the die.

In this chapter, we first review the floating-point representation and the basic operations on floating-point numbers, including addition, multiplication, fused multiply-add and division. In particular, we demonstrate different approaches to implement division, namely the digit recurrence method and the Newton-Raphson method. The first method uses only adders and shifters and produce one digit of quotient per iteration. The second method, on the other hand, uses a floating-point multiplier and requires less iterations. We then analyze the power dissipation in the implementation of these operations for different approaches in the design of floating-point units. To compare the energy consumption between different design choices, we evaluate different FPUs' configurations when executing realistic workloads. Since the consumed power is mostly dissipated in the form of heat, we also compare the different designs in terms of thermal distribution, namely peak temperature and thermal gradient. Finally, we discuss floating-point unit design for the next generation processors.

2 Floating-Point Representation

A floating-point representation is used to represent real numbers in a finite number of bits. Since the set of real numbers is infinite, it is only possible to exactly represent a subset of real numbers in the floating-point representation. The rest of the real numbers either fall outside the range of representation (overflow or underflow), or they are approximated by other floating-point numbers (roundoff). The most used representation is sign-and-magnitude, in which case a floating-point number x is represented by (S_x, M_x, E_x):

$$x = (-1)^{S_x} \times M_x \times b^{E_x} \tag{1}$$

where $S_x \in \{0, 1\}$ is the sign, M_x denotes the magnitude of the significand, b is a constant called the base and E_x is the exponent.

A floating-point representation system involves many parameters and historically many floating-point processors were designed using a variety of representation systems. To avoid incompatibilities between different systems, the IEEE Floating-point Standard 754 was developed, which is followed by most floating-point processors today. The latest version of the standard (IEEE 754-2008 [5]) defines the arithmetic formats of binary and decimal floating-point numbers as well as the operations that perform on these numbers. We briefly summarize the main parts of the IEEE Standard 754 for binary numbers in this section.

2.1 Formats

The magnitude of the significand M_x is represented in radix 2 normalized form with one integer bit:

$$1.F$$

where F is called the fraction and the leading 1 is called the hidden bit. The exponent E_x is base 2 and in biased representation with

$$B = 2^{e-1} - 1$$

where e is the number of bits of the exponent field. When $E_x = 0$ and F is non-zero, the number is called *subnormal*, and the hidden bit in its representation is a 0. Consequently, the value of a normal floating-point number represented in the IEEE format can be obtained as:

$$x = (-1)^{S_x} \times 1.F_x \times b^{E_x - B} \tag{2}$$

The three components are packed into one word with the order of the fields in S, E and F. The IEEE standard 754 defines four binary floating-point formats [5]:

- binary16 (Half): S(1), E(5), F(10).
- binary32 (Single): S(1), E(8), F(23).
- binary64 (Double): S(1), E(11), F(52).
- binary128 (Quad): S(1), E(15), F(112).

2.2 Rounding Modes

The standard defines five rounding modes, divided in two categories:

- Round to nearest: Round to nearest, ties to even (default); Round to nearest, ties away from zero.
- Directed: Round toward 0 (truncated); Round toward $+\infty$; Round toward $-\infty$.

2.3 Operations

Required operations include:

- Numerical: add, subtract, multiply, divide, remainder, square root, fused multiply-add, etc.
- Conversions: floating to integer, binary to decimal (integer), binary to decimal (floating), etc.
- Miscellaneous: change formats, test and set condition flags, etc.

2.4 Exceptions

The standard defines five exceptions, each of which sets a corresponding status flag when raised and by default the computation continues.

- overflow (result is too large to be represented).
- underflow (result is too small to be represented).
- division by zero.
- inexact result (result is not an exact floating-point number).
- invalid operation (when a Not-A-Number result is produced).

In the following sections, we describe the algorithm and implementation for floating-point operations. In specific, the operations described are: add/subtract, multiply, fused multiply-add and divide. Of all these operations, division is the most complex and we will present several algorithms and implementations for the division operation.

For each operation, we first present a high level description of the steps to be performed in generic form. Then, a hardware implementation of the operation is given to illustrate the execution of different algorithms. We assume the operands and results are represented by the triplet (S, M, E) as previously described. To simplify the description of algorithms, let $M^* = (-1)^S M$ represent the signed significand.

3 Floating-Point Addition

The addition/subtraction is described by the following expression:

$$z = x \pm y$$

The high level description of this operation is composed of the following steps:

1. Add/subtract significands and set exponent.

$$M_z^* = \begin{cases} (M_x^* \pm (M_y^* \times b^{(E_y - E_x)})) \times b^{E_x} & if\ E_x \geq E_y \\ ((M_x^* \times b^{(E_x - E_y)}) \pm M_y^*) \times b^{E_y} & if\ E_x < E_y \end{cases}$$

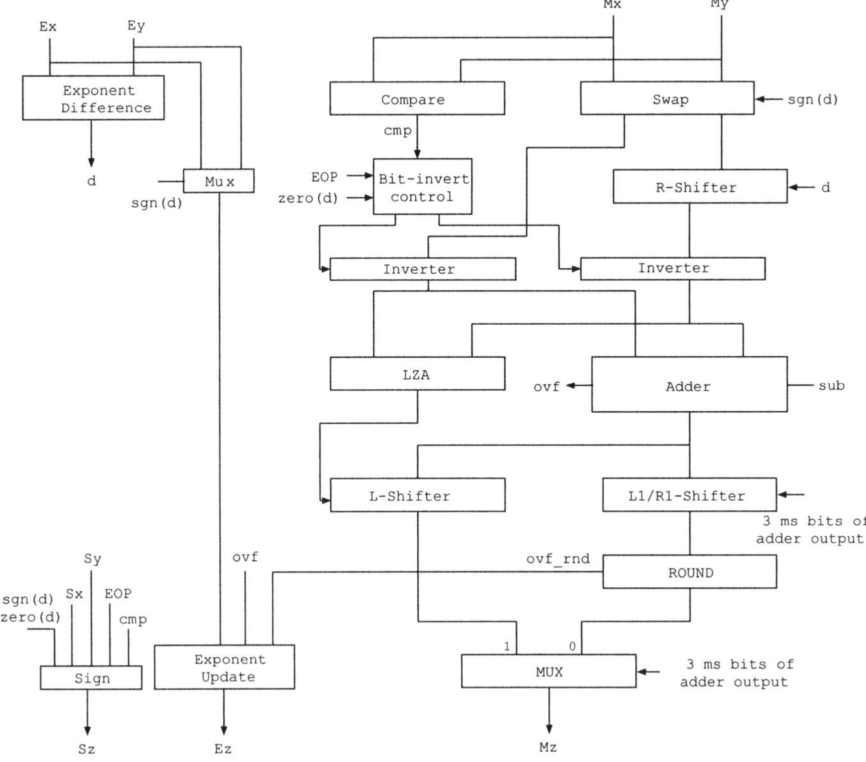

Fig. 1 Single path floating-point addition

$$E_z = max(E_x, E_y)$$

2. Normalize significand and update exponent.
3. Round, normalize and adjust exponent.
4. Set flags for special cases.

A single path implementation of the floating-point add operation is shown in Fig. 1. The figure is derived from [6], where a more detailed description of the unit is given. To avoid having two alignment shifters, the operands are swapped according to the sign of the exponent difference. A two's complement adder performs the sign-and-magnitude addition in step 1. When the effective operation is subtraction (determined by the operation and the signs of the operands), the smaller operand is complemented by bit-inversion plus carry-in to the adder. This is to avoid complementing the output of the adder when the result is negative. The leading zero anticipation (LZA) unit determines the position of the leading one in the result in parallel with the addition.

In the normalization step, two cases can occur. In the first case, the effective operation is subtraction and the output of the adder might have many leading zeros, which requires a massive left shift of the result and no roundup is necessary since the exponents difference is less than 2 and no initial massive right shift was performed. In the second case, the output of the adder contains only one leading zero or has an overflow due to addition. In this case, a shifting of only one position to the left or to the right is required and subsequently a roundup is necessary. The two cases can be designed into separate paths in order to reduce the latency in both paths [7].

4 Floating-Point Multiplication

The multiplication of two floating-point numbers x and y is defined as:

$$z = x \times y$$

The high level description of this operation is composed of the following steps:

1. Multiply significands and add exponents.

$$M_z^* = M_x^* \times M_y^*$$
$$E_z = E_x + E_y + B$$

2. Normalize M_z^* and update exponent.
3. Round.
4. Determine exception flags and special values.

The basic implementation of floating-point multiplication is shown in Fig. 2. For the sake of simplicity, we only show the data paths for the significands in block diagrams. Parallel multiplication (combinational) is a three steps computation [6]. We indicate with

$$z = x \times y$$

the product z ($n + m$ bits) of a n-bit operand x and a m-bit operand y.

1. First, m partial products

$$z_i = 2^i x \cdot y_i \qquad i = 0, \ldots, m-1$$

   are generated. Because $y_i = \{0, 1\}$, this step can be realized with a $n \times m$ array of AND-2 gates[1]
2. Then, the m partial products are reduced to 2 by an adder tree

$$\sum_{i=0}^{m-1} 2^i x \cdot y_i = z_s + z_c.$$

3. Finally, the carry-save product z_s, z_c is assimilated by a carry-propagate adder (CPA).

$$z = z_s + z_c.$$

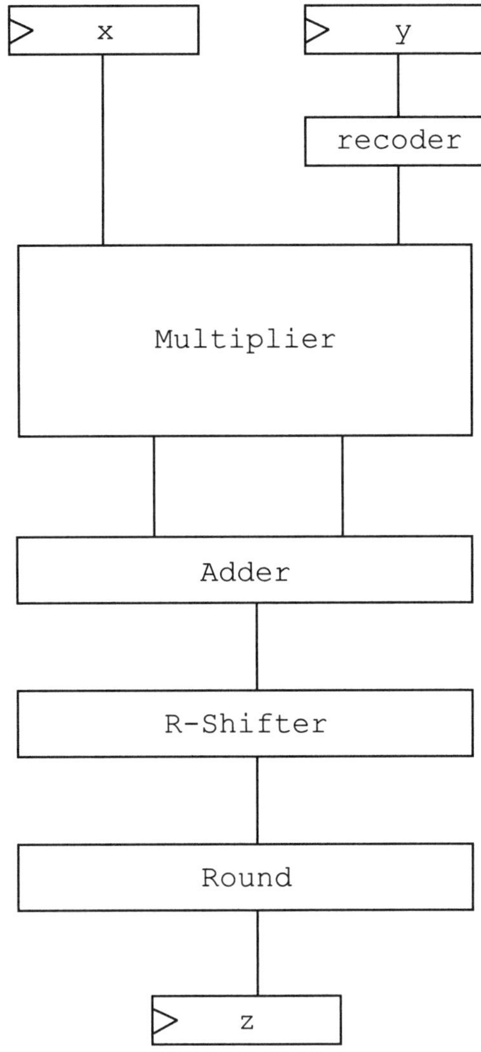

Fig. 2 Implementation of floating-point multiplication (significands only)

The delay in the adder tree and its area depend on the number of addends to be reduced (m : 2). By radix-4 recoding the multiplier y, often referred as Booth's recoding, the number of partial products is halved $\frac{m}{2}$. As a consequence, the multiplier's adder tree is smaller and faster. However, in terms of delay, the reduction in the adder tree is offset by a slower partial product generation, due to the recoding [6]. On the other hand, the reduction in area is significant, and the power dissipation is reduced as well

[1] Shifting (2^i) is done by hard-wiring the AND-2 array's output bits.

due to both the reduced capacitance (area) and the nodes' activity because sequences of 1's are recoded into sequences of 0's resulting in less transitions.

The significand of the product might have an overflow in which case it is necessary to shift the result one position to the right and increment the exponent. Finally, rounding is performed according to the specified mode.

5 Floating-Point Fused Multiply-Add

The fused multiply-add (FMA) operation is a three operand operation defined by the following expression:

$$z = a + b \times c$$

The high level description of this operation is composed of the following steps:

1. Multiply significands M_b^* and M_c^*, add exponents E_b and E_c, and determine the amount of alignment shift of a.
2. Add the product of $M_b^* \times M_c^*$ and the aligned M_a^*.
3. Normalize the adder output and update the result exponent.
4. Round.
5. Determine exception flags and special values.

The multiply-add operation is fundamental in many scientific and engineering applications. Many commercial processors include a FMA unit in the floating-point unit to perform double-precision floating point fused multiply-add operation as a single instruction. The main advantages of the fused implementation over the separate implementation of multiplication and addition are:

- The high occurrence of expressions of that type in scientific computation, and the consequent reduction in overhead to adjust the operands from the IEEE format to the machine internal representation (de-normalization, etc.).
- Improvement in precision, as the result of multiplication is added in full precision and the rounding is performed on $a + b \times c$.

The drawback is that if a large percentage of multiply and add cannot be fused, the overhead in delay and power is large especially for addition.

The architecture of an FMA unit for binary64 (double precision) significands, shown in Fig. 3, is derived from the basic scheme in [6] and [8]. Registers A, B and C contain the input operands and register Z contains the final result. To prevent shifting both a and the product of b and c, a is initially positioned two bits to the left of the most significant bit (MSB) of $b \times c$ so that only a right shift is needed to align a and the product. The zero bits inserted in the two least-significant (LS) positions are used as the guard and round bits when the result significand is a. The amount of shift depends on the difference between the exponents of a and $b \times c$. Moreover, a is conditionally inverted when the effective operation is subtraction.

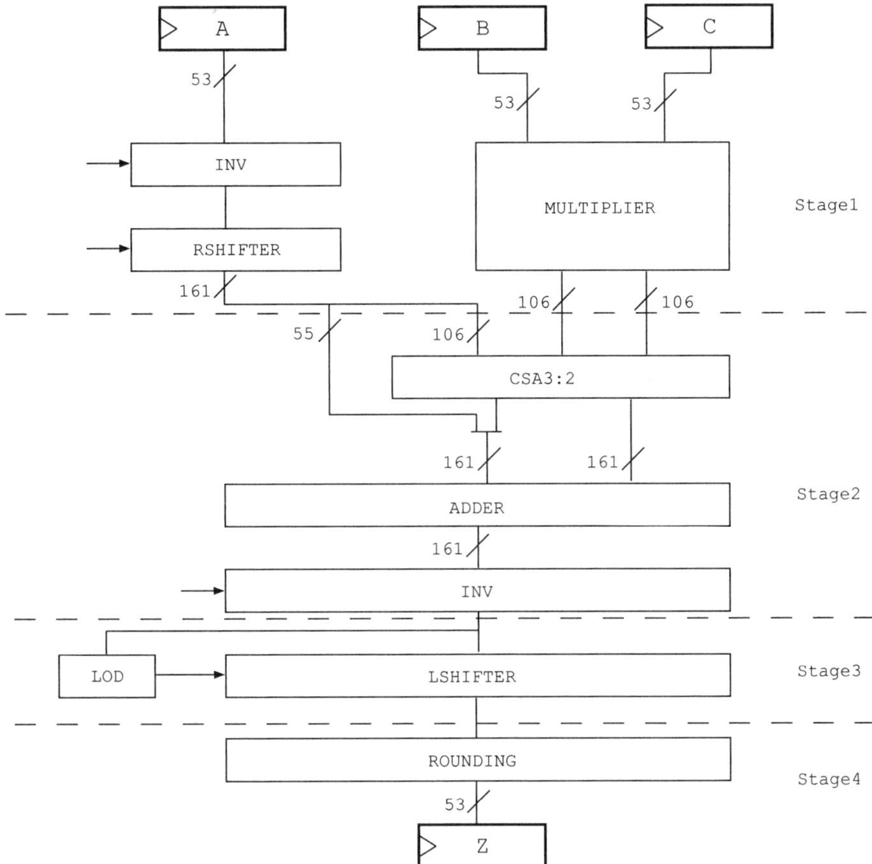

Fig. 3 Scheme of an FMA unit (significands only)

A Booth encoded tree multiplier computes the product of b and c and the result is output in carry-save format to be added with the shifted a. Since the product has 106 bits, only the 106 LSBs of the shifted a are needed in the carry-save adder (CSA). The 55 MSBs of the shifted a are concatenated with the sum of the CSA to form input to the adder. Since the carry in the output of the CSA has 106 bits, only one of the input to the adder has 161 bits.

Consequently, the leftmost 55 bits portion of the adder is implemented as an incrementer with the carry-out of the lower part as the increment input. The adder also performs end-around-carry adjustment for effective subtraction. As the result might be negative, an array of inverters is required at the output of the adder.

Once the result of the addition is obtained, the amount of normalization shift is determined by the leading one detector (LOD). No right shift for normalization is required due to the initial position of a.

To increase the throughput, the FMA unit is implemented in a four-stage pipeline. The position of the pipeline registers is indicated with dashed horizontal lines in Fig. 3.

The FMA unit can be used to perform floating-point addition by making $b = 1$ (or $c = 1$) and multiplication by making $a = 0$.

6 Floating-Point Division

The division operation is defined by the following expressions:

$$x = q \cdot d + rem$$

and

$$|rem| < |d| \cdot ulp \quad \text{and} \quad sign(rem) = sign(x)$$

where the *dividend* x and the *divisor* d are the operands and the results are the *quotient* q and the *remainder rem*.

The high-level description of the floating-point division algorithm is composed of the following steps:

1. Divide significands and subtract exponents.

$$M_q^* = M_x^* \,/\, M_d^*$$

$$E_q = E_x - E_d - B$$

2. Normalize M_q^* and update exponent accordingly.
3. Round.
4. Determine exception flags and special values.

Division is implemented in hardware in all general purpose CPUs and in most processors used in embedded systems. Several classes of algorithms exist to implement the division operation in hardware, the most used being the digit recurrence method, the multiplicative method and various approximation methods.

In the following, we briefly review these algorithms and implementations. Due to the differences in the algorithms, a comparison among their implementation in terms of performance and precision is sometimes hard to make. In Sect. 7.2, we will use power dissipation and energy consumption as metrics to compare among these different classes of algorithms.

6.1 Division by Digit Recurrence

The digit-recurrence algorithm [9] is a direct method to compute the quotient of the division

$$x = qd + rem$$

The radix-r digit-recurrence division algorithm is implemented by the residual recurrence

$$w[j+1] = rw[j] - q_{j+1}d \qquad j = 0, 1, \ldots, n$$

with the initial value $w[0] = x$. The quotient-digit q_{j+1}, normally in signed-digit format to simplify the selection function, provides $\log_2 r$ bits of the quotient at each iteration. The quotient-digit selection is

$$q_{j+1} = SEL(d_\delta, y) \qquad q_{j+1} \in [-a, a]$$

where d_δ is d truncated after the δ-th fractional bit and the estimated residual, $y = rw[j]_t$, is truncated after t fractional bits. Both δ and t depend on the radix and the redundancy (a). The residual $w[j]$ is normally kept in carry-save format to have a shorter cycle time.

The divider is completed by a on-the-fly convert-and-round unit [9] which converts the quotient digits q_{j+1} from the signed-digit to the conventional representation, and performs the rounding based on the sign of the remainder computed by a sign-zero detect (SZD) block. The conversion is done as the digits are produced and does not require a carry-propagate adder.

The digit-recurrence algorithm is quite a good choice for the hardware implementation because it provides a good compromise between latency, area and power and rounding is simple (the remainder is computed at each iteration). A radix-4 division scheme is implemented in Intel Pentium CPUs [10], in ARM processors [11] and in some IBM FPUs [12].

6.1.1 Radix-4 Division Algorithm

We now briefly summarize the algorithm for radix-4 with the quotient digit selected by comparison and speculative residual generation [11]. The radix-4 recurrence is

$$w[j+1] = 4w[j] - q_{j+1}d \qquad j = 0, 1, \ldots, n$$

with $q_{j+1} = \{-2, -1, 0, 1, 2\}$.

The quotient-digit q_{j+1} is determined by performing a comparison of the truncated residual $y = \widehat{4w[j]}$ (carry-save) with the four values (m_k) representing the boundaries to select the digit for the given d. That is,

$$\begin{aligned}
y \geq m_2 &\rightarrow q_{j+1} = 2 \\
m_1 \leq y < m_2 &\rightarrow q_{j+1} = 1 \\
m_0 \leq y < m_1 &\rightarrow q_{j+1} = 0 \\
m_{-1} \leq y < m_0 &\rightarrow q_{j+1} = -1 \\
y < m_{-1} &\rightarrow q_{j+1} = -2
\end{aligned}$$

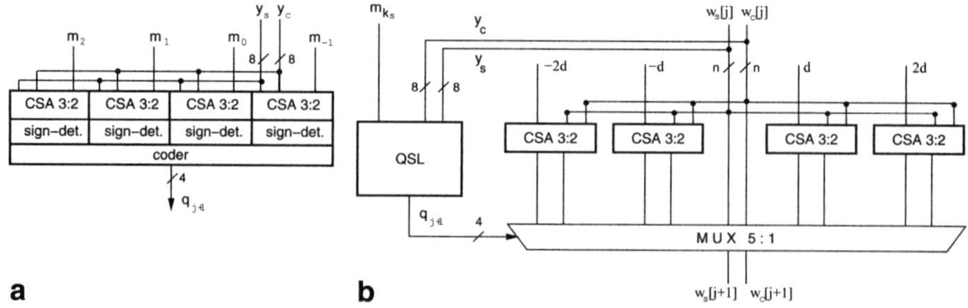

Fig. 4 **a** Selection by comparison (QSL). **b** Single radix-4 division stage

This selection can be implemented with a unit (QSL) similar to that depicted in Fig. 4a where four 8-bit comparators (sign-det.) are used to detect in which range y lies. The coder then encodes q_{j+1} in 1-out-4 code which is suitable to drive multiplexers.

In parallel, all partial remainders $w^k[j+1]$ are computed speculatively (Fig. 4b), and then one of them is selected once q_{j+1} is determined.

The critical path of the unit in Fig. 4 is

$$t_{REG} + t_{CSA}^{QSL} + t_{8b-CPA}^{QSL} + t_{buffer} + t_{MUX}$$

6.1.2 Intel Penryn Division Unit

The division unit implemented in the Intel Core2 (Penryn) family is sketched in Fig. 5 [10]. It implements IEEE binary32/binary64 compliant division, plus extended precision (64 bits significand) and integer division. The unit consists of three main parts: the pre-processing stage necessary to normalize integer operands to ensure convergence; the recurrence stage; and the post-processing stage where the rounding is performed.

The recurrence is composed of two cascaded radix-4 stages synchronized by a two-phase clock to form a radix-16 stage (4 bits of quotient computed) over a whole clock cycle. Each radix-4 stage is realized with a scheme similar to that of [11] shown in Fig. 4.

This scheme was selected by Intel because of the reduced logical depth. However, the speculation on the whole w-word (54 bits for [11], 68 bit for the Core2 format) is quite expensive in terms of area and power dissipation.

According to [10], a maximum of $6 + 15 = 21$ cycles are required to perform a division on binary64 (double-precision) operands.

6.1.3 Radix-16 by Overlapping Two Radix-4 Stages

An alternative to the Penryn solution, is to have a radix-16 divider obtained by overlapping (and not cascading) two radix-4 stages. In this scheme, the speculation

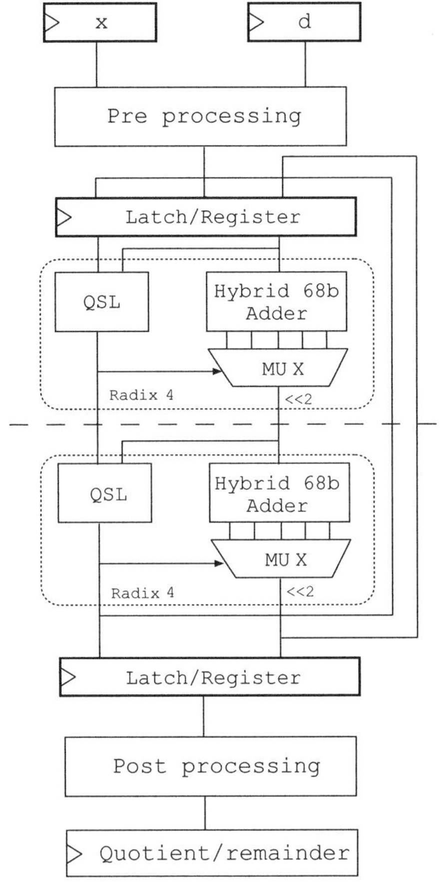

Fig. 5 Architecture of Penryn divider (significands only)

is applied to the narrower y-path as explained next. Examples of radix-16 dividers by radix-4 overlapping are reported in [9] and [13].

The radix-16 retimed recurrence, illustrated in Fig. 6a, is

$$v[j] = 16w[j-1] - q_{Hj}(4d)$$
$$w[j] = v[j] - q_{Lj}d$$

with $q_{Hj} \in \{-2,-1,0,1,2\}$, $q_{Lj} \in \{-2,-1,0,1,2\}$, and $w[0] = x$ (eventually shifted to ensure convergence). In Fig. 6a, the position of the registers is indicated with a dashed horizontal line. The recurrence is retimed (the selection function is accessed at the end of the cycle) to increase the time slack in the bits of the wide w-path (at right) so that these cells can be redesigned for low power [13].

The block QSL in Fig. 6b is the same as that of Fig. 4a. In this case, while q_H is computed, all five possible outcomes of q_L are computed speculatively. Therefore the computation of q_L is overlapped to that of q_H, and q_L is obtained with a small additional delay.

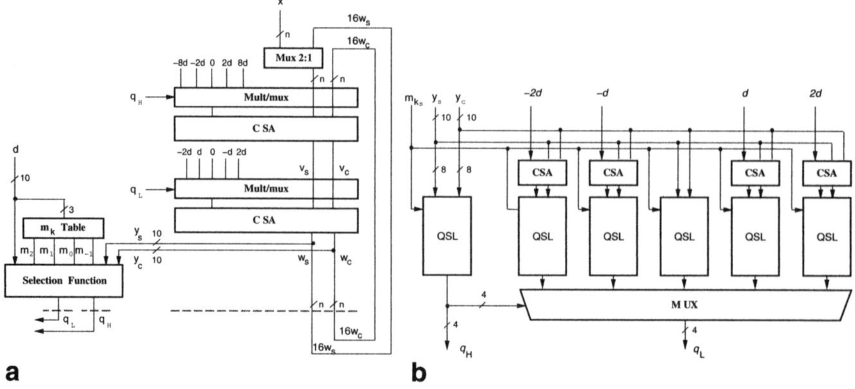

Fig. 6 **a** Recurrence radix-16. **b** Overlapped selection function

The total number of iterations to compute a binary64 division, including initialization and rounding, is 18.

6.2 *Division by Multiplication*

The quotient q of the division can also be computed by multiplication of the reciprocal of d and the dividend x

$$q = \frac{1}{d} \cdot x$$

This is implemented by the approximation of the reciprocal $R = 1/d$, followed by the multiplication $q = R \cdot x$.

By determining $R[0]$ as the first approximation of $1/d$, R can be approximated in m steps by the Newton-Raphson (NR) approximation [6]

$$R[j+1] = R[j](2 - R[j]d) \qquad j = 0, 1, \ldots, m$$

Each iteration requires two multiplications and one subtraction. The convergence is quadratic and the number of iterations m needed depends on the initial approximation $R[0]$, which is usually implemented by a look-up table.

Once $R[m]$ has been computed, the quotient is obtained by an additional multiplication $Q = R[m] \cdot x$. To have rounding compliant with IEEE standard, extra iterations are required to compute the remainder and perform the rounding according to the specified mode [6]:

- $rem = Qd - x$
- $q = ROUND(Q, rem, mode)$.

The NR algorithm for binary64 division ($m = 2$) with an initial approximation of 8 bits is summarized below.

```
R[0]  = LUT(d);
FOR i  := 0 TO 2 LOOP
    W = 2 - d *R[i];
    R[i+1] = R[i]  * W;
END LOOP;
Q = x * R[3];
rem = x - d * Q;
q = ROUND(Q,rem,mode);
```

Although division by iterative multiplication is expensive in power, it has been chosen to implement division in AMD processors [14], NVIDIA GPUs [15], and in Intel Itanium CPUs utilizing the FMA unit.

To implement the NR algorithm using the existing FMA instruction, the look-up table for the initial approximation has to be performed in software. Subsequently, the NR iterations can be executed directly in the FMA unit in Fig. 3. An extra clock cycle is required to forward the result from the output register to the input register between each FMA instruction. Thus, excluding the initial approximation a total of $8 \times 5 + 1 = 41$ cycles is required to implement division in software.

As a result, the latency of the software implementation is quite long. In the following, we illustrate how to implement the NR algorithm in hardware based on the FMA unit shown in Fig. 3. In order to achieve the initial approximation and implement the NR algorithm, the FMA unit in Fig. 3 needs to be augmented with a look-up table and several multiplexers and registers to bypass intermediate results. The implementation of the multiplicative method based on a FMA unit is shown in Fig. 7.

A look-up table, providing an 8-bit initial approximation is generated using the midpoint reciprocal method [16], of which the entries are the reciprocals of midpoints of the input intervals. The dividend x is stored in register B and divisor d in register C.

The first cycle is to obtain the initial approximation $R[0]$. After that, the operations performed in the 4-stage pipelined unit of Fig. 7 are the following (Stage 1 is abbreviated S1, etc.):

- S1 The initial approximation $R[0]$ is multiplied by d using the tree multiplier.
- S2 The product is subtracted from 2 to obtain $2 - R[0]d$. This is achieved by setting register A to the value of 2 in the previous stage. The result is stored in register W ($W[1] \leftarrow (2 - R[0]d)$).
- S1 $W[1]$ is multiplied by $R[0]$.
- S2 The new approximation $R[1] \leftarrow W[1]R[0]$ is stored in register R. The new approximated reciprocal has a precision of 16 bits.

The above four steps have to be repeated two more times to have $R[3]$ with the precision necessary for binary64 division.

Once the correct approximation of $1/d$ has been computed, another two iterations in the multiplier are required to compute:

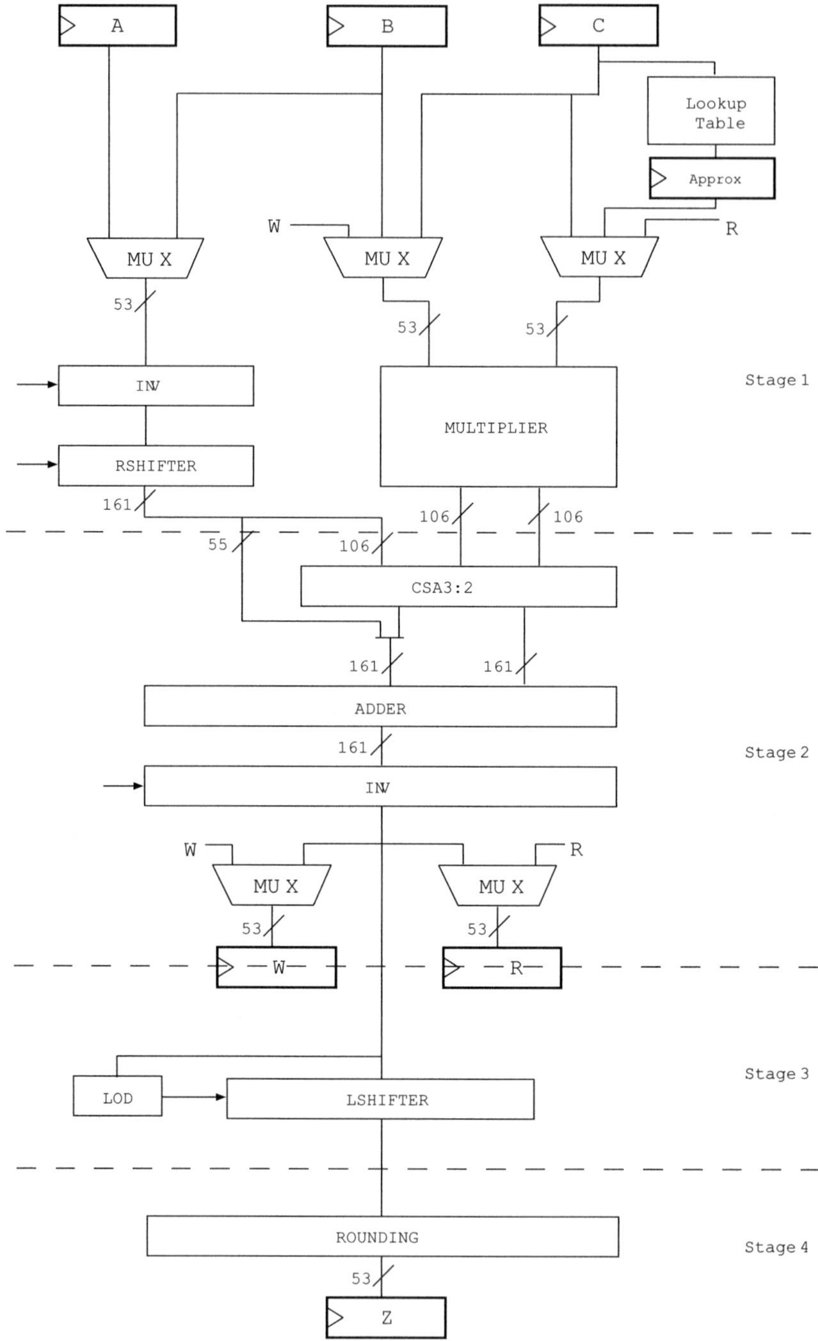

Fig. 7 Scheme of the modified FMA unit to support division

Table 1 Cycles for binary64 division in FMA unit

	Cycles
Initial approx. $R[0]$	1
Three NR iterations	$3 \times 4 = 12$
Non-rounded quotient $Q = x \cdot R[3]$	2
Remainder $rem = Q \cdot d - x$	2
Rounding	1
Total cycles	18

1. the non-rounded quotient: $Q = x \cdot R[3]$;
2. the remainder: $rem = Q \cdot d - x$ necessary for IEEE compliant rounding.

Finally, Q is rounded according to the remainder and the specified rounding mode

$$q = ROUND(Q, rem, mode)\,.$$

Summarizing, the number of clock cycles required for the implementation of the division algorithm with the unit of Fig. 7 is 18 as detailed in Table 1. The intermediate results are stored in denormalized format and consequently the normalization and rounding stages can be bypassed between iterations.

7 Energy dissipation in FP-units

7.1 Energy Metrics

At the algorithm level of design abstraction, a problem can usually be approached by different methods. For example, an application can be implemented in different ways with different timing and latency. When power is a primary design constraint, a common measure of the power and energy dissipation is required in order to evaluate and compare different algorithms.

Because the algorithms are in general different and the latency of the operations varies from case to case, it is convenient to have a measure of the energy dissipated to complete an operation. This energy-per-operation is given by

$$E_{op} = \int_{t_{op}} vi\ dt \qquad [J] \tag{3}$$

where t_{op} is the time elapsed to perform the operation. Operations are usually performed in more than one cycle (in n cycles) of clock period T_C and the expression of t_{op} is typically $t_{op} = T_C \times n$. By dividing the energy-per-operation by the number of cycles we obtain the energy-per-cycle

$$E_{pc} = \frac{E_{op}}{n} \qquad [J]. \tag{4}$$

This term is proportional to the average power dissipation that can be expressed in its equivalent forms:

$$P_{ave} = \frac{E_{pc}}{T_C} = E_{pc} f = \frac{E_{op}}{t_{op}} = V_{DD} I_{ave} \ [W] \tag{5}$$

where V_{DD} is the unit supply voltage and I_{ave} its average current. By rearranging (5) and substituting t_{op} we obtain

$$E_{op} = P_{ave} \times T_C \times n \qquad [J] \tag{6}$$

The term P_{ave} has an impact on the sizing of the power grid in the chip and on the die temperature gradient, while the term E_{op} impacts the electricity costs, and the battery lifetime in portable systems.

7.2 Implementation of the FP-Units

To analyze the impact on power dissipation of the different units and to evaluate the different approaches to division, we implemented the following units for binary64:

- **FPadd** is the floating-point add unit of Fig. 1.
- **FPmul** is the floating-point multiply unit of Fig. 2.
- **FMA** is the fused multiply-add unit of Fig. 7 modified to execute the Newton-Raphson (NR) division algorithm.
- **FMA-soft** is the fused multiply-add unit of Fig. 3 to execute the NR division algorithm in software.
- **FPdiv** (or **r16div**) is the radix-16 divide unit of Fig. 6 completed with convert-and-round unit and sign and exponent computation and update.

All units are synthesized, using a commercial 65 nm library of standard cells, to obtain the maximum speed.

Power estimation is based on randomly generated input vectors conformed to IEEE 754 binary64 format. The synthesis results are summarized in Table 2, where T_c is the minimum clock period, $Cycles$ is the number of clock cycles to finish an FP operation and $Latency$ is the total delay from applying inputs to obtaining results, that is $T_c \times Cycles$. The average power dissipation P_{ave} is normalized for all units at 1.3 GHz. The power dissipation data for the FMA unit are divided by operation.

As described in Sect. 5, the FMA unit has four pipeline stages. For the three operations: ADD, MUL and MA fused, the power was measured with the pipeline full to get the worst case power dissipation (P_{WC}) necessary to characterize the thermal behavior (Sect. 7.4) of the units. For division operations, being an iterative algorithm, a new instruction has to wait until the previous instruction finished execution and the power was measured per single operation (P_{sg}).

From the data of Table 2, it can be seen that an ADD operation in a FMA consumes much less power than a MUL operation but the latency is the same. For floating-point division, it is clear that the digit-recurrence approach (**r16div**) is much more

Table 2 Results of implementations

Unit	Area $[\mu m^2]$	T_c $[ns]$	Cycles	Latency $[ns]$	Oper.	P_{WC} $[mW]$	P_{sg} $[mW]$	E_{op} $[pJ]$
FPadd	16,461	0.75	3	2.25		39.3	18.9	42.5
FPmul	62,531	0.75	3	2.25		183.4	90.2	203.0
FMA	114,816	0.75	4	3.00	ADD	131.9	68.4	205.3
					MUL	266.8	110.9	332.8
					MA	290.7	119.5	358.5
			18	13.50	DIV		171.1	2309.6
FMA-soft	94,130	0.75	41	30.75	DIV		72.4	2226.6
FPdiv	14,054	0.75	18	13.50			27.0	365.0

P_{WC} and P_{sg} are average power measured at 1.3 GHz ($f = 1/T_C$).
P_{WC} is worst-case scenario with full pipeline.
P_{sg} is average power per single operation in pipeline

convenient in terms of latency, area and power dissipation. For example, with the same latency, FMA DIV consumes more than six times power than **FPdiv**.

In terms of energy per operation, the results in Table 2 show that in a FMA unit, the ratio of E_{op} between ADD and MUL is about $1/2$ and MA fused consumes slightly more than MUL operations. With the same latency, the energy per operation E_{op} is proportional to average power P_{sg}, thus implementing division in a FMA unit consumes much more energy than in **FPdiv**. On the other hand, although DIV operation in **FPdiv** has the lowest power consumption, the energy consumed in this unit is much larger than ADD operations due to the long latency in DIV operations. The latter observation motivates the optimization for power consumption in division.

The only argument in favor of the FMA DIV is that division is much less frequent than addition and multiplication, and a larger power dissipation for the operation can be tolerated. The software implementation of division in FMA has a even longer latency (as shown in Table 2), since each iteration has to go through all the pipeline stages and intermediate results have to be saved in the register file. The E_{op} for the hardware and software implementations of division in FMA is almost the same, but the former has a much shorter latency. Therefore in all the experiment results shown hereafter, we refer to the modified FMA with hardware support for division when comparing division by multiplication in a FMA and division by digit-recurrence.

7.3 *Energy Consumption in Floating-Point Workloads*

In this section, we evaluate the impact of the different floating-point operations, and their implementation, on the power dissipation of the whole FP-unit. We consider the instruction mix of the SPEC2006 floating-point benchmark suite profiled in [17] (Table 3). In [18], for the SPECfp92 benchmark suite, it is reported that the FP adder

Table 3 Instruction mix in floating-point SPEC2006

FP-operation	Profiled in [17] (%)	In FP-unit (%)
Add	19.4	52
Mul	17.4	47
Div	0.4	1
Total	37.2	100

consumes nearly 50 % of the multiply results which explains why fused multiply-add units are often used in modern processors.

Based on the implementation data in Table 2, we can obtain the clock cycle distribution for all FP operations (shown in Fig. 8) with the instruction mix in Table 3. Due to the much longer latency of DIV operation, the percentage of cycles spent in DIV operation is significantly larger than its percentage of instructions, which emphasizes the importance of optimizing DIV operation in terms of delay, power and energy consumption.

As for the power dissipation and energy consumption, we test two FP-unit configurations:

C1 composed of the stand-alone units FPadd, FPmul, FPdiv.
C2 where all the FP operations are executed in the FMA.

Moreover, for configuration C2, we consider what percent of multiplications can be fused with addition.

The comparison of configuration C1 and C2 is obtained by combining the values of E_{op} and P_{WC} of Table 2 and the mix of Table 3. In this way, we obtain the values of E_{op} and P_{WC} averaged on the frequency of the operations.

The results in Table 4 shows that:

1. the area of C1 is slightly smaller, but C1 dissipates significantly less power;
2. the implementation of FPadd and FPmul with stand-alone units is more power/energy efficient;

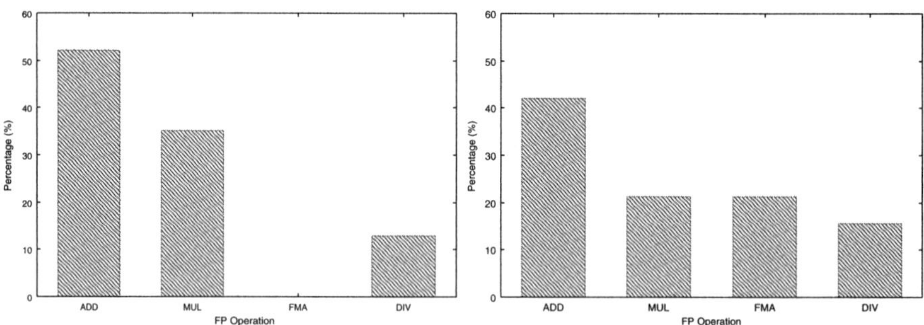

Fig. 8 Clock cycle distribution for all FP operations: not fused (*above*) and fused (*below*)

Table 4 Characteristics of the two configurations C1 and C2

Config.	Area	Fused	$\overline{E}_{op}\ (pJ)$						
	(μm^2)	%	Add	Mult	Fused	Div	AVE	Ratio	$\overline{E}_{div}/\overline{E}_{op}$
C1	93,046		22.1	95.4		3.7	121.2	1.00	0.03
C2	114,816	0	106.8	156.4	0.0	23.1	286.2	2.36	0.08
		25	93.6	132.9	47.7	26.2	300.4	2.48	0.09
		50	76.5	102.2	110.1	30.2	319.0	2.63	0.09
		75	53.1	60.4	195.2	35.7	344.3	2.84	0.10
		100	19.4	0.0	317.9	43.6	380.9	3.14	0.11
Ratio	0.81								
Config.		Fused	$\overline{P}_{WC}\ (mW)$						
		%	Add	Mult	Fused	Div	AVE	Ratio	$\overline{P}_{div}/\overline{P}_{WC}$
C1			20.4	86.2		0.3	106.9	1.00	0.003
C2		0	68.6	125.4	0.0	1.7	195.7	1.83	0.009
		25	60.2	106.6	38.7	1.9	207.4	1.94	0.009
		50	49.1	81.9	89.3	2.2	222.6	2.08	0.010
		75	34.1	48.4	158.2	2.6	243.4	2.28	0.011
		100	12.4	0.0	257.8	3.2	273.4	2.56	0.012

3. the most efficient workload for C2 is when all multiplicantions can be fused with addition (unrealistic scenario), and that for a 50 % of fused operations C1 is about twice more power efficient;
4. the impact of division operations on the energy of the whole unit is 3 % in C1 (digit–recurrence) and about 10 % in C2 (Newton-Raphson).

7.4 *Thermal Analysis*

Due to the large difference in power consumption between the two solutions, we also investigated the runtime thermal effect by performing steady-state thermal analysis.

To perform the thermal analysis, we use the model proposed in [19], which consists of a conventional RC model of the heat conduction paths around each thermal element. The differential equation modeling heat transfer according to the Fourier's law is solved by first transforming it into a difference equation, and using the electrical simulator SPICE to solve the equivalent RC circuit.

Figure 9 shows the RC equivalent model and the geometrical structure for a thermal cell. A circuit is meshed into three-dimensional thermal cells. The z direction is discretized into 9 layers and on each layer x and y directions are both discretized into 20 units which result in a grid of $20 \times 20 \times 9 = 3600$ cells in total. This provides us accurate temperature estimations at standard cell level with a reasonable

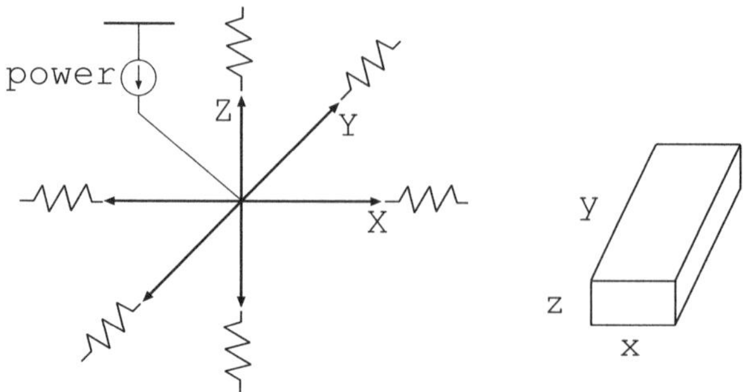

Fig. 9 Equivalent model of a thermal cell

simulation overhead. Cells inside the grid are connected to each other while cells on the boundary are connected to voltage sources that model the ambient temperature. In our thermal model, we adopted the thermal conductivities of different layers from [20]. The thermal map obtained for configuration C2 is shown in Fig. 10.

The thermal analysis results show that the peak temperature rise in C1 and C2 (FMA) are 8.5 and 28.8 °C, respectively. Due to the small circuit size, the temperature gradients (difference between maximum and minimum value) in the two circuits are insignificant.

It is obvious that C1 has a much lower temperature than C2 as for every instruction type, C1 consumes less power than C2. Furthermore, the components of C1, can remain in idle mode when the instruction is irrelevant to the component. The FMA, on the other hand, is used in all operations and all its parts have to participate in the execution.

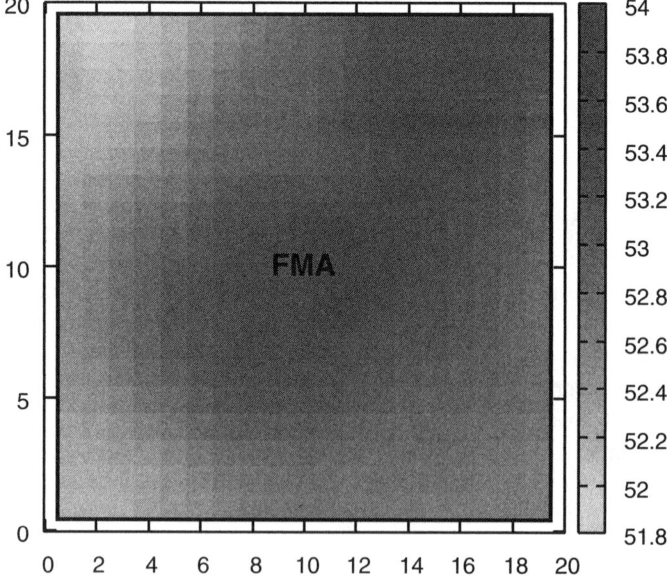

Fig. 10 Thermal map for configuration C2 (FMA). Temperature scale is in degree Celsius

8 Conclusions and Outlook on FP-Units

To get more floating-point operations completed in a given power envelope (higher FLOPS/W), it is necessary to have more power efficient FP-units. The trend to use fused multiply-add units (FMAs) in modern processors is justified by the presence of multiplications concatenated to additions in the benchmarks, but our analysis shows that this approach is less power efficient than having separate FPadd and FPmul units.

On the other hand, FMAs can increase the throughput (FLOPS) as two FP operations are merged in one, and, consequently, higher throughput can be traded-off with lower power consumption.

As the trend in supercomputers and in servers for scientific computations is a massive use of accelerators-in the form of GPUs or coprocessors, such as the Intel Xeon Phi-which are based on multicore architectures, it is advisable that operations such as division are implemented in the most efficient way to have low latency, low power dissipation and efficient use of chip area. Implementing division by multiplicative algorithms in the FMA is not power efficient, and the long latency of these algorithms might have an impact on the throughput as it keeps the FMA busy for several clock cycles.

Another way to have power efficient computation for applications requiring complex, but rather fixed, algorithms is to have application specific accelerators implemented on FPGAs. FPGA based accelerators allow efficient computation and flexibility for applications in which the market segment has not enough volume to justify the production of application specific processors on a chip (ASIC). FPGAs can exploit the parallelism of the algorithms, be re-configured (re-programmable) and can implement processors in non-common number systems, such as the decimal number system for financial applications [21].

Finally, lowering the power dissipation in FP-units reduces the temperature of the chip having a two-fold impact on costs: the cost of energy to perform the computation, and the cost of energy to cool down the servers.

References

1. S. Borkar, "Electronics beyond nano-scale CMOS," *Proc. of the 43rd ACM/IEEE Design Automation Conference*, pp. 807–808, 2006.
2. D. K. Schroder and J. A. Babcock, "Negative bias temperature instability: Road to cross in deep submicron silicon semiconductor manufacturing," *Journal of Applied Physics*, vol. 94, no. 1, pp. 1–18, July 2003.
3. X. Fan, W.-D. Weber, and L. A. Barroso, "Power Provisioning for a Warehouse-sized Computer," *Proc. of ACM International Symposium on Computer Architecture*, June 2007.
4. M. Cornea, "Precision, Accuracy, and Rounding Error Propagation in Exascale Computing," *Proc. of 21st IEEE Symposium on Computer Arithmetic*, pp. 231–234, Apr. 2013.
5. *IEEE Standard for Floating-Point Arithmetic*, IEEE Computer Society Std. 754, 2008.
6. M. D. Ercegovac and T. Lang, *Digital Arithmetic*. Morgan Kaufmann Publishers, 2004.
7. S. Oberman, G. Favor, and F. Weber, "AMD 3DNow! technology: architecture and implementations," *IEEE Micro*, vol. 19, no. 2, pp. 37–48, Mar./Apr. 1999.

8. T. Lang and J. Bruguera, "Floating-point multiply-add-fused with reduced latency," *IEEE Transactions on Computers*, vol. 53, no. 8, pp. 988–1003, Aug. 2004.
9. M. D. Ercegovac and T. Lang, *Division and Square Root: Digit Recurrence Algorithms and Implementations*. Kluwer Academic Publisher, 1994.
10. H. Baliga, N. Cooray, E. Gamsaragan, P. Smith, K. Yoon, J. Abel, and A. Valles, "Improvements in the Intel Core2 Penryn Processor Family Architecture and Microarchitecture," *Intel Technology Journal*, pp. 179–192, Oct. 2008.
11. N. Burgess and C. Hinds, "Design issues in radix-4 SRT square root and divide unit," *Conference Record of 35th Asilomar Conference on Signals, Systems and Computers*, vol. 2, pp. 1646–1650, 2001.
12. G. Gerwig, H. Wetter, E. Schwarz, and J. Haess, "High performance floating-point unit with 116 bit wide divider," *Proc. of 16th IEEE Symposium on Computer Arithmetic*, pp. 87–94, Jun. 2003.
13. A. Nannarelli and T. Lang, "Low-power division: comparison among implementations of radix 4, 8 and 16," *Proc. of 14th IEEE Symposium on Computer Arithmetic*, pp. 60–67, 1999.
14. S. Oberman, "Floating point division and square root algorithms and implementation in the AMD-K7 microprocessor," *Proc. of 14th IEEE Symposium on Computer Arithmetic*, pp. 106–115, 1999.
15. NVIDIA. "Fermi. NVIDIA's Next Generation CUDA Compute Architecture". Whitepaper. [Online]. Available: http://www.nvidia.com/content/PDF/fermi_white_papers/ NVIDIA_Fermi_Compute_Architecture_Whitepaper.pdf
16. D. DasSarma and D. Matula, "Measuring the accuracy of ROM reciprocal tables," *IEEE Transactions on Computers*, vol. 43, no. 8, pp. 932–940, Aug. 1994.
17. D. A. Patterson and J. L. Hennessy, *Computer Organization and Design-the hardware/software interface*, 4th ed.Morgan Kaufmann Publishers Inc., 2009.
18. S. Oberman and M. Flynn, "Design issues in division and other floating-point operations," *IEEE Transactions on Computers*, vol. 46, no. 2, pp. 154–161, Feb. 1997.
19. W. Liu, A. Calimera, A. Nannarelli, E. Macii, and M. Poncino, "On-chip Thermal Modeling Based on SPICE Simulation," *Proc. of 19th International Workshop on Power And Timing Modeling, Optimization and Simulation (PATMOS 2009)*, pp. 66–75, Sept. 2009.
20. T. Sato, J. Ichimiya, N. Ono, K. Hachiya, and M. Hashimoto, "On-chip thermal gradient analysis and temperature flattening for SoC design," *Proc. of the 2005 Asia and South Pacific Design Automation Conference (ASP-DAC)*, vol. 2, pp. 1074–1077, Jan. 2005.
21. A. Nannarelli, "FPGA Based Acceleration of Decimal Operations," in *Proc. of International Conference on ReConFigurable Computing and FPGA's*, Dec. 2011, pp. 146–151.

Providing Green Services in HPC Data Centers: A Methodology Based on Energy Estimation

Mohammed El Mehdi Diouri, Olivier Glück, Laurent Lefèvre and Jean-Christophe Mignot

1 Introduction

A supercomputer is an infrastructure built from an interconnection of computers capable of performing tasks in parallel in order to achieve very high performance. They are used in order to run scientific applications in various fields like the prediction of severe weather phenomena and seismic waves. To meet new scientific challenges, the HPC community has set a new performance objective for the end of the decade: Exascale. To achieve such performance (10^{18} FLoat Operations Per Second), an exascale supercomputer will gather several millions of CPU cores running up to a billion trends and will consume several megawatts. The energy consumption issue at the exascale becomes even more worrying when we know that we already reach energy consumptions higher than 17 MW at the petascale while the DARPA set to 20 MW the threshold for exascale supercomputers [2]. Hence, these systems that will be 30 times more performant than the current systems have to achieve an energy efficiency of 50 gigaFLOPS per watt while the current ones achieve between 2 and 3 gigaFLOPS per watt. As a consequence, reducing the energy consumption of high-performance computing infrastructures is a major challenge for the next years in order to be able to move to the exascale era.

M. E. M. Diouri (✉)
Institut supérieur du Génie Appliqué – Casablanca (IGA Casablanca), Casablanca, Morocco
e-mail: mehdi.diouri@iga-casablanca.ma

O. Glück · L. Lefèvre · J.-C. Mignot
INRIA Avalon team, LIP Laboratory, ENS Lyon, Lyon, France

O. Glück
e-mail: olivier.gluck@ens-lyon.fr

L. Lefèvre
e-mail: laurent.lefevre@ens-lyon.fr

J.-C. Mignot
e-mail: jean-christophe.mignot@ens-lyon.fr

S. U. Khan, A. Y. Zomaya (eds.), *Handbook on Data Centers*,
DOI 10.1007/978-1-4939-2092-1_9

While the future exascale applications are not yet designed and developed, services are highly studied. We call a service a software component that fulfills a given functionality for the successful execution of the application and that is expected to be necessary for all high performance computing applications. Besides being necessary for the successful execution of applications, their energy consumptions are expected to grow highly at the exascale. For these reasons, the goal of this chapter is to reduce the energy consumption of services instead of applications. In this chapter, we focus particularly on two services: fault tolerance and data broadcasting. Fault tolerance is an unavoidable service since it is anticipated that exascale systems will experience various kind of faults many times per day [5]. Furthermore, the applications executed on these systems will involve hundreds of exabytes [1]. There are several ways of implementing each of these two services.

We identify two classes of fault tolerance protocols: coordinated and uncoordinated protocols. These two protocols rely on checkpointing regularly (each checkpoint interval) the global state of the application in order to restart it in case of failure from the last checkpoint instead of re-executing the whole application. The problem of checkpointing is to ensure a global coherent state of the system. A global state is considered as coherent if it does not contain messages that are received but that were not sent. Coordinated protocols are currently the most used fault tolerance protocols in high performance computing applications. In order to ensure the global coherent state of the system, the coordinated protocol relies on a coordination that consists of synchronizing all the processes before checkpointing [20]. Coordination may result in a huge waste in terms of performances. Indeed in order to synchronize all the processes, it is necessary to wait for all the inflight messages to be transmitted and received. Moreover, in case of failure with the coordinated protocol, all the processes have to be restarted from the last checkpoint even if a single process has crashed. This results in a huge waste in terms of energy consumption since all the processes even the non-crashed ones have to redo all the computations and the communications from the last checkpoint. The uncoordinated protocol with message logging addresses this issue by restarting only the failed processes. Thus, the power consumption in recovery is supposed to be much smaller than for coordinated checkpointing. However, in order to ensure a global coherent state of the system, all message logging protocols need to log all messages sent by all processes during the whole execution and this impacts the performance [3]. Hence, in case of failure, the non-crashed processes send to the crashed ones the messages that they have logged.

As concern data broadcasting algorithms, two main MPI implementations exist: MPICH2[1] and OpenMPI[2]. Each implementation uses different data broadcasting algorithms. As a matter of fact, to broadcast a volume of data large enough (more than 128 Kbytes) to a number of professes large enough (more than 8 professes), OpenMPI uses the *Pipeline* algorithm with a configurable chunk size while MPICH2 *Scatter*s pieces of the broadcasted data and perform an *AllGather* in order to make

[1] MPICH2 : http://www.mcs.anl.gov/research/projects/mpich2/.
[2] OpenMPI : http://www.open-mpi.org/.

all the processes gather the remaining pieces of the broadcasted data. The *Scatter* algorithm used in MPICH2 is implemented using a binomial tree with a packet size equal to the total volume of data to broadcast divided by the total number of processes. Another promising approach for broadcasting data is to use hybrid programming by combining MPI for internode communications with OpenMP for intranode communications [23]. Indeed, MPI is optimized for architectures using distributed memory while OpenMP is more performant to program shared memories. With an hybrid data broadcasting algorithm, the root process uses MPI to broadcast the data to a single MPI process in each node. Then, each MPI process uses OpenMP to share the data broadcasted with all the others processes located in the same node. In this chapter, we focus on the two MPI broadcasting algorithms (*Scatter & AllGather* on the one hand and *Pipeline* on the other hand) and on the two hybrid algorithms that combine OpenMP to these two MPI algorithms.

Although some devices allow to measure the power and energy consumption of a service [9], measuring the energy consumption requires always to run the service at a large scale and this in all execution contexts. To reduce the number of measurements, we must be able to estimate accurately the energy consumption of a service, for any execution context and for any experimental platform. The advantage of such energy estimation is to evaluate the energy consumption of a service without pre-executing in each execution context, and in this order to be able to choose the version of the service that consumes less energy.

In order to adapt the energy estimations to the execution platform, we need to collect a set of power measurements of the nodes of the platform during the various operations that compose an application service. However, we learned from [10], that the nodes of a same cluster can have an heterogeneous idle power consumption while they have the same extra power consumption due to the execution of an giving operation. We deduce that we need to measure the idle power consumption of each node of a same cluster but we need to measure the extra power consumption due to an operation only for each type of node. Moreover, in order to estimate the energy consumption according to the execution platform, we also need to measure the execution time of each operation on this platform. However, we have shown in [10] that the nodes of a cluster are homogeneous in terms of performance. We deduce that we do not need to measure the execution time due to an operation for each type of node. In order to adapt the energy estimations to the execution context, our estimation approach is also based on a description of the parameters execution provided by the user.

In this chapter, we explain our estimation methodology from the identification of the operations found in a service to the energy estimation models of these operation, through a description of the calibration and the execution parameters that we need. We apply each step of our methodology to the fault tolerance protocols [11, 13] and to the data broadcasting algorithms [12] that we have presented. In Sect. 2, we identify the various operations for both studied services. Section 3 presents our methodology for calibrating the power consumption and the execution time of the identified operations. Section 4 shows how we estimate the energy consumption of the different operations by relying on the energy calibration and the different execution parameters. In Sect. 5, we evaluate the precision of the estimates for each of the two services by comparing then to the real energy measurements. In Sect. 6, we show how such energy estimations can be used in order to choose the energy-aware version of a service. Section 7 presents the conclusions of the chapter.

2 Identifying Operations in a Service

The first step of our methodology consists of identifying the various operations that we find in the different versions of a service. An operation is a task that the service may need to perform several times during the execution of an application. In Sects. 2.1 and 2.2, we identify the various operations of the services of fault tolerance and data broadcasting. In Sect. 2.3, we present the various parameters of which depends the energy consumption of the operations related to fault tolerance and data broadcasting.

2.1 *Fault Tolerance Case*

As described in Sect. 1, we study the two families of fault tolerance protocols: coordinated and uncoordinated protocols. For each of these two families, we distinguish two major phases: on the one hand, the checkpointing that occurs during a fault free execution (*i.e.,* without failure) of an application, and on the other hand, the recovery which occurs whenever a failure occurs. In our study, we focus on the checkpointing phase.

We consider an application using the fault tolerance protocol that is running on N nodes with p processes per node and where p is identical in the N nodes. In fault-tolerant protocols, we identify the following operations:

- Checkpointing: performed in both coordinated and uncoordinated protocols, it consists in storing a snapshot image of the current application state that can be later on used for restarting the execution in case of failure. In our study, we consider the system level checkpointing at the system level and not checkpointing at the application level. Such a choice is motivated by the fact that not all the applications embed global checkpointing and that we cannot select the optimal checkpointing interval with the applicative checkpointing. We consider the checkpointing provided in the Berkeley Lab Checkpoint/Restart library (BLCR), and available in the MPICH2 implementation. In checkpointing, the basic operation is to write a checkpoint of V_{data} size on a reliable media storage. For our study, we consider only the HDD since RAM is not reliable.
- Message logging: performed in uncoordinated protocols, it consists in saving on each sender process the messages sent on a specific storage medium (RAM, HDD, NFS, ...). In case of failure, thanks to message logging, only the crashed processes need to restart. In message logging, the basic operation is to write the

message of V_{data} size on a given media storage. For our study, we consider the RAM and the HDD.

- Coordination: performed in coordinated protocols, it consists in synchronizing the processes before taking the checkpoints. If some processes have inflight messages at the coordination time, all the other ones are actively polling until these messages are sent. This ensures that there will be no orphan messages: messages sent before taking the checkpoints but received after checkpointing. When there is no more inflight message, all the processes exchange a synchronization marker. In coordination, the basic operations are the active polling during the transmission of inflight messages of V_{data} and the synchronization of $N \times p$ processes that occurs when there is no more inflight message.

2.2 Data Broadcasting Case

We study the four data broadcasting algorithms that we described in Sect. 1, namely the two algorithms used in MPI *MPI/SAG* and *MPI/Pipeline* and two hybrid algorithms (MPI + OpenMP) *Hybrid/SAG* and *Hybrid/Pipeline*. In the two hybrid algorithms, the root process uses MPI to distribute the data to a master MPI process per node. Then, each MPI master process uses OpenMP to share the broadcasted data with all other processes that are in the same node: the routine used in OpenMP is *CopyPrivate*. We consider the broadcast of V_{data} among N nodes with p processes per node. We assume that the number p processes per node is identical in the N nodes.

Figure 1 presents the four broadcasting algorithms that we consider and shows the sizes of the messages exchanged between the different processes. In these four algorithms, we identify the following operations:

- *Scatter*: It consists of dividing a data into a number of smaller parts equal to the number of processes and sending a piece of data to each process using a binomial tree topology. It is used in MPI/SAG and Hybrid/SAG.
- *AllGather*: Given a set of elements distributed across all processes, *AllGather* will gather all of the elements to all the processes using a ring topology. It is used in MPI/SAG and Hybrid/SAG.
- *Pipeline*: It consists of splitting the source message into an arbitrary number of packets (called chunks) which are routed in a pipelined fashion. In Figure 1, the number of chunks is denoted by C. It is used in MPI/Pipeline and Hybrid/Pipeline.
- *CopyPrivate*: It consists of copying a data stored in a variable from one thread to the corresponding variables of all other threads within the same node. It is used in Hybrid/SAG and Hybrid/Pipeline.

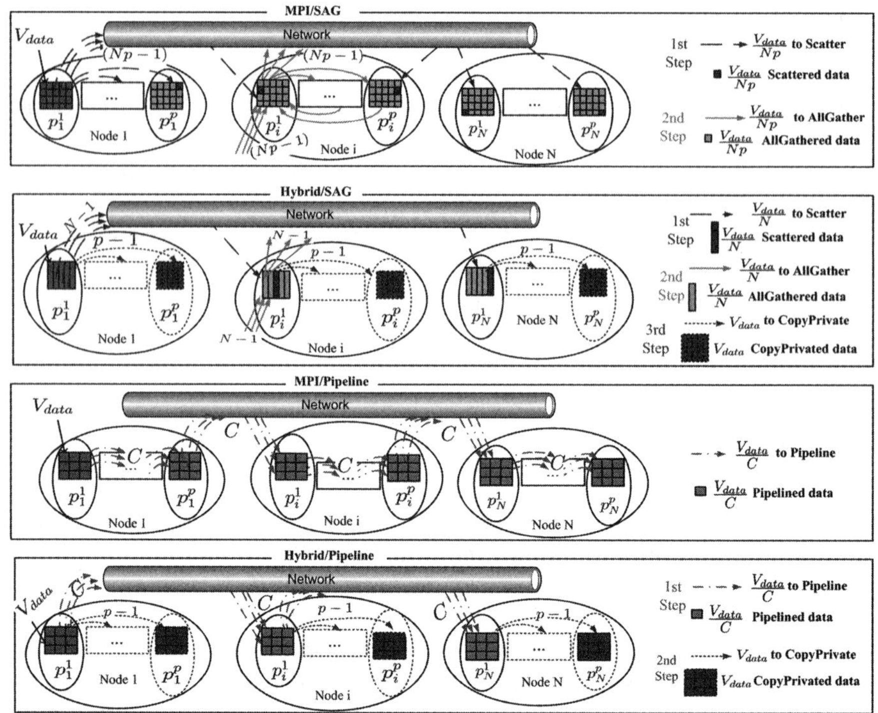

Fig. 1 Data broadcasting algorithms and associated parameters

2.3 Associated Parameters

In order to estimate the energy consumption of these operations, we need to take into account a large set of parameters. These operations are associated to parameters that depend not only on the protocols but also on the application features, and on the hardware used. Thus, in order to estimate accurately the energy consumption due to a specific implementation of a fault tolerance protocol, the estimator needs to take into consideration all the protocol parameters (checkpointing interval, checkpointing storage destination, etc.), all the application specifications (number of processes, number and size of messages exchanged, volume of data written/read by each process, etc.) and all the hardware parameters (number of cores per node, memory architecture, type of hard disk drives, etc.).

- service and application parameters: checkpointing interval, checkpointing storage destination, volume of data to broadcast, number of processes, number and size of messages exchanged between processes, type of storage media used (RAM, HDD, NFS, etc), volume of data written/read by each process, etc.
- hardware parameters: number of nodes, number of sockets per node, number of cores per socket, network topology, memory architecture, network technologies

(Infiniband, Gigabit Ethernet, proprietary solutions, etc), type of hard disk drives (SSD, SATA, SCSI, etc), etc.

We consider that a parameter is a variable of our estimator only if a variation of this parameter generates a significant variation of the energy consumption while all the other parameters are fixed. It is necessary to calibrate the execution platform by taking into account all the parameters to estimate the energy consumption.

3 Energy Calibration Methodology

Energy consumption depends strongly on the hardware used in the execution platform. For instance, the energy consumption of checkpointing depends on the checkpointing storage destination (SSD, SATA, SCSI, etc.), on the read and write speeds and on the access times to the resource. The goal of the calibration process is to gather energy knowledge of all the identified operations according to the hardware used in the supercomputer. To this end, we gather the information about the energy consumption of the operations by running a set of benchmarks allowing to collect at set of power measurements and execution times of the various operations. The goal of such calibration approach is to adapt to the supercomputer used, the energy evaluations computed from the theoretical estimation models, and this in order to make our energy estimations accurate on any supercomputer, regardless of specifications. Although this knowledge base has a significant size, it needs to be done only occasionally, for example when there is a change in the hardware (like a new hard disk drive).

To estimate the energy consumption of a node performing an operation *op*, we need to obtain the power consumption of the node during the execution of *op* and the execution time of this operation. We know from [10] that the nodes from a same cluster are homogeneous in terms of performance. Therefore, we do need to measure and estimate the execution time due to an operation only for each type of nodes. Thus, the energy $\xi_{op}^{Node_i}$ consumed by a node i performing an operation *op* is:

$$\xi_{op}^{Node_i} = \rho_{op}^{Node_i} \cdot t_{op}$$

Analogously, the energy consumption $\xi_{op}^{Switch_j}$ of a (*switch*) j during the operation *op* is:

$$\xi_{op}^{Switch_j} = \rho_{op}^{Switch_j} \cdot t_{op}$$

t_{op} is the time required to perform *op* by un type of nodes.
$\rho_{op}^{Node_i}$ is the power consumed by the node i during t_{op}.
$\rho_{op}^{Switch_j}$ is the power consumed by the switch j during t_{op}.

As a consequence, in order to calibration the energy consumption, we need a calibrator for the power consumption described in Sect. 3.1 and a calibrator for the execution time described in Sect. 3.2.

3.1 Calibration of the Power Consumption ρ_{op}

We showed in [10] that the power consumption of a node i performing an operation op is composed of a static part, $\rho_{idle}^{Node_i}$, which is the power consumption of the node i when it is idle and a dynamic part $\Delta\rho_{op}^{Node_i}$, which is the extra power cost related to the operation op. We have shown that $\rho_{idle}^{Node_i}$ can be different even for identical nodes from homogeneous clusters. Therefore, we measure $\rho_{idle}^{Node_i}$ for each node i. We also have shown in [10] that $\Delta\rho_{op}^{Node_i}$ is the same for identical nodes running the same operation op. Consequently, we measure $\Delta\rho_{op}^{Node_i}$, for each operation op, once for each type of nodes,

In [10], we have also highlighted that the number p of processes used per node may influence the power consumed by the node. Therefore, we need to measure $\rho_{op}^{Node_i}(p)$ for every operation op and for different values of p. Thus, the power consumption $\rho_{op}^{Node_i}(p)$ of a node i during an operation op using p processes of this node is:

$$\rho_{op}^{Node_i}(p) = \rho_{idle}^{Node_i} + \Delta\rho_{op}^{Node}(p)$$

Analogously, the power consumption $\rho_{op}^{Switch_j}$ of a switch j during the operation op is:

$$\rho_{op}^{Switch_j} = \rho_{idle}^{Switch_j} + \Delta\rho_{op}^{Switch}$$

$\rho_{idle}^{Node_i}$ (or $\rho_{idle}^{Switch_j}$) is the power consumption of a node i (or of a switch j) when it is idle (*i.e.,* switched on but executed nothing except the operating system) and $\Delta\rho_{op}^{Node_i}$ (or $\Delta\rho_{op}^{Switch_j}$) is the extra power consumption due to the execution of the operation op.

In order to compute $\Delta\rho_{op}^{Node}(p)$ (or $\Delta\rho_{op}^{Switch}$), we measure $\rho_{op}^{Node_i}(p)$ (or $\rho_{op}^{Switch_j}$) for a given node i (or a switch j) and subtract the static part of the power consumption which corresponds to the idle power consumption of the node i (or switch j). We measure $\rho_{op}^{Node_i}(p)$ (or $\rho_{op}^{Switch_j}$) by making the operation last a few seconds. Therefore, it is an *mean* extra power consumption because it is computed from the average of several power measurements (one every second).

Moreover, $\rho_{op}^{Node_i}(p)$ and so $\Delta\rho_{op}^{Node}(p)$ may vary depending on the number p of processes used by the node i. Therefore, we need to calculate $\Delta\rho_{op}^{Node}(p)$ and thus to measure $\rho_{op}^{Node_i}(p)$ for different values of p in order to be able to estimate $\Delta\rho_{op}^{Node}(p)$ for a number p of processes executing the operation op. To do this, we should be able to know how $\Delta\rho_{op}^{Node}(p)$ evolves according to p (and this for each type of nodes). We do not know such information a priori. To this end, we rely on four possible models presented in the table below:

Linear	$\Delta\rho_{op}^{Node}(p) = \alpha p + \beta$
Logarithmic	$\Delta\rho_{op}^{Node}(p) = \alpha ln(p) + \beta$
Power	$\Delta\rho_{op}^{Node}(p) = \beta p^{\alpha}$
Exponential	$\Delta\rho_{op}^{Node}(p) = \alpha^{p} + \beta$

For each type of nodes, we measure $\Delta\rho_{op}^{Node}(p)$ for five different numbers of processes:

- the smallest possible value p denoted p_{min}, which is equal to 1;
- the highest possible value p denoted p_{max}, which corresponds to the number of cores available in the node;
- the median value denoted p_2 which corresponds to half of the number of cores available in the node;
- the number p_1 which is located in the middle of the interval $[p_{min}; p_2]$;
- the number p_3 which is located in the middle of the interval $[p_2; p_{max}]$

Then, we determine thanks to the least squares method [24] the coefficients (α and β) of each of the four models according to the five measured values for $\Delta\rho_{op}^{Node}(p)$. We compute the coefficient of determination R^2 corresponding to each of the four adjusted models obtained with the least squares method. We consider $\Delta\rho_{op}^{Node}(p)$ evolves according to the adjusted model for which the coefficient of determination is the highest one (*i.e.,* that is to say, the closest to 1) .

For our measurements of $\Delta\rho_{op}^{Node}(p)$ (deduced from the measurements of $\rho_{op}^{Node_i}(p)$), we use an external wattmeter capable to provide us the mean power measurements with a sufficiently high frequency (1 Hz). We have shown in [9] that the OMEGAWATT wattmeter is a good candidate to collect such power measurements.

3.2 *Calibration of the Execution Time* t_{op}

The execution time t_{op} depends on one or many parameters according to the operation op. To take into account the possible effects of congestion, we consider that the number p of the same process node performing the same operation simultaneously op is a parameter to consider in our calibration of t_{op}. For example, this may occur if multiple processes on the same node try to write data simultaneously on the local hard drive.

To calibrate t_{op}, we need to measure the execution time by varying different parameters. To do this, we measure t_{op} for five values uniformly distributed between the minimum and maximum for each parameter (while fixing all the other parameters). The five values of each parameter are chosen similarly to what we have previously reported with the parameter p for the calibration of $\Delta\rho_{op}^{Node}(p)$.

We consider two cases:

1. "Known model" case: we know a model a model where t_{op} evolves with respect to the parameters. We know it from the literature, with the knowledge of the algorithm used in the operation op or resource requested by the operation op. In this case, we determine the coefficients of the theoretical model using the least squares method [24] based on the values of the five parameters.
2. "No model known" case: we do not know how t_{op} evolves with respect to the parameters. In this case, for each parameter, we proceed to the determination by the adjusted least squares method as presented for the calibration of $\Delta\rho_{op}^{Node}(p)$ relying on the four models (linear, logarithmic, exponential and power).

To measure t_{op}, we instrument the code of the algorithm or the protocol of the operation op, in order to obtain the corresponding execution time. To ensure that the calibration of the execution time is accurate, we realize each measurement 30 times and we compute the mean value of the 30 measurements.

3.2.1 Fault Tolerance Case

In this section, we describe the models used for the execution times of each operation of the fault tolerance protocols. For each operation op, t_{op} depends on different parameters.

We remind that the calibration of t_{op} is required for each type of nodes. In other words, we do not need to calibrate t_{op} on all nodes when they are all identical.

For each type of nodes, the time $t_{checkpointing}$ required for checkpointing a volume of data V_{data} is:

$$t_{checkpointing}(p, V_{data}) = t_{access}(p) + t_{transfer}(p, V_{data}) = t_{access}(p) + \frac{V_{data}}{r_{transfer}(p)}$$

Similarly, the time $t_{logging}$ required to log a message with a size equal to V_{data} is:

$$t_{logging}(V_{data}) = t_{access} + t_{transfer}(V_{data}) = t_{access} + \frac{V_{data}}{r_{transfer}}$$

p is the number of processes within the same node simultaneously trying to perform the checkpointing operation. t_{access} is the time required to access the storage media where the checkpoint will be saved or the message logged. $t_{transfer}$ is the time required to write data size V_{data} on the storage medium. $r_{transfer}$ is the transmission rate when writing on storage medium.

In the case of checkpointing, t_{access} and $r_{transfer}$ (and $t_{transfer}$) depend on the number p of processes per node since the p processes save their checkpoints simultaneously on the same storage media as the frequency of checkpoints writing is the same for all processes of an application.

A message is logged on a storage medium once it has been sent by a process of the node through the network interface used by the node. Thus, if several processes of the node try to send messages, there will be a traffic congestion at the network interface and the time for the current message will overlap the time of writing the message previously sent. In other words, this means that we consider that we can not find themselves in a situation where multiple messages are logged simultaneously by p processes of the node. Therefore, in the case of message logging, t_{access} and $r_{transfer}$ (and $t_{transfer}$) do not depend on the number p process per node.

As explained in Sect. 3.2, we measure $t_{checkpointing}$ considering both p and V_{data} parameters.

We know the theoretical model of $t_{checkpointing}$ based on V_{data} so for this parameter, we proceed to the determination of the coefficients of the theoretical model as explained in the case "with known model" (Sect. 3.2).

For p parameter, we do not have theoretical model giving $t_{checkpointing}$ based on p and therefore proceed as explained in the case of "no known model" (Sect. 3.2).

Regarding $t_{logging}$, it depends only on V_{data} and we have the theoretical model giving $t_{logging}$ depending on this parameter. So we proceed as explained in the "with known model" case.

We calibrate $t_{checkpointing}$ and $t_{logging}$ with respect to various storage media available on each node of the platform (RAM, local hard disk, flash SSD, etc..).

As we consider checkpointing at system-level, coordinated protocol requires a coordination between all processes.

The execution time for coordination between all processes is:

$$\begin{aligned} t_{coordination}(N, p, V_{data}) &= t_{polling}(V_{data}) + t_{synchro}(N, p) \\ &= \frac{V_{data}}{R_{transfer}} + t_{synchro}(N, p) \end{aligned}$$

p is the number of processes of the node i trying to perform coordination. $t_{synchro}(N, p)$ is the time required to exchange a marker synchronization between all processes. $t_{synchro}(N, p)$ depends on the number of nodes and the number of processes per node involved in the synchronization. We do not have a theoretical model for $t_{synchro}(N, p)$ neither in terms of N nor based on p. For the calibration, we proceed as explained in the "without known model" case (Sect. 3.2). $t_{polling}(V_{data})$ is the time required to finish transmitting the messages being transmitted at the time of coordination. In other words, $t_{polling}(V_{data})$ is equal to the time required to transfer the larger application message. $R_{transfer}$ is the transmission rate in the network infrastructure used for the platform.

Regarding the polling time, $t_{polling}(V_{data})$, we have a theoretical model giving $t_{polling}(V_{data})$. For the calibration, we proceed as in the "known model" case for V_{data} parameter.

3.2.2 Data Broadcasting Case

In this section, we describe the models of the execution times that we consider for each operation op.

In [25], the authors present theoretical models for the operations *Scatter*, *AllGather* and *Pipepline*. However, these models assume that there is only one process per node.

We adjusted the theoretical models presented in [25] in order to take into consideration the number p of processes per node (see Fig. 1).

Thus, the time required to perform a *Scatter*, a *AllGather* or a *Pipeline* with a volume of data V_{data} from N nodes with p processes per node is:

$$\begin{aligned} t_{Scatter}(N, p, V_{data}) &= t_{AllGather}(N, p, V_{data}) \\ &= (T_{Snet}(N, p) + \frac{V_{data}}{R_{net}(N, p)}) \cdot \frac{Np - 1}{Np} \\ t_{Pipeline}(N, p, V_{data}) &= (T_{Snet}(N, p) + \frac{V_{data}}{R_{net}(N, p)}) \cdot \frac{C + Np - 2}{C} \end{aligned}$$

$T_{Snet}(N, p)$ is the time needed to start the network link and $R_{net}(N, p)$ is the transfer rate to transmit a data volume V_{data}. C is the number of parts (of equal size) in which the data is divided into for *MPI/Pipeline*. It is equal to the total volume of data to be broadcasted divided by the size of each piece of data. Thus, the size of each piece and therefore C depends on the chosen implementation for the algorithm *MPI/Pipeline*.

Regarding $t_{Scatter}$, $t_{AllGather}$ and $t_{Pipeline}$, we have the theoretical models based on the parameters N, p and V_{data}. Therefore, for the calibration of the execution of these operations, we proceed as explained in the case "with known model" (Sect. 3.2).

The time required to copy data to all processes located on the same node is a function of the number of processes per node p and the volume of data to be copied V_{data}:

$$t_{CopyPrivate}(p, V_{data}) = t_{access}(p) + \frac{V_{data}}{r_{transfer}(p)}$$

t_{access} is the time required to access the RAM memory in which the data V_{data} is stored following an operation *AllGather* or *Pipeline*. $r_{transfer}$ is the transmission rate for a given copy in RAM. As explained in Sect. 3.2, we measure $t_{CopyPrivate}$ considering both p and V_{data} parameters. We have the theoretical model $t_{CopyPrivate}$ based on V_{data} so for this parameter, we proceed to the determination of the coefficients of the theoretical model as explained in the case "with known model" (Sect. 3.2). We do not have a theoretical model giving $t_{CopyPrivate}$ based on p and then proceed as explained in the "no known model" case (Sect. 3.2).

4 Energy Estimation Methodology

We have previously described how we realize the energy calibration. Once the calibration is done, the estimator is able to provide estimates of the energy consumed by the various operations identified for each studied service. Figure 2 shows the framework components related to the estimation of the energy consumed.

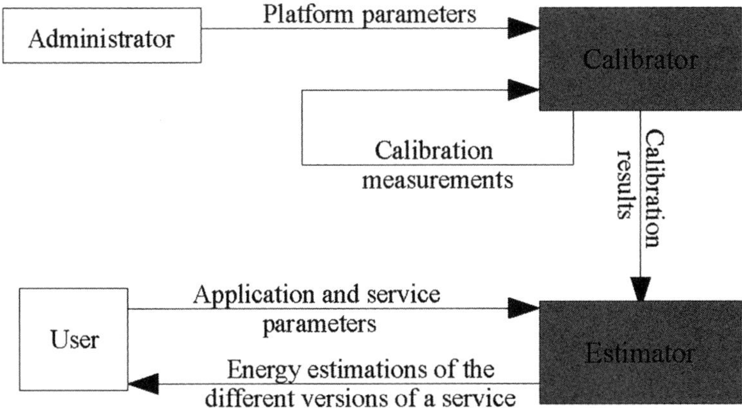

Fig. 2 Framework to estimate the energy consumption of services

We can now describe how to estimate the energy consumed by each of the identified operations. To this end, we rely on the parameters provided by the user and the data measured by our calibrator.

Once the administrator has provided the hardware settings of the platform, the calibrator performs the various steps required to build the knowledge base on the power consumption and the execution time of the various identified operations. Then, based on the calibration results and a description of the application (volume of data to broadcast, the application memory size, etc..) and runtime parameters (number of nodes used, number of processes per node, etc..) provided by the user, the estimator calculates the energy consumption of different versions of the service.

The parameters that we get from the user for the estimation depend on each operation to estimate. In case these parameters correspond to the values that we have measured during calibration, estimation directly uses these values to calculate the energy consumed by the operation. If this is not the case, that is to say, if there is a lack of measurement points in the calibrator, the estimator is uses the models created with the least squares method [24] during calibration. In Sects. 4.1 and 4.2, we show respectively how this method applies to the fault tolerance and data broadcasting services.

4.1 Fault Tolerance Case

This section describes how we estimate the energy consumed by each operation identified in the protocols of fault tolerance. For this, we show the necessary information: the parameters provided by the user and the data measured by our calibrator.

4.1.1 Checkpointing

To estimate the energy consumption of checkpointing, the estimator gets from the user the total memory size required by the application to run, the number of nodes N and the number p of processes per node.

From this information, the estimator calculates the average memory size V_{memory}^{mean} required by each process (total memory size divided by the number of processes). Then the estimator gets from the calibrator the extra power consumption $\Delta\rho_{checkpointing}(p)$ and the execution time $t_{checkpointing}(p, V_{memory}^{mean})$ depending on the models obtained by the least squares method in the step of the calibration.

It also gets the measurement $\rho_{idle}^{Node_i}$ for each node i. We denote respectively by $\xi_{checkpointing}^{Node_i}(p)$ and $\rho_{checkpointing}^{Node_i}(p)$ the energy consumption and the average power consumption of each node i performing checkpointing. The estimation of the energy consumption of a single checkpointing is given by:

$$\begin{aligned} E_{checkpointing} &= \sum_{i=1}^{N} \xi_{checkpointing}^{Node_i}(p) \\ &= \sum_{i=1}^{N} \rho_{checkpointing}^{Node_i}(p) \cdot t_{checkpointing}\left(p, V_{memory}^{mean}\right) \\ &= t_{checkpointing}\left(p, V_{memory}^{mean}\right) \cdot \sum_{i=1}^{N} \left(\rho_{idle}^{Node_i} + \Delta\rho_{checkpointing}(p)\right) \\ &= t_{checkpointing}\left(p, V_{memory}^{mean}\right) \cdot \left(N \cdot \Delta\rho_{checkpointing}(p) + \left(\sum_{i=1}^{N} \rho_{idle}^{Node_i}\right)\right) \end{aligned}$$

4.1.2 Message Logging

To estimate the energy consumption of message logging, the estimator gets from the user the number of nodes N, the number p of processes per node, the number and total size of all messages sent during the application that he wants to run.

With this information, the estimator calculates the average volume V_{data}^{mean} of data sent and therefore logged on each node (total size of all messages sent divided by the number of nodes N). Then, the estimator gets from the calibrator the extra power consumption $\Delta\rho_{logging}$ and the execution time $t_{logging}(p, V_{data}^{mean})$ depending on the models obtained with least squares method in the step of the calibration. It also receives the measurement of $\rho_{idle}^{Node_i}$ for each node i.

The estimation of the energy consumption of messages logging is given by:

$$\begin{aligned}
E_{logging} &= \sum_{i=1}^{N} \xi_{logging}^{Node_i}(p) \\
&= \sum_{i=1}^{N} \rho_{logging}^{Node_i}(p) \cdot t_{logging}\left(V_{data}^{mean}\right) \\
&= t_{logging}\left(V_{data}^{mean}\right) \cdot \sum_{i=1}^{N} \left(\rho_{idle}^{Node_i} + \Delta\rho_{logging}(p)\right) \\
&= t_{logging}\left(V_{data}^{mean}\right) \cdot \left(N \cdot \Delta\rho_{logging}(p) + \left(\sum_{i=1}^{N} \rho_{idle}^{Node_i}\right)\right)
\end{aligned}$$

4.1.3 Coordination

We remind that the coordination is divided into two phases: the active polling during the transmission of the inflight messages followed by the synchronization of all processes. To estimate the energy consumption of the coordination, the estimator calculates the average message size $V_{message}^{mean}$ as the total size of messages divided by the total number of messages exchanged. The estimator also uses the total number of nodes N and the number of processes per node p. Then the estimator gets from the calibrator the extra power consumption $\Delta\rho_{sync}(p)$ and the execution time $t_{sync}(N, p)$ depending on the models obtained with the least squares method in calibration step. It also receives the measurement $\rho_{idle}^{Node_i}$ for each node i. The estimation of the energy consumption $E_{synchro}$ of synchronization is given by:

$$\begin{aligned}
E_{synchro} &= \sum_{i=1}^{N} \xi_{synchro}^{Node_i}(N, p) \\
&= \sum_{i=1}^{N} \rho_{synchro}^{Node_i}(p) \cdot t_{synchro}(N, p) \\
&= t_{synchro}(N, p) \cdot \sum_{i=1}^{N} \left(\rho_{idle}^{Node_i} + \Delta\rho_{synchro}(p)\right) \\
&= t_{synchro}(N, p) \cdot \left(N \cdot \Delta\rho_{synchro}(p) + \left(\sum_{i=1}^{N} \rho_{idle}^{Node_i}\right)\right)
\end{aligned}$$

Regarding active polling, the estimator gets from the calibrator the extra power consumption $\Delta\rho_{polling}(p)$ and the execution time $t_{polling}(N, p, V_{message}^{mean})$ depending on the models obtained with least squares method in the calibration step. The estimation of the energy consumption of active polling is given by:

$$\begin{aligned} E_{polling} &= \sum_{i=1}^{N} \xi_{polling}^{Node_i}(N, p) \\ &= \sum_{i=1}^{N} \rho_{polling}^{Node_i}(p) \cdot t_{polling}\left(V_{message}^{mean}\right) \\ &= t_{polling}\left(V_{message}^{mean}\right) \cdot \sum_{i=1}^{N} \left(\rho_{idle}^{Node_i} + \Delta\rho_{polling}(p)\right) \\ &= t_{polling}\left(V_{message}^{mean}\right) \cdot \left(N \cdot \Delta\rho_{polling}(p) + \left(\sum_{i=1}^{N} \rho_{idle}^{Node_i}\right)\right) \end{aligned}$$

The estimator computes the energy consumption of coordination as follows:

$$E_{coordination} = E_{polling} + E_{synchro}$$

4.2 *Data Broadcasting Case*

This section describes how we estimate the energy consumed by each operation identified in the algorithms for data broadcasting. For this, we show the necessary information including: the parameters provided by the user and the data measured by our calibrator.

4.2.1 MPI/SAG and Hybrid/SAG

To estimate the energy consumption of the *MPI/SAG* and *Hybrid/SAG*, the estimator retrieves from the user the number of nodes N, the number p of processes per node, and the size of the data to be broadcasted V_{data}. With this information, the estimator then gets from the calibrator $\Delta\rho_{Scatter}(p)$, $\Delta\rho_{AllGather}(p)$, $\Delta\rho_{CopyPrivate}(p)$, $t_{Scatter}(N, p, V_{data})$, $t_{AllGather}(N, p, V_{data})$ and $t_{CopyPrivate}(p, V_{data})$ according to models obtained with least squares method in the calibration step. Moreover, the estimator retrieves from the calibrator the number M of network switches based on the physical description of the platform made by the administrator. The estimation of the energy consumption of the algorithm *MPI / SAG* is given by:

$$\begin{aligned} E_{MPI/SAG} &= \sum_{i=1}^{N} \xi_{MPI/SAG}^{Node_i} + \sum_{j=1}^{M} \xi_{MPI/SAG}^{Switch_j} \\ &= t_{Scatter}\left(N, p, V_{data}\right) \cdot \left(\sum_{i=1}^{N} \rho_{Scatter}^{Node_i}(p) + \sum_{j=1}^{M} \rho_{Scatter}^{Switch_j}\right) \\ &\quad + t_{AllGather}\left(N, p, V_{data}\right) \cdot \left(\sum_{i=1}^{N} \rho_{AllGather}^{Node_i}(p) + \sum_{j=1}^{M} \rho_{AllGather}^{Switch_j}\right) \end{aligned}$$

The estimation of the energy consumption of the algorithm *Hybrid/SAG* is given by:

$$
\begin{aligned}
E_{Hybrid/SAG} &= \sum_{i=1}^{N} \xi_{Hybrid/SAG}^{Node_i} + \sum_{j=1}^{M} \xi_{Hybrid/SAG}^{Switch_j} \\
&= t_{Scatter}(N,1) \cdot \left(\sum_{i=1}^{N} \rho_{Scatter}^{Node_i}(1) + \sum_{j=1}^{M} \rho_{Scatter}^{Switch_j} \right) \\
&\quad + t_{AllGather}(N,1) \cdot \left(\sum_{i=1}^{N} \rho_{AllGather}^{Node_i}(1) + \sum_{j=1}^{M} \rho_{AllGather}^{Switch_j} \right) \\
&\quad + t_{CopyPrivate}(p, V_{data}) \cdot \sum_{i=1}^{N} \left(\rho_{CopyPrivate}^{Node_i}(p) \right)
\end{aligned}
$$

4.2.2 MPI/Pipeline and Hybrid/Pipeline

Compared to the estimate of *MPI/SAG* and *Hybrid/SAG*, the estimator gets from the user an additional parameter which is the size of each piece (*chunk*) in order to estimate the energy consumption the *MPI/Pipeline* and *Hybrid/Pipeline*. This parameter is required for the determination of the performance of *Pipeline* since $t_{Pipeline}(N, p, V_{data})$ depends on the constant C. From the calibrator, the estimator gets $\Delta\rho_{Pipeline}(p)$ and $t_{Pipeline}(N, p, V_{data})$. The estimation of the energy consumption of the algorithm *MPI/Pipeline* is given by:

$$
\begin{aligned}
E_{MPI/Pipeline} &= \sum_{i=1}^{N} \xi_{MPI/Pipeline}^{Node_i} + \sum_{j=1}^{M} \xi_{MPI/Pipeline}^{Switch_j} \\
&= t_{Pipeline}(N, p, V_{data}) \cdot \left(\sum_{i=1}^{N} \rho_{Pipeline}^{Node_i}(p) + \sum_{j=1}^{M} \rho_{Pipeline}^{Switch_j} \right)
\end{aligned}
$$

The estimation of the energy consumption of the algorithm *Hybrid / Pipeline* is given by:

$$
\begin{aligned}
E_{Hybrid/Pipeline} &= \sum_{i=1}^{N} \xi_{Hybrid/Pipeline}^{Node_i} + \sum_{j=1}^{M} \xi_{Hybrid/Pipeline}^{Switch_j} \\
&= t_{Pipeline}(N,1) \cdot \left(\sum_{i=1}^{N} \rho_{Pipeline}^{Node_i}(1) + \sum_{j=1}^{M} \rho_{Pipeline}^{Switch_j} \right) \\
&\quad + \sum_{i=1}^{N} \left(t_{CopyPrivate}(p) \cdot \rho_{CopyPrivate}^{Node_i}(p) \right)
\end{aligned}
$$

5 Validation of the Estimations

To validate our estimations, we perform various real applications of high performance computing with different fault tolerance protocols or with different scenarios of data broadcasting on a homogeneous cluster of the experimental distributed platform for large-scale computing, Grid'5000 [4], then we compare the energy consumption actually measured to the energy consumption evaluated by our estimator. For the experiments to validate our estimations, we used a cluster of the Grid'5000 distributed platform. The cluster we used for our experiments offers 16 identical nodes Dell R720. Each node contains 2 Intel Xeon CPU 2.3 GHz, with 6 cores each; 32 GB of memory; a 10 Gigabit Ethernet network; a SCSI hard disk with a storage capacity of 598 GB. We monitor this cluster with an energy- sensing infrastructure of external wattmeters from the SME Omegawatt. This energy-sensing infrastructure, which was also used in [8], enables to get the instantaneous consumption in Watts, at each second for each monitored node [7]. Logs provided by the energy-sensing infrastructure are displayed lively and stored into a database, in order to enable users to get the power and the energy consumption of one or more nodes between a start date and an end date. We ran each experiment 30 times and computed the mean value over the 30 values. We use the same notations as in previous sections: N is the number of nodes, p is the number of processes and op denotes one of the identified operations (checkpointing, *Pipeline*, etc).

5.1 Calibration Results of the Platform

In this section, we present some of the calibration results on the considered platform according to the methodology described in Sect. 3. The considered platform is composed only of identical nodes: thus there is only one type of nodes. They are interconnected using a single network switch.

5.1.1 Calibrating the Power Consumption

First, we measure the idle power consumption ρ^i_{idle} for each node i of the experimental cluster. Figure 3 shows the idle power consumption of the 16 nodes belonging to the considered cluster. From this figure, even if the cluster is composed of homogeneous nodes, we notice the need to calibrate the electrical power when idle of each node.

For each identified operation op and for each of the cluster nodes, we calibrate with OMEGAWATT the average additional cost of electrical power due to the op operation, $\Delta\rho_{op}(p)$, as explained in Sect. 3.1. Since each node of the *Taurus* cluster has 12 processing cores, the five values of p we choose to calibrate $\Delta\rho_{op}(p)$ are 1, 4, 6, 9 and 12 processes per node. Figures 4 and 5 show the measurements $\Delta\rho_{op}(p)$ for the five values of p and for each operation op identified respectively in fault tolerance protocols and for data broadcasting algorithms.

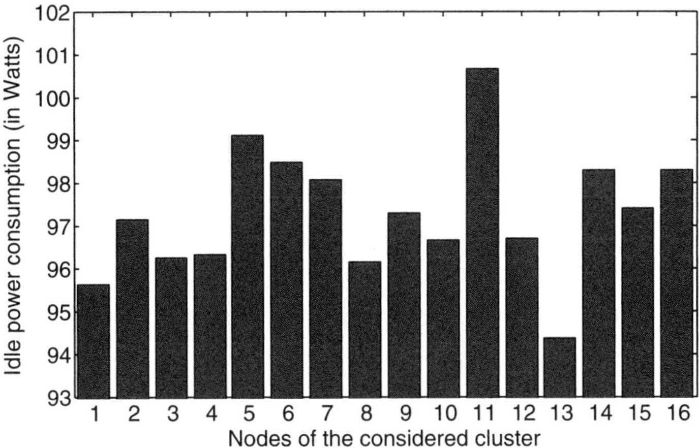

Fig. 3 Idle power consumption of the nodes of the cluster *Taurus*

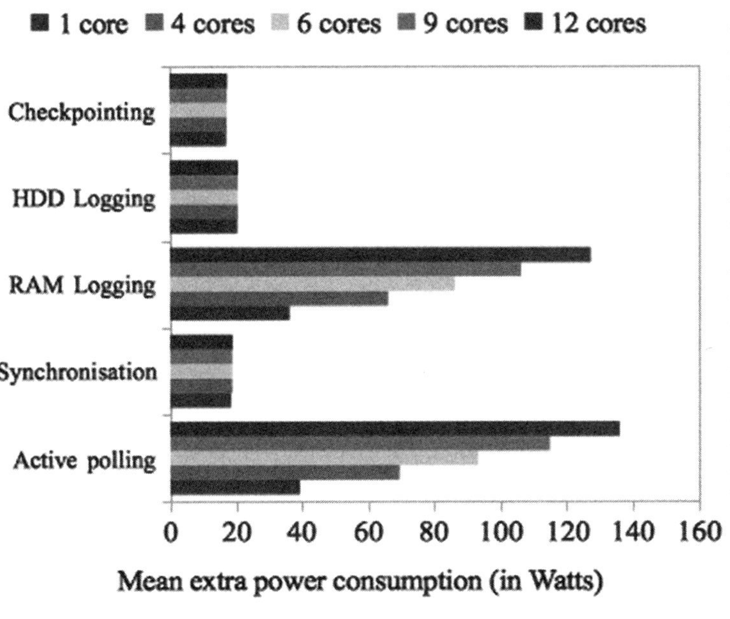

Fig. 4 Extra power consumption of operations related to fault tolerance protocols

We note in Figs. 4 and 5 that for some operations, $\Delta\rho_{op}(p)$ does vary depending on the number of cores per node that perform the same operation. For some operations, such as checkpointing $\Delta\rho_{op}(p)$ is almost a constant function of p. For $\Delta\rho op(p)$ of these operations, we obtain one of the four models of the calibrator (see

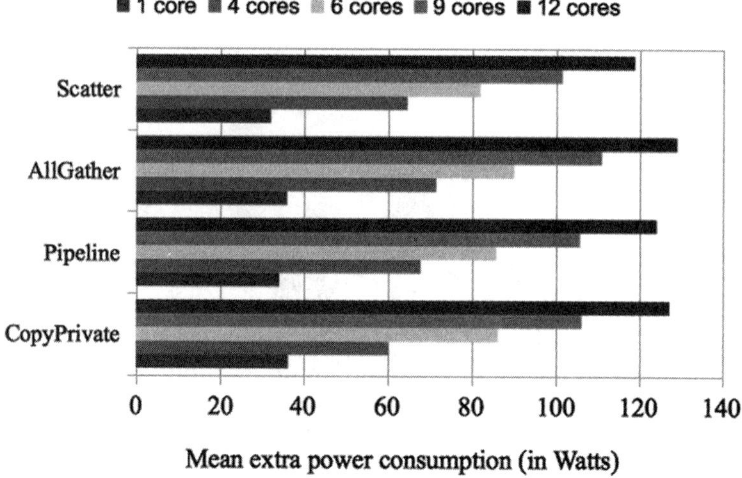

Fig. 5 Extra power consumption of operations related to data broadcasting algorithms

Sect. 3.1) with a coefficient α very close to 0 and a value of β very close to the constant value of $\Delta\rho_{op}(p)$ (*i.e.,* that is to say, quasi-stationary model). For example, the model of $\Delta\rho_{checkpointing}(p)$ adjusted by the least squares method for the five values of p is:

$$\Delta\rho_{checkpointing}(p) = 17.22 \cdot p^{0.0084271}$$

Although the fitted model is a power model, the very low coefficient α implies that $\Delta\rho_{checkpointing}(p)$ is a quasi-stationary function p. The coefficient of determination R^2 corresponding to this model is 0.976, which is very close to 1. For other operations such as *Pipeline*, $\Delta\rho op(p)$ increases with p. For example, the model of $\Delta\rho_{Pipeline}(p)$ obtained in the calibration is:

$$\Delta\rho_{Pipeline}(p) = 33.85 \cdot p^{0.518956}$$

The fact that α is very close to 0.5 means that $\Delta\rho_{Pipeline}(p)$ is almost expressed in terms of $\sqrt{p}$. The coefficient of determination R^2 corresponding to this model is 0.999, which is also very close to 1.

In addition, we measure the energy consumption when idle of 10 Gigabit Ethernet switch for 300 s followed by the electrical power during heavy network traffic for 300 s. To measure its electrical power when idle, we ensure that there is no network traffic by turning off all nodes that are interconnected by the network switch. To measure its electrical power during heavy network traffic, we run `iperf` in server mode on one of the nodes and `iperf` in client mode on all other interconnected nodes. Figure 6 shows the electrical measurements.

From Fig. 6, we note that the electric power switch remains almost constant throughout the duration of the experiment. In other words, the electrical power

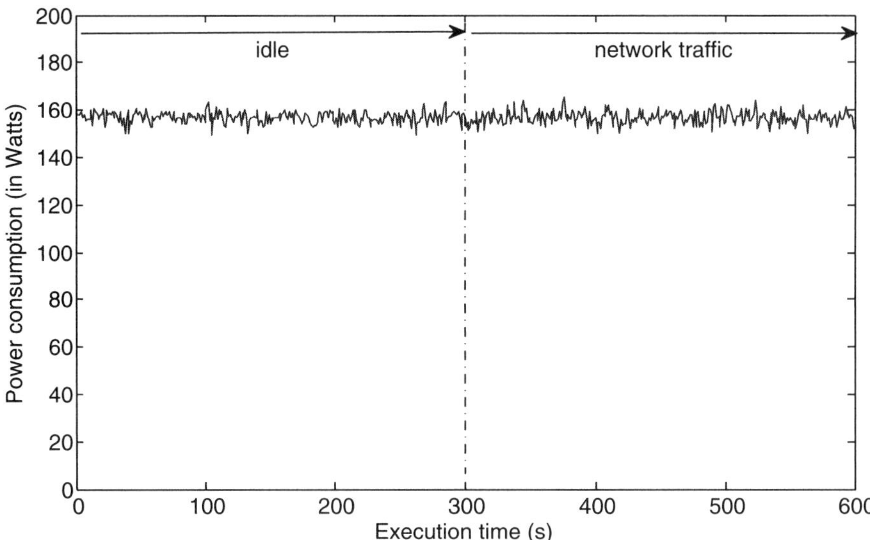

Fig. 6 Power consumption of the switch when idle for 300 s and with an intense network load for 300 s

network switch does not vary depending on the network traffic. This means that $\Delta\rho_{op}^{Switch}$ is (almost) equal to 0 for all operations $\left(\forall op, \rho_{op}^{Switch_j} = \rho_{idle}^{Switch_j}\right)$. A recent study [17, 19] confirms this fact in evaluating and demonstrating that the electrical power of multiple network devices is not affected by network traffic. That said, even if the electrical power of a network switch would depend on network traffic, our approach to calibration would allow to take into account in measuring $\Delta\rho_{op}^{Switch}$ for each operation op.

5.1.2 Calibration of the Execution Time

Based on the methodology presented in Sect. 3.2, we calibrate the execution time for each operation on each type of node of the experimental platform.

Fault Tolerance Case To calibrate the execution time of checkpointing on local hard drive, we consider a variable number of cores per node simultaneously checkpointing and we measure the time for different sizes of checkpoints V_{data} for one node of the experimental platform. Each node process saves a checkpoint with a size equal to V_{data}. In other words,when there are p processes that save checkpoints simultaneously a volume of $p \cdot V_{data}$ is saved on the local hard drive. Figure 7 shows the measured time for checkpointing on a node of the experimental platform. As explained in 3.2, we choose 1, 4, 6, 9 and 12 processes per node for the five values

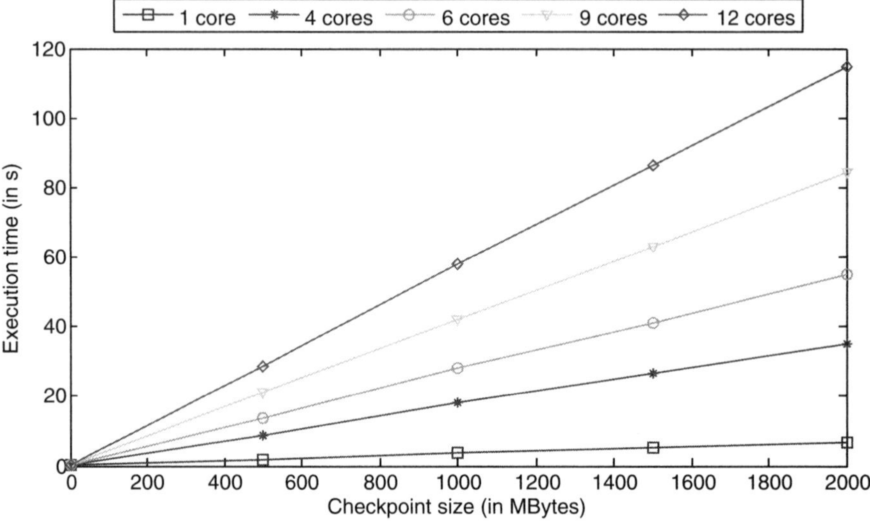

Fig. 7 Calibration of checkpointing on local hard drive

of p and 0 MB, 500 MB, 1000 MB, 1500 MB and 2000 MB for the five values of V_{data}. The choice of 2000 MB as the maximum size of checkpoint is motivated by the fact that each node has only 32 GB of memory that can be shared by 12 processing cores. For different values of p, Figure shows how evolves $t_{checkpointing}$ with respect to V_{data}.

First, we observe that the curves have a linear trend according to V_{data} for p fixed. For example, for $p = 4$, the model for $t_{checkpointing}$ adjusted by the least squares method from the five values of V_{data}:

$$t_{checkpointing}(4, V_{data}) = \frac{1}{0.56569 \cdot 10^9} \cdot V_{data} + 0.09433 \cdot 10^{-3}$$

We also note that for V_{data} fixed $t_{checkpointing}$ increases when p grows this is because of the congestion of the input-output generated by concurrent access by p process on local hard drive. For $V_{data} = 1000MB$ the model of $t_{checkpointing}$ adjusted by the least squares method from the five values of p:

$$t_{checkpointing}(p, 1000Mo) = 4.91359 \cdot p - 1.5026$$

If for example we want to estimate the time $t_{checkpointing}(3, 800MB)$, that is to say, for values of p and V_{data} which booth are not belonging to the five measured values, we calculate:

- on one side: $t_{checkpointing}(1, 800Mo)$, $t_{checkpointing}(4, 800Mo)$, $t_{checkpointing}(6, 800Mo)$, $t_{checkpointing}(9, 800Mo)$ et $t_{checkpointing}(12, 800Mo)$, respectively from the equations $t_{checkpointing}(1, V_{data})$, $t_{checkpointing}(4, V_{data})$, $t_{checkpointing}(6, V_{data})$, $t_{checkpointing}(9, V_{data})$ and $t_{checkpointing}(12, V_{data})$;
- on the other side: $t_{checkpointing}(3, 0Mo)$, $t_{checkpointing}(3, 500Mo)$, $t_{checkpointing}(3, 1000Mo)$, $t_{checkpointing}(3, 1500Mo)$, $t_{checkpointing}(3, 2000Mo)$, respectively from the equations $t_{checkpointing}(p, 0Mo)$, $t_{checkpointing}(p, 500Mo)$, $t_{checkpointing}(p, 1000Mo)$, $t_{checkpointing}(p, 1500Mo)$, $t_{checkpointing}(p, 2000Mo)$.

From the calculated values $t_{checkpointing}(1, 800Mo)$, $t_{checkpointing}(4, 800Mo)$, $t_{checkpointing}(6, 800Mo)$, $t_{checkpointing}(9, 800Mo)$ and $t_{checkpointing}(12, 800Mo)$, we determine by the least squares method, the model giving $t_{checkpointing}(p, 800Mo)$ as a function of p (as explained in Sect. 3.2) and calculate the determination coefficient R^2 corresponding to the adjusted model.

Similarly, from the values $t_{checkpointing}(3, 0Mo)$, $t_{checkpointing}(3, 500Mo)$, $t_{checkpointing}(3, 1000Mo)$, $t_{checkpointing}(3, 1500\ Mo)$, $t_{checkpointing}(3, 2000\ Mo)$, we determine the model giving $t_{checkpointing}(3, V_{data})$ as a function of V_{data} and calculate the determination coefficient R^2 corresponding to the thereby adjusted model. Then between $t_{checkpointing}(p, 800Mo)$ and $t_{checkpointing}(3, V_{data})$, we choose the model for which the determination coefficient is the closest to 1. Then we calculate $t_{checkpointing}(3, 800Mo)$ with the choosen model.

Figure 8 presents the execution time for message logging in RAM and on a HDD. To calibrate the execution time of message logging on memory or on disk, we measure the time for different message sizes V_{data} for one node of the experimental platform. The values choosen for V_{data} are 0 Ko, 500 Ko, 1000 Ko, 1500 Ko et 2000 Ko. As explained in Sect. 3.2.1, we do not need to calibrate $t_{logging}$ as a function of p because the processes do not write simultaneously the messages on the medium storage due to the contention during message sending. We measure the execution time when a single process ($p = 1$) of the node executes the message logging operation.

We observe that the curves have a linear trend and this as well for message logging on RAM on local hard drive. The message logging time on local hard drive is higher than the RAM one and this regardless of the size of the logged message. Similarly to checkpointing, we get the following adjusted models for $t_{logging}$:

$$\text{In RAM}: t_{logging}(V_{data}) = \frac{1}{4.4342 \cdot 10^9} \cdot V_{data} + 0.0426 \cdot 10^{-3}$$

$$\text{On the local HDD}: t_{logging}(V_{data}) = \frac{1}{1.0552 \cdot 10^9} \cdot V_{data} + 0.0858 \cdot 10^{-3}$$

Regarding coordination, we need to calibrate the time of the synchronization as well as the transfer time of a message.

To calibrate the synchronization time $t_{synchro}(N, p)$ of Np process, we measure this time for different values of N and for different values of p The measured values of p and N are chosen as explained in Sect. 3.2. In our case, the measured values of p are 1, 4, 6, 9 and 12 while the measured values for N are 1, 4, 8, 12 and 16. Figure 9 presents the synchronization time measured by the calibrator. For example,

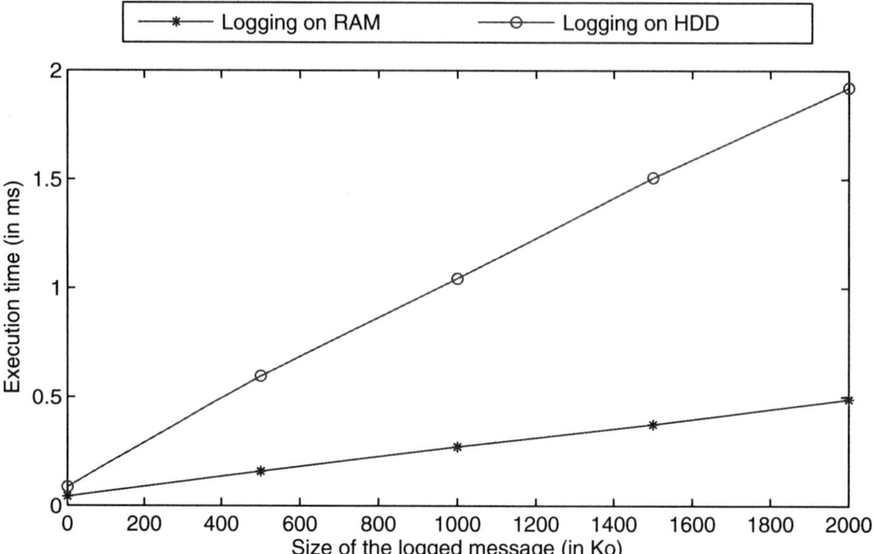

Fig. 8 Calibration of message logging on RAM and local disk

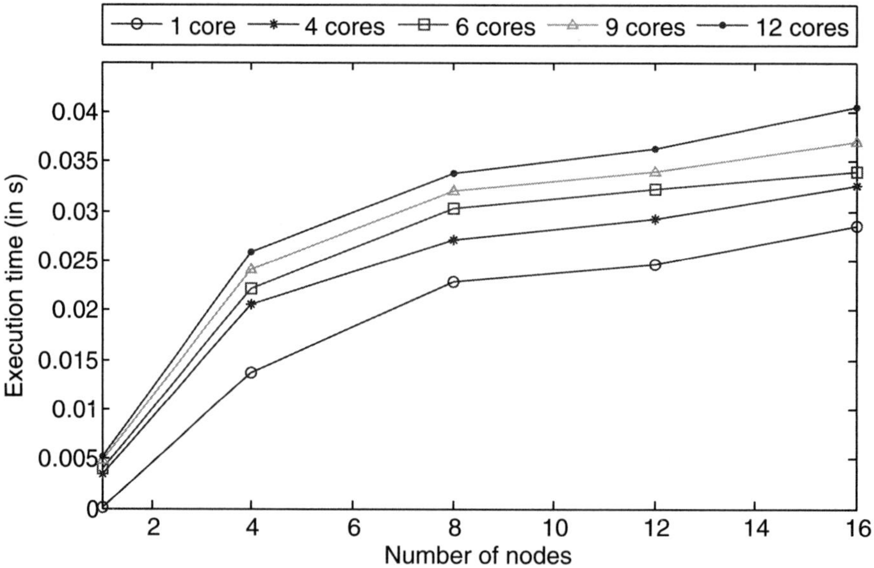

Fig. 9 Calibration of the synchronization time of the experimental platform

point 4 cores / 8 nodes is the time required to synchronize 32 processes 32 uniformly distributed over 8 nodes. First, we find that the time to synchronize processes located on the same node is lower than for processes located on different nodes. Indeed, it requires much less time to synchronize processes located on the same node than for processes located on different nodes. The transmission rate of the network is much lower than the transmission rate within a single node.

For example, for $p = 4$, the model for $t_{synchro}$ adjusted by the least squares method from the five values of N is:

$$t_{synchro}(N,4) = 0.0103757 \cdot ln(N) + 0.00445945$$

For $N = 8$, the model for $t_{synchro}$ adjusted by the least squares method from the five values of N is:

$$t_{synchro}(8,p) = 0.00443799 \cdot ln(p) + 0.02225942$$

If for example we want to estimate the time $t_{synchro}(N,p)$, that is to say, for values of N and p which booth are not belonging to the five measured values, then we proceed in a manner similar to that explained for $t_{checkpointing}(p, V_{data})$.

We calibrate the time needed to transfer a message (*i.e.,* the active polling occurring at the time of coordination) on the experimental platform by varying the of size V_{data} of the message to transfer. To do this, we measure the execution time $t_{polling}(V_{data})$ to transfer a message sent using *MPI_Send* by a process located on a given node to a process on a different node. In the general case, we must make this measurement for each pair of processes at different levels of the network hierarchy, ie for two processes that need to cross a single network switch, then for two processes that need to cross two network switches, etc. In our experimental platform, a single network switch interconnects all nodes so we only need to measure the time for a couple of processes on different nodes.

To calibrate the execution time to transfer a message over the network , we measure the time for different message sizes V_{data} for a couple of processes located on two separate nodes. The values chosen for V_{data} are 0 KB, 500 KB, 1000 KB, 1500 KB and 2000 KB. On Fig. 10, we present the calibration of the transfer time of a message.

The measured transfer time depends linearly on the size of the message transferred. Similarly to checkpointing, we get the following adjusted model for $t_{polling}$:

$$t_{polling}(V_{data}) = \frac{1}{0.60148 \cdot 10^9} \cdot V_{data} + 3.6222 \cdot 10^{-3}$$

Data Broadcasting Case To calibrate the execution times ($t_{Scatter}(N,p,V_{data})$, $t_{AllGather}(N,p,V_{data})$ and $t_{Pipeline}(N,p,V_{data})$), we measure this execution times for different values of N, de p and V_{data}. As explained in Sect. 3.2.2, since we know a theoretical model for these operations as a function of all the parameters (N, p et V_{data}), we determine the coefficients $T_{Snet}(N,p)$ et $R_{net}(N,p)$ that are present for the 3 models, using the least squares method.

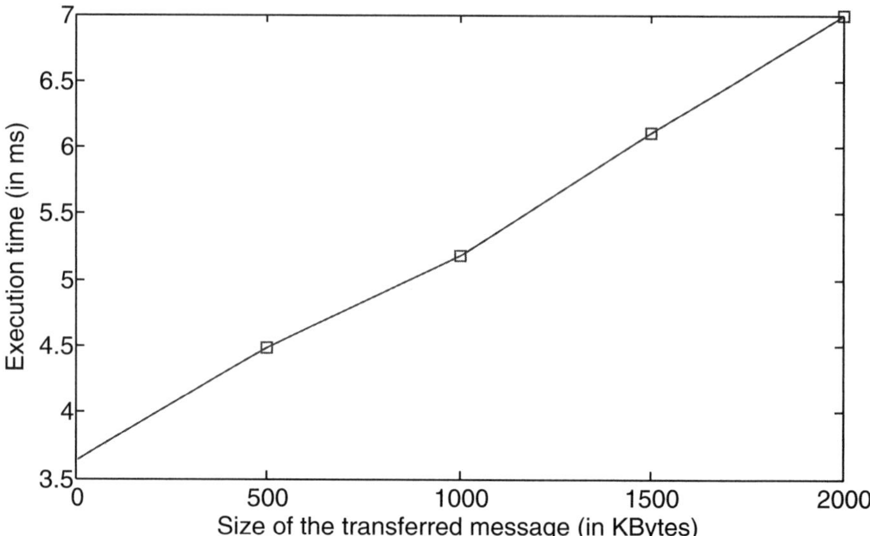

Fig. 10 Calibration of the active polling for the experimental platform

We present some of the results of the calibration of the execution times for operations *scatter* and *Pipeline* on Fig. 11 for a variable number of nodes (4, 8, 12, 16), for a process per node and for a data volume from 0 MB to 2000 MB through the values 500 MB, 1000 MB and 1500 MB (as explained in Sect. 3.2). In this figure, we have not shown the execution time measurements for the operation *AllGather* because the curves obtained for this operation are overlaid with those of the operation *Scatter* because the theoretical models are the same for these two operations (see Sect. 3.2.2).

In order not to overload the Fig. 11 (25×2 curves instead of 5×2 because two operations shown), we have presented the measurements for a fixed number of processes per node. The choice to present these measurements for a number $p = 1$ is motivated by the fact that we also use $t_{Scatter}(N, 1, V_{data})$, $t_{AllGather}(N, 1, V_{data})$ et $t_{Pipeline}(N, 1, V_{data})$ for the energy estimation of hybrid broadcast (which is not the case for $p > 1$).

For N and p fixed, Fig. 11 shows that $t_{Scatter}$ and $t_{Pipeline}$ evolve linearly with respect to V_{data}. In other words, $T_{Snet}(N, p)$ and $R_{net}(N, p)$ are constant if we consider a fixed number of nodes and a given number of processes per node. For a single fixed parameter p (or N), we determine $T_{Snet}(N, p)$ and $R_{net}(N, p)$ by applying the least squares method according on the other parameter N (or p respectively), similarly to what was presented for $t_{checkpointing}(p, V_{data})$. Thanks to these models we can estimate T_{Snet} et R_{net} for values of N and p given by the user when one of the two parameters is one of five values measured. When none of the two parameters were measured by our calibrator, we proceed in a manner similar to what is presented for checkpointing (see Sect. 5.1.2).

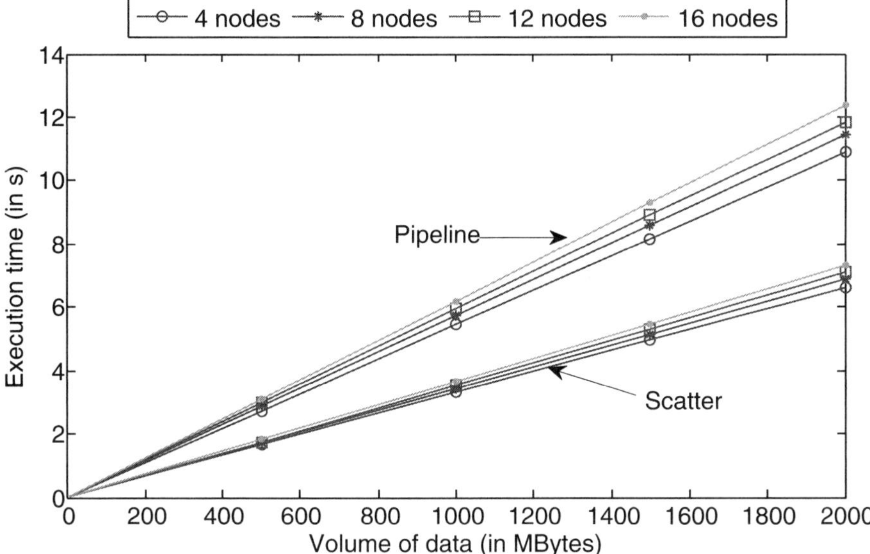

Fig. 11 Calibration of $t_{Scatter}(N, 1)$ et $t_{Pipeline}(N, 1)$

Regarding $t_{CopyPrivate}(p, V_{data})$, we calibrate in a manner similar as checkpointing (see Sect. 5.1.2) but with the RAM as support storage. For different numbers (1, 4, 6, 9, 12) of processes located on the same node, we measure $t_{CopyPrivate}(p, V_{data})$ for variable sizes of data (0 MB, 500 MB , 1000 MB, 1500 MB, 2000 MB).

Figure 12, presents the results of the calibration for $t_{CopyPrivate}(p, V_{data})$. $t_{CopyPrivate}(1, V_{data})$ is equal to $t_{logging}(V_{data})$ for writing a message in the RAM.

We need to take into account $t_{CopyPrivate}(p, V_{data})$ when we consider multiple processes per node with hybrid broadcasting algorithms because this time is not negligible. In addition, for a fixed volume of data , we see that $t_{CopyPrivate}(p, V_{data})$ remains almost unchanged when we consider an increasing number of processes per node. This is explained by the fact that the shared data is simultaneously without congestion in the RAM of the node.

5.2 *Accuracy of the Estimations*

In this section, we seek to compare the energy consumption achieved by our estimator once the calibration is made (but before executing the application) to the energy actually measured by the meters OmegaWatt during the execution of the application.

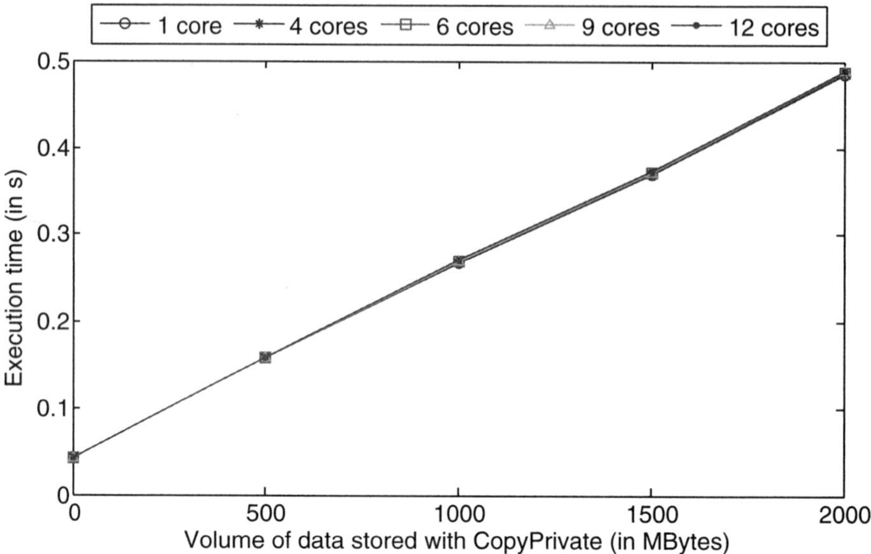

Fig. 12 Calibration of $t_{CopyPrivate}^{Node_i}(p)$

5.2.1 Fault Tolerance Case

We consider 4 HPC applications : CM1[3] with a 2400x2400x40 resolution and 3 NAS[4] of class D (SP, BT, et EP) executed on 144 processes (*i.e.,* 12 nodes with 12 cores per node) of the considered cluster.

With the infrastructure of external wattmeter OMEGAWATT, we measure for each application the energy consumption during the execution of the application with and without activation of the fault tolerance service. Specifically, we instrumented the source code implementations of the different protocols of fault tolerance in order to enable/disable each of the operations described above: checkpointing, message logging (on local disk or RAM disk) and coordination. Thus, we obtain the actual energy consumption for each operation. Each energy measurement is performed 30 times, and we consider the average values.

As concerns the uncoordinated protocol, we estimated and measured the energy consumption of all message logging. As concerns the coordinated protocol, we estimated and measured the energy consumption of a single checkpointing and therefore for one single coordination. To measure the energy consumption of a single checkpointing, we used a checkpoint interval greater than half of the application duration. Thus, the first (and only) checkpoint will occur in the second half of the application.

---

[3] Cloud Model 1: http://www.mmm.ucar.edu/people/bryan/cm1/.

[4] NAS: http://www.nas.nasa.gov/publications/npb.html.

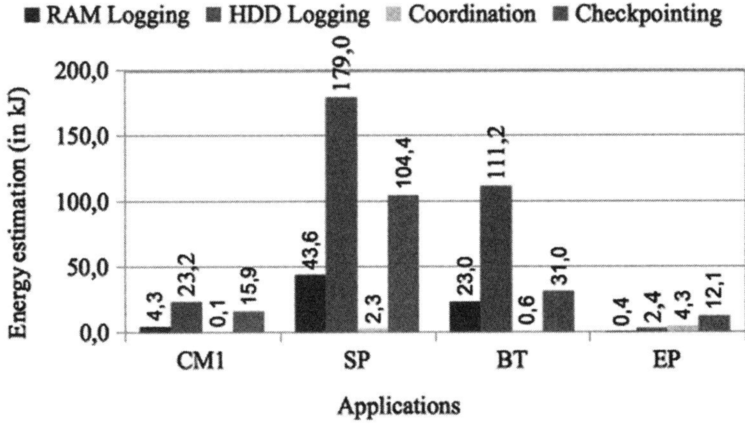

Fig. 13 Energy estimations (in kJ) of operations related to fault tolerance

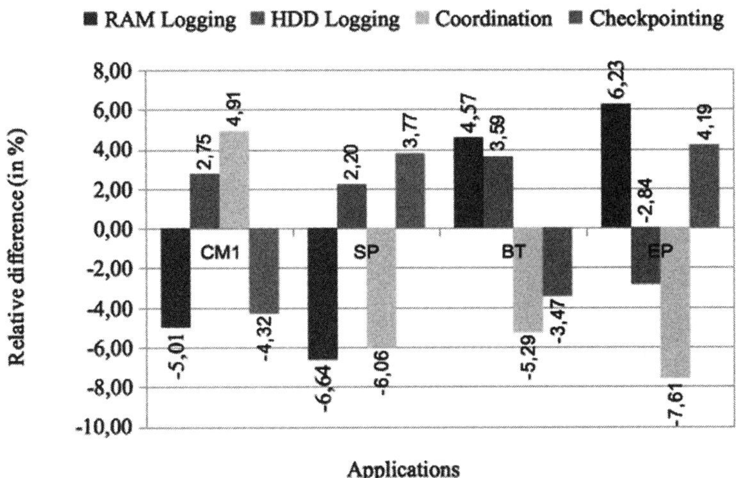

Fig. 14 Relative differences (in %) between the estimated and measured energy consumption of the operations related to fault tolerance

In Fig. 13, we show the energy estimations for different operations identified in the protocols of fault tolerance. In Fig. 14, we show the relative differences (in percent) between the estimated and the actual energy consumption. Figure 14 shows that the energy estimations provided in Fig. 13 are accurate. Indeed, the relative differences between the estimated and measured energy consumption is low. The worst estimate shows a gap of 7.6 % compared to the measured coordination with EP value. The average deviation of all tests is 4.9 %.

In comparison with message logging and checkpointing, we find that we estimate a little less coordination. This is due to the fact that this process takes much less time

Table 1 Execution contexts considered for the four data broadcasting applications

Name	Number of messages	Size of each message	Number of nodes	Number of processes per node
A	2000	1 MB	14	8
B	80	25 MB	16	1
C	4	500 MB	10	4
D	1	1.75 GB	6	12

than message logging. This is also due to the fact that this operation is evaluated from the estimated two sub-operations ($t_{polling}$ et $t_{synchro}$) which generates more inaccuracies in our estimation We will show in Sect. 6 how such energy estimations can reduce energy consumption related to protocols of fault tolerance when they are known before pre-executing the application.

5.2.2 Data Broadcasting Case

We consider four classes of data broadcasting applications involving the different execution contexts presented in Table 1.

For the four execution scenarios, we choose a large enough total volume of data for all broadcast messages so that each scenario execution lasts several seconds and so that the energy measurements are possible using the wattmeters OmegaWatt. The size parameters of the broadcast, the number of nodes and the number of processes per node messages are chosen such that:

- one parameter was not measured (applications B and D);
- two parameters were not measured (application C);
- three parameters were not measured (application A)

Each application (A, B, C, D) and for each algorithm of data broadcasting (*MPI/SAG*, *MPI/Pipeline*, *Hybrid/SAG*, *Hybrid/Pipeline*), we estimate the energy consumption by adding the energy consumed by the various operations identified in each algorithm. We estimate the energy of each operation by multiplying the energy associated with the data broadcasting of a single message by the number of application messages (A, B, C or D).

On the other hand, we measure the energy consumption for each algorithm and for each application. In our experiments, the operation *Pipeline* is performed with a fixed piece size of 128 KBytes in the implementation of the algorithm Each energy measurement is performed 30 times, and we consider the average values.

In Fig. 15, we show the energy estimations for the various broadcasting algorithms. In Fig. 16, we show the relative differences (in percent) between the estimated and the measured energy consumption. To get the energy estimation of a broadcast of a single message, we only need to divide the energy estimation displayed for all application messages (A, B or C) by the number of broadcasted messages.

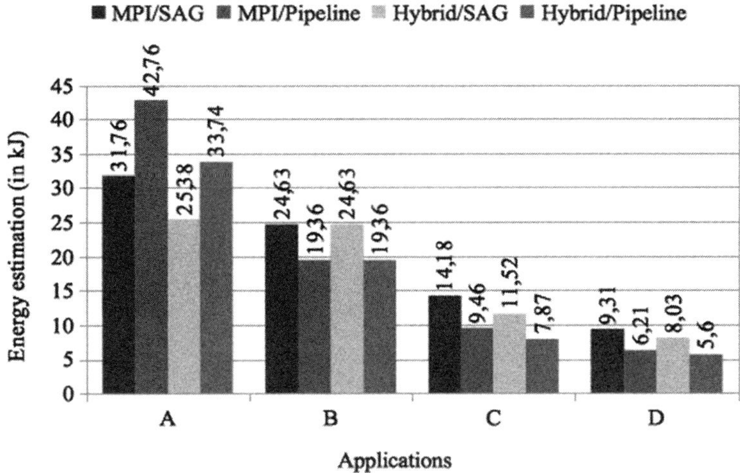

Fig. 15 Estimations of energy consumption (kJ) of the four broadcasting algorithms for the four applications

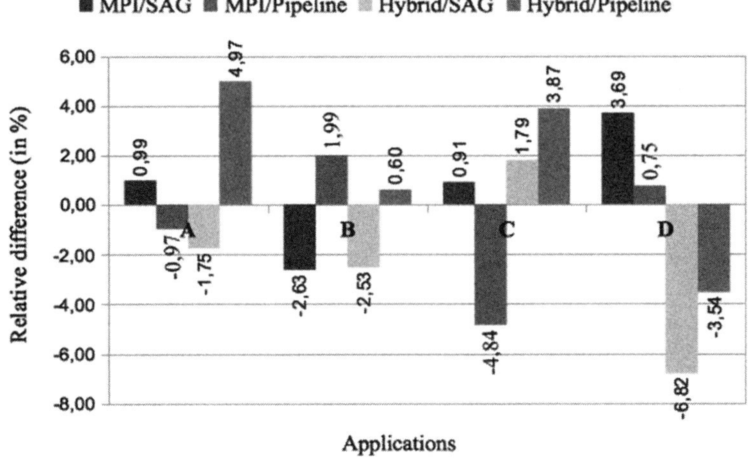

Fig. 16 Relative differences (in %) between measured and estimated energy consumption for the four broadcasting algorithms

Figure 16 shows that energy estimations provided in Fig. 15 are accurate. Indeed, the relative differences between the estimated and measured energy consumption are low. The worst estimation shows a difference of −6.82 % compared to the measured value for the algorithm *Hybrid/SAG* with D application. We will show in the next section how such energy estimations can reduce the energy consumption of data broadcasting if they are known in advance.

6 Energy-Aware Choice of Services for HPC applications

In this section, we show how we can rely on energy estimations in order to reduce the energy consumption of the services executed with high-performance computing applications. Several implementations may exist for a same service. The version of the service that consumes less energy may change depending on the considered application. The energy estimation that we are able to provide allows the users to choose the best version of service in terms of energy consumption according to the execution context. By making such choice, the user is able to reduce the energy consumption of the executed HPC services. We illustrate this in Sect. 6.1 in the case of fault tolerance protocols and in Sect. 6.2 in the case of data broadcasting algorithms.

6.1 Fault Tolerance Protocols

The two families of fault tolerance protocols that we considered are the coordinated and the uncoordinated protocols. We consider the 4 HPC applications that we studied in Sect. 5.2.1 : CM1 with a resolution of 2400 × 2400 × 40 and 3 NAS in Class D (SP, BT, and EP) running over 144 processes (i.e. 12 nodes with 12 cores per node). For each application and for each fault tolerance protocol, we estimate the energy consumption by considering the different operations that we have identified in Sect. 2. First, we highlight that the energy consumption of an fault tolerance operation depends highly on the application. Then, we show how the energy estimations of the different operations identified in Sect. 2.1 help the user in the choice of the fault tolerance protocol that consumes the less energy.

Figure 13 shows that energy consumption of the operations are not the same from one application to another. For instance, the energy consumption of RAM logging in SP is more than 10 times the one in CM1. This is because CM1 exchanges much less messages compared to SP. Another example is that checkpointing in CM1 is more than 20 times the one in EP. Indeed, the execution time of CM1 is much higher than EP so the number of checkpoints is more important in CM1. Moreover, the volume of data to checkpoint is more important in CM1 as it involves a more important volume of data in memory.

We can obtain the overall energy estimation of the entire fault-tolerant protocols by summing the energy consumptions of the operations considered in each protocol. For fault free uncoordinated checkpointing, we add the energy consumed by checkpointing to the energy consumption of message logging. For fault free coordinated checkpointing, we add the energy consumed by checkpointing to the energy consumption of coordinations.

Both of uncoordinated and coordinated protocols rely on checkpointing. To obtain a coherent global state, checkpointing is combined with message logging in uncoordinated protocols and with coordination in coordinated protocols. Therefore, to compare coordinated and uncoordinated protocols from an energy consumption point of view, we compare the energy cost of coordinations to message logging. In our experiments we considered message logging either in RAM or in HDD. Coordination will consume as much as there are still bulked messages that are being transferred at the moments of the processes synchronization. Message logging will consume as much as the number and the size of exchanged messages during the application are important.

Figure 13 shows that from one application to another the less energy consuming protocol is not always the same. In general, determining the less consuming protocol depends on the trade-off between the volume of logged data and the coordination cost. For BT, SP and CM1, the less energy consuming protocol is the coordinated protocol (Coordination values on Fig. 13 lower to the RAM and HDD logging values) since the volume of data to log for these applications is relatively important and leads to a higher energy consumption. Oppositely, the less energy consuming fault tolerance protocol for EP is the uncoordinated one.

These conclusions are specific to the case where there is only one checkpointing and so one coordination during the execution of these applications. If the user is interested in more reliability, and this specifically for the applications that last long (several hours), he should choose a smaller checkpoint interval and so a higher number of checkpointing and coordinations. This checkpoint interval can influence the choice of the fault tolerance protocol that consumes the less energy. Indeed, if for instance during the execution of SP, there are more than 19 checkpointing and therefore more than 19 coordinations, the energy consumption of coordinations will be higher than the one of RAM logging. As a consequence, as opposed to what we have seen previously, it would be better to use the uncoordinated protocol to reduce the energy consumption of fault tolerance.

This checkpoint interval can be selected by considering the models that define the optimal interval: the one that enables to maximize the reliability by minimizing the performance degradation [6, 26].

In case we use a higher number of processes for the execution of a same application, the energy consumption of coordination will be more important. However, the energy consumption of message logging may also increase since there may be more communications with an increased number of processes. Therefore, there will be more message to send and so more message to log.

Thus, by providing such energy estimations before executing the HPC application, we help the user to select the best fault tolerance protocol in terms of energy consumption depending on the number of checkpoints that he would like to perform during the execution of his application.

6.2 Data Broadcasting Algorithms

The four data broadcasting algorithms that we take into consideration are *MPI/SAG*, *MPI/Pipeline*, *Hybrid/SAG* and *Hybrid/Pipeline*. Our goal is to compare these four algorithms in terms of energy consumption when they are used with the data broadcasting applications that we have studied in Sect. 5.2.2.

For each application (A, B, C, D) presented in Sect. 5.2.2 and for each data broadcasting algorithm (*MPI/SAG*, *MPI/Pipeline*, *Hybrid/SAG* and *Hybrid/Pipeline*), we estimate the energy consumption by adding the energy consumptions of the different operations that we have identified in each algorithm. In our experiments, the *Pipeline* operation is performed with a chunk size fixed to 128 KBytes.

Figure 15 shows that from one application to another the less energy consuming algorithm is not always the same. First, we notice that each hybrid algorithm consumes less energy compared to its homologous version that uses MPI only, and this particularly when we consider several processes per node (applications A, C and D). In general, determining the less consuming algorithm depends on the trade-off between the volume of broadcasted data and the number of nodes involved. For A, the less energy consuming algorithm is *Hybrid/SAG* since the number of pipelined chunks is relatively low as the volume of data to broadcast is small. Oppositely, the less energy consuming broadcasting algorithm for B, C and D is *Hybrid/Pipeline* since the number of pipelined chunks starts to be high enough compared to the number of processes involved.

Thus, by providing such energy estimations before executing an application, we can select the best data broadcasting algorithm in terms of energy consumption. One may think that energy consumption of an algorithm is completely linked to its execution time. Our study shows that it is not true. Indeed, although Fig. 15 shows that the energy consumptions of hybrid algorithms are lower compared to the energy consumptions of MPI algorithms in applications A, C and D, the corresponding estimated execution times of the hybrid algorithms are slightly higher to the execution time of the MPI algorithms. Indeed, the power consumption of hybrid algorithms during MPI operations (*Scatter*, *AllGather* and *Pipeline*) is much lower than the one of MPI broadcasting algorithms (see Fig. 4). This is because in hybrid broadcasting, only one process is active during *Scatter*, *AllGather* and *Pipeline* while in pure MPI broadcasting algorithms, all the processes are active during MPI operations.

7 Conclusion

In this chapter, we presented our approach to accurately estimate the energy consumption of a given service. In particular, we applied our approach to fault tolerance protocols and algorithms for data broadcasting. Regarding fault tolerance, we focused on the phase without failure. We considered the case of coordinated protocols and uncoordinated protocols. Regarding the broadcasting of data, we considered two algorithms used in MPI (*MPI/SAG* et *MPI/Pipeline*) and two hybrid algorithms combining each of the two algorithms with OpenMP (respectively *Hybrid/SAG* and *Hybrid/Pipeline*).

Our estimation approach is to first identify the operations that we find in the different protocols or algorithms of the studied service. Then, in order to adapt our theoretical models to the specificities of the considered platform, we perform an energy calibration that consists in gathering a set of measurements of the electrical power and execution times of each of the identified operations. To calibrate the considered platform, the calibrator collects parameters describing the execution platform, such as the number of nodes or the number of cores per node. With this calibration, energy estimations that we provide can adapt to any platform. Once the calibration is complete, the estimator is based on the calibration results as well as a description of the execution context to provide an estimation of the energy consumption of the studied service.

We have shown in this chapter that our energy estimations are accurate for each operation whether for fault tolerance or for broadcasting. Indeed, comparing the energy measurements for each operation to energy estimations that we are able to provide, we have shown that the relative differences were small. Regarding fault tolerance, the relative differences between the estimates and energy measures are equal to 4.9 % on average and do not exceed 7.6 %. Regarding the broadcasting of data, the relative differences do not exceed 6.82 % for different execution contexts considered.

We described in the last section of this chapter, the way to use our estimations in order to consume less energy. By providing energy consumption estimations before the execution of the application, we showed that it is possible to choose the fault tolerance protocol or the broadcast algorithm which is consuming the less energy for a particular application in a given execution context.

We showed that the energy consumption of an application service was not always linked to the execution time. The impact of this is that users have the opportunity to make a choice between energy consumption and performance (execution time). The ability of the estimator to provide an estimation of the execution time gives the means to apply a realistic bi-criteria choice between energy consumption and performance. Providing a clear interface to help users with this choice is one of our short-time perspective.

Besides, thanks to the energy estimations, understanding the energy behaviour of the different versions of a service let us consider others solutions in order to reduce the energy consumption of a fault tolerance protocol or a data broadcasting algorithm. Indeed, by predicting the idle periods and the active polling periods, we would be able to apply some green levers such as slowing down resources (like DVFS [14, 15, 18]) or even shutting down [16, 22] some components if these idle or active polling periods are long enough [21].

Acknowledgment Experiments presented in this chapter were carried out using the Grid'5000 experimental testbed, being developed under the INRIA ALADDIN development action with support from CNRS, RENATER and several Universities as well as other funding bodies (see http://www.grid5000.fr).

References

1. Aloisio, G. and Fiore, S. (2009). Towards Exascale Distributed Data Management. *IJHPCA*, 23(4):398–400.
2. Bergman, K., Borkar, S., Campbell, D., and others (2008). ExaScale Computing Study: Technology Challenges in Achieving Exascale Systems. In *DARPA Information Processing Techniques Office*, page pp. 278, Washington, DC.
3. Bouteiller, A., Bosilca, G., and Dongarra, J. (2010). Redesigning the message logging model for high performance. *Concurrency and Computation: Practice and Experience*, 22(16):2196–2211.
4. Cappello, F., Caron, E., Daydé, M. J., Desprez, F., Jégou, Y., Primet, P. V.-B., Jeannot, E., Lanteri, S., Leduc, J., Melab, N., Mornet, G., Namyst, R., Quétier, B., and Richard., O. (2005). Grid'5000: A Large Scale, Reconfigurable, Controlable and Monitorable Grid Platform. In *IEEE/ACM Grid 2005, Seattle, Washington, USA*.
5. Cappello, F., Geist, A., Gropp, B., Kale, S., Kramer, B., and Snir, M. (2009). Toward exascale resilience. *International Journal of High Performance Computing Applications*, 23:374–388.
6. Daly, J. T. (2006). A higher order estimate of the optimum checkpoint interval for restart dumps. *Future Generation Comp. Syst.*, 22(3):303–312.
7. Dias de Assuncao, M., Gelas, J.-P., Lefèvre, L., and Orgerie, A.-C. (2010a). The green grid5000: Instrumenting a grid with energy sensors. In *5th International Workshop on Distributed Cooperative Laboratories: Instrumenting the Grid (INGRID 2010)*, Poznan, Poland.
8. Dias de Assuncao, M., Orgerie, A.-C., and Lefèvre, L. (2010b). An analysis of power consumption logs from a monitored grid site. In *IEEE/ACM International Conference on Green Computing and Communications (GreenCom-2010)*, pages 61–68, Hangzhou, China.
9. Diouri, M. E. M., Dolz, M. F., Glück, O., Lefèvre, L., Alonso, P., Catalán, S., Mayo, R., and Quintana-Ortí, E. S. (2013a). Solving some mysteries in power monitoring of servers: Take care of your wattmeters! In *Energy Efficiency in Large Scale Distributed Systems (EE-LSDS), Vienna, Austria, April*, 22–24 2013.
10. Diouri, M. E. M., Glück, O., and Lefèvre, L. (2013b). Your Cluster is not Power Homogeneous: Take Care when Designing Green Schedulers! In *4th IEEE International Green Computing Conference (IGCC), Arlington, VA USA*.
11. Diouri, M. E. M., Glück, O., Lefèvre, L., and Cappello, F. (2013c). ECOFIT: A Framework to Estimate Energy Consumption of Fault Tolerance protocols during HPC executions. In *13th IEEE/ACM International Symposium on Cluster, Cloud and Grid Computing (CCGrid)*, Delft, Netherlands.
12. Diouri, M. E. M., Glück, O., Lefèvre, L., and Mignot, J.-C. (2013d). Energy Estimation for MPI Broadcasting Algorithms in Large Scale HPC Systems. In *20th European MPI Users' Group Meeting on Recent Advances in Message Passing Interface (EuroMPI 2013), Madrid, Spain*.
13. Diouri, M. E. M., Tsafack Chetsa, G. L., Glück, O., Lefèvre, L., Pierson, J.-M., Stolf, P., and Da Costa, G. (2013e). Energy efficiency in high-performance computing with and without knowledge of applications and services. *International Journal of High Performance Computing Applications (IJHPCA)*, 27(3):232–243.
14. Etinski, M., Corbalan, J., Labarta, J., and Valero, M. (2010). Utilization driven power-aware parallel job scheduling. *Computer Science - Research and Development*, 25(3–4):207–216.
15. Freeh, V. W., Lowenthal, D. K., Pan, F., Kappiah, N., Springer, R., Rountree, B., and Femal, M. E. (2007). Analyzing the energy-time trade-off in high-performance computing applications. *IEEE Trans. Parallel Distrib. Syst.*, 18(6):835–848.
16. Hermenier, F., Loriant, N., and Menaud, J.-M. (2006). Power Management in Grid Computing with Xen. In *Frontiers of High Performance Computing and Networking - ISPA 2006 International Workshops, volume 4331 of Lecture Notes in Computer Science*, pages 407–416, Sorrento, Italy.

17. Hlavacs, H., Da Costa, G., and Pierson, J.-M. (2009). Energy consumption of residential and professional switches. In *IEEE CSE*.
18. Hotta, Y., Sato, M., Kimura, H., Matsuoka, S., Boku, T., and Takahashi, D. (2006). Profile-based optimization of power performance by using dynamic voltage scaling on a pc cluster. In *Proceedings of the 20th International in Parallel and Distributed Processing Symposium, IPDPS 2006*.
19. Mahadevan, P., Sharma, P., Banerjee, S., and Ranganathan, P. (2009). A power benchmarking framework for network devices. In *NETWORKING 2009 Conference*, Aachen, Germany, May 11–15, 2009., pages 795–808.
20. Netzer, R. H. B. and Xu, J. (1995). Necessary and sufficient conditions for consistent global snapshots. *IEEE Transactions on Parallel and Distributed Systems*, 6(2):165–169.
21. Orgerie, A.-C., Lefevre, L., and Gelas, J.-P. (2008). Save Watts in your Grid: Green Strategies for Energy-Aware Framework in Large Scale Distributed Systems. In *ICPADS 2008 : The 14th IEEE International Conference on Parallel and Distributed Systems*, Melbourne, Australia.
22. Pinheiro, E., Bianchini, R., Carrera, E. V., and Heath, T. (2001). Load balancing and unbalancing for power and performance in cluster-based systems. In *IN WORKSHOP ON COMPILERS AND OPERATING SYSTEMS FOR LOW POWER*.
23. Rabenseifner, R., Hager, G., and Jost, G. (2009). Hybrid mpi/openmp parallel programming on clusters of multi-core smp nodes. In *Parallel, Distributed and Network-based Processing, 2009 17th Euromicro International Conference on*, pages 427 –436.
24. Rao, C., Toutenburg, H., Fieger, A., Heumann, C., Nittner, T., and Scheid, S. (1999). Linear models: Least squares and alternatives. *Springer Series in Statistics*.
25. Wadsworth, D. M. and Chen, Z. (2008). Performance of MPI broadcast algorithms. In *IEEE IPDPS 2008, Miami*, Florida USA, April 14–18, *2008*, pages 1–7.
26. Young, J. W. (1974). A first order approximation to the optimum checkpoint interval. *Commun. ACM*, 17(9):530–531.

Part II
Networking

Network Virtualization in Data Centers: A Data Plane Perspective

Weirong Jiang and Viktor K. Prasanna

1 Introduction

A data center is a facility to provide large-scale computing resources. It consists of a great number of servers which are interconnected through high-bandwidth networks. While the servers in a data center have long been virtualized, the virtualization of data center networks has not attracted much attention until recently [1]. The data center network is virtualized to accommodate multiple tenants of the data center. By virtualizing the data center network, multiple virtual networks (VNs) on top of a shared physical network substrate can be created. Each virtual network is isolated from the physical network as well as from one another. This allows the network operators of each data center tenant to configure and manage their own virtual network flexibly and dynamically. The network infrastructure of a data center includes network links and network nodes (i.e. networking devices such as routers and switches). Hence data center network virtualization involves virtualizing the network links and/or the network nodes. Figure 1 shows an example of two virtual networks (VNs) created over the same physical data center network, where both network links and network nodes are virtualized.

Both network link virtualization and network node virtualization require supports from the networking devices such as routers and switches. A networking device is normally architected into two planes: control plane and data plane. While the control plane handles all kinds of control information such as routing protocol and administrative input, the data plane performs the heavy-duty job of processing every network packet. The volume of network traffic in a data center can be enormous. This

W. Jiang (✉)
Xilinx Research Labs, 2100 Logic Drive, San Jose, CA 95124, USA
e-mail: weirongj@acm.org

V. K. Prasanna
University of Southern California, 3740 McClintock Avenue, Los Angeles, CA 90089, USA
e-mail: prasanna@usc.edu

S. U. Khan, A. Y. Zomaya (eds.), *Handbook on Data Centers*,
DOI 10.1007/978-1-4939-2092-1_10

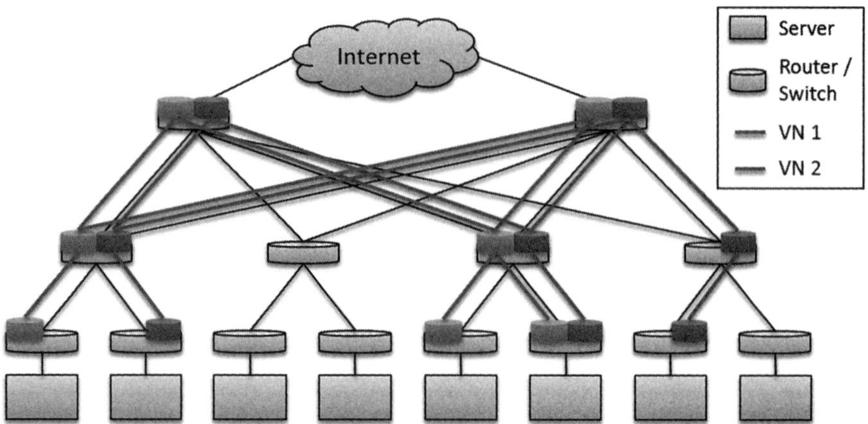

Fig. 1 Example of data center network virtualization

poses a great challenge on the data plane to meet performance requirements including high throughput and low power consumption, both of which are critical to modern data centers. While the control plane of today's networking devices is normally implemented as software, the data plane is usually built as dedicated hardware to achieve high performance. In this chapter we discuss network link virtualization and network node virtualization from the data plane perspective.

1.1 Network Link Virtualization

Network links carry the traffic in terms of network packets. The packets with the same source and destination endpoints comprise a flow. We can visualize a network link as a pipe carrying various flows. To enable multiple virtualize networks to share the same physical network link, we need to assign and classify the flows to different virtual networks. The traditional method is Virtual Local Area Network (VLAN) tagging. The packets tagged with the same VLAN ID belong to the same virtual network. Network nodes classify the flows by matching the VLAN ID. VLAN is limited to Layer 2 headers and cannot support finer-grained or more complicated flow definitions. For example, we may need a virtual network dedicated for HTTP traffic. Various solutions have been proposed to enable flexible flow matching in the data plane of the network. The most successful story to date is the OpenFlow technique [2] which defines flows using any combination of numerous packet headers.

1.2 Network Node Virtualization

Network nodes such as routers and switches connect multiple network links and forward packets from one port to another. Where a packet should be forwarded is determined by looking up the destination address of the packet against a forwarding table. The forwarding table is populated by the control plane based on the addressing and routing scheme. In a virtualized data center network, different virtual networks may employ different addressing and routing scheme. By virtualizing the network node, each virtual network gets its own virtual instance, called the virtual network node. Each virtual network node takes a slice of the physical resources and maintains a separate forwarding table based on the addressing and routing scheme of the associated virtual network. The size of a forwarding table can be large for a virtual network serving a large number of virtual machines. It may consume a large amount of memory and power to store and search the large flow tables. Thus the number of virtual networks is limited by the power budget and the available memory resources in the data plane. A promising solution is to consolidate the forwarding tables while isolating the traffic for different virtual networks.

1.3 Organization

The rest of the chapter is organized as follows. Section II presents the challenges and the solutions for flexible flow matching that enables fine-grained network link virtualization. Section III discusses the schemes for consolidating forwarding tables associated with different virtual networks. Section IV summarizes the chapter and provides our insights on the future research in this field.

2 Flexible Flow Matching for Network Link Virtualization

Due to the emerging network requirements such as user-level and fine-grained security, mobility and reconfigurability, fine-grained network link virtualization has been an essential feature for data center and cloud computing networks. Also major networking device vendors have recently initiated programs to open up their data plane to allow users to develop software extensions for proprietary hardware [3]. The key trend behind these moves is the so-called software-defined networking (SDN). Both fine-grained network virtualization and SDN require the underlying forwarding hardware to be flexible with rich flow definitions. The most successful story to date is the OpenFlow technique that employs flow tables with highly flexible flow definition at the data plane [2].

2.1 Background

The kernel operation in flexible forwarding hardware is matching each packet against a table of flow rules. The definition of a flow rule can be as flexible as users want.

Table 1 Example OpenFlow rule table

Rule	Ingress port	MAC Src	MAC Dst	Ether type	VLAN ID	VLAN priority	IP Src	IP Dst	IP Proto	IP ToS	Port Src	Port Dst
R1	*	00:01	20:01	*	*	*	0/0	0/0	*	*	*	*
R2	1	20:02	00:02	*	*	*	0/0	0/0	*	*	*	*
R3	*	*	00:ff	*	*	*	0/0	0/0	*	*	*	*
R4	*	20:0f	*	0 × 8100	10	5	0/0	0/0	*	*	*	*
R5	0	*	*	0 × 0800	*	*	0/0	64/2	*	*	*	*
R6	1	*	*	0 × 0800	*	*	32/3	192/2	TCP	*	1080	80
R7	*	*	*	0 × 0800	*	*	32/3	192/2	UDP	*	2000	6
R8	*	*	*	0 × 0800	*	*	128/3	192/3	*	*	1080	6
R9	3	00:ff	20:00	0 × 8100	1982	3	48/4	192/4	TCP	0	2000	80
R10	*	20:0f	00:aa	0 × 8100	1982	3	65	163	TCP	0	2001	80

Taking the OpenFlow 1.0 specification as an example, up to 12 header fields extracted from a packet can be used to define a flow. The 12 header fields supported in the OpenFlow 1.0 specification include the ingress port[1], 48-bit source/destination Ethernet (MAC) addresses, 16-bit Ethernet type, 12-bit VLAN ID, 3-bit VLAN priority, 32-bit source/destination IP addresses, 8-bit IP protocol, 6-bit IP Type of Service (ToS) bits, and 16-bit source/destination port numbers. Each field of a flow rule can be specified as an exact number or a wildcard. IP address fields can also be specified as a prefix. Table 1 shows a simplified example of OpenFlow rule table, where we assume MAC addresses to be 16-bit and IP addresses 8-bit. Let SA/DA denote the source/destination IP addresses and SP/DP the source/destination port numbers. We have following definitions:

- *Simple rule* is the flow rule where all the fields are specified as exact values, e.g. R10 in Table 1.
- *Complex rule* is the flow rule containing wildcards or prefixes, e.g. R1 ∼ 9 in Table 1.

A packet is considered matching a rule if and only if its header values match all the specified fields within that rule. If a packet matches multiple rules, the matching rule with the highest priority is used. In OpenFlow, a simple rule always has the highest priority. If a packet does not match any rule, the packet is forwarded to the centralized server (i.e., OpenFlow controller). The server determines how to handle it and may register a new rule in the switches. Hence dynamic rule updating needs to be supported.

[1] The width of the ingress port is determined by the number of ports of the networking device. For example, 6-bit ingress port indicates that the networking device has up to 63 ports.

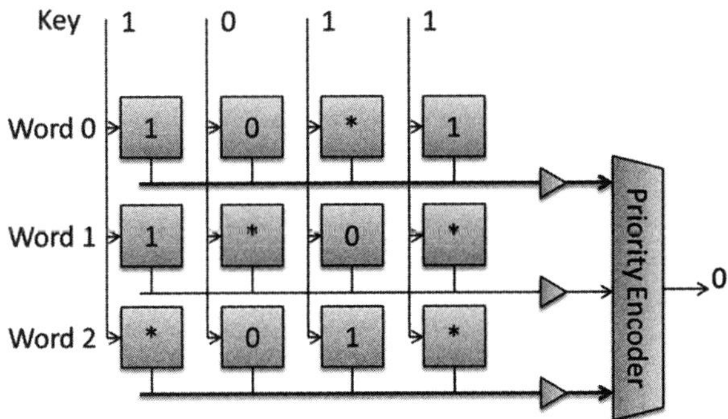

Fig. 2 A 3 × 4 bits TCAM

2.2 Existing Solutions

While OpenFlow technology is evolving, little attention has been paid on improving the performance of flexible flow matching. Most of the existing flexible forwarding hardware is focused on the functionality and simply adopts ternary content addressable memory (TCAM). TCAM is a specialized associative memory where each bit can be 0, 1, or "don't care" (i.e. "*"). For each input key, TCAM performs parallel search over all stored words and identifies the matching word(s) in a single clock cycle. A priority encoder is needed to obtain the index of the matching word with the highest priority. The word with lower index usually has the higher priority. Figure 2 shows a TCAM storing three 4-bit words: 10*1, 1*0*, and *01*. The input key 1011 matches both Word 0 and Word 2. Because Word 0 has a higher priority than Word 2, the output word index is 0.

TCAM has been widely used in the data plane of network infrastructure for various search functions. However, TCAMs are power-hungry, and do not scale well with respect to area and clock rate [4]. The power consumption per bit of TCAMs is 150 times that for static random access memories (SRAMs) [5]. On the other hand, field programmable gate array (FPGA) technology has become an attractive option for implementing real-time network processing engines [4]. A modern FPGA is a parallel assemblage of diverse programmable components, including millions of programmable logic gates, with a programmable interconnection network between these components. State-of-the-art SRAM-based FPGA devices such as Xilinx Virtex-7 [6] provide high clock rate, low power dissipation and large amounts of on-chip dual-port memory with configurable word width.

Naous et al. [7] implement an OpenFlow switch on NetFPGA which is a Xilinx Virtex-2 Pro 50 FPGA board tailored for network applications. A small TCAM is implemented on FPGA for complex rules. Due to the high cost to implement TCAM on FPGA, their design can support no more than few tens of complex rules. Though it is possible to use external TCAMs for large rule tables, high power consumption of TCAMs remains a big challenge. Luo et al. [8] propose using network processors to accelerate the OpenFlow switching. Similar to the software implementation of the OpenFlow switching, hashing is adopted for simple rules while linear search is performed on the complex rules. When the number of complex rules becomes large, using linear search leads to low throughput.

As the flexible forwarding hardware is still a concept under development, most of existing work focuses on the functionality rather than performance. Little work has been done in exploiting the algorithmic solutions and the power of state-of-the-art FPGA technology to achieve high-performance flexible flow matching. Few of existing schemes for OpenFlow-like flexible flow matching can support more than hundreds of complex rules while sustaining throughput above 10 Gbps in the worst case where packets are of minimum size i.e. 40 bytes.

2.3 Algorithmic Solution for Efficient Flexible Flow Matching

2.3.1 Motivations

Flexible flow matching can be viewed as an extension from the traditional five-field packet classification whose solutions have been extensively studied in the past decade. Comprehensive surveys for packet classification algorithms can be found in [5]. Among the existing packet classification solutions, decision-tree-based designs are considered the most scalable with respect to memory requirement [4, 5]. Traversal of the tree can be pipelined to achieve high throughput [4].

Decision-tree-based algorithms (e.g., HyperCuts [9]), take the geometric view of the packet classification problem. Each rule defines a hypercube in a d-dimensional space where d is the number of header fields considered for packet classification. Each packet defines a point in this d-dimensional space. The decision tree construction algorithm employs several heuristics to cut the space recursively into smaller subspaces. Each subspace ends up with fewer rules. The cutting process is performed until the number of rules contained by a subspace is small enough to allow a low-cost linear search to find the best matching rule. Such algorithms scale well and are suitable for rule sets where the rules have little overlap with each other. But they suffer from rule duplication which can result in $O(N^d)$ memory explosion in the worst case, where N denotes the number of rules. Moreover, the depth of a decision tree can be as large as $O(W)$, where W denotes the total number of bits per packet for lookup. Note that $d = 12$, $W > 237$ according to OpenFlow 1.0.

We aim to apply decision-tree-based algorithms to flexible flow matching while addressing their drawbacks, i.e. memory explosion and large tree depth. We observe that different complex rules in a flexible flow table may specify only a small number of fields while leaving other fields to be wildcards. This phenomenon is fundamentally due to the concept of flexible forwarding hardware which was proposed to support various applications on the same substrate. For example, both IP routing and Ethernet forwarding can be implemented in OpenFlow. IP routing will specify only the destination IP address field while Ethernet forwarding will use only the destination Ethernet address.

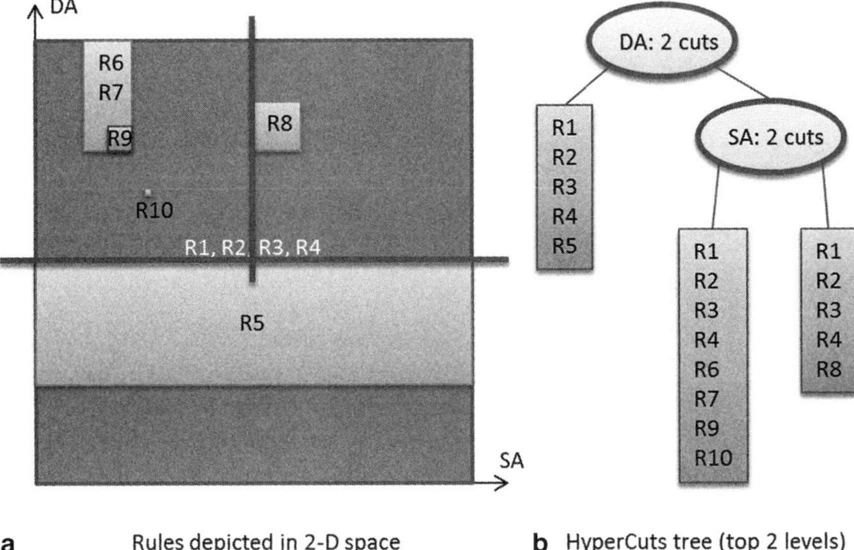

Fig. 3 HyperCuts for the example OpenFlow rules

The memory explosion for decision-tree-based algorithms in the worst case has been identified as a result of rule duplication [4]. A less specified field is more likely to cause rule duplication. Consider the example of OpenFlow table shown in Table 1. All the ten rules can be represented geometrically on a two-dimensional space as depicted in Fig. 3 where only the SA and DA fields are concerned. Decision-tree-based algorithms such as HyperCuts [9] cut the space recursively based on the values from SA and DA fields. As shown in Fig. 3, no matter how to cut the space, R1 ~ 4 will be duplicated to all children nodes. This is because their SA/DA fields are wildcards, i.e. not specified. Similarly, if we build the decision tree based on source/destination Ethernet addresses, R5 ~ 8 will be duplicated to all children nodes, no matter how the cutting is performed. The characteristics of real-life flexible flow table rules, i.e. sparse specified values with lots of wildcards, may cause severe memory explosion especially when the rule set becomes large.

An intuitive idea is to split a table of complex rules into multiple subsets. The rules within the same subset specify the same set of header fields. The number of header fields specified in a subset should be far smaller than that in the original rule table. For each rule subset, we build the decision tree based on the specified fields used by the rules within this subset. For instance, the example rule table can be partitioned into two subsets: one contains R1 ~ 4 and the other contains R5 ~ 10. We can use only source/destination Ethernet addresses to build the decision tree for the first subset while only SA/DA fields for the second subset. As a result, the rule duplication will be dramatically reduced. Meanwhile, since each decision tree after such partitioning employs a much smaller number of fields than the single decision tree without partitioning, we can expect considerable resource savings in hardware implementation. We call such an algorithm *decision forest*, which consists of multiple decision trees.

2.3.2 Algorithms

We develop the decision forest construction algorithms to achieve the following goals:

- Reduce the overall memory requirement.
- Bound the depth of each decision tree.
- Bound the number of decision trees.

Building a decision forest involves partitioning the rule set. Fong et al. [10] have shown that it can be computation-intensive for efficient rule set partitioning. Instead of performing the rule set partitioning and the decision tree construction in two phases, we combine them efficiently as shown in Algorithm 1. The rule set is partitioned dynamically during the construction of each decision tree. The function for building a decision tree, i.e. *BuildTree*, is shown in Algorithm 2. The parameter P bounds the number of decision trees in a decision forest. We have the rule set R_i to build the i-th tree whose construction process will split out the rule set R_{i+1}, $i=0,1,\ldots,P-1$. In other words, the rules in $R_i - R_{i+1}$ are actually stored in the i-th tree. The parameter *split* determines if the rest of the rule set will be partitioned. When building the last decision tree ($i=P-1$), *split* is turned to be FALSE so that all the remaining rules are used to construct the last tree. Other parameters include *depthBound* which bounds the depth of each decision tree, and *listSize* which is inherited from the original HyperCuts algorithm to determine the maximum number of rules allowed to be contained in a leaf node.

Algorithm 1: Building the decision forest

```
Input:    Rule set R
Input:    Parameters: listSize, depthBound, P
Output:   Decision forest: {T_i | i=0, 1, ..., P-1}

i ← 0, R_i ← R, split ← TRUE
While i < P do
    If i == P-1 then
            split ← FALSE
    {T_i, R_{i+1}} ← BuildTree(R_i, split, listSize, depthBound)
    i ← i + 1
```

Algorithm 2 is based on the original HyperCuts algorithm, where Lines $6\sim7$ and $12\sim14$ are the major changes. Lines $6\sim7$ are used to bound the depth of a tree. After determining the optimal cutting information (including the cutting fields and the number of cuts on these fields) for the current node (Lines $8\sim11$), we identify the rules which may be duplicated to the children nodes (shown as the *PotentialDuplicatedRule* function). These rules are then split out of the current rule set and pushed into the split-out rule set R_{remain}. The split-out rule set will be used to build the next decision tree(s). The rule duplication in the first $P-1$ trees will thus be reduced.

Algorithm 2: Building the decision tree and the split-out set

Input: Rule set R
Input: Parameters: *split*, *listSize*, *depthBound*, P
Output: Decision tree T=*nodeList* and the split-out rule set R_{remain}

Initialize the root node and push it into *nodeList*
While $nodeList \neq NULL$ do
    $n \leftarrow Pop(nodeList)$
    If $n.numRules < listSize$ then
        n is a leaf node. Continue.
    If $n.depth == depthBound$ then
        Assign to n the *listSize* most specified rules from *n.rules*. Push the rest of *n.rules* into R_{remain}. n is a leaf node. Continue.
    $n.numCuts = 1$
    For $f \in OptFields(n)$ do
        $numCuts[f] \leftarrow OptNumCuts(n,f)$
        $n.numCuts \leftarrow n.numCuts * numCuts[f]$
    If *split* is TRUE then
        $r \leftarrow PotentialDupRules(n,numCuts)$
        Push r into R_{remain}
    For $i \leftarrow 0$ to $2^{n.numCuts}$-1 do
        $n_i \leftarrow$ CreateChild(n,numCuts,i)
        Push n_i to *nodeList*

2.3.3 Architecture

To achieve line-rate throughput, we map the decision forest including P trees onto a parallel multi-pipeline architecture with P linear pipelines, as shown in Fig. 4. Each pipeline is used for traversing a decision tree as well as matching the rule lists attached to the leaf nodes. The pipeline stages for tree traversal are called the *tree stages* while those for rule list matching are called the *rule stages*. Each tree stage includes a memory block storing the tree nodes and the cutting logic which generates the memory access address based on the input packet header values. At the end of tree traversal, the index of the corresponding leaf node is retrieved to access the rule stages. Since a leaf node contains a list of *listSize* rules, we need *listSize* rule stages for matching these rules. All the leaf nodes of a tree have their rule lists mapped onto these *listSize* rule stages. Each rule stage includes a memory block storing the full content of rules and the matching logic which performs parallel matching on all header fields. Each incoming packet goes through all the P pipelines in parallel. A different subset of header fields of the packet may be used to traverse the trees in different pipelines. Each pipeline outputs a flow ID. The priority resolver picks the result with the highest priority in case of multiple flow matches among the outputs from the P pipelines. It takes $H + listSize$ clock cycles for each packet to go through the architecture, where H denotes the number of tree stages. To further improve the throughput, we exploit the dual-port on-chip RAMs provided by state-of-the-art FPGAs so that two packets are processed every clock cycle.

The size of the memory in each pipeline stage must be determined before hardware implementation. When simply mapping each level of the decision tree onto a separate stage, the memory distribution across stages can vary widely. Allocating memory

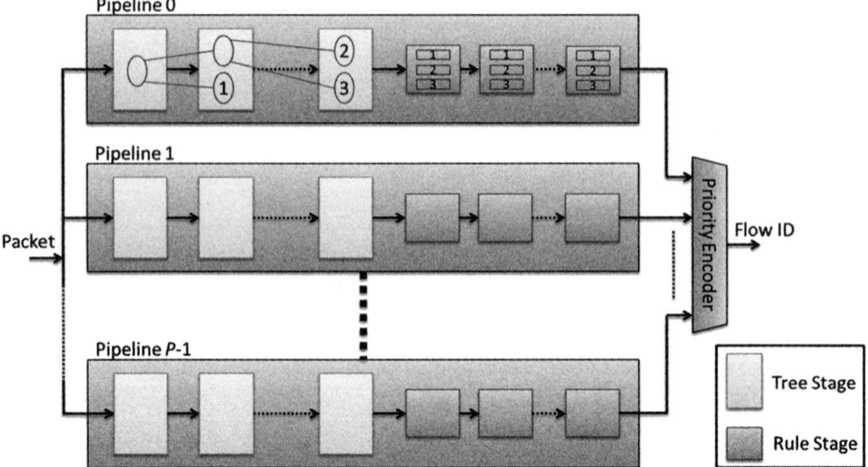

Fig. 4 Multi-pipeline architecture for searching the decision forest

with the maximum size for each stage results in large memory wastage [4]. We need to map tree nodes onto pipeline stages while balancing the memory distribution across stages. We adopt the heuristic that allows the nodes on the same level of the tree to be mapped onto different stages. This provides more flexibility to map the tree nodes. Only one constraint must be followed: If node A is an ancestor of node B in the tree, then A must be mapped to a stage preceding the stage where B is stored. Such a heuristic is enabled by adding an extra field to each tree node: the distance to the pipeline stage where the child node is stored. When a packet is passed through the pipeline, the distance value is decremented by one when it goes through every stage. When the distance value becomes zero, the child node's address is used to access the memory in that stage.

The architecture supports dynamic rule updates by inserting write bubbles into the pipelines. The new content of the memory is computed offline. When an update is initiated, a write bubble is inserted into the pipelines. Each write bubble is assigned an ID. There is one write bubble table in each stage, storing the update information associated with the write bubble ID. When a write bubble arrives at the stage prior to the stage to be updated, the write bubble uses its ID to look up the write bubble table and retrieves: (1) the memory address to be updated in the next stage, (2) the new content for that memory location, and (3) a write enable bit. If the write enable bit is set, the write bubble will use the new content to update the memory location in the next stage. Since the architecture is linear, all packets preceding or following the write bubble can perform their operations while the write bubble performs an update.

2.4 *Performance Evaluation*

We have conducted extensive experiments to evaluate the performance of the decision forest solution including the algorithms and the FPGA implementation of the architecture.

2.4.1 Experimental Setup

Due to the lack of large-scale real-life flexible flow rules, we generate synthetic 12-tuple OpenFlow-like rules to examine the effectiveness of the decision forest solution. Each rule is composed of 12 header fields that follow the OpenFlow 1.0 specification [2]. We use 6-bit field for the ingress port and randomly set each field value. Concretely, we generate each rule as follows:

a. Each field is randomly set as a wildcard. When the field is not set as a wildcard, the following steps are executed.
b. For source/destination IP address fields, the prefix length is set randomly from between 1 and 32, and then the value is set randomly from its possible values.
c. For other fields, the value is set randomly from its possible values.

In this way, we generate four OpenFlow-like 12-tuple rule sets with 100, 200, 500, and 1K rules, each of which is independent of the others. Note that our generated rule sets include many impractical rules because each field value is set at random. But we argue that the lower bound of the performance of the decision forest scheme is approximated by using such randomly generated rule sets which do not match well the characteristics observed in real-life flexible flow table rules. Better performance may be expected for large sets of real-life flexible flow rules which however are not available to date.

2.4.2 Algorithm Evaluation

To evaluate the performance of the decision forest algorithms, we use following performance metrics:

- *Average memory requirement* (bytes) per rule: It is computed as the total memory requirement of a decision forest divided by the total number of rules for building the forest.
- *Tree depth*: It is defined as the maximum directed distance from the tree root to a leaf node. For a decision forest including multiple trees, we consider the maximum tree depth among these trees. A smaller tree depth leads to shorter pipelines and thus lower latency.
- *Number of cutting fields* (denoted N_{cf}) for building a decision tree: The N_{cf} of a decision forest is defined as the maximum N_{cf} among the trees in the forest. Using a smaller number of cutting fields results in less hardware for implementing cutting logic and smaller memory for storing cutting formation of each node.

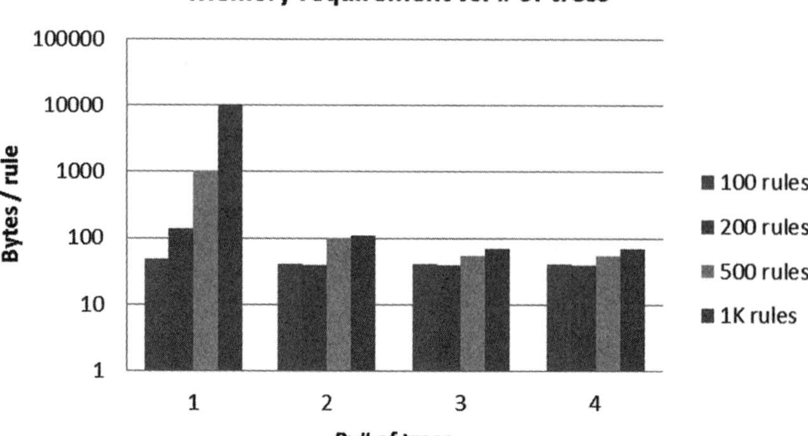

Fig. 5 Average memory requirement with increasing the number of trees

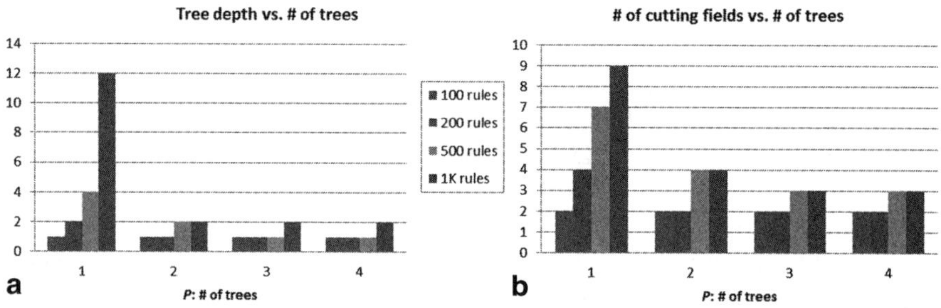

Fig. 6 **a** Tree depth and **b** N_{cf}, with increasing the number of trees

We set *listSize* = 64, *depthBound* = 16, and vary the number of trees $P = 1, 2, 3, 4$. Figure 5 shows the average memory requirement per rule, where logarithmic plot is used for the Y axis. In the case of $P = 1$, we can observe memory explosion when the number of rules is increased from 100 to 1K. On the other hand, increasing P dramatically reduces the memory consumption, especially for the larger rule set. Almost 100-fold reduction in memory consumption is achieved for the 1K rules, when P is increased just from 1 to 2. With $P = 3$ or 4, the average memory requirement per rule remains on the same order of magnitude for different size of rule sets.

Figure 6 shows that the tree depth and the number of cutting fields are reduced by increasing P. With $P = 3$ or 4, 6-fold and 3-fold reductions are achieved, respectively, in the tree depth and the number of cutting fields, compared with the case ($P = 1$) using a single decision tree.

Table 2 Breakdown of a decision forest ($P = 4$)

Trees	# of rules	# of tree nodes	Memory (bytes/rule)	Tree depth	# of cutting fields
Tree #1	712	545	78.70	2	3
Tree #2	184	265	84.70	2	5
Tree #3	65	17	41.78	1	2
Tree #4	39	9	45.23	1	2
Overall	1000	836	76.10	2	5

Table 3 Resource utilization

	Available	Used	Utilization (%)
# of slices	30,720	11,720	38
# of block RAMs	456	256	56
# of I/O pins	960	303	31

2.4.3 Hardware Implementation

To implement the decision forest for 1K rules in hardware, we examine the performance results of each tree in a decision forest. Table 2 shows the breakdown with $P = 4$, *listSize* = 32, *depthBound* = 4. We map the above decision forest onto the 4-pipeline architecture. The design is implemented on FPGA using Xilinx ISE 10.1 development tools. The target device is Virtex-5 XC5VFX200T with -2 speed grade. Post place and route results show that the design achieves a clock frequency of 125 MHz. The resulting throughput is 40 Gbps for minimum size (40 bytes) packets. Table 3 summarizes the resource utilization of the design. To the best of our knowledge, this is among the first single-chip hardware designs for flexible flow matching to achieve over 10 Gbps throughput while supporting 1K complex rules.

3 Resource Consolidation in Network Node Virtualization

Network node virtualization requires the networking device to maintain multiple forwarding tables for different virtual networks. Two general approaches for network node virtualization exist in literature. One is the isolation approach which stores a separate forwarding table for each virtual network. The other is the consolidation approach where all the forwarding tables are merged into a single one. The main limitation in virtualizing a network node is scalability. By scalability, we refer to the number of virtual networks supported on the available physical resources. Unless there are abundant resources, techniques to reduce the resource requirement should be considered in order to improve the scalability. The isolation approach is easier for management but requires more resources than the consolidation approach. We focus on the consolidation approach which has recently attracted growing interests from the community.

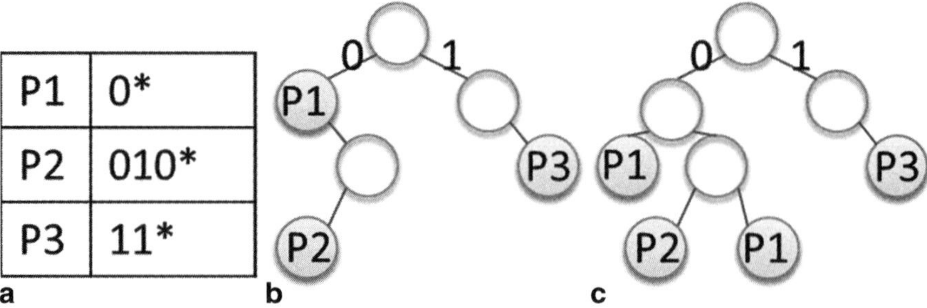

Fig. 7 **a** Prefix entries **b** Uni-bit trie **c** Leaf-pushed trie

3.1 Background

Several networking device manufacturers have introduced network node virtualization in their routers and switches to support up to hundreds of virtual networks. Cisco Systems [11] proposes a software and hardware virtualized router. Juniper Networks [12] achieves router virtualization by instantiating multiple router instances on a single hardware router to enforce security and isolation among virtual routers. The amount of physical resources limits the number of virtual network nodes that can be supported. For example, in a Juniper Networks router running logical router services, only up to 16 virtual routers are supported [13].

The primary function of routers and switches is to forward packets, where the destination address (e.g., IP address) extracted from each packet is looked up in the forwarding table. The entries in the forwarding table are specified using prefixes. The kernel of packet forwarding is IP lookup i.e. longest prefix matching. The most common data structure for IP lookup is some form of trie [13]. A trie is a binary tree, where a prefix is represented by a node. The value of the prefix corresponds to the path from the root of the tree to the node representing the prefix. The branching decisions are made based on the consecutive bits in the prefix. A trie is called a uni-bit trie if only one bit at a time is used to make branching decisions. The prefix entries in Fig. 7a correspond to the uni-bit trie shown in Fig. 7b. Each trie node contains two fields: the represented prefix and the pointer to the child nodes. By using the optimization called leaf pushing [14], each node needs only one field: either the pointer to the next-hop address or the pointer to the child nodes. Figure 7c shows the leaf-pushed uni-bit trie that is derived from Fig. 7 and 7b.

Given a uni-bit trie, IP lookup is performed by traversing the trie according to the bits in the IP address. When a leaf is reached, the last seen prefix along the path to the leaf is the longest matching prefix for the IP address. The time to look up a uni-bit trie is equal to the prefix length.

3.2 Existing Solutions

In [15] Fu and Rexford present a memory-efficient data structure for IP lookup in a virtualized router. They achieve significant memory saving by using a shared data structure. Their algorithm performs well when the different forwarding tables have similar structures. Otherwise, the memory requirement increases significantly. Song et al. [13] propose a novel approach to increase the overlap among the multiple tries that are to be merged. They introduce braiding bits at each node which allows swapping the left and the right branches for each node. This however increases the complexity for construction and update. Even though memory efficiency is claimed, the complexity of this algorithm may make it less appealing to real networking environments.

3.3 Efficient Algorithm for Resource Consolidation

We take the consolidation approach for network node virtualization by using a shared trie data structure. We propose a potential scheme to realize high memory efficiency. This method exploits the address space allocation of data center networks to minimize the memory requirement of the data plane.

3.3.1 Motivations

Each virtual network employs its own addressing scheme. Their addresses are most likely to be defined in a specific range. Packet forwarding in such networks is performed within a relatively small range in the address space of either IPv4 or IPv6. When such a network is mapped on to a trie, the range of IP addresses are located at a specific branch of the trie starting at the root. This is illustrated in Fig. 8 where the forwarding tables corresponding to three virtual networks A, B and C are mapped on to three uni-bit tries. It can be seen that the prefixes in a particular network have a common portion. We call this common portion as *common prefix*. The common prefixes of the three example tries are listed in Table 4. The structure and the size of the virtual network forwarding tables can be significantly different from one another. But we argue that the most significant difference is reflected at the common prefix while the structure of the sub-trie below the split node intends to be similar. This is because network operators intend to segment the network hierarchically in a similar way.

We use an example to show why we need a more efficient algorithm than the existing solutions for network node virtualization in a data center. Consider the two tries A and C in Fig. 8. These two tries reside in the two branches of the full trie[2]

---

[2] A full trie is defined as a complete binary trie covering all possible prefixes.

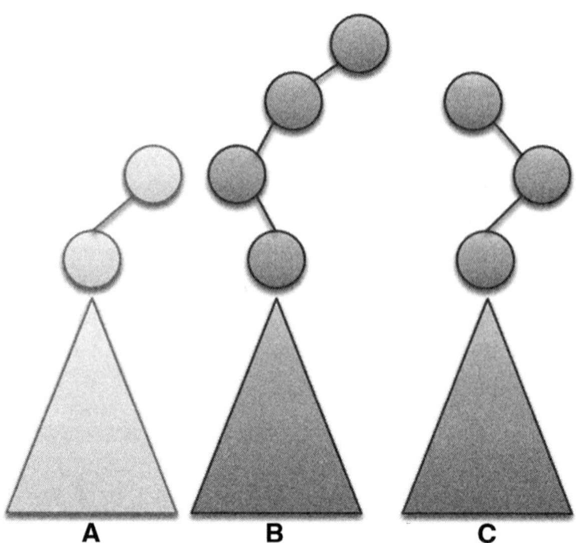

Fig. 8 Example tries for three different forwarding tables

Table 4 Common prefixes for the three example tries

Forwarding table	Common prefix	Common prefix length
A	0*	1
B	001*	3
C	10*	2

starting from the root: A on the left sub-trie and C on the right sub-trie. For simplicity, let us assume that we merge A on to C. As shown in Fig. 9a, merging them using the simple overlaying approach of [15] will create a new set of leaf and non-leaf nodes corresponding to A. This may result in a merged trie with a larger memory requirement than the sum of the memory requirements of the two tries. The memory requirement becomes larger because each leaf node in the merged trie has to store next hop information for both forwarding tables. While trie braiding [13] may result in maximum overlap between the two tries, the memory requirement at each node increase. This is because all the non-leaf nodes have to store braiding bits for each forwarding table. Also trie braiding achieves the best memory efficiency when the multiple forwarding tables have the same length of common prefixes. As shown in Fig. 9b, trie braiding does not work well for our specific example where the lengths of the common prefixes of different forwarding tables are different.

Instead we propose a scheme as illustrated in Fig. 9c. The key idea is to use the common prefix to identify different virtual networks while merging the forwarding tables at lower levels. We define the first node that has both left and right branches as the *split node*. For example, the trie corresponding to B in Fig. 8 starts to split at the sub-trie starting at 001*. Actually the prefix represented by the split node is

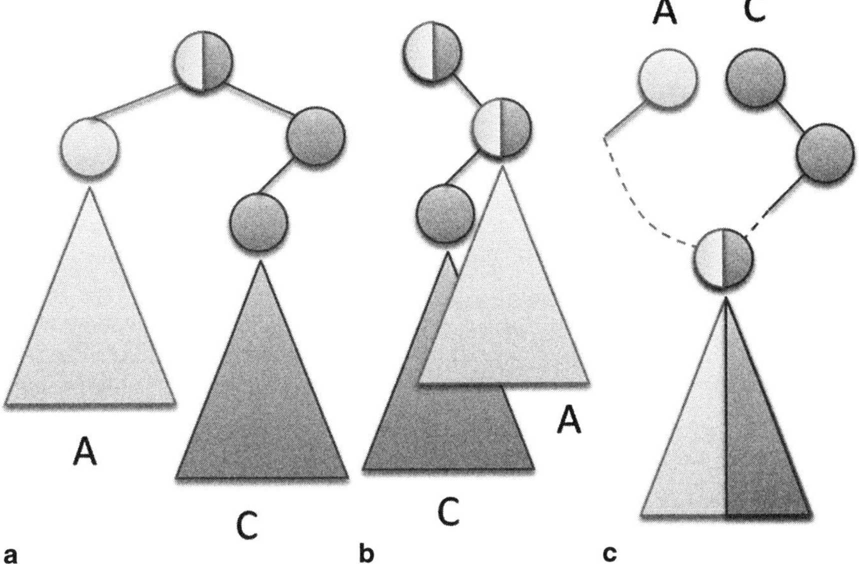

Fig. 9 Merging A and C **a** using [15] **b** using [13] **c** using our scheme

the common prefix. As shown in Fig. 9c, the amount of overlap between the tries A and C using our scheme is much higher than that of Figs. 9a, 9b. This is because we merge the tries at the split node rather than the top root node. Also we do not require any extra information to traverse the trie. As a result, our approach could require less memory than [13, 15].

3.3.2 Trie Merging

In order to take advantage of the characteristics of forwarding tables in virtualized data center networks, we use an effective trie merging algorithm to reduce the overall memory requirement without increasing the lookup complexity. Initially, we build a uni-bit trie for each virtual network forwarding table. Then we execute a recursive algorithm to find the split node at which the uni-bit trie starts its first split. After the split node is found, we truncate the constructed uni-bit trie at the split node. Then, we merge this truncated trie on to a merged trie. The merged trie is the trie that holds forwarding information for all the considered virtual networks (initially, the merged trie is an empty trie). The same process is repeated for all the other tries, where the merged trie is augmented. Once all the forwarding tables are merged this way, we do a leaf push to bring all the forwarding information down to the leaf nodes.

In addition to the above process, the information about the split nodes of all the tries is stored as a lookup table at the root node. This table is used for the initial lookup. When a packet arrives, the virtual network ID associated with the packet will be used to access the table to retrieve the necessary information to start traversing the merged trie.

In the original trie for a given virtual forwarding table, we may encounter a case where there might be an IP prefix above the split node. For example, the default gateway of a router is specified for the prefix 0.0.0.0/0 (default gateway) so that if there is no match for an incoming packet, this information will be used for packet forwarding. If such a case occurs, we can store that next hop information in a separate table. This information can be carried along with the packet and if there is no other match, the default next hop information can be used to forward the packet.

We name our approach *multiroot*, where multiple logical roots (i.e. the split nodes) are mapped to the root of the merged trie. Note that for a set of core routing tables, our algorithm will perform exactly like [15]. This is because the IP address range for a core routing table is fairly wide, and it might span over the entire IP address range. Hence there is no common prefix and the root node becomes the split node.

3.3.3 Lookup Process

The lookup process in our scheme is very similar to the existing trie-based IP lookup schemes proposed in the literature. The major differences in our scheme appear in the initial lookup stage and when accessing a leaf node. When a packet arrives, its destination IP and the Virtual Network Identifier (VNID, such as VLAN ID) are extracted. The VNID is used to access the initial lookup table to find the corresponding common prefix (CP) and the table index. The table index is an internal mapping for VNID to simplify the lookup process. When this information is looked up in the table, the destination IP address is compared with the common prefix corresponding to that forwarding table. In our scheme, the root of the merged trie might not be the root of the original trie corresponding to the forwarding table. Therefore, we cannot start the lookup process at the first bit of the incoming IP address. We use the length of the common prefix to give an offset to the index of the IP address bit where we start the traversal within the merged trie. Take Fig. 9c as an example. All packets belonging to A start traversing the merged trie using the second bit of the IP address while those belonging to C use the third bit of their IP addresses to start the traversal.

3.3.4 Traffic Isolation

In network virtualization, security is of utmost importance. Even though the hardware is shared among multiple virtual network nodes, packets that belong to a specific virtual network should not interfere with traffic from any other virtual network. The isolation and the consolidation approaches for network node virtualization enforce the security in two different ways. The former uses hard isolation by implementing multiple separate router instances, whereas the latter uses soft isolation in the data structure. As we take the consolidation approach, we use the leaf node data structure to isolate traffic from different virtual networks. There is an entry for each forwarding

Table 5 Theoretical comparison

Method	Memory requirement	Execution time
Simple overlaying [15]	$\Omega(M * N_{max})$	$O(N \log N)$
Trie braiding [13]	$\Omega(M * N_{max})$	$O(N^2)$
Our scheme (multiroot)	$\Omega(M * N_{max})$	$O(MN)$

table in every leaf node. Such an entry contains the forwarding information related to a specific forwarding table. When a leaf node is reached, the next hop information will be accessed using the table index. On the other hand, the non-leaf nodes (which are used to traverse the trie to locate the leaf node) are common to all the virtual networks.

3.4 *Analysis and Evaluation*

Our scheme is evaluated with respect to the memory requirement and the execution time to construct the data structure. The execution time is critical for quick updates. We also analyze the impact of common prefix length to the overall performance of our algorithm.

3.4.1 Theoretical Comparison

The existing network node virtualization methods exhibit similar construction procedure. Table 5 compares these methods with respect to the memory requirement and the execution time. Here, N is the average number of nodes in a trie and M is the number of virtual network nodes. N_{max} is the maximum trie size (in terms of the number of trie nodes) among the multiple tries that are being merged.

It should be noted that N_{max} in our scheme is smaller than that in simple overlaying or trie braiding. This is due to the common prefix extraction in our scheme, which reduces the number of prefix bits to be used to build the tries. Overall our scheme achieves high memory efficiency and fast execution time.

3.4.2 Experimental Setup

We focus on data center networks. But we do not have access to any data center network forwarding tables. Partitioning existing core routing tables may result in unrealistic forwarding tables. To overcome these problems, we use FRuG [16], a synthetic rule generator, to generate forwarding tables. FRuG provides us the flexibility to generate close-to-real synthetic forwarding tables with different structures and various common prefixes. We generate 16 forwarding tables, each having 100K prefixes to illustrate the performance of our algorithm for a worst case scenario in

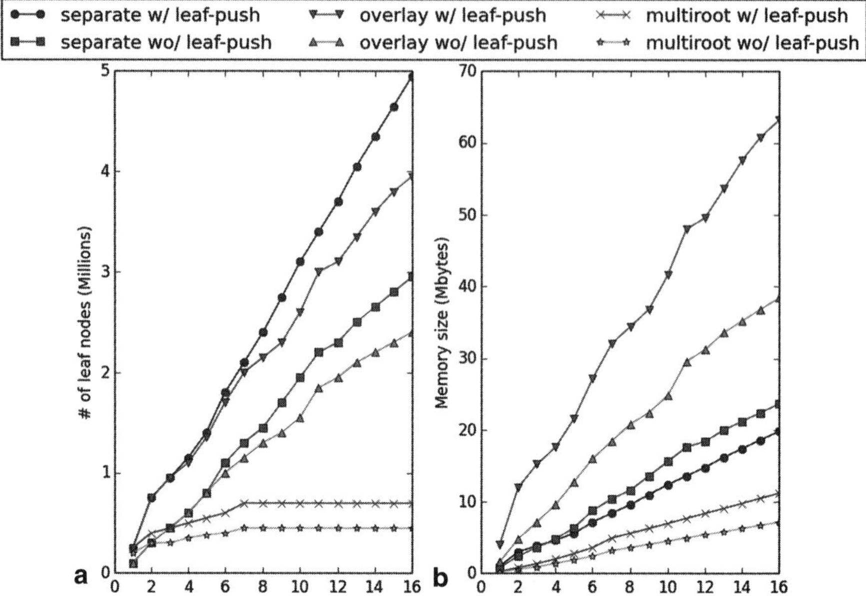

Fig. 10 Scalability comparison

data center networks. These tables are generated in such a way that each forwarding table follows the structure of one of the core routing tables in real life listed in [17].

3.4.3 Scalability

First, the experiments are conducted using forwarding tables with random common prefixes. The prefix lengths range from 2 to 5. Figure 10a illustrates the total number of leaf nodes in the merged trie. It shows that as we increase the number of forwarding tables, the total number of nodes starts to saturate. This is because the merged trie becomes dense and complete with more tries to be merged. Therefore, adding more tries does not necessarily result in more nodes. Figure 10b depicts the total memory requirement of the merged trie including the leaf and non-leaf nodes. The size of a leaf node is calculated as $M*H$ where M is the number of virtual networks and H is the bit width of each next hop field. In our analysis we consider each non-leaf node to be of 32-bit size (2*16-bit pointers) and $H = 6$ for leaf nodes. These results show that our solution requires much less memory compared with the existing methods. We achieve 6 fold and 5 fold memory reduction compared with the isolation approach and the simple overlaying [15], respectively.

While Fig. 10 is based on random common prefix, Fig. 11 compares the memory requirement for the common prefix with random length and with controlled length. L corresponds to the length of the common prefix. As shown in Fig. 11, the memory requirement of the merged trie can be further reduced significantly by merging the forwarding tables with the same length common prefix.

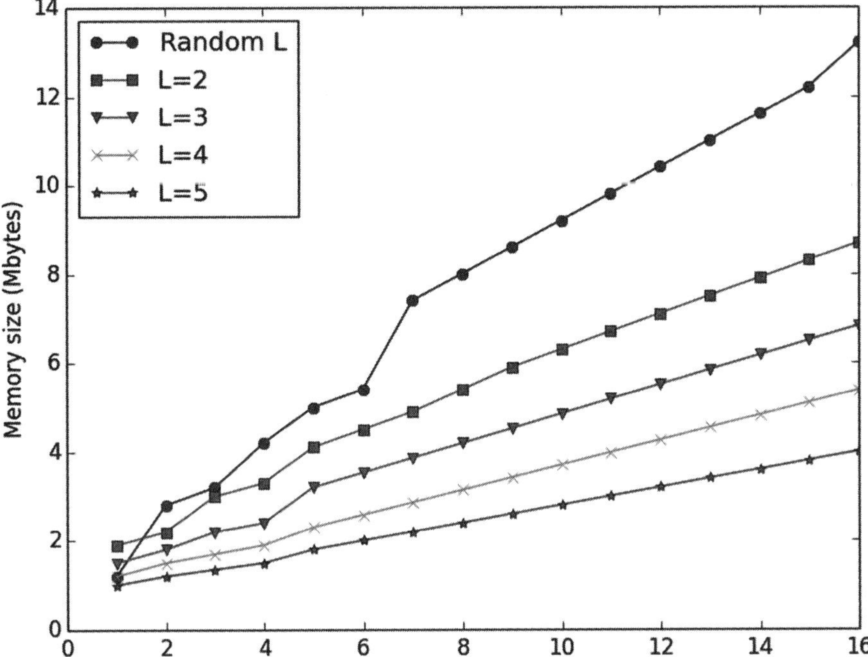

Fig. 11 Memory requirement with various common prefix lengths

3.4.4 Execution Time

In network node virtualization, adding a new virtual network to the existing virtualized network node should be done fairly quickly. Forwarding table updates should be performed in a timely fashion. Therefore the time overhead involved in these operations should be minimized. As shown in Table 5, our scheme has a lower time complexity compared to the existing approaches. The simplicity of our merging process reduces the time taken to reconstruct the merged forwarding data structure. Even though our experiments are conducted for fairly large forwarding tables, our algorithm completes the merging process for the 16 forwarding tables in 3.2s on a Quad-core AMD Opteron processor running at 2.00 GHz.

4 Summary and Discussion

In this chapter we discuss two issues associated with data plane for data center network virtualization. The first issue is flexible flow matching which enables fine-grained network link virtualization. The second issue is resource consolidation in network node virtualization to achieve high scalability.

Little research has been done so far on high-performance flexible flow matching. We present the *decision forest*, a parallel architecture to address the performance challenges for flexible flow matching. We develop a framework to partition a set of complex flow rules into multiple subsets and build each rule subset into a depth-bounded decision tree. The partitioning scheme is designed so that both the overall memory requirement and the number of packet header fields for constructing the decision trees are reduced. We implement the design on FPGA, a reconfigurable device. Experimental results demonstrate the effectiveness of the decision forest. The FPGA design supports 1K OpenFlow-like complex rules and sustains 40 Gbps throughput for minimum size (40 bytes) packets.

While some recent work has been done on merging multiple forwarding tables for network node virtualization, none of them consider the characteristics of data center networks. We present a novel approach called *multiroot* for efficient resource consolidation for network node virtualization in data centers. Our scheme improves the memory efficiency by merging multiple forwarding tables in an efficient way that exploits the address space allocation in data center networks. The key idea is to merge sub-tries from the optimal nodes rather than to merge the entire tries from the top root node. Experimental results show that our scheme outperforms existing solutions with respect to both memory efficiency and execution time.

In this chapter, we study the data plane problems associated with network link virtualization and network node virtualization, separately. It can be interesting to study both problems in an integrated context. For example, a virtual network assigned to a data center tenant may maintain a forwarding table with flexible flow definitions so as to slice the network into multiple overlays. Then it becomes a challenge in efficient merging of multiple forwarding tables, each of which may employ a different flow definition.

The techniques for data center network virtualization are still evolving. A lot of issues remain open in data plane to support efficient data center network virtualization. Neither flexible flow matching nor resource consolidation has been completely solved in our schemes. For example, the decision forest supports no more than 1K complex 12-tuple rules. The latest OpenFlow specification requires the data plane to support more than 20 header fields which account for over 600 bits to match. Another challenge in research is the lack of benchmarks and real-life data sets. Neither real-life flexible flow rule tables nor data center network forwarding tables are publicly available to date. While FRuG [16] represents an early attempt to solve the challenge, substantial collaboration between academia and industry are needed.

Acknowledgement This work is partly supported by the United States National Science Foundation under grant CCF-1116781. Equipment grant from Xilinx Inc. is gratefully acknowledged.

References

1. M. F. Bari, R. Boutaba, R. Esteves, L. Z. Granville, M. Podlesny, M. G. Rabbani, Q. Zhang and M. F. Zhani, "Data Center Network Virtualization: A Survey," *IEEE Communications Surveys & Tutorials*, vol. 15, no. 2, pp. 909–928, 2013.
2. OpenFlow, "Enabling Innovation in Your Network," [Online]. Available: www.openflow.org.

3. J. C. Mogul, P. Yalag, J. Tourrilhes, R. Mcgeer, S. Banerjee, T. Connors and P. Sharma, "API design challenges for open router platforms on proprietary hardware," in *Proceedings of the ACM Workshop on Hot Topics in Networks (HotNets)*, 2008.
4. W. Jiang and V. K. Prasanna, "Scalable Packet Classification on FPGA," *IEEE Transactions on Very Large Scale Integration (VLSI) Systems*, vol. 20, no. 9, pp. 1668–1680, 2012.
5. D. E. Taylor, "Survey and taxonomy of packet classification techniques," *ACM Comput. Surv.*, vol. 37, no. 3, p. 238–275, 2005.
6. "Virtex-7 FPGA Family," Xilinx, [Online]. Available: http://www.xilinx.com/products/silicon-devices/fpga/virtex-7/index.htm.
7. J. Naous, D. Erickson, G. A. Covington, G. Appenzeller and N. McKeown, "Implementing an OpenFlow Switch on the NetFPGA Platform," in *Proceedings of the 4th ACM/IEEE Symposium on Architectures for Networking and Communications Systems (ANCS '08)*, 2008.
8. Y. Luo, P. Cascon, E. Murray and J. Ortega, "Accelerating OpenFlow Switching with Network Processors," in *Proceedings of ACM/IEEE ANCS*, 2009.
9. S. Singh, F. Baboescu, G. Varghese and J. Wang, "Packet classification using multidimensional cutting," in *Proceedings of ACM SIGCOMM*, 2003.
10. J. Fong, X. Wang, Y. Qi, J. Li and W. Jiang, "ParaSplit: A Scalable Architecture on FPGA for Terabit Packet Classification," in *Proceedings of IEEE HOTI*, 2012.
11. Cisco Systems, Inc, [Online]. Available: www.cisco.com.
12. Juniper Networks, Inc., [Online]. Available: www.juniper.net.
13. H. Song, M. Kodialam, F. Hao and T. V. Lakshman, "Building Scalable Virtual Routers with Trie Braiding," in *Proceedings of IEEE Infocom*, 2010.
14. V. Srinivasan and G. Varghese, "Fast address lookups using controlled prefix expansion," *ACM Trans. Comput. Syst.*, vol. 17, pp. 1–40, 1999.
15. J. Fu and J. Rexford, "Efficient IP-Address Lookup with a Shared Forwarding Table for Multiple Virtual Routers," in *Proceedings of ACM CoNext*, 2008.
16. T. Ganegedara, W. Jiang and V. K. Prasanna, "FRuG: A Benchmark for Packet Forwarding in Future Networks," in *Proceedings of IEEE IPCCC*, 2010.
17. Routing Information Service (RIS), [Online]. Available: http://www.ripe.net/ris/.

Optical Data Center Networks: Architecture, Performance, and Energy Efficiency

Yuichi Ohsita and Masayuki Murata

1 Introduction

Online services such as cloud computing have recently become popular and the amounts of data that need to be processed by such online services are increasing. Large data centers with hundreds of thousands of servers have been built to handle such large amounts of data.

Large amounts of data are stored in large data centers in the memories or storage of numerous servers by using distributed file systems such as the Google File System [1]. Such large amounts of data are then handled by distributed computing frameworks such as MapReduce [2]. The distributed file systems or distributed computing frameworks require servers within a data center to communicate. Thus, the data center network plays an important role in data centers and affects their performance.

The data center network should provide communication with a sufficiently high bandwidth between communicating server pairs to prevent the network from becoming a bottleneck for data centers. The lack of bandwidth between servers may prevent communication between servers, and increases the time to obtain the required data. This degrades the performance of data centers. However, traditional data center networks, which are constructed with tree topologies, cannot provide communication with a sufficiently large bandwidth between servers because the root of tree topologies becomes a bottleneck and the number of hops between servers increases as the number of servers at data centers increases.

Another serious problem at data centers is energy consumption. The energy consumed by data centers increases as the amounts of data they handle rise. The energy consumed by data center networks is a non-negligible fraction of the total energy

Y. Ohsita (✉) · M. Murata
Graduate School of Information Science and Technology,
Osaka University, Suita, Japan
e-mail: y-ohsita@ist.osaka-u.ac.jp

M. Murata
e-mail: murata@ist.osaka-u.ac.jp

S. U. Khan, A. Y. Zomaya (eds.), *Handbook on Data Centers*,
DOI 10.1007/978-1-4939-2092-1_11

consumed by data centers [3] but increases as the networks grow. Thus, the energy consumed by data center networks should be reduced to reduce the energy consumed by large data centers.

There has been much research to construct data centers with sufficient performance or limited energy consumption [4–12]. For example, Al-Fares et al. proposed a topology called *FatTree* [4] that provided a sufficient bandwidth between all server pairs. The FatTree was a tree with multiple root nodes. Each node in this topology used half of its ports to connect it to the nodes of the upper layer, and the other half of its ports to connect it to the nodes of the lower layer. Another topology to provide sufficient bandwidth was proposed by Kim et al. [5] that was called the *flattened butterfly*, which provided enough bandwidth between servers and decreased the number of hops between servers by using nodes with large numbers of ports instead of constructing a tree.

Even though many data center network structures constructed of electronic switches have been proposed, it is difficult to achieve both sufficient bandwidth and small energy consumption by only using electronic switches. Electronic switches with many ports that provide communication with large bandwidths consume large amounts of energy. Although electronic switches with few ports consume less energy than switches with many ports, we need many switches to connect all servers in large data centers if we construct data center networks by using switches with few ports.

One approach to supplying sufficient bandwidth with lower energy consumption is through *optical data center networks* that use optical switches, which consume much less energy than electronic switches and provide communication with large bandwidths between their ports. There are two kinds of optical switches: *optical packet* and *optical circuit switches*. Optical packet switches relay packets constructed of optical signals without converting the signals to electronic signals. The destination port at each optical packet switch is determined based on the labels of the packets. Furthermore, multiple packets from different input ports share the same output port of each optical packet switch by waiting for the output port to become free at the buffer when the output port is busy. Optical circuit switches connect each input port with one of the output ports based on the configuration. The output ports of the optical circuit switch cannot be shared by multiple flows from different input ports unlike optical packet switches. However, the energy consumed by optical circuit switches is much less than that by optical packet switches because they do not require label processing.

This chapter introduces two approaches using optical switches. The first is aimed at providing large bandwidths between all servers by using optical packet switches because many ports are required to immediately provide communication between all servers if packets from different input ports cannot share the same output port. The second approach is aimed at minimizing energy consumption by using optical circuit switches because they consume much less energy. Optical circuit switches in this approach are deployed at the core of data centers and packet switches are connected to the optical circuit switches. The connections of packet switches can be changed in this network by configuring the optical circuit switches. Thus, we provide a sufficient bandwidth with small energy consumption by configuring the connections between packet switches to satisfy the current requirements and shutting down unused ports of packet switches.

The rest of this chapter is organized as follows. Sect. 2 overviews the optical switch architectures and Sect. 3 introduces the approach to providing a large bandwidth to all-to-all communication by using optical packet switches. Sect. 4 introduces an approach to achieving lower energy consumption and Sect. 5 is the conclusion.

2 Optical Switches Used in Optical Data Center Networks

Two kinds of switches, i.e., optical packet and optical circuit switches, are used in optical data center networks. This section overviews their architectures.

2.1 *Optical Packet Switches*

Optical packet switches relay optical packets constructed of optical signals without converting the signals to electronic signals. All optical packets have labels that indicate their destination. Optical packet switches receiving optical packets relay the packets to output ports based on the labels.

Figure 4 has a model of an optical packet switch, which is constructed of label processors, controllers, switching fabrics, and buffers. The label processors in the optical packet switch identify the labels of the optical packets. The controller then determines the destination ports for the optical packets and configures the switching fabrics based on the labels of the incoming optical packets. After the switching fabrics are configured, the incoming optical packets are relayed to output ports.

Multiple packets to the same output port may arrive simultaneously in packet switches. Buffers are deployed in optical packet switches to avoid packet loss.

The buffers may be constructed with fiber delay lines (FDLs) or electronic memories. The optical packets in FDL-based buffers can be stored without converting them into electronic packets. However, it is difficult to construct large buffers. However, large buffers can easily be implemented for electronic buffers, although optical packets must be converted into electronic packets before the packets are stored (Fig. 1).

The switching fabrics relay incoming optical packets to the required destination ports without converting optical signals into electronic signals. The switching fabrics can be constructed with arrayed waveguide grating routers (AWGRs)[13] or broadcast and select switching (B&S) [14]. AWGR is a passive switching fabric where the output port of the input signal depends on the wavelength of the input signal. Thus, the packets in the switching fabric constructed with AWGR are relayed to the destination port by changing the wavelength of the input signal according to the required output port. Wavelength converters or tunable lasers are deployed at all input ports of AWGR to change the wavelength. B&S is based on the wavelength-division

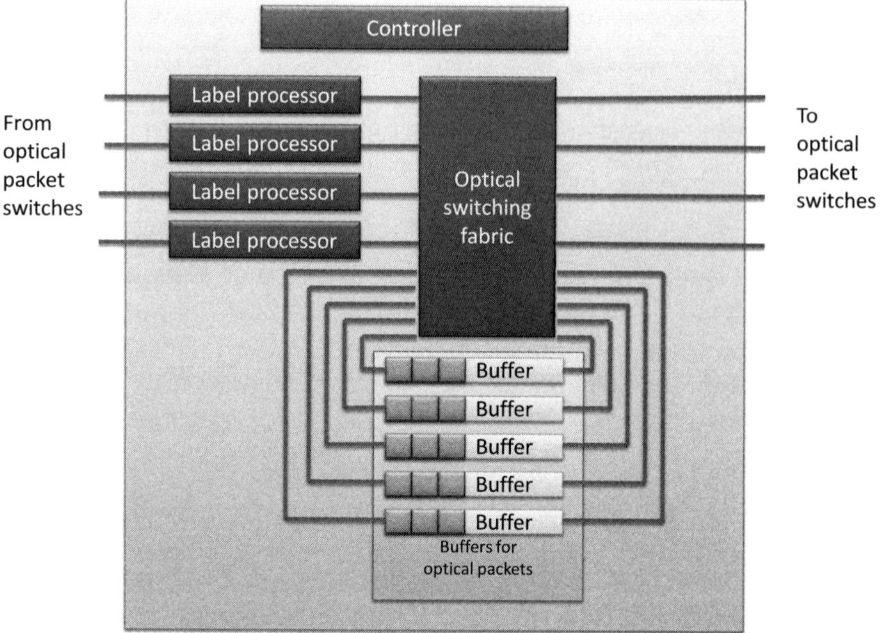

Fig. 1 Model of optical packet switch

multiplexing (WDM) star coupler. The input signals in B&S are broadcast through a splitter to all the output ports. Then, the optical signals are relayed to their output ports by setting each output port to select the signal corresponding to the port. Both types of switching fabrics can change the destination port immediately by setting the wavelength of the input signal or setting the selector. Thus, we can change the configuration for the switching fabrics each time a packet arrives.

Optical packet switches provide large bandwidths between their ports with less energy consumption, compared with electronic packet switches because they relay optical packets without converting optical signals into electronic signals. Thus, one approach to providing large bandwidths with less energy consumption is to use optical packet switches.

2.2 *Optical Circuit Switches*

Optical circuit switches relay input optical signals to output ports based on the configuration.

One of the most popular optical circuit switches is the micro-electronic-mechanical systems (MEMS) based optical circuit switch outlined in Fig. 2 [15] in which micro-mirrors are deployed. The input optical signals are reflected by the

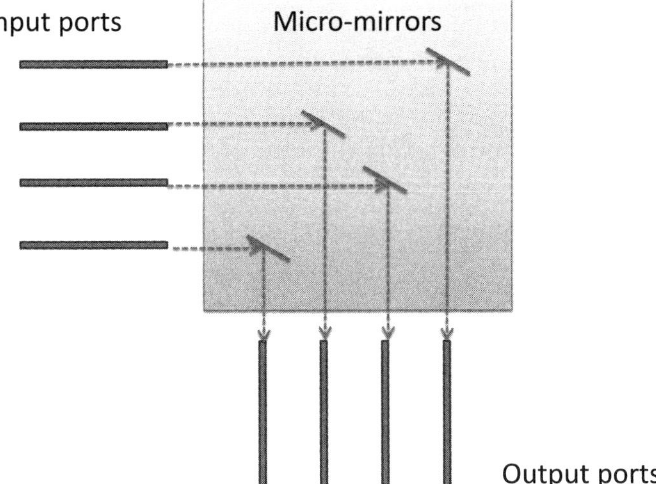

Fig. 2 MEMS optical circuit switch

micro-mirrors to output ports. Tiny motors are attached to the mirrors, whose angles can be changed. The output ports of input signals can be changed by changing the angles of the micro-mirrors.

A MEMS optical circuit switch can be configured with commands from a remote node. The configuration commands indicate the output ports that correspond to input ports. The controller within the MEMS optical switch controls the tiny motors that set the angles of the micro-mirrors so that the input signals are reflected to corresponding output ports.

The optical circuit switch only consumes a little energy because it only reflects the optical signals to their output ports with the micro-mirrors. However, the change in the angles of the mirrors takes a certain time. Thus, MEMS based optical circuit switches cannot be used as the switching fabrics of optical packet switches.

One of the most important applications of optical circuit switches is in the construction of virtual networks, where the core network is constructed of these switches. Packet switches are then connected to the ports of the optical circuit switches. By configuring the optical circuit switches, lightpaths called *optical paths* are established between the packet switches. The set of optical paths and packet switches forms a virtual network. The virtual network can be changed by reconfiguring the optical circuit switches based on the current amount of traffic.

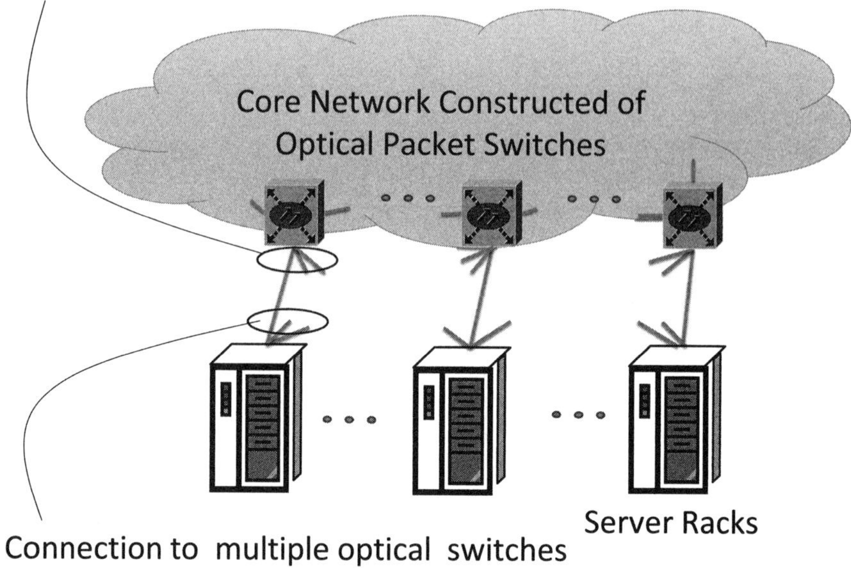

Fig. 3 Data center network using optical packet switches

3 Approach 1: Optical Data Center Networks to Provide Large Bandwidth for All-to-All Communication

Numerous servers cooperate in data centers that handle large amounts of data. The data center networks should accommodate all-to-all communication to enable cooperation between any server pairs. In addition, the lack of bandwidth between communications may increase the time to obtain the required data from other servers. Thus, data center networks should provide large bandwidth communication between all server pairs.

All-to-all communication requires packet switches, because circuit switch networks cannot accommodate all-to-all communications since multiple flows from different input ports cannot share the same output ports of circuit switches. This section discusses a network structure that uses optical packet switches to provide large bandwidths between all server pairs based on our research [16].

Optical packet switches provide large bandwidths with less energy consumption. Several optical packet switch architectures for data centers have been proposed [17–19]. Some of them are optical packet switches with many ports [18, 19]. However, networks using optical packet switches with many ports are vulnerable to failure by these switches because most of the traffic between servers traverses the switches.

This section introduces a network structure using optical packet switches with few ports that can provide sufficient bandwidths between all server pairs even when failures occur. Optical packet switches are used in this network structure to construct the core network of a data center. Figure 3 outlines the structure for a data center network using optical packet switches that are used to construct the core network so as to use the large bandwidth of these switches.

We deployed a top of rack (ToR) switch in each server rack similarly to that in conventional data centers. All servers in a server rack are connected to one ToR switch. The ToR switches are connected to the core network by connecting them to optical packet switches. Each optical packet switch is connected to multiple ToR switches, and it aggregates traffic from them to efficiently use the large bandwidth between optical packet switches. Each ToR switch is also connected to multiple optical packet switches to retain connectivity even when optical packet switches fail.

The packets from a server rack in this network are converted into optical packets at the first optical packet switch connected to the source server rack. Then, the optical packets are relayed in the core network constructed of optical packet switches. Finally, the optical packets are converted into electronic packets at the optical packet switches connected to the destination server rack, and are relayed to the destination server rack. Each ToR switch in this network only relays the electronic packets from or to the corresponding server rack, and does not relay the packets from or to the other server racks.

The details on a network structure suitable for data center networks using optical packet switches are discussed in the rest of this section.

3.1 Optical Packet Switches with Large Bandwidth

One approach to providing a large bandwidth between server racks is to use optical packet switches with a large bandwidth. This subsection introduces an optical packet switch architecture that provides a large bandwidth between its ports, which is used as an example of optical packet switches.

A large bandwidth is provided in optical networks by using multiple wavelengths. Using multi-wavelength packets is one approach to attaining multiple wavelengths. Multi-wavelength packets are constructed by dividing a packet into multiple wavelength signals. We can provide a large bandwidth to each port by using multiple wavelengths (Fig. 4).

Urata et al. [17] proposed and implemented an optical packet switch based on multi-wavelength packet technology. Figure 1 outlines an optical packet switch architecture. Optical packets constructed of multiple wavelengths are relayed in this architecture between optical packet switches. The optical packets from other optical packet switches are demultiplexed into the optical signals of each wavelength. Then, the optical signals are relayed to the destination port after label processing and multiplexed into optical packets. The optical packets are stored in the shared buffer constructed with a complementary metal oxide semiconductor (CMOS) after serial-to-parallel conversion in the case of collisions. We then try to relay the packets again after parallel-to-serial conversion.

This optical packet switch also has electronic ports in the data center, which can be used to connect them to the ToR switches. Packets from ToR switches are

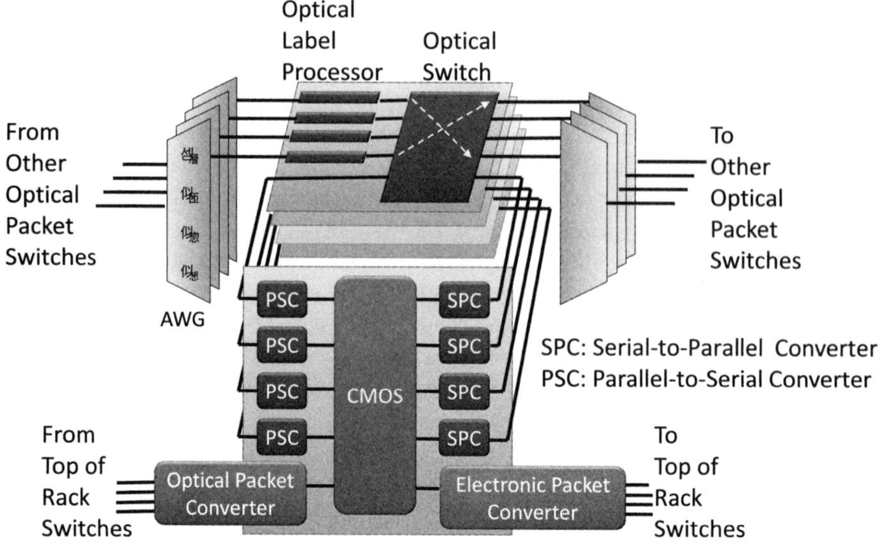

Fig. 4 Optical electronic packet switch

aggregated to the optical packets and stored in the shared buffer. Then, the packets are relayed after parallel-to-serial conversion. Optical packets whose destination is the ToR switches connected to the optical packet switch are also stored in the shared buffer. The packets are then sent to the destination ToR switches after the optical packets are demultiplexed into packets to each ToR switch.

3.2 Data Center Network Structure Using Optical Packet Switches

This section introduces a network structure that satisfies (1) the efficient use of links between optical packet switches, whose bandwidths are much larger than those of the ports of ToR switches, by aggregating traffic from multiple ToR switches and (2) the connectivity between all servers even when optical packet switches fail by connecting each ToR switch to multiple optical packet switches.

We divided the data center network in our topology into multiple groups. We avoided long links between optical packet switches and ToR switches by connecting each ToR switch to optical packet switches belonging to the same group. We respectively denote the number of ToR switches in each group, the number of optical packet switches in each group, and the number of groups as $N_{\mathrm{in}}^{\mathrm{tor}}$, $N_{\mathrm{in}}^{\mathrm{opt}}$, and G. Each optical packet uses P_{in} ports to connect optical packet switches belonging to the same group, and P_{gr} ports to connect optical packet switches belonging to other groups. We also denote the number of servers connected to each ToR switch as $P_{\mathrm{tor}}^{\mathrm{svr}}$. The number of

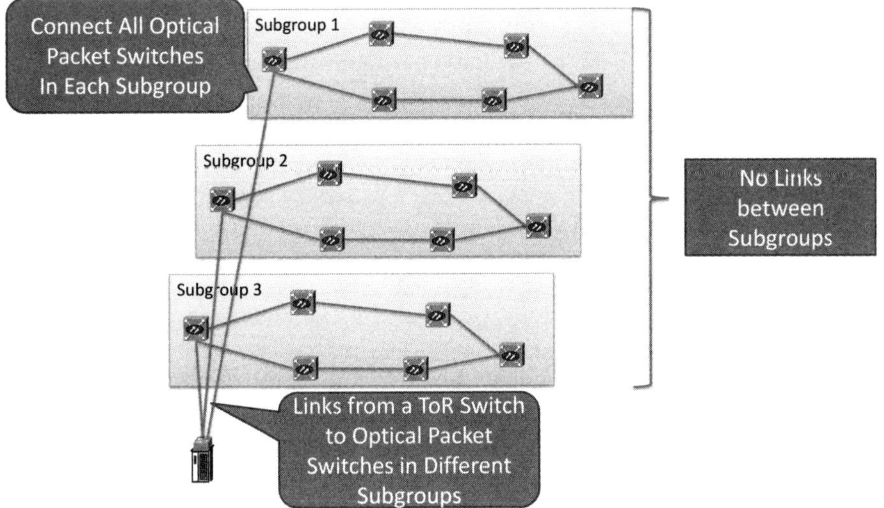

Fig. 5 Connection within group

ToR switches connected to each optical packet switch is denoted as $P_{\text{tor}}^{\text{opt}}$, and the number of optical packet switches connected to each ToR switch is denoted as $P_{\text{opt}}^{\text{tor}}$.

We also divide each group into $P_{\text{opt}}^{\text{tor}}$ subgroups. Each ToR switch is connected to optical packet switches belonging to different subgroups. All P_{in} ports of each optical packet switch are used to connect optical packet switches belonging to the same subgroup. No links are constructed between optical packet switches belonging to different subgroups as seen in Fig. 5.

We have $P_{\text{opt}}^{\text{tor}}$ distinct paths between all ToR switch pairs in this topology. Thus, we can retain the connectivity between all ToR switch pairs even when optical packet switches fail.

In addition, this topology effectively uses the ports of optical packet switches. The set of ToR switches connected to each subgroup is the same. Thus, no links between optical packet switches belonging to different subgroups are required. We greatly reduce the number of hops between ToR switches and optical packet switches by using all P_{in} ports of each switch to connect optical packet switches of the same subgroup.

We assign unique IDs to the groups, the subgroups in each group, and the optical packet switches in each subgroup. We respectively denote the group ID, the subgroup ID, and the optical packet switch ID of optical packet switch s as $D^{\text{gr}}(s)$, $D^{\text{sub}}(s)$, and $D^{\text{opt}}(s)$.

The rest of this subsection explains details on the connection within a group and that between groups.

3.2.1 Connection Within Group

We first connect optical packet switches belonging to the same subgroups. We then connect each ToR switch to $P^{\mathrm{tor}}_{\mathrm{opt}}$ optical packet switches belonging to different subgroups.

Optical packet switches belonging to the same subgroup are connected through the following steps. First, we construct a ring topology by connecting optical packet switches of the nearest optical packet switch IDs. Then, we add links between optical packet switches S_1 and S_2 if constraint

$$D^{\mathrm{opt}}(S_2) = \lfloor D^{\mathrm{opt}}(S_1) + i N_{\mathrm{sub}}/(P_{\mathrm{in}} - 1) \rfloor \bmod N_{\mathrm{sub}} \tag{1}$$

is satisfied, where N_{sub} is the number of optical packet switches belonging to each subgroup and i is a positive integer. If the optical packet switch, S_2, satisfying Eq. (1) does not have enough ports to enable connection within a group, we connect S_1 to the optical packet switch that has sufficient ports and has an optical packet switch ID close to S_2.

3.2.2 Connection Between Groups

We connect groups by adding links between optical packet switches belonging to different groups. The number of links used to connect a group to other groups is $N^{\mathrm{opt}}_{\mathrm{in}} P_{\mathrm{gr}}$. If $N^{\mathrm{opt}}_{\mathrm{in}} P_{\mathrm{gr}} \geq G - 1$, we can add links between all group pairs. We have assumed that we can add links between all group pairs in this subsection.

We select optical packet switches on both ends of links between groups to connect them. We select optical packet switch S_1 as the optical packet switch to be connected to the Kth link between groups $D^{\mathrm{gr}}(S_1)$ and $D^{\mathrm{gr}}(S_2)$ if the constraint

$$D^{\mathrm{in}}(S_1) = \begin{cases} \left\lfloor \frac{D^{\mathrm{gr}}(S_2)+K(G-1)}{P_{\mathrm{gr}}} \right\rfloor & (D^{\mathrm{gr}}(S_1) \geq D^{\mathrm{gr}}(S_2)) \\ \left\lfloor \frac{D^{\mathrm{gr}}(S_2)+K(G-1)-1}{P_{\mathrm{gr}}} \right\rfloor & (\text{Otherwise}) \end{cases}, \tag{2}$$

is satisfied, where D^{in} is the number defined by

$$D^{\mathrm{in}}(S_1) = D^{\mathrm{sub}}(S_1) \frac{N^{\mathrm{opt}}_{\mathrm{in}}}{P^{\mathrm{tor}}_{\mathrm{opt}}} + D^{\mathrm{opt}}(S_1).$$

3.2.3 Routing in Topology

We can calculate routes from ToR switches to optical packet switches and from optical packet switches to ToR switches in our topology by using the IDs assigned to optical packet switches, $[D^{\mathrm{gr}}(s), D^{\mathrm{sub}}(s), and D^{\mathrm{opt}}(s)]$ without exchanging any route information.

Routes from ToR Switches to Optical Packet Switches The routes from the ToR switch in group $D^{\mathrm{gr}}(s)$ to optical packet switch d are calculated through the following steps.

If the destination optical packet switch, d, belongs to $D^{\mathrm{gr}}(s)$, the source ToR switch first sends the packet to the optical packet switch that is directly connected to the source ToR switch and belongs to the same subgroup as destination optical packet switch d, (i.e., subgroup $D^{\mathrm{sub}}(d)$). The intermediate optical packet switch selects the next hop by calculating $H(d,a)$, defined by Eq. (3), for all neighboring optical packet switches a.

$$H(d,a) = |D^{\mathrm{opt}}(d) - D^{\mathrm{opt}}(a)| \tag{3}$$

The optical packet switch, a, having the smallest $H(d,a)$ is close to destination optical packet switch d. Thus, we select the optical packet switch having the smallest $H(d,a)$ as the next-hop optical packet switch. If there are multiple optical packet switches having the smallest $H(d,a)$, we regard all optical packet switches having the smallest $H(d,a)$ as candidates for the next hop, and balance the load by selecting the next-hop optical packet switch randomly from the candidates.

If the destination optical packet switch does not belong to $D^{\mathrm{gr}}(s)$, we first select the intermediate optical packet switch in group $D^{\mathrm{gr}}(s)$ having a link to an optical packet switch belonging to subgroup $D^{\mathrm{sub}}(d)$ in group $D^{\mathrm{gr}}(d)$. An intermediate optical packet switch is selected through the following steps. First, we calculate the range of $\tilde{k}$ in Eq. (2) where the $\tilde{k}$th link between groups $D^{\mathrm{gr}}(s)$ and $D^{\mathrm{gr}}(d)$ is connected to optical packet switches belonging to subgroup $D^{\mathrm{sub}}(d)$ in group $D^{\mathrm{gr}}(d)$ by solving the inequation

$$D^{\mathrm{sub}}(d)\frac{N_{\mathrm{in}}^{\mathrm{opt}}}{P_{\mathrm{opt}}^{\mathrm{tor}}} \leq \tilde{D}^{\mathrm{in}}(d) < (D^{\mathrm{sub}}(d)+1)\frac{N_{\mathrm{in}}^{\mathrm{opt}}}{P_{\mathrm{opt}}^{\mathrm{tor}}}, \tag{4}$$

where

$$\tilde{D}^{\mathrm{in}}(d) = \begin{cases} \left\lfloor \frac{D^{\mathrm{gr}}(s)+\tilde{k}(G-1)}{P_{\mathrm{gr}}} \right\rfloor & (D^{\mathrm{gr}}(d) \geq D^{\mathrm{gr}}(s)) \\ \left\lfloor \frac{D^{\mathrm{gr}}(s)+\tilde{k}(G-1)-1}{P_{\mathrm{gr}}} \right\rfloor & (\text{Otherwise}) \end{cases}$$

Then, we identify the optical packet switch, s', connected to an optical packet switch belonging to subgroup $D^{\mathrm{sub}}(d)$ in group $D^{\mathrm{gr}}(d)$ by substituting $\tilde{k}$ for K, s' for S_1, and d for S_2 in Eq. (2).

After the intermediate optical packet switch is selected, we calculate the routes from the source ToR switch to the intermediate optical packet switch and from the intermediate optical packet switch to the destination optical packet switch by using the same steps as those for destination optical packet switch d that belongs to $D^{\mathrm{gr}}(s)$.

Routes from optical packet switches to ToR switches If the destination ToR switch belongs to the same group as the source optical packet switch, we first select intermediate optical packet switch d^{opt} that belongs to the same subgroup as the source optical packet switches and this is directly connected to the destination ToR switch.

We have assumed that each optical packet switch knows the connections between all optical packet switches and all ToR switches within its group in this subsection. Thus, each optical packet switch can calculate d^{opt}. Then, we calculate the routes from the source optical packet switch to intermediate optical packet switch d^{opt} by using $H(d^{\mathrm{opt}}, a)$ in the same manner as that for routes from the ToR switch to the optical packet switch.

If the destination ToR switch does not belong to the same group as the source optical packet switch, we select an intermediate optical packet switch having a link to the group of the destination ToR switch. The intermediate optical packet switches having a link to the group of the destination ToR switch are obtained with Eq. (2). We then calculate the routes from the source optical packet switch to the intermediate optical packet switch, and from the intermediate optical packet switch to the destination ToR switch by using the same steps as where the destination ToR switch belonged to the same group as the source optical packet switch.

Routes between ToR Switches We can calculate routes between ToR switches by selecting an intermediate optical packet switch and calculating the routes from the source ToR switch to the intermediate optical packet switch and from the intermediate optical packet switch to the destination ToR switch. We can avoid large hop counts between ToR switches by selecting an intermediate optical packet switch at the end of a link between the group of the source ToR switch and the group of the destination ToR switch based on Eq. (2).

Handling Failures If optical packet switch S_1 cannot find any suitable next-hop optical packet switches for destination d because of failures, it returns the packet to the previous-hop optical packet switch, S_2. S_2 knows that S_1 has no suitable paths to destination d by receiving the returned packet. Thus, S_2 removes S_1 from the candidates of next-hop optical packet switches to d, and relays the packet to one of the other candidates. If S_2 also cannot find any suitable next-hop optical packet switches after S_1 is removed from the candidates, S_2 also returns the packet to the previous hop of S_2. All optical packet switches can remove switches with no suitable routes to d from their candidates for the next-hop switch to d by continuing these steps.

3.3 Parameter Settings

Our topology has three kinds of parameters, P_{gr}, P_{in}, and connection between the ToR switches and the optical packet switches. We set these parameters so that our topology could accommodate any traffic without limiting the bandwidth between servers.

We have assumed each server has a link with 1 Gbps connected to a ToR switch in this subsection. The bandwidth of each link between optical packet switches is B^{opt} Gbps.

We assumed that traffic was balanced by Valiant Load Balancing (VLB) [20] when setting the parameters. We selected the intermediate nodes randomly in VLB, regardless of the destination, to avoid traffic from concentrating on certain links even when certain node pairs had a large volume of traffic.

We selected an intermediate optical packet switch randomly with a probability of $\frac{1}{N_{\text{in}}^{\text{opt}} G}$ by applying VLB to this topology. Traffic was then sent via the intermediate optical packet switch that was selected. The volume of traffic from a ToR switch to an optical packet switch, $T_{\text{tor,opt}}$, and the volume of traffic from an optical packet switch to a ToR switch, $T_{\text{opt,tor}}$, satisfied two conditions by applying VLB:

$$T_{\text{tor,opt}} \leq \frac{P_{\text{tor}}^{\text{svr}}}{N_{\text{in}}^{\text{opt}} G} and \tag{5}$$

$$T_{\text{opt,tor}} \leq \frac{P_{\text{tor}}^{\text{svr}}}{N_{\text{in}}^{\text{opt}} G}. \tag{6}$$

Thus, we set the parameters of our topology to accommodate the traffic of $T_{\text{tor,opt}}^{\max} = T_{\text{opt,tor}}^{\max} = \frac{P_{\text{tor}}^{\text{svr}}}{N_{\text{in}}^{\text{opt}} G}$ between all ToR switch and optical packet switch pairs.

3.3.1 Parameters for Connection Between Groups

The total traffic sent between a certain group pair, T^{gr}, is constrained by applying VLB by

$$T^{\text{gr}} \leq (T_{\text{tor,opt}}^{\max} + T_{\text{opt,tor}}^{\max}) N_{\text{in}}^{\text{opt}} N_{\text{in}}^{\text{tor}}.$$

We have $\frac{P_{\text{gr}} N_{\text{in}}^{\text{opt}}}{G-1}$ bidirectional links between each group pair whose bandwidths are B^{opt} Gbps. Thus, we set P_{gr} to satisfy the following condition to avoid congestion on the links between groups:

$$\frac{2B^{\text{opt}} P_{\text{gr}} N_{\text{in}}^{\text{opt}}}{G-1} \geq (T_{\text{tor,opt}}^{\max} + T_{\text{opt,tor}}^{\max}) N_{\text{in}}^{\text{opt}} N_{\text{in}}^{\text{tor}}. \tag{7}$$

3.3.2 Parameters for Connection within Group

We denote the amount of traffic on link l as X_l and the set of links between optical packet switches within a group as L. We also denote the set of traffic from a ToR switch to an optical packet switch as $F_{\text{opt}}^{\text{tor}}$, and the set of traffic from an optical packet switch to a ToR switch as $F_{\text{tor}}^{\text{opt}}$.

The total amount of traffic traversing the links within a certain group, $\sum_{l \in L} X_l$, satisfies the following condition:

$$\sum_{l \in L} X_l \leq \sum_{i \in F_{\mathrm{opt}}^{\mathrm{tor}}} M_i T_{\mathrm{tor,opt}}^{\max} + \sum_{i \in F_{\mathrm{tor}}^{\mathrm{opt}}} M_i T_{\mathrm{opt,tor}}^{\max},$$

where M_i is the number of links within the group passed by traffic i.

We have $\frac{P_{\mathrm{in}} N_{\mathrm{in}}^{\mathrm{opt}}}{2}$ bidirectional links between optical packet switches within a group. Thus, the total bandwidth of the links within a group is $B^{\mathrm{opt}} P_{\mathrm{in}} N_{\mathrm{in}}^{\mathrm{opt}}$. Therefore, Eq. (8) should be satisfied to provide sufficient bandwidth between all ToR switches.

$$B^{\mathrm{opt}} P_{\mathrm{in}} N_{\mathrm{in}}^{\mathrm{opt}} \geq \sum_{i \in F_{\mathrm{opt}}^{\mathrm{tor}}} M_i T_{\mathrm{tor,opt}}^{\max} + \sum_{i \in F_{\mathrm{tor}}^{\mathrm{opt}}} M_i T_{\mathrm{opt,tor}}^{\max} \tag{8}$$

Eq. (8) indicates that one approach to providing enough bandwidth between ToR switches is to reduce the average number of hops between ToR switches and optical packet switches. Thus, we connect ToR switches to optical packet switches to minimize the average number of hops between them. Then, we check whether the condition in Eq. (8) is satisfied. If the condition in Eq. (8) is not satisfied, we add more links between optical packet switches within the group.

We set parameter P_{in} and the connections between ToR switches and optical packet switches in five steps.

Step 1 Initialize P_{in} to two.
Step 2 Construct the topology between optical packet switches including both intra- and inter-group connections based on the current parameter, P_{in}.
Step 3 Connect ToR switches to optical packet switches so that the average number of hops between ToR switches and optical packet switches is minimized.
Step 4 Check whether Eq. (8) is satisfied for all groups. If Eq. (8) is satisfied, go to Step 5. Otherwise, go back to Step 2 after incrementing P_{in} by one.
Step 5 End.

We need to minimize the average number of hops between ToR switches and optical packet switches in Step 2. However, it is difficult to obtain optimal connections between ToR switches and optical packet switches from all possible solutions. We selected one optical packet switch to be connected to a certain ToR switch to minimize the average number of hops from the ToR switch to all optical packet switches in each step, instead of finding the optimal solution from all possible solutions. We connected all ToR switches to optical packet switches by continuing this step.

3.4 Evaluation

3.4.1 Topologies

We evaluated our topology by comparing it with the topologies in Table 1, which is explained in this subsection.

Table 1 Topologies used in our evaluation

	# of Servers	# of optical Packet SWs	# of links Between optical SWs
Our topology	2400	24	48
Full torus	2400	24	48
Parallel torus	2400	24	48
FatTree (3 layers)	2400	20	32
FatTree (4 layers)	2400	56	140
Switch-based DCell	2400	30	60

Our Topology We set the number of optical packet switches connected to one ToR switch, $P_{\mathrm{opt}}^{\mathrm{tor}}$, to two in our evaluation, and the number of ToR switches connected to one optical packet switch, $P_{\mathrm{tor}}^{\mathrm{opt}}$, to 10. Each ToR switch was connected to 20 servers within a rack. We set the number of optical packet switches within a group, $N_{\mathrm{in}}^{\mathrm{opt}}$, to six, and the number of groups, G, to four. Thus, there were 24 optical packet switches in our topology. We set the parameters, P_{group} and P_{in}, according to the steps described in Sect. 3.3 and set B^{opt} to 100 Gbps. As a result, P_{group} and P_{in} were set to two.

Full Torus We constructed the torus topology using the same number of optical packet switches and the same number of links as those in our topology. Each optical packet switch in our topology had four ports in this evaluation. Thus, we also used optical packet switches with four ports in the full torus topology, and we connected optical packet switches as a 4×6 torus. We connected each ToR switch to two optical packet switches and each optical packet switch to ten ToR switches, which was similar to our topology.

Parallel Torus We constructed $P_{\mathrm{opt}}^{\mathrm{tor}}$ torus topologies without links between the different torus topologies. We connected each ToR switch to optical packet switches in the different torus topologies. We use the same number of optical packet switches and the same number of links as our topology. That is, we used 24 optical packet switches with four ports in this evaluation, and constructed two 3×4 torus topologies. We connected each ToR switch to two optical packet switches and each optical packet switch to ten ToR switches, which was similar to our topology.

FatTree We constructed the FatTree topology using optical packet switches with four ports with the method proposed by Al-Fares et al. [4]. This topology was a tree topology with multiple roots, where half the ports of an optical packet switch were used to connect it to nodes in the upper layer and the other half of the ports of an optical packet switch were used to connect it to nodes in the lower layer.

Although the method proposed by Al-Fares et al. [4] was used to construct a 3-layer FatTree, which was constructed of root switches and pods containing two layers of switches, we could construct higher-layer FatTree topologies. The k-layer

FatTree constructed of optical packet switches with four ports included $(2k-1)2^{k-1}$ optical packet switches.

We constructed two kinds of FatTree topologies for our evaluation, i.e., 3- and 4-layer FatTree topologies using optical packet switches with four ports. We only connected the ToR switches to optical packet switches in the lowest layer. We connected the same number of ToR switches as that in our topology to both topologies. We set the number of optical packet switches connected to each ToR switch to two. There were 30 ToR switches connected to each optical packet switch and 15 each for the 3- and 4-layer FatTree topologies.

Switch-based DCell DCell represent a topology for data center networks proposed by Guo et al. [6]. Since the original DCell was constructed by directly connecting server ports, we modified them for use in connections between optical packet switches. We called the modified version of the DCell *switch-based DCell.*

High-layer DCell are constructed from low-layer DCells in switch-based DCell. We denote the number of optical packet switches in one layer-k DCell as N_k^{DCell}. Switch-based DCell are constructed in the following steps. First, layer-0 DCell is constructed by adding links between all pairs of N_0^{DCell} optical packet switches. Then, layer-k DCell is constructed from $N_{k-1}^{\mathrm{DCell}}+1$ layer-$k-1$ DCells so that each layer-$k-1$ DCell is connected to all other layer-$k-1$ DCells with one link.

We constructed layer-1 switch-based DCell with $N_0^{\mathrm{DCell}}=5$ in our evaluation. Thus, there were 30 optical packet switches and five ports per optical packet switch, which were more than those in our topology. We connected the same number of ToR switches as those in our topology and set two as the number of optical packet switches connected to each ToR switch. Thus, there were eight ToR switches connected to one optical packet switch. We clarified that our topology could accommodate more traffic than switch-based DCell by comparing our topology with this topology, even though switch-based DCell had more links.

3.4.2 Properties of Topologies

We compared the topologies with two metrics.

Edge Betweenness The edge betweenness of link l, C_l is defined by

$$C_l = \sum_{s,d\in V,l\in L} \frac{|F_{s,l,d}|}{|F_{s,d}|},$$

where V is the set of nodes that are the source or destination nodes of traffic, L is the set of links, $F_{s,l,d}$ is the set of the shortest paths from nodes s to d passing link l, and $F_{s,d}$ is the set of the shortest paths from nodes s to d. The edge betweenness indicates the expected amount of traffic passing the link. Thus, a topology that has a large edge betweenness is easily congested. We calculated the maximum edge betweenness for traffic between ToR switches in our evaluation.

Minimum Cut The minimum cut indicates the smallest number of link failures to make the source node unable to reach the destination node. We calculated the

Table 2 Properties of topologies

	Edge betweenness	Minimum cut
Our topology	1000	2
Full torus	1600	2
Parallel torus	1200	2
FatTree (3 layer)	2700	2
FatTree (4 layer)	1575	2
Switch-based DCell	2065	2

minimum cut for all ToR switch pairs in our evaluation. Each ToR switch in all the topologies we used in our evaluation was connected to two optical packet switches. Thus, there were at most two minimal cuts.

Table 2 summarizes the results, where there are two minimum cuts for all topologies. That is, all server pairs could communicate even when one link failed in all the topologies.

The FatTree topologies have large edge betweenness regardless of the number of layers. Even though the 4-layer FatTree, especially, uses more than double the optical packet switches and links between optical packet switches compared with other topologies, its edge betweenness is larger than that in our topology and the parallel torus, and is similar to that in the full torus. This is caused by the large average number of hops between ToR switches. A large amount of traffic in the FatTree topologies passes root optical packet switches, which causes the large average number of hops. The large average number of hops leads to the large expected amount of traffic passing links.

The switch-based DCell also has large edge betweenness, even though the switch-based DCell used in our evaluations had more links than those in our topology and torus topologies. This is because the switch-based DCell had only one link between each layer-0 DCell pair. We connected many layer-0 DCell pairs in the switch-based DCell by limiting the number of links between each layer-0 DCell pair to one. This greatly reduced the number of hops between optical packet switches. One link between each layer-0 DCell pair, however, cannot provide enough bandwidth.

The parallel torus had smaller edge betweenness than the full torus. This was caused by the close connections between optical packet switches connected to different ToR switches. The parallel torus had more links between optical packet switches connected to different ToR switches instead of connecting optical packet switches coupled to the same ToR switch, while the full torus had links between optical packet switches connected to the same ToR switch. This close connection between optical packet switches coupled to different ToR switches greatly decreased the number of links passed by traffic between ToR switches, and reduced the amount of traffic between ToR switches passing each link.

Our topology had the smallest edge betweenness of the topologies used in our evaluation. Our topology used more links between optical packet switches connected

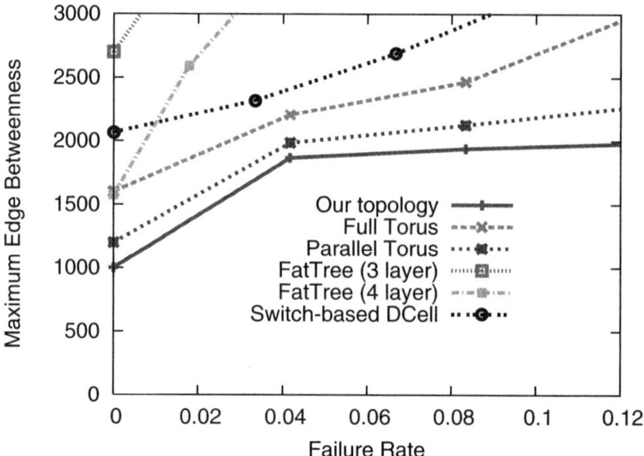

Fig. 6 Edge betweenness in case of failure

to different ToR switches instead of connecting optical packet switches connected to the same ToR switches, which was similar to the parallel torus. In addition, the parameters in our topology were set according to the steps described in Sect. 3.3, which aimed at avoiding concentrations of traffic on certain links. As a result, the parameters for our topology were set to greatly reduce the maximum edge betweenness.

We also compared the maximum edge betweenness when randomly selected optical packet switches failed by generating 100 patterns of random failures and calculating the average maximum edge betweenness for cases where all servers could mutually communicate. We compared the possibility that congestion occurred when some optical packet switches failed by using this metric. Figure 6 plots the results, where the horizontal axis indicates the failure rates for optical packet switches and the vertical axis indicates the maximum edge betweenness.

As we can see from Fig. 6, the maximum edge betweennesses of the FatTree topologies increase faster than those of the other topologies as the failure rate increases. Because there are many shortest paths between ToR switches passing each link in the FatTree topologies, the failure of each optical packet switch affects many ToR switch pairs. Moreover, the paths between the ToR switch pairs affected by failure also pass many links. As a result, failure in each optical packet switch has a large impact on the edge betweennesses of many links.

Figure 6 also indicates that our topology has the smallest edge betweenness even when some optical packet switches fail. As previously discussed, our topology has the smallest edge betweenness when there are no failures. In addition, because there are few shortest paths between ToR switches passing each link and few average number of hops between ToR switches, failure in each optical packet switch only affects few paths between ToR switch pairs and few links, unlike that in the FatTree topologies. As a result, the edge betweenness with our method remains the smallest even when some optical packet switches fail.

We have also confirmed that the edge betweenness of our topology does not increase even when the failure rate reaches more than 0.12. However, there is the probability that ToR switch pairs being unable to communicate will rise as the failure rate increases in our topology. Any optical packet switch in our topology has important links that connect different groups. Thus, as the failure rate increases, the number of redundant paths between groups decreases. Finally, when the number of paths between groups becomes zero due to failures, the ToR switches belonging to different groups become unable to communicate. However, as listed in Table 2 no topologies are more robust to the worst cases of failures than our topology. In addition, we can make our topology more robust to failures by setting $P_{\mathrm{opt}}^{\mathrm{tor}}$ to a large value.

3.4.3 Maximum Link Load

We define the link load as the total volume of traffic passing the link, and we compare the maximum link load without limiting the total volume of traffic passing each link in this subsection. We generated two kinds of traffic in this evaluation.

Uniform random Traffic was generated between all server pairs. We added traffic, whose volume was randomly generated, between randomly selected server pairs until the network interface cards of all servers had no remaining bandwidth.

Certain SW pair All the servers connected to the same ToR switch communicated with the servers connected to a certain ToR switch.

We randomly generated 20 patterns of traffic and calculated the maximum link load for each type of traffic.

Routes of traffic between ToR switches were calculated with two policies in our evaluation.

ECMP Traffic between ToR switches was equally divided among all shortest paths.

VLB One intermediate optical packet switch was randomly selected regardless of the destination. The traffic was then sent from the source ToR switch to the selected intermediate optical packet switch, and from the intermediate optical packet switch to the destination ToR switch.

We generated random failure in optical packet switches and investigated the maximum link utilization when all servers could communicate in this evaluation, which was similar to the scenario in Fig. 6. Figure 7 plots the results, where the horizontal axis indicates the failure rate for optical packet switches and the vertical axis indicates the maximum link loads.

Figures 7a and b indicate that our topology has the smallest link loads for uniform random traffic regardless of routing. Link loads for uniform random traffic are proportional to edge betweennesses. Thus, our topology that has the smallest edge betweenness shown in Fig. 6 has the smallest link loads.

Our topology using ECMP has much larger link loads than the parallel torus for certain switch pair traffic. This is caused by the number of distinct shortest paths.

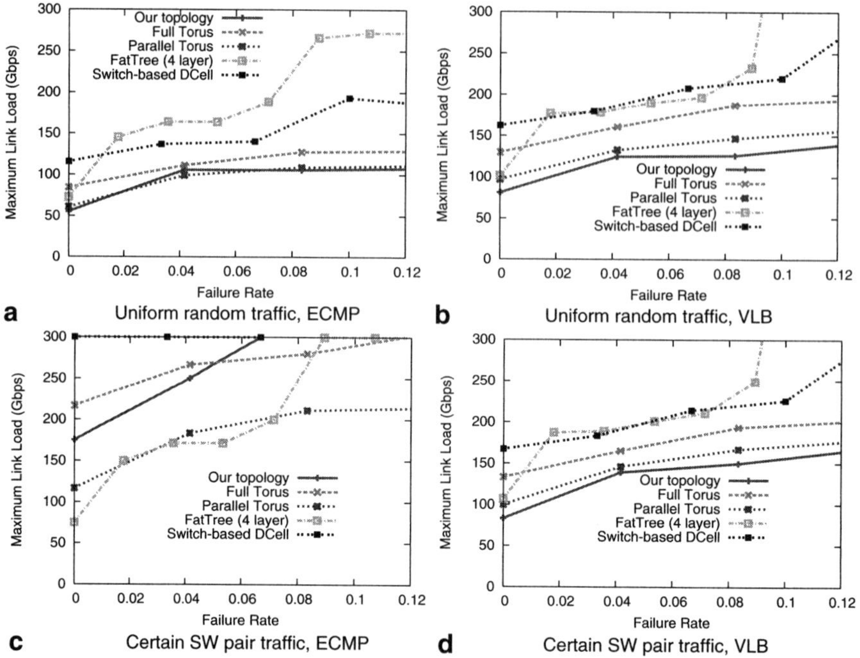

Fig. 7 Maximum link load

While the torus has many distinct shortest paths, there are few distinct shortest paths in our topology, which causes traffic to concentrate on certain links.

However, our topology achieves the smallest link loads even for certain switch pair traffic by calculating routes with VLB. This is because the parameters of our topology are set to avoid traffic from concentrating on certain links when the routes are calculated with VLB. As seen in Fig. 7, of all pairs of topologies and routing methods used in our evaluation, only the 4-layer FatTree topology using ECMP achieves slightly smaller link loads than those with our topology when there are no failures. The 4-layer FatTree, however, uses more than double the optical packet switches and links between optical packet switches in our topology. In addition, similar to the edge betweenness in Fig. 6, the link loads of the 4-layer FatTree increase rapidly as the failure rate increases. Therefore, our topology is the most suitable topology for accommodating traffic between ToR switches when some optical packet switches fail.

4 Approach 2: Networks to Achieve Low Energy Consumption

This section focuses on the energy consumption by data center networks, while the previous section focused on the bandwidth provided between servers. Energy consumption by data center networks is a non-negligible fraction of the total energy consumed by data centers [3] but increases as the size of the networks increases. Thus, the energy consumed by data center networks should be decreased to reduce the energy consumed by large data centers.

One approach to reducing energy consumption is to construct data center networks that accommodate traffic within the center with limited energy. Networks that are suitable for data centers depend on the applications and current loads of data centers. Networks should provide a large bandwidth between all servers related to applications where servers exchange large amounts of data. A small bandwidth, on the other hand, is sufficient where servers only exchange a small amount of data for applications and a network structure with only a small number of devices is preferable to reduce energy consumption.

However, the traffic demand at data centers changes over time [21]. Moreover, servers related to applications that suddenly become popular may start exchanging large amounts of data. Additional servers related to applications may be implemented to handle the abruptly increased demand for applications. As a result, data center networks are no longer suitable for current applications and loads. Although we can avoid the lack of bandwidth or large delays by constructing redundant networks, this approach consumes large amounts of energy.

Therefore, network structures based on current traffic demand within data centers need to be reconfigured to accommodate current traffic with only limited amounts of energy consumption. Networks using optical circuit switches enable network structures to be reconfigured, where the core of data center networks is constructed by using optical circuit switches and optical fibers. Electronic packet switches, deployed in each server rack, are then connected to the core network by connecting them to optical circuit switches. An optical path is established between two packet switches by configuring the optical circuit switches along the route between the electronic switches. A set of optical paths and electronic packet switches forms a virtual network. Traffic between electronic switches is carried over a virtual network.

The energy consumed by the data center network in this network is minimized by minimizing the number of ports of electronic switches used in the virtual network and shutting down unused ports because the energy consumed by electronic switches is much larger than that by optical circuit switches. When there are changes in demand, we need to maintain a sufficiently large bandwidth, short delays between servers, and low energy consumption by reconfiguring the virtual network.

Dynamic reconfiguration of virtual networks constructed over optical networks has also been discussed by many researchers [22–26]. However, most of them have aimed at optimizing virtual networks using monitored or estimated traffic demand, and their research has not been applied to data center networks, where traffic can change within a second [27], because their calculation time has been too long for large data centers.

Therefore, we have proposed a method of reconfiguring a virtual network that is suitable for a large data center network [28]. This section introduces this proposed method [28], in which traffic changes within short periods are handled by load balancing [20] over the virtual network. We designed the virtual network to achieve a sufficiently large bandwidth and small delay with low energy consumption by considering load balancing. The virtual network is reconfigured if the current virtual network cannot satisfy current demand. Our method reconfigures the virtual network by setting the parameters of the topology to avoid long calculation times at large data centers. We introduce a topology called the *Generalized Flattened Butterfly (GFB)* that we used to configure the virtual network. We also introduce a method of setting the parameters to match the current conditions.

4.1 Overview

This section introduces the virtual network configured over a data center network constructed of optical circuit and the electronic switches, where the core of the data center network is constructed by using optical circuit switches and optical fibers. Each ToR switch, which is an electronic packet switch, is connected to the core of the data center by connecting it to one of the ports of the optical circuit switches. An optical path is established between two electronic switches by configuring the optical packet switches along the route between the electronic switches. A set of optical paths and ToR switches forms a virtual network, where each optical path is regarded as a link and each ToR switch is regarded as a node in the virtual network. Traffic between electronic switches is carried over the virtual network. The energy consumed by the data center network can be minimized by minimizing the number of ports of electronic switches used in the virtual network and shutting down unused ports because the energy consumed by electronic switches is much greater than that by optical circuit switches.

The virtual network can easily be reconfigured by adding or deleting optical paths if the current virtual network is no longer suitable. We maintain a sufficiently large bandwidth, short delays between servers, and low energy consumption by reconfiguring the virtual network when there are changes in demand. Moreover, servers related to applications that suddenly become popular may start exchanging large amounts of data at the data center. A method of reconfiguring the virtual network is needed to handle such changes in traffic within a short period. However, existing methods of reconfiguring virtual networks [23, 24] cannot be applied to large data center networks because these methods require long calculation times to optimize virtual networks in large data center networks (Fig. 8).

Therefore, we introduce the method to reconfigure the virtual network by setting parameters of a topology so as to avoid large calculation time in a data center. As the topology used in the virtual network configuration, We introduce the topology called *Generalized Flattened Butterfly (GFB)*. We also introduce a method to set the parameters so as to suit the current condition.

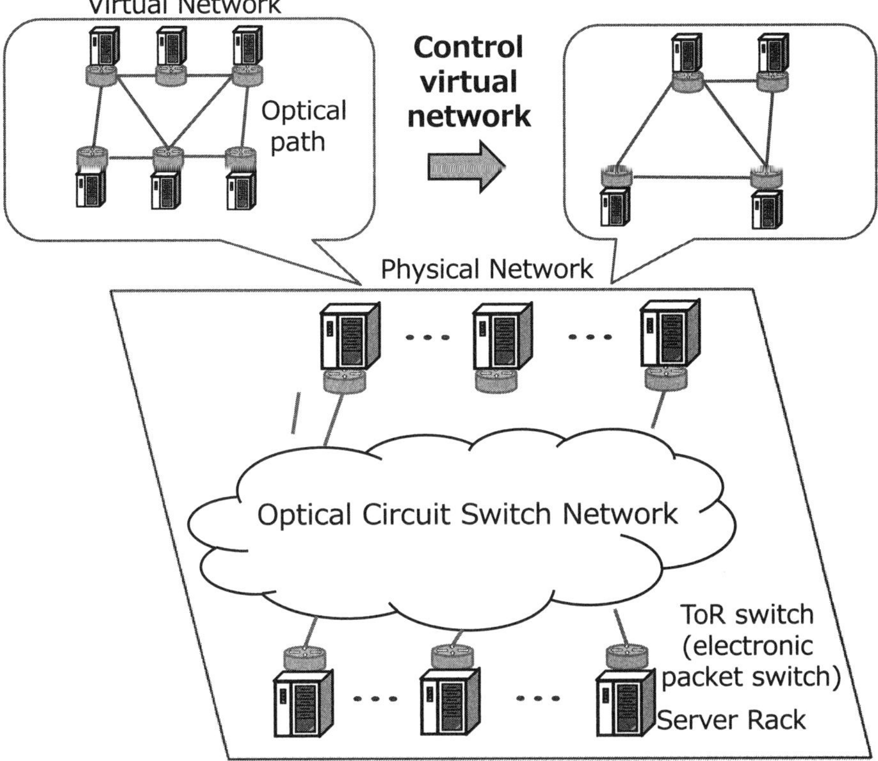

Fig. 8 Data center network using optical circuit switches

4.2 *Virtual Network Topologies Suitable for Optical Data Center Networks*

Because it is difficult to obtain the optimal topology for large data center networks within a short time, our method of reconfiguring the virtual network constructs the network by setting the parameters of its topology, which is suitable for data center networks, instead of calculating the optimal topology that achieves a sufficiently large bandwidth, short delays between servers, and low energy consumption.

This subsection discusses the requirements for the topology used in our reconfiguration of the virtual network. We then introduce the new topology that can be used to construct various data center networks by setting parameters.

4.2.1 Requirements

The virtual network should satisfy three main requirements.

Low Energy Consumption The energy consumed by the network is a non-negligible fraction of the total energy consumed by the data center. The energy consumed by the data center network should be reduced to decrease the energy consumed by the data center.

Most of the energy in the data center network discussed in this section is consumed by ToR switches because the energy consumed by optical circuit switches is much less than that by ToR switches. The energy consumed by ToR switches can be reduced by shutting down their unused ports. Thus, the energy consumed by the data center network can be minimized by constructing the virtual network with the smallest number of ports for ToR switches.

Large Bandwidth between Servers Large amounts of data are exchanged between servers in some applications such as distributed file systems. The bandwidth provided between servers is important for such applications because the lack of bandwidth increases the time required to transport data. Therefore, the virtual network should provide sufficient bandwidth between servers.

Short Delays between Servers Data centers handle large amounts of data by using distributed computing frameworks in which a large number of servers communicate. If the delays between servers are long, it takes a long time to obtain the required data from other servers, which degrades the performance of data centers. Thus, delay should be kept sufficiently short for the applications of data centers.

The delays between servers are difficult to forecast when constructing a virtual network, because they are affected by traffic load. We kept the delay short between servers by constructing a virtual network that could provide sufficient bandwidth and decreasing the number of hops between servers.

4.2.2 Existing Network Structures for Data Centers

We will introduce the existing network structures for data centers as the candidate topologies used in virtual networks before presenting our network topology.

FatTree One of the most popular network structures for data centers is the topology called *FatTree* proposed by Al-Fares et al. [4] that uses switches with few ports. FatTree is a tree topology constructed of multiple roots and multiple pods containing the multi-layer switches outlined in Fig. 9.

Each pod is regarded as a virtual switch with many ports constructed by using multiple switches that have few ports. Pods are constructed with a *butterfly topology*, where each switch uses half of its ports to connect it to switches in the upper layer, and the other half of its ports to connect it to switches in the lower layer. The switches in the lowest layer are connected to servers.

Although the method proposed by Al-Fares et al. [4] is used to construct a 3-layer FatTree, which is constructed of root switches and pods containing two layers of switches, we can construct higher-layer FatTree topologies. The k-layer FatTree constructed of switches with n ports includes $(2k-1)\frac{n}{2}^{k-1}$ switches.

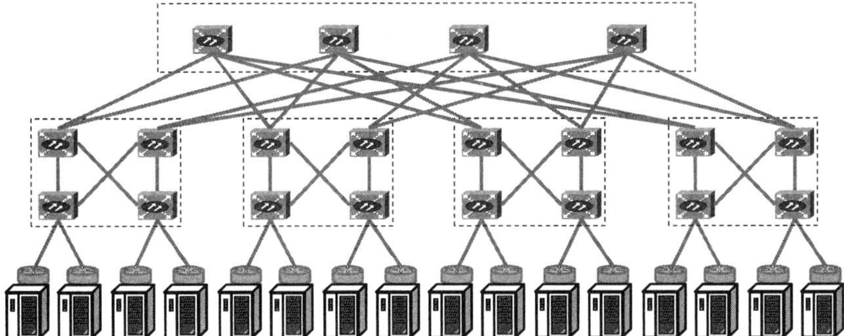

Fig. 9 FatTree

The number of links from the lower-layer switches in FatTree equals the number of links to the upper-layer switches at each switch. That is, the total bandwidth from a switch to the upper layer equals that from the lower layer to a switch. Therefore, no switches become bottlenecks, and we can provide a sufficiently large bandwidth between all servers.

However, FatTree is not suitable for virtual networks constructed of ToR switches. The switches in the upper layer are not connected to servers in FatTree. This means that ToR switches that are not connected to servers should be powered on, which consumes large amounts of energy.

Flattened Butterfly Kim et al. [5] proposed a data center network topology called the *flattened butterfly*, which was constructed by *flattening* the butterfly topology in Fig. 10. We combined the switches in each row of the butterfly topology into a single switch.

The flattened butterfly provides a sufficiently large bandwidth between all servers with lower energy consumption than FatTree [4]. In addition, all switches in the flattened butterfly are connected to servers. Thus, all ToR switches that are not connected to any working servers can be shut down if the flattened butterfly is constructed as a virtual network, unlike FatTree. However, the flattened butterfly requires switches with a large number of ports to construct a large data center network. Thus, the flattened butterfly is not preferred when there is little traffic demand.

DCell Guo et al. proposed a data center network comprised of *DCell* that was constructed from a small number of switches and servers with multiple ports, as shown in Fig. 11 [6]. The DCell uses a recursively-defined structure and the level-0 DCell is constructed by connecting one switch with n ports to n servers, and the level-k DCell is constructed by connecting servers belonging to different level-$k-1$ DCells.

DCell reduces the number of switches required to construct a large data center network by directly connecting server ports. However, DCell was not used as the topology for the virtual network introduced in this section, which was constructed of ToR switches.

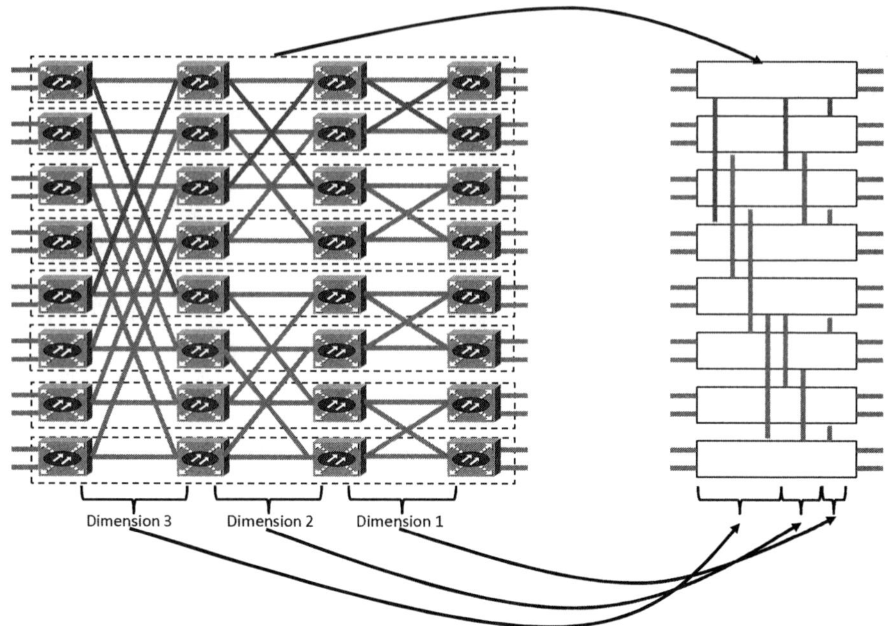

Fig. 10 Flattened butterfly

Therefore, we introduce a topology called *switch-based DCell*, where the level-0 DCells are replaced with a fully-connected network constructed of switches, as shown in Fig. 12. Similar to DCell, the switch-based DCell can construct a large data center network by using switches with few ports. That is, the switch-based DCell achieves low energy consumption. However, the switch-based DCell cannot provide large bandwidth between all servers, because they have only one link between lower-level DCells (Fig. 13).

4.2.3 Generalized Flattened Butterfly

As discussed above, the flattened butterfly [5] provides sufficient bandwidth between all server pairs but requires large amounts of energy. DCell [6] can be used to construct a topology that includes many servers using few ports but cannot provide sufficient bandwidth.

This subsection introduces a topology called the *Generalized Flattened Butterfly (GFB)*, in which the number of required ports, the maximum number of hops, and the bandwidth provided between servers can be changed by setting the parameters. GFB is constructed hierarchically and the upper-layer GFB is constructed by connecting multiple lower-layer GFBs. GFB has three parameters.

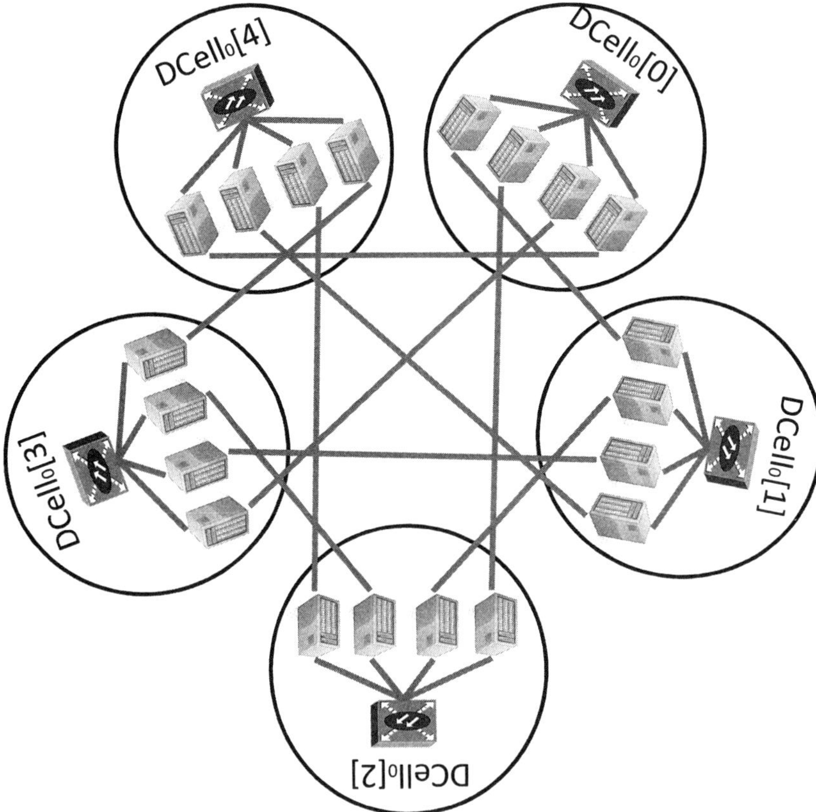

Fig. 11 DCell

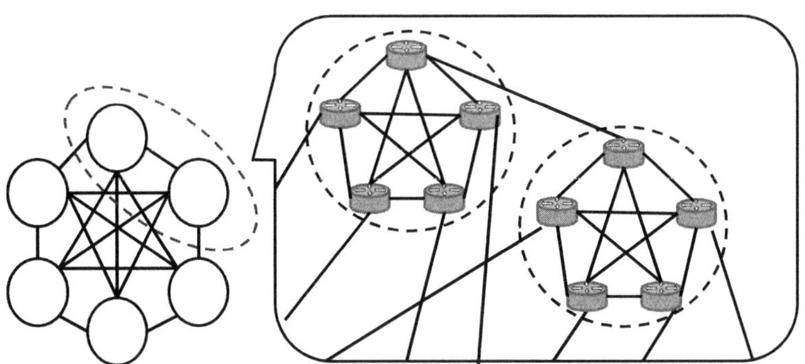

Fig. 12 Switch-based DCell

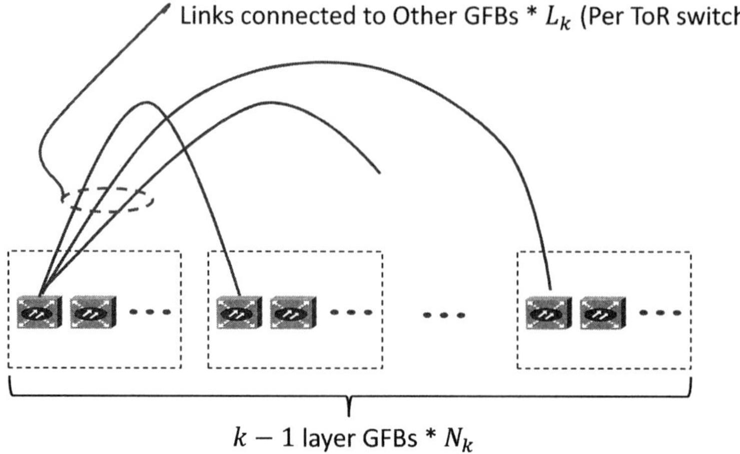

Fig. 13 Generalized flattened butterfly

- Number of layers: k
- Number of links per node used to construct layer-k GFB: L_k
- Number of layer-$k-1$ GFBs used to construct layer-k GFB: N_k

We can construct various topologies including the flattened butterfly and switch-based DCell by setting these parameters.

Steps to Construct Generalized Flattened Butterfly The layer-k GFB is constructed in two steps.

Step I Construct the connections between the layer-$k-1$ GFBs.
Step II Select the switches connected to the links between each layer-$k-1$ GFB pair.

We use the IDs assigned to the GFBs of each layer in these steps. A switch can be identified by the set of IDs of the GFBs the switch belongs to. We denote the ID of the layer-k GFB switch s belongs to as $D_k^{GFB}(s)$. We define the ID of switch s in the layer-k GFB by

$$D_k^{sw}(s) = \sum_{1 \le i \le k} \left(D_i^{GFB}(s) \prod_{j=1}^{i-1} N_j \right).$$

Connections between layer-$k-1$ GFBs We construct the connections between layer-$k-1$ GFBs in four steps.

Step I.I Calculate the number of links used to connect one layer-$k-1$ GFB to the other layer-$k-1$ GFBs, L_k^{GFB}, with

$$L_k^{GFB} = L_k \prod_{i=1}^{k-1} N_i \tag{9}$$

Step I.II If L_k^{GFB} is larger than $(N_k - 1)$, we can connect all layer-$k-1$ GFB pairs. Otherwise, construct the ring topology by connecting the GFBs having the nearest ID.

Step I.III Calculate the number of residual links $L_k^{'GFB}$ that can be used to connect one layer-$k-1$ GFB to the other layer-$k-1$ GFBs with

$$L_k^{'GFB} = L_k^{GFB} - \bar{L}_k^{GFB}, \tag{10}$$

where $\bar{L}_k^{GFB}$ is the number of links per layer-$k-1$ GFB constructed in Step I.II.

Step I.IV Check whether layer-$k-1$ GFBs have residual links to be used to connect layer-$k-1$ GFBs. If yes, connect the GFB of ID $D_{k-1}^{GFB}(a)$ to the GFB of ID $D_{k-1}^{GFB}(b)$ where the following equation is satisfied.

$$D_{k-1}^{GFB}(b) = (D_{k-1}^{GFB}(a) + \lceil p_k \rceil + C\lfloor p_k \rfloor) \bmod N_k. \tag{11}$$

C is the integer value and p_k is the value that defines the distance of connected layer-$k-1$ GFBs, and is calculated as

$$p_k = \frac{N_k}{L_k^{'gfb} + 1}.. \tag{12}$$

The links in the GFBs are connected at an equal distance of the ID for the layer-k GFBs to minimize the maximum number of hops between the layer-k GFBs.

Selection of switches used to connect layer-$k-1$ GFBs After the connections between layer-$k-1$ GFBs are constructed, we select switches that are used to connect layer-$k-1$ GFB pairs. The switch $D^{sw}(s)$ included in the GFB of ID $D_{k-1}^{GFB}(a)$ is connected to the GFB of ID $D_{k-1}^{GFB}(b)$ when the following condition is satisfied.

$$D^{sw}(s) = D_{k-1}^{gfb}(b) + \left\lfloor \frac{Cn_{D_{k-1}^{gfb}(a)}}{l_{(D_{k-1}^{gfb}(a), D_{k-1}^{gfb}(b))}} \right\rfloor$$

where C is an integer value, $n_{D_{k-1}^{gfb}(a)}$ is the number of switches in the GFB of ID $D_{k-1}^{gfb}(a)$, and $l_{(D_{k-1}^{gfb}(a), D_{k-1}^{gfb}(b))}$ is the number of links to be constructed between the GFBs of IDs $D_{k-1}^{gfb}(a)$ and $D_{k-1}^{gfb}(b)$. By connecting switches using the above condition, the intervals of switches connected to the same GFB become constant, and we can avoid the large number of hops from a switch to the other GFBs.

Properties of GFBs The maximum number of hops in GFBs or the number of paths passing each link can be calculated from the parameters described below.

Maximum Number of Hops The maximum number of hops between switches in the layer-k GFB, H_k, is calculated as

$$H_k = (h_k + 1)H_{k-1} + h_k, \tag{13}$$

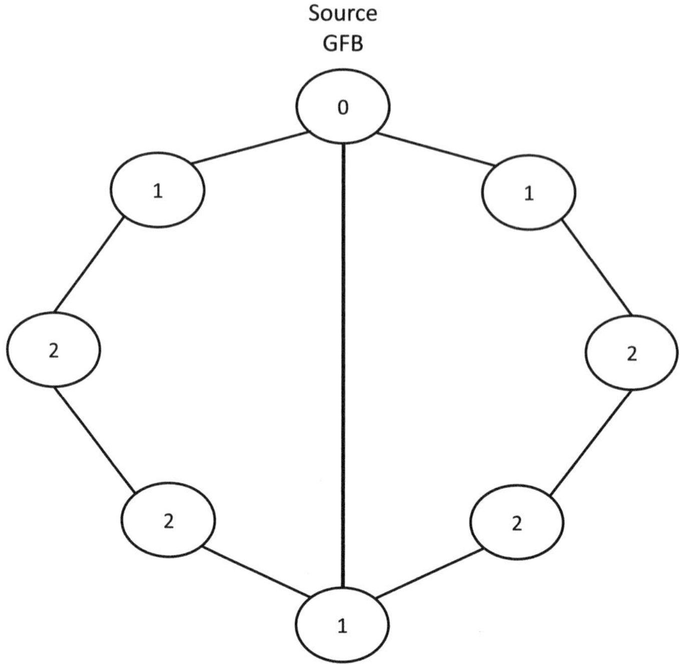

Fig. 14 Example of number of hops in topology constructed of low-layer GFBs ($L_k^{'GFB} = 1$)

where h_k is the most links between layer-$k-1$ GFBs passed by the traffic between layer-$k-1$ GFBs. H_k is obtained by calculating h_k. The rest of this paragraph discusses how h_k is calculated from the parameters of GFBs.

If L_k^{gfb} defined by Eq. (9) is larger than $M_k(N_k - 1)$, we add links between all pairs of layer-$k-1$ GFBs. Thus, $h_k = 1$.

If L_k^{GFB} is smaller than $N_k - 1$ and $L_k^{'GFB}$ defined in Eq. (10) is zero, the connections between layer-$k-1$ GFBs form a ring topology. In this case, h_k is $\lceil \frac{N_k}{2} \rceil$.

If L_k^{GFB} is smaller than $(N_k - 1)$ and $L_k^{'GFB}$ is a positive value, we add links to the GFBs satisfying Eq. (11). In this case, we discuss the calculation of h_k by dividing the topology constructed of layer-$k-1$ GFBs into modules so that each module includes the GFB whose ID is within the range from Cp_k to $(C+1)p_k$ where C is a integer variable and p_k is defined by Eq. (12). Then, we calculate the maximum number of hops from the source GFB whose ID is zero. Since all low-layer GFBs play the same role in high-layer GFBs, h_k is calculated by calculating the maximum number of hops from a GFB whose ID is zero.

If $L_k^{'GFB}$ is one, the topology constructed of layer $k-1$ GFBs is divided into the two modules in Fig. 14. Here, each module becomes a ring topology with p_k nodes. That is, h_k is $\lceil \frac{p_k}{2} \rceil$ in this case.

If $L_k^{'GFB}$ is more than one, the topology constructed of layer $k-1$ GFBs is divided into more than two modules. Here, at least one module does not include the source GFB, and the module without the source GFB includes a GFB whose number of hops

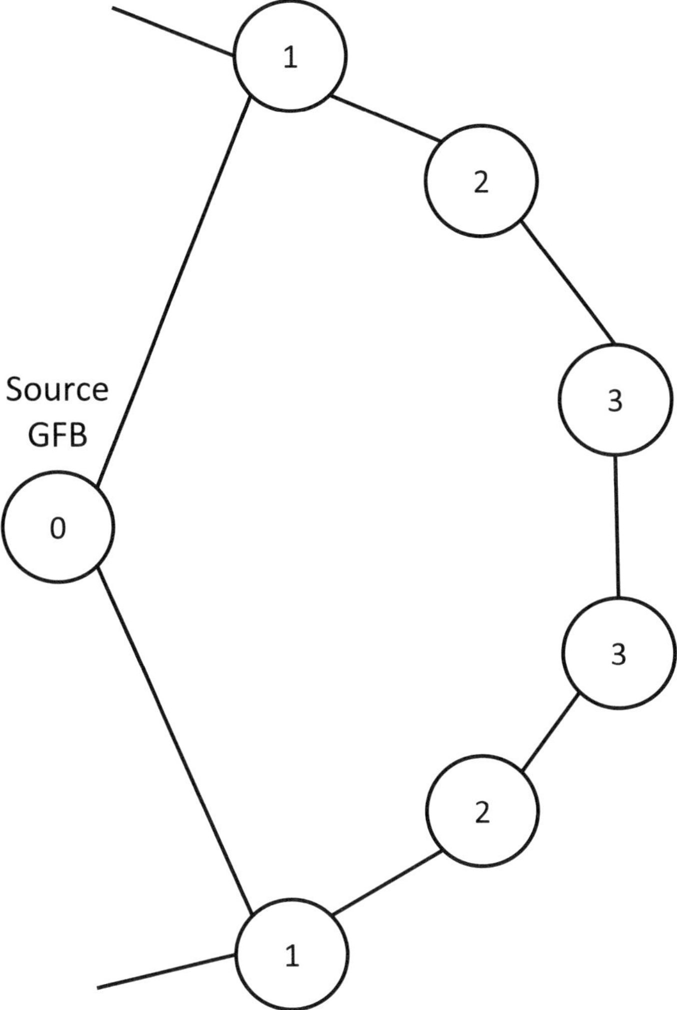

Fig. 15 Example of number of hops in topology constructed of low-layer GFBs ($L_k^{'GFB} > 1$)

from the source GFB is the largest. The modules without the source GFB includes $\lfloor p_k \rfloor$ GFBs. The GFBs at both edges of the module are connected to the source GFB, as shown in Fig. 15, where the source GFB and the GFB included in the modules form a ring topology with $\lfloor p_k \rfloor + 2$ nodes. Thus, h_k is $\lceil \frac{\lfloor p_k \rfloor + 2}{2} \rceil$ in this case.

Summarizing the above discussion, h_k is calculated as

$$h_k = \begin{cases} 1 & (L_k^{GFB} \geq (N_k - 1)) \\ \lceil \frac{p_k}{2} \rceil & (L_k^{GFB} < (N_k - 1) \text{ and } L_k^{'GFB} = 0, 1) \\ \lceil \frac{\lfloor p_k \rfloor + 2}{2} \rceil & (Otherwise) \end{cases} \quad (14)$$

h_k is defined by constructing the layer-k GFB in Step. I, and h_k can be calculated with the parameters. If the links are added between all pairs of layer-$k-1$ GFBs, h_k is 1. Otherwise, the links are added to the layer-$k-1$ GFB satisfying Eq. (11). In this case, h_k is defined by p_k.

Number of Flows through a Link The number of layer-$k-1$ GFB pairs whose traffic passes link l between layer-$k-1$ GFBs x_l^k is obtained by calculating the number of flows passing the link in an abstracted topology where the layer-$k-1$ GFB is regarded as a single node. Then, by multiplying it with the number of flows passing layer-$k-1$ GFB pairs, we obtain the number of flows passing each link. Since all layer-k GFBs play the same role, the number of flows passing between the layer-$k-1$ GFB pair is independent of the IDs of the GFBs.

Thus, the number of flows passing link l between layer-$k-1$ GFBs X_l^k is obtained as

$$X_l^k = F_k x_l^k,$$

where F_k is the number of flows from a layer-$k-1$ GFB to another layer-$k-1$ GFB. After this, we will calculate x_l^k and F_k.

We will first calculate x_l^k. There are two kinds of links in the abstracted topology, where a lower-layer GFB is regarded as a single node. The first is a link on the ring topology (we will call this link a *ring link* after this), and the second is a link added to give the ring topology a shortcut (we will call this link a *shortcut link* after this).

Since all layer-$k-1$ GFBs play the same role in the layer-k GFB, the number of flows passing each ring link is independent of the GFBs connected to the link. Similarly, the number of flows passing each shortcut link is also independent of the GFBs connected to the link. Therefore,

$$x_l^k = \begin{cases} \dfrac{M_k^{\text{ring}}}{2\prod_{i=1}^{k} N_i} & (l \text{ is a ring link}) \\ \dfrac{M_k^{\text{shortcut}}}{(L_k-2)\prod_{i=1}^{k} N_i} & (l \text{ is a shortcut link}) \end{cases}. \tag{15}$$

where M_k^{ring} is the total ring links passed by the traffic between layer-$k-1$ GFB pairs, and M_k^{shortcut} is the total shortcut links passed by the traffic between layer-$k-1$ GFB pairs. $2\prod_{i=1}^{k} N_i$ is the number of ring links between layer-$k-1$ GFBs, and $(L_k-2)\prod_{i=1}^{k} N_i$ is the number of shortcut links between layer-$k-1$ GFBs.

The traffic between layer-$k-1$ GFBs passes at most one shortcut link because the interval of the IDs of the GFBs connected to a certain GFB is constant. $2h_k\prod_{i=1}^{k} N_i$ flows do not pass the shortcut link. Thus,

$$M^{\text{shortcut}} = \prod_{i=1}^{k} N_i(\prod_{i=1}^{k} N_i - 1) - 2h_k \prod_{i=1}^{k} N_i.$$

In addition, M^{ring} is obtained by subtracting M^{shortcut} from the total number of links passed by the traffic between layer-$k-1$ GFBs:

$$M^{\text{ring}} = \sum_{i=1}^{h_k} i s_k(i) - M^{shortcut},$$

where $s_k(i)$ is the number of layer $k-1$ GFB pairs whose traffic passes i links in the abstracted topology.

$s_k(i)$ is obtained as follows. $s_k(1)$ is the same value as the number of links in the layer-k GFB. That is,

$$s_k(1) = \begin{cases} N_k(N_k - 1) & (L_k^{GFB} \geq (N_k - 1)) \\ N_k L_k \prod_{i=1}^{k-1} N_i & \text{(otherwise)} \end{cases}. \tag{16}$$

$s_k(i)$ for $i > 1$ is calculated by dividing the topology constructed of layer-$k-1$ GFBs into groups similar to the case of calculating h_k. By dividing the topology, $s_k(i)$ is calculated by the sum of the number of layer-$k-1$ GFBs that are i hops away from the source layer-$k-1$ GFB in each group. Thus, $s_k(i)$ is calculated as

$$s_k(i) = N_k \sum_{m_j \in M} U_{(k,m_j)}(i), \tag{17}$$

where $U_{(k,m_j)}(i)$ is the number of layer-$k-1$ GFBs that are i hops away from the source layer-$k-1$ GFB in group m_j. Since GFBs are included in each group, the source GFB and the GFBs directly connected to it form a ring topology:

$$U_{(k,m_j)}(i) = \begin{cases} 0 & \left(i > \left\lceil \frac{m_j+2}{2} \right\rceil\right) \\ 1 & \left(i = \left\lceil \frac{m_j+2}{2} \right\rceil \text{ and } |m_j| \text{ is odd}\right). \\ 2 & (Otherwise) \end{cases} \tag{18}$$

We calculate the number of flows between each layer-$k-1$ GFB pair, F_k. The number of flows between each layer-$k-1$ GFB pair is independent of the IDs of the source or destination GFB. Thus, we calculate the number of flows passing between layer-$k-1$ GFBs s and d, $F_k^{s \to d}$.

$F_k^{s \to d}$ is calculated as

$$\begin{aligned} F_k^{s \to d} &= f_k^{s \to s \to d \to d} + \sum_{n \in G} f_k^{n \to s \to d \to d} \\ &+ \sum_{n \in G} f_k^{s \to s \to d \to n} + \sum_{n_1, n_2 \in G} f_k^{n_1 \to s \to d \to n_2}, \end{aligned} \tag{19}$$

where $f^{a \to b \to c \to d}$ is the number of flows whose source and destination switches belong to the layer-$k-1$ GFBs, a and d, and that traverse the layer $k-1$ GFBs, b and c. G is the set of switches that do not belong to the layer-k GFB including the layer-$k-1$ GFBs, s and d.

$f_k^{s \to s \to d \to d}$ is calculated by the product of the number of switches included in the layer-$k-1$ GFB s and that included in the layer-$k-1$ GFB d. That is,

$$f_k^{s \to s \to d \to d} = \prod_{i=1}^{k-1} (N_i)^2. \tag{20}$$

$\sum_{n \in G} f_k^{s \to s \to d \to n}$ indicates the number of flows from the layer-$k-1$ GFB s to the outside of the layer k GFB via the layer $k-1$ GFB d. Because all layer-$k-1$ GFBs play the same role in the GFB, $\sum_{n \in G} f_k^{s \to s \to d \to n}$ is calculated by dividing the number of flows whose source switches belong to the layer-$k-1$ GFB s and destination switches belong to the different layer-k GFB by the number of layer-$k-1$ GFBs in the layer-k GFB.

$$\sum_{n \in G} f_k^{s \to s \to d \to n} = \frac{(\prod_{i=1}^{k-1} N_i)(\prod_{i=1}^{K_{\max}} N_i - \prod_{i=1}^{k} N_i)}{N_k}. \tag{21}$$

Similarly, $\sum_{n \in G} f_k^{n \to s \to d \to d}$ is calculated as

$$\sum_{n \in G} f_k^{n \to s \to d \to d} = \frac{(\prod_{i=1}^{k-1} N_i)(\prod_{i=1}^{K_{\max}} N_i - \prod_{i=1}^{k} N_i)}{N_k}. \tag{22}$$

$\sum_{n_1, n_2 \in G} f_k^{n_1 \to s \to d \to n_2}$ indicates the number of flows that arrive from outside the layer-k GFB via the layer-$k-1$ GFB s and go outside the layer-k GFB via the layer-$k-1$ GFB d. The number of flows arriving from outside the layer-k GFB via the layer-$k-1$ GFB s is the total flows on links that connect switches in the layer-$k-1$ GFB s and the switches outside the layer-k GFB, which is calculated as

$$\prod_{j=1}^{k-1} N_j \sum_{i=k+1}^{K} (X_l^i L_i). \tag{23}$$

We obtain the number of flows that arrive from outside the layer-k GFB via the layer-$k-1$ GFB s and that are sent to the layer-$k-1$ GFB d by dividing Eq. (23) by the number of layer-$k-1$ GFBs in the layer-k GFB. This calculated value includes flows whose destination switches belong to the layer-$k-1$ GFB d, whose number is $\sum_{n_1 \in G} f_k^{n_1 \to s \to d \to d}$. Therefore, $\sum_{n_1, n_2 \in G} f_k^{n1 \to s \to d \to n_2}$ is calculated as

$$\sum_{n_1, n_2 \in G} f_k^{n_1 \to s \to d \to n_2} = \frac{\prod_{j=1}^{k-1} N_j \sum_{i=k+1}^{K} (X_l^i L_i)}{N_k} - \sum_{n_1 \in G} f_k^{n_1 \to s \to d \to d}. \tag{24}$$

4.3 Control of Virtual Network Topology to Achieve Low Energy Consumption

4.3.1 Outline

The virtual network is constructed with our method to minimize the number of used ports taking two kinds of requirements into consideration, i.e., bandwidths and delays between servers.

One approach to providing sufficient bandwidths between servers is to construct the virtual network so that it can accommodate current traffic demand between servers. However, traffic at data center can change within a second [27]. Thus, if the virtual network is optimized for the current traffic demand, it may need to be reconfigured every second. However, the calculation time to optimize virtual networks for current traffic demand becomes too long at large data centers. Therefore, a virtual network cannot be constructed to accommodate current traffic demand between servers.

The traffic changes within short periods are handled in our method by load balancing [20] over the virtual network. We also designed the virtual network to achieve a sufficiently large bandwidth and short delay with small energy consumption, by taking load balancing into account.

We used a load balancing technique called *Valiant Load Balancing (VLB)* [20]. We randomly selected intermediate nodes in VLB regardless of the destination to avoid traffic from concentrating on certain links even when a certain node pair had a large amount of traffic. Then, traffic was sent from the source node to an intermediate node and from the intermediate node to the destination node. The amount of traffic between each ToR switch pair T is calculated by applying VLB as

$$T \leq \frac{T^{SWto} + T^{SWfrom}}{N_{\mathrm{all}}}. \tag{25}$$

The T^{SWto} in this equation is the maximum amount of traffic to a ToR switch, T^{SWfrom} is the maximum amount of traffic from a ToR switch, and N_{all} is the number of ToR switches in the virtual network. Thus, we can provide sufficient bandwidth by making the number of flows passing a link less than a threshold, which is calculated by dividing the capacity of an optical path by the amount of traffic between each switch pair calculated with Eq.(25).

Delay is also hard to forecast when designing the virtual network. We avoided overly long delay by providing enough bandwidth and making the maximum number of hops less than the threshold, which is discussed in this section.

4.3.2 Control of Topology to Satisfy Requirements

This subsection introduces a method of setting the parameters of the GFB to minimize the number of used ports and satisfy the requirements for bandwidth and the maximum number of hops between servers.

The number of switches connected in the virtual network, N_{all}, the acceptable maximum number of hops, H_{max}, the maximum amount of traffic from a ToR switch, T^{SWfrom}, and the maximum amount of traffic to a ToR switch, T^{SWto} are given. Our method sets the parameters according to the following steps.

First, we calculate the candidates for the number of layers. Because we cannot make the maximum number of hops of the GFB less than the case where $h_k = 1$ in Eq. (13) for all layers, the number of layers $K_{\max}$ must satisfy the condition below to make the maximum number of hops less than H_{max}.

$$2^{K_{\max}} - 1 \leq H_{max} \tag{26}$$

We consider all $K_{\max}$ that satisfy the above condition as candidates for the number of layers. We set suitable parameters according to two steps for each candidate.

Step 1 Set the parameters by considering the acceptable number of hops.
Step 2 Modify the parameters to provide a sufficient bandwidth.

Then, we construct a topology that uses the fewest virtual links for the candidates. The details of the two steps are described in the following.

Parameter Settings Considering Acceptable Number of Hops We set parameters N_k and L_k to make the maximum hops less than H_{max}. We set N_k to $\prod_{i=1}^{k-1} N_i + 1$ for $1 < k < K_{\max}$ in these steps to reduce the number of variables. By doing so, h_k becomes one even when $L_k = 1$.

$N_{K_{\max}}$ must satisfy

$$N_{K_{\max}} = \left\lceil \frac{N_{\text{all}}}{\prod_{i=1}^{k-1} N_i} \right\rceil \tag{27}$$

to connect N_{all} switches. We also set $L_{K_{\max}}$ in these steps so that $h_{K_{\max}}$ becomes one to reduce the number of variables.

$L_{K_{\max}}$ should satisfy

$$L_{K_{\max}} = \left\lceil \frac{N_{K_{\max}}}{\prod_{i=1}^{k-1} N_i} \right\rceil \tag{28}$$

to make $h_{K_{\max}}$ one. h_1 must satisfy

$$h_1 \leq \left\lceil \frac{H_{\max} + 1}{2^{K-1}} - 1 \right\rceil. \tag{29}$$

to make the maximum number of hops less than H_{max} according to Eq. (14). L_1 should satisfy

$$L_1 = \begin{cases} N_1 - 1 & (h_1 = 1) \\ 2 & (h_1 \geq \lfloor \frac{N_1}{2} \rfloor) \\ \lfloor \frac{N_1}{2h_1} + 1 \rfloor & (Otherwise) \end{cases} \tag{30}$$

to satisfy Eq. (29). All N_k $(k > 1)$ and L_k are calculated with N_1 under the above condition. Our objective in setting the parameters is to minimize the number of used ports for ToR switches. That is, we minimize $\sum_{1 \leq k \leq K} L_k$. Since $\sum_{1 \leq k \leq K} L_k$ is the convex function of N_1, we find N_1 that minimizes $\sum_{1 \leq k \leq K_{\max}} L_k$ by incrementing N_1 as long as $\sum_{1 \leq k \leq K_{\max}} L_k$ decreases.

Parameter Modifications to Provide Sufficient Bandwidth If the GFB with the parameters set in Step 1 cannot provide sufficient bandwidth, we add the links to a layer where sufficient bandwidth cannot be provided. To detect the lack of bandwidth, we check whether condition

$$TX_l^k \leq B \tag{31}$$

is satisfied for each layer k to detect the lack of bandwidth, where B is the bandwidth of one link, and T is calculated with Eq. (25). If Eq. (31) is not satisfied, we add L_k until it is satisfied.

4.4 Evaluation

We investigated the number of ports needed for ToR switches to satisfy the requirements. All topologies in this comparison included 420 ToR switches. We compared the topology constructed with our method with the four topologies of FatTree, torus, switch-based DCell [6], and the Flattened Butterfly [5]. Unlike the FatTree topology proposed by Al-Fares et al. [4], we assumed that traffic was not only generated from switches in the lowest layer but also from switches in the upper layer in the FatTree used in this evaluation, since powering up additional switches consumes more energy. The parameters of each topology in our evaluation were set to minimize the number of ports required by the topology under the constraint that it could provide sufficient bandwidth and the maximum number of hops was less than H_{max}.

We assumed that the number of wavelengths on optical fibers was sufficient in this evaluation. We set the bandwidth of one optical path to 10 Gbps.

We compared the number of virtual optical paths per ToR switch needed to satisfy the requirements by changing the maximum amount of traffic from or to each ToR switch. We set the acceptable maximum number of hops to a sufficiently large value in this comparison. That is, the bandwidth provided to each ToR switch was the main requirement for the virtual network.

Figure 16 presents the results, where the horizontal axis indicates the maximum amount of traffic from or to ToR switches that was required to be accommodated, and the vertical axis indicates the number of virtual links per ToR switch that was needed to satisfy the requirements. As we can see from this figure, the switch-based DCell could not accommodate more than 1 Gbps of traffic, and FatTree and torus could not accommodate more than 6 Gbps of traffic per ToR switch. The link between level-0 DCells became a bottleneck in switch-based DCell, which could not be solved by setting the parameters. We could not construct FatTree topologies having more links than the 3-layer FatTree. Thus, FatTree could not accommodate more traffic than that in the 3-layer FatTree. Similarly, torus could not accommodate more traffic than that in the torus with the largest dimensions of those constructed with 420 switches.

Although the flattened butterfly could accommodate a large amount of traffic, it required many virtual links. Fig. 16 indicates that our method used the least virtual

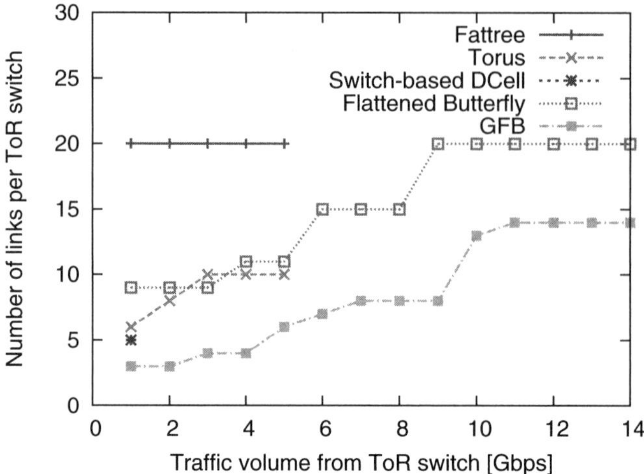

Fig. 16 Number of virtual links required to accommodate traffic from ToR switches

links to accommodate traffic regardless of the amount of traffic. This was because our method of setting the parameters of GFB only added links that were necessary to accommodate traffic. Therefore, the topology constructed with our method satisfied the requirement for bandwidth with the lowest energy consumption.

We also compared the number of virtual links per ToR switch needed to satisfy the requirements by changing the acceptable maximum number of hops. We assumed that the capacity of each virtual link was sufficient in this comparison. That is, the acceptable maximum number of hops was the only requirement for the virtual network.

Figure 17 plots the results, where the horizontal axis indicates the maximum number of hops, and the vertical axis indicates the number of virtual links per ToR switch needed to satisfy the requirements. As shown in this figure, the flattened butterfly required a large number of virtual links even if there were many acceptable maximum numbers of hops.

Conversely, the switch-based DCell and FatTree could construct a topology by using few links when there were many acceptable maximum numbers of hops. However, these topologies required many virtual links when there were few acceptable maximum numbers of hops. The maximum number of hops in these topologies was defined by the number of layers. These topologies required few layers when there were few acceptable maximum numbers of hops. However, fewer layers required many links in these topologies constructed of 420 switches. The torus constructed of 420 switches could not accommodate fewer than seven acceptable numbers of hops.

The topology constructed with our method used the least virtual links to satisfy the requirements as can be seen from Fig. 17 in all cases of the acceptable maximum number of hops. This is because our method of setting the parameters of GFB only added links that were necessary to achieve the maximum number of hops. Therefore,

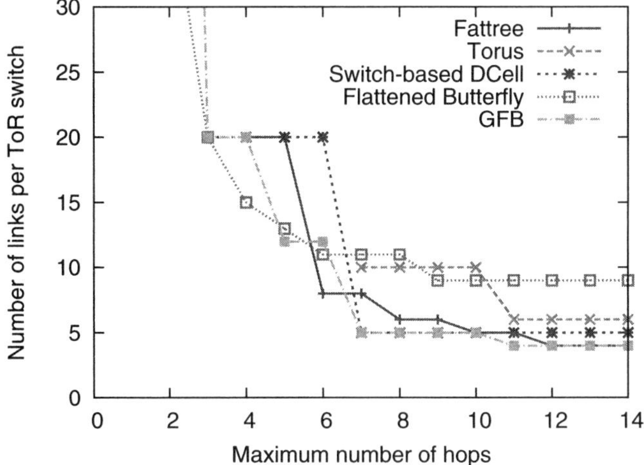

Fig. 17 Number of virtual links required to make maximum number of hops less than target value

the topology constructed with our method satisfied the requirement for the maximum number of hops with the lowest energy consumption.

5 Conclusion

Optical data center networks represent one approach to constructing data center networks that provide large bandwidths between servers with low energy consumption. We can use two types of optical switches in optical data center networks, i.e., optical packet and optical circuit switches.

This chapter introduced two approaches of our research [16, 28]. The first was introduced as an application of optical packet switches that focused on the bandwidth provided to all-to-all communication. We deployed optical packet switches in this approach and constructed a network structure that efficiently used the large bandwidth provided by these switches.

The second approach was introduced as an application of optical circuit switches and was aimed at minimizing energy consumption, where the switches were placed at the core of a data center network. ToR switches were connected to one of the ports of the optical circuit switches. The virtual network constructed of ToR switches was constructed by setting the optical circuit switches and the topology of the virtual network was changed by reconfiguring the switches. We reduced energy consumption by setting the optical circuit switches to minimize the number of ports required by packet switches. We also introduced a method of calculating suitable settings for the optical circuit switches in this approach.

One future direction is to combine these two approaches. Although optical packet switches consume less energy than electronic packet switches, they consume more energy than the optical circuit switches because they require label processing and buffers. Thus, our method of constructing a virtual network of optical packet switches by setting optical circuit switches could accommodate more traffic with less energy consumption, which was similar to the second approach explained in this chapter.

Acknowledgements The research introduced in this chapter was done as part of the "Research & Development of Basic Technologies for High Performance Opto-electronic Hybrid Packet Routers" supported by the National Institute of Information and Communications Technology (NICT).

References

1. S. Ghemawat, H. Gobioff, and S. Leung, "The google file system," in *Proceeding of ACM SIGOPS Operating Systems Review*, vol. 37, pp. 29–43, ACM, Dec. 2003.
2. J. Dean and S. Ghemawat, "Mapreduce: Simplified data processing on large clusters," *Communications of the ACM*, vol. 51, no. 1, pp. 107–113, 2008.
3. D. Abts, M. Marty, P. Wells, P. Klausler, and H. Liu, "Energy proportional datacenter networks," *ACM SIGARCH Computer Architecture News*, vol. 38, pp. 338–347, June 2010.
4. M. Al-Fares, A. Loukissas, and A. Vahdat, "A scalable, commodity data center network architecture," in *Proceedings of ACM SIGCOMM*, vol. 38, pp. 63–74, Aug. 2008.
5. J. Kim, W. Dally, and D. Abts "Flattened butterfly: a cost-efficient topology for highradix networks," in *Proceedings of ISCA*, vol. 35, pp. 126–137, June 2007.
6. C. Guo, H. Wu, K. Tan, L. Shi, Y. Zhang, and S. Lu "DCell: A scalable and fault-tolerant network structure for data centers," *ACM SIGCOMM Computer Communication Review*, vol. 38, pp. 75–86, Aug. 2008.
7. C. Guo, G. Lu, D. Li, H. Wu, X. Zhang, Y. Shi, C. Tian, Y. Zhang, and S. Lu "BCube: A high performance, server-centric network architecture for modular data centers," *ACM SIGCOMM Computer Communication Review*, vol. 39, pp. 63–74, Aug. 2009.
8. D. Guo, T. Chen, D. Li, Y. Liu, X. Liu, and G. Chen "BCN: expansible network structures for data centers using hierarchical compound graphs," in *Proceedings of INFOCOM*, pp. 61–65, Apr. 2011.
9. D. Li, C. Guo, H. Wu, K. Tan, Y. Zhang, S. Lu, and J. Wu "Scalable and cost-effective interconnection of data-center servers using dual server ports," *IEEE/ACM Transactions on Networking*, vol. 19, pp. 102–114, Feb. 2011.
10. Y. Liao, D. Yin, and L. Gao, "Dpillar: Scalable dual-port server interconnection for data center networks," in *Proceedings of ICCCN*, pp. 1–6, Aug. 2010.
11. A. Greenberg, J. Hamilton, N. Jain, S. Kandula, C. Kim, P. Lahiri, D. Maltz, P. Patel, and S. Sengupta, "VL2: A scalable and flexible data center network," *ACM SIGCOMM Computer Communication Review*, vol. 39, pp. 51–62, Aug. 2009.
12. R. N. Mysore, A. Pamboris, N. Farrington, N. Huang, P. Miri, S. Radhakrishnan, V. Subramanya, and A. Vahdat, "PortLand: A scalable fault-tolerant layer 2 data center network fabric," *ACM SIGCOMM Computer Communication Review*, vol. 39, pp. 39–50, Aug. 2009.
13. K. A. McGreer, "Arrayed waveguide gratings for wavelength routing," *IEEE Communications Magazine*, vol. 36, Dec. 1998.
14. B. Li, Y. Quin, X. R. Cao, K. M. Sivaligam, and Y. Danziger, "Photonic packet switching: Architecture and performance," *Optical Networks Magazine*, vol. 2, pp. 27–39, Jan. 2001.
15. P. Beebe, J. M. Ballantyne, and M. F. Tung, "An introduction to mems optical switches." http://courses.cit.cornell.edu/engrwords/final_reports/Tung_MF_issue_1.pdf, Dec. 2001.
16. Y. Ohsita and M. Murata, "Data center network topologies using optical packet switches," in *Proceedings of DCPerf*, pp. 57–64, June 2012.

17. R. Urata, T. Nakahara, H. Takenouchi, T. Segawa, H. Ishikawa, A. Ohki, H. Sugiyama, S. Nishihara, and R. Takahashi, "4x4 optical packet switching of asynchronous burst optical packets with a prototype, 4x4 label processing and switching sub-system," *Optics Express*, vol. 18, pp. 15283–15288, July 2010.
18. H. J. Chao and K. Xi, "Bufferless optical clos switches for data centers," in *Proceedings of OFC*, Mar. 2011.
19. K. Xi, Y. H. Kao, M. Yang, and H. J. Chao, "Petabit optical switch for data center networks." Technical Report, Polytechnic Institute of New York University, http://eeweb.poly.edu/chao/publications/petasw.pdf.
20. M. Kodialam, T. V. Lakshman, and S. Sengupta, "Efficient and robust routing of highly variable traffic," in *Proceedings of HotNets*, Nov. 2004.
21. T. Benson, A. Akella, and D. A. Maltz "Network traffic characteristics of data centers in the wild," in *Proceedings of Internet Measurement Conference*, Nov. 2010.
22. Y. Zhang, M. Roughan, N. Duffield, and A. Greenberg, "Fast accurate computation of large-scale IP traffic matrices from link loads," in *Proceedings of ACM SIGMETRICS Performance Evaluation Review*, vol. 31, pp. 206–217, June 2003.
23. Y. Ohsita, T. Miyamura, S. Arakawa, S. Ata, E. Oki, K. Shiomoto, and M. Murata "Gradually reconfiguring virtual network topologies based on estimated traffic matrices," *IEEE/ACM Transactions on Networking*, vol. 18, pp. 177–189, Feb. 2010.
24. M. Zhang, C. Yi, B. Liu, and B. Zhang, "GreenTE: power-aware traffic engineering," in *Proceedings of ICNP*, pp. 21–30, Oct. 2010.
25. A. Singla, A. Singh, K. Ramachandran, L. Xu, and Y. Zhang, "Proteus: a topology malleable data center network," in *Proceedings of ACM SIGCOMM Workshop on Hot Topics in Networks*, pp. 8–13, Oct. 2010.
26. N. Farrington, G. Porter, S. Radhakrishnan, H. Bazzaz, V. Subramanya, Y. Fainman, G. Papen, and A. Vahdat, "Helios: a hybrid electrical/optical switch architecture for modular data centers," in *Proceedings of ACM SIGCOMM Computer Communication Review*, pp. 339–350, Oct. 2010.
27. T. Benson, A. Anand, A. Akella, and M. Zhang, "MicroTE: Fine Grained Traffic Engineering for Data Centers," in *Proceedings of ACM CoNEXT*, pp. 1–12, Dec. 2011.
28. Y. Tarutani, Y. Ohsita, and M. Murata, "A virtual network to achieve low energy consumption in optical large-scale datacenter," in *Proceedings of IEEE International Conference on Communication Systems (ICCS 2012)*, Nov. 2012.

Scalable Network Communication Using Unreliable RDMA

Ryan E. Grant, Mohammad J. Rashti, Pavan Balaji and Ahmad Afsahi

1 Introduction

High-performance interconnects play a pivotal and essential role in the performance and functionality of modern large-scale computational systems, including datacenters and high-performance computing (HPC) architectures. Commercial datacenter applications require that a large number of small independent tasks be performed rapidly in parallel with upper bounds on individual task delays. This emphasis on large numbers of tasks and limited intertask dependencies leads to such computing being known as *capacity computing*. The term high-performance computing traditionally refers to large-scale applications running exclusively on a large system. This type of computing is referred to as *capability computing*. It requires that the system coordinate a small number of applications over a large number of resources (e.g.,

Sandia is a multiprogram laboratory operated by Sandia Corporation, a Lockheed Martin Company, for the United States Department of Energy's National Nuclear Security Administration under contract DE-AC04-94AL85000.

R. E. Grant (✉)
Scalable System Software Department, Sandia National Laboratories, MS-1319, P.O. Box 5800, Albuquerque, NM 87185-1319, USA
e-mail: regrant@sandia.gov

M. J. Rashti
RNET Technologies Inc., 240 W. Elmwood Dr., Dayton, OH 45459-4248, USA
e-mail: mrashti@rnet-tech.com

P. Balaji
Mathematics and Computer Science Division, Argonne National Laboratory, Bldg. 240, Rm. 3146, Argonne, IL, 60439, USA
e-mail: balaji@mcs.anl.gov

A. Afsahi
Department of Electrical and Computer Engineering, Queen's University, 19 Union Street, Walter Light Hall, Kingston, ON K7L 3N6, Canada
e-mail: ahmad.afsahi@queensu.ca

S. U. Khan, A. Y. Zomaya (eds.), *Handbook on Data Centers*,
DOI 10.1007/978-1-4939-2092-1_12

nodes). Capability computing is generally most useful for computation-intensive scientific applications. Cloud computing has begun to move HPC applications from dedicated machines into the cloud, where they can take advantage of commercial datacenter infrastructure.

Datacenter computing with commercial applications does not require as many interprocess connections as high-performance computing does. The reason for this is the significantly smaller number of interprocess dependencies in capacity computing than in HPC. Many of the scalability issues that exist in HPC are not present in capacity computing. For example, capacity computing has no requirements to have many nodes coordinate their computation. However, the overhead associated with managing many small tasks and potentially many client connections can still be significant. Some datacenter applications are much more tolerant of network packet loss than HPC applications. Applications such as audio and video streaming can tolerate a small amount of data loss with little noticeable effect. Datacenters benefit from networks that reduce overhead; freeing more resources for processing tasks and thus increasing system capacity.

High-performance computing offers several major challenges. Since a large system must be made to cooperate on a small number of jobs, synchronization between processes and parallelism of individual tasks is nontrivial. Unlike a system handling many independent jobs, the dependencies between processes create bottlenecks where the progress of the whole system can be impeded by a single straggling process. In addition, because the individual processes are working cooperatively, contention for shared resources can be a major issue, since certain execution phases will require that processes need shared resources at the same time. Another major issue is scalability. As the number of nodes and processes involved in a given job grows, the number of potential communication partners does as well. Hence, networks designed to handle thousands to tens of thousands of communication partners may not be suited to handle hundreds of thousands or millions of partners.

Designing high-performance networks that meet the needs of traditional datacenter and HPC applications is challenging. Although the requirements of both application spaces may seem to be noncomplementary, design solutions have been developed that are applicable to both. Traditional datacenter computing and HPC datacenters are both concerned with the fast and efficient movement of large amounts of data. In HPC this communication happens between computational peers, whereas in datacenters communication happens between small groups of nodes in the datacenter and out to clients over a wide area network (WAN).

1.1 The Significance of Data Communication

In parallel applications running over multinode distributed systems, data is frequently communicated among different application components over high-performance interconnects. Data movement is a key factor in the performance of datacenter services. It has become even more important with the emergence of the big-data paradigm, virtualization, and increased demand for distributed elastic storage and computing services (i.e., cloud computing). In many scientific applications, communication of intermediate computing results among processing elements accounts for a significant portion of the overall execution time. With the ever-increasing computing capabilities in modern distributed systems, the scalability of internode communication architecture has become critical for application performance in extreme-scale systems.

Several features of an interconnect are essential for the scalability and efficiency of distributed applications. The resource requirements of the hardware as well as the runtime and communication software are important for scalability. The complexity and synchronization requirements of communication protocols, as well as operating system (OS) involvement in the process of data movement, can also affect the ability of an application to efficiently utilize the available computing resources of a large-scale machine.

In modern communication subsystems, several techniques have been employed to provide efficient and scalable interconnects. Nonblocking communication is a technique that substantially reduces resource blockage by liberating essential computing resources. For example, asynchronous placement and capture of data can significantly reduce the utilization of a data server's processing power for data movement, increasing its capability for handling other relevant tasks or its capacity to serve more clients. Nonblocking network application programming interfaces (APIs) allow for computation to "overlap" communication, so that the CPU can continue doing useful work while it waits for a network operation to complete. This ability is especially useful when the communication tasks can be offloaded to networking hardware. Such asynchronous communication should not be confused with the term used to denote data transmissions without an external clock signal. OS bypass is another technique that reduces resource usage and prevents the OS from becoming a bottleneck to data movement, thus reducing communication latency and improving capacity. Virtual machine monitor (VMM) bypass is a concept analogous to OS bypass that is used for overcoming VMM-related communication bottlenecks in cloud datacenters.

Remote direct memory access (RDMA) leverages many of the performance-enhancing networking techniques and is the de facto communication technology in modern high-performance interconnects. RDMA refers to the capability to directly access a remote memory location for different purposes, including read, update, or atomic operations. RDMA operations are usually performed without the intervention of the host software in actual data transfer, and DMA technologies at both the data source and data sink are utilized to perform the bulk transfer of data. RDMA combines some of the above techniques to provide a low-overhead asynchronous substrate that has significantly contributed to increased efficiency of data movement in clusters and datacenters over the past decade. While RDMA can now be seen as the communication centerpiece in many data-processing systems, some applications still have not been able to utilize such technologies. This chapter discusses how RDMA networks can be made accessible to a larger application space and the changes needed in modern RDMA interconnects to allow this ability.

1.2 Datacenter Computing and RDMA

Datacenter computing serves commercial computing applications. As pointed out earlier, unlike HPC, datacenter systems are referred to as "capacity" systems because the workload in a commercial datacenter is typically made up of many independent tasks. The individual systems making up the datacenter are not required to work cooperatively on a single large problem; they can operate independently on their small-scale tasks with no need for global synchronization. If one considers a typical web-serving datacenter, a small amount of internode cooperation may occur, for example, a web server communicating with a database server for dynamic data. However, there is never a requirement for all the web servers to communicate with each other in order to synchronize their execution.

The "capacity" nature of commercial workloads means that they typically have many independent tasks that need to be completed within a deadline. The speed at which each individual task can be accomplished is important to meeting this deadline. Therefore, throughput is the most desired attribute of these systems, with the constraint that tasks must be completed within a given deadline.

Traditional non-offloaded networks create an additional CPU load for network processing and introduce additional latency through kernel-based software networking stacks, which could adversely affect both the deadline and throughput requirements of a datacenter application. RDMA offload-capable networks alleviate these overheads by bypassing the kernel and providing network processing on the RDMA-capable network interface card (RNIC).

Despite obvious benefits of RDMA, connection-based RDMA networks still impose overhead for storing the data associated with many connections, with each client using one or more unique connections to the system. Although not restrictive on small scale systems, these resource requirements impose huge limitations on the capacity of large scale systems, in addition to potential adverse effects on the performance of individual applications. Datagram-based approaches resolve this issue by using only a limited number of system resources to serve a large number of clients. This capability is accomplished by keeping minimal or no state information about the communication with the client. Such capability will certainly improve the capacity of datacenter servers.

1.3 High-Performance Computing and RDMA

HPC utilizes supercomputers to solve a single large problem in as little time as possible. Applications of HPC range from fundamental science simulations to computationally complex commercial applications such as oil and gas discovery and pharmaceutical research. HPC applications typically use very large datasets and can communicate large amounts of data through both shared memory and network communications. They require the coordination of many parallel tasks (e.g., threads or processes) and hence require low-latency interconnection links for synchronization and small-message transmission. Many HPC applications follow the bulk synchronous parallel program (BSP) model [1]. In this model, the computer comprises many processors connected by a communication network. Program progress is made in *supersteps* where, for each superstep, individual processes perform a combination of computation and communication (both sending and receiving messages). The superstep concludes with a global synchronization of all the processes to ensure that each has successfully completed its assigned superstep tasks.

Modern HPC applications significantly benefit from RDMA-capable networks. Low overhead and asynchronous nature of RDMA-based communication play a significant role in improving the performance, scalability, and resource efficiency of HPC systems. RDMA offload-capable networks constitute the majority of interconnects in the world's top supercomputers. Therefore, improving the capability and scalability of RDMA-enabled networks will undoubtedly improve the performance of a wide range of HPC applications.

1.4 RDMA and the Current Unreliable Datagram Network Transports

Unreliable network transports are an important mechanism for providing high-performance and low-latency communications. Unreliable transports (such as UDP) operate over many networks including Ethernet, one of the most ubiquitous computer networks in use today. Ethernet's wide adoption has made such networks available at lower cost than that of competing high-speed networks. Compatibility and cost lead to Ethernet's domination despite Ethernet's lack of cutting-edge network performance. However, Ethernet is not the only network that supports unreliable transports. Other high-speed networks such as InfiniBand provide unreliable transports as well (e.g. unreliable datagrams and unreliable connections).

Despite limited existing support for unreliable remote memory access in networks such as InfiniBand, none of the unreliable network transports available today support one-sided RDMA operations. InfiniBand's support is limited to two-sided send/recv operations over its unreliable datagram transport, and there is no specification for one-sided put and get operations. iWARP, which defines RDMA functionality of IP networks, provides operations only over reliable transports such as TCP and SCTP. Other network specifications such as Portals (used in Cray interconnects) do not currently support unreliable transports, although the Portals network specification [2] is connection agnostic.

Because of the large existing Ethernet infrastructure, it is desirable to upgrade existing networks piece by piece rather than overhauling networks by complete networking infrastructure replacement. Therefore, a high-speed Ethernet networking solution that can leverage the performance benefits of advanced networks such as operating system bypass, RDMA, and offloading for both TCP and UDP traffic is attractive.

In this chapter, we provide an overview of the RDMA technology and address the use of RDMA over unreliable transports, particularly over Ethernet. We discuss adding one-sided RDMA capability to unreliable datagrams and extending iWARP as a platform for utilizing RDMA with unreliable transports. We also present a software implementation of an unreliable datagram-based RDMA solution and compare its performance with that of a traditional connection-based solution.

2 Overview of RDMA Technology

Remote direct memory access is typically accomplished via two paradigms that rely on a queue-pair based communication model. The first one, where data placement is handled by the target node, is referred to as "channel" semantics and uses operations called Send and Receive (recv). Prior to the RDMA operation beginning, the target node places an entry into its receive queue that indicates the location in memory where a matching incoming message should be placed as well as the length of the available buffer and information used to match the recv entry to an incoming send. When an RDMA operation is to be performed, the initiator node posts a send request into its send queue, making the RDMA layer send the request along with the data to the target node. The target node matches the incoming send request with an entry in the target node's recv queue. It then places the incoming message directly into the memory location specified by the recv queue entry.

The second paradigm for RDMA provides memory semantics. This type of RDMA is closely related to how local memory accesses are described, with the exception that the target of the operation is a remote node. The operations used in memory semantics are Write and Read (also referred to as Put and Get). Prior to any operation being performed, the target node must expose its memory to any node that it allows RDMA operations from. Whether this information is broadcast as an advertisement to many nodes or exchanged only with a single node by request is implementation dependent. Once the information on the target's memory (the address of the start of the region and its size) is available, the initiator can compose the entire message with information about where the data is to be placed on the target node. When the target node receives the message, all the node needs to do is write the message into the appropriate memory location.

Because of the differences in how the paradigms work, send/recv is sometimes referred to as two-sided communication, since it requires both sides to actively participate in the communication. RDMA Write/Read is referred to as one-sided communication, since the actual message transmission does not require the target node to post any a priori information about the incoming data to its recv queue.

The following provides a brief overview of two of today's major RDMA-based interconnection standards: iWARP Ethernet and InfiniBand.

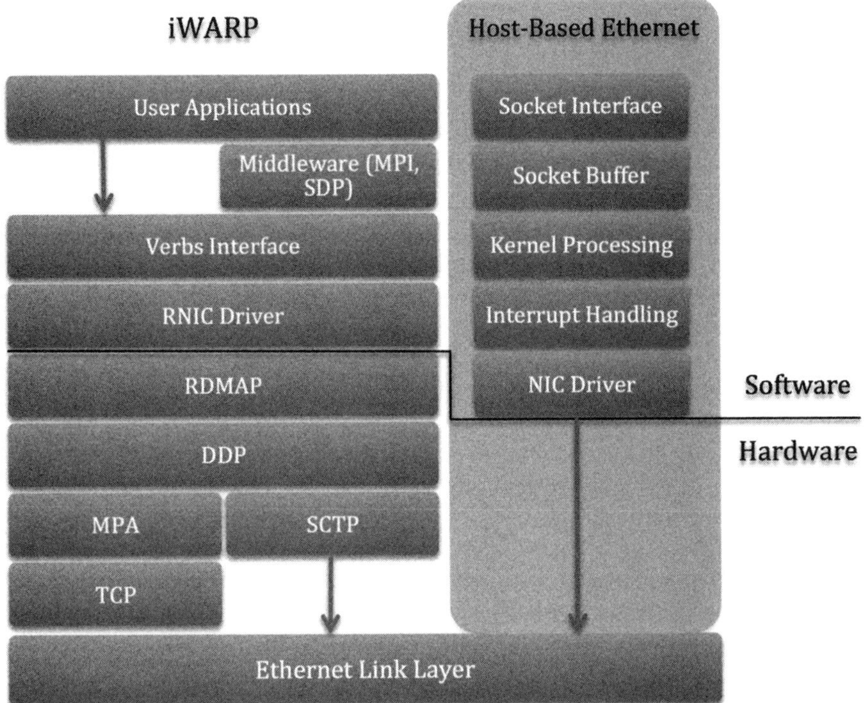

Fig. 1 iWARP stack compared with host-based TCP/IP

2.1 Overview of the iWARP Standard

The specifications for iWARP, or Internet Wide-Area RDMA Protocol, were first proposed in 2002 by the RDMA Consortium [3] to the Internet Engineering Task Force [4]. Utilizing a queue pair design and a communication semantic called verbs, it reflected many of the basic design choices used in Virtual Interface Architecture (VIA) based [5] architectures for high-speed communication. Much like the VIA architecture, the communication stack is intended to completely bypass the operating system, avoiding extra memory copies of data as well as any context switching that would be required when utilizing the OS communication stack. An iWARP adapter provides fully offloaded iWARP stack processing and therefore also requires fully offloaded lower-layer networking transport processing. Hence, hardware for iWARP stack processing as well as a full TCP Offload Engine (TOE) is needed in order to provide all the benefits of OS bypass and zero-copy operation. An overview of the iWARP stack compared with the traditional kernel communication stack is illustrated in Fig. 1.

The iWARP stack comprises three layers, with a fourth layer added for operation over a TCP transport layer. The layers are the verbs layer [6], the Remote Direct

Memory Access Protocol (RDMAP) layer [7], the Direct Data Placement (DDP) layer [8], and, if operating over TCP, the Marker PDU Alignment (MPA) layer [9].

The verbs layer is the direct middleware or application interface to the networking stack; it directly interfaces with the RDMAP layer. The RDMAP layer services all the communication primitives (send, recv, RDMA Write, and RDMA Read). Verbs requests to the RDMAP layer must take the form of a work request (WR) for a given queue pair. Although all requests from the RDMAP layer must be passed in order to the DDP layer, the individual WRs are processed asynchronously on the RNIC. The completion of a given WR typically results in the creation of a completion queue (CQ) entry. Some operations (e.g., send with solicited event) can trigger an event, with the event being an interrupt to signal completion. Most RDMAP operations require only a single operation from the RNIC, sending data to an indicated target or, in the case of RDMA Write, a specific memory location at a given target node. Others, such as RDMA Read, require that state information about the ongoing operation be kept, since they involve a multistep operation. In the case of an RDMA Read, the initiating node requests certain data from the target node, which serves this data back to the node that initiated the request.

The DDP layer [8] is responsible for transferring data to and from the RNIC to the specified user-level buffers. The goal is to do so without any additional memory copies (zero copies from reception to buffer placement). The DDP layer achieves this through two methods: tagged (RDMA Write/Read) and untagged (send/recv).

The tagged method specifies directly in the data header where the data is to be placed, via a steering tag (STag), offset value, and the length of the incoming data. For this model to work, the source node must be aware of the valid areas of memory at the target node in order to create a valid message header. This requires that prior to the data transfer taking place, the nodes exchange data on the STags and lengths of valid memory locations. Headers are checked upon arrival to ensure that the placement occurs at valid addresses only.

The untagged model does not require knowledge of valid locations in memory at the target node. Data buffers at the target nodes are handled via posted recv requests that specify the location in memory to place the data. All incoming data is matched to the posted recv requests. Unmatched (e.g., unexpected) data results in an error being passed to the Upper Layer Protocol (ULP), since no buffer is available for reception.

The MPA layer [9] is the lowest layer of the iWARP stack. The DDP layer is message-based; therefore, in order to protect DDP messages from being unrecoverable as a result of fragmentation that may occur on the network (called middlebox fragmentation), the MPA layer was developed. This layer overcomes the middlebox fragmentation issues for stream-based transport protocols by inserting markers into the data to be transmitted, in order to point to the correct DDP header for that data, should it become fragmented. This strategy requires modifying the outgoing data to insert the markers and removing the markers from the incoming data before passing it to the DDP layer. The MPA layer is required only for operation over stream-based transports such as TCP. Message-based transports such as SCTP and UDP do not require this costly operation because their message boundaries are clear.

Existing solutions to high-speed Ethernet have taken advantage of several offloading technologies, with many offering abilities such as stateless offloading, performing segmentation on the network adapter and calculating checksums. Solutions for stateful protocol offloading also exist, typically referred to as TCP offload engines, or TOEs. Existing iWARP-compatible network adapters offer stateless and stateful offloading capabilities in addition to RDMA capabilities, enabling them to perform zero-copy-based communications as well as bypassing the operating system, providing greater CPU availability, increased network throughput, and decreased latency [10]. These capabilities are available in existing iWARP networking hardware for TCP-based traffic.

2.2 *Overview of the InfiniBand Standard*

The InfiniBand Architecture Specifications [11] detail the requirements of InfiniBand (IB) compliant hardware. InfiniBand, like iWARP, uses a verbs programming interface. Unlike iWARP, IB networks require software for subnet management, which is responsible for bootstrapping the network and is required whenever any changes are made to the network or its settings. Existing IB solutions also support high bandwidths with an EDR 4X InfiniBand fabric having a bandwidth of 100 Gb/s (an actual data rate of $\sim$97 Gb/s once encoding is taken into account).

InfiniBand has several different transports: Reliable Connection (RC), Unreliable Connection (UC), Reliable Datagram (RD), and Unreliable Datagram (UD). For UD, InfiniBand supports only send/recv messaging; write/read support is not available. No current hardware implementations support RD, although with reliable datagrams both send/recv and write/read can be supported. Throughout this chapter, we will use the same notions of RC and UD for reliable connection-based (TCP-based) and unreliable connection-less (UDP-based) communication in iWARP as well.

In the rest of this chapter we discuss the support for RDMA over unreliable transports in high-performance interconnects such as iWARP Ethernet. Section 3 discusses the benefits of an unreliable underlying transport for RDMA transfers, and particularly for iWARP and Ethernet. Section 4 outlines a proposed solution for supporting both channel (two-sided) and memory (one-sided) communication semantics of RDMA in high-performance RDMA-enabled networks, with a focus on iWARP. In particular, after a discussion in Sect. 4.1 of iWARP's historical development and related technologies, we turn in Sect. 4.2 to a discussion of the proposed design for extending the iWARP standard for datagram-based RDMA support. Section 5 describes our prototype implementation of iWARP over UDP, followed in Sect. 6 by our experimental results, including the evaluation of send/recv and RDMA Write communication for traditional commercial datacenters.

3 The Case for RDMA over Unreliable Transports

In this section we discuss the reasons for needing unreliable connectionless RDMA and the benefits it can offer.

3.1 Importance of Unreliable Connectionless RDMA

Current high-speed networks provide high levels of performance for reliable transports in LAN and/or WAN environments. The limitations of connection-based transports such as nonscalable resource requirements and complex hardware protocols (e.g. TOEs) could represent a problem for future systems. Local computing systems will have increased node and core counts, and wide-area systems will serve more clients per node than current hardware is capable of handling. The existing iWARP RNICs are limited by the complexity of processing required for TCP streams and the behavior of TCP itself. Additional overhead is imposed the MPA layer handling, which is a result of the mismatch between message layers and stream transports.

Current high-performance networks do not allow for RDMA Write/Read operations to occur over a connectionless and/or unreliable transport. Moreover, they typically require that the underlying networking layers provide a reliable, in-order delivery of data for such operations (InfiniBand's unreliable datagrams allow unreliable send/recvs only). This requirement brings limitations compared with a design that allows for unreliable data transmission and does not require the storage and manipulation of data associated with ongoing network connections. For very large systems or for systems serving a very large number of clients, the administration and storage of the data associated with these many individual connections can be onerous. Existing high-speed networking technologies such as InfiniBand have acknowledged such limitations in connection-based networks by introducing technologies such as eXtended Reliable Connections (XRC) [12]. These technologies help mitigate the overhead incurred as a result of connections. Connectionless approaches eliminate this concern to a greater extent, by not incurring the overhead of connections in the first place.

A datagram-based solution offers flexibility to application developers in adapting the overhead of the network directly to their individual needs. For example, for a streaming application that requires time-dependent data, a reliable data stream may not be required because any data that had to be retransmitted would no longer be current enough to be of use. Alternatively, an application could require reliable communication but not need the flow control capabilities of TCP [13]. The reduced complexity of a datagram approach provides for faster communication and lower overhead than either of the current iWARP transports, TCP and Stream Control Transmission Protocol [14]. A datagram-based iWARP can also reduce the complexity of the networking stack by not requiring use of the Marker PDU Alignment layer (MPA) [9], since middle-box fragmentation is not an issue when going over datagram transports. This feature is helpful in enhancing performance as well as reducing the complexity of datagram-iWARP hardware solutions, leading to more scalability and efficiency. Moreover, datagram-enabled iWARP allows for the use of a much wider variety of applications than does traditional iWARP. In particular, it allows wide-area datacenter datagram-base applications to use iWARP.

Applications that can traditionally make use of datagram-based communication, such as streaming media or real-time financial trading applications, are part of the application set poised to make up ~ 90 % of all Internet consumer traffic [15]. VOIP and streaming media applications are typically built on top of protocols such as RTP [16], which can utilize UDP (datagrams) as a lower layer. Such applications have large throughput requirements, and therefore a hardware networking solution that is RDMA capable can reduce the burden on the CPU in transferring such large amounts of data to and from memory, thus freeing up the CPU for other important tasks. This solution translates into potentially reduced CPU requirements for a system, thereby providing both a power savings and a cost savings in initial costs as well as upkeep. Alternatively, this solution can result in increased system throughput, thereby increasing overall system efficiency.

The use of datagrams in HPC communication middleware can also provide performance and scalability improvements for traditional scientific applications. Such applications are able to leverage RDMA capabilities by using connection-based communication. However, RDMA Write operations over a datagram transport are not currently unavailable.

3.2 Benefits of RDMA over Unreliable Datagrams for iWARP

iWARP is an excellent high-performance networking candidate for unreliable get/put operations. The scalability of iWARP can be enhanced through the use of non-connection-based transport provisioning. While the main benefits of providing unreliable RDMA are scalability and better performance for data-loss-tolerant applications, some additional side benefits are also realizable. First, the removal of the MPA layer can provide increased performance. The reduced complexity of connectionless transports also provides the benefit of easy adoption of datagram-iWARP into existing iWARP silicon. Alternatively, datagram-iWARP provides the opportunity for a datagram offloaded iWARP solution while offering onloaded TCP processing. Such a solution would be much less expensive to produce, given its lack of a TOE and MPA processing, and hence would be potentially attractive for commercial datacenters, where applications that can tolerate loss can benefit the most from our proposed extensions. Furthermore, datagram-iWARP provides the opportunity to support broadcast and multicast operations that could be useful in a variety of applications including media streaming and HPC for collective operations. Since the iWARP RNIC is compatible with the OpenFabrics verbs API [17], methods developed for iWARP can also be easily adapted for use with other verbs-compatible interfaces such as InfiniBand.

iWARP's WAN support is also an important feature for leveraging datagram support, since many datagram-based applications require WAN capabilities. Applications such as online gaming use unreliable datagram transports such as UDP. The overhead imposed by reliable transports is detrimental to such applications because individual data is time dependent and limited to short time periods. Examples of time-dependent data include the current heading and location of a moving object or a single frame of a video. Another important application for on-time delivery is high-frequency trading systems for automated financial market trading. Receiving time-sensitive information after more recent information has arrived is unnecessary. In the case of packet loss, TCP and SCTP both guarantee in-order delivery of data. Therefore, network jitter is experienced because there is a delay in delivering data that has already arrived but is blocked by an incomplete message at the head of the receive queue. For a datagram transport such as UDP, data is delivered as it arrives, and this approach helps reduce network jitter, particularly for time-dependent data transmissions. These applications can make use of datagram-iWARP, enhancing their performance by using OS bypass, offloading, and zero-copy techniques. These performance-enhancing networking features are unavailable for such applications in traditional iWARP.

Scalability is a prime concern of high-performance networks. The communication channel isolation of connection-based transports is excellent for flow control and reliability of individual data streams, but it limits the resource sharing that can occur between the connections. Connections must have a current state, and communication over a connection can occur only to a single destination point. Offloading RNICs enables individual systems to have the capacity to serve more clients and therefore incur more connection-based transport overhead. Datagram transports do not have to keep state information as a connection-based transport does. For connection-based iWARP implementations, state information is kept in local memory on the RNIC. Alternatively it can be kept in system memory, in which case it must be accessed through slower requests over the system buses. Reducing the complexity of the networking protocols through the use of unreliable transports such as UDP will help reduce message latency and close the latency gap between iWARP and other high-speed interconnects. Offloading the network processing onto the RNIC will reduce the CPU requirements of providing high-throughput traffic for datagram applications by lessening the data movement responsibilities of the CPU. This will enable businesses to concentrate their infrastructure investments directly into networking technologies and capacity, thereby reducing the amount spent on CPU hardware while still supporting high bandwidth and low latency. This is a common benefit of WAN-capable RNICs that can support both WAN- and LAN-based applications. Another performance advantage of using datagram-iWARP is the existence of message boundaries. Unlike stream-based connections, having message boundaries recognized by the lower layer protocol (LLP) avoids having to mark packets in order to determine message boundaries later. Packet marking can be a high overhead activity and can be expensive to implement in hardware [9]. Therefore, avoiding it in datagram-based iWARP is a significant advantage over TCP. SCTP does allow message boundaries like UDP; however, it provides even more features than those in

TCP and consequently is more complicated. SCTP is also not as mature as either TCP or UDP and therefore does not have as much application support nor as long a history of performance tuning as does either TCP or UDP.

The elements of datagram-iWARP design that improve performance also have a beneficial effect on the implementation cost. Adding datagram-iWARP functionality to existing iWARP RNICs would be relatively easy and inexpensive, and the most likely form in which datagram-iWARP can be leveraged for real devices. The creation of a datagram-iWARP-only RNIC would be less complex and less expensive than a TCP-based iWARP RNIC. Because of the reduced silicon size, datagram-iWARP could also be used to create a highly parallelized RNIC capable of handling many simultaneous requests by multiplicating the communications stack processing pipelines. This would help increase network throughput by helping systems with many cores avoid delays due to resource availability conflicts.

4 RDMA over Unreliable Datagrams

Given the potential benefits of a datagram-based unreliable RDMA iWARP, it is important to examine how such a scheme can be designed. The RDMA Write-Record technique introduced in [18] is, to our knowledge, the first RDMA Write operation that can work over datagram-based unreliable transports. This solution seeks to provide RDMA functionality to a complete subset of applications (those using datagrams) that previously was not able to utilize such high-performance networking technology. In addition, it seeks to improve existing high-performance networks by offering a scalable (unreliable) communication option that can provide good performance while allowing for application-level reliability provisioning depending on runtime system characteristics and individual application needs.

With these capabilities, it has the potential to bring advanced networking performance and scalability to an entirely new area of applications that run over wide-area networks served by Ethernet. RDMA Write-Record can also be of potential use on other high-speed networking architectures, such as InfiniBand.

To increase the adoption of the datagram-iWARP among data-centric applications, we require a matching application interface that could reduce the amount of application development rework and ideally enable such applications to seamlessly run on the new transport. The verbs networking API is not immediately compatible with traditional socket-based network interfaces. Since the onerous task of rewriting a wide spectrum of datacenter application code to use a verb-based interface is neither desirable nor practical, a middleware layer is required to translate sockets-based applications to use the verbs interface. The Sockets Direct Protocol (SDP) [19] is an example of such layer that is used for TCP-based iWARP (as well as InfiniBand RC) to provide a socket-based interface to the verbs-based RDMA stack. SDP translates traditional TCP-based socket interface into the set of RDMA communication verbs. SDP, however, provides support only for applications using connection-based communication (e.g. TCP), not for datagram-based applications. In addition, since

SDP is designed specifically for stream-based protocols, it is not easily adaptable to use datagram-based protocols. Rsockets [20] is a new alternative to SDP, providing a user-level sockets-compatible interface for OpenFabrics stack-compatible RDMA methods. However, since OpenFabrics-based stacks do not support datagram RDMA Write/Read, a custom socket interface needed to be designed that could translate datagram-based networking calls to use the datagram-iWARP verbs. Such interface enables applications to utilize datagram-based RDMA communication without requiring any software rewrite.

4.1 Related Work and Development History

The current leading solution for improving existing Ethernet directly without the use of a new upper-layer networking stack is Converged Enhanced Ethernet (CEE) [21–26]. CEE consists of a set of compatible RFCs that are designed to add advanced functionality to the Ethernet standard. The flagship feature of CEE is error-free transmission channel for applications, offering capabilities such as priority flow control, shortest path bridging, enhanced transmission selection, and congestion notification. Given the availability of such a hardware-managed reliable channel, RDMA over Converged Enhanced Ethernet (RoCE) [27] has been designed to take advantage of the reliabile channel to provide RDMA operations. RoCE uses InfiniBand's RDMA stack directly, using Ethernet as the link/physical layer. This leads to the most highlighted disadvantage of RoCE, which is the lack of support for IP routing, making it applicable only to LANs and WANs supporting non-IP traffic. RoCE works only over the transport modes provided with InfiniBand (no unreliable RDMA Write/Read), and the potential addition of datagram-based RDMA Write-Record operation to InfiniBand can boost RoCE's functionality by making RDMA capable of working over unreliable transports for both channel and memory semantics (i.e., send/recv and RDMA Write/Read). Other alternatives to iWARP have been suggested in order to improve Ethernet. Open-MX allows for Myrinet Express networking traffic to function over Ethernet hardware (MXoE) [28].

Datagram-iWARP was first proposed in [29] using the UDP transport for send/recv based iWARP. The results demonstrated runtime improvements of up to 40 % over connection-oriented iWARP (with TCP) for MPI-based HPC applications, particularly for applications that perform a great deal of communication. Datagram-iWARP can also provide as much as a 30 % improvement in memory usage for moderately sized HPC clusters. RDMA Write-Record was first proposed and analyzed in [18] along with the iWARP socket interface for commercial datacenter applications. In this chapter an integrated view of the aforementioned work is presented. The rest of the chapter elaborates on the details of the datagram-iWARP methodology.

4.2 iWARP Extension Methodology

Datagram-iWARP represents a significant shift in the overall design of iWARP, since the current standard is based entirely on reliable connection-based transports that provide ordered data delivery. In contrast, datagram-iWARP is designed for an unreliable, nonordered network fabric. This requires a change in the existing assumptions made in the iWARP standard regarding the LLP data delivery and the ULP assumptions. We first outline the changes required to enable both two-sided (channel-based) and one-sided (memory-based) methods of communication to run on unreliable datagram-based transports. These include changes to the iWARP standard and behavioral requirements. The design of the RDMA Write-Record and its significant differences from a traditional iWARP RDMA Write operation then are discussed.

4.3 iWARP Design Changes

Adding support for connection-less communication mandates a number of transport-level changes that need to be performed in iWARP standard. In this section we briefly refer to some of these requirements. A high-level overview of the changes required for datagram-iWARP is shown in Fig. 2.

1) Verbs-Level Changes In a connectionless UDP flow, the source IP address and port of the incoming packets need to be reported to the application receiving the data in order to perform necessary flow matching. Therefore, either modification to existing verbs or introduction of new datagram-based verbs is required to pass along the required data structures for datagram-based traffic. Since existing OpenFabrics (OF) [17] verbs specification allows for datagram-based traffic (currently as send/recv datagram traffic InfiniBand), the underlying iWARP-specific verbs can be redesigned or added to be compatible with OF verbs. Additional changes to verbs for polling the completion queue are also required, as are changes to verbs for creating and modifying queue pairs for datagram-based communication.

Since datagram QPs require different initialization information from that of connected QPs, alterations of the QP creation or modification verbs are needed to accept datagram-related inputs that are different from those of existing connection-based verbs.

Communication transmission verbs (send and RDMA Write-Record) must be altered so that they receive a valid destination (IP address) and port with every operation, since they cannot rely on past behavior to determine the destination for a given data transmission. For connection-based communication, this is not an issue because data over a given connection always flows to the same destination. Likewise, the verbs for receiving data must be altered such that they deliver the address and port of the source of the transmission. For recv, this alteration is relatively straightforward,

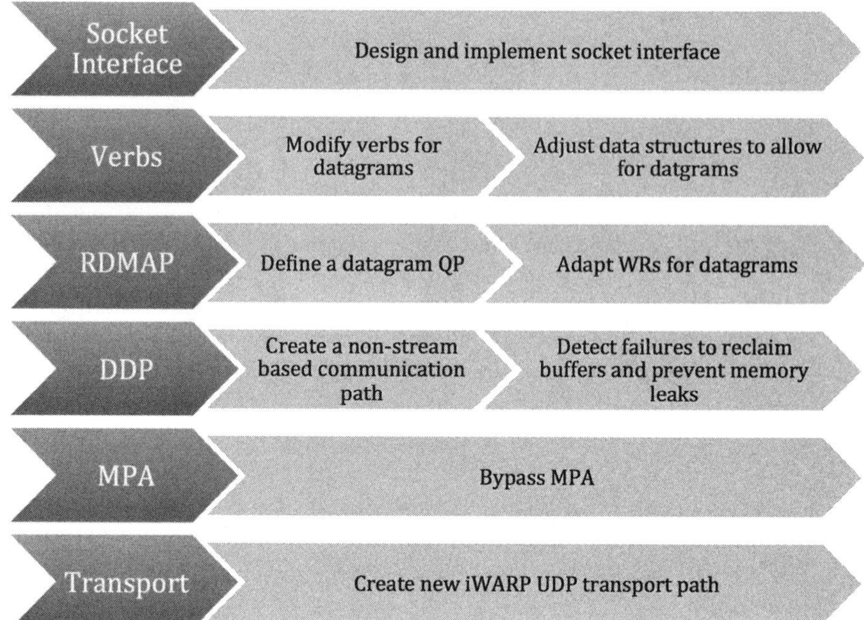

Fig. 2 Changes for datagram-iWARP

through the addition of data structures to relay this information. However, this is not done in the recv verb itself, but rather as a common change to both recv and RDMA Write-Record, by reporting the source address and port in the completion queue element passed back to the requesting process after a completion queue poll request.

2) Required Changes to the Core iWARP Layers One of the major conceptual changes is in the notion of work completion. In a reliable connection iWARP model, a work request completion is assured as soon as its delivery can be guaranteed, which essentially occurs upon passing the data to the underlying transport layer. For an unreliable transport however, such assurances can never be provided by the transport. Therefore, for datagram-iWARP, a work request completion is defined to occur when the data is passed to the LLP for delivery, without the expectation that it is guaranteed to be delivered.

The lack of guaranteed delivery for UD-iWARP requires that the polling for the completion of a recv operation as well as the completion of an RDMA Write-Record operation have a timeout period associated with the polling request. Otherwise, an infinite polling loop may result, which would cause an application to fail.

The RDMAP specifications in the iWARP standard [6] Sect. 5.1 states that LLPs must provide reliable in-order delivery of messages. The DDP standard [8], Sect. 5, item 3, states that the LLP must reliably deliver all packets. These specifications must be changed to allow for the operation of datagram-iWARP.

The DDP standard, Sect. 5, item 8, states that errors that occur on an LLP stream must result in the stream being marked as erroneous and that further traffic over the marked stream is not permitted. Likewise, the RDMAP specification requires an abortive teardown of an entire communication stream should an error occur on that stream. Such requirements need to be relaxed for these layers of the datagram-iWARP. For a datagram QP, the error must be reported to the corresponding upper layer; but no teardown is required, and traffic can continue to flow across the same QP without transition into an error state. This approach is necessary because of the possibility of the QP concurrently communicating with multiple targets. In addition, recovery from an error occurring over an unreliable transport might be possible at the ULP.

3) iWARP Behavioral Adaptations The iWARP standard has been designed for an LLP that offers guaranteed in-order delivery of data. In order to be compatible with an unreliable transport, the behavior of the communication stack needs to be adapted. In many cases, the applications utilizing datagram-iWARP must be aware of its unreliable nature and have provisions to deal with data loss and reordering. This is the case for applications that stream media over a lossy network. Another example is some implementations of HPC application middleware such as the Message Passing Interface (MPI), which communicate over an extremely low-loss-rate LAN where simple and low-overhead middleware-level reliability measures can effectively provide the required delivery assurance of data.

Connection establishment and teardown procedures are not required for datagram-iWARP. For datagram-iWARP, a QP can be transitioned into an RTS state after it is created and an LLP (i.e. UDP) socket is assigned to it. Since this setup can be accomplished without any transmissions to the target of the datagram QP, the other parameters used for the QP, such as the use of a cyclic redundancy check (CRC), must be configured by the ULP. Doing so also eliminates the requirement that the ULP configure both sides of the QP (target and source) at the same time.

The other set of changes is related to the error management behavior of the iWARP stack. For datagram-iWARP, errors must be tolerated, by reporting errors as they happen to the upper layer, but not causing a complete teardown of the QP. Consequently, the application may decide on the course of action to take upon notification that a communication error has occurred.

This approach requires that an error message detailing the error (and, in the case of a RDMA Write or Read communication, the message sequence number) be sent to the message source. The error message is locally placed in an error queue as opposed to the termination queue used in traditional iWARP.

Message segmentation and the requirement for marking of messages are significantly different for datagram-iWARP. Since the proposed LLP, UDP, has a maximum message size of 64 kB and since UDP delivers the message in its entirety, message segmentation is not needed in the iWARP stack. The application layer is responsible for ensuring that data transfers larger than 64 kB are segmented and reassembled properly.

The MPA layer and its marking of packets in order to facilitate reassembly are also not needed. Once datagram packets are segmented into maximum transmission unit (MTU)-sized frames by the IP layer they are not permitted to be further segmented by network hardware along the transmission route (unlike TCP traffic). Therefore, there is no need to perform the costly activity of marking the packets so they can be reassembled correctly at the target system.

4) Optional iWARP Changes Datagram-iWARP always requires the use of a CRC when sending messages. This requirement allows for the creation of datagram QPs without any communication between the source and target nodes. It also ensures that no CRC usage conflicts occur over a datagram QP, given that it can communicate with several other systems. Since the proposed LLP for the UD transport, UDP, does not require the use of a checksum and since its checksum is inferior to a CRC, a CRC check must be performed in the iWARP stack such that data transmission errors are identifiable.

4.4 RDMA Write-Record

In this section we define a one-sided RDMA Write operation over an unreliable transport. None of the existing interconnects supporting one-sided RDMA define support for such operations over unreliable datagrams. Because of the data-source-oriented nature of one-sided RDMA operations, the target host is not aware that an incoming transmission may be occurring. Over a network that provides guaranteed in-order delivery, the source can be assured that data passed to the lower layer networking protocols will arrive properly at the target node. Such a guarantee cannot be made for an unreliable datagram-based transport.

Current specifications require the use of a subsequent two-sided (send/recv) message after an RDMA Write operation has completed, in order to notify the target node that the data has been successfully placed in its local memory [3]. An optimization of this basic method allows the target node to determine whether the data is valid, not by an additional message, but by the modification of a given bit in memory once the operation is complete. Both methods rely on reliable transport. With an unreliable transport, the notification message or the message segment containing the notification bit could be dropped. Alternatively, the notification message could arrive, but the message itself might not; this situation would cause significant problems because the memory region would not contain correct data and most likely would cause application failure or invalidation of application results. Consequently, the traditional RDMA model does not adapt well to an unreliable network environment.

Therefore, in order to facilitate operation of RDMA over unreliable datagrams, a new method called RDMA Write-Record [18] can be used. It uses the iWARP tagged-model data structure while changing the behavior at the source and target nodes. At the source, the RDMA operation is not followed by any additional send/recv based notification in order to notify the target host. The data is sent to the target, but the data

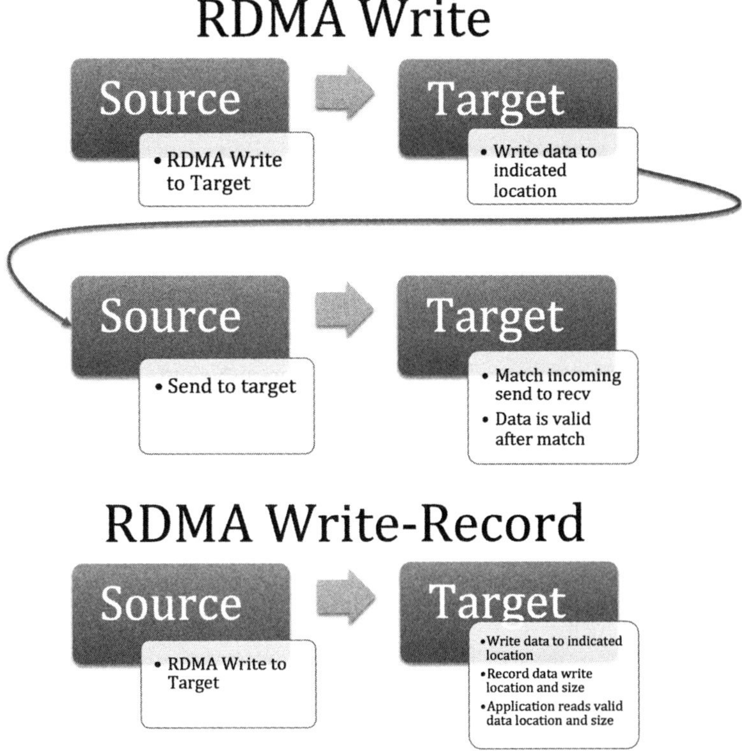

Fig. 3 Comparison of RDMA Write over RC and RDMA Write-Record over UD

source cannot assume that the data will be received. If such information is required at the data source, it must be supplied by using a reliable LLP or by notification messages at an upper layer. At the target node, upon receiving an RDMA message, the target RDMA stack places it in the destination memory as it would for a connection-based message; but for a UD message it also records that the memory write took place, the location of the transaction, and the length of the valid data placed. This information is then accessible as a completion notification through the completion queue. This method does not separate the notification from the actual data arrival, and therefore the notification is recorded if and only after the data is placed.

Since this method uses existing data structures and queues already present in traditional iWARP, its implementation is relatively lightweight. The design is illustrated in Fig. 3 where it is compared with a traditional RDMA Write operation. This design also has the benefit that it is easily compatible with socket semantics and makes an interface for socket-based applications easier to implement. The reason is that a completion notification exists for an RDMA operation, so it can be easily polled for via the completion queue and the requisite information passed back to the polling application.

In order to have adaptable design that supports different computing paradigms, two possible methods for creating CQ entries for RDMA Write-Record operations are proposed. These methods allow applications to use either a message-based or a memory-based networking semantic:

1. Message-based semantics: Each RDMA Write-Record operation, when completed at the data sink, inserts an individual entry in the completion queue. When the application polls for a completion, each completion entry can be interpreted as a separate message received, which is essential to making RDMA Write-Record applicable to existing message-based applications such as those utilizing sockets.
2. Memory-based semantics: In this case, the application views the data transfers as simply memory accesses rather than network messages. In this paradigm, the application needs to know what areas of memory are valid. For such ability, individual notifications for every Write-record operation are unnecessary. The CQ entry method that is used for the message-based semantics approach is still valid for this approach (thereby reducing overall device complexity); however, we create a validity map for any requesting application by simply traversing the CQ and aggregating the CQ elements into a single validity map structure. This approach provides additional application interface flexibility while only marginally increasing device complexity and not requiring a QP to be declared as using only one of the two methods. The manner in which completions are to be reported can be specified in the call to the CQ polling verb. We note that this method does not remove the requirement for synchronization between communicators before the target memory is reused. The state of the memory may be undefined if multiple writes are performed to a single memory location without synchronization. However, such synchronization is out of the scope of the iWARP, as the underlying transport fabric, and needs to be addressed through application / middleware layers. For example, in an HPC setting where sharing of such memory windows is common for one-sided operations, middleware such as MPI may provide adequate reliability and synchronization, so that using unreliable iWARP for scalability and performance is desirable.

As a final note on the introduction of the RDMA Write-Record operation, we emphasize that this operation differs from all types of the existing RDMA operations. One might find similarities with send/recv operations such as send with solicited event, however, in such case a recv must be posted before the specified event can be triggered upon reception of the data; this process also triggers an event to occur immediately upon reception (an interrupt). RDMA Write-Record, on the contrary, reports that an operation has occurred only when the upper layer software requests information concerning completed operations. Similarly, RDMA Write-Record varies from the InfiniBand RDMA Write with immediate operation, since in the InfiniBand case a recv is required to be posted in order to receive the immediate data. Conversely, RDMA Write-Record acts as a truly one-sided operation where a posted recv operation at the target is not required.

4.5 *Packet Loss Design Considerations*

The memory-semantic-based RDMA Write-Record paradigm implies the requirement for supporting unlimited message sizes at the iWARP layer, messages larger than 64 kB, as well as support for partial-message placement in the target memory. For networks that are relatively error free, this may imply some performance benefits due to avoiding segmentation costs for applications when passing large messages to the iWARP stack. However, segmentation will still happen at lower layers, for example, over traditional Ethernet networks, at the IP layer, where messages are segmented into MTU-size segments (usually a 1500-byte MTU, although a 9000-byte jumbo frame is also possible on some networks).

UDP relies on the IP layer for segmenting a 64 kB message into several MTU-compatible messages and reassembling that message at the target node. Messages larger than 64 kB cannot be handled by this mechanism in UDP and therefore must be segmented at the iWARP level before being passed to UDP. This approach results in several smaller message segments in iWARP that must be reassembled at the target node into a single-larger iWARP message. Here is where the partial-message placement support added to iWARP for RDMA Write-Record comes into play. It allows for the received portions of a message to be placed into memory as they arrive; the resulting memory validity map reflects the missing portions of the overall larger message, in case some segments are lost during the transfer. This approach is appropriate for applications that can handle some packet loss, such as online gaming, VOIP, and streaming media applications that can make use of large messages. Other applications such as streaming video applications can also handle packet loss as well as invalid data and can make use of partial-message placement.

For some applications, a single packet loss for a large, multisegment application message would normally result in a complete message loss. Partial-message delivery is not a requirement of RDMA Write-Record. It is detailed here for the subset of applications that can handle packet loss, as well as for future Ethernet networks, such as the changes proposed for CEE [27], where error-free streams for datagrams can be provided. Such network channels could make use of iWARP hardware segmentation and reassembly of large messages.

5 Datagram-iWARP Software Implementation

To evaluate the proposed datagram extension to the iWARP standard, we have developed a software implementation of datagram-iWARP based on a TCP-based iWARP implementation. This allows for a direct comparison between the UD and RC modes of datagram iWARP, using a publicly available RC implementation. Figure 4 shows an overview of this implementation including the additional upper-layer interfaces we have added, the iWARP socket interface, and modification to the existing OF verbs interface in order to make it compatible with more OF verbs middleware and

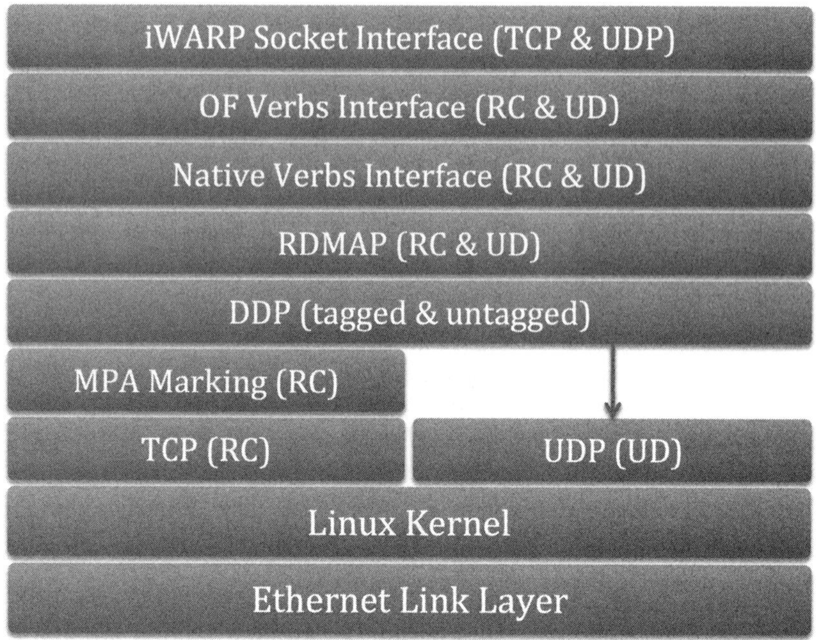

Fig. 4 Software implementation of datagram-iWARP

applications. The changes required to the verbs, RDMAP, and DDP layers, as described in Sect. 4.3, are reflected in the layers of the stack indicating both UD and RC support. While this implementation is fully capable of operating over a reliable UDP transport, all the testing and results of the implementation were performed over unreliable UDP.

The software implementation of iWARP that was used as the code-base for datagram-iWARP originated from a project by the Ohio Supercomputer Center (OSC) [30]. The OSC iWARP project implements both user-space [31] and kernel-space [32] implementations of iWARP, with the user version being used for datagram-iWARP. Other software iWARP solutions have also been developed; for example, SoftiWARP, a project from IBM Zurich [33], is integrated into the OpenFabrics Enterprise Distribution stack [17]. All these projects have implemented only the traditional iWARP stack, however, and no support for datagrams or design alternatives compatible with datagrams is presented.

Modifications to the existing incomplete OF verbs interface on top of the native software iWARP verbs were necessary because of the base RC iWARP implementation used for the design. OF verbs were originally designed for InfiniBand (OpenIB verbs) but are now also used to support iWARP hardware in a unified driver. A socket interface was also added in order to facilitate the use of socket-based applications without needing to rewrite the existing networking code.

The datagram-iWARP implementation has been extended to allow for IP-level broadcast operations. Using the existing IP broadcast provisions available for Ethernet networks, we adapted the datagram-iWARP implementation to demonstrate the bandwidth achievable using IP-level broadcasting. Send/recv is supported for broadcast operations, since the receiver-managed data placement is useful for ensuring correct delivery to multiple target nodes.

5.1 *iWARP Socket Interface*

The compatibility of iWARP to operate with existing software without the need for a long and expensive networking code rewrite is desirable. Existing iWARP implementations can operate over the Socket Direct Protocol or the Rsockets interface, which allows sockets-based applications (non-datagram) to run over iWARP verbs. However, SDP was designed specifically for TCP-based applications and does not support datagram-based applications. Rsockets supports both TCP and (recently) UDP but does not support the use of RDMA Write-Record. Therefore, an interface was needed to allow datagram-based applications to take advantage of datagram-iWARP without having to rewrite networking code in order to use verbs. A lightweight interface was developed to facilitate both RC-based and UD-based socket networking code to work with iWARP. This interface is not as complex or full featured as SDP or Rsockets but provides a baseline with which the performance of RC and UD socket-based applications can be compared.

The iWARP socket interface is loaded in the same manner as SDP/Rsockets. When running an application, the interface is dynamically preloaded and uses networking calls that override the existing operating system network calls, passing them to the iWARP networking stack. The iWARP socket interface operates by allowing for both TCP and UDP-based iWARP sockets to be opened, using the relevant iWARP lower-layer protocol. It limits sockets to having only one QP associated with them, so that socket-based applications do not have multiple streams over a single socket.

6 Experimental Results and Analysis

In this section, we briefly describe our evaluation of the proposed technology through our prototype implementation of unreliable datagram-based iWARP. We use microbenchmarks to evaluate individual features and the core functionality of datagram-based RDMA compared with traditional connection-based RDMA. We also utilize the technology in some datacenter applications in order to realize the potential benefits in a real environment.

We conducted our tests on a cluster of four nodes, each with two quad-core 2 GHz AMD Opteron processors, 8 GB RAM, and a NetEffect 10-Gigabit Ethernet (10GE) card connected through a Fujitsu 10GE switch. The cards were operated in native

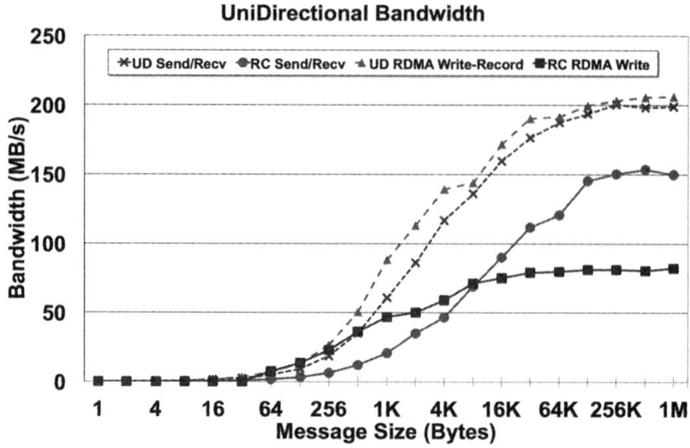

Fig. 5 Unidirectional verbs bandwidth

Ethernet mode. The nodes run Fedora 12 (kernel 2.6.31). All tests were run by using a software iWARP implementation over sockets on Ethernet hardware.

6.1 Verbs-Layer Microbenchmarks

Microbenchmark results are presented for both send/recv and RDMA Write-Record as compared with the connection-based traditional iWARP operations. These microbenchmark tests were performed on one pair of nodes.

Unidirectional bandwidths of the different iWARP methods are illustrated in Fig. 5. Both UD-based methods outperform their RC equivalents by a wide margin. This result is especially observable at larger message sizes, which is particularly significant for RDMA Write-Record compared with RC RDMA Write. UD RDMA Write-Record also enjoys a visible advantage over UD send/recv at 1 kB message sizes.

6.2 Send/Recv Broadcast

Since broadcast/multicast is a potentially useful feature in several classes of data center applications (such as media streaming applications), we evaluated the addition of such a feature to datagram iWARP.

In a verbs-level bandwidth evaluation benchmark over a network with IP-level hardware broadcasting operations, the source node delivers data to multiple targets via IP broadcast. With no overhead, the bandwidth is expected to scale linearly with an increasing number of targets. As can be observed in Fig. 6, for two and

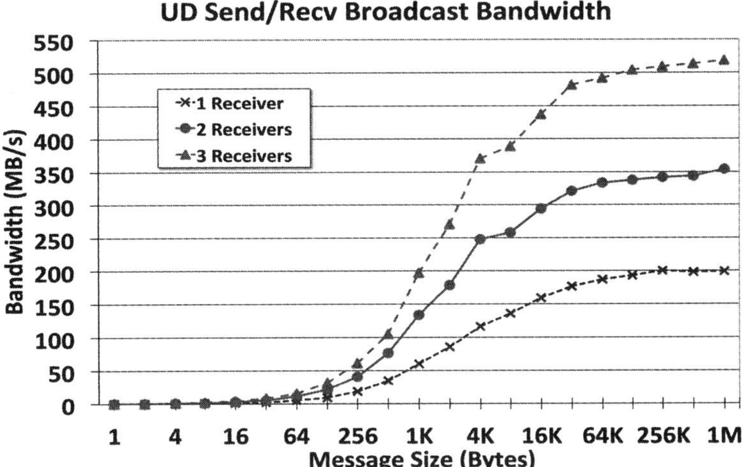

Fig. 6 Send/recv broadcast aggregate bandwidth

three receivers the bandwidth scales well for small and medium messages. Greater overhead is observed for large messages as a result of copying overhead by the broadcasting switch, leading to nonideal scaling. This small overhead (in the order of a few percent) is preferable to having the software iWARP stack replicate packets and send them to multiple targets. The slight change in the bandwidth curve slope at the 8 kB message size is due to exceeding the network MTU and consequently IP-level message segmentation, which slightly affects the efficiency of the network processing.

6.3 Packet Loss and Performance

In a LAN environment, where the probability of data loss is very low, the implications of using unreliable transports are not fully understood. Therefore, after analyzing results in LAN conditions we studied datagram-iWARP in a WAN-like environment, where the presence of packet loss can affect the overall observable bandwidth. In this test, we chose the packet loss percentage rates based on real-world network packet loss records as discussed in [34]. A loss rate of 0.1 % is similar to that of intra-U.S. web traffic, while a 0.5 % loss rate is in line with expectations of loss between a western European-U.S. transmission. Larger packet loss rates of 1–5 % approximate traffic in Africa and parts of Asia, with African nations typically having rates on the higher ends of the scale.

The bandwidth of UD send/recv datagram-iWARP under various packet loss conditions are illustrated in Figs. 7 and 8. Since the loss of a single segment of a message

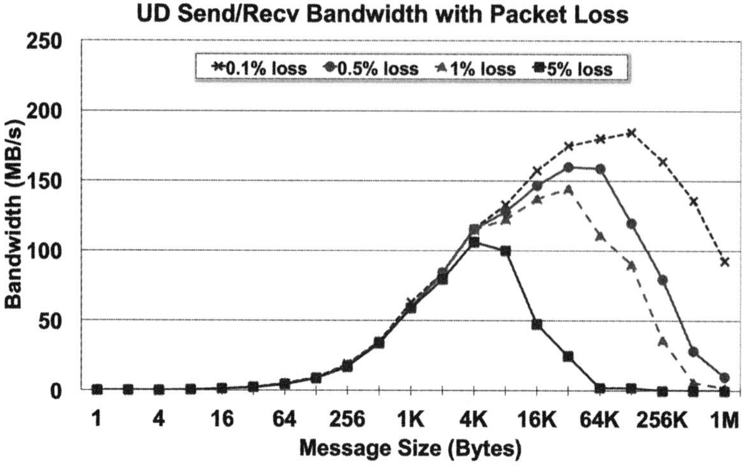

Fig. 7 UD send/recv bandwidth with packet loss

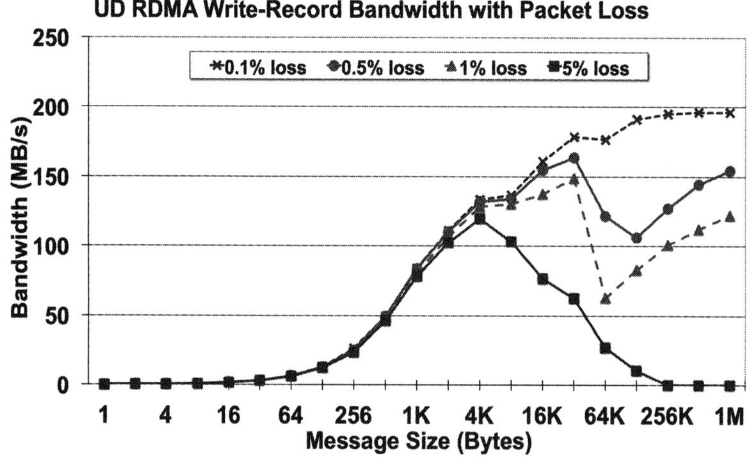

Fig. 8 UD RDMA Write-Record bandwidth with packet loss

causes the loss of the entire message, it will cause observable losses in throughput for medium-sized messages and catastrophic loss in performance for large messages.

As mentioned in Sect. 4.5 the MTU for a UDP datagram is 64 kB. This means that once the 64 kB packet size is exceeded, the chance of an iWARP message being dropped increases by the number of packets that must be sent for the message. Examining Fig. 8, one can see the effect that this has on the performance for RDMA Write-Record. A key observation is that for larger message sizes, the bandwidth does not drop as much as that for send/recv in Fig. 7. This is due not to any inherent differences between send/recv and RDMA Write-Record but rather to the fact that

RDMA Write-Record has been adapted to perform partial placement of messages, where it places all the packets of a message that do arrive, and does not drop a large message when one or more packets in that message are lost.

Although no technical limitation prevents partial placement of messages for send/recv, the partial placement of large messages for RDMA Write-Record is easy to accomplish. The reason is that the partial-placement feature of RDMA Write-Record also allows for memory validity mapping-type behavior. All the segments of a message that arrive can be delivered regardless of whether they are sequential. The partial-placement method allows large message sizes to be of use in the most typical packet loss scenarios for Internet traffic in the majority of the world. It is most useful for applications that can tolerate partial loss of messages, where retransmission of such losses can be skipped in most cases. Without a partial-placement feature, large message sizes are generally impractical for use over lossy networks.

6.4 Datacenter Application Results

Datagram-iWARP has obvious applications in datacenters running commercial applications. To assess the advantages of using datagram-iWARP for video streaming, we used VideoLan's VLC, a popular streaming media application [35]. For the performance of datagram-iWARP for systems where client scalability is of concern, we used SIPp [36], a framework for load-testing Session Initiation Protocol (SIP) servers.

1) Datacenter Performance Results Datagram-iWARP's performance with streaming media applications was tested by using VLC's UDP streaming mode and was compared with an RC-compatible mode (HTTP-based). The relative performance of the two modes was compared by using a network-intensive portion of the application, the buffering of a new media stream. Figure 9 illustrates the performance of the different iWARP modes during buffering. UD provides a 74.1 % reduction in buffering time versus RC. This increase in throughput for the system is a direct result not only of the UD iWARP stack but also of the advantages that UDP streaming has over the HTTP-based method in VLC. The UDP method benefits from lower overhead than the HTTP approach and therefore demonstrates that making UD transports available for future datacenters could provide benefits at multiple layers.

RDMA Write-Record does not demonstrate any performance benefits compared with send/recv in Fig. 9. This result is due to the software socket interface, which is required to provide buffers for incoming messages for Write-Record. Each time a buffer is consumed, the socket interface must advertise the remote buffer location. The socket interface has not been performance tuned to minimize the overhead involved in this process. For a hardware implementation, a rotating group of prenegotiated buffers could be used that would help reduce this overhead. For the software implementation, this approach results in similar performance to that of send/recv (essentially the prenegotiated buffers are like posted recvs).

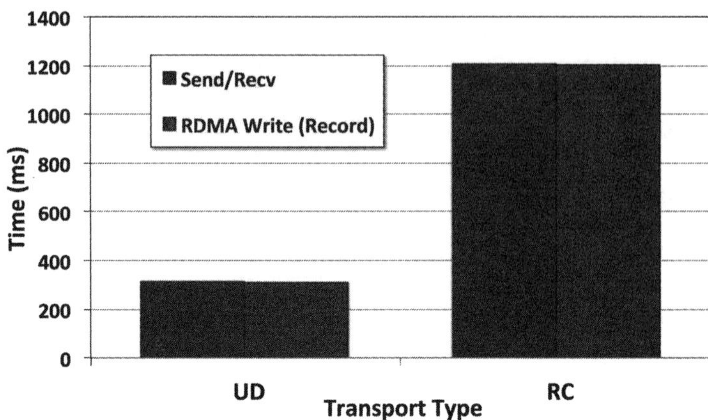

Fig. 9 VLC UD streaming vs. RC-based HTTP streaming

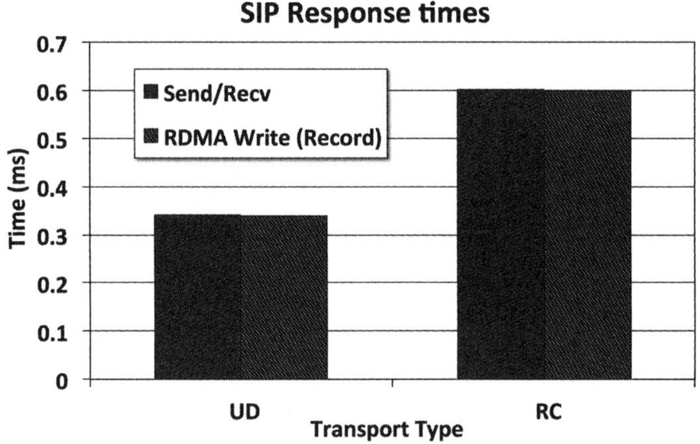

Fig. 10 SIP response times

SIPp was used to determine the benefits of datagram-iWARP for IP telephony. Under a light server load, the request response averages seen in Fig. 10 were obtained. The UD-iWARP response time is 43.19 % better than that of RC-iWARP for send/recv and 43.24 % better for RDMA Write (Record).

2) SIPp Memory Usage Results Datagram-iWARP is expected to have memory usage benefits over a connected iWARP solution because of the lack of client connections and their application-level memory usage implications. The memory footprint of SIPp was examined in order to quantify these benefits. Figure 11 was compiled by using the sum of the SIPp application memory usage and the Linux slab buffer

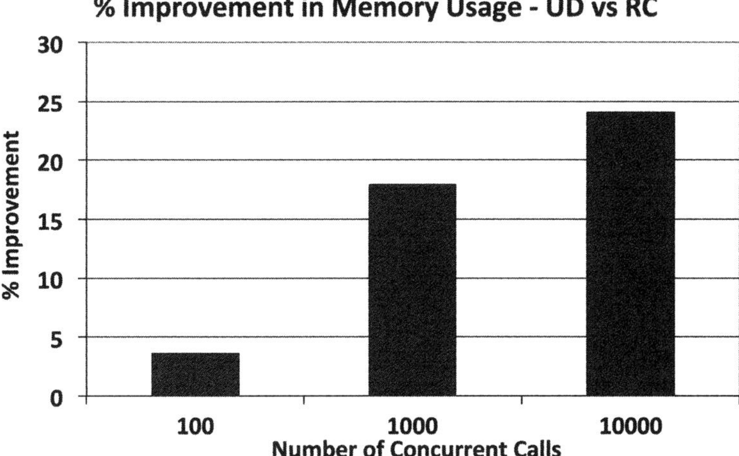

Fig. 11 SIP improvement in memory usage using send/recv datagram-iWARP over traditional iWARP

used for the sockets. The figure shows that a memory savings of 24.1 % is possible for a server with 10,000 clients.

We can compare this result to the maximum possible theoretical benefit based solely on the iWARP socket size. The maximum benefit is 28.1 % over RC. This shows only a 4 % difference between the measured and maximum possible benefit. The difference can be accounted for through the application's memory usage associated with the datagram sockets. For SIPp, even without a connection, some state must be stored for each of the clients. The true memory usage benefits of using a datagram-based RDMA approach will be realized with high-end datacenter hardware serving millions of clients.

The datacenter applications tested demonstrate that applications that are data loss tolerant and UDP-compatible can take great advantage of send/recv and RDMA Write-Record in datagram-iWARP. The packet loss tolerance features of datagram-iWARP make it an effective transport for datagram-based WAN communications. It is also compatible with the sockets network programming paradigm. Datagram-iWARP also demonstrates scalability benefits. Thus, three key concerns of future datacenter and WAN networks are addressed: performance, ability to operate on lossy networks, and scalability. In addition, datagram-iWARP provides backwards compatibility with existing Ethernet networks and an easy transition from existing hardware to future generation hardware with the ability to provide a staged upgrade path.

7 Summary

This chapter introduced the concepts of RDMA and discussed some of the most popular RDMA interconnects available. The differences between the needs of scientific HPC and commercial datacenter applications were discussed, along with how the UDP transport can be of benefit, especially in datacenters. The chapter illustrated the critical design requirements and resulting real-world performance of a proposed next-generation design for supporting RDMA over unreliable transports, in particular for Ethernet technology. It explored the challenges facing the current iWARP standard and addressed them by extending the standard to the datagram domain. The result is a full-featured datagram-iWARP design. This design includes the first method capable of providing one-sided RDMA communications over unreliable datagrams, called RDMA Write-Record. Such functionality can be extended for use on other RDMA-enabled networks.

As a proof of concept, a datagram-iWARP stack was implemented and supplemented with an OF verbs interface for standard verbs compatibility and a socket interface for providing iWARP functionality to existing socket-based applications. The results indicate that datagram-iWARP can provide superior performance and scalability to traditional iWARP.

Packet loss design considerations were discussed, with a solution for partial data placement to help throughput in typical Internet packet loss situations. The semantics of RDMA Write-Record were discussed as a good match for packet loss design and the partial data placement scheme was demonstrated to provide benefits in terms of bandwidth.

Additionally, the benefits of datagram-iWARP in commercial datacenters were assessed. Improved performance and scalability were shown, demonstrating that datacenters could benefit from adopting datagram-iWARP hardware. Further, it was noted that commercial datacenter applications are capable of leveraging all the benefits of RDMA Write-Record and datagram-iWARP, since they can tolerate some data loss. Providing application developers with tools such as RDMA Write-Record will enable applications that can tolerate data loss to benefit from reduced reliability constraints and greater scalability.

Acknowledgments This work was supported in part by the Natural Sciences and Engineering Research Council of Canada Grant #RGPIN/238964-2011; Canada Foundation for Innovation and Ontario Innovation Trust Grant #7154; U.S. Department of Energy, Office of Science, Advanced Scientific Computing Research, under Contract DE-AC02-06CH11357; and the National Science Foundation Grant #0702182.

References

1. L. G. Valiant, "A bridging model for parallel computation," *Commun. ACM*, vol. 33, no. 8, pp. 103–111, Aug. 1990. [Online]. Available: http://doi.acm.org/10.1145/79173.79181
2. B.W. Barrett, R. Brightwell, R.E. Grant, S. Hemmert, K. Pedretti, K. Wheeler, K.D. Underwood, R. Riesen, A.B. MacCabe, T. Hudson, The Portals 4.0.2 Networking Programming Interface, Sandia National Laboratories, October 2014, Tech. Rep. SAND2014-19568
3. RDMA Consortium, "July 2013." [Online]. Available: http://www.rdmaconsortium.org
4. Internet Engineering Taskforce. July 2013. [Online]. Available: www.ietf.org
5. D. Dunning, G. Regnier, G. McAlpine, D. Cameron, B. Shubert, F. Berry, A. M. Merritt, E. Gronke, and C. Dodd, "The virtual interface architecture," *Micro, IEEE*, vol. 18, no. 2, pp. 66–76, 1998.
6. J. Hilland, P. Culley, J. Pinkerton, and R. Recio, "RDMA protocol verbs specification," *RDMAC Consortium Draft Specification draft-hilland-iwarp-verbsv1. 0-RDMAC*, 2003.
7. R. Recio, P. Culley, D. Garcia, J. Hilland, and B. Metzler, "An RDMA protocol specification," IETF Internet-draft draft-ietf-rddp-rdmap-03. txt (work in progress), Tech. Rep., 2005.
8. H. Shah, J. Pinkerton, R. Recio, and P. Culley, "Direct data placement over reliable transports (version 1.0)," *RDMA Consortium, October*, 2002.
9. P. Culley, U. Elzur, R. Recio, S. Baily *et al.*, "Marker PDU aligned framing for TCP specification (version 1.0)," *RDMA Consortium, October*, 2002.
10. B. Hauser, "iWARP ethernet: eliminating overhead in data center designs," *NetEffect Inc. White paper*, 2006.
11. InfiniBand Trade Association. InfiniBand architecture specification, release 1.2.1, nov. 2007.
12. InfiniBand Trade Association. InfiniBand architecture specification, release 1.2.1, annex A14: Extended Reliable Connected Transport Service, mar. 2009.
13. G. Huston, "TCP performance," *The Internet Protocol Journal*, vol. 3, no. 2, pp. 2–24, 2000.
14. R. Stewart, Q. Xie, K. Morneault, C. Sharp, H. Schwarzbauer, T. Taylor, I. Rytina, M. Kalla, L. Zhang, and V. Paxson, "RFC 4960: Stream control transmission protocol," *Network Working Group*, 2007.
15. Cisco, VNI, "Hyperconnectivity and the approaching zettabyte era," *White paper*, 2013.
16. H. Schulzrinne, S. Casner, R. Frederick, and V. Jacobson, "RFC 3550," *RTP: a transport protocol for real-time applications*, vol. 7, 2003.
17. OpenFabrics Alliance, "July 2013." [Online]. Available: http://www.openfabrics.org
18. Ryan E. Grant, Mohammad J. Rashti, Pavan Balaji, and Ahmad Afsahi, "Remote Direct Memory Access over Datagrams", U.S. Patent #8903935, December 12, 2014.
19. J. Pinkerton, E. Deleganes, and M. Krause, "Sockets direct protocol (SDP) for iWARP over TCP (v1. 0)," *RDMA Consortium*, 2003.
20. S. Hefty. (2013) Rsockets. Intel Corporation. [Online]. Available: https://www.openfabrics.org/ofa-documents/doc_download/495-rsockets.html
21. IEEE. IEEE standard for local and metropolitan area networks - virtual bridged local area networks – amendment: priority-based flow control - 802.1qbb. [Online]. Available: http://www.ieee802.org/1/pages/802.1bb.html
22. IEEE. IEEE standard for local and metropolitan area networks - virtual bridged local area networks – amendment 10: congestion notification - 802.1qau. [Online]. Available: http://www.ieee802.org/1/pages/802.1au.html
23. IEEE. IEEE standard for local and metropolitan area networks - virtual bridged local area networks – amendment: enhanced transmission selection - 802.1qaz. [Online]. Available: http://www.ieee802.org/1/pages/802.1az.html
24. IEEE. IEEE standard for station and media access control connectivity - 802.1ab. [Online]. Available: http://www.ieee802.org/1/pages/802.1ab.html
25. INCITS technical committee T11. ANSI standard FC-BB-5 - fibre channel over ethernet (FCoE). [Online]. Available: http://www.t11.org/ftp/t11/pub/fc/bb-5/09-056v5.pdf

26. R. Perlman, "Introduction to TRILL," *The Internet Protocol Journal*, vol. 4, no. 3, pp. 2–20, 2011.
27. D. Cohen, T. Talpey, A. Kanevsky, U. Cummings, M. Krause, R. Recio, D. Crupnicoff, L. Dickman, and P. Grun, "Remote direct memory access over the converged enhanced ethernet fabric: Evaluating the options," in *Proceedings of the 17th IEEE Symposium on High Performance Interconnects (HOTI)*. IEEE, 2009, pp. 123–130.
28. B. Goglin, "Design and implementation of open-mx: High-performance message passing over generic ethernet hardware," in *Proceedings of the 22nd IEEE International Parallel and Distributed Processing Symposium (IPDPS)*. IEEE, 2008, pp. 1–7.
29. M. J. Rashti, R. E. Grant, P. Balaji, and A. Afsahi, "iWARP redefined: Scalable connectionless communication over high-speed ethernet," in *Proceedings of the 2010 International Conference on High Performance Computing (HiPC)*. IEEE, 2010, pp. 1–10.
30. Ohio Supercomputing Center, "Software implementation and testing of iWARP protocol," 2013. [Online]. Available: http://www.osc.edu/research/network_file/projects/iwarp/iwarp_main.shtml
31. D. Dalessandro, A. Devulapalli, and P. Wyckoff, "Design and implementation of the iWARP protocol in software," in *Proceedings of the 17th IASTED International Conference on Parallel and Distributed Computing and Systems, Phoenix, AZ*, 2005.
32. D. Dalessandro, A. Devulapalli, and P. Wyckoff, "iWARP protocol kernel space software implementation," in *Proceedings of the 20th International Parallel and Distributed Processing Symposium (IPDPS)*, 2006.
33. B. Metzler, P. Frey, and A. Trivedi, "A software iWARP driver for OpenFabrics," in *Proceedings of the OpenFabrics Alliance 2010 Sonoma Workshop*, 2010.
34. W. Matthews and L. Cottrell, "The PingER project: active internet performance monitoring for the HENP community," *Communications Magazine, IEEE*, vol. 38, no. 5, pp. 130–136, 2000.
35. VideoLan Project, "VLC media player, May 2013." [Online]. Available: http://www.videolan.org/vlc/
36. R. Gayraud, O. Jacques, and C. Wright, "SIPp: traffic generator for the SIP protocol," 2013.

Packet Classification on Multi-core Platforms

Yun R. Qu, Shijie Zhou and Viktor K. Prasanna

Supported by U.S. National Science Foundation under grant CCF-1320211.

1 Introduction

Internet routers perform *packet classification* on incoming packets for various network services such as network security and Quality of Service (QoS) routing. All the incoming packets need to be examined against predefined rules in the router; packets are filtered out for security reasons or forwarded to specific ports during this process. Another well-known name for packet classification is *packet filtering*. As shown in Fig. 1, *packet filters* or *firewall* is also used to refer to the hardware or software-based network security system performing packet classification.

Many hardware-based approaches have been proposed to enhance the performance of packet classification. One of the most popular methods for packet classification is to use Ternary Content Addressable Memories (TCAMs) [3]. TCAMs are not scalable and require a lot of power [4]. Recent work has explored the use of Field-Programmable Gate Arrays (FPGAs) [5]. These designs can achieve very high throughput for moderate-size rule set, but they also suffer long processing latency when external memory has to be used for large rule sets.

Use of software accelerators and virtual machines for classification is a new trend [6]. However, both the growing size of the rule set and the increasing bandwidth of the Internet make memory access a critical bottleneck for high-performance packet classification. State-of-the-art multi-core optimized microarchitectures [7, 8] deliver

Y. R. Qu (✉) · S. Zhou · V. K. Prasanna
Ming Hsieh Department of Electrical Engineering,
University of Southern California, Los Angeles, CA 90089
e-mail: yunqu@usc.edu

S. Zhou
e-mail: shijiezh@usc.edu

V. K. Prasanna
e-mail: prasanna@usc.edu

S. U. Khan, A. Y. Zomaya (eds.), *Handbook on Data Centers*,
DOI 10.1007/978-1-4939-2092-1_13

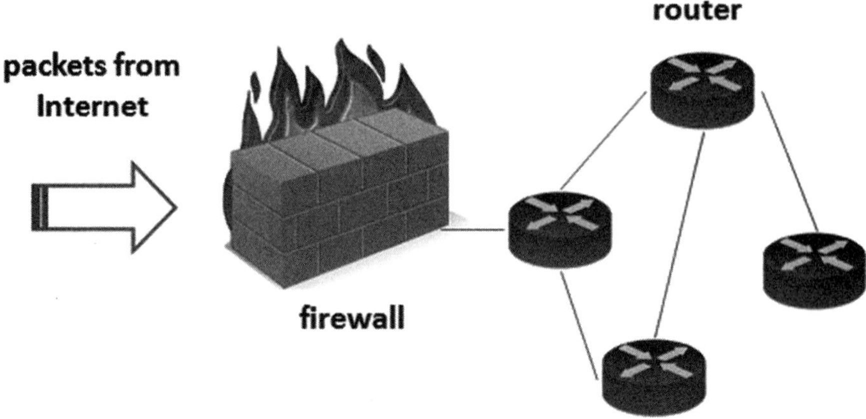

Fig. 1 Network firewall performing packet classification

a number of new and innovative features that can improve memory access performance. The increasing gap between processor speed and memory speed was bridged by caches and instruction level parallelism (ILP) techniques [9]. For cache hits, latencies scale with reductions in cycle time. A cache hit typically introduces a latency of two or three clock cycles. The cache misses are overlapped with other misses as well as useful computation using ILP. These features make multi-core processors an attractive platform for low-latency network applications. Efficient parallel algorithms are also needed on multi-core processors to improve the performance of network applications.

In this chapter, we focus on various approaches and their performance for packet classification on multi-core processors. Specifically, we conduct a thorough comparison for various implementations with respect to throughput and latency. The rest of the chapter is organized as follows. Section 2 formally describes the background; Sect. 3 covers the details of various algorithms. Section 4 summarizes performance results, and Sect. 5 concludes this chapter.

2 Background

2.1 Multi-field Packet Classification

Multi-field packet classification problem [10] requires five fields to be examined against the rules: 32-bit source IP address (SIP), 32-bit destination IP address (DIP), 16-bit source port number (SP), 16-bit destination port number (DP), and 8-bit transport layer protocol (PROT). In SIP and DIP fields, Longest Prefix Match (LPM) is performed on the packet header. In SP and DP fields, the matching criterion of a rule is a range match. The PROT field only requires exact value to be matched. We

Table 1 Example rule set

ID	SA	DA	SP	DP	PROT	PRI	ACT
1	192.77.88.155/20	119.106.158.230/20	0–65535	6888–6888	0x06	1	Act 0
2	175.77.88.6/32	36.174.239.222/32	0–65535	1704–1704	0x06	2	Act 1
3	175.77.88.4/32	36.174.239.222/32	0–65535	177–177	0x06	3	Act 0
4	95.105.143.51/32	*	0–65535	1521–1521	0x06	4	Act 2
5	95.105.143.51/32	204.13.218.182/32	0–65535	0–65535	0x01	4	Act 3
6	152.175.65.32/28	248.116.141.0/28	80–80	123–123	0x11	5	Act 5
7	17.21.12.0/23	224.0.0.0/5	0–65535	0–65535	0x00	6	Act 4
8	233.117.49.48/28	233.117.49.32/28	750–750	*	0x11	7	Act 3

denote this problem as the classic five-field packet classification problem in this chapter. We define the field requiring prefix match as *prefix match field*; similarly, *range match field* and *exact match field* can also be defined.

Each packet classification engine maintains a rule set. In this rule set, each rule has a rule ID, the matching criteria for each field, an associated action (ACT), and/or priority (PRI). For an incoming packet, a match is reported if all the five fields match a particular rule in the rule set. Once the matching rule is found for the incoming packet, the action associated with that rule is performed on the packet. We show an example rule set consisting of eight rules in Table 1; the typical rule set size (denoted as N) ranges from 50 to 1K [10]. "Note a "*" (don't care) in a field indicates the corresponding rule matches any value of the packet header in this field."

For example, a packet with the header (SA: 17.21.12.5, DA: 224.5.3.0, SP: 100, DP: 5, PROT: 0x00) matches the rule with ID 7 (Rule 7). This is because its SA matches the corresponding SA field of Rule 7 (17.21.12.0/23) following LPM criterion, and its DA matches the DA field of Rule 7 (224.0.0.0/5); also, its SP and DP fall into the corresponding ranges specified by Rule 7, while it also has the same protocol type as Rule 7. Since there is a match between the header of the incoming packet and a rule (Rule 7), the associated action (Act 4) is performed on this packet.

A packet can match multiple rules. If only the highest priority one needs to be reported, it is called *best-match* packet classification [11]; if all the matching rules need to be reported, it is a *multi-match* packet classification. In data center networks, since a single packet can be duplicated and distributed into different nodes in the network, a match can correspond to multiple actions; hence packet classification engines usually perform multi-match packet classification.

2.2 *Related Work*

Packet classification has a history originated back in 1990s. It was first used in network security systems. Nowadays, packet classification has become a kernel function in high-speed routers, data center networking, and other novel network architectures. For example, the emerging Software Defined Networking (SDN) performs *OpenFlow table lookup* [1, 2], which is similar as the classic multi-field packet classification mechanism. We refer it to a newer version of packet classification: *OpenFlow packet classification*. Compared to the classic five-field packet classification, OpenFlow packet classification requires a larger number of fields ($12 \sim 15$) to be examined.

In this chapter, we use the following two metrics to measure the performance of a packet classification engine:

a. Throughput: total number of packets processed per second
b. Latency: average processing time used for a single packet

Due to the rapid growth of the Internet traffic, there is a demand to design high speed routers capable of processing millions of packets per second. However, the growing rule set size and complex matching criteria make packet classification one of the fundamental challenges in future Internet. As can be seen, the challenges of multi-field packet classification include:

- Scalability: supporting a large number of fields, a large number of rules.
- Performance: sustaining high throughput with low processing latency.

Most of packet classification algorithms on general purpose processors fall into two categories: *decision-tree based* and *decomposition based* approaches.

2.3 Multi-core Processor

A multi-core processor consists of a small number of independent processor cores, each having access to its designated cache as well as shared caches. Each core also has access to large but much slower main memory. Modern multi-core processors explore multi-socket implementation, where all the cores are separated into groups (sockets); communication between different sockets is realized by technologies such as Quick Path Interconnect (QPI) [8].

Figure 2 shows an example of the architecture of a state-of-the-art multi-core processor. It has two sockets. Each socket has four cores, each having 32 KB L1 cache and 256 KB L2 cache. All the four cores in the same socket share a 20 MB L3 cache, and a DDR3 main memory.

Multi-core processors are widely used in many applications, including high-performance computing, networking, and digital signal processing. The general trend in processor development has moved from dual-core to octo-core chips [8]. The number of cores continues to increase to tens or even hundreds. Besides the large number of cores on-chip, multi-core chips also employ multithreading techniques. Each core of the state-of-the-art multi-core processors is usually capable of concurrently handling several threads; this offers more potential to achieve high performance for applications.

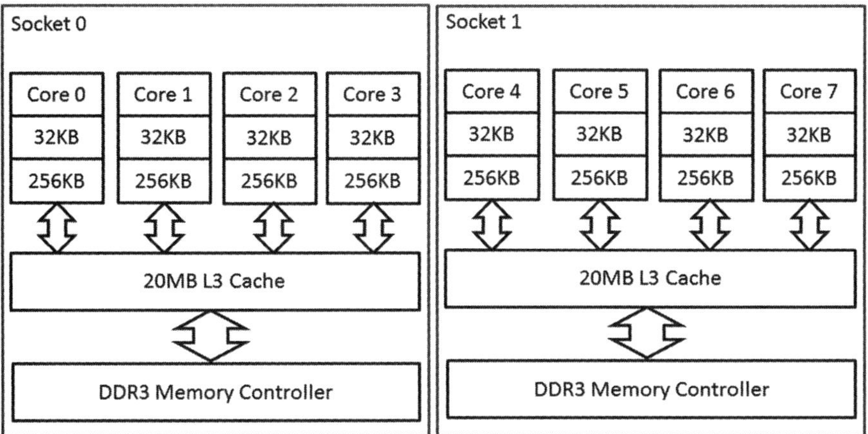

Fig. 2 Multi-core architecture

The improvement of processing performance of a multi-core processor, compared to the single-core processor, is highly dependent on the parallelism in the algorithms. In the best case, the speed-up factor can be as much as the number of the cores; however, designing an efficient parallel algorithm to achieve such high performance is very challenging.

On the other hand, similar to the traditional single-core processor, the performance per core for multi-core processors is bottlenecked by the number of memory accesses, especially the main memory accesses. Although the memory hierarchy of the multi-core process employs fast cache accesses, as the data size grows, it is not possible to fit all the data in the first few levels of cache. Hence expensive main memory accesses have to be used, where a single main memory access, typically to DDR3 DRAM, takes over hundreds of nanoseconds.

As a consequence, many applications are not easily accelerated on state-of-the-art multi-core processors. To achieve high performance, researchers have been exploring efficient algorithms that reduce the number of memory accesses.

3 Decision-Tree Based Approaches

3.1 Algorithms

The most well-known decision-tree based algorithms are HiCuts [12] and HyperCuts [13] algorithms. The idea of decision-tree based algorithms is that each rule defines a sub-region in the multi-dimensional space and a packet header is viewed as a point in that space. The sub-region which the packet header belongs to is located by cutting the space into smaller sub-regions recursively.

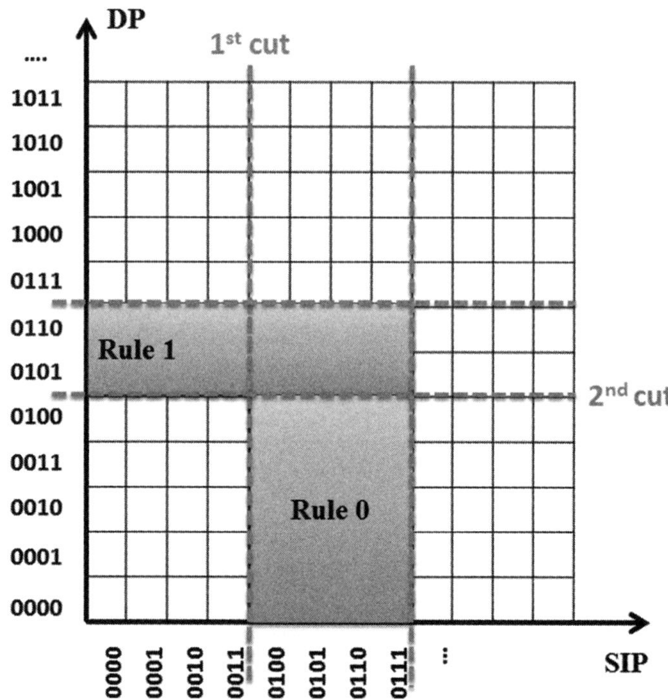

Fig. 3 Two-dimensional space formed by SIP and DP fields (in this example, Rule 0 has SIP: 01*, DP: 0–2; Rule 1 has SIP: 0*, DP: 5–6)

To distinguish different rules from each other, multiple cuts in this space are needed. Similar to the binary search in a 1-dimensinal space, cutting in multi-dimensional space leads to efficient search time.

We show an example of HiCuts in Fig. 3. Suppose both the SIP and DP fields of the packet header are 4-bit wide, Fig. 3 shows the 2-dimensional space formed by these two fields. Rule 0 and Rule 1 cover only part of the 2-dimensional space; "blank" regions denote the space where there is no match (NM) between the input and any of the rules. Without loss of generality, we assume the first cut is performed in SIP field.

- The first cut (vertical cut in x-axis between 0011 and 0100) in SIP field results in two smaller sub-regions. The left sub-region only contains Rule 1, while the right sub-region contains Rule 0 and Rule 1.
- The second cut (horizontal cut in y-axis between 0100 and 0101) in DP field results in a total number of four sub-regions. This cut distinguishes Rule 0 from Rule 1 in DP field.
- The cuts are performed alternatively in horizontal and vertical directions as shown in Fig. 3.

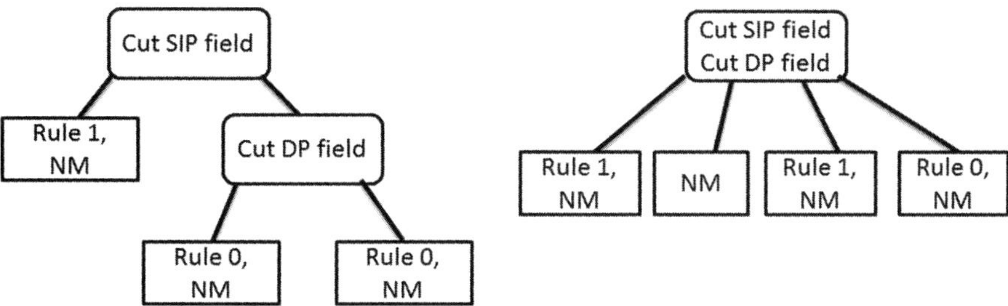

Fig. 4 Decision-tree for HiCuts (*left*) and HyperCuts (*right*) based on two cuts in Fig. 3

We show the resulting decision-tree of HiCuts after the first two cuts in Fig. 4 (left). As can be seen, each cut is represented as an internal node, while the rule IDs are stored in the leaf nodes. Overlapping rules can lead to a leaf node containing multiple rule IDs; in that case, a linear search is performed in a leaf node to get the final matching results.

It is possible to continuously cut the sub-regions so that each leaf node contains fewer rules. For example, a cut between Rule 1 and no-match sub-regions in Fig. 4 (left) further leads to two leaf nodes: one storing only Rule 1, and the other indicating there is no match between the input and any of the rules. However, more cuttings result in a deeper tree, and also an increase of memory consumption. In this case, tradeoffs have to be made between space and time [12].

HyperCuts is similar to HiCuts, except that HyperCuts allows cutting on multiple fields per step. This results in a fatter and shorter decision tree, as shown in Fig. 4 (right). Each internal node cuts the original space into smaller sub-regions based on the information from multiple fields. In the example as shown in Fig. 4, cuts are performed both in SIP and DP fields in an internal tree node; the resulting decision-tree is a four-way tree.

3.2 Challenges and Prior Work

Decision-tree based approaches face the following challenges:

- Cutting may lead to more sub-regions than the number of rules. For example, the two cuts in Fig. 3 results in four disjoint pieces (Rule 1 is cut into two pieces in the first cut). This means, depending on the rule set, decision-tree based approaches may require more memory than the memory storing the rule set.
- Searching in a large tree is hard to parallelize and it requires too many memory accesses. Therefore it is relatively challenging to achieve high performance using efficient search tree structure on state-of-the-art multi-core processors.
- It is challenging to use multiple trees or parallel threads to improve latency performance in decision-tree based approaches. For example, processing multiple packets concurrently improves throughput, but not latency.

A software-based router is implemented in [21] using HyperCuts on an eight-core Xeon processor. A throughput of up to 46 million Packets Per Second (MPPS) is achieved for five-field classification with the help of a TCAM. The classification time for each packet is measured by the total processing latency divided by all the packets processed; this is not an accurate metric. Moreover, it makes simple assumptions on accessibility of TCAM. In [18], a decision-tree-based solution using hashing is proposed on a four-core 2.8 GHz AMD Opteron system. It supports a five-field rule set consisting of 30 K rules. Although the throughput per core is 3.6 $\times$ the throughput per core achieved by HyperCuts, no processing latency is explicitly given, and no comparison is made with FPGA-based approaches.

In general, it is still a challenging research topic to achieve high performance using decision-tree based approaches on multi-core processors.

4 Decomposition-Based Approaches

4.1 Overview

The basic idea of decomposition-based approaches is: search each packet header field independently, get the partial results, and then merge them to produce the final packet header match. We show an example for decomposition-based approach in Fig. 5. In this example, a packet header having six fields is split; each field is compared against the corresponding field of the rule set. Since the comparison result in each field only suggests a match between the packet header and the rule set in this particular field, we denote this compassion result in a field as the "partial result". A final stage of merging is required to combine all the partial results into the final result.

For instance, suppose a packet has the header (SA: 95.105.143.51, DA: 39.240.26.229, SP: 100, DP: 1521, PROT: 0x06) and it is to be matched against rules in Table 1. In decomposition-based approaches, the SA field of the packet header is searched in the SA field of the rule set (Table 1). Notice the search only in this field indicates the packet header potentially matches both Rule 4 and 5. We may have multiple matching candidates (in this case, Rule 4 and 5), because these search operations are performed independently in each field.

Similarly, the DA field of the packet header is compared to the DA field of the rules in the rule set. This process is called *search* phase in each field and the matching candidates after a search is called *partial results*. Search phase is performed in all the packet header fields, until all the partial results are generated. We will introduce some search techniques later. To simplify the search phase, *preprocessing* of the rule set is usually required.

After the search phase, we notice the packet header (SA: 95.105.143.51, DA: 39.240.26.229, SP: 100, DP: 1521, PROT: 0x06) matches Rule 4 and 5 in SA field, Rule 4 in DA field, Rule 1 $\sim$ 5 in SP field, Rule 4 in DP field, and Rule 1 $\sim$ 4 in PROT field. These are not the final results since a packet is reported only if its five header fields all match a particular rule in the rule set. However, since we already have the

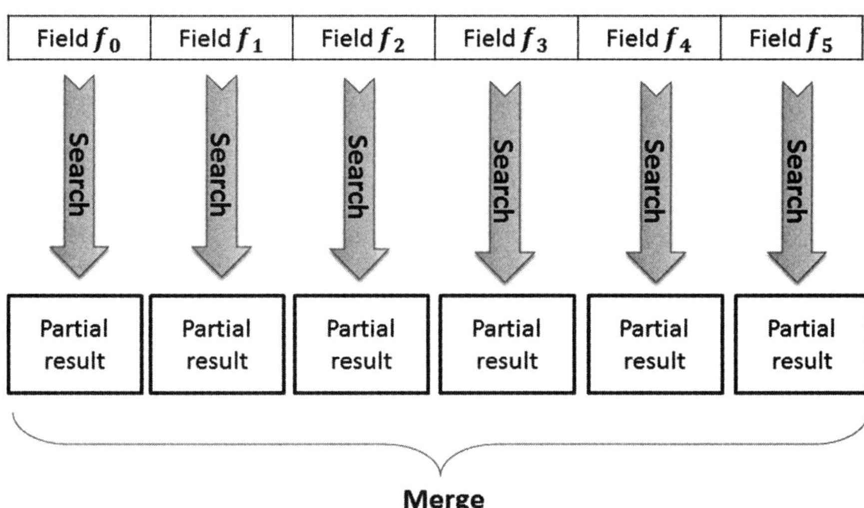

Fig. 5 Decomposition-based approach

partial results in all the fields, it is relatively simple to generate the final result. This process is called *merge* phase; we will introduce the merge algorithms later in this section.

In the example mentioned above, since the packet header only satisfies all the criteria specified by Rule 4 in all the fields, the packet header is considered to be matching Rule 4 only.

As can be seen, decomposition based approaches contain three phases. The first phase is to preprocess the rule set. The second phase is to *search* each field individually against the rule set. The third phase is to *merge* the partial results from all the fields.

4.2 Challenges and Prior Work

The key challenge of these algorithms is to parallelize individual search processes and handle merge process efficiently. For example, one of the decomposition based approaches is the Bit Vector (BV) approach [17]. The BV approach is a specific technique in which the lookup on each field returns an N-bit vector. Each bit in the bit vector corresponds to a rule. A bit is set to "1" only if the input matches the corresponding rule in this field. A bit-wise logical AND operation gathers the matches from all fields in parallel.

The BV-based approaches can achieve 100 Gbps throughput on FPGA [19], but the rule set size they support is typically small (less than 10 K rules). Also, for port

Table 2 Various search methods

Search method	Typical usage	Example	Data structure after preprocessing
Linear	Any field	Any	Linked-lists
Range-tree	Prefix/range match field	SA/DA, SP/DP	Range-tree
Hashing	Exact match field	PROT	Hash table

number fields (SP and DP), since BV-based approaches usually require rules to be represented in ternary strings, they suffers from range expansion when converting ranges into prefixes [20].

Some recent work has proposed to use multi-core processors for packet classification. For example, the implementation using HyperSplit [14], a decision-tree based algorithm, achieves a throughput of more than 6 Gbps on the Octeon 3860 multi-core platform. However, the decomposition based approaches on state-of-the-art multi-core processors have not been well studied and evaluated.

4.3 Preprocessing

The preprocess phase of the rule set depends on the algorithms used in the search phase. Typical search techniques in a packet header field include linear search, range-tree search [16], and hashing. Linear search can be applied to all the fields, regardless of the matching criteria. Range-tree search can be applied to fields requiring prefix match (SA and DA fields) or range match (SP and DP fields); hashing is usually applied to PROT field requiring exact match. We show various search methods and their corresponding data structures after preprocessing in Table 2.

First of all, before using any of the preprocessing techniques, let us take a close look at Table 1. We denote the rules before any preprocess the *original rules*. Notice that if we focus only on the SA field, Rule 4 and Rule 5 have exactly the same value in this field. As a result, out of eight original rules, we only have seven unique values in this field; we define such unique values as *unique rules* (URs) in this field. For example, in the PROT field of Table 1, we have four unique rules. This is one of the advantages for decomposition-based approaches: the number of unique rules in each field can be less than the total number of rules; this means lookup can be performed in a much smaller rule set [15].

i. **Linked-lists and range-tree**

After rules are projected onto a set of URs, we show an example of getting a set of linked-lists or a balanced binary range-tree in Fig. 6. As can be seen, if the URs require prefix match, they are first converted into a set of ranges. Then all the ranges are projected onto the same space to form a set of non-overlapping "subranges". In Fig. 6, for example, four URs (corresponding to six original rules) are translated into five subranges: a, b, c, d, and e. Each subrange is then "linked" to a set of IDs of the

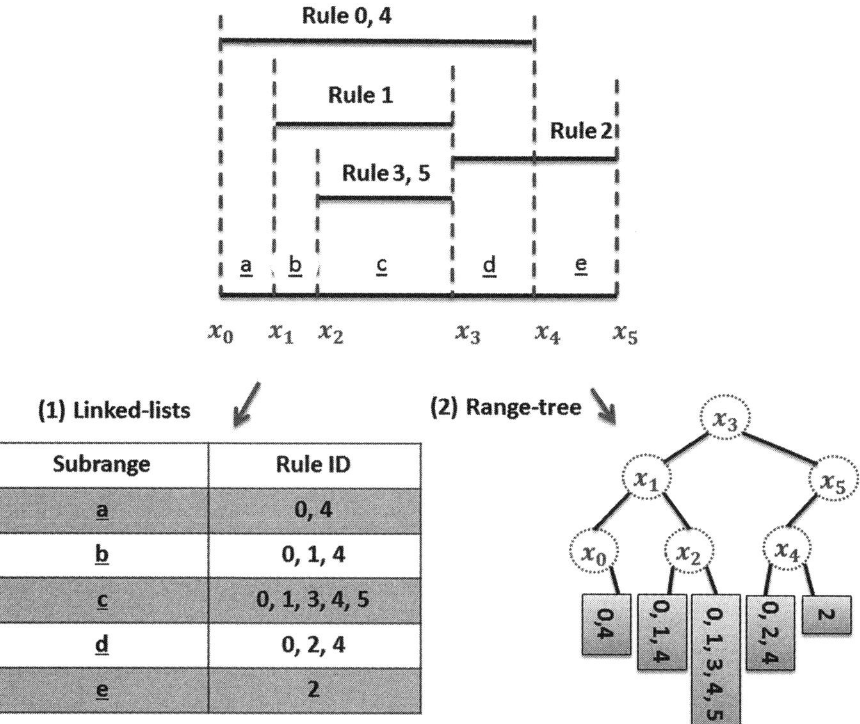

Subrange	Rule ID
a	0, 4
b	0, 1, 4
c	0, 1, 3, 4, 5
d	0, 2, 4
e	2

Fig. 6 Preprocessing 4 unique rules to (*1*) linked-lists or (*2*) a balanced binary range-tree

original rules, indicating a match between the input and a set of original rules. The linked-lists shown in Fig. 6 can be used directly in linear search.

If range-tree search is to be performed, a further conversion from subranges to range-tree is needed. The balanced binary range-tree is constructed of subrange boundaries, as shown Fig. 6. Notice that each leaf node corresponds to a non-overlapping subrange and a set of candidate URs that potentially match the input packet header.

ii. **Hash table**

Notice the search phase for an exact match field can be reinterpreted as: given a set of exact values stored in an array and an input integer, locate the exact value matching the integer. This problem can be efficiently solved by perfect hashing; without loss of generality, we use Cuckoo hashing [22] to reduce the number of memory accesses in the exact match fields.

We show an example of exact match field in Table 3. Notice that in most cases, exact match field has a "default" rule: "*", which means it matches any input. In Table 3, we only have five URs in this exact match field; however, the total number of classification rules can be much larger than five, as discussed before. Obviously,

Table 3 An example of exact match field

PROT field	UR 0	UR 1	UR 2	UR 3	UR 4
Linear	0x06	0x01	0x11	0x00	*

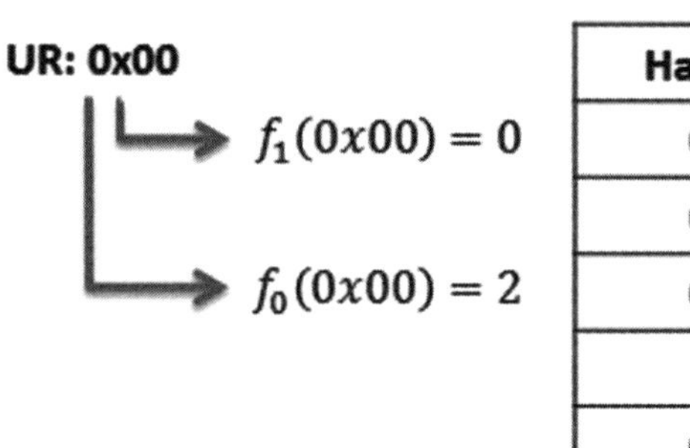

Fig. 7 Storing URs into a hash table

the corresponding IDs of the original rules should also be kept for each UR, usually in the form of Bit Vector (see Sect. 4.5).

In Fig. 7, we show an example of constructing a hash table for the URs in Table 3. Based on Cuckoo hashing, we use M hash functions to construct the hash table. Each entry in the hash table $H(k)$ is a (hash key, hash value) pair. Let us use $U_j(j = 0, 1, \ldots, j-1)$ to denote the hash key, and the corresponding hash value is given by $f_m(U_j)$ for some $m \in \{0, 1, \ldots, M-1\}$. The hash values are used directly as the memory index. For each exact match rule, at most M attempts are made to store this rule in the hash table. If all the memory locations indexed by $f_m(U_j)$ are occupied, $\forall_m = 0, 1, \ldots, M-1$, the hash table size has to be enlarged, and the M hash functions are chosen again.

For example, in Fig. 7, suppose we already stored key 0x06, 0xl1 and 0x01 in the hash table. When a UR 0x00 has to be inserted, we use $f_0(0x00) = 2$ as a possible index to store this UR; however, since the entry with index 2 is already taken by the key 0xl1, we need to make a second attempt. Using $f_0(0x00) = 0$ as the index, the UR 0x00 can be stored in the corresponding position in the hash table. In this way, multiple URs can be stored in the hash table using M hash functions. If the index computed by a hash function $f_m(U_j)$ is used by another entry in the hash table, another attempt using $f_{m+1}(U_j)$ as the index is needed, unless $m = M$, where we have to enlarge the hash table size or choose the M hash functions again.

4.4 Searching

i. **Linear search**

For linear search, every subrange needs to be examined against the input packet header bits. For example, in Fig. 6, all the five subranges (their six boundaries) have to be one-by-one compared against the input bits for a possible match. The reasons of doing linear search on the subrange boundaries rather than on the URs or in the field of the original rules are:

- Usually the number of URs is significantly less than the number of original rules, while the number of subrange boundaries is slightly greater than the number of URs. Namely, if we denote the total number of original rules as N, the number of URs in the field k as $q^{(\mathrm{k})}$, and the number of subrange boundaries as L, we have $q^{(k)} << N$ and $q^{(k)} < L$.
- Subranges are non-overlapping and their boundaries can be sorted in the preprocess phase. Each search corresponds to at most 1 comparison. For instance, finding a 3-bit number in sorted non-overlapping subranges [0, 2), [2, 3) and [3, 7) requires at most two comparisons (to two and three). However, if the original rules or URs are to be searched, the number of comparisons needed is usually larger. A singe comparison between a UR represented by a range and the input requires at least two comparisons (one for lowerbound and another for upperbound).

ii. **Range-tree search**

Binary search can be performed on a balanced binary range-tree. For binary trees, each tree node stores a subrange boundary as a key and compares the input with this key. The outcome of the comparison determines whether the left child or the right child of this node has to be searched next. As can be seen in Fig. 6, the search phase continues until a leaf node is reached; a set of rule IDs are returned as the partial result after this binary search.

iii. **Hashing**

For exact match field, hashing techniques can be applied. Suppose we have already constructed a hash table using the approach discussed before, as shown in Fig. 7. Now given an input, we need to efficiently identify which rule the input matches.

As shown in Fig. 8, let us consider the input 0x00. We use the same hash functions as discussed before to compute the first possible index $f_0(0x00) = 2$. As can be seen, the first attempt returns a hash key (0x11) which does not match the input. So a second attempt has to be made; in the second attempt, $f_1(0x00) = 0$ is used and its corresponding hash key matches the input. Hence the hash value "0" is used as the memory index to extract partial results (the corresponding IDs of the original rules).

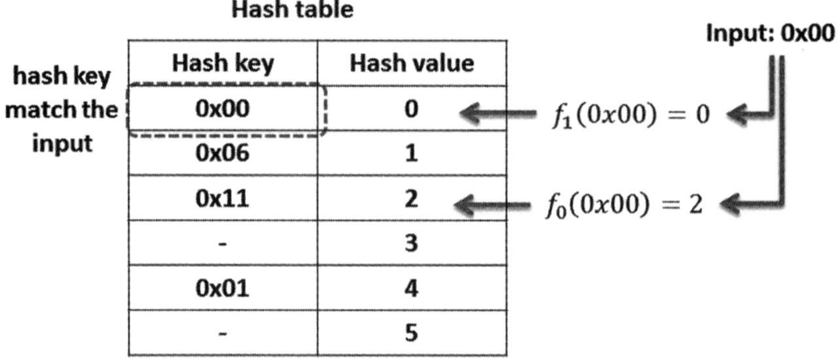

Fig. 8 Storing URs into a hash table

4.5 Merging

No matter how the searching is performed in each field, the partial results are represented by a list of original rule IDs. The most efficient way of recording the partial results is the Bit Vector (BV) [23]. BV-based approaches use one bit for each rule, and totally N bits for N rules. If the input matches a rule, then the corresponding bit is set to "1"; otherwise it is set to "0".

For example, consider a rule set having four rules and two fields W_0 and W_1, as shown in Fig. 9. We construct a 4-bit vector for W_0 and another 4-bit vector for W_1 Suppose in the W_0 field, we only have rule R_1 matches the input; the corresponding BV is then "0100", whose MSB corresponds to R_0 and its LSB corresponds to R_3. Similarly, in the W_1 field, the corresponding BV is "1110", indicating R_0, R_1 and R_2 all match the input in this field.

Recall that an input packet is considered as "matching" a rule only if the packet header matches all the fields of a rule in the predefined rule set. After we get the partial results from all the fields, a bitwise logical AND operation can be performed on all the BVs to form the final classification result. As shown in Fig. 9, ANDing BV "0100" and BV "1110" results in a BV "0100", indicating the only rule R_1 matches the input in both of the two fields. This is also the final result of the packet classification.

5 Performance Evaluation and Summary of Results

5.1 Experimental Setup

Since the decomposition-based approaches are not platform-dependent, we conducted the experiments on a 2 × AMD Opteron 6278 processor and a 2 × Intel Xeon E5-2470 processor. The AMD processor has 16 cores, each running at 2.4

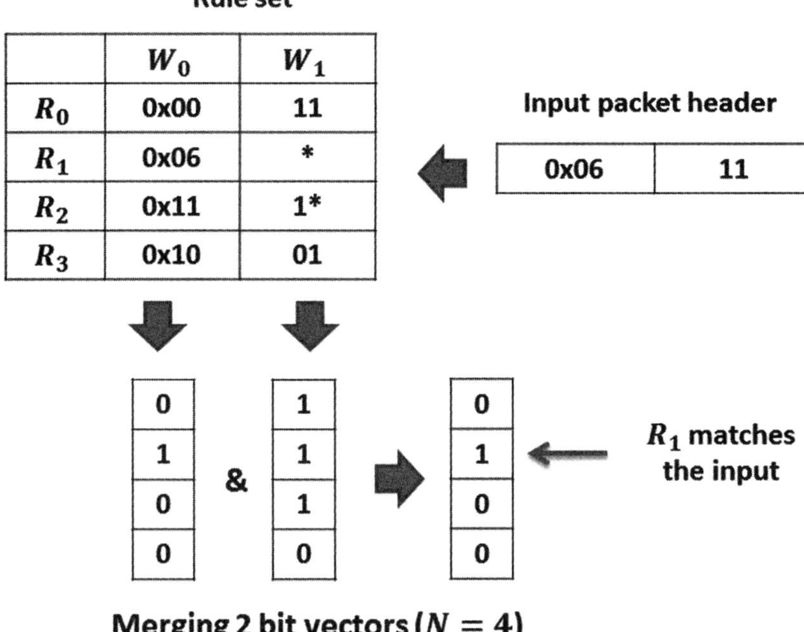

Fig. 9 Storing URs into a hash table

GHz. Each core is integrated with a 16 KB L1 data cache and a 2 MB L2 cache. A 6 MB L3 cache is shared among all 16 cores. The processor has access to 64 GB DDR3-1600 main memory through an integrated memory controller running at 2 GHz. The Intel processor has 16 cores, each running at 2.3 GHz. Each core has a 32 KB L1 data cache and a 256 KB L2 cache. All 16 cores share a 20 MB L3 cache. This processor has access to 48 GB DDR3-1600 main memory.

We implemented all the approaches using Pthreads on openSUSE 12.2. We used *perf*, a performance analysis tool in Linux, to monitor the hardware and software events such as the number of executed instructions, the number of cache misses and the number of context switches.

We generated synthetic rule sets using the same methodology as in [18]. We varied the rule set size from 1 to 64K to study the scalability of our approaches. We used processing latency and overall throughput as the main performance metrics. We also examined the relation between the number of threads per core and context switch frequency to study their impact on the overall performance.

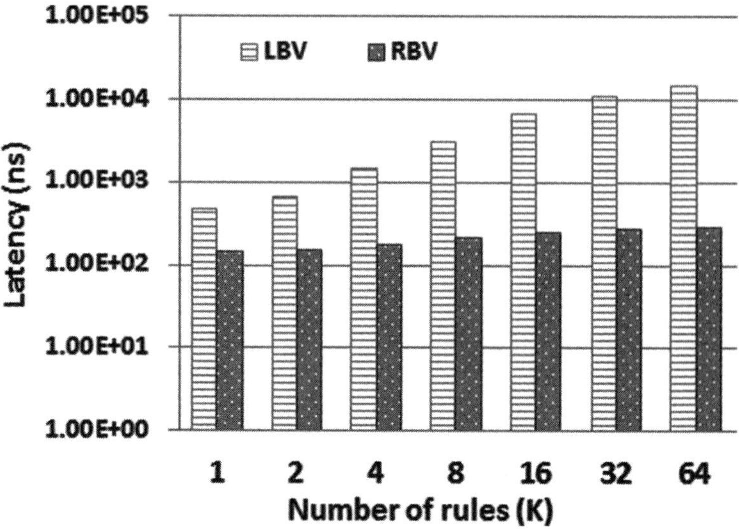

Fig. 10 Latency on the AMD processor

Table 4 Various implementations

Implementation	Searching	Merging
LBV	Linear search only	ANDing BVs
RBV	Prefix/range match field: range-tree Exact match field: hashing	ANDing BVs

5.2 *Latency*

It is shown in [23] that in the search phase of decomposition-based approaches on multi-core processors, it is better to allocate each core a single packet. An alternative way is to assign a single packet header to multiple cores, where each core deals with one or more packet header fields; in that case, the merge phase requires access to partial results from different cores. Since a large amount of data move between cores, this is less efficient than allocating each core an independent packet. Hence we assign each core a packet to improve the overall performance. In this case, we have five parallel search threads and one merge thread per core; this configuration is used for all our implementations (except in Sect. 5.5).

Figure 10 shows the latency performance on the AMD processor, while Fig. 11 shows the latency performance on the Intel processor. We show the detail of various implementations in Table 4.

We have following observations:

- Using range-tree and hashing techniques for individual field search significantly reduces the latency. Let N denote the total number of the rules; in the worst case,

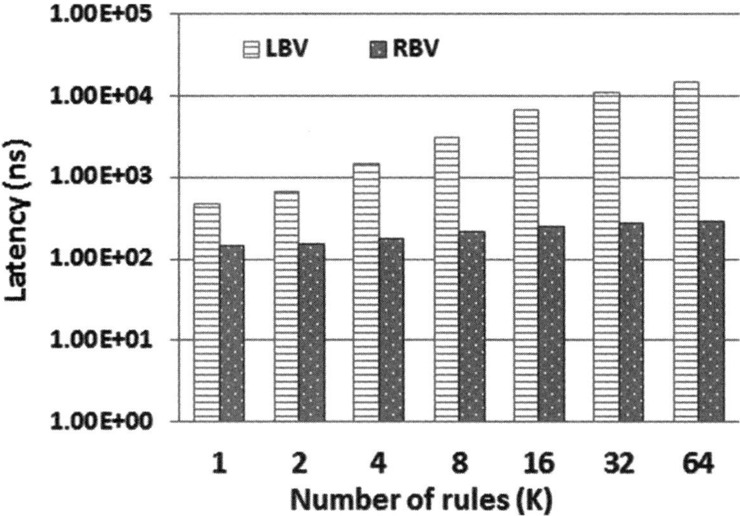

Fig. 11 Latency on the Intel processor

the linear search requires $0(N)$ memory accesses, while the range-tree search and hashing only require $0(\log N)$ and $0(1)$ memory accesses, respectively.

- As the number of rules increases, the latency of LBV also increases. This is because for larger rule sets, the linear search has to examine a large number of URs, which incur a large amount of latency.
- The processing latency of RBV, however, shows a small variation even for large rule sets. Note the time complexity for range-tree search is $0(\log N)$, while the time complexity for linear search is $0(N)$. For a balanced binary range-tree, when the rule set size doubles, one more tree level has to be searched, while linear search requires double the amount of search time.

We show the breakdown of the processing latency in terms of search and merge latencies in Fig. 12. As can be seen, the search latency contributes more to the total classification latency (around $70 \sim 80\,\%$ of the total latency) than the merge latency, since search operations are more complex compared to the merge operations. We have similar observations on both the AMD and Intel processors.

5.3 Throughput

We show the throughput performance on the AMD and Intel processors in Figs. 13 and 14, respectively. We observe consistent performance with the latency performance. The RBV implementation achieves $5.2 \sim 11.5$ Gbps throughput on the AMD processor, and achieves $4.8 \sim 10.6$ Gbps throughput on the Intel processor. The performance on the Intel processor is slightly worse than the performance on

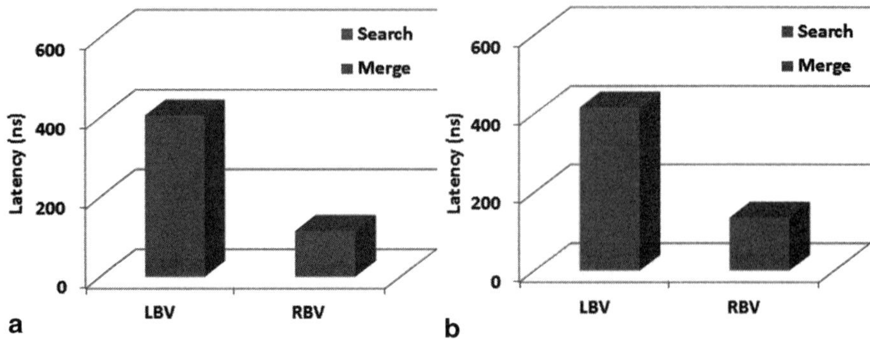

Fig. 12 Breakdown of the latency per packet (1K rule set, five threads per core). **a** On the AMD processor. **b** On the Intel processor

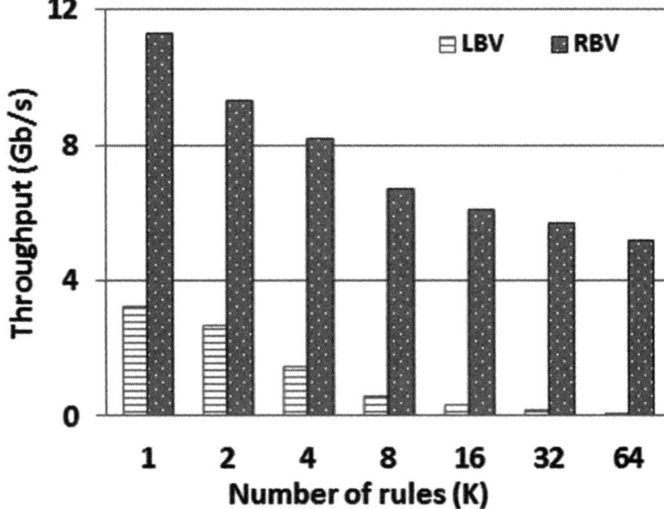

Fig. 13 Throughput on the AMD processor

the AMD processor. Note that the Intel processor has a lower clock rate and smaller cache size.

5.4 Cache Performance

To explore further why the performance deteriorates as the rule size grows, we measure the cache performance. We show the number of cache misses per 1K packets on the AMD processor in Fig. 15. As can be seen, the overall performance is consistent with the cache performance on the multi-core processors. As the size of

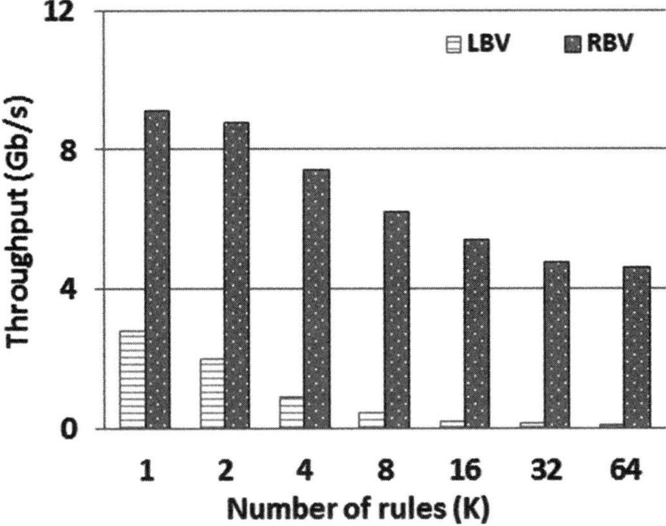

Fig. 14 Throughput on the Intel processor

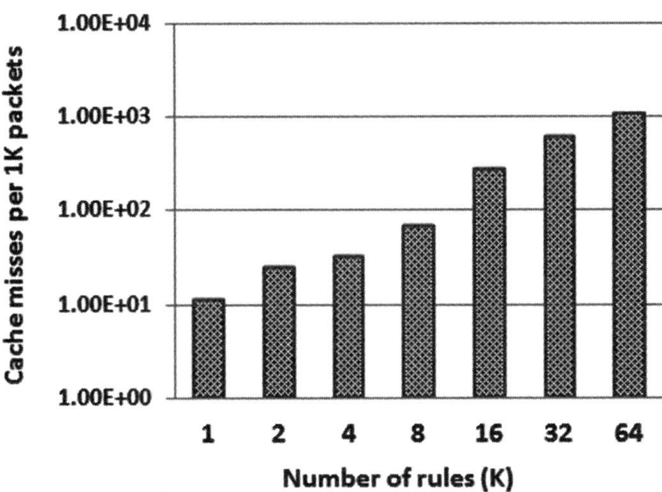

Fig. 15 Number of L2 cache misses per 1K packets on the AMD processor (RBV)

the rule set increases, it is more difficult to fit all the rules in the cache; a cache miss requires data to be read from a farther-level cache or the main memory, hence the performance degrades with respect to both throughput and latency. Similar results can also be seen for LBV and the implementations on the Intel processor.

5.5 Impact of the Number of Threads

A more thorough research [23] shows that the number of threads per core also has an impact on the performance. Our results indicate:

- The performance of RBV goes up as the number of threads per core increases from one to six. Once the number of threads per core exceeds six, the throughput begins to degrade.
- For LBV, the performance keeps increasing until we have ten threads per core.

Reasons for performance degradation include saturated resource consumption of each core and the extra amount of overhead brought by the context switch mechanism. When the number of threads increases, context switches happen more frequently, and switching from one thread to another requires a large amount of time to save and restore states.

5.6 Comparison with Existing Approaches

According to the target platform, packet classification can be categorized into hardware-based approaches and software-based approaches. In the hardware-based approaches, efficient architectures are exploited on FPGA. In software-based approaches, parallel algorithms are explored on multi/many-core General Purpose Processor (GPPs). Most of the algorithms used in hardware and software fall into two categories: decision-tree based and decomposition based algorithms. In this section, we compare existing solutions for packet classification. We consider the advantages and weakness of the algorithms, performance with respect to overall throughput and the scalability to support large rule set.

a. Approaches on Multi-core GPPs

We introduced Decision-tree-based approaches in Sect. 3. Most of the existing solutions along with their enhanced versions fall into this category. For example, in [21], HyperCuts is used to implement a software-based router on an eight-core Xeon processor. With the help of TCAM, a 15 Gbps throughput can be achieved for a rule set consisting of five-field 9K rules. However, it makes simple assumptions on accessibility of TCAM. In [14], a decision-tree-based solution using HyperSplit achieves more than 6 Gbps on the Octeon 3860 multi-core processor against 10K rules. However, the performance of decision-tree-based approaches is highly dependent on the statistical features of the rule set.

There are very few works using decomposition-based approaches on multi-core processors [23]. Similar to decision-tree-based approaches, for implementations on multi-core processors, it is challenging to achieve high throughput. The major advantage of implementations on multi-core processors is the scalability of the rule set size; in [23], a five-field rule set consisting of up to 32K rules can be supported on state-of-the-art multi-core processors.

Table 5 Comparison summary

Platform	Approach	Pros	Cons	Throughput (Gbps)	No. of rules (K)
FPGA	Decision-tree-based [24]	Pipelined architecture	Exponential rule set expansion	80	10
	Decomposition-based [25, 26]	Parallel search, exploiting massive bandwidth	Large amount of pseudo rules, limited on-chip memory	100 ~ 400	1
Multi-core	Decision-tree-based [21]	Efficient tree search	Challenging to support high throughput	15	9
	Decomposition-based [23]	Parallelism, large number of rules	5.2 ~ 11.5	1 ~ 64	

b. Approaches on FPGA

In [24], a decision-tree-based approach is implemented on FPGA. The decision tree is constructed by the nodes which cut the search space based on one or more fields of the rule. Then the decision tree is mapped into a deeply pipelined architecture on FPGA. The design can store either 10K 5-field rule or 1K 12-field rule in on-chip memory of a single FPGA and sustain 80 and 40 Gbps, respectively. However, the deep pipeline introduces long latency and the clock rate cannot be easily sustained for large rule sets due to linear alignment of BRAM on FPGA.

In decomposition-based approaches on FPGA [25], each packet header field is searched in the rule set individually, and the partial results are mapped to a rule ID by a perfect hash function. The approach achieves a throughput of 100 Gbps for 100 rules. However, the crossproducting of the partial results can potentially expand the original rule set to a much larger one. Hence external memory has to be used to store both the real and pseudo rules. The slow access rate and long access latency remains the bottlenecks of the approach.

In [26], a TCAM-like approach is implemented on FPGA, which can achieve over 400 Gbps throughput. This approach splits each field into strides of bit vectors. A bitwise AND operation is operated on the bit vectors of all the fields to gather the matches. Although this approach can achieve a very high throughput, it can only support at most 1K five-field rules due to limited on-chip resources.

c. Summary of Comparison

We summarize the comparisons between various approaches in Table 5. The main advantage of using multi-core processors is the capability of supporting very large rule sets. Although the approaches on FPGA can employ external memory to support large rule set, the access latency to external memory becomes a new performance bottleneck.

6 Conclusion

In this chapter, we introduced two important packet classification techniques: decision-tree-based and decomposition-based packet classification approaches on state-of-the-art multi-core processors. There are other classification techniques that do not belong to either one of the aforementioned two; a complete scope of all the packet classification algorithms is beyond the scope of this chapter.

In this chapter, we focused on the decomposition-based approaches, since exploring parallelism on multi-core platforms is a natural trend for the future Internet infrastructure [6]. For decomposition-based approaches, we explored the impact of various parameters on the performance with respect to latency and throughput. We also examined the cache performance and conducted experiments on various platforms. Our experimental results show that range-tree search is much faster than linear search and is influenced less by the size of the rule set.

A more interesting future work is to target OpenFlow packet classification where more packet fields are required to be examined.

References

1. J. Naous, D. Erickson, G. A. Covington, G. Appenzeller, and N. McKeown, "Implementing an OpenFlow Switch on the NetFPGA Platform", in Proc. of the 4th ACM/IEEE Symposium on Architectures for Networking and Communications Systems, ser. ANCS '08, (2008) 1–9.
2. G. Brebner, "Softly Defined Networking", in Proc. of the 8th ACM/IEEE Symp. on Architectures for Networking and Communications Systems, ser. ANCS '12, (2012) 1–2.
3. F. Yu, R. H. Katz, and T. V. Lakshman, "Efficient Multimatch Packet Classification and Lookup with TCAM", IEEE Micro, vol. 25, no. 1 (2005) 50–59.
4. W. Jiang, Q. Wang, and V. K. Prasanna, "Beyond TCAMs: an SRAM based parallel multi-pipeline architecture for terabit IP lookup", in Proc. IEEE INFOCOM (2008) 1786–1794.
5. G. S. Jedhe, A. Ramamoorthy, and K. Varghese, "A Scalable High Throughput Firewall in FPGA", in Proc. of IEEE Symposium on Field Programmable Custom Computing Machines (FCCM), (2008) 802–807.
6. T. Koponen, "Software is the Future of Networking," in Proc. of the 8th ACM/IEEE Symp. on Architectures for Networking and Communications Systems (ANCS), 2012, pp. 135–136.
7. "AMD Multi-Core Processors," http://www.computerpoweruser.com/articles/archive/c0604/29c04/29c04.pdf. 8.
8. "Intel Multi-Core Processors: Making the Move to Quad-Core and Beyond," http://www.cse.ohio-state.edu/~panda/775/slides/intel_quad_core_06.pdf.
9. "Multicore Computing- the state of the art," http://eprints.sics.se/3546/1/SMI-MulticoreReport-2008.pdf.
10. P. Gupta and N. McKeown, "Packet classification on multiple fields", In Proceedings of the conference on Applications, technologies, architectures, and protocols for computer communication, SIGCOMM (1999) 147–160.
11. W. Jiang and V. K. Prasanna, "A FPGA-based Parallel Architecture for Scalable High-Speed Packet Classification," in 20th IEEE International Conference on Application-Specific Systems, Architectures and Processors (ASAP), (2009) 24–31.
12. P. Gupta and N. McKeown, "Packet Classification using Hierarchical Intelligent Cuttings", IEEE Symposium on High Performance Interconnects (HotI) (1999).

13. S. Singh, F. Baboescu, G. Varghese and J. Wang, "Packet Classification using Multidimensional Cutting", ACM SIGCOMM (2003) 213–224.
14. D. Liu, B. Hua, X. Hu and X. Tang. "High-performance Packet Classification Algorithm for Any-core and Multithreaded Network Processor." in Proc. CASES, (2006).
15. D. E. Taylor and J. S. Turner, "Scalable Packet Classification using Distributed Crossproducing of Field Labels," in Proc. IEEE INFOCOM, (2005) 269–280.
16. P. Zhong, "An IPv6 Address Lookup Algorithm based on Recursive Balanced Multi-way Range Trees with Efficient Search and Update", in Proc. of international conference on Computer Science and Service System (CSSS), ser. CSSS '11, (2011) 2059–2063.
17. T. V. Lakshman, "High-Speed Policy-based Packet Forwarding Using Efficient Multi-dimensional Range Matching", ACM SIGCOMM (1998) 203–214.
18. F. Pong, N.-F. Tzeng, and N.-F. Tzeng, "HaRP: Rapid Packet Classification via Hashing Round-Down Prefixes", IEEE Transactions on Parallel and Distributed Systems, vol. 22, no. 7, (2011) 1105–1119.
19. W. Jiang and V. K. Prasanna, "Field-split Parallel Architecture for High Performance Muti-match Packet Classification using FPGAs," in Proc. of the 21st Annual Symp. on Parallelism in Algorithms and Arch. (SPAA), 2009, pp. 188–196.
20. V. Srinivasan, G. Varghese, S. Suri, and M. Waldvogel, "Fast and Scalable Layer Four Switching," in Proc. ACM SIGCOMM, 1998, pp. 191–202.
21. Y. Ma, S. Banerjee, S. Lu, and C. Estan, "Leveraging Parallelism for Multi-dimensional Packet Classification on Software Routers," SIGMETRICS Perform. Eval. Rev., vol. 38, no. 1, pp. 227–238, 2010.
22. R. Pagh and F. F. Rodler, Cuckoo Hashing. Springer, 2001.
23. S. Zhou, Y. R. Qu, and V. K. Prasanna, "Multi-core Implementation of Decomposition-based Packet Classification Algorithms", in Proc. of the 12th International Conference on Parallel Computing Technologies (PaCT'13), pp. 105–119.
24. W. Jiang and V. K. Prasanna, "Scalable Packet Classification on FPGA," IEEE Trans. VLSI Syst., vol. 20, no. 9, pp. 1668–1680, 2012.
25. V. Pus, J. Korenek, and J. Korenek, "Fast and Scalable Packet Classification using Perfect Hash Functions," in Proceedings of the ACM/SIGDA international symposium on Field Programmable Gate Arrays (FPGA), 2009, pp. 229–236.
26. T. Ganegedara and V. K. Prasanna, "StrideBV: Single chip 400G+Packet Classification," 13th IEEE International Conference on High Performance Switching and Routing (HPSR '12), June 2012, pp. 1–6.

Optical Interconnects for Data Center Networks

Khurram Aziz and Mohsin Fayyaz

1 Introduction

Traditional data center networks built with copper wires and electronic elements suffer from various problems. These include high energy consumption due to the wired architecture, high latency due to extra hops adding to the routing delay, fixed throughput of links, and very limited configurability. Data center networks built with optical fibers and optical components would solve all of these problems but they suffer from issues of their own including higher cost, immaturity of optical components, lack of optical buffers and complexity of design. It is clear however, that optical interconnects will replace their electronic counterparts in all data center network architectures due to their superior properties.

Over the past several years, data center network architectures have come a long way with several optical and electro-optical architectures employing optical interconnects being proposed in the literature. This chapter presents a detailed survey of these architectures with a brief discussion about their performance.

The rest of this chapter is organized as follows. Section 1 discusses the need for optical interconnects in data center networks. Section 2 presents an overview of the commonly used optical components in data center networks. Section 3 presents the various optical data center network architectures proposed in literature, grouped into several categories depending on what parameter they best optimize. The later sections build on the information gained from Sect. 3. Section 4 briefly discusses the data center traffic characteristics, Sect. 5 discusses the energy requirements for data center networks while Sect. 6 discusses the routing characteristics and issues of data center networks based on optical interconnects.

K. Aziz (✉)
Department of Electrical Engineering, COMSATS Institute of Information Technology, Abbottabad, Pakistan
e-mail: khurram@ieee.org

M. Fayyaz
e-mail: mohsinf@ciit.net.pk

S. U. Khan, A.Y. Zomaya (eds.), *Handbook on Data Centers*,
DOI 10.1007/978-1-4939-2092-1_14

2 Need for Optical Interconnects in Data Center Networks

Cloud based services demand high network performance. Scaling electronic switched data center network will prove to be costly and complex. In addition, electronic data center architectures have high energy consumption due to wired architecture, high latency due to large port count, fixed throughput of links and very limited reconfigurability [1]. Traditional data center networks are based a hierarchical architecture. Servers are arranged in the form of racks and are connected through top of rack (ToR) switches. Many ToRs are connected through aggregate or core switches which result in large port count requirements for the switch and contributes to high energy expenses.

Data center networks can be broadly classified into network centric design and server centric design. In network centric design, servers are connected through high end switches that have a huge port count. In server centric design, servers are computing units and they also participate in load balancing and packet forwarding. Optical interconnects are a suitable candidates for both server centric and network centric data center networking. Simple 2×2 switches can be used with traditional topologies like hypercube to create shortcuts for heavy flows. Network adaptively creates paths to reduce the impact of intermediate nodes. In this way additional benefits of high connectivity, small diameter, fault tolerance, simple control and routing are achieved. Thus a combination of traditional data center topologies and optical interconnects give the best of both worlds [10].

The deployment of optical switching in data center networks can bring some key benefits to data center networks. In addition to the long reach and high bandwidth provided by the optical fiber compared to its copper counter parts, optical fibers and switches are transparent to signal bit rate, are energy efficient, provide very high capacity, and are flexible to protocols and network upgrades [13].

The key benefits of employing optical interconnects in data center networks include:

i) Emerging applications requirement fulfillment
ii) Dealing with traffic heterogeneity
iii) Energy conservation
iv) Fast switching transition
v) Wavelength multiplexing and parallelism
vi) Reconfigurability
vii) Hop count reduction
viii) Contention resolution
ix) Optically connected memory

i) Emerging applications requirement fulfillment

Emerging applications utilizing video streaming (Youtube, Google Video..), search engines (Google, Bing), social networking (Facebook, Orkut), Email (Gmail, Hotmail), cloud computing services and Geo data (Google Earth) put a high demand for computing, network scalability and bandwidth on current data centers. Most data centers networks are based on electronic switches. Shifting these data centers' architecture to the optical domain will address most of the above mentioned issues in these applications.

ii) Dealing with traffic heterogeneity
The reconfigurable nature of optical interconnects in data center networks make them suitable for dealing with traffic heterogeneity that may be introduced when data centers handle a variety of the above-mentioned applications. Photonic resources exist that are reconfigurable in nature in terms of path establishment and bandwidth scalability [1]. Hence, longer traffic flows can be configured with optical circuit switching while shorter flows can be configured with optical packet switching.

It is also possible to combine the benefits of both electronic switching and optical switching with optical interconnects. Micro Electro-Mechanical Switches (MEMS) are commonly employed for wavelength switching in circuit switched networks. MEMS are power efficient but are not fast enough to handle bursty data. They are also not bandwidth efficient due to fixed and coarse granularity of wavelength that they are built to handle. They are most suitable for inter data center networking where there are long flows. Electronic switching on the other hand handles bursty traffic well. Hence for shorter traffic flows between pods, electronic switching gives best results [8].

iii) Energy conservation
Wavelength division multiplexing (WDM) technologies reduce cabling and optical fibers have low power consumption [13]. Detailed discussion about energy conservation follows in the later section of this chapter.

iv) Fast switching transition times
Combining the wavelength selective and broadband behavior of some optical devices like micro rings and Semiconductor Optical Amplifier (SOA) switch it is possible to achieve fast switching transition, low driving voltages and high extinction ratios. Such devices are compact in size and are capable of routing messages in nanosecond-scale switching speed. They also have flexibility of selecting different wavelengths for switching ports [3].

v) Wavelength multiplexing and parallelism
It is now possible to have reconfigurable optical packet and circuit switched architectures that have wavelength control at various levels of granularity. [12].

If a ToRs which wants to communicate with another ToR at w times the speed of a single port then it will need to use w ports. Each of these w ports will have a unique wavelength. WDM enables these w wavelengths to be multiplexed in to one optical fiber. A wavelength selective switch (WSS) can split these w wavelengths to the appropriate MEMS port. This makes it possible to set up a $w \times$ *line speed* connection from device A to B at runtime. A fiber cannot carry two channels in same direction over same wavelength. To use all the available wavelengths, each ToR is assigned a unique wavelength across the ports. Same wavelength is used to receive the traffic as well. Full port usage is made possible through the use of bi-directional traffic. Optical circulators between ToR and MEMS make it possible to detect multiple optical channels using a single photo detector. This offers high energy efficiency by reducing the number of communication links [6].

Optical interconnects in data center networks are also able to benefit from multiple input multiple output (MIMO) based on orthogonal frequency division multiplexing (OFDM). OFDM allows high spectral efficiency as multiple channels are spectrally overlapped i.e. they are orthogonal over one symbol period [7].

Most of today's systems also suffer from I/O limitations and head of line blocking problem. These problems are overcome by optical interconnects which exploit high capacity wavelength division multiplexing and parallelism [4].

vi) Hop count reduction

Hop count in a network has an impact on the energy consumption and latency of the network. Multi-hop network topologies like Fattree and Butterfly are based on store and forward mechanism due to large port count, which causes huge energy consumption and latencies. Network endpoint sockets are scaling exponentially, putting more burden on the underlying communication fabric. Smaller diameter networks are possible using high radix routers. High bandwidth and low power photonic switching devices can be arranged to function like a high radix router with lower complexity and power [2]. High radix switch design reduces the overall hop count of the network.

vii) Reconfigurability

Optical switches are now possible that can be reconfigured in a few nanoseconds time regardless of the port count. This minimizes the end to end latency. Contention resolution in optical switches can be supported by electronic buffers for example [5]. Optical interconnects are able to achieve flexible topology by exploiting reconfigurable property of MEMS. N ToRs can be connected to a single N port MEMS. Every ToR can be connected to any other ToR at any instant. If ToR graph is connected through MEMS, then hop by hop switching of such circuits can achieve a network wide connectivity [6].

viii) Contention resolution

Different optical devices can be used to reduce or remove contention in optical networks like Reflective Semiconductor Optical Amplifiers (RSOA). RSOA is a widely used active optical component. It acts as a mutual exclusion element. Mutual exclusion is widely used in distributed computing. With this principle shared resource is used without any incorrect operation. When two nodes use a shared resource and if resource is requested by both nodes, the mutex element (for example RSOA in this case) grants access to the shared resource to any one of the nodes. If two requests arrive at the same time to RSOA both requestors get the $P_{tot}/2$ reflected power and none of the grant detectors at either source node trigger. This is the case when no one gets a grant. RSOA is also able to scale well for N requestors. RSOA based techniques can be implied to achieve distributed contention resolution protocols [11].

It is also possible to combine strictly non-blocking optical architectures like Spanke with wavelength division multiplexing. Thus apart from the non-blocking ability, it also supports N WDM channels for each port. Furthermore, if the architecture is made modular, each module operates independently to switch WDM packets. This combination leads to several advantages which include output contention resolution, control and configuration time becomes independent of port count and scaling just means a linear increase of components and energy consumption [14].

Lost or buffered packets limit the maximum load of the network. In optical interconnects such architectures are possible that are non-blocking and re-arrangeable and have a very low reconfiguration time. Switching time is reduced to few nanoseconds regardless of the number of input or output nodes. The hierarchical structure of data center network topology creates bottleneck at rack and cluster level. The end to end latency within racks should be less than 1 ms. The control and implementation of an optical switch that has thousands of ports is challenging. Optical switches flatten the network as they scale to thousands of ports, thus providing high connectivity, large bandwidth and low latency meeting all the demands of data center networking. Optical switches are able to achieve the latency of less than 1 microsecond regardless of the port count and input load [9]. Lack of optical buffers in optical interconnects make them reliant on deflection routing or optical delay lines. A hybrid of both optical and electrical technology is also a viable solution [4].

ix) Optically connected memory
Large scale systems suffer from memory scalability issues. A high end system requires high memory capacity and bandwidth while simultaneously demanding low latency and energy efficiency of interconnects. Making the interface between processors and memory devices optically connected significantly increases the amount of data available to the data center server processors so that they operate at their maximum speeds [15].

3 Optical Components in Data Centers

Major optical components used in optical data center networks include

- Semiconductor optical amplifier(SOA)
- Silicon ring resonators
- Arrayed waveguide grating (AWGS)
- Wavelength selective switch (WSS)
- MEMS switch(Optical Crossbar, Optical Switching Matrix(OSM))
- Circulators
- Optical multiplexers/de-multiplexers.

3.1 Semiconductor Optical Amplifier (SOA)

Optical amplifiers amplify optical signals. In the simplest form, they consist of a p-n junction. They are electronically pumped and are small in size. Some the properties of SOAs include nonlinearity, fast transition time, moderate polarization dependence, low gain and high noise. Four main nonlinear operations that can be conducted on SOAs include four wave mixing, wavelength conversion, cross phase modulation and cross gain modulation. They can operate on low power lasers. Nonlinear nature of SOAs makes them suitable candidates for optical signal processing techniques like wavelength conversion and all optical switching. SOAs have the ability to amplify signals ranging from several Mbps to beyond 40 Gbps. This nature of SOAs makes them useful for a range of protocols with different data rates.

For amplification purposes, linear gain over 3dB power dynamic range is preferred. If they are operated outside this region it causes distortion at high output powers. Their gain is also dependent on the operating wavelength. Operation of SOA in gain compression region produces chirp in the amplified signal. Chirp is the frequency variation of the signal whereas gain compression is the reduction in differential gain due to the nature of device transfer function. Another important parameter of an SOA is the Noise Figure. Noise figure is the amount of degradation of signal to noise ratio during the amplification process. At 3dB, typical SOAs exhibit 80 nm optical gain bandwidth. Centering the gain peak is required in order to operate SOAs in low loss transmission window of fibers [30].

3.2 Silicon Micro Ring Resonator

The working of a silicon micro ring resonator is shown in Fig. 1 below. Terminal 1 takes in multiple wavelengths which are partially coupled to coupler 1. The signal in the ring is also coupled to the straight waveguide through coupler 2.

Resonant condition for wavelength λ_i is

$$n_{eff} L = m\lambda_i$$

Coupling wavelength λ_i gets enhanced and all others get suppressed. This resonant wavelength λ_i is output to terminal 2 while the rest are output to Terminal 4.

3.3 Arrayed Waveguide Grating

An arrayed waveguide grating (AWG) is based on an array of waveguides. It has both imaging and dispersive properties. AWGs image the signal from single input waveguide to multiple waveguides in such a way that different wavelength signals are imaged on to several different waveguides as shown in Fig. 2.

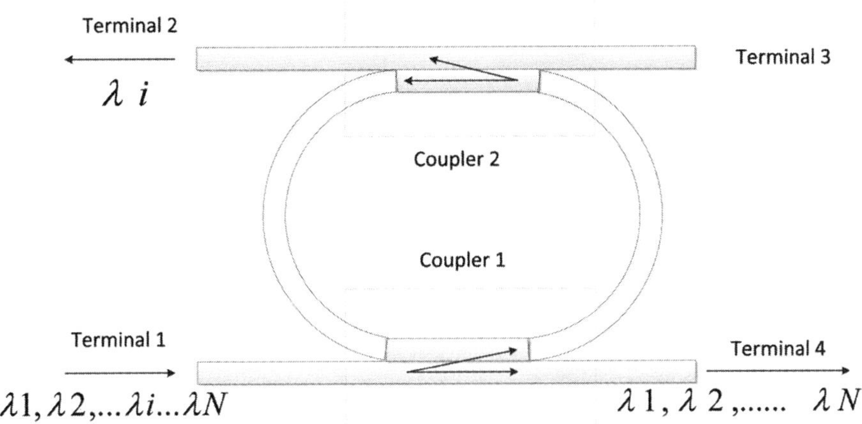

Fig. 1 Silicon micro-ring resonator

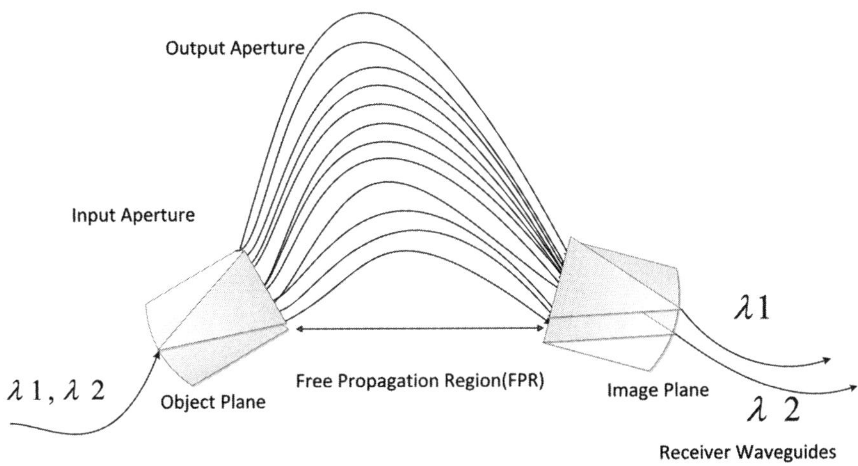

Fig. 2 Arrayed waveguide grating

Technologies used to realize AWGs include silicon on silicon and Indium Phosphide based semiconductor technology. AWGs are used to achieve a wide variety of functionalities in WDM networks. AWGs are compact and high functionality devices. When a light beam traveling in the transmitter waveguide enters the free propagation region, it becomes divergent and does not remain confined. When it arrives at the input aperture it is coupled to a large number of waveguides in the Free Propagation Region (FPR) and travels to the output aperture. Arrayed waveguide length is such

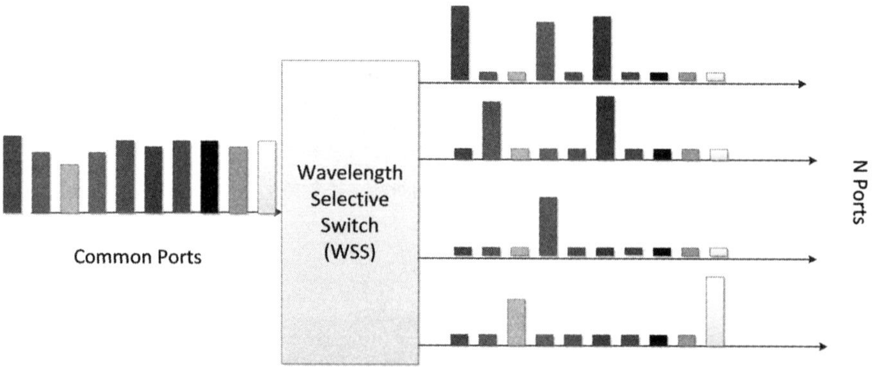

Fig. 3 Wavelength selective switch

that the difference in the length of individual waveguides is integer multiple of central wavelength. In this way a divergent beam is made convergent with equal phase and amplitude distribution. Spatial separation of different wavelengths is obtained by linear increase in length of waveguide array [31].

3.4 Wavelength Selective Switch

A wavelength selective switch can block, route or attenuate all densely multiplexed (DWDM) wavelengths within a node in network.

Wavelength switching or routing can be changed dynamically by an electronic communication control interface. As can be seen from Fig. 3, a WDM signal arrives at the input port and different wavelengths are routed on different output ports. This is done by attenuating some wavelengths while not changing the magnitude of other wavelengths.

Internal architecture of a simple MEMS based wavelength switch is shown in Fig. 4.

The light beam is collimated with lens and is de-multiplexed with the grating. The direction of beam after grating depends upon the central wavelength of the beam. The diffracted beams pass through lens again and are directed to a MEMS device. The MEMS device either changes amplitude (attenuates) or changes the direction of the beams. Reflective MEMS device is coupled to the fiber again [32].

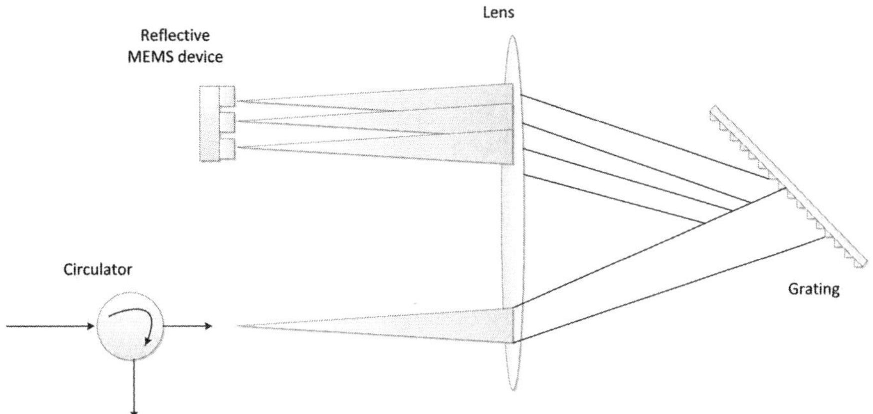

Fig. 4 MEMS based wavelength switch

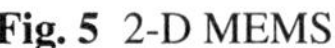

Fig. 5 2-D MEMS

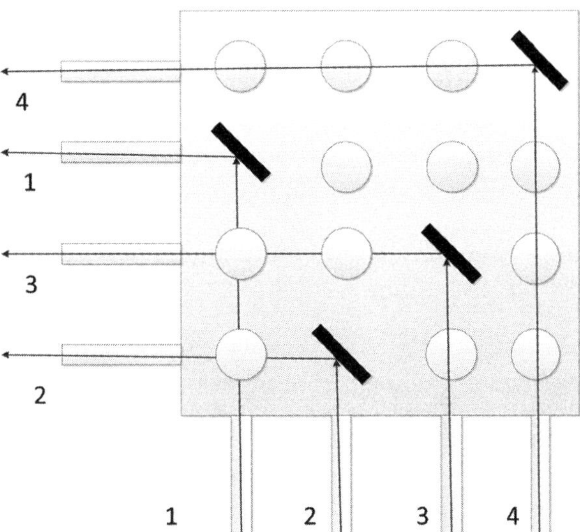

3.5 *MEMS Switch(Optical Switching Matrix, Optical Crossbar)*

These switches use moveable mirrors to redirect the light and achieve the switching functionality. These switches are low loss, low crosstalk and are economical to build. They rotate along axis and their two states are referred to as lying or standing. This architecture is very simple and non blocking in nature. N $\times$ N switch requires N^2 micro mirrors. As can be seen from Fig. 5, different wavelengths on input ports can be routed to different output ports [33].

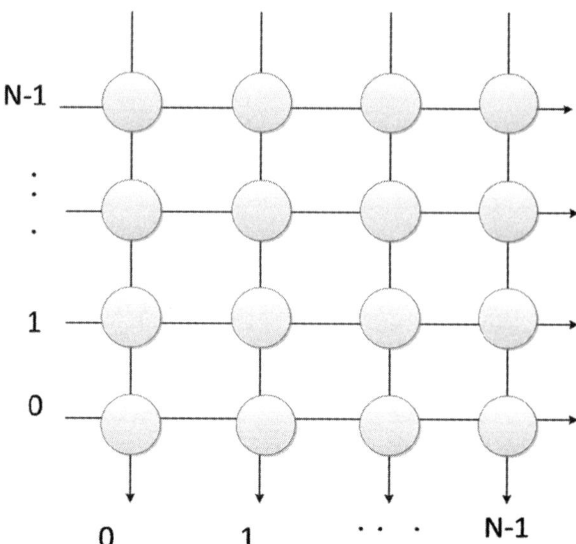

Fig. 6 Optical crossbar switch

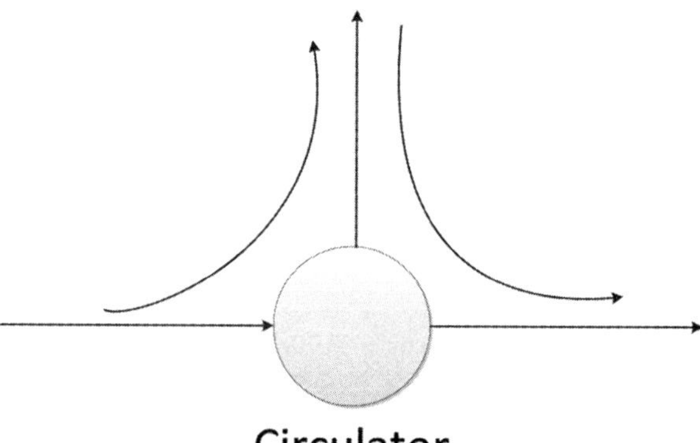

Fig. 7 Optical circulator

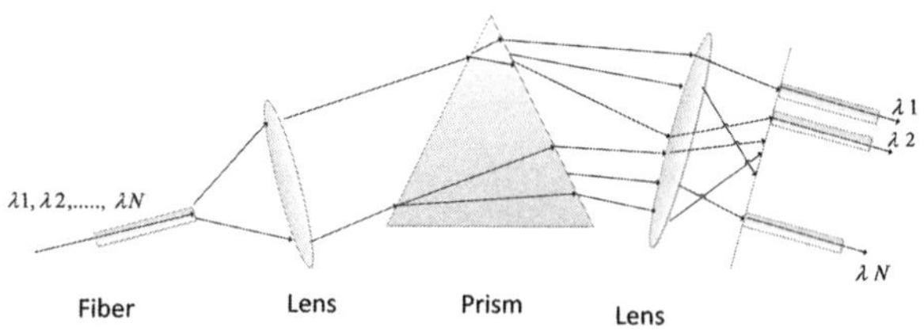

Fig. 8 Optical de-mulitplexer

A traditional optical crossbar is shown in Fig. 6. Optical crossbars can route wavelength from any input port to any output port. There is a wide range of possibilities to achieve this functionality. MEMS switch also presents the same functionality. There are also different ways to arrange these switching elements.

3.6 Circulators

Optical circulator is a multi port passive component. It transmits light waves from one port to the next sequential port with maximum intensity as shown in Fig. 7. It also blocks light from one port to previous port. Circulators are based on the Faraday's affect.

3.7 Optical Multiplexer and De-multiplexer

The functionality of optical de multiplexer is to separate the frequency components of light beam. Passive de-multiplexers are made up of prism, grating and spectral filters. The working of optical de-multiplexer using these passive components is shown in Fig. 8. When collimated beam of light falls on the surface of prism, each frequency component is refracted differently. The lens on the other side focuses each frequency at a different location where it is picked up by the receiver fiber.

Same optical components can be used to achieve the functionality of multiplexer by working in reverse direction. A multiplexer will combine several wavelengths of light in to a single beam.

4 Optical Interconnects in Data Center Networks and their Performance

Optical interconnects in DCNs can be categorized in to following main classes:

- Reconfigurable architectures
- Low latency architectures
- Low power consumption architectures
- Scaling link bandwidth architectures
- High radix switch architectures

These architectures are discussed in detail in the following sections.

4.1 Reconfigurable Architectures

Several reconfigurable architectures proposed in the literature and their performance is discussed in this section. The focus of this section is on reconfigurability of the data center network architectures.

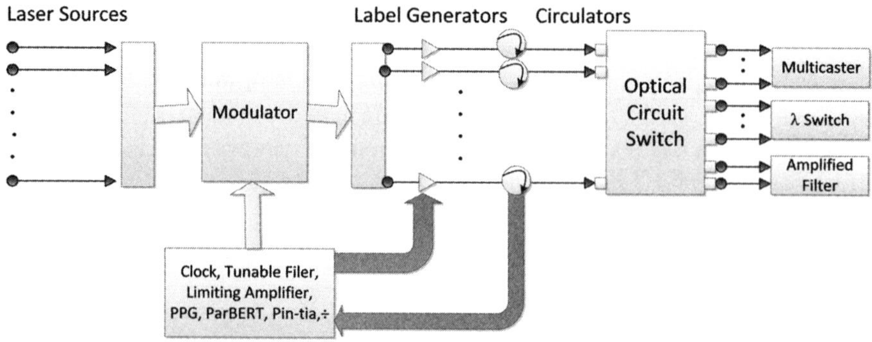

Fig. 9 Re-configurable data center architecture proposed by Wang et al

4.1.1 An Enhanced Optically Connected Network Architecture

Wang et al. [1] propose a reconfigurable architecture that supports a wider class of bandwidth intensive traffic patterns. This is shown in Fig. 9. The photonic resources are allocated on demand to optimize the communication between various applications of the data center. Data intensive applications become bottle necked when there is inter rack communication. Each photonic subsystem is treated as a physical resource that can be allocated on demand. The architectures that are dependent on MEMS switch are reliant on the stability of traffic. Traffic heterogeneity is an unavoidable property of data centers. The proposed architecture is adaptable to traffic. A central controller manages the resources. The controller either accepts explicit requests for resources and allocates based on demand estimation. Photonic capabilities are efficiently utilized. This architecture is modular which suits the incremental nature of data center expansion.

Performance Average output power of the optical signals is quite satisfactory for detection with open eye diagrams for all the output signals. Optical multicasting and optical local area network were successful.

4.1.2 OSA, a Novel Optical Switching Architecture for DCNs

Presented by Chen et. al. [6], this architecture dynamically changes its topology and link capacities to adapt to dynamic traffic patterns. This architecture is shown in Fig. 10. It introduces circulators to efficiently utilize the expensive optical ports, which doubles the usage of MEMS ports. This scheme gives a non blocking network which outperforms the hybrid structures. Its achieves 60 % non blocking all to all communication. The flexibility is achieved by using optical technologies like wavelength division multiplexing (WDM), wavelength selective switching (WSS), optical switching matrix (OSM), optical circulators and optical transceiver. OSM usually use MEMS switch. Flexibility of topology is achieved by using the re-configurability

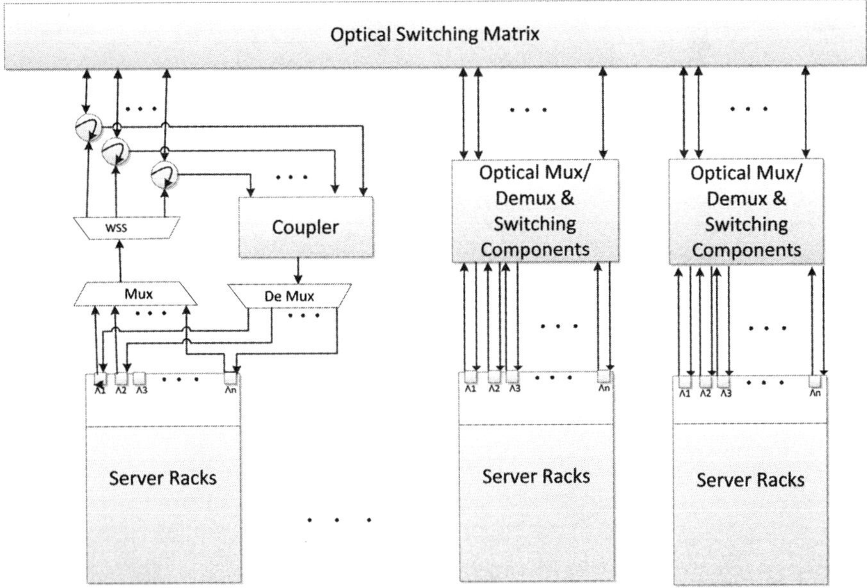

Fig. 10 OSA, a novel optical switching architecture for DCNs

of MEMS. WSS and MEMS configurations are decided by a central manager, which estimates traffic demands and calculates appropriate configurations of MEMS and WSS. The steps of this algorithm are:

a. Estimate traffic demand
b. Compute the topology (MEMS configuration)
c. Compute routes (ToR routing configuration)
d. Compute wavelength assignment (WSS configuration)

Performance Average bisection bandwidth for over 100 traffic instances was tested. This architecture delivered high bisection bandwidth (60–100 %) for synthetic and real traffic patterns as it adaptively adjusts topology and link capacity for various traffic patterns.

4.1.3 Wavelength-reconfigurable optical packet and circuit switched platform for DCNs

A wavelength reconfigurable platform for optical packet and circuit switched data centers is proposed by Zhang et al. [12] as shown in Fig. 11. A 2 × 2 switch is made up of Wavelength Selective Switches (WSSs) for wavelength selection and an optical circuit and packet switched platform. Packet and circuit switched ports are managed by FPGA.

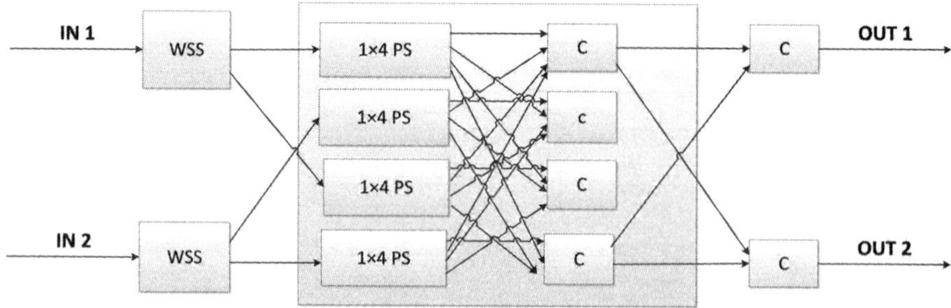

Fig. 11 Wavelength-reconfigurable optical packet and circuit switched platform for DCNs

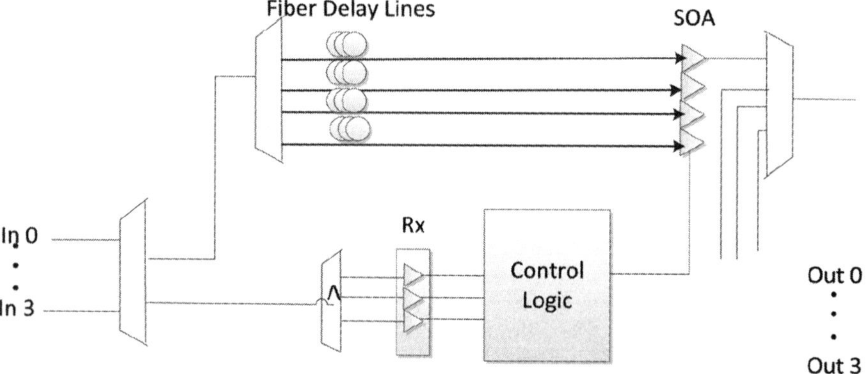

Fig. 12 Next-generation optically-interconnected high-performance data center architecture

Performance First tested thing is the correct routing of payloads through the switch fabric. The eye diagrams of channels from input and output are recorded which show a satisfactory eye opening. An error free transmission is achieved with BER less than 10^{-12}.

4.1.4 Next-Generation Optically-Interconnected High-Performance Data Centers

A next-generation reconfigurable optically interconnected high-performance architecture is presented by Zhang et al. [17] and shown in Fig. 12. The detailed implementation of Optical Network Interface Card (O-NIC) is also presented. This photonic platform consists of two sub-systems, first is a reconfigurable network that supports a variety of switch functionalities, second is an optical network interface that bridges the constraints of photonic network with the existing network protocols.

Performance The performance of switch is measured by recording BERs of for wavelengths evenly spaced across the spectrum of twenty five payload channels.

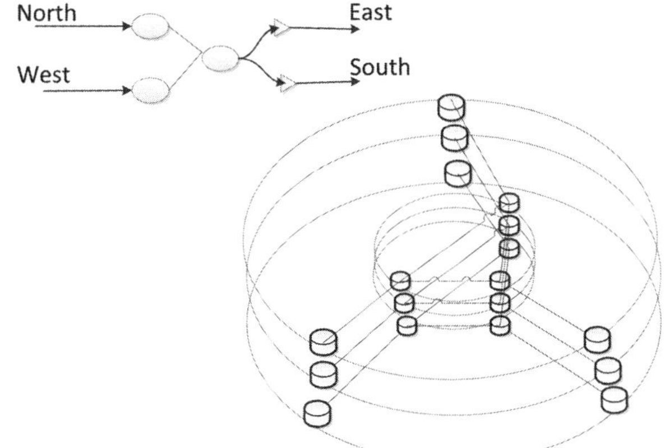

Fig. 13 The data vortex optical packet switched interconnection network

An error free propagation is observed and power penalties are ranging between 0.7 dB and 1.5 dB. Multi wavelength packets suffer from significant distortion due to carrier density variation, saturation affects and inter-channel non linear affects in SOA. Higher power penalties at 1530 nm and 1550 nm are due to high noise figure of SOA and EDFA used for switching and amplification.

4.1.5 The Data Vortex Optical Packet Switched Interconnection Network

The proposed architecture uses both, the packet switched and circuit switched traffic through its configurability. The configurability is based on SOAs. Input data is organized into multiple channels. The whole topology is made up of 2×2 switch elements that are organized as a fully directed and connected graph as shown in Fig. 13. The routing nodes are fully distributed and require no arbitration. The proposed model is scalable. Its main drawbacks include its complexity and its non deterministic latency [28].

Performance This architecture allows non blocking operations. No centralized arbitration is needed for routing the packets and packets are routed in the network in a distributed manner. It has a modular structure and thus efficiently scales to large number of nodes.

4.1.6 Proteus: A Topology Malleable Data Center Network

This architecture is based on Wavelength Selective Switches (WSS) and a MEMS based optical switching matrix as shown in Fig. 14. The operation is dependent on the traffic flows. For large flows MEMS switch is used while for short flows multi-hop communication is used. ToR switch flows are multiplexed and sent to WSS. A point to point connection is established between ToR switches. The MEMS configuration

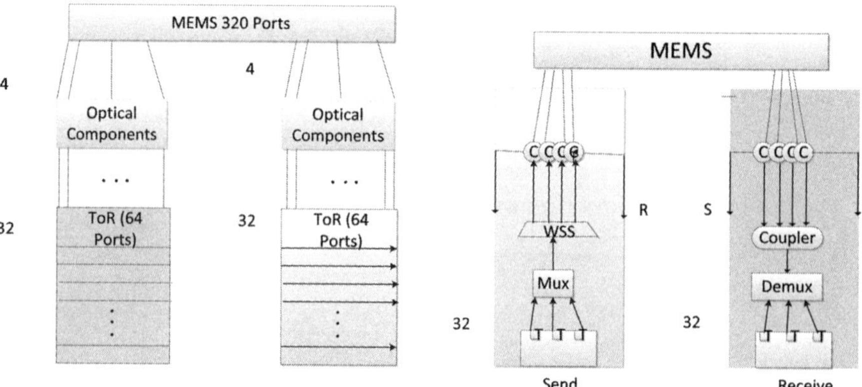

Fig. 14 Proteus architecture and send/receive structure for ToR

decides which ToR switches to connect. This scheme ensures that all ToR switches are connected when performing reconfiguration [21].

Performance This architecture works best when there are high volumes of ToR-ToR connections. At each ToR the packet undergoes OEO conversion. It also achieves more than 50 % power saving.

4.1.7 A Hybrid Optical Packet and Wavelength Selective Switch for High-Performance DCNs

A hybrid optical packet and wavelength selective switching platform is proposed by Xu et al. [3]. The proposed architecture is based on two components—silicon micro rings and semiconductor optical amplifiers (SOA). Using the wavelength selective nature of micro rings and broadband behavior of SOA switch, it is possible to achieve fast switching transitions, low driving voltages and high extinction ratio. Optical interconnects are characterized by high bandwidth-distance product, bit rate transparency and low power consumption.

Reconfigurable optical packet and circuit switch is made up of micro ring resonators, SOAs, PIN detectors, electronic control logic and other passive elements as shown in Fig. 15. System can utilize such packet switching for small random bursty traffic. Wavelength channels can be dynamically added if additional throughput is required. The number of wavelength channels for packet switching can be dynamically adjusted. Wavelength Selective Switch (WSS) reduces the number of ports, thus minimizing the SOA power consumption. Both switching elements optical packet and wavelength selective are controlled by electronic control logic.

Performance Amplitude shift keyed ASK signal is tesed for both packet switching mode and circuit switching mode. There is a 0.2 dB power penalty observed in both packet switching and circuit switching for ASK signal. Error free propagation is

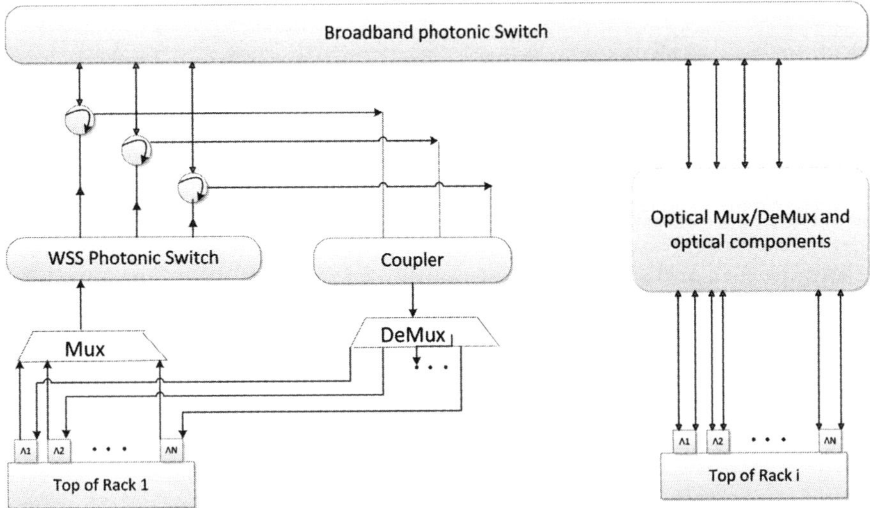

Fig. 15 Optical packet/circuit and wavelength/space hybrid switching platform

observed with BER less than 10^{-12} for all cases verifying format transparency and error free propagation.

4.2 Power Saving Architectures

In this section, we look at those data center network architectures whose main goal is to reduce the power consumption.

4.2.1 VCSEL Based Energy Efficient and Bandwidth Reconfigurable Architecture

The main advantage of proposed architecture is that its power consumption can be adjusted according to the traffic load. Transmitters are based on the Vertical-cavity surface-emitting laser (VCSEL) as shown in Fig. 16. The current of VCSEL is adjusted according to the traffic load. The bit rate can be scaled down when traffic is low giving a significant power saving. The proposed architecture is also reconfigurable. A controller controls the crossbar switch and a specific VCSEL laser is allocated to a specific node [25].

Performance The power consumption of this architecture is adjusted based on the traffic load. The latency of packets ranges between 1 and 2 ms.

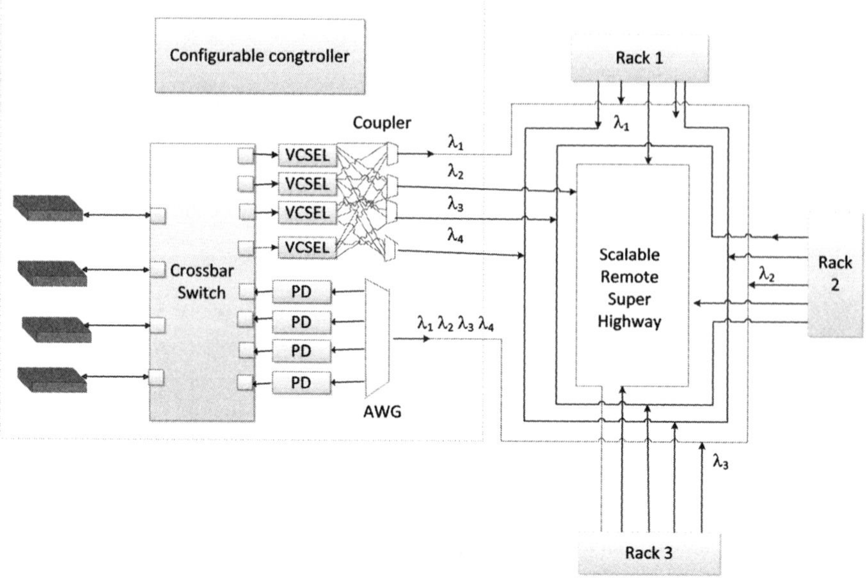

Fig. 16 VCSEL based energy efficient and bandwidth reconfigurable architecture

4.2.2 A Wavelength Striped, Packet Switched, Optical Interconnection Network

This architecture proposed by Shacham et al. [27] reduces the power consumption and the component cost. It is based on bi-directional SOAs. It uses 2×2 SOA based switch which can be used to build a big switch very easily using the tree topology as shown in Fig. 17. A connection between any of N ports can be established in nanoseconds.

Performance The achieved BERs are less than $^{10^{-12}}$. This scheme scales well to large number of nodes with reduced number of modules and thus reduced power. The total number of nodes is constrained by the total latency required and congestion management.

4.2.3 SPRINT: Scalable Photonic Switching Fabric for HIGH PERFORMANCE COMPUTING

The proposed architecture [2] can scale a large number of cores with photonic switching implemented using silicon micro ring resonators (MRRs). MRRs are low energy high bandwidth devices that can be arranged to function like high radix routers. As the platform scale increases the interconnection network becomes the bottleneck thus reducing the overall throughput. The drawback of using optics is that the cost of optics is very large compared to electrical counterparts. Arrayed waveguide grating

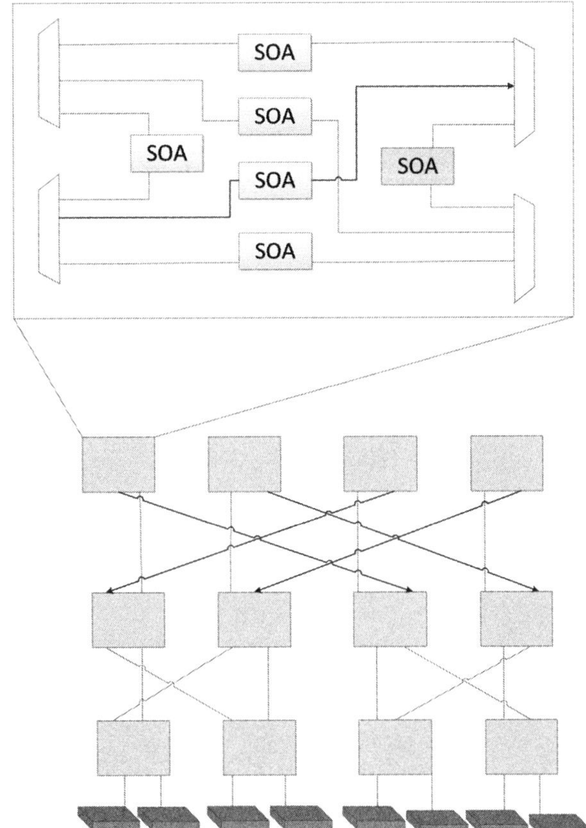

Fig. 17 A wavelength striped, packet switched, optical interconnection network based on SOAs

(AWG) is designed using MRRs. A 4 × 4 non blocking switch is designed that allows 4 links to interact as if they were 16 physical links as shown in Fig. 18. There is no need for electrical packet switching within this optical switch. Using this 4 × 4 switch the design can be extended to 256 and 1024 cores. Two designs—single MRR and dual MRR—are also evaluated which have area and power implications. The results indicate that AWG optical crossbar using low power MRRs can be used to scale the network design for data centers and high performance computing (HPC). Four cores are combined to form a group, core groups are combined to from clusters and clusters are combined to form the system domain.

The single ring AWG consists of three to four sets of MRRs in vertical and horizontal directions. Estimated area for each set is 60 μm × 90 μm. A 64 wavelength AWG consists of 16 MRRs. At each set the horizontal and vertical lengths will increase 16 fold. This makes 64 wavelength AWG area to be 1440 μm × 960 μm.

A double ring AWG is made up of eight sets. Each set has 32 MRRs thus a total of 256 MRRs. Each double MRR has a dimension of 15 μm × 20 μm. This also

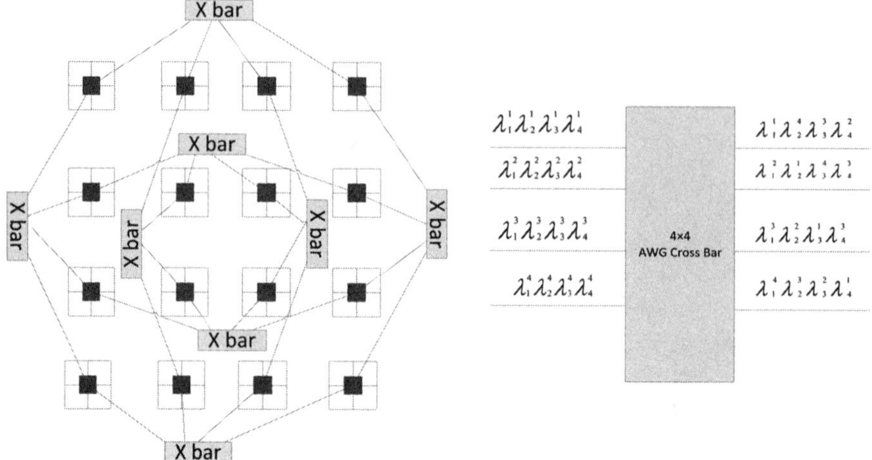

Fig. 18 SPRINT: Scalable Photonic Switching Fabric for High Performance Computing

includes 5 μm spacing between each double MRR. This makes the AWG height to be 80 μm and length to be 720 μm.

Single ring AWG has less optical loss than double ring AWG. Major contributor to single ring AWG is the bending loss as multiple bends are required to accommodate the 16 wavelength switching. So single ring AWG should be used when optical loss needs to be minimized and double ring AWG to be used when area occupied needs to be minimized.

Performance Results of this architecture include power consumption and throughput. Power consumption results show that this is a good alternative to the electrical networks. Throughput results also show a dramatic improvement over comparable networks [2].

4.3 *Low Latency Architectures*

In this section, we focus on architectures that focus on reducing latency.

4.3.1 DOS: A Scalable Optical Switch for Data Centers

In this architecture, the main advantage is that the latency is independent of the number of ports. This architecture provides the contention resolution in wavelength domain. It uses Arrayed Waveguide Grating Router (AWGR) for contention resolution. The optical interconnect consists of AWGR, tunable wavelength converter (TWC) and a shared buffer as shown in Fig. 19. The optical network is configured

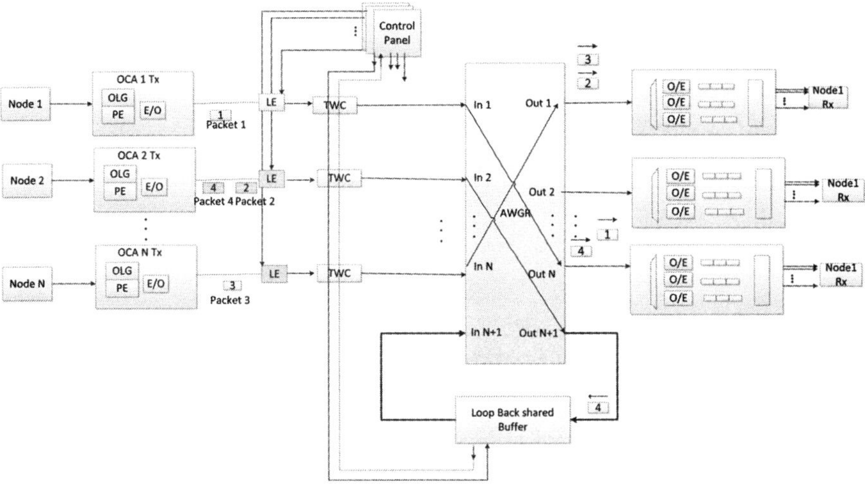

Fig. 19 DOS: A scalable optical switch for data centers

by a control plane which controls the TWC and label extractors. Optical label is interpreted by the control plane to determine the destination address and accordingly configure the TWC. By avoiding store and forward paradigm and optical parallelism, optical switch greatly reduces the average switching latency. The requests which are not possibly resolvable by AWGR are buffered and looped back through TWCs [4, 20].

Performance Experimental results show an error free propagation. A low latency of 118.2 ns in this contention free architecture is observed. System power consumption is below 200 pJ per bit by using discrete components. Average switching latency can be kept below 150 ns in the case of full contention.

4.3.2 Scalable Optical Packet Switch Architecture for Low Latency and High Load

A non-blocking switch architecture is proposed by Calabretta et al. [9] with a contention resolution subsystem. It supports highly distributed control and reduces the switching time to few nanoseconds regardless of the number of input/output nodes. The switch operates completely in optical domain. The focus is on synchronous operation i.e., system is time slotted and operates in discrete time. On each time step on which the input state of switch changes a new connection map is constructed. Controller can use several algorithms. Contention resolution block (CRB) is used for Contention resolution. CRB uses wavelength conversion done by Wavelength Selectors (WSs) and fixed wavelength converters (FWCs) as shown in Fig. 20.

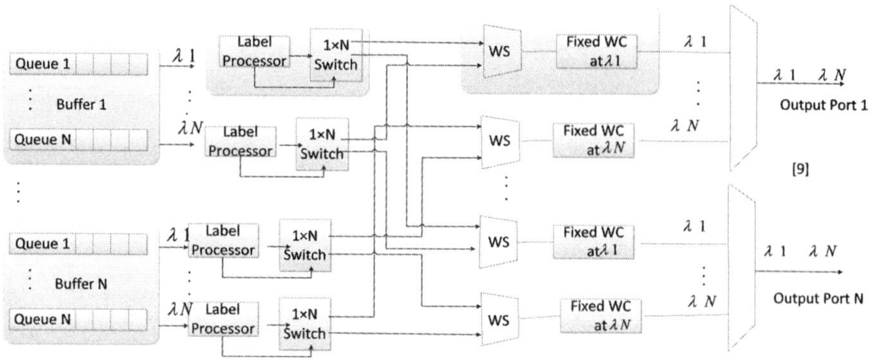

Fig. 20 Scalable optical packet switch architecture for low latency and high load

Performance Latency, packet loss and throughput of the architecture is investigated. For a 50 m link length, the round trip time is 546.67 ns. End to end latency increases linearly with up to 70 % load, and then it quickly approaches 1 microsecond. Packet loss also increases with the load increment. Throughput decreases with the load increment. This behavior occurs as contention probability increases with the port count.

4.3.3 AWGR Based Data Center Switches Using RSOA-based Optical Mutual Exclusion

In this architecture reflective semiconductor optical amplifier (RSOA) and arrayed waveguide grating router (AWGR) are used as mutual exclusion elements as shown in Fig. 21. Mutual exclusion refers to independent and concurrent nodes sharing the resources without incorrect operation [11].

Performance Throughput and latency results are discussed for this architecture. For throughput this architecture performs better than other comparable architectures like Distributed Loop Back buffer (DLB). Packet latency performance of this architecture is in between DLB and Flattened Butterfly (FBF) architecture. The performance of this switch is not directly impacted by number of nodes of the switch, but it does get impacted by the number of contending nodes for a particular output.

4.3.4 A Petabit Photonic Packet Switch (P^3S)

A low latency switch is proposed by Chao et al. [22] that reduces the latency. It uses buffers for congestion management. This architecture adopts a three stage network. Each stage consists of AWGRs for the routing of packets. TWCs are used to guide the laser through each stage of the network as shown in Fig. 22.

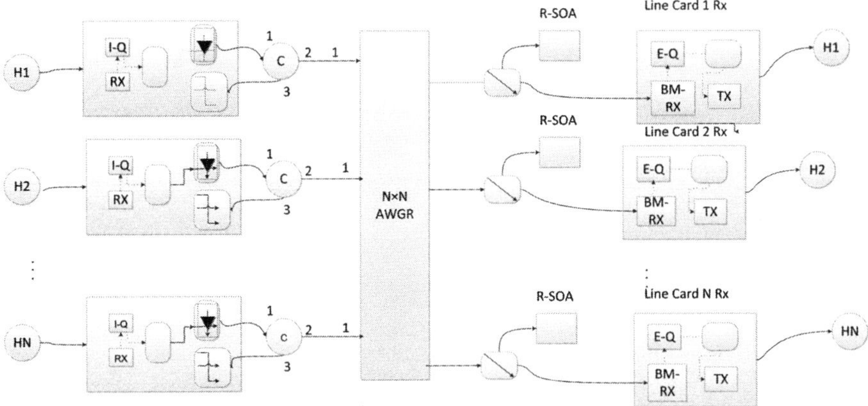

Fig. 21 AWGR based data center switches

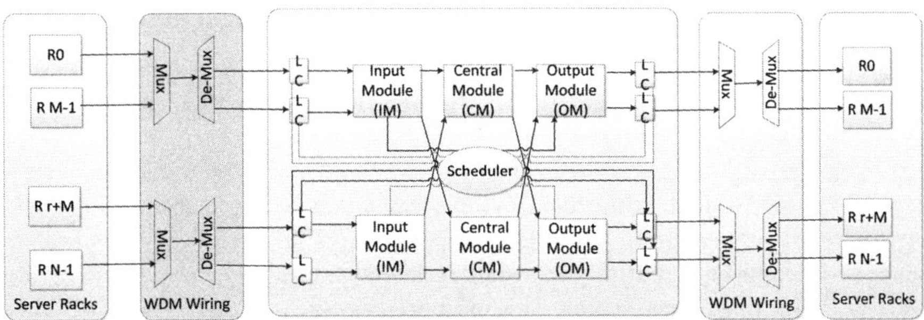

Fig. 22 A petabit photonic packet switch (P^3S)

Performance Average latency, queue length and throughput measurements about the architecture are made. The results show excellent scalability of the design.

4.3.5 Optical Interconnection Networks: The OSMOSIS Project

The proposed architecture is a low latency system. It is based on broadcast and select architecture using wavelength and space division multiplexing. It is made up of two stages as shown in Fig. 23. The first stage multiplexes all wavelengths on a common WDM line which is broadcast to all modules of the second stage. The second stage used semiconductor optical amplifiers (SOAs) that act as fiber selector gates to forward the wavelength to output [23].

Performance Improvements in throughput and latency are achieved by using two receivers per switch output port. Delay is also measured with increasing throughput. Delay increases with the increased throughput.

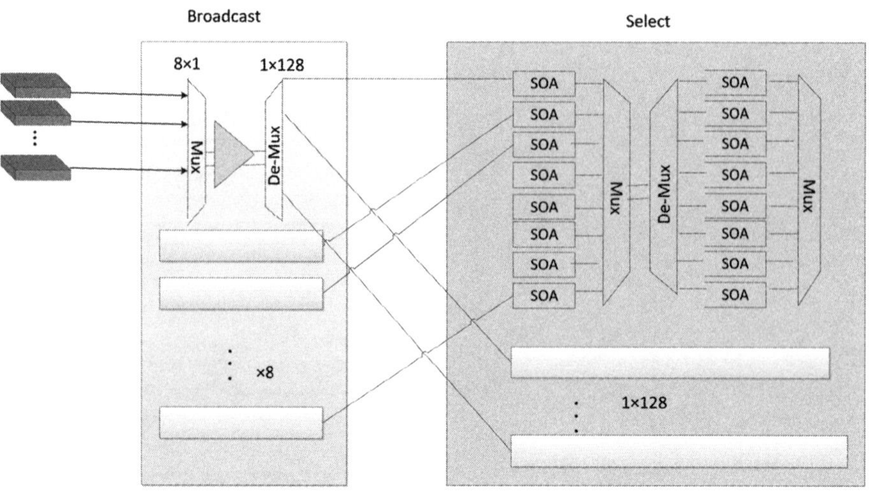

Fig. 23 Architecture of the Optical Interconnection Networks: The OSMOSIS project

4.3.6 A Scalable Optical Multi-Plane Interconnection Architecture

This scheme proposed by Liboiron-Ladouceur et al. [24]achieves low latency for networks that have high utilization. However main drawback of scheme is that it is expensive and increases the power consumption due to the use of SOAs. It utilizes space wavelength switching. Space switch architecture needs one fixed laser per port and a non-blocking optical switch. The non-blocking optical switch is made of SOAs that establish the connection for each time slot. SOAs are used for the connection establishment. Switching of the packets is done using a$1 \times M$ space switch that is made up of an array of SOAs in the form of a tree structure as shown in Fig. 24.

Performance This scheme is able to achieve low latency even under high network utilization. However, the main drawback of this architecture is that it uses SOA arrays that are expensive and increase overall power consumption.

4.3.7 Low Latency and Large Port Count OPS for Data Center Network Interconnects

A modular Optical packet switch architecture is proposed in [14] with a highly distributed control for inter cluster communication. This is shown in Fig. 25. The configuration time and latency of the switch are independent of port count. Scaling the port count linearly increases the number of components and energy consumption. The modular architecture allows the overall performance to be evaluated by testing a single optical module [14].

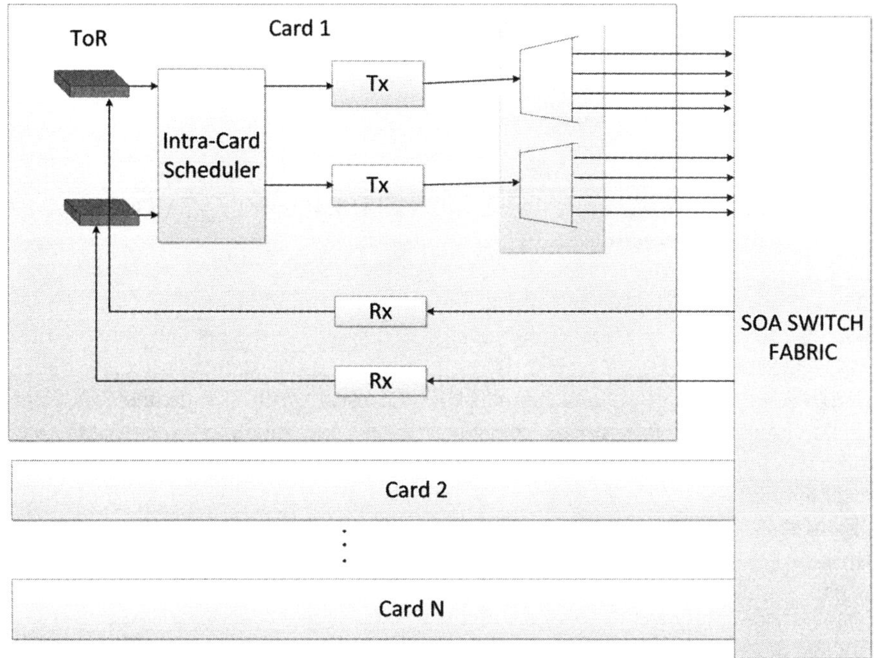

Fig. 24 A scalable optical multi-plane interconnection architecture

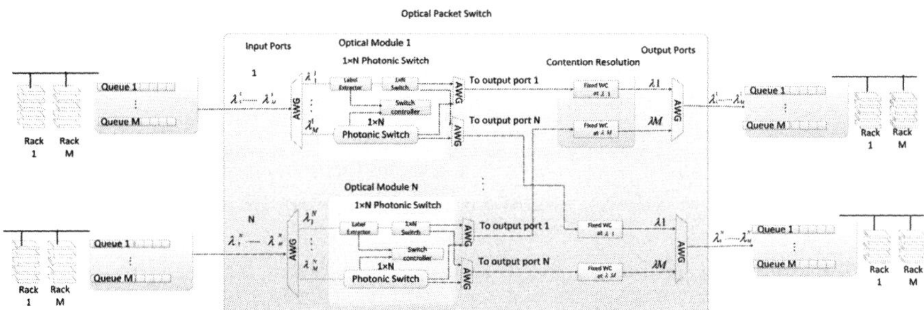

Fig. 25 Low latency and large port count OPS

Performance Experimental results are validated for 8 × 8 optical packet switched module. Forwarding operation of 8 input WDM channels at 40 Gbps to 8 output ports is performed. Error free propagation with 1.6 dB average power penalty is observed. Energy consumption is 76.5 pJ per bit.

4.4 *Link Bandwidth Scaling Architectures*

The bandwidth of the links can be scaled by exploiting orthogonal frequency division multiplexing and wavelength division multiplexing.

4.4.1 Data Center Network Based on Flexible Bandwidth MIMO OFDM Optical Interconnects

This architecture is based on optical multiple input multiple output (MIMO) orthogonal frequency division multiplexing (OFDM). Passive optical switch called cyclic arrayed waveguide grating is used. Parallel signal detection technology is used to detect multiple optical channels simultaneously with a single photo detector. This architecture reduces power consumption and operation cost. Intra data center communication is within the racks of a data center as opposed to inter data center communication which involves communication over long distances. In Optical OFDM the OFDM signal is generated electrically and then modulated to an optical carrier. It achieves high spectral efficiency by parallel transmission of spectrally overlapped low rate frequency domain components that are mathematically orthogonal over one symbol period.

Parallel signal detection is a key technology where a common photo detector simultaneously detects multiple OFDM signals from different sources. There should be no contention among OFDM sub carriers and wavelengths.

The basic optical component of the proposed architecture [7] is cyclic arrayed waveguide grating (CAWG) as shown in Fig. 26. It routes N different input ports to N different output ports in a cyclic fashion thus avoiding wavelength contention. The DCN interconnect that communicates with ToR switches is simply N $\times$ N passive CAWG. This reduces the power consumption at the interconnect level. This scheme also ensures that switched signals take exactly one hop, therefore the latency is also very low. Furthermore MIMO OFDM DCN scheme also uses WDM-OFDM ToR switches.

Components in this architecture can be divided into three categories. First are opto-electronic components including optical transmitters, receivers and transceivers. Second are electronic components which are used in ToR switches and aggregate switches. Third are the passive optical components including CAWG and optical couplers.

Performance A comparison is made between MIMO OFDM DCN and conventional electrical DCN. Power consumption increases linearly with the network scale and MIMO OFDM DCN scheme offers significant power saving. There is 50.2 % power reduction than its electrical counterparts.

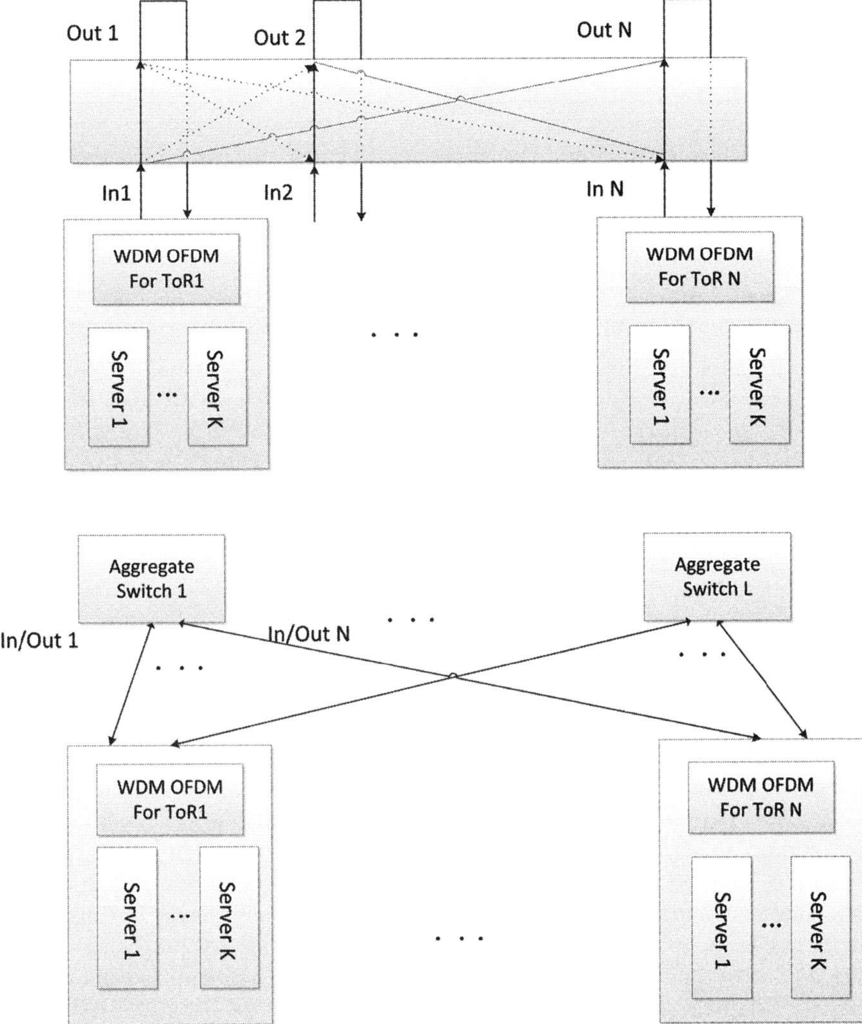

Fig. 26 Flexible bandwidth MIMO OFDM optical interconnects

4.4.2 Photonic Terabit Routers Employing WDM

The proposed architecture in the IRIS project [26] uses wavelength division multiplexing. All optical conversion based AWGRs are used. It is a three stage architecture which is non-blocking. First stage consists of wavelength space switches, second stage consists of time switches that have optical time buffers and the third stage consists of space switch. The internal architectures of first and third space switches are different as shown in Fig. 27. Control plan complexity is reduced and the optical random access memory is also eliminated.

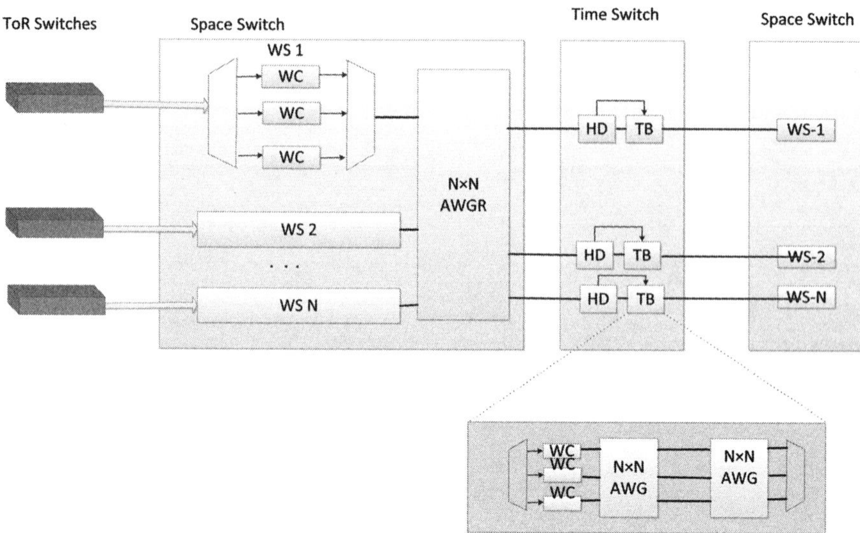

Fig. 27 A three-stage photonic terabit router employing WDM

4.5 High Radix Switch Design

IT equipment in the data center network typically consumes about 33 % of the total data center power. One solution to reduce the power consumption is to reduce the number of switches. Increasing the radix reduces the number of switches and thus the number of hops a packet must travel from source to destination. A combination of fewer hops and number of switches reduces the communication latency, component cost and power. For example, energy required by data center to send signal across the link is only a few Pico Joules per bit, however if the same signal has to cross a switch, the overhead shoots to 120 Pico Joules per bit. Binkert et al. [18] propose a scheme that uses silicon ring resonators. The ring can be used as modulator, wavelength specific switch or a drop filter.

Performance Over 100,000 port networks can be constructed using photonic I/O while consuming only one-third of the power of equivalent all-electronic networks. Employing photonic components within the switch can further reduce consumed power by about 50 % [18].

5 Data center traffic characteristics

The traffic within the DCN is highly unpredictable. Most of the resources of DCN remain underutilized. 1500 server Microsoft DCN shows that there are only few ToRs that are hot and most traffic passes through only a few ToRs. Thus for a few thousand servers DCN, the uniform capacity of the network is an overkill. Reconfigurable networks provide a solution to this that is cost affective [7].

Data Center Networks may also suffer from oversubscription. In oversubscribed networks intelligent work placement algorithms are needed. There are various workload placement algorithms, some are based on flexible bandwidth allocation, some work on periodic backup process and others work on dynamic deletion and creation of virtual machine instances. Flexible bandwidth allocation is more effective for data intensive DCNs where more bandwidth for data fetch is adjusted dynamically based on demand. In back creation procedure multiple copies of data are placed on different servers so that the traffic can be distributed uniformly across the network. Virtual machine is an environment that includes application layer, platform and operating system [6].

A data center consists of multiple racks and each rack accommodates multiple servers. The servers within the rack are connected through ToR. When a user request is generated it is forwarded to the front end of the data center. In front end switches load balance devices are used to route the request to appropriate rack and server.

There has been an exponential increase in the internet traffic due to emerging consumer applications like video streaming, social networking and cloud computing etc. Each application type poses different traffic characteristics constraints on the Data Center Network. User applications are data intensive and demand a high degree of interaction between the servers within the data centers. This interaction challenge demands for the high bandwidth and low latency interconnects within the data center. Even if the demand to fetch data from servers increase, the interconnects within a data center are required to show a reasonable latency. As the processing cores on a single chip are increasing the demand for communication requirements between racks is also increasing [19].

The traffic of data centers can be studied on the basis of the following constraints

- Applications
- Traffic flow locality
- Traffic flow size and duration.
- Concurrent traffic flows
- Link utilization

Applications The traffic could be related to applications like HTTP, HTTPS or LDAP. The traffic of a university campus data center for example, could be HTTP; however the traffic of cloud computing data center could be HTTPS, HTTP or LDAP.

Traffic Flow Locality It can be divided into two categories i.e. intra rack traffic and inter rack traffic. If the traffic is between the servers within the rack it is intra rack traffic, however if it is between servers of different racks it is inter rack traffic. Intra rack traffic is between 10 and 40 % and the inter rack traffic fluctuates between 10 and 80 %.The traffic flow locality has an impact on the design of network.

Traffic Flow Size and Duration Active connection between two or more server is the traffic flow. If the duration of traffic flow is large then reconfigurable interconnects can be used, as the reconfiguration over head can be compared to the flow duration, optical circuit switching is suitable if traffic flow is long and packet switching in feasible if the traffic flow is small.

Concurrent Traffic Flows If the optical interconnect supports a separate connection for each concurrent flow, then it presents a significant advantage

Link Utilization Link utilization inside rack is low and between racks it is high. So network can be designed specific to the link utilization profile.

6 Energy Requirements for Data Center Networks

Energy efficiency is one of the challenging issues of data center design. The energy consumption of data center is mostly due to network topology and the infrastructure.

Data centers based on electronic switches consumes a lot of electrical power. Compared to electronic switches the optical interconnects consume much less energy. All optical networks provide 75 % energy saving. In order to face the power consumption requirements in data centers, new schemes must be developed to conserve energy. Apart from using energy efficient architectures, such optical components can also be used which consume less energy like semiconductor optical amplifiers [19]. Other similar components are silicon micro ring resonators. These are low power components that can be arranged similar to a high radix router. Micro ring resonators require 26 μW/ring in heating to keep them working [2]. By combining the wavelength selective behavior of micro ring and broadband behavior of semiconductor optical amplifier it is possible to achieve very low driving voltages [3].

It is estimated by Gartner Group that energy consumed by data center accounts for 10 % of the operational expenses and it may increase to 50 % in a few years. Energy consumed by computing is not the only source of energy consumption. The heat generated also needs a cooling system that is also a source of energy consumption. If the data center temperature in not in safe range this may reduce the reliability of hardware.

The workload of data center changes on weekly and hourly basis. Average workload of data center consumes only 30 % of data center resources. Energy can be conserved by putting the 70 % of data center resources in sleep mode. To achieve this energy efficiency, central coordination and energy aware workload scheduling techniques are needed.

Two main sources of energy consumption in data center network are the server processing units and the network infrastructure. Network infrastructure includes communication links, switching and aggregation elements. Network infrastructure accounts for more than 30 % of total data center energy. In optical domain the communication links, switching and aggregation elements are all very much green compared to their electrical counterparts.

The energy consumption of communication infrastructure can be reduced by reducing the operational frequency, communication speed and the input voltage for switching elements and transceivers. Slowing down the communication infrastructure needs to be done very carefully by analyzing the demand of the user application. If the demand of application is not analyzed carefully this may result in serious bottlenecks in the network. Energy saving of up to 75 % can be achieved using this policy. This result is further improved if optical interconnects are used with this policy [29].

Traditional model of store and forward mechanism of packets increases the power consumption. This is due to fact that each node has to consume energy to store the packet in buffer, retrieving it, and forwarding it using the transmit power. Optical interconnects reduce or eliminate this store and forward mechanism thus providing a significant energy saving.

As the number of servers in the Data Center Network increases the number of line cards increases proportionally and so does the wiring length and complexity increases. Increases wiring also consumes more energy due to the resistance of wire which causes heat dissipation. On the other hand optical interconnects replace traditional wires with the optical fiber which consumes much less energy to sustain the signal across the medium [22].

Traditional electronic data center networks dissipate a large amount of heat. Apart from signal propagation, the cooling system of a data center is a significant contributor to the energy consumption. Cooling requirements of optical interconnects are much less than traditional networks. Underutilization of the network resources also wastes energy. In data starved networks the utilization is as low as 30 %. Thus 70 % unutilized resources of the data center network are a source of energy consumption [1]. Proposing efficient work placement algorithms can save energy. If patterns of data center network usage is known, the unutilized components can be put to sleep mode.

Energy consumption can also be reduced by reducing the number of network devices. As most of components of network devices are common, the only solution to reducing the number of network devices is through a high radix switch design. Using a high radix switch design such as a 100,000 port interconnection can save energy consumption per bit by a factor of 6 [18].

Multiple input multiple output MIMO orthogonal frequency division multiplexing OFDM is able to achieve energy consumption by reducing the number of fibers used. Parallel signals are detected simultaneously [7]. Similarly, wavelength division multiplexing WDM also reduces the energy consumption significantly [11]. Both schemes—MIMO OFDM and WDM—tend to reduce the amount of cabling in the data center network.

7 Routing in Data Centers

Routing of packets in optical interconnects reduces the latency significantly. Optical interconnects are distance immune. The latency is reduced due to less or no hop count between ToRs, the scalable bandwidth of optical links and eliminating the store and forward mechanism of packets. In traditional wired Data Center Networks there is a large hop count between ToRs due to various layers of access and aggregate switches. The bandwidth of electrical links is capped by the upper limit, however in optical case the bandwidth of the optical can be scaled through various techniques like wavelength division multiplexing (WDM), space wavelength multiplexing, and optical orthogonal frequency division multiplexing OFDM. Each of these techniques enhances the link capacity [22].

Optical OFDM enhances the capacity of multimode fiber (MMF). This technique provides larger data rates over large distances without addition of extra components. In MMF there are different optical modes that travel at different group velocities resulting in differential modal delay. This results in inter symbol interference if traditional single carrier is used. In OFDM there are orthogonal channels which suffer frequency selective attenuation. The use of coded OFDM can deal with the intermodal dispersion [16].

Key technology that enables OFDM is parallel signal detection. A common photo detector is able to detect signals from multiple sources. There should be no contention among OFDM wavelengths and sub carriers. As the transmission distance is short so there is no need for any guard band between sub carrier groups [7].

In WDM multiple wavelengths can be transmitted simultaneously on single mode fiber which makes optical circuit switching possible. Network is reconfigured dynamically without rerouting of cables manually [13].

WDM systems resolve the limitations if I/O and head of line blocking problem [18]. In head of line blocking problem when a packet get struck, it will also block the path for subsequent packets even though the destinations are free. This problem limits the throughput to 60 % [4].

Store and forward mechanism introduces transmission and propagation delay in traditional electrical platform. In all optical packet switching the optical packets are not stored and are delivered to the destination by employing contention resolution techniques in optical domain.

Most of the optical interconnects flatten the traditional hierarchical electronic network by replacing the whole network by a single optical switch. The idea of flattening the network is welcomed commercially by Juniper Networks.

Contention resolution is a big issue in Optical Data Center Networks. Traditionally, a small loop back buffer is used to temporarily store the contending packet and then retransmitting it. This loop back buffer is a bottleneck and a major hurdle to the scalability of the network in relation to data rate and port count. It is also a contributor to latency, design complexity and power consumption. By employing contention resolution techniques the inter stage loop back buffering can be eliminated in optical interconnects. AWRG has non-blocking switching characteristics [11].

Reconfigurable routing is possible in optical networks. Most prominent candidates for reconfigurable routing are the MEMS switches. Reconfigurable routing gives its best benefits for large traffic flows. Reconfigurable circuit switching is possible whose benefits surpass the reconfiguration overhead time. However circuit switching is not feasible for routing small traffic flows among ToRs and the reconfiguration time may exceed the benefits of circuit switching for a small flow [12]. Another drawback of MEMS based switches is their reconfiguration time [3].

The heterogeneity of traffic is unavoidable in data center networks. It is hence a good approach to use a hybrid optical network which has the benefits of both the optical packet switching and optical circuit switches. Longer flows adopt circuit switching whereas small flows adopt packet switching [1].

An exascale system has nearly 100,000 10 teraflops nodes. Such a scaling puts a large pressure on the interconnects. This problem is best addressed by high radix switch design. Optical interconnects also facilitate high radix switch design. High radix switch design reduces the hop count. A combination of reduced hop count and number of switches contributes to lower latency, energy consumption and reduced component cost.

The constraints of electrical signaling make the electronic switches an unsuitable candidate for high radix switch design. In electrical switches there is a tradeoff between port bandwidth and switch radix. A chip has limited number of pins, so it is impossible to increase the port count indefinitely without reducing the bandwidth [18].

References

1. Howard Wang et al, *Optically Interconnected Data Center Architecture for Bandwidth Intensive Energy Efficient Networking*, IEEE 14th International Conference on Transparent Optical Networks (ICTON), 2012.pp. 1–4. (doi10.1109/ICTON.2012.6253873)
2. Brian Neel et al, *SPRINT: Scalable Photonic Switching Fabric for High Performance computing*, IEEE/OSA Journal of Optical Communications and Networking, vol:4, No: 9, pp. A38–A47. (doi: 10.1364/JOCN.4.000A38).
3. Lin Xu et al, A hybrid optical packet and wavelength selective switching platform for high performance data center networks, OPTICS EXPRESS 2011, vol. 19, No. 24, pp. 24258–24267. (doi: 10.1364/OE.19.024258)
4. Roberto Proietti et al, *40 Gb/s 8 × 8 Low Latancy Optical Switch for Data Centers*, Optical Fiber Communication Conference. (doi: 10.1364/OFC.2011.OMV4)
5. S Di Lucente et al, *Study of the performance of an optical packet switch architecture with highly distributed control in data center environment*, 16th International Conference on Optical Network Design and Modeling, 2012, pp. 1–6. (doi: 10.1109/ONDM.2012.6210266)
6. Kai Chen et al, OSA: *An optical switching architecture for Data Center Networks with Unprecedented Flexibility,* Proceedings of the 9th USENIX conference on Networked Systems Design and Implementation 2012, pp. 18–18.
7. Philip N Ji et al, "*Energy Efficient Data Center Network based on Flexible Bandwidth MIMO OFDM Optical Interconnect*", IEEE 4th International Conference on cloud Computing Technology and Science, 2012, pp. 699–704. (doi: 10.1109/CloudCom.2012.6427601)
8. Li Mei Peng et al, *Cube based Intra Data Center Networks with LOBS-HC, IEEE International Conference on Communications (ICC),* 2011, pp. 1–6. (doi:10.1109/icc.2011.5962754)

9. Nicola Calabretta et al, *Scalable Optical Packet Switch architecture for low latency and High Load Computer Communication Networks*, 13th International Conference on Transparent Optical Networks, 2011, pp. 1–4. (doi: 10.1109/ICTON.2011.5971139)
10. Henggang Cui et al, *Optically Cross-Braced Hypercube: a reconfigurable physical layer for interconnects and server centric Data Centers*, Optical Fiber Communication Conference and Exposition, 2012, pp. 1–2. (doi: 10.1364/OFC.2012.OW3J.1)
11. Roberto Proitti et al, *Scalable and Distributed Contention Resolution in AWGR based Data Center Switches Using RSOA-based Optical Mutual Exclusion*, IEEE Journal of Selected Topics in Quantum Electronics, 2011, vol. 19, No. 2, pp. 3600111,3600111, March-April 2013. (doi: 10.1109/JSTQE.2012.2209113)
12. Wenjia Zhang et al, *Experimental demonstration of wavelength reconfigurable optical packet and circuit switched platform for Data Center Networks*, IEEE Optical interconnects conference, 2012, pp. 123–124. (doi: 10.1109/OIC.2012.6224415)
13. Lei Xu et al, *Optically Interconnected Data Center Networks*, OFC/NFOEC Technical Digest 2012. (doi: 10.1364/OFC.2012.OW3J.3)
14. Nicola Calabretta et al, *Experimental assessment of Low Latency and Large Port Count OPS for Data Center Network interconnects*, 14th International Conference on Transparent Optical Networks, 2012, pp. 1–4. (doi: 10.1109/ICTON.2012.6254381)
15. Daniel Brunina et al, *Building Data Centers With Optically Connected Memory*, Journal of Optical Communications and Networking, 2011, vol. 3, No: 8, pp. A40–A48. (doi: 10.1364/JOCN.3.000A40)
16. Yannis Benlachtar et al, *Optical OFDM for Data Center*, 12th International Conference on Transparent Optical Networks, 2010, pp:1–4. (doi: 10.1109/ICTON.2010.5549137)
17. Wenjia Zhang et al, *Next-Generation Optically-Interconnected High-Performance Data Centers*, IEEE Journal of Lightwave Technology, 2012, vol. 30, No. 24, pp. 3836–3844. (doi:10.1109/JLT.2012.2212696)
18. Nathan Binkert et al, *Optical high radix switch design*, IEEE Computer Society 2012, vol. 32 No. 3, pp. 100–109. (doi: 10.1109/MM.2012.24)
19. Christoforos Kachris et al, *A survey on Optical Interconnects for Data Centers*, IEEE Communications Surveys and Tutorials 2011, vol. 14, No. 4, pp. 1021–1036. (doi:10.1109/SURV.2011.122111.00069)
20. X. Ye et al, *DOS: A scalable optical switch for Data Centers*, ACM/IEEE symposium on Architectures for Networking and Communication Systems, 2010, pp. 1–12.
21. Ankit Singla et al, Proteus*: A topology Malleable Data Center Network*, Hotnets-IX Proceedings of the 9th ACM SIGCOMM Workshop on Hot Topics in Networks, 2010, No: 08. (doi:10.1145/1868447.1868455)
22. H.J Chao et al, *A petabit photonic packet switch* (P^3S), INFOCOM 2003. Twenty-Second Annual Joint Conference of the IEEE Computer and Communications. IEEE Societies, vol. 1, pp. 775–785. (doi: 10.1109/INFCOM.2003.1208727)
23. R. Luijten et al, *Optical Interconnection Networks: The OSMOSIS project*, IEEE 17th Annual Meeting of Lasers and Optoelectronics society, 2004, vol. 2, pp. 563–564. (doi:10.1109/LEOS.2004.1363363)
24. O. Liboiron-Ladouceur et al, *Energy efficient design of a scalable optical multiplane interconnection architecture*, IEEE journal of selected topics in Quantum computing, 2010, vol. 17, No. 2, pp. 377–383. (doi:10.1109/JSTQE.2010.2049733)
25. A. K Kodi et al, *Energy efficient and bandwidth reconfigurable photonic networks for high performance computing systems*, IEEE journal of selected topics in Quantum Electronics, 2010, vol. 17, No. 2, pp. 384–395.(doi:10.1109/JSTQE.2010.2051419)
26. J. Gripp. Et et al, Photonic Terabit Routers: The IRIS project, Optical fiber communication conference, OSA 2010, pp. 1–3
27. A Shacham et al, *An experimental validation of a wavelength striped, packet switched, optical interconnection network*, Journal of Lightwave Technology, 2009, vol. 27, No. 7, pp. 841–850. (doi:10.1109/JLT.2008.928541)

28. O. Liboiron et al, *The data vortex optical packet switched network,* Journal of Lightwave Technology, 2008, vol. 26, No. 13, pp. 1777–1789. (doi:10.1109/JLT.2007.913739)
29. Dzmitry Kliazovich et al, *Green Cloud: A packet level simulator for energy aware cloud computing data centers*, Springer Journal of Super Computing, 2010, vol. 62, pp. 1263–1283, (doi: 10.1007/s11227-010-0504-1)
30. http://www.kamelian.com/techarticles/App_Note_No_0001.pdf, last accessed Oct 31, 2013
31. Xaveer J. M. Leijtens et al, Arrayed Waveguide Gratings, http://alexandria.tue.nl/openaccess/Metis203741.pdf, last accessed Oct 31, 2013
32. http://www.fiberoptics4sale.com/wordpress/what-is-wavelength-selective-switchwss/, last accessed Oct 31, 2013
33. Gangxiang Shen et al, *A novel rearrangeable non-blocking architecture for 2D MEMS optical space switches*, Optical Networks Magzine 2012, vol. 3, Issue, 7, pp. 70–79

TCP Congestion Control in Data Center Networks

Rohit P. Tahiliani, Mohit P. Tahiliani and K. Chandra Sekaran

1 Introduction

Internet over the past few years has transformed from an experimental system into a gigantic and decentralized source of information. Data centers form the backbone of the Internet and host diverse applications ranging from social networking to web search and web hosting to advertisements. Data Centers are mainly classified into two types [1]: the ones that aim to provide *on-line services* to users, e.g., Google, Facebook and Yahoo, and others that aim to provide *resources* to users e.g., Amazon Elastic Compute Cloud (EC2) and Microsoft Azure.

Transmission Control Protocol (TCP) is one of the most dominant transport protocols, widely used by a large variety of Internet applications and also constitutes majority of the traffic in both types of Data Centers [2]. It has been the workhorse of the Internet ever since its inception. The success of the Internet, in fact, can be partly attributed to the congestion control mechanisms implemented in TCP. Though the scale of the Internet and its usage increased exponentially in recent past, TCP has evolved to keep up with the changing network conditions and has proven to be scalable and robust.

Data Center environment, however, is largely different than that of the Internet e.g., the Round Trip Time (RTT) in Data Center Networks can be as less as 250 μs in the absence of queuing [3]. The reason is that Data Center Networks are well designed and layered to achieve high-bandwidth and low-latency. Moreover, the nature of traffic in Data Center Networks largely varies from that of the Internet traffic. Traffic in Data Center Networks is classified mainly into three types [2]: (i) Mice traffic - the

R. P. Tahiliani (✉)
Department of Computer Science and Engineering, NMAM Institute of Technology, Nitte, Karnataka, 574110 India

M. P. Tahiliani · K. C. Sekaran
Department of Computer Science and Engineering, NITK, Surathkal Karnataka, 575025 India
e-mail: tahiliani@nitk.ac.in

K. C. Sekaran
e-mail: kchnitk@gmail.com

S. U. Khan, A. Y. Zomaya (eds.), *Handbook on Data Centers*,
DOI 10.1007/978-1-4939-2092-1_15

Table 1 Data Center Traffic: applications and performance requirements

Traffic Type	Examples	Requirements
Mice traffic (< 100 KB)	Google Search, Facebook	Short response times
Cat traffic (100 KB-5 MB)	Picasa, YouTube, Facebook photos	Low latency
Elephant traffic (> 5 MB)	Software updates, Video On-demand	High throughput

queries form the mice traffic (e.g. Google search, Facebook updates, etc). Majority of the traffic in a data center network is query traffic and its data transmission volume is usually less. (ii) Cat traffic - the control state and co-ordination messages form the cat traffic (e.g. small and medium sized file downloads, etc) and (iii) Elephant traffic - the large updates form the elephant traffic (e.g. anti-virus updates, movie downloads, etc). The different traffic types in Data Center Networks, their applications and performance requirements are summarized in Table 1.

Thus, bursty query traffic, delay sensitive cat traffic and throughput sensitive elephant traffic co-exist in Data Center Networks. Therefore, the three basic requirements of the data center transport are high burst tolerance, low latency and high throughput [2]. The state-of-the-art TCP fails to satisfy these requirements together within the time boundaries because of impairments such as TCP Incast [3], TCP Outcast [4], Queue build-up [2], Buffer pressure [2] and Pseudo-Congestion Effect [5] which are discussed further in this chapter.

Recently, a few TCP variants have been proposed for Data Center Networks. The major goal of these TCP Variants is to overcome the above mentioned impairments and improve the performance of TCP in Data Center Networks. This chapter presents the background and the causes of each of the above mentioned impairments, followed by a comparative study of TCP variants that aim to overcome these impairments. Although a few other transport protocols have also been proposed for Data Center Networks, we restrict the scope of this chapter to TCP variants because TCP is the most widely deployed transport protocol in modern operating systems.

2 TCP Impairments in Data Center Networks

Although TCP constantly evolved over a period of three decades, the diversity in the characteristics of present and next generation networks and a variety of application requirements have posed several challenges to TCP congestion control mechanisms. As a result, the shortcomings in the fundamental design of TCP have become increasingly apparent. In this section, we mainly focus on the challenges faced by the state-of-the-art TCP in DCNs.

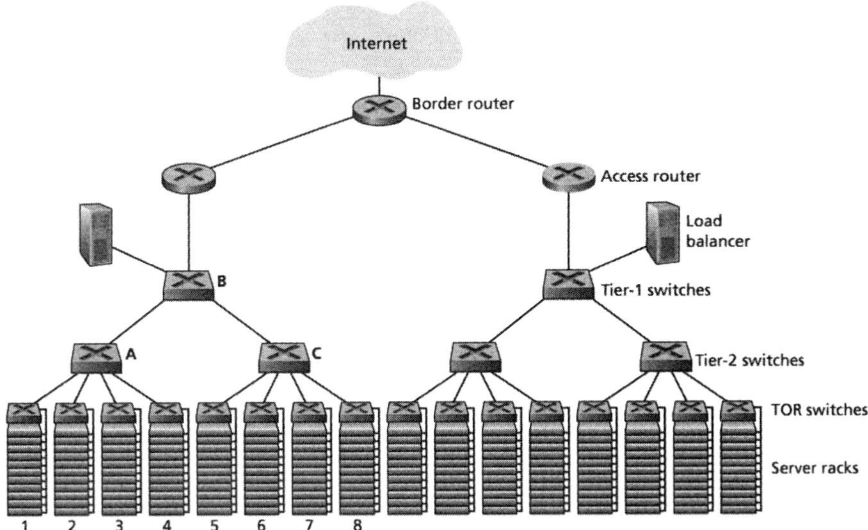

Fig. 1 Partition/aggregate application structure [8]

2.1 TCP Incast

TCP Incast has been defined as the pathological behavior of TCP that results in gross under-utilization of the link capacity in various many-to-one communication patterns [6], e.g. partition/aggregate application pattern as shown in Fig. 3. Such patterns are the foundation of numerous large scale applications like web search, MapReduce, social network content composition, advertisement selection, etc [2, 7]. As a result, TCP Incast problem widely exists in today's data center scenarios such as distributed storage systems, data-intensive scalable computing systems and partition/aggregate workflows [1].

In many-to-one communication patterns, an aggregator issues data requests to multiple worker nodes. The worker nodes upon receiving the request, concurrently transmit a large amount of data to the aggregator (see Fig. 2). The data from all the worker nodes traverse a bottleneck link in many-to-one fashion. The probability that all the worker nodes send the reply at the same time is high because of the tight time bounds. Therefore, the packets from these nodes happen to overflow the buffers of Top of the Rack (ToR) switches and thus, lead to packet losses. This phenomenon is known as *synchronized mice collide* [2]. Moreover, no worker node can transmit the next data block until all the worker nodes finish transmitting the current data block. Such a transmission is termed as *barrier synchronized transmission* [7].

Under such constraints, as the number of concurrent worker nodes increases, the perceived application level throughput at the aggregator decreases due to a large number of packet losses. The lost packets are retransmitted only after the Retransmit TimeOut (RTO), which is generally in the order of few *milliseconds*. As mentioned

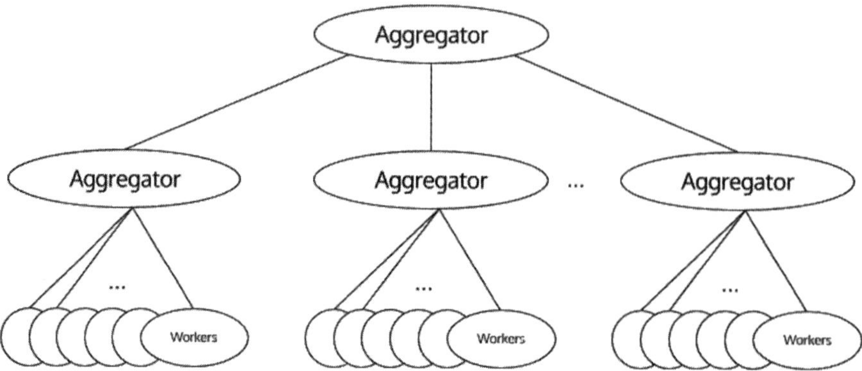

Fig. 2 Many-to-one communication pattern

earlier, mice traffic requires short response time and is highly delay sensitive. Frequent timeouts resulting out of Incast significantly degrade the performance of mice traffic as the lost packets are retransmitted after a few *milliseconds*.

It must be noted that *Fast Retransmit* mechanism may not be applicable to mice traffic applications since the data transmission volume of such traffic is quite less and hence, there are very few packets in the entire flow. As a result, the sender (or worker node) may not get sufficient duplicate acknowledgements (*dupacks*) to trigger a Fast Retransmit.

Mitigating TCP Incast: A lot of solutions, ranging from application layer solutions to transport layer solutions and link layer solutions have been proposed recently to overcome the TCP Incast problem. A few solutions suggest *revision of TCP*, others recommend to *replace TCP* while some seek solutions from layers other than the transport layer to solve this problem [1]. Ren et al. [9] provides a detailed analysis and summary of all such solutions.

2.2 *TCP Outcast*

When a large set of flows and a small set of flows arrive at two different input ports of a switch and compete for the same bottleneck output port, the small set of flows lose out on their throughput share significantly. This phenomenon has been termed as TCP Outcast [4] and mainly occurs in Data Center switches that employ drop-tail queues. Drop-tail queues lead to consecutive packet drops from one port and hence, cause frequent TCP timeouts. This property of drop-tail queues is termed as *Port Blackout* [4] and it significantly affects the performance of small flows because frequent timeouts lead to high latencies and thus, poor quality response times. Figure 3 shows an example scenario of port blackout where A and B are input ports whereas

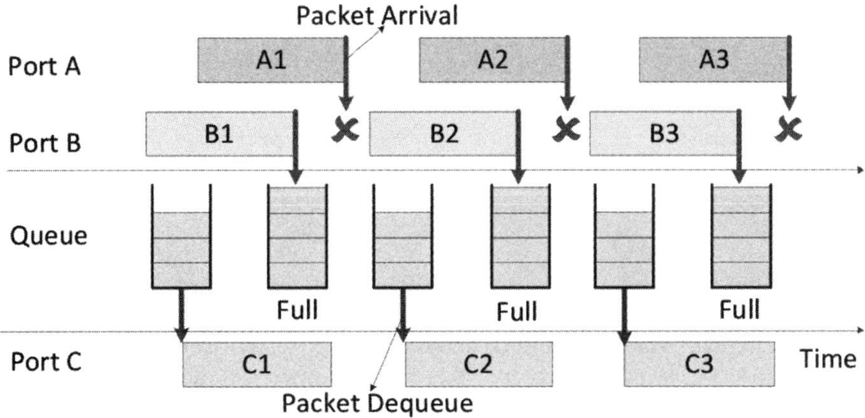

Fig. 3 Example scenario of Port Blackout [4]

C is the common output port. The figure shows that packets arriving at Port B are successfully queued whereas those arriving at Port A are dropped consecutively.

It is well known that the throughput of a TCP flow is inversely proportional to the RTT of that flow. This behavior of TCP leads to RTT-bias i.e., flows with low RTT achieve larger share of bandwidth than the flows with high RTT. However, it has been observed that due to TCP Outcast problem in Data Center Networks, TCP exhibits *Inverse RTT-bias* [4] i.e., flows with low RTT are outcasted by flows with high RTT.

The two main factors that cause TCP Outcast are: (i) the usage of drop-tail queues in switches and (ii) many-to-one communication pattern which leads to a large set of flows and a small set of flows arriving at two different input ports and competing for the same bottleneck output port. Both these factors are quite common in Data Center Networks since majority of the switches employ drop-tail queues and many-to-one communication pattern is the foundation of many cloud applications.

Mitigating TCP Outcast: One possible approach to mitigate TCP Outcast is to use queuing mechanisms other than drop-tail e.g., Random Early Detection (RED) [10], Stochastic Fair Queue (SFQ) [4], etc. Another possible approach is to minimize the buffer occupancy at the switches by designing efficient TCP congestion control laws at the end hosts.

2.3 Queue Buildup

Due to the diverse nature of cloud applications, mice traffic, cat traffic and elephant traffic co-exist in a Data Center Network. The long lasting and greedy nature of elephant traffic drives the network to the point of extreme congestion and overflows the bottleneck buffers. Thus, when both mice traffic and elephant traffic traverse through the same route, the performance of mice traffic is significantly affected due to the presence of the elephant traffic [2].

Following are two ways in which the performance of mice traffic is degraded due to the presence of elephant traffic [2]: (i) since most of the buffer is occupied by elephant traffic, there is a high probability that the packets of mice traffic get dropped. The impact of this situation is similar to that of TCP Incast because the performance of mice traffic is largely affected by frequent packet losses and hence, the timeouts. (ii) the packets of mice traffic, even when none are lost, suffer from increased queuing delay as they are in queue behind the packets of elephant traffic. This problem is termed as Queue build-up.

Mitigating Queue Buildup: Queue build-up problem can be solved only by minimizing the queue occupancy in the Data Center Network switches. Most of the existing TCP variants employ reactive approach towards congestion control i.e., they do not reduce the sending rate unless a packet loss is encountered, and hence, fail to minimize the queue occupancy. A proactive approach instead, is desired to minimize the queue occupancy and overcome the problem of queue build-up.

2.4 *Buffer Pressure*

Buffer pressure is yet another impairment caused by the long lasting and greedy nature of elephant traffic. When both mice traffic and elephant traffic co-exist on the same route, most of the buffer space is occupied by packets from the elephant traffic. This leaves a very little room to accommodate the burst of mice traffic packets arising out of many-to-one communication pattern. The result is that large number of packets from mice traffic are lost, leading to poor performance. Moreover, majority of the traffic in Data Center Networks is bursty [2] and hence, packets of mice traffic get dropped frequently because the elephant traffic lasts for a longer time and keeps most of the buffer space occupied.

Mitigating Buffer Pressure: Like Queue build-up, Buffer pressure problem too can be solved by minimizing the buffer occupancy in the switches.

2.5 *Pseudo-Congestion Effect*

Virtualization is one of the key technologies driving the success of Cloud Computing applications. Modern Data Centers adopt Virtual Machines (VMs) to offer on-demand cloud services and remote access. These data centers are known as *virtualized data centers* [1, 5]. Though there are several advantages of virtualization like efficient server utilization, service isolation and low system maintenance cost [1], it significantly affects the environment where our traditional protocols (e.g., TCP,

Table 2 TCP Impairments in Data Center Networks and their causes

TCP impairment	Causes
TCP Incast	Shallow buffers in switches and Bursts of mice traffic resulting from many-to-one communication pattern
TCP Outcast	Usage of tail-drop mechanism in switches
Queue Buildup	Persistently full queues in switches due to elephant traffic
Buffer Pressure	Persistently full queues in switches due to elephant traffic and Bursty nature of mice traffic
Pseudo-Congestion Effect	Hypervisor scheduling latency

UDP) work. The recent study of Amazon EC2 Data Center reveals that virtualization dramatically deteriorates the performance of TCP and UDP in terms of both, throughput and end to end delay [1]. Throughput becomes unstable and the end to end delay becomes quite large even if the network load is less [1].

When more number of VMs are running on the same physical machine, the hypervisor scheduling latency increases the waiting time for each VM to obtain an access to the processor. Hypervisor scheduling latency varies from microseconds to several hundred milliseconds [5], leading to unpredictable network delays (i.e., RTT) and thus, affecting the throughput stability and largely increasing the end to end delay. Moreover, hypervisor scheduling latency can be so high that it may lead to RTO at the VM sender. Once RTO occurs, VM sender assumes that the network is heavily congested and significantly brings down the sending rate. We term this effect as *pseudo-congestion effect* because the congestion sensed by the VM sender is actually *pseudo-congestion* [5].

Mitigating Pseudo-congestion Effect There are generally two possible approaches to address the above mentioned problem. One is to design efficient schedulers for hypervisor so that the scheduling latency can be minimized. Another approach is to modify TCP such that it can intelligently detect the *pseudo-congestion* and react accordingly.

2.6 Summary: TCP Impairments and Causes

We briefly summarize the TCP impairments discussed in the above subsections and mention the causes for the same in Table 2.

3 TCP Variants for Data Center Networks

3.1 TCP with Fine Grained RTO (FG-RTO) [3]

The default value of minimum RTO in TCP is generally in the order of milliseconds (around 200 ms). This value of RTO is suitable for Internet like scenarios where the average RTT is in order of hundreds of milliseconds. However, it is significantly larger than the average RTT in data centers which is in the order of a few microseconds. Large number of packet losses due to TCP Incast, TCP Outcast, Queue build-up, Buffer pressure and pseudo-congestion effect result in frequent timeouts and in turn, lead to missed deadlines and significant degradation in the performance of TCP. Phanishayee et al. [3] show that reducing the minimum RTO from 200 ms to 200 μs significantly alleviates the problems of TCP in Data Center Networks and improves the overall throughput by several orders of magnitude.

Advantages: The major advantage of this approach is that it requires minimum modification to the traditional working of TCP and thus, can be easily deployed.

Shortcomings: The real time deployment of fine grained timers is a challenging issue because the present operating systems lack the high-resolution timers required for such low RTO values. Moreover, FG-RTOs may be not suitable for servers that communicate to clients through the Internet. Apart from the implementations issues of fine grained timers, it must be noted that this approach of eliminating drawbacks of TCP in Data Center Networks is a *reactive* approach. It tries to reduce the impact of a packet loss rather than *avoiding* the packet loss in the first place. Thus, although this approach significantly improves the network performance by reducing post-packet-loss delay, it does not alleviate the TCP Incast problem for loss-sensitive applications.

3.2 TCP with FG-RTO + Delayed ACKs Disabled [3]

Delayed ACKs are mainly used for reducing the overhead of ACKs on the reverse path. When delayed ACKs are enabled, the receiver sends only one ACK for every two data packets received. If only one packet is received, the receiver waits for delayed ACK timeout period before sending an ACK. This timeout period is usually 40 ms. This scenario may lead to spurious retransmissions if FG-RTO timers (as explained in the previous section) are deployed. The reason is that receiver waits for 40 ms before sending an ACK for the received packet and by that time, FG-RTO which is in order of few microseconds, expires and forces the sender to retransmit the packet. Thus, either the delayed ACK timeout period must be reduced to a few microseconds or must be completely disabled while using FG-RTOs to avoid such spurious retransmissions. This approach further enhances the TCP throughput in Data Center Networks.

Advantages: It has been shown in [3] that reducing the delayed ACK timeout period to 200 μs while using FG-RTO achieves far better throughput than the throughput obtained when delayed ACKs are enabled. Moreover, completely disabling the delayed ACKs while using FG-RTO further improves the overall TCP throughput.

Shortcomings: The shortcomings of this approach are exactly similar to that of TCP with FG-RTO because this approach is an undesired side-effect of the previous approach.

3.3 *DCTCP: Data Center TCP [2]*

Additive Increase Multiplicative Decrease (AIMD) is the cornerstone of TCP congestion control algorithms. When an acknowledgement (ACK) is received in AIMD phase, the congestion window (*cwnd*) is increased as shown in (1). This is known as Additive Increase phase of the AIMD algorithm.

$$cwnd = cwnd + \frac{1}{cwnd} \tag{1}$$

When congestion is detected either through *dupacks* or Selective Acknowledgement (SACK), *cwnd* is updated as shown in (2). This is known as Multiplicative Decrease phase of the AIMD algorithm.

$$cwnd = \frac{cwnd}{2} \tag{2}$$

DCTCP employs an efficient multiplicative decrease mechanism which reduces the *cwnd* based on the *amount of congestion* in the network rather than reducing it by half. DCTCP leverages Explicit Congestion Notification (ECN) mechanism [11] to extract multi-bit feedback on the *amount of congestion* in the network from the single bit stream of ECN marks. The next subsection describes the working of ECN in brief:

3.3.1 Explicit Congestion Notification (ECN)

Explicit Congestion Notification (ECN) [11] is one of the most popular congestion signaling mechanisms in communication networks. It is widely deployed in a large variety of operating systems at end hosts, modern Internet routers and used by a variety of transport protocols. Moreover, it has been noticed that the use of ECN in the Internet has increased by three folds in the last few years.

As shown in Fig. 4 and 5, ECN uses two bits in the IP header, namely ECN Capable Transport (ECT) and Congestion Experienced (CE), and two bits in the TCP header, namely Congestion Window Reduced (CWR) and ECN Echo (ECE), for signaling congestion to the end-hosts. ECN is an industry standard and its detailed mechanism is described in RFC 3168. Table 3 and 4 show the ECN codepoints in the TCP header and the IP header respectively and Fig. 6 shows in brief, the steps involved in the working of ECN mechanism.

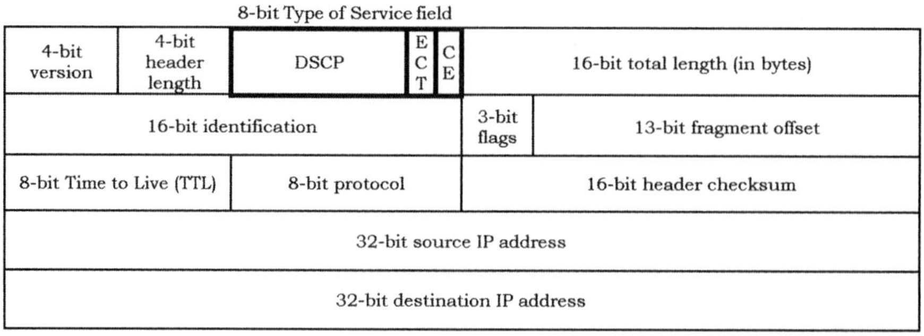

8-bit Type of Service field

4-bit version	4-bit header length	DSCP	ECT	CE	16-bit total length (in bytes)
16-bit identification				3-bit flags	13-bit fragment offset
8-bit Time to Live (TTL)	8-bit protocol				16-bit header checksum
32-bit source IP address					
32-bit destination IP address					

Fig. 4 ECN bits in IP header

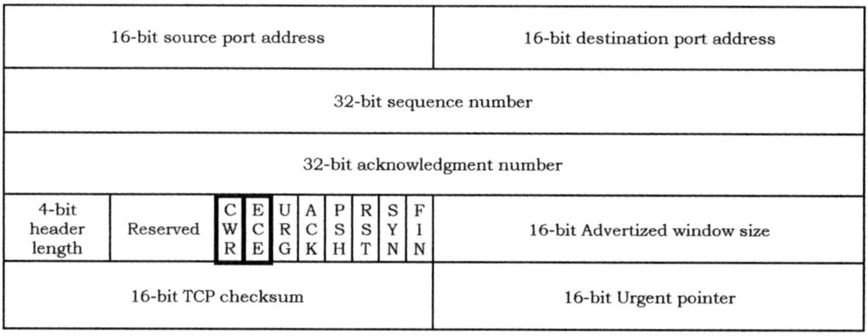

16-bit source port address										16-bit destination port address
32-bit sequence number										
32-bit acknowledgment number										
4-bit header length	Reserved	CWR	ECE	URG	ACK	PSH	RST	SYN	FIN	16-bit Advertized window size
16-bit TCP checksum										16-bit Urgent pointer

Fig. 5 ECN bits in TCP header

Table 3 ECN codepoints in the TCP header

Codepoint	CWR bit value	ECE bit value
Non ECN-set up	0	0
ECN-Echo	0	1
CWR	1	0
ECN-set up	1	1

Table 4 ECN codepoints in the IP header

Codepoint	ECT bit value	CE bit value
Non-ECT	0	0
ECT(1)	0	1
ECT(0)	1	0
CE	1	1

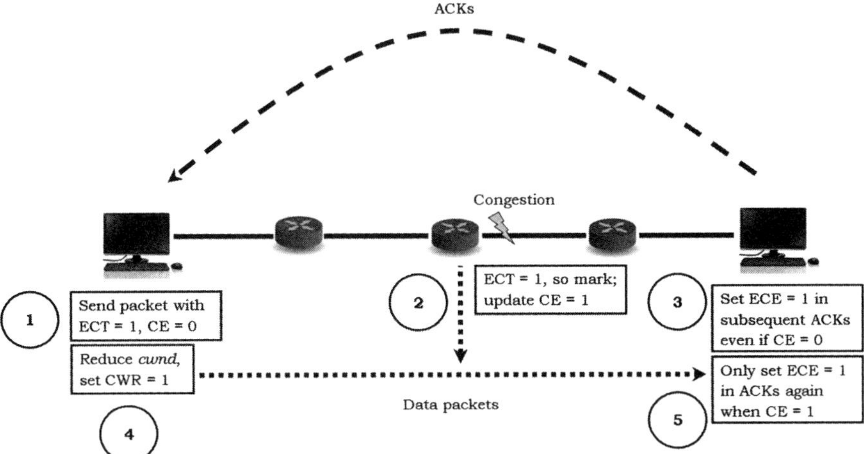

Fig. 6 Explicit congestion notification

As described in RFC 3168—the sender and the receiver must negotiate the use of ECN during the three-way handshake (See Fig. 7). If both are ECN capable, the sender marks every outgoing data packet with either ECT(1) codepoint or ECT(0) codepoint. This serves as an indication to the router that both sender and receiver are ECN capable. Whenever congestion builds up, the router marks the data packet by replacing ECT(1) or ECT(0) codepoint by the CE codepoint. When the receiver receives a marked packet with CE codepoint, it infers congestion and hence, marks *a series of* outgoing acknowledgments (ACKs) with ECE codepoint until the sender acknowledges with CWR codepoint (See Fig. 6).

The major observation here is that, even if the router marks just one data packet, the receiver continues to mark ACKs with ECE until it receives confirmation from the sender (See Step 3 of Fig. 6). This is to ensure the reliability of congestion notification; because even if the first marked ACK is lost, other marked ACKs would notify the sender about congestion. Note that this basic working of ECN aims to only notify the sender about congestion. It is not designed to provide the additional information about the *amount of congestion* to the sender.

At the receiver, counting number of packets marked by the router provides fairly accurate information about the *amount of congestion* in the network. However, conveying this information to the sender by using ECN is a complex task. One of the possible ways is to enable the sender to count the number of marked ACKs it receives from the receiver. The limitation, however, is that even if router marks just one data packet, receiver sends *a series of* marked ACKs. Hence, the number of marked ACKs counted by the sender would be much higher than the number of packets actually marked by the router. This would lead to incorrect estimation of the *amount of congestion* in the network.

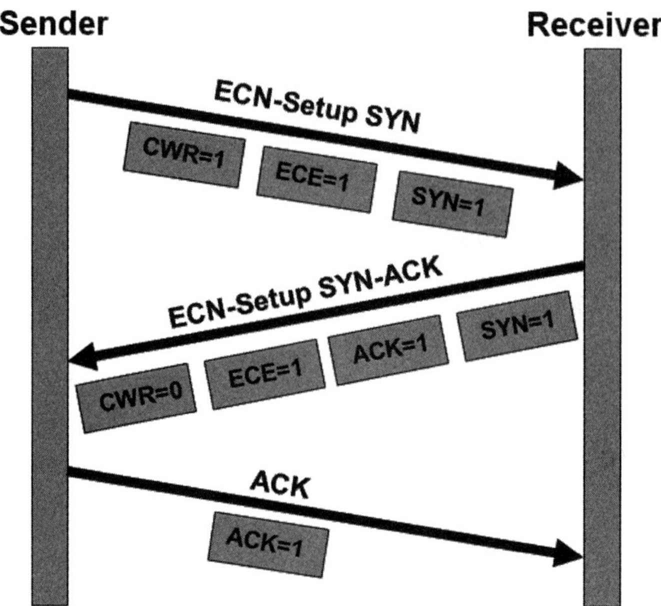

Fig. 7 ECN negotiation

To overcome this limitation, DCTCP modifies the basic mechanism of ECN. Unlike TCP receiver which sends *a series of* marked ACKs, DCTCP receiver sends a marked ACK *only when* it receives a marked packet from the router i.e., it sets ECE codepoint in the outgoing ACK *only when* it receives a packet with CE codepoint. Thus, the DCTCP sender obtains an accurate number of packets marked by the router by simply counting the number of marked ACKs it receives. Note that this modification to the original ECN mechanism reduces the reliability because if a marked ACK is lost, sender remains unaware of the congestion and does not reduce the sending rate. However, since Data Center Networks are privately controlled networks, the possibility that an ACK gets lost is negligible.

On receiving the congestion notification via ECN, the *cwnd* in DCTCP is reduced as shown in (3).

$$cwnd = cwnd \times (1 - \frac{\alpha}{2}) \tag{3}$$

where α ($0 \leqslant \alpha \leqslant 1$) is an estimate of the fraction of packets that are marked and is calculated as shown in (4). F in (4) is the fraction of packets that are marked in the previous *cwnd* and g ($0 < g < 1$) is the exponential weighted moving average constant. Thus, when congestion is low (α is near 0), *cwnd* is reduced slightly and when congestion is high (α is near 1), *cwnd* is reduced by half, just like traditional TCP.

$$\alpha = (1 - g) \times \alpha + g \times F \tag{4}$$

The major goal of DCTCP algorithm is to achieve low latency (desirable for mice traffic), high throughput (desirable for elephant traffic) and high burst tolerance (to avoid packet losses due to Incast). DCTCP achieves these goals by reacting to the *amount of congestion* rather than halving the *cwnd*. DCTCP uses a marking scheme at switches that sets the Congestion Experienced (CE) codepoint [11] in packets as soon as the buffer occupancy exceeds a fixed pre-determined threshold, K (17 % as mentioned in [12]). The DCTCP source reacts by reducing the window by a factor that depends on the fraction of marked packets: the larger the fraction, the bigger the decrease factor.

Advantages: DCTCP is a novel TCP variant which alleviates TCP Incast, Queue-up and Buffer pressure problems in Data Center Networks. It requires minor modifications to the original design of TCP and ECN to achieve these performance benefits. DCTCP employs a *proactive* behavior i.e., it tries to avoid packet loss. It has been shown in [2] that when DCTCP uses FG-RTO, it further reduces the impact of TCP Incast and also improves the scalability of DCTCP. The stability, convergence and fairness properties of DCTCP [12] make it a suitable solution for implementation in Data Center Networks. Moreover, DCTCP is already implemented in latest versions of Microsoft Windows Server operating system.

Shortcomings: The performance of DCTCP falls back to that of TCP when the degree of Incast increases beyond 35 i.e., if there are more than 35 worker nodes sending data to the same aggregator, DCTCP fails to avoid Incast and its performance falls back to that of the traditional TCP. However, authors show that dynamic buffer allocation at the switch and usage of FG-RTO can scale DCTCP's performance to handle up to 40 worker nodes in parallel.

Although DCTCP uses a simple queue management mechanism at the switch, it is ambiguous whether DCTCP can alleviate the problem of TCP Outcast. Similarly, DCTCP does not address the problem of pseudo-congestion effect in Virtualized Data Centers. DCTCP utilizes minimum buffer space in the switches, which in fact, is a desirable property to avoid TCP Outcast. However, experimental studies are required to conclude whether DCTCP can mitigate the problems of TCP Outcast and pseudo-congestion effect.

3.4 ICTCP: Incast Congestion Control for TCP [7]

Like DCTCP, the main idea of ICTCP is to *avoid* packet losses due to congestion rather than recovering from the packet losses. It is well known that a TCP sender can send a minimum of advertised window (*rwnd*) and congestion window (*cwnd*) (i.e. *min(rwnd, cwnd)*). ICTCP leverages this property and efficiently varies the *rwnd* to avoid TCP Incast. The major contributions of ICTCP are: (a) The available bandwidth is used as a quota to co-ordinate the *rwnd* increase of all connections. (b) Per flow congestion control is performed independently and (c) *rwnd* is adjusted based on the ratio of difference between expected throughput and measured throughput over expected throughput. Moreover, live RTT is used for the throughput estimation.

Advantages: Unlike DCTCP, ICTCP does not require any modifications at the sender side (i.e. worker nodes) or network elements such as routers, switches, etc. Instead, ICTCP requires modification only at the receiver side (i.e. an aggregator). This approach is adopted to retain the backward compatibility and make the algorithm general enough to handle the Incast congestion in future high-bandwidth, low-latency networks.

Shortcomings: Although it has been shown in [7] that ICTCP achieves almost zero timeouts and high throughput, the scalability of ICTCP is a major concern i.e., the capability to handle Incast congestion when there are extremely large number of worker nodes since ICTCP employs per flow congestion control. Another limitation of ICTCP is that it assumes that both the sender and the receiver are under the same switch, which might not be the case always. Moreover, it is not known how much buffer space is utilized by ICTCP. Thus, it is difficult to conclude whether ICTCP can alleviate Queue build-up, Buffer pressure and TCP Outcast problems. Like DCTCP, ICTCP too does not address the problem of pseudo-congestion effect in Virtualized Data Centers.

3.5 *IA-TCP: Incast Avoidance Algorithm for TCP [13]*

Unlike DCTCP and ICTCP which use window based congestion control, IA-TCP uses rate based congestion control algorithm to control the total number of packets injected in the network. The motivation behind selecting rate based congestion control mechanism is that window based congestion control mechanisms in Data Center Networks have limitations in terms of scalability i.e., number of senders in parallel.

The main idea of IA-TCP is to limit the total number of outstanding data packets in the network so that it does not exceed the bandwidth-delay product (BDP). IA-TCP employs ACK regulation at the receiver and like ICTCP, leverages the advertised window (*rwnd*) field of the TCP header to regulate the *cwnd* of every worker node. The minimum *rwnd* is set to 1 packet. However, if large number of worker nodes send packets with respect to a minimum *rwnd* of 1 packet, the total number of outstanding packets in the network may exceed the link capacity. In such scenarios, IA-TCP adds delay, Δ, to the ACK packet to ensure that the aggregate data rate does not exceed the link capacity. Moreover, IA-TCP also uses delay, Δ, to avoid the synchronization among the worker nodes while sending the data.

Advantages: Like ICTCP, IA-TCP also requires modification only at the receiver side (i.e. an aggregator) and does not require any modifications at the sender or network elements. IA-TCP achieves high throughput and significantly improves the query completion time. Moreover, the scalability of IA-TCP is clearly demonstrated by configuring up to 96 worker nodes sending data in parallel.

Shortcomings: Similar to the problem of ICTCP, it is not clear how much buffer space is utilized by IA-TCP. Thus, experimental studies are required to conclude whether IA-TCP can mitigate Queue build-up, Buffer pressure and TCP Outcast problems. Like DCTCP and ICTCP, studies are required in Virtualized Data Center environments to analyze the performance of IA-TCP with respect to the problem of pseudo-congestion effect.

3.6 D^{2}TCP: Deadline-aware Datacenter TCP [14]

D^{2}TCP is a novel TCP-based transport protocol which is specifically designed to handle high burst situations. Unlike other TCP variants which are deadline-agnostic, D^{2}TCP is deadline-aware. D^{2}TCP uses a *distributed* and *reactive* approach for bandwidth allocation and employs a novel deadline-aware *congestion avoidance* algorithm which uses ECN feedback and deadlines to vary the sender's *cwnd* via a gamma-correction function [14].

D^{2}TCP does not maintain per flow information and instead, inherits the distributed and reactive nature of TCP while adding deadline-awareness to it. Similarly, D^{2}TCP employs its congestion avoidance algorithm by adding deadline-awareness to DCTCP. The main idea, thus, is that far-deadline flows back-off aggressively and the near-deadline flows back-off only a little or not at all.

Advantages: The novelty of D^{2}TCP lies in the fact that it is deadline-aware and reduces the fraction of missed deadlines up to 75 % as compared to DCTCP. In addition, since it is designed upon DCTCP, it avoids TCP Incast, Queue build-up and has high burst tolerance.

Shortcomings: The shortcomings of D^{2}TCP are exactly similar to those of DCTCP: scalability and whether it is robust against TCP Outcast as well as pseudo-congestion effect. However, since it is deadline-aware, it would be interesting to analyze the robustness of D^{2}TCP against the pseudo-congestion effect in Virtualized Data Centers.

3.7 TCP-FITDC [15]

TCP-FITDC is an adaptive delay-based mechanism to *prevent* the problem of TCP Incast. Like D^{2}TCP, TCP-FITDC is also a DCTCP-based TCP variant which benefits from the novel ideas of DCTCP. Apart from utilizing ECN as an indicator of network buffer occupancy and buffer overflow, TCP-FITDC also monitors changes in the queueing delay to estimate variations in the available bandwidth.

If there is no marked ACK received during the RTT, TCP-FITDC infers that the queue length in the switch is below the marking threshold and hence, increases the *cwnd* to improve the throughput. If marked ACKs are received during the RTT, *cwnd* is decreased to control the queue length. TCP-FITDC maintains two separate variables called rtt_1 and rtt_2 for unmarked ACKs and marked ACKs respectively. By analyzing the difference between these two types of ACKs, TCP-FITDC gets more accurate estimate of the network conditions. The *cwnd* is then reasonably decreased to maintain low queue length.

Advantages: TCP-FITDC gets a better estimate of the network conditions by coupling the information received via ECN and the information obtained by monitoring the RTT. Thus, it has better scalability than DCTCP and scales up to 45 worker nodes in parallel. It avoids TCP Incast, Queue build-up and has high burst tolerance because it is built upon DCTCP.

Shortcomings: The shortcomings of TCP-FITDC are similar to those of DCTCP, except that it improves the scalability of DCTCP. Unlike D^2TCP, TCP-FITDC is deadline-agnostic and like all above mentioned TCP variants, it does not address TCP Outcast and pseudo-congestion effect problems.

3.8 TDCTCP [16]

TDCTCP attempts to improvise the working of DCTCP (and so, is DCTCP-based) by making three modifications. First, unlike DCTCP, TDCTCP not only decreases, but also increases the *cwnd* based on the *amount of congestion* in the network i.e, instead of increasing the *cwnd* as shown in (1), TDCTCP increases the *cwnd* as shown in (5). Thus, when the network is lightly loaded, the increment in *cwnd* is high; and vice-versa.

$$cwnd = cwnd * \left(1 + \frac{1}{1 + \frac{\alpha}{2}}\right) \tag{5}$$

Second, TDCTCP resets the value of α after every delayed ACK timeout. This is done to ensure that α does not carry the stale information about the network conditions, because if the stale value of α is high, it restricts the *cwnd* increment and thereby degrades the overall throughput. Lastly, TDCTCP employs an efficient approach to dynamically calculate the delayed ACK timeout with a goal to achieve better fairness.

Advantages: TDCTCP achieves 26–37 % better throughput and 15–20 % better fairness than DCTCP in a wide variety of scenarios ranging from single bottleneck topologies to multi-bottleneck topologies and varying buffer sizes. Moreover, it achieves better throughput and fairness even at very low values of K i.e., the ECN marking threshold at the switch. However, there is a slight increase in the delay and queue length while using TDCTCP as compared to DCTCP.

Shortcomings: Although TDCTCP improves throughput and fairness, it does not address the scalability challenges faced by DCTCP. Like most of other TCP variants discussed, TDCTCP too is deadline agnostic and does not alleviate the problems of TCP Outcast and pseudo-congestion effect.

3.9 TCP with Guarantee Important Packets (GIP) [17]

TCP with GIP mainly aims to improve the network performance in terms of goodput by minimizing the total number of timeouts. Timeouts lead to dramatic degradation in the network performance and affect the user perceived delay. The authors of TCP with GIP focus on avoiding mainly two types of timeouts in the network: (i) the timeouts caused due to the loss of full window of packets. These types of timeouts are termed as Full window Loss TimeOuts (FLoss-TOs) and (ii) the timeouts caused due to the lack of ACKs. These types of timeouts are termed as Lack of ACKs TimeOuts (LAck-TOs).

FLoss-TOs generally occur when the total data sent by all the worker nodes exceeds the available bandwidth in the network and thus, a few unlucky flows end up loosing all the packets of the window. On the other hand, LAck-TOs mainly occur when the transmission is *barrier synchronized transmission*. In such transmissions, the aggregator will not request the worker nodes to transmit the next stripe units until all the worker nodes finish sending their current ones. If a few packets get dropped at the end of the stripe unit, they cannot be recovered until the RTO fires because there may not be sufficient *dupacks* to trigger Fast Retransmit.

TCP with GIP introduces *flags* in the interface between the application layer and the transport layer. These *flags* indicate whether the running application follows many-to-one communication pattern or not. If the running application follows such a communication pattern, TCP with GIP redundantly transmits the last packet of the stripe unit at most three times and each worker node decreases its initial *cwnd* at the head of the stripe unit. On the other hand, if the running application does not follow many-to-one communication pattern, TCP with GIP works like a standard TCP.

Advantages: TCP with GIP achieves almost zero timeouts and higher goodput in a wide variety of scenarios including with and without the background UDP traffic. Moreover, the scalability of TCP with GIP is much more than any other TCP variant discussed above i.e., it scales well up to 150 worker nodes in parallel.

Shortcomings: TCP with GIP does not address the queue occupancy problem resulting out of the presence of elephant traffic. As a result, the Queue Buildup, Buffer pressure and TCP Outcast problems remain unsolved because all these problems arise due to the lack of the buffer space in the switches. Although timeouts are eliminated by TCP with GIP, but flows may miss the specified deadlines because of queueing delay. Moreover, the hypervisor scheduling latency is not taken into consideration and thus, the problem of pseudo-congestion effect also remains open. Note that high latencies introduced by hypervisor scheduling algorithm may also prevent flows from meeting the specified deadlines.

3.10 PVTCP: Para Virtualized TCP [5]

PVTCP proposes an efficient solution to the problem of pseudo-congestion effect. This approach does not require any changes to be done in the hypervisor. Instead, the basic working of TCP is modified to accept the latencies introduced by the hypervisor scheduler. An efficient approach is suggested to capture the *actual* picture of every packet transmission involving the hypervisor-introduced-latencies and then determine RTO more accurately to filter out pseudo-congestion effect.

Whenever the hypervisor introduces scheduling latency, sudden spikes can be observed during the regular measurements of RTT. PVTCP detects these sudden spikes and filters out the negative impact of these spikes by proper RTT measurement and RTO management. While calculating average RTT, PVTCP ignores the measurement of a particular RTT if a spike is observed in that RTT.

Advantages: PVTCP solves the problem of pseudo-congestion effect without requiring any changes in the hypervisor. By detecting the unusual spikes, accurately measuring RTT and proper management of RTO, PVTCP enhances the performance of Virtualized Data Centers.

Shortcomings: The scalability of PVTCP is ambiguous and thus, whether it can solve TCP Incast effectively or not is unclear. The queue occupancy while using PVTCP is not taken into consideration which may further lead to problems such Queue build-up, Buffer pressure, TCP Outcast and missed deadlines.

4 Summary: TCP Variants for DCNs

Table 5 summarizes the comparative study of TCP variants proposed for Data Center Networks. Apart from the novelty of the proposed TCP variant, the table also highlights the deployment complexity of each protocol. The protocols which require modifications in sender, receiver and switch are considered as hard to deploy. The ones which require modification only at the sender or receiver are considered as easy to deploy. Data Center Networks, however, are privately controlled and managed networks and thus, the former ones may also be treated as easy to deploy.

Apart from the above mentioned parameters, the summary also includes which problems amongst TCP Incast, TCP Outcast, Queue build-up, Buffer pressure and pseudo-congestion effect are alleviated by each TCP variant. The details regarding the tools used/approach of implementation adopted by the authors are also listed.

Table 5 Summary of TCP Variants proposed for Data Center Networks

TCP Variants proposed for Data Center Networks	Modifies Sender	Modifies Receiver	Modifies Switch	Solves TCP Incast	Solves TCP Outcast	Solves Queue build-up	Solves Buffer pressure	Is Deadline Aware	Detects pseudo-congestion	Implementation
TCP with FG-RTO	✓	x	x	✓	x	x	x	x	x	Testbed and *ns*-2
TCP with FG-RTO + Delayed ACKs disabled	✓	x	x	✓	x	x	x	x	x	Testbed and *ns*-2
DCTCP	✓	✓	✓	✓	x	✓	✓	x	x	Testbed and *ns*-2
ICTCP	x	✓	x	✓	x	x	x	x	x	Testbed
IA-TCP	x	✓	x	✓	x	x	x	x	x	*ns*-2
D^2TCP	✓	✓	✓	✓	x	✓	✓	✓	x	Testbed and *ns*-3
TCP-FITDC	✓	✓	✓	✓	x	✓	✓	x	x	Modeling and *ns*-2
TDCTCP	✓	✓	✓	✓	x	✓	✓	x	x	OMNeT++
TCP with GIP	x	✓	x	✓	x	x	x	x	x	Testbed and *ns*-2
PVTCP	✓	✓	x	✓	x	x	x	x	✓	Testbed

5 Open Issues

Although several modifications have been proposed to the original design of TCP, there is an acute need to further optimize the performance of TCP in DCNs. A few open issues are listed below:

- Except D^{2}TCP, all other TCP variants are deadline-agnostic. Meeting deadlines is the most important requirement in DCNs. Missed deadlines may lead to violations of Service Level Agreements (SLAs) and thus, incur high cost to the organization.
- Most of the Data Centers today employ virtualization for efficient resource utilization. Hypervisor scheduling latency ranges from microseconds to hundreds of milliseconds and hence, may hinder in successful completion of flows within the specified deadline. While making modifications to hypervisor is one viable solution, designing an efficient TCP which is deadline-aware and automatically tolerates hypervisor scheduling latency is a preferred solution.
- A convincing solution to TCP Outcast problem is unavailable. An optimal solution to overcome TCP Outcast must ensure minimal buffer occupancy at the switch. Since RED is implemented in most of the modern switches - it can be used to control the buffer occupancy. The parameter sensitivity of RED, however, poses further challenges and complicates the problem.

6 Concluding Remarks

Data Centers in the present scenario house a plethora of Internet applications. These applications are diverse in nature and have various performance requirements. Majority of these applications use many-to-one communication pattern to gain performance efficiency. TCP, which has been a mature transport protocol of Internet since past several decades, suffers from performance impairments such as TCP Incast, TCP Outcast, Queue build-up, Buffer pressure and Pseudo-congestion effect in Data Center Networks.

This chapter described each of the above mentioned impairment in brief along with the causes and possible approaches to mitigate them. Moreover, it presents a comparative study of TCP variants which have been specifically designed for Data Center Networks and the advantages and shortcomings of each TCP variant are highlighted.

References

1. J. Zhang, F. Ren, and C. Lin "Survey on Transport Control in Data Center Networks," *IEEE Network*, vol. 27, no. 4, pp. 22–26, 2013.
2. M. Alizadeh, A. Greenberg, D. A. Maltz, J. Padhye, P. Patel, B. Prabhakar, S. Sengupta, and M. Sridharan "Data Center TCP (DCTCP)," *SIGCOMM Computer Communications*

Review, vol. 40, no. 4, pp. 63–74, Aug. 2010. [Online]. Available: http://doi.acm.org/10.1145/1851275.1851192
3. V. Vasudevan, A. Phanishayee, H. Shah, E. Krevat, D. G. Andersen, G. R. Ganger, G. A. Gibson, and B. Mueller "Safe and effective Fine-grained TCP Retransmissions for Datacenter Communication," *SIGCOMM Computer Communications Review*, vol. 39, no. 4, pp. 303–314, Aug. 2009. [Online]. Available: http://doi.acm.org/10.1145/1594977.1592604
4. P. Prakash, A. Dixit, Y. C. Hu, and R. Kompella "The TCP Outcast Problem: Exposing Unfairness in Data Center Networks," in *Proceedings of the 9th USENIX Conference on Networked Systems Design and Implementation*, ser. NSDI'12. Berkeley, CA, USA: USENIX Association, 2012, pp. 30–30. [Online]. Available: http://dl.acm.org/citation.cfm?id=2228298.2228339
5. L. Cheng, C.-L. Wang, and F. C. M. Lau "PVTCP: Towards Practical and Effective Congestion Control in Virtualized Datacenters," in *21st IEEE International Conference on Network Protocols*, ser. ICNP 2013. IEEE, 2013.
6. Y. Chen, R. Griffith, J. Liu, R. H. Katz, and A. D. Joseph "Understanding TCP Incast Throughput Collapse in Datacenter Networks," in *Proceedings of the 1st ACM workshop on Research on Enterprise Networking*, ser. WREN '09. New York, NY, USA: ACM, 2009, pp. 73–82. [Online]. Available: http://doi.acm.org/10.1145/1592681.1592693
7. H. Wu, Z. Feng, C. Guo, and Y. Zhang "ICTCP: Incast Congestion Control for TCP in Data Center Networks," in *Proceedings of the 6th International Conference*, ser. Co-NEXT '10. New York, NY, USA: ACM, 2010, pp. 13:1–13:12. [Online]. Available: http://doi.acm.org/10.1145/1921168.1921186
8. J. F. Kurose and K. W. Ross, *Computer Networking: A Top Down Approach*.Addison-Wesley, 6th ed., 02/2012, ISBN-13: 978-0-13-285620-1, 2012.
9. Y. Ren, Y. Zhao, P. Liu, K. Dou, and J. Li "A survey on TCP Incast in Data Center Networks," *International Journal of Communication Systems*, pp. n/a–n/a, 2012. [Online]. Available: http://dx.doi.org/10.1002/dac.2402
10. S. Floyd and V. Jacobson, "Random Early Detection Gateways for Congestion Avoidance," *IEEE/ACM Transactions on Networking*, vol. 1, pp. 397–413, August 1993. [Online]. Available: http://dx.doi.org/10.1109/90.251892
11. K. K. Ramakrishnan and S. Floyd, "The Addition of Explicit Congestion Notification (ECN) to IP," 2001, rFC 3168.
12. M. Alizadeh, A. Javanmard, and B. Prabhakar "Analysis of DCTCP: Stability, Convergence and Fairness," in *Proceedings of the ACM SIGMETRICS, Joint International Conference on Measurement and Modeling of Computer Systems*, ser. SIGMETRICS '11. New York, NY, USA: ACM, 2011, pp. 73–84. [Online]. Available: http://doi.acm.org/10.1145/1993744.1993753
13. J. Hwang, J. Yoo, and N. Choi "IA-TCP: A Rate Based Incast-Avoidance Algorithm for TCP in Data Center Networks," *ICC 2012*, 2012.
14. B. Vamanan, J. Hasan, and T. Vijaykumar "Deadline-aware Datacenter TCP (D^2TCP)," *SIGCOMM Computer Communications Review*, vol. 42, no. 4, pp. 115–126, Aug. 2012. [Online]. Available: http://doi.acm.org/10.1145/2377677.2377709
15. J. Wen, W. Zhao, J. Zhang, and J. Wang "TCP-FITDC: An Adaptive Approach to TCP Incast Avoidance for Data Center Applications," in *Proceedings of the 2013 International Conference on Computing, Networking and Communications (ICNC)*, ser. ICNC '13. Washington, DC, USA: IEEE Computer Society, 2013, pp. 1048–1052. [Online]. Available: http://dx.doi.org/10.1109/ICCNC.2013.6504236
16. T. Das and K. M. Sivalingam, "TCP Improvements for Data Center Networks," in *Communication Systems and Networks (COMSNETS), 2013 Fifth International Conference on*. IEEE, 2013, pp. 1–10.
17. J. Zhang, F. Ren, L. Tang, and C. Lin "Taming TCP Incast Througput Collapse in Data Center Networks," in *21st IEEE International Conference on Network Protocols*, ser. ICNP 2013. IEEE, 2013.

Routing Techniques in Data Center Networks

Shaista Habib, Fawaz S. Bokhari and Samee U. Khan

1 Introduction

Data Centers are the core of cloud computing as they consist of thousands of computers which are interconnected in a way to provide cloud services. A Data Center Network (DCNs) can be defined as centralized infrastructure providing several large scale computing and diversified network services like video streaming and cloud computing to its subscribed users [1]. With the proliferation of internet applications, the demand for DCNs are increasing as they provide efficient platform for data storage to such applications. Figure 1 shows the block diagram of data center networks, where end users get services from data centers via the internet. In order to make data centers quick and cost effective, dynamic resources allocation is provided by assigning services to any server or machine in the network. Similarly, the performance isolation between services should also be managed in DCNs. This involves lot of server to server communication and huge amount of traffic is routed among servers in a data center network. Other than the economical (cost-effective) and technical motivations behind the deployment of data center networks, they are designed to facilitate its users.

Data centers have some unique characteristics that make them different from other networks such as the internet or LAN. First of all, data centers are designed to deliver large scale computing services and data intensive communications such as web searching, email and network storage. These types of services and applications

S. Habib (✉) · F. S. Bokhari
University of the Punjab (P.U.C.I.T), Lahore, Pakistan
e-mail: h_shaista@yahoo.com

F. S. Bokhari
e-mail: fawaz@pucit.edu.pk

S. U. Khan
North Dakota State University, Fargo, USA
e-mail: samee.khan@ndsu.edu

S. U. Khan, A. Y. Zomaya (eds.), Handbook on Data Centers,
DOI 10.1007/978-1-4939-2092-1_16

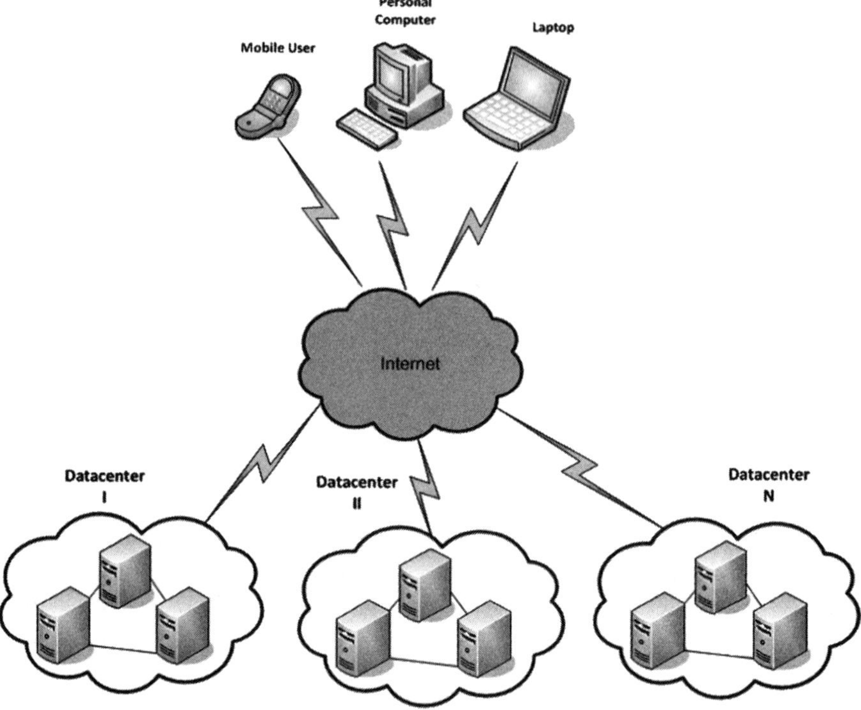

Fig. 1 Block Diagram of Data Center Network

demand high bandwidth to transfer data among distributed components in a data center. For this purpose, data centers have usually high bisection bandwidth to avoid hot spots between any pair of machines in the network. Moreover, data centers have thousands of servers which are densely interconnected and normally have 1:1 oversubscription ratio among the links. Since, data centers are built using commodity hardware and their failure is also a natural situation, there is a requirement of agility so that any server can service any demand. All these failure should be transparent to the client. Without agility a specific number of servers are fixed for a specific application. If number of request increase or decrease, resources becomes overloaded or underutilize respectively.

Routing protocols typically determine how communications between routers take place. They determine the best routes and share network information with their neighbors. Since, data traffic flows inside, outside, and within the servers, routing protocols are therefore, needed to route and forward data among servers in the network. Similarly, in order to explore services of data centers infrastructure, efficient routing protocols are required which are different from the traditional internet routing protocols such as OSPF and BGP. It is found that these traditional internet protocols are not suitable for DCNs due to its unique characteristics. Recently, many routing and forwarding schemes have been proposed for data centers [6–11,15,16,17–20].

One of the major objectives of data center networks from routing perspective is to connect several data center servers. Due to multiple interconnections, data center becomes efficient and fault tolerant. Three communication patterns are used in data centers i.e., one-to-one, one-to-all and all-to-all communication. *One-to-one* communication or traffic pattern is the most typical form of communication in data centers. For example, a large file is transferred from one location to another machine. Similarly, *one-to-all* type of communication involves replicating chunks of data from one machine/server to several servers to ensure data reliability in data center networks. Many applications such as Google File System (GFS) [2] and CloudStore [3] works in this fashion. Lastly, *all-to-all* traffic in which, every server transmits data to every other server. This type of traffic pattern is generated by MapReduce [4] type jobs.

Most existing data centers implement a tree like network configuration, which suffers from network scalability, high cost and single point of failure [9]. However, there has been research on designing efficient data center structures and topologies which take into account the characteristics of DCNs. These proposed structures fall into two categories namely *switch-centric* and *server-centric* data centric structures. As the names suggest, in switch-centric data center structure, the routing and forwarding is performed by the switches in a data center, whereas in server-centric, servers are also involved in forwarding and somewhat routing of the data in the DCN.

Generally, routing scheme determine routes between any two servers with short latency but in DCN, this requires further optimization in many parameters like minimum latency, maximum reliability, maximum throughput and energy, etc. Also if any link gets down or fail then a routing protocol should route traffic using alternate path without interrupting running application. All these optimizations come under traffic engineering (TE) problem. In data centers, multiple paths exist among servers, therefore, multipath routing can be used which helps in load balancing and fault tolerance [10]. There has some work done in designing efficient multipath routing algorithms for data center networks [10, 11]. To take advantage of multiple paths Equal Cost Management Protocol (ECMP) and Valiant Load Balancing (VLB) techniques are used. ECMP performs static load splitting among flows [11], where as VLB randomly selects any intermediate switch that will forward incoming traffic towards the destination. Since, data centers typically consist of huge amount of servers, and each device consumes power, therefore, a lot of research has also been done on designing efficient power saving routing protocols [9, 13]. In DCN, efficient multicasting routing schemes [5, 20] are also required in instances when query searchers are redirected to multiple indexing servers [12] which reduces cost of repeated transmission task. Similarly, in Content Distribution Networks (CDNs), where request routing is used [15, 16], several challenges exists such as routing of the client requests to a suitable surrogate server for serving and the chosen surrogate server should be nearest to the client and have minimum load.

Within the scope of this chapter, we focus on the problem of routing and data forwarding in the context of data center networks (DCNs). The rest of the chapter is

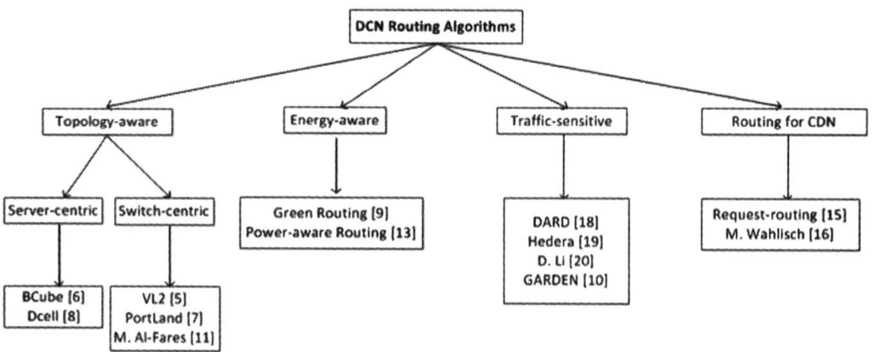

Fig. 2 Classification of data center routing schemes

organized as follows. We provide a comprehensive review of some of the recent well-known routing and data forwarding techniques in DCNs and classify these routing protocols according to their most prominent attributes in Sect. 2 together with the objectives and limitations of each of the approach. In Sect. 3, we discuss the open issues and challenges in the design of such routing schemes, followed by the chapter conclusion in Sect. 4.

2 Classification of Routing Schemes in Data Centers

In literature, several routing protocols have been proposed for data center networks [5–11,21–23] and they can be classified based on different criteria and approached that is used in those schemes. The criteria that we have used for classification is based on four parameters. The first parameter is *topology-aware* routing schemes in which we further classify the approaches based on a particular data center structure design i.e. switch-centric or server-centric approaches. The second parameter for classification is *energy-aware* which means that the main objective of the routing algorithm is to conserve energy or to reduce power consumption in data centers. Our third classification parameter represents the routing algorithms which are designed for either unicast, or multicast routing in data centers. All of these routing schemes in this classification have employed some type of traffic engineering approach and hence we call it *traffic-sensitive* routing approaches. There are some routing schemes defined for content distribution networks (CDNs) which are data centers in essence but are given different name because of their peculiar functionality of distributing content across multiple networks. We have classified such routing schemes on our fourth parameter i.e. *routing for CDNs*. The classification tree of all these protocols (which are discussed in this chapter) is shown in Fig. 2. Note that our classification based on these parameters may not be disjoint from each other and therefore, a particular traffic-sensitive routing scheme may overlap with another scheme belonging to a different category.

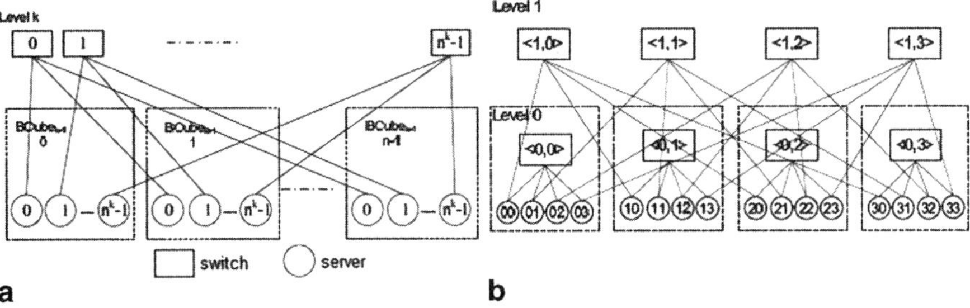

Fig. 3 **a** A Bcube$_k$ is constructed from n BCube$_{k\text{-}1}$ and n^k n-port switches. **b** A BCube1 with $n=4$ [6]

2.1 Topology-Aware Routing

The term topology means physical layout of a network. Routing algorithms which take into account the physical layout of the network for calculating routing paths and forwarding can be classified as topology-aware routing. We further classify these topology-aware routing protocols into *server-centric* and *switch-centric* routing approaches. Briefly, in server centric approach, the servers in a typical data center network act as relaying nodes in multi-hop communication. BCube [6] and Dcell [8] fall under this category. On the contrary, in switch centric approach, switches in data center networks act as relay nodes. Portland [7], VL2 [5] and M. Al-Fares [11] lie in this category. Next, we will study some of the existing topology-aware routing schemes in data center networks.

2.1.1 Server-Centric Approach

BCube In [6], the authors have proposed a novel data center network architecture called BCube which is designed specifically for shipping-container based modular data centers (MDC). MDC are ad-hoc data centers consisting of thousands of servers which are interconnected through switches that are readily deployable on a shipping-container. The authors have argued that existing conventional routing protocols like OSPF [26] does not scale well and lacks the load balancing feature which is necessary in a multi-path BCube network architecture. They have designed a novel BCube source routing (BSR) algorithm for MDC which takes into account the particular network structure of a BCube architecture. The BCube source routing provides maximum utilization of the available capacity, automatically load balance the traffic and scale up to thousands of routers. The rationale behind the selection of source routing for MDC is that the intermediate routers do not have to route the traffic, they just have to forward the data being sent to them and instead, the source will do all the routing decisions. The proposed Bcube structure is shown in Fig. 3.

In BSR routing algorithm, when a source obtains k + 1 parallel paths towards a destination, it sends probe packets on these paths and if any path is unavailable due to link failure, the source runs Breath First Search (BFS) algorithm [27] to determine alternate path. The failed link and the existing parallel paths are removed from the BCube graph and then BFS is applied. The intermediate servers upon receiving a packet determine the next hop and if the incoming link bandwidth is smaller than the bandwidth value mention in probe packet then it updates the probe bandwidth value and send probe response containing updated value. If next hop is not available, then a message about path failure is sent back to the source node. Similarly, if network condition changes due to path failure, the source perform alternate path adoption.

The authors have also considered the possibility of external communication in their proposed BSR routing algorithm. They have introduced the concept of *aggregator* and *gateway* nodes which are used to communicate with external networks. An aggregator is an ordinary layer-2 switch with 10G uplinks. Any Bcube server that connects with the aggregator node acts as gateway node. When there is a packet for some external network from internal server then it will be passed through gateway node. Gateway node performs the address mapping either manually or dynamically. When a packet arrives on a gateway node, it removes the Bcube packet header and sends the packet to the external network. The paths from the external network to the internal servers can be erected in the same manner.

The authors in [8] have proposed a novel network structure for data center networking called *DCell*. Their proposed structure is fault tolerant and can perform very well in situations of very few link or node failures. They have also proposed a distributed routing algorithm which is fault tolerant and provides high data center throughput and is scalable also. The authors have implemented *DCell* on a real test bed and performed comparative evaluations with shortest-path first (SPF) routing. It is found that *DCell* is resilient to node and link failures and provides high aggregate network throughput.

2.1.2 Switch-centric Approach

VL2 In [5], the authors propose a flexible data center network architecture called VL2 consisting of low cost ASIC switches which are arranged into a Clos topology [28] and is scalable in nature. The proposed network architecture provides uniform high capacity among servers, and layer-2 forwarding and routing semantics. For the provision of these services, VL2 uses flat addressing so that service instances can be placed anywhere in the network. In addition to that, the proposed scheme implements routing and forwarding with the help of a Valiant Load Balancing [29, 30] mechanism to distribute traffic evenly across network paths and an end-system based address resolution feature which facilitates scalability, without introducing too much complexity at the network control plane.

For routing and forwarding in VL2, two types of IP addresses are used i.e., location specific IP addresses (LA) and application specific IP addresses (AA). AA remains same even after migration of a virtual machine (VM) within a data center, but LA addresses change. In order to provide load balancing and hot spot free forwarding, VL2 implements two mechanisms. One is the Valiant Load Balancing (VLB) and the

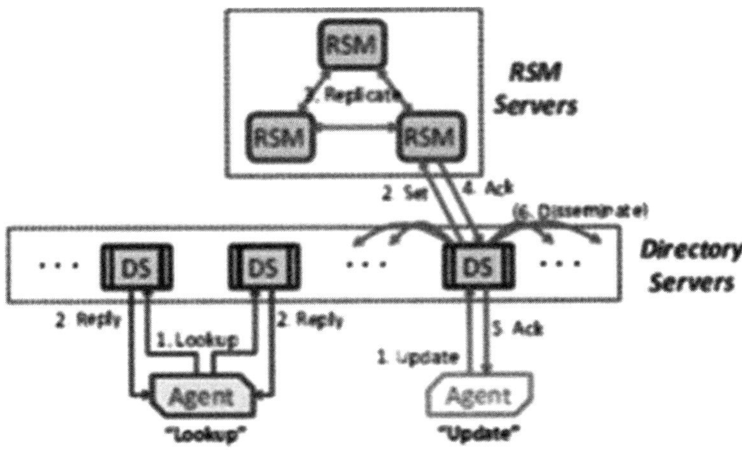

Fig. 4 VL2 Directory System Architecture [5]

second is Equal Cost Management Protocol (ECMP). VLB distributes traffic among intermediate nodes and ECMP distributes load among equal cost paths. VL2 agents running on each server implement VLB using encapsulation by sending traffic to intermediate nodes by selecting randomly chosen paths. Upon reception, an intermediate node de-encapsulates and sends the packet to the Top of Rack (ToR) switch for further forwarding.

For address resolution, the conventional address resolution protocol (ARP) is replaced by a Directory system to handle broadcast and multicast traffic. Directory system consists of directory servers (DS), who keep AA-LA mappings. Each DS performs three tasks, i.e., lookup, updates for AA-LA mapping and third one is reactive cache update under live VM migration. DS helps to trace physical location of application. When any two servers want to communication with each other, AA addresses are used and the source queries to DS for AA-LA mapping of the destination server. After address resolution is done, the packet can be forwarded to the destination server. Since, directory servers provide mapping functions, therefore, they can enforce access control policies. Figure 4 shows the proposed VL2 directory system architecture.

One of the limitations of VL2 is that it does not guarantee absolute bandwidth between any servers, which is a requirement for many real time applications. Moreover, the proposed architecture is highly dependent on Clos topology [28] which requires that switches implement OSPF, ECMP, and IP-in-IP encapsulation, which limits its deployment [1].

Portland In [7], the authors have proposed a scalable fault-tolerant layer-2 data center network fabric. Existing layer-2 and layer-3 network protocols face some limitations e.g., lack of scalability, difficult management, inflexible communication, or limited support for virtual machine migration. Some of these limitations are inherent from Ethernet/IP style protocols when trying to support arbitrary topologies in data center networks. However, today's DCNs are mostly managed by a single logical network fabric with a known baseline topology and growth model. The authors have therefore taken this particular characteristics of DCN into consideration and designed Portland; a scalable, fault tolerant layer-2 routing and forwarding protocol for such networks.

PortLand implements a centralized process called Fabric Manager which contains the network configuration in terms of its topology and does the ARP resolution function for forwarding. In order to achieve efficient routing and forwarding, the authors have proposed a Pseudo MAC (PMAC) addressing scheme. Each host in the Portland architecture is assigned a unique PMAC address which is used to encode the physical location of the host in the network topology and an actual MAC address (AMAC). When a source node wants to send some data to a destination node, the ingress switch upon receiving a packet looks for the source MAC address in the packet and checks its local table for AMAC to PMAC mapping. If no such entry is found, it will be created and then will be sent to the fabric manager for further processing of the ARP request. Similarly, when the packet reaches a particular egress node, it performs the necessary PMAC to AMAC mapping and then it is eventually sent to the destination node. In this way, the fabric manager in PortLand helps to reduce broadcast ARP traffic at the ethernet layer. Figure 5 shows how an ARP request is processed in PortLand architecture. The authors have called this ARP resolution mechanism as proxy ARP. When a packet is received by an edge switch in step 1, it creates an Address Resolution Protocol (ARP) request for IP to MAC mapping and sends this request to fabric manager to lookup its PMAC in step 2. The fabric manager then returns the PMAC address to the edge switch as shown in step 3 of Fig. 5. Finally, in step 4, the edge switch creates an ARP reply and sends it to the source node for further routing.

One of the limitations of PortLand is that it requires a multi-rooted fat-tree topology thus making it not applicable to other commonly used DCN topologies. Also, since, a single fabric manager is used to resolve address resolution protocol requests, there is a greater chance of malicious attacks on the fabric manager. Similarly, in PortLand, every edge switch should have at least half of its ports connected to servers [1]. The authors have shown through implementation and simulation results that PortLand holds promise for supporting a plug-and-play large-scale data center networks.

A Scalable, Commodity Data Center Network Architecture The authors of [11] have proposed a scalable data center network architecture consisting of commodity machines (switches and routers). The proposed architecture enables any arbitrary host to communicate with another host in a data center network by fully utilizing its local network interface bandwidth. Moreover, it makes use of commodity switches and machines to build large scale data centers which helps to reduce cost. The authors have also claimed that their proposed architecture is backward compatible without any modification to hosts running Ethernet and IP protocols. It is assumed that the networking devices are arranged in a fat-tree topology in the proposed architecture. A novel addressing scheme is proposed based on the fat-tree topology in which private

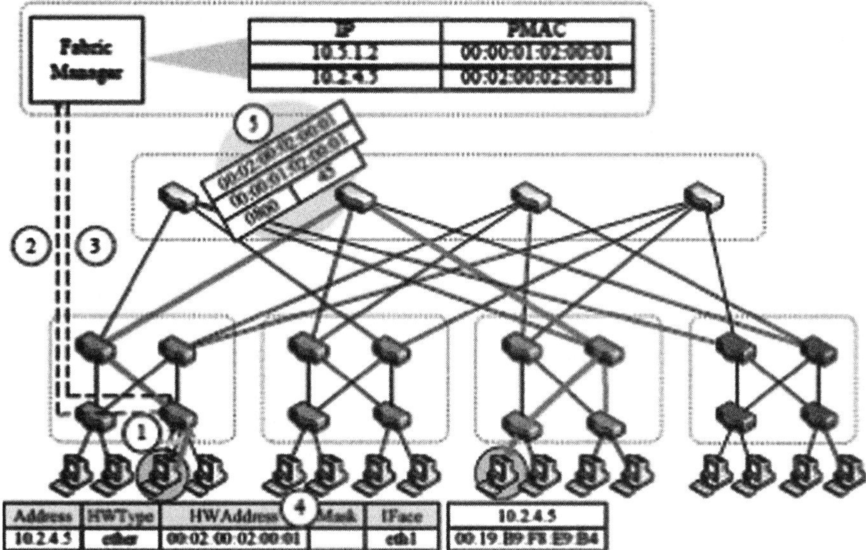

Fig. 5 Proxy ARP [7]

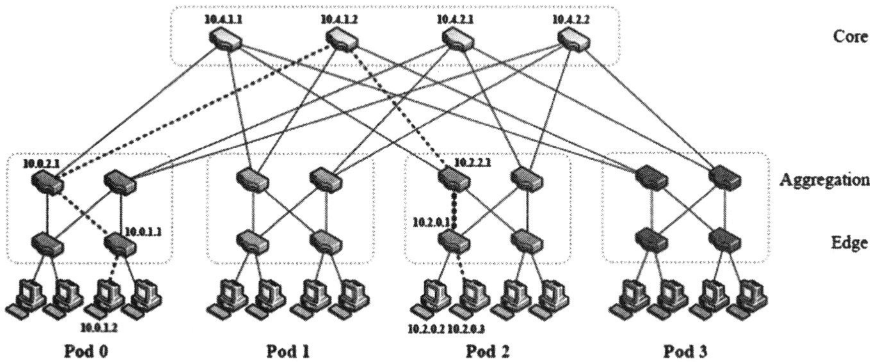

Fig. 6 Simple fat-tree topology. Using the two-level routing tables, packets from source 10.0.1.2 to destination 10.2.0.3 would take the dashed path [11]

addresses are assigned in the network and the format of the addressing scheme is *10.pod.switch.1* where *pod* determines the pod number and *switch* determines the switch number within a specific pod of a fat-tree topology. Figure 6 shows the addressing scheme in a fat-tree topology of a DCN.

In order to achieve maximum bisection bandwidth, which requires evenly distribution of traffic among the core switches of a DCN, the authors have extended an existing OSPF-ECMP [31] routing algorithm. They have introduced a two level prefix lookup, in which the primary table prefix points to the secondary table suffix. For a lookup query, If there is no entry against any prefix in the primary table then searching will be terminated, otherwise, longest suffix matching is done in the secondary table. It is possible that more than one prefix can point to a single suffix.

In order to understand the routing process in this architecture, let us consider an example. A source host 10.0.1.2 wants to send a packet to let us say this destination address 10.2.0.3 as shown in Fig. 6. First of all, the gateway switch near to the source will do the packet lookup with the /0 first-level prefix. It will then forward the packet based on the host ID byte according to the secondary table for that prefix. In that table, the packet matches the 0.0.0.3/8 suffix, which points to port 2 and switch 10.0.2.1. Switch 10.0.2.1 will also follow the same steps and forwards the packet on port 3 which is attached to core switch 10.4.1.1. The core switch upon lookup will forward the packet to the destination pod 2. After the packet has successfully reached the destination subnet, standard switching techniques are employed to deliver the packet to the destination host 10.2.0.3. The authors have performed extensive simulations and evaluations to validate their proposed scheme. They have implemented their proposed routing scheme including the two level lookup in the NetFPGA [32] using content-addressable memory (CAM) [17] which is quite efficient in searching.

Table 1 summarizes all the above mentioned topology-aware routing and forwarding schemes in data center networks. It states the objective of each algorithm, the procedures that are used in obtaining an efficient routing algorithm and the limitations of each scheme.

2.2 *Energy-Aware Routing*

Data centers consists of huge amount of devices. Each device consumes power. For this purpose efficient power saving routing protocols are needed so that energy can be saved by keeping those devices in sleep mode which are not participating in communication but without effecting DCN performance. For this purpose, in this section, we will explore some of the existing energy-aware routing protocols in data center networks.

2.2.1 Green Routing

In [9, 24], the authors propose energy aware routing which helps to save energy without compromising on routing performance. The energy consumed by power-hungry devices becomes a great trouble for many data centers owners. In the proposed concept such idle devices can be shut down. The authors have proposed a heuristic algorithm which first manipulates the network throughput by routing on all switches; this routing is called basic routing. Here throughput is taken as performance metric. Now one by one it removes switches until network throughput decreases to a predefine threshold value, and then it stops removing switches anymore. In one iteration, the algorithm removes one device and calculate the network throughput and compare it with the threshold value. If the network throughput is not less than the threshold, it

Table 1 Summary of the topology-aware routing approaches in DCNs

Techniques	Objective	Methodology	Limitations
BCube [6]	Design of network architecture for modular data centers (MDC)	Routing and forwarding to achieve load balancing, fault tolerance and graceful degradation	Control overhead, scalability issue, wiring cost
Dcell [8]	Design of a fault tolerant physical network infrastructure	DCell architecture, fault-tolerant distributed routing protocol	Wiring cost, not scalable because construction of complete graph at each level
VL2 [5]	Design of a network architecture for forwarding and routing that requires minimal or no data center hardware change	Routing with VLB and ECMP to provide load balancing, flat addressing for service portability	Not scalable due to flat addressing, does not guarantee absolute bandwidth, dependent on Clos topology
PortLand [7]	Scalable and fault tolerant layer-2 routing and forwarding protocol	Ethernet compatible routing, forwarding and address resolution protocol, loop free forwarding	Scalability due to layer-2 forwarding, only applicable for fat-tree topology, single point of failure because of fabric manager
M. Al-Fares [11]	Design of a DCN infrastructure to interconnect commodity switches in fat-tree topology	Extension of existing routing protocol, novel IP address assignment, two-level route lookups for multipath routing, fault tolerance scheme	Wiring overhead because of fat-tree topology

updates the topology and goes to the next iteration otherwise the algorithm terminates. In the end, the algorithm puts all those devices in sleep mode which are not part of the updated topology.

The proposed heuristic routing algorithm contains three modules. The first module is the Route Generation (RG) module and the second module is the Throughput Computation (TC) module. The third module is the Switch Elimination (SE) module. Now, we will explain each of the these modules proposed in [9].

Route Generation This module generates route for a selected traffic matrix at which the network throughput is maximum. The input inserted in this module is (G, T), where G is the data center topology and T is traffic matrix. RG() function returns all possible paths for T then select only that path at which the network gives maximum throughput. The data centre network is too large sometimes that it is not feasible to determine all paths. This also increases the complexity of the algorithm because of such large scale data centers. Moreover, we can take advantage of the topological structure of the data center. Similarly, for a specific flow, if several paths exists, the algorithm will select only that path on which least overlapping flows exists. If two or more flows have the same number of overlapping flows then select the one which have least number of hop count.

Throughput Computation This module computes the throughput of a network in a given topology. The authors have used a max-min fairness model to add up the throughputs of all flows.

Switch Elimination In this module, the algorithm will remove the switch which is carrying the least amount of traffic in a single iteration using a greedy approach. The algorithm works by selecting any active switch carrying least traffic load and removing it from the topology. The updated topology is then passed through the second round. It is also important to note that the eliminating switch should not be the critical one which means that by removing it would disconnect the topology. The authors have tested their proposed algorithm on BCube [6] data center architecture and found that at low network load significant amount of energy is saved. However, when network load increases, energy saving is decreased due to the fact that now almost all switches carry too much traffic load and it becomes difficult to remove all of them.

2.2.2 Power-Aware Routing

The authors have proposed a throughput guaranteed power aware routing algorithm in [13]. The objective of the proposed algorithm is to reduce power consumed by networking devices in densely connected data centers. The idle links or devices which are not involved in routing are put in sleep mode or shut down in order to save power. The authors have proposed a model of throughput guaranteed power aware routing and proved the time complexity of the algorithm to be NP-hard by reducing it to the famous 0–1 knapsack problem [25], The proposed model is a combination of general power consumption model and simplified model. They have called their proposed model a multi-granularity power saving strategy. Through this strategy, the power consumption of network devices can be reduced at both the device level and the component level. The propose heuristic algorithm consists of four steps. In step 1: basic routing is performed. Step 2 gradually removes the switches in the network and updates the topology without affecting the network throughput. In step 3, the algorithm keeps on removing links until network throughput start decreasing. The last step 4, puts all the devices in sleep mode which are not included in final topology.

In addition to the basic working of the proposed algorithm, it has five modules. These are the route generation (RG), throughput computation (TC), switch elimination (SE), link elimination (LE) and the reliability adaptation (RA) module. The relationship among these five modules is shown in Fig. 7.

The input to the algorithm is the quadruplet (*G*, *T*, *PR*, and *RP*). Where *G* denotes data center network topology, *T* denotes the traffic matrix. *PR* is the performance

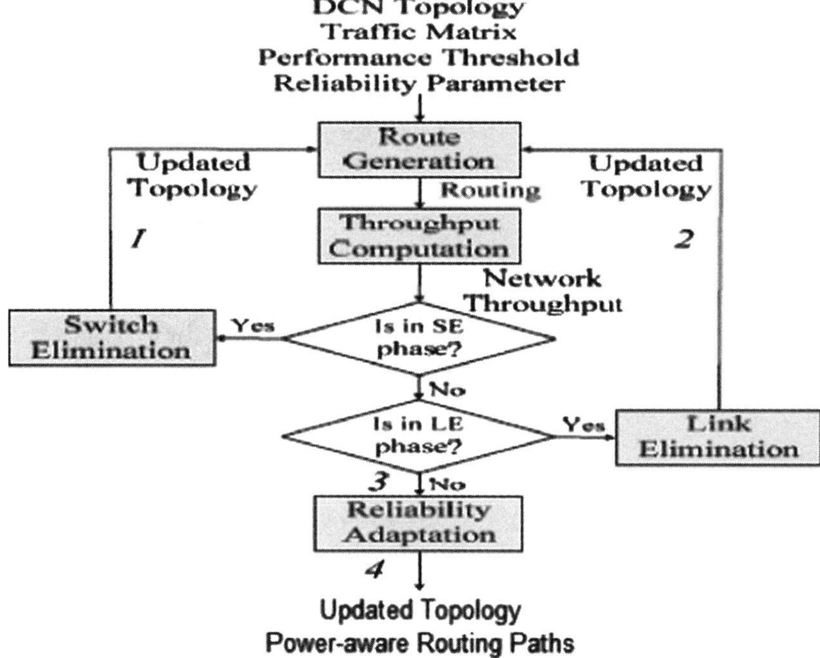

Fig. 7 Relationship among five modules of the proposed algorithm in [13]

threshold percentage and *RP* is the reliability parameter. As shown in Fig. 7, switch elimination module is initialized first followed by the link elimination module until the network performance in terms of the throughput starts to degrade. After this, the RA module takes the updated topology and adds some more links and network devices and gives updated topology.

Table 2 compares the energy-aware routing schemes surveyed above based on their objectives, the procedures that are used in obtaining an efficient routing algorithm and the limitations of each technique.

2.3 *Traffic-sensitive Routing*

Our third parameter for classification of existing routing schemes in data center networks is based on the type of traffic, a particular routing scheme is intended for. For example, whether a particular algorithm is designed for unicast, or multicast traffic or does it support multipath communication. We classify such routing algorithms as traffic-sensitive routing schemes. Note that all of the surveyed routing schemes in this classification have applied some form of traffic engineering to achieve their design objectives. [10,18–21] are the algorithms that fall in this category and we will explore them in this section.

Table 2 Comparison of Energy-aware routing schemes in data center networks

Techniques	Objective	Methodology	Limitations
Green Routing [9]	Reducing energy consumption in high density data center networks, without effecting network performance	Model formulation for energy-aware routing problem. Heuristic routing algorithm	Scalability issue, centralized approach, network disconnection
Power-aware Routing [13]	Minimize network power to provide routing services	Throughput guaranteed routing, max-min fairness model to calculate throughput	Single point of failure due to centralized approach, network disconnection

2.3.1 DARD

The authors of [18] have proposed a novel routing architecture called Distributed Adaptive Routing for Datacenter (DARD), which tries to fully utilize the bisection bandwidth typically common in DCNs in order to avoid any hot spots between any pair of hosts in the network. The authors have argued that the classical TCP protocol cannot dynamically select paths according to the current traffic load in the network which means that it is not adaptive to the network traffic load. Therefore, dynamic path selection mechanisms are adopted which have their own pros and cons and the authors have discussed those issues in [18]. However, to overcome the limitations of existing path selection mechanisms, the authors have proposed an end-to-end distributed path selection algorithm (DARD) that runs on the end hosts rather than on the switches and is load-sensitive to the dynamic traffic in the network. The proposed DARD architecture helps to move traffic from overloaded paths to the under loaded paths efficiently.

An overview of the proposed DARD architecture is shown in Fig. 8. As it can be seen in the figure, the proposed system consists of three components. One is the elephant flow detector which detects the if there is any elephant flows exists in the network. The second component is the path state monitor which monitors the traffic load on each link by periodically querying the switches and the third component is the path selector whose job is to move flows from overloaded paths to the under loaded paths.

Because of the distributed nature of the DARD algorithm, no end system or network device has complete network information and each end system has limited/local information based on which it selects paths. A hierarchical addressing scheme is used to uniquely identify end-to-end path with a pair of source and destination nodes. DARD adopts different paths to carry flows towards a destination. The purpose of doing so is to maximize utilization of link capacities between any pair of servers, and to avoid retransmissions caused by packet reordering. There are two paths in DARD one is the *Uphill Path* that is a partial path encoded by source address to keep entries

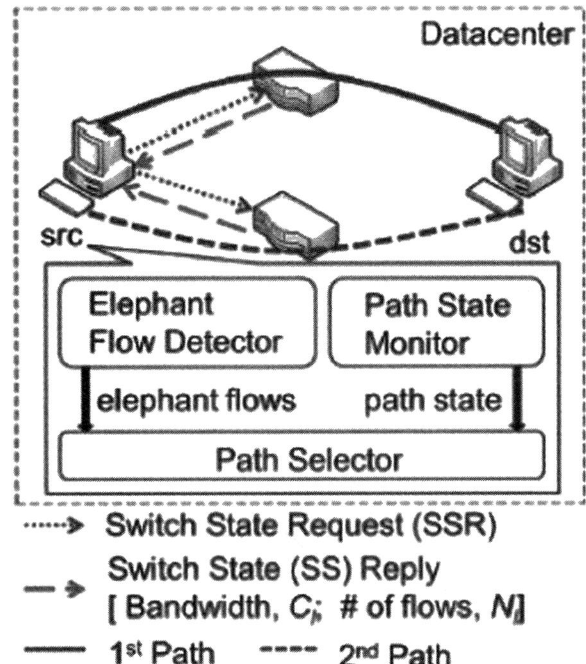

Fig. 8 DARD's system overview [18]

allocated from upstream nodes and the second is the *Downhill Path* that is a partial path encoded by destination address. The idea of splitting a path is taken from [33].

In DARD, when a packet is received by a switch, it notes the destination address and runs the longest prefix matching algorithm. If a match is found, it will forward the packet to the corresponding downstream switch, otherwise, the switch finds the source address in table and forwards the packet back to the corresponding switch in upward direction. The source adds source and destination address in the packet header and the intermediate switches view the header and forward the packet accordingly. When the destination node receives the packet, it de-encapsulates and passes the packet to the upper layer protocols. DARD notifies every end host about the traffic load in the network. Then each host will select appropriate paths for its outbound elephant flows according to the network conditions. Existing routing protocols can lead to oscillations and instability. The reason for this is that different resources move flows to under-utilized paths in a synchronized manner. In DARD, path oscillation is prevented by introducing a fixed span and a random span of time interval between two adjacent flow movements of the same end host.

The authors have performed extensive simulations and have implemented their prototype in a real test-bed. The authors have show that DARD is a scalable routing algorithm with reduced control overhead and provides efficient utilization of the bisection bandwidth in a data center network. The results have also shown that the DARD achieves fairness among elephant flows in the network.

```
GLOBAL-FIRST-FIT(f: flow)
    if f.assigned then
        return old path assignment for f
    foreach p ∈ P_src→dst do
        if p.used + f.rate < p.capacity then
            p.used ← p.used + f.rate
            return p
    else
        h = HASH(f)
        return p = P_src→dst(h)
```

a

```
SIMULATED-ANNEALING(n : iteration count)
    s ← INIT-STATE()
    e ← E(s)
    s_B ← s, e_B ← e
    T_0 ← n
    for T ← T_0 ... 0 do
        s_N ← NEIGHBOR(s)
        e_N ← E(s_N)
        if e_N < e_B then
            s_B ← s_N, e_B ← e_N
        if P(e, e_N, T) > RAND() then
            s ← s_N, e ← e_N
    return s_B
```

b

Fig. 9 **a** Pseudocode for global first fit. **b** Pseudocode for simulated annealing algorithm [19]

2.3.2 Hedera

In [19], the authors have proposed a centralized scalable and dynamic scheduler for flow routing to efficiently utilize aggregate network resources in data center networks. They called the flow scheduler as *Hedera*. The main objective of Hedera is therefore, to maximize the utilization of bisection bandwidth with least scheduling overhead in data center networks. Hedera achieves this objective by implementing a global view of the routing and traffic information. The scheduler works by collecting flow information from switches to determine the non-conflicting paths for flows, and then instructs the switches to forward traffic respectively. The scheduler measures the bandwidth utilization of each flow and when a flow demand is beyond a particular predefine threshold value, it is moved to an alternate route which fulfills the flow demand. The proposed scheduling system works by first detecting large flows at the edge switches and then it computes alternate paths by running placement algorithms (practical heuristic algorithms) to fulfill the bandwidth demand, followed by installing these paths on the switches.

Since, scheduling flows running in a network with the constraint that it should not exceed the capacity of any link in the network is called the multi-commodity-flow problem which is an NP-complete problem [34], the authors have proposed two practical heuristic algorithms for flow scheduling that can be applied to many topologies in DCNs. The first heuristic they have used for flow scheduling is the *Global First Fit* algorithm which is shown in Fig. 9a. With the possibility of having multiple paths between same source and destination pairs, when a new flow is detected, the scheduler greedily assigns the first path that accommodates the demands. This way, the algorithm does not guarantee that it will schedule all the flows but it performs relatively well in real scenarios. The second heuristic flow scheduling algorithm is based on *Simulated Annealing* shown in Fig. 9b. It is a probabilistic approach for efficiently computing paths for flows. The input to the algorithm is the set of flows to be scheduled and the flow demands. The algorithm searches through the solution space to find a near optimal solution to the scheduling problem. There is an energy function E that defines the energy in the current state. In each iteration, we move to another state with an acceptance probability P, which is a function of the current temperature and the energies of current and neighboring states. The algorithm terminates when the temperature hits zero.

The authors have implemented Hedera on a real test-bed and have also performed simulations to validate their findings. They have found that Simulate Annealing always outperforms Global First Fit flow scheduling algorithm and provides near optimal solution to the flow placement problem.

2.3.3 ESM: Multicast Routing for Data Centers

In [20], the authors have explored the design of an efficient and scalable multicast routing protocol for DCNs. Because of multiple interconnections among switches in a data center network, the traditional multicast routing protocols are inefficient in the formation of multicast trees in such networks. Moreover, switches in data centers have limited space to maintain multicast routing information which makes the design of a multicast routing protocols more challenging. To resolve these issues, the authors propose ESM which constructs efficient multicast trees using source-to-receiver expansion approach rather than the traditional receiver-driven multicast routing approach. To reduce the routing table size, ESM, combines In-packet Bloom Filters and in-switch entries to make it a scalable multicast routing protocol.

For the construction of multicast tree for data forwarding, ESM implements a multicast manager. The manager builds the multicast tree using source-to-receiver expansion approach which basically removes many unnecessary switches in the formed multicast tree. The problem of computing efficient multicast tree can be translated into the problem of generating Steiner tree in general graphs and Steiner tree problem in BCube has been proven to NP-Hard by the authors in [35]. They have also shown that the heuristic algorithms to solve Steiner tree problem has a high computational complexity and therefore, they have designed an approximate algorithm to generate multicast tree for routing in data center networks.

ESM, aggregates multiple multicast routing entries into a single entry by using in-packet bloom filters. Actually, the tree information is encoded into the in-packet bloom filter instead of the usual multicast routing entries in the switch. However, there is a drawback in using in-packet bloom filters that is the network bandwidth is wasted. Therefore, the authors have combined both the in-packet bloom filter and the traditional in-switch entries to achieve a scalable multicast routing in DCNs.

The authors have tested ESM's source-to-receiver approach and compared it with the traditional receiver-driven approach for multicast routing. They have found that ESM saves network traffic for tree building by 50 % and also doubles the application throughput when compared with receiver-driven approach. Similarly, they have found that combining in-packet bloom filter with in-switch entries significantly reduces the number of routing entries inside the switches.

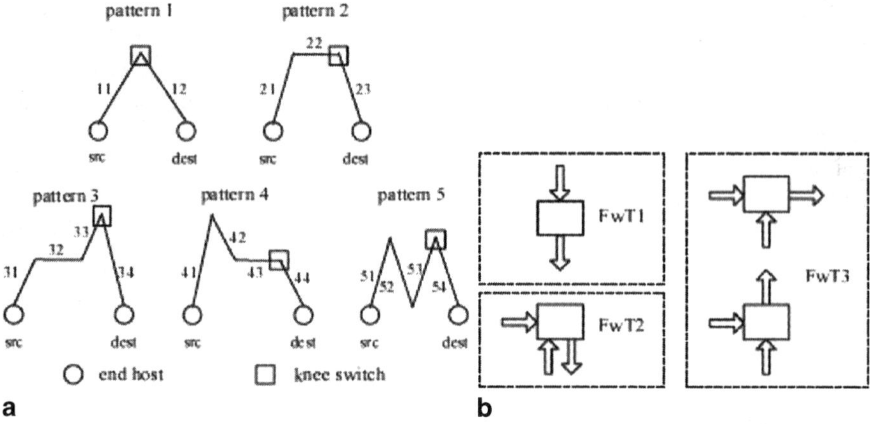

Fig. 10 **a** Knee switch example. **b** Forwarding model [10]

2.3.4 GARDEN

The authors of [10] have proposed a Generic Addressing and Routing for Data Center Networks (GARDEN) protocol. There are many routing and forwarding schemes which are designed for a particular network topology in a data center such as VL2 [5], BCube [6], Portland [7], and DCell [8]. The main idea in this paper is to design a forwarding scheme for data center networks which is not dependent on any particular underlying topology. In the proposed system, a multi-rooted tree is first constructed, so that the addressing of the nodes in the tree can be defined. Then a downward path from each node to the root is constructed, and a locator is assigned to each path. These locators are actually assigned to each host or switch in the network based on the newly created multi-rooted tree topology. As the name suggests, locators encode the location of a particular node in the tree. This effectively means that the locators are used for forwarding instead of the usual MAC addressing. This significantly reduces the forwarding states in the routing tables on each switch in the network as locators are addressed according to the hierarchical structure of the multi-rooted tree.

The authors have proposed a forwarding model to populate the forwarding tables in the switches. They have tried to explain their proposed forwarding model with the help of an illustrative example as in Fig. 10a. A single path from the source node to the destination may consists of many steps (segments). As it can be seen in Fig. 10a, some segment has an upward direction (*upward e.g. 21*), some have downward direction (*downward e.g. 23*) and some have straight direction i.e. *horizontal* (e.g. 32). The first switch towards the last downward path segment is called a *knee* and the highest switch among all switches is called a *peak* switch. Based on this example, the authors have applied two traffic constraints. The first constraint is that the traffic should change the direction from *upward* to *downward* only once and the second constraint is that the *knee* switch must be among the *peak* switches in a path. Now,

Table 3 Comparison of the four traffic-sensitive routing techniques in data centers

Techniques	Objective	Methodology	Limitations
DARD [16]	Design of load sensitive adaptive routing	Routing algorithm based on simulated annealing, provable convergence to Nash equilibrium	Requires modification at the end hosts
Hedera [19]	Maximize aggregate network utilization (bisection bandwidth)	Global flow scheduler, two heuristic flow scheduling algorithm: simulated annealing and global first fit	Not scalable as it is centralized
D. Li [20]	Design of efficient multicast tree for routing with minimum switches	In-pocket bloom filter and in-switch routing for scalable multicast routing, node based bloom filter to encode tree	Complexity, centralized, ignore control overhead
GARDEN [10]	Design of generic addressing, routing and forwarding protocol for DCNs	Central controller to form multi-rooted tree, addressing, forwarding model to reduce states, load balancing, and fault-tolerance	Scalability, traffic load not considered, offline solution

the switch traffic can be divided into three forwarding classes based on the above mentioned two constraints i.e., FwT1, FwT2 and FwT3. Figure 10b shows the block diagram of the proposed forwarding model.

In order to implement and evaluate GARDEN, the authors have built different sized testbeds of fat tree topologies. and compared the GARDEN forwarding states for different topologies, They have shown that GARDEN provide good fault tolerance and improved aggregate network throughput using different scheduling methods.

We compare the four algorithms surveyed above based on their objectives, the methodology that are used to provide efficient routing and forwarding, and the limitations of each approach in Table 3.

2.4 *Routing for Content Distribution Networks (CDN)*

In this section, we will study some of the existing routing schemes specifically designed for content distribution/delivery networks (CDNs). We have classified the routing schemes in data center networks based on this parameter. A CDN is an overlay network over multiple data center networks located in geographically dispersed locations and are accessible via internet [14]. The primary purpose of CDNs is to provide content to its subscribed users by maintaining many surrogate or replica

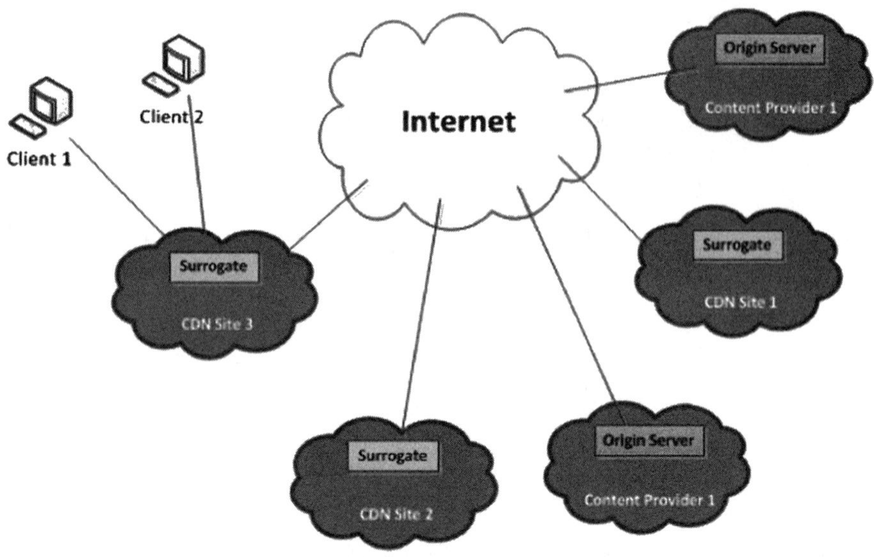

Fig. 11 A typical content distribution network (CDN)

servers on locations which are geographically close to the end user. A typical content distribution network is shown in Fig. 11. The protocols that have been proposed for content distribution networks are mostly propriety in nature and are not standardized as of yet. However, the IETF [36] has initiated a working group called Content Delivery Networks Interconnections (CDNI [37]) to standardize the protocols in the domain of CDNs.

In addition to routing schemes in CDNs, we will also survey some of the routing schemes in data centers that do not completely lie in any of the above mentioned classification parameters and therefore are also discussed in this section.

2.4.1 Request-Routing in CDNs

In [15], the authors have described CDN based request routing techniques that have been implemented by researchers. A typical CDN contains many surrogate servers, a distribution system, a request-routing system, and accounting systems. Surrogate servers are not the origin servers which are the servers that actually own the contents, instead, their responsibility are just to deliver the content closest to the end user.

When a client request for content, it reaches to the geographically nearest located surrogate server which maps the request to deliver the content information. This mapping is done by the *Distribution system* who interacts with the request-routing and accounting system for determining the availability of the requested content in different surrogate servers and the volume of content distribution respectively. The request routing system [38, 39] consisting of request routers directs the client request to the appropriate surrogate server by taking into consideration the network conditions and the load on surrogate server.

The authors have discussed many solutions to the request routing problem in CDNs. The first is the use of DNS servers for request routing in CDNs. In such routing, the client sends a name based query to the local Domain Name Server (DNS). If the name exists, the DNS returns the address of the corresponding surrogate server that is nearest to the client location. If look up fails, the local DNS forwards the request to the DNS root. The DNS root server returns the address of the authoritative DNS server. The authoritative DNS will return address of the surrogate server nearby to the client. Then the client will get the content from the appointed surrogate server. But, this type of request routing has limitations as mentioned in [15] such as, the inherent scalability issue in DNS multi-level redirections. The second solution could be implemented at the transport-layer in which the requesting client's IP address and port number will be checked and a session is established with corresponding surrogate server. Similarly, application-level request routing solutions also exist which are normally divided into two types i.e. header inspection and content modification. In header inspection, Uniform Resource Locator (URL) of the requested content is used for taking routing decisions. In the content modification method, the content provider can modify the references of the embedded items belonging to a particular content item, and the client can fetch the embedded item from the best designated surrogate server.

In [16] M. Wahlisch et. al., have done a measurement study on the efficiency of content centric routing under various offered load. It is found that service efficiency decreases due to the exhaustion of memory and processing resources due to the excessive state allocations in such CDN based routing protocols.

2.4.2 Symbiotic Routing

The authors of [22] have proposed symbiotic routing in data center networks. In traditional data center network architecture, one server is associated with other servers using network devices like switches and routers, however, in *CamCube* which is a DCN topology proposed in [40], one server is attached with other servers directly and forms a 3D torus topology which uses Content Addressable Network (CAN). In CamCube architecture, each service implements its own routing protocol which results in better performance at the service level. Each application will have a unique id and on every server, many application instances can be run. When a packet is received from any link, the kernel will be responsible for its delivery and inform via call-back to the service running in the user space about the source link on which the packet was received. The authors have used a single queue mechanism per link. When a packet is transmitted on a link, all the services will be polled to determine to which service this packet belongs and the packet is queued to that service queue. The routing service in the proposed architecture performs the key-based routing. Keys in the key-based routing determine the location of the server in the form of coordinate space. Using key-based routing, all packets are directed towards the server, who is responsible for the destination key.

The authors have evaluated individual services and concluded that by applying their proposed optimized protocols, better application-level performance can be achieved and network load can be reduced.

2.4.3 fs-PGBR: A Scalable and Delay Sensitive Cloud Routing Protocol

In [23], the authors propose an efficient routing protocol for accessing resources with minimum delay in the cloud. The proposed algorithm is called fast search-Parameterized Gradient Based Routing (fs-PGBR), which is fully distributed, scalable and provides minimum latency during resource discovery. The proposed algorithm is an extension of PGBR [41], the idea of which is inspired from *chemotaxis* (a bacteria motility process). In order to solve the typical scalability issue in existing PGBR problem, the authors have modified the hop count formula to make it scalable. With the modifications, each node compares the shortest path of its neighbours and evaluates it against the maximum and minimum weight of its neighbours. It is concluded that fs-PGBR is scalable its resource discovery delays are minimum due to its shortest path adoption policy.

2.5 *Summary of All Routing and Forwarding Techniques*

We have provided the survey of existing routing and forwarding techniques in data center networks in the previous section and summarized them based on their objectives, methodologies and limitations. We now present an overall summary of all the routing approaches mentioned above in Table 4, which shows the comparison of these schemes based on six characteristics which are briefly described as follows:

Implementation How the routing and forwarding is performed in the proposed algorithm i.e. in a centralized or distributed fashion?

L2/L3 Is the proposed routing and forwarding performed at Layer-2 (ethernet) or Layer-3 (network) or Layer 2.5 (shim layer) of the TCP/IP stack in a data center network?

Server/Switch Centric Who is involved for forwarding in a proposed routing technique in a data center i.e. switches or servers?

Traffic What type of traffic/communication does the proposed routing protocol support i.e. unicast, or multicast?

H/W Change Does the algorithm requires any switch or server modifications?

Table 4 Summary of characteristics of all routing approaches in data center networks

Characteristics	Implementation	L2/L3	Server/Switch centric	Traffic	H/W change	Topology
BCube [6]	Distributed	L3	Server-centric	Unicast	No	Hypercube
Dcell [8]	Distributed	L3	Server-centric	Unicast	No	Custom
VL2 [5]	Distributed	L2.5	Switch-centric	Unicast	No	Clos
PortLand [7]	Centralized	L2	Switch-centric	Multicast	Yes	Fat-tree
M. Al-Fares [11]	Centralized	L3	Switch-centric	Unicast	Yes	Fat-tree
Green Routing [9]	Centralized	L3	Switch-centric	Unicast	No	Generic
Power-aware Routing [13]	Centralized	L3	Switch-centric	Unicast	Yes	Generic
DARD [16]	Distributed	L3	Server-centric	Unicast	No	Generic
Hedera [19]	Centralized	L3	Switch-centric	Unicast	Yes	Fat-tree
D. Li [20]	Centralized	L3	Switch-centric	Multicast	Yes	Generic
GARDEN [10]	Centralized	L3	Switch-centric	Unicast	No	Generic
Request-routing [15]	Centralized	L3/4/5	Switch-centric	Unicast	No	Generic
Symbiotic routing [22]	Distributed	L3	Server-centric	Multicast	Yes	Generic
Fs-PGBR [23]	Distributed	L3	Switch-centric	Unicast	No	Generic

Topology Does the routing scheme takes into account the underlying physical network topology for generating paths and for forwarding? Is it designed for a particular data center structure or is independent of any DCN topology i.e. generic.

3 Open Issues and Challenges

In spite of a reasonable amount of research in the late literature, there are still some challenges and open issues that need to be addressed in designing efficient data forwarding and routing schemes particularly in DCNs. Below, we outline what we believe some of these challenges and open issues are.

Topology Independent Routing As explained in Sect. 2.1, most of the existing routing and forwarding schemes are designed for a particular data center network topology. There is a need to design efficient routing protocols for generic DCN topologies. Very few routing schemes have addressed this issue of topology independent routing and therefore, we believe that more research needs to be done in this area.

Load-balancing and Reliability More robust and efficient routing schemes are required to intelligently balance the load among multiple paths between a source and destination server in a data center network. Although, the existing schemes do partially solve the load balancing issue in DCNs, but a complete solution in terms of a routing and forwarding algorithm which is able to perform load balancing and also provides fault-tolerance is still one of the major challenges of routing in DCNs.

Energy-efficiency Data centers consume lots of energy, therefore, energy conservation or power saving in data centers is a hot research field nowadays. Although, some work has been done in designing efficient power-aware or energy-aware routing algorithms and we have gone through some of them in this chapter, but the concept of energy conservation in data centers is new and it is still progressing and evolving to its maturity.

Lack of Secure Routing Protocols There is a growing need to design secure routing protocols to protect both the control plane and the data plane of a routing infrastructure in a data center network. Securing control plane means that the topology discovery mechanism should be protected and data plane security relates to protecting the flow data from tampering or modification. The control plane security of a data center routing infrastructure can be made possible by the fact that most of the existing data center routing schemes are tightly coupled with the underlying physical network interconnections and topologies. However, to the best of our knowledge, there is no secure routing protocol particularly designed for data centers with the focus of securing data plane of a DCN routing infrastructure. We believe the reason for the lack of research in secure routing is because data centers as a whole is still not fully matured yet and therefore, a lot of research potential still exists in the field of routing data and control plane security in DCNs.

4 Conclusions

In this chapter, we have discussed the problem of data forwarding and routing in data center networks. We have defined various types of communication patterns that are commonly used in DCNs and have provided different data center structures and topologies that most of the existing routing techniques in DCNs make use of. We then presented a survey of some of the existing forwarding and routing techniques in DCNs by classifying them using parameters such as energy, traffic and topology aware upon which many routing schemes are based. We discussed the rationale behind using each of these parameters for our classification. We have also tabulated a comparative summary of all the discussed routing schemes based on six characteristics. Finally, we discussed some of the challenges and open issues in designing efficient routing schemes in DCNs.

References

1. Md. Faizul Bari, Raouf Boutaba, "Data Center Network Virtualization: A Survey", Communications Surveys & Tutorials, IEEE (Volume:15, Issue: 2).
2. S. Ghemawat, H. Gobioff, and S. Leung. "The Google File System", in SOSP, 2003.
3. CloudStore. Higher Performance Scalable Storage. http://kosmosfs.sourceforge.net/.
4. J. Dean and S. Ghemawat, "MapReduce: Simplified Data Processing on Large Clusters," in Proc. USENIX OSDI, December 2004.
5. A. Greenberg, J. Hamilton, N. Jain and etc., "VL2: A Scalable and Flexible Data Center Network", In Proceedings of ACM SIGCOMM'09, Aug 2009.
6. C. Guo, G. Lu, D. Li, H. Wu, X. Zhang, Y. Shi, C. Tian, Y. Zhang, and S. Lu, "BCube: A High Performance, Server-centric Network Architecture f or Modular Data Centers," in Proc. ACM SIGCOMM, August 2009.
7. R. Mysore, A. Pamboris, N. Farrington and etc., "PortLand: A Scalable Fault-Tolerant Layer 2 Data Center Network Fabric", In Proceedings of ACM SIGCOMM'09, Aug 2009.
8. C. Guo, H. Wu, K. Tan, L. Shi, Y. Zhang, and S. Lu, "Dcell: A scalable and fault-tolerant network structure for data centers," SIGCOMM 2008.
9. Yunfei Shang, Dan Li, Mingwei Xu, "Green Routing in Data Center Network: Modeling and Algorithm Design". Proceedings of the first ACM SIGCOMM workshop on Green networking.
10. Yan Hu, Ming Zhu, Yong Xia, Kai Chen "GARDEN: Generic Addressing and Routing for Data Center Networks", Cloud Computing (CLOUD), 2012 IEEE 5th International Conference.
11. M. Al-Fares, A. Loukissas, and A. Vahdat, "A Scalable, Commodity Data Center Network Architecture," in Proc. ACM SIGCOMM, August 2008.
12. Kai Chen; Chengchen Hu; Xin Zhang; Kai Zheng; Yan Chen; Vasilakos, A.V.," Survey on routing in data centers: insights and future directions", Network, IEEE (Volume:25, Issue: 4)
13. Mingwei Xu, Yunfei Shang, Dan Li, Xin Wang, "Greening Data Center Networks with Throughput-guaranteed Power-aware Routing", Computer Networks Volume 57, Issue 15, 29 October 2013, Pages 2880–2899.
14. http://en.wikipedia.org/wiki/Content_delivery_network.
15. Md. Humayun Kabir, Eric G. Manning, Gholamali C. Shoja, "Request-Routing Trends and Techniques in Content Distribution Network".
16. Matthias, Thomas, M.vahlenkamp, "Bulk of interest: performance measurement of content-centric routing", ACM SIGCOMM Computer Communication Review—Special october issue SIGCOMM '12.
17. L. Chisvin and R. J. Duckworth. "Content-Addressable and Associative Memory: Alternatives to the Ubiquitous RAM". Computer, 22(7):51–64, 1989.
18. Wu, Xin; Yang, Xiaowei, "DARD: Distributed Adaptive Routing for Datacenter Networks", Distributed Computing Systems (ICDCS), 2012 IEEE 32nd International Conference
19. M. Al-Fares, S. Radhakrishnan, B. Raghavan, N. Huang, and A. Vahdat. "Hedera: Dynamic flow scheduling for data center networks". In Proceedings of the 7th ACM/USENIX Symposium on Networked Systems Design and Implementation (NSDI), San Jose, CA, Apr. 2010.
20. Dan Li, Jiangwei Yu, Junbiao Yu, Jianping Wu, "Exploring efficient and scalable multicast routing in future data center networks", INFOCOM, 2011 Proceedings IEEE.
21. Schlansker, Turner, Tourrilhes, Karp, "Ensemble routing for datacenter networks", Architectures for Networking and Communications Systems (ANCS), 2010 ACM.
22. H. Abu-Libdeh et al., "Symbiotic Routing in Future Data Centers," SIGCOMM, 2010.
23. Julien Mineraud, Sasitharan Balasubramaniam, Jussi Kangasharju and William Donnelly, "Fs-PGBR: a scalable and delay sensitive cloud routing protocol", ACM SIGCOMM Computer Communication Special october issue SIGCOM, Volume 42 Issue 4, October 2012.
24. Yunfei Shang, Dan Li, Mingwei Xu, "Energy-aware Routing in Data Center Network". Proceedings of the first ACM SIGCOMM workshop on Green networking.
25. R.M.Karp. Reducibility Among Combinatorial Problems, in R.E.Miller and J.W. Thatcher (Eds.), Complexity of Computer Computations. Plenum Press, New York, 1972.

26. J. Moy. OSPF: Anatomy of an Internet Routing Protocol. Addison-Wesley, 2000.
27. http://en.wikipedia.org/wiki/Breadth-first_search
28. W. J. Dally and B. Towles. Principles and Practices of Interconnection Networks. Morgan Kaufmann Publishers, 2004.
29. M. Kodialam, T. V. Lakshman, and S. Sengupta. "Efficient and Robust Routing of Highly Variable Traffic". In HotNets, 2004.
30. R. Zhang-Shen and N. McKeown. "Designing a Predictable Internet Backbone Network". In HotNets, 2004.
31. D. Thaler and C. Hopps. Multipath Issues in Unicast and Multicast Next-Hop Selection. RFC 2991, Internet Engineering Task Force, 2000.
32. J. Lockwood, N. McKeown, G. Watson, G. Gibb, P. Hartke, J. Naous, R. Raghuraman, and J. Luo. "NetFPGA–An Open Platform for Gigabit-rate Network Switching and Routing". In IEEE International Conference on Microelectronic Systems Education, 2007.
33. X. Yang. "Nira: a new internet routing architecture". In FDNA '03: Proceedings of the ACM SIGCOMM workshop on Future directions in network architecture, pages 301–312, New York, NY, USA, 2003. ACM.
34. EVEN, S., ITAI, A., AND SHAMIR, A. "On the Complexity of Timetable and Multicommodity Flow Problems". SIAM Journal on Computing 5, 4 (1976), 691–703.
35. D. Li, J. Yu, J. Yu, and J. Wu, "Exploring efficient and scalable multicast routing in future data center networks," in Proc. IEEE INFOCOM, Apr. 2011, pp. 1368–1376.
36. http://www.ietf.org/
37. http://datatracker.ietf.org/wg/cdni/
38. M. Day, B. Cain, and G. Tomlinson, "A Model for CDN Peering", http://www.contentalliance.org/docs/draft-daycdnp-model-03.html (work in progress), November 2000.
39. B. Cain, F. Douglis, M. Green, M. Hofmann, R. Nair, D. Potter, and O. Spatscheck, "Known CDN Request-Routing Mechanisms", http://www.contentalliance.org/docs/draft-caincdnp-known-req-route-00.html (work in progress), November 2000.
40. P. Costa, A. Donnelly, G. O'Shea, and A. Rowstron. "CamCube: A Key-based Data Center". Technical Report MSR TR-2010–74, Microsoft Research, 2010.
41. S. Balasubramaniam, J. Mineraud, P. Mcdonagh, P. Perry, L. Murphy, W. Donnelly, and D. Botvich. "An Evaluation of Parameterized Gradient Based Routing With QoE Monitoring for Multiple IPTV Providers". IEEE Transactions on Broadcasting, 57(2):183–194, 2011.

Part III
Cloud Computing

Auditing for Data Integrity and Reliability in Cloud Storage

Bingwei Liu and Yu Chen

1 Introduction

As a new computing paradigm, cloud computing has enhanced the data storage centers with multiple attractive features including on-demand scalability of highly available and reliable pooled computing resources, secure access to metered services from nearly anywhere, and displacement of data and services from inside to outside the organization. Due to the low cost of storage services provided in the cloud, compared with purchasing and maintaining storage infrastructure, it is attractive to companies and individuals to outsource applications and data storage to public cloud computing services.

Outsourcing data to remote data centers that are based on cloud servers is a rapidly growing trend. It alleviates the burden of local data storage and maintenance. Security and privacy, however, have been the major concerns that make potential users reluctant to migrate important and sensitive data to the cloud. The fact that data owners no longer possess their data physically forces service providers and researchers to reconsider data security policies in the storage cloud. On one hand, evidences such as data transmission logs can prevent disputation among users and service providers [8–11]; on the other hand, the service providers need to convince users that their data stored in the cloud is tamper free and crash free, and that their data can be retrieved anytime when needed. Traditional cryptographic methods cannot meet these new challenges in the new paradigm of cloud storage environments. Downloading the entire data set to verify its integrity is not practical due to constraints of the communication network and the massive amount of data.

B. Liu (✉)
Department of Electrical and Computer Engineering, Binghamton University,
State University of New York, Binghamton, New York, USA
e-mail: bliu@binghamton.edu

Y. Chen
Department of Electrical and Computer Engineering, Binghamton University,
State University of New York, Binghamton, New York, USA
e-mail: ychen@binghamton.edu

S. U. Khan, A. Y. Zomaya (eds.), *Handbook on Data Centers*,
DOI 10.1007/978-1-4939-2092-1_17

Integrity and reliability of data in the cloud are not inherently assured. On the one hand, cloud service providers themselves face the same threats that traditional distributed systems need to handle. On the other hand, cloud service providers have incentive to hide data loss or to discard parts of user data without informing the user, since they aim at making profit and need to maintain their reputation. Trusted third party (TTP) based auditing is promising to solve this dilemma. Therefore, a customized auditing scheme is desired, which is expected to keep track of accesses and operations on stored data in the cloud. The recorded information is also essential for digital forensics or disputation resolving. In this chapter, we will discuss the rationale and technologies that are potentially capable of meeting this important challenge in the storage cloud.

There are technologies to verify the retrievability of a large file F in its entirety on a remote server [1, 3, 12]. Juels and Kaliski [12] have developed Proof of Retrievability (POR), a new cryptographic building block. The POR protocol encrypts a large file and randomly embeds randomly-valued check blocks, called sentinels. To protect against corruption by the prover of a small portion of F, they also employed error-correcting codes in the POR scheme. The tradeoff of these sentinel-based schemes is that preprocessing is required before uploading the file to remote storage. Because sentinels must be indistinguishable from regular file blocks, POR can only be applied to encrypted files and has a limited number of queries that are decided prior to outsourcing.

A Provable Data Possession (PDP) scheme [1] allows a user to efficiently, frequently, and securely verify that the server possesses the original data without retrieving the entire data file and provides probabilistic guarantees of possession. The server can only access small portions of the file when generating the proof of its possession of the file. The client stores a small amount of metadata to verify the server's proof.

However, PORs and PDPs mainly focus on static, archival storage. Considering dynamic operations in which the stored data set will be updated, such as inserting, modifying, or deleting, these schemes need to be extended accordingly. Dynamic Provable Data Possession (DPDP) schemes [7] aim to verify file possession under these situations.

Due to constraints at user side such as limited computing resources, researchers also seek solutions that migrate the auditing task to a third party auditor (TPA). This approach will significantly reduce users' computing burden. However, new challenges appear. Privacy protection of users' data against external auditors becomes a major issue. Privacy-preserving public auditing has attracted a lot of attention from the cloud security research community.

The rest of the chapter is organized as follows. The basics of information auditing are introduced in Sect. 2. Section 3 discusses the principles of POR and PDP schemes and illustrates several typical implementations. Section 4 presents recent reported efforts considering privacy-preservation in cloud storage. Section 5 discusses several open questions and indicates potential research directions in the future. Finally, we conclude this chapter in Sect. 6.

2 Information Auditing: Objective and Approaches

The past decades have witnessed the rapid development of information technologies and systems. Such an evolution has made system architecture very complex. Information auditing plays the central role in effective management since it is critical to any organization to obtain a good understanding of information storage, transmission, and manipulation. As more and more components have been introduced, the focus and definition of information auditing are expanded. In this section, a definition of information auditing is given first. Then, three typical approaches are discussed that actually reflect the particular view of an auditor focusing on an organization.

2.1 Definition of Information Auditing

In past 30 years, the application of information auditing has been extended from identifying formal information sources, which emphasizes document management, to monitoring the information manipulations on the organizational level. As an independent, objective assurance and consulting activity, information auditing helps to add value and improve operations. It provides clients and service providers information for internal control, risk management, and so on. Defined by the ASLIB Knowledge & Information Management Group [6], information auditing is:

> A systematic examination of information use, resources and flows, with a verification by reference to both people and existing documents, in order to establish the extent to which they are contributing to an organization's objectives.

According to this definition, information auditing could include one or more of the following objectives:

- Identifying control requirements
- Supporting vender selection
- Reviewing vendor management
- Assessing data migration
- Assessing project management
- Reviewing/assessing/testing control flow
- Logging digital footprints for forensics

Corresponding to the objectives, the following are questions an information auditing system is expected to address:

- **Data**: What information does this system store, transfer, or manipulate?
- **Function**: How does the system work? What has done to the data?
- **Infrastructure**: Where are the system components and how are they connected?

- **User**: Who launches the work? What is the work flow model?
- **Time**: When do events happen? How are they scheduled?
- **Motivation**: Why are functions executed? What are the goals and strategies?

2.2 *Three Approaches of Information Auditing*

Considering the complexity of today's information systems, an IT manager may be interested in certain components of an organization instead of all components. To allow more dimensions to auditing, an auditor can adopt particular views against an organization and variant approaches can be taken. Three approaches are suggested by researchers: strategic-oriented, process-oriented, and resource-oriented. The strategic-oriented approach focuses on the routines by which an organization achieves its strategic objectives under the constraints of available information resources. The expected output in this dimension would be an information strategy for the organization. Typically a strategic-oriented information auditing system should consider the following questions:

- **Goal**: What is this system for?
- **Approaches**: How can we achieve this goal?
- **Resources**: What information/infrastructure resources do we have/use?
- **Constraints**: Is there any resource/performance gap/constraints?
- **Essential Concerns**: What are the most essential concerns?

The process-oriented approach focuses on a process, which is a sequence of activities the system takes to achieve the expected outcomes. Processes reflect system characteristics and reveal how information flows and how functions cooperate. There are four main types of processes: core processes, support processes, management processes, and business network processes [14]. The key output would include processed-based mapping and information flow/resources analysis. Typical questions [6] a process oriented information auditing system will answer include:

- **Activities**: What do we do?
- **Approaches**: How do we do it?
- **Attestation**: How can we prove we do what we say we will do?
- **Resources**: What information resources do we use and require?
- **Facilities/Tools**: What systems do we use?
- **Concerns**: What problems do we experience?

The resource-oriented approach aims at identifying, classifying, and evaluating information resources. Instead of associating resources with a strategic goal or an operational process, the major purpose of the resource-oriented approach is to allow auditors to manage or categorize resources according to strategic importance or according to their ability to support critical processes. Questions [6] that resource-oriented information auditing systems should address include:

- **Identification**: What are the information resources?
- **Utilization**: How are the information resources used?
- **Management**: How does the system manage and maintain them?
- **Policy**: What are the regulations of utility?
- **Priority**: Which are the most critical information resources? Which are useless?

Considering the properties and security expectations of cloud storage, process-oriented auditing is the most suitable candidate among the three approaches. Information collected in this dimension will provide sufficient evidence for both digital forensics and the reputation estimation of the data storage service provider. However, there is no reported effort that tries to develop such a process-oriented auditing system for cloud computing services. We hope this book chapter can inspire more activities in this important area.

3 Auditing for Data Integrity in Distributed Systems

One of the important applications of distributed storage service systems is to store large research data files that are not frequently accessed but cannot be reproduced because the devices that collected the data are unavailable or because of the expense. A client might choose to store such data in remote a storage system provided by trusted professional services. Usually the data files will be replicated in case one of the storage servers is unavailable because of maintenance or disk damage. For each server that possesses the client's data files, both the client and the service provider need to assure that each file is retrievable in its entirety whenever the client needs. It is not practical to download the entire file to verify its integrity when dealing with large archive file. Users (data owners) need to be assured that their outsourced data is not deleted or modified at the server, without having to download the entire data file.

In this section, we investigate the general categories of strategies for auditing data integrity in distributed systems. Then three popular schemes are discussed in more details.

3.1 Strategies of Auditing Data Integrity

A straightforward way to assure the integrity of our data is to utilize message authentication code (MAC). The client, who wants to use the storage service of a server, calculates a short MAC for each file block, save them in local storage and upload his data to the server. To increase the security of data, the client can choose to encrypt the data before MAC calculation. When the client needs to check the server's possession of the data, he simply asks the server to calculate MACs for all file blocks and send them as a proof of possession. The obvious problem of this simple strategy is the huge overheads of computation and communication. In order to make sure the entirety of data, the server needs to access all file blocks, executing expense computation when the file is large. The communication cost to transmit all MACs is also unacceptable.

More practical strategies in auditing data integrity can be divide into two categories:

1. **Sentinel Embedding**. The strategy is to utilize sentinels produced by the client to secure data integrity. Sentinels are created by a one-way function. By appending the predefined number of sentinels to the encoded file and permuting the resulting file, the client is able to check fixed number of sentinels during each challenge period to the server. Since the server has no knowledge about the position of these sentinels, it cannot modify any block of the client's data without being detected in one or several challenges that could ask for the entirety of any block.
2. **Random Sampling Authenticators**. The other way to audit the integrity of data is based on authenticators. An authenticator is produced for each file block before uploading data to the server. The client only stores some metadata, such as cryptographic keys and functions, and uploads his data along with authenticators to the server. The key of this strategy is the algorithm that we use to calculate the authenticators. This algorithm should be able to verify the integrity in an aggregating way so that the proof from the server will not be proportional to the number of blocks that we want to check.

Juels and Kaliski [12] proposed the Proof of Retrievability (POR) based on sentinel embedding. Although it is a strong protocol for data integrity, there is one inevitable problem. The number of sentinels is predefined, causing a fix number of challenges. This is unacceptable in some applications. In order to obtain higher confidence of data integrity, the entire file needs to be retrieved to embed more sentinels.

Ateniese et al. [1] on the other hand suggested random sampling in their Provable Data Possession scheme. The rest schemes that we shall discuss in this chapter [17, 18, 20] are all constructed under similar idea, with various choices of authenticator algorithms for specific purposes, such as privacy preservation or dynamic data operations etc.

One problem of random sampling schemes is that they cannot assure in 100 % confidence. Ateniese et al. [1] suggested that checking 460 blocks in each challenge is able to achieve 99 % confidence to detect server misbehavior if 1 % of data is changed. This seems good enough for most applications, but still needs to be improved for more flexible storage services.

In the following subsections, we focus on POR [12], PDP [1] and Compact POR [17] for distributed storage system. Next section further discusses the challenges in Cloud storage services and efforts [18, 20] to solve them.

3.2 *Proof of Retrievability*

Juels and Kaliski [12] proposed a cryptographic building block known as a proof of retrievability (POR) for archived files. POR enables a user (Verifier) to determine

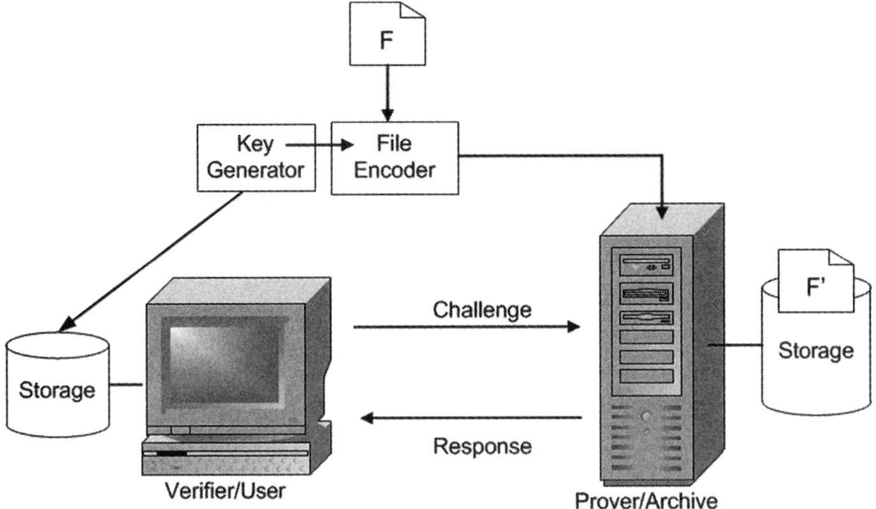

Fig. 1 Schematic of a POR System [12]

that an archive (Prover) "possesses" a file or data object F. A successfully executed POR assures a Verifier that the Prover presents a protocol interface through which the Verifier can retrieve F in its entirety.

Figure 1 shows the schematic of a POR. Two parties are involved in this model: the archive server as the Prover and the owner of the archived file or the user as the Verifier.

At the Verifier side, a key generation algorithm and an encoding algorithm are used to preprocess the file F. The key generation algorithm produces a key to encode the file F. This key should be independent of F and is stored by the Verifier. The encoding algorithm transforms raw file F into encoded file $\tilde{F}$ by randomly embedding a set of randomly-valued check blocks called sentinels.

After storing the encoded file into the Prover, the Verifier challenges the Prover by specifying the positions of a collection of sentinels and asking the Prover to return the associated sentinel values. If the Prover has modified or deleted a substantial portion of F, then with high probability it will also have suppressed a number of sentinels. It is therefore unlikely to respond correctly to the Verifier. To protect against corruption by the Prover of a small portion of F, a POR scheme also employs error-correcting codes.

A POR system (PORSYS) consists of six algorithms: **keygen**, **encode**, **extract**, **challenge**, **verify** and **respond**. Table 1 summarizes inputs, outputs of these algorithms and provides a brief description of each algorithm.

Definition 1 *Algorithm*: An algorithm with n inputs and m outputs is denoted as

$$\mathcal{A}(input_1, \cdots, input_n) \rightarrow (output1, \cdots, output_m)$$

where $\mathcal{A}$ is the name of the algorithm.

Table 1 Six Algorithms of a POR System [12]

Role	Algorithm	Description
Verifier	**keygen**$[\pi] \rightarrow \kappa$	Generate a secret key κ, could be a public/private key pair. For security concern, this key can be decomposed into multiple keys
	encode$(F; \kappa, \alpha)[\pi] \rightarrow \tilde{F}_\eta)$	Encode the original file with κ into $\tilde{F}_\eta$, where η denotes the unique file id (handle) of $\tilde{F}$ in the file system
	extract$(\eta; \kappa, \alpha)[\pi] \rightarrow F$	Extract the original file F by a sequence of challenges to the Prover
	challenge$(\eta; \kappa, \alpha)[\pi] \rightarrow c$	Take as input the file handle η, secret key κ, and state α. Output a challenge value c
	verify$((r, \eta); \kappa, \alpha) \rightarrow b \in \{0, 1\}$	Determine whether the receiver response r is valid to challenge c. If success, output 1, otherwise output 0
Prover	**respond**$(c, \eta) \rightarrow r$	Generate a response to a challenge c

In these algorithms, α denotes a persistent state during a Verifier invocation, and π denotes the full collection of system parameters. π should at least include the security parameter j. In particular, we can also include the length, formatting, encoding of files and challenge/response sizes in π.

The **encode** algorithm is the core of this system since all operations and data for verification are accomplished in this algorithm. The basic steps include error correction, encryption, sentinel creation and permutation.

Figure 2 shows the file structure changes in POR system. Suppose F with a message-authentication code (MAC) has b blocks, denoted as: $F[1], \cdots, F[b]$. It is divided into s chunks, each has k l-bit blocks. Thus we can view it as an $s \times k$ matrix, where each element is a block. For simplicity, the error-correcting code (ECC) also operates over l-bit symbols and sentinels, and l-bit values computed by a one-way function have l-bit length. This basic scheme adopts an efficient (n, k, d)-error correcting code with even-valued d. This code has the ability to correct up to $d/2$ errors. After applying ECC to F, each chunk is expanded to n blocks, resulting in a new file $F' = (F'[1], F'[2], \cdots, F'[b'])$, where the number of blocks is $b' = bn/k$. The encryption step applies a symmetric-key cipher E to F', yielding file F''.

A sentinel is created by a suitable one-way function f, taking as input the key generated by **keygen** and the index of this sentinel. Suppose we have s sentinels. These sentinels are appended to F'', yielding F''' with $b' + s$ blocks. Finally, in the **encode** algorithm, we apply a permutation function to F''', obtaining the output file $\tilde{F}$, where $\tilde{F}[i] = F'''[g(\kappa, i)]$.

The auditing procedure involves **challenge**, **response** and **verify** algorithms.The Verifier use a state α to track the state of each challenge. For simplicity, we let the

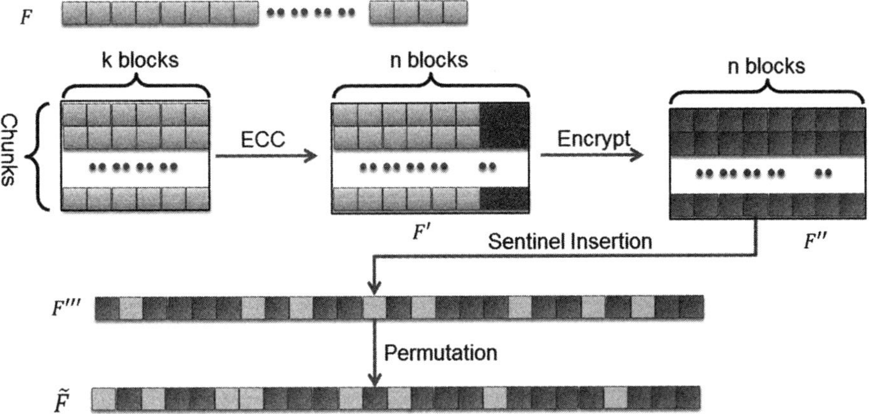

Fig. 2 File blocks changes in the encode step of sentinel POR system

Verifier state α initially be 1, incrementing it by q during each challenge[1]. The current value of α indicates that in last challenge phase the client requested sentinel position from $\alpha - q$ to $\alpha - 1$. The positions of sentinels that the Verifier wants to check are simply generated by the permeation function with two inputs: the secret key and the position of sentinel before applying permutation, that is $b' + \alpha$. The Prover then send the Verifier requested blocks (in the Prover's point of view). The **Verify** uses the one-way function f to calculate all sentinels that are being checked and compare with the Prover's response. In this way, the Verifier can detect the Prover's misbehavior in a relatively low cost of checking a small number of sentinels.

The overhead of POR mainly includes the storage for error-correcting code and sentinels, as well as computation of error-correcting code and permutation operations. Several optimization can be done to improve POR's performance. For example the length of response can be further hashed to a compact fixed length proof and the challenge can also be compressed by passing a seed to the Prover instead of all index of sentinel blocks. However, the major problem of this scheme is the limited number of challenge once the sentinel embedded file is upload to the prover.

3.3 *Provable Data Possession*

PDP was first proposed by Ateniese et al. [1]. Earlier solutions for verifying a server retaining a file need either expensive redundancy or access to the entire file. The PDP model provides probabilistic proof of the possession of a file with the server accessing small portions of the file when generating the proof. The client only stores fixed size

[1] In [12], the state α is not clearly defined. This interpretation of α is based on the σ in challenge function in Sect. 3.1 of [12].

of metadata and consumes a constant bandwidth. The challenge and the response are also small (168 and 148 bytes respectively). This subsection will introduce the PDP scheme in detail, including the definition and two enhanced versions of PDP algorithms: S-PDP and E-PDP.

3.3.1 Preliminaries

The PDP schemes are based on RSA algorithm. Readers are referred to [16] for more information about RSA algorithm.

First we choose two safe primes p and q that are large enough. Let $N = pq$, all exponentiations are calculated modulo N. We denoted $\mathbb{Z}_N = \{0, 1, \cdots, N-1\}$ and $\mathbb{Z}_N^*$ is the set of all numbers in $\mathbb{Z}_N$ that are relatively prime to N. That is

$$\mathbb{Z}_N^* = \{a \in \mathbb{Z}_N : gcd(a, N) = 1\}$$

Definition 2 *Quadratic Residue*: An integer $a \in \mathbb{Z}_N$ is a quadratic residue (mod N) if $x^2 \equiv a \bmod N$ has a solution. Let QR_N be the set of all quadratic residues of $\mathbb{Z}_N$ and g be a generator of QR_N.

In the RSA algorithm [16], a large integer d is randomly chosen such that it is relatively prime to $(p-1)(q-1)$. The other integer e is computed so that

$$ed \equiv 1 \mod (p-1)(q-1).$$

When using RSA-based algorithm for verification, there is a slight difference in choosing e and d in that

$$ed \equiv 1 \mod \frac{p-1}{2} \cdot \frac{q-1}{2}.$$

Definition 3 *Sets of Binary Numbers*: The set of all binary numbers with length n is denoted by $\{0, 1\}^n$. Specifically, $\{0, 1\}^*$ is the set of arbitrary length of binary numbers.

When we want to randomly choose a number k from a set S, we use the notation $k \overset{R}{\leftarrow} S$. For example, $k \overset{R}{\leftarrow} \{0, 1\}^\kappa$ means k is a number randomly chosen from the set of all κ-bit binary numbers.

Definition 4 Homomorphic Verifiable Tags (HVTs): An HVT is a pair of values (T_{m_i}, W_i). W_i is a random value obtained from the index i. T_{m_i} will be store on the server.

As building blocks of PDP, HVTs have the properties of unforgeable and blockless verification. The PDP scheme use HVTs as the verification metadata of file blocks.

Finally, we introduce four cryptographic functions:

- $h : \{0, 1\}^* \rightarrow QR_N$ is a secure deterministic hash-and-encode function.
- $H : \{0, 1\}^* \rightarrow \mathbb{Z}_N$ is a cryptographic hash function.

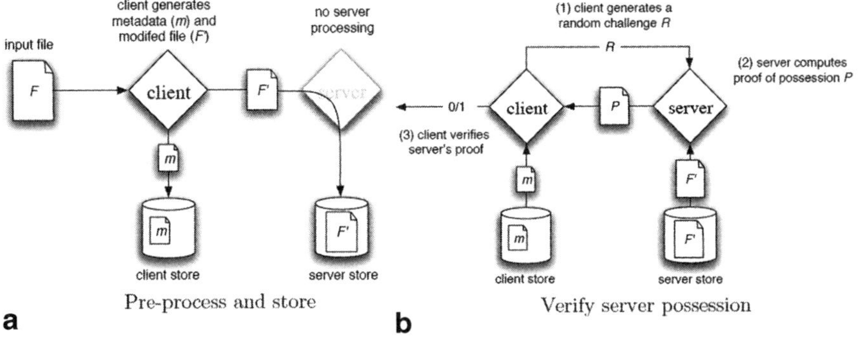

Fig. 3 Protocol for provable data possession [1]

- $f : \{0,1\}^{\kappa} \times \{0,1\}^{\log_2 n} \rightarrow \{0,1\}^{\ell}$ is a pseudo-random function (PRF). Specifically, we use $f_k(x)$ to denote $f(k,x)$.
- $\pi : \{0,1\}^{\kappa} \times \{0,1\}^{\log_2 n} \rightarrow \{0,1\}^{\log_2 n}$ is a pseudo-random permutation (PRP).

A hash function often takes as input an concatenation of two binary strings. We use $s_1||s_2$ to denote the concatenation of s_1 and s_2.

3.3.2 Defining the PDP Protocol

The PDP protocol involves a client, denoted as $\mathcal{C}$, who wants to store a large file in a remote server and a server, denoted as $\mathcal{S}$, who provides storage services. Fig. 3 depicts the PDP protocol in [1].

The PDP protocol consists of four polynomial-time algorithms: **KeyGen**, **TagBlcok**, **GenProof** and **CheckProof**. Table 2 summarizes these algorithms. Among them, **KeyGen**, **TagBlcok** and **CheckProof** are executed on the client side. The server need only to run the **GenProof** algorithm to generate a proof that it is possessing the client's file upon receiving a challenge from the client.

A file $\mathbf{F}$ is divided into n blocks, that is $\mathbf{F} = (m_1, \cdots, m_n)$. If not explicitly stated, the letter n always means the number of blocks in file $\mathbf{F}$. At the beginning of the setup phase, $\mathbf{F}$ is pre-processed by the client $\mathcal{C}$ into a new file $\mathbf{F}'$. This process could include encrypting the file and generating a tag for each file block. The client then uploads $\mathbf{F}'$ to the server $\mathcal{S}$. To verify whether the server is storing the entire file, the client then periodically generates a challenge and sends it to the server. Upon receiving a challenge, the server computes a proof of possession as a response to this challenge and sends back to the client. Finally, the client can check the server's response and verifies whether the server possesses the correct file.

A PDP protocol consist of two phases: the Setup Phase and the Challenge Phase.

- **Setup Phase**:
  The Setup Phase at the client side includes generation of necessary keys (public key and private key), calculation of a tag for each file block, transmission of

Table 2 Four Algorithms of a PDP System [1]

Role	Function	Description
Client	**KeyGen**$(1^\kappa) \rightarrow (pk, sk)$	Generate a secret key pair (pk, sk), taking as input a secret parameter κ
	TagBlock$(pk, sk, m) \rightarrow T_m$	Generate the verification metadata T_m for the input file block m
	CheckProof$(pk, sk, chal, \mathcal{P})) \rightarrow \{0, 1\}$	Validate a proof of possession $\mathcal{P}$. IF $\mathcal{P}$ is a correct proof of possession, output 1, else output 0
Server	**GenProof**$(pk, \mathbf{F}, chal, \Sigma) \rightarrow \mathcal{P}$	Generate a proof of possession $\mathcal{P}$ for given challenge $chal$

the processed file to the server and finally deletion local copy of the file. These operations are all executed on the client side. In particular, this phase consists of the following steps:
1. **KeyGen**$(1^\kappa) \rightarrow (pk, sk)$ generate secret keys.
2. Apply **TagBlock** to each file block $m_i, i = 1, \cdots, n$, resulting in n tags T_{m_i}.
3. Send $\{pk, \mathbf{F}, \Sigma = (T_{m_1}, \cdots, T_{m_n})\}$ to $\mathcal{S}$.
4. Delete $\mathbf{F}$ and Σ in local storage.

- **Challenge Phase**:
  In the Challenge Phase we use a challenge-response style to verify the integrity of the client's file. The challenge message specifies a predefined number of blocks, with their indices. The server needs to prove it is possessing all these blocks by calculating a proof $\mathcal{P}$ using all block data. The necessary steps of this phase are:
  1. $\mathcal{C}$ generates a challenge **chal**, specifying the set of blocks that it wants $\mathcal{S}$ to prove that it possesses these blocks.
  2. $\mathcal{C}$ sends **chal** to $\mathcal{S}$.
  3. $\mathcal{S}$ runs **GenProof** to get the proof of possession $\mathcal{P}$
  4. $\mathcal{S}$ sends $\mathcal{P}$ to $\mathcal{C}$.

Considering the tradeoff between security and efficiency, Ateniese et al.[1] introduced a secure PDP scheme (S-PDP), which has a strong data possession guarantee, as well as an efficient PDP scheme (E-PDP), providing better efficiency by means of a weaker data possession guarantee. The next two subsections will discuss these two schemes in detail.

3.3.3 The Secure PDP Scheme (S-PDP)

This section provides the construction of the Secure PDP Scheme (S-PDP) [1], including implementation of each algorithm and the two phases. The S-PDP is able to assure that the server possesses all blocks that are specified in the challenge message.

The key generation algorithm produce the public key $pk = (N, g)$ and the secret key $sk = (e, d, v)$. The RSA modulus N is the product of two distinct large primes p and q. Let g be a generator of QR_N. The public key pk is then formed by N

and g. Among the three integers in the secret key v is randomly chosen from $\{0,1\}^{\kappa}$. d is used to generate tags (authenticators) in **TagBlock** algorithm and e is used in **CheckProof** algorithm. pk and sk should be stored on the client side.

For each block of data, m_i, the **TagBlock** algorithm calculates a tag using the data as a number and its index. An index related number W_i is first generated by concatenating v with the index i, denoted as $W_i = v||i$. The tag of this block T_{m_i} is then computed as

$$T_{m_i} = (h(W_i) \cdot g^m)^d \mod N.$$

After getting all tags $\Sigma = (T_{m_1}, \cdots, T_{m_n})$, the client sends them to the server together with the original file F and the public key pk. That is $\{pk, \mathbf{F}, \Sigma\}$ are sent to the server. The client then deletes F and Σ on its local storage. This finishes the Setup Phase.

In the Challenge Phase, a challenge **chal** $= (c, k_1, k_2, g_s)$ is generated as follows. First of all, we randomly choose three integers: k_1,k_2 and s. k_1 and k_2 are selected from $\{0,1\}^{\kappa}$, serving as keys for the pseudo-random permutation π and the pseudo-random function f respectively. The last number s belongs to $\mathbb{Z}_N^*$ and is used to mask the generator g. The challenge message is then formed as (c, k_1, k_2, g_s), where c is the number of blocks that each challenge will pick and $g_s = g^s \mod N$.

At the server side, there is only one algorithm **GenProof** that is executed upon receiving a challenge requested by the client. For each number j from 1 to c, an index $i_j = \pi_{k_1}(j)$ and a mask $a_j = f_{k_2}(j)$ are calculated. The file blocks that the client want to check are then indicated by $\{i_1, i_2, \cdots, i_c\}$. Finally, two numbers T and ρ are computed as the proof for this challenge:

$$T = \prod_{j=1}^{c} T_{m_{i_j}}^{a_j}, \rho = H(g_s^{a_1 m_{i_1} + \cdots + a_c m_{i_c}} \mod N).$$

The motivation for putting the coefficients a_j in the challenge phase is to strengthen the guarantee that $\mathcal{S}$ possesses each block queried by the client. In each challenge phase, there is a randomly chosen key for calculation of these coefficients. $\mathcal{S}$ cannot store combinations of the original blocks to save storage cost. Since the proof of possession has a constant length regardless the number of blocks being requested, this scheme can maintain constant communication cost in the challenge phase.

Once T and ρ are ready, the server response the client with its proof to **chal**, $\mathcal{P} = (T, \rho)$. The client runs **GenProof** to check the correctness of the proof. First, it computes i_j, a_j and W_{i_j} as the server did. Then $\tau = T^e$ is divided by $h(W_{i_j})^{a_j}$ for each j from 1 to c. This actually results in

$$\tau = g^{a_1 m_{i_1} + \cdots + a_c m_{i_c}} \mod N.$$

If the proof is valid, the following equation should be true:

$$H(\tau^s \mod N) = \rho.$$

Table 3 Comparison between S-PDP and E-PDP [1]

Algorithm	S-PDP	E-PDP
GenProof	$a_j = f_{k_2}(j)$	delete
	$T = \left(\prod_{j=1}^{c} T_{i_j}^{a_j}\right) \bmod N$	$T = \prod_{j=1}^{c} T_{i_j} \bmod N$
	$\rho = H(g_s^{a_1 m_{i_1} + \cdots + a_c m_{i_c}} \bmod N)$	$\rho = H(g_s^{m_{i_1} + \cdots + m_{i_c}} \bmod N)$
CheckProof	$i_j = \pi_{k_1}(j), W_{i_j} = v\|\|i_j, a_j = f_{k_2}(j)$	$i_j = \pi_{k_1}(j), W_{i_j} = v\|\|i_j$
	$\tau = (\tau / h(W_{i_j})^{a_j}) \bmod N$	$\tau = [T^e/(h(w_{i_1}) \cdots h(w_{i_c}))] \bmod \mathrm{N}$

3.3.4 The Efficient PDP Scheme (E-PDP)

The Efficient Provable Data Possession (E-PDP) [1] scheme achieved a higher performance at the cost of a weaker guarantee by eliminating all coefficients a_j in the GenProof and CheckProof algorithms. Table 3 shows a comparison between S-PDP and E-PDP. As all coefficients $a_j = 1$, the E-PDP scheme reduces the expensive exponential computation. However, the server $\mathcal{S}$ can only possess the sum of the blocks $m_{i_1}, \cdots, m_{i_c}$ for a challenge. In order to completely pass the challenge phase every time, the server $\mathcal{S}$ needs to compute every combination of c blocks out of n blocks, that is $\binom{n}{c}$. The client can choose values of n and c such that make it impractical for the server to simply store all sums.

3.4 Compact Proof of Retrievability

Because of the predefined number of sentinels, the sentinel-based POR scheme has a limited number of possible challenges. Based on Juels and Kaliski's work [12], Shacham and Waters [17] introduced two new schemes that achieved public and private verifiability, with compete proof of security.

Shacham and Waters [17] proposed two new proof of retrievability schemes, with private and public verifiability respectively. The main advantage of SW PORs is the unlimited number of queries. The private verification scheme, based on pseudo-random functions (PRFs), has the shortest response of any POR scheme (20 bytes) with the cost of a longer query. The second scheme used short signatures introduced by Boneh, Lynn and Shacham (BLS) [5] to verify the authentication of data in remote servers, hence assuring public verifiability is secure. At an 80-bit security level, this scheme has the shortest query (20 bytes) and response (40 bytes)of any POR scheme.

An important contribution of [17] is that it provided a complete security proof for both schemes. Interested readers can consult this paper for more details.

3.4.1 System Model

Shacham and Waters's system model has similar functions with Juels and Kaliski's POR description in [12], but modules were redefined and more details were added. In this model, key generation and verification procedures no longer maintain any state. In addition, Shacham and Waters's protocols [17] allow challenge and response to be arbitrary.

There are two parties in this system, the Verifier and the Prover. The Verifier could be the data owner itself or a third party auditor. The prover is the storage server. Similar to POR, a file M with size b is divided into n blocks, each further split into s sectors. Thus, we can refer to a sector as m_{ij}, $1 \leq i \leq n, 1 \leq j \leq s$. M can also be treated as an $n \times s$ matrix $\{m_{ij}\}$. Each sector is an element of $\mathbb{Z}_p = \{0, 1, \cdots, p-1\}$, where p is a large enough prime number.

Our description of these algorithms in the following sections about public and private verification will be slightly different with [12] in input and output parameters. We redefine part of these algorithms to maintain internal consistency. We hereby ignore all input and output parameters since they vary in the public scheme and the private scheme and focus on the functionalities of these algorithms. There are four algorithms in the system model:

- **KeyGen**. The key generation algorithm. The Verifier runs this algorithm to generate necessary private keys (for private verification) or key pairs (for public verification) for other algorithms.
- **Store**. The file processing algorithm. The Verifier runs this algorithm to produce the processed file M^* and a file tag t. The processed file M^* is stored in the server. The file tag t is saved in the data owner's local storage or at the server side depends on the desire of service agreement.
- **Prove**. The proving algorithm. The Prover runs this algorithm to generate a proof of retrievability according the index indicated in the challenge message sent by the Verifier.
- **Verify**. The verifying algorithm. If it is a third party auditor who is running this algorithm, the file tag need to be retrieved and verified first. Then the challenge message is produced. When the prover sends the response message back, the Verify algorithm continues to verify the response. After running this algorithm, the Verifier will know whether the file is being stored on the server and can be retrieved as needed. If the algorithm fails, it outputs 0. The Verifier can then use an extractor algorithm to attempt to recover the file.

This scheme still works in two phases. In the setup phase the data owner runs **KeyGen**and **Store** to process the file and upload the resulting file to the server (Prover). The challenge phase involves the Verifier (data owner or TPA), running **Verify**, and the Prover, running **Prove**.

3.4.2 Private Verification Construction

The construction of a private verification consists of the implementation of the four functions that are discussed in Sect. 3.4.1.

Since this scheme is for private verification purposes, we only need a secret key $sk = (k_{enc}, k_{mac})$ in the **KeyGen** algorithm, where k_{enc} is an encryption key and k_{mac} is an MAC key.

In the **Store** algorithm, the file M is preprocessed with an erasure code before applying the following operations. This erasure code should be able to recover the file even the Prover erases some portion of the file. The processed file is denoted as M^* as a part of output. This file is divided into n blocks, each has s sections. Hence we can write $M^* = \{m_{ij}\}, (i = 1, \cdots, n; j = 1, \cdots, s)$. To detect any modification to the file by the Prover, the Verifier (data owner) calculate an authenticators

$$\sigma_i = f_{k_{prf}}(i) + \sum_{j=1}^{s} \alpha_j m_{ij}, i = 1, \cdots, n$$

for each block, where α_j are randomly chosen from $\mathbb{Z}_p$. These authenticators provide strong assurance that the Prover cannot forge any one of them since it has no knowledge of the PRF as well as the PRF key. A file tag t is also computed to include PRF key and α_j. These numbers are concatenated and encrypted first. Then the encrpyted bit string is appended to the number of blocks n, forming an initial tag t_0. Finally, the file tag is produced by appending the MAC of t_0, keyed with k_{mac}, with itself. The Verifier only stores sk and outsources the processed file M^*, authenticators $\{\sigma_i\}_{i=1}^{n}$ and the file tag t to the Prover.

In the **Verify** algorithm, the Verifier sends an l-element query $Q = \{(i, v_i)\}$, $1 \leq i \leq n$, specifying the index of blocks that are to be verified, to the Prover. In each pair of (i, v_i), i is a random index of file block and v_i is randomly chosen from B, a subset of $\mathbb{Z}_p$. For simplicity, we can let $B = \mathbb{Z}_p$. There are totally l pairs in Q, suggesting that the set of all i, $I = \{i : (i, v_i) \in Q\}$, has the size l.

In the **Prove** algorithm, the Prover uses v_i as coefficients when calculating a proof for a specific Q. This also prevents the Prover from using a previously calculated proof. The response $r = \{\mu_1, \cdots, \mu_s, \sigma\}$ is produced as follows. For each $1 \leq j \leq s$, we let $\mu_j = \sum_{i \in I} v_i m_{ij}$. The last number σ is simply the sum of all products of v_i and σ_i, that is $\sigma = \sum_{i \in I} v_i \sigma_i$. Now the Prover have all it needs for the response r and send it to the Verifier.

Back to the **Verify** algorithm, upon receiving r, the Verifier checks if

$$\sigma = \sum_{i \in I} v_i f_{k_{prf}}(i) + \sum_{j=1}^{s} \alpha_j \mu_j$$

is true. If it is, there is a high probability that M is retrievable.

As shown in the protocol, this private verification scheme has less computation overhead than PDP since multiplication is the only operation except addition.

3.4.3 Public Verification Construction

A public verifiable POR scheme allows anyone who has the public key of the data owner to query the Prover and verify the return response. With this protocol, user offloads the verification task to a trusted third party auditor. The public verification scheme in [17] used BLS signatures [5] for authentication values instead of utilizing PRF.

In this public verification construction, Shacham and Waters employ bilinear map in the verify algorithm V. We briefly introduce bilinear map here. Interested readers are referred to [4, 5] for more details. Let G_1, G_2 and G_T be multiplicative cyclic groups of the same prime order p. g_1 is a generator of G_1 and g_2 is a generator of G_2. A bilinear map is a map $e : G_1 \times G_2 \rightarrow G_T$ with the following properties:

1. Bilinear: for all $u, v \in G$ and $a, b \in \mathbb{Z}$, $e(u^a, v^b) = e(u, v)^{ab}$.
2. Non-degenerate: $e(g_1, g_2) \neq 1$

In this scheme, we let $G = G_1 = G_2$.

The **KeyGen** algorithm now needs a public and private key pair. At first a signing key pair (spk, ssk) is generated. The public key $pk = (v, spk)$ and the secret key $sk = (\alpha, ssk)$, where α is randomly chosen from $\mathbb{Z}_p$ and $v = g^\alpha$.

The file tag t contains a $name \in \mathbb{Z}_p$ and s randomly chosen $u_k \in G, k = 1, \cdots, s$ as well as the number of blocks n. These data are concatenated, resulting $t_0 = name||n||u_1|| \cdots ||u_s$, and appended by its signature keyed with ssk. The final file tag is $t = t_0||SSig_{ssk}(t_0)$. u_k is also used for authenticator calculation. For each i from 1 to n, the authenticator of block m_i is $\sigma_i = (H(name||i) \cdot \prod_{j=1}^{s} u_j^{m_{ij}})^\alpha$, where $H : \{0, 1\}^* \rightarrow G$ be the BLS hash [5]. After calculating the above information, the user then sends the erasure coded file M^* together with $\{\sigma_i\}$ to the Prover.

The challenge phase is almost the same as the private verification scheme except that σ in the response message is changed to the sum of $\sigma_i^{v_i}$ instead of $v_i\sigma_i$ before. Of course, the verification equation need to be modified since bilinear group is used here. The user now check whether the following condition is held:

$$e(\sigma, g) = e\Big(\prod_{(i,v_i)\in Q} H(name||i)^{v_i} \cdot \prod_{j=1}^{s} u_j^{\mu_j}, v \Big).$$

4 Auditing in Cloud Storage Platform

Cloud computing is migrating traditional computing services to remote cloud service providers. Cloud storage has advantages such as high flexibility, ultimately low price and relatively high data security for a wide spectrum of users. However, not all problems with traditional distributed storage are solved by cloud computing. Security is still the major concern, even though the cloud providers all claim they can protect our data in more secure way than the users can do themselves.

This section analyzes changes brought by cloud computing in data storage and introduces researchers' attempt to solve these problems.

4.1 Challenges

Although there are reported efforts in information auditing for distributed systems, the special features in the cloud storage platforms necessitate customized design due to new challenges. This subsection briefly lists the problems that need to be considered.

1. **Dynamic Data Operations**. Clients in cloud services might not have files of large size such as the original PDP and PoR schemes assume, but the number of files is greater, and the flexibility requirements are stronger. Files in the cloud storage will be changed more frequently. Modification, deletion and insertion need to be considered in the design of storage system.
2. **Public Verifiability**. Computing devices that cloud clients have might not be powerful enough to accomplish the computational task of integrity auditing of their own data in cloud storage. Meanwhile, these clients' end devices might have multiple tasks to do, which cannot allow limited computing resources consumed by this single task. It is desired to offload the verification procedure to a third party auditor (TPA), which has sufficient computing resources and expertise in data auditing. It is expected to make the verification protocol a public verifiable one.
3. **Privacy Preserving**. This seems to conflict with the public verifiability requirement at first glance. How can the TPA execute auditing protocol and yet not be trusted? Studies in preserving privacy in using TPAs for auditing purposes showed that it is feasible and practical to design a verification protocol for untrusted TPAs. Using this protocol, file blocks should not be retrieved in order to verify the integrity of files.
4. **Computational Efficiency**. A cloud client can be a portable device like PDA or smartphone which usually has weaker computation ability and limited communication bandwidth. Data auditing protocols in cloud storage should try to reduce both computation and bandwidth as much as possible.
5. **Multiple Files**. A cloud client's storage request could consist of large number of files instead of a single large file.
6. **Batch auditing**. A cloud server can be audited by a TPA for thousands of users' files. In this case, if aggregating multiple proofs as a single message to the TPA is applicable, the communication burden of the protocol could be significantly reduced to an acceptable level.

In the following subsections, we shall discuss more schemes trying to tackle some of these problems.

4.2 Public Verifiability

There is a variant of PDP scheme [1] that can support public verifiability. A PDP scheme with public verifiability property allows anyone to challenge the server for the possession of the specific file as long as they have the client's public key.

To support public verifiability, the following changes are made to the S-PDP protocol:

1. Besides N and g, the Client should make e public.
2. A PRF $\omega : \{0,1\}^{\kappa} \times \{0,1\}^{\log_2 n} \rightarrow \{0,1\}^{\ell}$ is used to generate W_i by randomly choose a v from $\{0,1\}^{\kappa}$ as a key. That is, $W_i = \omega_v(i)$.
3. The client makes v public after the Setup phase.
4. The challenge **chal** in GenProof and CheckProof no longer contains g_s or s.
5. In GenProof, the server computes $M = a_1 m_{i_1} + \cdots + a_c m_{i_c}$ instead of ρ and returns $\mathcal{V} = (T, M)$.
6. In CheckProof, the client checks $g^M = \tau$ and $|M| < \lambda/2$.

In sect. 3.4, we also saw a public verifiable POR scheme.

4.3 Dynamic Data Operations Support

The PDP scheme [1] did not employ dynamic data operations like modification, deletion and insertion due to the original motivation to verify integrity of archive files, which will not involve many dynamic operations. Similarly, the POR scheme [12] cannot support data dynamics due to the verification mechanism of embedding pre-computed sentinels. However, these operations are vital features for cloud storage services.

Ateniese et al. [2] propose a dynamic version of PDP scheme . The extended scheme achieved higher efficiency because it only relied on symmetry-key cryptography. But the number of queries was limited, hence, the scheme cannot support fully dynamic data operations. Erway et al. [7] introduced a formal framework for dynamic provable data possession (DPDP) . Their first scheme utilized authenticated skip list data structure to authenticate tag information of blocks, thereby eliminating the index information in tags. They also provided an alternative RSA tree based construction, which improved the detection probability at the cost of an increased Server computation burden.

Wang et al.[20] extended the Compact POR in Sect. 3.4 to support both public verifiability and data dynamics in cloud storage. We'll focus on this model to discuss dynamic data operation support. Table 4 shows the six algorithms in [20].

In cloud data storage, Clients could be portable devices that have limited computation ability. A third party auditor is necessary for the verification procedure. The system we shall consider in this section includes three entities: Client, Cloud Storage Server and the prover, a Third Part Auditor (TPA). The TPA is trusted and unbiased while the Server is untrusted. Privacy preserving is not considered in [20].

Table 4 Algorithms of Extended POR System [20]

Role	Algorithm	Description
Client	**KeyGen**$(1^\kappa) \to (pk, sk)$	This algorithm is the same as in SW's model in sect. 3.4. It generates a secret key pair (pk, sk), taking as input a secret parameter κ
	SigGen$(sk, F) \to (\Phi, sig_{sk}(H(R)))$	Generates the signature set $\Phi = \{\sigma_i\}$ on file blocks $\{m_i\}$ and sign the root R of a Merkel hash tree $sig_{sk}(H(R))$
	VerifyUpdate$(pk, update, \mathcal{P}_{update}) \to \{(1, sig_{sk}(H(R'))), 0\}$	Verifies the update operation
Client/TPA	**VerifyProof**$(pk, chal, \mathcal{P}) \to \{0, 1\}$	Validate a proof $\mathcal{P}$. IF $\mathcal{P}$ is correct, output 1, else output 0
Server	**GenProof**$(pk, \mathbf{F}, \mathbf{chal}, \Sigma) \to \mathcal{P}$	Generates a proof $\mathcal{P}$ for given challenge **chal**
	ExecUpdate$(F, \Phi, update) \to (F', \Phi', \mathcal{P}_{update})$	According to the type of "update" request from the Client, this algorithm executes the corresponding update operation and outputs the updated file F', signatures Φ' and proof $\mathcal{P}$

The Client encodes the raw file $\tilde{F}$ into $\mathbf{F}$ using Reed-Solomon codes. The file $\mathbf{F}$ consists of n blocks $m_1, \cdots, m_n, m_i \in \mathbb{Z}_p$, where p is a large prime. $e : G \times G \to G_T$ is a bilinear map, with a hash function $H : \{0,1\}^* \to G$ serving as a random oracle. g is the the generator of G. h is a cryptographic hash function.

In order to accomplish dynamic dada operation, the well studied Merkle hash tree (MHT) [13] is a good choice to assure the value and positions of data blocks. A MHT is a binary tree with all data blocks as leaf nodes. A parent node is the hash of the concatenation of its two children. This procedure continues until reach a common root node. All dynamic operations will result in a update of MHT by recalculating every node that is in the path from affected blocks to the root. The sibling data that is needed for a recalculation is called auxiliary information.

The verification of data relies on BLS hash [5] and bilinear map as in [17]. The differences here reside in file tag, authenticators calculation, components of a proof and the verification of a proof. The file tag is shorter in that it only include the concatenated name, number of blocks and a random value for authentication purpose. A proof includes four parts: a block data related value, a authenticator related value, the auxiliary information set and the signature of the MHT root's BLS hash. When computing an authenticator, the **SigGen** algorithm no longer take into account the index or name as in [1] or the public scheme of [17]. It is simply $\sigma_i = (H(m_i) \cdot u^{m_i})^\alpha$, where u is a random value and α is a part of the secret key. The bilinear map e is used twice in **VerifyProof**, one to authenticate MHT root and the other to verify the rest of the proof.

We now consider three types of dynamic data operations: Modification, Insertion and Deletion. The advantage of MHT lies in the convenience in modifying the structure of the tree, hence embedding dynamic data operations into the scheme.

- Modification. The Client wants to replace a block block. First, it computes the signature for new block and a update message is sent to the Server. The Server runs **ExecUpdate** to update the block. The update procedure includes replacement of new block, new authenticator and the leaf node in MHT. The root is then updated. A proof of update message must be sent to the Client so that he can know whether the update is valid. This message includes all auxiliary information, the old signature of the hash of the root and the new root. The Client generate old root using auxiliary information and uses bilinear map to check whether the signature is valid. If it is true, new MHT root is computed and compared with the one transmitted back by the Server. The modification is valid if and only if it passes all these tests.
- Insertion. The Client wants to insert block m^* after block m_i. It generates signature σ^* and sends the Server a update message. The Server runs **ExecUpdate** to execute a insertion operation, storing new block, inserting new authenticator, generating new root, and sends a update message to the Client like it did in the modification operation. The Client also need to verify this operation according to the message it receive.
- Deletion. Inverse operation to insertion. Similar to modification and insertion.

This scheme efficiently solves the dynamic operation problem but has the drawback that the messages exchanged between the Client and the Server is proportional to the number of file blocks.

4.4 Privacy Preserving

Although auditing storage data through a third party auditor, who has expertise in auditing and powerful computing capabilities, has many advantages to the client, the auditing procedure has the possibility to reveal user data to the TPA. Previous schemes [1, 17, 20] for data verification do not consider the privacy protection issue when offloading the verification job to the TPA. They all assumed that the TPA is trusted and will not try to look into user's data when verifying the integrity of data. A privacy-preserving public auditing scheme was proposed for cloud storage in [19]. Based on a homomorphic linear authenticator, integrated with random masking, the proposed scheme is able to preserve data privacy when TPA audits the stored data in the server.

There are three entities in this system: The user, the Cloud Server and the TPA. Since dynamic data operations were not considered in this scheme, only four algorithms (**KeyGen**, **SigGen**, **GenProof** and **VerifyProof**) are needed in this protocol, without the two algorithms for update purpose in Sect. 4.3. Still, we have two phases in the system: Setup Phase and Audit Phase. The mathematical integrity assurance technique is still a bilinear map $e : G_1 \times G_2 \rightarrow G_T$ as in Sect. 3.4.3. These groups should be different groups but has the same order. The server has knowledge about G_1, G_T and $\mathbb{Z}_p$ (all file blocks are elements of this group).

Like all other public key cryptosystem, the **KeyGen** algorithm needs a public-private key pair. A pair of signing keys (spk, ssk) is generated for the verification of file tag, which includes the identifier of the file.

The secrete key sk includes the secret signing key ssk and a random integer chosen from $\mathbb{Z}_p$. The public key $pk = (spk, g, g^x, u, e(u, v))$ on the other hand includes more values. g is a generator of G_2, u is an element of G_1 and $e(u, v) \in G_T$ is the image of u and v under the bilinear map e.

The **SigGen** algorithm calculates the file tag and authenticators in a different way. The file tag in [19] is shorter than previous schemes [17, 20], only the identifier of the file is included. This identifier, denoted as $name$, is also an element of $\mathbb{Z}_p$. The signing key pair is generated just for verification of $name$. An authenticator the block m_i is $\sigma_i = (H(name||i) \cdot u^{m_i})^x$. The hash function $H : \{0, 1\}^* \rightarrow G_1$ maps a bit string into G_1, which means all authenticators will fall into G_1. The set of the authenticators and the file tag are sent to the server. This finishes the setup phase.

During the audit phase, the file tag is retrieved and verified by the TPA. If t is valid, the file name is recovered. The challenge **chal** is generated in the same way as in [17]. Upon receiving **chal**, the server runs **GenProof** to calculate the proof that it possesses the requested file blocks. There are three components in the response to **chal**: μ, σ, and R. σ is the aggregation of all authenticators that are indicated by **chal**. Each authenticator σ_i is raised to the power v_i and their product is the value of σ. The other two values are related to a random number r from $\mathbb{Z}_p$. R is the result of raising the image of u, v to the power r. A number μ' is directly calculated from all indicated blocks. This value is highly related to the file. To hide it from the TPA, the server uses r and the hash value of R. The final component $\mu = r + h(R)\mu'$ is obtained. The TPA runs **VerifyProof** to validate the response. If $R \cdot e(\sigma^{\gamma} g) = e((\prod_{i \in I} H(W_i)^{v_i})^{\gamma} \cdot u^{\mu}, v)$ is true, the response is a valid one. The audit procedure is then accomplished.

Data dynamic operations can also be supported by adapting this scheme using MHT as in [20].

4.5 Multiple Verifications

Since the cloud server is accessed by multiple users, the possibility that many clients request verification for different files or one client requests verifying multiple files. These requests should be treated in different way and hence need different auditing schemes. For example, multiple clients have different key pairs, whereas one single client requesting multiple verifications has the same key pair. Most schemes that claim to be able to support batch auditing belong to the first category.

Both [19] and [20] have the extension to support multiple verifications thanks to the aggregation property of bilinear signature schemes [4]. [20] uses auxiliary information in a proof, hence has relatively long proof message for multiple clients batch auditing. Only σ in each proof can be aggregated in one value. [19] aggregates by multiply all R's.

Batch auditing can reduce the computation cost on TPA since the K responses are aggregated into one. But the practical efficiency still needs to be verified by further experiments.

5 Open Questions

In this chapter, we provided an overview of general issues on information auditing to clarify the major goal of information auditing. Then we discussed two popular protocols that audits for data integrity in distributed data storage: PDP and POR. They were proposed almost at the same time to address different security concerns. PDP provides a high probability guarantee that a system possesses a file with high efficiency in computation and communication. POR and Compact POR allow a stronger guarantee of retrievability with the cost of more complex algorithms. Most schemes discussed in this chapter came with security proof in the original research papers, in which interested readers can find proof details and mathematical analysis. However, there is not sufficient study on efficiency and performance.

Since it is still a new research area in Cloud storage, we anticipate more new schemas will come out in the academic community, trying to resolve different challenges from various perspective. When evaluating a scheme, it usually includes the following metrics:

- Server computation overhead for a proof in each storage node.
- Server communication overhead when transmit computing results to form the final proof.
- Client computation overhead for authenticators, error-correcting code and verification algorithm.
- Communication cost between any two parties of the client, the server and TPA.
- Client storage for necessary metadata.
- Server misbehavior detection probability.

Compared to traditional information auditing, there are still numerous open problems in cloud data security auditing. The number one impending issue is the lack of standardization and consistency in auditing development efforts due to the heterogenity in infrastructure, platforms, software, and policy. While a "silver bullet" is highly desired, the diversity in auditing and assurance practices in cloud computing makes it extremely challenging to find a one-for-all solution. Essentially, in terms of data security oriented auditing, a thorough study is expected on balancing the tradeoffs among confidentiality, integrity, availability and usability.

From the cloud service providers' point of view, allowing external auditing implies more components such as transparency, responsibility, assurance, and remediation [15]. To accommodate these central components, a cloud service provider is required to:

- Set up policies that are consistent with external auditing criteria.
- Provide transparency to clients/users.
- Allow external auditing.
- Support remediation, such as accident management and compliant handling.
- Enable legal mechanisms that support prospective and retrospective accountability.

6 Conclusions

An efficient auditing system is critical to establish accountability for cloud users who do not have physical possession of their data. Existance of a trustworthy third party audits enable users to check the data integrity, track suspicious activities, obtain evidence for forensics, and evaluate service providers' behaviors when needed. This chapter provides our readers fundamental understanding of cloud auditing technologies. We expect to witness development of standard framework for cloud auditing and efforts at cloud service providers to make their policies and mechanisms more auditable and accountable.

References

1. Ateniese, G., Burns, R., Curtmola, R., Herring, J., Kissner, L., Peterson, Z., Song, D.: Provable data possession at untrusted stores. In: Proceedings of the 14th ACM conference on Computer and communications security, CCS '07, pp. 598–609. ACM, New York, NY, USA (2007). DOI 10.1145/1315245.1315318. URL http://doi.acm.org/10.1145/1315245.1315318
2. Ateniese, G., Di Pietro, R., Mancini, L.V., Tsudik, G.: Scalable and efficient provable data possession. In: Proceedings of the 4th international conference on Security and privacy in communication networks, SecureComm '08, pp. 9:1–9:10. ACM, New York, NY, USA (2008). DOI 10.1145/1460877.1460889. URL http://doi.acm.org/10.1145/1460877.1460889
3. Ateniese, G., Kamara, S., Katz, J.: Proofs of Storage from Homomorphic Identification Protocols. In: M. Matsui (ed.) Advances in Cryptology - ASIACRYPT 2009, *Lecture Notes in Computer Science*, vol. 5912, chap. 19, pp. 319–333. Springer Berlin / Heidelberg, Berlin, Heidelberg (2009). DOI 10.1007/978-3-642-10366-7_19. URL http://dx.doi.org/10.1007/978-3-642-10366-7_19
4. Boneh, D., Gentry, C., Lynn, B., Shacham, H.: Aggregate and verifiably encrypted signatures from bilinear maps. Advances in Cryptology-EUROCRYPT 2003 pp. 641–641 (2003)
5. Boneh, D., Lynn, B., Shacham, H.: Short signatures from the weil pairing. Journal of Cryptology **17**, 297–319 (2004). URL http://dx.doi.org/10.1007/s00145-004-0314-9. 10.1007/s00145-004-0314–9
6. Buchanan, S., Gibb, F.: The information audit: Role and scope. International journal of information management **27**(3), 159–172 (2007)
7. Erway, C., Küpçü, A., Papamanthou, C., Tamassia, R.: Dynamic provable data possession. In: Proceedings of the 16th ACM conference on Computer and communications security, CCS '09, pp. 213–222. ACM, New York, NY, USA (2009). DOI 10.1145/1653662.1653688. URL http://doi.acm.org/10.1145/1653662.1653688
8. Feng, J., Chen, Y.: A fair non–repudiation framework for data integrity in cloud storage services. International Journal of Cloud Computing **2**(1), 20–47 (2013)

9. Feng, J., Chen, Y., Liu, P.: Bridging the missing link of cloud data storage security in aws. In: Consumer Communications and Networking Conference (CCNC), 2010 7th IEEE, pp. 1–2. IEEE (2010)
10. Feng, J., Chen, Y., Summerville, D., Ku, W.S., Su, Z.: Enhancing cloud storage security against roll-back attacks with a new fair multi-party non-repudiation protocol. In: Consumer Communications and Networking Conference (CCNC), 2011 IEEE, pp. 521–522. IEEE (2011)
11. Feng, J., Chen, Y., Summerville, D.H.: A fair multi-party non-repudiation scheme for storage clouds. In: Collaboration Technologies and Systems (CTS), 2011 International Conference on, pp. 457–465. IEEE (2011)
12. Juels, A., Kaliski Jr., B.S.: Pors: proofs of retrievability for large files. In: Proceedings of the 14th ACM conference on Computer and communications security, CCS '07, pp. 584–597. ACM, New York, NY, USA (2007). DOI 10.1145/1315245.1315317. URL http://doi.acm.org/10.1145/1315245.1315317
13. Merkle, R.: Protocols for public key cryptosystems. In: IEEE Symposium on Security and privacy, vol. 1109, pp. 122–134 (1980)
14. Ould, M.A.: Business Processes: Modeling and Analysis for Re-engineering and Improvement. Wiley, Chichester (1995)
15. Pearson, S.: Toward accountability in the cloud. Internet Computing, IEEE **15**(4), 64 –69 (2011). DOI 10.1109/MIC.2011.98
16. Rivest, R.L., Shamir, A., Adleman, L.: A method for obtaining digital signatures and public-key cryptosystems. Commun. ACM **21**(2), 120–126 (1978). DOI 10.1145/359340.359342. URL http://doi.acm.org/10.1145/359340.359342
17. Shacham, H., Waters, B.: Compact Proofs of Retrievability Advances in Cryptology - ASIACRYPT 2008. In: J. Pieprzyk (ed.) Advances in Cryptology - ASIACRYPT 2008, *Lecture Notes in Computer Science*, vol. 5350, chap. 7, pp. 90–107. Springer Berlin / Heidelberg, Berlin, Heidelberg (2008). DOI 10.1007/978-3-540-89255-7_7. URL http://dx.doi.org/10.1007/978-3-540-89255-7_7
18. Wang, C., Chow, S., Wang, Q., Ren, K., Lou, W.: Privacy-preserving public auditing for secure cloud storage. Computers, IEEE Transactions on **PP**(99), 1 (2011). DOI 10.1109/TC.2011.245
19. Wang, C., Wang, Q., Ren, K., Lou, W.: Privacy-preserving public auditing for data storage security in cloud computing. In: INFOCOM, 2010 Proceedings IEEE, pp. 1–9 (2010). DOI 10.1109/INFCOM.2010.5462173
20. Wang, Q., Wang, C., Ren, K., Lou, W., Li, J.: Enabling public auditability and data dynamics for storage security in cloud computing. Parallel and Distributed Systems, IEEE Transactions on **22**(5), 847 –859 (2011). DOI 10.1109/TPDS.2010.183

I/O and File Systems for Data-Intensive Applications

Yanlong Yin, Hui Jin and Xian-He Sun

1 Parallel File Systems vs. Data-Intensive File Systems: A Comparison

Large-scale parallel computing increasingly plays important roles on accelerating scientific advances, providing versatile internet services, and many other knowledge discoveries. During the evolution of parallel computing, it forms two major camps: high-performance computing (or Supercomputing) and cloud computing. HPC is computing-oriented and the typical applications are scientific simulation, numerical computation, and etc. They rely on low-latency networks for message passing and use parallel programming paradigms such as MPI to enable parallelism [1]. Cloud computing is usually data-processing-oriented and the typical framework is designed for large-scale batch data processing.

These two camps of parallel computing adopts two different types of file systems to manage their data, namely parallel file systems (PFS) and distributed file systems or data-intensive file systems (DFS).

A parallel I/O system used by HPC typically consists of several layers from top to bottom: applications, high level I/O Library, I/O middleware, parallel file systems, and the underlying storage devices. Parallel file systems currently serve as general-purpose file systems to support I/O operations of HPC applications. A parallel file system manages large numbers of storage nodes and disks to form a large storage space. Any data request accessing this storage space will be fulfilled using multiple

Y. Yin (✉) · X.-H. Sun
Department of Computer Science, Illinois Institute of Technology, Chicago, IL 60616, USA
e-mail: yyin2@iit.edu

X.-H. Sun
e-mail: sun@iit.edu

H. Jin
Parallel Execution Group, Oracle Corporation, Redwood City, CA 94065, USA
e-mail: hui.x.jin@oracle.com

S. U. Khan, A. Y. Zomaya (eds.), *Handbook on Data Centers*,
DOI 10.1007/978-1-4939-2092-1_18

underlying storage devices in parallel. To achieve this parallel access, files saved in parallel file systems are partitioned into small data stripes that are distributed over multiple storage nodes or devices. Representative examples of parallel file systems include IBM GPFS [2], Oracle Lustre [3, 4], and OrangeFS/PVFS2 [5, 6]. HPC applications access PFS via either POSIX interface or MPI-IO, a subset of the MPI-2 specification [7] that enables performance optimizations such as Collective I/O [8].

Data-intensive file systems are specialized file systems for data-intensive computing frameworks such as MapReduce [9]. Leading data-intensive file systems include Google file system (GFS) [10], Hadoop file system (HDFS) [11], and Kosmos file system (KFS) [12]. Data-intensive file systems usually come with interfaces to interact with general HPC applications. For Java based HDFS, libHDFS can be used as the programming interface to support MPI applications [13]. The Kosmos file system offers a native interface to support HPC applications [12]. POSIX imposes many hard consistency requirements that are not needed for MapReduce applications and are not natively supported for data-intensive file systems. MPI-IO [8] was designed on general-purpose file systems and its access to data-intensive file systems is currently not supported as well.

Parallel file systems and data-intensive file systems share similar high-level designs. They are both cluster file systems that are deployed on a bunch of nodes. Both of them divide a file into multiple pieces (stripes or blocks/chunks), which are distributed onto multiple I/O servers. However, there are many differences between them. Governed by the CAP theorem [14], the design goals of distributed systems can be only two out of the three of Consistency, Availability, and Partition-tolerance. Parallel file systems choose "CA" since they assume general HPC applications as ACID (Atomicity, Consistency, Isolation, and Durability). Data-intensive file systems are designed to support MapReduce jobs with batch processing pattern on top of commodity hardware and choose "AP" as the design goals. The difference between "CA" and "AP" leads to distinct design decisions on several key components.

Data Layout Data layout means the method how the data stripes and blocks are placed and distributed over all the available storage nodes. In modern parallel file systems [6], data are typically distributed over multiple storage nodes in a round-robin fashion, to take advantage of parallel access. This round-robin data layout is most widely used because it can provide acceptable I/O performance for many scenarios. Parallel file systems provide more than one data layout methods to advanced users for choosing optimal layout configurations. We name three most popularly adopted data layout methods as 1-DH, 1-DV, and 2-D data layout [15]. 1-DH data layout is the simple striping method and distributes data across all storage nodes. 1-DV data layout refers to the policy that data to be accessed by each I/O client process is stored on one storage node. 2-D data layout refers to the policy in which data to be accessed by each process is stored on a subset (called storage group) of storage nodes. Given a file or directory, the data layout is predefined by default or by user's customization, and all the new data append to that file will follow the predefined policy. However, in distributed file system, there is no predefined data layout. While a task generating data, it asks the Namenode (the metadata server) to create a new data block with a fixed block size, and then the generated data will fill into that block. When the data write reaches the end of the block (i.e. the block is full), a new block is created, and so on. The selection of the new block's location is not following any of the above mentioned 1-DV, 1-DH, or 2-D layouts. Instead, the location is either the local node or some random remote node. And this policy favors the co-locality between the task and its data.

Data Locality Parallel file systems are designed for typical HPC architecture that separates I/O nodes from compute nodes. File system server processes are deployed on I/O nodes and client processes are deployed on compute nodes. Client processes see server processes as symmetric and data locality is not considered by PFS. On the other hand, the deployment of data-intensive file systems calls for the existence of local disk on each compute node. The client processes of data-intensive file systems should be co-located with server processes to gain high data locality and better I/O performance.

Data Partition Granularity As we mentioned, both of parallel file systems and distributed file systems partition their data into data stripes and data blocks. However, the sizes of their stripes/blocks are largely different. For parallel file system, the default stripe size is relatively small, for example, 64 KB for OrangeFS. For distributed file system, the default block size is large, for example, 64 MB for HDFS. This design differences are determined by their different data locality designs. Parallel file systems choose small stripe size because they want to benefit each request with parallel access; if the size is too large, then handling the request may just involve very few nodes resulting low parallel degree. Distributed file systems have to adopt large block size, because they intend to let each task have all its data on the same local node. If the block size is too small, a task accessing a large chunk of data may have to access remote nodes.

Concurrency and Locking One data chunk is exclusively used by one process/task in MapReduce applications. As such, concurrency is not supported well by data-intensive file systems. Concurrent write operation to one shared file is not supported by HDFS. KFS supports shared file write by placing an exclusive lock on each chunk. All the processes accessing the same chunk compete for the lock to perform I/O operations. Parallel file systems are designed to support POSIX interface and concurrency is inherently supported. GPFS and Lustre leverage more complex distributed locking mechanisms to mitigate the impact of caching. For example, GPFS employs a distributed token-based locking mechanism to maintain coherent caches across compute nodes [16].

Parallel file systems and distributed file systems are also different on some special features and optimizations.

File Caching Client-side cache is an effective approach to improving the bandwidth, especially for small I/O requests. However, the adoption of cache also threatens the data consistency. Data-intensive file systems employ cache for better performance since consistency is not the top design goal. The client accumulates the write requests in memory until its size reaches the chunk size (usually 64 MB) or the file is closed, which triggers the write operation to I/O servers. To guarantee consistency and durability, PVFS drops client side cache. GPFS and Lustre support file caching but depend on sophisticated distributed locking mechanism to assure the consistency [17].

Fault Tolerance Parallel file systems do not have native fault tolerance support inherently and usually rely on hardware level mechanism like RAID for fault tolerance. Failures could occur frequently for data-intensive file systems that assume commodity hardware at scale. As such, chunk-level replication is adopted to support the fault tolerance of data-intensive file systems.

Due to these design differences, parallel file systems and data-intensive file systems serve the applications in their own camp very well but not well enough for the applications from the other camp. However, there are occasions that require the cross-camp system integrations. One important occasion is the "HPC in Cloud." The following section will present our work on the I/O optimization that lets data-intensive file systems support HPC applications efficiently. In particular, we consider the scenario of running HPC applications on top of MapReduce file systems, and propose solutions to bridge the semantic gap between the two. A chunk-aware I/O strategy is introduced to enable efficient N-1 data access such as checkpointing for data-intensive file systems. Some early-stage progress of this work is previously published in [18].

2 Chunk-Aware I/O: Enabling HPC on Data-Intensive File Systems

2.1 Motivation

The advent of Cloud computing has revolutionary impacts on every aspect of technical computing, as well as high-performance computing. Amazon EC2, the leading Infrastructure-as-a-Service (IaaS) cloud computing platform, has hit the Top 500 supercomputer list since 2010 [19]. More and more HPC scientists see Cloud as a promising alternative to classical supercomputers, due to the feature of elasticity, the flexible pay-as-you-go pricing model of Cloud computing. There have been numerous efforts that investigate the potential of "HPC in the Cloud" computing paradigm [20–23].

Classical MPI based HPC applications usually rely on parallel file systems (PFS) for data manipulation. However, PFS is designed based on the premise of dedicated, highly reliable hardware with fast network connectivity, which makes it unrealistic to be deployed in Cloud environment that assumes inexpensive commodity hardware.

Performance is another concern of utilizing PFS as checkpointing storage for traditional HPC systems. As discussed in Chap. 3, PFS are usually deployed on dedicated I/O servers that are separated from computing nodes. The number of compute nodes is often one or two magnitude greater than the number of I/O servers [24, 25]. Furthermore, the communication between the compute nodes and I/O servers rely on a single or a handful of network links that are built during the installation [26]. The inherent centralization design of PFS significantly limits its support of data-intensive parallel checkpointing that comes with overwhelming I/O workloads.

Data-intensive distributed file systems are storage systems specially designed for applications running under MapReduce frameworks [9–12]. Sharing the same assumptions, MapReduce is designed to be well coupled with Clouds. As a consequence, data-intensive file systems are a natural choice to manage the storage media for HPC applications running in the Cloud.

Due to the merits of scalability, fault tolerance, and data locality, the data-intensive file systems were recently recognized as a promising alternative to offload the workload from the stretched traditional parallel file systems in traditional HPC frameworks [26].

Unfortunately, data-intensive file systems are not designed with HPC semantics in mind, and few HPC applications can benefit from them directly even if they are not consistency constrained. Many HPC applications are either not supported or cannot perform well on data-intensive distributed file systems.

N-1 (N to 1) is a widely used data access pattern for parallel applications such as checkpointing and logging [27]. The N processes usually issue requests to different regions of one shared file, which leads to non-sequential data access, unbalanced data distribution, and violates the data locality. All these factors make N-1 based HPC applications not usable on data-intensive file systems.

We have set up an experimental environment to compare the write performance of N-1 and N-N (N to N) data access patterns on two data-intensive file systems, Hadoop distributed file system (HDFS) [11] and Kosmos file system (KFS) [12]. We added components to the IOR benchmark to access data-intensive file systems. We utilize the API provided by libHDFS to access HDFS. However, the N-1 write is not supported by HDFS since libHDFS currently only allows 'hdfsSeek' in read only mode [13]. On the other hand, KFS, a C++ based data-intensive file system, supports the N-1 data access by allowing concurrent non-sequential writes to one chunk [12].

Figure 1 compares the performance of N-1 and N-N performance on 16 I/O nodes (chunk servers) selected from SCS cluster. The chunk (block) size is 64 MB for both file systems. We have 16 processes in each run to issue strided I/O requests. The N-1 curve presents unstable performance with different request sizes. Smaller request sizes lead to more contention in the shared chunk and more performance degradation. The problem is common for HPC applications as often the request size is much smaller than the chunk size of data-intensive file systems (64 MB or higher) [28].

In recognition of the semantic gap between HPC applications and data-intensive file systems, the objective of this research is to bridge the gap and facilitate efficient shared data access of HPC applications to data-intensive file systems.

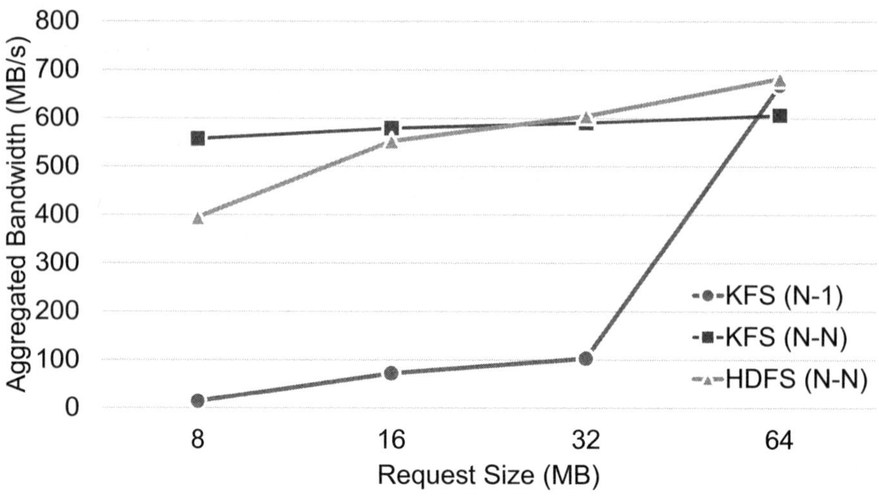

Fig. 1 Performance comparison of N-1 and N-N (write)

The contribution of this study is three-fold:

1. CHAIO, a chunk-aware I/O strategy to enable efficient N-1 data access patterns on data-intensive distributed file systems, is introduced. CHAIO reorganizes data from different processes to avoid contention and achieve sequential data access.
2. An aggregator selection algorithm is proposed to decide a process that issues the I/O requests on behalf of the conflicting processes to balance the I/O workload distribution and regain the data locality.
3. CHAIO is prototyped over the Kosmos file system. Extensive experiments have been carried out to verify the benefit of CHAIO and its potential in fostering scalability.

2.2 *Chunk-Aware I/O Design*

Data Access Patterns Data access patterns in HPC applications like checkpointing can be classified as either N-N or N-1 [27]. In N-N data access pattern, each process accesses an independent file with no interference with other processes. Figure 2a demonstrates N-N data access pattern and how it is handled by data-intensive file systems. We assume the chunk size and request size as 64 MB and 40 MB, respectively, which means a chunk is composed of 1.6 requests. Each compute node has one process, and we have four nodes host the data-intensive file system.

Each process issues three I/O requests, which are marked by logical block number (LB#) to reflect its position in the file. The file view layer in the Fig. 2 shows the mapping between the requests and their positions in the file. Based on the data access

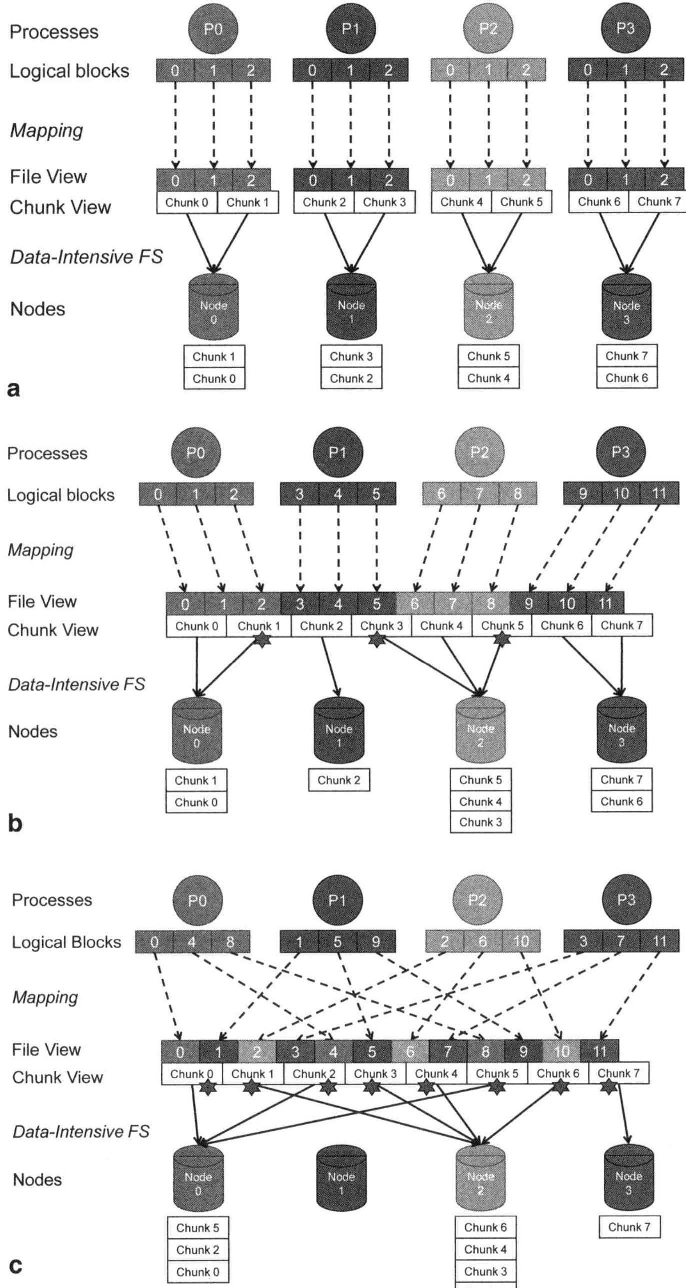

Fig. 2 Data access patterns **a** N-N, **b** N-1 segmented, **c** N-1 strided

information and chunk size, the requests are translated into chunks by the data-intensive file system, which are distributed onto the nodes with the consideration of data locality.

In the N-N data access case of Fig. 2a, each process accesses an individual file and does not incur contention. The I/O workload is evenly distributed such that each node holds two chunks. The downside of the N-N data access pattern, however, is that it involves more files and requires extra cost in metadata management, which is unwanted for data-intensive file systems because of the centralized metadata management.

N-N data access pattern is the ideal case to avoid contention. However, most HPC applications have the processes cooperate with each other and adopt N-1 data access pattern in practice. The processes access different regions of one shared file in N-1 data access. Depending on the layout of regions, N-1 data access can be further classified into two categories: N-1 segmented and N-1 strided [27].

In N-1 segmented data access pattern, each process accesses a contiguous region of the shared file. Figure 2b illustrates N-1 segmented data access pattern and how it is handled by the data-intensive file system. The request size is determined by HPC applications and does not match the chunk size well. The requests from multiple processes could be allocated to one chunk and lead to contention. We term a chunk as conflict chunk if it is accessed by multiple requests. In Fig. 2b we have three conflict chunks with ID 1, 3, and 5.

Conflict chunks degrade the I/O performance because of the following reasons:

1. The file system alternates among different requests on the conflict chunk, which violates the sequential data access assumption of data-intensive file systems.
2. The conflict chunk is composed of requests from multiple compute nodes and only one node is selected to host the chunk. Data locality is not achieved for the requests from other compute nodes. For example, for chunk 3 of Fig. 2b, the request from P1 (LB# 5) is not a local data access.
3. The chunk placement is decided by the first request with the consideration of data locality. This mechanism results in unbalanced data distribution. In Fig. 2b we can observe that three chunks (3, 4, and 5) are allocated onto node 2 while node 1 only has one chunk. It is more critical for data-intensive file systems to balance the chunk distribution since the chunk size is normally sized 64 MB or higher, which is magnitudes higher than the strip size (usually 64KB) of PFS.

In the N-1 strided data access pattern, each process issues I/O requests to the file system in an interleaved manner. As illustrated in Fig. 2c, strided data access has a higher probability to incur conflict chunks and has greater impact in degrading the performance. Actually, all the 8 chunks have contention in the case shown in Fig. 2c. The data locality and balanced data distribution will be further deteriorated as well. Figure 2c demonstrates the worst case that node 2 has 4 chunks, while no chunk is allocated to node 1.

In practice, N-1 strided is a more common data access pattern than N-1 segmented for HPC applications such as checkpointing [27].

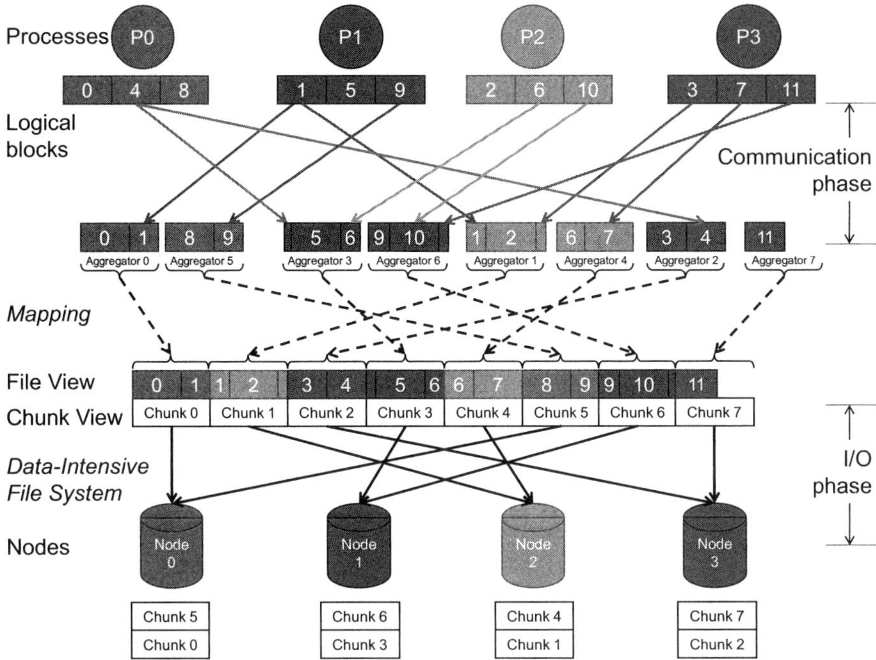

Fig. 3 N-1 Strided write with CHAIO

This study is motivated by the performance issues with the N-1 data access on data-intensive file systems. The proposed new CHAIO strategy rearranges the I/O requests to eliminate conflict chunks, achieve data locality, and balance data distribution.

CHAIO Design The basic idea of CHAIO is to reorganize the I/O requests such that each chunk is accessed by one process to eliminate contention. Figure 3 shows how CHAIO handles the scenario shown in Fig. 2c.

We add a communication phase to exchange data among processes. One process is selected as an aggregator process for each conflict chunk. The aggregator collects data from the non-aggregator processes accessing the same conflict chunk and issues the I/O requests to the file system. From the perspective of data-intensive file systems, each chunk is accessed by the aggregator process only. Even though CHAIO introduces slight message passing overhead, it improves the performance significantly by removing the contention and marshaling the I/O requests.

With CHAIO, each chunk has only one aggregator process that acts as the file system client to issue the I/O request, as shown in the I/O phase of Fig. 3. The data locality is assured since the file system by default allocates the chunk to the node where the aggregator process resides.

The N-1 read of CHAIO is performed in the reverse order. The aggregator first gets data from the file system and distributes the data to the corresponding processes.

Table 1 CHAIO aggregator selection algorithm

```
Definition:
A chunk is allocated if its aggregator process has been decided.
Chunk c is serviced by node s if there is at least one I/O request from s to c.
Terminology:
C is the collection of the unallocated chunks.
a(s) is the number of chunks that have been allocated to node s.
n(s) is the number of unallocated chunks that are served by node s.
g(c) is the number of nodes that service chunk c.
p is the number of conflict chunks.
q is the number of nodes.
f(c) = s means we select a process on node s as the aggregator process for chunk c.
Algorithm:
Initialize C, n(s), g(c), p and q based on the data access info, chunk size and the process distribution.
a(s) = 0
threshold = ⌈p/q⌉
while (size(C)>0) do
        find the node s with min(a(s)+n(s)) and satisfies a(s)≤threshold.
        if s == null then
                increase threshold by 1
                continue
        end if
        for each chunk serviced by node s, find the chunk c with min(g(c)).
        f(c) = s
        a(s) = a(s)+1
        for each node that services c do
                remove c from its chunk list
                set n(s) = n(s)-1
        end for
        remove c from C.
end while
```

Next we introduce the aggregator selection algorithm that balances the chunk distribution among the nodes. The aggregator selection algorithm takes the data access pattern, chunk size and process distribution as input. The output of the algorithm is the decision of the aggregator process for each chunk.

The pseudo code of the aggregator selection algorithm is listed in Table 1. The node with less conflict chunks has higher priority to be selected to host the aggregator process. If multiple chunks are serviced by the selected node, the chunk with the least service nodes is selected. The algorithm is a greedy algorithm that is biased toward the node or chunk with least matching options.

The algorithm sets a threshold to limit the aggregator processes on each node and guarantee balanced I/O workload distribution. A node will not be selected to run more aggregator processes if it is already fully loaded with the threshold number of aggregator processes. The threshold value will be increased if there is no eligible node but not all the chunks have been allocated yet.

Multiple processes from the selected node may access the same conflict chunk in a multicore architecture. The process with the largest I/O request size to minimize the message passing overhead will be selected in this case. Figure 3 shows an example where each node has two chunks with the assistance of the aggregator selection algorithm.

2.3 Chunk-Aware I/O Implementation

CHAIO can be implemented either inside the application code or in the I/O middleware layer such as MPI-IO. CHAIO takes the data access information, chunk size and the process distribution information as input. The data access information can be obtained from the application or from MPI primitives, i.e., MPI_File_get_view. The data-intensive file system needs to expose the chunk size information to CHAIO, which is trivial to implement. We also need to know the process distribution information that indicates the mapping between processes and nodes, usually in a round-robin or interleaved manner. We can obtain this information easily from the job scheduler.

Each process first captures the aforementioned input and carries out the aggregator selection algorithm. The output of the algorithm is organized in a hash table data structure which stores the chunk ID and the corresponding rank ID for the aggregator process.

When a process carries out an I/O request, it first calculates the chunk ID of the I/O request and checks the hash table derived from the aggregator selection algorithm. If the chunk ID of the I/O request matches one entry in the hash table, it means the I/O request is involved in a conflict chunk and we need to take action. If the rank ID of the process matches the aggregator process ID from the hash table, the process will receive data from other processes and then issue the I/O request of the entire chunk to the file system. If the process is not selected as the aggregator, it simply sends the data to the aggregator process.

We use non-blocking send for the non-aggregator processes so that the following I/O requests are not blocked by the message passing. Blocked receive is adopted by the aggregator to guarantee that the process is carrying out one I/O request at a time.

2.4 Chunk-Aware I/O Analysis

There are potential alternative solutions to the problem of N-1 data access besides the CHAIO approach. The straightforward solution is to adopt methodologies such that one I/O request generates one individual chunk in the data-intensive file systems. To implement the idea, we can adapt the chunk size to the I/O request size in the file system.

The primary concern with this approach is the metadata management overhead it introduces to the file system. The number of chunks is equal to the number of I/O requests, which could be significant considering small request sizes from HPC applications [28]. On the other hand, the file system namespace and file BlockMap of the data-intensive file system is kept in the memory of the centralized metadata server (Namenode). A large number of chunks could overwhelm the centralized metadata management of the data-intensive file system and degrade I/O performance.

CHAIO aggregates multiple I/O requests of one chunk to form sequential data access and does not increase the metadata management overhead to the file system.

CHAIO is implemented at either the application level or I/O middle-ware level and does not introduce complexity to the data-intensive file system.

Data-intensive applications usually adopt multiple replicas of one chunk to achieve fault tolerance. The data locality and balanced data distribution are not concerned by non-primary replicas since they select nodes randomly to store the data. Multiple replicas do not obscure the advantages of CHAIO for read operations. The I/O request returns after reading one replica from the file system and CHAIO does help to alleviate contention in this scenario. Furthermore, the performance of the primary (first) replica is improved by CHAIO for write operations. The performance of the first replica is usually more critical than others since it concerns the application elapsed time. It is a widely used optimization technique for replica based file systems to return to the application after the first replica is completed and process the rest of the replicas in parallel with the applications [29].

2.5 *CHAIO Performance*

2.5.1 Experiment Setup

We have carried out experiments on the SCS cluster. One computer node of the cluster is dedicated as the job submission node and the metadata server of the Kosmos file system. The experiments were tested with Open MPI v1.4.

KFS is utilized as the underlying data-intensive file system in the experiments. We use IOR-2.10.2 from Lawrence Livermore National Laboratory as the benchmark to evaluate the performance [30]. We have added a KFS interface to the IOR benchmark to enable data access to the Kosmos file system. The KFS interface was implemented with the methodology similar to other interfaces of IOR such as POSIX. We implement CHAIO in IOR benchmark and compare its performance with the original IOR benchmark. We set the chunk size at 64 MB in the experiments and each chunk has one replica by default.

2.5.2 Performance with Different Request Sizes

We keep the number of nodes fixed at 32 in Fig. 4 and study the performance with different I/O request sizes. We fix the size of the shared file at 32 GB and each process issues 16 interleaved I/O requests to implement N-1 strided data access. The number of processes is varied accordingly with different request sizes. We run each setting 10 times in the experiments, get the mean and standard deviation of the aggregated bandwidth and plot them in the figure. The standard deviation is reflected by the error bars.

The write performance is illustrated in Fig. 4a. A smaller size of I/O requests means more contention in conflict chunks and leaves more opportunity for performance improvement in CHAIO. Actually, when the request size is 4 MB, it is not

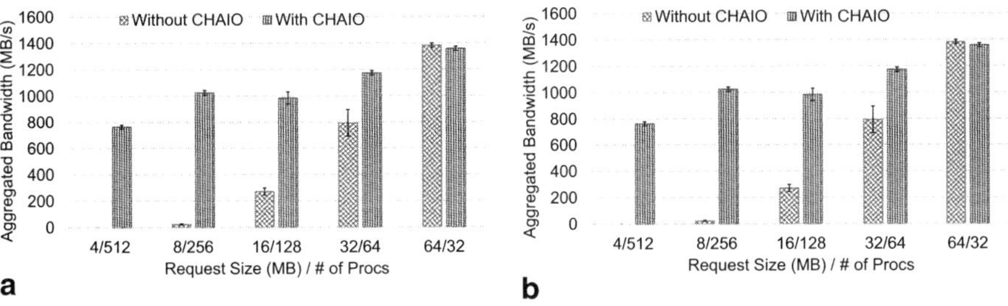

Fig. 4 CHAIO performance with different request sizes **a** write, **b** read

possible to have successful N-1 data access by the existing approach due to the overwhelming contention on the conflict chunks. We were able to get successful data access for request size of 8 MB but the performance was still very poor (22.92 MB/s). CHAIO achieved a write bandwidth of 983.52 MB/s for 16 MB request size, which is three times higher than 270.6 MB/s, the bandwidth achieved by the existing approach.

When the request size is 64 MB, CHAIO does not show advantage in bandwidth performance. Since the request size is equal to the chunk size, there is no contention on the chunks and the benefit of CHAIO cannot be observed.

When the request size is 4 MB, CHAIO shows less bandwidth than in the case with larger request sizes. There are possibly two factors leading to the performance degradation. First, a smaller request size needs more data exchange in the communication phase. Furthermore, each 8-core node is overloaded with 16 processes when the request size is 4 MB and could considerably harm the overall I/O performance. Our later analysis in sub-sect. 2.5.5 shows that the impact of the small request size incurs little overhead and we can attribute the performance degradation to the second factor.

Figure 4b compares the read bandwidth of CHAIO and the existing approach. It is easy to observe the advantage of CHAIO over the existing approach. The bandwidth of reads more than doubles the existing approach when the request size is 16 MB or less.

2.5.3 Performance with Two Replicas

We set the number of replicas as two and demonstrate its performance with different request sizes in this set of tests, and the results are shown in Fig. 5.

Though the advantage of CHAIO is less significant for more replicas, as discussed in sub-sect. 2.5.2, it still presents satisfactory write performance improvement as shown in Fig. 5a. CHAIO achieved a write bandwidth of 685.7 MB/s, which doubled the existing approach, 328.9 MB/s.

We also observe that both CHAIO and the existing approach have bandwidth degradation for two replicas than the one replica case of Fig. 4a. For example, in the

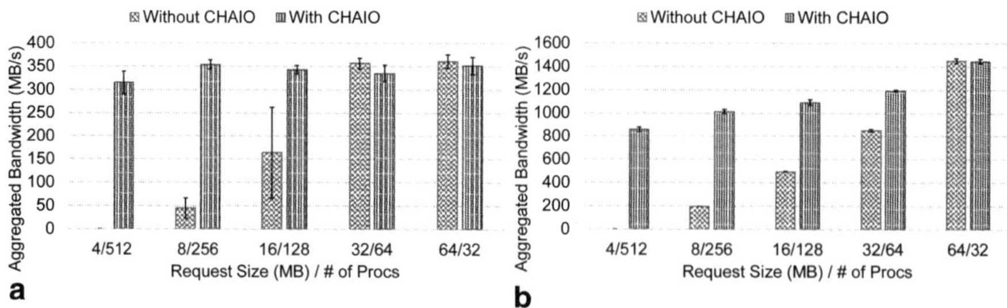

Fig. 5 CHAIO performance with two replicas **a** write, **b** read

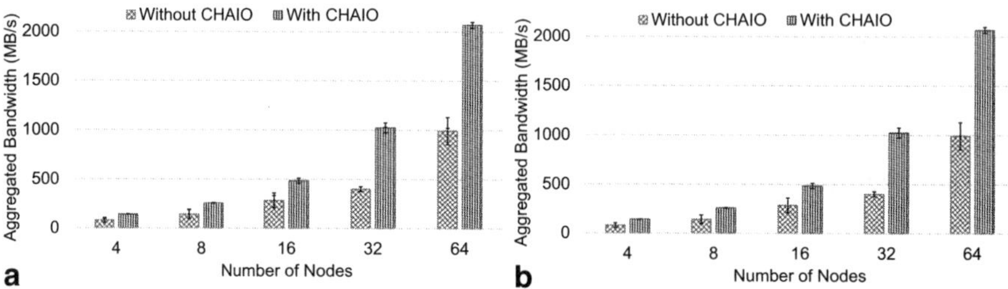

Fig. 6 CHAIO performance with different number of nodes **a** write, **b** read

contention-free case with 64 MB request size, the write bandwidth of two replicas is about 700 MB/s, which is considerably lower than the 1300 MB/s bandwidth in the case of one replica. A detailed study reveals that the performance degradation is due to node-level contention. When the number of replicas increases to two or more, each node not only services the first chunk, but the non-primary copies will also compete for the node and incur node-level contention. This study focuses on the chunk-level contention problem caused by N-1 data access. The node-level contention problem is on our roadmap for future studies.

As illustrated in Fig. 5b, read performance is not impacted by multiple replicas and CHAIO outperforms the existing approach consistently.

2.5.4 Performance with Different Number of Nodes

In Fig. 6 we vary the number of data nodes from 4 to 64 and observe its impact on the performance. For each node we spawn 4 MPI processes to carry out the I/O requests. Each process issues 16 interleaved I/O requests to one shared file to implement N-1 strided writes. The I/O request size is kept at 16 MB.

The write bandwidth is presented in Fig. 6a. The bandwidth of CHAIO is two-fold higher than that of the existing approach.

The existing approach also exhibits more variance in bandwidth than CHAIO, which is caused by the unbalanced chunk distribution. In the existing approach, the

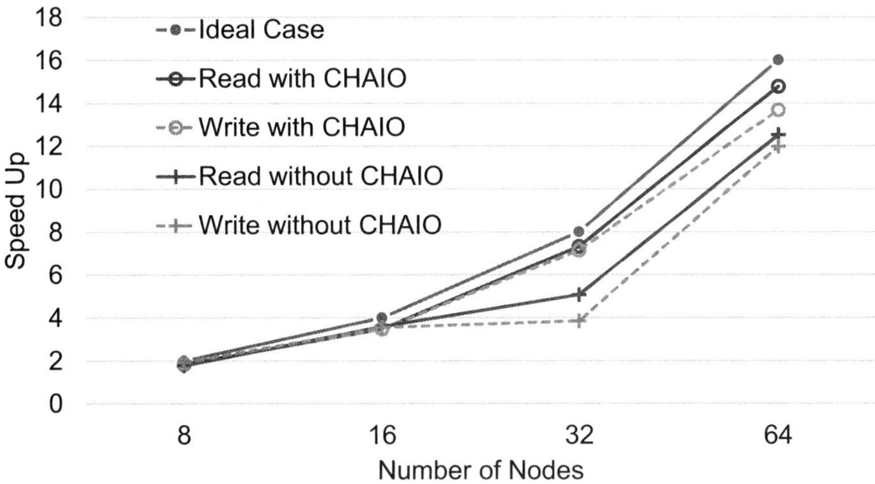

Fig. 7 Scalability analysis

conflict chunk selects the node based on the first I/O request coming into the file system, which results in uncertainties in the chunk distribution and a large variance in bandwidth.

Figure 6b compares the read bandwidth of the CHAIO approach with the existing approach and confirms the advantage of CHAIO as well. CHAIO achieved a bandwidth two times higher than the existing approach for all cases.

We use the performance with 4 nodes as the baseline and plot the speedup for each scenario in Fig. 7. CHAIO performs well in terms of scalability for both write and read operations, which is close to the ideal speedup case. We can observe that the existing approach achieved speedup as well; however, it is not stable. For example, there is no much improvement between 16 nodes and 32 nodes for the existing approach. A detailed study reveals that due to the uneven chunk distribution, a small set of the nodes is constantly selected as the chunk servers and hurts the scalability. The CHAIO approach reduces the access contention and regains the access locality by rearranging requests, and achieves better and stable scalability.

2.5.5 Overhead Analysis in Large-Scale Computing Environments

We have shown the performance improvement of CHAIO in terms of both bandwidth and scalability in previous sub-sections. We have performed tests to evaluate the potential of CHAIO in large-scale computing environment as well. To achieve that, we have measured the communication phase cost of CHAIO on the ANL SiCortex test bed.

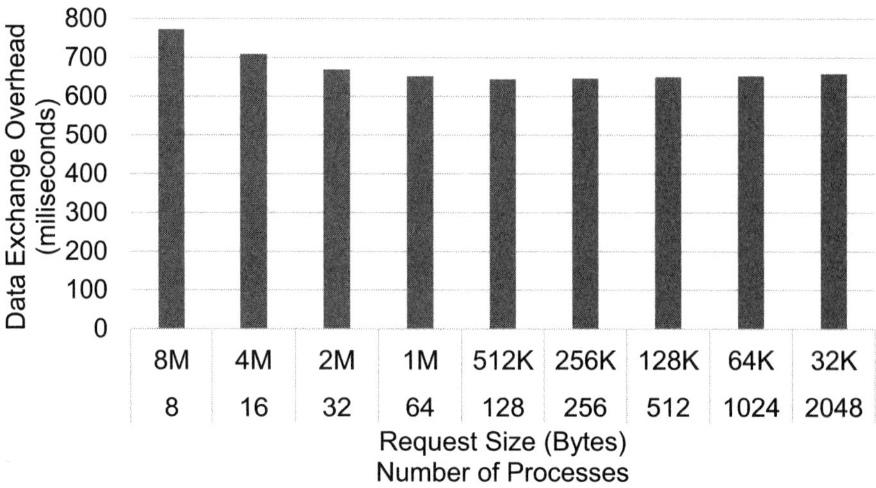

Fig. 8 Data exchange overhead (1 conflict chunk)

Data exchange cost is the primary, if not only, overhead introduced by CHAIO and is used as the metric to study the CHAIO overhead[1]. Figure 8 reports the measure data exchange cost of one conflict chunk with different request sizes and number of processes. It can be observed that the data exchange cost is kept less than 1 s for all cases which is a minor overhead. The data exchange overhead is increased by 0.15 s when reducing the request size from 8 MB to 16KB, which is still trivial compared with the I/O performance CHAIO helps to improve. The experimental tests confirm that CHAIO improves the I/O bandwidth considerably with only introducing a minor communication overhead.

We keep the number of processes at 2048, vary the request size from 1 to 32 MB, and demonstrate the data exchange overhead in Fig. 9. In this set of experiments, we have irregular request sizes that cannot divide the default block size of 64 MB perfectly, e.g., 3, 5, 10 MB, etc. Such irregular sized requests result in irregular conflict patterns and impose challenges to the aggregator selection algorithm. As reported in Fig. 9, we do not observe much performance deviation of irregular request sizes compared to the regular ones. This infers that CHAIO is applicable to various request sizes with trivial data exchange cost.

[1] We eliminated the I/O phase in the SiCortex experiments and only measured the communication phase cost for overhead analysis. The lack of local disk and the job scheduler of SiCortex make it impractical, if not impossible, to deploy Kosmos file system on SiCortex.

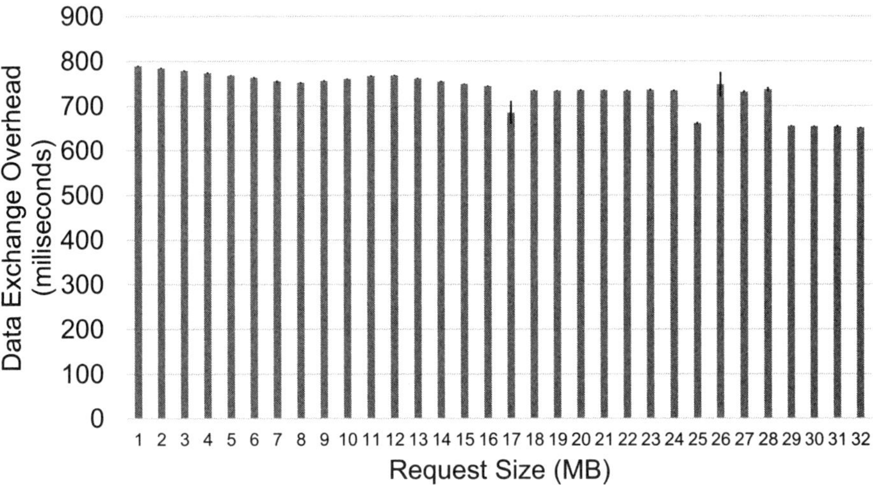

Fig. 9 Data exchange overhead (2048 processes)

2.5.6 Load Balance

The primary objective of the aggregator selection algorithm is load balance. CHAIO should balance the number of chunks allocated to each data node to gain better parallelism and performance. As such, we evaluate the load balance of CHAIO and compare it with the existing approach. We first calculate the ideal data layout that conflict chunks are evenly distributed to all the involved data nodes. The data layout of CHAIO and the existing approach are also calculated. We next compute the Manhattan distances between the ideal data layout and the two scenarios (with/without CHAIO), which is used as the metrics of unbalance [31]. Figure 10 reports the degree of unbalance by varying request size from 1 to 32 MB with 2048 processes on SiCortex. CHAIO performs significantly better in balancing the workload among each data node than the existing approach. In particular, we observe perfect load balance (0 in y-axis) for 17 out of the 32 samples, including 11 irregular requests that cannot divide 64 MB perfectly.

3 Related Works

Merging HPC architecture and MapReduce framework has great practical importance and many researchers have put efforts on this topic. In this section, we introduce some other researcher's efforts, which fall into two categories: (1) HPC on data-intensive file systems and (2) N-1 data access and its handling.

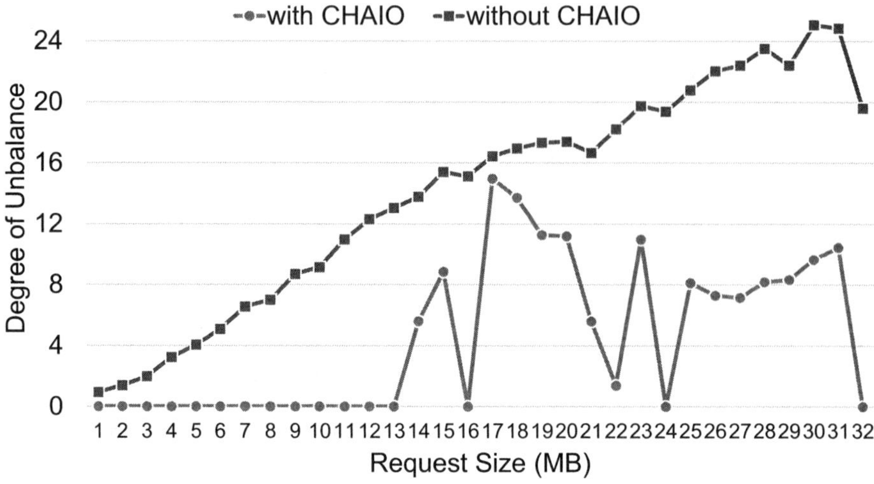

Fig. 10 Load balance evaluation (2048 Processes)

3.1 HPC on Data-Intensive File Systems

There is an increasing interest in incorporating the emerging data-intensive file systems with the HPC framework. The pioneering works exploited the merge of the two from different perspectives.

In [26], the authors introduced VisIO to utilize HDFS as the storage for large-scale interactive visualization applications. VisIO provides a mechanism for using non-POSIX distributed file system to provide linear scaling of I/O bandwidth. The application targeted by VisIO is N-N read, which is naturally supported by data-intensive distributed file systems.

In [28], the authors studied the requirements of migrating data from HPC storage system to data-intensive frameworks such as MapReduce, and proposed MRAP to bridge semantic gaps. MRAP extends MapReduce to eliminate multiple scans and also reduces the number of pre-processing MapReduce programs.

In [32], the authors examined both HPC and Hadoop workloads on two representative file systems, PVFS and KFS. Their study confirmed the performance degradation of N-1 data access pattern on the Kosmos file system.

In [33], the authors compared PVFS and HDFS, and enhanced PVFS to match the HDFS-specific optimizations. Unmodified Hadoop applications can store and access data in PVFS using the proposed non-intrusive shim layer that implements several optimizations, including prefetching data, emulating replication and relaxed consistency, to make PVFS performance comparable to HDFS.

Nevertheless, all these existing works acknowledged the concurrency issue on data-intensive file systems but did little to overcome it. The proposed CHAIO is motivated by the observation that some HPC applications with concurrent I/O access cannot perform well even they are not consistency constrained. CHAIO extends the scope of HPC applications supported by data-intensive file systems, and improves the overall I/O performance of HPC systems as a consequence.

Several efforts have also been made by industry to enable HPC applications with MapReduce. Hamster is an initiative from Hadoop community that aims at supporting MPI applications in a Hadoop cluster [34]. Apache Mesos is a cluster manager that provides resource sharing across different applications, e.g., Hadoop, MPI, Hypertable, and Spark [35, 36]. These efforts offer feasibility of running MPI applications with Hadoop, but do little to handle the data manipulation issue under this computing scenario, which is complemented by CHAIO.

MapR is a data-intensive application platform from EMC to support Hadoop framework. MapR has a functionality of Direct Access NFS to enable the capability of accessing MapR file system with the POSIX interface [37]. This technique is implemented at file system layer, which does not consider the data exchange among parallel processes and, therefore, eliminates possible performance improvement opportunities. The idea of CHAIO can be extended to support MapR to deliver both capability and efficiency to parallel applications.

3.2 N-1 Data Access and its Handling

Modern PFS either leverages distributed locking protocols to achieve consistency for N-1 shared data access (GPFS and Lustre), or does not support POSIX semantics for concurrent writes and relies applications to solve conflicting operations (PVFS). The lock-based solution imposes considerable overhead and several works have been conducted to address this concern.

Collective I/O merges the requests of different processes with interleaved data access patterns and forms a contiguous file region, which is further divided evenly into non-overlapping, contiguous sub-regions denoted as file domains [8]. Each file domain is assigned an aggregator process that issues the I/O requests on behalf of the rest of the processes in that file domain. Collective I/O does not take underlying file system into consideration when deciding file domains and cannot eliminate the conflict chunks. It is still possible that two aggregator processes concurrently access one shared chunk in Collective I/O. Users can customize the collective buffer size on each aggregator process by setting parameter cb_buffer_size but that does not solve the problem.

In [17], the authors proposed to partition files based on the underlying locking protocols such that the file domains are aligned to locking boundaries. Data shipping was introduced by GPFS to bind each file block to a single I/O agent that acts as the delegator [16].

PLFS is a virtual parallel log structured file system that sits between parallel file systems and applications and transforms the N-1 data access into N-N pattern [27]. PLFS currently supports parallel file systems such as GPFS and PanFS. Extra efforts need to be taken to adapt PLFS to support data-intensive file systems because the underlying N-N data access potentially imposes more overhead to the metadata management, which is unwanted for data-intensive file systems due to the centralized metadata server.

BlobSeer is a storage system that supports efficient, fine-grain access under heavy concurrency. In [31], the authors demonstrated the potential of BlobSeer in substituting HDFS to enable efficient MapReduce applications. BlobSeer adopts versioning instead of locking protocols to handle the concurrency issue.

While demonstrating their success on the N-1 data access of PFS, the ideas of these works can also be applied to data-intensive file systems to alleviate the problem. However, unique features of data-intensive file systems require additional efforts for a complete solution. The selection of the aggregator process is actually a key factor in determining the overall performance of N-1 access on data-intensive file system, especially when the requests from different processes are irregular with varied sizes. However, the selection of aggregator process is not covered by existing PFS optimization techniques since the client processes are usually independent of server processes.

4 Summary

In this chapter, we introduced and studied two types of most popular cluster file systems: parallel file systems that serve HPC and data-intensive file systems that serve distributed cloud computing. We introduce their differences in categorized details. Due to these design differences, parallel file systems and data-intensive file systems serve the applications in their own camp very well but not well enough for the applications from the other camp.

We have identified three factors that degrade the performance of N-1 data access on data-intensive file systems: non-sequential data access, uneven chunk distribution, and the violation of data locality. A chunk-aware I/O (CHAIO) strategy was proposed to address these issues and overcome the challenge. The CHAIO introduced an data communication phase that collects data from multiple processes and issues the I/O requests to the file system to achieve sequential data access and data locality. An aggregator selection algorithm has been proposed to balance the chunk distribution among nodes. CHAIO can be implemented at either the application level or the I/O middle-ware level and does not introduce complexity to the underlying file systems.

We have prototyped the CHAIO idea and conducted experiments with the IOR benchmark over the Kosmos file system. Experimental results show that CHAIO improves both the write and read performance significantly. The overhead analysis showed that CHAIO introduces little overhead for small request sizes and has a real potential for large-scale computing environment. The performance gain of CHAIO is robust to different requests size. The aggregator selection algorithm works efficiently for load balance.

References

1. "The Message Passing Interface (MPI) standard" [Online]. Available: http://www.mcs.anl.gov/research/projects/mpi/.
2. F. Schmuck and R. Haskin, "GPFS: A Shared-disk FileSystem for Large Computing Clusters," in *Proceedings of the 1st USENIX Conference on File and*, 2002.
3. "Lustre File Systems Website," [Online]. Available: http://wiki.lustre.org/index.php/Main_Page.
4. P. J. Braam., "The Lustre Storage Architecture," [Online]. Available: http://www.lustre.org/documentation.html.
5. "OrangeFS Website," [Online]. Available: orangefs.org.
6. Carns, P.H., Ligon, W.B. III, and Ross, R.B., "PVFS: A Parallel File System for Linux Clusters," in *Proceedings of the 4th Annual Linux Showcase and Conference*, 2000.
7. "MPI-2: Extensions to the Message-Passing Interface," [Online]. Available: http://www.mpi-forum.org/docs/mpi-20-html/mpi2-report.html.
8. R. Thakur, W. Gropp, and E. Lusk, "Data Sieving and Collective I/O in ROMIO," in *FRONTIERS '99: Proceedings of the 7th Symposium on the Frontiers of Massively Parallel Computation*, 1999.
9. Dean, Jeffrey, and Ghemawat, Sanjay, "MapReduce: Simplified Data Processing on Large Clusters," in *Sixth Symposium on Operating System Design and Implementation*, 2004.
10. Sanjay Ghemawat, Howard Gobioff, and Shun-Tak Leung, "The Google File System," in *19th ACM Symposium on Operating Systems Principles*, 2003.
11. "Hadoop Distribute Filesystem Website," [Online]. Available: http://hadoop.apache.org/hdfs/.
12. "Kosmos Distributed Filesystem " [Online]. Available: http://code.google.com/p/kosmosfs/.
13. "libHDFS Source Code " [Online]. Available: http://github.com/apache/hadoop-hdfs/blob/trunk/src/c++/libhdfs/hdfs.h.
14. Brewer, E, "PODC Keynote Presentation," 2000. [Online]. Available: http://www.cs.berkeley.edu/~brewer/cs262b-2004/PODC-keynote.pdf.
15. H. Song, Y. Yin, Y. Chen, and X.-H. Sun, "A Cost-Intelligent Application-Specific Data Layout Scheme for Parallel File Systems," in *Proc. of the 20th International ACM Symposium on High Performance Distributed Computing*, 2011.
16. Prost, J.-P.; Treumann, R.; Hedges, R.; Jia, B.; Koniges, A., "MPI-IO/GPFS, an Optimized Implemetation of MPI-IO on top of GPFS," in *Proc. of the International Conference for High Performance Computing, Networks, Storage and Analysis (Supercomputing)*, 2001.
17. Liao, Wei-keng, and Choudhary, Alok, "Dynamically Adapting File Domain Partitioning Methods for Collective I/O Based On Underlying Parallel File System Locking Protocols," in *International Conference for High Performance Computing, Networking, Storage and Analysis, SC 2008*, 2008.
18. H. Jin, J. Ji, X.-H. Sun, Y. Chen and R. Thakur, "CHAIO: Enabling HPC Applications on Data-Intensive File Systems," in *41st International Conference on Parallel Processing*, 2012.
19. "TOP500 Supercomputer Sites" [Online]. Available: http://www.top500.org/.
20. "Magellan Project: A Cloud for Science," [Online]. Available: http://magellan.alcf.anl.gov/.
21. Walker, E., "Benchmarking Amazon EC2 for High-Performance Scientific Computing," *Usenix Login*, 2008.
22. He, Q.; Zhou, S.; Kobler, B.; Duffy, D.; McGlynn, T., "Case Study for Running HPC Applications in Public Clouds," in *Proc. of 1st Workshop on Scientific Cloud Computing (ScienceCloud)*, 2010.
23. "HPC in the Cloud," [Online]. Available: http://www.hpcinthecloud.com/.
24. Moody, A.; Bronevetsky, G.; Mohror, K.; Supinski, B. R., "Design, Modeling and Evaluation of a Scalable Multi-Level Checkpointing System," in *Proc. of the International Conference for High Performance Computing, Networks, Storage and Analysis (Supercomputing)*, 2010.
25. Oldfield, R.; Ward, L.; Riesen, R.; Riesen, A.; Widener, P.; Widener, T., "Lightweight I/O for Scientific Applications," in *Proc. of IEEE Cluster Computing (Cluster)*, 2006.

26. C. Mitchell, J. Ahrensy and J. Wang, "VisIO: Enabling Interactive Visualization of Ultra-Scale, Time Series Data via High-Bandwidth Distirburted I/O Systems," in *IEEE International Parallel & Distributed Processing Symposium*, 2011.
27. Bent John and Gibson Garth and Grider Gary and McClelland Ben and Nowoczynski Paul and Nunez James and Polte Milo and Wingate Meghan, "PLFS: A Checkpoint Filesystem for Parallel Applications," in *Proceedings of the Conference on High Performance Computing Networking, Storage and Analysis*, 2009.
28. Sehrish Saba and Mackey Grant and Wang Jun and Bent John, "MRAP: a Novel Mapreduce-based Framework to Support HPC Analytics Applications with Access Patterns," in *Proceedings of the 19th ACM International Symposium on High Performance Distributed Computing*, 2010.
29. Al-Kiswany, S.; Ripeanu, M.; Vazhkudai, S. S.; Gharaibeh, A., "stdchk: A Checkpoint Storage System for Desktop Grid Computing," in *Proc. of The 28th International Conference on Distributed Computing Systems (ICDCS)*, 2008.
30. "IOR HPC Benchmark," [Online]. Available: http://sourceforge.net/projects/ior-sio/.
31. B. Nicolae, G. Antoniu, L. Bougé, D. Moise and A. Carpen-Amarie, "BlobSeer: Next-Generation Data Management for Large Scale Infrastructures," *Journal of Parallel and Distributed Computing*, vol. 2, pp. 169–184, 2011.
32. M.-E. Esteban, G. Maya, M. Carlos, J. Bent and S. Brandt, "Mixing Hadoop and HPC Workloads on Parallel," in *the 2009 ACM Petascale Data Storage Workshop (PDSW 09)*, 2009.
33. W. Tantisiriroj, S. Patil, G. Gibson, S. W. Son, S. J. Lang and R. B. Ross, "On the Duality of Data-Intensive File System Design: Reconciling HDFS and PVFS," in *International Conference for High Performance Computing, Networking, Storage and Analysis (SC)*, 2011.
34. "Hamster: Hadoop And Mpi on the same cluSTER," [Online]. Available: http://issues.apache.org/jira/browse/MAPREDUCE–2911.
35. "Apache Mesos" [Online]. Available: http://mesos.apache.org/.
36. B. Hindman, A. Konwinski, M. Zaharia, A. Ghodsi, A. D. Joseph, R. Katz, S. Shenker and I. Stoica, "Mesos: A Platform for Fine-Grained Resource Sharing in the Data Center," in *the 8th USENIX conference on Networked systems design and implementation*, 2011.
37. "MapR Direct Access NFS" [Online]. Available: http://www.mapr.com/products/only-with-mapr/direct-access-nfs.

Cloud Resource Pricing Under Tenant Rationality

Xin Jin and Yu-Kwong Kwok

1 Introduction

Infrastructure-as-a-Service (IaaS) clouds such as Amazon Web Services [1], Windows Azure [15], and Google App Engine [6] operate at Internet-scale. In this novel business model, cloud resources including network bandwidth, CPUs, and memory are packaged into virtual instances for rent. Consequently, tenant users can conveniently build and host web applications in cloud with no requirements of dedicated infrastructure deployment, and achieve cost and risk reduction. At the same time, cloud providers obtain profit gains thanks to economies of scale [2]. In such a cloud market, cloud resource pricing fundamentally determines cloud revenue, cloud resource allocation, and tenant demand dynamics.

To achieve revenue maximization, the cloud provider optimally adapts pricing decisions to tenant demand responses. Notably, prices decrease at demand troughs but increase at demand peaks. There is no doubt that data centers possess dominant control over resource prices. Nonetheless, a tenant user can strategically adjust the amount of resource procurements so as to maximize its own surplus (i.e., tenant utility minus dollar cost). Indeed, the strong substitutability of resources offered by different cloud providers and the existence of elastic and delay-tolerant demands render it feasible for tenants to judiciously adjust the optimal demand levels. In general, the pricing dynamics and demand patterns critically depend on elastic tenant demands and strategic cloud pricing decisions. In this chapter, we consider a monopoly cloud market by formulating a competitive market among tenants. In such a cloud market, a natural but critical question arises: How to achieve cloud revenue maximization in cloud resource pricing under optimal tenant demand responses?

X. Jin (✉)
Yahoo Inc., Sunnyvale, CA, USA
e-mail: tojinxin@gmail.com

Y.-K. Kwok
The University of Hong Kong, Pokfulam, Hong Kong

S. U. Khan, A. Y. Zomaya (eds.), *Handbook on Data Centers*,
DOI 10.1007/978-1-4939-2092-1_19

To this end, we propose a Stackelberg game to tractably analyze the strategic interactions between the cloud provider and tenant users for optimal cloud resource pricing in Sect. 2. Specifically, we first build a general game model to realistically capture strategic cloud resource pricing and optimal demand adjustments of tenant consumers. Tenant surplus is used to model tenant rationality, while cloud providers aim at revenue maximization in pricing. We then analytically perform equilibrium analysis for both non-uniform pricing and uniform pricing in Sect. 3. In Nash equilibrium, different price rates are offered based on the capacity provision of the cloud provider. We also reveal hidden effects for both pricing schemes. Namely, tenants with low demands may be crowded out of the system. Moreover, non-uniform pricing with price differentiation is cheating proof.

To quantify the efficiency of different pricing schemes, Price of Anarchy is used to measure the effectiveness of our strackelberg strategies in Sect. 4. Interestingly, uniform pricing can achieve social welfare maximization, though non-uniform pricing attains higher cloud revenue via price differentiation. The role of brokers in pricing is discussed based on the notion of Nash bargaining solution (NBS) in Sect. 5. In particular, tenants with low demands reserve resources from brokers for pricing discounts, which mitigates the hidden effects and improves social welfare. Finally, to empirically evaluate the analyses in Sect. 6, we conduct extensive simulations driven by 40 GB of realistic workload traces from Google [7].

Our models provide an insightful abstraction of strategic interactions in the cloud market. Moreover, hierarchical cloud resource pricing improves tenant surplus and thus social welfare in a practical interrelated market with the consideration of brokers. We believe, however, that our analyses will inspire numerous studies on the impact of strategic interactions on pricing and systems as brought up here, as well as other issues that we do not address, such as price competition among cloud providers.

2 The Game Model

We consider a cloud market with a large number of data centers and tenant users. Data centers offer cloud resources in bundle of virtual instances as *sellers*. Tenant users as *buyers* dynamically access virtual instances.

2.1 User Model and Virtual Instances Pricing

Throughout this chapter, we adopt the pay-as-you-go model, which offers guaranteed services [14]. The pay-as-you-go pricing (a.k.a., usage-based pricing) is widely used in realistic cloud resource markets, which has been adopted by Amazon EC2 [1], etc. In particular, the cloud sells virtual instances at a fixed price p. This regular price is updated infrequently, and tenants are charged only for what they use. That is, a tenant that accesses x instances pays $p \cdot x$ dollars per unit time.

More generally, sellers may provide price differentiation to absorb more of tenant surplus and transform this surplus into revenues. Intrinsic to our user model is a tenant's willingness to pay, which is captured by parameter θ_i, for tenant i, $i \in \{1, \cdots, N\}$. Parameter θ_i is not a decision variable, but reflects tenants' inherent valuation of cloud resources. That is, higher θ_i implies higher valuation of the cloud resources. Parameter θ_i realistically captures factors such as the availability of substitutes, and the inherent tenant demand. For instance, a tenant with higher inherent demand may value cloud resources more to satisfy their own service requirements. Thus, instead of fixed charges for all cloud users, we consider the most general scenario in which the unit price $p_i(\theta_i)$ is a function of the valuation coefficient θ_i. For ease of exposition, we use p_i to represent $p_i(\theta_i)$ in the remainder of this chapter.

2.2 *Modeling Cloud Revenue and Tenant Surplus*

Geographically distributed data centers possibly reside in multiple cloud providers with different pricing strategies. We consider the cloud market of a typical data center with N tenant customers. Denote by $\mathcal{N} = \{1, \cdots, N\}$ the set of tenants, where $N = |\mathcal{N}|$. We assume the capacity of the data center is Q, i.e., the aggregate demand of virtual instances from all tenant users should be no larger than Q. That is,

$$\sum_{i=1}^{N} d_i \leq Q, \tag{1}$$

where d_i represents the demand (i.e., usage level) of tenant i.

In this chapter, the interactions among tenant customers and a local data center is modeled as a two-stage Stackelberg game. Stackelberg game is a strategic game with a leader and several followers. The leader moves first and the followers move subsequently to compete on certain resources. In our cloud market, as the leader, the data center first announces the instance prices in Stage I. In Stage II, given current pricing and resource allocation policies determined by the data center, tenants make usage decisions by maximizing their own surplus.

2.2.1 Stage I: Cloud Revenue Maximization

Thus, in Stage I, the data center's objective is to maximize its revenue obtained from selling virtual instances to tenants. The revenue of the data center can be easily obtained:

$$\Pi(\mathbf{p}, \mathbf{d}) \triangleq \sum_{i=1}^{N} p_i \cdot d_i. \tag{2}$$

where $\mathbf{p}$ is the price vector with $\mathbf{p} = [p_1, \cdots, p_N]^T$, and $\mathbf{d}$ is a vector of usage levels with $\mathbf{d} = [d_1, \cdots, d_N]^T$. Note that $\forall i$, d_i is a function of p_i and θ_i under our game theoretic formulation. That is, the usage level of tenant i is dependent on the associated cloud resource p_i and its valuation coefficient θ_i. The data center needs to find the optimal pricing policy by maximizing its aggregate revenue. This can be achieved by solving the following optimization problem, constrained by the total resource supply.

Problem 1 For a data center with aggregate resource supply Q, its optimal pricing policy is to maximize its aggregate revenue:

$$\begin{aligned} &\max_{\mathbf{p}\succeq 0} \quad \Pi(\mathbf{p}, \mathbf{d}) \\ &\text{subject to} \quad \sum_{i=1}^{N} d_i \leq Q. \end{aligned} \tag{3}$$

2.2.2 Stage II: Tenant Surplus Maximization

At the tenants' side, we adopt constant price elasticity of demand (or simply *elasticity*) to develop utility of heterogeneous tenant customers. Denote by k the constant elasticity, which represents the ratio between the percentage change of demand and the percentage change of price. By assuming linear relationship between θ_i and tenant i's utility, we can define a concave utility function:

$$U_i(d_i) \triangleq \theta_i \cdot \ln(1 + k \cdot d_i), \tag{4}$$

which also reflects the law of diminishing return. That is, user satisfaction saturates with the increase of demand.

Tenant surplus is utility minus dollar cost of cloud resource usage.

Given p_i which is under the control of the data center, tenant i's surplus is thus:

$$\pi_i(p_i, d_i) \triangleq U_i(d_i) - p_i \cdot d_i. \tag{5}$$

Given the data center's pricing choices, tenant i's optimal demand level is to maximize its surplus. This problem can be mathematically formulated as a convex maximization problem.

Problem 2 Given the pricing strategy p_i for type-i tenants, the optimal demand level of tenant i is obtained by maximizing its surplus:

$$\max_{d_i \geq 0} \quad \pi_i(p_i, d_i). \tag{6}$$

2.3 *Stackelberg Equilibrium*

In our proposed Stackelberg game for cloud resource pricing and allocation, the Stackelberg equilibrium (i.e., SE) is defined as follows.

Definition 1 Let $\mathbf{p}^*$ represent a solution for Problem 2.2.1, and the optimal d_i^* be a solution for Problem 2.2.2 for tenant i. Then, $(\mathbf{p}^*, \mathbf{d}^*)$ is a SE for the proposed Stackelberg game if and only if

$$\Pi(\mathbf{p}^*, \mathbf{d}^*) \geq \Pi(\mathbf{p}, \mathbf{d}^*), \qquad \text{and}$$
$$\pi_i(d_i^*) \geq \pi_i(d_i) \tag{7}$$

for any $(\mathbf{p}, \mathbf{d})$ satisfying $\mathbf{p} \geq 0$ and $\mathbf{d} \geq 0$.

Stackelberg equilibrium is a subgame perfect equilibrium. At the equilibrium, both the data center and tenant customers have no incentives to unilaterally change their pricing and usage choices.

3 Usage-Based Cloud Resource Pricing

In this section, we apply backward induction to perform equilibrium analysis, by considering the following two pricing schemes: *non-uniform pricing* and *uniform pricing*, followed by a comparison of the two schemes.

3.1 Non-Uniform Pricing

We first consider non-uniform pricing, in which tenants are charged based on different unit rates.

3.1.1 Stage II: Tenant Surplus Maximization

The tenant surplus maximization in Problem 2.2.2 can be easily formulated as:

$$\max_{d_i \geq 0} \quad \theta_i \cdot \ln(1 + k \cdot d_i) - p_i \cdot d_i. \tag{8}$$

This is a typical convex optimization problem, the solution of which can be easily solved by the Karush-Kuhn-Tucker (KKT) conditions. Therefore, the optimal solution can be easily obtained in the following lemma. We omit the detailed proof.

Lemma 1 *For a given price p_i, the optimal demand level of tenant i (i.e., the optimal solution to Problem 2.2.2) is:*

$$d_i^* = \left(\frac{\theta_i}{p_i} - \frac{1}{k}\right)^+, \tag{9}$$

where $(x)^+ = \max(x, 0)$.

From the above Lemma, we can see that tenants will not use any cloud resource if the price p_i is too high. That is, tenant i will not be admitted to the cloud system (i.e., $d_i = 0$), when p_i is large enough.

3.1.2 Stage I: Cloud Pricing Choices

Substitute Eq. 2 and 9 into Problem 2.2.1. The optimization problem is formulated as:

$$\max_{\mathbf{p}\succeq 0} \quad \sum_{i=1}^{N} p_i \cdot \left(\frac{\theta_i}{p_i} - \frac{1}{k}\right)^{+}$$

$$\text{subject to} \quad \sum_{i=1}^{N} \left(\frac{\theta_i}{p_i} - \frac{1}{k}\right)^{+} \leq Q. \tag{10}$$

For $\forall\, i \in \{1, \cdots, N\}$, we define the indicator function:

$$I_i = \begin{cases} 1 & \text{if } p_i < k \cdot \theta_i \\ 0 & \text{otherwise.} \end{cases} \tag{11}$$

Then, the above problem can be transformed into

$$\max_{\mathbf{p},\mathbf{I}\succeq 0} \quad \sum_{i=1}^{N} p_i \cdot \left(\frac{\theta_i}{p_i} - \frac{1}{k}\right) \cdot I_i$$

$$\text{subject to} \quad \sum_{i=1}^{N} \left(\frac{\theta_i}{p_i} - \frac{1}{k}\right) \cdot I_i \leq Q, \tag{12}$$

where $\mathbf{I} \triangleq [I_1, \cdots, I_N]^T$. The above problem is not convex due to indicator functions.

However, an important property of the above problem is that it is convex for any given $\mathbf{I}$. Therefore, we first assume that Q is large enough so that all tenants are admitted to the system. Under this assumption, we have the following equivalent convex optimization problem:

Problem 3 Under the assumption of large enough Q, Problem 2.2.1 can be transformed into:

$$\min_{\mathbf{p}\succeq 0} \quad \sum_{i=1}^{N} \frac{p_i}{k}$$

$$\text{subject to} \quad \sum_{i=1}^{N} \frac{\theta_i}{p_i} \leq Q + \frac{N}{k}. \tag{13}$$

The optimal solution of the above convex optimization problem is given by the following lemma.

Lemma 2 *Under the assumption of large enough Q, the data center's optimal pricing strategy for tenant i (i.e., the optimal solution to Problem 3.1.2) is:*

$$p_i^* = \frac{\sum_{j=1}^{N} \sqrt{\theta_j}}{Q + \frac{N}{K}} \cdot \sqrt{\theta_i}, \forall i \in \{1, \cdots, N\}. \tag{14}$$

Proof The Lagrangian associated with Problem 3.1.2 is

$$\mathcal{L}(\mathbf{p}, \lambda, \mu) = \sum_{i=1}^{N} \frac{p_i}{k} + \lambda \cdot \left(\sum_{i=1}^{N} \frac{\theta_i}{p_i} - Q - \frac{N}{k} \right) - \sum_{i=1}^{N} \mu_i \cdot p_i, \tag{15}$$

where λ and μ respectively are non-negative Lagrangian multipliers associated with the constraints $\sum_{i=1}^{N} \frac{\theta_i}{p_i} \leq Q + \frac{N}{k}$ and $p_i \geq 0$.

The KKT conditions are:

$$\frac{\partial \mathcal{L}(\mathbf{p}, \lambda, \mu)}{\partial p_i} = 0, \ \forall i, \tag{16}$$

$$\lambda \cdot \left(\sum_{i=1}^{N} \frac{\theta_i}{p_i} - Q - \frac{N}{k} \right) = 0, \tag{17}$$

$$\sum_{i=1}^{N} \mu_i \cdot p_i = 0, \ \forall i, \tag{18}$$

$$\lambda \geq 0, \ \mu_i \geq 0, \ p_i \geq 0, \tag{19}$$

$$\sum_{i=1}^{N} \frac{\theta_i}{p_i} - Q - \frac{N}{k} \leq 0 \tag{20}$$

From condition 16, we have

$$\frac{\partial \mathcal{L}(\mathbf{p}, \lambda, \mu)}{\partial p_i} = \frac{i}{k} - \frac{\lambda \cdot \theta_i}{p_i^2} - \mu_i = 0, \ \forall i, \tag{21}$$

which yields

$$p_i = \sqrt{\frac{\lambda \cdot \theta_i}{1/k - \mu_i}}, \ \forall i. \tag{22}$$

Firstly, we have

$$\mu_i = 0. \tag{23}$$

Otherwise, from condition 18, we have $p_i = 0$. This contradicts to condition 20. Therefore,

$$p_i = \sqrt{k\lambda \cdot \theta_i}, \ \forall i. \tag{24}$$

Secondly, we have

$$\sum_{i=1}^{N} \frac{\theta_i}{p_i} - Q - \frac{N}{k} = 0. \tag{25}$$

Otherwise, from condition 17, we have $\lambda = 0$. From condition 24, $p_i = 0$ as a result. This again contradicts condition 20.

Substitute Eq. 24 into 25. We can obtain

$$\sqrt{k\lambda} = \frac{\sum_{j=1}^{N} \sqrt{\theta_j}}{Q + N/k}. \tag{26}$$

Then, we have

$$p_i = \frac{\sum_{j=1}^{N} \sqrt{\theta_j}}{Q + \frac{N}{k}} \cdot \sqrt{\theta_i}. \tag{27}$$

■

Then, we get the following lemma.

Lemma 3 *The sufficient and necessary condition that the solution given by Lemma 2 is also the optimal solution to Problem 2.2.1 is given by:*

$$Q > \frac{\sum_{j=1}^{N} \sqrt{\theta_j}}{\min_i k \cdot \sqrt{\theta_i}} - \frac{N}{k}. \tag{28}$$

Proof First, we consider the sufficient condition. If all the indicators are equal to 1, that is, $p_i < k \cdot \theta_i$, $\forall i$, the solution given by Lemma 2 is the optimal solution of Problem 2.2.1. Substitute the solution in Lemma 2 into all these conditions, we have

$$Q > \frac{\sum_{j=1}^{N} \sqrt{\theta_j}}{Q + \frac{N}{k}} \cdot \sqrt{\theta_i},\ \forall i, \tag{29}$$

which can be expressed as

$$Q > \frac{\sum_{j=1}^{N} \min_i \sqrt{\theta_j}}{Q + \frac{N}{k}} \cdot \sqrt{\theta_i}. \tag{30}$$

Second, we consider the necessary condition. For ease of exposition, we assume that $\theta_1 > \theta_2 > \cdots > \theta_N$. Then, the condition in Lemma 3 becomes

$$Q > T_N, \tag{31}$$

where $T_N = \frac{\sum_{j=1}^{N} \sqrt{\theta_j}}{k \cdot \sqrt{\theta_N}} - \frac{N}{k}$.

Now, suppose that $T_r < Q \leq T_{r+1}$, $r \leq N - 1$. Suppose that the optimal solution in Lemma 2 is still the optimal solution. Then, $I_i = 1$, $\forall i \leq r$, and $I_i = 0$, $\forall r + 1 < i \leq N - 1$. Then, Problem 2 becomes

$$\max_{\mathbf{p} \succeq 0} \sum_{i=1}^{r} p_i \cdot \left(\frac{\theta_i}{p_i} - \frac{1}{k} \right)$$

$$\text{subject to} \quad \sum_{i=1}^{r}\left(\frac{\theta_i}{p_i} - \frac{1}{k}\right) \le Q. \tag{32}$$

This convex optimization problem has the same structure as Problem 3.1.2. Thus, the optimal solution to the above problem is given by:

$$p_i^* = \frac{\sum_{j=1}^{r}\sqrt{\theta_j}}{Q + \frac{r}{K}} \cdot \sqrt{\theta_i}, \forall i \in \{1, \cdots, r\}, \tag{33}$$

which is different from the optimal solution given in Lemma 2. This gives a contradiction. Thus, it is also a necessary condition. ■

From the proof of the above lemma, we can obtain the following theorem about the optimal solution of Problem 2.2.1.

Theorem 1 *Under the assumption that* $\theta_1 > \cdots > \theta_N$*, the optimal solution to Problem 2.2.1 is given by*

$$\mathbf{p}^* = \begin{cases} s_N \cdot [\sqrt{\theta_1}, \cdots, \sqrt{\theta_N}]^T & \text{if } Q > T_N \\ s_{N-1} \cdot [\sqrt{\theta_1}, \cdots, \sqrt{\theta_{N-1}}, \infty]^T & \text{if } T_N > Q > T_{N-1} \\ \vdots & \vdots \\ s_r \cdot [\sqrt{\theta_1}, \cdots, \sqrt{\theta_r}, \infty, \cdots, \infty]^T & \text{if } T_{r+1} > Q > T_r \\ \vdots & \vdots \\ s_1 \cdot [\sqrt{\theta_1}, \infty, \cdots, \infty]^T & \text{if } Q > T_1 \\ s_0 \cdot [\infty, \cdots, \infty]^T & \text{otherwise} \end{cases} \tag{34}$$

where $s_r = \frac{\sum_{j=1}^{r}\sqrt{\theta_j}}{Q+\frac{r}{k}}$, $s_0 = 1$, *and* $T_r = \frac{\sum_{j=1}^{r}\sqrt{\theta_j}}{k\cdot\sqrt{\theta_r}} - \frac{r}{k}$, $\forall r \in \{1, \cdots, N\}$.

The above theorem implies an economic observation: the data center maximizes its revenue via price differentiation. That is, tenants with higher willingness to pay are charged with higher unit prices. Therefore, a very important question is whether tenants possess incentives to cheat about their willingness to pay, and receive a lower unit price. This is answered in Theorem 2.

Theorem 2 *The price differentiation scheme is cheating proof. That is, tenants do not have incentives to cheat about their valuation types.*

Proof Assume that $\theta_1 > \cdots > \theta_N$. Suppose that the maximized surplus of a type-i tenant is π_i^*, when the tenant truthfully reveals its valuation parameter θ_i. Then, from Theorem 1, we have

$$\pi_i^* = \pi_i(p_i^*, d_i^*), \tag{35}$$

where $d_i^* = \left(\frac{\theta_i}{p_i^*} - \frac{1}{k}\right)^+$.

Tenants have no incentives to reveal a higher valuation parameter, which will lead to higher unit resource prices. Suppose that the maximized surplus of a type-i tenant is π_i', when the tenant claims that its valuation parameter is θ_l, $\forall l > i$. Then, from Lemma 1, we have

$$\pi_i' = \pi_i(p_j^*, d_i'), \tag{36}$$

where $d_i' = \left(\frac{\theta_i}{p_l^*} - \frac{1}{k}\right)^+$.

Therefore, the necessary and sufficient condition that type-i tenants truthfully reveal their valuation parameters is

$$\pi_i^* \geq \pi_i'. \tag{37}$$

From Theorem 1, we can derive π_i^* and π_i'.

1) We first assume that the resource supply Q is large enough. That is, $Q > T_N$. All tenants are admitted. Then, condition 37 becomes

$$\frac{1}{2} \cdot \theta_i \cdot \ln \frac{\theta_i}{\theta_l} + \frac{1}{k} \cdot \frac{\sum_{j=1}^{N} \sqrt{\theta_j}}{Q + \frac{N}{k}} \cdot \left(\sqrt{\theta_i} - \sqrt{\theta_l}\right) \geq 0, \tag{38}$$

which is obviously satisfied because $\theta_i > \theta_l$, $\forall l > i$.

2) Second, we assume that $T_{r+1} > Q > T_r$. Tenants will not have incentives to cheat for $l > r$, because

$$\pi_i'|_{T_{r+1}>Q>T_r} = 0,\ l > r,\ \forall i. \tag{39}$$

Then, we have two cases: a) $l \geq i$, and $l \leq r$; b) $i < r$, and $l > r$. In case a), type i tenants have no incentives to cheat. The proof is similar to the case $Q > T_N$. In case b), condition 37 becomes

$$\pi_i^*|_{T_{r+1}>Q>T_r} \geq 0,\ \forall i < r. \tag{40}$$

Thus, in case b), tenants also have no incentives to cheat. ■

3.2 *Uniform Pricing*

Next, we consider uniform pricing, in which heterogeneous tenants are charged based on the same unit rates $p_i = \overline{p}$, $\forall i \in \{1, \cdots, N\}$.

3.2.1 Stage II: Tenant Surplus Maximization

The tenant surplus maximization in Problem 2.2.2 can be easily transformed into:

$$\max_{d_i \geq 0} \quad \theta_i \cdot \ln(1 + k \cdot \overline{d}_i) - \overline{p} \cdot \overline{d}_i. \tag{41}$$

This is a typical convex optimization problem. Similar to Lemma 1, we can obtain the following lemma on the optimal demand level of tenants. We omit the detailed proof.

Lemma 4 *For a given uniform price $\overline{p}$, the optimal demand level of tenant i (i.e., the optimal solution to Problem 2.2.2) is:*

$$\overline{d}_i^* = \left(\frac{\theta_i}{\overline{p}} - \frac{1}{k}\right)^+, \forall i \in \{1, \cdots, N\}, \tag{42}$$

where $(x)^+ = \max(x, 0)$.

3.2.2 Stage I: Cloud Pricing Choices

Similar to non-uniform pricing, we first assume that Q is large enough so that all tenants are admitted to the system. Under this assumption, similar to Problem 3.1.2, we get the following equivalent convex optimization problem:

Problem 4 Under the assumption of large enough Q, Problem 2.2.1 can be transformed into:

$$\min_{\overline{p} \geq 0} \quad \frac{N \cdot \overline{p}}{k}$$

$$\text{subject to} \quad \sum_{i=1}^{N} \frac{\theta_i}{\overline{p}} \leq Q + \frac{N}{k}. \tag{43}$$

The optimal solution of the above optimization problem is thus given by the following lemma.

Lemma 5 *Under the assumption of large enough Q and uniform pricing, the data center's optimal pricing strategy for tenant i (i.e., the optimal solution to Problem 3.2.2) is:*

$$\overline{p}^* = \frac{\sum_{j=1}^{N} \theta_j}{Q + \frac{N}{K}}, \forall i \in \{1, \cdots, N\}. \tag{44}$$

Similar to Lemma 3, the condition on which the above optimal solution is also the optimal solution to the original problem given by Problem 2.2.1 is given by the following lemma, under uniform pricing.

Lemma 6 *Under uniform pricing, the sufficient and necessary condition that the solution given by Lemma 2 is also the optimal solution to Problem 2.2.1 is given by:*

$$Q > \frac{\sum_{j=1}^{N} \theta_j}{\min_i k \cdot \theta_i} - \frac{N}{k}. \tag{45}$$

Similar to Theorem 1, we can obtain the following theorem about the optimal solution of Problem 2.2.1.

Theorem 3 *Under uniform pricing and the assumption that $\theta_1 > \cdots > \theta_N$, the optimal solution to Problem 2.2.1 is given by*

$$\overline{\mathbf{p}}^* = \begin{cases} s'_N \cdot [1, \cdots, 1]^T & \text{if } Q > T'_N \\ s'_{N-1} \cdot [1, \cdots, 1, \infty]^T & \text{if } T'_N > Q > T'_{N-1} \\ \vdots & \vdots \\ s'_r \cdot [1, \cdots, 1, \infty, \cdots, \infty]^T & \text{if } T'_{r+1} > Q > T'_r \\ \vdots & \vdots \\ s'_1 \cdot [1, \infty, \cdots, \infty]^T & \text{if } Q > T'_1 \\ s'_0 \cdot [\infty, \cdots, \infty]^T & \text{otherwise} \end{cases} \tag{46}$$

where $s'_r = \frac{\sum_{j=1}^{r} \theta_j}{Q + \frac{r}{k}}$, $s'_0 = 1$, *and* $T'_r = \frac{\sum_{j=1}^{r} \theta_j}{k \cdot \theta_r} - \frac{r}{k}$, $\forall r \in \{1, \cdots, N\}$.

4 The Effectiveness of Stackelberg Strategies

It is well known that agent selfishness may deteriorate the network performance. The ratio of the maximized total network utility over the total network utility incurred by selfish behaviors is usually used to measure such performance degradation. This ratio is called Price of Anarchy (PoA). In this section, we obtain results on PoA to scrutinize the impact of agent selfishness under the cloud computing environment.

4.1 Centralized Aggregate Network Utility Maximization

The aggregate utility of the entire network $U_t(\mathbf{d})$ is the aggregate surplus of all tenants, plus revenue of the data center:

$$\begin{aligned} U_t(\mathbf{d}) &= \Pi(\mathbf{p}, \mathbf{d}) + \sum_{i=1}^{N} \pi_i(p_i, d_i) \\ &= \sum_{i=1}^{N} U_i(d_i) \\ &= \sum_{i=1}^{N} \theta_i \cdot \ln(1 + k \cdot d_i). \end{aligned} \tag{47}$$

Problem 5 For a data center with aggregate resource supply Q, the centralized optimal solution to maximize the total network utility is given by solving the following problem:

$$\max_{\mathbf{d}\succeq 0} \quad U_t(\mathbf{d}) = \sum_{i=1}^{N} \theta_i \cdot \ln(1 + k \cdot d_i)$$

$$\text{subject to} \quad \sum_{i=1}^{N} d_i \leq Q. \tag{48}$$

The optimal solution of the above convex optimization problem is given by the following theorem.

Theorem 4 *The optimal demand level of tenant i to maximize the total network utility is (i.e., the centralized optimal solution to Problem 4.1) is:*

$$d_i^c = \frac{\theta_i}{\sum_{j=1}^{N} \theta_j} \cdot \left(Q + \frac{N}{k}\right) - \frac{1}{k}, \ \forall i \in \{1, \cdots, N\}, \tag{49}$$

which yields the maximized total network utility:

$$U_t^c = \sum_{i=1}^{N} \theta_i \cdot \ln\left(\frac{k \cdot \theta_i}{s'_N}\right), \tag{50}$$

where $s'_N = \frac{\sum_{j=1}^{N} \theta_j}{Q + \frac{N}{k}}$.

Proof The Lagrangian associated with Problem 4.1 is

$$\mathcal{L}(\mathbf{d}, \alpha, \beta) = \sum_{i=1}^{N} \theta_i \cdot \ln(1 + k \cdot d_i) + \alpha \cdot \left(\sum_{i=1}^{N} d_i - Q\right) - \sum_{i=1}^{N} \beta_i \cdot d_i, \tag{51}$$

where α and β_i respectively are non-negative Lagrangian multipliers associated with the constraints $\sum_{i=1}^{N} d_i \leq Q$ and $d_i \geq 0$.

The KKT conditions are:

$$\frac{\partial \mathcal{L}(\mathbf{d}, \alpha, \beta)}{\partial d_i} = 0, \ \forall i, \tag{52}$$

$$\alpha \cdot \left(\sum_{i=1}^{N} d_i - Q\right) = 0, \tag{53}$$

$$\sum_{i=1}^{N} \beta_i \cdot d_i = 0, \ \forall i, \tag{54}$$

$$\alpha \geq 0, \ \beta_i \geq 0, \ d_i \geq 0, \tag{55}$$

$$\sum_{i=1}^{N} d_i - Q \leq 0. \tag{56}$$

From condition 52, we have

$$\frac{\partial \mathcal{L}(\mathbf{d}, \alpha, \beta)}{\partial d_i} = \frac{k \cdot \theta_i}{1 + k \cdot d_i} + \alpha - \beta_i = 0, \ \forall i, \tag{57}$$

which yields

$$d_i = \frac{\theta_i}{\beta_i - \alpha} - \frac{1}{k}, \ \forall i. \tag{58}$$

Firstly, we have

$$\beta_i = 0. \tag{59}$$

Otherwise, from condition 54, we have $d_i = 0$. This contradicts to condition 56. Therefore,

$$d_i = -\frac{\theta_i}{\alpha} - \frac{1}{k}, \ \forall i. \tag{60}$$

Secondly, we have

$$\sum_{i=1}^{N} d_i - Q = 0. \tag{61}$$

Otherwise, from condition 53, we have $\alpha = 0$. From condition 60, $d_i = -\infty$ as a result. This contradicts condition $d_i \geq 0$.

Substitute Eq. 60 into 61. We can obtain

$$\alpha = -\frac{\sum_{j=1}^{N} \theta_j}{Q + \frac{N}{k}}. \tag{62}$$

Then, we have the optimal solution

$$d_i^c = \frac{\theta_i}{s'_N} - \frac{1}{k}, \tag{63}$$

where $s'_N = \frac{\sum_{j=1}^{N} \theta_j}{Q + \frac{N}{k}}$.

Therefore, the maximized total network utility is

$$U_t^c = \sum_{i=1}^{N} \theta_i \cdot \ln(1 + k \cdot d_i^c). \tag{64}$$

Substituting the optimal solution into the above equation, we have:

$$U_t^c = \sum_{i=1}^{N} \theta_i \cdot \ln\left(\frac{k \cdot \theta_i}{s'_N}\right). \tag{65}$$

■

Moreover, from Lemma 4 and Lemma 5, we can obtain the following proposition.

Proposition 1 *Centralized resource allocation is equivalent to the resource allocation under uniform pricing when the capacity of the data center is large enough.*

4.2 Total Network Utility Under Selfish Interactions

Under selfish interactions between tenants and data centers, we also have two cases: non-uniform pricing and uniform pricing. We first consider the scenario of non-uniform pricing.

Lemma 7 *Under the assumption that $\theta_1 > \cdots > \theta_N$ and non-uniform pricing, the total network utility, incurred by selfish behaviors of the data center and tenants, is given by*

$$U_t^s = \begin{cases} \sum_{i=1}^{N} \theta_i \cdot \ln\left(\frac{k \cdot \sqrt{\theta_i}}{s_N}\right) & \text{if } Q > T_N \\ \sum_{i=1}^{N-1} \theta_i \cdot \ln\left(\frac{k \cdot \sqrt{\theta_i}}{s_{N-1}}\right) & \text{if } T_N > Q > T_{N-1} \\ \vdots & \vdots \\ \sum_{i=1}^{r} \theta_i \cdot \ln\left(\frac{k \cdot \sqrt{\theta_i}}{s_r}\right) & \text{if } T_{r+1} > Q > T_r \\ \vdots & \vdots \\ \theta_1 \cdot \ln\left(\frac{k \cdot \sqrt{\theta_1}}{s_1}\right) & \text{if } Q > T_1 \\ 0 & \text{otherwise} \end{cases} \tag{66}$$

where $s_r = \frac{\sum_{j=1}^{r} \sqrt{\theta_j}}{Q + \frac{r}{k}}$, $s_0 = 1$, *and* $T_r = \frac{\sum_{j=1}^{r} \sqrt{\theta_j}}{k \cdot \sqrt{\theta_r}} - \frac{r}{k}$, $\forall r \in \{1, \cdots, N\}$.

Proof From Theorem 1 and Lemma 1, we have:

$$\mathbf{d}^* = \begin{cases} \frac{1}{s_N} \cdot [\sqrt{\theta_1}, \cdots, \sqrt{\theta_N}]^T - \frac{1}{k} & \text{if } Q > T_N \\ \frac{1}{s_{N-1}} \cdot [\sqrt{\theta_1}, \cdots, \sqrt{\theta_{N-1}}, 0]^T - \frac{1}{k} & \text{if } T_N > Q > T_{N-1} \\ \vdots & \vdots \\ \frac{1}{s_r} \cdot [\sqrt{\theta_1}, \cdots, \sqrt{\theta_r}, 0, \cdots, 0]^T - \frac{1}{k} & \text{if } T_{r+1} > Q > T_r \\ \vdots & \vdots \\ \frac{1}{s_1} \cdot [\sqrt{\theta_1}, 0, \cdots, 0]^T - \frac{1}{k} & \text{if } Q > T_1 \\ \frac{1}{s_0} \cdot [0, \cdots, 0]^T & \text{otherwise.} \end{cases} \tag{67}$$

Substituting it into $U_t^s = U_t(\mathbf{d}^*) = \sum_{i=1}^N \theta_i \cdot \ln(1 + k \cdot d_i^*)$, we obtain the results in the lemma. ■

We then consider the scenario of uniform pricing.

Lemma 7 *Under the assumption that* $\theta_1 > \cdots > \theta_N$ *and uniform pricing, the total network utility, incurred by selfish behaviors of the data center and tenants, is given by*

$$\overline{U}_t^s = \begin{cases} \sum_{i=1}^N \theta_i \cdot \ln\left(\frac{k \cdot \theta_i}{s'_N}\right) & \textit{if } Q > T'_N \\ \sum_{i=1}^{N-1} \theta_i \cdot \ln\left(\frac{k \cdot \theta_i}{s'_{N-1}}\right) & \textit{if } T'_N > Q > T'_{N-1} \\ \vdots & \vdots \\ \sum_{i=1}^r \theta_i \cdot \ln\left(\frac{k \cdot \theta_i}{s'_r}\right) & \textit{if } T'_{r+1} > Q > T'_r \\ \vdots & \vdots \\ \theta_1 \cdot \ln\left(\frac{k \cdot \theta_1}{s'_1}\right) & \textit{if } Q > T'_1 \\ 0 & \textit{otherwise} \end{cases} \tag{68}$$

where $s'_r = \frac{\sum_{j=1}^r \theta_j}{Q + \frac{r}{k}}$, $s'_0 = 1$, *and* $T'_r = \frac{\sum_{j=1}^r \theta_j}{k \cdot \theta_r} - \frac{r}{k}$, $\forall r \in \{1, \cdots, N\}$.

Proof From Theorem 3 and Lemma 4, we have:

$$\overline{\mathbf{d}}^* = \begin{cases} \frac{1}{s'_N} \cdot [\theta_1, \cdots, \theta_N]^T - \frac{1}{k} & \text{if } Q > T'_N \\ \frac{1}{s'_{N-1}} \cdot [\theta_1, \cdots, \theta_{N-1}, 0]^T - \frac{1}{k} & \text{if } T'_N > Q > T'_{N-1} \\ \vdots & \vdots \\ \frac{1}{s'_r} \cdot [\theta_1, \cdots, \theta_r, 0, \cdots, 0]^T - \frac{1}{k} & \text{if } T'_{r+1} > Q > T'_r \\ \vdots & \vdots \\ \frac{1}{s'_1} \cdot [\theta_1, 0, \cdots, 0]^T - \frac{1}{k} & \text{if } Q > T'_1 \\ \frac{1}{s'_0} \cdot [0, \cdots, 0]^T & \text{otherwise.} \end{cases} \tag{69}$$

Substituting it into $\overline{U}_t^s = U_t(\mathbf{d}^*) = \sum_{i=1}^N \theta_i \cdot \ln(1 + k \cdot d_i^*)$, we obtain the results in the lemma. ■

4.3 Asymptotic Analysis of Price of Anarchy

From the definition of PoA, we obtain the formulation of PoA in the following.

Definition 2 i) The PoA under non-uniform pricing is given by:

$$\text{PoA}_{nu} = \frac{U_t^c}{U_t^s}. \tag{70}$$

ii) The PoA under uniform pricing is given by:

$$\text{PoA}_u = \frac{\overline{U}_t^c}{U_t^s}. \tag{71}$$

Theorem 5 *Suppose that the capacity of the data center under consideration is large enough. From Proposition 1, we can easily get the PoA under uniform pricing:*

$$PoA_u = 1. \tag{72}$$

For non-uniform pricing, we resort to asymptotic analysis. That is, there exist a large number of tenants. We also assume that the valuation parameter θ_j follows uniform distribution with $\theta_j \in (0, 1), \forall j \in \{1, \cdots, N\}$. Then, we have the following theorem.

Theorem 6 *Suppose that the capacity of the data center under consideration is large enough, and that there are a large number of tenant consumers. If the valuation parameter θ_j follows uniform distribution, then we have*

$$PoA_{nu} = \frac{\ln k - \ln X - \frac{1}{2} - \ln 2 + \ln 3}{\ln k - \ln X - \frac{1}{4} + \ln 2}, \tag{73}$$

where $X = \frac{N}{Q + \frac{N}{k}}$.

Proof sketch. We first have

$$\begin{aligned} U_t^s &= N \cdot \mathbf{E}\left[\theta_i \cdot \ln\left(\frac{k \cdot \sqrt{\theta_i}}{s_N}\right)\right] \\ &= N \cdot \left(\frac{1}{2}\ln k - \frac{1}{8} - \frac{1}{2}\ln s_N\right), \end{aligned} \tag{74}$$

where $s_N = \frac{N \cdot \mathbf{E}\left[\sqrt{\theta_j}\right]}{Q + \frac{N}{k}} = \frac{2}{3} \cdot \frac{N}{Q + \frac{N}{k}}$. Similarly, we can derive U_t^c and S_N'. ■

5 Broker Resource Pricing

Cloud brokers are widely discussed to leverage demand correlation among different tenants for resource multiplexing when procuring resources from the cloud. However, pricing strategies taken by brokers are still largely unexplored. In this section, we tackle this challenge by introducing brokers in our framework of differentiated cloud resource pricing. Denote by p_i^b the unit price charged by the broker for tenant i with valuation parameter θ_i. In this section, suppose that the data center capacity is large enough to admit all tenants, and that θ_i follows uniform distribution with $\theta_i \in (0, 1)$.

Lemma 9 *If and only if $\theta_i \leq \left(\frac{p_i^b}{s_N}\right)^2$, the optimal strategy for tenant i is to procure resources from the data center; otherwise (i.e., $\theta_i > \left(\frac{p_i^b}{s_N}\right)^2$), the optimal strategy is to obtain resources from the broker.*

Proof When procuring resources from the broker, the tenant surplus of a typical tenant i is given by:

$$\pi_i^b = \theta_i \cdot \ln\left(1 + k \cdot d_i^b\right) - p_i^b \cdot d_i^b, \tag{75}$$

where the demand level of tenant i is

$$d_i^b = \left(\frac{\theta_i}{p_i^b} - \frac{1}{k}\right)^+. \tag{76}$$

Therefore, the tenant surplus is

$$\pi_i^b = \begin{cases} 0 & \text{if } \theta_i < \frac{p_i^b}{k} \\ \theta_i \cdot \ln\left(\frac{k\theta_i}{p_i^b}\right) - \theta_i + \frac{p_i^b}{k} & \text{otherwise.} \end{cases} \tag{77}$$

if tenant i obtains resources from the broker.

For clarity, we assume that the capacity of the data center is large enough to admit all tenants for resource procurement. Then, if tenant i obtains resources from the cloud directly, then the surplus of tenant i is

$$\pi_i^* = \theta_i \cdot \ln\left(k \cdot \frac{\sqrt{\theta_i}}{s_N}\right) - \theta_i + \frac{\sqrt{\theta_i}}{k} \cdot s_N, \tag{78}$$

and the optimal price charged by the data center is

$$p_i^* = s_N \cdot \sqrt{\theta_i} < k\sqrt{\theta_i \cdot \theta_N} < k \cdot \theta_i. \tag{79}$$

Then, tenant i procures resources from the broker if and only if

$$\Delta\pi_i = \pi_i^b - \pi_i^* > 0. \tag{80}$$

Case 1: $\theta_i < \frac{p_i^b}{k}$. We have $\Delta\pi_i = -\pi_i^* < 0$. Then, tenant i procures resources from the data center.

Case 2: $\frac{p_i^b}{k} < \theta_i < \left(\frac{p_i^b}{s_N}\right)^2$. That is, $p_i^* < p_i^b < k \cdot \theta_i$. Because π_i^b is a decreasing function of p_i^b when $p_i^b < k \cdot \theta_i$, $\Delta\pi_i < 0$. That is, tenant i procures resources from the data center.

Case 3: $\theta_i > \left(\frac{p_i^b}{s_N}\right)^2$. That is, $p_i^b < p_i^* < k \cdot \theta_i$. Because π_i^b is a decreasing function of p_i^b when $p_i^b < k \cdot \theta_i$, $\Delta\pi_i > 0$. That is, tenant i procures resources from the broker. ■

The broker obtains revenue by selling resources to tenants and at the same time shares part of the revenue with the data center by negotiation. Denote by γ the fraction of revenue shared with the data center. Assuming uniform distribution of θ_i, from the above lemma, we can obtain that the broker revenue obtained from selling resources to tenants with high valuation parameters is

$$\begin{aligned}\Delta\Pi_b(\gamma, \mathbf{p}^b) &= (1-\gamma) \cdot \int_{\left(\frac{p_i^b}{s_N}\right)^2}^{1} p_i^b \cdot d_i^b d\theta_i \\ &= (1-\gamma) \cdot \int_{\left(\frac{p_i^b}{s_N}\right)^2}^{1} \left(\theta_i - \frac{p_i^b}{k}\right) d\theta_i \end{aligned} \tag{81}$$

The revenue increase of the data center is the revenue increase shared by the broker minus the revenue loss from tenants with high valuation parameters, due to the competition from the broker. Then, we have the revenue increase of the data center:

$$\begin{aligned}\Delta\Pi(\gamma, \mathbf{p}^b) &= \gamma \cdot \int_{\left(\frac{p_i^b}{s_N}\right)^2}^{1} p_i^b \cdot d_i^b d\theta_i - \int_{\left(\frac{p_i^b}{s_N}\right)^2}^{1} p_i^* \cdot d_i^* d\theta_i \\ &= \gamma \cdot \int_{\left(\frac{p_i^b}{s_N}\right)^2}^{1} \left(\theta_i - \frac{p_i^b}{k}\right) d\theta_i \\ &\quad - \int_{\left(\frac{p_i^b}{s_N}\right)^2}^{1} \left(\theta_i - \frac{s_N \cdot \sqrt{\theta_i}}{k}\right) d\theta_i. \end{aligned} \tag{82}$$

The broker bargains with the cloud provider on γ and p_i^b in broker resource pricing. Therefore, we use the Nash bargaining solution (NBS) to solve the problem of broker resource pricing. NBS is Pareto efficient and promotes fairness for revenue sharing between the data center and the broker.

Problem 6 The broker pricing under the Nash bargaining framework is to solve the following optimization problem:

$$\max_{\mathbf{p}^b \succeq 0, 0<\gamma<1} \quad \Delta\Pi_b(\gamma, \mathbf{p}^b) \cdot \Delta\Pi(\gamma, \mathbf{p}^b). \tag{83}$$

To obtain insights into broker resource pricing and the revenue sharing between the data center and

Theorem 7 *The optimal broking pricing strategy is given by:*

$$p^{b*} = \frac{s_N}{\sqrt{5}} = \frac{2\sqrt{5}}{15} \cdot \frac{N}{Q + \frac{N}{k}}. \tag{84}$$

We omit the proof here. The broker pricing problem can be solved by first deriving the optimal value for γ given a specific p^b, and then obtaining the optimal value of p^b.

6 Performance Evaluation

We present our results on performance evaluation in this section.

6.1 Setup

We augment our evaluation with empirical data on workload traces from Google production clusters [8].

The workload traces are used to derive the distribution of tenant valuation coefficient θ. Considering the fact that uniform pricing is commonly adopted by practical cloud systems in the pay-as-you-go model, we use the market price of \$0.06 for a small virtual instance in July, 2013. The amount of resources (i.e., the number of CPU cores) requested by tenants is obtained from the traces. Due to the normalization of CPU numbers in the Google traces, we assume the smallest requested amount is 1 core to scale the demand. From the optimal demand $d_i = (\frac{\theta_i}{p} - \frac{1}{k})^+$, we derive the distribution of the valuation parameters.

We use $k = 0.5$ by default. In the following, we first use Google traces to get resource prices and demands.

6.2 Economic Implications of Cloud Resource Pricing

Figure 1a and b show the impact of resource supply on cloud resource prices and tenant demands. An inverse relationship exists between supply and prices, and uniform prices fall between the prices charged to tenants with high resource valuation and low resource valuation.

Importantly, under non-uniform pricing, tenants with higher demands are charged with higher unit resource prices. This contradicts the industry practice of volume discounts, which partially explains the adoption of uniform pricing by the industry. In particular, there are no available prices for tenants with $\theta = 0.05$ in that tenants with low resource valuation are crowded out of the system. This phenomenon will

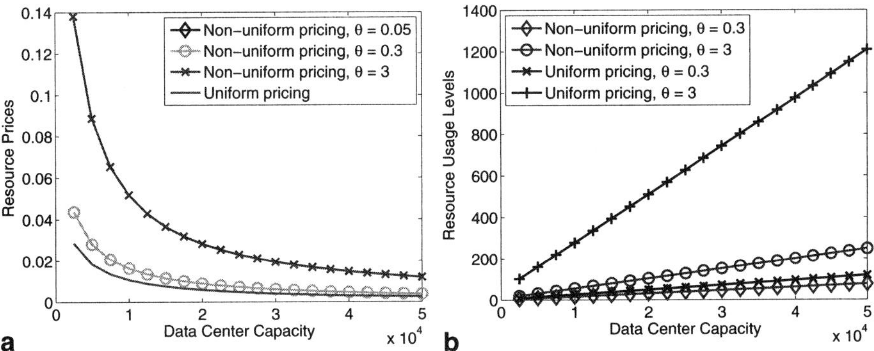

Fig. 1 Economic implications at equilibrium. **a** Resource prices for tenants. **b** Tenant demand levels

be elaborated in detail later. However, from the Google workload traces, we can only obtain large enough observable demands in that tenants with low demands are crowded out of the system with no demand records in the trace. Therefore, to compare tenant surplus, revenue, social welfare and hidden effects, we use uniform distribution with $\theta_i \in (0, 1)$.

6.3 Social Welfare Tradeoffs, and Hidden Effects

Figure 2a and b compare tenant surplus and cloud revenue under non-uniform and uniform pricing. It is observed that non-uniform pricing can achieve higher cloud revenue by exploiting more tenant surplus via price differentiation.

From Fig. 2c and the definition of PoA,

we learn that there exists a tradeoff between cloud revenue maximization and social welfare optimization. The revenue differences are small compared with the gap in PoA, which may also partially explain the industry adoption of uniform pricing. Hidden effects as implied in the parable of broken windows refer to unintended consequences, including unexpected benefits or more likely adverse effects, in social sciences and especially economics [5, 12]. We also observe such unintended consequences in the cloud system. That is, tenants with low valuation coefficients may be crowded out of the system due to low resource supply, as shown in Fig. 2d. This is true for both non-uniform and uniform pricing. Moreover, such hidden effects are more serious under uniform pricing for oblivious prices charged to tenants with different resource valuation types.

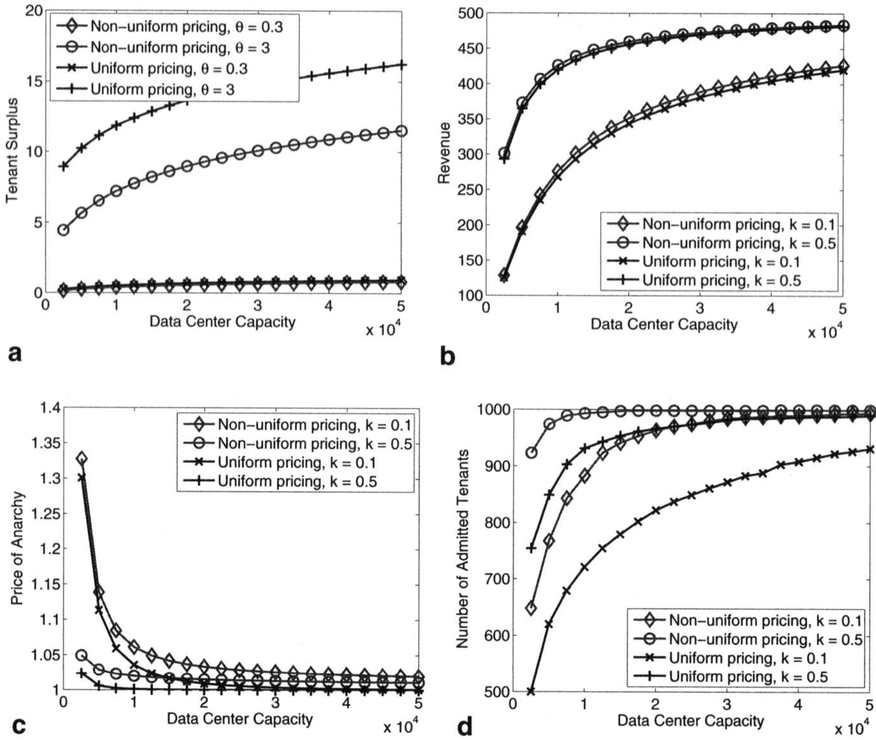

Fig. 2 Revenue-efficiency tradeoffs and hidden effects. **a** Tenant surplus. **b** Data center revenue. **c** Price of anarchy. **d** Hidden effects as exemplified by rejected admission of tenants in the cloud system

7 Related Work

Due to the critical role of cloud resource pricing, there already exist some studies on pricing of cloud resources. Wang et al. [10] pinpoint the importance of pricing in cloud system design. Wang et al. [14] discuss optimal resource capacity segmentaion between the pay-as-you-go pricing model and spot pricing with the objective of revenue maximization. Niu et al. [3, 4] propose a pricing scheme to multiplex bandwidth demands of a cloud based VoD system. Mihailescu et al. [11] consider dynamic pricing in federated clouds. Kantere et al. [13] explore optimal service pricing in cloud cache services. Most recently, Xu et al. [9] argue cloud revenue maximization by proposing centralized optimization solutions.

8 Concluding Remarks

In this chapter, we explore optimal cloud resource pricing by considering the strategic interactions and optimal responses of both cloud providers and tenant users. We propose a Stackelberg game to model such strategic cloud resource pricing. We then conduct equilibrium analysis by considering both non-uniform and uniform pricing, and explore the degradation of system performance via Price of Anarchy analysis. The results revealed insightful observations for practical pricing scheme design. In the future, we would like to extend our general model to the scenario of price competition among different cloud providers.

References

1. Amazon EC2, 2013. http://aws.amazon.com/ec2/.
2. Byung Chul Tak, Bhuvan Urgaonkar, and Anand Sivasubramaniam. Cloudy with a Chance of Cost Savings. *IEEE Transactions on Parallel and Distributed Systems*, 24(6):1223–1233, June 2013.
3. Di Niu, Chen Feng, and Baochun Li. A Theory of Cloud Bandwidth Pricing for Video-on-Demand Providers. In *Proc. of INFOCOM*, March 2012.
4. Di Niu, Chen Feng, and Baochun Li. Pricing Cloud Bandwidth Reservations under Demand Uncertainty. In *Proc. of SIGMETRICS*, June 2012.
5. Frederic Bastiat. What is seen and what is not seen. *Selected Essays on Political Economy*, 1995.
6. Google App Engine, 2013. https://appengine.google.com/start.
7. Google Cluster Data, 2013. https://code.google.com/p/googleclusterdata/.
8. Google Cluster Data, 2013. http://code.google.com/p/googleclusterdata/wiki/ClusterData2011_1.
9. Hong Xu and Baochun Li. A Study of Pricing for Cloud Resources. *ACM SIGMETRICS Performance Evaluation Review, Special Issue on Cloud Computing*, March 2013.
10. Hongyi Wang, Qingfeng Jing, Rishan Chen, Bingsheng He. Distributed Systems Meet Economics, Zhengping Qian, and Lidong Zhou: Pricing in the Cloud. In *Proc. of USENIX HotCloud*, June 2010.
11. Marian Mihailescu and Yong Meng Teo. Dynamic Resource Pricing on Federated Clouds. In *Proc. of 10th IEEE/ACM International Conference on Cluster, Cloud and Grid Computing*, 2010.
12. Mark Skidmore and Hideki Toyaz. Does Natural Disasters Promote Long-run Growth? *Economic Inquiry*, 40(4):664–687, October 2002.
13. Verena Kantere, Debabrata Dash, Gregory Francois, Sofia Kyriakopoulou, and Anastasia Ailamaki. Optimal Service Pricing for a Cloud Cache. *IEEE Transactions on Knowledge and Data Engineering*, 23(9):1345–1358, September 2011.
14. Wei Wang, Baochun Li, and Ben Liang. Towards Optimal Capacity Segmentation with Hybrid Cloud Pricing. In *Proc. of ICDCS*, June 2012.
15. Windows Azure Pricing Calculator, 2013. http://www.windowsazure.com/en-us/pricing/calculator/.

Online Resource Management for Carbon-Neutral Cloud Computing

Kishwar Ahmed, Shaolei Ren, Yuxiong He and Athanasios V. Vasilakos

1 Introduction

The rapid growth of Internet and cloud services in recent years has contributed to the dramatic increase in the number and scale of data centers, resulting in huge brown energy consumption (e.g., electricity) and carbon emissions. This growth in electricity consumption raises serious concerns for data centers: increase in annual operational expenditure by significant amount for large data centers [2], and detrimental effect on environment due to data centers' dependence on coal or other carbon-intensive sources that produce huge carbon footprints [3]. In recent years, data center operators have been increasingly pressured to reduce the net carbon footprint to zero i.e., *carbon neutrality*. While some initial efforts have been made to achieve carbon neutrality for data centers [4–6], they require accurate prediction of long-term future information (e.g., workloads, renewable energy availability) that is very difficult, if not impossible to obtain in practice. In this chapter, we propose an efficient online resource management solution to minimize data center's operational cost (defined as

This chapter is mainly based on the authors' prior research [1].

S. Ren (✉) · K. Ahmed
Florida International University, Miami, USA
e-mail: sren@fiu.edu

K. Ahmed
e-mail: kahme006@fiu.edu

Y. He
Microsoft Research, Redmond, USA
e-mail: yuxhe@microsoft.com

A. V. Vasilakos
National Technical University of Athens, Athens, Greece
e-mail: vasilako@ath.forthnet.gr

S. U. Khan, A. Y. Zomaya (eds.), *Handbook on Data Centers*,
DOI 10.1007/978-1-4939-2092-1_20

a weighted sum of electricity cost and delay cost) while achieving carbon neutrality without requiring long-term future information.

1.1 Background

Data centers, with recent enormous growth in size and scale containing tens of thousands of servers, consume a huge amount of electricity. According to recent studies, the combined electricity consumption of global data centers amounts to 623 billion kWh annually and would rank 5th in the world if the data center were a country [7]. Data centers in the U.S. consume approximately 2% of the national electricity supply, which has been growing at rate of 12 % annually [7]. The statistics is alarming for data centers of large companies with thousands of servers: consuming significant amount of electricity to operate and costing millions of dollars. For example, annual electricity consumption of Microsoft data center servers is over 600 GWh, costing more than $36 M, while servers of Google consume more than 1120 GWh to operate annually and costs over $67 M [8]. As a significant portion of this electricity is produced by coal or other carbon-intensive sources, it is often labelled as "brown energy" and the growing trend of data center electricity consumption has raised serious concerns about its carbon footprint. With the release of large amount of carbon dioxide and other greenhouse gases into the atmosphere, there have been severe environmental impacts of carbon emissions recently, such as altering global patterns of temperature, rainfall, and consequently creation of drought, flood etc. In view of this huge carbon footprint [2, 9], large data centers are now increasingly urged to find effective solutions to reduce the carbon emission for sustainable computing and achieve an overall net zero carbon footprint i.e., carbon neutrality [10, 11].

Besides creating a huge carbon footprint, increase in data center energy consumption contributes to the rise in annual operational expenditure by millions of dollars for large-scale data centers [2]. Many megawatts of electricity are required to power large-scale geographically distributed data centers, and companies like Google and Microsoft spend a large portion of their overall operational costs on electricity bills. Extensive research has been done to increase energy efficiency in data centers [12–15], thereby reducing the energy cost. While recent studies have attempted to minimize the operational cost of data centers subject to capping the brown energy consumption [4], they require a priori knowledge of long-term future information and do not explicitly address carbon neutrality. Realizing that carbon neutrality has become increasingly important and that accurate prediction of long-term future information is practically infeasible, we seek to develop an *online* resource management solution to minimize the operational cost of data centers while achieving carbon neutrality. Following the paper [1], we outline the proposed solution along with key findings in this chapter.

1.2 Carbon Neutrality: Benefits and Challenges

In addition to achieving highly coveted sustainability, carbon neutrality provides other significant benefits in various forms such as reduction of tax, lower-cost contracts with the power utility companies, favourable accreditation, and even business promotion. Consequently, several companies such as Facebook, Google and Microsoft have set capping their carbon usage and achieving net zero carbon emission as their long-term strategic goals [16–18]. While achieving carbon neutrality is an attractive option, there are challenging issues that need to be resolved. In particular, data centers need to budget electricity usage over a long timescale such that the "unknown" future brown energy consumption can be completely offset by limited renewables (e.g., generated by on-/off-site projects or obtained by purchasing renewable energy credits, or RECs). In other words, carefully "budgeting the energy usage" over a long term (which we refer to as *energy budgeting*) is crucial for being carbon neutral, and meanwhile neither operational costs nor quality of service can be considerably compromised. While the existing *power* budgeting technique that allocates the peak power to servers given the current workload (e.g., [19]) might seem to be applicable for this purpose, energy budgeting fundamentally differs from power budgeting and faces a significant challenge: data center operator needs to decide its energy usage in an online manner that cannot possibly foresee the far future time-varying workloads or intermittent renewable energy availability.

To sum up, there are several challenges that must be addressed to efficiently manage data center resources for carbon neutrality:

- Given a limited and even possibly unknown budget of renewables, how to efficiently distribute the budget throughout the budgeting period (e.g., one year) such that the overall net carbon footprint is zero (i.e., achieve carbon neutrality)?
- How to decide the energy usage without knowing the long-term future information and in the presence of time-varying workload or intermittent renewable energy availability?
- Amid quest for carbon neutrality, how to ensure that the two primary concerns of data center operations (i.e., operational energy cost and quality of service, or QoS) are not significantly compromised?

1.3 Current Research and Limitations

Before presenting our online resource management solution for carbon-neutral data centers, we first discuss the existing work from the following point of views.

- **Cost minimization:** With the objective of reducing the operating cost in data center, much research has been done recently to identify methods of cost cutting while ensuring the QoS at the same time. Finding a balance between energy cost of data center and performance loss through dynamically provisioning the workload has been the primary focus of many recent studies [15, 20–25]. Other approaches that are complementary to dynamic capacity provisioning include utilizing storage devices to reduce the operational cost of data center, exploiting the spatio-temporal variation of electricity price [26–29], and using batteries to reduce the peak power usage from the power grid [30, 31]. The recent studies [32, 33] have focused on operational cost minimization of data centers considering multi sources of energy (e.g., grid energy, on-site power generation etc). Moreover, optimization of data center operation through response time minimization [34] has also been widely explored and surveyed. In [20], the peak power budget is allocated to homogeneous servers to minimize the total response time based on a queueing-theoretic model; [19] studies a problem similar to [20] but in the context of virtual system.

Several prior studies have focused on the potential of reducing the energy cost in data centers by exploring geographical and temporal variations in electricity prices. For example, the problem of reducing the electricity cost by utilizing time-varying electricity prices in geographically distributed data centers has been extensively studied [2, 35, 36]. Electricity cost can be further reduced when the advantage of geographical load balancing is combined with the dynamic capacity provisioning approach [13, 37, 38]. Other approaches related to geographical load balancing include utilizing renewable energy for brown energy reduction in data centers. For example, the work [12] schedules workloads to data centers with more renewable energy to lower the brown energy consumption. [39] explores a three-way tradeoff between electricity cost, carbon emissions and response times in geo-distributed data centers. Nevertheless, none of these studies have addressed the long-term carbon neutrality.

- **Capping and reducing brown energy:** The increasing pressure on data center to cap the surging brown energy has motivated the research community to identify methods to reduce brown energy consumption and ultimately to achieve net zero carbon footprint. Much research has been done to optimize the usage of renewable energy (e.g., [40–43]) to green the data centers. Some works reduce the brown energy usage via "following the renewables" by scheduling the workloads to data centers with more green energies [39, 44], but none of these studies have explicitly considered carbon neutrality, which is becoming increasingly important for large scale data center operators [10, 17]. Although there have been studies on achieving energy capping [4–6], these works require long-term prediction of future information, which is infeasible in practice. Moreover, these studies only use empirical evaluations without providing any performance guarantees. By contrast, our solution provides provable analytical guarantees and also bounds the potential deviation from long-term carbon neutrality even in the worst case.

1.4 Contributions

To minimize the operational cost while achieving carbon neutrality, we introduce a provably-efficient online resource management algorithm, COCA (optimizing for COst minimization and CArbon neutrality), to control the electricity usage for

Table 1 List of notations

Notation	Description
$\lambda_i(t)$	Workloads processed by server i
$x_i(t)$	Service rate of server i
$r(t)$	on-site renewable energy
$f(t)$	off-site renewable energy
$u(t)$	Electricity price
$p(t)$	Server power consumption
$e(t)$	Electricity cost
$d(t)$	Delay cost
V	Cost-delay parameter
Z	Total RECs
$q(t)$	Carbon deficit queue

minimizing data center operational cost while satisfying carbon neutrality without requiring long-term future information. By using COCA, each server autonomously adjusts its speed as well as power states and optimally decides the amount of workloads to process. Our main contributions are summarized as follows:

- We present an online resource management algorithm, COCA, to minimize the operational cost of a data center in the presence of time-varying workloads and intermittent renewable energy supplies. We simultaneously address two critical costs for a data center, i.e., energy cost and delay cost, so that the operational cost can be optimized without considerably compromising user experiences. It is rigorously proved that COCA is efficient in terms of minimizing the parametrized operational cost compared to the optimal offline algorithm with T-step lookahead information, while approximately satisfying the carbon neutrality constraint.
- We conduct a trace-based simulation to empirically validate COCA. The results show that COCA can reduce the operational cost and achieve close-to-minimum value while satisfying the carbon neutrality. When compared with state-of-the-art prediction-based methods, it is shown that COCA can reduce the cost by approximately 25 %. Since a large data center contains thousands of servers, millions of dollars can be saved using COCA. Moreover, COCA is robust against various factors such as inaccurate knowledge of workload arrivals.

2 Model

The system model is described in this section. Our focus is on dynamically distributing the workload and determining processing speed of each server in a data center to minimize the operational cost of the system, incorporating both the energy cost and delay cost. Key notations are summarized in Table 1.

2.1 Some Assumptions

Following are some of the assumptions that are used throughout the chapter.

- **Time model:** We consider a discrete-time model, and divide the entire budgeting period into K time slots of equal length denoted by $t = 0, 1, \cdots, K - 1$. We assume the length of prediction window is one hour, where the operator can update its capacity provisioning and load distribution decisions at the beginning of each hour. Moreover, we consider that the data center operator can accurately predict the hour-ahead information (including workload arrival rate, renewable energy supply, and electricity price), which can be easily obtained in practice. Note that, whenever longer-term prediction is feasible, our model is extendible to such prediction window size (e.g., day-ahead prediction).
- **Workload type:** In this chapter, our focus is on *delay sensitive* interactive workloads, while *delay tolerant* batch jobs can easily be maintained through separate batch job queue [36]. Moreover, M/G/1/PS (Memoryless/General/1/Processor-Sharing) queueing model is used as analytical tool for quantifying delay cost incurred by workload arrival at the data center. Note that, often the workload inter-arrival time may not be exponentially distributed and the scheduling policy may not be processor-sharing. However, since it is difficult to analyze non-exponential workload inter-arrival time and/or with general scheduling policies, the M/G/1/PS model has been adopted as a reasonable estimation for actual service process and utilized for modelling queueing delay [12, 20, 37, 45].

2.2 Energy Sources

To accomplish carbon neutrality, data centers normally depend on a combination of approaches, such as generation of on-site and off-site energy (alternatively known as renewable energy) and purchase of renewable energy credits (RECs), to offset brown energy usage. The available on-site renewable energy during time t is denoted by $r(t) \in [0, r_{max}]$, which may follow an arbitrary trajectory throughout the budgeting period. Moreover, the available off-site renewable energy generated via Power Purchase Agreement (PPA) for time t is denoted by $f(t) \in [0, f_{max}]$, while the fixed amount of RECs available throughout the entire budgeting period is represented as Z. In the following, these three types of renewable energy are described in detail:

- **On-site renewable energy:** An intuitive and natural way to offset the brown energy can be to produce green energy directly from on-site projects or utility companies. On-site renewable energy systems are usually installed on or adjacent to data center and supplies green energy directly to power the data center. For example, solar photovoltaic (PV) panels and wind turbines have been widely installed for providing on-site renewable energy sources [46], although they are highly dependent on weather conditions and exhibit an intermittent nature. Moreover, the location of data center may not be suitable for generation of sufficient green energy to satisfy the data center requirement. In spite of these difficulties, a substantial amount of energy required for the data center functioning can be supplied by the on-site renewable energy [17]. Note that energy storage device can be an effective option to store the unused renewable energy [27] and hence may further reduce the electricity cost. However, since these storage devices are quite expensive and charging/discharging of batteries is not the main focus of our study, we exclude energy storage devices while concentrating on capacity provisioning and load distribution instead.

- **Off-site renewable energy:** Data centers resort to off-site renewable energy to offset a significant portion of brown energy [17, 18]. For example, PPA has become an important and widely-used type of off-site renewable energy, where companies make contract with several renewable energy developers so that the generated renewable energy will be directly utilized to offset the brown energy usage of data centers. An increasing number of companies (e.g., Google) have signed contract with several renewable energy plants to buy renewable electricity [3, 10]. However, since off-site renewable energy becomes undifferentiated due to the mixture with other types of energy after entering the grid, it is necessary for data centers to draw electricity and pay utility companies for accountability reasons.

Another off-site renewable energy option available to data center is REC (a.k.a. green tag), which is purchased by data centers to fund green power projects. The objective here is to offset data center's brown energy usage, mitigating the environmental impact of their energy usage. We assume that the REC purchasing decision has been made prior to a budgeting period, although various approaches to purchasing RECs such as dynamic purchasing in real time can also be accommodated. Although some companies may buy RECs at the end of a budgeting period, it is not pertinent here, because the amount of available RECs in our work can be considered as *desired* REC usage, while the remaining brown energy may still be offset by purchasing additional RECs at the end of a budgeting period [10, 11].

2.3 Data Center

We consider one data center that has N servers. We measure the processing speed in terms of the *service rate*, i.e., how many jobs can be processed on average in a unit time on average, and interchangeably use "processing speeds" and "service rates" wherever applicable.

In general, we consider that server i can choose its speed x_i out of a finite set $\mathcal{S}_i = \{s_{i,0}, s_{i,1}, \cdots, s_{i,K_i}\}$, where $s_{i,0} = 0$ represents zero-speed (e.g., sleep or shut down) and K_i is the total number of positive processing speeds available to server i. Next, assuming that the servers consume a negligible power under the zero-speed mode, we express the average power consumption of server i as:

$$p_i(\lambda_i, x_i) = \begin{cases} p_{i,s} + p_{i,c}(x_i) \cdot \frac{\lambda_i}{x_i}, & \text{if } x_i > 0, \\ 0, & \text{if } x_i = 0, \end{cases} \tag{1}$$

where λ_i is the workload arrival rate distributed to server i, $p_{i,s}$ is the static power regardless of the workloads as long as server i is turned on, and $p_{i,c}(x_i)$ is the computing power incurred only when server i is processing workloads at a speed of x_i.

We focus on the server power consumption, while the power consumption of non-IT parts of the data center such as cooling and power supply system is captured using the factor of power usage effectiveness (PUE), which is the ratio of the total data center power consumption to IT power consumption measuring data center's energy efficiency [12]. Currently, the PUE factor ranges from as high as 2.0 in most enterprise data centers [47], to a PUE of 1.1 at a few state-of-the-art facilities [48]. Thus, the total server power consumption at time t is given by:

$$p(\vec{\lambda}(t), \vec{x}(t)) = \sum_{i=1}^{N} p_i(\lambda_i(t), x_i(t)), \tag{2}$$

where $\vec{\lambda}(t) = (\lambda_1(t), \cdots, \lambda_N(t))$ and $\vec{x}(t) = (x_1(t), \cdots, x_N(t))$ are the load distribution and capacity provisioning decisions for time t, respectively.

- **Electricity cost.** We denote the electricity price at time t by $u(t)$ which is known to the data center no later than the beginning of time t and may change over time/locations. We also assume that the electricity price remain fixed during a time period t but may change *arbitrarily* over time and at different locations. We can express the incurred electricity cost (measured in dollars per hour) as:

$$e(\vec{\lambda}(t), \vec{x}(t)) = u(t) \cdot \left[p(\vec{\lambda}(t), \vec{x}(t)) - r(t) \right]^{+}, \tag{3}$$

  where $[\,\cdot\,] = \max\{\cdot, 0\}$ indicates that no electricity will be drawn from the power grid if on-site renewable energy is already sufficient. Although Eq. (3) represents a linear electricity cost function for a data center at time t, the model can also incorporate other electricity cost function such as nonlinear convex functions. For example, data center may be charged at a higher rate when consuming more power. Moreover, large data centers may have contract with utility companies for a constant electricity price, when in such case, $u(t)$ becomes constant throughout the budgeting period.

2.4 Workload

We denote by $\lambda(t)$ the total arrival rate of workloads in the data center during time t. As assumed in prior work [4, 12, 49], the value of $\lambda(t)$ is accurately available at the beginning of each time slot t, while our simulation results further demonstrate the robustness of COCA against inaccurate knowledge of workload arrival rates. Moreover, the time-varying arrival of workload follows a non-stationary distribution and is bounded by the limit $[0, \lambda_{\max}]$, where $\lambda_{\max}$ is the maximum possible arrival rate. The workloads first arrive at a load distributor before they are distributed to servers for processing. We denote the workload arrival rate distributed to server i at time t by $\lambda_i(t) \geq 0$.

- **Delay cost.** To quantify the overall data center delay performance, we introduce the notion of *delay cost* capturing the delay-induced revenue loss and/or user dissatisfaction [12]. Based on the M/G/1/PS queueing model, multiplying the arrival rate of workload with the average response time for each server, the queueing delay cost can be calculated as:

$$d(\vec{\lambda}(t), \vec{x}(t)) = \sum_{i=1}^{N} d_i(\lambda_i(t), x_i(t)) = \sum_{i=1}^{N} \frac{\lambda_i(t)}{x_i(t) - \lambda_i(t)}, \tag{4}$$

in which we ignore the network delay cost between the load distributor and servers.

3 Problem Formulation

In this section, the optimization objective and constraints are first specified. Then, the offline problem formulation for capacity provisioning and load distribution is presented.

3.1 Objective and Constraints

The objective of COCA is to minimize the long-term operational cost subject to carbon neutrality and a set of other constraints introduced as follows.

- **Objective.** A data center incurs various types of costs such as cost for building data centers, installing renewal generators and so on. However, in this chapter, the objective is to minimize operational cost, ignoring capital costs. Note that since electricity cost takes up a significant portion of operational cost and delay cost is important due to its impact on user experiences [12, 15], we consider these two costs in our study. A parameterized cost function representing a combination of energy cost and delay cost is utilized to represent overall operational cost:

$$g(\vec{\lambda}(t), \vec{x}(t)) = e(\vec{\lambda}(t), \vec{x}(t)) + \beta \cdot d(\vec{\lambda}(t), \vec{x}(t)), \tag{5}$$

where $\beta \geq 0$ is a weighting parameter that converts the delay cost to energy cost [12, 34]. In special cases, when β reduces to zero, the data center only minimizes the electricity, whereas when β goes to infinity the data center minimizes the delay cost, ignoring electricity cost. Since data center operates for a long time period, we focus on minimizing the long-term average cost $\bar{g}$ under a particular control policy over a sufficiently large but finite budgeting period (e.g., a year or a month, depending on the actual budgeting) [4, 50]:

$$\bar{g} = \frac{1}{K} \sum_{t=0}^{K-1} g\left(\vec{\lambda}(t), \vec{x}(t)\right), \tag{6}$$

where K is the total number of time slots over the entire budgeting period.

- **Constraints.** To achieve carbon neutrality, a data center has to offset its electricity usage by the off-site renewable energy and RECs [17, 18]. In the following, the long-term carbon neutrality constraint is specified:

$$\frac{1}{K} \sum_{t=0}^{K-1} \left[p\left(\vec{\lambda}(t), \vec{x}(t)\right) - r(t)\right]^{+} \leq \frac{\alpha}{K} \cdot \left[\sum_{t=0}^{K-1} f(t) + Z\right], \tag{7}$$

where $f(t)$ represents the off-site renewable energy generated from PPAs at time t, while Z denotes the amount of available RECs throughout budgeting period. Moreover, $\alpha \geq 0$ is the desired capping constraint of electricity usage over the entire budgeting period relative to the total off-site renewable energy plus RECs. With less α, the data center achieves carbon neutrality more aggressively. Note that our study can also address the scenario in which part of the electricity is produced by green energy sources: by multiplying the electricity usage with a certain factor that indicates the percentage of "brown" electricity, (7) specifies the constraint on the actual brown electricity usage. Other constraints that need to be satisfied are introduced in the following:

$$0 \leq \lambda_i(t) \leq \theta \cdot x_i(t), \ \forall i, t, \tag{8}$$

$$\sum_{i=1}^{N} \lambda_i(t) = \lambda(t), \ \forall t. \tag{9}$$

Constraint (8) represents that the service demand can not exceed the service capacity for server i at time t, where $\theta \in (0, 1)$ is the maximum utilization constraint for each server. Constraint (9) avoids workload dropping. Naturally, server i can only select one of its supported service rates, i.e.,

$$x_i(t) \in \mathcal{S}_i = \{s_{i,0}, s_{i,1}, \cdots, s_{i,K_i}\}, \ \forall i, t. \tag{10}$$

Note that additional constraints, such as peak power and maximum delay, can also be incorporated with little impact on our online algorithm.

3.2 *Offline Problem Formulation*

Given the objective function and set of constraints, the optimal offline problem formulation for capacity provisioning and load distribution is presented as follows:

$$\mathbf{P1}: \quad \min_{\mathcal{D}} \bar{g} = \frac{1}{K} \sum_{t=0}^{K-1} g\left(\vec{\lambda}(t), \vec{x}(t)\right) \tag{11}$$

$$s.t., \quad \text{constraints } (7), (8), (9), (10), \tag{12}$$

where $\mathcal{D}$ represents a sequence of decisions, i.e., $\vec{\lambda}(t)$ and $\vec{x}(t)$, for $t = 0, 1, \cdots, K-1$, which we need to optimize. However, there is one significant challenge that needs to be addressed to derive the optimal solution to **P1**: solving **P1** requires complete offline information (i.e., workload arrivals, renewable energy supplies, and electricity prices) over the entire budgeting period, which may be only possible in an idealized scenario but practically infeasible. Moreover, the problem may be further complicated due to the bursty and unpredictable nature of workloads arrivals in data center [51], making online algorithms necessary to solve **P1**.

4 Algorithm for Cost Optimization and Carbon Neutrality

In this section, we first outline our online algorithm, COCA, and then prove that it is efficient with respect to cost minimization compared to the optimal offline algorithm with T-step lookahead information. COCA only uses online information and allows the data center operator to adaptively adjust the tradeoff between cost saving and how much the electricity consumption *potentially* exceeds the desired carbon neutrality constraint.

4.1 *Carbon Deficit Queue*

The long-term carbon neutrality constraint couples the online decisions across different time slots, and the current decisions will impact the future decisions (e.g., higher electricity usage at one time slot may lead to a lower remaining carbon budget for future use). Hence, it is challenging to make online decisions while satisfying the long-term carbon neutrality constraint. To decouple the decisions for different time slots and hence enable an online algorithm, we leverage the recently-developed Lyapunov optimization technique [52] and introduce a virtual carbon deficit queue, $q(t)$. In particular, assuming $q(0) = 0$, we write the carbon deficit queue dynamics in the following:

$$q(t+1) = \left\{ q(t) + \left[p\left(\vec{\lambda}(t), \vec{x}(t)\right) - r(t) \right]^{+} - \alpha \cdot f(t) - z \right\}^{+}, \tag{13}$$

Algorithm 1 COCA

1: Input $\lambda(t)$, $r(t)$, $f(t)$, $u(t)$, at the beginning of each time $t = 0, 1, \cdots, K-1$
2: **if** $t = rT, \forall r = 0, 1, \cdots, R-1$ **then**
3: $\quad q(t) \leftarrow 0$ and $V \leftarrow V_r$
4: **end if**
5: **P2:** Choose $\vec{\lambda}(t)$ and $\vec{\mathbf{x}}(t)$ subject to (8)(9)(10) to minimize

$$V \cdot g\left(\vec{\lambda}(t), \vec{\mathbf{x}}(t)\right) + q(t) \cdot \left[p\left(\vec{\lambda}(t), \vec{\mathbf{x}}(t)\right) - r(t)\right]^{+} \tag{14}$$

6: Update $q(t)$ according to (13).

where the arrival rate for the queue is the electricity usage at time t while the departure rate is denoted by the average allocated budget. Moreover, $z = \frac{\alpha}{K} \cdot Z$ denotes the average electricity budget per time slot. Note that introduction of the carbon deficit queue helps us decouple the decisions across different time slots, therefore replacing the long-term carbon neutrality constraint. Clearly, a larger queue length represents a greater electricity consumption than the total off-site renewable energy plus RECs provided, hence it needs to reduce electricity usage to achieve carbon neutrality.

4.2 *Optimizing for Cost Minimization and Carbon Neutrality*

In this subsection, we discuss COCA for a single data center that can dynamically determine the workload distribution and perform server speed selection. COCA only uses online information and allows the data center operator to adaptively adjust the tradeoff between cost saving and how much the electricity consumption exceeds the desired carbon neutrality constraint. Algorithm 1 outlines the online algorithm for a single data center.

4.2.1 Working Principle of COCA

COCA only requires the currently available information (i.e., $\lambda(t)$, $r(t)$, $f(t)$, $u(t)$) as the inputs and hence the algorithm is purely online. $V_0, V_1, \cdots, V_{R-1}$ denote a sequence of positive control parameters and are used to dynamically adjust the tradeoff between cost minimization and electricity usage over R frames, each having T time slots. Lines 2–4 reset the carbon deficit queue at the beginning of each frame r, such that the cost-carbon parameter V can be adjusted and the carbon deficit in a new time frame will not be affected by its value resulting from the previous time frame. Line 5 defines an optimization problem based on online information: minimizing the original cost scaled by V plus $q(t) \cdot [p(\vec{\lambda}(t), \vec{\mathrm{x}}(t)) - r(t)]^{+}$. Its intuition is as follows. By considering the perturbing term $q(t) \cdot [p(\vec{\lambda}(t), \vec{\mathrm{x}}(t)) - r(t)]^{+}$, the data center operator places a higher weight on the electricity usage when making resource management decision: the weighting factor for the electricity usage is scaled by V plus the carbon queue length $q(t)$, whereas the weighting factor for delay cost is only scaled by V. As a consequence, when $q(t)$ increases (i.e., the current electricity usage further exceeds the supplied renewable energy and RECs), minimizing the electricity usage is more critical for the data center operator due to the carbon neutrality constraint. In essence, the carbon deficit queue maintained without foreseeing the future guides the data center decisions towards carbon neutrality, thereby enabling online decisions.

4.2.2 Distributed Implementation

In practical systems, distributed solutions are desired such that each server can make autonomous decisions. Here, we develop a distributed algorithm, called GSD (Gibbs Sampling-based Distributed optimization), based on a variation of Gibbs sampling presented in Algorithm 2.

GSD is a distributed algorithm working as follows: at each iteration, a randomly selected server first autonomously updates its speed, and then the servers decide their optimal load distribution decisions (also distributedly), after which the servers communicate decisions to each other. Line 3 in GSD, i.e., minimizing (15), can be solved efficiently using any distributed optimization techniques (see [53] for a solution based on dual decomposition). Note that during the iterations, servers do not need to actually adjust their speeds or load distribution decisions, which is only needed after the completion of the algorithm. In line 7, to randomly select a server, we can assign each server with a random timer to "compete" for the updating opportunity, like in random channel access in wireless networks. Each server i, for $i = 1, 2, \cdots, N$, maintains $x_i^*(t)$ as its current processing speed, while exploring (possibly) new processing speed $x_i^e(t)$ to avoid being trapped in a locally optimal solution. We use $\vec{\lambda}^*(t)$ and $\vec{\lambda}^e(t)$ to denote the optimal load distribution decisions corresponding to $x_i^*(t)$ and $x_i^e(t)$, respectively. The parameter $\delta > 0$, referred to as *temperature* [54], is used to control exploration versus exploitation (i.e., the degree of randomness).

Now, we briefly describe the performance analysis of GSD:

- **Complexity:** Despite being distributed, GSD still incurs a worst-case complexity that is exponential in the number of servers (although in practice a reasonably good solution is often quickly identified). In practice, the computational complexity of GSD can be effectively reduced by making capacity provisioning decisions on a *group* basis: changing speed selections for a whole group of (homogeneous) servers in batch.
- **Accuracy:** Theorem 1, whose proof can be found in [1], formally shows that GSD can solve the optimization problem **P2** with an arbitrarily high probability as the temperature $\delta \to \infty$.

Theorem 1 *As $\delta > 0$ increases, GSD converges with a higher probability to the globally optimal solution that minimizes (14) subject to* (8), (9), (10). *When $\delta \to \infty$, Algorithm 2 converges to the globally optimal solution with a probability of 1.*

Theorem 1 indicates that using Algorithm 2, the servers can select the optimal processing speeds distributedly with an arbitrarily high probability.

Algorithm 2 GSD: Distributed Optimization for **P2**

1: Initialization: servers choose feasible values and set $\vec{x}^*(t) \leftarrow \vec{x}^e(t)$, $\vec{\lambda}^*(t) \leftarrow \vec{\lambda}^e(t)$, $\tilde{g}^* \leftarrow \infty$
2: **if** $\lambda(t) \leq \gamma \cdot \sum_{i=1}^{N} x_i^e(t)$ **then**
3: Obtain $\vec{\lambda}^e(t)$ by minimizing over $\vec{\lambda}(t)$

$$V \cdot g\left(\vec{\lambda}(t), \vec{x}^e(t)\right) + q(t)\left[p\left(\vec{\lambda}(t), \vec{x}^e(t)\right) - r(t)\right]^+, \tag{15}$$

subject to (8)(9), and set $\tilde{g}^e$ to the minimum value of (15)
4: $u \leftarrow \frac{\exp\left(\frac{\delta}{\tilde{g}^e}\right)}{\exp\left(\frac{\delta}{\tilde{g}^*}\right)+\exp\left(\frac{\delta}{\tilde{g}^e}\right)}$
5: With a probability of u: servers set $\vec{x}^*(t) \leftarrow \vec{x}^e(t)$, $\vec{\lambda}^*(t) \leftarrow \vec{\lambda}^e(t)$ and $\tilde{g}^* \leftarrow \tilde{g}^e$; with a probability of $1-u$: servers set $\vec{x}^e(t) \leftarrow \vec{x}^*(t)$
6: **end if**
7: Randomly select a server i; server i randomly selects a processing speed $x_i'(t) \in \mathcal{S}_i$ and sets $x_i^e(t) \leftarrow x_i'(t)$
8: Return $\vec{x}^*(t)$ and $\vec{\lambda}^*(t)$ if the stopping criterion is satisfied; otherwise, go to Line 3

4.3 Performance Analysis

This subsection presents the performance analysis of COCA in Theorem 2, whose proof is provided in [1]. First, we describe an optimal algorithm with T-step lookahead information, which we use as the benchmark to compare COCA with.

***T*-step lookahead algorithm.** We now present an offline algorithm with T-step lookahead information, which has full knowledge of the data center states and workload arrivals up to next T time steps. The entire budgeting period is divided into R frames, each containing $T \geq 1$ times slots such that $K = RT$. Then, the cost minimization problem over the r-th frame, for $r = 0, 1, \cdots, R-1$ can be formulated as:

$$\mathbf{P3}: \min_{\vec{\lambda}(t), \vec{x}(t)} \frac{1}{T} \sum_{t=rT}^{(r+1)T-1} g\left(\vec{\lambda}(t), \vec{x}(t)\right) \tag{16}$$

$$s.t., \text{ constraints (8), (9), (10),} \tag{17}$$

$$\sum_{t=rT}^{(r+1)T-1} [p\left(\vec{\lambda}(t), \vec{x}(t)\right) - r(t)]^+ \leq \alpha \cdot f_r. \tag{18}$$

Essentially, **P3** defines a family of offline problems parametrized by the lookahead information window size T. In **P3**, we denote the minimum of $\frac{1}{T}\sum_{t=rT}^{(r+1)T-1} g\left(\vec{\lambda}(t), \vec{x}(t)\right)$ by G_r^*, for $r = 0, 1, \cdots, R$, which is achievable considering all the actions including those that are chosen with the perfect information of data center states and workload arrivals over the entire frame. Next, to ensure the existence of at least one feasible solution to **P2**, we make the two assumptions that are very mild in practice.

Boundedness Assumption The workload arrival rate $\lambda(t)$, electricity price $u(t)$, as well as renewable energy supplies $r(t)$ and $f(t)$ are finite, for $t = 0, 1, \cdots, K-1$.

Feasibility Assumption For the r-th frame, where $r = 0, 1, \cdots, R-1$, there exists at least one sequence of capacity provisioning and load distribution decisions that satisfy the constraints of **P2**.

The boundedness assumption, combined with (8), ensures that the cost function is finite, while the feasibility assumption guarantees that there is at least one sequence of feasible decisions to solve **P2**. We denote the long-term average minimum cost by the optimal T-step lookahead algorithm by $\frac{1}{R}\sum_{r=0}^{R-1} G_r^*$. Later, we shall show that our online algorithm can achieve a cost close to this value. Moreover, $f_r = \sum_{t=rT}^{(r+1)T-1} f(t) + \frac{Z}{R}$ represents the total available off-site renewable energy on the r-th frame plus the total RECs over the R frames. Note that (18) is a *stronger* version of the original carbon neutrality constraint in (7), as it requires the satisfaction of carbon neutrality for every T time slots. Nonetheless, if T is sufficiently large, (18) will be almost equivalent to (7), and the oracle *approximately* solves the original problem **P1** [52].

Theorem 2 *Suppose that boundedness and feasibility assumptions are satisfied. Then, for any $T \in \mathbb{Z}^+$ and $R \in \mathbb{Z}^+$ such that $K = RT$, the following statements hold.*

a. *The carbon neutrality constraint is approximately satisfied with a bounded deviation:*

$$\frac{1}{K}\sum_{t=0}^{K-1}\left[p\left(\vec{\lambda}(t), \vec{x}(t)\right) - r(t)\right]^+ \leq \frac{\alpha}{K}\cdot\left[\sum_{t=0}^{K-1} f(t) + Z\right] + \frac{\sum_{r=0}^{R-1}\sqrt{C(T) + V_r\left(G_r^* - g_{\min}\right)}}{R\sqrt{T}}, \tag{19}$$

b. *The average cost $\bar{g}^*$ achieved by COCA satisfies:*

$$\bar{g}^* \leq \frac{1}{R}\sum_{r=0}^{R-1} G_r^* + \frac{C(T)}{R}\cdot\sum_{r=0}^{R-1}\frac{1}{V_r}, \tag{20}$$

where $C(T) = B + D(T-1)$ with B and D being finite constants (more details on B and D can be found in [1]), and $g_{\min}$ is the minimum hourly cost that can be achieved by any feasible decisions throughout the budgeting period.

It follows from Theorem 2 that, COCA is $O(1/V)$-optimal with respect to the average cost against the optimal T-step lookahead policy, while the carbon neutrality constraint is bounded by $O(V)$. With a larger V, the cost is closer to the infimum, while the deviation from the carbon neutrality constraint can be larger, and vice versa. Thus, by appropriately tuning the cost-carbon parameter V, we can achieve a desired tradeoff between cost and carbon neutrality. Moreover, since both the boundedness and feasibility assumptions are mild, the established bounds are applicable for almost all practical scenarios.

5 Simulation

To validate our analysis, this section presents trace-based simulation studies of a large data center and performance evaluation of COCA. First, we present the data sets, and then we present the following sets of simulations.

- Efficiency of COCA: We show that under different settings, COCA provides a satisfactory performance in terms of average cost while satisfying the carbon neutrality constraint.
- Comparison with prediction-based method: We compare COCA with the state-of-the-art prediction-based method and show that COCA reduces the average cost by more than 25 % while satisfying the desired carbon neutrality better.

5.1 Data Sets

A single data center is considered with peak server power of 50 MW. The model assumes delay sensitive workloads and considers the server power consumption, ignoring the cooling power or server power for delay tolerant batch jobs. Due to the practical difficulty in implementing COCA in a real system, only event-based simulations with real-world trace data are considered to validate the analysis, which is a common approach in the literature [12, 37].

The budgeting period in our study is one year, and the default total allowed electricity usage is 1.43×10^5 MWh (i.e., 92 % of 1.55×10^5 MWh , where 1.55×10^5 MWh is the electricity usage of carbon-unaware algorithm). Among the 1.43×10^5 MWh renewable budget, off-site renewable energy and RECs contribute 40 and 60 %, respectively. The weighting parameter converting the delay to monetary cost is $\beta = 10$. Since the data center capacity provisioning and load distribution are updated hourly, all the energy consumption and electricity cost throughout this section are hourly values unless otherwise stated.

- **Workload information:** We consider "mice-type" synthetic workloads (e.g., web requests), and use real-world trace to drive our simulation. The service time of an individual request follows an exponential distribution with a mean of 100 ms (when the server is running at its full speed), which may not perfectly capture a real system but suffices for our evaluation purpose. The workload arrival processes for simulations are taken from two real-world traces. The first set of workload trace is taken from the server I/O usage log of Florida International University (FIU, a large public university in the U.S. with over 50,000 students). The traced period

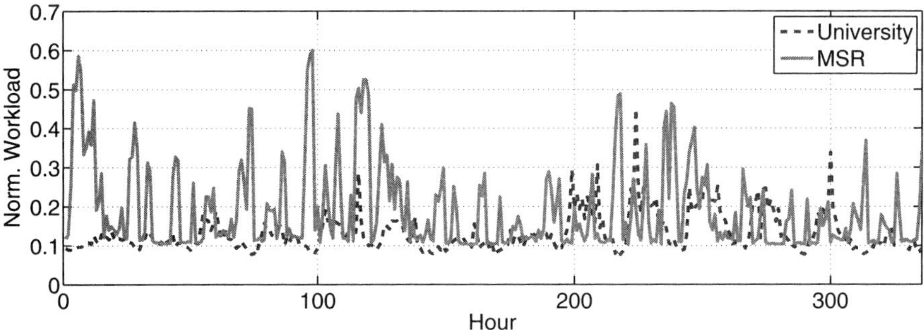

Fig. 1 Workload trace. Normalized with respect to the maximum data center capacity

was taken from January 1 to December 31, 2012 and was profiled. Fig. 1 shows a snapshot of such trace (for first week of the month of July, 2012) , normalized with respect to the maximum data center capacity. For sensitivity studies, we also use workload trace for Microsoft Research (MSR) first shown in [15] and repeat the trace for one year by adding random noises of up to $\pm 40\,\%$. The I/O trace for MSR is taken from 6 RAID volumes at MSR Cambridge, and the traced period is 1 week starting from 5PM (GMT) on February 22, 2007 [15]. Figure 1 shows the one week MSR workload trace, normalized with respect to the maximum data center capacity.

- **Server:** There are approximately 216 K servers in the data center and each server has a maximum power of 231 W. Powerpack [55] is used to measure the power consumption of a server with a quad-core AMD Opteron 2380 processor that supports four different speeds via DVFS. Specifically, if turned on, each server has an idle power of 140 W and supports the following four different processing speeds/powers: 0.8 GHz (184 W), 1.3 GHz (194 W), 1.8 GHz (208 W), and 2.5 GHz (231 W). The normalized service rate for each server is 10 jobs per second, i.e., when running at the maximum speed, it is assumed that each server can process 10 requests per second on average.
- **Electricity price and renewable energy:** The hourly electricity price for Mountain View, California, is obtained from [56]. Note that it is assumed that the data center participates in a real-time electricity market [2, 12, 26, 37]. The hourly renewable energies (generated through solar panels and wind turbines) for the city of Mountain View as well as the state of California during the year of 2012 are obtained from [56], and was scaled proportionally such that on-site renewable accounts for approximately 20 % of the total energy consumption.

Since the access to commercial data centers is unavailable, we obtain the trace data from various sources, but it captures the variation of workloads, renewable energy supplies and electricity prices over the budgeting period. Thus, it serves the purpose of evaluating the performance and benefits of COCA.

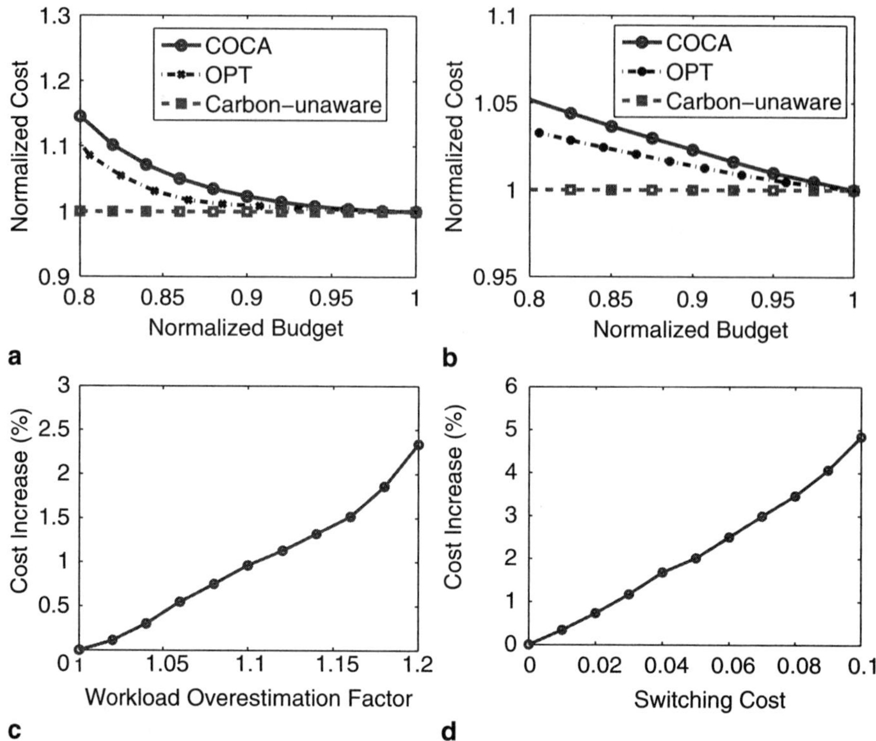

Fig. 2 Efficiency of COCA. **a** Average cost given different carbon budgets. **b** Average cost given different carbon budgets for MSR workloads. **c** Cost increase with workload overestimation. **d** Cost increase with switching cost

5.2 Results

We drive the event-based simulation using the above synthetic trace data. The power consumption and delay are recorded as outputs of the simulation. Next, we present the simulation results as follows.

5.2.1 Efficiency of COCA

We perform the following simulation studies to demonstrate the efficiency of COCA. For all the cases, we appropriately choose V such that carbon neutrality is satisfied.

- **Impact of carbon budget:** We now show in Fig. 2a the impact of carbon budget (i.e., off-site renewables plus RECs) on the cost. According to our simulation settings, the carbon-unaware algorithm consumes 1.55×10^5 MWh electricity energy over a year, which we normalize to 1. It can be seen that given a 85 % carbon budget, which is equivalent to using only 85 % of electricity, COCA only exceeds the capping-unaware algorithm by approximately 5 % in terms of the average cost, while still being able to satisfy carbon neutrality (which is clearly violated by the capping-unaware algorithm). As a comparison, we also show the optimal offline algorithm (OPT), which has the complete offline information and minimizes the operational cost under carbon neutrality. COCA works remarkably well even compared to OPT. This demonstrates that, with careful energy budgeting, the long-term energy consumption can be significantly reduced while still keeping the operational cost low (even compared to the optimal offline algorithm and the carbon-unaware algorithm). Note that, if a higher carbon budget (e.g., 1.05) is used, COCA will be almost the same as the capping-unaware algorithm without using up the budget, because our optimization objective incorporates both electricity and delay costs and using excessive electricity will increase the total cost (although decreasing the delay cost).

- **Different workload trace:** We now consider the MSR workload trace as illustrated in Fig. 1. Fig. 2b shows the normalized cost achieved by COCA, OPT and the carbon-unaware algorithm under different normalized carbon budgets. It delivers the same message as Fig. 2a and demonstrates that COCA works well with different workload traces.
- **Workload overestimation:** Since in general, it is difficult to always predict the hour-ahead workload arrivals accurately, a conservative approach can be to overestimate the workload and therefore keep more server turned on to cope with unexpected traffic spikes. Note that, workload overestimation can also be considered as imperfect modeling of server's service rate. In Fig. 2c, we show the percent increase in average cost caused by overestimated workload for COCA. We see that total cost increases by less then 2.5 % even when we overestimate the workloads by 20 %. This is because although workload overestimation may turn on more servers and incur a higher electricity cost at some time slots, the delay cost will be decreased. The result confirms that COCA is robust against workload prediction error. The reason behind choosing 20 % is that, according to prior research [57], the prediction error is typically within 20 % for hour-ahead prediction.
- **Switching cost:** We study the performance of COCA when the server toggling cost (such as energy/time waste, "wear and tear", and other risks) is added to the power cost. We incorporate all these factors related to server toggling cost and use switching cost as the combined cost measured in terms of energy consumption. We normalize the switching cost (obtained by turning on/off one server) with respect to the maximum hourly energy consumption of a single sever (i.e., 0.231 KWh). In Fig. 2d, we see that the average operational cost increases by less than 5 % even when the switching cost of one server takes 10 % of its maximum hourly energy consumption (i.e., 0.0231 KWh). This shows that performance of COCA is not significantly degraded even when the server switching cost is considered.

We further note that with different combinations of off-site renewables and RECs (but with the same total amount), COCA achieves almost the same cost (less than 1 % change), indicating that COCA is not sensitive to renewable energy portfolios, but

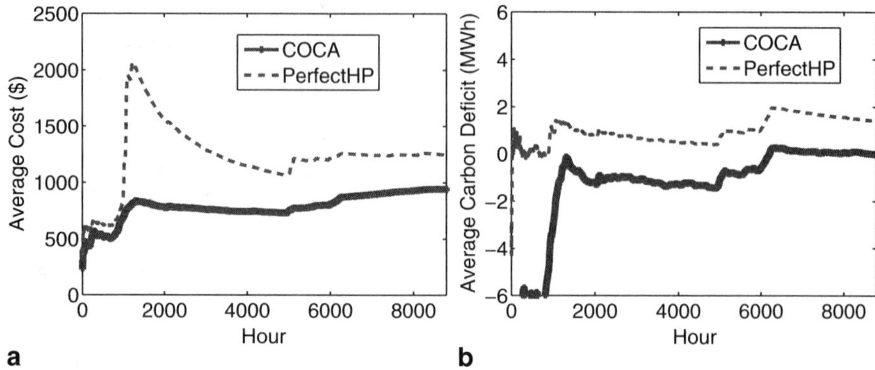

Fig. 3 Comparing COCA with PerfectHP. **a** Average cost. **b** Average carbon deficit

rather mainly depends on the total budget (as shown in Fig. 2a). Other studies such as different server settings are also performed, demonstrating that COCA provides a satisfactory performance in various scenarios and pointing to its applicability in real systems. These results are omitted due to space limitations.

5.2.2 Comparison with Prediction-Based Method

This subsection compares COCA with the best known relevant solution—prediction-based method studied in [4, 5]. However, since none of the existing prediction-based methods have considered both the nonlinear delay cost and intermittent off-site renewable energy supplies, we incorporate these factors by considering a heuristic variation as follows.

Perfect hourly prediction heuristic (PerfectHP) The data center operator leverages 48-hour-ahead prediction of hourly workloads and allocates the carbon budget (RECs plus off-site renewables, but not including the on-site renewables) in proportion to the hourly workloads. The operator minimizes the cost subject to the allocated hourly carbon budget; if no feasible solution exists for a particular hour (e.g., workload spikes), the operator will minimize the cost without considering the allocated carbon budget. We consider 48-hour-ahead prediction in the comparison, because prediction beyond 48 hours will typically exhibit large errors [12], especially for solar and wind energy supplies that are commonly used for data centers but highly subject to weather condition.

Figure 3 shows the comparison between COCA and the prediction-based PerfectHP in terms of the average hourly cost and carbon deficit.[1] Figure 3 demonstrates that COCA is more cost-effective compared to the prediction-based method with a

[1] The average at time t in Fig. 3 is obtained by summing up all the values from time 0 to time t and then dividing the sum by $t+1$.

cost saving of more than 25 % over one year. COCA achieves the benefits because it can still focus on cost minimization even though the workload spikes and carbon neutrality is *temporarily* violated, since the carbon deficit queue only penalizes the data center for overusing electricity in later time slots while guaranteeing a bounded deviation from the carbon neutrality. On the other hand, without foreseeing the long-term future, short-term prediction-based PerfectHP may over-allocate the carbon budget at inappropriate time slots and thus have to set a stringent budget for certain time slots when the workload is high, thereby significantly increasing the delay cost. In addition to cost saving, COCA also satisfies the desired carbon neutrality constraint better, as shown in Fig. 3b.

6 Extension to Geographic Load Balancing

Large cloud providers have become increasingly interested in geo-distributed data centers, as geographical load balancing helps reduce the latency as well as guarantee the service availability. Moreover, geo-distributed data centers can potentially provide additional benefits in minimizing energy cost [13, 37]. Now, we extend COCA to incorporate advantages of geo-distributed data centers and show the effect of geographic load balancing on carbon neutrality. The new online resource management algorithm is called COCA-GLB (COCA - Geographic Load Balancing). We first compare COCA-GLB with a benchmark algorithm called COST (that minimizes the cost without considering the carbon neutrality constraint) as well as with *non-GLB* (which does not consider the data center location). Then, we investigate COCA-GLB for three different budget levels of 97, 95 and 93 % of carbon-unaware brown energy consumption indicated by COCA-GLB(H), COCA-GLB(M) and COCA-GLB(L) respectively. By default, COCA-GLB refers to COCA-GLB(M).

Figure 4a shows the workload distribution among the four data centers (Mountain View - CA, Prineville - OR, Dallas - TX, and Forest City - NC) for three different algorithms. We see that COST sends relatively more workloads to the data center in TX and less to the data center in CA than COCA-GLB. It is because the data center in TX has lower electricity price and hence, cost-minimizing COST sends more workload to the data center in TX to take advantage of the lower power cost. On the other hand, to satisfy the carbon neutrality constraint, COCA-GLB sends relatively more workloads to the data center in CA which has a lower *power per unit service capacity* [2], thereby resulting in an overall lower energy consumption. Similar pattern can also be observed in Fig. 4b where lower renewable budget causes COCA-GLB to give more emphasis on energy saving, which in turn increases workloads for the data center in CA.

[2] Lower power per unit service capacity indicates less energy for the same amount of workload served.

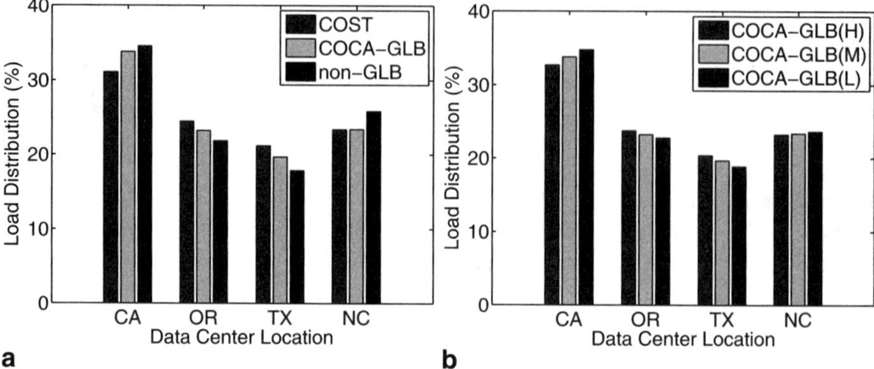

Fig. 4 Workload distribution among data centers. **a** Different algorithms result in different workload distributions due to difference in electricity price. **b** Lower budget for COCA-GLB shifts the workload from data center with lower electricity price (TX) to data center with lower power consumption for unit service capacity (CA)

7 Conclusions

In this chapter, we studied the long-term carbon neutrality and proposed a provably-efficient distributed online algorithm, called COCA, to control the electricity usage for minimizing the data center operational cost while satisfying carbon neutrality. Compared to the existing research that addresses carbon neutrality based on prediction of future information, COCA makes online decisions without requiring long-term future information. Leveraging Lyapunov optimization technique, it was rigorously proved that COCA achieves a close-to-minimum operational cost compared to the optimal offline algorithm with future information, while bounding the potential violation of carbon neutrality constraint. A trace-based simulation study was also presented in the chapter, complementing the analysis and showing that COCA can reduce the average cost compared to the state-of-the-art prediction-based method while resulting in a smaller carbon footprint. Moreover, an online geographic load balancing algorithm, COCA-GLB, was also introduced which dynamically dispatches workloads to distributed data centers for minimizing the operational cost while satisfying carbon neutrality.

References

1. S. Ren and Y. He, "Coca: Online distributed resource management for cost minimization and carbon neutrality in data centers," in *SuperComputing*, 2013.
2. A. Qureshi, R. Weber, H. Balakrishnan, J. Guttag, and B. Maggs "Cutting the electric bill for internet-scale systems," in *SIGCOMM*, 2009.
3. "How clean is your cloud?" Greenpeace, April 2012.

4. K. Le, R. Bianchini, T. D. Nguyen, O. Bilgir, and M. Martonosi "Capping the brown energy consumption of internet services at low cost," in *IGCC*, 2010.
5. C. Ren, D. Wang, B. Urgaonkar, and A. Sivasubramaniam "Carbon-aware energy capacity planning for datacenters," in *MASCOTS*, 2012.
6. N. Deng, C. Stewart, D. Gmach, and M. F. Arlitt "Policy and mechanism for carbon-aware cloud applications," in *NOMS*, 2012.
7. "How dirty is your data? a look at the energy choices that power cloud computing," Greenpeace, 2011.
8. A. Qureshi "Power-demand routing in massive geo-distributed systems," Ph.D. dissertation, MIT, 2010.
9. "Electricity from coal, http://www.powerscorecard.org."
10. "Google. google's green ppas: What, how, and why." April 2011.
11. "Microsoft. becoming carbon neutral: How microsoft is striving to become leaner, greener, and more accountable."
12. Z. Liu, M. Lin, A. Wierman, S. H. Low, and L. L. Andrew "Greening geographical load balancing," in *SIGMETRICS*, 2011.
13. N. Buchbinder, N. Jain, and I. Menache "Online job migration for reducing the electricity bill in the cloud," in *IFIP Networking*, 2011.
14. Z. Liu, M. Lin, A. Wierman, S. H. Low, and L. L. Andrew "Geographical load balancing with renewables," *SIGMETRICS Perform. Eval. Rev.*, vol. 39, no. 3, pp. 62–66, Dec. 2011.
15. M. Lin, A. Wierman, L. L. H. Andrew, and E. Thereska "Dynamic right-sizing for power-proportional data centers," in *IEEE Infocom*, 2011.
16. "Facebook statement: Facebook and greenpeace collaboration on clean and renewable energy, http://www.greenpeace.org."
17. Google, "Google's green ppas: What, how, and why."
18. Microsoft, "Becoming carbon neutral: How microsoft is striving to become leaner, greener, and more accountable."
19. H. Lim, A. Kansal, and J. Liu "Power budgeting for virtualized data centers," in *USENIX ATC*, 2011.
20. A. Gandhi, M. Harchol-Balter, R. Das, and C. Lefurgy "Optimal power allocation in server farms," in *SIGMETRICS*, 2009.
21. D. Meisner, B. T. Gold, and T. F. Wenisch "The powernap server architecture," *ACM Trans. Comput. Syst.*, vol. 29, no. 1, pp. 3:1–3:24, Feb. 2011.
22. Y. Chen, A. Das, W. Qin, A. Sivasubramaniam, Q. Wang, and N. Gautam "Managing server energy and operational costs in hosting centers," *SIGMETRICS Perform. Eval. Rev.*, vol. 33, no. 1, pp. 303–314, Jun. 2005.
23. R. Urgaonkar, U. Kozat, K. Igarashi, and M. Neely "Dynamic resource allocation and power management in virtualized data centers," in *NOMS*, 2010.
24. B. Guenter, N. Jain, and C. Williams "Managing cost, performance and reliability tradeoffs for energy-aware server provisioning," in *IEEE Infocom*, 2011.
25. T. Lu, M. chen, and L. Andrew "Simple and effective dynamic provisioning for power-proportional data centers," *IEEE Transactions on Parallel and Distributed Systems*, vol. 24, no. 6, pp. 1161–1171, 2013.
26. R. Urgaonkar, B. Urgaonkar, M. J. Neely, and A. Sivasubramaniam "Optimal power cost management using stored energy in data centers," in *SIGMETRICS*, 2011.
27. D. Wang, C. Ren, A. Sivasubramaniam, B. Urgaonkar, and H. K. Fathy "Energy storage in datacenters: What, where and how much?" in *SIGMETRICS*, 2012.
28. V. Kontorinis, L. Zhang, B. Aksanli, J. Sampson, H. Homayoun, E. Pettis, D. Tullsen, and T. Simunic Rosing "Managing distributed ups energy for effective power capping in data centers," in *ISCA*, 2012.
29. Y. Guo and Y. Fang, "Electricity cost saving strategy in data centers by using energy storage," *IEEE Transactions on Parallel and Distributed Systems*, vol. 24, no. 6, pp. 1149–1160, 2013.
30. S. Govindan, D. Wang, A. Sivasubramaniam, and B. Urgaonkar "Leveraging stored energy for handling power emergencies in aggressively provisioned datacenters," in *ASPLOS*, 2012.

31. S. Govindan, A. Sivasubramaniam, and B. Urgaonkar "Benefits and limitations of tapping into stored energy for datacenters," *SIGARCH Comput. Archit. News*, vol. 39, no. 3, pp. 341–352, Jun. 2011.
32. W. Deng, F. Liu, H. Jin, C. Wu, and X. Liu "Multigreen: cost-minimizing multi-source datacenter power supply with online control," in *e-Energy*, 2013.
33. J. Tu, L. Lu, M. Chen, and R. K. Sitaraman "Dynamic provisioning in next-generation data centers with on-site power production," in *e-Energy*, 2013.
34. M. Lin, Z. Liu, A. Wierman, and L. L. H. Andrew "Online algorithms for geographical load balancing," in *IGCC*, 2012.
35. N. U. Prabhu, *Foundations of Queueing Theory*. Kluwer Academic Publishers, 1997.
36. S. Ren, Y. He, and F. Xu "Provably-efficient job scheduling for energy and fairness in geographically distributed data centers," in *ICDCS*, 2012.
37. L. Rao, X. Liu, L. Xie, and W. Liu "Reducing electricity cost: Optimization of distributed internet data centers in a multi-electricity-market environment," in *IEEE Infocom*, 2010.
38. M. A. Adnan, R. Sugihara, and R. K. Gupta "Energy efficient geographical load balancing via dynamic deferral of workload," in *Cloud*, 2012.
39. P. X. Gao, A. R. Curtis, B. Wong, and S. Keshav "It's not easy being green," *SIGCOMM Comput. Commun. Rev.*, vol. 42, no. 4, pp. 211–222, Aug. 2012.
40. I. Goiri, K. Le, M. Haque, R. Beauchea, T. Nguyen, J. Guitart, J. Torres, and R. Bianchini "Greenslot: Scheduling energy consumption in green datacenters," in *Super Computing*, 2011.
41. A. Krioukov, C. Goebel, S. Alspaugh, Y. Chen, D. E. Culler, and R. H. Katz "Integrating renewable energy using data analytics systems: Challenges and opportunities." *IEEE Data Eng. Bull.*, vol. 34, no. 1, pp. 3–11, 2011.
42. C. Li, A. Qouneh, and T. Li "iswitch: coordinating and optimizing renewable energy powered server clusters," *SIGARCH Comput. Archit. News*, vol. 40, no. 3, Jun. 2012.
43. Y. Zhang, Y. Wang, and X. Wang "Greenware: greening cloud-scale data centers to maximize the use of renewable energy," in *Middleware*, 2011.
44. C. Chen, B. He, X. Tang, C. Chen, and Y. Liu "Green databases through integration of renewable energy." in *CIDR*, 2013.
45. M. Harchol-Balter "The effect of heavy-tailed job size distributions on computer system design," in *Applications of Heavy Tailed Distributions in Economics*, 1999.
46. P. Costello and R. Rathi, "Data center energy efficiency, renewable energy and carbon offset investment best practices," RealEnergyWriters.com, January 2012.
47. J. Mogul "Improving energy efficiency for networked applications," *ANCS*, 2007.
48. Google, "http://www.google.com/green/bigpicture."
49. N. Deng, C. Stewart, D. Gmach, M. Arlitt, and J. Kelley "Adaptive green hosting," in *ICAC*, 2012.
50. Y. Zhang, Y. Wang, and X. Wang "Electricity bill capping for cloud-scale data centers that impact the power markets," in *ICPP*, 2012.
51. A. Gandhi, M. Harchol-Balter, R. Raghunathan, and M. A. Kozuch "Autoscale: Dynamic, robust capacity management for multi-tier data centers," *ACM Trans. Comput. Syst.*, vol. 30, no. 4, pp. 14:1–14:26, Nov. 2012.
52. M. J. Neely, *Stochastic Network Optimization with Application to Communication and Queueing Systems*. Morgan & Claypool, 2010.
53. S. Boyd and L. Vandenberghe, *Convex Optimization*. Cambridge University Press, 2004.
54. C. Robert and G. Casella, *Monte Carlo Statistical Methods*. New York: Springer-Verlag, 2004.
55. R. Ge, X. Feng, S. Song, H.-C. Chang, D. Li, and K. W. Cameron "Powerpack: Energy profiling and analysis of high-performance systems and applications," *IEEE Trans. Parallel and Dist. Systems*, vol. 21, no. 5, pp. 658–671, May 2010.
56. "California ISO, `http://www.caiso.com/`."
57. Z. Liu, Y. Chen, C. Bash, A. Wierman, D. Gmach, Z. Wang, M. Marwah, and C. Hyser "Renewable and cooling aware workload management for sustainable data centers," in *SIGMETRICS*, 2012.

A Big Picture of Integrity Verification of Big Data in Cloud Computing

Chang Liu, Rajiv Ranjan, Xuyun Zhang, Chi Yang and Jinjun Chen

1 Introduction

Big data is attracting more and more interests from numerous industries. A few examples are oil and gas mining, scientific research (biology, chemistry, physics), online social networks (Twitter, Facebook), multimedia data, and business transactions. With mountains of data collected from increasingly efficient data collecting devices as well as stored on fast-growing storage hardware, people are keen to find solutions to store and process the data more efficiently, and to discover more values from the mass at the same time. When referring to big data research problems, people often brings the 4 v's—volume, velocity, variety, and value. These pose various brand-new challenges to computer scientists nowadays.

The recently emerged cloud computing, known to be the latest development in data center technology, parallel distributed systems and service computing, is widely considered as the most promising technological backbone for solving big data problems [1]. The pay-as-you-go payment model of cloud can cut into the investments by enabling zero expense in setting up and maintaining of expensive computational and storage hardware, as well as provide on-the-fly problem solving. The services

C. Liu (✉) · X. Zhang · C. Yang · J. Chen
Faculty of Engineering and IT, University of Technology Sydney, Sydney, Australia
e-mail: changliu.it@gmail.com

X. Zhang
e-mail: xyzhanggz@gmail.com

C. Yang
e-mail: chiyangit@gmail.com

J. Chen
e-mail: jinjun.chen@gmail.com

R. Ranjan
Computational Informatics, CSIRO, Marsfield, Australia
e-mail: rranjans@gmail.com

S. U. Khan, A. Y. Zomaya (eds.), *Handbook on Data Centers*,
DOI 10.1007/978-1-4939-2092-1_21

cloud can provide, ranging from SaaS (Software-as-a-Service), PaaS (Platform-as-a-Service), and IaaS (Infrastructure-as-a-Service), can offer solutions for big data problems from any level. Cloud also offers elasticity and scalability which can result in further saving of costs in many practical applications involving fast-updating dynamic data. To date, large amounts of business data of numerous big companies have been moved into and managed by clouds such as Amazon AWS, IBM SmartCloud and Microsoft Azure.

Despite those stand-out advantages of cloud, there are still strong concerns regarding service qualities, especially data security. In fact, data security has been frequently raised as one of the top concerns in using cloud. In this new model, user datasets are entirely outsourced to the cloud service provider (CSP), which means they are no longer stored and managed locally. As CSPs cannot be deemed completely trusted, this fact brings several new issues. To name a few, first, when applied in cloud environments, many traditional security approaches will stop being either effective or efficient especially when handling big data tasks. Second, not only CSPs need to deploy their own security mechanisms (mostly conventional), but the clients also need their own verification mechanisms, no matter how secure the server-side security mechanisms claimed to be; the verifications may not bring additional security risks and must be efficient in computation, communication and storage in order to work in correlation with cloud and big data. Third, as the storage server is only semi-trusted, the client may be deceived by deliberately manipulated responses. All these new requirements have made the problem very challenging and therefore started to attract computer science researchers' interest in recent years.

Main dimensions in data security include confidentiality, integrity and availability. In this chapter, we will focus on data integrity. Integrity verification and protection is an active research area; numerous research problems belong to this area have been studied intensively in the past. As a result, the integrity of data storage can now be effectively verified in traditional systems through the deployments of Reed-Solomon code, checksums, trapdoor hash functions, message authentication code (MAC), digital signatures, and so on. However, as stated above, the data owner (cloud user) still needs a method to verify their data stored remotely on a semi-trusted cloud server, no matter how secure the cloud claim to be. A straightforward approach is to retrieve and download from the server all the data the client wanted to verify. Unfortunately, when data size is large, it is very inefficient in the sense of both time consumption and communication overheads. To address this problem, scientists are developing schemes mainly based on traditional digital signatures to help users verify the integrity of their data without having to retrieve them, which they term as provable data possession (PDP) or proofs of retrievability (POR). In this book chapter, we will provide an analysis to some latest research on this problem, as well as providing a look into the future, to eventually form a big picture for this research topic.

The rest of this chapter is organized as follows. Section 2 gives some motivating examples regarding security and privacy in big data application and cloud computing. Section 3 analyzes the research problem and propose a lifecycle of integrity verification over big data in cloud computing. Section 4 shows some representative

Table 1 Acronyms/abbreviations

AAI	Auxiliary authentication information
BLS	Boneh-Lynn-Shacham signature scheme
CSS	Cloud storage server
HLA	Homomorphic linear authenticator
HVT	Homomorphic verifiable tag
MAC	Message authentication code
MHT	Merkle hash tree
PDP	Provable data possession
POR	Proof of retreivability
TPA	Third-party auditor

approaches and their analyses. Section 5 provides a brief overview of other schemes in the field. Section 6 provides conclusions and points out future work.

For the convenience of readers, we list some frequently-used acronyms in Table 1.

2 Motivating Examples

Big data and cloud computing is receiving more and more interest from both industry and academia nowadays. They have been recently listed as important strategies by Australian Government [2, 3]. To address big data problems, cloud computing is believed to be the most potent platform. In Australia, big companies such as Vodafone Mobile and News Corporation are already moving their business data and its processing tasks to Amazon cloud—Amazon Web Services (AWS) [4]. Email systems of many Australian universities are using cloud as the backbone. To tackle the large amount of data in scientific applications, CERN, for example, is already putting the processing of petabytes of data into cloud computing [5]. There has also been a lot of research regarding scientific cloud computing, such as in [6–8]. For big data applications within cloud computing, data security is a problem that should always be properly addressed. In fact, data security is one of the biggest reasons why people are reluctant in using cloud [9–11]. Therefore, more effective and efficient security mechanisms are direly in need to help people establish their confidence in all-around cloud usage.

Data integrity is always an important part in data security, and there is no exception for cloud data [12]. As stated in Sect. 1, client-side verification is as important as server-side protection. As data in most big data applications are dynamic in nature, we will focus on verification of dynamic data. A large proportion of the updates are very small but very frequent. For example, in 2010 Twitter is producing every day up to 12 terabytes of data, composed of tweets with a size of 140 characters maximum [13]. Business transactions and loggings are also good examples. The dataset in these big data applications are very large in size and requires heavy-scale processing

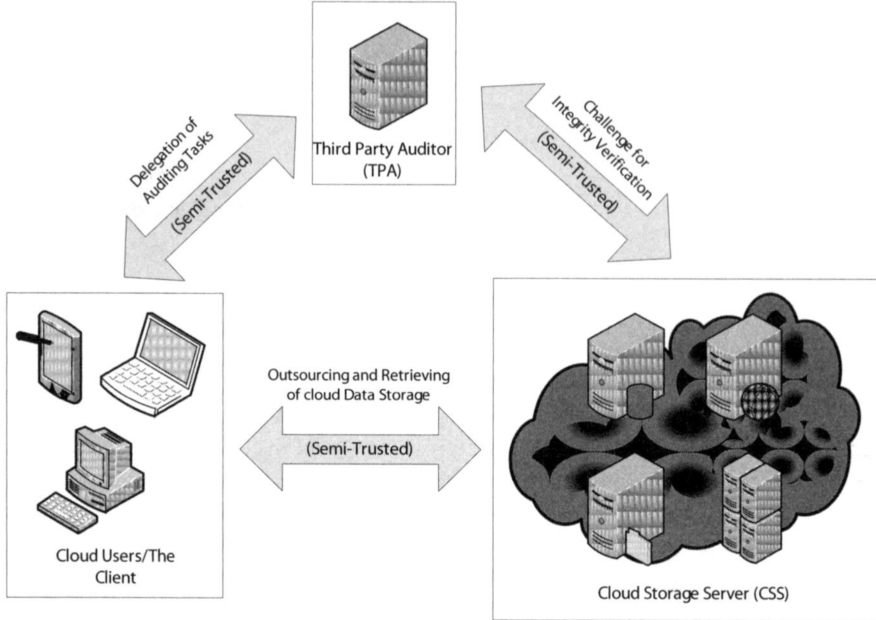

Fig. 1 Relations between the participating parties

capabilities. Therefore, the requirements are not only in security, but also in good efficiency.

3 Problem Analysis—Framework and Lifecycle

There are 3 participating parties in an integrity verification scheme: client, CSS and TPA. The client stores her data on CSS, while TPA's objective is to verify the integrity of client's data stored on CSS. Although the three forms a robust and efficient triangle, each of the two parties are only semi-trusted by each other as shown in Fig. 1. New security exploits may appear while verifying data integrity, which is why we need a good framework to address this problem systematically. The main lifecycle of a remote integrity verification scheme with support for dynamic data updates can be analyzed in the following steps:

Setup and data upload → Authorization for TPA → Challenge for integrity proof → Proof verification → Updated data upload → Updated metadata upload → Verification of updated data

The relationship and order of these steps are illustrated in Fig. 2. We now analyze in detail how these steps work and why they are essential to integrity verification of cloud data storage.

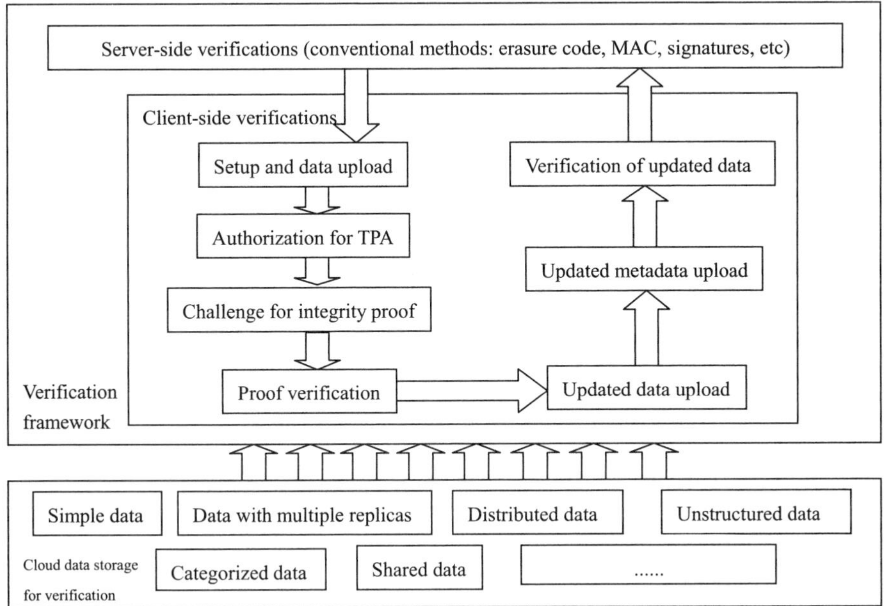

Fig. 2 A brief overview of integrity verification over big data in cloud computing—lifecycle and research topics

Setup and data upload: In cloud, user data is stored remotely on CSS. In order to verify the data without retrieving them, the client will need to prepare verification metadata, namely homomorphic linear authenticator (HLA) or homomorphic verifiable tag (HVT), based on homomorphic signatures [14]. Then, these metadata will be uploaded and stored alongside with the original datasets. These tags are computed from the original data; they must be small in size in comparison to the original dataset for practical use.

Authorization for TPA: This step is not required in a two-party scenario where clients verify their data for themselves, but it is important when users require a semi-trusted TPA to verify the data on their behalf. If a third party can infinitely ask for integrity proofs over a certain piece of data, there will always be security risks in existence such as plaintext extraction.

Challenge and verification of data storage: This step is where the main requirement—integrity verification—to be fulfilled. The client will send a challenge message to the server, and server will compute a response over the pre-stored data (HLA) and the challenge message. The client can then verify the response to find out whether the data is intact. The scheme has public verifiability if this verification can be done without the client's secret key. If the data storage is static, the whole process would have been ended here. However, as discussed earlier, data are always dynamic in many big data contexts (often denoted as velocity, one of the four v's). In these scenarios, we will need the rest of the steps to complete the lifecycle.

Data update: Occurs in dynamic data contexts. The client needs to perform updates to some of the cloud data storage. The updates could be roughly categorized in insert, delete and modification; if the data is stored in blocks with varied size for efficiency reasons, there will be more types of updates to address.

Metadata update: In order to keep the data storage stay verifiable without retrieving all the data stored and/or re-running the entire setup phase, the client will need to update the verification metadata (HLA or HVT's), according with the existing keys.

Verification of updated data: This is also an essential step in dynamic data context. As the CSS is not completely trusted, the client needs to verify the data update process to see if the updating of both user data and verification metadata have been performed successfully in order to ensure the updated data can still be verified correctly in the future.

We will show in the next section how each steps in this lifecycle was developed and evolved by analyzing some representative approaches in this research area.

4 Representative Approaches and Analysis

We first introduce the basic idea behind the designs as well as some common notations. The file m is stored in the form of a number of blocks, denoted as m_i. Each of the block is accompanied with a tag called HLA/HVT denoted as T_i, computed with the client's secret key. Therefore CSS cannot compute T_i (or more frequently denoted as σ_i) from m_i. The client will choose a random set of m_i, send over the coordinates, and ask for proofs. CSS will compute a proof based on the tags T_i according to m_i. Due to homomorphism of the tags, the client will still be able to verify the proof with the same private key used for tag computation.

4.1 Preliminaries

We now introduce some preliminaries laid as foundation stones for our research area. HLA or HVT is evolved from digital signatures; current methods in verifiable updates utilized authenticated data structures. Therefore, we will introduce here two standard signature schemes (RSA and BLS) and one authenticated data structure (MHT) involved in representative approaches.

4.1.1 RSA Signature

The RSA signature is classic and one of the earliest signature schemes under the scope of public-key cryptography. While the textbook version is not semantically secure and not resilient to existential forgery attacks, there is a large body of research work on its improvements later on, and eventually makes it a robust signature scheme. For example, a basic improvement is to use *h(m)* instead of m where h is a one-way hash function.

The setup is based on an integer $N = pq$ where p and q are two large primes, and two integers d and e where $ed = 1 \bmod N$; d is kept as the secret key and e is the public key. The signature σ of a message m is computed as $\sigma = m^d \bmod N$. Along with m, the signature can be verified through verifying whether $m = \sigma^e \bmod N$ holds.

4.1.2 Bilinear Pairing and BLS Signature

BLS signature is proposed by Boneh, Lynn and Shacham [15] in 2004. In addition to the basic soundness of digital signature, this scheme has a greatly reduced signature length, but also increased overheads due to the computationally expensive paring operations.

Assume a group G is a gap Diffie-Hellman (GDH) group with prime order p. A bilinear map is a map constructed as $e : G \times G \rightarrow G_T$, where G_T is a multiplicative cyclic group with prime order[1]. A usable e should have the following properties: bilinearity— $\forall m, n \in G \Rightarrow e(m^a, n^b) = e(m, n)^{ab}$; non-degeneracy— $\forall m \in G, m \neq 0 \Rightarrow e(m, n) \neq 1$; and computability—$e$ should be efficiently computable. For simplicity, we will use this symmetric bilinear map in our scheme description. Alternatively, the more efficient asymmetric bilinear map in the form of $e : G \times G \rightarrow G_T$ may also be applied, as was pointed out in [15].

Based on a bilinear map $e : G \times G \rightarrow G_T$, a basic BLS signature scheme works as follows. Keys are computed as $y = g^x$ where $g \in G, x$ is secret key and $\{g, y\}$ is public key. Signature σ for a message m is computed as $\sigma = (h(m))^x$. People can then verify this signature through verifying whether $e(\sigma, g) = e(h(m), y)$.

4.1.3 Merkle Hash Tree

The Merkle Hash Tree (MHT) [16] is an authenticated data structure which has been intensively studied in the past and later utilized to support verification of dynamic data updates. Similar to a binary tree, each node N will have a maximum of 2 child nodes. Information contained in one node N in a MHT T is H—a hash value. T is constructed as follows. For a leaf node LN based on a message m_i, we have $H = h(m_i), r_{LN} = s_i$; A parent node of $N_1 = \{H_1, r_{N1}\}$ and $N_2 = \{H_2, r_{N2}\}$ is constructed as $N_p = \{h(H_1 \parallel H_2)\}$ where H_1 and H_2 are information contained in N_1 and N_2 respectively. A leaf node m_i's AAI Ω_i is defined as a set of hash values chosen from every of its upper level so that the root value R can be computed through $\{m_i, \Omega_i\}$. For example, for the MHT demonstrated in Fig. 3, m_1's AAI $\Omega_1 = \{h(m_2), h(e), h(b)\}$.

---

[1] For simplicity, we only discuss symmetric pairing here, although specific asymmetric parings could also be applied for better efficiency.

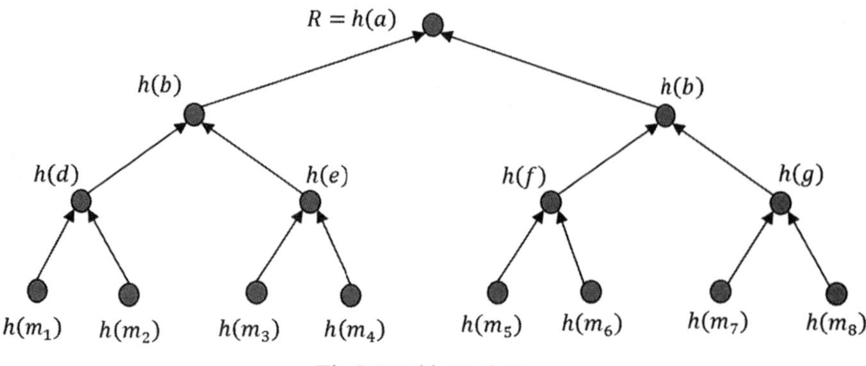

Fig. 3 Merkle hash tree

4.2 Representative Schemes

Now we start to introduce and analyze some representative schemes. Note that all computations are within the cyclic group $\mathbb{Z}_p$ or $\mathbb{Z}_N$.

4.2.1 PDP

Proposed by Ateniese et al. in 2007, PDP (provable data possession) can provide authors with efficient verification over their outsourced data storage [17, 18]. It is the first scheme to provide blockless verification and public verifiability at the same time.

The tag construction is based on RSA signature, therefore all computations are modulo N by default. Let N, e, d be defined as the same as in RSA signature, g is a generator of QR_N, and v is a random secret value; $\{N, g\}$ is the public key and $\{d, v\}$ is the secret key. The tag is computed as $\sigma_i = (h(v \parallel i)g^{mi})^d$. To challenge CSS, the client sends the indices (or, coordinates) of the blocks they want to verify, and correspondingly chooses a set of coefficients a_i, as well as a $g_s = g^s \bmod N$ where s is a random number, and send them to CSS along with the indices. To prove data integrity, CSS will compute $\sigma = \Pi_i \sigma_i{}^{ai}$, along with a value $p = H(\Pi_i, g_s{}^{a_i m_i})$, and send back $\{\sigma, p\}$ as the proof. To verify this proof, the client (or TPA) will compute $\tau = \frac{\sigma^e}{\Pi_i h(v \| i)^{a_i}}$, then verify if $p = H(\tau^s \bmod N)$.

The authors have also proposed a light version called E-PDP, in contrast to the formal S-PDP scheme, for better efficiency. The basic idea is to throw away the coefficients a_i. However, the light version was later proved not secure under the compact POR model. However, as a milestone in this research area, a lot of settings continued to be used by the following work. Mixing in random coefficients is one of the example. Another example is that the paper proposed a probability analysis and found that only a constant small number of blocks are to be verified, if the client needs to have 95 % or even 99 % confidence in that the integrity of the entire file is good. This analysis also became a default setting in the following schemes.

4.2.2 Compact POR

Compact POR is proposed by Shacham et al. in 2008 [19]. Compared to original POR, the authors provided an improved rigorous security proof.

They proposed first a construction for private verification. In this case, data can only be verified with the secret key, therefore no other party can verify it except for the client. The metadata HVT is computed as $\sigma_i = f_k(i) + am_i$, where $f_k()$ is a pseudo-random function (PRF). α and the PRF key k is kept as secret key. When the server is challenged with a set of block coordinates and a set of corresponding public coefficients v_i (same definition as α_i in PDP above), it will compute $\sigma = \sum_i v_i \sigma_i$ and $\mu = \sum_i v_i m_i$ to return $\{\sigma, \mu\}$ as the proof. Upon receiving the proof, the client can simply verify if $\alpha = \alpha\mu + \sum_i (v_i f_k(i))$. The scheme is efficient because it admits short response length and fast computation.

The other construction with public verification is even more impressive compared to schemes at that time. It is the first BLS-based scheme that supports public verification. Due to the shortened length of BLS signature, the proof size is also greatly reduced compared to RSA-based schemes. Similar to BLS signature, the tag construction is based on a bilinear map $e : G \times G \rightarrow G_T$ where G is a group of prime order p. Two generators g and u of $\mathbb{Z}_p$ are chosen to be the public key, as another value $v = g^\alpha$ where α is the secret key for the client. The tag is computed as $\sigma_i = (H(i)u^{mi})^\alpha$, same as the one with private verification, a set of coefficients v_i is also chosen with the designated block coordinates. When challenged, the proof $\{\sigma, \mu\}$ is computed as $\sigma = \Pi_i \sigma_i{}^{v_i}$ and $\mu = \sum_i v_i m_i$. The client can then verify the data integrity through verifying if $e(\sigma, g) = e(\Pi_i (H(i)^{v_i}) \cdot u^\mu, v)$.

Another great contribution of this work is the rigorous security framework it provided. In their model, a verification scheme is secure only when it is secure against an arbitrary adversary with a polynomial extraction algorithm to reveal the message from the integrity proof. To prove the security, they also defined a series of interactive games under the random oracle model. Compared to the previous security frameworks in first PDP and first POR schemes, the adversary defined in this framework is stronger and stateless, and the definition of extraction algorithm (therefore the overall soundness) is stronger. Also, their framework suits perfectly with the public verification, and even multi-replica storage and multi-prover scenarios. To date, this model is considered the strongest and is very frequently used to prove the security of newly-proposed verification algorithms.

4.2.3 DPDP

DPDP (Dynamic PDP), proposed in 2009, is the first integrity verification scheme to support full data dynamics [20]. It is from here that the processes in integrity verification schemes started to form a lifecycle. They utilized another authenticated data structure—rank-based skip list—for verification of updates. A rank-based skip list is similar to MHT in the sense that they will both incur a logarithm amount of operations when an update occurs. All types of updates—insert, delete and modification—are supported for the first time. This design is essentially carried on by all the following schemes with dynamic data support. However, public verifiability was not supported by the scheme, and there was no follow-up work to fill in the blank. Therefore, we will only give a brief introduction here. The readers can refer to the next subsection to see how data dynamics is supported with an authenticated data structure such as MHT.

4.2.4 Public Auditing of Dynamic Data

As the DPDP scheme did not provide support for public verifiability, Wang et al. proposed a new scheme that can support both dynamic data and public verifiability at the same time [21]. They term the latter as 'public auditability', as the verification is often done by a sole-duty third-party auditor (TPA).

A MHT is utilized to verify the updates where the root R is critical authentication information. The tree structure is constructed on blocks, and the structure is stored along with the verification metadata. Compared to compact POR, they compute the tags using $H(m_i)$ instead of $H(i)$ in order to support dynamic data, otherwise all tags of the following blocks must be changed upon each one insert or delete update, which will be very inefficient. Aside from this, the tag construction and verification are similar: $\sigma_i = (H(m_i)u^{m_i})^\alpha$. The proof is also computed as $\sigma = \Pi_i {\sigma_i}^{v_i} \mu = \sum_i v_i m_i$. While the verification is to verify whether $e(\sigma, g) = e(\Pi_i(H(m_i)^{v_i})u^\mu, v)$, TPA will first verify $H(R)$'s signature to ensure the MHT is correct at server side.

To verify data updates, the client will first generate the tag for new block: $\sigma_i' = (H(m_i')u^{m_i'})^\alpha$, then upload it to CSS along with the update request. CSS will update the metadata as requested, and send back R' along with the old block $H(m_i)$, the AAI Ω_i (note Ω_i will stay unchanged if m_i is the only block that has changed) and the client-signed old MHT root $H(R)$. The client can then verify the signed $H(R)$ to ensure CSS has not manipulated it, then it can verify R' with m_i' and Ω_i to see if the update of data and metadata was correct.

There was also a follow-up work to improve this scheme for privacy preserving public auditing [22]. When computing integrity proof, they added a random masking technique to prevent the part of original file being extracted from several integrity proofs over this specific part of data.

4.2.5 Authorized Auditing with Fine-Grained Data Updates

Although the above schemes have already supported dynamic data and public verification/ auditability, they only support insert/delete/modification with blocks with a fixed size, which are later termed as 'coarse-grained updates'. Lack of support of fine-grained updates, i.e., arbitrary-length updates, especially small updates, will cause functionality and efficiency problems. Liu et al. [23] proposed a public auditing scheme with support of fine-grained updates over variable-sized file blocks. In addition, an authentication process between the client and TPA is also proposed to prevent TPA from endless challenges, thereby cut the possibility of attacks over multiple challenges (like the one in [24]) from source.

Similar to previous work, this scheme is also based on BLS signature. Unlike previous schemes which are based on evenly distributed file blocks, here the file blocks are of variable size, with an upper bound of $s_{\max}$ sectors per block. The tag construction is $\sigma_i = (H(m_i)\Pi_{j=1}^{s_i} u_j^{m_{ij}})^{\alpha}$ where $u_j \in U, U = \{u_k \in \mathbb{Z}_p\}, k \in [1, s_{\max}]$ is chosen according to $s_{\max}$. To challenge CSS, TPA must first obtain authorization from client to be eligible for auditing. The client will compute $sig_{AUTH} = Sig_{ssk}(AUTH \parallel t \parallel VID)$, which is a signature with client's secret key where *VID* is the verifier ID and *AUTH* is a message shared secretly earlier between client and CSS. In this case, only the client can generate this signature and only the CSS (other than the client herself) will be able to verify sig_{AUTH}. After CSS has finished verifying sig_{AUTH}, it will compute the proof $P = \{\sigma, \{\mu_k\}_{k\in[1,w]}, \{H(m_i), \Omega_i\}_{i\in I}, sig\}$ where $\sigma = \Pi_i \sigma_i{}^{v_i}$ and $\mu_k = \sum_{i\in I} v_i m_{ik}$, then send P back to TPA. TPA will then verify the proof through verifying whether $e(sig, g) = e(H(R), v)$ and $e(\sigma, g) = e(\omega, \upsilon)$, where $\omega = \Pi_{i\in I} H(m_i)^{v_i} \Pi_{k\in[1,w]} u_k^{\mu_k}$.

For support in fine-grained updates, 5 types of necessary updating operations including $PM, M, D, JandSP$ are analyzed; a theorem was provided to illustrate that all updates can be divided into this 5 basic operations. For more efficient verification of fine-grained updates, a modified verification scheme for PM operations (which was the majority of the operations in many occasions found through analysis) is also provided, where only the modified part of the new block, instead of a whole block, is needed to retrieved and transferred back to the client for tag re-computation. Experimental results have also demonstrated some significant efficiency improvements.

5 Other Related Work

Other than the ones stated in the previous section, a great amount of work has also been proposed in recent years to address the research problem of integrity verification and public auditing of cloud data and other outsourced data storage. The concept of POR is proposed in 2007 by Juels et al. [25], but the security framework was not complete and it only suits for static data storage like library and archives. After PDP, Ateniese et al. also proposed an improvement they call Scalable PDP [26] to support dynamic data verification. Alas, only partial data dynamics is supported, i.e., only limited types of data updates is supported. Therefore, this scheme is not suitable for practical use. Curtmola et al. proposed a verification for multi-replica cloud storage, which is named MR-PDP [27]. This is also a practical solution, because cloud will constantly keep a number of replicas of user data in the aim of availability. Ateniese et al. also proposed a framework to transfer homomorphic identification protocols into integrity verification schemes [28].

There is also some work proposed in the most recent years. Based on previous work and the recent developments of big data and cloud, they can be the more practical solution for specific cloud environments and applications. As mentioned before, [23] is a good example. For big enterprises, data migration is a big problem in the adoption process of cloud, because the different security/control levels in data and the heavy cost in migration itself. Therefore, hybrid cloud has been a more practical solution; enterprises will keep relatively static and security-sensitive data on private cloud, and put all services into the cloud. Zhu et al. proposed a PDP scheme for Hybrid Cloud [29] for verification of data stored in separated domains. As cloud data sharing becomes a hot topic, Wang et al. worked on secure data verification of shared data storage [30] and also with efficient user management [31]. Zhang et al. proposed a scheme with a new data structure called update tree [32]. Without conventional authenticated data structures such as MHT, the proposed scheme has a constant proof size and support fully data dynamics. However, the scheme does not support public verification/auditing at the moment.

6 Conclusions and Future Work

As we can see from the above, the topic of integrity verification of big data in cloud computing is a flourishing area that is attracting more and more research interest and there is still lots of research currently ongoing in this area. Cloud and big data is a fast-developing topic. Therefore, even though existing research has already achieved some amazing goals, we are confident that integrity verification mechanisms will also continue evolving along with the development of cloud and big data applications to meet emerging new requirements and address new security challenges. For future developments, we are particularly interested in looking at the following aspects.

Efficiency Due to high efficiency demands in big data processing overall, efficiency is one of the most important factors in designing of new techniques related to big data and cloud. In integrity verification/ data auditing, the main costs can come from every aspects, including storage, computation, and communication, and they can all affect the total cost-efficiency due to the pay-as-you-go model in cloud computing.

Security Security is always a problem between spear and shield; that is, attack and defend. Although the current formalizations and security model seemed very rigorous and potent, new exploits can always exist, especially with dynamic data streams and varying user groups. Finding the security holes and fixing them can be a long-lasting game.

Scalability/elasticity As the cloud is a parallel distributed computing system in nature, scalability is one of the key factors as well. Programming models for parallel and distributed systems, such as MapReduce, are attracting attentions from a great number of cloud computing researchers. Some of the latest work in integrity verification is already considering how to work well with MapReduce for better parallel processing [29]. On the other hand, elasticity is one of a biggest reason why big companies are moving their business, especially service-related business, to the cloud [4]. User demands vary all the time, and it would be a waste of money to purchase hardware that can handle the demands at peak times. The advent of cloud solved this problem—cloud allows their clients to deploy their applications on a highly elastic platform whose capabilities can be scaled up and down on-the-fly, and the cost is based solely on usage. Therefore, an integrity verification mechanism that has the same level of scalability and elasticity will be highly resourceful for big data applications in a cloud environment.

References

1. Michael Armbrust, Armando Fox, Rean Griffith, Anthony D. Joseph, Randy Katz, Andy Konwinski, Gunho Lee, David Patterson, Ariel Rabkin, Ion Stoica and Matei Zaharia, “A View of Cloud Computing,” *Communications of the ACM*, vol. 53, no. 4, pp. 50–58, 2010.
2. Australian Government Department of Finance and Deregulation, “Big Data Strategy—Issues Paper,” 2013; http://agimo.gov.au/files/2013/03/Big-Data-Strategy-Issues-Paper1.pdf.
3. Australian Government Department of Finance and Deregulation, “Cloud Computing Strategic Direction Paper: Opportunities and Applicability for Use by the Australian Government,” 2011; http://agimo.gov.au/files/2012/04/final_cloud_computing_strategy_version_1.pdf.
4. Customer Presentations on Amazon Summit Australia, Sydney, 2012, available: http://aws.amazon.com/apac/awssummit-au/, accessed on: 25 March, 2013,
5. Nick Heath, “Cern: Cloud Computing Joins Hunt for Origins of the Universe,” 2012, available: http://www.techrepublic.com/blog/european-technology/cern-cloud-computing-joins-hunt-for-origins-of-the-universe/262, accessed on: 25 March, 2013,
6. Christian Vecchiola, Rodrigo N. Calheiros, Dileban Karunamoorthy and Rajkumar Buyya, “Deadline-driven Provisioning of Resources for Scientific Applications in Hybrid Clouds with Aneka,” *Future Generation Computer Systems*, vol. 28, no. 1, pp. 58–65, 2012.
7. Lizhe Wang, Marcel Kunze, Jie Tao and Gregor von Laszewski, “Towards Building A Cloud for Scientific Applications,” *Advances in Engineering Software*, vol. 42, no. 9, pp. 714–722, 2011.
8. Lizhe Wang, Jie Tao, M. Kunze, A.C. Castellanos, D. Kramer and W. Karl, “Scientific Cloud Computing: Early Definition and Experience,” in *Proceedings of the 10th IEEE International Conference on High Performance Computing and Communications (HPCC ’08)* pp. 825–830, 2008.
9. Stephen E. Schmidt, “Security and Privacy in the AWS Cloud,” Presentation on Amazon Summit Australia, 17 May 2012, Sydney, 2012, available: http://aws.amazon.com/apac/awssummit-au/, accessed on: 25 March, 2013,
10. Jinhui Yao, Shiping Chen, Surya Nepal, David Levy and John Zic, “TrustStore: Making Amazon S3 Trustworthy with Services Composition,” in *Proceedings of the 10th IEEE/ACM*

International Conference on Cluster, Cloud and Grid Computing (CCGRID '10), pp. 600–605, 2010.

11. Dimitrios Zissis and Dimitrios Lekkas, "Addressing Cloud Computing Security Issues," *Future Generation Computer Systems*, vol. 28, no. 3, pp. 583–592, 2011.
12. Surya Nepal, Shiping Chen, Jinhui Yao and Danan Thilakanathan, "DIaaS: Data Integrity as a Service in the Cloud," in *Proceedings of the 4th International Conference on Cloud Computing (IEEE CLOUD '11)*, pp. 308–315, 2011.
13. Erica Naone, "What Twitter Learns from All Those Tweets," Technology Review, 28 September, 2010, available: http://www.technologyreview.com/view/420968/what-twitter-learns-from-all-those-tweets/, accessed on: 25 March, 2013,
14. Robert Johnson, David Molnar, Dawn Song and David Wagner, "Homomorphic Signature Schemes," Topics in Cryptology—CT-RSA 2002, Lecture Notes in Computer Science, vol. 2271, pp. 244–262, 2002.
15. Dan Boneh, Hovav Shacham and Ben Lynn, "Short Signatures from the Weil Pairing," *Journal of Cryptology*, vol. 17, no. 4, pp. 297–319, 2004.
16. Ralph C. Merkle, "A Digital Signature Based on a Conventional Encryption Function," in Proceedings of A Conference on the Theory and Applications of Cryptographic Techniques on Advances in Cryptology (CRYPTO '87), pp. 369-378, 1987.
17. Giuseppe Ateniese, Randal Burns Johns, Reza Curtmola, Joseph Herring, Lea Kissner, Zachary Peterson and Dawn Song, "Provable Data Possession at Untrusted Stores," in *Proceedings of the 14th ACM Conference on Computer and Communications Security (CCS '07)*, pp. 598–609 2007.
18. Giuseppe Ateniese, Randal Burns, Reza Curtmola, Joseph Herring, Osama Khan, Lea Kissner, Zachary Peterson and Dawn Song, "Remote Data Checking Using Provable Data Possession," ACM Transactions on Information and System Security, vol. 14, no. 1, pp. Article 12, 2011.
19. Hovav Shacham and Brent Waters, "Compact Proofs of Retrievability," in Proceedings of the 14th International Conference on the Theory and Application of Cryptology and Information Security (ASIACRYPT '08), pp. 90–107 2008.
20. Chris Erway, Alptekin Küpçü, Charalampos Papamanthou and Roberto Tamassia, "Dynamic Provable Data Possession," in *Proceedings of the 16th ACM Conference on Computer and Communications Security (CCS'09)*, pp. 213–222, 2009.
21. Qian Wang, Cong Wang, Kui Ren, Wenjing Lou and Jin Li, "Enabling Public Auditability and Data Dynamics for Storage Security in Cloud Computing," *IEEE Transactions on Parallel and Distributed Systems*, vol. 22, no. 5, pp. 847–859, 2011.
22. Cong Wang, S.M. Chow, Qian, Kui Ren and Wenjing Lou, "Privacy-Preserving Public Auditing for Secure Cloud Storage," IEEE Transactions on Computers, In Press, 2011.
23. Chang Liu, Jinjun Chen, Laurence T. Yang, Xuyun Zhang, Chi Yang, Rajiv Ranjan and Kotagiri Ramamohanarao, "Authorized Public Auditing of Dynamic Big Data Storage on Cloud with Efficient Verifiable Fine-grained Updates," IEEE Transactions on Parallel and Distributed Systems, in press, 2013.
24. Cong Wang, S.M. Chow, Qian, Kui Ren and Wenjing Lou, "Privacy-Preserving Public Auditing for Secure Cloud Storage," *IEEE Transactions on Computers*, vol. 62, no. 2, pp. 362–375, 2013.
25. Ari Juels and Jr. B. S. Kaliski, "PORs: Proofs of Retrievability for Large Files," in *Proceedings of the 14th ACM Conference on Computer and Communications Security (CCS '07)*, pp. 584–597, 2007.
26. Giuseppe Ateniese, Roberto Di Pietro, Luigi V. Mancini and Gene Tsudik, "Scalable and Efficient Provable Data Possession," in *Proceedings of the 4th International Conference on Security and Privacy in Communication Netowrks (SecureComm '08)*, pp. 1–10, 2008.
27. Reza Curtmola, Osama Khan, Randal C. Burns and Giuseppe Ateniese:, "MR-PDP: Multiple-Replica Provable Data Possession.," in *Proceedings of the 28th IEEE International Conference on Distributed Computing Systems (ICDCS '08)*, pp. 411–420, 2008.
28. Giuseppe Ateniese, Seny Kamara and Jonathan Katz, "Proofs of Storage from Homomorphic Identification Protocols," in Proceedings of the 15th International Conference on the Theory

and Application of Cryptology and Information Security (ASIACRYPT '09), pp. 319–333, 2009.

29. Yan Zhu, Hongxin Hu, Gail-Joon Ahn and Mengyang Yu, "Cooperative Provable Data Possession for Integrity Verification in Multi-Cloud Storage," *IEEE Transactions on Parallel and Distributed Systems*, vol. 23, no. 12, pp. 2231–2244, 2012.
30. Boyang Wang, Sherman S.M. Chow, Ming Li and Hui Li, "Storing Shared Data on the Cloud via Security-Mediator," in *33rd IEEE International Conference on Distributed Computing Systems (ICDCS '13)*, 2013.
31. Boyang Wang, Baochun Li and Hui Li, "Public Auditing for Shared Data with Efficient User Revocation in the Cloud," in *Proceedings of the 32nd Annual IEEE International Conference on Computer Communications (INFOCOM'13)*, pp. 2904–2912, 2013.
32. Yihua Zhang and Marina Blanton, "Efficient Dynamic Provable Possession of Remote Data via Update Trees," IACR Cryptology ePrint Archive, Report 2012/291, 2012.

An Out-of-Core Task-based Middleware for Data-Intensive Scientific Computing

Erik Saule, Hasan Metin Aktulga, Chao Yang, Esmond G. Ng and Ümit V. Çatalyürek

1 Introduction

Petascale scientific computing, next-generation telescopes, high-throughput experiments, data-oriented business technologies and the Internet have been driving a rapid growth in data acquisition and generation. Analysis of large-scale datasets is likely to bring new breakthroughs in the academic and industrial world. These analyses typically require the use of large computer systems, such as those that can be found in data centers or high performance computing (HPC) facilities.

While the computing power of large computer systems that can enable timely and scalable data analysis has been increasing steadily for decades, their memory capacities have not been able to keep pace [1], see Fig. 1. As we move towards the future, this gap is anticipated to widen even further. The main reason for this trend is that it is not possible to meet the storage capacity and power consumption requirements of future machines using the DRAM technology. Non-volatile memory (NVM) solutions, on the other hand, feature much higher storage densities and lower

E. Saule (✉)
Department of Computer Science, University of North Carolina at Charlotte,
Charlotte, NC 28223, USA
e-mail: esaule@uncc.edu

H. M. Aktulga · C. Yang · E. G. Ng
Computational Research Division, Lawrence Berkeley National Laboratory,
Berkeley, CA 94720, USA
e-mail: hmaktulga@lbl.gov

C. Yang
e-mail: cyang@lbl.gov

E. G. Ng
e-mail: egng@lbl.gov

Ü. V. Çatalyürek
Department of Biomedical Informatics, The Ohio State University,
Columbus, OH 43210, USA
e-mail: umit@bmi.osu.edu

S. U. Khan, A. Y. Zomaya (eds.), *Handbook on Data Centers*,
DOI 10.1007/978-1-4939-2092-1_22

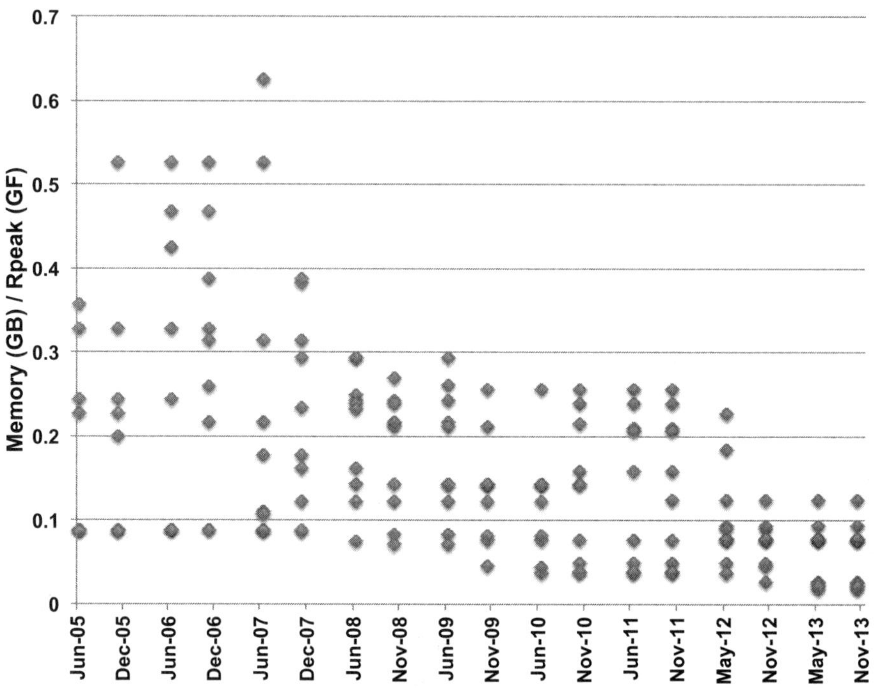

Fig. 1 Memory in gigabytes per gigaflop of computing power for the leading 10 supercomputers on the TOP500 list over the years

power requirements compared to DRAM. Therefore NVM technology will be one of the key enablers for future high end computing architectures.

In datacenters, NVM storages are experiencing a fast adoption rate due to the high bandwidth and low latency advantages that they provide over the traditional disk-based storage systems in the management and analysis of large datasets. Several NVM storage solutions from companies such as Fusion-IO, OCZ Technologies, HP and Seagate already exist in the marketplace. Initially, these NVM storages were used as mere disk replacements and they were connected to the compute resources through low performance interfaces such as SCSI or SATA. Nowadays, we increasingly see high performance NVM storages being connected through the PCI Express bus. As the technology improves, it is anticipated that NVM storages will take their place as a new layer in the memory hierarchy for datacenter systems [2].

NVM storages are already incorporated in today's HPC architectures that are designed to tackle challenging data-intensive problems. For example, the Gordon computer at the San Diego Supercomputing Center (SDSC) houses 300 TBs of flash memory storage in addition to 64 TBs of DRAM space. Trinity (Los Alamos) and NERSC-8 systems, which are planned for operation in 2015, will use flash memory based storages at a much greater scale. Their total flash memory storage capacity is expected to be on the order of 5–10 PBs, which corresponds to about 2–3 times the total DRAM capacity on those systems. An important use case for the flash memory storage on the Trinity and NERSC-8 systems will be to provide a fast workspace for data-intensive applications. As we move towards the exascale era, NVM storages which are currently seen as fast disk alternatives only will be introduced as a new layer into the memory hierarchy in HPC systems as well. Non-Volatile Random Access Memory (NVRAM) is a main component of exascale computer architecture designs by AMD and IBM as part of their efforts in the DOE's Fast Forward program, [3, 4].

The drastic changes in system architecture will require rethinking systems software as well. Specifically, with improvements in hardware performance, software efficiency will become the next bottleneck. Scalable and efficient analytics on large computer systems require advanced parallel programming skills. However, most computational scientists and data scientist are not parallel programming experts. Besides the need for carefully organizing communication and computations in large scale applications, the need to manage data stored on NVM devices emerges in current architectures designed for data-intensive computing. This adds considerable complexity in code design and development.

Our vision is to increase the programmer productivity while still ensuring good performance and scalability by enabling the separation of computation and data movement. In our approach, the programmer can focus on the computational operations that he/she wants to apply to the sets of data and delegates the chore of data movement to the task-based data-flow middleware, DOoC (**D**istributed **O**ut-**o**f-**C**ore), that we describe in this chapter. DOoC is a runtime environment that determines and executes optimal data movement policies for systems with deep memory/storage hierarchies. Conceptually, in DOoC the entire computation is represented as a Directed Acyclic Graph (DAG), where an operation on a dataset corresponds to a vertex, the input data for the computational task is represented as an incoming edge to that vertex and the resulting data is represented as an outgoing edge of the vertex. Our runtime environment carefully considers the characteristics of the underlying memory/storage subsystem and the needs of the data-intensive applications that it supports to enable efficient execution of large-scale computations. The overall goal of our work is to provide an easy-to-use high-level application interface for data-intensive workloads, while providing efficient and scalable execution by orchestrating pipelined execution of computation, communication and I/O.

We have designed and implemented DOoC to be a generic middleware that can be used in a wide spectrum of applications in fields as diverse as graph mining, bioinformatics and scientific computing. A customizable frontend allows the application developer to interact with the DOoC framework through a simple programming interface. In this chapter, after giving an overview of the DOoC framework, we introduce the Linear Algebra Frontend (LAF) which is developed to enable the implementation of iterative numerical methods using DOoC. We present a case study on the implementation of a block eigensolver for the solution of large-scale eigenvalue problems arising in nuclear structure computations. We give detailed performance and scalability analysis for the resulting distributed out-of-core eigensolver on an experimental testbed equipped with NVM storages. We conclude our chapter with a discussion on the future work planned.

2 Related Work

One can draw similarities between our approach and other approaches that use directed acyclic graphs (DAG) to model computational dependencies. In the classical DAG scheduling [5], the complete task graph is generated before scheduling. However, in our system the task graph is generated dynamically on-the-fly. Two other middlewares are similar to our effort: StarPU [6] and PaRSEC [7]. They both have been recently used for sparse linear algebra [8].

StarPU [6] is a task-based middleware like DOoC. It has been used for both dense and sparse linear algebra. It is designed to take advantage of multicore systems with accelerators and has been ported to support multiple architectures such as CUDA devices, OpenCL devices, the IBM Cell processor and multicore CPUs. StarPU has no support for out-of-core processing. It also allows multiple copies of a data item to exist on multiple devices as long as they are identical copies. Once a modification is made on one copy, the other existing copies must be deallocated. Two recent developments in StarPU are the composability of StarPU applications [9] and the support for distributed memory computing using MPI [10].

PaRSEC (previously known as DAGuE [7]) has originally been designed for in-core, dense linear algebra computations. Recently, it has been used to perform sparse linear algebra operations [8]. It supports both accelerators and distributed memory computing. The highlight of PaRSEC is the use of Parametrized Task Graph [11] to store the task graph in a compact form to reduce the scheduling overheads and synchronizations [12].

Out-of-core algorithms for sparse numerical linear algebra applications involving large matrices have been an attractive research topic, especially back in the 90's. Toledo gives an excellent survey of such algorithms [13]. More recently, out-of-core direct solvers on a single node have been investigated for symmetric [14, 15] or asymmetric matrices [16, 17]. A parallel (but still single node) out-of-core multi-frontal method has recently been developed [18] and recently improved to reduce the amount of I/O transfers [19]. Distributed out-of-core computations were considered to compute the steady state of Markov chains using Jacobi or Conjugate Gradient algorithms [20]. Also approximations to compute the Page Rank of a graph accessed from the disk has recently been proposed [21].

Another related area of work is the field of memory aware scheduling algorithms. Out-of-core computing relies on reusing available data as much as possible and minimizing the amount of data to transfer from the disk to perform the computation. Many works in scheduling are applicable to out-of-core algorithms. [22] studies the problem of scheduling independent tasks and DAGs onto a cluster to minimize both the makespan of the application and the memory consumption of the node with the most used memory. In this model, the assumption is that once memory is used it is never freed. This assumption can model either the cost of a reading from the disk or the space used on the disk by the tasks. This model is extended in [23] in the context of load balancing for file servers where the author investigates the use of replication of data items and their reallocation to better take the change in the load into account.

The previously described model uses memory as an abstract concept. Some other models attach actual piece of data to the computations and focus on assigning the data to a compute node in order to minimize the cost of off-node data accesses [24].

Other related scheduling problems are concerned with the execution of a task graph under memory pressure where data is deallocated once it is no longer used and the goal of the scheduler is to execute the application using the least amount of memory. This problem has historically been solved to schedule the execution of binary arithmetic trees in compilers with unitary space cost to minimize the amount of used registers [25]. Most of the work in the area is concerned with trees since it has been shown that the problem is NP-Complete on DAGs [26]. The problem of scheduling non-binary in-trees with arbitrary cost has been solved in polynomial time [27]. There also have been interest in the case where multiple chains need to be computed and a cache is available to store the result of some tasks removing the need to compute it. Unfortunately, this problem has also been proved NP-Complete, but some polynomial time approximation algorithms have been proposed for it [28].

Most of the work on memory pressured scheduling only consider the problem of minimizing the memory requirement in a sequential setting. But if the problem can not be solved in memory, then it becomes important to try to minimize the amount of I/O performed to compute the final solution. This problem is shown to be NP-Complete and heuristics have been proposed and tested on instances coming from multifrontal methods [29]. Also, the trade-off of memory and execution time of the execution of an in-tree on a parallel machine have recently been investigated in [30].

During the last decade there has been little interest in distributed memory out-of-core numerical linear algebra algorithms. We argue that the main reason has been the poor performance of these algorithms due to the high latency and low bandwidth associated with traditional disk-based storage systems. At this point, the emergence of clusters equipped with non-volatile NAND-flash memory based solid state drives (SSD) presents unique opportunities and this is exactly what we explore in this chapter.

3 An Out-of-Core Task-based Middleware

DOoC (**D**istributed **O**ut-**o**f-**C**ore) is a recently developed generalized middleware for distributed out-of-core computation and data analysis [31]. DOoC runs on top of DataCutter [32], which itself is a distributed, coarse-grain data-flow middleware. We have built our framework on top of DataCutter instead of directly implementing using MPI (or any other low-level library that enables distributed-memory programming), because the programing model of DataCutter naturally enables the separation of the computations from the data movements and provides an efficient runtime system that orchestrates pipelined executions with computation and communication overlapping.

Figure 2 depicts the architectural overview of our proposed framework, which is composed of DOoC and LAF (**L**inear **A**lgebra **F**rontend) [33]. DOoC provides efficient execution of task graphs with given input and output data dependencies. In

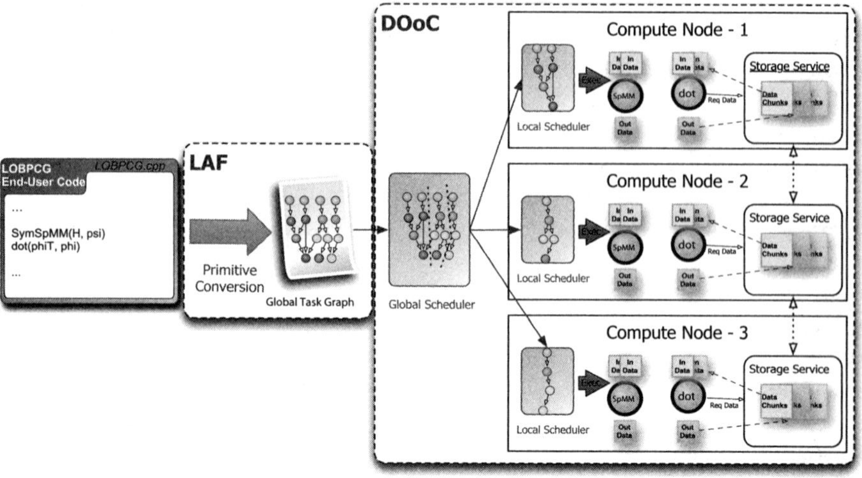

Fig. 2 Schematic overview of our framework Distributed Out-of-Core (DOoC) with Linear Algebra Frontend (LAF)

DOoC, task graphs and task codes need to be generated manually by the application developer. Since our focus in this chapter is on iterative eigensolvers for large-scale sparse matrices, we have designed and developed LAF, which we describe in more detail in Sect. 4. LAF customizes our framework for linear algebra computations by providing a high-level interface to application developers. It acts as a frontend that translates basic linear algebra primitives into global task graphs that can be executed by DOoC.

DOoC is composed of two parts: (i) a hierarchical scheduler responsible for ordering and triggering the execution of tasks, and (ii) a storage service responsible for managing the memory as a resource and handling transfers of data, which is either the input for local computational tasks or the output of them. Data transfer in the context of a distributed out-of-core computation involves reading from or writing to the permanent storage system, or communicating with other compute nodes.

In DataCutter, which serves as the distributed data storage layer for DOoC, the *immutable object* paradigm is adopted. In immutable object paradigm, a given memory location can only be written once and can not be read before being written. This removes race conditions and the need for distributed memory coherency protocols (which are major concerns in similar systems with mutable objects such as Global Array [34]).

Below, we describe each component of the proposed framework in more detail.

3.1 Global and Local Schedulers

Within the scheduler, the application is represented as a set of tasks. Each task is annotated with the set of data it needs (input data) and the set of data it generates (output data). These annotations are used to generate a partial ordering between the tasks (such as the one presented in Fig. 2). An efficient partial ordering is achieved by the use of hash tables, where for each data the mapping of which tasks use it as input and which tasks produce it as output is kept.

Each individual task is sequentially executed on a single computing node. The tasks are created on the global scheduler. The global scheduler is responsible for assigning these tasks to the local schedulers on compute nodes for processing, as well as tracking the completions of those tasks. It assigns a task to a local scheduler only when all the input data of the task have been generated or will be generated as a result of executing the tasks already assigned to that particular local scheduler. Among all the compute nodes, the global scheduler allocates a task onto the node where most of the input data is already located at. This is a heuristic aimed at minimizing the data movement required for starting to process tasks. Alternatively, a task assignment can be forced to a different node by the application programmer, too.

The local scheduler obtains regularly (default every 100 ms) from the storage service (which we describe in the next subsection) the list of data that is available on the local memory. Based on this information, the local scheduler decides which tasks among those assigned to itself are ready for execution. The scheduler triggers the execution of a ready task as soon as a computation thread becomes idle. There are as many computation threads as the number of cores on a compute node. The output data from executing a task, which will serve as the input data for a subsequent task, resides in the compute node's memory until it is consumed.

Another key responsibility of the local scheduler is to enable the pipelined execution of computation, communication and I/O. It achieves this by sending prefetching requests to the storage service. The local scheduler first queries the storage service to learn the amount of memory space available for prefetching. As long as there is space available and there are tasks that are waiting for input data to be executed, the local scheduler determines the data to be prefetched by using the greedy algorithm presented in Algorithm 1 to order tasks.

This greedy algorithm orders the tasks in the local scheduler's list based on the amount of additional input data that needs to be brought into the local memory to make each task ready for execution. The task which requires the least amount of additional input data is ordered first, and the prefetching requests for its input data are issued. Those input data are added to the list of available data, and the algorithm continues to determine the next task for prefetching. Note that, data will be actually available after it has been prefetched by the storage service. Prefetching is paused when there is no more memory space available. The prefetched data is consumed when ready tasks are executed. As soon as enough memory space becomes available, prefetching is reinstantiated.

Algorithm 1: Task Ordering Algorithm.

```
AVAILDATA ← storage().getAvailData()
AVAILMEM ← storage().getAvailMem()
TASKS ← global().getSchedulableTasks()
OUTOFMEM ← False
PREFETCHLIST ← ∅
while not OUTOFMEM and not TASK.empty() do
    for t ∈ TASKS do
        TOFETCH ← input_data(t) - AVAILDATA
        cost_t ← Σ_{d∈ToFetch} size_of (d)
    end
    t* = argmin_{t∈Tasks} cost_t
    TOADD ← input_data(t*) - PREFETCHLIST
    for d ∈ TOADD do
        if size_of (d) > AVAILMEM then
            OUTOFMEM ← True
        end
        else
            PREFETCHLIST ← PREFETCHLIST ∪ {d}
            AVAILDATA ← AVAILDATA ∪ {d}
            AVAILMEM ← AVAILMEM - size_of(d)
        end
    end
    TASKS ← TASKS − {t*}
end
return PREFETCHLIST
```

3.2 Storage Service

The storage service is responsible for managing the local memory, managing the data transfer to/from the permanent storage system and handling the communication between compute nodes.

When the storage service starts, it queries the permanent storage system through its file system and makes a list of the data stored there. This information is reported to the global scheduler. In addition, the storage service provides functions to declare new data objects and to destroy ones that are no longer necessary. In DOoC, declaring a new data object does not actually induce memory allocation, it just induces the creation of appropriate meta-data. The memory allocation is done when the newly created data object is accessed for the first time.

The way DOoC handles an access to a data object differs based on whether it is a read access or a write access. In a read access, if the data object is currently not in that node's local memory, it may be stored either on the permanent storage system or on the memory of another node. If the data is stored on permanent storage system, it is simply read from there. Otherwise, it needs to be communicated from the hosting node. The storage service randomly queries other nodes until it locates the one where the data object is stored. Once the data is located, a *hint* is created to speedup the querying process in subsequent accesses to the same data object.

Write access to a data object is only possible if the data object resides in local memory. Notice that because the data objects in DOoC are immutable, they are only written once. Therefore there is no need for a complex coherency protocol. All data access operations are performed asynchronously to be able to process multiple requests simultaneously. However, after a certain number of simultaneous requests (default: 50) within a node, subsequent ones are queued.

A deallocation procedure is triggered when there is no more memory available on a compute node. The input data that are necessary for executing the tasks currently scheduled on the cores of that compute node, as well as any data object that cannot be reobtained are excluded from deallocation. A data object cannot be reobtained, if it was created on the node itself. Such data objects must be kept until they are written to the permanent storage system or they are explicitly deallocated by the application programmer. On the other hand, a data object can be reobtained if it was read from the permanent storage system, or communicated from another node. Such data objects are eligible for deallocation along with remaining data objects that do not fit into any of the categories above. The storage service frees data objects eligible for deallocation according to the Least Recently Used policy.

4 Linear Algebra Frontend (LAF)

The Linear Algebra Frontend (LAF) is a C++ library which works with objects of different data types including dense and sparse matrices, (dense) vectors, and scalars. Objects are persistent, and can be partitioned into chunks and distributed in the system. Each object is identified by a string that gives it a unique name. Each object is considered immutable, similar to objects in functional programming. Hence it is generated once and is never overwritten. New objects can be generated from the stored data, and also as a result of computation using provided primitives.

When an object is no longer needed, the associated memory needs to be deallocated within the system. This is triggered upon the destruction of the object in the frontend which can be explicit or automatic when the program exits the scope an object was declared in.

Currently supported primitives are listed in Table 1. Although not comprehensive, these operations are sufficient to implement various numerical methods for the solution of linear systems or eigenvalue problems that are widely used in scientific computing. The Conjugate Gradients, LOBPCG, Lanczos, and Page-Rank algorithms are among the examples that can be implemented using the primitives that currently exist in LAF.

Some of these primitives (such as dot product, MM and MV) require a reduction phase when the data are partitioned into multiple chunks. The reduction operation can be implemented using a static reduction tree. Since the summation operation required for these reductions are commutative, it does not matter in which order the different chunks are added up. So the reduction is first performed locally on each node and then globally on the destination node to reduce communication overheads. In order to prevent the accumulation of intermediate results on a node (which may be very costly in terms of memory space), local reduction tasks are implemented to listen on scheduling events. When the number of intermediate results associated

Table 1 Primitives that are currently available in LAF. A, B and C are matrices, y, x and w are vectors, and a and b are scalars

Primitives	Operation
Primitives that creates Matrix	
MM, (Sym)SpMM	$C = AB$
addM	$C = A + B$
axpyM	$C = aA + b$
randomM	$C = random()$
Primitives that creates Vector	
MV, (Sym)SpMV	$y = Ax$
addV	$y = x + w$
axpyV	$y = ax + b$
Primitives that creates scalar	
dot	$a =< x, y >$

with a reduction operation reaches a threshold (default: 5), a local reduction task is dynamically created.

5 A Case Study: Block Iterative Eigensolver Using DOoC+LAF

In this section, we present a case study using our DOoC+LAF framework. We give the implementation details of a block eigensolver for the solution of large-scale eigenvalue problems arising in nuclear structure computations.

5.1 *Eigenvalue Problem in the Configuration Interaction Approach*

The eigenvalue problem arises in nuclear structure calculations because the nuclear wave functions Ψ are solutions of the many-body Schrödinger's equation:

$$H\psi = E\psi \tag{1}$$

$$H = \sum_{i<j} \frac{(p_i - p_j)^2}{2mA} + \sum_{i<j} V_{ij} + \sum_{i<j<k} V_{ijk} + \ldots \tag{2}$$

In the Configuration Interaction (CI) approach, both the wave functions ψ and the Hamiltonian H are expanded in a finite basis of Slater determinant of single-particle states (anti-symmetrized product of single-particle states). Each element of this basis is referred to as a many-body basis state. The representation of H under this basis expansion is a sparse symmetric matrix $\hat{H}$. Thus, in CI calculations, Schrödinger's

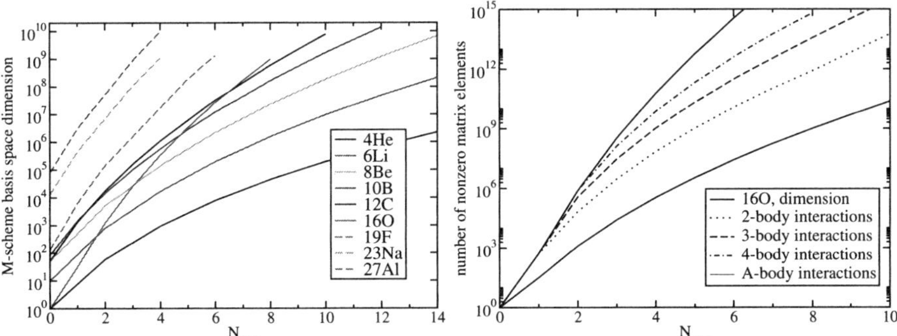

Fig. 3 The dimension and the number of non-zero matrix elements of the various nuclear Hamiltonian matrices

equation becomes a finite-dimensional eigenvalue problem, where we seek the lowest eigenvalues (energies) and their associated eigenvectors (wave functions). Many-body basis state i corresponds to the ith row and column of the Hamiltonian matrix. The total number of many-body states or the dimension of $\hat{H}$ in our adopted harmonic oscillator (HO) basis, which we denote by n, is controlled by the number of particles A, the truncation parameter $N_{\max}$, and the maximum number of HO quanta above the minimum for a given nucleus (see Fig. 3). Higher $N_{\max}$ values yield more accurate results for the same nucleus, but at the expense of an exponential growth in the dimension of $\hat{H}$. The sparsity of $\hat{H}$ is determined by the interaction potential used which can be a 2-body, 3-body or even a higher order interaction. The approach described above is implemented in the MFDn (Many Fermion Dynamics nuclei) code, which is a state-of-the-art CI code to study the properties of light nuclei with high precision [35–37]. In MFDn, a round-robin distribution of the many-body basis states to the processors is used to ensure a uniform distribution of the nonzero matrix elements in the $\hat{H}$ matrix. This way load imbalances among processors is reduced significantly [38].

In order to find the lowest nev number of eigenvalues and eigenvectors of $\hat{H}$, we use the locally optimal block preconditioned conjugate gradient (LOBPCG) algorithm [39]. As mentioned above, in this paper we are focused on the efficient execution of a single LOBPCG iteration in our out-of-core approach, rather than how fast the LOBPCG algorithm converges for a given nuclear structure calculation. Therefore, for simplicity of presentation, we take the preconditioning matrix M to be the identity matrix. Algorithm 2 gives the pseudocode for a simplified version of the LOBPCG algorithm, assuming $M = I$.

Algorithm 2: Pseudocode of the LOBPCG algorithm for the eigenproblem of the form $\hat{H}\Psi = E\Psi$. The preconditioning matrix M is assumed to be the identity matrix.

Input: $\hat{H}$, matrix of dimensions $n \times n$
Input: Ψ_0, a block of vectors of dimensions $n \times nev$
Output: Ψ and E such that $\|\hat{H}\Psi - \Psi E\|_F$ is small.
Orthonormalize the columns of Ψ_0
$P_0 \leftarrow 0$
for i = *0, 1, . . . , until convergence* **do**
  $E_i \leftarrow {\Psi_i}^T \Psi_i / {\Psi_i}^T \hat{H} \Psi_i$
  $R_i \leftarrow \Psi_i - E_i \hat{H} \Psi_i$
  Use Rayleigh-Ritz method on the span$\{\Psi_i, R_i, P_i\}$
  $\Psi_{i+1} \leftarrow \underset{Y \in span\{\Psi_i, R_i, P_i\}}{\text{argmin}} \quad Y^T Y / Y^T \hat{H} Y$
  $P_{i+1} \leftarrow \Psi_i$
  Check convergence
end

5.2 Implementation Using 1D partitioning

Our first implementation of the out-of-core eigensolver is an implementation of the LOBPCG algorithm given in Algorithm 2 using the linear algebra primitives of the DOoC+LAF framework and using a one dimensional partitioning of the matrix. In this scheme, the matrix is cut into p bands of equal size $\frac{n}{p}$, and each band is of length $\frac{n}{2}$. The allocation of the parts of the matrix to each node is depicted in Fig. 4a. The implementation is composed of two main parts: symmetric SpMM computations, followed by two inner products. Each matrix block $\hat{H}_{ij}$ stored on the permanent storage system essentially corresponds to a task, which we denote by SymSpMM(i, j). The input data of SymSpMM(i, j) are Ψ_i and Ψ_j subvectors. The 1D decomposition of the matrix $\hat{H}$ is ensured by having the compute node p create the subvector blocks $\Psi_{rs_p}, \Psi_{rs_p+1}, \ldots, \Psi_{re_p}$ for the initial guess Ψ using the DOoC+LAF primitive randomM. As mentioned above, the global scheduler assigns each task to the compute node which stores the most amount of input data required for that task. Consequently, all tasks SymSpMM(i, j), where $rs_p \le i \le re_p$ and $1 \le j \le n_b$, would be scheduled to the compute node p, essentially resulting in a load balanced 1D decomposition of the SpMM operation.

As a result of executing the task SymSpMM(i, j) on node p, two intermediate output vector blocks of $\hat{H}\Psi_i'$ and $\hat{H}\Psi_j'$ are produced. $\hat{H}\Psi_i'$ is consumed by a local reduction task denoted by addV($\hat{H}\Psi_i, \hat{H}\Psi_i'$) on node p. Similarly, $\hat{H}\Psi_j'$ is consumed by the task addV($\hat{H}\Psi_j, \hat{H}\Psi_j'$). However, note that $\hat{H}\Psi_j$ is stored on node k such that $rs_k \le j \le re_k$. Assuming that $k \ne p$, the intermediate result vectors $\hat{H}\Psi_j'$ first need to be communicated to node k for the execution of the task addV($\hat{H}\Psi_j, \hat{H}\Psi_j'$).

Lemma Assume that on node p, the difference between the sizes of the smallest and largest matrix blocks, as measured by the space required to store a block in Compressed Sparse Column (CSC) format, is less than the size of any vector block Ψ_i, for $1 \le i \le n_b$. Then Algorithm 1 orders the set of tasks on node p

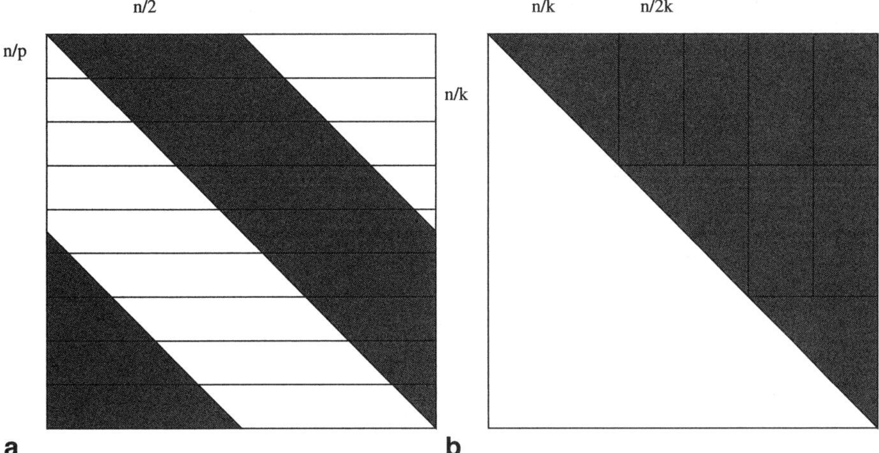

Fig. 4 Different partitioning of the matrix $\hat{H}$ on the processors. Notice that since the matrix is symetric, only half of it needs to be stored. **a** 1D decomposition on p nodes. **b** 2D decomposition on $p = k^2$ nodes

$\{\texttt{SymSpMM}(i, j) \mid rs_p \leq i \leq re_p \wedge 1 \leq j \leq n_b\}$ such that they are executed in a column-major order.

Proof Without loss of generality, let SymSpMM(rs_p, j) be the first task executed on node p for some j. Then the subvector Ψ_j is the only input data on the local memory of node p, besides the locally stored subvectors Ψ_i for $rs_p \leq i \leq re_p$. Additional input data required to execute other tasks associated with the matrix blocks in the jth column is the matrix block itself only. However, to execute a task corresponding to a matrix block in a column $c \neq j$, both the matrix block and the subvector Ψ_c would be needed. Hence, the tasks of the jth column would be ordered by Algorithm 1 before the tasks in any other column. This leads to a column-major processing of matrix-blocks. □

As a result, our out-of-core implementation using the DOoC+LAF framework is able to execute the computations related to the solution of the eigenvalue problem in a way that reduces the communication overheads. It is a natural result of the task ordering algorithm, and the pipelined execution of computation, communication and I/O operations in the DOoC+LAF framework. Since no explicit effort is required to achieve this, a significant burden on the application programmer is removed.

After the symmetric SpMM computations are completed, two inner products of the form $Y^T Y$ and $Y^T \hat{H} Y$, where $Y = \text{span}\{\Psi, R, P\}$ and $\hat{H} Y = \text{span}\{\hat{H}\Psi, \hat{H} R, \hat{H} P\}$, need to be performed. Vector blocks R and P, and consequently Y, are also partitioned according to the partitioning of Ψ and $\hat{H}\Psi$. Hence these inner products are performed on node k, for $k = 1, 2, \ldots, n_p$, as a set of tasks denoted by dot(Y_i, Y_i) and dot(Y_i, $\hat{H} Y_i$), where $rs_k \leq i \leq re_k$. The local inner products are reduced on node 1. Then all computing nodes estimate the Rayleigh quotients. Once the estimates for eigenvalues E and eigenvectors Ψ are obtained, the computation continues with the next iteration.

5.3 Implementation Using a 2D Partitioning

The 1D partitioning scheme, shown in Fig. 4a, requires that each node touches $n(\frac{1}{2} + \frac{1}{p})$ row/column. When the number of nodes p increases, the volume of communication will be proportional to the problem dimension, i.e., with $\frac{n}{2}$. This indicates a potential scalability bottleneck, as the number of nodes and problem dimensions increase together in order to solve larger problems.

One can partition the upper triangle of the matrix in two dimensions (2D) by using horizontal and vertical bands. Because the matrix is symmetric, a classical checkerboard partitioning would make the nodes responsible for the diagonal blocks processing half the non zeros of the other nodes. Therefore, we propose to split the non diagonal blocks in two so as to remove this problem. Such a partitioning is depicted in Fig. 4b and requires a number of nodes which is a square number $p = k^2$. Diagonal nodes touch only $\frac{n}{k}$ row/columns since the rows one touches are the same as the columns it touches. Meanwhile the non diagonal nodes touch $\frac{3n}{2k}$ row/columns. Since $p = k^2$, the communication volume will behave like $\frac{3n}{2\sqrt{p}}$ and is much better than the number of node increases than the 1D decomposition.

Notice also that improving the communication volumes is not the only interest of this 2D decomposition. Indeed when a processor touches a row or a column, not only it will perform communications, but also it needs to store the partial results. So a 2D decomposition will be necessary to allow to scale the computation to larger problems in terms of size of the matrix or number of vectors.

In term of implementation within the DOoC+LAF framework, there is no difference between a 1D decomposition of the work and a 2D decomposition of the work. It is sufficient to place the blocks of the matrix on the computing nodes that will process them. The framework will automatically add the appropriate communications.

6 Experiments

Experiments are run on an experimental SSD testbed on the Carver cluster at NERSC. The testbed is composed of 48 nodes: 40 computational nodes and 8 I/O nodes. Each node is equipped with two Intel Xeon X5550 processors clocked at 2.67 GHz (4 cores each, hyper-threading is disabled) and 24 GB of DDR3 memory. Each node runs on Red Hat 5.5 with Linux kernel 2.6.18-238.12.1.el5. Nodes are interconnected by 4X QDR InfiniBand technology, providing 32 Gb/s of point-to-point bandwidth for high-performance message passing and I/O. Our codes are compiled with GCC 4.5.2. The InfiniBand interconnect is leveraged through the use of the MVAPICH 1.2

Table 2 General information on the testcase

	$N_{max} = 8$
Matrix Dimension (n)	159.9×10^6
# Nonzero matrix elements	123.6×10^9
Total matrix size	920 GB
# Block row/columns (n_b)	87
Total number of matrix blocks	3828
Average size of a matrix block	246 MB

Table 3 General information on the vector block sizes

	$N_{max} = 8$
Number of eigenpairs (nev)	8
Size of a subvector block Ψ_i	58.8 MB
Total size of the vector block Ψ	5.1 GB
Total size of all 6 vector blocks	30.6 GB

library. Each I/O node is equipped with two SSD cards, Virident tachIOn 400 GB, connected through the PCI-express bus. Each card can deliver up to 1 GB/s sustained read bandwidth, leading to a peak bandwidth of 2 GB/s per I/O node, and 16 GB/s maximum I/O bandwidth from the permanent storage system to the compute nodes. I/O nodes are accessed by the compute nodes through the Global Parallel File System [40]. Data is streamed from the I/O nodes to the compute nodes using the 4X QDR InfiniBand interconnect as well.

Performance evaluation of our out-of-core implementation is done with the nuclear structure computations of the ^{10}B (5 protons, 5 neutrons) nucleus. The truncation parameter $N_{max} = 8$ is used. Some key properties of this testcase are summarized in Table 2. Since storage space is at premium for MFDn, matrix blocks are stored in single precision CSC format.

6.1 Practical Considerations

The number of eigenpairs to be computed is fixed at $nev = 8$ for our test-case. Table 3 gives detailed information regarding the sizes of vector blocks involved when $nev = 8$. The size of the entire Ψ vector block, which is also stored in single precision, is 5.1 GB for the $N_{max} = 8$ case. In the LOBPCG algorithm, 6 such vectors (Ψ, R, P from the previous iteration and $\hat{H}\Psi$, $\hat{H}R$, $\hat{H}P$ of the current iteration) need to be hosted on the volatile memory available to compute nodes. The total space required for this purpose would be 30.6 GB for the $N_{max} = 8$ case, respectively. On Carver, about 5 GB of the 24 GB memory on a compute node is reserved for the OS kernel, and the network file system (NFS). Since matrix blocks to be read are on the order of hundreds of MBs, and the messages to be communicated

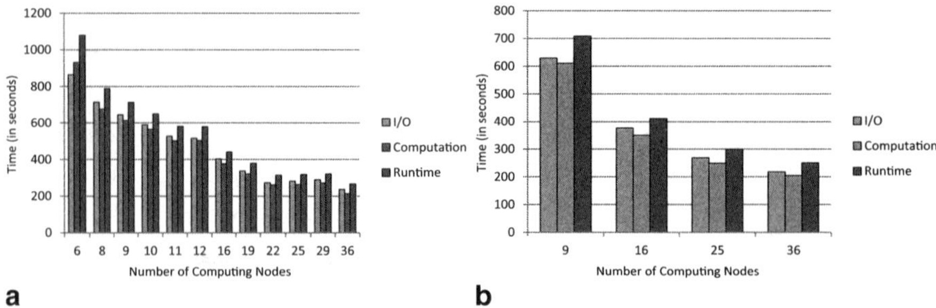

Fig. 5 Runtime of the application and time spent doing computations and I/O. The I/O and Computation mostly overlap. **a** 1D partitioning. **b** 2D partitioning

are on the order of tens of MBs (see the size of Ψ_i in Table 3), significant space is needed for the I/O and MPI buffers. As a result, only 15 GB out of the 24 GB memory on a compute node can be used by our out-of-core eigensolver. We choose to use at most 5 GB of the usable memory for hosting the vector blocks, and the remaining memory for processing the tasks. Therefore the minimum number of nodes required for $N_{\max} = 8$ computations is 6, respectively.

We create 8 computation threads (one for each core), which collectively work on the tasks assigned to a node. Since there are lots of I/O and communication operations involved in our out-of-core eigensolver, per iteration timings may fluctuate during execution. Therefore, we report the timings from the first 5 iterations of the LOBPCG algorithm for a reliable performance evaluation.

Since all the computing nodes share the same file system, each node will read its data in different directory so as to provide data partitioning.

6.2 Performance Results for $N_{\max} = 8$

Figure 5 presents the runtime obtained when executing the application on different number of nodes. The figure also presents the time taken by the computations and by the I/O separately. These times varies on all computation and I/O threads. The figure reports the maximum value of all threads but the average value is fairly close to the maximum.

The first remark is that the difference between the Runtime and the maximum of I/O and computation is fairly small. This indicates that the computations and I/O are fairly well overlapped and that the design of our middleware is sound. The runtime decreases with an increase of the number of computing nodes. Though, the runtime is fairly stable after 20 nodes. This comes from a saturation of the GPFS after 20 computing nodes which draws 16 GB/s, the peak performance for the I/O nodes. This shows that the traditional organization of the cluster with I/O nodes on one side and compute node on the other one is not a scalable setup for the data-intensive clusters.

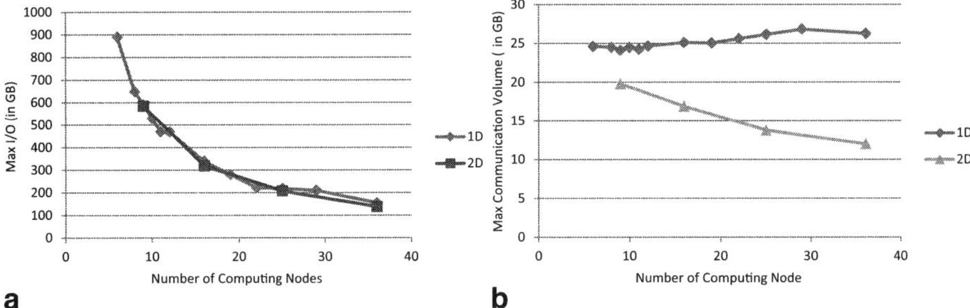

Fig. 6 Comparison of 1D and 2D partitioning. **a** Amount of GPFS I/O. **b** Amount of Inter node communication

Indeed, within a single I/O node, we get a bandwidth from the disks to the memory of about 2 GB/s. Yet to be able to reach such bandwidth from the application more than twice the amount of compute nodes are required.

We can see on Fig. 5 that there is little difference in total runtime between using 1D and 2D decomposition. The only existing difference is entirely explained by the difference of I/O performed by the 1D and 2D decomposition. The differences between 1D and 2D decompositions are presented in Fig. 6. One can see that the amount of GPFS I/O (Fig. 6a) is slightly lower for the 2D decomposition. Indeed, when 2D decomposition is used, less memory is used for storing the intermediate values of the multiplications which leaves more memory available for caching the data from the matrix (as explained in Sect. 5.3). Another interest of the 2D decomposition lies in the amount of communication performed by each node involved in the computation which is depicted in Fig. 6b. With a 1D decomposition, each processor transfers a whole Ψ vector at each iteration. Leading to a communication volume (per node) constant when the number of nodes increases. Meanwhile the communication volution when using a 2D decomposition decreases when the number of node increases. This confirms the analysis of Sect. 5.3 that 2D decomposition is more scalable than a 1D decomposition.

The DOoC+LAF runtime environment generates a detailed log file on each compute node for all the steps it takes during the execution of a code. The analysis of these log files can give important insights. One way to analyze how our out-of-core eigensolver performs is to look at the number of jobs in the local scheduler's queue versus execution time plot, as shown in Fig. 7a. Here we plot the first 3 iterations of the $N_{max} = 8$ case on 12 nodes with 1D decomposition. There are 87 rows of matrix blocks in this calculation, therefore 3 nodes (nodes 1, 3 and 4) are responsible for an extra row of matrix blocks compared to other compute nodes. This is reflected as a higher peak at the start of an iteration for those 3 nodes. The rise of the peak corresponds to the building and partitioning of the task graph part. The percentage of this part is again negligible compared to the total time per iteration. The fall of the peak means that the task graph is shrinking, because tasks are being executed. As

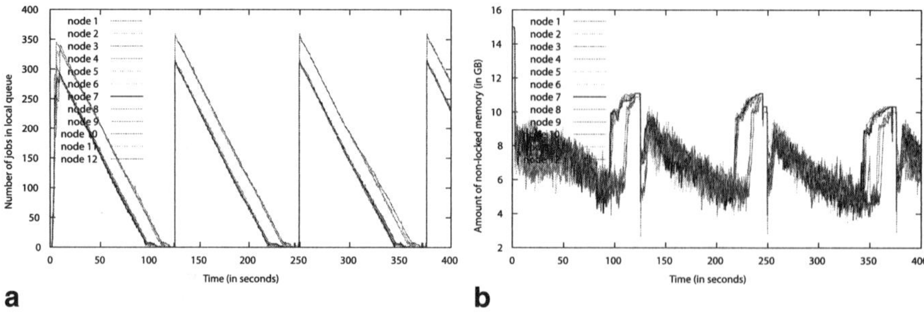

Fig. 7 Amount of free memory available and jobs in the local scheduler during an execution on 12 nodes with 1D decomposition

seen in the plot, the peak falls at a constant slope during the SpMM computations. This means that computation and I/O operations are overlapped efficiently, and the SpMM computations progress smoothly, without idling.

When using the DOoC+LAF framework, it is important to keep track of the amount of memory available. Because this memory is used to prefetch the data of the upcoming tasks. Here, the available memory is used to buffer the blocks of $\hat{H}$ from the file system and Ψ_i vectors from other nodes. If the available memory is low, the prefetching is no longer possible, the computation are sequentialized and the overlapping of I/Os, computations and communications might not be effective.

Figure 7b shows the amount of available memory as the execution progresses. At the start of an iteration, the local scheduler reserves memory space and issues prefetching requests for the initial batch of matrix blocks. This results in a sharp drop in the amount of memory available. As tasks associated with these matrix blocks are completed, the memory space that becomes available is filled in further with other matrix blocks. Once all the SpMM tasks are finished, we see a sudden jump at the amount of memory space available. This is because the inner product computations do not consume much memory. The slight load imbalance caused due to the higher number of tasks on 3 nodes, is reflected as a phase difference in this plot. Nodes 1, 3 and 4 finish their SpMM computations a little after other compute nodes, and the amount of memory available makes a peak slightly later on these nodes.

7 Conclusions

Adaption of NVM-based memory in future HPC architectures and data centers will only increase with time. Efficient use of such, multilevel memory hierarchies will require advanced parallel programming skils. Here, we presented an attempt to relieve such burden from programmer, by providing a domain specific frontend that uses already familiar Basic Linear Algebra Subprograms (BLAS)-like application interface and leverages a capable task-based runtime system that will take care of efficient orchestration of the execution of use applications. Specifically, we have presented early results of our out-of-core task-based runtime system, (**D**istributed **O**ut-**o**f-**C**ore (DOoC), together with a specialized frontend, Linear Algebra Frontend (LAF), which is developed to enable the implementation of iterative numerical methods using DOoC. Although our out-of-core runtime system generic and could work with any storage system, existance of high-bandwith, low-latency storage system that are based on non-volatile memory makes it feasible to execute larger problems that will not fit into physical RAM memory of the compute nodes. Our results shows that LAF+DOoC pushes the hardware limitations of the underlying testbed we have carried our experiments, while providing an extremely easy application interface.

We argue that in the future systems by co-locating SSD storages with computation [41], one can further optimize the out-of-core execution further. Our task-based runtime system DOoC is well positioned to take advantage of such hardware changes without requiring the rewrite of application program.

References

1. P. Kogge and J. Shalf, "Exascale computing trends: Adjusting to the new normal in computer architecture," *Computing in Science Engineering*, vol. PP, no. 99, pp. 1–1, 2013.
2. P. Ranganathan and J. Chang, "(Re)designing data-centric data centers," *Micro, IEEE*, vol. 32, no. 1, pp. 66–70, 2012.
3. E. Barragy, B. Brantley, S. Gurumurthi, M. Ignatowski, N. Jayasena, A. Lee, G. Loh, S. Manne, M. O'Connor, P. Popescu, S. Reinhardt, and M. Schulte, "Amd's fastforward extreme-scale computing processor and memory research," in *US DOE Exascale Research Conference*, Arlington, VA, USA, Oct. 2012.
4. R. Nair, J. Moreno, and D. Joseph, "Advanced memory concepts for exascale systems," in *US DOE Exascale Research Conference*, Arlington, VA, USA, Oct. 2012.
5. Y.-K. Kwok and I. Ahmad, "Static scheduling algorithms for allocating directed task graphs to multiprocessors," *ACM Comput. Surv.*, vol. 31, no. 4, pp. 406–471, Dec. 1999.
6. C. Augonnet, S. Thibault, R. Namyst, and P.-A. Wacrenier, "StarPU: A Unified Platform for Task Scheduling on Heterogeneous Multicore Architectures," *Concurrency and Computation: Practice and Experience, Special Issue: Euro-Par 2009*, vol. 23, pp. 187–198, Feb. 2011.
7. G. Bosilca, A. Bouteiller, A. Danalis, T. Hérault, P. Lemarinier, and J. Dongarra, "DAGuE: A generic distributed DAG engine for high performance computing," *Parallel Computing*, vol. 38, no. 1-2, pp. 37–51, 2012.
8. G. Bosilca, M. Faverge, X. Lacoste, I. Yamazaki, and P. Ramet, "Toward a supernodal sparse direct solver over DAG runtimes," in *Proceedings of PMAA'2012*, London, UK, Jun. 2012.
9. A.-E. Hugo, A. Guermouche, R. Namyst, and P.-A. Wacrenier, "Composing multiple StarPU applications over heterogeneous machines: a supervised approach," in *Third International Workshop on Accelerators and Hybrid Exascale Systems*, Boston, États-Unis, May 2013.
10. C. Augonnet, O. Aumage, N. Furmento, R. Namyst, and S. Thibault, "StarPU-MPI: Task Programming over Clusters of Machines Enhanced with Accelerators," in *EuroMPI 2012*, ser. LNCS, S. B. Jesper Larsson Träff and J. Dongarra, Eds., vol. 7490. Springer, Sep. 2012, poster Session.
11. M. Cosnard and M. Loi, "Automatic task graph genera tion techniques," *Parallel Processing Letters*, vol. 5, no. 4, p. 527–538, 1995.
12. M. Cosnard, E. Jeannot, and T. Yang, "Slc: Symbolic scheduling for executing parameterized task graphs on multiprocessors," in *Proc. ICPP*, 1999.

13. S. Toledo, "A survey of out-of-core algorithms in numerical linear algebra," in *External memory algorithms*, J. M. Abello and J. S. Vitter, Eds. Boston, MA, USA: American Mathematical Society, 1999, pp. 161–179.
14. J. K. Reid and J. A. Scott, "An out-of-core sparse cholesky solver," *ACM Trans. Math. Softw.*, vol. 36, no. 2, 2009.
15. V. Rotkin and S. Toledo, "The design and implementation of a new out-of-core sparse cholesky factorization method," *ACM Trans. Math. Softw.*, vol. 30, no. 1, pp. 19–46, 2004.
16. P. R. Amestoy, I. S. Duff, Y. Robert, F.-H. Rouet, and B. Ucar, "On computing inverse entries of a sparse matrix in an out-of-core environment," CERFACS, Tech. Rep. TR/PA/10/59, 2010.
17. J. A. Scott, "Scaling and pivoting in an out-of-core sparse direct solver," *ACM Trans. Math. Softw.*, vol. 37, no. 2, 2010.
18. E. Agullo, A. Guermouche, and J.-Y. L'Excellent, "A parallel out-of-core multifrontal method: Storage of factors on disk and analysis of models for an out-of-core active memory," *Parallel Computing, Special Issue on Parallel Matrix Algorithms*, no. 6–8, 2008.
19. E. Agullo, A. Guermouche, and J.-Y. L'Excellent, "Reducing the I/O Volume in Sparse Out-of-core Multifrontal Methods," *SIAM Journal on Scientific Computing*, no. 6, 2010.
20. W. J. Knottenbelt and P. G. Harrison, "Distributed disk-based solution techniques for large markov models," in *Proc. of Numerical Solution of Markov Chains*, 1999.
21. Y.-Y. Chen, Q. Gan, and T. Suel, "Local methods for estimating pagerank values," in *Proceedings of the thirteenth ACM international conference on Information and knowledge management*, ser. CIKM '04. New York, NY, USA: ACM, 2004, pp. 381–389.
22. E. Saule, P.-F. Dutot, and G. Mounié, "Scheduling With Storage Constraints," in *Proc of IPDPS'08*, Apr. 2008, conference, acceptance rate: 25.6%.
23. S. S. Tse, "Online bicriteria load balancing using object reallocation," *IEEE Transactions on Parallel and Distributed Systems*, vol. 20, no. 3, pp. 379–388, 2009.
24. Ü. V. Çatalyürek, K. Kaya, and B. Uçar, "Integrated data placement and task assignment for scientific workflows in clouds," in *The Fourth International Workshop on Data Intensive Distributed Computing (DIDC 2011), in conjunction with the 20th International Symposium on High Performance Distributed Computing (HPDC 2011)*, Jun 2011.
25. R. Sethi, "Pebble games for studying storage sharing." *Theor. Comput. Sci.*, vol. 19, pp. 69–84, 1982.
26. S. Biswas and S. Kannan, "Minimizing space usage in evaluation of expression trees," in *Foundations of Software Technology and Theoretical Computer Science*, ser. Lecture Notes in Computer Science, P. Thiagarajan, Ed. Springer Berlin Heidelberg, 1995, vol. 1026, pp. 377–390.
27. C.-C. Lam, D. Cociorva, G. Baumgartner, and P. Sadayappan, "Memory-optimal evaluation of expression trees involving large objects," in *High Performance Computing – HiPC'99*, ser. Lecture Notes in Computer Science, P. Banerjee, V. Prasanna, and B. Sinha, Eds. Springer Berlin Heidelberg, 1999, vol. 1745, pp. 103–110.
28. V. Rehn-Sonigo, D. Trystram, F. Wagner, H. Xu, and G. Zhang, "Offline scheduling of multi-threaded request streams on a caching server," in *IPDPS*, 2011, pp. 1167–1176.
29. M. Jacquelin, L. Marchal, Y. Robert, and B. Uçar, "On optimal tree traversals for sparse matrix factorization," in *Parallel Distributed Processing Symposium (IPDPS), 2011 IEEE International*, 2011, pp. 556–567.
30. L. Marchal, O. Sinnen, and F. Vivien, "Scheduling tree-shaped task graphs to minimize memory and makespan," INRIA, Rapport de recherche RR-8082, Oct. 2012.
31. Z. Zhou, E. Saule, H. M. Aktulga, C. Yang, E. G. Ng, P. Maris, J. P. Vary, and Ü. V. Çatalyürek, "An out-of-core dataflow middleware to reduce the cost of large scale iterative solvers," in *2012 International Conference on Parallel Processing (ICPP) Workshops, Fifth International Workshop on Parallel Programming Models and Systems Software for High-End Computing (P2S2)*, Sep 2012.
32. M. D. Beynon, T. Kurc, Ü. V. Çatalyürek, C. Chang, A. Sussman, and J. Saltz, "Distributed processing of very large datasets with DataCutter," *Parallel Computing*, vol. 27, no. 11, pp. 1457–1478, Oct. 2001.

33. Z. Zhou, E. Saule, H. M. Aktulga, C. Yang, E. G. Ng, P. Maris, J. P. Vary, and Ü. V. Çatalyürek, "An out-of-core eigensolver on SSD-equipped clusters," in *Proc. of IEEE Cluster*, Sep. 2012.
34. J. Nieplocha, B. Palmer, V. Tipparaju, M. Krishnan, H. Trease, and E. Apra, "Advances, applications and performance of the global arrays shared memory programming toolkit," *International Journal of High Performance Computing Applications*, vol. 20, pp. 203–231, 2006.
35. P. Maris, H. M. Aktulga, M. A. Caprio, Ü. V. Çatalyürek, E. G. Ng, D. Oryspayev, H. Potter, E. Saule, M. Sosonkina, J. P. Vary *et al.*, "Large-scale ab initio configuration interaction calculations for light nuclei," *Journal of Physics: Conference Series*, vol. 403, no. 1, p. 012019, 2012.
36. P. Maris, H. M. Aktulga, S. Binder, A. Calci, Ü. V. Çatalyürek, J. Langhammer, E. Ng, E. Saule, R. Roth, J. P. Vary, and C. Yang, "No-Core CI calculations for light nuclei with chiral 2- and 3-body forces," *Journal of Physics: Conference Series*, vol. 454, no. 1, p. 012063, 2013.
37. H. M. Aktulga, C. Yang, E. G. Ng, P. Maris, and J. P. Vary, "Improving the scalability of a symmetric iterative eigensolver for multi-core platforms," *Concurrency and Computation: Practice and Experience*, p. in press, 2013.
38. P. Sternberg, E. G. Ng, C. Yang, P. Maris, J. P. Vary, M. Sosonkina, and H. V. Le, "Accelerating configuration interaction calculations for nuclear structure," in *Proc. of SC08*, 2008.
39. A. V. Knyazev, "Toward the optimal preconditioned eigensolver: Locally optimal block preconditioned conjugate gradient method," *SIAM Journal on Scientific Computing*, vol. 23, no. 2, pp. 517–541, 2001.
40. F. B. Schmuck and R. L. Haskin, "GPFS: A shared-disk file system for large computing clusters," in *Proc. of FAST'02*, 2002, pp. 231–244.
41. M. Jung, E. H. W. III, W. Choi, J. Shalf, H. M. Aktulga, C. Yang, E. Saule, Ü. V. Çatalyürek, and M. Kandemir, "Exploring the future of out-of-core computing with compute-local non-volatile memory," in *Proc. of Conference on High Performance Computing Networking, Storage and Analysis (SC '13)*, Nov 2013.

Building Scalable Software for Data Centers: An Approach to Distributed Computing at Enterprise Level

Fernando Turrado García, Ana Lucila Sandoval Orozco and Luis Javier García Villalba

1 Introduction to Big Data Problems

Big data can be defined as a large collection of data that it is difficult to process due to its size or complexity. In 2001 Doug Laney, a META Group (now Gartner) analyst, published a research report defining 3 dimensions that characterize big-data problems: Volume, Variety and Velocity (also known as 3V's). The original report can be found at the garter site [1].

Volume refers to the amount of data to be handled. In astronomy science, the Large Synoptic Survey Telescope produces 30 TB of data per night of raw data. The total volume (data stored after processing) is about 100 PB [2].

Velocity refers to the performance (relative to the amount of input data) required in the solution. In many cases, the right answer delivered at the wrong time is a wrong answer. Consider a system that uses comments in a social network to provide traffic guidance [3].

Variety refers to the variety present in the sources, structures, formats and quality of the data to be processed. This factor, added to the previous ones, greatly increases the computational resources needed to solve the problem.

Nowadays social networks are a new data source to be analyzed and studied. In the following examples, we will present the benefits obtained from mining and analyzing those large datasets. In [4] the twitter mood (seen as the global mood of its users) was correlated to the stock market fluctuations. In [5], more than 100.000 tweets were analyzed in order to make predictions for the German federal elections in 2009. Similar work is carried out in [6] where the twitter stream is analyzed looking for linking text sentiment to public opinion time series in the U.S. In this case, more than

L. J. García Villalba (✉) · A. L. Sandoval Orozco · F. Turrado García
Group of Analysis, Security and Systems (GASS), Department of Software Engineering and Artificial Intelligence (DISIA), Faculty of Information Technology and Computer Science, Office 431,
Universidad Complutense de Madrid (UCM), Calle Profesor José García Santesmases 9, Ciudad Universitaria, 28040 Madrid, Spain
e-mail: javiergv@fdi.ucm.es

S. U. Khan, A. Y. Zomaya (eds.), *Handbook on Data Centers*,
DOI 10.1007/978-1-4939-2092-1_23

one billion (1.000.000.000) tweets were used. Another example of the twitter stream being used for social behavior is [7]. Golder and Macy used more than 500.000.000 tweets to identify diurnal and seasonal mood rhythms at individual level.

In these situations, where the data does not fit in a single machine, the hardware & software solutions must process those data sets in a parallel & distributed way. To achieve this particular goal, the new software stack has to be designed for working in a computing cluster instead of a single super-computer server. A computing cluster can be defined as a large collection of commodity hardware, examples of these clusters can be found at Amazon EC2, Twitter, Google and others. Inside an enterprise, this computing cluster can be managed as a private cloud provided from some internal or external providers. In [8], Armbrust and others argue that this approach, the construction and operation of extremely large-scale, commodity-computer data centers at low-cost locations, is the key necessary enabler for cloud computing. Another example of a commodity cluster can be found at [9].

In computer science, from a performance point of view, scalability is the ability of a system to handle a growing amount of input data without compromising performance or its ability to be enlarged to maintain it. In a real world environment, where the computing resources (processors, memory, storage, network bandwidth,···) are finite, the growth of the input data will cause the system performance limit (the point at which it decreases as the input data grows) to be reached at a given time. The methods for increasing system capabilities by incorporating additional resources are grouped into two broad categories: vertical or horizontal scaling.

Vertical scaling, or scale in, is done when one of the cluster nodes (or some of them) is enhanced by increasing its hardware resources: adding extra processors, memory··· By using this method, the node workload capability can be increased to its maximum. This maximum is established by the technical specifications of the server. Horizontal scaling, or scale out, is done when new nodes are added to the cluster. By applying this method, the system computational limits can be removed.

In 1994 B.C. Nueman published a paper [10] about how scaling affects distributed systems. In this paper, four main problems must be confronted while designing or building a distributed system: reliability, system load, administration and heterogeneity. It also presents how replication, caching and distribution of the services can help to solve those problems. In the final section, some useful guidelines are found:

- *For replication*: replicate important services, distribute the replicas and use loose consistency
- *For distribution*: Distribute across multiple servers, distribute evenly, exploit locally and bypass upper levels of hierarchies (in hierarchically organized systems).
- *For caching*: Cache frequently accessed data, consider access patterns when caching, establish a cache timeout, cache at multiple levels, look first locally and minimize the change frequency of extensively shared data.

At this point, a private cloud or similar infrastructure (based on a commodity cluster) is a recommended solution for providing a scalable hardware solution. This approach guarantees the vertical and horizontal scaling needed for solving these kind of problems. This chapter aims to provide an introduction on how to design and build (using existing frameworks) usable software in this environment.

2 Known Solutions at Design Phase: Overview of Design Patterns for Parallel & Distributed Computing

A design pattern is defined as a general reusable solution to a commonly occurring problem within a given context. Design patterns are built on top of expert knowledge of design methods, constraints and best practices. They represent a standard on design reuse and alleviate the accidental and inherent complexities of software design.

There are many papers and books published regarding this matter; we will now mention the books that are considered as "classic works", i.e. the ones that must be read to get introduced in this field of study:

- *Design Patterns*: Elements of reusable Object Oriented Software [11]. It is also known as the Gang of Four design patterns. It contains the definition for many of the basic patterns applied today:
  - *Singleton*: How to create a unique instance of an object in a system.
  - *Factory Method*: How to define an interface for creating an object but letting the subclasses decide which class to instantiate.
  - *Composite*: How to compose objects into tree structures to represent leaf-node hierarchies.
  - *Facade*: How to provide a unified interface to a system or subsystem.
  - *Proxy*: How to provide a surrogate or placeholder for another object.
  - *Observer*: How to define a one-to-many dependency between objects so that when one object changes state, all its dependents are notified and updated automatically.
  - And many others like Template Method, Strategy, Visitor, and Iterator...
- *Pattern oriented software architecture*: a system of patterns [12]. This book is the first one of a collection dedicated to design patterns. For example, it contains the following patterns:
  - *Layers*: How to decompose a system into groups of subtasks in which each group of subtasks is at the same level of abstraction.
  - *Pipes and filters*: How to structure a system that processes a stream of data.
  - *Broker*: How to structure distributed software systems with decoupled components that interact by remote service invocations.
  - *Master-Slave*: How to distribute work among several identical processes having a main process that computes the aggregated final result.
- *Analysis patterns*: reusable object models [13]. This book provides a collection of patterns applied to domains like trading, corporate financial analysis, planning and others.

Related to the topic of this chapter, building scalable software, the second volume of the Pattern Oriented Software Architectures [14] is dedicated to design patterns for concurrent and networked software. It contains the definition (amongst others) of the following patterns:

- *Reactor*: In event-driven applications, how to demultiplex and dispatch service requests that are delivered to an application from one or more clients.
- *Proactor*: In event-driven applications, how to efficiently demultiplex and dispatch service requests triggered by the completion of asynchronous operations.
- *Acceptor-Connector*: How to decouple the connection and initialization of cooperating peer services in a networked system from the processing they perform.
- *Half-Sync/Half-Async*: How to decouple asynchronous and synchronous service processing in concurrent systems, to simplify programming without reducing overall performance.
- *Leader/Followers*: This pattern provides an efficient concurrency model where multiple threads take turns sharing a set of event sources in order to process service requests that occur on the event sources.

However, in those books, the design patterns are basically described in an individual manner. Their relations or the tradeoffs to be considered when combining two of them are not analyzed in depth. So in a real world problem where several design patterns can be applied to solve it and many of them need to be combined, the software architects need to read and study them one by one, investing time and effort into understanding the tradeoffs so they can make the right choices.

Prior to the application of design patterns to software construction, in 1977 Alexander and others [15] introduced the concept of pattern language and applied it to the civil engineering sector. A pattern language can be viewed as a collection of design patterns and their relations applied to solve a concrete problem in a domain. The fifth book of the Pattern Oriented Software Architectures series [16] explains how this concept can be applied and adapted to software development.

The fourth book on the pattern oriented software architectures, labeled a pattern language for distributed computing [17], contains a pattern language for distributed computing and aimed to provide an overview of the state-of-the-art in some critical areas of distributed software systems. Another pattern language for parallel programming, called OPL, is shown at [18].

In this pattern language, the design patterns are grouped into five categories:

1. *Structural patterns*: In this category, the patterns describe the overall organization of the application and the way the computational elements that make up the application interact.
2. *Computational patterns*: These patterns describe the classes of computations that make up the application.
3. *Concurrent algorithm strategies*: These patterns define high-level strategies to exploit concurrency in a computation for execution on a parallel computer.
4. *Implementation strategies*: These are the structures that are realized in source code to support how the program itself is organized and common data structures specific to parallel programming.

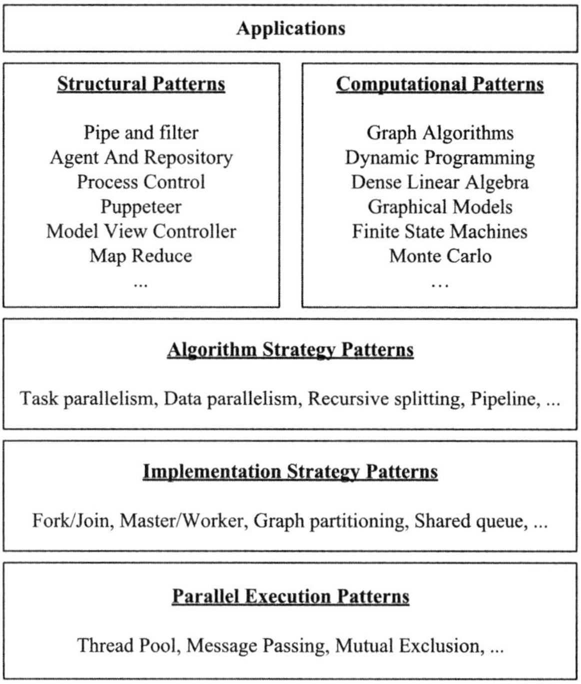

Fig. 1 OPL pattern language. Taken from [18]

Applications

Structural Patterns

Pipe and filter
Agent And Repository
Process Control
Puppeteer
Model View Controller
Map Reduce
...

Computational Patterns

Graph Algorithms
Dynamic Programming
Dense Linear Algebra
Graphical Models
Finite State Machines
Monte Carlo
...

Algorithm Strategy Patterns

Task parallelism, Data parallelism, Recursive splitting, Pipeline, ...

Implementation Strategy Patterns

Fork/Join, Master/Worker, Graph partitioning, Shared queue, ...

Parallel Execution Patterns

Thread Pool, Message Passing, Mutual Exclusion, ...

5. *Parallel execution patterns*: These are the approaches used to support the execution of a parallel algorithm. This includes strategies that advance a program counter and basic building blocks to support the coordination of concurrent tasks.

Figure 1 shows how these categories are organized into design layers. The first two categories, placed at the same design layer, are focused on the application internal structure and the algorithms to be implemented. Categories 3, 4 & 5 are focused on how the parallelism can be achieved.

Using the above pattern languages, the one described at [17] or the one defined at [18], a software architect can design a scalable software solution reusing the knowledge of experts in areas such as task scheduling, parallelization, computational methods ... But one of the major disadvantages of using patters is that only the design effort can be reused. All the details needed for an efficient implementation of those patterns, all the knowledge obtained by implementing the same pattern in different domains or the complexities found at the coding phase are not shared in the pattern specification.

The use of application frameworks alleviates the risks associated with those complexities and provides a way to reuse not only design concepts but also software components. The benefits of using frameworks can be read at [19]. In the next sections, we will focus on the structural pattern called Map Reduce [20] as it represents a design trend for developing applications in the big-data context and the Apache Hadoop framework (its most used implementation).

3 Introduction to MapReduce Programming Model

In 2004, two engineers from Google (Dean & Ghemawat) published a paper [20] entitled: "MapReduce: Simplified Data Processing on Large Clusters" that has changed the way that big volumes of data are processed. In this paper MapReduce is defined as a programming model in which the programs are written in a functional style and are automatically parallelized and executed on a large cluster of commodity servers.

In this programming model, all the data is viewed as pairs of key-value objects and the developers have to write two functions:

- *Map*: This function takes an input pair and produces a collection of output pairs. The underlying system groups all the intermediate values associated with the same key and passes them to the Reduce function.
- *Reduce*: This function takes an intermediate key and the set of valued for that key. It merges together these values trying to form a smaller set of values.

In their paper, Dean and Ghemawat presented an example of counting the number of occurrences of each word in a large collection of documents. Here is the pseudo-code they wrote:

- map(String key, String value):
  // key: document name, value: document contents
  for each word w in value:
    EmitIntermediate(w, "1");
- reduce(String key, Iterator values):
  // key: a word, values: a list of counts
  int result = 0;
  for each v in values:
    result + = ParseInt(v);
  Emit(AsString(result));

In [21] more examples, like matrix multiplication and relational algebra operations can be found. The typical execution of a MapReduce program, shown at Fig. 2, is as follows:

1. The input data (a file) is loaded into the system. This file is partitioned into a set of M splits. Then the MapReduce library starts up several copies of the program.
2. Each split is processed (Map function) by different nodes.
3. The Reduce function invocations are distributed by partitioning the intermediate key set into R parts.
4. The result of the MapReduce algorithm is stored in R output files.

The cluster nodes are organized using the Master/Slave pattern (described as Command in [11]), so the computing nodes are divided into two categories: Master node and Worker nodes. The Master node is in charge of maintaining the status of each Map and Reduce task, outlining the work to be done (at each worker) and keeping track of the files generated by the Map functions (with the Reduce task to which it is destined). The worker nodes are where the Map and Reduce functions are executed. All those nodes (up to several thousand [22, 23]) share the data (input, output

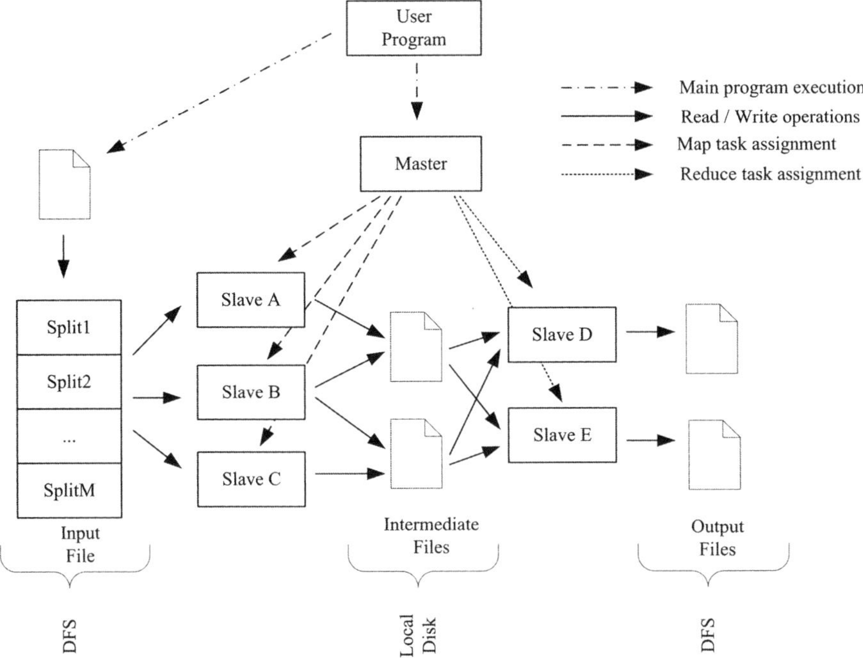

Fig. 2 MapReduce program execution overview

and temporary) using a distributed file system (DFS). Examples of distributed file systems are GFS (proprietary, created by Google) or Apache's HDFS.

These DFS are built on top of two premises: files can be enormous (up to a terabyte) and the files are rarely modified. It is assumed that once the file is created, it will be read several times but the update operations will not be frequent. In these conditions, commodity hardware and a large set of machines, the hardware errors will occur with relatively high frequency. So data replication is needed to prevent data losses and provide fault tolerance. To achieve this goal, the files are divided into smaller chunks (typically 64 megabytes in size) and those chunks are replicated several times at different nodes of the system. Also, another smaller file is generated for the data file, let us call it name file, and it contains the location of the different chunks that compose the original data file. This file, the name file, is also replicated, and a directory for the whole system is created (with all the name files). Fault tolerance is also available at task level, when a worker fails during the execution of an assigned task the Master node discover the error (using a heartbeat protocol) and reassigns the task to another worker.

So, in the simplest case, the software stack needed to run a MapReduce program in a node is composed of a Distributed File System and a MapReduce framework (or runtime implementation). Figure 3 shows how this software stack is organized.

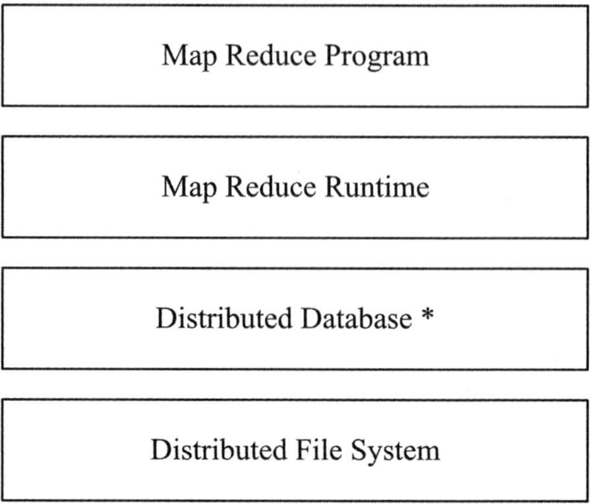

Fig. 3 Software stack, optional layers marked with*

On top of these two modules several add-ons can be applied to enhance the system:

a. A high level programming language that simplifies the writing ofMapReduce programs. Examples of these modules are: Hive [24], Pig [25] and Dryad/Linq [26].
b. Integration with distributed databases like MongoDB [27], Apache Hbase (based on Google's BigTable) [28] and Apache Cassandra [29].

The relationship between a MapReduce system and a distributed relational database has been discussed and analyzed in many ways. In 2009, Pavlo and others published a benchmark report [30] between a MapReduce framework (Hadoop) and two parallel databases (DBMS-X and Vertica). This report claimed that the MapReduce framework was several times slower than the parallel databases. Dean and Ghemawat argued that the conclusions of the comparison paper were based on flawed assumptions about MapReduce and overstated the benefit of parallel database systems [31].

In a later paper, Stonebraker, Pavlo and others [32] reviewed the arguments shown at [31] and stated that both systems where compatible if not complementary. MapReduce based systems excel at data manipulation (complex analytics, ETL tasks, $\cdots$) and the parallel database systems excel at efficient querying of massive datasets.

4 Overview of Apache Hadoop: A Framework for Distributed Computing

In the previous sections of this chapter we have introduced the so called big-data problems, where massive amounts of raw data are processed and analyzed, reusable design patterns for leveraging the design effort required for the software construction and a new programming model for processing massive data.

However, those theoretical concepts need to be applied to real world in order to build a concrete application or system. For application developers, one approach is to make their own implementations of all those modules mentioned before. Another approach is to use an enterprise application framework (its definition and the advantages of use can be read in [19]) and build the application on top of it.

In this case, we will show how the Apache Hadoop [33] framework provides the key components needed to build a MapReduce based application. Nowadays Hadoop is the open source reference implementation for the MapReduce programming model. Yahoo! has been its major contributor and it is adopted in a very large scale (see [34]).

In [35] is a list of companies or institutions that use this framework for production or educational uses. Some of them are:

- *AOL*: A cluster of 150 machines used for doing behavioral analysis and targeting.
- *EBay*: A cluster of 532 nodes, 5.3 PB. Used for search optimization and research.
- *Facebook*: 2 major clusters, one with 1100 machines and 12 PB of storage and another with 300 machines and 3 PB of storage, used for reporting, analytics and machine learning.
- *Last.fm*: A cluster of 100 nodes, used for charts calculation, royalty reporting and log analysis.

In this section we will provide a brief introduction to the core features of Hadoop and some of the more popular add-ons like Hive, Pig and HBase. The following sections are based on Apaches's official documentation, which is licensed under the Apache 2.0 license [46].

4.1 *Distributed File System: HDFS*

The Apache Hadoop framework provides its own implementation of a distributed file system. It is called HDFS and the full documentation can be found at [33].

HDFS has been designed and built under the following assumptions:

- On commodity hardware, failures are frequent and can't be managed as exceptions. On large installations, the non-trivial probability of failure makes the probability of some components to be offline near to 100
- Big data problems involve handling large data sets. A typical file in HDFS can be from gigabytes to terabytes in size.
- On large data sets, an application is more efficient if it is executed near the data it needs.

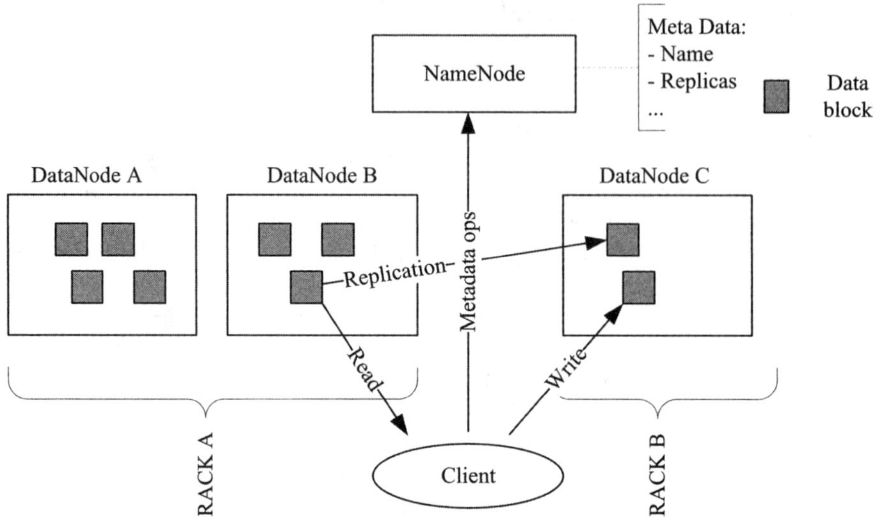

Fig. 4 Standard HDFS architecture

- Lesser data coherency requirements. Applications using HDFS are expected to use a write-once-read-many access model for files.
- Streaming data access to the files. Applications using HDFS are expected to be designed for batch processing rather than interactive use.
- HDFS must work on heterogeneous hardware and software.

In HDFS the Master node is called NameNode and is the node that manages the metadata of the cluster (optionally a secondary NameNode can be configured). The worker nodes are called DataNodes, and installing a task execution service in each one is common practice. Due to this configuration, the task execution framework is allowed to effectively schedule tasks on the nodes where data is already stored, reducing the bandwidth requirements for the cluster.

Therefore, a file is divided into smaller chunks that are distributed among the datanodes. In this implementation, the NameNode executes the operations such as opening, closing and renaming of files or directories. It is also in charge of determining the mapping of chunks to the datanodes. The DataNodes perform all chunk related operations: create, delete, read, write etc. Figure 4 shows an example of a standard HDFS architecture.

More documentation about HDFS configuration and its features, like replication, quotas or permissions is located at [36]. There are several (most of them) commercial alternatives to HDFS, we will enumerate some of the most popular:

- NetApp provides an improved version of HDFS in their "NetApp Open Solution for Hadoop" [37]

- MapR sells a Hadoop distribution with a proprietary DFS (MapR File System) [38].
- IBM has tuned its GPFS for Hadoop. [39]
- CFS, from DataStax(the Cassandra makers) is another open source alternative.

4.2 MapReduce Framework & API

In this section we will provide a brief introduction to the Apache Hadoop API and how it can be used to build user applications. As an example, we will implement the map reduce program that counts the words present in the files located at a given folder.

Typically a Hadoop based application is a collection of MapReduce jobs; each job can be viewed as a process that takes some input data (from files, tables, $\cdots$), transforms it and produces some output data (stored in files, tables, $\cdots$). The execution of these jobs is performed by two different services running on the cluster:

- *JobTracker*: The JobTracker process is responsible for distributing the software/configuration to the slave nodes. It also schedules and monitors the tasks, providing status and diagnostic information to the job client. Usually the JobTracker process is initiated in a different node from the NameNode due to performance reasons. Inside the cluster there is only one instance running, so this is a point of failure of the global system.
- *TaskTracker*: The TaskTracker process is responsible for executing the MapReduce tasks in each DataNode. There is one instance running in each DataNode.

Jobs are described using the class org.apache.hadoop.mapred.JobConf; in its instances the programmers can define all the parameters needed for the job execution: job name, map function, reduce function, input & output formats, $\cdots$

In our example, the Hadoop job implementation is follows defined in Algorithm 1.

Algorithm 1: An example of Hadoop Job implementation

```
package com.springer.datacenter.example;
import org.apache.hadoop.fs.Path;
import org.apache.hadoop.io.IntWritable;
import org.apache.hadoop.io.Text;
import org.apache.hadoop.mapred.FileInputFormat;
import org.apache.hadoop.mapred.FileOutputFormat;
import org.apache.hadoop.mapred.JobClient;
import org.apache.hadoop.mapred.JobConf;
import org.apache.hadoop.mapred.TextInputFormat;
import org.apache.hadoop.mapred.TextOutputFormat;
public class MapReduceTask {
    public static void main(String[] args) throws Exception {
        JobConftaskConfiguration = new JobConf(MapReduceTask.class);
        taskConfiguration.setJobName("wordcount");
        taskConfiguration.setOutputKeyClass(Text.class);
        taskConfiguration.setOutputValueClass(IntWritable.class);
        taskConfiguration.setMapperClass(MapFunction.class);
        taskConfiguration.setCombinerClass(ReduceFunction.class);
        taskConfiguration.setReducerClass(ReduceFunction.class);
        taskConfiguration.setInputFormat(TextInputFormat.class);
        taskConfiguration.setOutputFormat(TextOutputFormat.class);
        FileInputFormat.setInputPaths(taskConfiguration, new Path(args[0]));
        FileOutputFormat.setOutputPath(taskConfiguration, new Path(args[1]));
        JobClient.runJob(taskConfiguration);
    }
}
```

Input data, and the mechanisms needed to access it, are defined using subclasses of the interface org.apache.hadoop.mapred.InputFormat. The framework uses this instances to validate the specification of the data, split the data into logical shards (each of which is then assigned to an individual instance of the map function) and provide a mechanism to access the input records (an instance of org.apache.hadoop.mapred.RecordReader).

Output data, and the mechanisms needed to store it, are defined using subclasses of the interface org.apache.hadoop.mapred.OutputFormat. In an analogous case, the framework uses this instance to perform validations on the output data. The mechanism for storing the data is provided by an instance of org.apache.hadoop.mapred.RecordWriter.

Map functions are defined by implementing the Mapper interface from the package org.apache.hadoop.mapred. Most of the time, these implementations are passed (to the framework) using the job configuration object (a JobConf instance). One Mapper instance is created for each InputSplit generated by the InputFormat. The Mapper interface only defines one method called map that accepts the following parameters:

- Key of the input record to process.
- Value of the input record to process.
- An OutputCollector instance. This object provides a common way to collect the data generated by the function.
- A Reporter instance. This object provides a tool for reporting progress, updating counters or status information, ...

In our example the map function implementation is defined in the Algorithm 2.

Algorithm 2: Map function implementation

```
package com.springer.datacenter.example;

import java.io.IOException;
import java.util.StringTokenizer;

import org.apache.hadoop.io.IntWritable;
import org.apache.hadoop.io.LongWritable;
import org.apache.hadoop.io.Text;
import org.apache.hadoop.mapred.MapReduceBase;
import org.apache.hadoop.mapred.Mapper;
import org.apache.hadoop.mapred.OutputCollector;
import org.apache.hadoop.mapred.Reporter;
import java.io.IOException;
import java.util.StringTokenizer;

import org.apache.hadoop.io.IntWritable;
import org.apache.hadoop.io.LongWritable;
import org.apache.hadoop.io.Text;
import org.apache.hadoop.mapred.MapReduceBase;
import org.apache.hadoop.mapred.Mapper;
import org.apache.hadoop.mapred.OutputCollector;
import org.apache.hadoop.mapred.Reporter;

public class MapFunction extends MapReduceBase implements
Mapper<LongWritable, Text, Text, IntWritable> {
    private final static IntWritable one = new IntWritable(1);
    private Text intermediateWordKey = new Text();
    public void map(LongWritable key, Text value, OutputCollector<Text,
     IntWritable> output, Reporter reporter) throws IOException {
        String line = value.toString();
        StringTokenizer tokenizer = new StringTokenizer(line);
        while (tokenizer.hasMoreTokens()) {
            intermediateWordKey.set(tokenizer.nextToken());
            output.collect(intermediateWordKey, one);
        }
    }
}
```

Reduce functions are defined by implementing the Reducer interface from the package org.apache.hadoop.mapred. However, the reduce task is done in three phases: shuffle, sort and reduce. In the shuffle phase, the relevant data (for each reduce instance) generated by the mappers is fetched by HTTP. Once this phase is done, the resulting data is sorted because different mappers can produce the same keys. Finally, in the reduce phase as its name suggests the reduce function is applied to the sorted data. The Reducer interface only defines one method called reduce that accepts the following parameters:

- Key of the input record to process.
- An Iterator to access the collection of values mapped into the same key.
- An OutputCollector instance. As in the Mapper interface, this object provides a common way to collect the data generated by the function.
- A Reporter instance. As in the Mapper interface, this object provides a tool for reporting progress, updating counters or status information, $\cdots$

In our example the reduce function implementation is defined in the Algorithm 3.

Algorithm 3: Reduce function implementation

```
package com.springer.datacenter.example;

import java.io.IOException;
import java.util.Iterator;

import org.apache.hadoop.io.IntWritable;
import org.apache.hadoop.io.Text;
import org.apache.hadoop.mapred.MapReduceBase;
import org.apache.hadoop.mapred.OutputCollector;
import org.apache.hadoop.mapred.Reducer;
import org.apache.hadoop.mapred.Reporter;

public class ReduceFunction extends MapReduceBase implements Reducer
    <Text, IntWritable, Text, IntWritable> {
    public void reduce(Text key, Iterator<IntWritable> values,
            OutputCollector<Text, IntWritable> output, Reporter reporter) throws
            IOException {
        int totalCount = 0;
        while (values.hasNext()) {
            totalCount = totalCount + values.next().get();
        }
        output.collect(key, new IntWritable(totalCount));
    }
}
```

Once the job is configured, it is necessary to submit the job to the Hadoop cluster. This submission can be completed using a command from the Hadoop distribution.

4.3 Database Support: HBase

Apache Hbase is a distributed database for the Hadoop framework. It is based on the Bigtable system [40] built at Google by Chang, Dean and others. Bigtable was described [40] as "a distributed storage system for managing structured data that is designed to scale to a very large size: petabytes of data across thousands of commodity servers".

In this system a Bigtable (a table in a RDBMS) is a sparse, distributed, persistent multidimensional sorted map. The map is indexed by a row key, a column key and a timestamp. Each value stored in the map is an uninterpreted array of bytes. The following design considerations were made:

- The row keys could be arbitrary strings (up to 64 KB).
- Every operation (read/write) made to data under the same row key has to be atomic (with independence of the number of column keys involved).
- The data has to be maintained in lexicographic order by row key.
- The row range for a table has to be dynamically partitioned. Each row range was the unit of distribution and load balancing.
- Column keys were grouped into sets called column families. A column family is the basic unit of access control; all the data stored in a column family was intended to be of the same type. So, the column family must be created prior to the writing of the data.
- Each cell (value in a row key for a column key) can contain several versions of the same data. These versions are identified by the timestamps values (64-bit integers). The different values were stored in decreasing (timestamp value) order to optimize the access to the most recent one.

As a result of the above design choices, HBase can be viewed as a type of "NoSQL" database. However, HBase is closer to a "Data Store" than a "Data Base" system because it lacks many of the features usually found in an RDBMS. However, HBase has many features for supporting horizontal and vertical scaling:

- Strongly consistent read/write operations. This makes it very suitable for tasks such as high-speed counter aggregation.
- Automatic sharding of tables: HBase tables are distributed on the cluster via regions (row ranges), and regions are automatically split and re-distributed as your data grows.
- Automatic node (RegionServer) failover.
- Hadoop/HDFS Integration: HBase supports HDFS out of the box as its distributed file system.
- MapReduce: HBase supports massively parallelized processing via MapReduce.

But there are notorious differences with a standard RDBMS:

- HBase is not an ACID compliant database. However, it does guarantee certain specific properties. The description of these properties can be found at [41].
- HBase does not supports joins is a common way or as it expected from a RDBMS.

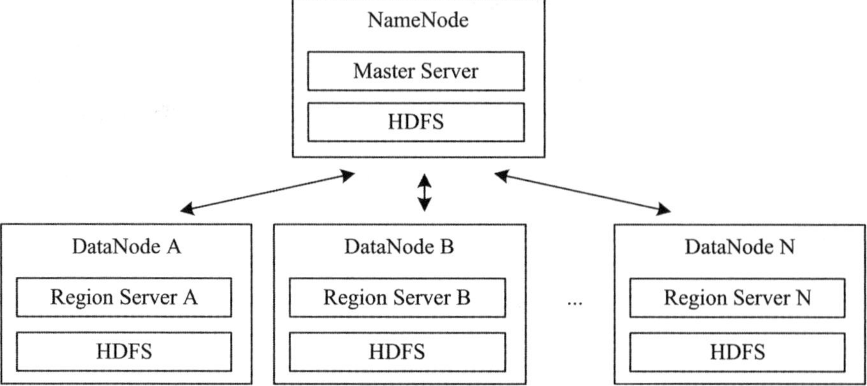

Fig. 5 The standard topology for HBase based systems

When installed in a distributed environment, the different nodes are grouped into two categories using the Master / Slave design pattern mentioned in previous sections. In HBase, HMaster is the implementation of the Master Server and it typically runs on the NameNode. This Master server is responsible for monitoring all RegionServer (slaves) instances in the cluster, and is the interface for all metadata changes. All this information is stored in two system tables called ROOT and META. A RegionServer is responsible for serving and managing regions which are the basic element of availability and distribution for tables. In a distributed cluster, a RegionServer usually runs on a DataNode. A standard HBase installation is showed in Fig. 5.

HBase provide the following operations for data manipulation:

- *Scan*: Allows iteration over the rows selected for a set of attributes given.
- *Get*: Returns attributes for a specified row.
- *Put*: Adds or updates rows in a table.
- *Delete*: Removes a row from a table.

The integration between the Hadoop framework and HBase is seamless. HBase tables are exposed as instances of the TableInputFormat class (which is a subclass of InputFormat from the Hadoop API) for reading and as instances of TableOutputFormat for writing the reduce data.

4.4 High Level Programming Language: Pig

Apache Pig is a tool for analyzing massive data sets built to execute on top of a Hadoop cluster. It consists of a high level programming language called Pig Latin and a compiler that translates the Pig Latin programs into MapReduce ones.

The main goal of this project is to provide a programming language that allows the user to simply define complex data analysis tasks. The details of how the parallelism is achieved are hidden from the user. So, the system can optimize the execution of those tasks automatically, allowing the user to focus on the data analysis rather than optimizing his programs. It also provides an extensible API that allows users to write their own Pig Latin functions.

The grammar of the Pig Latin language and the provided functions is located at [42]. Some of these functions are:

- *LOAD*: Loads data from the hadoop file system (by default). However, the users can define load functions to retrieve the data from other storage options. For example, the Pig standard distribution also includes a load function for Apache HBase.
- *STORE*: Saves the data selected to the hadoopfilesystem (by default). As in the LOAD functions, the users can provide their own store implementation. A function to write the data into Apache HBase is also provided.
- *FILTER*: Select records from a data set based on some condition given.
- *ORDER BY*: Sorts a data set based on one or more fields.
- *GROUP*: Groups the data in one or more sets.
- *FOREACH*: Generates data transformations based on fields.

Let us introduce an example of its use as the following problem: A system administrator needs to catalog the HTTP log of web servers. Assume all the log files are concatenated into one (and only one) big file. The file contains two fields: IP of the client and URL visited.

A Pig Latin program to count the different URL visited from each IP is shown in the Algorithm 4.

Algorithm 4: An example of a PIG program

```
1: A = LOAD '/springer/data/accessLog.log' USING PigStorage(',') AS
   (IP:chararray,URL:chararray);
2: B = GROUP A BY IP;
3: C = FOREACH B GENERATE (group,COUNT(A));
4: STORE C INTO '/resultsTest2' USING PigStorage(',');
```

4.5 Hive: Another Database Support & High Level Programming Language

Apache Hive is another tool for data summarization, data analysis and querying built on top of the Hadoop framework. Hive is compatible with HDFS, HBase and other Hadoop compatible file systems. It also provides tools to extract, transform and load data, methods to unify the structure of the stored data and a proprietary SQL-like language (called Hive QL) for query execution. It also has support for several kinds of indexes and allows joins in the queries.

But Hive is not designed to be used as a relational database, for example online transaction processing, real-time queries, row updates and so on··· It is focused on batch processing of large sets of permanent data.

From an architectural point of view, Hive is divided into three main components. The first one is called MetaStore and it is a metadata server implementation; it is used for holding all the information about the tables and the partitions in the system. The second one, called SerDe, is a collection of libraries of serializers and deserializers for several data formats; it also allows users to build their own implementations. The third, and last one, is the Query processor which is a framework for converting the Hive QL queries to a graph of MapReduce jobs and executing them.

Inside Hive the data is organized using the following structures ordered by size in descendant order:

- Database: Databases represent namespaces for data unit separation and naming conflicts prevention.
- Tables: Tables are homogeneous sets of data that share the same scheme.
- Partitions: Each table can be divided into several parts using a key; this key also determines how the data is stored. This is a similar concept to regions in HBase or table partitions in a RDBMS.
- Buckets (or Clusters): The data stored in each partition can also be grouped into a smaller entity (a bucket). The data is grouped using a hash function on some columns of the table.

As in a standard RDBMS, the Hive QL sentences differences two types of sentences: the ones expressed in its Data Definition Language (DDL) or the ones expressed in its Data Manipulation Language (DML). The full definition of those languages can be found at [43] and [44] respectively.

A DDL sentence for creating a table may be like the following is shown in the Algorithm 5.

Algorithm 5: Hive DDL example

```
CREATE TABLE springer.access_log(SERVER_DNS STRING, USER_IP
STRING, ACCESS_TIME STRING, URI STRING);
COMMENT 'This is the DDL Sample sentence';
PARTITIONED BY(SERVER_DNS STRING);
CLUSTERED BY(USER_IP) SORTED BY(ACCESS_TIME) INTO 32
BUCKETS;
```

Algorithm 6: Another Hive DDL example

```
CREATE INDEX URI_INDEX
ON TABLE springer.access_log (URI)'
PARTITIONED BY(SERVER_DNS STRING)
AS 'org.apache.hadoop.hive.ql.index.compact.CompactIndexHandler'
```

A simple Hive QL query will look like the written in the Algorithm 7.

Algorithm 7: An example of an HiveQL query

```
SELECT springer.access_log.*
FROM springer.access_log
WHERE ACCESS_TIME ≤ '2013-03-01' AND ACCESS_TIME ≤ '2013-03-31'
```

5 Conclusions

The MapReduce programming model is based on solid design principles and is compatible with other approaches for developing parallel and distributed software as shown at [18]. There are many others patterns for designing and developing such kind of applications [45] but those patters usually lack two of the major benefits of this approach:

- From the developer point of view, MapReduce users are unaware of the underlying parallelism, distribution and fault tolerance mechanisms. This simplifies the coding task and allows them to concentrate on building the application; the existence of a high quality framework also reduces the testing phase as the base layer (provided by the framework) does not need to be tested.
- From the systems administration point of view, this approach guarantees two features required in batch processing: horizontal scaling and fault tolerance. This features makes it possible to deploy such applications on a cluster built on top of commoditized hardware (which is a key factor in leveraging the data center costs).

The MapReduce programming model and its reference implementation (the Apache Hadoop framework) is going to become a de facto standard for processing massive data sets (if it isn't it already). Although it is not intended for general use, for example it probably will not replace distributed databases, this framework is being used in large data centers [22, 34] for building scalable software. Having such corporations supporting it, in addition to the fact that it is developed and maintained by a large community of developers, is a guarantee of its continuity in the long run.

Acknowledgment Part of the computations of this work were performed in EOLO, the HPC of Climate Change of the International Campus of Excellence of Moncloa, funded by MECD and MICINN.

References

1. Laney, D.: 3D Data Management Controlling-Data Volume, Velocity and Variety (February 2001)
2. LSST Corporation: LSST and Technology Innovation (2013)
3. Google Corporation: Waze Champs Meetup at Waze HQ. http://blog.waze.com/ (2013)

4. Bollen, J., Mao, H., Zeng, X.: Twitter Mood Predicts the Stock Market. Journal of Computational Science **2**(1) (2011) 1–8
5. Tumasjan, A., Sprenger, T.O., Sandner, P.G., Welpe, I.M.: Predicting Elections with Twitter: What 140 Characters Reveal About Political Sentiment. In: Proceedings of the Fourth International AAAI Conference on Weblogs and Social Media. (May 23–26 2010) 178–185
6. O'Connor, B., Balasubramanyan, R., Routedge, B., Smith, N.: From Tweets to Polls: Linking Text Sentiment to Public Opinion Time Series. In: Proceedings of the Fourth International AAAI Conference on Weblogs and Social Media. (May 23–26 2010) 122–129
7. Golder, S.A., Macy, M.W.: Diurnal and Seasonal Mood Vary with Work, Sleep, and Daylength Across Diverse Cultures. Science **333**(6051) (may 2002) 1878–1881
8. Armbrust, M., Fox, A., Griffith, R., Joseph, A.D., Katz, R., Konwinski, A., Lee, G., Patterson, D., Rabkin, A., Stoica, I., Zaharia, M.: A View of Cloud Computing. Communications of the ACM **53**(4) (April 2010) 50–58
9. Dorband, J.E., Raytheon, J.P., Ranawake, U.: "Commodity Computing Clusters at Goddard Space Flight Center". Online journal of space communication, School of Media Arts and Studies Scripps College of Communication, Ohio University (2013)
10. Neuman, B.C.: Scale in Distributed Systems. Readings in Distributed Computing Systems (1994) 463–489
11. Gamma, E., Helm, R., Johnson, R., Vlissides, J.: Design Patterns: Elements of reusable Object-Oriented Software. Addison-Wesley (1994)
12. Buschmann, F., Meunier, R., Rohnert, H., Sommerlad, P., Stal, M.: Pattern Oriented Software Architecture: A System of Patterns. Volume 1. J. Willey (1999)
13. Fowler, M.: Analysis Patterns: Reusable Object Models. 1 edn. Addison-Wesley Professional (1996)
14. Schmidt, D., Stal, M., Rohnert, H., Buschmann, F.: Pattern-oriented Software Architecture: Patterns for Concurrent and Networked Objects. Volume 2. J. Willey (2000)
15. S. Ishikawa, M.S.: Pattern Language: Towns, Buildings, Construction. Oxford University Press (1977)
16. Buschmann, F., Henney, K., Schmidt, D.C.: Pattern-Oriented Software Architecture: On Patterns and Pattern Languages. Volume 5. J. Willey (April 2007)
17. Buschmann, F., Henney, K., Schmidt, D.C.: Pattern-Oriented Software Architecture: A Pattern Language for Distributed Computing. Volume 4. J. Willey (2007)
18. OPL Working Group, B.U.: A Pattern Language for Parallel Programming ver2.0 (2013)
19. Fayad, M., Schmidt, D.C.: Object-Oriented Application Frameworks. Communications of the ACM **40**(10) (October 1997) 32–38
20. Dean, J., Ghemawat, S.: MapReduce: Simplified Data Processing on Large Clusters. In: Proceedings of the 6th Symposium on Operating System Design and Implemention, San Francisco, CA (December 2004) 1–13
21. Rajaraman, A., Ullman, J.D.: Mining of Massive Datasets. Cambridge University Press (2011)
22. Thusoo, A.: Hive - A Petabyte Scale Data Warehouse using Hadoop. https://www.facebook.com/note.php?note_id=89508453919 (2009)
23. O'Malley, O., Murthy, A.: Hadoop Sorts a Petabyte in 16.25 Hours and a Terabyte in 62 Seconds. http://developer.yahoo.com/blogs/hadoop/hadoop-sorts-petabyte-16-25-hours-terabyte-62-422.html (2013)
24. Apache Software Foundation: Hive™. http://hive.apache.org/ (2013)
25. Apache Software Foundation: Pig™. http://pig.apache.org/ (2013)
26. Microsoft Corporation: DryadLINQ™. http://research.microsoft.com/en-us/projects/dryadlinq/ (2013)
27. MongoDB Inc: MongoDB™. http://www.mongodb.org/ (2013)
28. Apache Software Foundation: HBase™. http://hbase.apache.org/ (2013)
29. Apache Software Foundation: Cassandra™. http://cassandra.apache.org/ (2013)
30. Pavlo, A., Paulson, E., Rasin, A., Abadi, D.J., DeWitt, D.J., Madden, S., Stonebraker, M.: A Comparison of Approaches to Large-Scale Data Analysis. In: Proceedings of the International Conference on Management of Data, New York, NY, USA, ACM (August 2009) 165–178

31. Dean, J., Ghemawat, S.: MapReduce: A Flexible Data Processing Tool. Communications of the ACM **53**(1) (January 2010) 72–77
32. Stonebraker, M., Abadi, D., DeWitt, D.J., Madden, S., Paulson, E., Pavlo, A., Rasin, A.: MapReduce and Parallel DBMSs: Friends or Foes? Communications of the ACM **53**(1) (January 2010) 64–71
33. Apache Software Foundation: Hadoop. http://hadoop.apache.org/ (2013)
34. Singh, S.: Hadoop at Yahoo!: More Than Ever Before. http://developer.yahoo.com/blogs/hadoop/hadoop-yahoo-more-ever-095826045.html/ (2013)
35. Apache Software Foundation: Hadoop Wiki. http://wiki.apache.org/hadoop/PoweredBy/ (2013)
36. Apache Software Foundation: HDFS Users Guide. http://hadoop.apache.org/docs/stable/hdfs_user_guide.html/ (2013)
37. NetApp Corporation: Open Solution for Hadoop. http://www.netapp.com/us/solutions/big-data/hadoop.aspx/ (2013)
38. MapR Technologies, Inc: MapR^{TM} Distribution for Apache Hadoop Advantages. http://www.mapr.com/products/why-mapr/ (2013)
39. Gupta, K., Jain, R., Koltsidas, I., Pucha, H., Sarkar, P., Seaman, M., Subhraveti, D.: GPFS-SNC: An Enterprise Storage Framework for Virtual-Machine Clouds. IBM Journal of Research and Development **55**(6) (December 2011) 2:1–2:10
40. Chang, F., Dean, J., Ghemawat, S., Hsieh, W.C., Wallach, D.A., Burrows, M., Chandra, T., Fikes, A., Gruber, R.E.: Bigtable: A Distributed Storage System for Structured Data. In: Proceedings of the 7th USENIX Symposium on Operating Systems Design and Implementation, Berkeley, CA, USA, USENIX Association (November 2006) 15–15
41. Apache Software Foundation: Apache HBase^{TM}. http://hbase.apache.org/acid-semantics.html/ (2013)
42. Apache Software Foundation: Pig Latin Basics. http://pig.apache.org/docs/r0.11.1/basic.html/ (2013)
43. Apache Software Foundation: LanguageManual DDL. LanguageManual+DDL/ (2013)
44. Apache Software Foundation: LanguageManual DML. https://cwiki.apache.org/confluence/display/Hive/LanguageManual+DML/ (2013)
45. Snir, M.: A Compilation of Parallel Patterns http://www.cs.uiuc.edu/homes/snir/PPP/ (2013)
46. Apache Software Foundation: Apache 2.0 license: http://www.apache.org/licenses/LICENSE-2.0.txt

Cloud Storage over Multiple Data Centers

Shuai Mu, Maomeng Su, Pin Gao, Yongwei Wu, Keqin Li and Albert Y. Zomaya

1 Introduction

Cloud storage has become a booming trend in the last few years. Individual developers, companies, organizations, and even governments have either taken steps or at least shown great interests in data migration from self-maintained infrastructure into cloud.

Cloud storage benefits consumers in many ways. A recent survey among over 600 cloud consumers [80] has shown that primary reasons for most clients in turning to cloud are: (1) to have highly reliable as well as available data storage services; (2) to reduce the capital cost of constructing their own datacenter and then maintaining it;

S. Mu (✉)
Department of Computer Science and Technology, Tsinghua National Laboratory for Information Science and Technology (TNLIST), Tsinghua University, Beijing, China
Research Institute of Tsinghua University in Shenzhen, Shenzhen, China
e-mail: msmummy@gmail.com

M. Su · P. Gao · Y. Wu
Tsinghua University, Beijing, China
e-mail: maomengsu19881010@gmail.com

P. Gao
e-mail: pin.gao2008@gmail.com

Y. Wu
e-mail: wuyw@tsinghua.edu.cn

K. Li
Department of Computer Science, State University of New York at New Paltz, New Paltz, USA
e-mail: lik@newpaltz.edu

A. Y. Zomaya
Centre for Distributed and High Performance Computing School of Information Technologies, The University of Sydney, Sydney, Australia
e-mail: albert.zomaya@sydney.edu.au

S. U. Khan, A. Y. Zomaya (eds.), *Handbook on Data Centers*,
DOI 10.1007/978-1-4939-2092-1_24

and (3) to provide high-quality and stable network connectivity to their customers. The prosperity of the cloud market has also inspired many companies to provide cloud storage services of different quality to vast variety of companies.

Reliability and availability are the most important issues in designing a cloud storage system. In cloud storage, data reliability often refers to that data is not lost, and availability refers to data accessibility. To major cloud storage providers, accidents of data loss seldom happen. But there was an accident that Microsoft once completely lost user data of T-Mobile cellphone users [81]. Because data loss accidents seldom happen, data availability is often a more important concern to most cloud storage consumers. Almost all cloud storage providers have suffered from temporary failures, lasting from hours to days. For example, Amazon failed to provide service in October, 2012, which caused many consumers such as Instgram to halt their service.

Data replication is an effective way to improve reliability and availability. Limited by cost, cloud providers usually use commodity hardware to store consumers' data. Replicating data into different machines can tolerate hardware failures, and replicating data into multiple data centers can tolerate failures of a datacenter, caused by earthquakes, storms and even wars.

To reduce cost of replication, data are usually divided into stripes instead of the original copy. There are many coding methods for this, originating from traditional storage research and practice. Erasure coding, which is widely used in the RAID systems, provides suitable features for data striping requirements in cloud storage environment.

Data consistency is also an important issue in cloud storage. As opposed to traditional storage systems which usually provide a strong consistency model, cloud storage often offers weaker consistency model such as eventual consistency. Some also propose other reduced consistency models such as session consistency, and fork-join consistency.

Privacy and security are very essential to some consumers. Consumers are often concerned about how their data are visible to cloud providers; whether administrators can see their data transparently. For other consumers who are less sensitive to privacy, they are more concerned about access control to data, because all the traffic finally leads to charges.

In spite of all the efforts of cloud storage providers, there is an emerging trend to build an integration layer on top of current cloud storages, also named "cloud-of-clouds". A cloud-of-cloud system makes use of current cloud storage infrastructure, but still provides a uniformed user interface to top-level application developers. It also targets the reliability, availability and security issues, but takes a different approach of using each cloud as a building block. The advantage of this approach is that it can tolerate performance fluctuation of single cloud storage, and can avoid potential risks of a provider's shutdown.

In the remainder of this chapter, we first review cloud storage architecture at a high level in Sect. 2. Section 3 describes common strategies used in data replication.

Section 4 gives a brief introduction on data striping. Section 5 introduces the consistency issue. Section 6 briefly highlights a new model of cloud-of-clouds. Section 7 discusses privacy and security issues in storage systems. Section 8 summarizes and suggests future directions to cloud storage research and practice.

2 Cloud Storage in a Nutshell

In this section we give an overview of cloud storage architecture and its key components. Cloud storage environments are usually complex systems mixed with many mature and new techniques. On a high level, cloud storage design and implementation consist of two parts: metadata and data, which will be investigated later.

2.1 Architecture

Early public clouds and most of today's private ones are built into a single datacenter, or several datacenters in nearby buildings. They are composed of hundreds or even thousands of commodity machines and storage devices, connected by high-speed networks. Besides large amounts of hardware, many other storage middleware such as distributed file systems are also necessary to provide storage service to consumers. The typical architecture of cloud storage usually includes storage devices, distributed file system, metadata service, frontend, and other components.

In practice, we find that the data models and library interfaces of different clouds are fairly similar; thus, we could support a minimal set to satisfy most users' needs. The data model shared by most services could be summarized as a "container-object" model, in which file objects are put into containers. Most services containers do not support nesting; i.e., users cannot create a sub-container in a container.

In the last decade, major cloud storage such as Amazon S3 [1] and Windows Azure Storage [2] have upgrade their service from running in separate datacenters to different data centers and different geographic regions. Compared to a single datacenter structure, running services in multiple data centers require a multitude of more resource management functions, such as, resource allocation, deployment, and migration.

An important feature of cloud storage is the ability to store and provide access to an immense amount of storage. Amazon S3 currently has a few hundred petabytes of raw storage in production, and it also has a few hundred more petabytes of raw storage based on customer demand [21]. Modern cloud storage architecture could be divided into three layers: storage service, metadata service, and front-end layer (Fig. 1).

- Metadata service—The metadata service is in charge of following functions: (a) handling high level interfaces and data structures; (b) managing a scalable namespace for consumers' objects; (c) storing object data into the storage service.

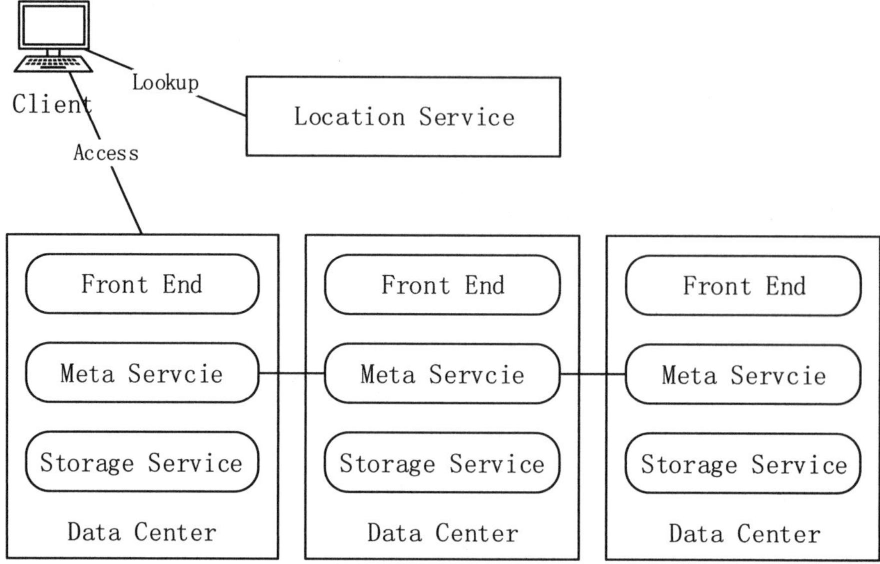

Fig. 1 Cloud storage architecutre over mutiple data centers

Metadata service holds the responsibility to achieve scalability by partitioning all of the data objects within a datacenter. This layer consists of many metadata servers, each of which serves for a range of different objects. Also, it should provide load balance among all the metadata servers to meet the traffic of requests.

- Storage service—This storage service is in charge of storing the actual data into disks and distributing and replicating the data across many servers to keep data reliable within a datacenter. The storage service can be thought of as a distributed file system. It holds files, which are stored as large storage chunks. It also understands how to store them, how to replicate them, and so on, but it does not understand higher level semantics of objects. The data is stored in the storage service, but it is accessed from the metadata service.
- Front-End (FE) layer—The front-end layer consists of a set of stateless servers that take incoming requests. Upon receiving a request, an FE looks up the account, authenticates and authorizes the request, then routes the request to a partition server in the metadata service. The system maintains a map that keeps track of the partition ranges and which metadata server is serving which partition. The FE servers cache the map and use it to determine which metadata server to forward each request to. The FE servers also file large objects directly from the storage service and cache frequently accessed data for efficiency.

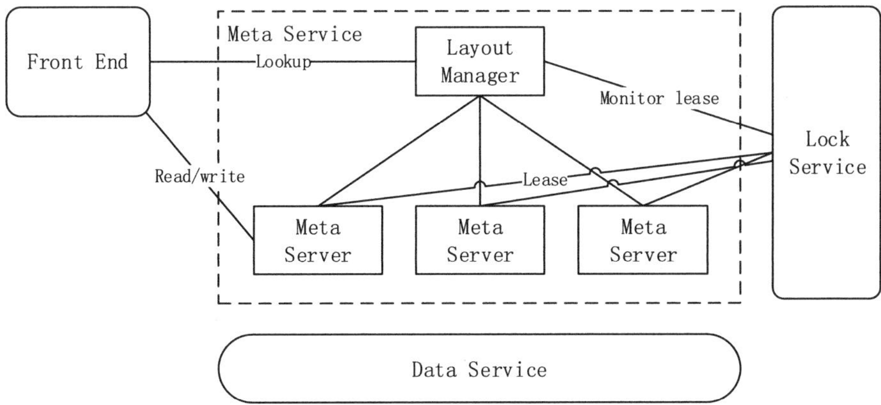

Fig. 2 Meta-layer architecture

2.2 *Metadata Service*

The metadata service contains three main architectural components: a layout manager, many meta-servers, and a reliable lock service (Fig. 2). The architecture is similar to Bigtable [5].

2.2.1 Layout Manager

A layout manager (LM) acts as a leader of the meta-service. It is responsible for dividing the whole metadata into ranges and assigning each meta-server to serve several ranges and then keeping track of the information. The LM stores this assignment in a local map. The LM must ensure that each range is assigned only to one active meta-server, and that two ranges do not overlap. It is also in charge of load balancing ranges among meta-servers. Each datacenter may have multiple instances of the LM running, but usually they function as reliable replications of each other. For this they need a Lock Service to maintain a lease for leader election.

2.2.2 Meta-Server

A meta-server (MS) is responsible for organizing and storing a certain set of ranges of metadata, which is assigned by LM. It also serves requests to those ranges. The MS stores all metadata into files persistently on disks and maintains a memory cache for efficiency. Meta-servers keep leases with the Lock Service, so that it is guaranteed that no two meta-servers can serve the same range at the same time.

If a MS fails, LM will assign a new MS to serve all ranges served by the failed MS. Based on the load, LM may choose a few MS rather than one to serve the ranges. LM firstly assigns a range to a MS, and then updates its local map which specifies

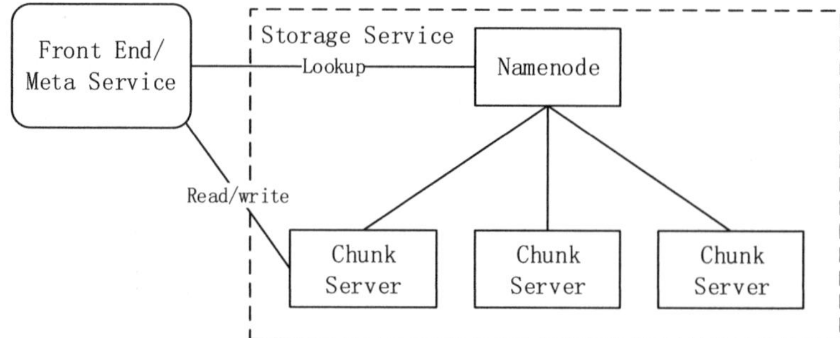

Fig. 3 Storage service architecture

which MS is serving each range. When a MS gets a new assignment from LM, it firstly acquires for the lease from Lock Service, and then starts serving the new range.

2.2.3 Lock Service

Lock Service (LS) is used by both of layout manager and meta-server. LS uses Paxos [16] protocol to do synchronous replication among several nodes to provide a reliable lock service. LM use LS for leader election; MS also maintains a lease with the LS to keep alive. Details of the LM leader election and the MS lease management are discussed here. We also do not go into the details of Paxos protocol. The architecture of lock service is similar to Chubby [18].

2.3 Storage Service

The two main architecture components of the storage service are the namenode and chunk server (Fig. 3). The storage service architecture is similar to GFS [4].

2.3.1 Namenode

The namenode can be considered as the leader of the storage service. It maintains file namespace, relationships between chunks and each file, and the chunk locations across the chunk servers. The namenode is off the critical path of client read and write requests. In addition, the namenode also monitors the health of the chunk servers periodically. Other functions of namenode include: lazy re-replication of chunks, garbage collection, and erasure code scheduling.

The namenode periodically checks the state of each chunk server. If the namenode finds that the replication number of a chunk is smaller than configuration, it will start a re-replication of the chunk. To achieve a balanced chunk replica placement, the namenode randomly chooses chunk server to store new chunk.

The namenode is not tracking any information about blocks. It remembers just files and chunks. The reason of this is that the total number of blocks is so huge that the namenode cannot efficiently store and index all of them. The only client of data service is the metadata service.

2.3.2 Chunk Servers

Each chunk server keeps the storage for many chunk replicas, which are assigned by the namenode. A chunk server machine has many large volume disks attached, to which it has complete access. A chunk server deals only with chunks and blocks, and it does not care about file namespace in the namenode. Internally on a chunk server, every chunk on disk is a file consisting of data blocks and their checksum. A chuck server also holds a map which specifies relationships between chunk and file. Each chunk server also keeps a view about the chunks it owns and the location of the peer replicas for a given chunk. This view is copied from namenode and is kept as a local cache by the chunk server. Under instructions from namenode, different chunk servers may talk to each other to replicate chunks, or to create new copies of an existing replica. When a chunk no longer stores any alive chunks, the namenode starts garbage collection to remove the dead chunks and free the space.

3 Replication Strategies

3.1 Introduction

Currently, more data-intensive applications are moving their large-scale datasets into cloud. To provide high availability and durability of storage services as well as improving performance and scalability of the whole system, data replication is adopted by many mature platforms [1, 2, 4, 6, 12] and research studies [7–10, 14, 30] in cloud computing and storage. Data replication is to keep several identical copies of a data object in different servers that may distribute across multiple racks, houses and region-scale or global-scale datacenters, which can tolerate different levels of failures such as facility outages or regional disasters [4, 10, 23, 30]. Replication strategy is now an indispensable feature in multiple datacenters [1, 2, 6–9, 12], which may be hundreds or thousands of miles away from each other, to completely replicate data objects of services, not only because wide-area disasters such as power outages or earthquakes may occur in one datacenter [10, 23], but also because replication across geographically distributed datacenters can mostly reduce latency and improve the whole throughput of the services in the cloud [6–9, 11].

Availability and durability is guaranteed as one data object is replicated on many servers across datacenters, thus in the presence of failing of a few number of components such as servers and network at any time [1, 4, 10, 23] or natural disasters occurring in one datacenter, the durable service of cloud storage won't be influenced because applications can normally access their data through servers containing replicas in other datacenters. Moreover, as each data object is replicated over multiple datacenters, it enables different applications to be served from the fastest datacenter or the datacenter with the lowest working load in parallel [1, 6, 9, 11, 31], thus providing high performance and throughput of the overall cloud storage system.

Common replication strategies can be divided into two categories: asynchronous replication and synchronous replication. They own distinct features and have different impacts on availability, performance, and throughput of the whole system. Besides, the cloud storage service should provide the upper applications with a consistent view of the data replicas especially during faults [6–9, 11], which requires that data copies among diverse datacenters should be consistent with each other. However, these two replication strategies bring in new challenges to replication synchronization, which finally will influence the consistency of data replicas over multiple datacenters.

Additionally, the placement of data replicas is also an important aspect of replication strategy in multiple datacenters as it highly determines the load distribution, storage capacity usage, energy consumption and access latency, and many current systems and studies [1, 4, 6, 10, 24, 26] adopt different policies for the placement of data replicas in the multiple-datacenter design on different demands.

In this section, we will present the main aspects and features of asynchronous replication, synchronous replication, and placement of replicas.

3.2 *Asynchronous Replication*

Figure 4a illustrates the working mechanism of asynchronous replication over multiple datacenters. As shown in Fig. 4, the external applications issue write requests to one datacenter, which could be a fixed one configured previously or a random one chosen by applications, and get a successful response if the write requests completely commit in this datacenter. The updated data will be eventually propagated to other datacenters in background in an asynchronous manner [1, 2, 12]. Asynchronous replication is especially useful when the network latency between datacenters is at a high cost as applications only need to commit their write requests in one fast datacenter and don't have to wait for the data to be replicated in each datacenter. Therefore, the overall performance and throughput for writes will be improved and systems with asynchronous replication can provide high scalability as they are decentralized. Now many prevailing systems such as Cassandra [12], Dynamo [1], and PNUTS [14] are using asynchronous replication.

However, asynchronous replication presents a big challenge to consistency, since replicas may have conflicting changes with each other, that is, the view of all the replicas over multiple datacenters has the probability to be inconsistent at some time. Figure 5 presents a simple scenario that will cause inconsistency among replicas. Assume there are three datacenters A, B and C, and all of them hold data replica d. When a write request for d from application P is issued to A and successfully commits in A, A will response to P and then replicates the updated data d_1 to B and C. However,

D_l for one datacenter

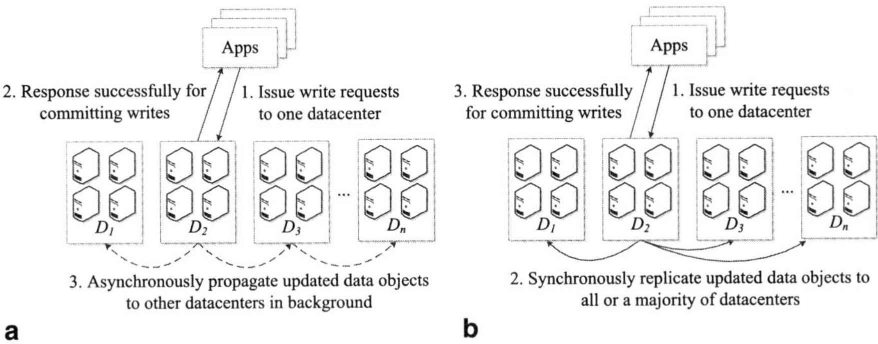

Fig. 4 The working mechanism of asynchronous replication and synchronous replication over multiple datacenters. **a** for asynchronous replication and **b** for synchronous replication

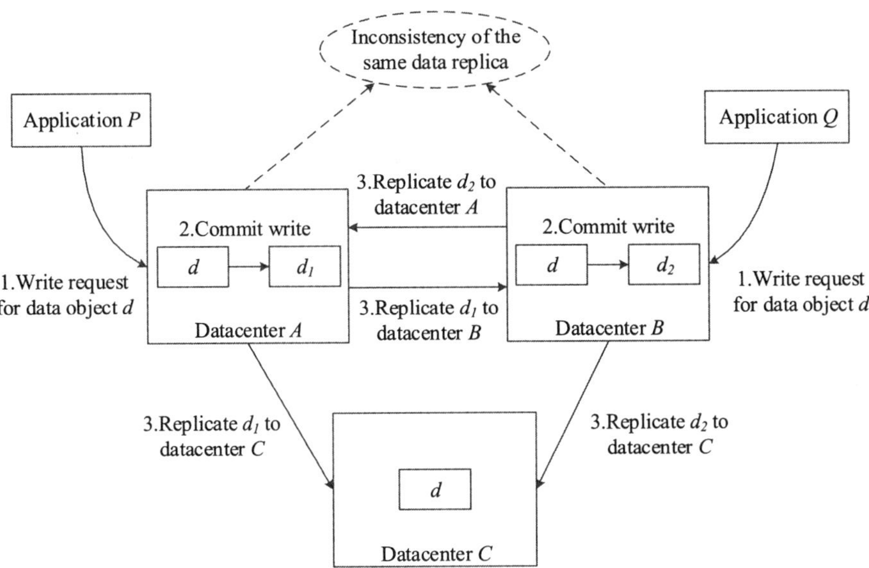

Fig. 5 A scenario that causes inconsistent view of data replicas among datacenters under asynchronous replication

at the same time, another write request for d from application Q is issued to C. As C hasn't gotten to know the update of d in A, it normally accepts and processes this write request and then d in C turns into d_2 and will be replicated to A and B. As a result, there are now two different versions of the same data replica, and the system steps into an insistent state which means that a subsequent read may get two different data objects.

As there are also other factors such as server or network failure that will cause inconsistency in asynchronous replication over multiple datacenters [1, 11], a few researches have been addressing this challenge of asynchronous replication. Eventual consistency [21] model is one scheme that is widely adopted by many studies and widespread distributed systems [1, 12–14]. Eventual consistency model allows the whole system to be inconsistent temporarily but eventually, the conflicted data objects will merge into one singe data object and the view of the data replicas across multiple datacenters will become consistent at last. The process of merging conflicted data objects is critical in eventual consistency model and the merging decision can be made by the write timestamp [12, 21], a chosen master [13, 14] or even the applications [1].

Under asynchronous replication, a read request may get a stale data object from some datacenters, which will decline the performance of current reads and complicates application development. However, whether this circumstance is adverse depends on the applications. If applications such as search engine and shopping carts allow weaker consistency at reading or demand high quality of writing experience [1, 12], asynchronous replication won't bring negative impacts to these applications.

3.3 Synchronous Replication

In contrast to asynchronous replication, synchronous replication requires that the updated data objects of write requests must be synchronously replicated to all or a majority of datacenters before applications get a successful response from the datacenter accepting the requests in the cloud, as presented in Fig. 4b. This synchronous replication mechanism can effectively guarantee a consistent view of cross-datacenter replicated data and it enables developers to build distributed systems that can provide strong consistency and a set of ACID semantics like transactions, which, compared with that in loosely consistent asynchronous replication, simplifies application building for the wide-area usage for the reason that applications can make use of serializable semantic properties while are free from write conflicts and system crashes [6, 9, 11, 20, 25].

The key point of synchronous replication is to keep states of replicas across different datacenters the same. A simple and intuitive way to realize this is to use synchronous master/slave mechanism [4, 6, 11]. The master waits for the writes to be fully committed in slaves before acknowledging to applications and is responsible for failure detection of the system. Another method to maintain consistent and up-to-date replicas among datacenters is to use Paxos [16], which is a fault-tolerant and optimal consensus algorithm for RSM [15] in a decentralized way. Paxos works well when a majority of datacenters are alive and at current, many system services adopt Paxos [2, 6, 11, 17, 18, 20] as their underlying synchronous replication algorithm.

However, no matter which method is used, the overall throughput and performance of the services based on synchronous replication will be constrained when the communication latencies between datacenters are at high expense [7, 9, 11] and scalability is limited by strong consistency to certain extent. As a result, many researches put forward mechanisms to help improve the throughput and scalability of the whole system while not destroying the availability and consistency for applications. These mechanisms include reasonable partitioning of data [2, 6, 11], efficient use of concurrent control [7, 11, 31] and adopting combined consistent models [7–9, 22, 25].

3.4 *Placement of Replicas*

As cloud storage now holds enormous amount (usually petabytes) of data sets from large-scale applications, how to place data replicas across multiple datacenters also becomes a very important aspect in replication strategy as it is closely related to load balance, storage capacity usage, energy consumption and access latency of the whole system [6, 10, 19, 24]. It is essential for both efficiently utilizing available datacenter resources and maximizing performance of the system.

Unbalanced replica placement will cause over-provisioning capacity and skewed utilization of some datacenters [26]. One way to address this issue is to choose a master or use partition layers to decide in which datacenter each data replica is placed [2, 4, 6]. This requires the master or partition layers to record the up-to-date load information of every datacenter so that they won't make unbalanced replica placement policies and can immediately decide to migrate data between datacenters to balance load. Another way is to use a decentralized method, as presented in Fig. 6. We can form datacenters as a ring, each responsible for a range of keys. A data object can get its key through hash functions such as consistent hash and locate a datacenter according to its key. Then, replicas of this data object could be placed in this datacenter and its successive ones, similar to [1, 12]. In this way, there is no need to maintain a master to keep information of each datacenter and if the hash functions could evenly distribute the keys, load balance can be achieved automatically.

Furthermore, as datacenters now consumes about 1.5 % of the world's total energy and a big fraction of it does come from the consumption of storage in them [28, 29], the number of datacenters to place the data replicas should also be considered carefully. If the number of datacenters to hold replicas increases, the storage capacity of the whole system will accordingly decease and the energy consumption improves [24, 26, 27] as those datacenters will contain large amounts of replicated data objects in storage. In addition, placing data replicas in a high number of datacenters enables applications to survive wide-area disasters that will cause a few datacenter failures and thus, this can provide high availability for applications at the expense of storage capacity and energy consumption [2, 6, 7, 11, 25]. Moreover, when the number of datacenters to place replicas is large, applications can have a low access latency based on geographic locality, i.e., they can communicate with datacenters that are faster or have less working load [6, 7, 9]. Hence, system developers have to consider the trade-off between these features for the placement strategy of data replicas across multiple datacenters when they are building geographically distributed services for applications.

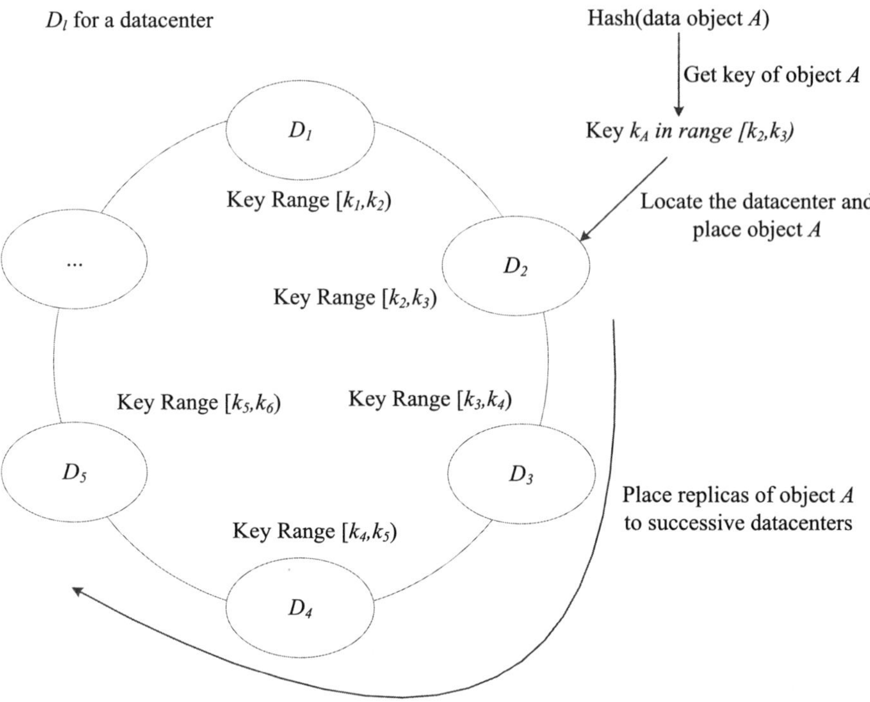

Fig. 6 The decentralized method to place data replicas across multiple datacenters

4 Data Striping Methods

4.1 Introduction

The main purpose of a storage system is to make data persistent, so reliability and availability should be top priority concern for storage systems. Actually, there are a variety of factors that may cause storage system unavailable. For example, if a server fails, the storage system is unable to provide storage services. Some physical damage to a hard disk will result in the loss of data stored. Therefore, it is indispensable for storage systems to introduce some techniques to make them reliable.

A lot of research work has been done in recent years to improve the availability and reliability of storage systems. The main idea is to generate some redundant information of every data block and distribute them on different machines. When one server becomes outage, another server that holds the redundant data can replace the role of the broken server. During this time, the storage system can still provide storage service. When one data block is broken, then other redundant data blocks will restore the broken one. Thus, the availability and reliability is improved. Generally, redundant data can be presented in two ways: one is using full data backup mechanism, called full replication; the other is erasure code.

Full replication, also known as multi-copy method, is to store multiple replicates of data on separate disks, in order to make the data redundant. This method does not involve specialized encoding and decoding algorithms, and it has better fault-tolerance performance. But full replication has lower storage efficiency. Storage efficiency is the sum of effective capacity and free capacity divided by raw capacity. When storing N copies of replica, the disk utilization is only $1/N$. For relatively large storage systems, full replication brings extra storage overhead, resulting in high storage cost.

Along with the increase of the data that a storage system holds, a full replication method has been difficult to adapt to mass storage system for redundant mechanism in disk utilization and fault tolerance requirements. Therefore, erasure code is becoming a better solution for mass storage.

4.2 Erasure Code Types

Erasure code is derived from communication field. At first, it is mainly used to solve error detection and correction problems in data transmission. Afterwards, erasure code gradually applied to improve the reliability of storage systems. Thus, erasure code has been improved and promoted according to the characters of storage system. The main idea of erasure code is that the original data can be divided into k data fragments, and according to the k data fragments, m redundant fragments can be computed according some coding theory. The original data can be reconstructed by any of the $m + k$ fragments. There are many advantages of erasure code, the foremost of these is the high storage efficiency compared with the mirroring method.

There are many types of erasure code. Reed-Solomon code [52] is an MDS code that can meet any number of data disks and redundant disk number. MDS code (maximum distance separable code) is a kind of code that can achieve the theoretically optimal storage utilization. The main idea of Reed Solomon code is to visualize the data encoded as a polynomial. Symbols in data are viewed as coefficients of a polynomial over a finite field. Reed Solomon code is a type of horizontal codes. Horizontal code has the property that data fragments and redundant fragments are stored separately. That is to say, each stripe is neither data stripe nor redundant stripe. Reed Solomon codes are usually divided into two categories: one is Vandermonde Reed Solomon code, and the other is Cauchy Reed Solomon code [53]. The difference between these two categories of Reed Solomon codes is that they are using different generation matrix. For Vandermonde Reed Solomon code, the generation matrix is Vandermonde matrix, and multiplication on Galois filed is needed which is very complex. For Cauchy Reed Solomon code, it is Cauchy matrix, and every operation is XOR operation, which is coding efficient. Figure 7 shows the Encoding principle for Reed-Solomon codes.

Compared with Reed Solomon Codes, Array Code [54] is totally based on XOR operation. Due to the efficient of encoding, decoding, updating and reconstruction,

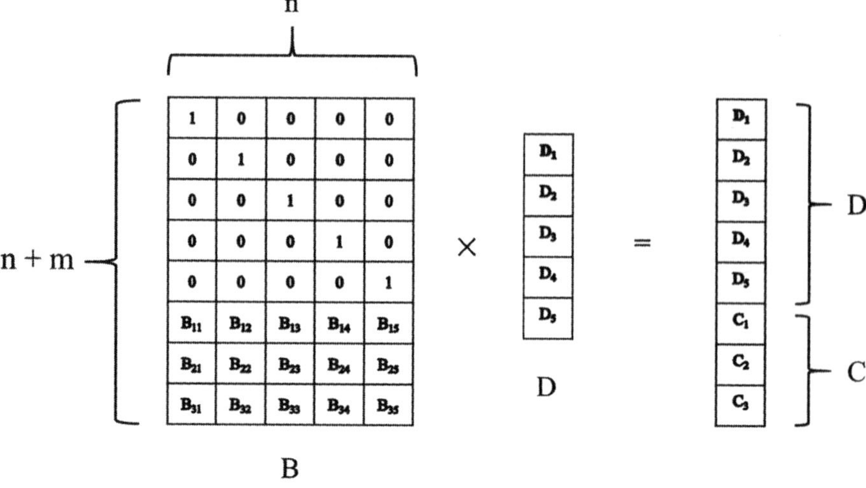

Fig. 7 Reed-Solomon codes

Array code is widely used. Array code can be categorized as two types due to the placement of data fragment and redundant fragment.

Horizontal parity array codes make data fragments and redundant fragments stored on different disks. By doing this, Horizontal parity array codes have better scalability. But most of it can just hold 2 disk failures. It has a drawback on updating data. Every time updating one data block will result in at least one read and one write operation on redundant disk. EVENODD code [55] is one kind of Horizontal parity array codes that used widely.

Vertical parity array codes make data fragment and redundant fragment stored in the same stripe. Because of this design, the efficiency of data update operation will be improved. However, the balance of vertical parity array code leads to a strong interdependency between the disks, which also led to its poor scalability. XCODE [56] is a kind of vertical parity array code, which has theoretically optimum efficiency on data update and reconstruction operation.

4.3 Erasure Codes in Data Centers

In traditional storage systems such as early GFS and Windows Azure Storage, to ensure the reliability, triplication has been favored because of its ease of implementation. But triplication makes the stored data triple, and storage overhead is a major concern. So many system designers are considering erasure coding as an alternative. Most distributed file systems (GFS, HDFS, Windows Azure) create an append-only write workload for large block size. So data update performance is not a concern.

Using erasure code in distributed file systems, data reconstruction is a major concern. For one data of k data fragment and m redundant fragment, when any one

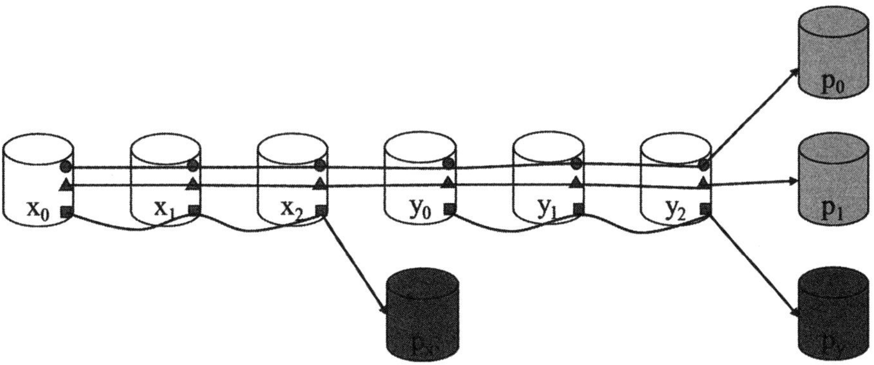

Fig. 8 LRC codes

of that fragment is broken or lost, to repair that broken fragment, k fragments size of network bandwidth will be needed. So some researchers found that the traditional erasure code does not fit distributed file system very well. In order to improve the performance of data repair, there are two ways.

One is reading from fewer fragments. In Windows Azure Storage System, a new set of code called Local Reconstruction Codes (LRC) [57] is adopted. The main idea of LRC is to reduce the number of fragments required to reconstruct the unavailable data fragment. To reduce the number of fragments needed, LRC introduced local parity and global parity. As Fig. 8 shows below, x_0, x_1, x_2, y_0, y_1, y_2 are data fragments, p_x is the parity fragment of x_0, x_1, x_2. p_y is the parity fragment of y_0,y_1,y_2. p_0 and p_1 are global parity fragments. p_x and p_y are called local parity. p_0 and p_1 are global parity. When reconstructing x_0, instead of reading p_0 or p_1 and other 5 data fragment, it is more efficient to read x_1, x_2 and p_x to compute x_0. As we can see LRC is not a MDS, but it can greatly reduce the cost of data reconstruction.

Another way to improve reconstruction performance is to read more fragments but less data size from each. Regenerating codes [58] provide optimal recovery bandwidth among storage nodes. When reconstructing fragments, it does not just transmit the existing fragments, but sends a liner combination of fragments. By doing this, the recovery data size to send will be reduced. Rotated Reed-Solomon codes [59] and RDOR [60] improve reconstruction performance in a similar way.

5 Consistency Models

5.1 *Introduction*

Constructing a globally distributed system requires many trade-offs between availability, consistency, and scalability. Cloud storages are designed to serve for a large amount of internet-scale applications and platforms simultaneously, which is often named as infrastructure service. To meet operational requirements, cloud storage must be designed and implemented as highly available and scalable, in order to serve consumers requests from all over the world.

One of the key challenges in build cloud storage is to provide a consistency guarantee to all client requests [63]. Cloud storage is a large distributed system deployed world-widely. It has to process millions of requests every hour. All the low-probability accidents in normal systems are often to happen in the datacenters of cloud storage. So all these problems must be taken care of in the design of the system. To guarantee consistent performance and high availability, replication techniques are often used in cloud storage. Although replication solves many problems, it has its costs. Different client requests may see inconsistent states of many replicas. To solve this problem, cloud storage must define a consistency model that all requests to replicas of the same data must follow.

Like many widespread distributed systems, cloud storage such as Amazon S3 often provides a weak consistency model called eventual consistency. Different clients may see different orders of updates to the same data object. Some cloud storage like Windows Azure also provides strong consistency that guarantees linearizability of every update from different clients. Details will be discussed in the following sub-sections.

5.2 *Strong Consistency*

Strong consistency is the most programmer-friendly consistency model. When a client commits an update, every other client would see the update in subsequent operations. Strong consistency can help achieve transparency of a distributed system. When developer uses a storage system with strong consistency, it appears like the system is a single component instead of many collaborating sub-components mixed together.

However, this approach has been proved as difficult to achieve since the middle of last century, in the database area for the first time. Databases are also systems with heavy use of data replications. Many of such database systems were design to shut down completely when it cannot satisfy this consistency because of node failures. But this is not acceptable for cloud systems, which is so large that small failures are happening every minute.

Strong consistency has its weak points, one of which is that it lowers system availability. In the end of last century, with large-scale Internet systems growing up, designs of consistency model are rethought. Engineers and researchers began to reconsider the tradeoff between system availability and data consistency. In the year of 2000, CAP theorem was introduced [61]. The theorem states that for three properties of shared-data systems—data consistency, system availability, and tolerance to network partition—only two can be achieved at any given time.

It is worth noting, that the concept of consistency in cloud storage is different to that in transactional storage systems such as databases. The common ACID property (atomicity, consistency, isolation, durability) defined in databases is a different kind of consistency guarantee. In ACID, consistency means that the database is in a consistent state when a transaction is finished. No go-between situation is allowed.

5.3 Weak Consistency

According the CAP theory, a system can achieve both consistency and availability, if it does not tolerate network partitions. There many techniques which make this work, one of which is to use transaction protocols like two phase commit. The condition for this is that both client and server of the storage systems must be in the same administrative environment. If partition happens and client cannot observe this, the transaction protocol would fail. However, network partitions are very common in large distributed systems, and as the system scale goes up, the chances of network partition would increase. This is one reason why one cannot achieve consistency and availability at the same time. The CAP theory provides two choices for developers: (1) sticking to strong consistency and allowing system goes unavailable under partitions (2) using relaxed consistency [65] so that system is still available under network partitions.

No matter what kind of consistency model the system uses, it requires that application developers are fully aware of the consistency model. Strong consistency is usually the easiest option for client developer. The only problem the developers have to deal with is to tolerate the unavailable situation that might happen to the system. If the system takes relaxed consistency and offers high availability, it may always accept client requests, but client developers have to remember that a write may get its delays and a read may not return the newest write. Then developers have to write the application in a way so that it can tolerant the delay update and stale read. There are many applications that can be design compatible for such relaxed consistency model and work fine.

There are two ways of looking at consistency. One is from the developer/client point of view: how they observe data updates. The other is from the server side: how updates flow through the system and what guarantees systems can give with respect to updates.

Let's show consistency models using examples. Suppose we have a storage system which we treat as a black box. To judge its consistency model we have several clients issuing requests to the system. Assume they are client A, client B, client C. All three clients issue both read and write requests to the system. The three clients are independent and irrelevant. They could run on different machines, processes, or threads. The consistency model of the system can be defined by how and when observers (in this case the clients A, B, or C) see updates made to a data object in the storage systems. Assume client A has made an update to a data object:

- Strong consistency. After the update completes, any subsequent access (from any of A, B, or C) will return the updated value.

- Weak consistency. The system does not guarantee that subsequent accesses will return the updated value. A number of conditions need to be met before the value will be returned. The period between the update and the moment when it is guaranteed that any observer will always see the updated value is dubbed the inconsistency window.

There are many kinds of weak consistency; we list some of the most common ones as below.

- Causal consistency [66]. If client A has communicated to client B that it has updated a data item, a subsequent access by client B will return the updated value, and a write is guaranteed to supersede the earlier write. Access by client C that has no causal relationship to client A is subject to the normal eventual consistency rules.
- Eventual consistency [62]. This is a specific form of weak consistency; the storage system guarantees that if no new updates are made to the object, eventually all accesses will return the last updated value. If no failures occur, the maximum size of the inconsistency window can be determined based on factors such as communication delays, the load on the system, and the number of replicas involved in the replication scheme. The most popular system that implements eventual consistency is the domain name system. Updates to a name are distributed according to a configured pattern and in combination with time-controlled caches; eventually, all clients will see the update [64].
- Read-your-writes consistency. This is an important model where client A, after having updated a data item, always accesses the updated value and never sees an older value. This is a special case of the causal consistency model.
- Session consistency. This is a practical version of the previous model, where a client accesses the storage system in the context of a session. As long as the session exists, the system guarantees read-your-writes consistency. If the session terminates because of a certain failure scenario, a new session must be created and the guarantees do not overlap the sessions.
- Monotonic read consistency. If a client has seen a particular value for the object, any subsequent accesses will never return any previous values.
- Monotonic write consistency. In this case, the system guarantees to serialize the writes by the same client. Systems that do not guarantee this level of consistency are notoriously difficult to program.

These consistency models are not exclusive and independent. Some of the above can be combined together. For example, the monotonic read consistency can be combined with session-level consistency. The combination of the both consistencies is very practical for developers in a cloud storage system with eventual consistency. These two properties make it much easier for application developers to build up their apps. They also allow the storage system to keep a relax consistency and provide high availability. As you can see from these consistency models, quite a few different circumstances are possible. Applications need to choose whether or not one can deal with the consequences of particular consistency.

6 Cloud of Multiple Clouds

6.1 Introduction

Although cloud storage providers claim that their products are cost saving, trouble-free, worldwide 24/7 available and reliable, reality shows that (1) such services are sometimes not available to all customers; and (2) customers may experience vastly different accessibility patterns from different geographical locations. Furthermore, there is also a small chance that clients may not even be able to retrieve their data from a cloud provider at all, which usually occurs due to network partitioning and/or temporary failure of cloud provider. For example, authors of [67] reported that this may also cause major cloud service providers to fail providing services for hours or days sometimes. Although cloud providers sign Service Level Agreements (SLA) with their clients to ensure availability of their services, users have complained that these SLAs are sometimes too tricky to break. Moreover, even when a SLA is violated, the compensation is only a minor discount for the payment and not to cover a customer's loss resulted by the violated SLA.

Global access experience can be considered as one specifically important issue of availability. In current major cloud storages, users are asked to create region-specific accounts/containers before putting their data blobs/objects into them. The storage provider then stores data blobs/objects into a datacenter in the selected locations; some providers may also create cross-region replicas solely for backup and disaster recovery. A typical result of such topology is an observation where users may experience vastly different services based on the network condition between clients and the datacenter holding their required data. Data loss and/or corruption are other important potential threats to users' data should it be stored on a single cloud provider only. Although users of major cloud storage providers have rarely reported data loss and/or corruption, prevention of such problems is not 100 % guaranteed either. Medium to small sized cloud providers may provide a more volatile situation to their customers as they are also in danger of bankruptcy as well.

In this section, we present a system named μLibCloud to address the two aforementioned problems of cloud customers; i.e., (1) availability of data as a whole and (2) different quality of services for different customers accessing data from different locations on the globe. μLibCloud is designed and implemented to automatically and transparently stripe data into multiple clouds—similar to RAID's principle in storing local data. μLibCloud is developed based on Apache libCloud project [3], and evaluated through global-wide experiments.

Our main contributions include: (1) to conduct global-wide experiments to show how several possible factors may affect availability and/or global accessibility of cloud storage services; (2) to use erasure codes based on observations. We then design and implement μLibCloud using erasure code to run benchmarks accessing several commercial clouds from different places in the world. The system proved the effectiveness of our method.

Fig. 9 Layer abstraction of cloud storage

Applications
Library
REST/SOAP
Cloud Storage

6.2 Architecture

Using a "cloud-of-cloud" rationale [68], μLibCloud is to improve availability and global access experience of data. Here the first challenge is how to efficiently and simultaneously use multiple cloud services. They follow different concepts and offer different ways to access their services. As shown in Fig. 9, cloud storage providers usually provide REST/SOAP web service interface to developers along with their libraries for different programming languages for developers to further facilitate building cloud applications. To concurrently use multiple cloud storages, two options are available. The first option is to set up proxy among cloud storages and applications. In this case, all data requests need to go through this proxy. To store data in cloud storages, this proxy receives original data from client, divides the data into several shares, and sends each share to different clouds using different libraries. To retrieve data, it fetches data shares from each cloud, rebuilds the data, and sends it back to clients. The second option—more complicated—is to integrate the support for multiple cloud storages directly into a new client library—replacing original ones. In this case, client applications only use this newly provided library to connect to all clouds. The main difference between these two options is the transparency in the second option to spread/collect data to/from multiple clouds.

The first choice is more straightforward in design; it uses a single layer for extra work, keeps the client neat and clean, includes many original libraries when implemented, and is usually run on independent servers. It also brings more complexity to system developers to maintain extra servers and their proper functioning. The second choice, on the other hand, benefits developers by providing them a unique tool; this approach also reduces security risk because developers do not need to put their secret keys on the proxy. It however also leads to other challenges on how to design and implement such systems; e.g., how multiple clients can coordinate with each other without extra servers. Furthermore, the client library must be efficient and resource saving because it needs to be run along with application codes.

In the design of μLibCloud, we chose to practice the second option so that it will have lesser of a burden on application developers. We also assume that consumers who choose to use cloud storage rather than to build their own infrastructure would not want to set up another server to make everything work. Figure 10 shows the basic architecture of μLibCloud with a single client; this figure also shows how μLibCloud serves upper-level users, while hiding most of development complexities of such systems.

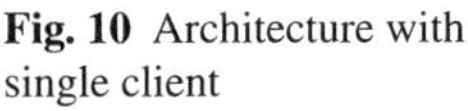

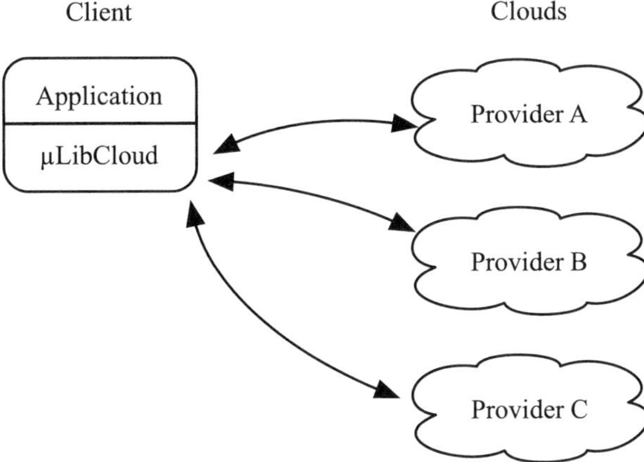

Fig. 10 Architecture with single client

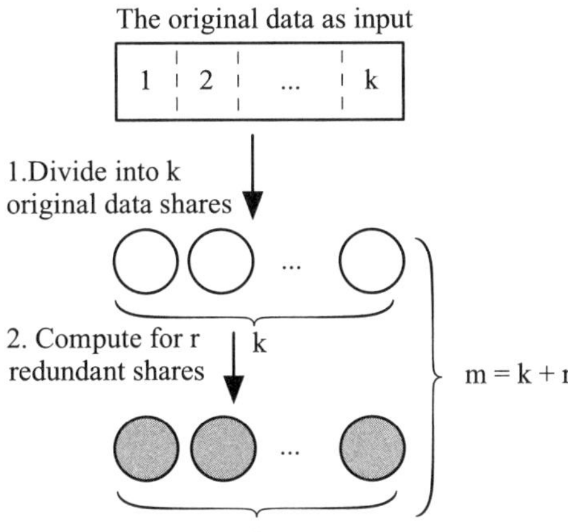

Fig. 11 Principle of erasure coding

6.3 *Data Striping*

As described before, data is first encoded into several original and redundant shares, and then stored on different providers. Through this redundancy, data not only is protected against possible failures of particular providers—high availability, but also tolerates the instability of individual clouds and provides consistent performance.

Among many possible choices for data encoding [69], we choose the most widely used erasure code [70] that is widely used in both storage hardware [71] and distributed systems [72]. Here, coding efficiency is a major concern because all the data striping algorithm work is performed at clients' side; i.e., large overheads that could decrease performance of applications is strongly unacceptable.

Figure 11 shows principles of erasure coding. As can be seen, data is first divided into k equal-sized fragments called original data shares. Then, r parity fragments with the same size as original data shares are computed and called redundant data

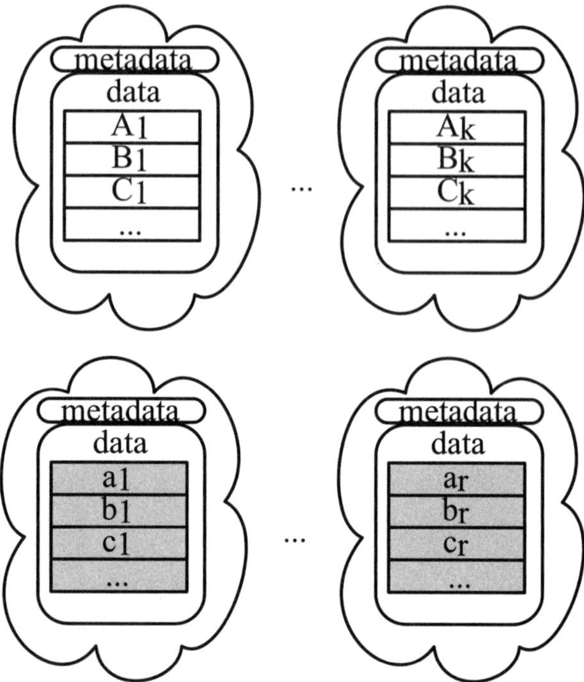

Fig. 12 Data stripes stored in each cloud

shares. This will generate a total of $m = k + r$ equal-sized shares. The erasure code algorithm guarantees any arbitrary k shares—out of total m shares—is sufficient enough to reconstruct the original data. Both k and r are positive values and are predefined by each user.

Here, we also define redundancy rate as $R = m/k$ to reflect the amount of storage overhead for storing data. For example, if $k = 1$, $m = 2$; then, $R = m/k = 200\,\%$. It means that each data takes twice of its original size when stored: one original and one replica. In this case, each piece is enough to reconstruct the original data—like RAID 1 (mirroring). If $k = 4$, $m = 5$ (like RAID 5); then, $R = m/k = 120\,\%$. It means that we need extra 20 % of storage to store any data. In this case, every four pieces—out of all available five pieces—are enough to reconstruct the original data.

In practice, we do not simply just divide an object into several parts and encode them, but the original data is first divided into several chunks, and then erasure coding is performed on each chunk; default chunk's size is usually 64 KB (Fig. 12). There are two benefits in splitting data into several chunks: (1) computation of erasure coding can be parallelized, and (2) reading and writing of file data—such as video and audio—can also be easily supported.

6.4 Retrieving Strategy

If a developer divides data into (*k*, *m*) shares, among all *m* parts of data the client library only needs *k* parts to reconstruct the original data. Although retrieving all parts of data could avoid the potential risks of failures, it is unnecessary in most cases. It also wastes more bandwidth and costs more money. Here, although retrieving *k* data shares to recover the data is enough, selecting the best possible *k* shares can be tricky. In μLibCloud we offer the following three data fetching strategies.

1. Efficient: Users want to use the *k* most available clouds to retrieve data pieces. Here, to determine which ones are faster, μLibCloud dynamically measures their download speed. When retrieving an object, all metadata files are downloaded first and their link speed is recorded. Upon that, *k* fastest clouds to fetch data are selected. During downloading the main data, μLibCloud keeps recording the download speed to compute its average. The larger data is, the more accurate network estimation would be.
2. Economical: If application is mainly run in the background—like a backup program storing data into clouds [73]—, users can tolerate spending more time. In such cases, economical cost is more important than speed. μLibCloud also offers a cost-saving mode, in which it will select k providers with lowest prices.
3. Custom: We also offer an option, allowing developers to set priorities of their own. This may be preferable in case that they are using computing and storage resources provided by the same provider. For example, if a developer is deploying applications into EC2 and use storage of S3, it would be reasonable that s/he wants to use S3 as the first choice.

6.5 Mutual Exclusion

When there is more than one client in the system, they must be able to coordinate with each in certain ways to avoid conflicts. Such conflicts can result in not only client read failures, but also inconsistent states and/or even data loss. For example, if two clients concurrently write to the same data file without any locking, they may write to each other's share and produce problems. In the worst case scenario, if the provider takes an eventual consistency model (like Amazon S3), all unordered writes would succeed although only the later ones become effective. As a result, it would be very probable that a client succeeds modifying several data shares, while the other client succeeds in the rest of data shares; both clients would return successful, while data inconsistency has already occurred! The following options are among the most suitable one for our needs.

1. Setting up a central lock server such as ZooKeeper [74] to coordinate all writes. This approach is easy and correct for a system like μLibCloud, yet with certain flaws. Firstly, with this approach clients need to maintain another system, which violates goals and principles of using clouds for simplicity in the first place.

Secondly, coordinators like ZooKeeper usually have throughput issues because of their leader-follower architecture, especially in internet-scale situation. Although this can be reduced by manually partitioning data onto multiple groups of ZooKeeper systems, this would still make the system extremely complex.

2. Running a client-client agreement protocol. Here, instead of deploying an additional central lock service, agreement protocols such as Paxos [75] handles the situation. This approach eliminates the trouble of bringing a lock service, but requires clients to be able to communicate with each other. In this case, frequent membership changes can seriously damage system performance. In fact, this approach is almost the same—in logic—as the first option if each client runs with a ZooKeeper member deployed to the same machine.
3. Manipulating lock files on each cloud storage. Instead of setting up an additional lock server or running an agreement protocol among clients, there is another approach more suitable to this situation. Each client creates empty files on each cloud as lock-files; this is called mutual exclusion in the area of distributed algorithms [76]. This option is more difficult to achieve because each cloud is purely an object storage that offers neither computing ability, nor a common compare and swap (cas) semantics usually used in fulfilling lock services.

In order to achieve mutual exclusion without introducing new bottlenecks, we introduced Algorithm 1 based on the third option. This algorithm is an improved version of another algorithm formerly designed by Bessani [77].

Algorithm 1 Mutual exclusion at client side

```
input: id, and providers[1, 2, .., m]
output: success or failure
for i from 1 to m do
  create a file named lock_id on provider[i]
end for
count = 0
for i from 1 to m do
  list all lock files on provider[i]
  if found any lock files created by other client then
    count = count + 1
  end if
end for
if count ≥ m/2 then
  for i from 1 to m do
    delete lock_id from provider[k]
  end for
else
  {critical section}          // lock succeeds
end if
```

Following comments is worth noting about this algorithms.

1. The algorithm is fault tolerate to possible failures of less than *m*/2 providers. In case a client fails and stops during any step, we add a timestamp t_{create} to the name of each lock file. Thus, when a client lists a file name with the $t_{create} + t_{delta} < t_{now}$, s/he can confidently deletes the expiring lock. To maintain correctness, we must choose a t_{delta} large enough to cover the entire operation time when created; it must also be able to tolerate possible time differences among clients.
2. To be correct, the algorithm requires each cloud to have an appropriate consistency model. To be specific, after each 'create' command all lists must see the creation. However, several major cloud providers, such as Amazon S3, employ an eventual consistency model [78]. It means the writes are not visible to reads immediately, and if one client detects a change, it does not imply other clients can also detect it. To tolerate eventual consistency, the client may need to wait for another time period, after each write to make sure it can be seen by all clients too. The time period is set by observation to model time delays among clients [64].

Amazon S3 recently releases an enhanced consistency model to most of its cloud storages, namely "read-after-write" consistency to ensure that for newly created objects, the write (not overwrite) can be seen immediately. Our algorithm (Algorithm 1) employs this feature in its locking system; this is why Algorithm 1 creates new lock files instead of writing to the old ones.

3. The algorithm is obstruction-free [79]; i.e., it is still possible—although very rare—that no client can progress. This flaw could be tolerated because most applications tend to have many more reads than writes—where only very few writes require mutual exclusion.

7 Privacy and Security of Storage System

7.1 *Introduction*

In the last few years, cloud computing has enabled more and more customers (such as companies or developers) to run their applications on the remote servers with elastic storage capacity and computing resources required on demand. The proliferation of cloud computing encourages customers to store and keep their data in the cloud instead of maintaining local data storage [32–34, 38, 39]. However, a key factor that may hinder the process of data migration from local storage to the cloud is the potential privacy and security concerns inside clouds [33, 34]. As customers don't own and manage remote servers directly by themselves, any malicious applications or administrators in the cloud can get access to, abuse or even damage the data of normal customers' applications. This phenomenon is especially adverse to the confidentiality of sensitive data objects of customers such as banks or financial companies. Under this circumstance, datacenters in the cloud must maintain strong protections on the privacy and security of data objects against untrustworthy applications, servers and administrators during the process of data storing and accessing [34, 36, 39, 46].

To guarantee data privacy and security in storage system of datacenters in the cloud, several basic solutions such as data access control [38–41], data isolation [36, 37, 42, 46, 47] and cryptographic techniques [35, 40, 43–45] have been proposed by researchers. All these solutions are intended to meet different requirements of data privacy and security and to make even the most privacy and security demanding applications to migrate their sensitive data into cloud with no concerns. In this section, combined with our experience of building privacy and security policies in datacenters in the cloud, we will present how these mechanisms can be used in a real world.

7.2 Fine-Grained Data Access Control

Data access control is highly related to the privacy and security provided to applications when they are accessing the data [33, 38, 39, 41]. Applications, if not allowed, don't have the authority to access the data of others. Besides, each application may have its own access control policies to maintain the data privacy and security among its users. For instance, one application may require that only its administrators can have the authority to modify and delete its data and other common users can only read these data. Therefore, storage systems in datacenters must ensure strict and flexible data access control mechanisms for upper applications to secure the data object sets of every application.

Algorithm 2 **The procedure of storing a data object.**

Input: P_n, d_m

1. *Get* k_m *of* d_m
2. *Get* L_n *of* P_n
3. *Insert* k_m *into* L_n
4. *Keep* d_m *in physical storage*

Figure 13 illustrates the overview of a fine-grained access control mechanism on data object level in a datacenter. As presented in Fig. 13, there are two main data structures for the correct process of fine-grained data-object-level access control: a set of lists keeping the keys of data objects that belong to each application and a set of tables recording each application's access control policy. Every application owns its list of keys and access control policy table. When one application stores a data object into the datacenter, the storage system will allocate a globally unique key to this data object and add this key into the list of this application, which means this data object does belong to such application. Denote an application as P_n, the list of P_n as L_n, a data object as d_m and the key of data object d_m as k_m, then the process of storing data in this mechanism could be summarized as Algorithm 2.

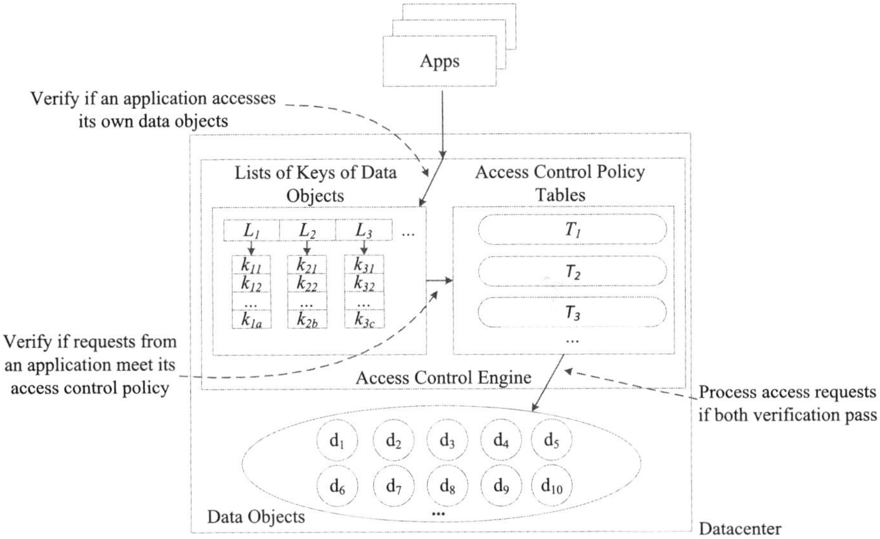

Fig. 13 The overview of the fine-grained access control mechanism on the data object level

When an access request for a data object issued from an application arrives at the datacenter, the storage system will first get the key of the data object and verifies if this key is in this application's list. Storage system will forbid the application to access this data object if the verification fails. This procedure ensures that data objects of one application are isolated from the other applications against illegal intrusion. Moreover, if the verification passes, the system will further check if this access request meets the requirements listed in the access control policies table of the application. This will prevent unauthorized application users from abusing operations on data of this application that may potentially damage these data. Applications can set and modify their access control policies according to their own demands and the policy information are recorded in their access control policy tables respectively. The access request is accepted and processed only after the check in the access control policy table successes. Denote an access request as R_p and access control policies table of application P_n as T_n, then the procedure to process an access request can be illustrate as Algorithm 3.

With Algorithm 2 and Algorithm 3, the data privacy and security could be achieved across applications through fine-grained data-object-level access control mechanism without impacting the normal usage of data by authorized users of each application. Furthermore, as these two data structures (lists and tables) that are used by the access control mechanism could keep a consistent view across multiple datacenters using replication strategy presented in Sect. 3, the privacy and security of data could be easily guaranteed through this fine-grained data-object-level data access control mechanism among multiple datacenters in the cloud.

Algorithm 3 The procedure of processing an access request on a data object.

Input: P_n, R_p, d_m

Get k_m *of* d_m
Get L_n *of* P_n
if $k_m \in L_n$ **then**
  Get T_n *of* P_n
  if R_p *meets requirements in* T_n **then**
    Process R_p
  else
    Refuse to process R_p
  end if
else
  Refuse to process R_p
end if

7.3 *Security on Storage Server*

Under fine-grained data-object-level access control mechanism, the privacy and security of applications' data could be protected against external untrusted users and applications. However, data stored in the storages servers of datacenters are still prone to abuse or compromise by untrusted processes running in these servers or malicious administrators of datacenters that can get the whole authority of the OS [36, 37, 51]. To address this issue, most studies [36, 37, 42, 46, 47, 51] use virtual-machine-based protection mechanisms to isolate applications' data kept in hardware (memories and disks) of storage servers from operating systems and other processes, and to authenticate the integrity of these data. This protection ensures that even operating systems carry out the overall task of managing data they cannot read or modify them. With this guarantee, even though malicious administrators or untrusted processes get the authority of OS, they have no access to abusing or damaging the data stored in the hardware. When trusted applications request to get their data, this mechanism would make sure that these applications will be presented with a normal view of their original data, hiding the complex underlying details of protection. Hence, the privacy and security of applications' data can be maintained in storage servers of datacenters in the cloud.

Figure 14 characterizes the architecture of the privacy and security protection mechanism in storage servers. The key component, as shown in Fig. 14, to protect the privacy and security of applications' data in hardware is the virtual machine monitor (VMM). The VMM could monitor the process/OS interactions such as system calls [36, 42] and directly manage the hardware to isolate memories and disks from operating systems [37, 42, 46], which makes it possible to prevent the data privacy and security against malicious processes or administrators that can get the authority to control operating systems.

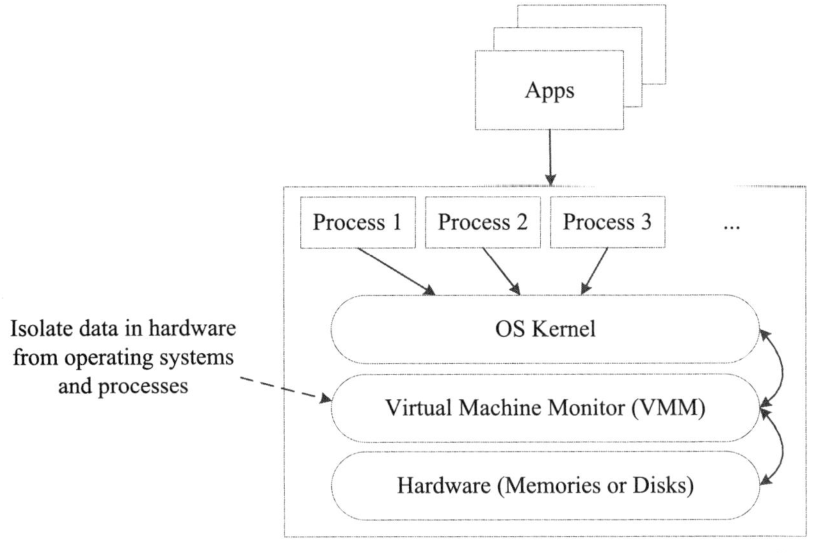

Fig. 14 The architecture of the privacy and security protection mechanism in a storage server

Generally, each process owns its independent virtual memory address space and is associated with a page table that maps the virtual memory address into the physical memory address [48] to use memory. The page tables of processes and the operations of address mappings are managed by the OS and thus, it has the authority to access the memory address space of all processes running on it. As applications' requests are served by specific processes in storage servers of datacenters, once malicious processes or administrators steal the operating system's authority, they can easily access the data of other normal processes through their page tables and threaten the privacy and security of applications' data. To address this challenge, VMM could protect the page tables of each process and complete the operations of memory address mappings instead of operating system [48, 51]. The OS can only access its kernel memory space through its own page table, without interleaving with other processes. However, even though the OS doesn't know the distribution of processes' virtual memory in the physical memory, malicious processes or administrators could also access the physical memory through OS [48, 49] and analyze or tamper the data in the memory [42, 50]. As a result, VMM is responsible for keeping the data in the physical memory in an encrypted and integrated view [49]. When a process requests to put data into memory, VMM will detect this request, encrypt the data and then put the encrypted data into the memory. If one trusted process requests to get its data in the memory, VMM will first authenticate the integrity of the encrypted data and then decrypt them before returning the original clear data to this process, which doesn't have to cover this middle process and just utilizes memory as normal. To complete the encrypting and decrypting procedures mentioned above, VMM holds a specific zone of memory that is secure enough against the attacks from operating systems and processes. Consequently, when processes are serving applications' requests in memory of storage servers, the privacy and security of their data in memory can be strongly protected.

As most of applications' data will be stored into disks of storage servers in datacenters, it is also critical to guarantee the privacy and security of data in disks [36, 42, 51] not only because untrusted processes and administrators that get the authority of OS can directly access data in disks through I/O operations, but administrators could fetch disks manually. As a result, data in disks must also be stored in an encrypted view so that even some processes or administrators get control on the disks of storage servers, they have no way to abuse or compromise the data stored in them. VMM also has the responsibility for data encryption/decryption when processes interact with disks through the OS. When a process wants to write its data into disks, it will use a system call sys_write [48] and passes the data to the operating system, which will execute the operations to really write data to disks. VMM will detect this system call from the process and obtain the data before passing to the operating system. Then VMM encrypts these data and calculate the checksum of the encrypted data for future integrity verification. After this procedure, VMM will transfer the encrypted data to the operating system that will normally write these data into disks. Similarly, when one process requests to get its data from disks, it will issue a system call sys_read to the operating system to fetch these data. VMM will also detect this system call and wait for the operating system to complete the read operations of the encrypted data. Then VMM authenticates the integrity of the encrypted data, decrypts them and return plain data to the process. All the underlying details of encryption/decryption are still hidden from the processes and to the operating system, although it manages the data during the operations of read and write, it only views data after encryption and can't threaten the privacy and security of the original data objects.

With these virtual-machine-based mechanisms, the data of applications can be kept in storage servers of datacenters without concerns of being abused or compromised by malicious processes or administrators in the datacenters. As data privacy and security can be achieved in hardware of each storage server, datacenters in the cloud have the ability to provide high privacy and security for applications to move their large sets of data into cloud and freely access their data on demands.

8 Conclusion and Future Directions

In this chapter we mainly discussed the architecture of modern cloud storage and several key techniques used in building such systems. Cloud storage systems are typically large distributed systems composed of thousands of machines and network devices over many datacenters across multiple continents. Cloud storage and cloud computing are the very mixture of modern storage and network technology. To build and maintain such systems calls for large amount of efforts from numerous developers and maintainers. Although we have discussed about replication, data striping, data consistency, security and some other issues, there are still much more of the iceberg we have not touched. Many conventional techniques in traditional storage techniques applied in cloud storage have the potentiality to evolve, such as the example we give about cloud-of-clouds, which arise from the traditional RAID system. To summarize, cloud storage is a valued area in both practice and research, and the goal of this chapter is to provide a glimpse into it when it grows into a global scale.

References

1. Varia, Jinesh. "Cloud architectures." White Paper of Amazon, jineshvaria. s3. amazonaws. com/public/cloudarchitectures-varia. pdf (2008).
2. Brad Calder, Ju Wang, Aaron Ogus, Niranjan Nilakantan, Arild Skjolsvold, Sam McKelvie, Yikang Xu, Shashwat Srivastav, Jiesheng Wu, Huseyin Simitci, Jaidev Haridas, Chakravarthy Uddaraju, Hemal Khatri, Andrew Edwards, Vaman Bedekar, Shane Mainali, Rafay Abbasi, Arpit Agarwal, Mian Fahim ul Haq, Muhammad Ikram ul Haq, Deepali Bhardwaj, Sowmya Dayanand, Anitha Aduzsumilli, Marvin McNett, Sriram Sankaran, Kavitha Manivannan, Leonidas Rigas. Windows Azure Storage: a highly available cloud storage service with strong consistency. Proceedings of the Twenty-Third ACM Symposium on Operating Systems Principles (SOSP'11), pages 143–157, 2011.
3. G. DeCandia, D. Hastorun, M. Jampani, G. Kakulapati, A. Lakshman, A. Pilchin, S. Sivasubramanian, P. Vosshall, W. Vogels. Dynamo: amazon's highly available key-value store. Proceedings of twenty-first ACM SIGOPS Symposium on Operating Systems Principles (SOSP' 07), pages 205–220, 2007.
4. Sanjay Ghemawat, Howard Gobioff, Shun-Tak Leung. The Google file system. Proceedings of the nineteenth ACM Symposium on Operating Systems Principles (SOSP' 03), pages 29–43, 2003.
5. Chang, Fay, et al. "Bigtable: A distributed storage system for structured data." ACM Transactions on Computer Systems (TOCS) 26.2 (2008): 4.
6. James C. Corbett, Jeffrey Dean, Michael Epstein, Andrew Fikes, Christopher Frost, J. J. Furman, Sanjay Ghemawat, Andrey Gubarev, Christopher Heiser, Peter Hochschild, Wilson Hsieh, Sebastian Kanthak, Eugene Kogan, Hongyi Li, Alexander Lloyd, Sergey Melnik, David Mwaura, David Nagle, Sean Quinlan, Rajesh Rao, Lindsay Rolig, Yasushi Saito, Michal Szymaniak, Christopher Taylor, Ruth Wang, Dale Woodford, D. Woodford. Spanner: Google's globally-distributed database. Proceedings of the 10th USENIX conference on Operating Systems Design and Implementation (OSDI' 12), pages 251–264, 2012.
7. Yair Sovran, Russell Power, Marcos K. Aguilera, Jinyang Li. Transactional storage for geo-replicated systems. Proceedings of the Twenty-Third ACM Symposium on Operating Systems Principles (SOSP'11), pages 385–400, 2011.
8. Wyatt Lloyd, Michael J. Freedman, Michael Kaminsky, David G. Andersen. Don't settle for eventual: scalable causal consistency for wide-area storage with COPS. Proceedings of the Twenty-Third ACM Symposium on Operating Systems Principles (SOSP'11), pages 401–416, 2011.
9. Cheng Li, Daniel Porto, Allen Clement, Johannes Gehrke, Nuno Preguica, Rodrigo Rodrigues. Making geo-replicated systems fast as possible, consistent when necessary. Proceedings of the 10th USENIX conference on Operating Systems Design and Implementation (OSDI'12), pages 265–278, 2012.
10. Luiz André Barroso, Urs Hölzle. The Datacenter as a Computer: An Introduction to the Design of Warehouse-Scale Machines. Morgan & Claypool Publishers, DOI: 10.2200/S00193ED1V01Y200905CAC006, 2009.

11. Jason Baker, Chris Bond, James C. Corbett, JJ Furman, Andrey Khorlin, James Larson, Jean-Michel Leon, Yawei Li, Alexander Floyd, Vadim Yushprakh. Megastore: Providing Scalable, Highly Available Storage for Interactive Services. In 5th Conference on Innovative Data Systems Research, pages 223–234, 2011.
12. Avinash Lakshman, Prashant Malik. Cassandra: a decentralized structured storage system. ACM SIGOPS Operating Systems Review, 44(2), pages 35–40, 2010.
13. D. B. Terry, M. M. Theimer, Karin Petersen, A. J. Demers, M. J. Spreitzer, C. H. Hauser. Managing update conflicts in Bayou, a weakly connected replicated storage system. Proceedings of the fifteenth ACM Symposium on Operating Systems Principles (SOSP'95), pages 172–182, 1995.
14. Brian F. Cooper, Raghu Ramakrishnan, Utkarsh Srivastava, Adam Silberstein, Philip Bohannon, Hans-Arno, Nick Puz, Daniel Weaver, Ramana Yerneni. PNUTS: Yahoo!'s hosted data serving platform. Proceedings of the VLDB Endowment, 1(2), pages 1277–1288, 2008.
15. Fred B. Schneider. Implementing fault-tolerant services using the state machine approach: a tutorial. ACM Computing Surveys (CSUR), 22(4), pages 299–319, 1990.
16. Leslie Lamport. Paxos made simple. ACM SIGACT News Distributed Computing Column, 32(4), pages 18–25, 2001.
17. Tushar D. Chandra, Robert Griesemer, Joshua Redstone. Paxos made live: an engineering perspective. Proceedings of the twenty-sixth annual ACM Symposium on Principles of Distributed Computing, pages 398–407, 2007.
18. Mike Burrows. The Chubby lock service for loosely-coupled distributed systems. Proceedings of the 7th symposium on Operating Systems Design and Implementation (OSDI'06), pages 335–350, 2006.
19. Jeff Dean. Designs, Lessons, and Advice from Building Large Distributed Systems. Keynote from LADIS, 2009.
20. Stacy Patterson, Aaron J. Elmore, Faisal Nawab, Divyakant Agrawal, Amr El Abbadi. Serializability, not serial: concurrency control and availability in multi-datacenter datastores. Proceedings of the VLDB Endowment, 5(11), PAGES 1459–1470, 2012.
21. Werner Vogels. Eventually consistent. Communications of the ACM—Rural engineering development, 52(1), pages 40–44, 2009.
22. Lisa Glendenning, Ivan Beschastnikh, Arvind Krishnamurthy, Thomas Anderson. Scalable consistency in Scatter. Proceedings of the Twenty-Third ACM Symposium on Operating Systems Principles (SOSP'11), pages 15–28, 2011.
23. Daniel Ford, François Labelle, Florentina I. Popovici, Murray Stokely, Van-Anh Truong, Luiz Barroso, Carrie Grimes, Sean Quinlan. Availability in globally distributed storage systems. Proceedings of the 9th USENIX conference on Operating Systems Design and Implementation (OSDI'10), No. 1–7, 2010.
24. Sage A. Weil, Scott A. Brandt, Ethan L. Miller, Carlos Maltzahn. CRUSH: controlled, scalable, decentralized placement of replicated data. Proceedings of the 2006 ACM/IEEE conference on Supercomputing, 2006.
25. Wyatt Lloyd, Michael J. Freedman, Michael Kaminsky, David G. Andersen. Stronger Semantics for Low-Latency Geo-Replicated Storage. Proceedings of the 10th USENIX Symposium on Networked Systems Design and Implementation (NSDI'13), 2013.
26. Sharad Agarwal, John Dunagan, Navendu Jain, Stefan Saroiu, Alec Wolman, Harbinder Bhogan. Volley: automated data placement for geo-distributed cloud services. Proceedings of the 7th USENIX conference on Networked Systems Design and Implementation (NSDI'10), 2010.
27. Anton Beloglazov, Rajkumar Buyya. Energy Efficient Resource Management in Virtualized Cloud Data Centers. Proceedings of the 2010 10th IEEE/ACM International Conference on Cluster, Cloud and Grid Computing, pages 826–831, 2010.
28. Zhichao Li, Kevin M. Greenan, Andrew W. Leung, Erez Zadok. Power Consumption in Enterprise-Scale Backup Storage Systems. Proceedings of the Tenth USENIX Conference on File and Storage Technologies (FAST '12), pages 65–71, 2012.

29. J. G. Koomey. Growth in data center electricity use 2005 to 2010. Technical report, Standord University, 2011.
30. Yi Lin, Bettina Kemm, Marta Patiño-Martínez, Ricardo Jiménez-Peris. Middleware based data replication providing snapshot isolation. Proceedings of the 2005 ACM SIGMOD international conference on Management of data, pages 419–430, 2005.
31. Daniel Peng, Frank Dabek. Large-scale incremental processing using distributed transactions and notifications. Proceedings of the 9th USENIX conference on Operating Systems Design and Implementation, 2010.
32. Cong Wang, Qian Wang, and Kui Ren, Wenjing Lou. Privacy-Preserving Public Auditing for Data Storage Security in Cloud Computing. Proceedings of IEEE INFOCOM, 2010.
33. S. Subashini, V. Kavitha. A survey on security issues in service delivery models of cloud computing. Journal of Network and Computer Applications, 34(1), pages 1–11, 2011.
34. H. Takabi, J.B.D. Joshi, G. Ahn. Security and Privacy Challenges in Cloud Computing Environments. IEEE Security and Privacy, 8(6), pages 24–31, 2010.
35. Kevin D. Bowers, Ari Juels, Alina Oprea. HAIL: a high-availability and integrity layer for cloud storage. Proceedings of the 16th ACM conference on Computer and Communications Security (CCS'09), pages 187–198, 2009.
36. Fengzhe Zhang, Jin Chen, Haibo Chen, Binyu Zang. CloudVisor: retrofitting protection of virtual machines in multi-tenant cloud with nested virtualization. Proceedings of the Twenty-Third ACM Symposium on Operating Systems Principles (SOSP'11), pages 203–216, 2011.
37. Xiaoxin Chen, Tal Garfinkel, E. Christopher Lewis, Pratap Subrahmanyam, Carl A. Waldspurger, Dan Boneh, Jeffrey Dwoskin, Dan R.K. Ports. Overshadow: a virtualization-based approach to retrofitting protection in commodity operating systems. Proceedings of the 13th international conference on Architectural Support for Programming Languages and Operating Systems, pages 2–13, 2008.
38. Wassim Itani, Ayman Kayssi, Ali Chehab. Privacy as a Service: Privacy-Aware Data Storage and Processing in Cloud Computing Architectures. Proceedings of Eighth IEEE International Conference on Dependable, Autonomic and Secure Computing, pages 711–716, 2009.
39. Shucheng Yu, Cong Wang, Kui Ren, Wenjing Lou. Achieving Secure, Scalable, and Fine-grained Data Access Control in Cloud Computing. Proceedings of IEEE INFOCOM, 2010.
40. Vipul Goyal, Omkant Pandey, Amit Sahai, Brent Waters. Attribute-based encryption for fine-grained access control of encrypted data. Proceedings of the 13th ACM conference on Computer and Communications Security (CCS'06), pages 89–98, 2006.
41. Myong H. Kang, Joon S. Park, Judith N. Froscher. Access control mechanisms for inter-organizational workflow. Proceedings of the sixth ACM symposium on Access Control Models and Technologies, pages 66–74, 2001.
42. H. Chen, F. Zhang, C. Chen, Z. Yang, R. Chen, B. Zang, P. Yew, and W. Mao. Tamper-resistant execution in an untrusted operating system using a virtual machine monitor. Parallel Processing Institute Technical Report, Number: FDUPPITR-2007-0801, Fudan University, 2007.
43. Lein Harn, Hung-Yu Lin. A cryptographic key generation scheme for multilevel data security. Computer & Security, 9(6), pages 539–546, 1990.
44. Christian Cachin, Klaus Kursawe, Anna Lysyanskaya, Reto Strobl. Asynchronous verifiable secret sharing and proactive cryptosystems. Proceedings of the 9th ACM conference on Computer and Communications Security (CCS'02), pages 88–97, 2002.
45. Phillip Rogaway. Bucket hashing and its application to fast message authentication. CRYPTO, volume 963 of LNCS, pages 29–42, 1995.
46. David Lie, Chandramohan A. Thekkath, Mark Horowitz. Implementing an untrusted operating system on trusted hardware. Proceedings of the nineteenth ACM Symposium on Operating Systems Principles (SOSP'03), pages 178–192, 2003.
47. Stephen T. Jones, Andrea C. Arpaci-Dusseau, Remzi H. Arpaci-Dusseau. Geiger: monitoring the buffer cache in a virtual machine environment. Proceedings of the 12th international conference on Architectural Support for Programming Languages and Operating Systems, pages 14–24, 2006.

48. Abraham Silberschatz, Peter Baer Galvin, Greg Gagne. Operating System Concepts. John Wiley & Sons, 2009.
49. Guillaume Duc, Ronan Keryell. CryptoPage: an Efficient Secure Architecture with Memory Encryption, Integrity and Information Leakage Protection. Proceedings of the 22nd Annual Computer Security Applications Conference (ACSAC'06), pages 483–492, 2006.
50. David Lie, Chandramohan Thekkath, Mark Mitchell, Patrick Lincoln, Dan Boneh, John Mitchell, Mark Horowitz. Architectural support for copy and tamper resistant software. ACM SIGPLAN Notices, 35(11), pages 168–177, 2000.
51. Hou Qinghua, Wu Yongwei, Zheng Weimin, Yang Guangwen. A Method on Protection of User Data Privacy in Cloud Storage Platform. Journal of Computer Research and Development, 48(7), pages 1146–1154, 2011.
52. Reed I S, Solomon G. Polynomial codes over certain finite fields [J]. Journal of the Society for Industrial & Applied Mathematics, 1960, 8(2): 300–304.
53. Roth R M, Lempel A. On MDS codes via Cauchy matrices [J]. Information Theory, IEEE Transactions on, 1989, 35(6): 1314–1319.
54. Blaum M, Farrell P, Tilborg H. Array Codes [M]. Amsterdam, Netherlands: Elsevier Science B V, 1998.
55. Blaum M, Brady J, Bruck J, et al. EVENODD: An efficient scheme for tolerating double disk failures in RAID architectures[J]. Computers, IEEE Transactions on, 1995, 44(2): 192–202.
56. Xu L, Bruck J. X-code: MDS array codes with optimal encoding[J]. Information Theory, IEEE Transactions on, 1999, 45(1): 272–276.
57. Huang, Cheng, et al. "Erasure coding in windows azure storage." USENIX ATC. 2012.
58. Dimakis A G, Godfrey P B, Wu Y, et al. Network coding for distributed storage systems[J]. Information Theory, IEEE Transactions on, 2010, 56(9): 4539–4551.
59. Khan, Osama, et al. "Rethinking erasure codes for cloud file systems: Minimizing I/O for recovery and degraded reads." Proc. of USENIX FAST. 2012.
60. Xiang, Liping, et al. "Optimal recovery of single disk failure in RDP code storage systems." ACM SIGMETRICS Performance Evaluation Review. Vol. 38. No. 1. ACM, 2010.
61. Brewer, Eric A. "Towards robust distributed systems." PODC. 2000.
62. Vogels, Werner. "Eventually consistent." Communications of the ACM 52.1 (2009): 40–44.
63. Birman, Kenneth P. "Consistency in Distributed Systems." Guide to Reliable Distributed Systems. Springer London, 2012. 457–470.
64. Bermbach, David, and Stefan Tai. "Eventual consistency: How soon is eventual? An evaluation of Amazon S3's consistency behavior." Proceedings of the 6th Workshop on Middleware for Service Oriented Computing. ACM, 2011.
65. Zhou, Yuanyuan, et al. "Relaxed consistency and coherence granularity in DSM systems: A performance evaluation." ACM SIGPLAN Notices. Vol. 32. No. 7. ACM, 1997.
66. Adve, Sarita V., and Kourosh Gharachorloo. "Shared memory consistency models: A tutorial." computer 29.12 (1996): 66–76.
67. Serious cloud failures and disasters of 2011. http://www.cloudways.com/blog/cloud-failures-disastersof-2011/.
68. D. Bernstein, E. Ludvigson, K. Sankar, S. Diamond, and M. Morrow, "Blueprint for the intercloud—protocols and formats for cloud computing interoperability," Internet and Web Applications and Services, International Conference on, vol. 0, pp. 328–336, 2009.
69. R. G. Dimakis, P. B. Godfrey, Y. Wu, M. O. Wainwright, and K. Ramch, "Network coding for distributed storage systems," in In Proc. of IEEE INFOCOM, 2007.
70. L. Rizzo, "Effective erasure codes for reliable computer communication protocols," SIGCOMM Comput. Commun. Rev.,vol. 27, no. 2, pp. 24–36, Apr. 1997. [Online]. Available: http://doi.acm.org/10.1145/263876.263881
71. H. P. Anvin. The mathematics of raid-6. http://kernel.org/pub/linux/kernel/people/hpa/raid6.pdf.
72. H. Weatherspoon and J. Kubiatowicz, "Erasure coding vs. replication: A quantitative comparison," Peer-to-Peer Systems, pp. 328–337, 2002.
73. M. Vrable, S. Savage, and G. M. Voelker, "Cumulus: Filesystem backup to the cloud," Trans. Storage, vol. 5, no. 4, pp. 14:1–14:28, Dec. 2009. [Online]. Available: http://doi.acm.org/10.1145/1629080.1629084

74. P. Hunt, M. Konar, F. P. Junqueira, and B. Reed, "Zookeeper: waitfree coordination for internet-scale systems," in Proceedings of the 2010 USENIX conference on USENIX annual technical conference, ser. USENIXATC'10. Berkeley, CA, USA: USENIX Association, 2010, pp. 11–11.
75. L. Lamport, "Paxos made simple," ACM SIGACT News, vol. 32, no. 4, pp. 18–25, 2001.
76. N. A. Lynch, Distributed algorithms. Morgan Kaufmann, 1996.
77. A. Bessani, M. Correia, B. Quaresma, F. Andr´e, and P. Sousa, "Depsky: dependable and secure storage in a cloud-of-clouds," in Proceedings of the sixth conference on Computer systems, ser. EuroSys'11. New York, NY, USA: ACM, 2011, pp. 31–46.
78. W. Vogels, "Eventually consistent," Communications of the ACM, vol. 52, no. 1, pp. 40–44, 2009.
79. M. Herlihy, V. Luchangco, and M. Moir, "Obstruction-free synchronization: Double-ended queues as an example," in Distributed Computing Systems, 2003. Proceedings. 23rd International Conference on. IEEE, 2003, pp. 522–529.
80. Csc cloud usage index. http://www.csc.com/.
81. D. Ionescu. (Oct. 2009) Microsoft red-faced after massive sidekick data loss. pcworld.

Part IV
Hardware

Realizing Accelerated Cost-Effective Distributed RAID

Aleksandr Khasymski, M. Mustafa Rafique, Ali R. Butt, Sudharshan S. Vazhkudai and Dimitrios S. Nikolopoulos

1 Introduction

The deluge of data from scientific instruments (SNS [1], LHC [2]), experiments (DZero [3]) and observations (SDSS [4]) will soon surpass the ability of storage systems to store and retrieve data in a reliable and cost-effective manner. While the capacity, performance and the mean time to failure (MTTF) of a single disk has been improving, large-scale storage systems and parallel file systems (PFS) can comprise tens of thousands of drives, thus bringing down the overall mean time to data loss (MTTDL) of the entire system to unacceptably low levels. For example, the Lustre-based Spider PFS of the Jaguar supercomputer (No. 3 machine on the Top500 [5] list) comprises 10,000+ disks [6]. An exaflop machine in 2018 is projected [7] to host hundreds of thousands of drives to support the desired I/O throughput. The "law of large numbers" in this case only reiterates that failure will be a norm and not an exception. The reliability and robustness of the I/O system is crucial to large-scale applications that generate and analyze terabytes of data. Trends from commercial

A. Khasymski (✉) · A. R. Butt
Virginia Tech, Blacksburg, VA, 24061 USA
e-mail: khasymskia@cs.vt.edu

M. M. Rafique
IBM Research, Ballsbridge, Ireland
e-mail: mustafa.rafique@ie.ibm.com

A. R. Butt
e-mail: butta@cs.vt.edu

S. S. Vazhkudai
Oak Ridge National Laboratory, Oak Ridge, TN, 37831 USA
e-mail: vazhkudaiss@ornl.gov

D. S. Nikolopoulos
Queen's University of Belfast, Belfast, UK
e-mail: d.nikolopoulos@qub.ac.uk

S. U. Khan, A. Y. Zomaya (eds.), *Handbook on Data Centers*,
DOI 10.1007/978-1-4939-2092-1_25

and HPC centers suggest, that on average, 3–7 % of disks fail per year [8]. Thus, storage systems are a significant contributor to system failure.

To increase the fault tolerance of storage systems, disks in each storage server are usually combined in a RAID (Redundant Array of Independent Disks) array that provides some level of redundancy. For example, RAID-6 contains $k + m$ disks, with k data disks and $m = 2$ parity disks. The array can recover from simultaneous failure of up to m disks. Data reconstruction time (in the event of a failure) is proportional to the drive size, load and the number of drives in the RAID group, and is in the order of a few hours even for standard disk sizes. For example, the reconstruction time for a 2 GB disk is approximately 30 min [9]. During reconstruction, applications achieve degraded I/O rates, at best. This is only bound to get worse with large-scale storage systems.

Historically, RAID has been implemented in hardware because of its high throughput compared to a software based solution. Hardware RAID controllers, unlike the rest of the storage system, rely on proprietary hardware. These non-commodity parts are usually expensive, receive infrequent software upgrades, and can become a bottleneck, especially during degraded array reconstruction. The high cost of sophisticated embedded RAID controllers—a typical 16 TB RAID setup could easily cost in excess of $ 15000 [10]—also implies that such solutions are beyond the reach of mid-sized institutions, small-scale clusters, and storage systems with limited provisioning budgets. Even with supercomputing centers, where the cost of provisioning and operating a scalable, reliable storage system can run on the order of millions of dollars, there is a need to reconcile the storage cost against the FLOPS purchased, as machines are often ranked in terms of peak FLOPS. Thus, providing a desired level of reliability and redundancy for the storage system under a given budget constraint is always a challenge, be it in supercomputing centers, mid-sized or small-scale systems.

In recent years, GPUs from NVIDIA and AMD have shifted from closed peripherals used to render graphics images to inexpensive commodity parallel accelerators. They provide general-purpose APIs that can be used to accelerate many types of computations. While currently GPUs are mainly used in scientific workloads, recent studies have explored applying them to I/O workloads [11–15]. These efforts have shown that GPUs can be used effectively for parity computation using Reed-Solomon coding [16] as well as other I/O workloads, such as hashing [17].

Further, large-scale machines are beginning to be provisioned with GPUs. For example, the state-of-the-art Keeneland supercomputer [18], is a combination of Intel Nehalem and NVIDIA Tesla GPUs. Similarly, a planned 20 petaflop machine for 2012, Titan [19], will use a hybrid architecture, with each node featuring two 16-core AMD Opteron processors and two Tesla X2090 GPUs. Additionally, GPUs provide a cost-efficient solution compared to general purpose CPUs (GPPs), especially when GPUs are coupled with a few GPPs [20]. Thus, GPUs are quickly being adopted in a myriad of fields, ranging from scientific workload processing [21] to education in the developing world [22]. These architectures present opportunities to explore the utility of GPUs towards improving storage system reliability.

In this paper we propose a novel way to utilize low-cost GPUs in conjunction with a PFS to provide fault tolerance and end-to-end data integrity. We capitalize the resources provided by the PFS, such as striping individual files over multiple disks, with the computational power of a GPU to provide flexible and fast parity computation for encoding and rebuilding of degraded RAID arrays. We attain end-to-end data integrity by performing encoding and decoding at the compute node, where data is produced and consumed. We implement our client-driven, per-file RAID in the widely used Lustre PFS [23], which will facilitate wider adoption of our system.

We evaluate our system using a medium-scale cluster based on nodes with off-the-shelf, GPUs and show that our approach: provides a customizable interface for an application to tailor the RAID array parameters and provides default values to support legacy applications. The results demonstrate that leveraging GPUs for I/O support functions, i.e., RAID parity computation, is a feasible approach and can provide an efficient alternative to specialized-hardware-based solutions.

The rest of the paper is organized as follows. Section 2 provides of brief overview of Lustre, erasure codes and the CUDA programming environment. Section 4 summarizes related work. We present our implementation in Sect. 5 and experimental results in Sect. 6. We explore future work in Sect. 7 and conclude in Sect. 8.

2 Background

2.1 Rationale

A PFS typically provides fault tolerance at the storage backend. For example, the data drives on each storage server are arranged in a RAID-5 (or higher) configuration. An alternative of computing parity at the backend is client-driven, per-file RAID [24]. In this section, we highlight the potential benefits and drawbacks of this approach, and make the case that such a framework is a good candidate for GPU acceleration.

2.1.1 Backend vs. Client-driven Parity Generation

GPUs are becoming ubiquitous, with good performance and flexibility features. Modern HPC clusters and supercomputers are being equipped with GPUs. In such settings, a client-driven parity generation can utilize the GPU resources already available on the client machines. Such a trend will also be supported by emerging technologies such as the Intel Sandy Bridge chip, which supports an integrated GPU and CPU [25]. Furthermore, client-driven parity generation allows for unprecedented flexibility. For example, the parity computation power of the system is not constrained by the hardware at the backend, and can be changed dynamically with the number of clients.

Hardware RAID controllers typically require all disks in an array to be co-located on the same blade. This can result in data loss because all the drives in the array can fail simultaneously, due to power failure, over-heating, etc. A client driven per-file RAID system does not impose any spatial limitation on the locality of drives, allowing data to be spread across the system and not just one location. Furthermore, the ability of each client to generate parity opens the door for end-to-end data integrity checking. Typically, data has to pass through several network interconnects, and memory and storage hierarchies, all of which can introduce errors, albeit with very small probability. If absolute data integrity is required, the client can choose to obtain parity as part of a read operation and check consistency of the data on the fly.

2.1.2 Block-Based vs. Per-File RAID

When compared to block-based RAID, a per-file RAID scheme will allow each file or directory tree to have a desired fault tolerance level. For example, small files can use a simple RAID-1, while large ones can use the state-of-the-art RAID-6 code. In a block-based RAID, it is difficult or impossible to directly map any lost sectors back to higher-level file system data structures. In fact, it has been recently argued [26] that such factors will continue to diminish the utility of simple block-based RAID. In a conventional hardware RAID, a single RAID controller is responsible for parity coding. For a large array, that can mean hours until the array is rebuilt—during which time, an unrecoverable read error (URE) can occur, potentially causing the entire array to fail. Using a software RAID, rebuilding of the array can be done in parallel. A number of machines at the backend can be equipped with GPUs and rebuilding of separate files can be farmed out to different machines.

2.1.3 Hardware vs. Accelerated Software RAID

Direct performance comparison between a hardware RAID controller and a GPU is hard to quantify, given that the two carry very different hardware. In the context of our approach, however, such a performance comparison is not necessary. For one, unlike a hardware RAID, the GPU in our system resides on the client and as such its available throughput to the storage drives is limited to the network throughput of the client. We show that a GPU can sustain encoding throughput that exceeds available network bandwidth even if the client is connected over 10 Gbps interconnect. A CPU alone, however, is not enough to meet such performance requirements. Previous research has shown that unlike GPUs, conventional x86-based processors are slow in performing a large number of finite field multiplications—the majority of operations required for parity generation [16].

One area that a GPU has a clear advantage over a hardware RAID solution, however, is programmability. The best fault tolerance that a hardware RAID controller typically supports is a Reed-Solomon [27] implementation of RAID-6. In contrast, any number of coding techniques can be used in a software solution, including triple parity RAID or any implementation of RAID-6. The programmability of the GPUs, thus, provides a unique opportunity to exploit the advances in parity encoding, such as minimum density coding schemes like Blaum-Roth [28] and Liberation codes [29].

2.1.4 Discussion

Unprecedented flexibility and increased fault tolerance do not come for free, of course, because more data has to move over the client-server interconnects. Moreover, parity generation can be computationally expensive and thus a burden to the clients. We address the later by offloading the computationally expensive codes to a GPU and show that in doing so we introduce acceptable overheads to the client systems. While some increase in data traffic is unavoidable in a client-driven approach, modern PFSs, like Lustre, maintain large caches on the client side, which absorb a large portion of parity modification caused by frequent small writes. Thus, a large portion of parity updates never hit the interconnect. Even in the context of frequent large writes that exceed client caches, our system provides enough flexibility to address the increase in traffic. Applications can set their own operating point with respect to data reliability and I/O performance. For example, by switching a file to a RAID-5 from a RAID-6, an application achieves a 2× decrease in network traffic due to parity. Another approach is increasing the per-file RAID array size. For example, moving from a (8,2) to a (16,2) RAID-6 array, drops parity from 20 % to 11 % of overall data. Normally a hardware RAID array is not larger than 16 drives, because increasing its size would result in unacceptably long time to rebuild it. As per-file arrays can be rebuilt in parallel in a client-driven approach, an application can set its desired array size based on network traffic and GPUs available to rebuild the array.

2.2 *Enabling Technologies*

In the following, we describe the enabling technologies that are used in realizing our GPU accelerated software-based RAID-6 distributed PFS.

2.2.1 Erasure Codes

In recent years, RAID-6 systems have become increasingly important as they can tolerate a complete failure of one drive occurring in combination with a latent failure of a block on a second drive. Such a failure scenario would result in a permanent data loss on a RAID-5 system. Unlike RAID-1 through RAID-5, which provide exact data encoding techniques, RAID-6 is only a specification and as a consequence there are a number of available coding techniques. The recently introduced Liberation codes promise to become a standard for RAID-6.

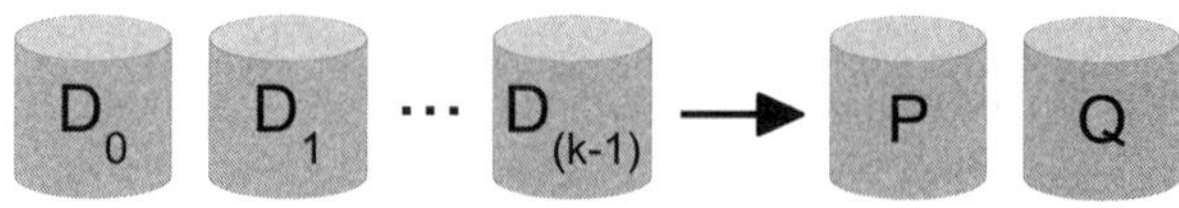

Fig. 1 Logical overview of a RAID-6 system

A RAID-6 system (Fig. 1) is composed of $k + 2$ data nodes and can tolerate the failure of any two devices. Devices D_0 through D_{k-1} can each store B bytes, whereas the remaining 2B bytes are in the P and Q coding devices. The P device is calculated to be the parity of all data devices, while the implementation of the Q device is left to the designer, with the sole constraint that it cannot hold more than B bytes and the resulting system must be able to recover from the failure of any two devices.

Liberation coding (Fig. 2) is similar to Cauchy Reed-Solomon coding [30]. The system splits each data device into w words and uses a $w(k + m) \times wk$ matrix to perform the encoding, where k and m represent the number of data and encoding devices respectively. For all RAID-6 techniques, the value for m is two. All operations are performed in Galois Field (2), where addition and multiplication are bitwise XOR and AND operations, respectively. The matrix is called a Binary Distribution Matrix (BDM) and each element is either one or zero. BDM is multiplied by the vector representing device bits, to produce a vector representing the data and encoding devices. The BDM is quite restricted as the top $k(w \times w)$ portion of the matrix is the identity, $D_{0,1}$ through $D_{0,k-1}$ are also identity matrices that produce the P device, and the bottom row can be customized as per rules laid out in [30].

The encoding matrices for the Liberation codes are shown to be optimal or close to optimal. The decoding matrix is produced by inverting the portion of the encoding matrix that corresponds to the data devices that are still active. However, the resulting matrix typically has far more 1s than optimal and in some cases it is more efficient to calculate a word in one of the failed devices from a previously computed product, rather than from the original BDM matrix by data vector product. To take advantage of this, a schedule is created from the BDM that does the least number of XORs. The optimized schedule produces a significant speedup for decoding. A schedule can also be used in the encoding process as it is a more compact representation of the operations than the BDM itself [29].

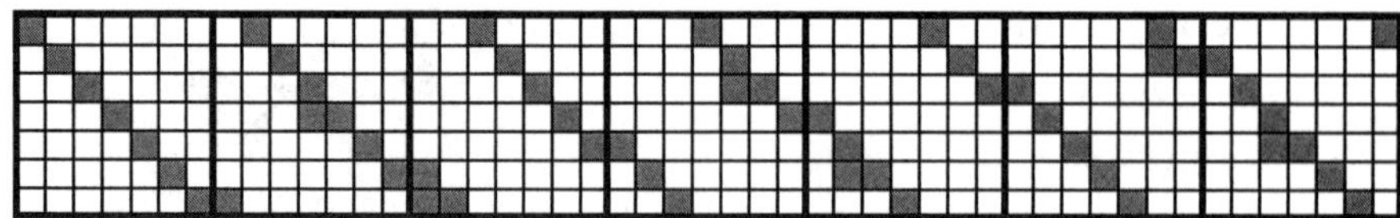

Fig. 2 *Bottom row* of BDM used to compute parity for the Q device for a system with 7 devices and word size of 7. *Gray* boxes represent a 1, *white* a 0

2.2.2 The Lustre Parallel File System

Lustre [23] is a storage architecture for Linux-based clusters and provides a POSIX-compliant UNIX file system interface. It is best known for powering seven of the ten largest HPC machines worldwide, with thousands of client systems, petabytes of storage and hundreds of gigabytes per second I/O throughput. Many HPC sites use Lustre as a site-wide global file system, serving dozens of clusters on an unprecedented scale, e.g., the Spider file system [31]. A Lustre file system comprises of the following key components: *Client*, *MDS* (MetaData Server) and *OSS* (Object Storage Server). Each OSS can be configured to host several *OST*s (Object Storage Target) that manage the storage devices.

The Lustre client that runs on the compute nodes of the cluster communicates with the MDS to obtain privileges and layout for a given file. Once file metadata has been received, the client is able to directly communicate with the OSTs that house the objects associated with the file. An important feature of the Lustre file system that we exploit in our design is its ability to store files in multiple same-sized objects striped over multiple OSTs. Moreover Lustre provides extensive management and recovery features that are useful in identifying the files affected in the event of an OST failure. Thus, Lustre provides some key building blocks to turn each file into its own RAID array.

Lustre also supports hot-swappable hard-drives on each OSS. In the case of a disk failure, a new disk can easily replace the failed disk. Upon a mount, the Lustre manager node detects and recreates the objects that were present in the failed disk. During the per-file RAID array rebuild process, our system restores the data in the lost objects, while reusing the objects that have not failed.

2.2.3 KGPU

Recent research has explored the potential of GPUs to accelerate computationally intense OS operations [32, 33]. The current state-of-the-art NVIDIA's and AMD's proprietary drivers do not support accessing the GPU from kernel space, therefore all these efforts rely on a userspace daemon to execute the GPU requests.

We use KGPU [32] in our implementation because of its service oriented approach and associated low latency. In contrast to the standard approach, where both data and kernel code are copied to the GPU before each execution, KGPU substantially decreases the latency of a GPU kernel launch by keeping the kernel alive even after it has completed its execution. KGPU incurs full latency only when a GPU kernel that provides a different service needs to be loaded. In our implementation, all RAID-6 array sizes are processed by a single kernel, therefore KGPU never incurs the extra latency of loading and unloading a kernel.

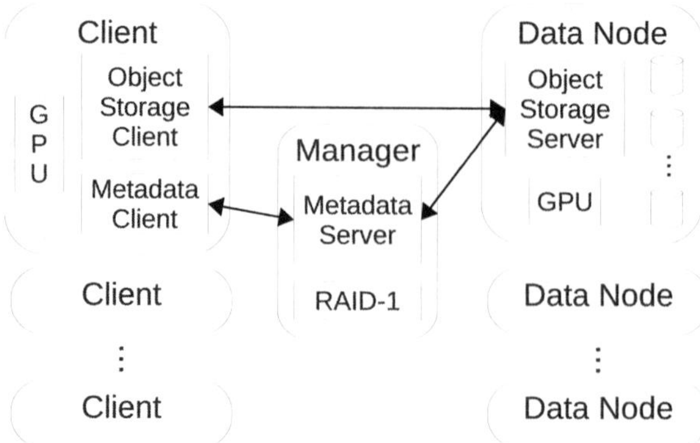

Fig. 3 High-level architecture of the GPU-enabled RAID system

3 Design

In this section, we present the design of our GPU-enabled RAID system and its realization within the Lustre PFS [23]. We also describe the use of KGPU [32], a GPU management framework, in our system.

3.1 System Overview

A high-level overview of the hardware and software components used in our system is shown in Fig. 3. The *Data Nodes* serve as the main storage components; the *Client* provides the user-side interface to the system; and the *Manager* directs and facilitates the interactions between all components. All system components are tightly integrated with the Lustre PFS. The clients typically run on the GPU-enabled compute nodes of the cluster. All or a subset of the Data nodes are equipped with a GPU to perform parity computation during a RAID array rebuild. This hardware addition is feasible on many deployments, since modern motherboards typically have a built-in PCI-Express (PCIe) slot. For the setups where installing a GPU on Data nodes is not an option, the array rebuild process can be offloaded to idle client machines. Each Data node runs an *Object Storage Server* (OSS), which provides file I/O services and network request handling for all the *Object Storage Targets* (OSTs). The OSTs manage the disk drives that store chunks of files called objects. A file in the Lustre PFS can be striped over any number of equally sized objects.

In our design, the Manager is equipped with a hardware or software RAID-1. The Lustre guidelines suggest using RAID-1 or RAID-1 + 0 for the disks on the Manager, which efficiently performs frequent updates on small metadata files. The Manager runs a MDS that only stores metadata (such as file names and layout, directories, and permissions), which generally accounts for only 1 % of the total storage capacity of the system [34]. This ensures that only a small number of disks are required to store the entire metadata in a typical deployment. Hence, equipping the Manager with a low-end RAID-1 controller with a small number of ports (or utilizing a software RAID) fits with our overall goal of achieving fault tolerance with minimal cost.

Each client node in our design is equipped with a programmable GPU that is used to accelerate the file encoding and decoding process. Each client node runs a *Metadata Client*, which communicates with the MDS at the Manager to serve all directory and file operations, such as opening and closing, on behalf of the client. Each client also runs an *Object Storage Client*, which interacts with the OSS at the Data node to read and write to the file objects in parallel. This enables the client to bypass the Manager for all subsequent read and write operations after opening a file and receiving its layout on the Data nodes.

We use the fault tolerant Manager to "bootstrap" the per-file RAID-6 arrays created by our system. If an OST device fails, the Manager identifies all the surviving objects of a given file, which are then used to reconstruct the lost objects.

3.2 RAID-enabled PFS Design

One of our key design objectives is to make our system compatible with Lustre so that it can be easily integrated with extant Lustre deployments. To this end, our first design choice is to keep the Lustre backend software infrastructure (Manager, OSS, etc.) intact and limit our software-level modifications to the client nodes.

One design obstacle for integrating parity acceleration on the client-side is that the NVIDIA CUDA toolkit is designed to run in user-space, while the Lustre client is implemented as a kernel module. One option is to augment `liblustre` [35], a user-space implementation of the Lustre client, to handle parity generation and storage. This approach decreases the number of context switches that are otherwise required to send data between the client module and the GPU. However, `liblustre` is not widely used in practice as it does not support many performance enhancing features of the kernel implementation, including client-side caching and the support for multi-threaded applications. Hence, we integrate all parity generation inside the Lustre client module and use KGPU to access the GPU directly from kernel space. We implement parity encoding and decoding as a service provided by the user-space component of KGPU.

Another challenge is to find the appropriate location to transparently store the extra parity information. One option is to create a separate "shadow" parity file for each file. This is promising, especially in Lustre, where a file can be striped over any collection of OSTs and by ensuring that the shadow file is stored on different OSTs from the OSTs containing the actual file contents, we can provide a complete RAID-6 array. Moreover, the data file and its attributes remains intact and can be accessed without modification. However, this approach doubles the number of files in the storage system and may introduce a bottleneck at the manager node. Additionally, updating the parity would require write locks on two different files simultaneously and would complicate the locking procedure. An alternative approach that incurs minimal bookkeeping overhead is to interleave data and parity in the same file.

However, utilizing this approach requires an effective mechanism to hide the parity from the user. To this end, we modify all file and inode operations that can expose the parity information, such as `write`, `read`, `seek`, `get` and `set` attribute. For operations such as `seek`, and `get/set` attribute, we perform a translation between the size of the actual file including the parity and size of the data. The bulk of the parity generation modifications are contained in the `write` call.

An important feature of our system that significantly decreases overheads when writing to small files is that as long as a file is smaller than a single object it is configured as a RAID-1. We achieve this by mirroring each write into the first parity object, while keeping the second one empty. If a write anywhere outside the first data object is submitted to the system it automatically locks all stripes and converts the file to a RAID-6 array. Note that while in the RAID-1 state no extra space is wasted as the empty blocks inside the second parity object are never written to disk. To maintain consistency we lock the parity object instead of the data, which ensures that a concurrent write to any portion of the stripe, would conflict with the current write and thus will be properly serialized. In this RAID-1 state the GPU is completely bypassed eliminating the expensive read-modify-write step.

3.3 Control Flow

We now describe the interactions between the different components of our system and how they come together to realize the flexible RAID-6 solution.

Figure 4 illustrates the control flow between different components of the system. The system is initialized by reading a configuration file, which specifies different architectural and RAID array specific parameters, such as available GPU memory, maximum supported file object size, and maximum number of disks a file can be striped over. These parameters are used to initialize global defaults, such as the coding bit matrix used in the default parity algorithm. Some of these parameters are passed on to the KGPU framework, which spawns a GPU management daemon, T_{manage}, that we later utilize to compute the parity. T_{manage} initializes its *request* containers and allocates their associated buffers. The daemon then waits for the jobs to be submitted to the request queue.

The Manager initializes the appropriate storage pools before the Lustre file system can be mounted. In Lustre, any OST can be assigned to a number of storage pools, which we use to define default RAID arrays in our system. Storage pools can be modified at runtime to support addition or removal of storage devices. Once Lustre is mounted on the client, the root directory is assigned to the default storage pool. Files and directories created under the root are recursively assigned the default pool. On creation, each file receives a randomized order in which to write to the OSTs in its pool, which ensures that parity is spreads around the OSTs. In addition to the given defaults, applications have full control to assign files and directory trees to any other pool using standard Lustre system calls.

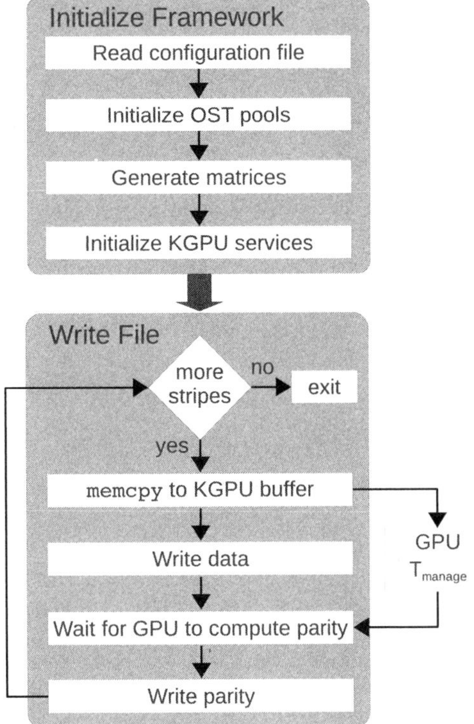

Fig. 4 Control flow in our GPU-enabled RAID system

The bulk of the operations are performed during a write. Lustre caches data on the client side and as a consequence most data writes are processed asynchronously. Synchronous I/O is triggered when the Lustre cache fills up. Lustre breaks down the write in a loop based on the object size. In each iteration, the client asks for a lock on the object and proceeds to update the object, releases the lock, and moves on to the next object. In order to ensure consistency of the parity during simultaneous writes to the same file stripe, we acquire a lock that spans all of objects in a stripe. Thus, we increase the granularity of Lustre's locking from an object to a stripe of objects. Note that we still allow multiple clients to be simultaneously reading and writing to the same file, as long as it is to a different stripe.

After acquiring the lock on a stripe, we copy the relevant portion of the write buffer to CUDA page-locked buffer previously initialized by KGPU and send a request to the KGPU module. The copy is required to maximize the PCIe bandwidth utilization ensured by the CUDA page-locked buffer. The request is then forwarded to our parity generation service implemented in the KGPU user-space daemon that interacts with the GPU to compute the parity for the buffer and return it to the caller. KGPU's call is asynchronous with the data write to the Lustre cache and for a full stripe write completes before it, thus hiding all the latencies associated with moving data to and from the GPU and computing parity. The only overhead exposed is due to the `memcpy` call and parity write. We quantify these latencies in our evaluation.

A read operation also acquires the lock in a loop. In the common case when only data is read, the read loop skips over the parity objects in each stripe of the file. However, a user can also read parity along with the data to ensure end-to-end integrity. In that case, locking is again done at the granularity of the stripe and data and parity is sent to the GPU for validation. If data corruption is detected the read call is restarted. However, if the call fails again an error is returned, as it indicates a permanent error in one of the system components.

3.4 Degraded Array Reconstruction

Unlike a conventional hardware RAID controller, our system is capable of utilizing multiple GPUs to reconstruct a degraded array. If a disk fails, it can be replaced manually or via a hot spare. The disk is formatted if necessary and assigned the same internal Lustre ID as the failed disk. When the new disk is mounted, the Manager recreates all the missing objects and relinks them to the file objects on the surviving disks. Next, the system requests a list of files that have been affected, and based on the location of the failed disks and the availability of GPUs, mounts a Lustre client on the machines to reconstruct the lost objects. The list of affected files is then split accordingly and forwarded to the reconstructing clients to rebuild the affected files in paralle

4 Implementation

We have implemented our system as described in Sect. 3 using 1272 lines of C/C++ and CUDA code. The implementation runs on Linux (kernel version 2.6.32) and is portable to CUDA-enabled GPUs. We based our parity generation implementation on the definition of Liberation Codes [29], which is provided in a freely available library, called *Jerasure* [36]. *Jerasure* provides a single threaded implementation for both Liberation and Blaum-Roth functionality.

Our analysis of Liberation and Blaum-Roth codes' single threaded implementation revealed that more than 95 % of the time is spent in the function that performs the XOR operations. We also noted that the same function is used for both encoding and decoding, with the only difference being the schedule. Furthermore, the work done in this function has the potential for both coarse and fine-grain parallelism, making it a good candidate for offloading to the GPU. Therefore in our implementation, we offload only XOR operations on the data to the GPU to maximize SIMD parallelism. Note that most of the other operations in the coding process, such as creating the BDM and converting it to a schedule are computed once and sent to the KGPU service at initialization. As these operations are at most quadratic in the number of drives, which are usually in the tens in a typical RAID array, the overhead for these tasks is negligible.

4.1 Basic GPU Implementation

As described earlier, a schedule is derived from the original BDM matrix while performing the XOR operations on the given data or while copying it between different devices. The schedule is a two dimensional array of integers of size $5 \times N$, where N is the number of operations that need to be performed for encoding. The operations defined are XOR or `memcpy`. The five integers in each tuple identify which words will be operated upon. The first two integers identify the id of the device and the word that will serve as source, while the next two identify the destination. The last integer is either 1 for XOR or 0 for `memcpy`. For example, the operation $< 00700 >$ can be interpreted as the first word of device 0 is to be copied over the first word of device 7. In the case of encoding, the schedule is used to compactly represent the BDM. In the offloaded function, the schedule is also flattened to a single dimensional array for easier copying of the data from the host to the GPU memory. Furthermore, since this data is relatively small and does not change during GPU kernel execution, it is copied directly to the GPU's constant memory to enable faster access by the GPU threads. The kernel iterates over all the operations in the given schedule and each thread performs an XOR or `memcpy` operation on the corresponding words (in 4 byte chunks) in parallel. Hence, the amount of parallelism exposed depends directly on the word size, which is determined by the size of each data object.

4.2 Optimizations

The main drawback of the basic GPU port is that it reveals only the fine-grain parallelism that is present within a scheduled operation. We analyzed data dependencies and found that entire operations can be done in parallel as well. Specifically, the schedule produces the $2w$ words of the coding devices of a RAID-6 array, where w is 8 and 16 for the Liberation and Blaum-Roth coding, respectively. All the operations associated with computing a single word in a coding device can be performed in parallel with the ones that encode the rest of the words. To exploit this, we modify the schedule and create our optimized port shown in Fig. 5.

We create a two dimensional grid of thread blocks and assign each of the $2w$ rows of blocks to perform the operations associated with one of the encoding words. We use an additional structure, `num_reads`, to store the number of operations needed to compute each of the $2w$ coding words. This enables the kernel to execute fewer iterations compared to the basic port, thus simultaneously reducing the work of each thread and exposing more parallelism.

```
__constant__ int d_schedule[8192];
__constant__ int d_num_reads[1024];
__constant__ long d_data[64];

__global__ void xor_gpu(int packetsize,
                        int blocksize, int k, int w) {
  int dest, source, i, y = (k + 2)*2*blockIdx.y;
  const unsigned int g_tid = blockIdx.x*blockDim.x +
                             threadIdx.x;
  int *dptr = (int *)((char *)d_data[d_schedule[y]] +
                        d_schedule[y+1]*packetsize);
  int *sptr = (int *)((char *)d_data[d_schedule[y+2]] +
                        d_schedule[y+3]*packetsize);
  dest = sptr[g_tid];
#pragma unroll
  for(i = 4; i < d_num_reads[blockIdx.y] * 2; i += 2) {
    sptr = (int *)((char *)d_data[d_schedule[y+i]] +
                    d_schedule[y+i+1]*packetsize);
    source = sptr[g_tid];
    dest = dest ^ source;
  }
  dptr[g_tid] = dest;
}
```

Fig. 5 GPU parity computation kernel

5 Evaluation

In this section, we present the evaluation of our GPU-enabled RAID system. We first describe our testbed, and then present the I/O measurements of our system. The goal is to show the impact of different design parameters and features, such as RAID stripe size and end-to-end integrity checking, on the overall system performance. Next, we evaluate the performance of RAID array reconstruction. Finally, we quantify performance under a real workload.

5.1 Experimental Setup

We have set up a Lustre cluster, consisting of one Manager node and three OSSs, each with six OSTs. The Lustre server machines are identical with four Opteron quad-cores each, and 64 GB of main memory. Additionally, each OSS has a GeForce 9500 GT GPU with 1 GB of graphics memory connected to an 8× PCIe slot. Our client machine has two Intel Xeon quad-cores, 48 GB of RAM and a Tesla C2070 GPU with 6 GB of GDDR memory. All the machines are connected using a dedicated

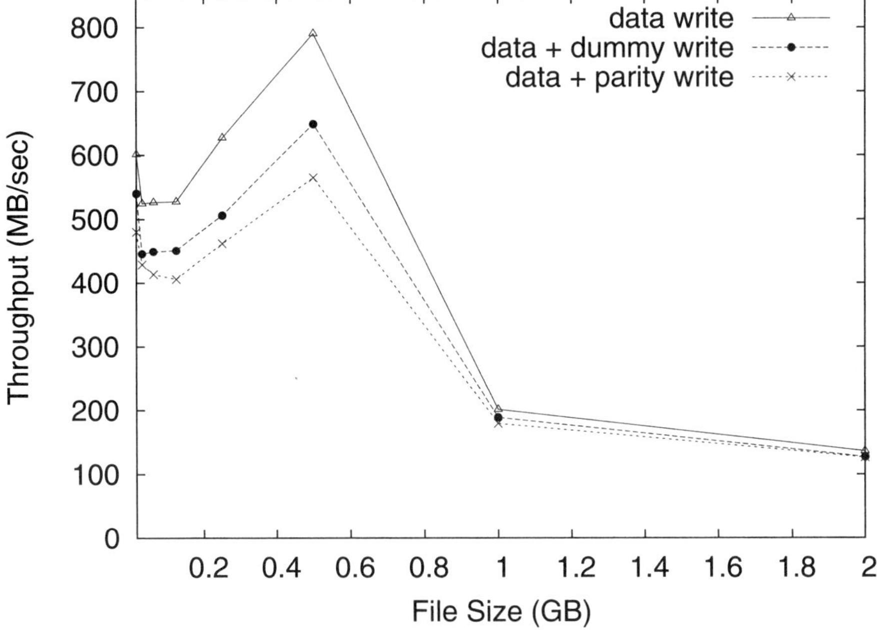

Fig. 6 Write throughput for a file striped over 16 OSTs + 2 parity OSTs

Gigabit switch. We use Lustre patched Linux 2.6.32 kernel, Lustre 1.8.5, and CUDA SDK 4.0.

5.2 *I/O Throughput Measurement*

5.2.1 Raw Throughput

We first measure the raw write throughput that our client machine can achieve. Figure 6 compares the throughput of writing a file striped over 16 OSTs with a stripe/block size of 1 MB, denoted as *data write*. The file size ranges between 16 MB and 2 GB. A Lustre client maintains a 32 MB local cache per OST, which is flushed periodically. If a write submitted by a client fits in the cache, the Lustre module returns from the write immediately after the write buffer is written to the cache. If there is no space left in the cache, it is flushed to the corresponding OST and the write returns after the write buffer has been written to the Lustre back-end. Since we are writing to 16 OSTs, the combined available cache is 512 MB, and consequently writes smaller than 512 MB exhibit throughput higher than the theoretical throughput of Gigabit Ethernet. The throughput of files larger than the cache quickly levels out to an effective available bandwidth of around 125 MB/s.

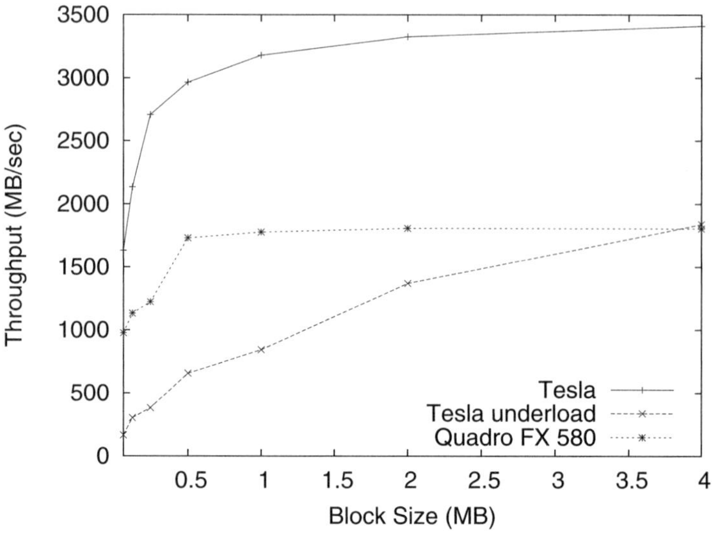

Fig. 7 GPU encoding throughput

In the Figure, *data + parity write* curve shows the throughput when our RAID encoding system is turned on. In this case, the same data as in the base case is striped over 16 OSTs and concurrently the parity is generated and written to the remaining 2 OSTs. As a point of reference we also include a *data + dummy write* curve, which generates the same traffic as the RAID encoding, without computing the parity.

Writes that fit into the Lustre cache exhibit a very high throughput and as a result the overhead of `memcpy`-ing data into KGPU buffers results in around 10 % overhead (difference between *data + dummy write* and parity *data + parity write*). The rest is attributed to copying the extra parity ($1/8^{th}$ of data in this case) to the Lustre cache. It is important to note that in this experiment all the overhead associated with parity generation remains completely hidden from the application.

5.2.2 Encoding Throughput

Next, we evaluate the parity encoding throughput delivered by the GPU. We measured throughput delivered by our high and low-end GPUs, which includes moving the data to and from the GPU's memory as well as actual parity computation on the GPU (Fig. 7). A low-end GPU can deliver encoding throughput around 1 GB/s for 512 KB files, which quickly increases to 1.7 GB/s for files large than 8 MB. The Tesla GPU delivers encoding rates of 1.6 GB/s for 512 KB files and in excess of 3 GB/s for files larger than 8 MB. Therefore, our system using a low-end GPU can generate parity faster than the speed at which Lustre commits the data to its caches. As parity is encoded asynchronously with the commit to cache, the overhead of generating it remains hidden.

We also include the encoding rates of the Tesla GPU, while it is under heavy load from an N-body simulation. We used the N-body simulation from the CUDA SDK and ran multiple iterations in the benchmark mode using the default number of objects based on the specifications of the Tesla GPU. Even under heavy load, the Tesla GPU delivers sufficient throughput for all but the smallest files, for which the encoding overheads are exposed to the system as increased latencies due to the heavy load. However, the performance of the simulation is unaffected by the parity generation kernels and remains constant at 484.650 single-precision GFLOP/s. This is because Tesla GPU can perform efficient context switches at the kernel boundary and asynchronous data transfer for different contexts can run simultaneously. Therefore the parity data can be transfered to the GPU, while the simulation kernel is running and vice versa. Moreover, the parity generation kernels complete 2-3$\times$ faster than the simulation kernels.

It is important to note that not all background loads on the GPU have the same effect on the parity generation. An iterative workload with kernel execution times in the order of milliseconds, such as the N-body simulation that can be rendered in real time, would not block the parity kernels and cause an unacceptable slowdowns to our system. However, for GPU kernels with execution times in seconds, alternative techniques such as "context funneling"[1] can be used to minimize the overhead. The down side is that the GPU workloads need to be modified, e.g., as a KGPU service. This enables even a long running kernel to run parity generation concurrently.

5.2.3 Impact of Number of Disks on Throughput

Next, we study the effect of number of OSTs that a file is striped over on the write throughput of our system. Figure 8 shows the baseline write throughput and the throughput of our system for writing a 256 MB file. When the file is striped over 6 OSTs or less, it cannot fit in the client caches and as a result, raw network bandwidth is exposed to the application. If the 256 MB file is striped over more than 6 drives, parity is cached and flushed after the write returns. As the write fits in the caches, throughput levels out. As the file size remains constant splitting and committing it to more caches becomes less efficient, which causes the slight dip in the throughput.

Striping does not have such an effect when writing a 1024 MB file as the file and its parity cannot be cached (Fig. 9). In the *data write* case, writing to more drives achieves better throughput because of efficient bandwidth utilization when the file is striped over all available OSTs. For the *data + parity write* case, decreasing the number of drives has the effect of decreasing the length of the RAID 6 array, e.g., striping data over four disks produces a (4,2) RAID 6 array where four objects in a stripe are data and the rest are parity, having a parity overhead of 50 %. Increasing the array length decreases the relative size of the parity, e.g., in a (16,2) RAID 6 array

[1] Context funneling uses advanced features of the Fermi architecture to execute concurrent kernels, which must be launched from the same context [37].

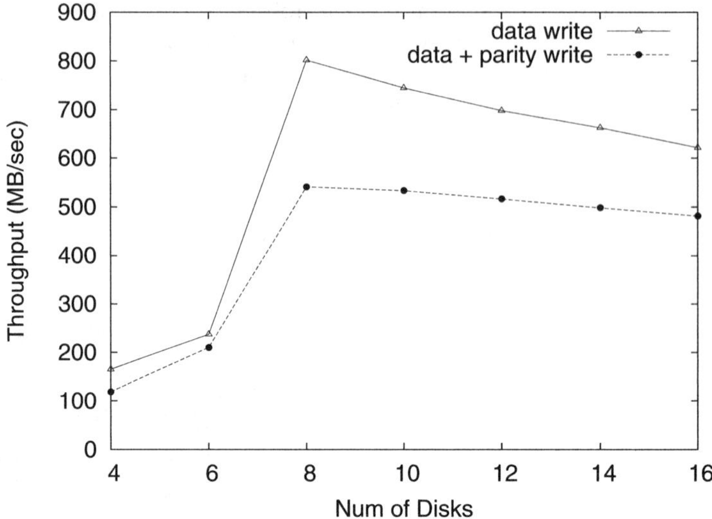

Fig. 8 Effect of number of disks on throughput (file size = 256 MB)

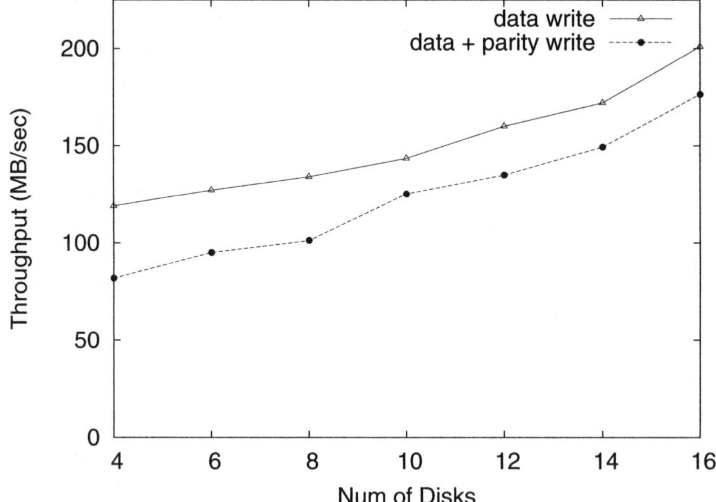

Fig. 9 Effect of number of disks on throughput (file size = 1024 MB)

parity is 12.5 %. Thus, for the *data + parity write* case, there is a linear increase in throughput available for data with the increase in the array length.

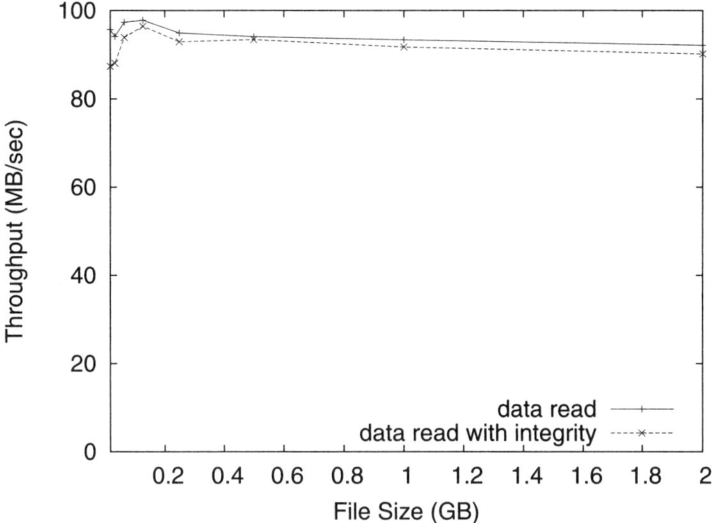

Fig. 10 Read throughput with end-to-end data integrity

5.2.4 End-to-End Data Integrity

One of the important features of our system is that it can provide end-to-end data integrity checks for the I/O operations. Figure 10 shows the achieved read throughput when the end-to-end integrity check is enabled. To ensure that data is not corrupted on the disk or on the network interconnects, one of the parity objects in each stripe is read along with the data. Both data and parity needs to be sent to the GPU for verification to successfully complete the `read` call. For files with single stripes, the synchronous call causes an overhead of 9 %. However, when reading files with more than one stripe, the parity check for each stripe is performed in parallel with the read of the next stripe, resulting in a negligible overhead of 2 %.

5.3 RAID Reconstruction Cost

It is critical to minimize the degraded RAID reconstruction time for maintaining the integrity of data, as the system is exposed to unrecoverable read errors during the reconstruction process. Table 1 shows the reconstruction time for rebuilding 20, 100, and 200 degraded files with a combined size of 5 GB, 25 GB, and 50 GB, respectively. Files are striped on all 18 OSTs (16 for data and 2 for parity). As the RAID arrays are defined per file, their rebuilding can be distributed between the available machines, resulting in a speedup of close to 2× when reconstructing for the 200 files case. During this test, we use all the machines in our setup, which results in the utilization

Table 1 RAID reconstruction time and normalized speedup with respect to 1 Node

Data Size (GB)	1 Node	2 Nodes		3 Nodes	
	Time (s)	Time (s)	Speedup	Time (s)	Speedup
5	46	30	1.53	25	1.84
25	237	151	1.57	125	1.90
50	493	345	1.43	258	1.91

of all the available network bandwidth. It is important to note that each of our low-end GPU achieves an effective reconstruction rate of 1.5 GB/s, therefore even higher speedup is possible, if network and disk throughput permit it.

These results show that our GPU-enabled RAID solution is feasible, and provides a configurable and flexible solution.

5.4 Impact on Applications

Next, we examine the performance of our system under load by a real-world application, the Data Cube (DC) NAS OpenMP [38] benchmark. DC performs a data-intensive operation known in data mining as the Data Cube Operator (DCO), which computes views of a dataset represented as a set of n tuples and involves O(log n) memory accesses per tuple.

Figure 11 shows the performance of DC executing on our client machine with varying number of threads. It is configured to write out the views to disk as they are computed, thus stressing both memory and the storage subsystem. At two threads the benchmark is actually CPU-bound, thus generating and writing out the extra parity for each view introduces a small additional slowdown of 2 %. Beyond four threads the benchmark becomes I/O-bound and as a result, the overheads due to parity produce a 5 and 10 % slowdown for four and eight threads, respectively. We also measured performance of our parity generation system under a heavy background GPU load produced by an N-body simulation application running on the GPU. When the DC benchmark is running with eight threads, the background job does not affect performance at all, because our system is able to schedule the workload for the eight threads more effectively. With fewer threads requesting parity, the system cannot obtain enough time on the GPU and as a result exposes parity generation overheads to a portion of the write operations, resulting in a slowdown of around 5 %.

6 Related Work

There are a number of parallel file systems that were built from the ground up to withstand failure, such as ZFS, Ceph, and Panasas [24, 39, 40]. ZFS maintains data integrity by using checksums for on-disk blocks, while Ceph relies on replication at

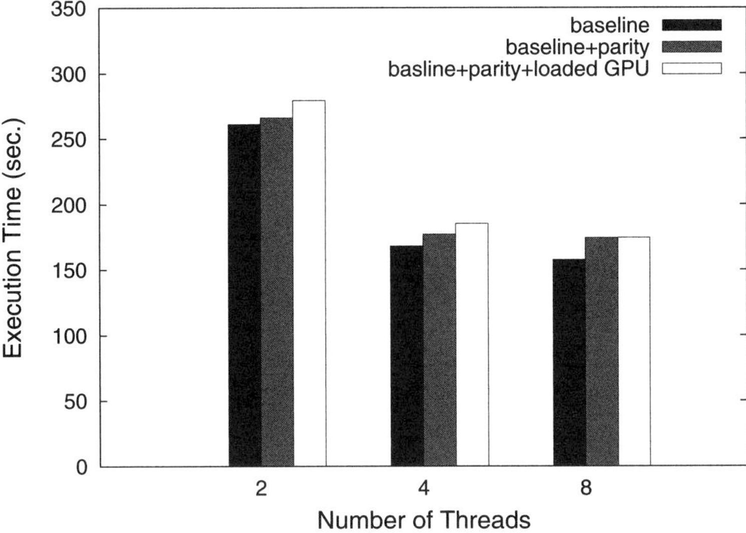

Fig. 11 Performance of NAS DC benchmark

the granularity of an entire object storage device, and thus both file systems cannot detect errors introduce during network transmission to the client. Panasas [24] offers a commercial solution with features similar to the system presented in this paper. However, it relies on the CPU to generate a per-file RAID-5 array. It does not support RAID-6, since generating the required parity without a hardware accelerator would result in high overheads.

Utilizing GPUs as commodity accelerators for general purpose applications has been on the rise [41]. stdchk [42] uses hashing to detect content similarity between two successive checkpoint images. Several efforts [43–45] have attempted to improve the performance of such hash computations by offloading them to the GPU. Similarly, GPUs have also been used to accelerate parity computation [46] and data encryption [47, 48] for storage systems.

The work most similar to ours is Gibraltar GPU based RAID [49], which focuses on accelerating Reed-Solomon [27] based parity codes to create a block based RAID in user-space. Gibraltar has several limitation, including the need to use the O_DIRECT flag in order to bypass the Linux buffer cache, which hurts performance. Additionally, Gibraltar cannot perform end-to-end integrity checks, parallel array rebuild of degraded arrays, or provide the same level of flexibility delivered by our per-file arrays tightly integrated with a PFS.

7 Conclusion

Fault tolerance on large-scale storage servers is largely based on proprietary, expensive, hardware-based solutions with limited flexibility and scalability. In this paper, we have presented a cost-effective alternative that uses commodity GPUs to implement RAID-6 in software, in conjunction with the Lustre PFS. Our solution leverages low-cost GPUs on the client and server nodes to accelerate minimum-density RAID-6 coding schemes. We have shown, through a prototype implementation, that our software-controlled parity computation scheme imposes acceptable overhead on application performance, and constitutes, overall, a feasible, low-cost, and efficient alternative to specialized hardware-based solutions.

As future work, we will extend our RAID solution to other accelerators in a heterogeneous setting, to expedite the encoding schemes. We will also explore the use of CPUs from emerging many-core nodes, those that cannot be fully utilized by a typical application, to compute the RAID encoding.

Acknowledgement This research was supported in part by the National Science Foundation under Grants CCF-0746832, CNS-1016793, and CNS-1016408, and used the resources of, the Oak Ridge Leadership Computing Facility, located in the National Center for Computational Sciences at ORNL, which is managed by UT Battelle, LLC for the U.S. DOE (under the contract No. DE-AC05-00OR22725).

References

1. "Spallation Neutron Source," http://www.sns.gov/, 2008.
2. Conseil Européen pour la Recherche Nucléaire (CERN), "LHC– the large hadron collider," July 2007, http://lhc.web.cern.ch/lhc/.
3. B. Abbott, A. Baranovski, M. Diesburg, G. Garzoglio, T. Kurca, and P. Mhashilkar, "Dzero data-intensive computing on the open science grid," *Journal of Physics: Conference Series*, vol. 119, 2008.
4. "Sloan digital sky survey," http://www.sdss.org, 2005.
5. "Top500 supercomputer sites," http://www.top500.org/.
6. S. Oral, F. Wang, D. Dillow, G. M. Shipman, R. Miller, and O. Drokin, "Efficient object storage journaling in a distributed parallel file system," in *USENIX Conference on File and Storage Technologies*, 2010, pp. 143–154.
7. Brooke Crothers, "DARPA 'exascale' supercomputer in the works," August 2010, http://news.cnet.com/8301-13924'_3-20013088-64.html.
8. Z. Zhang, C. Wang, S. S. Vazhkudai, X. Ma, G. Pike, J. Cobb, and F. Mueller, "Optimizing center performance through coordinated data staging, scheduling and recovery," in *Proceedings of Supercomputing 2007 (SC07): Int'l Conference on High Performance Computing, Networking, Storage and Analysis*, Jun. 2007.
9. M. D. R. Alex Osuna, Siebo Friesenborg, "Considerations for raid-6 availability and format/rebuild performance on the ds5000," 2009, document Number: REDP-4484-00.
10. B & H Foto & Electronics Corp., "Active Storage 16TB ActiveRAID Hard Drive Array," 2011, http://www.bhphotovideo.com/c/product/697437-REG/Active_Storage_AC16SFC02_16TB_ActiveRAID_Hard_Drive.html.
11. J. Michalakes and M. Vachharajani, "Gpu acceleration of numerical weather prediction," in *IEEE International Symposium on Parallel and Distributed Processing (IPDPS)*, april 2008, pp. 1–7.

12. C. Trapnell and M. C. Schatz, "Optimizing data intensive gpgpu computations for dna sequence alignment," *Parallel Comput.*, vol. 35, pp. 429–440, August 2009.
13. M. Fatica, "Accelerating linpack with cuda on heterogenous clusters," in *Proceedings of 2nd Workshop on General Purpose Processing on Graphics Processing Units*, ser. GPGPU-2. New York, NY, USA: ACM, 2009, pp. 46–51.
14. T. D. Hartley, U. Catalyurek, A. Ruiz, F. Igual, R. Mayo, and M. Ujaldon, "Biomedical image analysis on a cooperative cluster of gpus and multicores," in *Proceedings of the 22nd annual international conference on Supercomputing*, ser. ICS '08. New York, NY, USA: ACM, 2008, pp. 15–25.
15. M. M. Rafique, A. R. Butt, and D. S. Nikolopoulos, "A capabilities-aware framework for using computational accelerators in data-intensive computing," *J. Parallel Distrib. Comput.*, vol. 71, pp. 185–197, February 2011.
16. M. Curry, A. Skjellum, H. Ward, and R. Brightwell, "Arbitrary dimension reed-solomon coding and decoding for extended raid on gpus," in *Petascale Data Storage Workshop, 2008. PDSW '08. 3rd*, nov. 2008.
17. D. A. Alcantara, A. Sharf, F. Abbasinejad, S. Sengupta, M. Mitzenmacher, J. D. Owens, and N. Amenta, "Real-time parallel hashing on the gpu," *ACM Trans. Graph.*, vol. 28, pp. 154:1–154:9, December 2009.
18. G. I. of Technology, "Keenland," 2010, http://keeneland.gatech.edu/.
19. Damon Poeter, "Cray's Titan Supercomputer for ORNL Could Be World's Fastest," 2011, http://www.pcmag.com/article2/0,2817,2394515,00.asp.
20. M. M. Rafique, A. R. Butt, and D. S. Nikolopoulos, "Designing accelerator-based distributed systems for high performance," in *Proc. IEEE/ACM International Symposium on Cluster, Cloud and Grid Computing (CCGRID'2010)*, Melbourne, Australia, May. 2010.
21. M. A. Clark, "Qcd on gpus: cost effective supercomputing," 2009. [Online]. Available: http://lattice.github.com/quda/
22. NVIDIA Corporation, "Science & Education," 2011, http://www.nvidia.com/object/nvidia_userful_success.html.
23. Sun Microsystems, Inc., "Lustre file system - High-performance storage architecture and scalable cluster file system," 2007.
24. B. Welch, M. Unangst, Z. Abbasi, G. Gibson, B. Mueller, J. Small, J. Zelenka, and B. Zhou, "Scalable performance of the panasas parallel file system," in *Proceedings of the 6th USENIX Conference on File and Storage Technologies*, ser. FAST'08. Berkeley, CA, USA: USENIX Association, 2008, pp. 2:1–2:17.
25. Intel Corporation, "Intel Microarchitecture Codename Sandy Bridge," 2011, http://www.intel.com/technology/architecture-silicon/2ndgen/index.htm.
26. R. Appuswamy, D. C. van Moolenbroek, and A. S. Tanenbaum, "Block-level raid is dead," in *Proceedings of the 2nd USENIX conference on Hot topics in storage and file systems*, ser. HotStorage'10. Berkeley, CA, USA: USENIX Association, 2010, pp. 4–4.
27. I. S. Reed and G. Solomon, "Polynomial codes over certain finite fields," *Journal of the Society for Industrial and Applied Mathematics*, vol. 8, no. 2, pp. 300–304, 1960.
28. M. Blaum and R. Roth, "New array codes for multiple phased burst correction," *Information Theory, IEEE Transactions on*, vol. 39, no. 1, pp. 66–77, jan 1993.
29. J. S. Plank, "The raid-6 liberation codes," in *Proceedings of the 6th USENIX Conference on File and Storage Technologies (FAST)*. Berkeley, CA, USA: USENIX Association, 2008, pp. 7:1–7:14.
30. J. S. Plank and L. Xu, "Optimizing cauchy reed-solomon codes for fault-tolerant network storage applications," in *Proceedings of the Fifth IEEE International Symposium on Network Computing and Applications*. Washington, DC, USA: IEEE Computer Society, 2006, pp. 173–180.
31. G. M. Shipman, D. A. Dillow, S. Oral, and F. Wang, *The Spider center wide file system: From concept to reality*, 2009. [Online]. Available: http://www.nccs.gov/wp-content/uploads/2010/01/shipman_paper.pdf

32. KGPU, "KGPU: enabling GPU computing in Linux kernel," 2011, http://code.google.com/p/kgpu.
33. C. J. Rossbach, J. Currey, M. Silberstein, B. Ray, and E. Witchel, "PTask: Operating System Abstractions To Manage GPUs as Compute Devices," in *Proc. ACM SOSP*, 2011.
34. Oracle Corporation, "Lustre Documentation," 2011, http://wiki.lustre.org/index.php/Lustre_Documentation.
35. Sun Microsystems, Inc., "LibLustre How-To Guide," 2010, http://wiki.lustre.org/index.php/LibLustre_How-To_Guide.
36. J. S. Plank, S. Simmerman, and C. D. Schuman, "Jerasure: A library in C/C++ facilitating erasure coding for storage applications - Version 1.2," University of Tennessee, Tech. Rep. CS-08-627, August 2008.
37. L. Wang, M. Huang, and T. El-Ghazawi, "Towards efficient gpu sharing on multicore processors," in *Proceedings of the second international workshop on Performance modeling, benchmarking and simulation of high performance computing systems*, ser. PMBS '11. New York, NY, USA: ACM, 2011, pp. 23–24. [Online]. Available: http://doi.acm.org/10.1145/2088457.2088473
38. M. A. Frumkin and L. V. Shabanov, "Benchmarking memory performance with the data cube operator." in *ISCA PDCS'04*, 2004, pp. 165–171.
39. Y. Zhang, A. Rajimwale, A. C. Arpaci-Dusseau, and R. H. Arpaci-Dusseau, "End-to-end data integrity for file systems: a zfs case study," in *Proceedings of the 8th USENIX conference on File and storage technologies*, ser. FAST'10. Berkeley, CA, USA: USENIX Association, 2010, pp. 3–3. [Online]. Available: http://dl.acm.org/citation.cfm?id=1855511.1855514
40. S. A. Weil, S. A. Brandt, E. L. Miller, D. D. E. Long, and C. Maltzahn, "Ceph: a scalable, high-performance distributed file system," in *Proceedings of the 7th symposium on Operating systems design and implementation*, ser. OSDI '06. Berkeley, CA, USA: USENIX Association, 2006, pp. 307–320. [Online]. Available: http://dl.acm.org/citation.cfm?id=1298455.1298485
41. J. D. Owens, D. Luebke, N. Govindaraju, M. Harris, J. Krger, A. Lefohn, and T. J. Purcell, "A survey of general-purpose computation on graphics hardware," in *Eurographics 2005, State of the Art Reports*, Aug. 2005, pp. 21–51.
42. S. Al-Kiswany, M. Ripeanu, S. S. Vazhkudai, and A. Gharaibeh, "stdchk: A checkpoint storage system for desktop grid computing," in *Proceedings of the 2008 The 28th International Conference on Distributed Computing Systems*, ser. ICDCS '08. Washington, DC, USA: IEEE Computer Society, 2008, pp. 613–624. [Online]. Available: http://dx.doi.org/10.1109/ICDCS.2008.19
43. S. Al-Kiswany, A. Gharaibeh, E. Santos-Neto, G. Yuan, and M. Ripeanu, "Storegpu: exploiting graphics processing units to accelerate distributed storage systems," in *Proceedings of the 17th international symposium on High performance distributed computing*, ser. HPDC '08. New York, NY, USA: ACM, 2008, pp. 165–174.
44. A. Gharaibeh, S. Al-Kiswany, S. Gopalakrishnan, and M. Ripeanu, "A gpu accelerated storage system," in *Proceedings of the 19th ACM International Symposium on High Performance Distributed Computing*, ser. HPDC'10. New York, NY, USA: ACM, 2010, pp. 167–178.
45. S. Al-Kiswany, A. Gharaibeh, E. Santos-Neto, and M. Ripeanu, "On gpu's viability as a middleware accelerator," *Cluster Computing*, vol. 12, pp. 123–140, June 2009.
46. G. Falcão, L. Sousa, and V. Silva, "Massive parallel ldpc decoding on gpu," in *Proceedings of the 13th ACM SIGPLAN Symposium on Principles and practice of parallel programming*, ser. PPoPP '08. New York, NY, USA: ACM, 2008, pp. 83–90.
47. O. Harrison and J. Waldron, "Practical symmetric key cryptography on modern graphics hardware," in *Proceedings of the 17th conference on Security symposium*. Berkeley, CA, USA: USENIX Association, 2008, pp. 195–209.
48. A. Moss, D. Page, and N. P. Smart, "Toward acceleration of rsa using 3d graphics hardware," in *Proceedings of the 11th IMA international conference on Cryptography and coding*, ser. Cryptography and Coding'07. Berlin, Heidelberg: Springer-Verlag, 2007, pp. 364–383.
49. M. L. Curry, H. L. Ward, A. Skjellum, and R. Brightwell, "A lightweight, gpu-based software raid system," *Parallel Processing, International Conference on*, vol. 0, pp. 565–572, 2010.

Efficient Hardware-Supported Synchronization Mechanisms for Manycores

José L. Abellán, Juan Fernández and Manuel E. Acacio

1 Introduction

Data centers are evolving by hosting emerging parallel and distributed applications such as cloud computing, streaming video or social networking. These new applications demand not only more storage capacity or network bandwidth, but also higher performance and throughput thus requiring multicore processors as building blocks of every computational node or server [27].

As more and more cores are being integrated on chip, manycore architectures [17, 36, 38] have emerged as the next generation of multicores. Manycores are systems specially tailored to the exploitation of massive throughput by incorporating many simple and low-frequency computing units. This paradigm shift towards throughput-oriented servers leads to parallel workloads with ever more number of threads that need to communicate and synchronize among them, typically relying on a single shared memory domain per server.

In that context, conventional implementations of synchronization operations, such as barrier and locks, make use of shared variables which are atomically updated. In particular, when considering global barriers and highly-contended locks (i.e., a significant amount of threads requesting the lock at the same time), without the proper hardware support, typical software-based implementations cannot provide good scalability as the number of cores increases.

J. L. Abellán (✉)
Boston University, Boston, USA
e-mail: jabellan@bu.edu

J. Fernández
Intel-Labs Barcelona, Barcelona, Spain
e-mail: juan.fernandez@intel.com

M. E. Acacio
University of Murcia, Murcia, Spain
e-mail: meacacio@ditec.um.es

S. U. Khan, A. Y. Zomaya (eds.), *Handbook on Data Centers*,
DOI 10.1007/978-1-4939-2092-1_26

Regarding barrier implementations in software, as we will discuss in Sect. 3, the use of shared variables creates a performance bottleneck, and demands not only a significant amount of resources but also of energy consumption. In more depth, the cache coherence protocol must come into play to maintain memory consistency across all levels of the memory hierarchy. In turn, coherence activity translates into traffic injection in the interconnection network that may interfere with application-related traffic. On the other hand, the busy-waiting required to wait for the completion of the barrier synchronization on locally-cached shared variables has also significant implications on the energy consumed by the L1 caches.

As to the software implementations for highly-contended locks, as we will expose in Sect. 8, they are critical to performance since lock contention causes serialization. Therefore, an implementation based on the use of shared variables is not efficient enough due to the performance bottlenecks, the significant amount of resources and energy consumption that they entail, as explained above.

In this Chapter, we analyze and propose techniques to mitigate the problem of synchronization at server (manycore processor) level in datacenters. Particularly, we propose two different strategies that provide very efficient, scalable and lightweight hardware implementations for barriers and highly-contended locks. We implement our synchronization architectures using two different technologies. The first is a state-of-the-art full-custom technology, namely *G-Lines*, whilst the second is a cost-effective mainstream industrial toolflow with an advanced 45 nm technology, or *Standard* technology.

The rest of the Chapter is organized as follows. Section 2 introduces the state-of-the-art *G-Lines* technology that we also use to implement our proposals. From Sects. 3 to 7, we present and evaluate our *GBarrier* proposal for barrier synchronization in many-core CMPs. Next, from Sects. 8 to 12, we introduce and quantify the efficiency of our *GLock* proposal for highly-contended locks in many-core CMPs. Finally, we summarize the main conclusions of this Chapter in Sect. 13.

2 The *G-Lines* Technology

G-Lines have already been successfully integrated in a silicon substrate in order to enable speed-of-light point-to-point communications. Chang et al. [35] and Jose et al. [3] showed early point-to-point circuits allowing transmission-line, wave-like velocity for 10 mm of interconnect. Nonetheless, this initial implementation suffers from significant overheads in terms of power dissipation and die area. A great effort has been devoted to overcome such limitations. For instance, Ito et al. [15] extended *G-Lines* to support broadcast, multi-drop and bidirectional transmissions. This contribution enables both low-latency and multi-drop ability on a transmission line with low-power dissipation. However, their results still exhibit several integration density issues. Additionally, Ho et al. [33] and Mensink et al. [11] have shown that a capacitive feedforward method of global interconnect reduces both power dissipation and die area overheads. In particular, they achieve nearly single-cycle delay for long wires with voltage-mode signaling. As a result, every *G-Line* is basically a shared wire that broadcasts 1-bit messages (signals from now on) across one dimension of the chip in a single clock cycle. A practical use of *G-Lines* is presented by Krishna et al. [40] in the context of networks-on-chip (NoC). Krishna et al. leveraged *G-Lines* using multi-drop connectivity and the *S-CSMA* collision detection technique to enhance a flow control mechanism (EVC) in terms of latency and power dissipation. In particular, these *G-Lines* are used to broadcast the control signals of EVC in order to communicate the availability of free buffers and virtual channels much more accurately. Furthermore, the authors employ the *S-CSMA* technique to calculate how many virtual channels or free buffers are demanded at any time in order to grant requests accordingly.

As we will see, in this Chapter we also leverage this technology to deploy dedicated *G-Line*-based networks on chip, in order to implement barriers and highly-contended locks to overcome the previously described performance limitations in future manycore servers.

3 Hardware Barrier Synchronization

A barrier is a synchronization primitive that enables multiple processes or threads to wait in a particular point of execution, until all of them have reached it before any of them can continue. A typical example of its usage is utilizing barriers to separate the different phases commonly found in parallel applications [32]. By doing so, the programmer ensures that the second phase does not start until all processes or threads from the application have completed the first one.

In the context of systems that implement a shared-memory programming model [10], with the advent of manycore architectures, new challenges are arising to provide an efficient barrier implementation. This is mainly due to the fact that differently to classical multiprocessor applications which target coarse-grained parallelism, manycore applications tend to exploit fine-grained parallelism, and therefore, they may be highly sensitive to barrier performance [26].

Typical implementations of current software-based barriers (SW-barriers, from now on) rely on busy-waiting on shared variables which are atomically updated [20]. Nevertheless, the use of shared variables implies that the cache coherence protocol must come in on maintaining their consistency across all levels of the memory hierarchy. In turn, coherence activity translates into traffic injection in the interconnection network. As a result, an ever-growing amount of resources and energy may need to be devoted to support SW-barriers as the number of cores in manycore servers increases. On the other hand, busy-waiting on locally-cached shared variables has also significant implications on the energy consumed by the L1 caches.

As an example, Fig. 1 illustrates the potential performance losses suffered in the EM3D parallel application when using SW-barriers in future manycore servers (for details about the evaluation see Sect. 6). In particular, we present the results obtained for a sophisticated binary combining-tree barrier (which is considered one of the

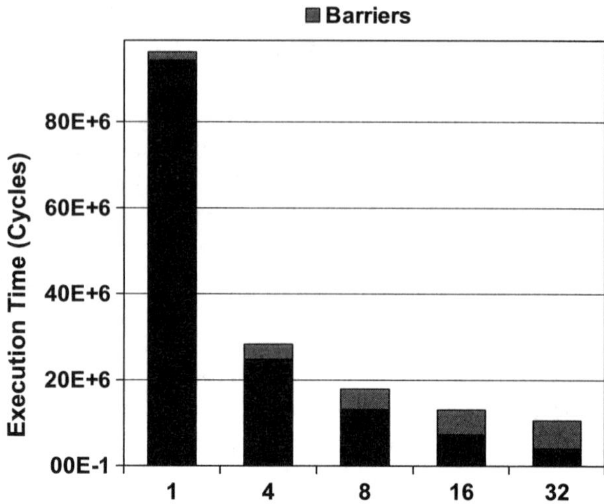

Fig. 1 Fraction of time due to barriers in EM3D

most efficient SW-barriers) as the number of cores is increased from 1 to 32. Each bar shows the fraction of the execution time due to barrier synchronization in orange color. As can be derived from the Figure, as the number of cores increases so does the fraction of the execution time due to barrier synchronization (up to 63 % for 32 cores), thereby considerably limiting scalability.

In this section, we describe and evaluate an efficient barrier synchronization mechanism specifically designed for manycore servers. Differently from SW-Barriers, our proposal, namely *GBarrier*, has been implemented entirely in hardware. To implement *GBarrier*, we have explored two different technologies. On the one hand, we make use of the state-of-the-art full-custom *G-Lines* technology and the *S-CSMA* technique explained in Sect. 2. In short, every *G-Line* enables almost speed-of-light 1-bit communications across one dimension of the entire chip, and the *S-CSMA* technique is employed to detect the number of simultaneous transmissions over a *G-Line*. On the other hand, we utilize the mainstream industrial toolflow with standard cells in an advanced 45 nm process (*Standard* technology from now on) in order to obtain a cost-effective implementation for our proposal at the expense of some negligible performance losses.

4 The *GBarrier* Synchronization Mechanism

In this section we present our proposal to build an efficient hardware infrastructure for barrier synchronization in the context of manycore servers. To do so, we start by describing the architecture of the dedicated on-chip network that our proposal entails. For simplicity, the explanation will be given assuming the *G-Lines* technology with the *S-CSMA* technique. As a case study, we choose a server with a 2D-mesh data interconnection network with R rows of C cores each (for a total of $N = R \times C$ cores),

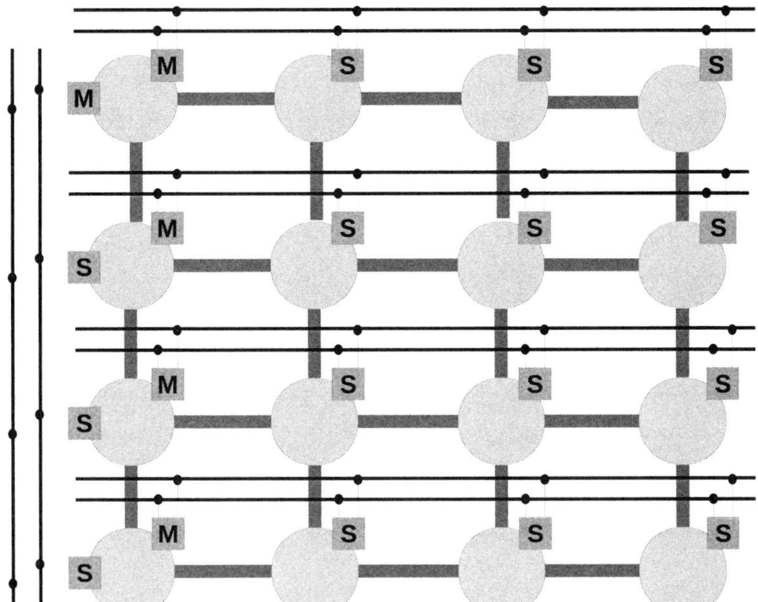

Fig. 2 *GBarrier* architecture for a 16-core server with a 2D-mesh network

although our proposal is not restricted to this topology. Next, we show how the *GBarrier* mechanism would operate and finally, we describe the interface for programmers.

4.1 Dedicated On-Chip Network Architecture

The *GBarrier* mechanism relies on a dedicated on-chip network as it can be observed in the example in Fig. 2. For simplicity, we concentrate on a version of the proposed network providing support for one barrier. As shown in Fig. 2, the *GBarrier* infrastructure is made up of two kind of components. *G-Lines* (horizontal and vertical finer black lines), that are used to transmit the signals required by the synchronization protocol; and controllers (M and S), that actually implement the synchronization protocol.

As discussed in Sect. 2, every *G-Line* is a wire that enables the transmission of one bit of information across one dimension of the chip in a single clock cycle. Our *G-Line*-based network employs two *G-Lines* per barrier for every row and two more for the first column. In this way, for any 2D-mesh layout with R rows and C columns, the total number of *G-Lines* per barrier that would be needed is equal to $2 \times (R + 1)$ (e.g., 10 *G-Lines* for the 16-core server assumed in the example).

In addition to the *G-Lines*, our proposal also incorporates a set of controllers in charge of the synchronization protocol required for a barrier synchronization. In particular, we distinguish two types of controllers: master and slave controllers (see M and S in Fig. 2, respectively). Each controller is attached to two *G-Lines*: one of them is used to transmit signals, whilst the other is employed to receive signals. More specifically, the *G-Line* used by the master controller to receive signals is the one used by the slave controllers to send signals, and vice versa. Moreover, the master controller is responsible for carrying out the count of signals transmitted from all slave controllers attached to the *G-Line*. To do so, the master controller contains a device that implements the *S-CSMA* technique. Recall that, this technique implements voltage amplitude sensing to determine the number of simultaneous transmitters over a particular *G-Line* at any given instant in time.

Finally, for design constraints [40] every *G-Line* can support up to six transmitters and one receiver as much, resulting in a server configuration with up to 7×7 cores. Note that, our *GBarrier* is not restricted to this number of cores and can be easily extended to operate with even larger core counts by means, for example, of a hierarchical tree-based placement of controllers.

4.2 Synchronization Protocol

The synchronization protocol implemented on top of the *G-Line*-based network previously described relies on the exchange of 1-bit messages (signals) between the master and slave controllers, and the use of the *S-CSMA* technique in the master controllers to count the number of signals transmitted across every *G-Line*. In our proposal, every barrier synchronization is carried out by using a two-phase protocol: the *account phase* and the *release phase*. The first phase starts when the first thread arrives at the barrier and finishes when the last one reaches the barrier. Then, the second phase, in which all threads participating in the barrier are commanded to resume execution, is initiated. The exact interplay among threads, *G-Lines* and controllers is detailed below with an example.

Without loss of generality, we assume that all cores execute the same barrier at the same time and we explain how the barrier synchronization would take place on a 2×2 mesh layout (see Fig. 3). We distinguish between horizontal and vertical controllers depending on the couple of *G-Lines* they are attached to. In this setting, there are four horizontal and two vertical *G-Lines*. Thus, there are two horizontal master controllers (see Mh in cores 0 and 2), two horizontal slave controllers (see Sh in cores 1 and 3), one vertical master controller (see Mv in core 0) and one vertical slave controller (Sv in core 2).

As shown in Fig. 3, each master controller employs a couple of hardware elements during a barrier synchronization. The first is the *ScntH* and *ScntV* counters required by the horizontal and vertical master controllers respectively. These counters keep track of the number of signals (obtained through the *S-CSMA* technique) received from the horizontal slaves (cores 1 or 3) or vertical slaves (just one in this case),

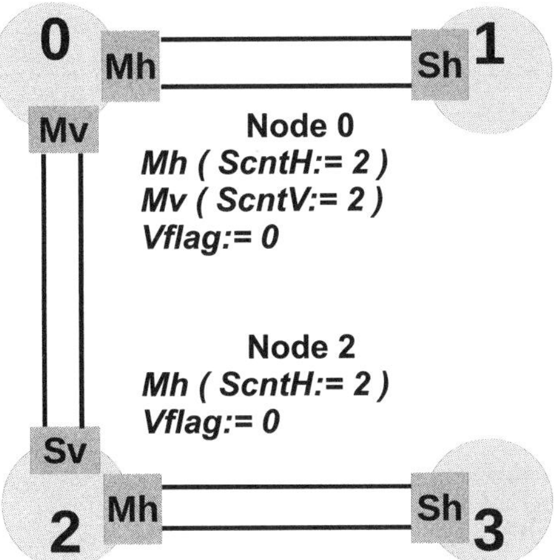

Fig. 3 *GBarrier* for a 4-core server with a 2D-mesh network showing initial state of registers and flags

and whether the server core the master controller is attached to has arrived at the barrier. The second element is the *Vflag* flag, which is used to establish a local synchronization between horizontal master controllers and the corresponding vertical controllers (master and slaves) located in the same core (Mh-Mv in core 0 and Mh-Sv in core 2). In particular, each *ScntH* counter is initialized with the number of slaves controllers in each row plus one to also account for the local core. *ScntH* is decremented every time a signal from a slave controller in its row is received through the corresponding *G-Line* (Sh in cores 1 and 3, in the example) and also when the local core arrives at the barrier. Once each *ScntH* counter reaches zero, the corresponding *Vflag* flag is set. Similarly, the initial value of the *ScntV* counter is the number of vertical slaves plus one, and is decremented on receiving a signal from a slave controller in its column (Sv in core 2, in the example) and when its local *Vflag* flag is set. It is worth noting that, an initial setup is required in order to initialize both *ScntH* and *ScntV* counters to their maximum values. In the example of Fig. 3, since all cores participate in the barrier, these counters will be initialized to two for both horizontal and vertical master controllers. From now on, these values will be referred to as *MAXH* and *MAXV* for the *ScntH* and *ScntV* counters, respectively.

Taking the initial setup shown in Fig. 3 as the starting point, Fig. 4 illustrates an example of how the barrier synchronization process would be performed. It is worth noting that we are assuming theoretical synchronization latencies that may not be reflected in the exact number of clock cycles required for the two physical *GBarrier* implementations (see Sect. 5.1).

At cycle 0, the *account phase* starts because all threads notify their arrival at the barrier. To do so, the horizontal slaves (Sh) signal, through their corresponding transmission lines, the arrival of cores 1 and 3 at the barrier. In turn, the horizontal masters

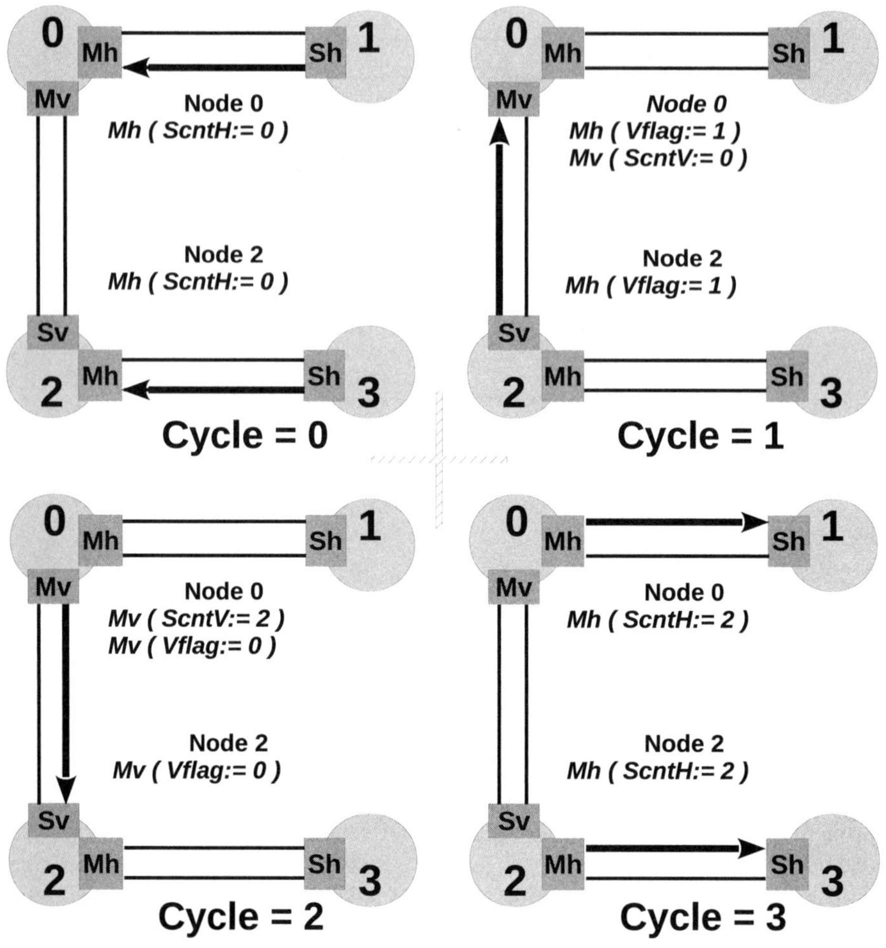

Fig. 4 Barrier synchronization under *GBarrier*

decrement their *ScntH* counters with the number of received signals (*ScntH:=1*). Besides, each *ScntH* counter is also decremented to reflect the fact that cores 0 and 2 have also arrived at the barrier (see *ScntH:=0* in the Figure). At cycle 1, once each horizontal master has detected that its local counter *ScntH* has reached zero, it sets its *Vflag* flag (*Vflag:=1*) in order to make the corresponding vertical slave (Sv) or master (Mv) controller to proceed with the vertical stage of the *account phase*. Then, the vertical slave (Sv) signals, through its corresponding transmission line, the arrival of cores 2 and 3 at the barrier and the vertical master (Mv) decrements its *ScntV* counter (*ScntV*:=1). Moreover, the *ScntV* counter is also decremented because the cores 0 and 1 have also arrived at the barrier and the *Vflag* flag was set (*ScntV*:=0). After the *ScntV* counter reaches zero, the *release phase* is initiated. To do so, at cycle 2, the vertical master unsets the *Vflag* flag (*Vflag:=0*), resets the local *ScntV* counter

```
GL_Barrier() {
        asm {
                # Arrival at the barrier
                  mov 1, bar_reg
                # Wait until all cores arrive
                  loop:
                        bnz bar_reg, loop
                # Resume execution
        }
}
```

Fig. 5 Encapsulating the *GBarrier* functionality into the *GL_Barrier* intrinsic

to its initial value (*ScntV:=2*) and signals the vertical slave, through the corresponding vertical *G-Line*. Upon reception of the signal, the vertical slave also resets the *Vflag* in order to make the horizontal masters to proceed with the horizontal stage of the *release phase*. At cycle 3, the horizontal masters initialize the local *ScntH* counters (*ScntH:=2*), and signal, through their corresponding horizontal *G-Lines*, the completion of the barrier synchronization to all waiting horizontal slaves. It is worth noting that all participating threads are spinning on a register until the whole process is completed as will be explained in Sect. 4.3.

4.3 Programmability Issues

The *GBarrier* mechanism proposed in this section is intended to be used by programmers in a transparent way. For that reason, as shown in Fig. 5, we propose to provide a special library-level barrier method (*GL_Barrier* in the Figure) that encapsulates the functionality of *GBarrier* and that could be used in parallel applications to deal with barrier operations. This barrier method uses a special 1-bit register, called *bar_reg*, to notify the arrival at the barrier by setting its value to one (see the *mov* instruction in Fig. 5). The *bar_reg* register needs as many bits as the number of *GBarriers* provided in hardware (one bit per barrier). In this way, several barrier operations involving different sets of cores (the threads in each set running one application) could take place simultaneously. In this way, the register file of each core must be augmented with the *bar_reg* register and the interplay between controllers and these registers must be enabled, switching on the controllers whenever the *bar_reg* registers are written, and resetting the registers and switching off the controllers once all controllers have finished the synchronization (cycle 3 in Fig. 4). In this way, the synchronization protocol explained in the previous section would be invoked as a result of the activation of the *bar_reg* register by a server core. Then, each core would enter in a loop waiting until the rest of cores have reached the barrier. Once all cores have set their corresponding *bar_reg* register and the synchronization protocol has been completed, all *bar_reg* registers are reset by the corresponding *GBarrier*'s controllers and then, all cores would leave the loop in order to resume execution.

5 Performance Implications

In this section, we analyze *GBarrier* to gain insight into its potential impact on performance. First, we start by discussing some considerations taken when using both *G-Lines* and *Standard* technologies to implement our proposal. Next, for both implementations, we show their potential contributions to performance in terms of some important raw statistics such as on-chip area overhead, power dissipation, maximum operating speed and minimum latencies to complete a barrier operation.

5.1 Implementation Technologies

5.1.1 *G-Lines* Technology

There were several reasons why we decided to use the *G-Lines* technology to develop our synchronization mechanism for barriers in manycore servers. First, the connectivity pattern utilized to deploy the dedicated *GBarrier*'s network (see Sect. 4.1) is based on long 1-bit single-dimension links which perfectly fit into the concept of *G-Lines*. Second, the promising results that could be achieved using this technology in terms of marginal area overhead and minimal power dissipation. Note that, according to the results reported in [40], that show negligible area overhead for a 392-*G-Line* network, the 32-core server system evaluated in this section (further details in Sect. 6.1) is made up of approximately one-20th of the latter number of *G-Lines*, thereby even lower implications for on-chip area would be obtained. This marginal area overhead will have also a negligible impact on power dissipation. Finally, the *GBarrier*'s synchronization protocol explained in Sect. 4.2 could take advantage of the extremely fast transmissions at 2.5 GHz that the use of the *G-Lines* technology would entail. In this way, we can directly adopt the same theoretical synchronization latencies for the gather and release phases explained in that section.

5.1.2 *Standard* Technology

The *GBarrier* architecture has also been implemented relying on the mainstream industrial synthesis toolflow with an STMicroelectronics 45 nm standard cell technology library. The main reason why we decided to employ this technology was to

Table 1 Raw statistics using *G-Lines* and *Standard* technologies for a single *GBarrier* in a 32-core server layout

	Frequency (MHz)	Latency (cycles)	Area (μm^2)	Power (mW)
G-Lines	2500	4	*Negligible*	26.4
Standard	670	14	5441	*Negligible*

provide a cost-effective implementation and to precisely quantify the performance losses due to the use of this interconnect-dominated nanoscale technology.

Since RC-based wires are very critical to performance degradation, we have implemented each *GBarrier*'s controller by separating the delay that signals take along the wires, from the effective computation that the controllers require to generate their output signals. Notice that, for small manycore servers, the critical path that limits the maximum operating speed in our *GBarrier* infrastructure is defined by the most complex controller (i.e., the master controller that samples signals from the highest number of slaves), but as the wire length increases for larger servers, the wires could represent such critical path. Consequently, separating wire delays from controllers delays become essential in order to achieve maximum clock speeds. In this way, by using this technology, we cannot directly assume the synchronization latencies achieved by using *G-Lines*, and a higher number of cycles will be required for the gather and release phases. In addition, to minimize the length of wires, we have situated the master controllers in the central column/row of the 2D-mesh topology, rather than the first column and first row as depicted in Fig. 2. Note that, in case of *G-Lines* technology this optimization would not be necessary since every *G-Line* is specially designed to implement one-cycle latency, one-bit transmissions across one dimension of the chip.

Finally, for a real characterization of our *GBarrier* proposal, our mechanism has been synthesized by defining non-routable obstructions. Such obstructions are placed to mimic the area of every core of the simulated system explained in Sect. 6.1. In this section, we assume that this area is equal to $550 \times 550\ \mu m^2$. Additionally, fences are defined to limit the area where the cells of each *GBarrier*'s controller can be placed. Such obstructions and fences also ensure minimum-length routing for the wires in order to reduce their impact on performance and area overhead as the wire length increases.

5.2 *Raw Performance Statistics*

Table 1 shows the main raw performance statistics obtained from the use of both technologies to implement *GBarrier*. In particular, we illustrate the maximum operating speed, the latencies of a barrier synchronization and also the area overhead and power that our proposal entails.

As we can see, the maximum operating speed achieved by the *G-Lines* technology is 3.7 times higher than for the *Standard* technology. Moreover, the number of clock cycles employed by the former technology to complete a barrier is 3.5 lower than that achieved by the latter technology. The reason is that every *GBarrier*'s controller and wire involved take a different clock cycle in the synchronization process. Besides, the internal communication using the *Vflag* flag between controllers located in the same core (e.g., Mh and Mv in Fig. 3) requires an extra clock cycle to achieve the maximum operating speed. Therefore, the superior efficiency of *G-Lines* technology reports roughly a thirteen times faster *GBarrier* implementation.

In addition, negligible overheads in terms of die area are reported for both technologies. First, regarding the *G-Lines* technology, as discussed above, our *GBarrier* infrastructure uses one-20th of the minimal area overhead reported in [40] and then, we can assume that its on-chip area is negligible. And second, for the *Standard* technology, an area overhead for *GBarrier* equal to 5441 μm^2 is required that corresponds to a negligible 0.06 % of the total area employed for the simulated 32-core server layout (remember that we assume that each core is 550×550 μm^2 in size).

The latter marginal on-chip overheads will introduce a negligible impact on power dissipation. To exemplify that, we estimate the power dissipated by the *G-Line*-based implementation. To do so, we employ the power dissipation parameters for a 65-nm CMOS process simulated in [40]: 0.6 mW per transmitter; 0.4 mW per receiver; and 2.4 mW per receiver that implement the *S-CSMA* technique. Moreover, according to [40] no static power is dissipated by the *G-Lines*.

To estimate the power dissipation, we must deal with the maximum number of transmitters and receivers in the system operating at once. From the synchronization protocol already explained and illustrated in Fig. 4, the worst case of power dissipation per clock cycle is when all cores initiate the gather phase at the same time. Therefore, for the simulated 32-core server in Sect. 6.1, and considering a 4×8-core 2D-mesh layout[1], there will be seven horizontal slaves per row (i.e., 28 transmitters) signaling the arrival at the barrier, and four horizontal master controllers that count the latter signals through the *S-CSMA* technique (i.e., four receivers). Hence, the total power estimated will be 26.4 mW (28×0.6 + 4×2.4). Utilizing CACTI [16], the magnitude of this dissipation is less than one-11st of the power dissipated per read port in the L1 caches simulated in this section (see Table 2).

As a conclusion, the above results suggest that the fastest technology is the most appropriate implementation to materialize *GBarrier*. Although synchronization delay would become the discriminating factor, we have also to consider the major drawback of using *G-Lines*: *The G-Lines technology is a full-custom technology that is not cost-effective in the embedded computing domain, hence not being within reach of a standard cell design methodology*. In consequence, it would be of paramount importance to determine the exact magnitude of such performance degradation when using the *Standard* technology. In case of being negligible, the slower technology

[1] For simplicity, we assume that 8 cores per row can be materialized in *G-Lines*. Recall that this technology is limited to 7 cores per row and, for example, a 6×6-core server layout must be considered instead to span the simulated 2D-mesh 32-core system.

Table 2 server baseline configuration

Number of cores	32
Core	3 GHz, in-order 2-way model
Cache line size	64 Bytes
L1 I/D-Cache	32 KB, 4-way, 2 cycle
L2 Cache (per core)	256 KB, 4-way, 12+4 cycles
Memory access time	400 cycles
Network configuration	2D-mesh
Network bandwidth	75 GB/s
Link width	75 bytes

would be the preferred *GBarrier* implementation. This experiment will be conducted in Sect. 6.3.1, by comparing synchronization timings of the two *GBarrier* implementations in comparison to the best SW-barrier implementation.

6 Evaluation

In this section we give details of our experimental methodology and performance results. We describe the simulation environment and the set of microbenchmarks and scientific applications that we have used in Sect. 6.1. The two SW-barrier implementations the *GBarrier* mechanism is compared with are presented in Sect. 6.2. Finally, the performance results are analyzed in Sect. 6.3 in terms of execution time, network traffic and energy consumption.

6.1 Experimental Setup

In order to support *GBarrier*, the Sim-PowerCMP [1] performance simulator has been extended. Sim-PowerCMP is a detailed architecture-level power-performance simulator for tiled-server architectures that also estimates energy consumption for the full server. Table 2 summarizes the values of the main configurable parameters assumed in this section. As can be seen, we have simulated a 32-core server with an aggressive 2D-mesh network built in a 45 nm process technology.

To evaluate the performance benefits derived from *GBarrier*, we have used one synthetic benchmark, three kernels and three scientific applications. First, the synthetic benchmark is intended to measure the latency of barriers themselves. Hence, it helps us provide some insight into the potential benefits that our *GBarrier* mechanism could provide. To do that, we follow the methodology described in [10]: performance is measured as average time per barrier over a 100,000-iterations loop of four consecutive barriers with no work or delays between them. Second, for the kernels we have employed three kernels from Livermore loops [13]. Following the recommendations

Table 3 Configuration of the benchmarks used in this section

Benchmark	Input size	#Barriers	Period
Synthetic	100,000 iterations	400,000	2568
Kernel 2	1024 elements, 1000 iterations	10,000	3103
Kernel 3	1024 elements, 1000 iterations	1000	4953
Kernel 6	1024 elements, 1000 iterations	1,022,000	4908
Unstructured	Mesh.2K, one time step	80	67,361
Ocean	258×258 ocean	364	205,206
EM3D	38,400 nodes, degree 2, 15 % remote, 25 steps	198	3673

given in [25], we focus on Kernels 2, 3 and 6. And third, we have considered three scientific applications: Unstructured, EM3D and Ocean. These applications were chosen since they present a non-negligible fraction of the total execution time due to barrier operations. We would like to point out that all experimental results reported in this section are for the parallel phase of all of the benchmarks under study.

We summarize the characteristics of the set of benchmarks used in Table 3. For each of them we account for the input size, the total number of barrier executions (*#Barriers*), and the estimated barrier period (the number of cycles on average between two consecutive barrier executions). The latter is calculated by dividing the total number of execution cycles into the total number of barrier executions in every case. Notice that, the barrier period is a simple metric that somehow quantifies the presence of barriers in every benchmark. For example, the Ocean application presents 364 barrier operations every 205,206 cycles on average (see Table 3). Consequently, from this high barrier period, we should not expect to obtain a significant fraction of the total execution time due to barriers. The latter result also limits the potential benefits that our *GBarrier* mechanism could provide. A more detailed analysis will be given below in Sect. 6.3.

6.2 Barrier Implementations

To quantify the benefits of our *GBarrier* mechanism, we consider that barriers found in the benchmarks previously described are implemented by using two SW-barrier implementations: a centralized sense-reversal barrier based on locks (or CSW), and a binary combining-tree or distributed barrier (DSW). On the one hand, in a CSW barrier, each core increments a centralized shared counter when it reaches the barrier, and spins until that counter indicates that all cores are present. On the other hand, in a DSW barrier, there are several shared counters distributed in a binary tree fashion. Thus all cores are divided into groups assigned to each leaf (variable) of the tree. Each core increments its leaf and spins. Once the last one arrives in the group, it continues up the tree to update the parent and so on towards the root. Finally, the release phase is similar but in the opposite direction (towards the leaves).

In general, the implementation of a barrier can be split into three typical stages: the notification stage (S1), when each core indicates its arrival at the barrier; the busy-wait stage (S2), to wait the arrival of the remaining cores; and the release stage (S3), in order to resume execution. At first glance, our *GBarrier* proposal should accelerate all the three stages because they are executed without involving any network transaction or coherence activity. Remember that, our mechanism operates just by means of a simple synchronization protocol implemented atop a dedicated lightweight on-chip network, taking only four cycles (the best-case scenario for a 7×7-core server and *G-Lines* technology) to perform a barrier operation among all threads or cores (see Sect. 4.2). However, we could identify two typical situations in which our proposal may entail negligible improvement. The first situation occurs when a parallel application contains a reduced number of barriers or a very high barrier period. This helped us to pick the most significant benchmarks for our evaluation (e.g., choosing Ocean among all of the applications from the SPLASH-2 benchmark suite). The second situation takes place when barrier latency is dominated by the stage S2. For instance, this fact may suggest that the application is under workload imbalance. We will take into consideration these conclusions when analyzing the performance results in the next section.

6.3 *Performance Results*

The evaluation of the *GBarrier* mechanism has been carried out taking into account the execution times achieved for the benchmarks shown in Table 3, as well as the amount of traffic in the interconnect and the energy-delay^2 product (ED^{2}P) metric for the full server.

6.3.1 Execution Time

First of all, we consider the implementation of the *GBarrier* that relies on the *G-Lines* technology. Figure 6 illustrates the execution times obtained for the synthetic benchmark under study depending on the number of cores in x-axis (from 4 to 32 cores). Notice that y-axis is in logarithmic scale. Remember that, the use of this benchmark allows us to measure the latency of barrier operations themselves. As we can observe, there are three lines depending on the three barrier implementations explained in Sect. 6.2: CSW, DSW and our *GBarrier* mechanism (GB).

From the results presented in Fig. 6, we can derive two main appreciations. First, the DSW implementation is much more efficient and scalable than the CSW barrier. It is due to the fact that the CSW implementation employs a centralized shared counter among all threads, which clearly becomes a bottleneck as the number of cores increases. In contrast, DSW significantly alleviates contention by using several shared counters distributed in a binary tree fashion. And second, it is clear that *GBarrier* highly outperforms the others in terms of execution time and scalability.

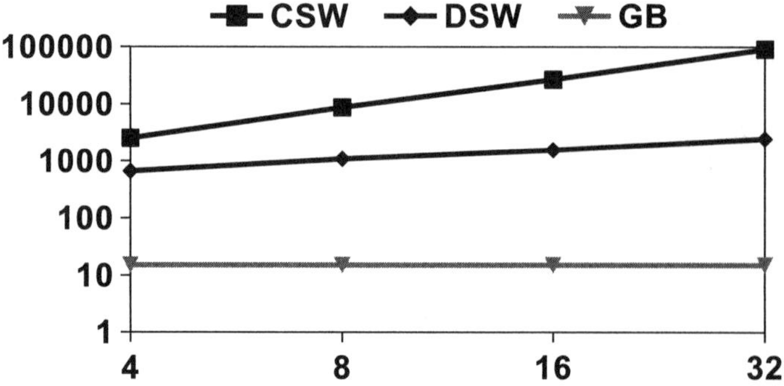

Fig. 6 Average times for three different barrier mechanisms

On the one hand, the *GBarrier* mechanism drastically reduces execution times of S1, S2 and S3 stages (up to four cycles for the best-case scenario). On the other hand, we deploy a dedicated *G-Line*-based network to implement barrier synchronizations thus removing any coherence activity or synchronization-related traffic in the interconnection network. We would like to point out that our *GBarrier* implementation suffers from a slight overhead in the times obtained (see 13 cycles in Fig. 6). It is due to the overhead introduced by the simulator when applications call our barrier implementation, because it must be accomplished through its application library.

Figure 7 shows the average normalized execution times over a 32-core server layout for the rest of applications under study. In particular, for Kernels 2, 3 and 6, and the scientific applications: Unstructured, Ocean and EM3D. Furthermore, we depict the breakdown of execution time depending on the best SW-barrier implementation (DSW) and our hardware barrier mechanism (GB). Execution time is further broken down into several categories: *Barrier* is the time spent on barriers (sum of the time taken in the S1, S2 and S3 stages explained above); *Write* and *Read* are the times spent on memory accesses; *Lock* is the time for lock synchronizations; finally, *Busy* is the time for computational work (e.g., arithmetic operations). In addition, we also illustrate the average times of all kernels and applications for each barrier implementation (see *AvgK* and *AvgA*).

Regarding the kernels results, we can see that our proposal involves a reduction in execution time of 54 % on average (see *AvgK*). In more depth, Kernels 2, 3 and 6 present reductions of 70, 46 and 47 %, respectively. The exact extent of the reduction in each case depends on the barrier period that each kernel has: 3103, 4953 and 4908 cycles, respectively (see Table 3). That is, the lower barrier period the higher performance efficiency. For that, Kernel 2 presents the highest reduction in execution time. Moreover, the reductions in execution time obtained also depend on the *Write* and *Read* categories, since our *GBarrier* mechanism operates without involving any memory-related instructions (e.g., see reduction of *Write* category for Kernel 6).

On other hand, the fraction of the execution time that barrier synchronization consumes is lower when scientific applications are considered. In these cases, most of

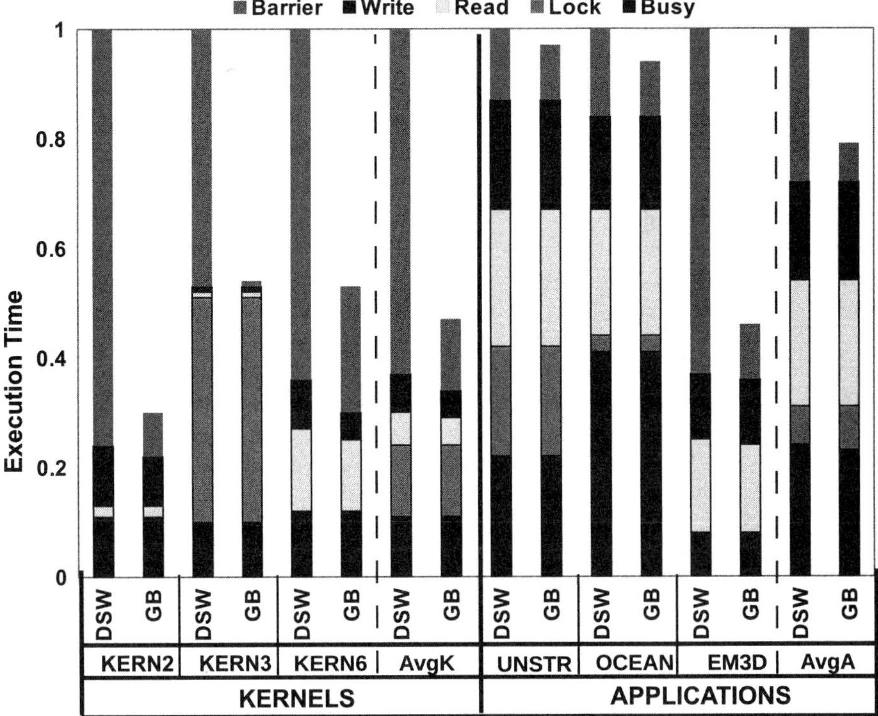

Fig. 7 Normalized execution time over a 32-core server

the time is spent on computations and memory accesses (*Busy*, *Write* and *Read* categories), resulting in lower barrier periods. As a result, lower reductions in execution time can be observed for Unstructured, Ocean and EM3D (21 % on average). In particular, worse results stem from the applications Unstructured and Ocean since they present a very high barrier period (67,361 and 205,206 cycles, respectively), which translates into reductions of only 3 and 6 % in the total execution time, respectively. The exception is EM3D, because it presents significant reductions in execution time (54 %) due to its very small barrier period (3673 cycles).

Table 4 shows the speedup results for the scientific applications (Ocean, Unstructured and EM3D) when scaling the number of cores parameter with the values 4, 8, 16 and 32. Moreover, we use two different barrier implementations: DSW in comparison to our *GBarrier* mechanism (GB). From the results shown in Table 4, we can extract two important observations. First, all of the benchmarks scale as the number of cores is increased. Second, the exact extent of speedups depends on the efficiency of the barrier implementation we are using. In this way, higher speedups are obtained when employing our *GBarrier* mechanism.

According to the discussion given at the end of Sect. 5.1.2, it would be of paramount importance determining whether the performance losses in terms of synchronization latency derived from the use of the *Standard* technology can be

Table 4 Speedups for the scientific applications

Benchmark	Barrier version	4	8	16	32
UNSTR	DSW	3.32	5.91	10.48	17.43
	GB	3.33	6.01	10.68	17.97
OCEAN	DSW	3.69	7.02	13.46	23.56
	GB	3.70	7.10	13.98	25.06
EM3D	DSW	3.36	5.38	7.32	9.13
	GB	3.42	6.12	10.55	16.82

considered negligible. To this end, Table 5 shows the normalized execution times with respect to those obtained when the DSW barrier is used, depending on the two kind of *GBarrier* implementations studied in this section: *G-Lines*[2] and *Standard* technologies. From the results shown in the table, it can be derived that average performance degradations of 6.3 and 4.3 % are reported when using the *Standard* technology for the kernels and scientific applications, respectively. These performance gap is very small if we take into account the significant average reductions in execution time of 48 and 16.6 % (kernels and scientific applications) achieved by the *Standard* technology in comparison to the most efficient SW-barrier implementation (DSW). Consequently, we can affirm that our *GBarrier* mechanism is not so dependent on a full-custom technology to provide extremely efficient barrier synchronizations.

Obviously, the performance gap between both technologies will be higher for greater server layouts due to the negative effects of the interconnect-dominated nanoscale *Standard* technology. However, the use of a very lightweight interconnection network, that features a hierarchical design, along with a very simple synchronization protocol help relieve such negative effects on performance making the *GBarrier* design really scalable. In particular, in [19], where we explore different hardware-based barrier layouts using *Standard* technology, impressive results are shown for a 64-core server layout when comparing performance against the best SW-barrier.

Finally, we also carried out a sensitivity analysis to evaluate the extent to which our proposal is affected by longer link latencies when considering the *G-Line*-based technology. To do so, we simulate several configurations of the *G-Line*-based network with varying latencies for the links and evaluate the impact that this has on performance. Several clock cycles may be necessary to transmit a signal across one dimension of the chip if, for example, we consider longer links that cannot support a propagation delay of a single clock cycle, or even if lower clock frequencies are required to integrate our *GBarrier* infrastructure in the manycore server. Figure 8 illustrates the normalized execution times when *G-Lines* take from 1 (results presented in Fig. 7) to 12 clock cycles (see z-axis in the Figure). As we can observe,

[2] Note that, the results for the implementation that uses *G-Lines* are the same as those presented in Fig. 7.

Table 5 Normalized execution times for *G-Lines* and *Standard* technologies

	KERN2	KERN3	KERN6	UNSTR	OCEAN	EM3D
G-Lines	0.30	0.54	0.53	0.97	0.94	0.46
Standard	0.39	0.61	0.56	0.99	0.96	0.55

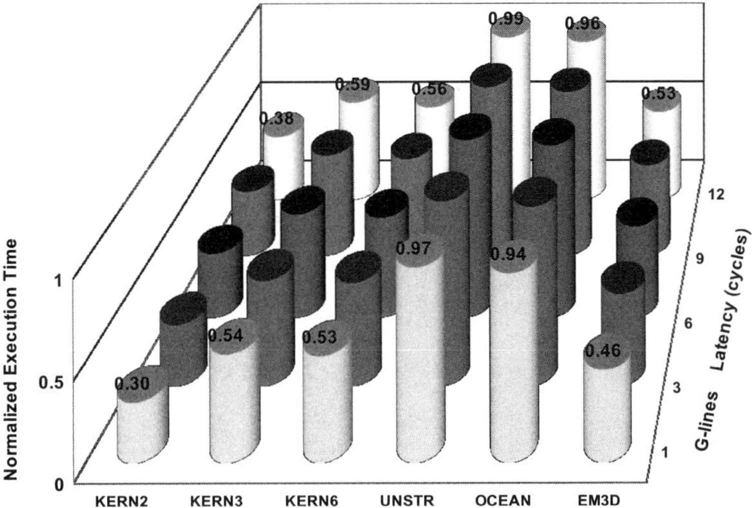

Fig. 8 Normalized execution times for the benchmarks depending on the latency of the *G-Lines* (a 32-core server is assumed)

very small performance losses are derived even when dealing with 12-cycle *G-Lines*. Particularly, performance degradations of just 5.3 and 3.6 % on average in the worst case are shown for the kernels and scientific applications, respectively. Note that the results observed for the 12-cycle case are slightly lower than those obtained for the *Standard* technology previously reported in Table 5. According to Sect. 5.1, the implementation based on *Standard* technology is roughly 13 times slower than the *G-Line*-based infrastructure, so that the former would be roughly equivalent to a 13-cycle *G-Line*-based implementation what explains such similarities.

6.3.2 Network Traffic

Our proposal does not generate any coherence messages on the main data network when performing barrier synchronizations. In the end, this translates into significant reductions in terms of network traffic. Figure 9 shows the total network traffic across the main data network. In particular, each bar plots the number of bytes transmitted through the interconnection network (the total number of bytes transmitted by all the switches of the interconnect) normalized with respect to the DSW case. Each bar is broken down into three categories: *Coherence* corresponds to the messages generated

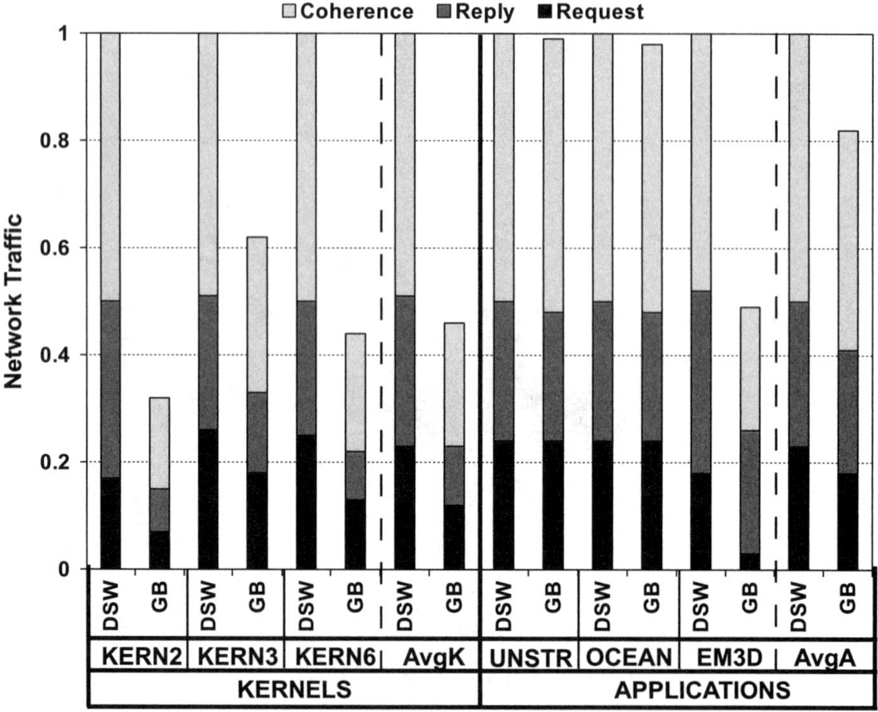

Fig. 9 Normalized network traffic over a 32-core server

by the cache coherence protocol (e.g., invalidations and Cache-to-Cache transfers); *Request* comprehends messages generated when load and store instructions miss in cache and must access a remote directory; and finally, *Reply* involves the messages with data.

For the kernels, important reductions in network traffic are achieved (53 % on average). In general, these reductions are directly related to the extents of the improvements in execution time previously reported. Moreover, since the simulated L2 cache is shared among the different processing cores, but it is physically distributed between them (see Sect. 6.1), some accesses to the L2 cache will be sent to the local slice while the rest will be serviced by remote slices. This will also affect the timings for lock acquisition and release operations. In contrast, since our *GBarrier* implementation skips the memory hierarchy we have not obtained such negative impact on network traffic. In particular, Kernel 2, 3 and 6 show important reductions of 68, 37 and 56%, respectively.

Finally, regarding the scientific applications, we can see a slight reduction in network traffic (see 18 % in *AvgA*). More specifically, the applications Unstructured, Ocean and EM3D present reductions of 1, 2 and 51 %, respectively. As before, there is a correlation between the fraction of the execution time devoted to barrier synchronization and the amount of network traffic that is saved. In this way, for

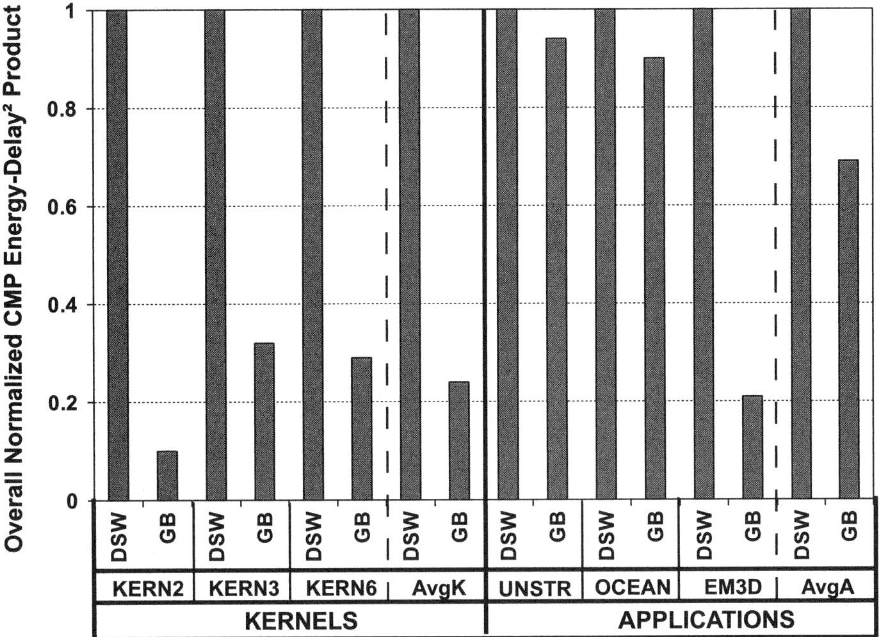

Fig. 10 Normalized ED^{2}P metric for the full server

Unstructured and Ocean we could expect more than 1 or 2 % reductions in network traffic, due to the 3 and 5 % reduction in execution time, respectively. However, we noticed that the latency of barriers for these benchmarks is dominated by the S2 stage and, as we mentioned, this implies workload imbalance. For the case of DSW, this stage involves negligible network traffic because, once shared variables are loaded in cache, busy-waiting is performed locally. As a result, our *GBarrier* mechanism reports a very low traffic reduction for both benchmarks. Finally, as we expected, EM3D presents a considerable reduction in network traffic (51 %) because of its very small barrier period.

6.3.3 Energy Efficiency

The use of our *GBarrier* mechanism leads to important reductions in execution time and network traffic, as explained above. In this section, we also quantify the benefits in energy efficiency that our proposal could entail. More specifically, we present in Fig. 10 the normalized energy-delay^2 product (ED^{2}P) metric for the full server. To account for the energy consumed by the *GBarrier* architecture (the *G-Lines*-based network described in Sect. 4.1), we extend the Sim-PowerCMP with the consumption model of *G-Lines* and controllers described in previous Sect. 5.2. As we conclude in that section, the power dissipation associated with our two technology-aware

GBarrier implementations is negligible, hence the power statistics presented in this section will be mainly due to the improvements in execution time and network traffic reported in previous sections.

As in the previous two sections, all results in Fig. 10 have been normalized with respect to the DSW case. As can be observed, important improvements in the ED^2P metric of the whole server are

achieved when applying our proposal. In particular, the *GBarrier* mechanism brings average improvements in ED^2P of 76 and 31 % for the kernels and scientific applications, respectively. Particularly, the Kernel 2, 3 and 6 show reductions of 90, 68 and 71 %, respectively. Additionally, reductions of 6, 10 and 79 % are achieved for Unstructured, Ocean and EM3D.

In general, the magnitude of these savings is directly related to the extents of the improvements in execution time and network traffic previously reported. We have found that when *GBarrier* is employed, the number of instructions executed per barrier operation is drastically reduced. Note that while DSW barrier must deal with a distributed shared counter in a binary tree fashion, *GBarrier* only needs a single assignment instruction on a register to notify the arrival at the barrier (see Sect. 4.3). Obviously, less instructions executed means less energy consumed in the server cores.

Moreover, since we reduce the latency to notify the arrival at the barrier (S1 stage), the busy-wait process (S2 stage) is also shortened with *GBarrier*. While busy-waiting, a server core repeatedly accesses the L1 cache to check the value of a shared variable. In this way, shorter busy-waiting implies less accesses to the L1 cache, and therefore, less energy consumed in this structure. Finally, given the fact that our proposal skips the memory hierarchy, we save all the energy derived from coherence activity when barriers are executed. In particular, we remove all of the L1 cache misses related to barrier operations and the corresponding messages transferred across the interconnect. This brings reductions in the energy consumed at the L2 cache banks and the interconnection network.

7 Related Work

To overcome the performance limitations imposed by SW-barriers, there have been proposed several hardware-based optimizations in the context of both traditional multiprocessors and, more recently, servers. In this section, we make an attempt at categorizing most of them in terms of the part of the system they improve or augment: memory-based approaches, network-based and global lines approaches.

Regarding memory-based approaches, Goodman et al. [24] proposed a set of efficient primitives for process synchronization based on the use of synchronization bits (syncbits). Syncbits are logically associated with every block in memory to provide a simple mechanism for mutual exclusion.

Sampson et al. [25] presented barrier filters, a mechanism to implement fast barrier synchronization on servers. The key idea is that they ensure that all threads arriving at a barrier require an unavailable cache line to proceed. Then, the barrier filter starves their requests until they all have arrived. Monchiero et al. [29] proposed a hardware module to optimize busy-waiting synchronization in servers. This module is integrated in the memory controller, namely the Synchronization-operation Buffer (SB). The SB manages locally the polling on shared variables, avoiding traffic in the network and memory accesses.

Differently from these previous approaches, our proposal decouples completely barrier synchronization from any kind of memory-related activity.

Regarding network-based approaches, Hsu and Yew [41] proposed a multistage shuffle-exchange network to efficiently handle synchronization traffic of SW-barriers by combining packets in the switches in order to relieve hot-spot congestion from the network.

For example, the network architecture of the Connection Machine CM-5 [8] contains a dedicated network (control network) to perform synchronizations of an entire set of servers through specific messages interchanged between outgoing and incoming FIFO queues at the network interface level. In addition, the Blue Gene/L [31] also contains a dedicated interconnection network for barrier synchronization. Sartori and Kumar pointed in [26] that although a dedicated interconnection network manages barrier operations efficiently, its integration in future manycore servers may not be a feasible solution due to the large on-chip area and power dissipation that it could entail. They propose three barrier implementations, that are hybrid of software and hardware aimed at achieving closer approximation to the performance of a dedicated interconnection network but at a fraction of the cost.

Differently from any of the above proposals, *GBarrier* operates independently of the main data network, thus removing all synchronization-related traffic. Moreover, we use a very reduced number of state-of-the-art global links that introduce negligible area overhead and power dissipation.

Finally, as to global lines approaches, Cyclops [7] is a highly parallel server-and-memory system on a chip (32-quad-core server architecture). This architecture implements a fast barrier operation through a special purpose register (SPR). It is actually implemented as a wired OR for all the threads on the chip. Each thread writes its SPR independently, and it reads the ORed values of all the threads' SPRs. The register has eight bits which provides four distinct barriers (2 bits per barrier). One of the bits holds the state of the *current* barrier cycle whilst the other holds the state of the *next* barrier cycle. All threads participating in the barrier initially set their *current* barrier bit to 1. The threads not participating in the barrier leave both bits set to 0. Then, when a thread reaches the barrier it writes 0 to the *current* bit, thereby removing its contribution to the *current* barrier cycle, and one to the *next* bit. Hence, the barrier is completed when all *current* bits become 0. Furthermore, the use of the *current* and *next* bits are interchanged after each execution of the barrier. To communicate the SPRs' values, Cyclops employs a dedicated 16-bit bus which enables the completion of a barrier operation among all threads in only a few dozens of cycles [42].

In contrast, rather than buses, our proposal communicates signals through a more scalable on-chip network based on 1-bit width links deployed in a hierarchical layout.

Moreover, as aforementioned, our accounting process is distributed, hence more scalable.

TLSync [22] is a sophisticated design for barrier synchronization that provides very efficient barriers although being fully dependent on non-standard technology, namely *Transmission Lines* [5]. In particular, the process of synchronization is performed by allocating different radio-frequency bands from the high-frequency part of the spectrum per barrier, thereby allowing multiple groups of threads to be concurrently synchronized very quickly. While this is a very efficient hardware design, a successful implementation is restricted to leading-edge technology thus not being within reach of a standard cell design methodology. In contrast, in light of the impressive performance results shown in this section, a cost-effective implementation of *GBarrier* is also feasible.

8 Hardware Lock Synchronization

Lock synchronization in parallel applications has long been devised to ensure that a block of code manipulating a shared data structure, namely critical section (CS), is executed by only one process or thread at a time (i.e., the lock owner), thereby guaranteeing mutual exclusion among processes or threads and preserving the integrity of the shared data [12].

In shared-memory parallel systems, this kind of synchronization mechanism commonly comprises a pair of operations. First, the *lock* operation that a thread utilizes before executing the CS to request the lock ownership. And second, once the thread becomes the lock owner and executes the CS, the *unlock* operation, that is executed straight afterwards the CS in order to release the lock ownership, so that another thread can become the next lock owner.

Typical software-based implementations for *lock/unlock* rely on a combination of memory operations on shared variables that involve special instructions such as *LL/SC*, or atomic read-modify-write instructions like *test&set*. Nonetheless, the use of shared variables for lock synchronization has two important implications for performance and scalability, especially in future manycore servers. First, the cache coherence protocol must come into play in order to maintain the consistency of shared variables across all levels of the memory hierarchy. Coherence activity translates into traffic injection in the interconnection network. As a result, an ever-growing amount of resources may need to be devoted to support lock synchronization as the core count increases. Moreover, lock acquisition and release operations timing is deeply affected by the performance and scalability of the cache coherence protocol especially under the presence of highly-contended locks. Second, lock contention has long been recognized as a key impediment to performance and scalability since it causes serialization [30]. Consequently, the longer the idle time spent on lock acquisition and release operations, the larger the parallel efficiency reduction.

As an evidence, we show in Fig. 11 the potential benefits to performance when lock synchronizations do not involve the cache coherence protocol and have zero

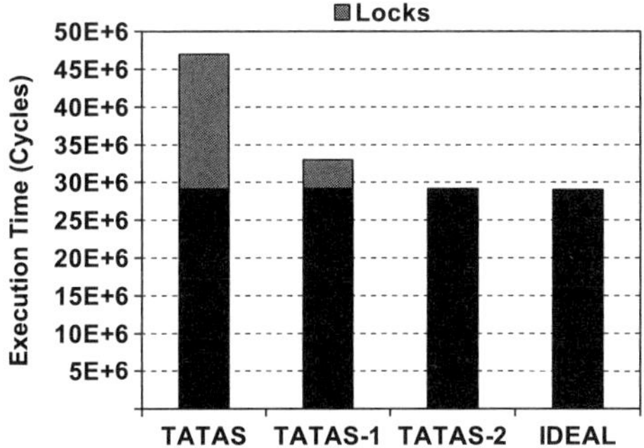

Fig. 11 Potential benefits for Raytrace when using *ideal locks*

latency. To do so, the Raytrace application from the SPLASH-2 benchmark suite [37] is run by using distinct lock implementations (for details of the evaluation see Sect. 11). In each case, we highlight in orange color the fraction of the execution time due to the locks. Shared-memory-based locks use `test-and-test&set` (see TATAS bar in Fig. 11). In turn, ideal locks (see IDEAL bar in Fig. 11) do not deal with the cache coherence protocol to eliminate any inherited performance or scalability side-effects. Besides, lock acquisition and release operations take a single clock cycle each to minimize serialization due to contention. As expected, ideal locks clearly outperform shared-memory-based locks since the lock acquisition and release operations account for a significant fraction of the execution time in Raytrace. However, a post-mortem analysis of Raytrace lock usage reveals that only 2 out of its 34 locks are highly-contended. In this sense, if all the locks other than the highly-contended ones are implemented using regular shared-memory-based locks, a reduction in the execution time similar to that of ideal locks is obtained (see TATAS-1 and TATAS-2 bars[3] in Fig. 11). The latter result suggests that only highly-contended locks can truly benefit from a more efficient lock implementation.

In this section, we present and evaluate a new lock synchronization mechanism aimed at accelerating highly-contended locks. Our proposal, namely *GLock*, is a lightweight on-chip network infrastructure devoted to implement a very simple token-based message-passing protocol providing extremely efficient execution for highly-contended locks. As with the *GBarrier* mechanism presented in Chap. 3, we have explored two different technologies to implement *GLock*. On the one hand, we make use of the state-of-the-art full-custom *G-Lines* technology introduced in Sect. 2, that enables almost speed-of-light 1-bit communications across one dimension of the entire chip. On the other hand, we employ the mainstream industrial toolflow with standard cells in an advanced 45 nm technology in order to obtain a cost-effective implementation for our proposal at the expense of some negligible performance loss.

[3] TATAS-X means that one (X = 1) or two (X = 2) of the highly-contended locks have been implemented as ideal locks.

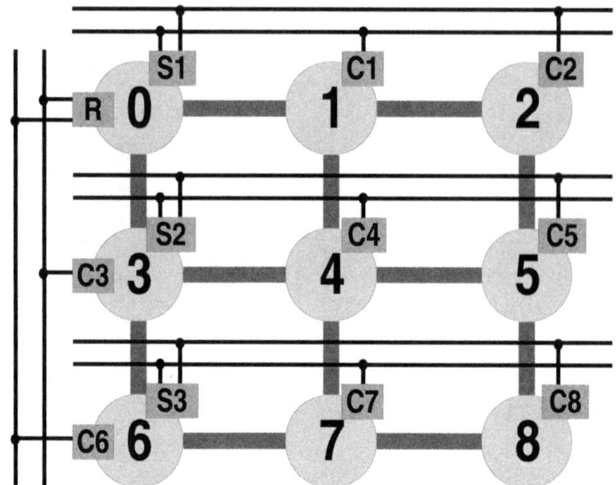

Fig. 12 *GLock* architecture for a 9-core server with a 2D-mesh network

9 The *GLock* Synchronization Mechanism

In this section, we present our proposal to build an efficient synchronization mechanism for highly-contended locks in manycore servers. To do so, we will focus on describing the hardware components required and the synchronization protocol employed, rather than going into any technical aspects of the two implementation technologies used (further details in Sect. 10.1). In more depth, we start by describing the dedicated on-chip network that our proposal entails. As a case study, we choose a server with a 2D-mesh data interconnection network with R rows of C cores each (for a total of $N = R \times C$ cores), although our proposal is not restricted to this topology. Next, we show how the *GLock* mechanism would operate and finally, the interface for programmers.

9.1 Dedicated On-Chip Network Architecture

The *GLock* mechanism proposed in this section relies on a dedicated on-chip network as can be observed in the example in Fig. 12. For simplicity, we concentrate on a version of the proposed network providing support for one lock. As we can see, the network is made up of two kind of components. *Links* (horizontal and vertical finer black lines), that are used to transmit the signals required by the synchronization protocol; and controllers (R, Sx and Cx), that actually implement the synchronization protocol.

Every *link* is simply a wire that enables the transmission of one bit of information across one dimension of the chip, employing one *link* per transmitter and lock.

Every *link* will be used to request the associated lock and grant lock acquisitions. In this way, for any 2D-mesh layout the total number of *links* per lock that would be needed is equal to $N - 1$, where N is the number of cores of the server (e.g., eight *links* for the 9-core server shown in Fig. 12). It is worth noting that our proposal is aimed at providing this kind of hardware support just for a very limited number of locks, enabling the opportunity to deal with very efficient highly-contended lock synchronizations with marginal area overhead (see Sect. 10.1).

In addition to the *links*, our proposal also incorporates a set of controllers. In particular, we distinguish two types of controllers: the *local controllers* (Cx in Fig. 12) and the *lock managers* (R and Sx in Fig. 12). The *local controllers* send and receive signals to and from their corresponding *lock managers* through their dedicated *links* (e.g., $C1$ sends and receives signals to/from $S1$). The exception is when the *local controller* is located in the same core as its associated *lock manager*. In this case, the functionality of the *local controller* is encapsulated in the *lock manager*, and communication is performed locally by means of a flag. For example, $S1$ monitors not only signals from *local controllers* one and two ($C1$, $C2$) through their corresponding *links*, but also from the local core through an internal flag (for clarity, this flag is not shown in Fig. 12).

The *lock managers* control lock ownership by monitoring signals from either *links* (remote cores) or the flags (local core). Besides, *lock managers* are divided into two groups: primary and secondary *lock managers*. Secondary *lock managers* (Sx) are responsible for monitoring signals from their corresponding *local controllers*, whereas the primary *lock manager* (R) is responsible for monitoring signals from the secondary ones. Primary and secondary *lock managers* communicate with each other by means of the vertical *links* shown in Fig. 12.

Finally, to have a clear understanding of our proposal, we represent the architecture described above as the hierarchy shown in Fig. 13. In particular, the dedicated network that our proposal is based on can be represented as a three-level hierarchy. The root of the hierarchy is the primary *lock manager*. The secondary *lock managers* would be located at the intermediate nodes. Finally, the leaves of the hierarchy would be the server cores (with the *local controllers*). All elements are connected using *links* (continuous lines) or locally by means of an internal flag (dashed lines). The flags (*fx* and *fSx*) store the signals sent by the controllers to the corresponding *lock manager* (primary and secondary). In this way, we need flags not only to store the signals sent between Sx and the *local controllers* (one flag per Cx controller: $f1$ for $C1$, $f2$ for $C2$, etc.), but also to store the signals transmitted between R and Sx (one flag per Sx controller: $fS1$ for $S1$, $fS2$ for $S2$, etc.).

9.2 Synchronization Protocol

The synchronization protocol implemented on top of the network previously described is based on the exchange of 1-bit messages (signals) between the *local controllers* and the *lock managers*. More specifically, the protocol uses three types

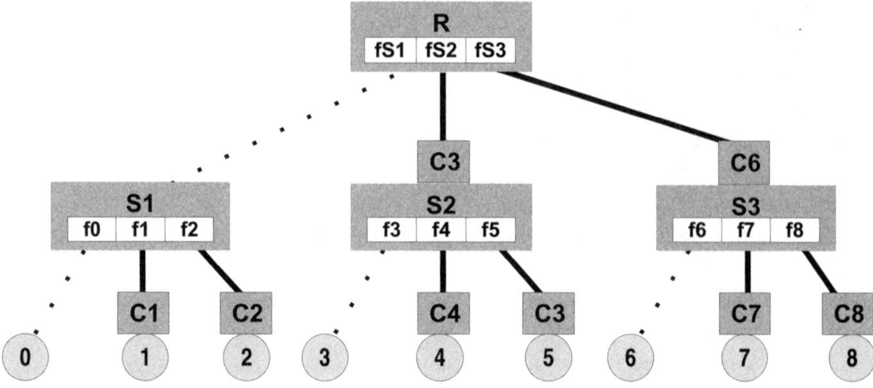

Fig. 13 Logical view of the *link*-based network for a 9-core server with a 2D-mesh network

of signals to perform a lock synchronization. The REQ and REL signals, which are sent from the *local controllers* to their corresponding *lock manager* to ask for the lock and to release the lock, respectively; and the TOKEN signal which is sent from a *lock manager* to a particular *local controller* to grant access to a lock. In addition, these signals are also transmitted between primary and secondary *lock managers* in a lock synchronization. In particular, the secondary *lock managers* ask for the lock by sending the REQ signal to the primary *lock manager* and receive authorization from the latter through the TOKEN signal. Similarly, after the lock is released, a secondary *lock manager* notifies the primary one by means of the REL signal.

Lock managers (both the primary and secondary ones) use a round-robin strategy to grant the lock among those server cores which are competing for becoming the next owner. Let's assume that all of the cores in Fig. 13 send the REQ signal to their corresponding secondary *lock manager* at the same time. In this case, the TOKEN signal granting the lock would be received by Core0 first; then, once Core0 has released the lock, Core1 would become the next holder; and so on, until Core8 is reached. Next, the process would start again from Core0 if there are additional pending lock requests. Since the *GLock* mechanism is aimed at accelerating highly-contended locks we do not expect that the election of the strategy to grant the lock in these situations will have any impact on performance. However, this is a key design point to ensure the fairness expected from a lock implementation [10]. The latter is the reason why we use the round-robin strategy.

As an example of how the synchronization protocol works, Fig. 14 presents the case where the nine cores of the server depicted in Fig. 12 try to get access to the lock at the same time. To clarify the explanation, the arrows in the Figure mark the sense of the transmissions. Moreover, each arrow is labeled with the cycle in which communication occurs, starting with cycle 1. It is worth noting that we are assuming theoretical synchronization latencies that may not be reflected in the exact number of clock cycles required for the two physical *GLock* implementations (see

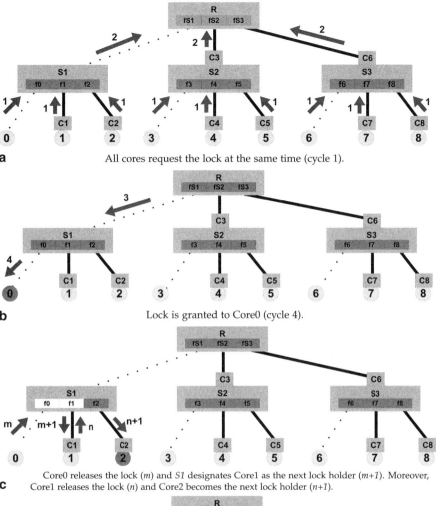

a All cores request the lock at the same time (cycle 1).

b Lock is granted to Core0 (cycle 4).

c Core0 releases the lock (*m*) and *S1* designates Core1 as the next lock holder (*m+1*). Moreover, Core1 releases the lock (*n*) and Core2 becomes the next lock holder (*n+1*).

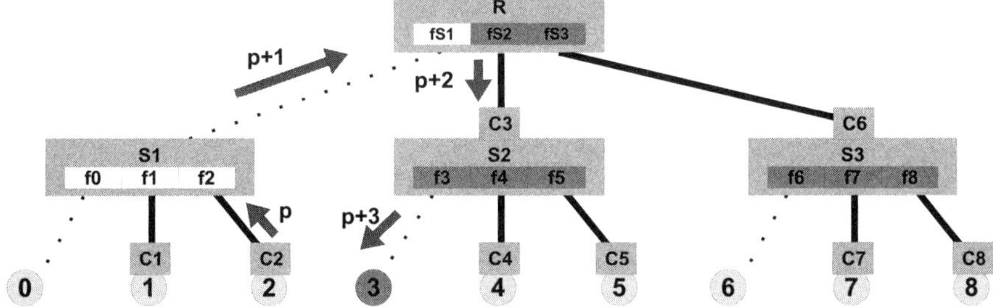

d Core2 releases the lock (*p*) and *S2* designates Core3 as the next lock holder (*p+3*).

Fig. 14 Example of lock synchronization under the *GLock* mechanism

Sect. 10.1). Finally, we highlight with dark gray the flags that are written and the core that acquires the lock in each case.

At cycle 0, all cores try to get the lock (see Fig. 14a). To do so, every *local controller* (Cx in the Figure) sends the REQ signal at cycle 1 to the corresponding secondary *lock manager* (Sx in the Figure). As a result, all fx flags would be written, and each Cx would be busy-waiting until the TOKEN signal is received. At cycle 2, once each Sx detects that at least one of its fi flags has been written, REQ signals towards the primary *lock manager* (R in the Figure) are sent in order to write the corresponding fSx flags. At this moment, R must make a decision about the secondary *lock manager* that will be granted the lock ownership. This process is shown in Fig. 14b. In this case, R would choose $S1$ by following the round-robin scheduling policy already discussed and would send the TOKEN signal at cycle 3. At cycle 4 and based on the round-robin policy, $S1$ chooses Core0 and sends the TOKEN signal granting access to the lock.

Figure 14c shows the scenario in which an Sx can grant the lock ownership without involving any additional notifications to R. More specifically, once Core0 releases the lock at cycle m, its controller sends the REL signal (by writing to the local $f0$ flag, as we mentioned) to $S1$. Next, at cycle $m + 1$, $S1$ grants the lock ownership (by means of the TOKEN signal) to the next core by following the round-robin policy from the active fx flags. In this case, Core1 becomes the new lock holder. In the same way, Core2 would be granted the lock in cycle $n + 1$ ($m < n$). Finally, in Fig. 14d we illustrate the scenario when an Si finishes its scheduling because either it has reached the last active fx or there are no more pending local requests for the lock. In this case, Si must send the REL signal towards R, which will choose another available Sj *lock manager* from those that activated the fSx flags. In the Figure, $S1$ sends the REL signal to R at cycle $p + 1$ ($n < p$), which following the round-robin policy grants the lock to $S2$. Finally, $S2$ sends the TOKEN signal giving access to the lock to Core3 at cycle $p + 3$.

9.3 Programmability Issues

The *GLock* mechanism proposed in this section is intended to be used by programmers in a transparent way. For that, as shown in Fig. 15, we propose to provide special library-level lock and unlock methods (`GL_Lock` and `GL_Unlock` in the Figure) that encapsulate the functionality of *GLock* and that could be used in parallel applications to deal with contended locks. This synchronization method uses a couple of special 1-bit registers added to each server core. First, the `lock_req` register that is used to request the lock and wait for lock acquisition. Second, the `lock_rel` register that is used to release the lock[4].

---

[4] Note that all pairs of flags (one per lock) could be grouped in each core using one special lock register.

```
GL_Lock() {
    asm {
        # Arrival at the CS: set lock_req
        mov 1, lock_req

        # Busy-wait until lock_req is reset
        loop:
                bnz lock_req, loop
    }
}

GL_Unlock() {
    asm {
        # Release lock: set lock_rel
        mov 1, lock_rel
    }
}
```

Fig. 15 Encapsulating the *GLock* functionality into the lock/unlock library-level intrinsics

As a result of the activation of the `lock_req` register by a server core, the synchronization protocol explained in the previous section would be invoked. In particular, the corresponding fx flag is activated by the *local controller*, and the secondary and primary *lock managers* start with delivering the lock ownership (granting the token). Straight afterwards, the server core enters in a loop waiting for the lock ownership (see Fig. 15). Next, once the lock is granted, the `lock_req` register is reset by the *local controller*, and the core can resume to execute the corresponding critical section protected by the lock. Once the critical section is executed, the server core sets the `lock_rel` register that will be used to release the lock. In consequence, the *local controller* would deactivate the fx flag and the `lock_rel` register would be reset as well.

The `lock_req` and `lock_rel` registers need as many bits as the number of *GLocks* provided in hardware (one bit per contended lock). In this way, several lock operations involving different sets of cores (the threads in each set running one application) could take place simultaneously. To this end, the register file of each core must be augmented with both registers and the interplay between controllers and them must be enabled, switching on the controllers whenever the `lock_req` registers are written, and switching off the controllers once all `lock_rel` registers are reset and all controllers have unset all the fx and fSx flags.

As pointed out through this section, our *GLock* mechanism is aimed at accelerating highly-contended locks. Obviously, the programmer is responsible for identifying locks of this kind and using the `GL_Lock` and `GL_Unlock` methods previously described for them. In the literature, there have been proposed several heuristics to detect contended locks in those cases in which it could be a tedious or difficult task. As an example, Tallent et al. [30] have recently proposed strategies for gaining insight into performance losses due to lock contention. Their goal was to understand where a parallel program needed improving.

As a final observation, the programmability of our *GLock* proposal is orthogonal to the utilization of any optimizations to harness the commented process of busy-waiting to conduct some other useful work while the lock ownership is not granted yet. For example, similar to *try locks* [28], upon a thread requests the lock the thread could execute some alternative code, or as in [14, 21], it could be involved some special queuing and scheduling kernel functions in order to deschedule the waiting thread allowing another one to make progress until the lock is eventually granted. Nevertheless, the implementation of these other approaches does not fall within the scope of this Chapter.

10 Performance Implications

In this section, we analyze *GLock* to determine its potential impact on performance. For that, we start by describing the two types of technologies employed to implement our *GLock* infrastructure. Next, for both implementations, we show their potential contributions to performance in terms of some important raw statistics such as on-chip area overhead, power dissipation, maximum operating speed and minimum latencies for acquiring and releasing a lock.

10.1 Implementation Technologies

10.1.1 *G-Lines* Technology

As discussed in the previous section, there were several reasons why we decided to use this technology to develop our synchronization mechanism for highly-contended locks in manycore servers. First, the connectivity pattern utilized to deploy the dedicated *GLock*'s network (see Sect. 9.1) is based on long 1-bit single-dimension *links* which perfectly fit into the concept of *G-Lines*. Second, according to the results reported in [40], that show negligible area overhead for a 392-*G-Line* network, the 32-core server system evaluated in this section (further details in Sect. 11.1) is made up of one-12th of the latter number of *G-Lines*, thereby even lower implications for on-chip area will be obtained. This marginal area overhead will have also a negligible impact on power dissipation. Finally, the *GLock*'s synchronization protocol explained in Sect. 9.2 could take advantage of the extremely fast transmissions at 2.5 GHz that the use of the *G-Lines* technology would entail. In this way, we can directly adopt the same theoretical synchronization latencies for acquiring and releasing a lock explained in that section.

10.1.2 *Standard* Technology

The *GLock* architecture has also been implemented relying on the mainstream industrial synthesis toolflow with an STMicroelectronics 45 nm standard cell technology library.

While this standard design methodology leads to cost-effective implementations in the embedded computing domain, low-latency communications for the *GLock*'s *links* are non-trivial to materialize. First, *links* have to be synthesized as RC-based wires[5] that are fully exposed to the effects of technology scaling. More specifically, the RC propagation delay of every wire will degrade as feature sizes shrink, making *links* increasingly slow. For this reason, this technology is also known as an interconnect-dominated nanoscale technology. And second, the propagation delay also affects the internal *GLock*'s logic thus reducing its maximum operating speed.

As for *GBarrier* in the previous section, it is worth noting that our mechanism has been synthesized by ensuring minimum wire lengths by situating lock managers in the central row/column of the 2D-mesh layout depicted in Fig. 12. In addition, we define non-routable obstructions that are placed to mimic the area of every core ($550\times550\ \mu m^2$) of the simulated system explained in Sect. 11.1. Additionally, fences are defined to limit the area where the cells of each *GLock*'s controller can be placed. Such obstructions and fences also ensure minimum-length routing for the *links* in order to reduce their impact on performance and area overhead as the wire length increases.

Due to the fact that RC-based *links* are very critical to performance degradation, we have implemented each *GLock*'s controller by separating the delay that signals take along the wires, from the effective computation that the controllers require to generate their output signals. Notice that, for small manycore servers, the critical path that limits the maximum operating speed in our *GLock* infrastructure is defined by the most complex controller (i.e., the *lock manager* which communicates with a higher number of controllers), but as the wire length increases for larger servers, the wires could represent such a critical path. Consequently, separating wire delays from controllers delays becomes essential in order to achieve maximum clock speeds. In this way, by using this technology, wc cannot directly assume the theoretical synchronization latencies explained in Sect. 9.2, and a higher number of cycles will be required to acquire and release the lock.

[5] We use the terms *links* and wires interchangeably.

Table 6 Raw statistics using *G-Lines* and *Standard* technologies for a single *GLock* in a 32-core server layout

	Freq. (MHz)	Latency (cycles)	Area (μm^2)	Power (*mW*)
G-Lines	2500	Acquire: 4 (worst), 2 (best) Release: 1	*Negligible*	28
Standard	714	Acquire: 9 (worst), 5 (best) Release: 3	6269	*Negligible*

10.2 Raw Performance Statistics

Table 6 shows the main raw performance statistics obtained from the use of both technologies to implement *GLock*. In particular, we illustrate the maximum operating speed, the latencies of the lock acquisition and release (assuming that the lock is free) and also the area overhead with an estimation of power dissipation that our proposal entails.

The maximum operating speed achieved by the *G-Lines* technology is 3.5 times higher than for the *Standard* technology. Moreover, the number of clock cycles employed by the former technology to acquire and release a lock is half of those achieved by the latter technology. The reason is that every *GLock*'s controller and *link* involved take a different clock cycle in the synchronization process. Therefore, the superior efficiency of *G-Lines* technology reports roughly an eight times faster *GLock* implementation.

Due to the very lightweight infrastructure deployed to implement *GLock*, negligible overheads in terms of die area are obtained for both technologies. Regarding the *G-Lines* technology, as aforementioned, our *GLock* infrastructure requires one-12th of the number of *G-Lines* reported in [40] thus leading to even lower implications for on-chip area. Moreover, as to the *Standard* technology, an area overhead for *GLock* equal to 6269 μm^2 is reported that corresponds to a negligible 0.07 % of the total area employed for the simulated 32-core server layout (remember that we assume that each core is 550×550 μm^2 in size).

The latter marginal on-chip overhead must also lead to a negligible impact on power dissipation. We demonstrate this by estimating the power dissipation for a worst-case scenario in which the maximum number of *GLock*'s transmitters and receivers are operating at once. As an example, we detailed the power estimation considering the *G-Lines*-based implementation for *GLock*. According to the *GLock*'s synchronization protocol already described (see Fig. 14), this situation arises when all cores request the lock ownership at the same time. In this way, for the simulated environment described later in Sect. 11.1, where we considered a 4×8-core server[6],

[6] For simplicity, we assume that 8 cores per row can be materialized in *G-Lines*. Recall that this technology is limited to 7 cores per row and, for example, a 6×6-core server layout must be considered instead to span the simulated 2D-mesh 32-core system.

there will be a total of seven local controllers per row (i.e., 28 transmitters) transmitting the 28 REQ signals towards the corresponding four secondary lock managers, which in turn store those signals in the corresponding fX flags (i.e., 28 receivers are required). For the power estimation, we assume the same power dissipation parameters for a 65-nm CMOS process simulated in [40]: 0.6 mW per transmitter; and 0.4 mW per receiver. Moreover, according to [40] no static power is dissipated by the *G-Lines*. Hence, for the number of transmitters and receivers discussed before, the total power estimated is 28 mW ($28 \times 0.6 + 28 \times 0.4$). It is worth noting that, utilizing CACTI [16], the magnitude of this dissipation is less than one-10th of the power dissipated per read port in the L1 caches simulated in this section (see Table 7).

As a conclusion of this section, the above results suggest that the fastest technology is the most appropriate implementation to materialize *GLock*. Although synchronization delay would become the discriminating factor, we have also to take into account that the *G-Lines* technology is not within reach of a standard cell design methodology. In consequence, it would be of paramount importance to determine the exact magnitude of such performance degradation when using the *Standard* technology. In case of being negligible, the slower technology would be the preferred *GLock* implementation. This experiment will be conducted in Sect. 11.4.1, by comparing synchronization timings of the two *GLock* implementations in comparison to the best software-based implementation for highly-contended locks.

11 Evaluation

In this section we give details of our experimental methodology and performance results. For that, the raw performance statistics already discussed in Sect. 10 have been integrated into the simulation environment described in Sect. 11.1. In the latter section, we also describe the sort of benchmarks and their main characteristics utilized to evaluate *GLock*, and a post-mortem analysis is carried out in Sect. 11.2 to precisely quantify the exact degree of contention of locks in every benchmark. Moreover, Sect. 11.3 describes the most efficient software implementation for highly-contended locks that *GLock* is compared against. Finally, Sect. 11.4 shows performance results in terms of execution time, network traffic and energy consumption.

11.1 Experimental Setup

As *GLock* has been specifically tailored to work in the context of manycore servers, we have integrated our proposal into the Sim-PowerCMP performance simulator as for our *GBarriers* proposal already presented in this Chapter. In particular, Table 7 shows the values of the main configurable parameters assumed in this section. In short, we have simulated a 32-core server architecture with an aggressive 2D-mesh network built in a 45 nm process technology.

Table 7 server baseline configuration

Number of cores	32
Core	3 GHz, in-order 2-way model
Cache line size	64 Bytes
L1 I/D-Cache	32 KB, 4-way, 2 cycles
L2 Cache (per core)	256 KB, 4-way, 12+4 cycles
Memory access time	400 cycles
Network configuration	2D-mesh
Network bandwidth	75 GB/s
Link width	75 bytes

To evaluate the performance benefits derived from *GLock*, five microbenchmarks and three real applications are used. On the one hand, the microbenchmarks that we have employed are: SCTR, MCTR, DBLL, PRCO and ACTR [34]. They were chosen because of exhibiting different highly-contended access patterns to shared data that can be commonly found in parallel applications. On the other hand, regarding real applications, we have considered Qsort sorting algorithm as well as two programs belonging to the SPLASH-2 benchmark suite [37]: Ocean and Raytrace. These applications were chosen since they present a significant lock synchronization overhead due to the existence of highly-contended locks[7]. In fact, these locks are accessed following similar patterns to those of the microbenchmarks. We summarize the characteristics of the microbenchmarks and applications used in this section in Table 8. For each of them we account for the input size, the total number of different locks, the number of these locks that are highly-contended (H-C Locks), and point out the highly-contended lock access patterns in terms of the microbenchmarks they are similar to.

It is important to note that only contended locks are implemented using the *GLock* mechanism. For the rest of the locks, we rely on a straightforward implementation called *Simple Lock*, that atomically toggles a boolean flag to acquire and release the lock (further details in Sect. 12), that is enhanced with the `test-and-test&set` optimization. This includes the locks used in the applications' library of our simulator to implement barriers. Apart from not being application-level, these locks do not exhibit high contention levels since our simulator provides applications with an efficient tree barrier implementation (up to two threads requesting every lock). In this way, barriers are not affected by our proposal. Finally, all experimental results reported in this section are for the parallel phase of all of the benchmarks previously described.

[7] In this Chapter, highly-contended locks are those locks accessed by all threads simultaneously or very close in time.

Table 8 Configuration of the benchmarks and lock-related characteristics

Benchmark	Input size	Locks	H-C Locks	Access pattern
SCTR	1000 iterations	1	1	–
MCTR	1000 iterations	1	1	–
DBLL	1000 iterations	1	1	–
PRCO	1000 iterations	1	1	–
ACTR	1000 iterations	2	2	–
RAYTR	teapot	34	2	SCTR
OCEAN	258×258 ocean	3	1	SCTR
QSORT	16,384 elements	1	1	PRCO

11.2 Post-mortem Analysis of Benchmarks

To determine the contention of locks, we performed a post-mortem analysis of the benchmarks under study where locks use the *Simple Lock* algorithm enhanced with the `test-and-test&set` optimization. Every time a core tries to acquire a lock, we register the number of concurrent requesters (group of acquiring cores or grAC ranging from 1 to 32) on a cycle-by-cycle basis until the lock is granted to the core. In this way, we can precisely compute each lock's contention rate as the number of cycles where the number of concurrent requesters is equal to each grAC divided by the total amount of cycles where the number of concurrent requesters belongs to the range [1,32]. That is, the lock's contention rate (*LCR*) of a particular lock (*Lock*) for each grAC ($i \in [1,32]$) would be defined by Eq. 1.

$$LCR_{grAC_i} = \frac{Cycles(Lock, grAC_i)}{\sum_{g=1}^{32} Cycles(Lock, grAC_g)} \tag{1}$$

In Fig. 16, the lock's contention rate for all of the benchmarks (x-axis) is shown. In particular, we show the lock's contention rate (y-axis) for all of the possible values of grAC (z-axis). Moreover, we decompose the results for each benchmark on a per-lock basis[8]. To do that, we assume that Eq. 2 is satisfied and redefine Eq. 1 as Eq. 3. That is, every lock's contention rate has also been estimated depending on the amount of clock cycles it uses. From this, we can easily identify in Fig. 16 those locks that present high contention, and those that although exhibiting high contention are executed during a negligible amount of clock cycles. Due to their very low impact on execution time, the latter kind of locks would be implemented by using the *Simple Lock* algorithm enhanced with the `test-and-test&set` optimization.

$$LCR_{benchmark} = \sum_{i=1}^{Locks} \sum_{j=1}^{32} L_i CR_{grAC_j} = 1 \tag{2}$$

[8] Although Raytrace has 34 locks, we only include the results for the two most highly-contended locks (RAYTR-L1 and RAYTR-L2) and aggregate the rest (RAYTR-LR).

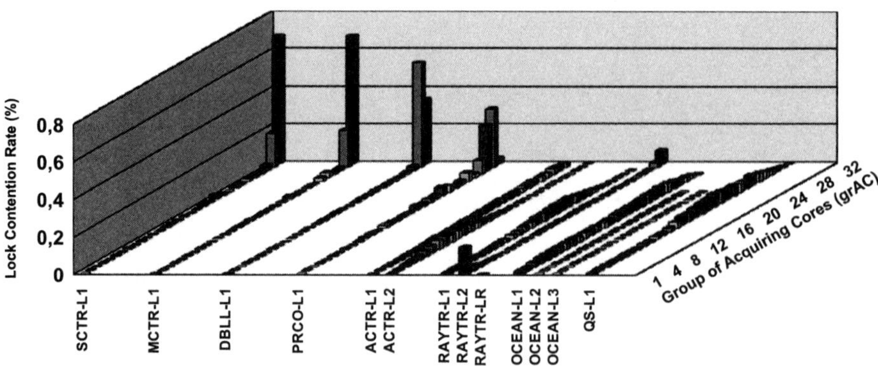

Fig. 16 Lock contention rate

$$L_i CR_{grAC_j} = \frac{Cycles(Lock_i, grAC_j)}{\sum_{l=1}^{Locks} \sum_{g=1}^{32} Cycles(Lock_l, grAC_g)} \quad (3)$$

As expected, the microbenchmarks exhibit a very high lock's contention rate when grAC is close to the total number of cores. The exception is the ACTR microbenchmark which presents a moderate homogeneous level of contention across all the grAC range. This is mainly due to the barrier synchronization interleaved between the two lock acquisition operations. The real applications also report a behavior similar to that of the ACTR microbenchmark. In this case the reason is their much coarser granularity which spreads the acquire operations throughout the parallel phase. Finally, it is worth noting that Ocean and Raytrace just have one and two highly-contended locks, respectively.

11.3 Lock Implementations

To fairly quantify the benefits of our *GLock* mechanism, we consider the case that highly-contended locks found in the benchmarks previously described are implemented by using *MCS Locks*. As we will explain in Sect. 12, *MCS Locks* are one of the most efficient software algorithms for lock synchronization. In particular, *MCS Locks* gracefully manage high-contention situations by having a distributed queue of waiting lock requesters. On the other hand, for the rest of locks (non-contended ones), we employ the *Simple Lock* algorithm enhanced with the `test-and-test&set` optimization due to it has been shown to lead to lower latencies when threads try to acquire a lock without competition. Finally, since the number of highly-contended locks is commonly very small in real applications (up to two in the applications evaluated in this section), we assume that two *GLocks* are provided at hardware level.

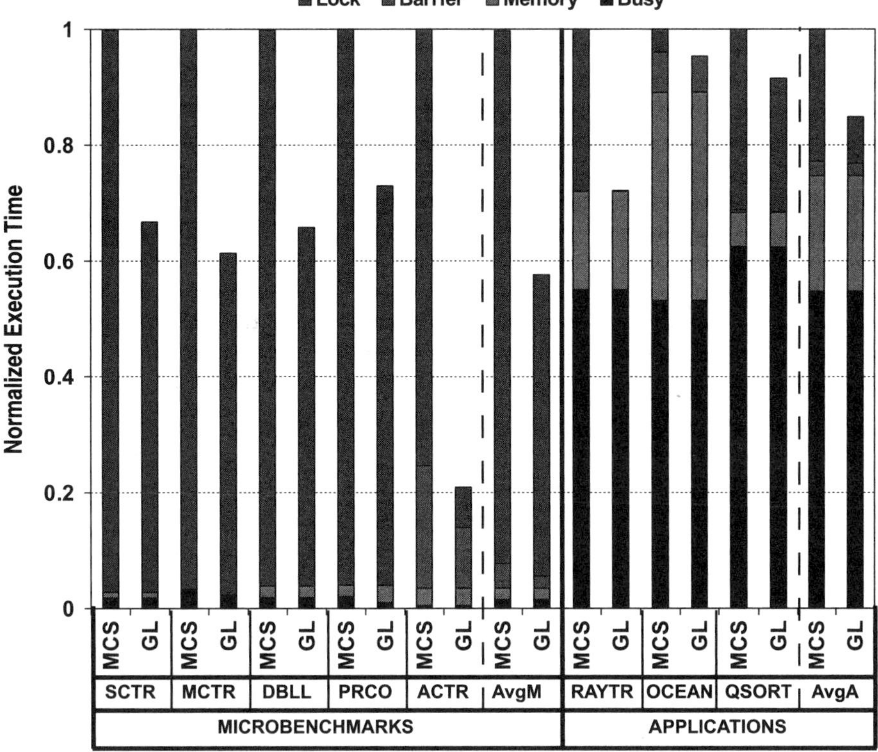

Fig. 17 Normalized execution time

11.4 Performance Results

The evaluation of the two *GLock* implementations presented in Sect. 10.1 has been carried out taking into account the execution times achieved for the benchmarks shown in Table 8, as well as the amount of traffic in the interconnect and the energy-delay^2 product (ED^{2}P) metric for the full server.

11.4.1 Execution Time

First of all, we consider the implementation of the *GLock* that relies on the *G-Lines* technology. Figure 17 shows the execution times that are obtained for the set of benchmarks under study when either *GLock* or *MCS Locks* are employed for the highly-contended locks (GL bars and MCS bars respectively). In particular, execution times have been normalized with respect to those obtained when *MCS Locks* are used. Additionally, each bar shows the fraction of the execution time due to

Table 9 Speedups for the real applications

Benchmark	Lock Version	4	8	16	32
RAYTR	MCS	3.91	7.53	13.61	20.69
	GL	3.93	7.97	15.67	28.78
OCEAN	MCS	3.70	7.12	13.48	23.62
	GL	3.80	7.32	13.93	25.66
QSORT	MCS	3.67	6.49	9.68	11.38
	GL	3.69	6.55	9.92	12.40

lock and barrier synchronizations (*Lock* and *Barrier* categories respectively), memory accesses (*Memory* category) and computation (*Busy* category). Finally, average execution times are shown in separate bars for the microbenchmarks (*AvgM*) and applications (*AvgA*).

Regarding the microbenchmarks, we can observe that our proposal presents an average reduction of 42 % in execution time (see *AvgM*). The exact extent of the reduction in each case depends on both: the number of highly-contended locks that each microbenchmark has (see Table 8), and also the contention rates exhibited by each lock (see Fig. 16). In particular, our proposal is applied in SCTR, MCTR, DBLL and PRCO to their single contented lock, resulting in reductions of 33, 39, 34, 25 % in execution time, respectively. On the other hand, two contended locks are found in the ACTR microbenchmark, which increases the benefits of our proposal (reductions of 81 % are obtained). This high reduction is also explained since ACTR presents a much lower contention rate. In particular, in Fig. 16 we can observe that SCTR, MCTR, DBLL and PRCO present a contention rate close to 80 % when considering grACs higher than 20 cores. In contrast, ACTR presents an aggregate contention of only 20 % for the same grACs. As we mentioned, *MCS Locks* become inefficient for the low contention case, which accentuates even more the differences between *MCS Locks* and our proposal.

A more in depth analysis reveals that the former reductions come from two kind of effects that the *GLock* mechanism has. First, the time taken to acquire and release the lock is drastically reduced as derived from the improvements shown in the *Lock* category. And second, the fact that our proposal removes from the main data network all extra coherence traffic that a shared-memory-based lock implementation would introduce, also has an effect on the *Barrier* category for the ACTR microbenchmark.

On other hand, the fraction of the execution time that lock synchronization consumes is lower when real applications are considered. In these cases, most of the time is spent on computations and memory accesses (*Busy* and *Memory* categories). This explains the lower reductions in execution time observed for Raytrace, Ocean and Qsort (14 % on average). Moreover, since Qsort presents higher contention rates than Raytrace (aggregate contentions of 60 and 29 %, respectively, for grACs higher than 20 cores), the *MCS Locks* become more efficient which translates into lower performance differences between *MCS Locks* and the *GLock* mechanism.

Table 9 shows speedup results for the real applications (Raytrace, Ocean and Qsort) when scaling the number of cores parameter with the values 4, 8, 16 and

Table 10 Normalized execution times for *G-Lines* and *Standard* technologies

	SCTR	MCTR	DBLL	PRCO	ACTR	RAYTR	OCEAN	QSORT
G-Lines	0.67	0.61	0.66	0.73	0.19	0.72	0.95	0.92
Standard	0.68	0.63	0.68	0.75	0.20	0.74	0.96	0.93

32. Moreover, we use two different lock implementations for the high contention case: *MCS Locks* (MCS) and our *GLock* mechanism (GL). From the results shown in Table 9, we can extract two important observations. First, all of the benchmarks scale as the number of cores is increased. Second, the exact extent of the speedups depends on the efficiency of the lock implementation we are using. In this way, higher speedups are obtained when employing our *GLock* mechanism which are even very close to ideal speedups in the case of Raytrace.

According to the discussion given at the end of Sect. 10.1.2, it would be of paramount importance determining whether the performance losses in terms of synchronization latency derived from the use of the *Standard* technology can be considered negligible. To this end, Table 10 shows the normalized execution times with respect to those obtained when *MCS Locks* are used, depending on the two kind of *GLock* implementations studied in this section: *G-Lines*[9] and *Standard* technologies. As we can see, very small performance degradations of 1.6 and 1.3 % on average are shown for the microbenchmarks and real applications, respectively. In consequence, we can affirm that our *GLock* mechanism is not so dependent on a full-custom technology to provide extremely efficient synchronizations for highly-contended locks.

Finally, we also carried out a sensitivity analysis to evaluate the extent to which our proposal is affected by longer *link* latencies. To do so, we simulate several configurations of the *G-Line*-based network with varying latencies for the *links* and evaluate the impact that this has on performance. Several clock cycles may be necessary to transmit a signal across one dimension of the chip if, for example, we consider longer *links* that cannot support a propagation delay of a single clock cycle, or even if lower clock frequencies are required to integrate our *GLock* infrastructure in the manycore server. Figure 18 illustrates the normalized execution times when *G-Lines* take from 1 (results presented in Fig. 17) to 10 clock cycles (see z-axis in the figure). As we can observe, negligible performance losses are derived even when dealing with 10-cycle *G-Lines*. Particularly, performance degradations of just 1.8 and 1.6 % on average in the worst case are shown for the microbenchmarks and real applications, respectively. Note that the results observed for the 10-cycle case are very similar to those obtained for the *Standard* technology previously reported. According to Sect. 10.1, since *Standard*-based implementation is roughly eight times

[9] Note that, the results for the implementation that uses *G-Lines* are the same as those presented in Fig. 7.

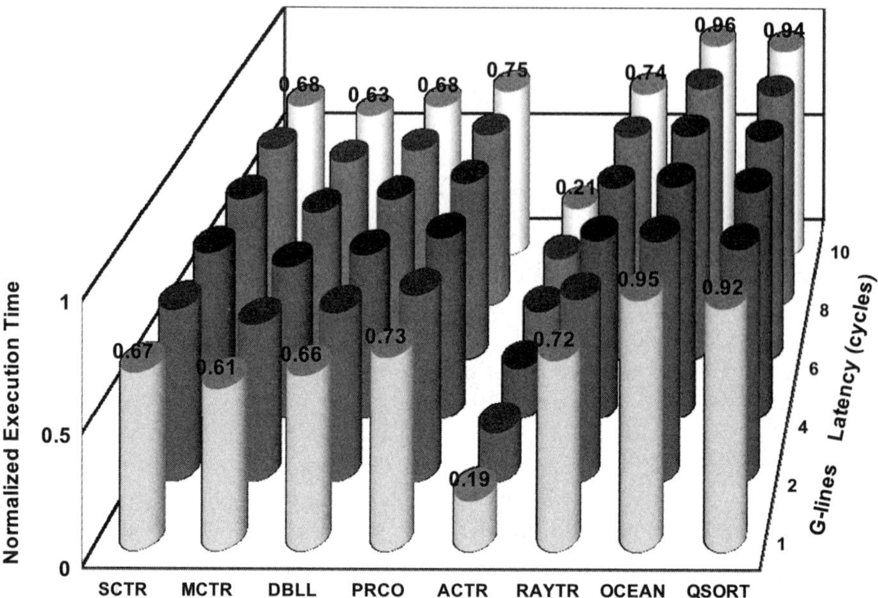

Fig. 18 Normalized execution times of benchmarks depending on *G-Lines* latency running on a 32-core server

slower than a *G-Lines*-based infrastructure, the former would be equivalent to an 8-cycle *G-Line*-based implementation of the *GLock* mechanism, which explains such similarities.

11.4.2 Network Traffic

Our proposal does not generate any coherence messages on the main data network when performing lock synchronizations for any of the two *GLock* implementations. At the end, this translates into the same significant reductions in terms of network traffic. Figure 19 shows the total network traffic across the main data network. In particular, each bar plots the number of bytes transmitted through the interconnection network (the total number of bytes transmitted by all the switches of the interconnect) normalized with respect to the *MCS* case. Each bar is broken down into three categories: *Coherence* corresponds to the messages generated by the cache coherence protocol (e.g., invalidations and Cache-to-Cache transfers); *Request* comprehends messages generated when load and store instructions miss in cache and must access a remote directory; and finally, *Reply* involves the messages with data.

For the microbenchmarks, important reductions in network traffic are achieved (76 % on average). In general, these reductions are directly related to the extents of the improvements in execution time previously reported. Moreover, since the simulated L2 cache is shared among the different processing cores, but it is physically

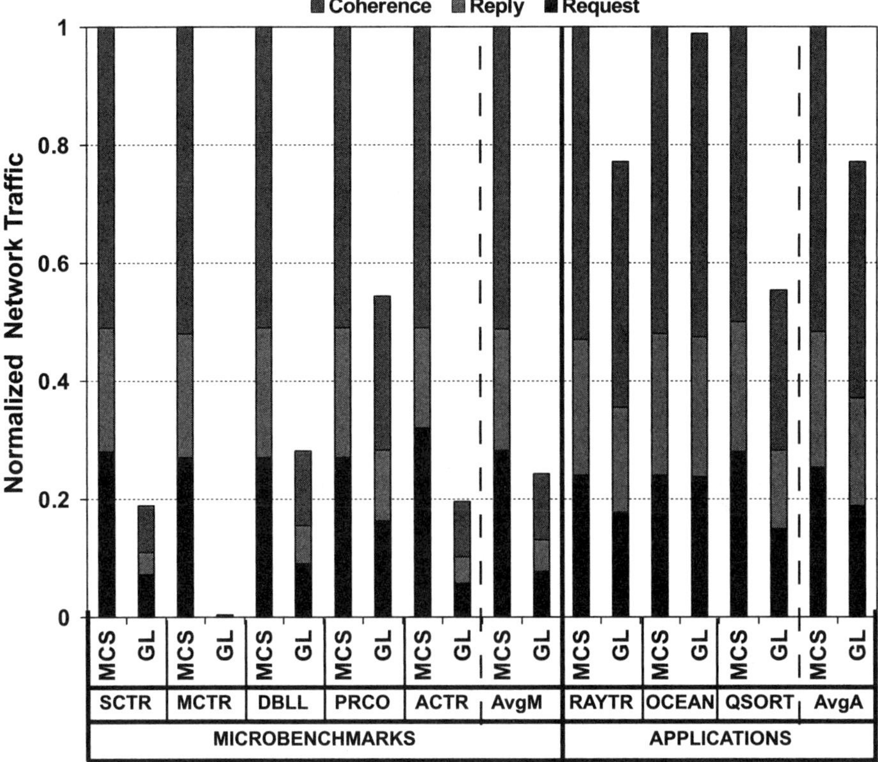

Fig. 19 Normalized network traffic

distributed between them (see Sect. 11.1), some accesses to the L2 cache will be sent to the local slice while the rest will be serviced by remote slices. This will also affect to lock acquisition and release operations timings. In contrast, since our *GLock* proposal skips the memory hierarchy we have not obtained such negative impact on network traffic. In particular, SCTR, MCTR, DBLL and PRCO show reductions of 81, 99, 72 and 46 %, respectively. This is due to the fact that almost all network traffic of these microbenchmarks is due to lock synchronizations. The exception is ACTR, where the barrier used in between the two phases also generates network traffic. However, since the barrier time is approximately 20 % of the lock time (see *Barrier* and *Lock* categories in Fig. 17), a reduction of 80 % in network traffic is obtained.

Finally, regarding the real applications, we can see an average reduction of 23 % in network traffic (see *AvgA* in Fig. 19). More specifically, the applications Raytrace, Ocean and Qsort present reductions of 23, 1 and 45 %, respectively. As before, there is a correlation between the fraction of the execution time devoted to lock synchronization and the amount of network traffic that is saved. For instance, Ocean

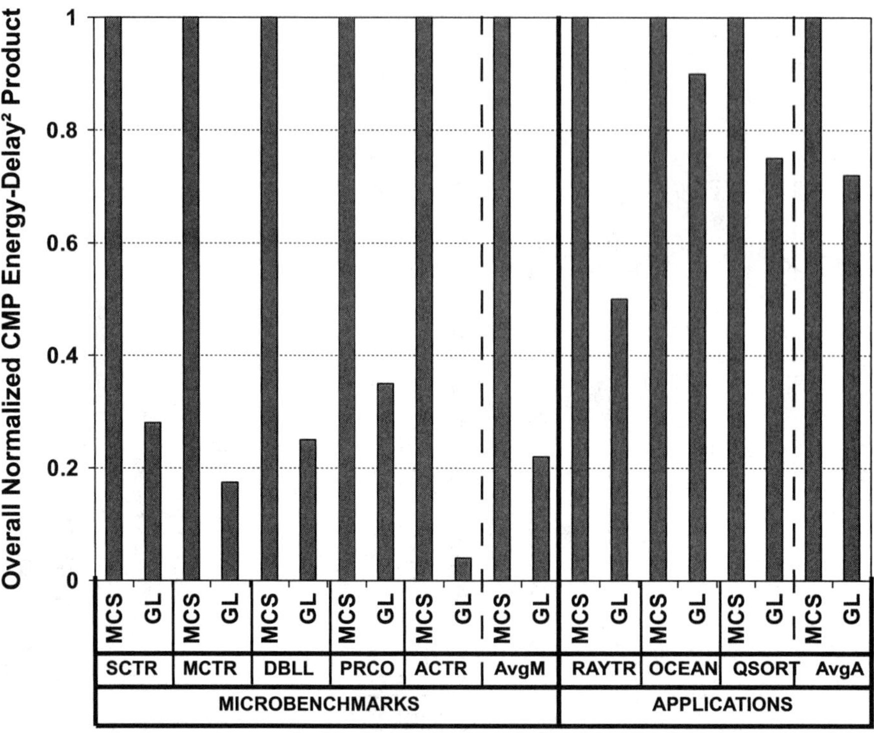

Fig. 20 Normalized energy-delay^2 product (ED^{2}P) metric for the full server

presents the lowest reduction in network traffic since less than 5 % of its execution time (see Fig. 17) is spent on locks.

11.4.3 Energy Efficiency

Finally, we also consider the benefits in energy efficiency that our proposal entails. More specifically, we present in Fig. 20 the normalized energy-delay^2 product (ED^{2}P) metric for the full server. To account for the energy consumed by the *GLock* architecture (the *G-Lines*-based network described in Sect. 9.1), we extend the Sim-PowerCMP with the consumption model of *G-Lines* and controllers described in previous Sect. 10.2. According to our discussion in Sect. 10.1.2, the power dissipation associated with our two technology-aware *GLock* implementations is negligible, hence the power statistics presented in this section will be mainly due to the improvements in execution time and network traffic reported in previous sections.

As in the previous two sections, all results in Fig. 20 have been normalized with respect to the *MCS* case. As can be observed, important improvements in the ED^{2}P metric of the whole server are achieved when applying our proposal. In particular, the *GLock* mechanism brings average improvements in ED^{2}P of 78 and 28 % for the microbenchmarks and real applications, respectively. The SCTR, MCTR, DBLL, PRCO and ACTR microbenchmarks show reductions of 72, 83, 75, 65 and 96 %, respectively. Additionally, reductions of 50, 10 and 25 % are achieved for Raytrace, Ocean and Qsort.

In general, as commented above, the magnitude of these savings is directly related to the extents of the improvements in execution time and network traffic previously reported. We have found that when the *GLock* mechanism is employed, the number of instructions executed per lock acquisition and release operation is drastically reduced. Note that while *MCS Locks* must deal with a distributed queue of waiting threads requesting the lock, *GLock* only needs two assignment instructions on two registers to notify the arrival to the lock and the subsequent release operation (see Sect. 9.3). Obviously, less instructions executed means less energy consumed in the server cores.

Moreover, since we reduce the latency of lock acquisitions, the busy-wait process is also shortened with *GLock*. While busy-waiting, a server core repeatedly access the L1 cache to check the value of a shared variable. In this way, shorter busy-waiting implies less accesses to the L1 cache, and therefore, less energy consumed in this structure. Finally, given the fact that our proposal skips the memory hierarchy, we save all the energy derived from coherence activity when locks are executed. In particular, we remove all of the L1 cache misses related to lock operations and the corresponding messages transferred across the interconnect. This brings reductions in the energy consumed at the L2 cache banks and the interconnection network.

12 Related Work

Performance degradation of software-based schemes for lock/unlock operations in parallel machines has long been recognized as a key impediment to scalability and high performance as the server/core count increases. For that reason, in the literature there have long been devised some architectural extensions that go from simple hardware support, such as improved network/memory controllers, to those proposals that integrate sophisticated interconnection networks for conveying synchronization traffic.

A comprehensive description of the major proposals for lock/unlock operations at both software and hardware levels are described below. To this end, we firstly give a review of some well-known software-based implementations exposing their main performance bottlenecks in order to understand why hardware support becomes essential. Secondly, we expose the most relevant hardware-based schemes by comparing them against our *GLock* proposal.

The simplest software-based synchronization algorithms rely on atomic read-modify-write instructions, such as `test&set`, `fetch&operation`, `swap` or `compare&swap`, to implement the lock and unlock synchronization primitives [10]. For instance, *Simple Lock* repeatedly tries to acquire the lock by toggling a boolean flag from false to true with a `test&set` instruction. Next, the lock is released by simply toggling the flag back from true to false. The main drawback of this algorithm is the continuous generation of cache-coherence network traffic while busy waiting for lock acquisition. To ameliorate this problem two optimizations, namely `test-and-test&set` and exponential back-off, have been proposed. The former issues standard loads that hit on the local cache while busy waiting for lock acquisition. Hence, the `test&set` is only issued when the lock appears to be free thus reducing cache-coherence network traffic. The latter inserts a delay between consecutive attempts to acquire the lock in order to reduce contention. Anderson [39] found that exponential back-off is the most effective form of delay. Nevertheless, as contention increases these improvements are not enough to guarantee scalability especially for highly-contended locks.

More elaborated algorithms such as *Ticket Lock*, *Array-based Lock* and *MCS Lock* provide more scalable and fair lock implementations at the expense of increased storage cost and higher latency for the low contention case [10]. The *Ticket Lock* algorithm consists of a pair of counters, a *ticket* counter and a *now-serving* counter. To acquire a lock a thread gets its turn by issuing a `fetch&increment` on the ticket counter and then busy waits until the now-serving counter equals its ticket. To release the lock a thread simply increases the now-serving counter. *Array-based Lock* just replaces the now-serving counter by an array of locations. The idea behind *MCS Locks* [20] is similar to that of *Array-based Locks*. An *MCS Lock* builds a distributed queue of waiting threads requesting the lock. In this way, each thread busy waits on a unique, locally accessible flag rather than competing for a single counter. *MCS Locks* are considered the most efficient software algorithm for lock synchronization [18, 20]. In all three cases, cache-coherence network traffic is reduced because only one thread actually attempts to obtain the lock when it is released by the previous owner.

In general, simple algorithms tend to be fast under low contention and inefficient when contention is high. In contrast, sophisticated algorithms specifically designed to deal with contention usually incur a non-negligible overhead when there is little contention. For this reason, a number of hybrid approaches have been proposed. *Reactive Lock* [4] is a library-based adaptive approach that chooses the best synchronization algorithm under different levels of contention. This technique switches between *Simple Lock* and *MCS Lock* for the low and high contention cases, respectively. *Smart Lock* [18] uses heuristics and machine learning to choose the most appropriate algorithm following a specific user-defined goal in terms of performance, energy consumption or problem-specific criteria.

A completely different software approach that is not based on atomic read-modify-write instructions called *MP-Locks* is presented in [6]. With *MP-Locks* synchronization operations are implemented using message passing, over the main data network, and embedded kernel lock managers. This approach comes in three different flavors, namely centralized, distributed and reactive that are differentiated from each other in how the lock managers control lock ownership. A comparison between *MP-Locks* and *MCS Locks* reports significant performance and scalability gains at the expense of increased software complexity and limited portability. A similar idea, proposed in the context of distributed systems, called *Token-based Locks* appears in [9]. In this case, the right to acquire a lock is represented by a token which is unique in the whole system. Threads willing to acquire a lock must wait for token arrival and release the token upon critical section completion.

Remote Core Locking [23] (RCL) is an efficient software-based implementation specially designed for highly-contended locks. RCL replaces lock acquisitions by remote procedure calls (RPCs) to a dedicated server core in order to exploit cache locality. The reason is that when a CS accesses shared data that has recently been accessed by another core, there will result in cache misses. So, the idea is to avoid these cache misses. RCL entails a tool that transforms CS code to be executed as an RPC as well as a runtime for Linux OS that includes the RCL code. The implementation of RCL is based on an array of requesting cores (clients) cached in the server core, and devoted to establish an interaction with the server that quits when the server executes the CS. For a 48-core machine significant performance improvements are shown. Nevertheless, the efficiency of this software implementation for highly-contended locks may be hampered by higher core counts, since dedicating an entire core to implement a CS is a centralized approach that may lead to potential performance bottlenecks as the number of clients increases. Differently, our *GLock* proposal does not dedicated cores to execute a CS and is based on a scalable and distributed infrastructure to implement highly-contended locks. In addition, our proposal neither injects synchronization-related traffic into the main interconnect nor uses the memory system, thereby not interfering with QoS of parallel applications.

Hardware support for lock synchronization has also been the target of a number of proposals. Queue-On-Lock-Bit (QOLB) [2] is based on a distributed queue of waiting threads requesting the lock. Unlike *MCS Locks*, in QOLB the queue is implemented entirely in hardware at the cache controller level.

The Synchronization-operation Buffer (SB) [29] is a hardware module which augments the memory controller to queue and manage lock operations issued by the threads. QOLB reports non-negligible performance gains when compared to *MCS Locks*. In general, all of the hardware-supported solutions require modifications at some level of the memory hierarchy. In contrast, our proposal, namely *GLock*, completely decouples lock synchronization from any kind of memory-related activity, by deploying a dedicated lightweight on-chip network infrastructure to implement a simple synchronization protocol aimed at accelerating highly-contended locks.

13 Conclusions

In this Chapter, we have identified two fundamental performance bottlenecks in manycore processors in the context of servers belonging to datacenters: highly contended synchronization in barriers and locks as the number of servers' cores increases.

We have proposed two distinct and complementary hardware-based solutions to overcome such performance bottlenecks. Moreover, we have also considered the use of non-digital technology to help us break such limitations obtaining superior efficiency and scalability. To do so, we have leveraged the full-custom state-of-the-art *G-Lines* technology, although we have also explored the efficiency of our proposals using a current standard cell design methodology.

The first of our proposals (namely *GBarrier*) is aimed to overcome performance limitations of barrier operations in manycore servers. *GBarrier* is a novel hardware-based barrier mechanism specifically designed to enable efficient barriers by removing all performance limitations of software-based barrier implementations, and even in all hardware-barrier mechanisms to date. In particular, our *GBarrier* mechanism consists of two main components: First, a very lightweight dedicated on-chip network that could be deployed in a hierarchical layout for scalability. The second is a simple and very fast synchronization protocol implemented atop the previous infrastructure. The reason why our proposal is much more efficient is that differently to software approaches based on the use of atomic read-modify-write instructions operating on shared-memory positions, *GBarrier* does not have any influence on the memory system, hence saving traffic and energy. More specifically, we have avoided all coherence activity, barrier-related network traffic and the involved energy consumption, that software approaches introduce and that restrict scalability.

To evaluate *GBarrier*, we have considered two implementations of our infrastructure by leveraging *G-Lines* and *Standard* technologies. Our study in terms of raw performance statistics reveals that differences in on-chip area overhead and power dissipation can be considered negligible between both technologies, although, as expected, the former technology reports the minimum synchronization latency, whereas the latter leads to a cost-effective implementation. We integrate both *GBarrier* implementations into a detailed execution-driven simulator (Sim-PowerCMP) of a 32-core server running a set of benchmarks: kernels and scientific applications. From this study, both *GBarrier* implementations report very similar reductions in execution time, thus not making our proposal so dependent on a full-custom technology to achieve extremely efficient synchronization in manycore servers. In particular, for the kernels and the scientific applications under study our proposal brings average reductions of 54 and 21 %, respectively, in total execution time, resulting in improved scalability for the applications. We also have obtained reductions of 53 and 18 %, respectively, in network traffic. The reason is that our proposal does not rely on shared memory positions and the cache coherence protocol saves a significant amount of messages on the main interconnection network. Finally, all these gains lead to improvements of 76 and 31 %, respectively, in the energy-delay^2 product (ED^{2}P) metric for the full server.

Finally, regarding lock synchronization, we have identified that contention is a key constraint to performance and scalability when there is a significant amount of threads willing to access into the same CS at once. To achieve a fair, very efficient and scalable solution for locks that are highly contended, we have proposed *GLock*. *GLock* is based on a dedicated on-chip network and relies on a simple token-based messaging-protocol. Due to the fact that lock mechanisms only exhibit problems in high-contention scenarios, our proposal could be combined with a software-based implementation for low contention (e.g. *Simple Locks* enhanced with the well-known `test-and-test&set` optimization). Moreover, a deep analysis of some relevant benchmarks discloses a reduced number of highly-contended locks in most cases, so that replication of the *GLock*'s resources is not expected to be a constraint. As for *GBarrier*, to evaluate *GLock*, we have made use of *G-Lines* and *Standard* technology coming to the very same conclusions in terms of raw performance: negligible on-chip area overhead and power dissipation in both technologies, cost-effective implementation for *Standard*, and reduced synchronization latency for the *G-Lines* technology. We integrate both *GLock* implementations into Sim-PowerCMP, and discuss synchronization efficiency results as compared to the most efficient software-based lock implementation. To do so, we have simulated a 32-core server with a 2D-mesh data network and employ a set of microbenchmarks and real applications. Both *GLock* implementations report very similar reductions in execution time, hence not making our proposal so dependent on a full-custom technology. From our evaluation, significant average reductions for the microbenchmarks and the real applications respectively are achieved: 42 and 14 % in execution time; 76 and 23 % in network traffic; and 78 and 28 % in the ED^2P metric for the full server.

Acknowledgements This work was supported by the Spanish MINECO, as well as European Commission FEDER funds, under grant TIN2012-38341-C04-03. This work was done while Juan Fernández was a member of the Computer Engineering Department of the University of Murcia.

References

1. A. Flores, J. L. Aragón and M. E. Acacio. Sim-PowerCMP: A Detailed Simulator for Energy Consumption Analysis in Future Embedded CMP Architectures. In *Proceedings of the 21st International Conference on Advanced Information Networking and Applications Workshops*, 2007.
2. A. Kägi, D. Burger and J. R. Goodman. Efficient Synchronization: Let Them Eat QOLB. In *Proceedings of the 24th International on Computer Architecture*, 1997.
3. A. P. Jose and K. L. Shepard. Distributed Loss-Compensation Techniques for Energy-Efficient Low-Latency On-Chip Communications. *IEEE Journal of Solid State Circuits*, 42(6):1415–1424, 2007.
4. B-H. Lim and A. Agarwal. Reactive Synchronization Algorithms for Multiprocessors. *ACM SIGPLAN Notices*, 29(11):25–35, 1994.
5. B. M. Beckmann and D. A. Wood. TLC: Transmission Line Caches. In *Proceedings of the 36th Annual IEEE/ACM International Symposium on Microarchitecture*, 2011.
6. C-C. Kuo, J. B. Carter and R. Kuramkote. MP-LOCKs: Replacing H/W Synchronization Primitives with Message Passing. In *Proceedings of the 5th International Symposium on High-Performance Computer Architecture*, 1999.
7. C. Cascaval, J. G. CastaÃos, L. Ceze, M. Denneau, M. Gupta, D. Lieber, J. E. Moreira, K. Strauss and H. S. Warren. Evaluation of a Multithreaded Architecture for Cellular Computing. In *Proceedings of the 8th International Symposium on High-Performance Computer Architecture*, 2002.
8. C. E. Leiserson, Z. S. Abuhamdeh, D. C. Douglas, C. R. Feynman, M. N. Ganmukhi, J. V. Hill, W. D. Hillis, B. C. Kuszmaul, M. A. St. Pierre, D. S. Wells, M. C. Wong, S. W. Yang and R. Zak. The Network Architecture of the Connection Machine CM-5. In *Proceedings of the ACM Symposium on Parallel Algorithms and Architectures*, 1992.
9. C. Wagner and F. Mueller. Token-based Read/Write-Locks for Distributed Mutual Exclusion. In *Proceedings of the 6th International Euro-Par Conference on Parallel Processing*, 2000.

10. D. E. Culler, J. P. Singh and A. Gupta. *Parallel Computer Architecture: A Hardware/Software Approach*. Morgan Kaufmann, 1998.
11. E. Mensink, D. Schinkel, E. Klumperink, E. Tuijl and B. Nauta. A 0.28pf/b 2gb/s/ch Transceiver in 90nm CMOS for 10mm On-Chip Interconnects. In *Proceedings of the IEEE Solid-State Circuits Conference*, 2007.
12. E. W. Dijkstra. Solution of a Problem in Concurrent Programming Control. *Communications of the ACM*, 8(9):569, 1965.
13. F. H. McMahon. Livermore Fortran Kernels: A Computer Test of Numerical Performance Range. Technical Report UCRL-53745, Lawrence Livermore National Laboratory, 1986. http://www.netlib.org/benchmark/livermorec.
14. H. Franke, R. Russell and M. Kirkwood. Fuss, Futexes and Furwocks: Fast Userlevel Locking in Linux. In *Proceedings of the Ottawa Linux Symposium*, 2002.
15. H. Ito, M. Kimura, K. Miyashita, T. Ishii, K. Okada and K. Masu. A Bidirectional-and Multi-Drop-Transmission-Line Interconnect for Multipoint-to-Multipoint On-Chip Communications. *IEEE Journal of Solid State Circuits*, 43(4):1020–1029, 2008.
16. HP Labs. CACTI, 2012. http://www.hpl.hp.com/research/cacti/.
17. Intel Labs. Single-chip Cloud Computer, 2009. http://techresearch.intel.com/ articles/Tera-Scale/1826.htm.
18. J. Eastep, D. Wingate, M. D. Santambrogio and A. Agarwal. Smartlocks: Self-Aware Synchronization through Lock Acquisition Scheduling. In *Proceedings of the 7th IEEE/ACM International Conference on Autonomic Computing and Communications*, 2009.
19. J. L. Abellán, J. Fernández, M. E. Acacio, D. Bertozzi, D. Bortolotti, A. Marongiu and L. Benini. Design of a Collective Communication Infrastructure for Barrier Synchronization in Cluster-Based Nanoscale MPSoCs. In *Proceedings of the Design, Automation & Test in Europe Conference & Exhibition*, 2012.
20. J. M. Mellor-Crummey and M. L. Scott. Algorithms for Scalable Synchronization on Shared-Memory Multiprocessors. *ACM Transactions on Computer Systems*, 9(1):21–65, 1991.
21. J. Mauro, R. McDougall. *Solaris Internals: Core Kernel Components*. Sun Microsystem Press, 2001.
22. J. Oh, M. Prvulovic and A. Zajic. TLSync: Support for Multiple Fast Barriers Using On-Chip Transmission Lines. In *Proceedings of the 38th International Symposium on Computer Architecture*, 2011.
23. J. P. Lozi, G. Thomas, J. Lawall and G. Muller. Efficient Locking for Multicore Architectures. Technical Report RR-7779, INRIA, 2011.
24. J. R. Goodman, M. K. Vernon and P. J. Woest. Efficient Synchronization Primitives for Large-Scale Cache-Coherent Multiprocessors. In *Proceedings of the 3rd International Conference on Architectural Support for Programming Languages and Operating Systems*, 1989.
25. J. Sampson, R. González, J. F. Collard, N. P. Jouppi, M. Schlansker and B. Calder. Exploiting Fine-Grained Data Parallelism with Chip Multiprocessors and Fast Barriers. In *Proceedings of the 39th Annual IEEE/ACM International Symposium on Microarchitecture*, 2006.
26. J. Sartori and R. Kumar. Low-Overhead, High-Speed Multi-core Barrier Synchronization. In *Proceedings of the 5th International Conference on High Performance Embedded Architectures and Compilers*, 2010.
27. L. Barroso and Urs Hölzle. *The Datacenter as a Computer. An Introduction to the Design of Warehouse-Scale Machines*. Morgan and Claypool Publishers, 2009.
28. M. L. Scott and W. N. Scherer. Scalable Queue-Based Spin Locks with Timeout. In *Proceedings of the 8th ACM SIGPLAN Symposium on Principles and Practice of Parallel Programming*, 2001.
29. M. Monchiero, G. Palermo, C. Silvano and O. Villa. An Efficient Synchronization Technique for Multiprocessor Systems on-Chip. *ACM SIGARCH Computer Architecture News*, 34(1):33–40, 2006.
30. N. R. Tallent, J. M. Mellor-Crummey and A. Porterfield. Analyzing Lock Contention in Multithreaded Applications. In *Proceedings of the 15th ACM SIGPLAN Symposium on Principles and Practice of Parallel Programming*, 2010.

31. P. Coteus, H. R. Bickford, T. M. Cipolla, P. G. Crumley, A. Gara, S. A. Hall, G. V. Kopcsay, A. P. Lanzetta, L. S. Mok, R. Rand, R. Swetz, T. Takken, P. La Rocca, C. Marroquin, P. R. Germann and M.J. Jeanson. Packaging the Blue Gene/L Supercomputer. *IBM Journal of Research and Development*, 49(2):213–248, 2005.
32. P. Tang and P. C. Yew. Processor Self-Scheduling for Multiple-Nested Parallel Loops. In *Proceedings of the the International Conference on Parallel Processing*, 1986.
33. R. Ho, T. Ono, R. D. Hopkins, A. Chow, J. Schauer, F. Y. Liu and R. Drost. High-Speed and Low-Energy Capacitively-Driven On-Chip Wires. *IEEE Journal of Solid State Circuits*, 43(1):52–60, 2008.
34. R. Rajwar and J. R. Goodman. Transactional Lock-free Execution of Lock-based Programs. In *Proceedings of the 10th Annual Conference on Architectural Support for Programming Languages and Operating Systems*, 2002.
35. R. T. Chang, N. Talwalkar, P. Yue and S. S. Wong. Near Speed-of-Light Signaling over On-Chip Electrical Interconnects. *IEEE Journal of Solid-State Circuits*, 38(5):834–838, 2003.
36. S. Bell et al. TILE64 - Processor: A 64-Core SoC with Mesh Interconnect. In *Proceedings of the International Solid-State Circuits Conference Digest of Technical Papers*, 2008.
37. S. C. Woo, M. Ohara, E. Torrie, J. P. Singh and A. Gupta. The SPLASH-2 programs: Characterization and Methodological Considerations. In *Proceedings of the 22nd International Symposium on Computer Architecture*, 1995.
38. S. D. Sherlekar. Intel Many Integrated Core (MIC) Architecture. In *Proceedings of the IEEE International Conference on Parallel and Distributed Systems*, 2012.
39. T. E. Anderson. The Performance Implications of Spin-Waiting Alternatives for Shared Memory Multiprocessors. In *Proceedings of the Intel Conference on Parallel Processing*, 1989.
40. T. Krishna, A. Kumar, L-S. Peh, J. Postman, P. Chiang and M. Erez. Express Virtual Channels with Capacitively Driven Global Links. *IEEE Micro*, 29(4):48–61, 2009.
41. W. T.-Y. Hsu and P.-C. Yew. An Effective Synchronization Network for Hot-Spot Accesses. *ACM Transactions on Computer Systems*, 10(3):167–189, 1992.
42. Z. Hu, J. del Cuvillo, W. Zhu and G. R. Gao. Optimization of Dense Matrix Multiplication on IBM Cyclops-64: Challenges and Experiences. In *Proceedings of the 12th International European Conference on Parallel and Distributed Computing*, 2006.

Hardware Approaches to Transactional Memory in Chip Multiprocessors

J. Rubén Titos-Gil and Manuel E. Acacio

1 Introduction

Multicores are nowadays at the heart of almost every computational system, from the smartphone in our pocket, to the server-class machines in datacenters that provide us with a myriad of cloud services. With the advent of chip multiprocessors, the shift to mainstream parallel architectures is inevitable, and both programmers and architects are presented with immense opportunities and enormous challenges. Despite the fact that multiprocessor systems have existed for a long time, multi-threaded programming has not been much of a focus. Instead, multiprocessors were of interest only to the small community of high-performance computing (HPC), and so was parallel programming, which was mostly ignored by software vendors, and not widely investigated nor taught. As a matter of fact, most software development over time has been predicated on single-core hardware, and the collective knowledge of software developers across organizations has been based primarily on single processor platforms.

Now that the *free lunch* is over [76], programmers must change the way they create applications to fully leverage multicore hardware. At every layer of the computing stack, whether the targeted platform is a handheld device or a warehouse-scale computer, programmers are being pushed towards unfamiliar programming models in order to deliver parallel software that takes advantage of the newly available computational resources and meets the demands of the end user. In the context of datacenters, the task is even more daunting because of the massive scale and complex architecture of these systems where efficient exploitation of parallelism is paramount at every level. Ideally, parallel software developed for these large-scale clusters should be able

J. R. Titos-Gil (✉)
Chalmers University of Technology, Gothenburg, Sweden
e-mail: ruben.titos@chalmers.se

M. E. Acacio
Universidad de Murcia, Murcia, Spain
e-mail: meacacio@ditec.um.es

S. U. Khan, A. Y. Zomaya (eds.), *Handbook on Data Centers*,
DOI 10.1007/978-1-4939-2092-1_27

to harness the potential of their multicore building blocks, while improving aspects that impact the total cost of ownership such as energy efficiency, server utilization, code maintainability or programmer productivity. Hybrid programming models that use shared memory for intra-node parallelism and message passing for inter-node communication are a good example of how programmers exploit these large-scale systems with multi-core processors. New programming models keep appearing in today's datacenters as a result of the wide spectrum of applications and their diverse characteristics. On the one hand, traditional HPC datacenters usually run scientific workloads that have long, computationally-intensive jobs, often as a single binary exclusively executed on a large number of nodes, where synchronization and communication abounds. On the other hand, Internet services exhibit ample parallelism given their large data sets of relatively independent records (e.g. web pages) and the thousands of independent requests received per second. In either environment, the programmer's job is to find the most appropriate way to efficiently exploit the parallelism that is inherent to the problem, maintaining high productivity while producing correct code that is easily verifiable and composable.

Many applications that run in today's datacenters have very strong requirements in terms of response time. This is particularly true for those online services that provide an almost instantaneous reply to the user, such as a web search engine. While the work required to process a user's request can be rather easily partitioned across different nodes in independent units of data, each task that executes in a single node generally performs a substantial amount of computation due to very large data sets. This alleviates the overheads inherently imposed by the communication and synchronization of hundreds or thousands of parallel tasks amongst different nodes of the datacenter. Given the considerable extent of the job performed by each task, the algorithms executed at the task level may also be subject to parallelization in order to speedup the task and reduce overall latency. This could improve utilization of the datacenter too, addressing the important trade-off between keeping machines busy and response times low. Since each task is mapped to a single computing node where all processing cores share the same address space, the intuitive abstraction of shared memory may simplify programmer's job of turning a monolithic task into a parallel, multi-threaded program.

A fundamental problem that all programmers face when writing multi-threaded code is the difficulty of simultaneously achieving both high efficiency and productivity/correctness. Designing a parallel algorithm involves orchestrating the concurrent execution of the parts to improve performance while at the same time guaranteeing correctness. Complex and hard-to-find, software defects unique to multi-threaded applications such as race conditions and deadlocks can quickly derail a software project [18]. Software engineering tools have yet to simplify the programming for these shared-memory architectures in order to make the new hardware resources accessible to the average programmer. In order to avert a software crisis, developers must adapt and improve such tools to make them better suited for parallel multicore software development [83]. The reality is that software has not matured enough to take advantage of the number of cores that are already available in today's systems, and the vast majority of applications are still single-threaded [30]. The rise of multicores has brought such problem of effective concurrent programming to the forefront of computing research. To help alleviating this problem, *Transactional Memory* (TM) has been proposed as a concurrency control mechanism that aims to simplify concurrent programming with reasonable scalability.

This chapter examines the state-of-the-art of Transactional Memory, paying special attention to its hardware implementations (*Hardware Transactional Memory* or HTM). Recent inclusion of HTM support in commodity multicore processors (Intel's Transactional Synchronization Extensions [91]) and commercial mainframes (IBM's Transaction Execution Facility [40]) has converted TM into a reality for current and future datacenters.

The remainder of this chapter is organized as follows. In Sect. 2 we delve into the problems that traditional parallel programming with locks has and discuss how TM can alleviate them. Subsequently, we present the fundamentals of TM in Sect. 3 and the hardware mechanisms TM requires in Sect. 4. Next, we describe the programming interfaces of the hardware TM support provided by the new Intel processors (Sect. 5), and present a brief performance analysis of it (Sect. 6). Section 7 summarizes the most relevant proposals found in the HTM research literature. The main conclusions of this chapter are summarized in Sect. 8.

2 Why Transactional Memory Is Going Mainstream

Concurrent programming is a far more challenging task than sequential programming: A parallel program is undoubtedly more difficult to design, write, and debug than its sequential counterpart. Orchestrating the concurrent execution of the parts to improve performance while at the same time guaranteeing correctness is by no means an easy task. Designing parallel algorithms requires restructuring code and data in often counter-intuitive ways, so that it can be split into parallel tasks. Balancing the workload among the available processors, or communicating and managing shared data between different processors are some of the many factors that make parallel programming a complicated endeavour. Programmers need to reason carefully about possible interactions of their threads when running concurrently, and not doing so may result in programs that are incorrect, perform poorly, or both. To add insult to injury, parallel programs are very hard to debug due to the combinatorial explosion of possible execution orderings: Parallel programs often produce non-deterministic results, making it harder to prove programs correct, and their bugs are often elusive and notoriously difficult to find and fix, because of the difficulty to reproduce the exact same execution (i.e. interleaving of threads, etc.) that leads to a race.

In the context of shared memory architectures where concurrent tasks process shared data, guaranteeing correctness while maintaining efficiency and productivity is a key challenge. Parallel thread execution requires synchronization for accessing shared data. Programmers are responsible for ensuring that concurrent accesses to shared data structures are correct, and often rely on mutual exclusion mechanisms to protect these critical sections, so that no more than one thread can simultaneously enter the same critical section and access the same shared data.

2.1 The Drawbacks of Lock-Based Synchronization

Traditional multi-threaded programming models use low-level primitives such as locks to guarantee mutual exclusion. Unfortunately, the complexity of lock-based synchronization makes parallel programming an error prone task, particularly when fine-grained locks are used to extract more parallelism. At one end, heroic programmers seeking performance try to minimize the amount of shared resources (data) that are protected by the same lock, so that different threads accessing different data do not have to serialize their execution unnecessarily, thus enabling maximum concurrency. However, the use of fine-grain locks adds more programming complexity, since programmers must be careful to acquire them in a fixed, predetermined order so as to avoid deadlocks. At the other end, common programmers seeking productivity (correctness) choose to reduce the complexity of reasoning, i.e. likelihood of deadlock, by using fewer locks with coarser granularity where each lock is responsible for protecting larger critical section. This naturally comes at the cost of sacrificing performance, when threads without true data races contend for the same lock. Though programmers can also include deadlock detection mechanisms in their programs, to try and recover from deadlocks, this alternative also adds substantial complexity.

As if deadlocks were not enough, locking brings about other undesired situations like priority inversion (when a high priority thread is unable to acquire a lock because a lower priority thread is holding it), convoying (when a lock holder is de-scheduled from execution, impeding others to progress) and lack of fault tolerance (when a lock holder modifies data and then crashes, causing the whole program to fail). Furthermore, locking breaks the abstraction principle, as programmers using a module need to be aware of the locks it uses, to ensure that the program still follows the predetermined locking order that prevents deadlock. Therefore, locks jeopardize the code composability property, as two individually correct modules can deadlock when combined together.

2.2 The Transactional Abstraction

The trade-off between programming ease and performance imposed by locks remains one of the key challenges to programmers and computer architects of the multicore era. Transactional Memory (TM) [34, 36] has been proposed as a conceptually simpler programming model that can help boost developer productivity by eliminating the complex task of reasoning about the intricacies of safe fine-grained locking. TM inherits the concept of *transaction* from the database community, and applies it to the domain of shared-memory programming in an attempt to simplify the task of thread synchronization. Transactions in the multi-threaded programming world are blocks of code that are guaranteed to be executed atomically and in isolation with respect to all other code. At a high level, the programmer or compiler annotates sections of the code as atomic blocks or transactions. The underlying system then executes these transactions speculatively in an attempt to exploit as much concurrency as possible.

TM systems generally employ an optimistic approach to concurrency control in order to let multiple transactions execute in parallel, while still preserving the properties of atomicity and isolation. Therefore, the TM system attempts to make best use of available concurrency in the application while guaranteeing correctness. By using transactions to safely access shared data, programmers need not reason about the safety of interleavings or the possibility of deadlocks to write correct multi-threaded code. Hence, TM addresses the performance-productivity trade-off by not discouraging programmers from using coarse-grain synchronization, since the underlying system can potentially achieve performance comparable to fine-grained locks by executing transactions speculatively. In addition to addressing such critical trade-off, TM tries to solve other limitations of lock-based synchronization. Transactional code is robust in the face of both hardware and software failures, as the system can always rollback the speculative updates to its pre-transactional state in case a thread crashes inside a transaction. Unlike locks, transactions are composable, and they can be safely nested without any risk of deadlocks [6].

2.3 High-Performance Transactional Memory

Transactions are a promising abstraction that could ease parallel programming and make it more accessible to the common programmer. Transactional semantics can be entirely supported in software, hardware, or using a combination of both. According to this, we can classify TM systems into software transactional memory (STM), hardware transactional memory (HTM), and hybrid transactional memory systems.

STM implementations [26, 35, 48, 72] allow running transactional workloads on existing systems without requiring special hardware support, providing a great degree of flexibility at little cost. Unfortunately, implementing the necessary mechanisms entirely in software imposes too high an overhead and thus STM systems do not fare well against traditional lock-based approaches when performance is important. For this new paradigm to be a viable alternative to locks, the key mechanisms that provide transactional semantics must be implemented at the architectural level.

Hybrid TM systems [4, 12, 23, 43, 74, 77] attempt to combine both the speed and flexibility by using simple hardware to accelerate performance-critical operations of an STM implementation. In this way, hybrid implementations of TM rely on some kind of software intervention to execute transactions, though they minimize the overheads of providing transactional semantics in comparison to a software-only solution. Hybrid TM models use the STM as a backup to handle situations where the hardware cannot execute the transaction successfully [34].

Transactional semantics can also be supported largely in hardware [1, 14, 16, 31, 50, 51, 89], allowing for good performance with varying degrees of complexity, which change considerably from one HTM proposal to another depending on what kind of transactions the TM system is capable of committing without resorting to fall-back mechanisms. Simple HTM schemes [17, 22, 36] adopt a "best-effort" solution that cannot not guarantee that all transactions will eventually commit successfully using hardware support alone, mostly because of the limitations imposed by the hardware structures involved. More sophisticated HTM proposals [1, 31, 51] address this limitation in transaction size, guaranteeing that certain "bounded" transactions can be entirely executed in hardware. These proposals typically behave in the same way as best-effort ones as long as hardware structures are sufficient, and then fall back to additional hardware mechanisms to maintain transactional properties on resource overflow. However, neither bounded nor best-effort solutions can commit transactions that encounter events that are too complicated to handle in hardware, like context switches, page faults, I/O, exceptions or interrupts [37], and in such circumstances the transaction is invariably aborted. Even more elaborated HTM schemes have been designed [1, 64] to handle all transactions in hardware, ensuring that the same transaction is not indefinitely aborted because of its size, duration or other events it may encounter. Unfortunately, the complexity of these "unbounded" HTM designs makes them too costly for processor manufacturers to consider them in practice.

2.4 *Industrial Adoption of Hardware Transactional Memory*

In the early 2000s, Transmeta was the first company to implement a form of transactional memory in its x86-compatible Crusoe microprocessor, though this hardware only meant to support aggressive speculative optimizations in its dynamic binary translation system [24].

More recently, Azul Systems included HTM support in its Vega systems [22], a specialized appliance designed to massively scale the usable compute resources available to Java applications. However, the HTM support was only used to accelerate Java locks and not exposed to programmers.

Sun Microsystems was the first general-purpose processor manufacturer that ventured to introduce support for transactions in a chip multiprocessor. In 2007, the company announced that its high-end Rock processor would have support for both transactional memory and speculative multithreading [17]. Unfortunately, Sun cancelled the project in 2009, and Rock chips never made it to the market, though some prototypes were distributed for research purposes.

Around the time the Rock project was cancelled, AMD proposed the Advanced Synchronization Facility (ASF) [21], a set of instruction extensions to the x86 architecture that provide limited support for lock-free data structures and transactional memory. To date, it is unknown whether any future AMD products will implement ASF.

In mid 2011, IBM revealed that its BlueGene/Q compute chip would feature transactional memory support. The custom design was a system-on-a-chip that integrated 18 PowerPC cores with memory and networking subsystems [33]. Cores shared a multiversioned L2 cache which supports transactional memory and speculative multithreading. With the lessons learned from BlueGene/Q, IBM began to ship the IBM zEnterprise EC12 system in the fall of 2012, less than a year after the first BlueGene/Q system made its debut in the Top500 list. The zEC12 processor introduces the Transactional Execution Facility [40], which extends the z/Architecture used on IBM mainframes with transactional memory support. The zEC12 has given IBM the distinction of becoming the first company to deliver commercial chips with this technology [27].

In early 2012, Intel announced that its new *Haswell* microarchitecture would implement hardware transactional memory through a set of new instructions called *Transactional Synchronization Extensions* (TSX) [61, 91]. Shortly after, Intel's TSX specification was released, describing how TM is exposed to programmers, but withholding details on the actual implementation. In mid-2013, Intel began shipping processors based on its 4th-generation Core microarchitecture, making the Core i3/i5/i7 and Xeon v3 processor families the first chips with TM support that are available in the consumer and server markets. The adoption of transactional memory by mainstream, commodity x86 processors culminates a two decade journey of active academic research. Section 7 provides a good overview of the contributions have brought the industry here.

3 Fundamentals of Transactional Memory

Transactional Memory (TM) [34, 36] has been proposed as an easier-to-use programming model that can help developers build scalable shared-memory data structures, relieving them from the burdens imposed by fine-grained locking. Under the TM model, the programmer declares *what* regions of the code must appear to execute atomically and in isolation (called *transactions*), leaving the burden of *how* to provide such properties to the underlying levels. The TM system then executes optimistically transactions, stalling or aborting them whenever real run-time data races (called *conflicts*) occur amongst concurrent transactions. The TM programming model thus replaces explicit synchronization mechanisms like locking with a more declarative approach whose aim is to decouple performance pursuit from programming productivity. The transactional abstraction is provided at the programming language level through a new construct, e.g. `atomic`, employed by programmers to delimit accesses to shared data—i.e. *critical sections*—thus structuring their parallel code into *atomic blocks* or transactions. A transaction is said to *commit* when it completes its execution successfully—confirming its speculative updates to shared memory—while it is *aborted* or *squashed* when some condition occurs—e.g. a conflict with a concurrent transaction—that impedes its completion with success. To guarantee race-free execution of a transactional multi-threaded application, TM implementations must satisfy two basic properties, namely atomicity and isolation, which are inherited from the database domain.

The *atomicity* property dictates that a transaction is either executed to completion or not executed at all. If the transaction successfully commits, all of its speculative changes are made globally visible at once. Otherwise, if the transaction aborts, all its tentative updates are discarded in order to revert the system to its pre-transactional state, as if the transaction had never executed. To the outside world, this means that a transaction appears as an indivisible operation that cannot be partially executed. On its part, the *isolation* property requires that the intermediate (speculative) state of a partially completed transaction must remain hidden from other code. By satisfying these properties, transactions appear to execute in some serial global order, i.e. committed transactions are never observed by different processors as executed in different orders. To provide these properties, the TM system must implement two basic mechanisms, namely data versioning and conflict management. The policy and implementation of these two mechanisms constitutes the two fundamental dimensions of the TM design space.

Version management handles the simultaneous storage of both speculative data (new values that will become visible if the transaction commits) and pre-transactional data (old values retained if the transaction aborts). Only one of the two values can be stored *in-situ*, i.e. in the corresponding memory address, while the other needs to be placed somewhere else. The data versioning policy dictates how the system handles the storage of both versions, and it constitutes a major design point of the system. Depending on which value, old or new, gets to stay "in place" during the course of the transaction, the data version management policy can be classified as eager or lazy. Lazy versioning keeps old values *in-situ* until the commit phase, buffering speculative updates "on the side" in the meantime. In contrast, an eager approach to versioning uses a per-thread *transaction log* to backup the old value of a memory location prior to each write, and then updates the memory location with the new value.

When two concurrent transactions access the same memory location, and at least one of the accesses is a write operation, we say that there is a *conflict* or *race* between them. TM systems implement a *conflict management* mechanism to detect and resolve such conflicts. For this purpose, the data read and written by each transaction must be tracked. The set of data addresses that a transaction modifies during its execution is known as *write set*. Similarly, the *read set* refers to the group of memory locations read by the transaction. In these terms, a conflict between two concurrent transactions happens when a transaction's write set overlaps with other concurrent transactions' read or write sets. Depending on the meta-data information used for transactional book-keeping, conflict detection can take place at different levels of granularity, from objects, to cache lines to word or even byte-level addresses.

4 Hardware Mechanisms for Transactional Memory

HTM systems must identify memory locations for transactional accesses, manage the read-sets and write-sets of the transactions, detect and resolve data conflicts, manage architectural register state, and commit or abort transactions [34].

4.1 ISA Extensions

Identifying transactional boundaries is accomplished by extending the instruction set architecture (ISA). All HTM implementations introduce a pair of new instructions, i.e. "begin transaction" and "commit transaction", to delimit the scope of a transaction. On the one hand, the execution of the "begin transaction" instruction causes the processor to enter into "transactional mode" (usually setting some bit in the status register) and perform some common actions related to the initialization of the basic transactional mechanisms, like checkpointing the architectural registers to a shadow register file. The architectural registers and memory combined form the precise state of the processor, and therefore the register state also needs to be restored to a known precise state in case of abort. The operation of creating a shadow copy of the architectural registers at the start of a transaction is rather straightforward and can often be performed in a single cycle. On the other hand, the "commit transaction" instruction attempts to confirm the speculative updates of the transaction by publishing them to the rest of the system, and it returns the processor to non-transactional state if successful, discarding the register checkpoint.

The most straightforward step to identify transactional accesses is to leverage these two instructions that mark the beginning and end of a transaction, so that all the loads and store instructions executed while in transactional mode are implicitly considered transactional. This is the approach that most modern HTM proposals follow, including the Haswell microarchitecture [91], and the failed Rock processor [17]. Another option is to further augment the ISA with explicit "transactional load" and "transactional store" instructions, separated from their conventional counterparts. Though allowing a transaction to contain both transactional and non-transactional accesses may complicate things, this provides increased flexibility and may aid programmers to reduce the pressure on the underlying TM mechanisms, as non-transactional accesses do not participate in data versioning nor conflict detection. The original HTM proposal by Herlihy and Moss [36] as well as the AMD Advanced Synchronization Facility (ASF) [21] are explicitly transactional designs.

Some proposed HTMs also include an "abort transaction" instruction to explicitly roll back the tentative work of transaction. This is an example of flexible design that may enable TM hardware to be applied toward solving problems beyond guaranteeing mutual exclusion during the execution of critical regions. Programmers using hardware transactions may find useful the ability to explicitly rollback execution upon a certain condition, which need not necessarily be a conflict with other transaction.

4.2 Transactional Book-Keeping

HTM systems must track a transaction's read and write set in order to detect data races amongst concurrent transactions. Many HTMs extend the cache line metadata kept at the private cache level, with two new bits that record, respectively, whether the line has been speculatively read (SR) and/or speculatively modified (SM) during the ongoing transaction [31, 51, 82]. Such designs also support the capability to clear all the read bits in the data cache instantaneously, an action that is performed when the transaction commits or aborts. The private caches serve as a natural place to track a transaction's read and write sets, enabling low overhead tracking, although they also constrain the granularity of conflict detection to that of a cache line.

All HTM systems that leverage the private level cache to perform transactional book-keeping are susceptible of transactional *overflows* due to the cache's limited capacity or associativity. Best effort designs would automatically abort the transaction if a cache line whose SR or SM bit is set is replaced, while bounded schemes would resort to some safety net in order to keep tracking read and write sets and detecting conflicts in the presence of spilled lines.

An alternative scheme of transactional book-keeping which does not leverage the private level cache is to use Bloom filters to conservatively summarize a transaction's data accesses using "address signatures" [14, 89]. The main disadvantage of hash encoding is that false positives may signal spurious conflicts, this is, the signature may indicate that an address belongs to the transaction read and write sets when in fact it does not.

4.3 Data Versioning

Besides keeping read and write set metadata, private caches are the natural place to buffer speculative values, since they are on the access path for the local processor and thus can automatically forward the latest transactional update to subsequent loads without special search. Write-back caches can be modified to behave as write buffers that support versioning of speculative data. Depending on the implementation, one or multiple versions of a speculatively modified cache line may be allowed.

For HTM systems with eager version management [51, 89], caches need no changes as they do not really have any notion of speculative writes. All writes go to the memory hierarchy in the same way, whether they occur inside or outside a transaction. It is then the responsibility of the coherence protocol to detect accesses to speculatively written data, and ensure no other threads or transactions observe it. Unlike lazy systems, evictions of speculatively written data from the private caches are tolerated, and they need no special treatment from the point of view of the versioning mechanism. However, specialized hardware is required to fill this virtualized log with the old value of each memory location that is being updated inside a transaction. The contents of the log are simply discarded on commit, by resetting the log pointer to its initial position. On an abort, a software handler walks the log restoring the original values into memory.

4.4 Conflict Detection and Resolution

HTM proposals leverage coherence mechanisms for conflict detection. Invalidation-based, cache coherence protocols allow HTM implementations to detect conflicts among concurrently running transactions at the granularity of cache lines. While unnecessary transactional conflicts may arise as a result of false sharing, for most transactional workloads this choice of granularity represents a good trade-off between design cost and performance.

More important than the granularity of the detection is the design decision of *when* conflicts with concurrent transactions must be detected. Strategies for conflict detection and resolution vary depending on when a processor examines the bookkeeping information of its read and write sets. In systems with eager detection—sometimes also referred to as pessimistic—conflicts are detected as soon as they happen, i.e. on every individual memory access. In the opposite approach, called lazy or optimistic conflict detection, this check is delayed until transaction commit, and the resolution is generally a committer-wins scheme.

The coherence protocol already provides mechanisms to locate the copies of a requested cache line, and thus the detection of transactional conflicts can be achieved with straightforward extensions. In snooping-based protocols, all caches observe all coherence traffic for all lines, allowing cache controllers to check for conflicts whenever a request is observed on the bus. In directory-based protocols, cache controllers only observe the coherence traffic corresponding to the lines that are currently privately cached. In a typical MESI directory protocol, a local store to a shared (S) line results in a write miss, since the protocol ensures that no cache can have permissions to write the data at this point. An exclusive request is sent to the directory, which in turn forwards invalidation messages to the current sharers of the line (except maybe the requester). The sharers are then able to check whether the requested address belongs to their read set—by checking the SR bit in cache, the read signature, etc.—and appropriately detect a *write-read* conflict. Similarly, a load (store) miss to a line that is remotely cached in modified (M) or exclusive (E) state results in a read (write) request forwarded from the directory to the cache that has the latest copy of the data, which then checks its write-set (read- and write-set) metadata to determine if a *read-write* (*write-write*) conflict exists.

Once an HTM system detects a conflict, it must determine how to resolve it. The conflict resolution policy constitutes another design dimension in HTM by dictating which transaction wins the conflict and is granted access to the data. The loser transaction can stall its execution, or it can be aborted: The alternatives change depending on when the conflict is detected.

In HTMs with eager conflict detection, there are several policies for resolution: requester wins, requester aborts, or requester stalls using a scheme of conservative deadlock avoidance. The implementation of the requester wins policy is straightforward: The cache or caches that detect a conflict simply trigger abort and yield to the requester. If the conflicted data was not speculatively modified (write-read conflict), the loser responds with the appropriate invalidation acknowledgement or data message. Otherwise, the response may be delayed until the data is conveniently restored. The main drawback of this policy is that it can produce livelock scenarios. The opposite option is to abort the requester. This is accomplished by augmenting the coherence protocol with negative acknowledgements (*nack*) messages, so that a cache controller that detects a conflict responds to a forwarded request or invalidation with a *nack* message. On reception of a *nack* response, the requester knows it has lost the conflict and can take the appropriate actions. The simplest alternative is to trigger its own abort, but this can also result in livelock. A less draconian, livelock-free solution is to stall the transaction and periodically retry the conflicting memory access until a positive response (different from the *nack*) is received. In this case, cyclic dependencies amongst transactions can bring the system to a deadlock, and so the system must have a way out such possible cycles. LogTM [51] uses a simple timestamp-based scheme to conservatively detect cycles, aborting the youngest transaction to break the possible cycle.

HTM systems with lazy conflict detection must resolve conflicts when a committer seeks to commit a transaction that conflicts with one or more other transactions. The resolution policy in this scenario can abort all others, or else stall or abort the committer. In general, lazy HTMs follow a committer wins policy [10, 32] that favours forward progress and is both deadlock- and livelock-free. Unfortunately, the committer wins policy does not guarantee fairness and can result in starvation for some transactions.

4.5 Transaction Commit

The execution of the "commit transaction" instruction attempts to make the transaction's tentative changes permanent and visible to other processors instantaneously. Such publication is in itself a task that must occur atomically and without interference from other processors. For most HTMs, publishing speculative updates means obtaining exclusive ownership for all cache lines in the write set, and then releasing isolation over both transactional sets at once.

The implementation of the commit instruction is a straightforward operation in eager HTMs, since writes were performed in place and therefore all write set lines held in cache have write (exclusive) permissions. As for lazy HTMs, the requirement of en-masse publication of speculative updates to shared memory at commit time poses more challenges when multiple speculative versions of the same data can coexist. Commits in this case are non-trivial because each SM line must be located and its coherence permission upgraded while every other copy in remote caches gets invalidated. On the other hand, lazily-versioned HTMs that allow at most one speculative version easily provide local commits since speculative writes are performed only when the protocol has obtained exclusive ownership (write permissions) for the line. To simultaneously support both local commits and aborts, the coherence protocol must be able to tolerate silent replacements of exclusively owned lines, and it must be adapted to ensure the consistent version of the data is always written back to the shared levels of the memory hierarchy before the first speculative write.

4.6 Transaction Abort

A hardware transaction may be implicitly aborted by the conflict resolution mechanism, or the abort can be explicitly triggered from the program via an "abort transaction" instruction. Aborting a transaction means discarding all its tentative changes and return the state of the processor to the exact same state it was right before the transaction began. Book-keeping information (SM and SR bits, signatures, etc.) must always be cleared on abort, and the last step of the abort process is the restoration of the architectural registers using the checkpoint that was saved in the shadow register file at the beginning of the transaction.

Implementing the abort functionality is quite simple in lazy HTMs, since speculative writes were performed "on the side" (in private structures local to the core) and therefore the shared memory still contains consistent, pre-transactional values. Aborts are cheap since silent invalidations of shared-state lines are generally supported by the protocol, and thus lazy HTM systems can quickly discard the speculative state, by extending the cache design with conditional gang-invalidation of lines whose SM bit is set.

Eager HTMs, on the other hand, must restore each cache line in the write set with the pre-transactional value that was backed up in the transaction log. The log unroll is generally done in software, by trapping to an abort handler that accesses the log base and pointer registers, and walks the log in reverse direction—those entries that were added last must be processed first. No transactional conflicts should arise during this process, as the coherence protocol ensures that the lines that belong to the write set of the aborting transaction are isolated and cannot belong to any other transaction. Because aborting is a slow process in eager HTMs, isolation over the read set is usually released as soon as the abort is triggered, as it is safe for other transactions to access it while the log is unrolled.

5 Intel TSX: TM Support in Mainstream Processors

Almost a decade after the research community regained interest in hardware implementations of TM, the world's largest semiconductor company adopted these ideas for a commercial product. The fourth generation of the Intel Core microarchitecture, commonly known by its code name *Haswell*, implements the basic mechanisms to provide programmers with a best-effort yet fast implementation of the transactional abstraction. Intel began shipping Haswell-based processors in 2013, making the Core i3/i5/i7 and Xeon v3 processor families the first chips with TM support that are available in the consumer and server markets. Given Intel's market share on mobile, desktop and servers platforms, Haswell is an important milestone towards the expansion of transactions as a synchronization primitive for multi-threaded applications.

Following the *tick* of Ivy Bridge, which shrunk the Sandy Bridge microarchitecture to the 22-nm process technology, Haswell's *tock* extends the instruction set architecture in a number of ways, from which the *Transactional Synchronization Extensions* (TSX) are certainly one of the most prominent novel features. Through the new TSX instructions, Haswell offers programmers two interfaces to exploit its ability to use optimistic concurrency in thread synchronization: *Hardware Lock*

Elision (HLE) is meant to accelerate conventional lock-based programs while maintaining legacy compatibility, while *Restricted Transactional Memory* (RTM) allows programmers to explicitly start, commit and abort transactions, thus providing a natural way of implementing transactions as a synchronization abstraction. Regardless of the TSX interface used, the same underlying hardware mechanisms are involved in the transactional execution.

5.1 Hardware Lock Elision

Hardware Lock Elision (HLE) is a legacy-compatible ISA extension aimed at extracting more thread-level parallelism from conventional lock-based programs, by using speculation to allow concurrent execution of critical sections protected by the same *mutex*. HLE comes in the form of two instruction prefixes, `XACQUIRE` and `XRELEASE`, which act as hints to delimit the boundaries of a critical section. If the processor supports TSX, each of these prefixes modifies the behaviour of the instructions that are typically used, respectively, to acquire and release a lock variable; otherwise, the prefixes are ignored and the processor executes the code without entering transactional execution, making HLE-ready binaries backwards compatible.

When the `XACQUIRE` hint is used in conjunction with the atomic instruction that attempts to acquire a free lock (e.g. `cmpxchg`), it alters its usual behaviour and prevents (*elides*) the associated write of the "busy" value. Instead, the processor enters transactional execution, adds the address of the lock to its read set and proceeds to execute the critical section speculatively. Because the globally visible value of the lock remains unchanged (i.e. "free"), other threads can read it without causing a data conflict and also enter the critical section protected by the lock. While in transactional execution, each processor leverages coherence traffic to monitor memory accesses, detecting data conflicts and rolling back as necessary.

Similarly, the `XRELEASE` prefix is paired with the store instruction that releases the lock, so that the associated write of the "free value" is again avoided. Instead, the processor attempts to commit the transactional execution. In this way, as long as threads do not perform any conflicting operations on each other's data, they can concurrently execute the critical section without unnecessary serialization due to a coarse grain lock.

If speculation fails, the processor will rollback and re-execute the critical section without using lock elision. Mutual exclusion in the re-execution of the critical section is automatically ensured, because the address of an elided lock is always added to the read set of the transaction, and thus non-transactional writes associated to lock acquisition will always cause data conflicts with all other threads that may be eliding the same lock at that time, which will be forced to rollback and also retry without elision. Therefore, code that makes use of HLE maintains the same forward progress guarantees as the underlying lock-based execution.

5.2 Restricted Transactional Memory

Unlike HLE, Restricted Transactional Memory (RTM) gives up backwards compatibility to provide programmers with a more flexible interface for transactional execution. It introduces new instructions to define transaction boundaries, `XBEGIN` and `XEND`, as well as to explicitly abort a transaction from software, `XABORT`. Transactional nesting is supported in TSX by means of flattening: the nesting level is incremented by `XBEGIN` and decremented by `XEND`, and commit is only attempted when the nesting level goes to zero.

Programmers must provide an alternative code path to the `XBEGIN` instruction, where control is transferred to in case the transaction aborts, after the processor has discarded all speculative updates, restored architectural state to appear as if the speculation never occurred, and resume execution non-transactionally. After an abort, the `EAX` register is used to communicate its cause (explicit, data conflict, internal buffer overflow, faults, etc.) to the fallback routine, as well as the 8-bit immediate taken as argument by the `XABORT` instruction. In this way, programmers may freely use the fallback path in different ways to decide the most profitable course of action, manage contention, etc. It is important to remark that according to the TSX specification, the HTM implementation is best-effort, as there are no guarantees as to whether an RTM transaction will ever successfully commit. Thus, the fallback code is entirely responsible for guaranteeing forward progress.

Figure 1 shows a simple implementation of the fallback path, which attempts to retry a transaction a number of times before acquiring a global lock to execute the transaction in serial irrevocability. This implementation shares similarities with that found in GCC's *libitm* library, since version 4.7.0. As we can see, `serial_lock` is read after the transaction has successfully started so that, when a thread enters serial irrevocable mode by acquiring the lock, it automatically causes the abort of all other running transactions due to a conflict on the lock variable, achieving a similar behaviour to what the HLE interface provides.

6 Analysing Intel TSX Performance on Haswell

In this section, we present a brief performance analysis of the Intel TSX extensions, with the purpose of shedding light into the benefits of hardware support for transactions. For this evaluation, we use a benchmark from the STAMP suite (*Stanford Transactional Applications for Multi-Processing* [13]). STAMP benchmarks are extensively used in the TM research literature. Unlike other benchmarks (e.g. SPLASH-2 [88]), the STAMP applications have been developed from scratch using coarse grain transactions, in an attempt to capture the features of future transactional workloads.

For the sake of simplicity, we focus exclusively on the application *intruder*, whose use of transactions we can see in the code snippet shown in Fig. 2. This benchmark emulates a signature-based network intrusion detection system in which packets

```
[...]
#include <immintrin.h>
#include <xmmintrin.h>
#include "spinlock.h"


void beginTransaction() {
 int nretries = 0;
 while(1) {
    /* Wait for serial irrevocable transaction to complete */
    while (!arch_read_can_lock(&serial_lock)) _mm_pause();
   ++nretries;
   unsigned status = _xbegin();
   if(status == _XBEGIN_STARTED) {
     if (!arch_read_can_lock(&serial_lock)){
     /* Started transaction but someone is executing the transaction section non-speculatively
       * (acquired fall-back lock) -> abort */
         _xabort(0xff);
     } else {
         return; /* Successfully started transaction */
     }
   }
   /* This is the beginning of the fallback path (abort handler) */
   if((status & _XABORT_EXPLICIT) &&
        _XABORT_CODE(status)==0xff &&
        !(status & _XABORT_NESTED))    {
     while (!arch_read_can_lock(&serial_lock)) _mm_pause();  /* Wait until the lock is free */
   }
   if(nretries >= MAX_RETRIES) break;     /* Too many retries, take the fallback lock */
 }
   arch_write_lock(&serial_lock);    /* Execute atomic region in serial irrevocable mode */
}

void endTransaction() {
   if (!arch_read_can_lock(&serial_lock)){      /* We were in serial irrevocable mode */
       arch_write_unlock(&serial_lock);
   } else  { /* Regular transaction, commit. */
       _xend();
   }
}
```

Fig. 1 A possible implementation of the fallback for Intel RTM

are processed in parallel and go through three phases: capture, reassembly, and detection. Transactions are used to synchronize access to the shared data structures used in the capture and reassembly phases, respectively, a simple FIFO queue and a self-balancing tree. We can see how the resulting code is simple as that of coarse-grain locks, effectively easing the task of the programmer, as opposed to the use of fine-grain locking on the queue and tree data structures.

We pick *intruder* because it exhibits several interesting characteristics. First, it comprises several transactions that access different data structures. On the one hand, its first and third transactions are used to extract an element at the head of a queue, and thus have small read and write set sizes, since it basically consist of a read-modify-write operation of the head pointer. On the other hand, its main transaction has medium-sized transactional sets—in the order of a few tens of cache lines—since it carries out most of the processing of the packet (reassembly) by traversing the tree structure. Despite its coarse granularity, its main transaction can be accommodated by the Haswell hardware without constantly causing capacity aborts. Last but not

```
void
processPackets (void* argPtr)
{
 [...]
   while (1) {
        TM_BEGIN();
        bytes = TMSTREAM_GETPACKET(streamPtr);
        TM_END();
        if (!bytes) {
            break;   /* No more packets to process */
        }
        [...]
        TM_BEGIN();
        error = TMDECODER_PROCESS(decoderPtr, bytes, (PACKET_HEADER_LENGTH + packetPtr->length));
        TM_END();
        [...]
        TM_BEGIN();
        data = TMDECODER_GETCOMPLETE(decoderPtr, &decodedFlowId);
        TM_END();
        if (data) {
            error_t error = PDETECTOR_PROCESS(detectorPtr, data);
            [...]
        }
    }
}
```

Fig. 2 Example of coarse grain transactions in *intruder*

least, intruder exhibits high levels of contention that are desirable to evaluate TM performance in less favourable conditions.

Our experiments with TSX are performed on a 3.4 GHz quad-core Intel Core i7-4770 processor with 16 GB of main memory, running Linux kernel 3.11. Each core has support for two SMT threads, but we choose to disable hyperthreading from the BIOS, in order to dedicate all available resources for speculative buffering (e.g. L1 data cache) to a single thread per core. Each core has an eight-way, 32 KB L1 data cache. Given the four hardware contexts available, we run the program with one, two and four threads. We pin one thread to each core using *pthread affinity*. The benchmark is compiled with GCC v.4.8.1, using the O3 optimization level. Since version 4.8, GCC supports the Intel RTM intrinsics, built-in functions and code generation by including the `<immintrin.h>` header and enabling the `-mrtm` flag. The `begin_transaction` and `end_transaction` functions shown in Fig. 1 are used to implement the fallback-path. The read-write spinlock implementation from `linux-3.11/arch/x86/include/asm/spinlock.h` is used to enforce serialization. Transactions are allowed to retry up to eight times before resorting to serialization via the fallback lock. For each configuration, a minimum of 20 executions are averaged to derive statistically meaningful results. Our experiments use the large input size recommended for non-simulator runs [13]: 256K traffic flows are analyzed, 10 % of which have attacks injected, where each flow has a maximum of 128 packets.

To observe the relative performance gain achieved by TSX, we consider in this experiment other two synchronization schemes that do not make use of the hardware support for transactions. We run a lock-based version of the benchmark in which transactions are implemented as critical sections protected through a *single global lock* (SGL). Additionally, we compare TSX performance against a software TM

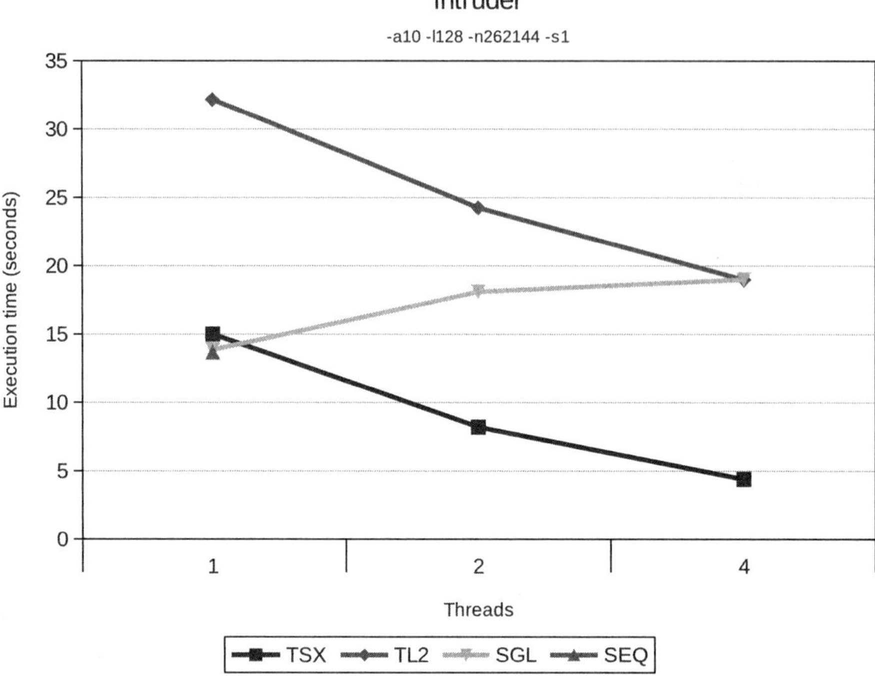

Fig. 3 Performance comparison of TSX versus TL2 and single global lock

system called *Transactional Locking II* (TL2) [26], which is distributed with STAMP. We also include a sequential flavour of the benchmark (SEQ), which is stripped of all synchronization.

Figure 3 shows the execution time of the three synchronization schemes implemented by the underlying library. The plot shows execution time (in seconds) for each of the synchronization flavours considered, and runs with one to four threads. As we can see in Fig. 3, *intruder* achieves good scalability through optimistic concurrency in spite of the coarse grain transactions used. Both hardware (TSX) and software (TL2) implementations of TM scale significantly better than a single global lock. In particular, we see how TSX is able to bring the execution time from 15 s with a single thread, down to around 4.5 when running 4 threads. As opposed to HTM and STM solutions, adding more threads in the SGL scheme does not speedup execution but rather causes a slight performance degradation: The global lock precludes all concurrency in the application, and adding more threads only makes things worse by increasing the contention on the lock variable. Given that the use of coarse grain transactions entails a similar complexity to that of single global lock, this performance comparison between SGL and two TM implementations confirms that, from the point of view of the programmer, transactions are indeed able to keep up its promise of achieving better scalability than coarse grain locks, with the same programming effort.

```
 Intel(r) Performance Counter Monitor: Intel(r) Transactional Synchronization Extensions Monitoring Utility
 Copyright (c) 2013 Intel Corporation
Num logical cores: 4
Num sockets: 1
Threads per core: 1
[...]
Detected Intel(R) Core(TM) i7-4770 CPU @ 3.40GHz "Intel(r) microarchitecture codename Haswell"
Update every 0 seconds
 Executing "./intruder -a10 -l128 -n262144 -s1 -t4" command:
Percent attack  = 10
Max data length = 128
Num flow        = 262144
Random seed     = 1
Num attack      = 29395
Elapsed time    = 4.451121 seconds
Num found       = 29395
Core | IPC  | Instructions | Cycles  | Transactional Cycles | Aborted Cycles   | #RTM  |  Cycles/Transaction
   0   0.52        11 G       22 G         13 G (61.25%)        5569 M (24.84%) 7779 K    1765
   1   0.21      3236 M       15 G         13 G (88.90%)        3825 M (24.56%) 7177 K    1929
   2   0.21      3203 M       15 G         13 G (89.08%)        3763 M (24.17%) 7103 K    1952
   3   0.22      3348 M       15 G         13 G (88.72%)        4457 M (28.63%) 7336 K    1882
-----------------------------------------------------------------------------------------------------
   *   0.31        21 G       69 G         55 G (79.93%)          17 G (25.48%)   29 M    1879
Event0: RTM_RETIRED.START Number of times an RTM execution started.
Event1: RTM_RETIRED.ABORTED_MISC1 Number of times an RTM execution aborted due to various- memory events
Event2: TX_MEM.ABORT_CONFLICT Number of times a transactional abort was signaled due to a data conflict.
Event3: RTM_RETIRED.COMMIT Number of times an RTM execution successfully committed
Core | Event0  | Event1  | Event2  | Event3
   0   7087 K    619 K     470 K     5869 K
   1   7093 K    666 K     510 K     5842 K
   2   7039 K    618 K     463 K     5835 K
   3   7023 K    624 K     472 K     5800 K
----------------------------------------------
   *     28 M    2528 K    1916 K      23 M
```

Fig. 4 PCM output for *intruder* (four threads), showing TSX performance event counts

Furthermore, the numbers obtained by TSX demonstrate that hardware implementations of transactional semantics are necessary to dramatically reduce the substantial performance overheads seen in software-only solutions, as a result of the instrumentation on every memory access within a transaction which is required to track transactional reads and writes and be able to detect conflicts. Hardware TM implementations can exhibit their full potential in those cases where most transactions are appropriately sized to avoid overflowing the hardware buffering capacity, as it is the case of *intruder* in our Haswell-based setup. Other STAMP benchmarks with larger transaction footprints may not be as well suited and exhibit substantially higher capacity-induced aborts [91].

In spite of the good speedup achieved by TSX with four threads (3.3 times faster than sequential), the scalability of *intruder* starts to deviate more and more from the ideal. This is a direct result of the increasing level of contention seen in the application: several threads attempting to capture the same packet from the FIFO queue, concurrent accesses to the dictionary while the tree is rebalanced, etc. Using the Intel Performance Counter Monitor (PCM), an open source tool, we can monitor TSX performance events in order to obtain relevant information about the execution of the program, such as the amount of contention (e.g. number of aborts due to data conflicts, `TX_MEM.ABORT_CONFLICT`). Figure 4 shows the output of PCM when running the benchmark *intruder* with four threads using TSX. The number of conflict-induced aborts may increase quickly with contention, particularly in HTM implementations that resolve conflicts using a *requester wins* policy. Though details about its implementation have not been disclosed at the time of this writing, it is likely that Intel has adopted such simple yet livelock-prone conflict resolution strategy in Haswell, with the intent of keeping the changes in its coherence protocol to a minimum. In any case, the TSX specification clearly places on the fallback path the responsibility of providing forward progress when it detects that a transaction has failed too many times. As commented earlier, in our implementation of the abort handler all running transactions are automatically *killed* when the fallback lock is acquired due to a data conflict, thus adding to the number of contention-induced aborts.

7 An Overview of Hardware TM Research

Research in HTM design has been very active since the introduction of multicores in mainstream computing. In the early 1990s, Herlihy and Moss introduced Transactional Memory [36] as a hardware alternative to lock-based synchronization. Their main idea was to generalize the LL/SC primitives in order to perform atomic accesses not to one but to several independent memory locations, thus eliminating the need for protecting critical sections with lock variables. Almost a decade later, architects began to recover their interest in transactions at a hardware level. Rajwar and Goodman's Transactional Lock Removal (TLR) [63] was the first to apply the concept of transaction to the execution of lock-protected critical sections, merging the idea of Speculative Lock Elision (SLE) [62] with a timestamp-based conflict resolution scheme.

The early proposal by Herlihy and Moss was revived ten years later by Hammond et al., who present Transactional Coherence and Consistency (TCC) [32] as a novel coherence and consistency model that uses continuous transactional execution. The novelty of TCC stems from its "all transactions, all the time" philosophy, where transactions are the basic unit of parallel work, synchronization, memory coherence and consistency. TCC's lazy approach contains speculative updates within private caches and lazily resolves races when a committing transaction broadcasts its write-set, employing a bus to serialize transaction commits.

In contrast to Stanford's TCC, Wisconsin's LogTM [51] explores the opposite corner of the HTM design space. Moore et al. take a more evolutionary approach to transactional memory in LogTM, combining transactional support with a conventional shared memory model that enables a more gradual change towards transactional systems. LogTM is a purely eager HTM system that leverages a standard coherence protocol to perform conflict detection on individual memory requests, and makes commits fast by storing old values to a per-thread log in cacheable virtual memory, which is unrolled by a software handler in case of abort. Unlike TCC, LogTM can tolerate evictions of transactional data from caches thanks to the log, and enables conflict detection on evicted blocks through an elegant extension to the coherence protocol.

LogTM has been subsequently refined. Moravan et al. [52] introduce support for nested transactions, enabling both closed nesting with partial aborts and open nesting [55]. Open nesting is a programming language construct motivated by performance, which can improve concurrency by relaxing the atomicity guarantee. When an open nested transaction commits, the TM system releases its read and written data so that other transactions can access them without generating conflicts. Thanks to open nesting, otherwise-offending transactions can access the exposed data after the nested transaction commits, while the outer transaction still runs. This can enhance the degree of concurrency achieved by the flattening scheme found in LogTM, which enforces isolation until the outermost transaction commits. In [3], Baek et al. propose FanTM, a design that uses address signatures in hardware [14] to efficiently support transaction nesting.

Later on, Yen et al. [89] decouple transactional support from caches, removing read and write bits used for transactional book-keeping, and replacing them with *hash signatures*. This latest improvement, called LogTM-SE (*Signature Edition*), borrows the concept of Bloom filters [5] to conservatively encode a transaction's read and write set metadata. The idea of applying hash encoding towards conflict detection/thread disambiguation was first introduced into the realm of TM by Ceze et al. in [14] and [15]. The use of hash signatures for transactional book-keeping has been further explored by several authors. In [69], Sanchez et al. examine different signature organizations and hashing schemes to achieve hardware-efficient and accurate TM signatures. Quislant et al. have also studied signature organizations, basing their works in LogTM-SE. In [59], they show that locality can be exploited in order to reduce the number of bits inserted in the filter for those addresses nearby located, and reducing the number of false conflicts. More recently, the authors have studied multiset signature designs [60] which record both the read and write sets in the same Bloom filter. Yen et al. developed Notary [90], which introduces a privatization interface that allows the programmer to explicitly declare shared and private heap memory allocation, which can be used to reduce the signature size as well as the number of false conflicts arising from private memory accesses. Sanyal et al. exploit the same concept in [70], proposing a scheme that dynamically identifies thread-local variables and excludes them from the commit set, both reducing the pressure on the versioning mechanisms and improving the scalability of such phase in lazy HTMs.

In the context of signature-based eager HTM systems, Titos-Gil et al. have proposed a scheme of conflict detection at the directory level [29] that is not only capable of dealing with contention more efficiently than LogTM-SE, but can also minimize the performance degradation caused by false positives. Their solution moves transactional bookkeeping from caches to the directory, introducing separate hardware module that acts as conflict controller and works independently of the coherence controller, leaving the protocol largely unmodified.

In FASTM [46], Lupon et al. extend LogTM with a coherence protocol that enables fast abort recovery in an otherwise eager HTM, by leveraging the private cache to buffer speculative state, effectively avoiding traps to software handlers that perform log unroll as long as speculatively modified data does not overflow the private cache level. LogTM's approximation of making commits fast has also inspired OneTM [8]

[7], which uses a cache to reduce the frequency with which transactions overflow on chip resources, and proposes a simple irrevocable execution switch to handle such overflows as well as context switches, I/O or system calls inside transactions, at the cost of limited concurrency.

Bobba et al. propose TokenTM [11], another unbounded HTM design that uses the abstraction of tokens [49] to precisely track conflicts on an unbounded number of memory blocks and it handles both paging, thread migration and context switching, but incurs high state overhead. In [41], Jafri et al. improve on TokenTM and propose LiteTM, a design that maintains the same virtualization properties of TokenTM while greatly reducing the state overhead, and without sacrifying much performance. Support for transactions of unlimited duration, size and nesting depth has also been considered by proposals such as UTM [1, 44] or VTM [64], which focus on hardware schemes that provide virtualization of transactions. However, both achieve this goal by introducing large amounts of complexity in the processor and the memory subsystem. On its part, XTM [20] implements transaction virtualization support in software, using virtual memory and operating at page granularity. A similar approach is taken by Chuang et al. [19] in PTM, a page-based, hardware-supported TM design that combines transaction bookkeeping with the virtual memory system to support transactions of unbounded size, as well as to handle context switches and exceptions.

While it is not an issue for eager systems like LogTM, parallelism at commit is important for lazy systems when running applications with low contention but a large number of transactions. Transactions that do not conflict should ideally be able to commit simultaneously. The very nature of lazy conflict resolution protocols makes it difficult since only actions taken at commit time permit discovery of data races among transactions. Simple lazy schemes like the ones employing a global commit token [10] or a bus [31] do not permit such parallelism. The reason for limited parallelism at commit time is that the committing transaction has no knowledge of which other concurrently running transactions must abort to preserve atomicity. The TCC design [31] was later extended to scalable DSM architectures using directory based coherence. This proposal is called Scalable TCC (STCC) [16], and it employs selective locking of directory banks to avoid arbitration delays and thereby improve commit throughput. Pugsley et al. [58] improve over STCC by proposing even more scalable commit algorithms that reduce the number of network messages, remove the need for a centralized agent, and tackle deadlocks, livelocks and starvation scenarios.

Another approach to improve the scalability of the commit process in lazy systems has been explored by EazyHTM [82]. Tomic et al. record the information pertaining to potential conflicts, which is readily available from coherence messages during the lifetime of any transaction, and use this information at commit time to allow true commit parallelism. All potentially conflicting transactions that must be aborted would be known, and committers that have not seen races can commit in a truly parallel fashion.

Pi-TM [53] builds upon the ideas explored by EazyHTM, and leverages the concept of pessimistic self-invalidation to enable parallel lazy commits without affecting the execution in the common case. Negi et al. develop an early conflict detection—lazy conflict resolution HTM design with modest extensions to existing protocols, which uses information regarding conflicts and performs pessimistic invalidation of potentially conflicting lines on commit and abort, enabling fast common-case execution.

FlexTM [75] also provides lazy conflict resolution by recording conflicts as they happen, using this information to enable distributed commits. Unlike EazyHTM, Shriraman et al. choose to do so in software, sacrifying progress guarantees to gain greater parallelism. Performance costs associated with software intervention and software verification challenges without watertight forward progress guarantees could limit the value of this approach. EazyHTM, on the other hand, provides parallel lazy commits in hardware and ensures forward progress, but trades off common-case performance to achieve it. FlexTM allows flexibility in policy but it does so by implementing critical policy managers in software. It provides a significant improvement in speed over software TM implementations by proposing the use of alert-on-update hardware, but the considerable cost of software intervention renders a comparison with pure HTMs moot. In the context of HTM, Shriraman and Dwarkadas [73] have also analyzed the interplay between conflict resolution time and contention management policy. They show that both policy decisions have a considerable impact on the ability to exploit available parallelism and demonstrate that conflict resolution time has the dominant effect on performance, corroborating that lazy HTMs are able to uncover more parallelism than eager approaches.

With DynTM [47], Lupon et al. introduce a cache coherence protocol that allows transactions in a multi-threaded application run either eagerly or lazily based on some heuristics like prior behavior of transactions, at the cost of adding extra complexity at level of the coherence controller. Recognizing the fact that contention is more a property of data rather than that of an atomic code block, ZEBRA [79] chooses a different dimension when combining eager and lazy policies into a HTM design, allowing per-cache-line selection of versioning and conflict resolution policies. While DynTM selects policies at the level of transactions, ZEBRA is a data-centric design which works at the same granularity of the underlying coherence infrastructure—i.e. cache lines—and therefore introduces less complexity into existing protocols. This hybrid design is able to track closely or exceed the performance of the best performing policy for a given workload, bringing together the benefits of parallel commits (inherent in traditional eager HTMs) and good optimistic concurrency without deadlock avoidance mechanisms (inherent in lazy HTMs), with little increase in complexity.

LV* [54], a proposal that utilizes snoopy coherence, allows programmer control over policy in hardware but with the constraint that all transactions in an application must use the same policy at any given time. The requirement of programmer-assisted policy change is a drawback too since the same phase of an application can exhibit different behavior with varying datasets.

The mitigation of the performance penalty associated with transaction aborts has been of interest to the HTM community. Waliullah and Stenstrom study the utility of intermediate checkpoints in lazy HTM systems [85, 86], as a means to reduce the amount of work that is discarded on abort. In their scheme transactions record conflicting addresses upon abort, and use this historical information to insert a checkpoint before a memory reference predicted as conflicting is executed. If the transaction is squashed, it is rolled back to the checkpoint associated with the first conflicting access, rather than all the way back to the beginning. Reducing the penalty of abort was also considered by Armejach et al. [2], who propose a reconfigurable private level data cache to improve the efficiency of the version management mechanism in both eager and lazy HTMs.

Titos et al. have also analysed how the lack of effective techniques for store management results in a quick degradation in the performance of eager HTM systems with increasing contention and, thus, lends credence to the belief that eager designs do not perform as well as their lazy counterparts when conflicts abound [80]. The authors present two simple ways to improve handling of speculative stores which yield substantial improvements in execution time when running applications with high contention, allowing eager designs to exceed the performance of lazy ones.

The applications of data forwarding and value prediction for conflict resolution have also been explored in the context of eager HTM systems. Pant et al. [56, 57] observe that shared-conflicting data is often updated in a predictable manner by different transactions, and propose the use of value prediction in order to capture this predictability and increase overall concurrency by satisfying loads from conflicting transactions with predicted values, instead of stalling. In DATM [66], Ramadan et al. investigate the advantages of value forwarding for speculative resolution of true data conflicts amongst concurrent transactions. DATM is an eager system that discovers and tracks the data dependencies amongst concurrent transactions, allowing writer transactions to proceed in the presence of other conflicting transactional accessors, and reader transactions to obtain uncommitted data produced by a concurrent transaction, while still enforcing a legal serialized order that preserves consistency.

Hardware TM systems can suffer a series of pathological behaviours that negatively affect performance. Bobba et al. explore HTM design space, identifying how some of these undesirable scenarios [10] affect each kind of system depending on the choice of policies for version and conflict management. Some pathologies such as starvation have been further analysed and resolved in other subsequent works [87]. Other pathologies that affect HTM performance have been the topic of several studies. Volos et al. [84] investigate the interaction of transactional memory implementations and lock-based code, and discover other problematic scenarios that may arise in these circumstances. False sharing, another undesired situation that may arise in multi-threaded codes, becomes even a bigger problem when it occurs in conjunction with hardware transactional memory [51] due to the detection of conflicts at a cache line granularity. Tabba et al. [78] propose a mechanism that takes the concepts of coherence decoupling [39] and value prediction, and combines them to mitigate the effects of coherence conflicts in transactions. The granularity of conflict detection in HTM has also been the subject of the works by Khan et al. [42], whose HTM proposal is able to detect conflicts at the level of objects—instead of cache lines—which leads to a novel commit scheme as well as an elegant solution to the problem of version management virtualization.

Another kind of pathological behaviour affecting HTM performance happens when concurrent operations on data structures that are not semantically conflicting—such as two insertions in two different buckets of a hash table—result in conflicting transactions because of updates on auxiliary program data—e.g. the *size* field. Inspired by instruction replay-based mechanisms [25], Blundell at al. propose RetCon [9], a hardware mechanism that eliminates the performance impact of such spureous transactional conflicts. RetCon tracks the relationship between input and output values symbolically and uses this information to transparently repair the output state of a transaction at commit.

Ramadan et al. have examined the architectural features necessary to support HTM in the Linux kernel for the x86 architecture [65, 68]. They propose MetaTM, an HTM model that contains features that enable efficient and correct interrupt handling for an x86-like architecture. Using TxLinux—a Linux kernel modified to use transactions in place of locking primitives in several key subsystems—they quantify the effect of architectural design decisions on the performance of such a large transactional workload. TxLinux, based on the Linux 2.4 kernel and thus characterized by its simple, coarse-grained synchronization structure, is used by Hoffman et al. in [38] to show that a minimal subset of TM features supported in hardware can simplify synchronization, provide comparable performance to fine-grained locking and handle overflows. The challenge of operating system (OS) support in HTM is also addressed Wang et al. [71] and Tomic et al. [81]. DTM [71] proposes a hardware-based solution that fully decouples transaction processing from caches, while HTM-OS [81] leverages the existing OS virtual memory mechanisms to support unbounded transaction sizes and provide transaction execution speed that does not decrease when transaction grows. A related challenge that has been addressed in the HTM literature is the support of input/output operations within transactions: Lui et al. [45] analyse this problem and propose an HTM system that supports I/O within transactions by means of partial commits, using commit-locks and blocking/waking-up of transactional threads.

The applicability of hardware transactional memory (HTM) has also been considered in the context of dynamic memory management. Dragojevic et al. [28] demonstrate that HTM can be used to simplify and streamline memory reclamation for practical concurrent data structures. The use of HTM to aid lightweight dynamic language runtimes in evolving more capable and robust execution models while maintaining native code compatibility has been studied too. Using a modified Linux kernel and a Python interpreter, Riley at al. [67] explore the lack of thread safety in native extension modules and use features found in an HTM implementation to address several issues that impede to the effective deployment of dynamic languages on current and future multicore and multiprocessor system.

8 Conclusions

Following the recent inclusion of hardware support for Transactional Memory in commodity multicore processors [91] and commercial mainframes [40], the time has come for architects and programmers of datacenters to ponder the new opportunities that may unfold in the coming years. This chapter examines the state-of-the-art of Transactional Memory, paying special attention to its hardware implementations (*Hardware Transactional Memory* or HTM). Transactions not only address one the key challenges of the multicore era, i.e. the trade-off between programming ease and performance, but also bring about other important benefits such as better code composability and fault tolerance. For these reasons, parallel software developed for large-scale clusters may also find in Transactional Memory an attractive programming model to unlock the full potential of the multicore processors that power a datacenter, while improving aspects that impact the total cost of ownership such as server utilization or code maintainability.

Acknowledgements This work was supported by the Spanish MINECO, as well as European Commission FEDER funds, under grant TIN2012-38341-C04-03.

References

1. C. Scott Ananian, Krste Asanovic, Bradley C. Kuszmaul, Charles E. Leiserson, and Sean Lie. Unbounded transactional memory. In *Proceedings of the 11th Symposium on High-Performance Computer Architecture*, pages 316–327, 2005.
2. Adria Armejach, Azam Seydi, Rubén Titos-Gil, Ibrahim Hur, Adrián Cristal, Osman Unsal, and Mateo Valero. Using a reconfigurable l1 data cache for efficient version management in hardware transactional memory. In *Proceedings of the 20th International Conference on Parallel Architectures and Compilation Techniques*, 2011.
3. Woongki Baek, Nathan Bronson, Christos Kozyrakis, and Kunle Olukotun. Making nested parallel transactions practical using lightweight hardware support. In *Proceedings of the 24th International Conference of Supercomputing*, pages 61–71, 2010.
4. Lee Baugh, Naveen Neelakantam, and Craig Zilles. Using hardware memory protection to build a high-performance, strongly atomic hybrid transactional memory. In *Proceedings of the 35th International Symposium on Computer Architecture*, pages 115–126. 2008.
5. Burton H. Bloom. Space/time trade-offs in hash coding with allowable errors. *Communications of the ACM*, 13:422–426, 1970.
6. Colin Blundell, E Christopher Lewis, and Milo Martin. Subtleties of transactional memory atomicity semantics. *Computer Architecture Letters*, 5(2), 2006.
7. Colin Blundell, E Christopher Lewis, and Milo M. K. Martin. Unrestricted transactional memory: Supporting i/o and system calls within transactions. Technical Report CIS-06-09, Department of Computer and Information Science, University of Pennsylvania, 2006.
8. Colin Blundell, Joe Devietti, E Christopher Lewis, and Milo Martin. Making the fast case common and the uncommon case simple in unbounded transactional memory. In *Proceedings of the 34th International Symposium on Computer Architecture*, pages 24–34, 2007.
9. Colin Blundell, Arun Raghavan, and Milo M.K. Martin. RETCON: transactional repair without replay. In *Proceedings of the 37th International Symposium on Computer Architecture*, pages 258–269, 2010.

10. Jayaram Bobba, Kevin E. Moore, Luke Yen, Haris Volos, Mark D. Hill, Michael M. Swift, and David A. Wood. Performance pathologies in hardware transactional memory. In *Proceedings of the 34th International Symposium on Computer Architecture*, pages 81–91, 2007.
11. Jayaram Bobba, Neelam Goyal, Mark D. Hill, Michael M. Swift, and David A. Wood. TokenTM: Efficient execution of large transactions with hardware transactional memory. In *Proceedings of the 35th International Symposium on Computer Architecture*, pages 81–91, 2008.
12. Chi Cao Minh, Martin Trautmann, JaeWoong Chung, Austen McDonald, Nathan Bronson, Jared Casper, Christos Kozyrakis, and Kunle Olukotun. An effective hybrid transactional memory system with strong isolation guarantees. In *Proceedings of the 34th International Symposium on Computer Architecture*, pages 69–80, 2007.
13. Chi Cao Minh, JaeWoong Chung, Christos Kozyrakis, and Kunle Olukotun. STAMP: Stanford transactional applications for multi-processing. In *Proceedings of the IEEE Intl. Symposium on Workload Characterization*, pages 35–46, 2008.
14. Luis Ceze, James Tuck, Calin Cascaval, and Josep Torrellas. Bulk disambiguation of speculative threads in multiprocessors. In *Proceedings of the 33rd International Symposium on Computer Architecture*, pages 227–238, 2006.
15. Luis Ceze, James Tuck, Pablo Montesinos, and Josep Torrellas. BulkSC: bulk enforcement of sequential consistency. In *Proceedings of the 34th International Symposium on Computer Architecture*, pages 278–289, 2007.
16. Hassan Chafi, Jared Casper, Brian D. Carlstrom, Austen McDonald, Chi Cao Minh, Woongki Baek, Christos Kozyrakis, and Kunle Olukotun. A scalable, non-blocking approach to transactional memory. In *Proceedings of the 13th Symposium on High-Performance Computer Architecture*, pages 97–108, 2007.
17. Shailender Chaudhry, Robert Cypher, Magnus Ekman, Martin Karlsson, Anders Landin, Sherman Yip, Håkan Zeffer, and Marc Tremblay. Rock: A high-performance Sparc CMT processor. *IEEE Micro*, 29(2):6–16, 2009.
18. Ben Chelf. Ensuring code quality in multi-threaded applications. http://www.coverity.com/library/pdf/coverity_multi-threaded_whitepaper.pdf.
19. Weihaw Chuang, Satish Narayanasamy, Ganesh Venkatesh, Jack Sampson, Michael Van Biesbrouck, Gilles Pokam, Brad Calder, and Osvaldo Colavin. Unbounded page-based transactional memory. In *Proceedings of the 12th International Symposium on Architectural Support for Programming Language and Operating Systems*, pages 347–358, 2006.
20. JaeWoong Chung, Chi Cao Minh, Austen McDonald, Travis Skare, Hassan Chafi, Brian D. Carlstrom, Christos Kozyrakis, and Kunle Olukotun. Tradeoffs in transactional memory virtualization. In *Proceedings of the 12th International Symposium on Architectural Support for Programming Language and Operating Systems*, pages 371–381, 2006.
21. Jaewoong Chung, Luke Yen, Stephan Diestelhorst, Martin Pohlack, Michael Hohmuth, David Christie, and Dan Grossman. Asf: Amd64 extension for lock-free data structures and transactional memory. In *Proceedings of the 43rd International Symposium on Microarchitecture*, pages 39–50, 2010.
22. Cliff Click. Azul's experiences with hardware transactional memory, 2009. http://sss.cs.purdue.edu/projects/tm/tmw2010/talks/Click-2010_TMW.pdf.
23. Peter Damron, Alexandra Fedorova, Yossi Lev, Victor Luchangco, Mark Moir, and Daniel Nussbaum. Hybrid transactional memory. In *Proceedings of the 12th International Symposium on Architectural Support for Programming Language and Operating Systems*, pages 336–346, 2006.
24. James C. Dehnert, Brian K. Grant, John P. Banning, Richard Johnson, Thomas Kistler, Alexander Klaiber, and Jim Mattson. The transmeta code morphing software: using speculation, recovery, and adaptive retranslation to address real-life challenges. In *Proceedings of the 1stInternational Symposium on Code Generation and Optimization (CGO)*.
25. Rajagopalan Desikan, Simha Sethumadhavan, Doug Burger, and Stephen W. Keckler. Scalable selective re-execution for edge architectures. In *Proceedings of the 11th International Symposium on Architectural Support for Programming Language and Operating Systems*, pages 120–132, 2004.

26. David Dice, Ori Shalev, and Nir Shavit. Transactional locking ii. In *Proceedings of the 19th Intl. Symposium on Distributed Computing*, 2006.
27. Ivan Dobos *et al.* IBM zEnterprise EC12 Technical Guide, February 2013. http://www.redbooks.ibm.com/redbooks/pdfs/sg248049.pdf.
28. Aleksandar Dragojević, Maurice Herlihy, Yossi Lev, and Mark Moir. On the power of hardware transactional memory to simplify memory management. In *Proceedings of the 30th International Symposium on Principles of Distributed Computing*, pages 99–108, 2011.
29. J. Rubén Titos Gil, Manuel E. Acacio, and José M. García. Efficient eager management of conflicts for scalable hardware transactional memory. *IEEE Transactions on Parallel and Distributed Systems*, 24(1):59–71, 2013.
30. Tom Groenfeldt. Software programmers lag behind hardware developments, 2011. http://blogs.forbes.com/tomgroenfeldt/2011/04/21/software-programmers-lag-be hind-hardware-developments/.
31. Lance Hammond, Brian D. Carlstrom, Vicky Wong, Mike Chen, Christos Kozyrakis, and Kunle Olukotun. Transactional coherence and consistency: Simplifying parallel hardware and software. *IEEE Micro*, 24(6), 2004.
32. Lance Hammond, Vicky Wong, Mike Chen, Brian D. Carlstrom, John D. Davis, Ben Hertzberg, Manohar K. Prabhu, Honggo Wijaya, Christos Kozyrakis, and Kunle Olukotun. Transactional memory coherence and consistency. In *Proceedings of the 31st International Symposium on Computer Architecture*, pages 102–113, 2004.
33. Ruud Haring, Martin Ohmacht, Thomas Fox, Michael Gschwind, David Satterfield, Krishnan Sugavanam, Paul Coteus, Philip Heidelberger, Matthias Blumrich, Robert Wisniewski, alan gara, George Chiu, Peter Boyle, Norman Chist, and Changhoan Kim. The ibm blue gene/q compute chip. *IEEE Micro*, 32(2):48–60, March 2012.
34. Tim Harris, James R. Larus, and Ravi Rajwar. *Transactional Memory, 2nd Edition*. Morgan & Claypool, 2010.
35. Maurice Herlihy, Victor Luchangco, Mark Moir, and III William N. Scherer. Software transactional memory for dynamic-sized data structures. In *Proceedings of the 22nd Symposium on Principles of Distributed Computing*, pages 92–101, 2003.
36. Maurice Herlihy and J. Eliot B. Moss. Transactional memory: Architectural support for lock-free data structures. In *Proceedings of the 20th International Symposium on Computer Architecture*, pages 289–300, 1993.
37. Owen S. Hofmann, Donald E. Porter, Christopher J. Rossbach, Hany E. Ramadan, and Emmett Witchel. Solving difficult HTM problems without difficult hardware. In *TRANSACT '07: 2nd Workshop on Transactional Computing*, 2007.
38. Owen S. Hofmann, Christopher J. Rossbach, and Emmett Witchel. Maximum benefit from a minimal HTM. In *Proceedings of the 14th International Symposium on Architectural Support for Programming Language and Operating Systems*, pages 145–156, 2009.
39. Jaehyuk Huh, Jichuan Chang, Doug Burger, and Gurindar S. Sohi. Coherence decoupling: making use of incoherence. In *Proceedings of the 11th International Symposium on Architectural Support for Programming Language and Operating Systems*, pages 97–106, 2004.
40. Christian Jacobi, Timothy Slegel, and Dan Greiner. Transactional memory architecture and implementation for IBM System z. In *Proceedings of the 45th International Symposium on Microarchitecture*, pages 25–36, 2012.
41. Syed Ali Raza Jafri, Mithuna Thottethodi, and T. N. Vijaykumar. LiteTM: Reducing transactional state overhead. In *Proceedings of the 16th Symposium on High-Performance Computer Architecture*, pages 1–12, 2010.
42. Behram Khan, Matthew Horsnell, Ian Rogers, Mikel Luján, Andrew Dinn, and Ian Watson. An object-aware hardware transactional memory. In *Proceedings of the 10th International Conference on High Performance Computing and Communications*, pages 93–102, 2008.
43. Sanjeev Kumar, Michael Chu, Christopher J. Hughes, Partha Kundu, and Anthony Nguyen. Hybrid transactional memory. In *Proceedings of the 11th Symposium on Principles and Practice of Parallel Programming*, pages 209–220, 2006.

44. Sean Lie. Hardware support for unbounded transactional memory. Master's thesis, 2004. Massachusetts Institute of Technology.
45. Yi Liu, Xin Zhang, He Li, Mingxiu Li, and Depei Qian. Hardware transactional memory supporting I/O operations within transactions. In *Proceedings of the 10th International Conference on High Performance Computing and Communications*, pages 85–92, 2008.
46. Marc Lupon, Grigorios Magklis, and Antonio González. FASTM: A log-based hardware transactional memory with fast abort recovery. In *Proceedings of the 18th International Conference on Parallel Architectures and Compilation Techniques*, pages 293–302, 2009.
47. Marc Lupon, Grigorios Magklis, and Antonio González. A dynamically adaptable hardware transactional memory. In *Proceedings of the 43rd International Symposium on Microarchitecture*, pages 27–38, 2010.
48. Virendra J. Marathe, William N. Scherer III, and Michael L. Scott. Adaptive software transactional memory. In *Proceedings of the 19th Intl. Symposium on Distributed Computing*, 2005.
49. Milo M.K. Martin. *Token Coherence*. PhD thesis, CS Dept., Univ. of Wisconsin-Madison, 2003.
50. Austen McDonald, JaeWoong Chung, D. Carlstrom Brian, Chi Cao Minh, Hassan Chafi, Christos Kozyrakis, and Kunle Olukotun. Architectural semantics for practical transactional memory. In *Proceedings of the 33rd International Symposium on Computer Architecture*, pages 53–65, 2006.
51. Kevin E. Moore, Jayaram Bobba, Michelle J. Moravan, Mark D. Hill, and David A. Wood. LogTM: Log-based transactional memory. In *Proceedings of the 12th Symposium on High-Performance Computer Architecture*, pages 254–265, 2006.
52. Michelle J. Moravan, Jayaram Bobba, Kevin E. Moore, Luke Yen, Mark D. Hill, Ben Liblit, Michael M. Swift, and David A. Wood. Supporting nested transactional memory in LogTM. In *Proceedings of the 12th International Symposium on Architectural Support for Programming Language and Operating Systems*, pages 359–370, 2006.
53. Anurag Negi, J. Rubén Titos Gil, Manuel E. Acacio, José M. García, and Per Stenström. Pi-tm: Pessimistic invalidation for scalable lazy hardware transactional memory. In *Proceedings of the 18th Symposium on High-Performance Computer Architecture*, pages 141–152, 2012.
54. Anurag Negi, M.M. Waliullah, and Per Stenstrom. LV*: A low complexity lazy versioning HTM infrastructure. In *Proceedings of the Intl. Conference on Embedded Computer Systems: Architectures, Modeling, and Simulation (IC-SAMOS 2010)*, pages 231–240, 2010.
55. Yang Ni, Vijay S. Menon, Ali-Reza Adl-Tabatabai, Antony L. Hosking, Richard L. Hudson, J. Eliot B. Moss, Bratin Saha, and Tatiana Shpeisman. Open nesting in software transactional memory. In *Proceedings 12th ACM SIGPLAN symposium on Principles and Practice of Parallel Programming*, pages 68–78, 2007.
56. Salil Pant and Gregory Byrd. Extending concurrency of transactional memory programs by using value prediction. In *Proceedings of the 6th ACM conference on Computing Frontiers*, pages 11–20, 2009.
57. Salil Pant and Gregory Byrd. Limited early value communication to improve performance of transactional memory. In *Proceedings of the 23rd International Conference of Supercomputing*, pages 421–429, 2009.
58. Seth H. Pugsley, Manu Awasthi, Niti Madan, Naveen Muralimanohar, and Rajeev Balasubramonian. Scalable and reliable communication for hardware transactional memory. In *Proceedings of the 17th International Conference on Parallel Architectures and Compilation Techniques*, pages 144–154, 2008.
59. Ricardo Quislant, Eladio Gutierrez, and Oscar Plata. Improving signatures by locality exploitation for transactional memory. In *Proceedings of the 18th International Conference on Parallel Architectures and Compilation Techniques*, pages 303–312, 2009.
60. Ricardo Quislant, Eladio Gutierrez, and Oscar. Plata. Multiset signatures for transactional memory. In *Proceedings of the 25th International Conference of Supercomputing*, pages 43–52, 2011.

61. Ravi Rajwar and Martin Dixon. Intel transactional synchronization extensions, 2012. Intel Developer Forum (IDF2012).
62. Ravi Rajwar and James R. Goodman. Speculative lock elision: Enabling highly concurrent multithreaded execution. In *Proceedings of the 34th International Symposium on Microarchitecture*, pages 294–305, 2001.
63. Ravi Rajwar and James R. Goodman. Transactional lock-free execution of lock-based programs. In *Proceedings of the 10th International Symposium on Architectural Support for Programming Language and Operating Systems*, pages 5–17, 2002.
64. Ravi Rajwar, Maurice Herlihy, and Konrad Lai. Virtualizing transactional memory. In *Proceedings of the 32nd International Symposium on Computer Architecture*, pages 494–505, 2005.
65. Hany E. Ramadan, Christopher J. Rossbach, Donald E. Porter, Owen S. Hofmann, Aditya Bhandari, and Emmett Witchel. MetaTM/TxLinux: transactional memory for an operating system. In *Proceedings of the 34th International Symposium on Computer Architecture*, pages 92–103, 2007.
66. Hany E. Ramadan, Christopher J. Rossbach, Owen S. Hofmann, and Emmett Witchel. Dependence-aware transactional memory. In *Proceedings of the 41st International Symposium on Microarchitecture*, pages 246–257, 2008.
67. Nicholas Riley and Craig Zilles. Hardware transactional memory support for lightweight dynamic language evolution. In *Dynamic Language Symposium*, 2006.
68. Christopher J. Rossbach, Hany E. Ramadan, Owen S. Hofmann, Donald E. Porter, Aditya Bhandari, and Emmett Witchel. TxLinux and MetaTM: transactional memory and the operating system. *Communications of the ACM*, 51:83–91, 2008.
69. Daniel Sanchez, Luke Yen, Mark D. Hill, and Karthikeyan Sankaralingam. Implementing signatures for transactional memory. In *Proceedings of the 40th International Symposium on Microarchitecture*, pages 123–133, 2007.
70. Sutirtha Sanyal, Adrián Cristal, Osman S. Unsal, Mateo Valero, and Sourav Roy. Dynamically filtering thread-local variables in lazy-lazy hardware transactional memory. In *Proceedings of the 11th International Conference on High Performance Computing and Communications*, pages 171–179, 2009.
71. Wang Shaogang, Dan Wu, Zhengbin Pang, and Xiaodong Yang. DTM: Decoupled hardware transactional memory to support unbounded transaction and operating system. In *Proceedings of the 38th International Conference on Parallel Processing*, pages 228–236, 2009.
72. Nir Shavit and Dan Touitou. Software transactional memory. In *Proceedings of the 14th ACM Symposium on Principles of Distributed Computing*, pages 204–213, 1995.
73. Arrvindh Shriraman and Sandhya Dwarkadas. Refereeing conflicts in hardware transactional memory. In *Proceedings of the 23rd International Conference of Supercomputing*, pages 136–146, 2009.
74. Arrvindh Shriraman, Virendra J. Marathe, Sandhya Dwarkadas, Michael L. Scott, David Eisenstat, Christopher Heriot, William N. Scherer III, and Michael F. Spear. Hardware acceleration of software transactional memory. In *Workshop on Languages, Compilers, and Hardware Support for Transactional Computing (TRANSACT)*, 2006.
75. Arrvindh Shriraman, Sandhya Dwarkadas, and Michael L. Scott. Flexible decoupled transactional memory support. In *Proceedings of the 35th International Symposium on Computer Architecture*, pages 139–150, 2008.
76. Herb Sutter. The free lunch is over: A fundamental turn toward concurrency in software. 30(3), 2005.
77. Fuad Tabba, Cong Wang, James R. Goodman, and Mark Moir. NZTM: Nonblocking, zero-indirection transactional memory. In *Workshop on Transactional Computing (TRANSACT)*, 2007.
78. Fuad Tabba, Andrew W. Hay, and James R. Goodman. Transactional conflict decoupling and value prediction. In *Proceedings of the 25th International Conference of Supercomputing*, pages 33–42, 2011.

79. Rubén Titos-Gil, Anurag Negi, Manuel E. Acacio, Jose M. Garcia, and Per Stenstrom. Zebra: A data-centric, hybrid-policy hardware transactional memory design. In *Proceedings of the 25th International Conference of Supercomputing*, pages 53–62, 2011.
80. Ruben Titos-Gil, Anurag Negi, Manuel E. Acacio, Jose M. Garcia, and Per Stenstrom. Eager beats lazy: Improving store management in eager hardware transactional memory. *IEEE Transactions on Parallel and Distributed Systems*, 99(PrePrints):1, 2012.
81. Sasa Tomic, Adrian Cristal, Osman Unsal, and Mateo Valero. Hardware transactional memory with operating system support, HTMOS. In *Proceedings of the 13th European Conference on Parallel Processing (Euro-Par)*, pages 8–17, 2007.
82. Sasa Tomic, Cristian Perfumo, Chinmay Kulkarni, Adria Armejach, Adrián Cristal, Osman Unsal, Tim Harris, and Mateo Valero. EazyHTM: Eager-lazy hardware transactional memory. In *Proceedings of the 42nd International Symposium on Microarchitecture*, pages 145–155, 2009.
83. Hans Vandierendonck and Tom Mens. Averting the next software crisis. *IEEE Computer*, 44:88–90, 2011.
84. Haris Volos, Neelam Goyal, and Michael M. Swift. Pathological interaction of locks with transactional memory. In *TRANSACT '08: 3rd Workshop on Transactional Computing*, 2008.
85. M. M. Waliullah. Efficient partial roll-backing mechanism for transactional memory systems. *Transactions on high-performance embedded architectures and compilers*, 3:256–274, 2011.
86. M.M. Waliullah and Per Stenstrom. Reducing roll-back overhead in transactional memory systems by checkpointing conflicting accesses. In *Proceedings of the 22nd International Parallel and Distributed Processing Symposium*. 2008.
87. M. M. Waliullah and Per Stenstrom. Schemes for avoiding starvation in transactional memory systems. *Concurrency and Computation: Practice and Experience*, 21:859–873, 2009.
88. Steven C. Woo, Moriyoshi Ohara, Evan Torrie, Jaswinder Pal Singh, and Anoop Gupta. The SPLASH-2 programs: Characterization and methodological considerations. In *Proceedings of the 22nd International Symposium on Computer Architecture*, pages 24–36, 1995.
89. Luke Yen, Jayaram Bobba, Michael R. Marty, Kevin E. Moore, Haris Volos, Mark D. Hill, Michael M. Swift, and David A. Wood. LogTM-SE: Decoupling hardware transactional memory from caches. In *Proceedings of the 13th Symposium on High-Performance Computer Architecture*, pages 261–272, 2007.
90. Luke Yen, Stark C. Draper, and Mark D. Hill. Notary: Hardware techniques to enhance signatures. In *Proceedings of the 41st International Symposium on Microarchitecture*, pages 234–245, 2008.
91. Richard Yoo, Christopher Hughes, Konrad Lai, and Ravi Rajwar. Performance evaluation of intel transactional synchronization extensions for high performance computing. In *Proceedings of the International Conference for High Performance Computing, Networking, Storage, and Analysis (SC)*, 2013.

Part V
Modeling and Simulation

Data Center Modeling and Simulation Using OMNeT++

Asad W. Malik and Samee U. Khan

Advances in cloud computing have rendered Service Oriented Architecture (SOA) eminent. In addition, SOA offers support for cloud computing solutions. Due to master worker paradigms, SOA is extensively adopted for cluster, grid and Cloud environment. The term Cloud computing is relatively new, compared to others. Cloud computing is define as; it is a pool of easily available, shared computing resources, including servers, services, storage and networks. Cloud comprises of computing servers arranged in racks and are connected with multiple tiers of switches to provide redundancy; this arrangement of equipment is termed as data centers. A single Cloud may comprise of multiple data centers connected through high speed communication links. Data centers have gained great publicity in recent years; however, the concepts of data center simulation models, communication protocols and analysis of data center traffic flow, remain relatively been less explored. It is important to understand how these systems work. Given the complexity of these systems, models and simulations are the best way to gain an insight into the workings of such systems. In this chapter, we provide a step by step tutorial for building traditional three tier data center simulation model using OMNeT+ +. The chapter is organized as follows: Section I presents core modeling and simulation concepts. In section II different architectures of data centers are discussed in detail. A step by step guide to modeling data center architectures in OMNeT++ is presented in Section III. Finally, Section IV concludes this chapter.

A. W. Malik (✉)
Department of Computing, NUST—School of Electrical Engineering and Computer Science, Islamabad, Pakistan
e-mail: Asad.malik@seecs.edu.pk

S. U. Khan
Department of Electrical and Computer Engineering, North Dekota State University—NDSU, Fargo, USA
e-mail: Same.khan@ndsu.edu

S. U. Khan, A. Y. Zomaya (eds.), *Handbook on Data Centers*,
DOI 10.1007/978-1-4939-2092-1_28

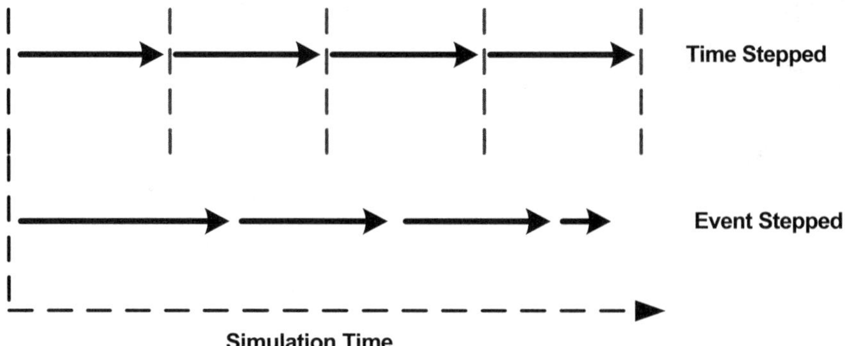

Fig. 1 Simulation model

1 Introduction to Modeling and Simulation (M&S) Methodology

Modeling and simulation of systems has aided in the understanding of system behavior since early 1990's. These concepts of modeling and simulations are being adopted in many fields, ranging from computing to architectural designs. Models are considered as abstractions of real entities; whereas simulation is used to analyze the behavior of these entities under a set of conditions, which may be otherwise difficult [1]. Many crucial decisions are based on models or simulations, especially in time critical applications. In addition, models prove to be cost effective and involve little or no risk when analyzed. Over the last few decades, modeling and simulation has become one of the most significant areas of research. Different techniques have been developed not only to study real time systems but also to improve system.

Modeling and Simulation can be further categorized as continuous or discrete event simulation. Within the simulation domain, continuous simulation is analyzing system response over a given time interval. These models are often used to compute numerical solutions of differential equations. Digital circuits are a common example of a continuous simulation models.

In discrete event simulation model, the system being simulated only changes state at discrete points in simulated time and are presented in chronological ordering of events [2]. Discrete Event Simulation (DES) manages time dependent events and is further categorized into Time and Event step simulation models; Time Step Simulation (TSS) is employed for events that require monitoring under fixed time intervals. The allocation of time interval depends on various factors, such as pending events in a process queue; whereas Event Step Simulation (EST) manages such events that are removed from the queue after having been processed according to the time stamp on each of these. Event step simulation eliminates the need for time step simulation. Figure 1 shows the differences between time and event step simulations. The event step simulation models are used in communication networks and flow networks [2, 3].

1.1 Parallel Discrete Event Simulation—PDES

A Parallel simulation or a Parallel Distributed Simulation (PDS) is an execution of a task over multiple processors or distributed machines. This reduces execution time substantially compared to sequential execution. However, the results obtained from parallel simulation must match against the results of sequential simulation. With their inherent ability to distribute simulation tasks, PDS overcome the limitations of traditional simulation techniques that employed a sequential process to execute an entire simulation with a single thread [4, 5]. Such technological advancements are leading towards more rapid adoption of parallel and distributed simulations and the emanated increased use of parallel simulation in engineering applications.

As discussed earlier, the simulators may be categorized as event based simulators as their time progression is based on pending events PDES events are generated at discrete time intervals depending on the nature of simulation. PDES is normally defined as a three step process: (i) receive events, (ii) process and perform operations on events (iii) update its local time and generate events for (other/own) processes or nodes. Each process must execute events in the order of the timestamps on them [5].

Rapid advancement in technology is increasing the necessity for simulators, since simulators are fast and cost effective tools to analyze systems, such as complex large scale distributed systems and cloud computing solutions. Cloud computing has become more renowned given its architecture and benefits for the end user [6]. A Cloud comprises of data centers, consisting of tens of thousands of computing servers and switches or routers connected via high speed networks with specialized cooling equipments. Multinational companies, such as Google, Facebook, Amazon and Microsoft, host their own distributed data centers [7]. These companies provide computational services and promote web based services. Clouds exclusively provide services such as infrastructures, platforms, and software [8]. User demand has lead the evolution of data center architecture. However, since efficient use of data centers in terms of computing power and heat energy dissipation, remains less explored; the maintenance costs of data centers remain high [9].

There has been a massive increase in the number of users switching to Cloud computing to host their data and applications [10, 11]. Due to this tremendous growth in the number of Cloud computing users, critical issues such as scalability, efficiency and security arise; researchers are developing different types of simulators to analyze and overcome these issues [7]. In this chapter, we demonstrate how to build a data center model using a discrete event simulator i.e. OMNeT++. OMNeT++ is an open source, discrete event simulator, developed by Andre Vergas [12]. The dexterity of OMNeT++ lies in its powerful graphical interface that makes the internal models completely visible to the end user. It uses packet transmission for communication between modules and nodes that is visible on the GUI. OMNeT++ simulation models are based on modules which can further be nested within other modules to form a complex network or system models.

OMNeT++ was developed in 2003, written in C++ and supported windows, Linux, UNIX and Mac OS X. OMNeT+ + allows the use of existing modules and frameworks that can easily be imported into other simulations, this further facilitates Cloud data center simulation. OMNeT++ provides features, such as plotting (plove), debugging through Valgrind utility [13]. Other network simulators commonly used for academic research are: NS-2 and NS-3. NS-2 was developed in 1996, written in C++ and mainly used for UNIX operating systems; whereas NS-3 was developed in 2008, written in C++, supported simulation models in C++ or Python languages. NS-3 was also targeted for UNIX based operating system.

In the next section, we briefly describe the traditional data center models. It is important to develop an understanding about these models before actual implementation.

2 Data Center Architectures

Over the last few decades, Cloud computing has become the most popular computing paradigm used all over the world. Clouds provide on demand computing access through networks. The basic objective of Cloud is to provide computing resources through its server pools. This approach is similar to master and worker example, where the master assigns scientific tasks to his workers. Furthermore, Cloud introduces a concept of virtualization that enables the sharing of resources among the users [14].

Cloud services are generally categorized as, Software as a Service (SaaS); Infrastructure as a Service (IaaS); and Platform as a Service (PaaS). For SaaS, application services are delivered over the network on demand basis. Microsoft, Google and Salesforce are some of the SaaS providers [15, 16]; whereas for IaaS, computation services and storage are provided over a network. For PaaS, software development is provided in the form of Application Programming Interfaces (API) [14].

The architecture of Cloud computing has evolved remarkably with the increase in its usage. Initially a two tier architecture were used to connect computing servers with other external networks. Figure 2 shows the two tier architecture connected with switches to provide mesh network connectivity. In data centers, computing servers are arranged in racks and connected with switches; these switches are further connected with layer L3 switches to provide complete redundant connectivity among computing servers and external networks. Two tier architecture designs only support a limited number of computing servers. Typically a two tier data center can hold up to 5500 nodes, depending on the types of switches being used [17].

The three tier architecture is employed to address scalability issues by supporting an additional layer of switches and routers. The three layers of switches are referred to as access, aggregate and core, shown in Fig. 3. BCube [18] and DCell [19] are two other advanced Cloud computing architectures. Dcell is designed for efficient interconnection and handling of exponential increase in the number of servers. In Dcell, computing servers are also used for routing purposes, thus providing fault tolerance mechanism and eliminates rack level failure. A four degree Dcell can support millions of servers without using any expensive core switches or routers.

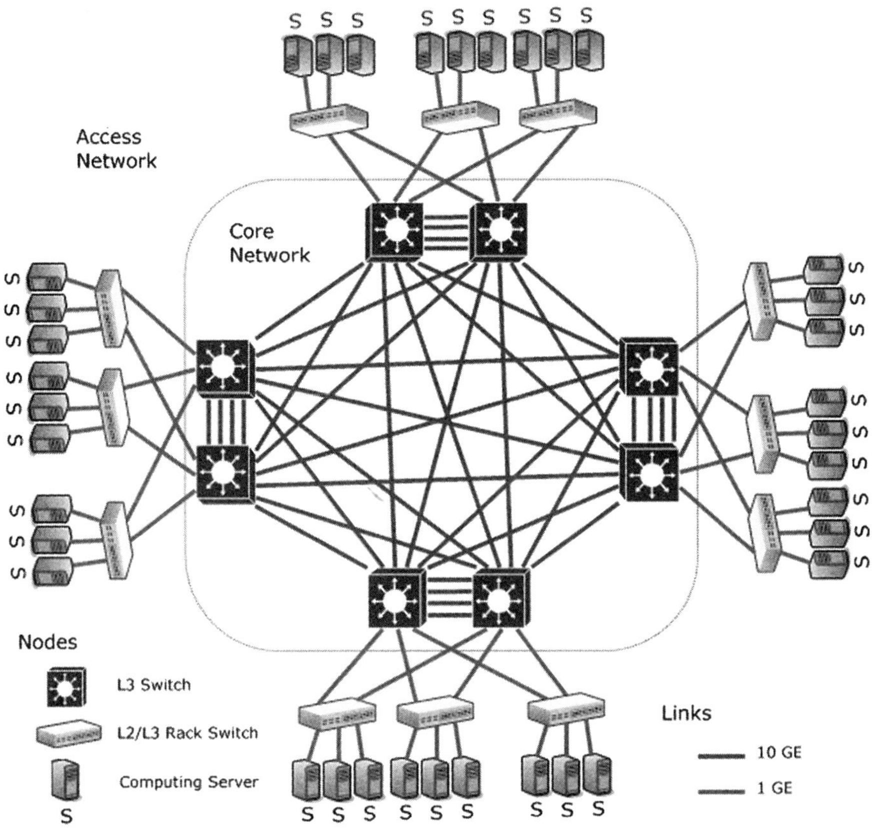

Fig. 2 Two tier architecture

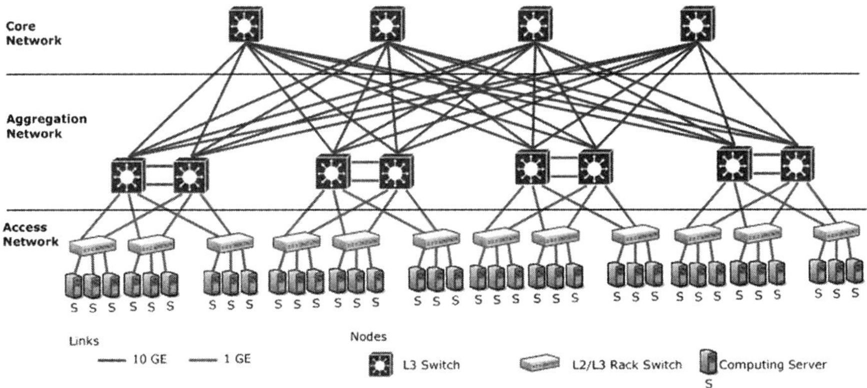

Fig. 3 Three tier architecture

Similarly, Bcube is based on a server centric architecture, where servers perform computational tasks and act as a relay to other servers. In this chapter, we focus on building traditional three tier data center simulation model.

In the next section, we discuss a step by step formulation of data center architecture model using OMNeT++.

3 Data Center Modeling Using OMNeT++

OMNeT++[1] is a discrete event simulator based on event step technique. OMNeT++ is chosen for this tutorial given its popularity among researchers, since it is open source, provides a rich graphical user interface and its cross-platform support. In OMNeT++ every simulation consists of modules and networks. These modules are interconnected using various communication links. Developers can model communication channels by varying its data rate, delay and bit error rate. OMNeT++ provides an extensive framework developed in C++ to build simulation models.

In OMNeT++, each simulation model consists of either simple or compound modules. Each simple module is supported by its own C++ class; whereas compound modules are collections of simple modules grouped together to achieve the required goals. We must first develop a simple two node simulation and then extend this simulation model to a more complex design based on compound modules that simulate three tier data center architectures.

3.1 Simple Two Node Simulation

We start with building a simple two node simulation, where each node sends a message to the other.

Open OMNeT++ editor, and select File → New → OMNeT++ Project: type the project name: "**demo**". Click next and choose an empty project then press the finish button.

Step-1: Add a simple module definition. Right click on "demo" (*OMNeT++ project*) and add a file with an extension ".ned" and name it "**node.ned**". Write the following lines of code (given in Table 1) in **node.ned** file.

Step-2: Create another ".ned" file that contains the definition of a network. The network consists of one or more communication nodes. The communication links between nodes must be defined within this file. Different properties for each communication link can also be specified. We name this file "NetSim.ned" and add the following lines of code (see Table 2).

In the code above, we have created a network named **SIMULATION**, consisting of two nodes (i.e. NodeA and NodeB) inherited from the simple module i.e. node.

[1] www.omnetpp.org.

Table 1 Simple module definition

Code
simple node
{
gates:
input in;
output out;
}

Table 2 Network definition

Code
Network SIMULATION
{
submodules:
NodeA: node;
NodeB: node;
connections:
NodeA.out → NodeB.in;
NodeA.in ← NodeB.out;
}

The communication link is established inside connection label, the output port of nodeA is connected with input port of node B, similarly the output port of node B is connected with input port of node A (arrow heads show the communication path between the nodes).

Step-3: Add a programming logic for each module. As discussed earlier, a network consists of simple modules which are connected via communication channels. A programming logic, or a behavior description of a node, must be implemented for each of the modules. This logic is implemented in C++ classes.

Now add a C++ class to model the desired functionality of the node. Right click on the OMNeT++ framework, select New → OMNeT++ class and add name as "**node**". This generates a header and a source file under the "src" directory. This class must be inherited from cSimpleModule and must provide definitions for the two virtual functions, *initialize*() and *handleMessage*(). These functions are invoked by OMNeT++ simulation kernel. Function *initialize*() is invoked only once and is used to set the initial values while *handleMessage*() is invoked every time the message is received at a node's gate or port. Prototypes of these functions are given in Table 3.

The node.cpp file contains the definition of the functions defined in a header file i.e. node.h. Write the following code (given in Table 4) in node.cpp file.

The *Define_Module*() macro function registers the "**node**" class by taking the class name as an argument (Table 4: line 2). OMNeT++ provides a built-in message

Table 3 Basic functions prototypes in node header file

Line No.	Code
1	Class node: public cSimpleModule
2	{
3	Protected:
4	virtual void initialize();
5	virtual void handleMessage(cMessage *msg);
6	}

Table 4 Basic function definition in node source file

Line No.	Code
1	#include "node.h"
2	Define_Module(node);
3	void node::initialize()
4	{
5	cMessage *msg = new cMessage("Message");
6	send(msg, "out");
7	}
8	void node::handleMessage(cMessage *msg)
9	{
10	send(msg, "out");
11	}

Table 5 Network configuration

Line No.	Code
1	[General]
2	network = NetSim

class for communication. The messages can be exchanged between the nodes by creating a pointer to this class (Table 4: line 5). The send() function is called in order to exchange packets or messages between the nodes. This function takes two arguments; first, the message to be sent, and second, the output gate name (Table 4: line 10.). When a message is received at a node, function *handleMessage*() is called.

Step-4: The next step is to add the "**.ini**" file. This file is used to tell the simulation kernel which networks it should simulate. Go to → New → Initialization File (ini) and name the file "**omnetpp**.ini". Write the code given in Table 5 in this file:

"***omnetpp.ini***" file is used for configuration and to instruct the kernel to load the network simulation model i.e. "NetSim" (Table 5: line 2).

Step-5: To build a simulation, right click on project → Build Project. On successful compilation, run the simulation by right clicking on project → RunAs → OMNeT++ Simulation. Figure 4 shows the simulation GUI containing the two

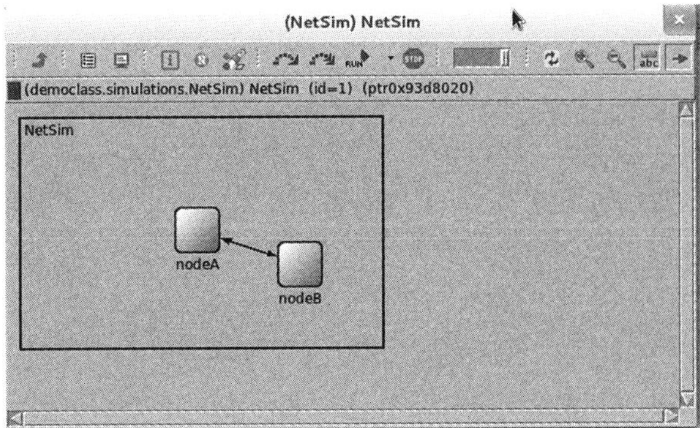

Fig. 4 Simulation GUI

nodes, nodeA and nodeB, connected through a bidirectional link. You can run this simulation model by pressing the RUN button on the toolbar.

We have now successfully created a two node simulation, where each node generates and exchanges messages with the other.

3.2 Advance Level Simulation

In realistic network simulation, each node is assigned an IP address and connected with routers or a switch. To model this scenario, we need to alter our demo project. OMNeT++ provides built-in modules that simulate networks with dynamic IP assignment. INET[2] is an open source module that can be used for this purpose. Download the INET project and import it into the OMNeT++ framework.

Step-1: A reference to INET project must be added to our "**demo**" simulation project to reuse its built-in functionality. Right click on the "**demo**" project and select Project References. This displays the list of open projects in OMNeT++ framework. Click on the INET project and press the **OK** button.

Step-2: The required changes in the "**NetSim.ned**" file are shown in Table 6.

INET contains protocol implementations that can be used in other simulation models. In Sect. 3.1, we connected NodeA and NodeB directly through defined gates or ports. Since we have now assigned IP addresses to the nodes, a router must be added to handle routing mechanism In addition, there may be inevitable communication delays between nodes that can be addressed through link properties available in INET project.

Step 1: Add an import statement at the top of ".ned" file to use built-in modules of INET (Table 6: lines 1–4).

Step 2: There are existing modules that implement the required network layers on the host and the router. Therefore nodes, nodeA and nodeB, are inherited from

[2] INET Module: http://inet.omnetpp.org.

Table 6 Network definition

Line No.	Code
1	import inet.nodes.inet.StandardHost;
2	import inet.nodes.inet.Router;
3	import inet.networklayer.autorouting.ipv4.FlatNetworkonfigurator;
4	import inet.nodes.ethernet.Eth10M
5	network NetSim
6	{
7	submodules:
8	NodeA: StandardHost
9	NodeB: StandardHost;
10	RouterC: Router;
11	Configurator: FlatNetworkConfigurator;
12	connections:
13	NodeA.ethg++ <−−>Eth10M <−−>RouterC.ethg++;
14	NodeB.ethg++ <−−>Eth10M <−−>RouterC.ethg++;
15	}

StandardHost (Table 6: lines 8–9). The StandardHost module is responsible for acquiring an IP address, resolving ARP (address resolution protocol) requests, and handles multiple applications at the top layer.

Step 3: Add a router to connect multiple nodes. To do this, add a new sub-module, RouterC, which is inherited from Router, an existing module in INET framework (Table 6: line 10). A separate module, FlatNetworkConfigurator, is available in the INET framework, which is responsible for assigning IP addresses to hosts and routers through its own sub-module Configurator (Table 6: line 11). This is the simplest IP assignment module implemented in INET.

Step 4: The links are defined under the connections tag with the "ethg" suffix, NodeA.ethg and NodeB.ethg; since the nodes are now inherited from StandardHost (Table 6: lines 13–14). In StandardHost and Router classes, the ports or gates are defined as bidirectional vectors. The import of inet.nodes.ethernet.Eth10M allows the following physical channel properties to be included:

- datarate $= 1e^{07}$
- delay $= 5e^{-08}$
- ber $= 0$

In addition to other modules similar to Eth10M, user defined modules may also be used (channel or link properties are assigned to each link, Table 6, lines 13–14)

Step 5: To build this simulation, the "**omnetpp.ini**" file must be altered. The changes are shown in Table 7.

Table 7 Advance network configuration

Code
[General]
network = NetSim
**.nodeA.numUdpApps = 1
**.nodeA.udpApp[*].typename = UDPBasicApp
**.nodeB.numUdpApps = 1
**.nodeA.udpApp[*].typename = UDPSink

The UDP application, running on the nodes generates UDP packets for the destination node. An explanation for the code given in Table 7:

**.nodeA.numUdpApps = 1
Only one UDP application is executed at the top of the application layer. This line of code declares the number of UDP applications that run at nodeA. Similarly we can use TCP applications
**.nodeA.udpApp[*].typename = UDPBasicApp
The UDPBasicApp is the class that runs the UDP application and contains the definitions of virtual functions declared in cSimpleModule. This class generates messages at regular time intervals to randomly selected destinations. The INET project contains the implementation of UDP (i.e. UDPBasicApp) and TCP (i.e. TCPBasicApp) classes
**.nodeB.numUdpApps = 1
This line of code declares the number of UDP applications that run at nodeB
**.nodeB.udpApp[*].typename = UDPSink
The UDPSink class contains the definitions of virtual functions declared in cSimpleModule. This class only receives messages from other nodes

Step 6: Compile the simulation, right click on demo project and select → Build Project. On successful compilation, run the simulation. Right click on demo project and select → RunAs → OMNeT++ Simulation. Certain parameters are required by UDPBasicApp and UDPSink class at this stage, such as *local port, destination port, send interval time* and *message length*. When prompted by OMNeT++, provide the values given in Table 8.

Figure 5 shows the complete network design using existing modules. IP addresses for each module are clearly shown on the GUI. These IP's are generated using the existing module "*FlatNetworkConfigurator*", as discussed earlier. Figure 6 shows

Table 8 Required parameters of UDP application

Parameters	Values
Local port	100
Destination port	100
Send interval (seconds)	10s
Message length (bytes)	10B

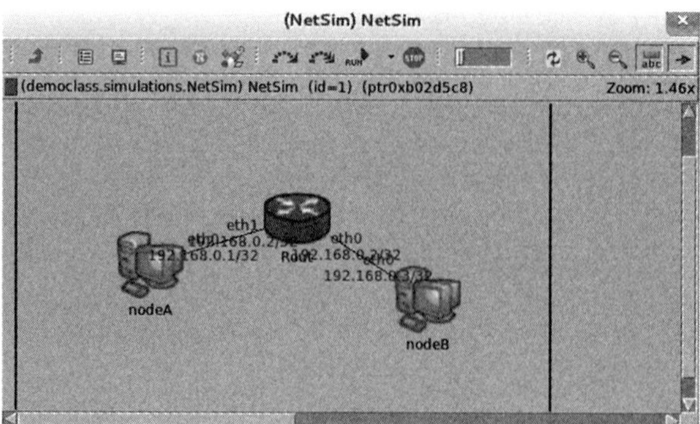

Fig. 5 Two node real network simulation model

the TCP/IP stack implementation within "*StandardHost*". These modules can be overwritten by the user.

3.3 *Data Center Simulation Model*

In the previous sections, we have discussed all the necessary steps to be followed to build simple and advance level simulations. In this section, we focus on building data center simulations using built-in modules available inside INET project. We require the INET framework for IP assignment, a TCP/IP stack at each node, and a router.

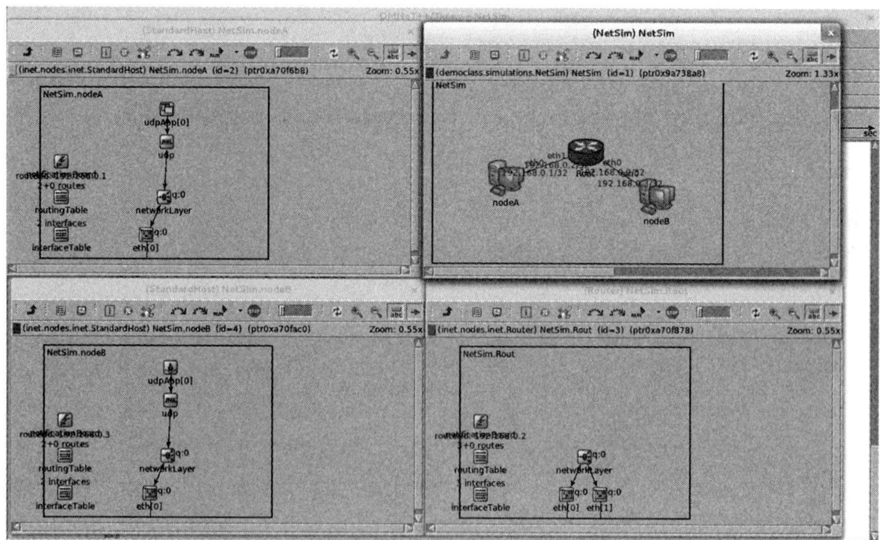

Fig. 6 Internal view of nodes and router

Table 9 Three tier network module definition

Line No.	Code
1	Module Rack
2	{
3	Parameter:
4	int N @prompt("Nodes per rack");
5	gates:
6	inout iogate[];
7	Submodules:
8	ComputingServer[N]: StandardHost;
9	AccessRouter: Router;
10	Connections:
11	for i = 0.. N-1{
12	AccessRouter.ethg++ <--> Eth10M <--> ComputingServer[i].ethg++; }
13	AccessRouter.ethg ++ <--> iogate++;
14	}

Step 1: The first step is to model the three tier simulation network in the ".ned" file, which is more complicated than the previously discussed examples. Simple modules are supported by a single class while compound modules are collections of several simple modules, as discussed earlier. Compound modules allow communication with other simple or compound modules through communication gates.

Create a new project and name it "**ThreeTierDC**". Add a new ".ned" file and name it "**NetworkDefination**.ned". As discussed in Sect. 2, three tier data centers consist of racks that hold servers and routers i.e. access, aggregate and core routers. In this simulation model, we build a compound module of racks that connects with aggregate and core routers through access router. Add the following lines of code (given in Table 9) in **NetworkDefination**.ned file:

Module on line 1 indicates that Rack is a compound module. The number of servers can vary on the racks and this number can either be fixed or provided by the user at run time.

Line 4, prompts the user to enter an integer value for variable '*N*'. The code on (Table 9) line 8, uses this '*N*' to declare the number of servers on each of the racks. The compound module defines bidirectional, vector in-out gates, shown on Table 9, line 6 of the code. The racks accommodate the servers and access routers (Table 9: lines 8–9).

The computation servers must be connected with access routers which in turn must be connected with compound module gates. The connection between the servers and access routers is established under the *connection* tab, with a loop (Table 9: lines 11).

Table 10 Three tier network definition

Line No.	Code
1	network ThreeTierDatacenter
2	{
3	Parameter:
4	int N= default(4);
5	int AGR= default(4);
6	int CR= default(2);
7	Submodules:
8	AGRouters[AGR]: Router;
9	CRouter[CR]: Router;
10	Racks[N]: Rack;
11	Configurator: FlatNetworkConfigurator;
12	Connections allowunconnected:
13	for i = 0..CR-1, for j= 0.. AGR-1{
14	CRouter[i].ethg++ <---> Eth100M <---> AGRouter[j].ethg++; }
15	for i = 0..1, for j = 0..1{
16	AGRouter[i].ethg++ <---> Eth100M <---> AGRouter[j].ethg++; }
17	for i = 2..3, for j = 2..3{
18	AGRouter[i].ethg++ <---> Eth100M <---> Racks[j].iogate++; }
19	}

The code on (Table 9) line 13, connects the access router to the compound module gates for external communication.

Step 2: Create a three tier data center model, with racks, and routers both aggregate and core. Add the following lines of code (see Table 10) to the "NetworkDefination.ned" file:

At this stage a network that connects all the modules together has been defined. For the sake of simplicity, the default number of racks and aggregate routers is hard-coded at four and two for the core switches (Table 10: lines 4–6).

The required modules, i.e. routers, racks and network configurator, are defined under the *Submodules* section (Table 10: lines 7–11).

In Table 10, line 12 of code, *Connections allowunconnected*, instructs the simulation kernel to allow this simulation model with unconnected gates to avoid the OMNeT++ kernel throwing an exception in case a gate was not connected.

Connections are established between the racks and routers, with nested loops (Table 10: lines 15–18). Code on line 18, connects the aggregate routers to the compound module rack through communication gates.

Table 11 Three tier network configuration settings

Line No.	Code
1	[General]
2	network = ThreeTierDatacenter
3	**.Racks[*].ComputingServer[*].numUdpApps = 1
4	**.Racks[*].ComputingServer[*].typename = UDPBasicApp
5	**.udpApp[*].localPort = 100
6	**.udpApp[*].destPort = 100
7	**.udpApp[*].messageLength = 1024B
8	**.udpApp[*].sendInterval = 1s

Step 3: The final step is to amend the "**omnetpp**.ini" file and specify the number of UDP applications and C++ source files (see Table 11).

In the previous example, "**omnetpp**.ini" file held the simulation network model. Modifying this file, lines 3–4, declare the number of UDP applications to run on each of the servers. In previous examples the user entered the values for local and destination ports, message length and send intervals. However, the default values for these variables are hard-coded in this example (Table 11: lines 5–8).

This model allows for the servers to generate tasks for a randomly selected destination node. This random node selection process is implemented in the UDPBasicApp class available in INET framework. Figure 7 and 8 shows the three tier simulation model and rack view within the OMNeT++ framework, the green colored

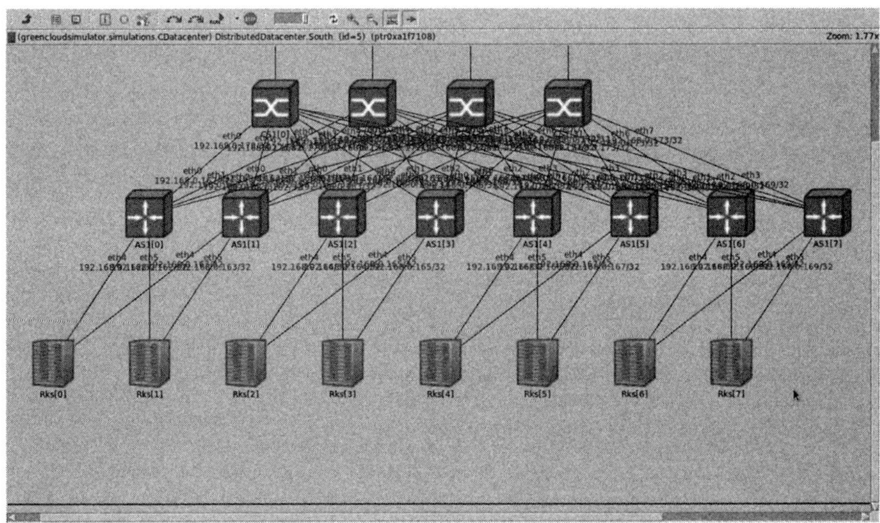

Fig. 7 Three tier data center simulation model

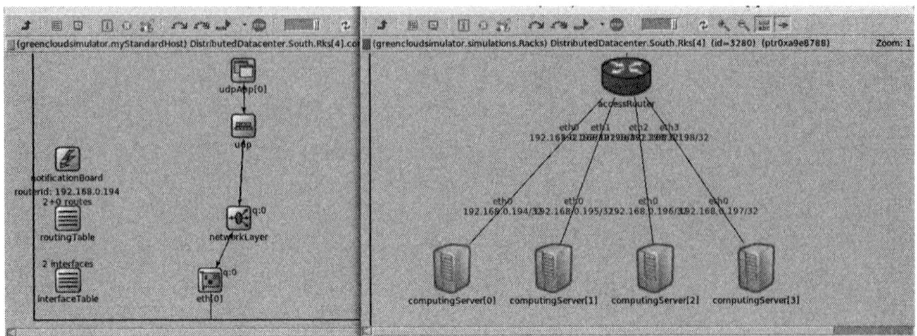

Fig. 8 Internal view of rack and computing server

overlapping texts depicts dynamically assigned IP's. The user can change the functionality of UDPBasicApp by defining its own class. Keeping parameters within the "omnetpp.ini" file allows the code to be used without recompilation.

4 Wrap Up

Cloud computing is a fascinating area of research. The importance of cloud computing has been highlighted with a sharp increase in its use. Technological advances and current user requirements are rendering cloud computing more desirable and inevitable to adopt. Along with these benefits, Cloud computing introduces intriguing research areas, including scalability, energy optimization and task scheduling.

It is important to understand how complex systems work. Given the complexity of these systems, models and simulations are the best way to gain an insight into the workings of such systems. Agile technology expansion is increasing the necessity for simulators, since simulators are fast and cost effective tools to analyze systems. In this chapter, we focused on modeling of data center architecture to facilitate researchers to perform in depth analysis on core issues instead of building basic modules. The aim of this chapter has been to provide an insight of building data center simulation models using OMNeT++ and served as a platform for researchers to build advance level data center architectures models using available components.

Acknowledgement The authors would like to thank Miss Maham Fatima Nasir (SCME-NUST) for her valuable support during write-up process.

References

1. A. Buss, and L. Jackson, Distributed simulation modeling: a comparison of HLA, CORBA, and RMI. Winter Simulation Conference (WSC'98), Washington, DC, USA, 1998, pp. 819–825.
2. R. M. Fujimoto, Parallel discrete event simulation. Communications of the ACM archive vol. 33 (1990) pp. 30–52.

3. A. Park, and R. M. Fujimoto, Efficient Master/Worker Parallel Discrete Event Simulation. ACM/IEEE/SCS 23rd Workshop on Principles of Advanced and Distributed Simulation, 2009. PADS '09, pp. 145–152.
4. R.M. Fujimoto, Parallel and Distributed Simulation Systems, Wiley interscience Publication, New York, 2000.
5. R.M. Fujimoto, A.W. Malik, and A.J. Park, Parallel and Distributed Simulation in the Cloud Magazine of the society for Modeling and Simulation (M&S), July 2010, pp. 1–10.
6. M.B. Mollah, K.R. Islam, and S.S. Islam, Next generation of computing through cloud computing technology. 25th IEEE Canadian Conference on Electrical & Computer Engineering (CCECE), 2012, pp. 1–6.
7. A. Vahdat, M. Al-Fares, N. Farrington, R.N. Mysore, G. Porter, and S. Radhakrishnan, Scale-Out Networking in the Data Center. Micro, IEEE vol. 30 (2010) pp. 29–41.
8. S. Akioka, and Y. Muraoka, HPC Benchmarks on Amazon EC2. 24th IEEE International Conference on Advanced Information Networking and Applications Workshops (WAINA), 2010, pp. 1029–1034.
9. B. Aksanli, J. Venkatesh, and T.S. Rosing, Using Datacenter Simulation to Evaluate Green Energy Integration. The Computer Journal vol. 45, pp. 56–64.
10. T. Ercan, Effective use of cloud computing in educational institutions. The Journal of Procedia—Social and Behavioral Sciences vol. 2 (2010) pp. 938–942.
11. P. Gupta, A. Seetharaman, and J.R. Raj, The usage and adoption of cloud computing by small and medium businesses. International Journal of Information Management vol. 33 (2013) pp. 861–874.
12. A. Varga, Using the OMNeT + + Discrete Event Simulation System in Education. IEEE Transactions on Education vol. 42 Issue 4. (1999) p. 372.
13. P. Vilhan, and J. Gajdos, ADEUS: Tool for Rapid Acceleration of Network Simulation in OMNeT++. 14th International Conference on Computer Modelling and Simulation (UKSim), 2012, pp. 591–595.
14. K. Bakshi, Considerations for cloud data centers: Framework, architecture and adoption. Aerospace Conference, 2011 IEEE, pp. 1–7.
15. S. Ming-Chien, Let's Walk Out of the Cloud. Fifth IEEE International Symposium on Service Oriented System Engineering (SOSE), 2010, pp. 5–5.
16. Y. Bo-Wen, T. Wen-Chih, C. An-Pin, and S. Ramandeep, Cloud Computing Architecture for Social Computing—A Comparison Study of Facebook and Google. International Conference on Advances in Social Networks Analysis and Mining (ASONAM), 2011, pp. 741–745.
17. Dzmitry Kliazovich, Pascal Bouvry, and S.U. Khan, GreenCloud: a packet-level simulator of energy-aware cloud computing data centers. The Journal of Supercomputing vol. 62 (2012) pp. 1263–1283.
18. C. Guo, G. Lu, D. Li, H. Wu, X. Zhang, Y. Shi, C. Tian, Y. Zhang, and S. Lu, BCube: a high performance, server-centric network architecture for modular data centers. SIGCOMM Comput. Commun. Rev. vol. 39 (2009) pp. 63–74.
19. C. Guo, H. Wu, K. Tan, L. Shi, Y. Zhang, and S. Lu, Dcell: a scalable and fault-tolerant network structure for data centers. Proceedings of ACM SIGCOMM conference on Data communication, ACM, Seattle, WA, USA, 2008.

Power-Thermal Modeling and Control of Energy-Efficient Servers and Datacenters

Jungsoo Kim, Mohamed M. Sabry, Martino Ruggiero and David Atienza

1 Introduction

This continuous growth in demand for computing has resulted in larger collections of servers machines, referred to as clusters or server farms, being hosted in denser datacenters thus having a higher computational and storage capability per occupied unit volume. While projections indicate a continued scaling of server density and manufacturing cost for another decade, the semiconductor manufacturing industry has already renounced following *Dennard scaling*[1] and almost reached the physical limits of voltage scaling in Complementary Metal-Oxide-Semiconductor (CMOS) technologies, which results in an energy-scalability wall that makes transistor power

J. Kim was also affiliated with ESL-EPFL during the period this research was developed.

[1] The scaling theory he and his colleagues formulated in 1974 postulated that MOSFETs continue to function as voltage-controlled switches while all key figures of merit (such as layout density, operating speed, and energy efficiency improve provided geometric dimensions, voltages, and doping concentrations) are consistently scaled to maintain the same electric field. This property underlies the achievement of Moore's Law and the evolution of microelectronics over the last few decades.

J. Kim (✉)
DMC Research Center, Samsung Electronics, Suwon, Republic of Korea
e-mail: jungsoo9.kim@samsung.com

M. M. Sabry · M. Ruggiero · D. Atienza
Embedded Systems Laboratory, EPFL, Lausanne, Switzerland
e-mail: mohamed.sabry@epfl.ch

M. Ruggiero
e-mail: martino.ruggiero@epfl.ch

D. Atienza
e-mail: david.atienza@epfl.ch

S. U. Khan, A. Y. Zomaya (eds.), *Handbook on Data Centers*,
DOI 10.1007/978-1-4939-2092-1_29

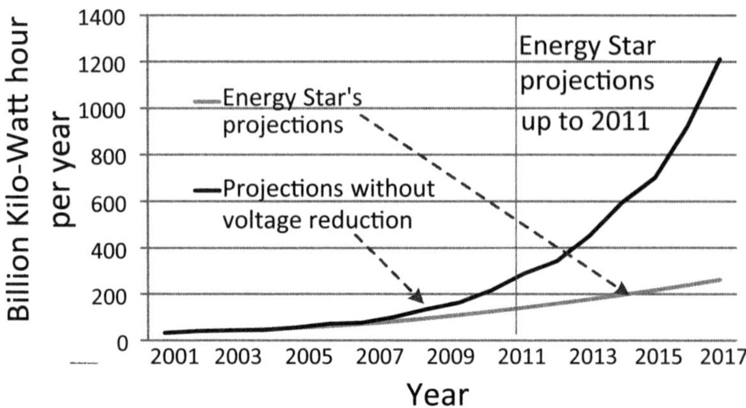

Fig. 1 Datacenters current energy use and projection [2]

consumption increase with further increases in density. At a large-scale, this "economic meltdown trend of Moore's law" for servers and datacenters [1], translates in a dramatic increase in computation and cooling electricity costs.

Energy-efficiency constraints have therefore become the dominant limiting factor for datacenters because their growing size and electrical power demands cannot be met with state-of-the-art design practices and their electricity bill is skyrocketing, as Fig. 1 shows. This figure depicts the *Energy Star*s [2] electricity usage measured and projected up to 2011. If we extrapolate these values linearly up to 2017, as voltages stop scaling down according to the current *International Technology Roadmap for Semiconductors* (ITRS) projections, the electricity use would exponentially increase. Moreover, the expected increase in energy prices would only exacerbate the cost of using datacenters. Thus, datacenter operation will require more money per year on energy costs than on IT equipment replacement. In 2007, datacenters in Western Europe consumed an estimated total of 56 terawatt-hours (TWh) of power per year. The European Union (EU) estimates that this figure is likely to reach 124 TWh by 2020 [2].

Power and thermal monitoring and control play a key role to reduce the power consumption of datacenters while maintaining the performance requirements and the maximum temperature constraints by manipulating multiple control knobs in the systems. As monitoring and control solutions are developed by being tightly coupled with hardware architecture and workload characteristics running on datacenters, we first revisit the datacenter structures (Sect.1.1) and the workload characteristics running on current datacenters (Sect. 1.2). Then, we present an energy efficiency figure of state-of-the-art datacenters (Sect. 1.3), which motivates us to develop effective power and thermal monitoring and control solutions by manipulating multiple control knobs to achieve further global/holistic energy savings in datacenters.

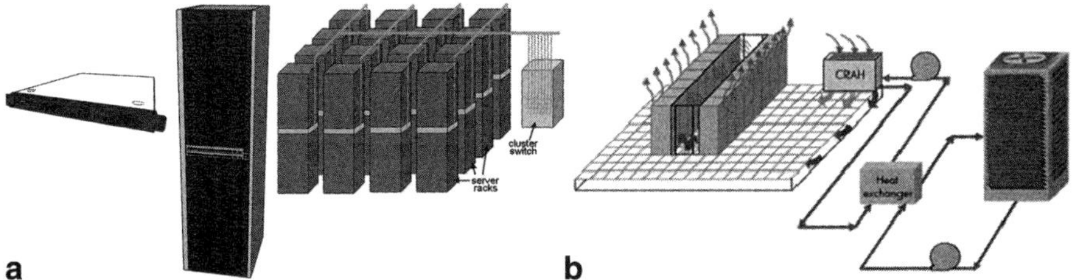

Fig. 2 Organization of datacenters: computing and cooling systems (**a**) and server organization (**b**) [3]

1.1 Overall Datacenter Architecture

A datacenter can largely be decomposed into three parts: (1) IT, i.e., aggregation of servers, (2) cooling, and (3) power distribution units. Servers are the key constituent of datacenters and produce a significant amount of heat as they provide the capability of data manipulation and processing. In a server room, there is a large number of servers to sustain performance requirements. Figure 2a shows an example of typical server organization in a server room with a typical 1 U[2]. Server are typically placed in 42 U racks such that the servers are interconnected with local rack Ethernet switch, and then, connected to cluster-level Ethernet switches, which can potentially span more than ten thousand individual servers [3].

Datacenter cooling systems are deployed to remove heat generated by the servers along with additional amount of heat inside a server room, which needs to be removed as well. Power is delivered to servers through power distribution units (PDUs) and stored in un-interruptible power supply (UPS) systems to cope with power blackout. In this chapter, we focus on IT and cooling parts of datacenters. As shown in Fig. 2b, in a typical datacenter, a cooling system consists of computer room air conditioning/handler (CRAC/CRAH) in a server room and heat exchanger (namely, chiller) and cooling tower outside the server room. CRAC/CRAH provides cold air, such that the air condition of server rooms maintains safe operating temperature and humidity through the exchange of hot air exhausted by servers in the room with cold air (or water) provided from a chiller. According to the *American Society of Heating, Refrigerating and Air-Conditioning* (ASHRAE) 2009 recommendation, it is recommended to maintain the server room air condition as follows:

- Temperature: 64.4–80.6 °F
- Humidity: 41.9 °F at dew point (DP) to 60 % RH and 59 °F DP.

However, these values are quite conservative as they are determined by assuming that servers in a server room are fully utilized, which rarely happens as will be explained

[2] A rack unit, **U** or **RU**, is a unit of measure to describe the height of rack-mount servers placed in 19-in. or a 23-in. rack, where 1U corresponds to 1.75 in. (44.45 mm) high.

in Sect. 1.2. Due to the over-provisioning of cooling capability to server rooms, huge amount of power are now wasted in datacenters, which motivates us to develop an efficient system control solution that adaptively adjusts cooling configurations along with existing power and thermal management solutions developed for servers to achieve further energy savings. The effective control solution is only obtained through accurate-yet-efficient monitoring of power consumption and temperature of multiple points of datacenters, which urges to develop an efficient monitoring system for datacenters.

1.2 Datacenter Workload Characteristics

Many types of applications are running on datacenters, ranging from high-performance computing (HPC) to large-scale services, e.g., web search, streaming service, etc. Recently, due to the big advancements on cloud service providers (e.g., Amazon, Microsoft, Google, etc.), it becomes easier to deploy large-scale services, which leads to the drastic increase on servers hosting large-scale applications. The common characteristics of the large-scale services are that they are unprecedentedly parallel as it uses big chunk of data by splitting into small chunk. Figure 3 illustrates the overall operation which manipulates big chunk of dataset. In [4], Ferdman et al., examined applications running on today's clouds and presented top six most commonly found applications as follows:

- *Data serving*: serving as the backing store for large-scale web applications, e.g., Facebook inbox, Google Earth, etc.
- *MapReduce*: large-scale data analysis by first performing filtering and transformation of the data (namely, *map* procedure) and then aggregate the results (namely, *reduce* procedure)
- *Media streaming*: streaming services by packetizing and transmitting media files ranging from megabytes to gigabytes
- *SAT solver*: large-scale computations for solving complex algorithms, e.g., symbolic execution
- *Web frontend*: web services which schedule independent client requests across a large number of stateless web servers
- *Web search*: web search engines such as those powering Google and Microsoft Bing, which indexes terabytes of data obtained from online sources.

Up to now, most of the control solutions have been developed by targeting HPC workload characteristics. However, the workload characteristics of such large-scale applications are quite different from traditional HPC applications in both macroscopic and microscopic scales [4], which mandates us to develop the control solutions for the large-scale applications.

In a macroscopic scale, the application, first, is user-interactive, thereby, the amount of required computing capacity is highly variable and fast-changing [6] due to the dependence with external factors, i.e., number of clients/queries, etc. The

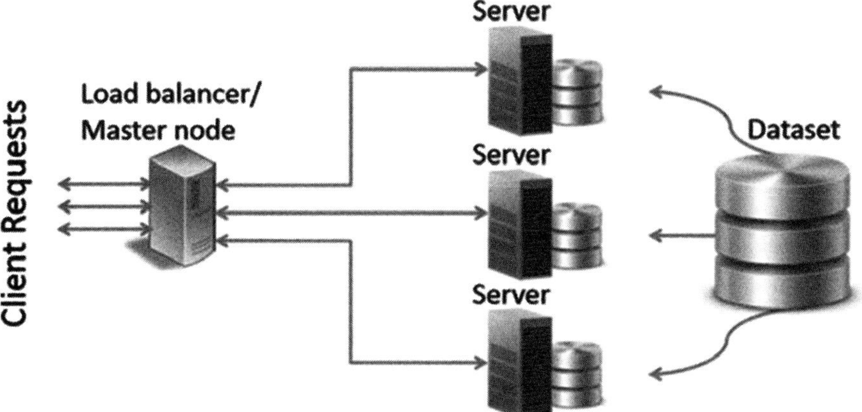

Fig. 3 An example of scale-out applications [5]

characteristics of the workload traffic are well analyzed in [7]. In the coarse-grained time interval (few tens of minutes to hours), the characteristics of users' requests are distinctly different over time while the global pattern has a strong correlation with adjacent time periods as well as the same period in different days. On the other hand, in the fine-grained time interval (less than few seconds), the characteristics of user requests depend on burstiness of traffic and arrival patterns and we can model the characteristics of users' request at the microscopic scale with (1) ON/OFF periods and (2) inter-arrival time between two consecutive requests during ON period. ON period is defined as the longest continual period during which all the request inter-arrival times are smaller than predefined value. Accordingly, OFF period is defined as a period between two on periods. As presented in [7], ON/OFF period and inter-arrival time are time-varying and uncertain while each of them forms lognormal distribution.

Second, the responsiveness (or latency) should come at the first criteria to be satisfied as the level of user satisfaction leads to the success of the business [10]. Third, the amount of required resources is usually far beyond the level that single server can sustain; thereby, massively parallel nodes are cooperatively working by forming a cluster architecture [8]. For instance, in a web search application, a big chunk of search index is divided into multiple smaller datasets, and then, allocated into multiple VMs (or servers) each of which is called a *index searching node (ISN)*. Once a query is arrived, each ISN independently searches matched data with the allocated dataset and a master node gathers the search results from multiple ISNs, then sends the results to clients. Due to the deployment of multiple nodes for a single application, such workload is called *scale-out* applications [4].

Microscopic-scale characteristics of the application are well studied in [4]. The following summarizes the four distinctive micro-architectural workload characteristics in the applications:

- High instruction cache miss rates
- Low instruction- and memory-level parallelism
- Large memory footprint far exceeding the capacity of on-chip caches
- Low on-chip and off-chip bandwidth requirements.

Due to the lack of the control solutions accounting for the distinctive workload characteristics of large-scale cloud application, in this chapter, we will present a power management solution optimized for the workload characteristics of the large-scale cloud applications.

1.3 Energy Efficiency of Datacenters

Due to the conservative cooling provision and lack of the consideration on workload characteristics, vast amount of energy is wasted in todays' datacenter. *Power usage efficiency* (PUE) is the most widely used metric to quantify the power efficiency of datacenters, which is defined as follows:

$$PUE = \frac{\text{Total power consumed by a datacenter}}{\text{Power consumed by servers}} \tag{1}$$

Thus, the lower, the better and it can ideally be reached to 1.0 According to *US Environmental Protection Agency* (EPA) report [2], the PUE of average datacenters around world amounts to 1.9, which means that for every watt of power consumed in the computing equipment, an additional 0.9 W of power is needed for cooling and power delivery. Figure 4a shows the breakdown of energy usage of typical datacenters (The PUE value amounts to $1/0.45 = 2.22$) when assuming $10 \sim 30\,\%$ IT load scenario [3]. Cooling system, comprised with chiller and CRAC/CRAH, consumes around 30 % of energy consumption while the power system spends additional 23 % of energy caused by uninterruptible power supply (UPS), power distribution unit (PDU) and AC-DC conversion losses. Other facility elements, e.g., humidifier, lighting, transformers, contribute around 2 % of total energy consumption. Such inefficiency corresponds to waste of money in the business sense. Figure 4b shows the monthly costs breakdown in a state-of-the-art datacenter assuming a 3-year server amortization and a 15-year infrastructure amortization [9]. This figure illustrates that, in less than three years, the accumulated cooling costs are higher than the actual server deployment costs, thus datacenters energy and thermal management is directly related to effective cooling and power delivery.

Among the various reasons contributing to the poor energy efficiency (e.g., voltage conversion loss in UPS, excessive cooling provision, etc.), the loss in the datacenter cooling facility caused by the over-provisioned cooling capability takes the most significant portion in the entire loss as it is adjusted to guarantee safe operating conditions of servers targeting the worst-case workload scenario which happens rarely. In order to improve the energy ineffectiveness, datacenter designers and a large set of recent search works in the literature have identified three key guidelines as follows:

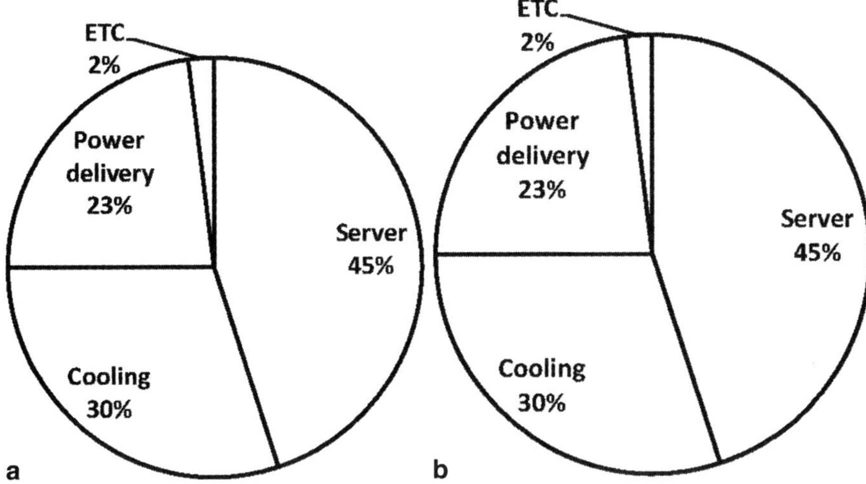

Fig. 4 **a** Breakdown of datacenter energy overheads [3]. **b** Datacenter costs breakdown assuming a 3-year servers and 15-year infrastructure amortization model [9]

- Fine-grained monitoring of PUE
- Server rack layout minimizing hot and cold air mixing by cold-aisle/hot-aisle layout, containment, duct, and analysis of computational fluid dynamics (CFD)
- Adjustment of thermostat of server room to the highest level where servers can be safely operated

However, there still exist huge gap until it reaches to its ideal value, i.e., 1.0, which necessitates the energy- and thermal-aware design in unprecedented ways. The main reason is that all these practices are still focused only on worst-case cooling scenarios designs without any holistic view that considers the dynamic cooling needs of the computing infrastructure at run-time. These results pose very drastic consequences in the design and modes of operation for next-generation datacenters.

1.4 Chapter Organization

In this chapter, we focus on presenting solutions to reduce the energy consumptions of servers and cooling systems through effective power and thermal control solutions based on accurate yet efficient power and temperature modeling and monitoring solutions. The rest of the chapter is organized as follows. Section 2 reviews state-of-the-art datacenters, especially focused on computing and cooling parts of datacenters to understand state-of-the-art technologies and figure out control knobs which are manipulated in control solutions. Section 3 shows approaches of modeling and monitoring power and temperature in servers as well as datacenters. Section 4 explains dynamic power and thermal management solutions for single servers, ranging from

Table 1 Server power breakdown [3]

Component	**Proportion** (%)
CPU	33
DRAM	30
Disk	10
Networking	5
Etc.	22

conventional air-convection cooled servers to liquid cooled ones. Section 5 explains power and thermal management solutions for large-scale computing server clusters in a datacenter. Section 6 explains the joint power and thermal management solutions for large-scale datacenters including both of computing and cooling power consumptions, especially targeting a hybrid cooling architecture which selectively uses free cooling according to required cooling capability. Section 7 summarizes the chapters.

2 State-of-the-Art in Datacenter Design

In this section, we explain state-of-the-art techniques to improve the energy efficiency of datacenters while meeting the temperature constraint, especially focusing on the two biggest energy consumers in datacenters, i.e., computing servers and datacenter cooling facility.

2.1 Computing Servers

1) Energy-Proportional Server Designs Server architectures have traditionally target performance optimization to support the ever-increasingly IT services demands and energy-efficiency has only become an important concern in the last five years. Due to the continuous technology scaling-driven performance improvement and the fact that single microprocessor architectures recently reached its performance limits [11], server designs have evolved since 2005 towards multi-cores architectures. A good example of this trend in state-of-the-art server designs is the HP DL980 blade server, which includes eight CPU sockets and each of them can support up to 10 cores [12]. Currently, the power consumed by servers takes more than 50 % of total power consumed by datacenters [3]. Table 1 shows the power breakdown of existing servers, which outlines that the largest portion of total power consumption in servers is taken by the CPU, but also DRAM memories must be considered as important blocks to develop power and thermal management strategies at server level.

In addition, future server designs trends by major server vendors, e.g., Sun Labs-Oracle, IBM, etc., show an evolution towards 3D-stacked technology integration programs [11], which enables the integration of a larger number of processing cores in very limited chip volumes and can significantly reduce the memory access latency by stacking memory layers on top of processing cores. Furthermore, 3D integration enables easier development of heterogeneous computing architectures because it is possible to integrate multiple memory types (e.g., 3D-stacked DRAM, phase change memory), and storage (e.g., solid-state disk) devices from different manufacturing processes, as in the EuroCloud server project [13]. However, as a side effect, power density is expected to significantly increase in 3D multi-core computing systems (i.e., up to 300 W/cm^3 [14]), which will make extremely difficult to properly dissipate the generated heat with current air-based cooling systems [15]. In particular, if free cooling is used, it will be a must to consider jointly the conception of the cooling and computing architecture.

One of the recent topics in server research is achieving energy-proportional components, which implies that computing systems should consume different amounts of active power according to their actual utilization. Nowadays, although servers are currently optimized to handle high-performance computation demands, most of the servers in a datacenter run at or below 40 % utilization during a significant part of the time, yet still draw almost full power during the process [16]. Therefore, latest server designs include many sensors (e.g., power, temperature, etc.) to accurately detect the current server utilization state [17]. Also, server components (i.e., processor, memory, and disk) now provide various operating states (e.g., active/idle/sleep/dormant) as well as various voltage and frequency (v/f) levels in processor and memory [18]. Therefore, recent works [19, 20] have shown the potential of developing energy proportionality in servers by exploiting the different power states and v/f levels according to the performance demand of local server utilization. Nonetheless, all these approaches focus on power consumption optimization of computing systems, thus they do not formally guarantee an optimal v/f point under thermal-induced power variations or can provide thermal damage prediction.

In order to reduce idle-time (leakage) power consumption, server processors provide nowadays hardware support for virtualization (e.g., AMD-V, Intel VT-x), which is a technique to enable increased physical server utilization by running applications from multiple OS instances in the so-called virtual machines (VMs) [21]. Moreover, on top of the hardware support, several virtualization software frameworks (e.g., Citrixs XenServer, Microsofts Hyper-V, VMWare ESXi, etc) have been recently developed to host multiple VMs with negligible performance degradation. Figure 5a illustrates the server virtualization. Recent improvements in the server virtualization techniques enable to run applications in a virtualized server within acceptable performance loss, i.e., $\sim 20\,\%$ for running CPU intensive workload [22] compared to running on a native system, while it is known to be degraded further when running memory- and disk-intensive workloads [23].

These various control options described above, i.e., power state, v/f level, VM placement, etc., give us great opportunities to achieve further power savings by fully utilizing the various control options while posing the challenges to develop an efficient control solution at the same time due to the large solution space, which necessitates us to develop an effective yet low-complexity control scheme.

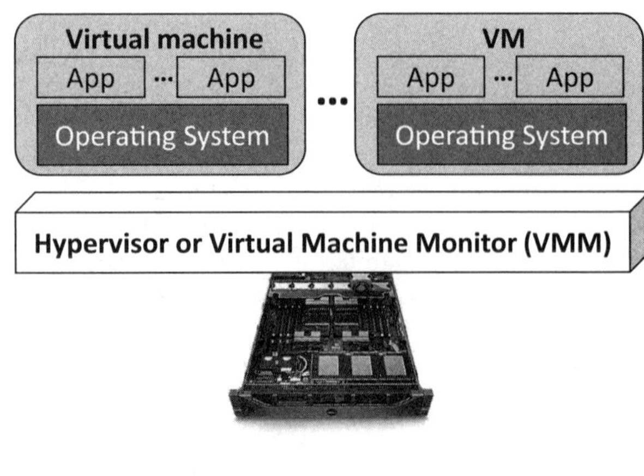

Fig. 5 Concept of server virtualization: hosting multiple VMs with the aid of hypervisor

Fig. 6 Hot- and cold-aisle isolation [9]

2.2 *Cooling Infrastructure*

In order to achieve energy-efficient datacenter cooling, various solutions have been presented. In this section, we address the three most widely used and effective solutions: (1) hot- and cold-aisle isolation, (2) closed-coupled cooling, and (3) free cooling. Then, we present how to utilize the cooling solutions more effectively to achieve further energy savings.

1) Hot- and Cold-Aisle Isolation Figure 6 shows a typical way of server room cooling. The cold air is provided by computer room air conditioning (CRAC) units through a raised floor, a steel grid resting on stanchions installed 2–4 ft. above the concrete floor. The cold air flows into racks through perforated tiles, and then, hot air is exhausted through a rear side of rack after absorbing heat generated by servers in the rack. One way of improving cooling efficiency is to prevent mixing the cold air provided from CRAC and hot air exhausted by servers. It is realized by a solution, so called *hot- and cold-aisle isolation*, which arranges server racks such that the intakes of cold air in server racks are faced each other, i.e., cold aisle, while preventing the mixture of hot air in different aisle side, i.e., hot aisle. The hot air is eventually drawn by the CRAC, and then, cold air is again provided to cold aisles by exchanging the heat with cold air (or water) provided from chillers.

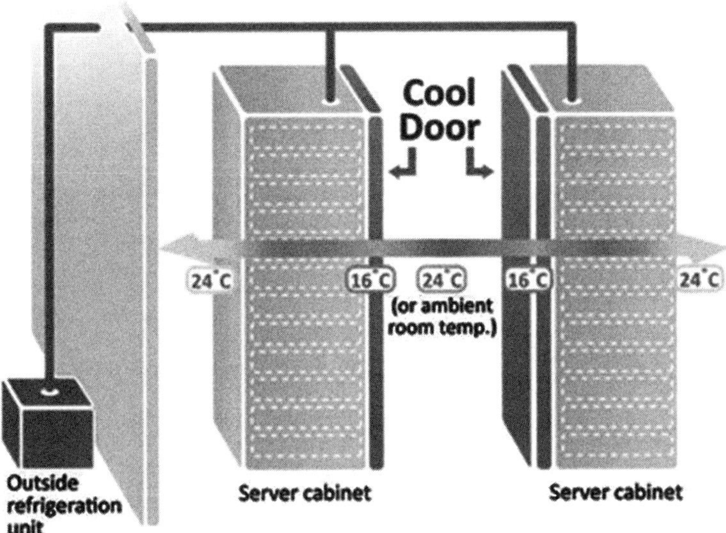

Fig. 7 In-rack cooling [24]

2) Closed-Coupled Cooling Closed-couple cooling solutions place cooling units more closely to computing units so as to remove any losses incurred throughout the delivery of cooling medium and quickly react to spatial temperature distribution. In this cooling solution, there are largely two classifications according to the granularity of computing cluster covered by single cooling unit, i.e., in-row and in-rack coolings. An in-row cooling adjusts cooling condition at every row according to the corresponding conditions while an in-row cooling adapts its cooling configuration according to operating condition at each rack. Figure 7 shows an example of an in-rack cooling solution where the cold air is directly fed into the front door of racks, namely, *CoolDoor* while the hot air is drawn by the CRAC with the same way in Sect. 2.2. The effectiveness of the solution is quite obvious in terms of the energy efficiency in that it can adjust only necessary parts instead of adjusting whole cooling configuration based on the worst-case scenario. It is reported that PUE of this cooling solution can reach down to $1.1 \sim 1.2$ [3]. However, the capital expenditure for the installation is quite high.

3) Free Cooling A recent approach to improve energy efficiency in datacenters is the concept of free cooling, which relies on the use of outside cold air and/or water for cooling instead of electricity. This is a promising architectural innovation for datacenter cooling infrastructure that can enable PUE to approach values near 1.0. Google has recently constructed two datacenters in Ireland and Belgium based on this concept and reports drastically improved PUE figures up to 1.09 [3].

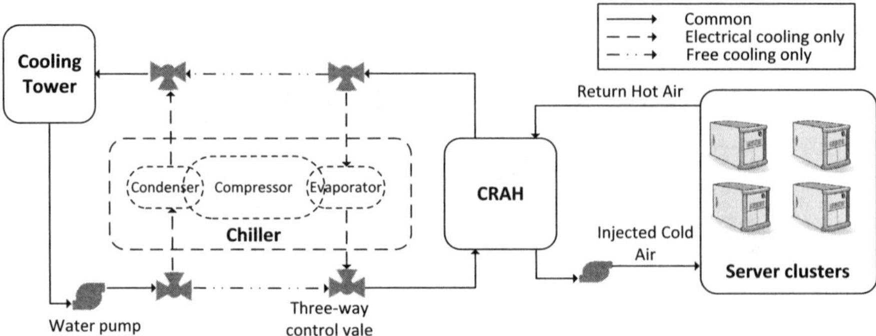

Fig. 8 Datacenter cooling architecture [43]

Despite the promising advantages on cooling-energy efficiency, the fundamental issue of free cooling is its limited applicability, as it can only be used in a very limited set of geographical locations because the cooling capability is tightly coupled with climate condition (e.g., temperature and humidity). Thus, it suffers from wide variations of cooling efficiency during the year, which translates in significantly high computing systems failure rates [25]. Hybrid cooling, which provisions back-up cooling infrastructure along with free cooling, is an intuitive solution to extend the usability of free cooling. Two main types of hybrid cooling architectures exist [26, 27]. The first architecture switches between free- and electricity-based cooling according to the outside temperature: if the outside temperature is lower than a certain threshold, free cooling is used; otherwise, chiller-based electricity cooling is employed as shown in Fig. 8. However, in real-life conditions, datacenters can use free cooling in very limited periods of the year and the average reported PUE is approximately 1.5. The second proposed architecture uses a cooperative hybrid cooling solution to increase the time free cooling is used. In this case, free cooling complements the chiller by pre-cooling hot return water with cold outside water before entering the chiller. This second architecture enables using free cooling, at least partially, for the entire year, and provides up to 50 % energy savings in cooling infrastructure (PUE$\simeq$1.25). However, it still suffers from significant higher failure rate than chiller-based solution due to lack of efficiency in the combined cooling scheme, which makes the current computing systems to operate at higher and variable temperatures. Moreover, due to the continuous increase in server power density, driven by the ever-increasing IT demand, the applicability of current free cooling will be even more limited in the future.

Figure 9 shows the variation of the power consumed by computing and cooling facilities as well as PUE measured for a datacenter equipped with hybrid cooling architecture deployed in Finland. As indicated the PUE line, PUE value varies 1.09 $\sim$ 1.60 and can be largely classified into three periods according to the PUE value. In this datacenter, free cooling is used only when the outside temperature is lower than 8 °C, which is set to very conservative value so as to cope with the worst-case

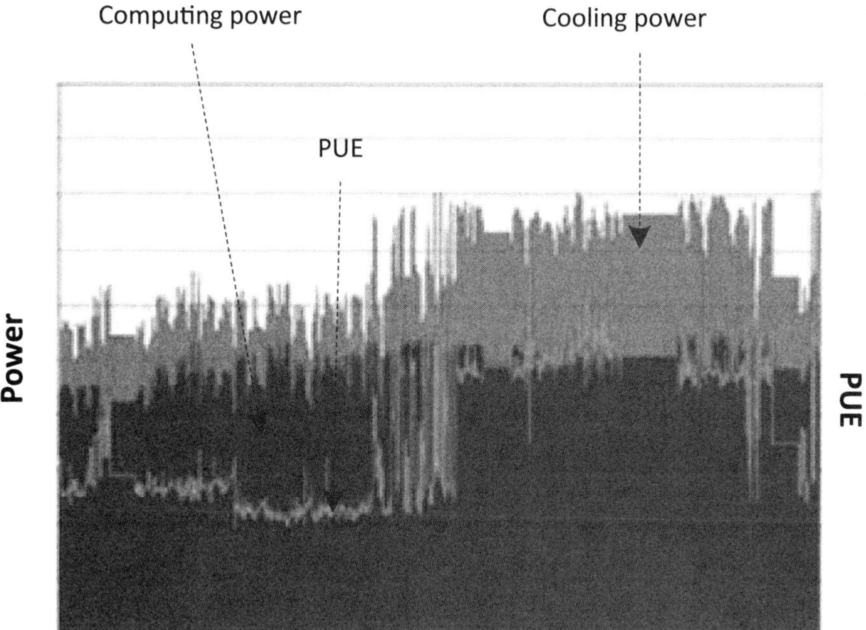

Fig. 9 Variations of power and PUE throughout a year [28] measured by a datacenter equipped with hybrid cooling architecture deployed in Finland

scenario. First, during the winter, PUE value is low as the free cooling is used for the most of the time period while it becomes increased during the summer as the electrical cooling is more frequently used as the temperature goes up.

Thus, free cooling as such cannot provide the ultimate solution to improve datacenter energy efficiency due to the limitation of the cooling capability and the dependency on outside temperature. In order to be generally applicable it must be combined in synergistic ways with innovative energy-proportional server design and cooling solutions, as well as holistic datacenter thermal control.

3 Power and Temperature Modeling and Monitoring

Accurate-yet-efficient modeling and monitoring on power and temperature of datacenters are necessary to develop control solutions for target systems. In this section, we first explain how we can model the power consumption and the temperature of existing servers and cooling facility in datacenters. Then, we address scalable and cost-effective power and temperature monitoring systems for large-scale datacenters.

3.1 Server Modeling

1) Power Modeling A server consists of various components, i.e., CPU, DRAM, disk, network interface (NIC), etc. As presented in Table 1, vast amount of the power is consumed by CPUs, memory, and disk, i.e., more than 70 %. Extensive works have been presented to accurately model power consumption of each component. *McPAT* is micro-architectural power model for chip multiprocessor (CMP), including in-order and out-of-order processor cores, networks-on-chips, shared caches, integrated memory controllers, and multiple-domain clocking, while tacking into account various process characteristics, e.g., bulk CMOS, SOI, and double-gate transistors, based on the forecast in the ITRS roadmap. The accuracy is validated using various processor implementations, i.e., Niagara, Niagara2, Alpha 21364, and Xeon Tulsa, whose errors range 10.84 $\sim$ 22.61 %, compared to the measured values. *DRAMSim* [29] and Micron's *System Power Calculator* [30] provide accurate and detailed timing and power models of various types of DRAM, e.g., DDR, DDR2. DDR3, Mobile LPDRAM, etc., accounting for the operations.

Although such accurate power models exist to model individual component of servers, it is difficult to use all such accurate models together due to the speed of the simulation. It becomes more exacerbated when we target to simulate the large number of servers in datacenters. Thus, high-level power models are widely used to track and estimate the power consumption of servers based on the observation that the power consumption for a given server is highly correlated with distinctive workload characteristics, e.g., CPU-, memory-, or disk-intensive, stressed on servers. To capture the relationship, various works have presented high-level power model which estimates the power consumption based on the utilizations [31–33]. Among them, Economous et al. [31] present a linear regression power model which estimates the server power consumption with respect to utilizations of CPU (u_{cpu}), memory (u_{mem}), and disk (u_{disk}), and network interface (u_{net}) as follows.

$$P_{server} = C_0 + C_1 u_{cpu} + C_2 u_{mem} + C_3 u_{disk} + C_4 u_{net} \tag{2}$$

where $\{C_0, C_1, C_2, C_3\}$ is a set of fitting parameters, which varies according to the target server system. This model is validated through two types of servers: (1) blade servers containing 2.2 GHz AMD Turion processor, 512 MB SDRAM, 40 GB HDD, 10/100 MBit Ethernet and (2) Itanium servers containing four Itanium2 chips, 1 GB DDR, 36 GB HDD, 10/100 MBit Ethernet. According to their evaluations, the errors are within 10 % in most of test cases using various benchmark suites, i.e., SPECcpu200, SPECjbb2000, SPECweb2005. Further evaluations for developing the high-level server power modeling have been conducted in [32] by comparing five different forms of power models as follows:

$$Type1 : P_{server} = C_0 \tag{3}$$

$$Type2 : P_{server} = C_0 + C_1 u_{cpu} \tag{4}$$

$$Type3 : P_{server} = C_0 + C_1 u_{cpu}^{r} \tag{5}$$

$$Type4: P_{server} = C_0 + C_1 u_{cpu} + C_2 u_{disk} \tag{6}$$

$$Type5: P_{server} = C_0 + C_1 u_{cpu} + C_2 u_{mem} + C_3 u_{disk} + C_4 u_{net} \tag{7}$$

Type 1 modes the power consumption in a static value. Type 2 and 3 model the power consumption with respect to CPU utilization, i.e., u_{cpu}, in linear and nonlinear manners, respectively. Type 3 and 4 add additional term to take into account the variations caused by disk (u_{disk}), memory (u_{mem}), and network (u_{net}). It concludes that Type 2 power model is enough for modeling CPU-intensive workload while Type 5 power model, using both of OS-reported component utilizations and CPU performance counters, is needed to cover broad workload characteristics, i.e., memory- and disk-intensive workloads, and aggressively power-managed servers.

In [33], Pedram et al. further enhance the accuracy of the power model by adjusting the fitting parameters according to various operating voltage and frequency and the number of active cores. It used Intel Xeon E5410 processor for the validation with various test cases, i.e., combination of the number of active cores and operating voltage and frequency level. Recently, *Joulemeter* is provided to automatically tune the parameters in power models by measuring battery usage in laptop or measuring power consumption in servers.

Fans also consume significant amount of power in servers. Indeed, it is well known that the fan power consumption has a cubic relationship with fan speed [34], as follows:

$$P_{fan} = C_0 + C_1 s_{fan}^3 \tag{8}$$

where $\{C_0, C_1\}$ is a set of fitting parameters and s_{fan} represents fan speed. Thus, lowering the fan speed enables us to reduce drastic amount of power consumption.

2) Temperature Modeling Accurate temperature models for servers are required to capture the temporal and spatial temperature variations. Especially, due to the high area and cost of placing thermal sensors in a silicon die as well as frequent failures of thermal sensors, the needs for the accurate temperature modeling becomes more important. Computational fluid dynamics (CFD) simulation is known to be a solution to develop accurate and complete 3D thermal map of servers by using numerical methods and algorithms to solve and analyze problems that involve fluid flows. In [35], Choi et al. present a CFD-based thermal modeling solution of servers by solving the governing transport equations shown in the following conservation law form:

$$\frac{\partial \rho \phi}{\partial t} + \frac{\partial \rho U_j \phi}{\phi \partial x_j} = \frac{\partial}{\partial x_j}\left(\Gamma_{phi,eff} \frac{\partial \phi}{\partial x_j}\right) + S_\phi \tag{9}$$

where ϕ is a general variable used for different context, e.g., mass, velocity, temperature, or turbulence properties; ρ is a fluid (air) density; t is a time for transient simulations; x_j is a coordinate x, y, or z direction when j is 1, 2, or 3, U_j is the velocity in each direction; Γ is the diffusion coefficient; S is the source for a particular variable such as the heat flux from a target system when the air temperature is ϕ. The four terms in Eq. (9) corresponds to transient, convection, diffusion, and source

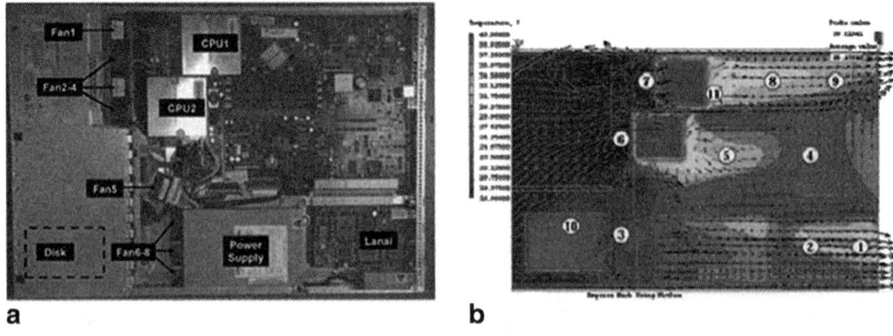

Fig. 10 Layout of IBM X335 server (**a**) and temperature map (**b**) [35]

parts of transport phenomenon at the spatial domain/extent. Figure 10a and b show pictures of IBM X335 server comprising of multiple components and its temperature map, respectively. As shown in Fig. 10, the spatial temperature variation can be accurately modeled. Despite the high accuracy of the CFD simulation, the simulation complexity is quite high because it does not have any closed-form solution for solving the differential equation in Eq. (9), which leads to adopt computer-based numerical procedures.

In [36], T. Heath et al. present a solution of constructing temperature map of servers while relieving the complexity of CFD simulation with negligible accuracy degradation, i.e., within 0.32 °C compared to CFD simulation. The simplification is achieved by abstracting heat- and air-flow with simplified graphs. Recently, a further simplified temperature model for servers has been presented in [38], especially targeting the CPU and memory sub-system of servers considering varied heat removal capability as a fan speed changes. It is developed by constructing thermal RC network of the system based on well-known duality between thermal and electrical phenomena [37], as shown in Fig. 11. In the RC network of CPU socket, P_j^c represents the power consumption of each core in a socket; R_l^c and R_v^c represent the lateral and vertical thermal resistance, respectively, where R_l^c is normally ignored as $R_v^c << R_l^c$; R_s^c and R_{ca}^c are thermal resistance of heat spreader and case-to-ambient (i.e., heat sink), respectively. C_j^c, C_s^c, and C_{ca}^c are thermal capacitances of die, heat spreader, and heat sink, respectively; T_{ja}^c represents the junction temperature which is used as an input to dynamic thermal management (DTM) units such that T_{ja}^c is lower than T_{max}. R_{ca}^c is the sum of the thermal resistances of heat sink and convective resistance, i.e., $R_{ca}^c = R_{hs}^c + R_{conv}^c$, where R_{conv} is changed according to the fan speed as follows:

$$R_{conv}^c \propto \frac{1}{A \cdot s_{fan}^{\alpha}} \tag{10}$$

where A is the effective area and α is a factor with a range of 0.8∼1.0.

In the RC network of memory part, P_{chip}^D is the power consumed in each DRAM chip; R_{chip}^D and C_{chip}^D are thermal resistance and capacitance of each chip; T_j^D is

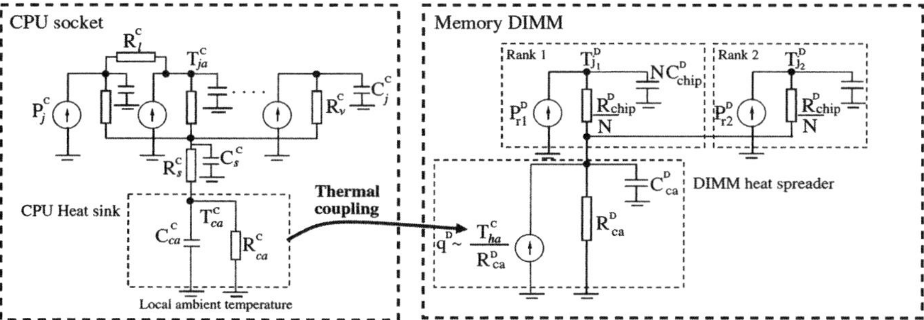

Fig. 11 RC network based temperature model [38]

the junction temperature of a DRAM chip; N is the number of ranks in a single DRAM chip. In addition, they observe that the temperature of DRAM is correlated with the temperature of CPU as the air inside a server flows from CPU to DRAM, thereby, air absorbing heat in CPU socket affects to the temperature of DRAM as it is equivalent to raising ambient temperature at DRAM. This phenomenon is called thermal coupling and modeled as follows:

$$q^D \propto \frac{T_{ha}^C}{R_{ca}^D} \tag{11}$$

where q^D is the dependent coupling heat source of the memory; T_{ha}^C is the heat sink sink temperature of the CPU; R_{ca}^D is the thermal resistance of the case to ambient of the memory DIMMs. This model is validated using Intel dual socket Xeon server, which shows a strong match between the actual measurement and the model within a 0.27 °C average error.

3.2 Datacenter Modeling

1) Computing Facility Basically, the temperature of servers in datacenter can be calculated using models in Sect. 3.1. However, for accurate temperature estimation for servers in a datacenter, we need to take into account interactions of generated heats among multiple servers in a server rack because servers are placed in a server rack in vertical direction and cold air flows from bottom to top of the server rack such that the heat generated at bottom is recirculated and affects to servers placed at upper side of the server rack. We call it *heat recirculation* in a datacenter. The amount of heat recirculation in a datacenter can be described by a cross-interference matrix, which is represented by $\Phi_{N \times N} = \{\phi_{i,j}\}$ where N is the number of servers in a server rack. $\phi_{i,j}$ indicates the contribution of the outlet heat rate of the i-th server in the inlet heat rate of the j-th one. Assuming Q_i^{out} and Q_j^{in} are, respectively, the outlet and inlet heat rates for the i-th and j-th server, the inlet heat rate for j-th server can be calculated as follows [39]:

$$Q_j^{in} = \sum_{i=1}^{N} \phi_{i,j} Q_{out}^i + Q_{amb} + P_j \tag{12}$$

where Q_{amb} represents the heat rate delivered from cold aisle of a server room and P_j denotes the power consumed by j-th server.

In the vector form, we can write this relationship as follows:

$$\boldsymbol{Q}_{in} = \boldsymbol{\Phi}^T \boldsymbol{Q}_{out} + \boldsymbol{Q}_s + \boldsymbol{P} \tag{13}$$

Based on the heat rate, we can calculate the temperature at each server within a server rack using temperature models in Sect. 3.1.

2) Cooling Facility The typical cooling facility consists of a cooling tower, a chiller, and CRAH (or CRAC) as explained in Sect. 1.1. The heat generated by servers in a server room is absorbed by cold air provided from CRAH, and then, drawn by CRAH. CRAH exchanges the heat drawn from the server room with cold water (or air) provided from a chiller based on refrigeration cycle. In [42], A. Qouneh et al. provide a comparative and quantitative analysis of cooling power as varying processor utilization and adjusting the server room temperature accordingly. For further analysis of the power consumption of the cooling facility, some models have been presented in [40, 41] which model the power consumption based on thermo-fluid principles.

However, based on our analysis of real datacenter setups of our industrial partners in this work, we have observed that an alternative procedure can be used, where PUE mainly depends on the temperature set-point of server room (T_{room}), outside temperature (T_{out}), and total power consumed by servers (P_{cl}). Moreover, T_{room} is the dominant factor compared to the others. Thus, we can simply characterize PUE with respect to T_{room}. Figure 12 shows PUE with respect to T_{room}. As shown in this figure, the PUE of electrical and free cooling ranges $1.53 \sim 1.83$ and $1.08 \sim 1.14$, respectively. Assuming that T_{room} is set to the highest temperature of which servers in active mode can satisfy the maximum temperature limit, i.e., T_{pm}^{max}, we can model PUE as a function of the power consumption of servers, i.e., P_{pm}. By matching the results shown in Fig. 12, we can approximate the PUE with a relatively simple form, namely:

$$PUE = a_1 P_{pm}^2 + a_2 P_{pm} + a_3 \tag{14}$$

where a_1, a_2, and a_3 are curve fitting parameters. In the case of electrical and free cooling, the sets we have obtained for $\{a_1, a_2, a_3\}$ are $\{3.32 \times 10^{-5}, -9.45e \times 10^{-4}, 1.30\}$ and $\{0, 0, 1.08\}$, respectively. Then, the maximum (average) root mean square (RMS) error amounts to 4.38 % (0.76 %) and 0.56 % (0.56 %), respectively.

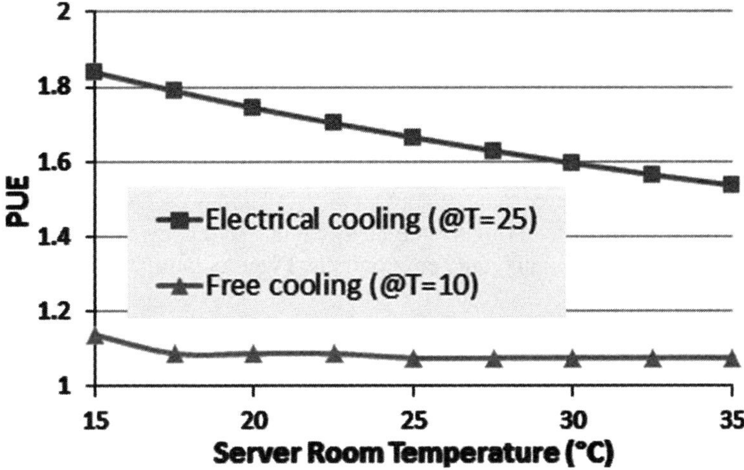

Fig. 12 Power usage effectiveness (PUE) in electrical and free cooling as power consumption of server varies [43]

Finally, the temperature of the server room, T_{room}, depends on CRAH efficiency, ϵ_{CRAH}, which is defined as follows [43]:

$$\epsilon_{CRAH} = \frac{T_{CRAH}^{air} - T_{room}}{T_{CRAH}^{air} - T_{CRAH}^{water}} \tag{15}$$

where T_{CRAH}^{air} represents temperatures of air exhausted from server room; T_{CRAH}^{water} is the temperature of chilled water flowing into the CRAH, which corresponds to the set-point of chiller and outside temperature when electrical and free cooling is used, respectively. Note that these values can be calculated using the procedure in [40], which depends on server power consumption, outside temperature, etc. Since ϵ_{CRAH} is always less than 1, T_{room} is always higher than T_{CARH}^{water}.

3.3 *Monitoring System for Datacenters*

Power management in datacenters is an area of increasing interest from several viewpoints as it is backed up by real concerns on energy usage and cost by modern computing systems. Data center computing applications and platforms have been typically designed without regard to power consumption. With increased awareness of energy cost, power consumption tracking and management is now an issue even for compute-intensive server clusters.

Datacenters ecosystem is facing an increasing need for decision support systems for datacenter management. Building and administration of datacenters are indeed evolving towards increasingly complex scenarios. IT infrastructure managers have to optimize the datacenter utilization and costs, under several constraints generated by heterogeneous and diverging technical challenges: customer requirements, infrastructure costs, energy costs, physical space available, etc.

Datacenters that have some energy measuring capabilities carry out those monitoring tasks through Data-Center Infrastructure Management (DCIM). This concept includes the integration of IT and Facility Management, with the aim of centralising monitoring, management and intelligent capacity planning of data centre systems. Capacity planning focuses primarily on energy but also on power, space, network, IT equipment, cabling, cooling and environmental factors (temperature and relative humidity).

Understanding total capacity of all factors ultimately gives the optimal position where equipment should be moved, added or changed for optimised use of the available capacity. It also directly indicates where potential capacity is still present but unused (stranded capacity). Currently, in many datacenters this task is carried out manually or through site audits. This is a tedious, time-consuming and labour-intensive process, with a high risk of human error. An advanced DCIM system automates and simplifies this process, benefiting to IT and facility staff, but also to the energy efficiency of the datacenter.

A DCIM system can in particular map and manage the complete power chain and hence the energy capacity of the datacenter. Starting at the power sources (grid power or alternative power sources) up to the outlets on a rack Power Distribution Unit (PDU) or even the components within the servers, including all devices in between, DCIM systems are essential to plan energy flows and perform trending and analysis. They bring full access to all available devices, from facility to IT, as well as life cycle management, support contracts, and logical and physical cable connections.

Whereas DCIM systems are usually a good fit for large datacenters, the needs of small to medium-size urban datacenters are not adequately met today. Existing systems are generally too complex, pricy, difficult to use and not modular enough for urban facilities. In addition, solutions offered on the market today are generally proprietary and tend to lock their users in to single vendors. Innovative DCIM support systems for datacenter management are thus needed. PMSM (i.e., Power Monitor System and Management) [44–56], developed at EPFL in cooperation with Credit Suisse [45], is an example of such an innovation.

4 Power and Thermal Managements of Servers

As the servers operating workloads are time-varying, the accompanying power consumption and thermal profile vary as well. In order to maintain controlled power consumption and thermal dissipations, run-time dynamic power and thermal management (DPM and DTM) mechanisms are required. These management schemes exploit the utilization of power and temperature-affecting control knobs that exist in different layers of abstraction of the system, to aid in power and thermal reduction. In addition, a fundamental challenge of any developed power or thermal management scheme is to have minimal, or preferably zero percent, performance degradation. If any management mechanism has a significant impact on the processing performance, it interferes with the architectural characteristics, hence considered a degrading rather than a managing element.

In this section, we explore the various power and thermal management mechanisms for server architectures. We first start by showing the state-of-the-art in power and thermal management solutions in Sect. 4.1. In Sect. 4.2, we explore our recent development in hierarchical power and thermal management schemes. Finally, we show our advances in power and thermal management in liquid-cooled server architectures.

4.1 Overview of CPU Power and Thermal Management Techniques

Power and thermal management solutions have been extensively existing in literature, which has been reflected in the various power and thermal management schemes [71, 72]. Nevertheless, we explore the recent works on power and thermal management in the state-of-the-art.

1) Temperature-Affecting Control Knobs As mentioned earlier, run-time management schemes utilize various control knobs that either reduce the causes of high heat generation, or increase the ability of the utilized cooling methodology. In the case of 3D MPSoCs, these control knobs are classified as follows.

a) *Workload Activity Knobs* At the software-level (application, system software, and OS), workloads can be altered and customized such that they can be thermally-aware. For example, task scheduling and task migration [73] have been extensively used to balance the workload on planar 2D MPSoCs [74]. Another example involves the intra-task instruction scheduling to prevent the processing element temperature from elevating to alarming values.
b) *Circuit Switching Activity Knobs* This class of control knobs affects the operating conditions of the processing element. These knobs may stall the processing element temporarily to reduce the heat generation, such as clock gating [75]. Alternatively, these knobs may reduce the operating speed of the processing element, which implies lower power consumption, hence lower heat generation, such as dynamic frequency scaling (DFS) or dynamic voltage and frequency scaling (DVFS) [75, 76].
c) *Thermal Package Control Knobs* The knobs at the thermal package level are responsible of changing the cooling capabilities, which is related to the injected fluid in the case of 3D MPSoCs with liquid cooling. For instance, the volumetric flow rate of the injected fluid can be varied by changing either the liquid pumping power [77], or varying the value of a flow-control valve [78].

2) Power and Thermal Management of Air-Cooled 2D and 3D MPSoCs Ogras et al. [79] proposes the control of power usage in processing elements (PEs) and routers by using model predictive control at design time, and Bogdan et al. [80] elaborate further this approach by considering both PEs and routers in the control scheme for voltage and frequency. However, they only consider power management and do not explore thermal control aspects. In fact, consolidating the power consumption in processing elements could undermine temperature issues while the power consumption is reduced. Thus, explicit thermal management schemes that include temperature as a key role in optimization or imposing temperature as a constraint are required for thermal balancing.

Initial research efforts have been focusing on combined power and thermal management by presenting a set of scheduling mechanisms for MPSoCs that perform temperature management at the system-level [81], using thread migration techniques to achieve temperature reduction in localized hot spots [75], or using a temperature-aware dynamic scheduling algorithm with negligible performance overhead [74]. These methods do not exploit history information and take reactive control actions based on the current thermal profile and frequency setting of the MPSoC.

However, recent works exploit history information to improve thermal management policies. Previous work [82] exploits a temperature forecast technique based on an auto-regressive moving average model. Another work proposes a novel technique that adapts the thermal management policy to the current workload characteristics [76], where the adaptation is done online exploiting information related to the workload history. Two recent approaches [83, 84] describe two methodologies to achieve thermal prediction by combining the information of thermal model, thermal sensors and power consumption statistical properties. These approaches rely on open-loop search or optimization where it is assumed that power can be estimated accurately.

More advanced solutions apply the concepts of *model-predictive control* (MPC) to turn the control from open-loop to closed-loop [87]. A chip-level power control algorithm based on optimal control theory is proposed [85], where the power consumption of the MPSoC is controlled to maintain the temperature of each core below a specified threshold. A recent work [86] proposes MPC utilization to solve the thermally-aware frequency assignment problem of a planar MPSoC.

However, most previous policies do not completely avoid hot-spots, but they simply reduce their frequency, because the interaction among the prediction method, the thermal behavior of the MPSoC and the frequency assignment of the MPSoC have not been addressed as a joint optimization problem.

In a similar vein, recent work considers dynamic thermal management for 3D MPSoCs. Previous work evaluates several policies for task migration and DVFS [88]. This previous work explores thermal profiles of adjacent processing elements being on the same vertical column (interlayer adjacent) or within the same layer (intralayer). Based on this analysis, a combined DVFS and a task migration policy, named *THERMOS*, is implemented. However, this work do not consider controlling the thermal packaging knobs, whether it is air or liquid cooling. Another work [89] integrates a thermally-aware task scheduler with DVFS on a two-tier 3D MPSoC with eight cores. A recent paper proposes a temperature-aware scheduling method specifically designed for air-cooled 3D MPSoCs [91]. This method takes into account the thermal heterogeneity among the different layers of the system, but there is no study on the effect of the thermal packaging control knobs as active thermal management parameters. The resulting temperatures obtained in these papers are significantly high (85–120 °C). These results imply that 3D MPSoCs are prone to high temperatures, and with increasing power densities conventional thermal management techniques and air-based cooling are incapable of controlling the temperature while preserving system performance.

3) Thermal Management of Liquid-Cooled 3D MPSoCs Prior liquid cooling work [90] evaluates existing thermal management policies on a 3D MPSoC with a fixed-flow rate setting, and also investigates the benefits of variable flow using a policy to increment or decrement the flow rate based on temperature measurements, but without considering pump energy consumption.

Thermal management methods for 3D MPSoCs using a variable-flow liquid cooling have been recently proposed [77]. These policies use experimentally-driven sets of rules to control the temperature profile of the 3D MPSoC while ensuring performance requirements to be satisfied. These approaches use a centralized control concept, which is inappropriate if the controlled parameters increase [92], as in the case of targeted 3D MPSoC designs with liquid cooling in this work.

Recently, Qian et al. explore the use of a cyber-physical approach 3D MPSoCs thermal management with inter-tier liquid cooling [93]. They construct their control mechanism with software-based thermal estimation and prediction. They use a non-uniform liquid flow in different microchannels to meet the cooling demands of different modules. They take their control decisions on software-based thermal estimation and prediction. They use a non-uniform liquid flow in different microchannels, to meet the cooling demands of different modules. However, they have not shown the overhead of their software-based thermal estimation. Moreover, they do not show the feasibility of having a non-uniform flow in different channels, as a physical implementation.

4.2 Run-Time Hierarchical Power and Thermal Management for Server Architectures

We have proposed another proactive management scheme that relies on model predictive controller (MPC) [94]. In this work, we have developed a thermal management scheme that controls task scheduling, DVFS, and the cooling infrastructure. In particular, we target the cooling infrastructure case of interlayer liquid cooled 3D MPSoC, where we can alter dynamically the injected liquid flow rate. At each time interval, a new set of workloads arrive, and the management scheme allocates these tasks to various cores and sets the corresponding flow rate such that the predicted peak temperature is reduced while minimizing the 3D MPSoC power consumption (cooling and computation power). Then for each processing element it applies MPC to the assigned workload such that the local predicted temperature is reduced while using the minimum computing energy possible via DVFS. The formulation of this problem is stated as follows:

$$J = \sum_{\tau=1}^{h} \left(\|\mathbf{R}\mathbf{p}_\tau\| + \|\mathbf{T}\mathbf{u}_\tau\| \right) \tag{16}$$

$$min \;\; J \tag{17}$$

$$subject\ to: f_{\min} \preceq \mathbf{f}_\tau \preceq \mathbf{f}_{\max} \;\; \forall\, \tau \tag{18}$$

$$\mathbf{x}_{\tau+1} = \mathbf{A}\mathbf{x}_\tau + \mathbf{B}\mathbf{p}_\tau \;\; \forall\, \tau \tag{19}$$

$$\tilde{\mathbf{C}}\mathbf{x}_{\tau+1} \preceq \mathbf{t}_{\max} \;\; \forall\, \tau \tag{20}$$

$$\mathbf{u}_\tau \succeq \mathbf{0} \;\; \forall\, \tau \tag{21}$$

$$\mathbf{u}_\tau = \mathbf{w}_\tau - \mathbf{f}_\tau \;\; \forall\, \tau \tag{22}$$

$$\mathbf{l}_\tau \succeq \mu \mathbf{f}_\tau^{\mathbf{2}} \;\; \forall\, \tau \tag{23}$$

$$-\,\mathbf{w} \preceq \mathbf{m}_{\tau+1} - \mathbf{m}_\tau \preceq \mathbf{w} \;\; \forall\, \tau \tag{24}$$

$$0 \preceq \mathbf{m}_\tau \preceq \mathbf{1} \;\; \forall\, \tau \tag{25}$$

$$\mathbf{p}_\tau = [\mathbf{l}_\tau ; \mathbf{m}_\tau] \;\; \forall\, \tau \tag{26}$$

where matrices **A**, **B** are related to the overall 3D MPSoC system description. These matrices represent the 3D MPSoC system using a coarse granularity of the thermal cells and where the sampling time of the resulting discrete-time system is T_{GC}. The horizon of this predictive policy is defined as h [87]. Then, the objective function J is expressed by a sum over the horizon.

In the cost function (Eq. (16)), the first term $\|\mathbf{R}\mathbf{p}_\tau\|$ is the norm of the power input vector p weighted by matrix **R**. Power consumption is generated here by two main sources. Vector p is a vector containing normalized power consumption data the p tiers and the pumping power. Matrix **R** contains the maximum value of the power consumption of the tiers and the cooling system. The second term $\|\mathbf{T}\mathbf{u}_\tau\|$ is the norm of the required workload, but not yet executed. To this end, the weight matrix **T** quantifies the importance that executing the required workload from the scheduler has in the optimization process. Then, Inequality (18) defines a range of working frequencies to be used, but this does not prevent from adding in the optimization problem a limitation on the number of allowed frequency values.

Equation (19) defines the evolution of the 3D MPSoC according to the present state and inputs. Equation (20) states that temperature constraints should be respected at all times and in all specified locations. Since the system cannot execute jobs that have not arrived, every entry of $\mathbf{u}_\tau$ has to be greater than or equal to 0 as stated by Eq. 21. The undone work at time τ, u_τ is defined by Eq. 22. Equation 23 defines the relation between the power vector **l** and the working frequencies. μ is a technology-dependent constant.

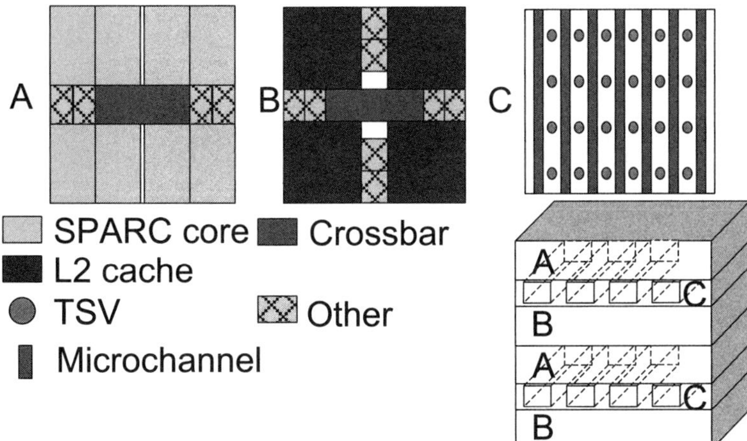

Fig. 13 Schematic diagram of the four-tier liquid-cooled 3D MPSoC used in the thermal evaluation of the proposed thermal management scheme

Then, Eqs. 24–25 define constraints on the liquid cooling management. The normalized pumping power value (**m**) scales, and any time instance τ, from 0 (no liquid injection) to 1 (power at the maximum pressure difference allowable), as shown in Eq. 25. Moreover, the maximum increment/decrement change in the pumping power value from time (τ) to ($\tau + 1$) is limited by a another normalized value **w**, as shown in Eq. 24, which models the mechanical dynamics of the pump.

Equation 26 defines formally the structure of vector **p**. Vector $\mathbf{l} \in \Re^{\mathbf{p}}$ is the power input vector, where p is the number of tiers of 3D MPSoC.

Finally, the control problem is formulated over an interval of h time steps, which starts at current time τ. Indeed the result of the optimization is an optimal sequence of future control moves (i.e., amount of workload to be executed in average for each tier of the 3D MPSoC which is stored in vector **f**). Then, only the first samples of such a sequence are applied to the target 3D MPSoC, while the remaining moves are discarded. Thus, at each next time step, a new optimal control problem based on new temperature measurements and required frequencies is solved over a shifted prediction horizon (e.g., the "receding-horizon" [87] mechanism), which represents a way of transforming an open-loop design methodology into a feedback one, as at every time step the input applied to the process depends on the most recent measurements.

To evaluate the effectiveness of this thermal control, we apply this management scheme on a four-tier 3D MPSoC based on the UltrsSPARC T1 MPSoC [112], which is shown in Fig. 13. In addition, we compare it against different state-of-the-art thermal management techniques, which are as follows:

- **Liquid cooling with LB (LC_LB)** [95]: It applies the maximum cooling flow rate, while the jobs are scheduled with load balancing policy (LB). LB balances the workload by moving threads from a core's queue to another if the difference in queue lengths is over a threshold.

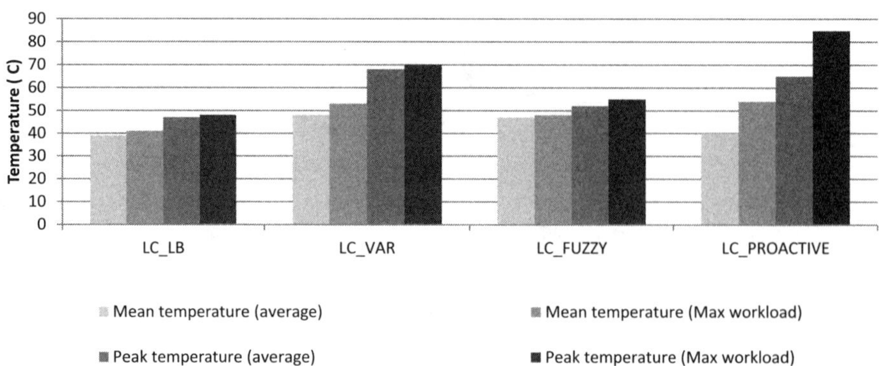

Fig. 14 Peak and average temperatures observed using all the policies, both for the average case across all workloads and maximum workload on four-tier 3D MPSoC [94]

- **LUT-based flow rate control with LB (LC_VAR)** [77]: It dynamically changes the flow rate based on the predicted maximum temperature, while the jobs are scheduled with LB.
- **Fuzzy-logic control (LC_FUZZY)** [96]: This mechanism utilized fuzzy logic in deriving thermal management mechanism that controls the variable liquid flow rate and DVFS.

In addition we refer to this management scheme as *LC_PROACTIVE* in the following paragraphs. In this evaluation of different thermal management policies, *LC_PROACTIVE* is compared with respect to the other management techniques mentioned above based on the:

- Maximum and average temperatures.
- Computational and cooling power consumption.

Thermal impact of all the policies on a four-tier 3D MPSoC (cf. Fig. 13) is shown in Fig. 14. This figure shows that LC_LB reduces the peak temperature to 47 °C, whereas LC_FUZZY and LC_VAR push the system into a higher peak of 52 and 67 °C, respectively, but still avoids any hot-spots. This is the similar case in LC_PROACTIVE, where the peak temperature reaches 84 °C. The alteration between the peak temperature comes from the fact that main target is to reduce the peak temperature to any value below 85 °C. However, since each technique has a different management policy, with different control elements, the peak and average temperatures are affected.

Figure 15 shows the total consumed power when running the various policies on the four-tier MPSoC with the average workload [94]. Energy consumption values are normalized with respect to the load balancing policy on the 3D-MPSoC with LC_LB. In this figure, LC_PROACTIVE manages to reduce the cooling power and the overall system power by 60 and 23 %, respectively, with respect to LC_LB. Moreover, LC_PROACTIVE even reduces the cooling energy more than LC_VAR and LC_FUZZY by 40 and 22 %, respectively.

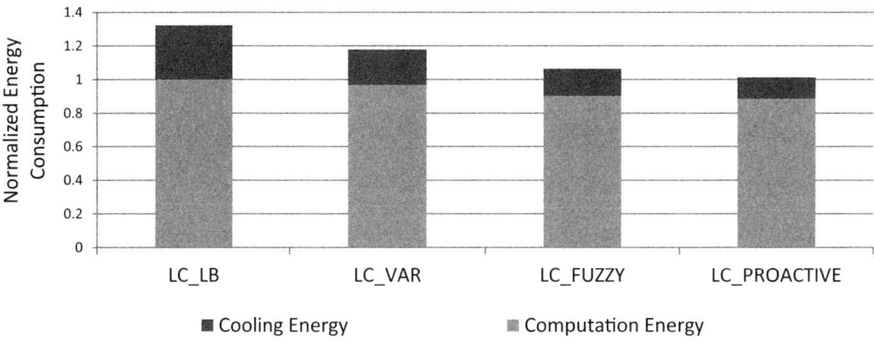

Fig. 15 The normalized energy consumption in the whole system (chip and cooling network) [94]

4.3 *Design-Time Power and Thermal Optimizations*

In addition to run-time management schemes, several works conduct power and thermal optimizations at design-time. In the case of MPSoCs, several approaches have been taken to optimize the power utilization and heat generation or dissipation. At the platform level, different modules can be designed to reduce the overall power density, hence heat generation, while preserving the system functionality. This approach has been taken recently in low-power (hence low temperature) processor designs such as ARM big.LITTLE processing architecture [97]. Another approach at the platform level is to reduce the operating power supply of the platform to near-threshold values [98]. Near-threshold computing allows the processing units to operate close to the voltage threshold value of the used transistor, hence reducing the overall power and thermal density.

In the case of 3D MPSoCs, recent work proposes multiple supply voltages utilization to optimize the voltage islands distribution in 3D MPSoCs [99]. In this work, a temperature-aware voltage island generation methodology is proposed that formulates this problem as a mixed-integer linear programming (MILP) problem. The main aim in this work is to minimize the thermal hotspots in 3D MPSoCs while keeping the performance and timing requirements satisfied. The interdependency between power and heat densities made it feasible to formulate this problem and achieve significant results.

Another work utilizes various microarchitectural techniques to control the thermal hotspots in 3D MPSoCs via thermal herding [100]. This technique explores different architectural disciplines by spitting several microarchitectural blocks between the different layers of 3D MPSoC to enhance the throughput while controlling the thermal hotspots such as, register file splitting. This splitting is based on general application trends and the significance of particular instructions or data locations to the execution flow.

Previous works have investigated the rearrangement of various hardware modules within the MPSoC to minimize the global thermal impact, which is also known in literature as *temperature-aware floorplanning*. Initial work on temperature-aware floorplanning [101] has shown its significant impact on reducing the peak temperature. This work has defined a metric called *thermal diffusion* that resembles the lateral heat dissipation. This metric has been used in an optimization problem to maximize the gains of *thermal diffusion*. Other similar works have proposed simulated annealing utilization [102] or genetic algorithms [103] to achieve temperature-aware floorplanning.

In the context of 3D MPSoCs, temperature-aware floorplanning has also been extended by including the interlayer thermal dissipation and interconnect characteristics [102, 104–106]. For example, initial work has been proposed [107] for temperature-aware microarchitectural floorplanning. The main objective in this work is to place the processing submodules of a single processor in several layers such that the wire lengths and the temperatures are minimized. To achieve this, a mixed integer linear programming (MILP) problem is formulated to minimize the weighted sum of performance, area and thermal-related aspects. Another work uses simulated annealing to minimize the temperature of 3D MPSoC via floorplanning [105] by considering the additional power consumption of the interconnects.

As for liquid-cooled 3D MPSoCs, Mizunuma et al. use their thermal model to explore floorplanning solutions to homogenize temperature distributions in this architecture [108]. The results in this work, which is further assisted by the observations in other work [96], show that in the case of liquid cooled 3D MPSoC, temperature-aware floorplanning follows the trend of placing more heat dissipating modules at the fluid inlet port, while lower heat dissipating modules at the outlet port. In other words, the optimal heat dissipation pattern for temperature-aware floorplanning would be monotonically decreasing from the distance of the fluid inlet port.

Our recent proposed framework, namely *GREENCOOL*, optimizes the active cooling path of microchannel-based iterlayer liquid cooled 3D MPSoCs to balance the thermal profile of the target 3D MPSoC while significantly reducing the active cooling energy demands [109]. This design-optimization methodology uses the concept of channel modulation, where we change the microchannel aspect ratio (channel width/channel height) to enhance the heat transfer capability from the target 3D MPSoC via changing the convective thermal resistance [110]. Using the conventional CMOS fabrication process for etching the channels, such as deep reactive iron etching [111], it is possible to modulate the width of the channel from inlet to outlet (and hence its aspect ratio) and create any kind of channel width profile, while keeping the height of the channels constant. Thus, channel width modulation requires only a change in the patterns on the masks used for etching channels amounting to minimal additional fabrication costs. To summarize, using careful design it is possible to modify the local channel aspect ratios so as to contain the pumping power while constraining the thermal gradients.

To understand how the channel width affects the change in temperature due to convection (ΔT_{conv}) in detail, an analysis is performed on a single microchannel shown in Fig. 16. We start by the following set of equations governing the Nusselt number (a dimensionless form of heat transfer coefficient), and the product of friction factor

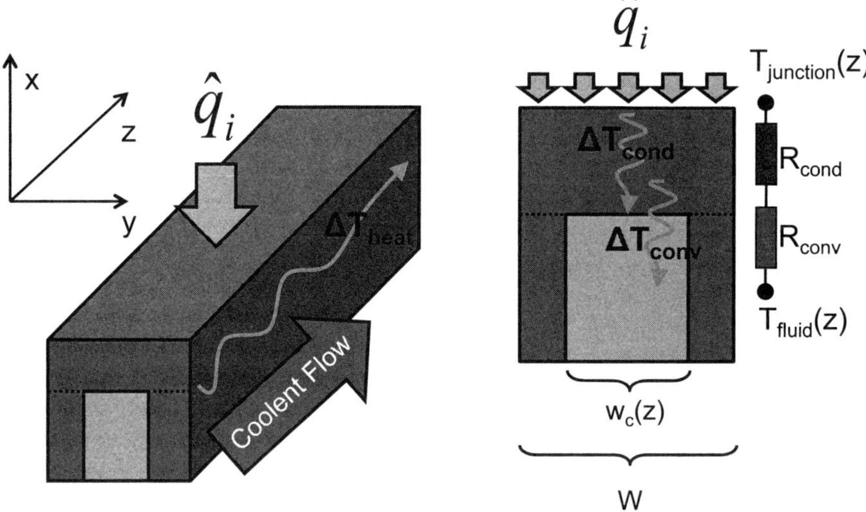

Fig. 16 Test structure: a single microchannel cooling a strip of an IC with uniform heat flux distribution. The figure shows both the 3D and the cross-sectional views [109]

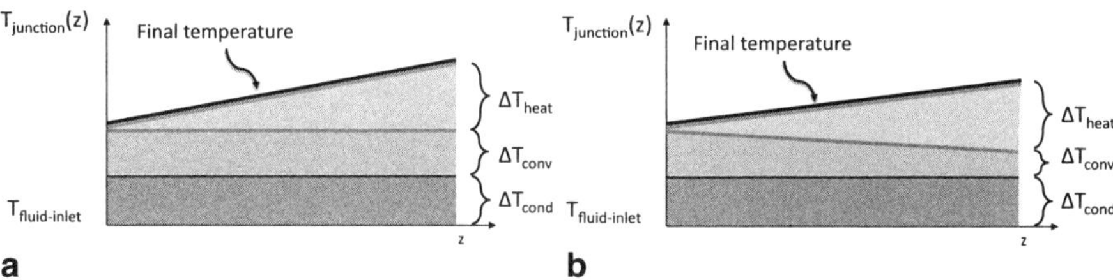

Fig. 17 R_{conv} as a function of the channel width for the structure in Fig. 16

and Reynold's number for microchannels, under fully developed conditions [110]:

$$Nu = 8.235 \cdot (1 - 2.0421AR + 3.0853AR^2 - 2.4765AR^3 + 1.0578AR^4 - 0.1861AR^5)$$

$$fr \cdot Re = 24 \cdot (1 - 1.3553AR + 1.9467AR^2 - 1.7012AR^3 + 0.9564AR^4 - 0.2537AR^5), \quad (27)$$

where AR is the aspect ratio reciprocal (height/width) of the channel. Using the Nusselt number, the heat transfer coefficient (a measure of the amount of heat transferred per unit area for one Kelvin difference in temperature between the fluid and the microchannel wall surface, expressed in W/m^2K) can be written as:

$$h = \frac{k_{coolant} \cdot Nu}{d_h} \quad (28)$$

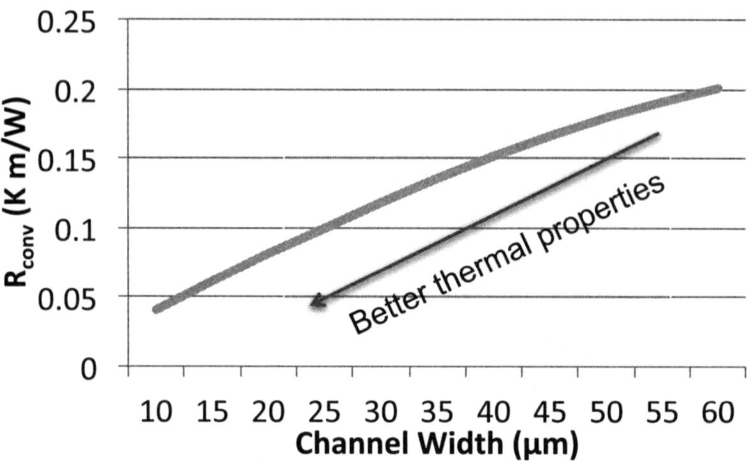

Fig. 18 Junction temperature distribution for the structure in Fig. 16. **a** With uniform non modulated channel width. **b** With modulated channel width to compensate for sensible heat absorption [109]

where $k_{coolant}$ is the thermal conductivity of the coolant and d_h is the hydraulic diameter of channel. The effective heat transfer coefficient as seen by the junction looking down the channel from the top can be written by projecting the heat transfer coefficient above from the side wall surfaces onto the top as follows:

$$h_{eff} = h\frac{2 * H_C + w_C}{W} \tag{29}$$

where H_C is the height and w_C is the width of the channel, and W is the total width of the structure as shown in Fig. 16. The convective resistance R_{conv} for this structure can be obtained as a reciprocal of this quantity. The R_{conv} for this structure is plotted as a function of w_C in Fig. 17, assuming water as the coolant, $H_C = 100\ \mu\text{m}$, $W = 100\ \mu\text{m}$ and varying w_C from 10 to 50 μm.

Figure 17 shows that the convective resistance (and also ΔT_{conv}) drops quickly as the channel width is reduced. Since the goal is to modify the convective resistance to compensate for ΔT_{heat}, it can be postulated that the channel width must no longer be a constant but instead should be a function of the distance along the channel $w_C(z)$. The width must be larger near the inlet where the fluid temperature is low and smaller near the outlet where the fluid temperature is high. Hence, theoretically, for the case of uniform heat flux, it is possible to lower the final thermal gradient by steadily modulating the channel width from inlet to outlet, as shown in Fig. 18b.

GREENCOOL uses this principle in formulating an optimal control problem to find the optimal channel width profile for each microchannel, from the fluid inlet to outlet ports. The target of this optimization is to minimize the peak temperature and thermal gradients of the 3D MPSoC, as well as reducing the energy needed by cooling. When applied various 3D MPSoC architectures, significant thermal gradient reductions as well as cooling power savings, with respect to worst-case designs.

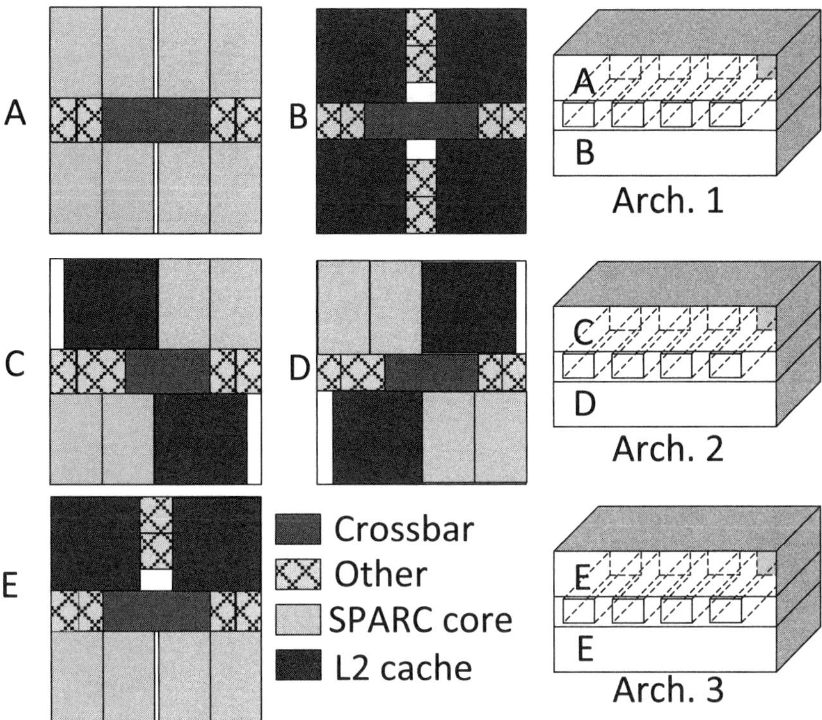

Fig. 19 Layout of the 3D-MPSoCs used in our experiments [113]

For instance, when *GREENCOOL* is applied to different architectural layouts of the UltraSPARC T1 Niagara MPSoC [112], a 31 % thermal gradient reduction is observed. Figure 19 shows the layout of the different two-dies 3D-MPSoCs used in this experiment. The dies are of size 1 cm × 1.1 cm and the heat flux densities range from 8 to 64 W/cm^2 in the two dies. Further details about the floorplan and power dissipations can be found in pervious works [77, 96, 112].

In this experiment, the worst-case (peak) power dissipation of the 3D-MPSoC functional elements [77, 96, 112] (obtained using measurements) are used in the optimization process. *GREENCOOL* achieves a thermal gradient reduction of 31 % (23 °C to 16 °C). When the peak heat flux levels were replaced by average values, this same optimal channel modulation configuration manages to reduce the thermal gradient by 21 % compared to the uniform channel width case. The thermal gradients obtained for the different cases and for various channel types are plotted in Fig. 20.

In another set of experiments to demonstrate the energy-efficiency of *GREENCOOL*, significant cooling energy savings that reach up to 80 % has been achieved [109]. Furthermore, *GREENCOOL* aids in developing efficient cooling layout in the cases where uniform cavity utilization is infeasible.

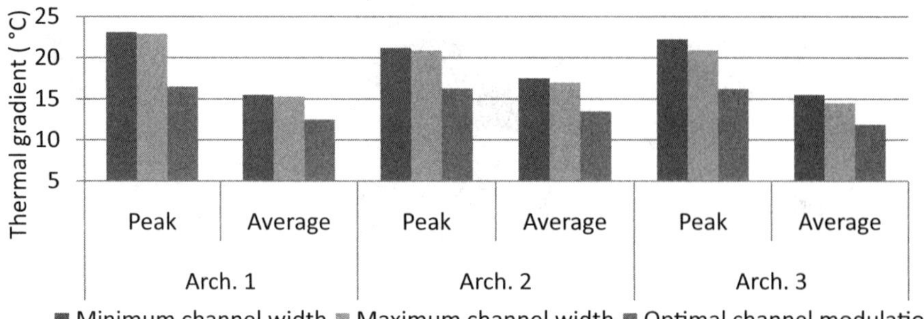

Fig. 20 Thermal gradients observed in the different 3D-MPSoC architectures dissipating peak and average level heat fluxes, using maximum, minimum and optimally modulated channel widths [113]

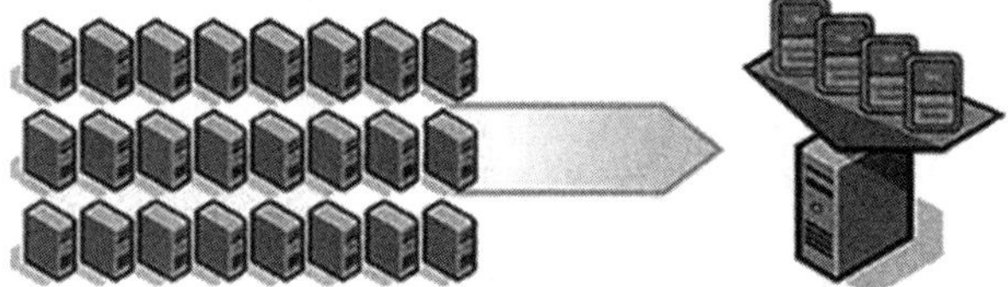

Fig. 21 Concept of server consolidation

5 Power and Thermal Managements for Server Clusters

5.1 Conventional Solution to Minimize Power Consumption for Server Clusters

In datacenters, servers are normally severely under-utilized, less than 30 % in more than 90 % of the total time [3]. In addition, as explained in Sect. 3, the power consumption of servers is not proportional to the utilization, i.e., the idle power consumes around 50 % of the peak power consumption. Due to the poor energy-proportionality, the power consumed by servers in datacenters can be reduced as we minimizes the number of active servers by packing workloads into the minimal number of active servers [46]. The technique is called *Server consolidation*. The key enabler to realize the solution is server virtualization, explained in Sect. 2.1, as it enables to migrate workloads easily by encapsulating workloads with a form of virtual machines (VMs) and run multiple VMs in a single physical server with the aid of hypervisor. Figure 21 shows the concept of the server consolidation in a virtualized server environment.

In the server consolidation, we need to take care such that the performance after the consolidation should not be degraded, or within an acceptable range. To achieve this goal, many works have developed the consolidation solutions such that the sum of the peak required utilization among co-located VMs does not exceed the server's capability [46]. However, as analyzed in many works [6], the peak utilization happens rarely and much higher than off-peak (e.g., 90th/95th/99th percentile) values. Thus, the server consolidation based on the peak value makes us to lose the opportunity for further power savings. To overcome the conservative solution, some works [6, 47] presents server consolidation solution which packs VMs into servers based on off-peak (e.g., 90th/95th/99th percentile) of server utilization.

The advantage obtained from the server consolidation is obvious in terms of power savings. However, it may cause unexpected performance degradation due to the conflict of using shared resources among co-located VMs, especially last-level cache (LLC) [48, 49]. Tickoo et al. [49] analyzed how the performance is degraded as VMs are allocated to share LLC with others using *SpecJBB* and *Sysbench* benchmark suites in order to evaluate the amount of the performance degradation caused by different cache usage characteristics of co-located VMs. The results show that sharing LLC between two copies of VMs both hosting SpecJBB leads to $\sim$30 % performance degradation while a case of sharing LLC between VMs hosting SpecJBB and Sysbench leads to $\sim$20 % degradation. In [50], Govindan et al. characterize the amount of interference with a set of parameters, i.e., effective number of used sets and ways. Then, it presents a solution to allocate VMs by accounting for the amount of the interference such that the performance interference becomes minimized while meeting the required performance requirement.

5.2 *Correlation-Aware Power and Temperature Management*

We can achieve further compact server consolidation by considering correlation among workload variation. In [51], Verma et al. found out that workloads running on datacenters are strongly correlated one another. In order to achieve further power savings while maintaining quality of service (QoS) level, correlations among VMs' workload have been exploited in recent works [51–54]. In [51], Verma et al. presented a clustering-based correlation-aware VM placement solution. To efficiently characterize the workload correlation, it first transforms utilization traces with a form of an envelope which is defined as a binary sequence which is '1' when CPU utilization is higher than a threshold value, e.g., 90th percentile, otherwise '0'. Second, it clusters VMs such that the envelops of VMs' CPU utilization included in different clusters do not overlap. Finally, it allocates VMs to physical servers such that VMs in different clusters are co-located in a single server so as to minimize the possibility when peaks are coincided. To meet the performance requirement after the consolidation, it allocates VMs based on their off-peak utilization demands (e.g., 90th percentile) while reserving a shared peak buffer to handle resource demand higher than the off-peak value for all co-located VMs. However, this approach is applicable only when the envelops of VMs are stationary and distinctively different one from another, thereby, producing multiple clusters. Hence, it does not work well with applications with non-stationary and fast-changing VM behaviors.

In [52], Meng et al. presented a joint-VM sizing technique that pairs two uncorrelated VMs into a *super-VM* and provision *super-VMs* by predicting the the aggregated workloads. However, once *super-VM*s are formed, this solution does not

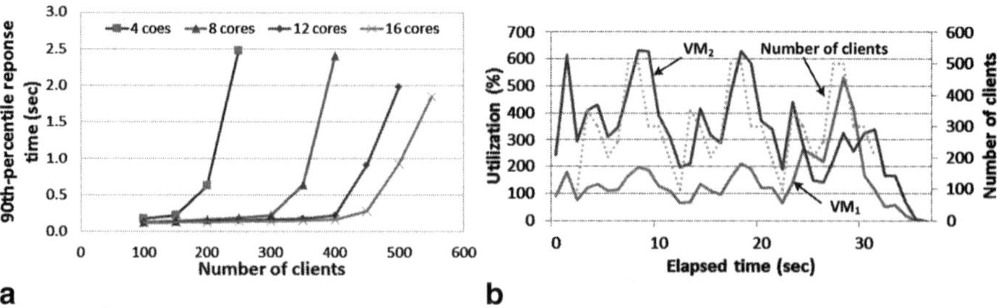

Fig. 22 **a** 90th-percentile response time (in seconds) with respect to the number of clients and allocated cores. **b** Variations of CPU utilization of two index searching nodes (ISNs) with respect the number of clients

consider the correlations of VMs within a same *super-VM* anymore. Thus, it may lose the chance of further power savings by leveraging time-varying correlations in scale-out applications. In [54], Halder et al. extends the scheme such that aggregated workload of multiple VMs can be utilized for VM placement. However, this solution can be applicable only when future servers' utilization is perfectly known.

However, all the correlation-aware VM placement solutions target conventional HPC application, thereby they cannot work well with scale-out applications whose workload characteristics are quite different, as we explained in Sect. 1.2. To overcome the drawbacks of existing solutions, we [56] developed a power management solution for datacenters hosting scale-out application, especially targeting following distinctive workload characteristics of scale-out applications. We used a websearch application in CloudSuite [4] as a proxy to characterize the workload characteristics of scale-out applications.

- **User-interactive and fast changing:** Owing to the user-interactive nature of scale-out applications, responsiveness, quantified in terms of latency, is the first criterion we need to satisfy when running the applications. Therefore, we should provision VMs in a conservative manner, based on the peak (or Nth percentile according to QoS requirement) resource demand of each VM. As the scale-out applications are commonly highly parallel, we can meet the required performance level for running VMs by assigning the right number of cores. Figure 22a demonstrates the 90th percentile response time of a websearch cluster with respect to the number of queries as we vary the number of allocated cores to host the websearch cluster from 4 to 16. Furthermore, the resource demand is time-varying and mostly lower than the provisioned amount of resources. However, as described in [6], due to the significant performance degradation caused by the long transition latency between power modes and fast changes of resource demands, dynamic power gating (turning on/off cores) cannot be applicable to such applications. Motivated by these observation, *it is required the solution allocating the right number of cores for each VM according to its peak (or off-peak depending on QoS level) resource demand to guarantee QoS levels to all VMs while scaling v/f level to achieve power savings.*

Table 2 Performance metrics of a web search application co-located with a VM running PARSEC benchmark: numbers in parenthesis show the case when a web search application is running alone

	IPC	L2 MPKI	L2 miss rate (%)
w/ Backshcoles	0.76 (0.75)	2.38 (2.40)	11.28 (11.57)
w/ Swaptions	0.75 (0.77)	2.32 (2.43)	11.02 (9.63)
w/ Facesim	0.70 (0.70)	2.41 (2.36)	11.41 (11.31)
w/ Canneal	0.76 (0.78)	2.46 (2.43)	11.76 (11.67)

The amount of required CPU utilization of websearch clusters is dynamically varied as the amount of user requests changes over time. Figure 22b shows the CPU utilization traces for two VMs with respects to the number of queries, each of which VM is an index serving node (ISN), in a single web search cluster to process queries requested from the varying number of clients. As shown in the figure, the CPU utilizations of both VMs are highly synchronized with the variation of the number of clients. Moreover, the loads between VMs in a cluster are not perfectly balanced because the CPU utilization depends on the amount of matched results corresponding to a user request. Thus, we can improve the resource utilization by sharing cores among multiple VMs, such that they can more flexibly use cores depending on their time-varying resource demands.

- **Negligible performance degradation caused by LLC conflict:** As analyzed in [4], the performance degradation caused by sharing caches is negligible because the required memory footprint is too large to be sustained by on-chip caches. Table 2 shows the measured performance metrics of a websearch application when it is allocated to share core and cache with various applications in PARSEC benchmark suite. We compared instruction per clock cycles (IPC), L2 miss-per-kilo-instruction (MPKI), and L2 miss ratio (percentage), which are obtained using Xenoprof patched for the AMD15h Bulldozer architecture [55]. The numbers in parenthesis show the cases before co-location. As presented, there are only negligible variations over all the metrics before and after the co-location, which correspond to a negligible performance degradation due to cores sharing. Motivated by these observations, *we can efficiently utilize multiple cores in a server by allocating co-located VMs to share the cores assuming that the performance degradation is negligible.*
- **High correlation among VMs:** As jobs are distributed to multiple VMs in a cluster, workloads of VMs within a same cluster are highly correlated compared to different clusters (or services). In Fig. 22b, we can observe the intra-cluster correlation between two VMs in a cluster, both of which are strongly synchronized with the variation of the number of clients. Thus, *the proposed solution takes into account the pervasive correlation in scale-out applications, i.e., within a cluster as well as among clusters, such that correlated VMs are not co-located.*

Figure 23 illustrates an example of demonstrating the effectiveness of the correlation-aware VM provisioning solution. Let's assume that we have two servers, $Server_1$, and $Server_2$, each of which consists of eight cores, and four

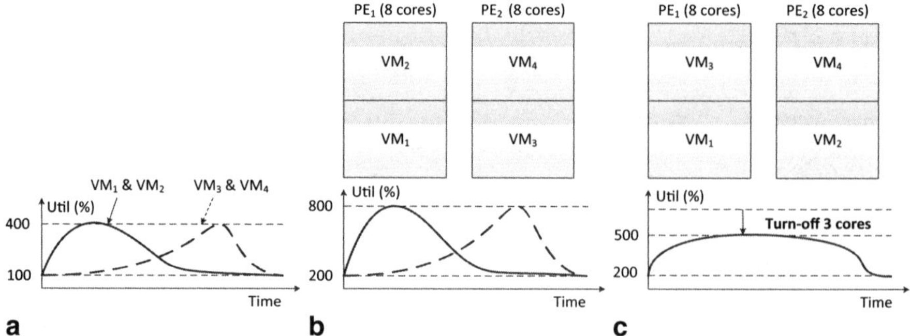

Fig. 23 A motivational example to show the effectiveness of considering correlation information: utilization traces (**a**); VM allocations (**b**) without considering correlation, and with considering correlation (**c**)

VMs, i.e., VM_1, VM_2, VM_3, and VM_4, where $\{VM_1, VM_2\}$ and $\{VM_3, VM_4\}$ are in $Service_1$ and $Service_2$, respectively. We assume that all the VMs have the same amount of the tail distribution on CPU utilization, and VMs in a same service are highly correlated (as load is quite well balanced among VMs) while VMs in different services are less correlated. Figure 23a shows an example of CPU utilization traces. If we do not take into account the correlation, we allocate sets of $\{VM_1, VM_2\}$ and $\{VM_3, VM_4\}$ into $Server_1$ and $Server_2$, respectively, as shown in Fig. 23b. In this case, the maximum CPU utilization amounts to 800 % per each server, thereby, all cores should be in active state. However, if we pair $\{VM_1, VM_3$ and $\{VM_2, VM_4\}$, as shown in Fig. 23c, the actual maximum utilization can be lowered down to 500 for both servers, thereby, we can turn-off (or idle low-power state) three cores per each server and/or lower v/f level without any quality degradation.

Based on the observations and motivations above, we presented a server consolidation solution in [56]. First, to efficiently capture correlation information, they present a low-complexity measure for evaluating workload correlation among co-located VMs, and then, developed VM allocation algorithm.

1) Efficient Correlation Measure for VM Allocation: *Pearson product-moment correlation coefficient*, or *Pearson's correlation*, is most widely used correlation measure to quantify the correlation of used CPU utilization among VMs [53]. It is calculated as the ratio of covariance of the two random variables to the product of their standard deviations. However, the overhead to calculate the metric for a certain time interval is high for a short time period due to the concentrated computation at the end of the time period, because it utilizes the average values of CPU utilization samples collected during each time period. In addition, Pearson's correlation is also partly inefficient because the value reflects correlation throughout the corresponding time interval because we only require the correlation at (off-)peak utilizations in VM placement. Equation (30) is presented in [56] as a new measure to quantify the correlation between two VMs to overcome the inefficiency of the conventional correlation metric.

$$Cost_{i,j}^{vm} = \frac{\hat{u}_{cpu}(VM_i) + \hat{u}_{cpu}(VM_j)}{\hat{u}_{cpu}(VM_i + VM_j)} \tag{30}$$

where $Cost_{i,j}^{vm}$ represents the newly defined correlation measure between VM_i and VM_j. $\hat{u}_{cpu}(VM_i)$ is a reference utilization of VM_i, which is either the peak or the Nth percentile value depending on QoS requirement. The numerator represents the worst-case peak CPU utilization when the peaks of two VMs coincide, while the denominator is an aggregated actual peak utilization when VM_i and VM_j are co-located into a same server. Thus, the higher $Cost_{i,j}^{vm}$, the lower correlation between VM_i and VM_j. Moreover, we can update the values at each sampling period of utilization. Thus, we can save memory space to store all samples as well as evenly distributing computational effort to measure the correlation across a certain time horizon. Using our new $Cost_{i,j}^{vm}$ function, we can model correlations among all VMs by constructing a 2-D matrix, namely, $\mathcal{M}_{cost}^{vm}$ where the (i,j)-th element corresponds to $Cost_{i,j}^{vm}$.

2) Correlation-Aware VM Allocation for Scale-Out Applications Based on the correlation metric in Eq. (30), we can minimize the correlation among co-located VMs in $Server_i$, i.e., $\mathbb{V}_i^{alloc} = \{VM_{i,1}, \cdots, VM_{i,n_i^{vm}}\}$ where n_i^{vm} is the number of VMs allocated to $Server_i$, by allocating VMs such that a weight sum of $Cost_{i,j}^{vm}$ is minimized while the sum of $\hat{u}_{cpu}(VM_{i,j})$ in the server does not exceed the total CPU capability of the server, i.e., Cap_i. The correlation of $Server_i$ is defined as follows:

$$\overline{Cost}_i^{server} = \sum_{j=1}^{n_i^{vm}} w_{i,j}^{vm} \cdot \left(\sum_{k=1\&k\neq j}^{N_{vm}} \frac{Cost_{j,k}^{vm}}{n_i^{vm} - 1} \right) \tag{31}$$

where $w_{i,j}^{vm}$ represents a weight of $VM_{i,j}$, defined as the ratio of $\hat{u}(VM_{i,j})$ to the sum of $\hat{u}(VM_{i,j})$'s of all co-located VMs in $Server_i$.

The problem is a well-known bin-packing problem [43]. In order to reduce the solution complexity within negligible solution quality degradation, Kim et al. present a heuristic based on a *First-Fit-Decreasing* where it first manipulates VMs having the highest utilization among unallocated VMs. Figure 24 shows a pseudo code to achieve this goal. In this algorithm, we periodically adjust VM allocation at every t_{period} based on the workload predictions. The algorithm largely consists of two phases: (1) *UPDATE* (lines 1 $\sim$ 8) and (2) *ALLOCATE* (lines 9 $\sim$ 18). In the *UPDATE* phase, we initialize parameters and update CPU utilization statistics. Then, we allocate VMs to servers in the *ALLOCATE* phase.

The *UPDATE* phase consists of five steps, namely:

- *Initialization*: a set of unallocated VMs ($\mathbb{V}^{unalloc}$), sets of allocated VMs ($\mathbb{V}_i^{alloc}$), remaining capacity (Rem_i) for all servers, and a correlation threshold (TH_{cost}) (lines 1 $\sim$ 4).
- *Prediction*: predict the workload based on history, as we previously prepared in [43] (line 5).

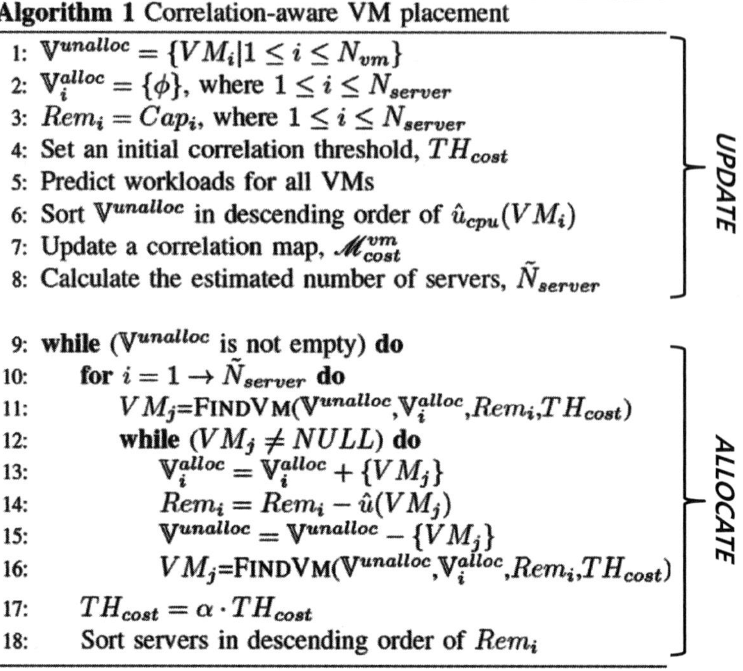

Fig. 24 The correlation-aware VM placement presented in [56]

- *Sorting*: we sort VMs in $\mathbb{V}^{unalloc}$ in descending order of predicted $\hat{u}_{cpu}(VM_i)$ to reduce the fragmentation of the bin-packing problem (line 6)
- *Update cost function*: update $\mathcal{M}^{vm}_{corr}$ by updating the $Cost^{vm}_{i,j}$ for all VM pairs (line 7)
- *Estimate the number of active server sets*: determine the number of estimated active servers, i.e., $\tilde{N}_{server}$, as presented in Eq. (32) (in line 8):

$$\tilde{N}_{server} = \frac{\sum_{i=1}^{N_{vm}} \tilde{\hat{u}}_{cpu}(VM_i)}{N_{core}} \tag{32}$$

  where $\tilde{\hat{u}}_{cpu}$ represents an estimate of $\hat{u}_{cpu}$. Then, $\tilde{N}_{server}$ is equal to the minimum number of servers to accommodate all VMs in $\mathbb{V}^{unalloc}$. We provision VMs to reduce the number of active servers while satisfying performance requirements.

Based on the update information and the predictions, we allocate VMs in *ALLOCATE* phase by iterating the procedure (in line 10 $\sim$ 18) until all VMs are allocated to $\tilde{N}_{server}$ servers (line 9).

- Select a server having the largest remaining CPU capability, i.e., Rem_i (line 10).
- Find a VM to be allocated into $Server_i$ which has the highest $\overline{Cost}_i^{server}$ with VMs in $\mathbb{V}_i^{alloc}$, while satisfying two conditions: (1) $\overline{Cost}_i^{server}$ should be larger than TH_{cost}; and (2) $\hat{u}_{cpu}(VM_i)$ should be less than or equal to Rem_i (line 11)

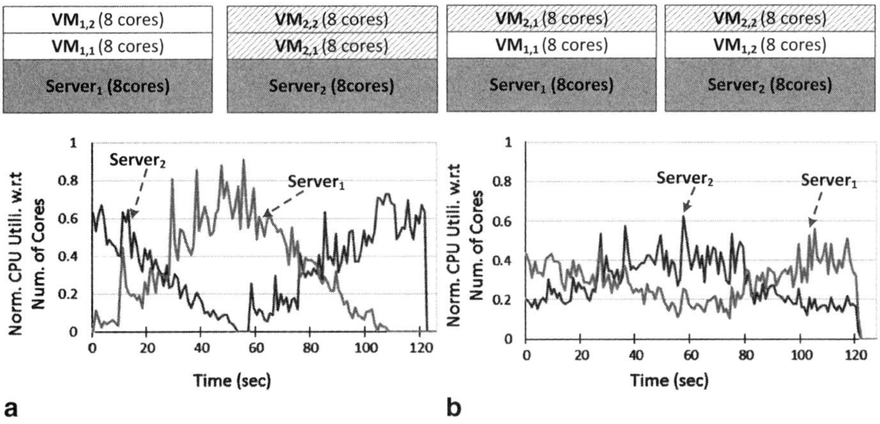

Fig. 25 CPU utilization traces: correlation-unaware (**a**) and correlation-aware (**b**) VM placements [56]

- Update $\mathbb{V}_i^{alloc}$, Rem_i, and $\mathbb{V}^{unalloc}$ accordingly in caase we find a VM (lines $12 \sim 15$)
- Iterate the procedure to find VMs to be allocated in $Server_i$ until until there is no VM left (lines $12 \sim 16$).
- If we have unallocated VMs at the end of the iteration, we need to repeat the procedure in lines $10 \sim 16$ with a degenerated TH_{cost} by a factor of α (line 17) along with a list of servers sorted in descending order of Rem_i (line 18)

3) Setting Voltage and Frequency Level Due to the correlation-aware VM allocation, the actual peak server utilization becomes much lower than the server's computing capability. Figure 25 shows the comparisons of CPU utilization traces when we allocate VMs in correlation-unaware and correlation-aware manner, respectively. Websearch benchmark is used in CloudSuite benchmark suite [4] and configured two websearch clusters each of which has two VMs, i.e., ISNs, in a single websearch cluster, and applied cosine and sin wave user request patterns to each cluster. In the figure, $VM_{i,j}$ represents j-th VM in i-th websearch cluster. As shown in Fig. 22b, workloads of VMs in a same cluster are highly correlated. Thus, a correlation-unaware VM placement solution allocates VMs in a same cluster into a same server as shown in Fig. 25a while the correlation-aware solution allocates VMs in different websearch clusters into a same server as shown in Fig. 25b. As illustrated in the figure, the correlation-aware VM placement solution leads to lowered peak CPU utilization, which enables to lower voltage and frequency (v/f) levels for further power savings.

However, we do not know exactly how much we can lower v/f level when multiple VMs are co-located into a server because $Cost_{i,j}^{vm}$ only captures the correlation between two VMs. An empirical solution to provide rough guideline to lower v/f is provided in [56] which utilizes $\overline{Cost_i}^{server}$ in Eq. (31). Figure 26 shows an empirical relationship representing possible v/f slowdown for servers with respect to

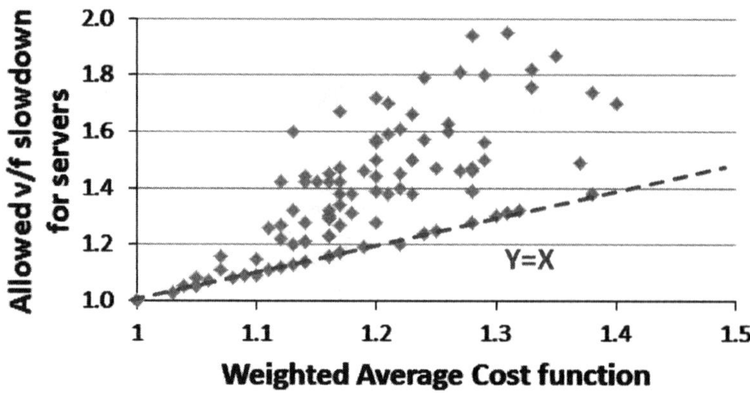

Fig. 26 Relationship between $\overline{Cost}_i^{server}$ in Eq. (31) and possible v/f scaling factor

$\overline{Cost}_i^{server}$. The dots are scattered while the red line, which is a form of $y = x$, shows the lower bound where we can safely lower v/f level without any performance degradation. Based on this relationship, we can determine the frequency level of $Server_i$, i.e., f_i as follows:

$$f_i = \left(\frac{1}{\overline{Cost}_i^{server}} \right) \cdot \left(\frac{\sum_{j=1}^{n_i^{vm}} \hat{u}_{cpu}(VM_{i,j})}{N_{core}^{server}} \right) \cdot f^{max} \tag{33}$$

where f^{max} is the maximum frequency level. f_i is set by lowering the worst-case peak required frequency level (i.e., the second parenthesis assuming the situation when peaks of VMs coincide) with a factor of $1/\overline{Cost}_i^{server}$.

Figure 27 shows 90th percentile response time of websearch benchmark in four different VM placement solutions and v/f levels.

- *Segregated*: allocate VMs into a server such that no VMs share cores
- *Shared-UnCorr*: allocate VMs to share cores without any consideration on their correlation
- *Shared-Corr (2.1G)*: correlation-aware VM allocation while running a server with 2.1 GHz
- *Shared-Corr (1.9G)*: correlation-aware VM allocation while running a server with 1.9 GHz

As shown in Fig. 27, allocating VMs to share cores provides better performance compared to the segregated allocation case. In addition, the correlation-aware VM allocation provides better response time compared to the correlation-unaware allocation scheme as it enables to reduce the actual CPU utilization, thereby the lowered utilization can be used to lower the v/f/ level without any performance degradation compared to the correlation-unaware solution.

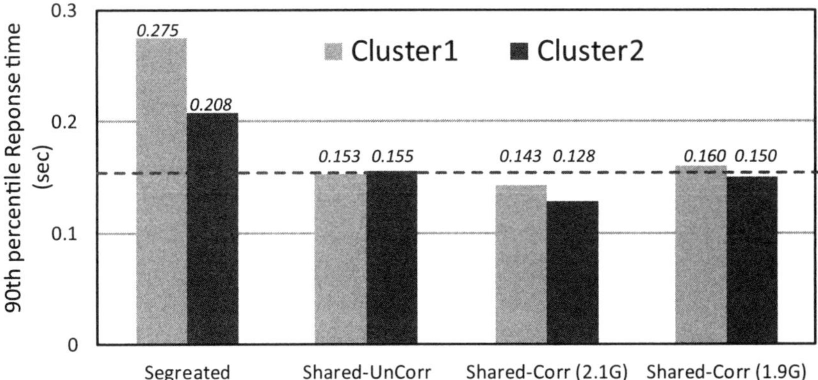

Fig. 27 90th percentile response time of $Cluster_1$ and $Cluster_2$ for three different VM allocations in [56]

4) Simulation Results: Effectiveness of the 3Correlation-Aware VM Placement For further validation of the correlation-aware VM placement, Kim et al. in [56] performed the evaluation using server utilization traces from an actual datacenter setup. It used CPU utilization traces of 40 VMs where each sample is collected at every 5 min for a day while synthesizing fine-grained utilization per 5 s with a lognormal random number generator [7], whose mean is the same as the sampled value for the corresponding 5-min sample. It targeted an Intel Xeon E5410 server configuration which consists of eight cores and two frequency levels (2.0 GHz and 2.3 GHz) and used the power model in [33] to compare the power consumption results among various solutions. It compares three different VM placement approaches as follows.

- *Best-Fit-Decreasing placement (BFD)*: a conventional best-fit-decreasing heuristic approach without taking into account correlation information
- *Peak clustering-based placement (PCP)* [51]: a correlation-aware VM allocation clustering VMs based on the envelopes of VMs' CPU utilization such that VMs coinciding their peaks of the envelopes are not allocated in a same server
- *Correlation-aware placement (CAP)* [56]: a correlation-aware VM allocation considering workload characteristics of scale-out applications manipulating a new correlation metric in Eq. (30).

Table 3a compares the power consumption and performance violations of the three approaches when we statically set the v/f level at the time of VM placement, i.e., $t_{period} = 1$ h. The power consumption results are normalized with respect to the power consumed by *BFD*, and the maximum violation shows the maximum per-period ratio of the number of over-utilized time instances (i.e., when the aggregated utilization among co-located VMs is beyond the CPU capacity of a corresponding server) to t_{period}, during the entire periods, i.e., 24 h. *CAP* provides up to 13.7 % power savings compared to *BFD* and *PCP*, while drastically reducing the number of the violations. It is noteworthy that *PCP* provides almost similar results with *BFD*

Table 3 Comparisons for static (**a**) and dynamic (**b**) v/f scaling

(a)

	Normalized power	Maximum violations (%)
BFD	1	18.2
PCP [51]	0.999	18.2
CAP	*0.863*	*2.6*
(b)		
BFD	1	20.3
PCP	0.997	20.3
CAP	*0.958*	*3.1*

because, due to high and fast-changing correlations among VMs in our utilization traces, *PCP* classifies VMs into only '1' cluster during the most of the time periods (22 out of 24 time periods). When the number of clusters is '1', *PCP* behaves exactly same with *BFD*. The power savings obtained by our proposed solution are due to the aggressive-yet-safe v/f settings utilizing the lowered actual peak resource demand, i.e., Eq. (33). Moreover, the proposed solution provides a drastic reduction of the violations (i.e., 15.6 %) compared to the other approaches. Note that we allocated VMs based on their peak utilizations, which were predicted from the their history. Despite the provision based on the peak utilization, we observed quality degradation over the three approaches due to the mis-predictions of the peak utilization, especially during abrupt workload changes. However, the proposed solution can statistically reduce the probability of the violation by co-locating uncorrelated VMs. Thus, the probability of joint under-predictions among the co-located VMs is drastically decreased.

Table 3b shows the comparisons for the simulated case of servers using dynamic v/f scaling for further investigation of the effectiveness of *CAP*. To prevent frequent oscillations of v/f level (which affects server reliability [70]), we performed the v/f scaling at every 12 samples (i.e., 1 min). As shown in Table 3b, the power savings become smaller compared to the static v/f scaling because the other approaches also adaptively scale v/f level according to the time-varying utilization demand. However, the amount of the violations is unacceptably high in the other approaches. Thus, more servers need to be activated to achieve the same QoS level obtained by the proposed solution, which leads to higher power consumption.

6 Power Minimization of Datacenters with Hybrid Cooling Architectures

The power consumption of datacenter can be further optimized as we jointly reduce the computing and cooling power consumption because the conventional computing power minimization solutions discussed in Sect. 5 usually require higher cooling capability due to the increased heat density of active servers by increasing actual

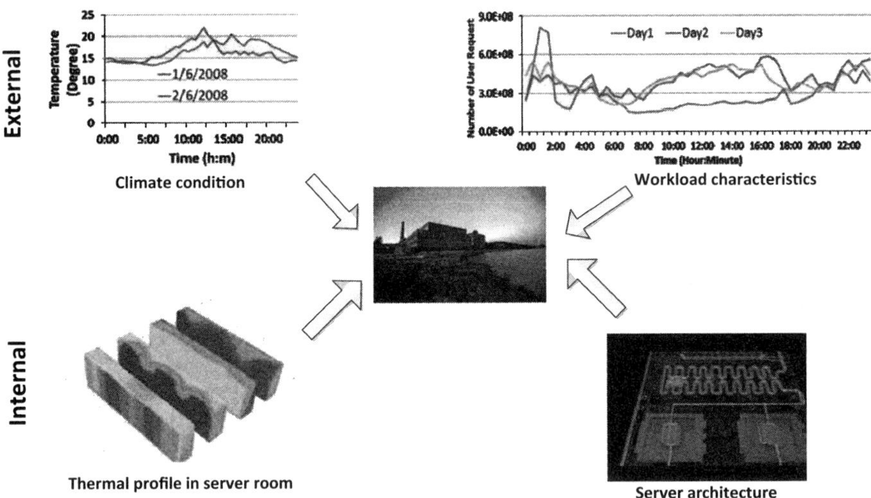

Fig. 28 Proposed solution overview [43]

CPU utilization. Especially, when it comes along with hybrid cooling solutions in a datacenter [26–28, 59–58], explained in Sect. 2.2, we need to revisit existing VM placement solutions [60–65] as it further reduces the chance of using free cooling as the solutions requires higher cooling capability due to the higher operating temperature of active servers. Motivated by this observation, Kim et al. present a joint power and thermal optimization solution for datacenters equipped with hybrid cooling architecutre to achieve further power savings while satisfying service-level agreement (SLA) requirements by extending the usability of free cooling for datacenters having a hybrid cooling architecture [43]. Figure 28 illustrates the solution overview explained in this section. The solution largely takes into account four input parameters as follows:

- Climate condition
- Workload characteristics
- Temperature profile in a server room
- Server cooling architecture

As the climate condition and workload characteristics are non-deterministic, the solution is implemented using a predictive control scheme utilizing predictions of the values. The temperature profile of a server room and the dependency between the server temperature and cooling solutions can be modeled using the solutions explained in Sect. 3.

6.1 Formal Problem Definition

To jointly minimize the computing and cooling power consumption of a datacenter equipped with hybrid cooling architecture, we need to determine the optimal pair of cooling mode, m_{co} (electrical vs. free cooling) and maximum power consumption of active servers (namely, *power capping*) based on four input parameters. In addition, switching cooling mode, i.e., turning on and off chillers, leads to overhead in terms of power and time. Thus, we jointly minimize the number of cooling mode switches along with the power consumption by judiciously considering the switching overhead into the objective function. Based on the requirements, the problem can be formulated as follows:

$$\text{Find} \quad \chi = \{m_{co}, [b_{i,j}]_{N_{pm} \times N_{vm}}\} \tag{34}$$

$$\text{Minimize} \quad J_{dc} = P_{cl} + P_{co} + O_{tr} \tag{35}$$

$$\text{Subject to} \quad T_{pm_i} \leq T_{pm}^{max}, \text{where } 1 \leq i \leq N_{pm} \tag{36}$$

$$Pr(t_{act} > t_{req}) \leq (1 - \beta) \tag{37}$$

The problem we are trying to tackle is two-fold, namely, determining both the (1) cooling mode and (2) VM placement such that the power consumption of datacenter, i.e., $P_{dc} = P_{cl} + P_{co}$ where P_{cl} and P_{co} represent the computing and cooling power consumption in a datacenter, and the overhead caused by cooling mode transition, i.e., O_{tr}, are jointly minimized while satisfying temperature and SLA requirements. m_{co} represents datacenter cooling mode: '1' when free cooling is selected, otherwise '0'; $b_{i,j}$ is a binary variable representing VM placement: '1' when vm_j is mapped into pm_i; N_{pm} and N_{vm} represent the number of servers and VMs, respectively; J_{dc} is an objective function consisting of power consumption of datacenter, i.e., $P_{dc} = P_{cl} + P_{co}$, and overhead caused by switching cooling mode, i.e., O_{tr}; T_{pm_i} and T_{pm}^{max} represent temperature of i-th server (or physical machine) and the maximum temperature constraint of servers, respectively. Then, t_{act} and t_{req} are actual and required execution time, respectively, and $Pr(t_{act} > t_{req})$ represents the probability when t_{act} is larger than t_{req}; β is SLA requirement.

As a matter of fact, this optimization problem can be translated into a bin-packing problem with variable bin size by exploiting the analogy between a bin and a server because, for a given bin size (analogy with threshold of server utilization), the power consumption is minimized when the number of bins (analogy with the number of active servers in which VMs are assigned) is minimized, i.e., server consolidation. Hence, the bin size, i.e., the threshold of server utilization, depends on m_{co} as well as T_{out}. However, due to the interdependency between m_{co} and $b_{i,j}$'s, the solution complexity is even higher than conventional bin-packing problem.

To reduce the solution complexity, we can solve this problem with a two-phase solution. First, we determine a power-optimal pair of $\{m_{co}, u_{pm}^{th}\}$ such that J_{dc} is minimized while satisfying temperature and performance requirements assuming that ideal VM consolidation[3] is applied, i.e., utilization of every active server equals to u_{pm}^{th} while others are '0'. Second, we assign VMs to servers such that the number of servers where VMs are allocated is minimized while total utilization of every server does not exceed u_{pm}^{th}. Moreover, in order to achieve further improvement by considering time-varying characteristics of T_{out} and the user requests, we iterate the optimization procedure at every predefined time interval, t_{opt}. Note that we can reuse server consolidation techniques explained in Sect. 5. Therefore, in this section, we simply focus on explaining the first step of this problem.

6.2 *Multi-objective Trade-offs Exploration Between Cooling Mode and Utilization Threshold*

We explore the best approach to determine the optimal pair of $\{m_{co}, u_{pm}^{th}\}$ which minimizes the multi-objective function, J_{dc}. Since external conditions, i.e., outside temperature and user requests, are time-varying, the optimal pair of $\{m_{co}, u_{pm}^{th}\}$ varies as well. Thus, we periodically adjust $\{m_{co}, u_{pm}^{th}\}$ based on the predictions of the external conditions and the predictive sequence of cooling mode transition. Assuming the ideal VM consolidation at a certain instant, we can approximate the problem as follows:

$$\text{Find} \quad \chi(k) = \{m_{co}(k), u_{pm}^{th}(k)\} \tag{38}$$

$$\text{Min} \quad J_{dc}(k) = \sum_{l=k}^{k+N_h-1} \alpha^{l-k}\left(\widetilde{P}_{cl}(l) + \widetilde{P}_{co}(l) + \widetilde{O}_{tr}(l)\right) \tag{39}$$

$$\text{s.t} \quad u_{pm}^{th}(l) \geq \frac{\hat{U}_{tot}(l)}{N_{pm}}, \forall l \in [k, k+N_h-1] \tag{40}$$

$$u_{pm}^{th}(l) \leq min\left(u_{pm}^{max}, u_{pm}^{temp,max}(l)\right), \forall l \tag{41}$$

where N_h is the number of time periods; α is a weighting factor, $0 \leq \alpha \leq 1$; $\widetilde{P}_{cl}(l)$, $\widetilde{P}_{co}(l)$, and $\widetilde{O}_{tr}(l)$ are predictions of P_{cl}, P_{co}, and O_{tr} at the l-th period, which are expressed as follows:

$$\widetilde{P}_{cl}(l) = \sum_{mode \in \{act,idle,sleep\}} \widetilde{N}_{pm}^{mode}(l)\widetilde{P}_{pm}^{mode}(l) \tag{42}$$

$$\widetilde{P}_{co}(l) = (PUE(u_{pm}^{th}(l)) - 1) \cdot \widetilde{P}_{cl}(l) \tag{43}$$

$$\widetilde{O}_{tr}(l) = w_{tr}^{co} \cdot (m_{co}(l) - m_{co}(l-1))^2 \tag{44}$$

---

[3] In order to reduce the solution complexity, we find the solution assuming that the ideal VM consolidation. The approach is optimistic in that the estimated power consumption is lower than actual scenario due to the fragmentation of the server utilization caused by different utilizations among VMs and fractional ratio of the obtained server utilization to VM utilization in actual scenario.

where $\widetilde{P}_{pm}^{mode}(l)$ is the estimated average power consumption of server at the l-th period when the operating mode of the server is active (i.e., $u_{pm} = u_{pm}^{th}(k)$ based on the assumption of ideal VM consolidation), idle, and sleep modes. $\widetilde{N}_{pm}^{mode}(l)$ is the corresponding number of servers. PUE is obtained using Eq. (14). $(m_{co}(l) - m_{co}(l-1))^2$ represents whether cooling mode is switched at the l-th period, and w_{tr}^{co} is a weighting factor which models the overhead caused by cooling mode transition. $\widetilde{N}_{pm}^{act}(l)$, $\widetilde{N}_{pm}^{idle}(l)$, and $\widetilde{N}_{pm}^{sleep}(l)$ are defined as follows:

$$\widetilde{N}_{pm}^{act}(l) = \frac{\widetilde{U}_{tot}(l)}{u_{pm}^{th}(l)} \tag{45}$$

$$\widetilde{N}_{pm}^{idle}(l) = \frac{\hat{U}_{tot}(l)}{u_{pm}^{th}(l)} - \widetilde{N}_{pm}^{act}(l) \tag{46}$$

$$\widetilde{N}_{pm}^{sleep}(l) = N_{pm} - (\widetilde{N}_{pm}^{act}(l) + \widetilde{N}_{pm}^{idle}(l)) \tag{47}$$

where N_{pm} is the number of servers; $\widetilde{U}_{tot}(l)$ is the prediction of average user requests normalized with respect to the maximum number of user requests processed by single server, i.e., $0 \leq \widetilde{U}_{tot}(l) \leq N_{pm}$; $\hat{U}_{tot}(l)$ is the normalized maximum[4] user requests which is characterized a priori based on extensive characterization.

The first constraint (Eq. (40)) represents the lower bound of $u_{pm}^{th}(l)$ which is determined such that $\hat{U}_{tot}(l)$ user requests can be processed while satisfying SLA requirement. The second constraint (Eq. (41)) represents the upper bound of $u_{pm}^{th}(l)$, which is determined by the minimum value between the utilization level where multiple VMs can run in a single server without acceptable performance loss (i.e., u_{pm}^{max}) and the highest utilization satisfying maximum temperature constraint, i.e., $u_{pm}^{temp,max}(l)$ which is obtained from temperature models in Sect. 3.1.

At the start of k-th period, we solve the optimization problem with two steps: (1) prediction of the external condition, i.e., $\widetilde{U}_{tot}$ and T_{out} for $[k, k + N_h - 1]$-th periods and (2) optimization to find $\{m_{co}(k), u_{pm}^{th}(k)\}$.

1) Temperature and Workload Prediction At the start of k-th period, we predict $T_{out}(l)$ and $\widetilde{U}_{tot}(l)$ where $k \leq l \leq (k + N_h - 1)$. Prediction of T_{out}'s can accurately be predicted by daily and weekly weather forecast. However, accurate prediction of $\widetilde{U}_{tot}$'s is not trivial due to uncertain and non-stationary characteristics of user requests. For accurate prediction, we adopt non-stationary Kalman filter [66], which outperforms other predictors especially when a prediction value is uncertain and non-stationary.

$\widetilde{U}_{tot}(k)$ is predicted based on the history of measured U_{tot} in past few periods as well as the history of the same period in past few days (or weeks). The predictions

---

[4] In this work, we target the SLA violation to be less than 5 %. Thus, we used 95th-percentile value instead of the maximum value to characterize the worst-case behavior of the corresponding period. Considering the correlation among VMs, we can use lower percentile values, e.g., 90-, 80-th percentile, etc., to reduce more power consumption while satisfying SLA requirement, as presented in [51]. Our optimization approach is directly applicable to these cases as well.

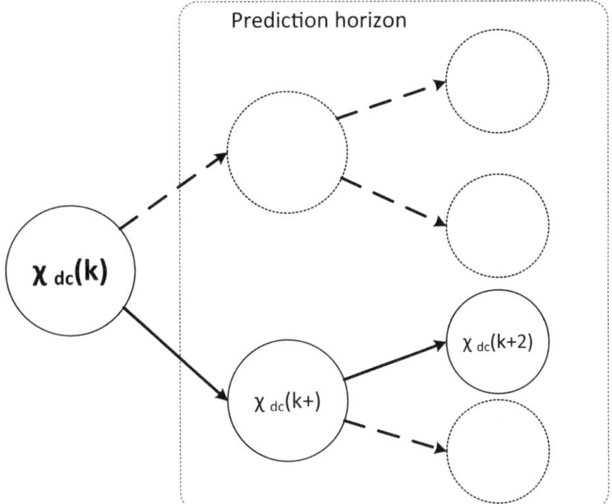

Fig. 29 An example of the predictive control scheme when $N_h = 3$

obtained from the former history is denoted as $\widetilde{U}^{(1)}_{tot}(k)$ while the other is denoted as $\widetilde{U}^{(2)}_{tot}(k)$. Then, we can obtain $\widetilde{U}_{tot}(k)$ by a weighted sum of $\widetilde{U}^{(1)}_{tot}(k)$ and $\widetilde{U}^{(2)}_{tot}(k)$ as follows:

$$\widetilde{U}_{tot}(k) = w^{(1)}_p \widetilde{U}^{(1)}_{tot}(k) + (1 - w^{(1)}_p)\widetilde{U}^{(2)}_{tot}(k) \tag{48}$$

where weight, $w^{(i)}_p(k)$ is weight factor.

2) Predictive Control Scheme To solve the multi-objective problem considering the uncertainty of T_{out} and $\widetilde{U}_{tot}$, we adopt receding horizon control scheme as shown in Fig. 29. At the start of the k-th period, we first predict $\widetilde{U}_{tot}$'s and T_{out}'s for $[k, k + N_h - 1]$-th periods as explained in Sect. 6.2. Second, we find the optimal utilization threshold corresponding to each cooling mode, i.e., $m_{co} = \{0, 1\}$, for $[k, k + N_h - 1]$-th periods, as follows. For a given cooling mode, we can express $\widetilde{P}_{dc}(k) = \widetilde{P}_{dc}(k) + \widetilde{P}_{cl}(k)$ as a continuous form with respect to $u^{th}_{pm}(k)$ using Eqs. (42)–(47). In addition, $\widetilde{P}_{dc}(k)$ is convex with respect to $u^{th}_{pm}(k)$ because, as $u^{th}_{pm}(k)$ increases, $\widetilde{P}_{cl}(k)$ is monotonically decreased (due to the decreased number of active servers) while $\widetilde{P}_{co}(k)$ increases because PUE is monotonically increased. Figure 30 shows the relationship of the power consumption with respect to the u^{th}_{pm}. When an electrical cooling is used, we can find an inflection point as the computing and the cooling power consumptions varies in opposite directions. When a free cooling is used, the total power consumption is usually decreased as the decrease of the computing power as u^{pm}_{th} increases is much larger than the increase of the cooling power. However, the cooling capability of the free cooling is limited, thereby, u^{pm}_{th} cannot be set too high.

Owing to the continuity and convexity of $\widetilde{P}_{dc}(k)$ with respect to $u^{th}_{pm}(k)$ for given $m_{co}(k)$, the unconstrained optimal solution of $u^{th}_{pm}(k)$ can be obtained by finding

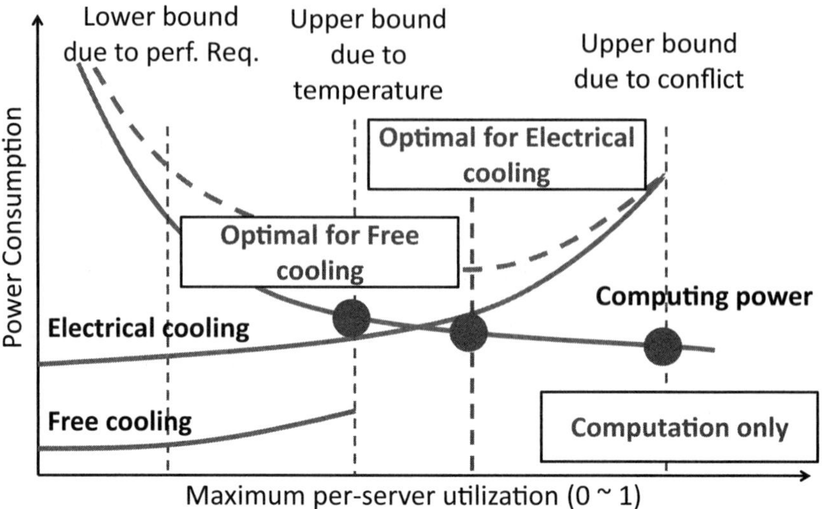

Fig. 30 Solution overview

value which satisfies following equation.

$$\text{Find } u_{pm}^{th}(k) \Longrightarrow \frac{\partial\big(P_{cl}(k) + P_{co}(k)\big)}{\partial u_{pm}^{th}(k)} = 0 \tag{49}$$

The root can be efficiently obtained by root-finding algorithms, e.g., Newton-Raphson method, binary search, etc. [67]. When $u_{pm}^{th}(k)$ satisfies the constraint, we directly set utilization threshold with $u_{pm}^{th}(k)$; otherwise, we set $u_{pm}^{th}(k)$ with lower-bound (Eq. (40)) and upper-bound (Eq. (41)) values so as to satisfy the constraint.

Third, with the pairs of $\{m_{co}, u_{pm}^{th}\}$'s and including the overhead caused by cooling mode transition, i.e., O_{tr}, we find the optimal sequence of cooling mode transition from k-th to $(k + N_h - 1)$-th periods, i.e., $\chi_{dc}(k) \rightarrow \chi_{dc}(k+1|k) \rightarrow \cdots \rightarrow \chi_{dc}(k + N_h - 1|k)$ where $\chi_{dc}(k + l|k)$ is the optimal solution at $(k + l)$-th period when $\chi_{dc}(k)$ is determined as the optimal solution at k-th period. Then, we select only $\chi_{dc}(k)$ and discard the other steps of the sequence. Finally, the entire process is repeated at the start of $(k + 1)$-th period with the updated predictions.

The complexity is $O(N_{pm}^{N_h-1})$. Despite the exponential complexity, the solution can normally be found in low overhead by confining the search range to the proximity of previous N_{pm}'s.

Table 4 Comparisons of power consumption and number of cooling mode transitions in May, June, and July

Period	FIXED-TEMP	P-ADAPTIVE	PT-ADAPTIVE
May 1 ~ May 4	1.00 / 7	0.781 / 8	0.784 / 6
June 1 ~ June 4	1.00 / 0	0.738 / 8	0.743 / 4
July 1 ~ July 4	1.00 / 0	0.879 / 29	0.898 / 13

6.3 Simulation Results

To evaluate the effectiveness of the joint optimization, we used CloudSim [68], an event-driven simulator providing toolkits to model behavior of cloud system components such as datacenters, virtual machines (VMs), and scheduling policies. We configured the target system with 100 servers and 100 VMs and used temperature data measured at EPFL in Lausanne, Switzerland from May 2008 to July 2008. To account for the overhead caused by VM migration, we assumed 100 s and 10 % as the migration time and performance degrdation, respectively. Then, we compared the following cooling mode decision solutions for datacenters:

- **FIXED-TEMP**: a conventional cooling mode decision scheme which uses free cooling only when T_{out} is lower than fixed pre-defined temperature, i.e., $T_{th} = 10\,°C$ [28], and sets u_{pm}^{th} to u_{pm}^{max}.
- **P-ADAPTIVE**: this is our first proposed scheme which adaptively adjusts the cooling mode and the utilization threshold such that only power consumption of datacenter is minimized.
- **PT-ADAPTIVE**: this is our second proposed scheme which jointly optimizes the power consumption and transition overhead caused by cooling mode transition with receding horizon control scheme.

To simply evaluate the effectiveness of the joint cooling mode decision scheme, we applied the same VM allocation solution based on the peak [46] for all the three comparisons above. Remind that these solutions are complementary with existing VM allocation and power management solutions.

Table 4 shows the comparisons in terms of power consumption and number of cooling mode transitions during the first four days in May, Jun, and July. The first column represents the simulated time period. The second to fourth columns show the normalized power consumption with respect to FIXED-TEMP and the number of cooling mode transitions in each month.

First, in May, PT-ADAPTIVE yields 21.6 % power savings compared to FIXED-TEMP. The reason for the improvement can be analyzed by observing the traces of cooling mode and utilization schedules presented in Fig. 31 where a and b depict the traces for FIXED-TEMP and PT-ADAPTIVE, respectively. X-axis represent data (month/date) and the left and right Y-axis are cooling mode/utilization and outside temperature, respectively. The temperature ranges 7 ~ 22 °C in May, thereby FIXED-TEMP uses the free cooling for short time period only when the outside

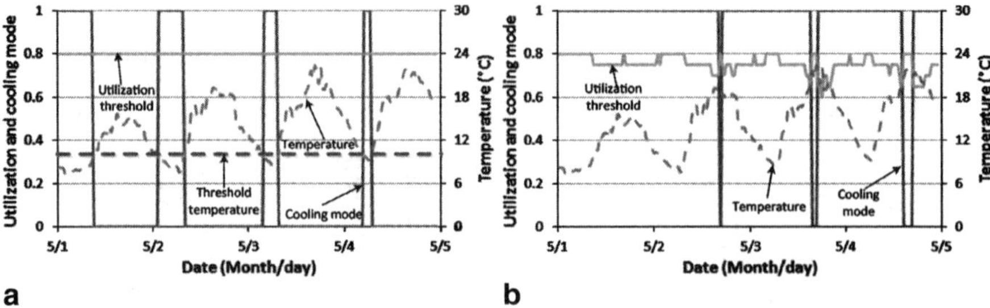

Fig. 31 Schedule of free mode and utilization threshold in May: MAX-UTIL (**a**) and PT-ADAPTIVE(**b**)

Table 5 Comparisons of power consumption and the number of cooling mode transitions as P_{static}/P_{tot} changes in June

P_{static}/P_{tot}	FIXED-TEMP	P-ADAPTIVE	PT-ADAPTIVE
0.3	1.00 / 0	0.722 / 2	0.722 / 2
0.5	1.00 / 0	0.738 / 8	0.743 / 4
0.7	1.00 / 0	0.852 / 24	0.878 /12

temperature is lower than the threshold value, i.e., 10 °C in this evaluation. On the contrary, PT-ADAPTIVE enables to use the free cooling for the longer time period as it dynamically adjusts the maximum power consumption of servers by capping the maximum server utilization according to the amount of demanding workload and the outside temperature.

The most highest power savings are observed in June, i.e., 25.7 %. The reason is that the outside temperature is always higher than 10 °C, which makes impossible to use the free cooling in FIXED-TEMP while PT-ADAPTIVE still decides to use the free cooling by lowering the maximum server power consumption. However, in July, the temperature is too high to use the free cooling while meeting the performance requirements despite capping the maximum server power consumption, which leads to rather smaller power savings, i.e., 10.2 %, compared to other months.

Compared to P-ADAPTIVE, PT-ADAPTIVE provides almost similar (or slightly less) power savings. However, PT-ADAPTIVE schedules the cooling mode such that the number of cooling mode transitions is drastically reduced by accounting for the overhead caused by the cooling mode transitions. Especially, in July, P-ADAPTIVE switches the cooling mode too often, i.e., around 7 times per day while PT-ADAPTIVE can reduce the number of transitions down to 3.25 times per day. Figure 32a and b show the traces of P-ADAPTIVE and PT-ADAPTIVE in July, respectively.

One important observation is that the effectiveness of PT-ADAPTIVE gets enhanced as the energy proportionality of server becomes improvement, which is the direction where server designers are now focusing on. Table 5 shows the normalized power consumption in June as the power-proportionality of servers, defined as the

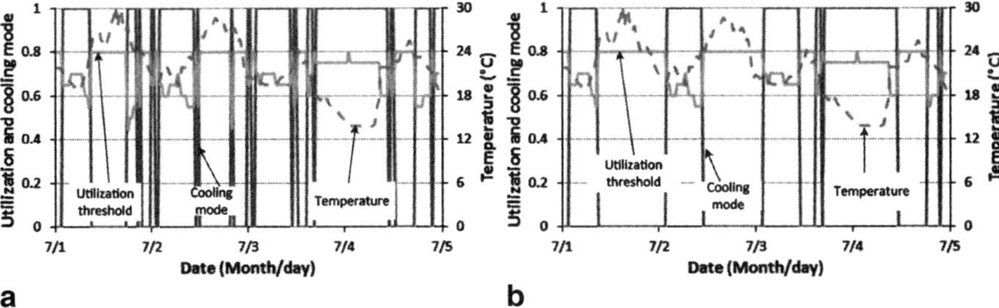

Fig. 32 Schedule of free mode and utilization threshold in July: P-ADAPTIVE (**a**) and PT-ADAPTIVE(**b**)

ratio of the static power to the total power consumption, i.e., P_{static}/P_{tot}, is at 0.3, 0.5, and 0.7. As shown in Table 5, PT-ADAPTIVE provides more power savings as P_{static}/P_{tot} is lowered. As a matter of fact, when P_{static}/P_{tot} is low, we can use free cooling for longer periods of time by lowering the server utilization threshold, thereby we have a smaller number of active servers. Furthermore, as state-of-the-art servers are designed to achieve higher energy-proportionality [69], these experiments demonstrate that PT-ADAPTIVE is able to provide even more power savings in possible future datacenter setups.

7 Conclusions

Recently, the energy-efficiency constraints have become the dominant limiting factor for datacenters due to their unprecedented increase of growing size and electrical power demands. In this chapter, we have explained the power and thermal modeling and control solutions which can play a key role to reduce the power consumption of datacenters considering time-varying workload characteristics while maintaining the performance requirements and the maximum temperature constraints. We have first explained simple-yet-accurate power and temperature models for computing servers, and then, extended the model to cover computing servers and cooling infrastructure of datacenters. Second, we have presented the power and thermal management solutions for servers manipulating various control knobs such as voltage and frequency of servers, workload allocation, and even cooling capability, especially, flow rate of liquid cooled servers). Finally, we have presented the solution to minimize the server clusters of datacenters by proposing a solution which judiciously allocates virtual machines to servers considering their correlation, and then, the joint optimization solution which enables to minimize the total energy consumption of datacenters with hybrid cooling architecture (including the computing servers and the cooling infrastructure of datacenters).

Acknowledgment This work has been partially supported by the Nano-Tera.ch TRANSCEND Strategic Action, the PMSM: CT Monitoring research grant for ESL-EPFL funded by Credit Suisse AG, an ERO Research Grant Donation from Oracle for ESL-EPFL, and the EC FP7 GreenDataNet STREP project (agreement No. 609000).

References

1. K. G. Brill, "The invisible crisis in the data center: The economic meltdown of Moore's law," *white paper, Uptime Institute*, 2007.
2. Energy Star Program, "EDA Report to Congress on Server and Data Center Energy Efficiency," 2007.
3. L. A. Barroso and U. Holzle. "The datacenter as a computer: An introduction to the design of warehouse-scale machines," *Synthesis Lectures on Computer Architecture* 4, no. 1 (2009): 1–108.
4. M. Ferdman, A. Adileh, O. Kocberber, S. Volos, M. Alisafaee, D. Jevdjic, C. Kaynak, A. D. Popescu, A. Ailamaki, and B. Falsafi. "Clearing the clouds: a study of emerging scale-out workloads on modern hardware," in *ACM SIGARCH Computer Architecture News*, vol. 40, no. 1, pp. 37–48. ACM, 2012.
5. A. Adileh, P. Lotfi-Kamran, S. Volos, S. Volos, and C. Kaynak, "CloudSuite on Flexus tutorial," in *international symposium on computer architecture (ISCA)* 2012.
6. D. Meisner, C. M. Sadler, L. A. Barroso, W.-D. Weber, and T. F. Wenisch, "Power management of online data-intensive services," in *Computer Architecture (ISCA), 2011 38th Annual International Symposium on*, pp. 319–330. IEEE, 2011.
7. T. Benson, A. Anand, A. Akella, and M. Zhang, "Understanding data center traffic characteristics," *ACM SIGCOMM Computer Communication Review 40*, no. 1 (2010): 92–99.
8. H. Goudarzi and M. Pedram, "Energy-efficient virtual machine replication and placement in a cloud computing system," in *Cloud Computing (CLOUD), 2012 IEEE 5th International Conference on*, pp. 750–757. IEEE, 2012.
9. 42U Datacenter Efficiency Consulting Corporation, "Data Center Energy Efficiency Calculator," http://www.42u.com/efficiency/energy-efficiency-calculator.htm, 2011.
10. E. Schurman and J. Brutlag, "The user and business impact of server delays, additional bytes, and HTTP chunking in web search," in *Presentation at the OReilly Velocity Web Performance and Operations Conference*, 2009.
11. G. H. Loh and Y. Xie, "3D stacked microprocessor: Are we there yet?," *Micro, IEEE 30*, no. 3 (2010): 60–64.
12. HP DL980, [online available] http://h18000.www1.hp.com/products/servers/platforms/ .
13. Eurocloud, [online avalable] http://www.eurocloudserver.com/ .
14. D. Meisner and T. F. Wenisch, "Does low-power design imply energy efficiency for data centers?," in *Proceedings of the 17th IEEE/ACM international symposium on Low-power electronics and design*, pp. 109–114. IEEE Press, 2011.
15. A. Coskun, J. Meng, D. Atienza, and M. M. Sabry, "Attaining single-chip, high-performance computing through 3D systems with active cooling," *Micro, IEEE 31*, no. 4 (2011): 63–75.
16. U. S. Deparment of Energy, "FEMP Best Practices Guide for Energy-Efficient Data Center Design," in 2011.
17. A. N. Nowroz, R. Cochran, and S. Reda, "Thermal monitoring of real processors: Techniques for sensor allocation and full characterization," in *Proceedings of the 47th Design Automation Conference*, pp. 56–61. ACM, 2010.
18. H. David, C. Fallin, E. Gorbatov, U. R. Hanebutte, and O. Mutlu, "Memory power management via dynamic voltage/frequency scaling," in *Proceedings of the 8th ACM international conference on Autonomic computing*, pp. 31–40. ACM, 2011.

19. R. Raghavendra, P. Ranganathan, V. Talwar, Z. Wang, and X. Zhu, "No power struggles: Coordinated multi-level power management for the data center," in *ACM SIGARCH Computer Architecture News*, vol. 36, no. 1, pp. 48–59. ACM, 2008.
20. X. Wang and Y. Wang, "Coordinating power control and performance management for virtualized server clusters," *Parallel and Distributed Systems, IEEE Transactions on 22*, no. 2 (2011): 245–259.
21. R. Uhlig, G. Neiger, D. Rodgers, A. L. Santoni, F. C. Martins, A. V. Anderson, S. M. Bennett, A. Kagi, F. H. Leung, and L. Smith, "Intel virtualization technology," *Computer 38*, no. 5 (2005): 48–56.
22. P. Muditha Perera and C. Keppitiyagama, "A performance comparison of hypervisors," in *Advances in ICT for Emerging Regions (ICTer), 2011 International Conference on*, pp. 120–120. IEEE, 2011.
23. N. Huber, M. Quast, M. Hauck, and S. Kounev, "Evaluating and Modeling Virtualization Performance Overhead for Cloud Environments," in *CLOSER*, pp. 563–573. 2011.
24. CoolDoor, [online available] http://www.cooldoor.com.au/html/specifications.html .
25. M. Pawlish and A. S. Varde, "Free cooling: A paradigm shift in data centers," in *Information and Automation for Sustainability (ICIAFs), 2010 5th International Conference on*, pp. 347–352. IEEE, 2010.
26. D. Garday, "Reducing data center energy consumption with wet side economizers," *White paper, Intel* (2007).
27. D. Atwood and J. G. Miner, "Reducing data center cost with an air economizer," *White Paper: Intel Corporation* (2008).
28. T. Lu, X. Lu, M. Remes, and M. Viljanen, "Investigation of air management and energy performance in a data center in Finland: Case study," *Energy and Buildings 43*, no. 12 (2011): 3360–3372.
29. D. Wang, B. Ganesh, N. Tuaycharoen, K. Baynes, A. Jaleel, and B. Jacob,"DRAMsim: a memory system simulator," *ACM SIGARCH Computer Architecture News 33*, no. 4 (2005): 100–107.
30. Micron's system power calculators, [online available] http://www.micron.com/products/support/power-calc.
31. D. Economou, S. Rivoire, C. Kozyrakis, and P. Ranganathan, "Full-system power analysis and modeling for server environments," in *Proceedings of Workshop on Modeling, Benchmarking, and Simulation*, pp. 70–77. 2006.
32. S. Rivoire, P. Ranganathan, and C. Kozyrakis, "A Comparison of High-Level Full-System Power Models," *HotPower* 8 (2008): 3–3.
33. M. Pedram and I, Hwang,"Power and performance modeling in a virtualized server system," in *Parallel Processing Workshops (ICPPW), 2010 39th International Conference on*, pp. 520–526. IEEE, 2010.
34. M. K. Patterson, "The effect of data center temperature on energy efficiency," in *Thermal and Thermomechanical Phenomena in Electronic Systems, 2008. ITHERM 2008. 11th Intersociety Conference on*, pp. 1167–1174. IEEE, 2008.
35. J. Choi, Y. Kim, A. Sivasubramanjam, J. Srebric, Q. Wang, and J. Lee, "A CFD-based tool for studying temperature in rack-mounted servers," *Computers, IEEE Transactions on 57*, no. 8 (2008): 1129–1142.
36. T. Heath, A. P. Centeno, P. George, L. Ramos, Y. Jaluria, and R. Bianchini, "Mercury and freon: temperature emulation and management for server systems," in *ACM SIGARCH Computer Architecture News*, vol. 34, no. 5, pp. 106–116. ACM, 2006.
37. W. Huang, S. Ghosh, S. Velusamy, K. Sankaranarayanan, K. Skadron, and M. R. Stan, "HotSpot: A compact thermal modeling methodology for early-stage VLSI design," *Very Large Scale Integration (VLSI) Systems, IEEE Transactions on 14*, no. 5 (2006): 501–513.
38. R. Ayoub, R. Nath, and T. Rosing, "JETC: Joint energy thermal and cooling management for memory and CPU subsystems in servers," in *High Performance Computer Architecture (HPCA), 2012 IEEE 18th International Symposium on*, pp. 1–12. IEEE, 2012.

39. E. Pakbaznia and M. Pedram, "Minimizing data center cooling and server power costs," in *Proceedings of the 14th ACM/IEEE international symposium on Low power electronics and design*, pp. 145–150. ACM, 2009.
40. D. C. Hwang., V. P. Manno, M. Hodes, and G. J. Chan, "Energy savings achievable through liquid cooling: A rack level case study," in *Thermal and Thermomechanical Phenomena in Electronic Systems (ITherm), 2010 12th IEEE Intersociety Conference on*, pp. 1–9. IEEE, 2010.
41. T. J. Breen, E. J. Walsh, J. Punch, A. J. Shah, and C. E. Bash, "From chip to cooling tower data center modeling: Part I influence of server inlet temperature and temperature rise across cabinet," in *Thermal and Thermomechanical Phenomena in Electronic Systems (ITherm), 2010 12th IEEE Intersociety Conference on*, pp. 1–10. IEEE, 2010.
42. A. Qouneh, C Li, and T. Li. "A quantitative analysis of cooling power in container-based data centers," in *Workload Characterization (IISWC), 2011 IEEE International Symposium on*, pp. 61–71. IEEE, 2011.
43. J. Kim, M. Ruggiero, and D. Atienza, "Free cooling-aware dynamic power management for green datacenters," in *High Performance Computing and Simulation (HPCS), 2012 International Conference on*, pp. 140–146. IEEE, 2012.
44. Smart data center energy monitoring: a thermal-aware design approach to 'Green IT', http://esl.epfl.ch/cms/op/edit/lang/en/pid/57400
45. Credit Suisse, https://www.credit-suisse.com/
46. E. Pakbaznia, *et al.*, "Minimizing data center cooling and server power costs," in *Proc. ISLPED*, 2009.
47. N. Bobroff, *et al.*, "Dynamic placement of virtual machines for managing sla violations," in *Proc. IM* 2007.
48. P. Padala, X. Zhu, Z.i Wang, S. Singhal, and K. G. Shin. "Performance evaluation of virtualization technologies for server consolidation," in *HP Labs Tec. Report*, 2007.
49. O. Tickoo, R. Iyer, R. Illikkal, and D. Newell, "Modeling virtual machine performance: challenges and approaches," in *ACM SIGMETRICS Performance Evaluation Review 37*, 2010.
50. S. Govindan, J. Liu, A. Kansal, and A. Sivasubramaniam. "Cuanta: quantifying effects of shared on-chip resource interference for consolidated virtual machines," in Proceedings of the 2nd ACM Symposium on Cloud Computing, p. 22. ACM, 2011.
51. A. Verma, *et al.*, "Server workload analysis for powr minimization using consolidation," in *Proc. USENIX*, 2009.
52. X. Meng, *et al.*, "Efficient resource provisioning in compute clouds via VM multiplexign," in *Proc. ICAC*, 2010.
53. M. Chen, *et al.*, "Effective VM sizing in virtualized data centers," in *Proc. IM*, 2011.
54. K. Halder, *et al.*, "Risk aware provisioning and resource aggregation based consolidation of virtual machines," in *Proc. Cloud*, 2012.
55. A. Menon, J. R. Santos, Y. Turner, G. J. Janakiraman, and W. Zwaenepoel, "Diagnosing performance overheads in the xen virtual machine environment," in *Proceedings of the 1st ACM/USENIX international conference on Virtual execution environments*, pp. 13–23. ACM, 2005.
56. J. Kim, M. Ruggiero, D. Atienza, and M. Lederberger, "Correlation-aware virtual machine allocation for energy-efficient datacenters," in Proc. *Conference on Design, Automation and Test in Europe (DATE)*, pp. 1345–1350, 2013.
57. M. K. Patterson, D. Atwood, and J. G. Miner, "Evaluation of air-side economizer use in a compute-intensive data center," *ASME*, 2009.
58. M. Pervila and J. Kangasharju,"Running servers around zero degrees," *ACM SIGCOMM Computer Communication Review 41*, no. 1 (2011): 96–101.
59. "Google data center," http://www.google.cim/about/datacenters/#.
60. P. Barham, B. Dragovic, K. Fraser, S. Hand, T. Harris, A. Ho, R. Neugebauer, I. Pratt, and A. Warfield, "Xen and the art of virtualization," *ACM SIGOPS Operating Systems Review 37*, no. 5 (2003): 164–177.

61. C. Clark, K. Fraser, S. Hand, J. G. Hansen, E. Jul, C. Limpach, I. Pratt, and A. Warfield, "Live migration of virtual machines," in *Proceedings of the 2nd conference on Symposium on Networked Systems Design and Implementation-Volume 2*, pp. 273–286. USENIX Association, 2005.
62. D. Kusic, J. O. Kephart, J. E. Hanson, N. Kandasamy, and G. Jiang, "Power and performance management of virtualized computing environments via lookahead control," *Cluster computing 12*, no. 1 (2009): 1–15.
63. G. Dhiman, G. Marchetti, and T. Rosing, "vGreen: a system for energy efficient computing in virtualized environments," in *Proceedings of the 14th ACM/IEEE international symposium on Low power electronics and design*, pp. 243–248. ACM, 2009.
64. J. Xu and J. A. Fortes, "Multi-objective virtual machine placement in virtualized data center environments," in *Green Computing and Communications (GreenCom), 2010 IEEE/ACM Int'l Conference on and Int'l Conference on Cyber, Physical and Social Computing (CPSCom)*, pp. 179–188. IEEE, 2010.
65. J.-W. Jang, M. Jeon, H.-S. Kim, H. Jo, J.-S. Kim, and S.l Maeng, "Energy reduction in consolidated servers through memory-aware virtual machine scheduling," *Computers, IEEE Transactions on 60*, no. 4 (2011): 552–564.
66. S.-Y. Bang, K. Bang, S. Yoon, and E.-Y. Chung, "Run-time adaptive workload estimation for dynamic voltage scaling," *Computer-Aided Design of Integrated Circuits and Systems, IEEE Transactions on 28*, no. 9 (2009): 1334–1347.
67. K. Madsen, "A root-finding algorithm based on Newton's method," *BIT Numerical Mathematics 13*, no. 1 (1973): 71–75.
68. R. Buyya, R. Ranjan, and R. N. Calheiros, "Modeling and simulation of scalable Cloud computing environments and the CloudSim toolkit: Challenges and opportunities," in *High Performance Computing and Simulation, 2009. HPCS'09. International Conference on*, pp. 1–11. IEEE, 2009.
69. D. Meisner, B. T. Gold, and T. F. Wenisch, "PowerNap: eliminating server idle power," in *ACM Sigplan Notices*, vol. 44, no. 3, pp. 205–216. ACM, 2009.
70. Y. Guo, D. Zhu, and H. Aydin, "Reliability-aware power management for parallel real-time applications with precedence constraints," in *Green Computing Conference and Workshops (IGCC), 2011 International*, pp. 1–8. IEEE, 2011.
71. J. Kong et al. Recent thermal management techniques for microprocessors. In *ACM Computing Surveys*, 44(3):13:1–13:42, 2012.
72. I. Koren and C. M. Krishna. Temperature-aware computing. In *Sustainable Computing: Informatics and Systems*, 1(1):46–56, 2011.
73. J. Choi et al. Thermal-aware task scheduling at the system software level. In *ISLPED*, 2007.
74. A. K. Coskun, T. Simunic Rosing, and K. Whisnant. Temperature aware task scheduling in MPSoCs. In *DATE*, pages 1659–1664, 2007.
75. J. Donald and M. Martonosi. Techniques for multicore thermal management: Classification and new exploration. In *ISCA*, pages 78–88, 2006.
76. A. K. Coskun et al. Temperature management in multiprocessor socs using online learning. In *DAC*, pages 890–893, 2008.
77. A. K. Coskun et al. Energy-efficient variable-flow liquid cooling in 3D stacked architectures. In *DATE*, pages 111–116, 2010.
78. Festo electric automation technology. http://www.festo-didactic.com/ov3/media/customers/1100/00966360001075223683.pdf.
79. Y. U. Ogras, R. Marculescu, D. Marculescu, and E. G. Jung. Design and management of voltage-frequency island partitioned networks-on-chip. *IEEE Transactions on VLSI*, 17(3):330–341, 2009.
80. P. Bogdan, S. Jian, R. Tornero, and R. Marculescu. An optimal control approach to power management for multi-voltage and frequency islands multiprocessor platforms under highly variable workloads. In *ISNoC*, pages 35–42, 2012.
81. W-L. Hung, Y. Xie, N. Vijaykrishnan, M. Kandemir, and M. J. Irwin. Thermal-aware task allocation and scheduling for embedded systems. In *DATE*, pages 898–899, 2005.

82. A. K. Coskun, T. Simunic Rosing, and K. Gross. Proactive Temperature Balancing for Low-Cost Thermal Management in MPSoCs. In *ICCAD*, pages 250–257, 2008.
83. R. J. Cochran et al. Consistent Runtime Thermal Prediction and Control Through Workload Phase Detection. In *DAC*, pages 62–67, 2010.
84. Y. Zhang et al. Adaptive and Autonomous Thermal Tracking for High Performance Computing Systems. In *DAC*, pages 68–73, 2010.
85. Y. Wang et al. Temperature-constrained power control for chip multiprocessors with online model estimation. In *ISCA*, pages 314–324, 2009.
86. F. Zanini et al. Online Convex Optimization-Based Algorithm For Thermal Management of MPSoCs. In *GLSVLSI*, pages 203–208, 2010.
87. A. Bemporad et al. The explicit linear quadratic regulator for constrained systems. *Automatica*, 38(1):3 –20, 2002.
88. C. Zhu et al. Three-dimensional chip-multiprocessor run-time thermal management. *IEEE Transactions on Computer-Aided Design of Integrated Circuits and Systems*, 27(8):1479–1492, August 2008.
89. X. Zhou et al. Thermal management for 3D processors via task scheduling. In *ICPP*, pages 115–122, 2008.
90. A. K. Coskun, J. Ayala, D. Atienza, T. Simunic Rosing. Modeling and Dynamic Management of 3D Multicore Systems with Liquid Cooling. In *VLSI-SoC*, pages 60–65, 2009.
91. A. K. Coskun et al. Dynamic thermal management in 3D multicore architectures. In *DATE*, pages 1410–1415, 2009.
92. T. Emi et al. Tape: Thermal-aware agent-based power economy for multi/many-core architectures. In *ICCAD*, pages 302 –309, 2009.
93. H. Qian et al. Cyber-physical thermal management of 3D multi-core cache-processor system with microfluidic cooling. *ASP Journal of Low Power Electronics*, 7(1):1–12, 2011.
94. F. Zanini, M. M. Sabry, D. Atienza, and G. De Micheli. Hierarchical thermal management policy for high-performance 3d systems with liquid cooling. *IEEE JETCAS*, 1(2):88–101, 2011.
95. F. Mulas et al. Thermal balancing policy for multiprocessor stream computing platforms. *IEEE Transactions on Computer-Aided Design of Integrated Circuits and Systems*, 28(12):1870–1882, 2009.
96. M. M. Sabry et al. Energy-Efficient Multi-Objective Thermal Control for Liquid-Cooled 3D Stacked Architectures. *IEEE Transactions on Computer-Aided Design of Integrated Circuits and Systems*, 30(12):1883–1896, 2011.
97. P. Greenalgh. Big.LITTLE Processing with ARM Cortex-A15 and Cortex-A7. www.arm.com/files/downloads/big.LITTLE_Final.pdf.
98. R. G. Dreslinski et al. Near-Threshold Computing: Reclaiming Moore's Law Through Energy Efficient Integrated Circuits. In *Proc. of the IEEE*, 98(2), 2010.
99. N. Xu et al. Thermal-Aware Post Layout Voltage-Island Generation for 3D ICs. In *Journal of Computer Science and Technology*, 28(4):671–681, 2013.
100. K. Puttaswamy and G. H. Loh. Thermal Herding: Microarchitecture Techniques for Controlling Hotspots in High-Performance 3D-Integrated Processors. In *HPCA*, pages 193–204, 2007.
101. Y. Han et al. Temperature Aware Floorplanning. In*Workshop on Temperature Aware Computing Systems*, 2005.
102. K. Sankaranarayanan, S. Velusamy, M. Stan, and K. Skadron. A Case for Thermal-Aware Floorplanning at the Microarchitectural Level. In*Journal of Instruction-Level Parallelism*, 8:1–16, 2005.
103. W-L. Hung et al. Thermal-Aware Floorplanning Using Genetic Algorithms. In*ISQED*, 2005.
104. J. Cong, J. Wei, and Y. Zhang. A Thermal-Driven Floorplanning Algorithm for 3D-ICs. In*ICCAD*, pages 306–313, 2004.
105. W.-L. Hung et al. Interconnect and Thermal-Aware Floorplanning for 3D Microprocessors. In*ISQED*, pages 98–104, 2006.

106. M. Healy et al. Multiobjective Microarchitectural Floorplanning for 2-D and 3-D ICs. In*IEEE Transactions on Computer-Aided Design of Integrated Circuits and Systems*, 26(1), 2007.
107. M. Ekpanyapong et al. Thermal-aware 3D Microarchitectural Floorplanning. Georgia Institute of Technology, 2004.
108. H. Mizunuma et al. Thermal Modeling and Analysis for 3D-ICs with Integrated Microchannel Cooling. In*IEEE Transactions on Computer-Aided Design of Integrated Circuits and Systems*, 30(9):1293–1306, 2011.
109. M. M. Sabry et al. Greencool: An energy-efficient liquid cooling design technique for 3d mpsocs via channel width modulation. *IEEE Transactions on Computer-Aided Design of Integrated Circuits and Systems*, 32(4):524–537, 2013.
110. R. Shah and A. London. *Laminar flow forced convection in ducts*. New York: Academic Press, 1978.
111. Y. Tan et al. Modeling and simulation of the lag effect in a deep reactive ion etching process. *Journal of Micromechanics and Microengineering*, 16, 2006.
112. A. Leon et al. A power-efficient high-throughput 32-thread SPARC processor. *ISSCC*, 42(1):7 – 16, 2007.
113. M. M. Sabry, A. Sridhar, and D. Atienza. Thermal balancing of liquid-cooled 3d-mpsocs using channel modulation. In *DATE*, 2012.

Thermal Modeling and Management of Storage Systems in Data Centers

Xunfei Jiang, Ji Zhang, Xiao Qin, Meikang Qiu, Minghua Jiang and Jifu Zhang

1 Introduction

Thermal modeling and management techniques have been widely investigated in recent years. Prior studies show thermal management could increase energy efficiency of data centers. The thermal impacts of CPUs on data storage have been extensively studied; however, disk thermal models are still in their infancy. In our study, we aim at building thermal models that take into account both CPUs and disks. We propose an approach to developing thermal models to estimate a data node's outlet temperature based on its CPU and disk activities. Integrating our thermal model into an energy consumption model, we can evaluate the total energy cost of a data center.

X. Jiang (✉)
Department of Computer Science,
Earlham College, Richmond, IN, USA
e-mail: jiangxu@earlham.edu

J. Zhang · X. Qin
Department of Computer Science and Software Engineering,
Auburn University, Auburn, AL, USA
e-mail: jzz0014@auburn.edu

X. Qin
e-mail: xqin@auburn.edu

M. Qiu
Computer Engineering, San Jose State University, San Jose, CA, USA
e-mail: meikang.qiu@sjsu.edu

M. Jiang
College of Mathematics and Computer Science, Wuhan Textile University, Wuhan, China
e-mail: jmh@wtu.edu.cn

J. Zhang
Taiyuan University of Science and Technology, Taiyuan, China
e-mail: jifuzh@sina.com

S. U. Khan, A. Y. Zomaya (eds.), *Handbook on Data Centers,*
DOI 10.1007/978-1-4939-2092-1_30

We apply our models to study the impact of various thermal management strategies on energy efficiency.

Energy consumption of data centers has dramatically increased in the past few years [1, 2]. Statistics show that computing cost coupled with cooling cost can take up to 25 % of the total energy cost of data centers [3]. In improving energy efficiency of data centers, much effort has been made to reduce the computing cost and cooling cost of storage systems [4]. For example, a wide range of workload distribution strategies have been proposed to reduce computing cost of data centers. Furthermore, recent studies demonstrate that cooling cost can be lowered by reducing outlet temperatures of storage nodes, balancing temperature distribution in data centers, or minimizing heat recirculation.

In many use cases, massive amount of data must be transferred through networks. These use cases include, but are not limited to, data backup (e.g., iDrive), file sharing (e.g., dropbox), video on demand (e.g., YouTube), and photo management (e.g., Flickr). The energy cost of data transmission activities becomes a key issue in achieving high energy efficiency of data centers. For example, the worldwide monthly active users of the social network Facebook increased from 100 million in the third quarter of 2008 to 1155 million in the second quarter of 2013 [5]. There are over 72 million links shared, 300 million photos uploaded, 2.5 billion status updated, and 2.7 billion likes and comments are made every day [6]. Transferring such a huge amount of data through the network inevitably results in high energy consumption. It is appealing and challenging to improve energy efficiency of data centers by reducing the energy consumption induced by data transmissions.

Contributions We propose a modeling approach to building thermal models used to estimate outlet temperatures of a data node. Our thermal models are the driving force behind the thermal-aware management strategies developed in this study. We make the following three contributions.

- First, we investigate the thermal profile of a storage server. The profiling data is collected by imposing CPU-intensive workload (i.e., Whetstone [7]) and I/O intensive workload (i.e., Postmark [8]). When the CPU and disk are running under various workload scenarios, we monitor CPU and disk temperatures in addition to the server's inlet and outlet temperatures.
- Second, we develop a thermal model to derive outlet temperatures from inlet temperatures, CPU and disk workload.
- Third, we propose two thermal management strategies aiming to reduce energy consumption of data storage systems.

Organization The rest of this chapter is organized as follows. The next section presents prior studies and related studies. In Sect. 3, we conduct experiments to investigate thermal behaviours of data nodes. Then, we introduce a thermal model for data nodes and validate the thermal models against real-world measurement acquired by temperature sensors. In Sect. 4, we propose new thermal-aware management strategies to save energy consumption of data storage systems. Section 5 shows the experimental results of our proposed strategies. Finally, Sect. 6 concludes the chapter.

2 Related Work

2.1 Efficient Data Centers

A study by DatacenterDynamics demonstrates that worldwide investment in data centers in 2012 had increased by 22.1 % up to US$ 105 billion compared with 2011; such an investment is going to grow by another 14.5 % to US$ 120 billion in 2013 [9]. Evidence shows that the energy consumption of data centers has increased rapidly for the past few years [1, 2].

Growing attention has been paid to building energy-efficient data centers [10, 11, 12]. One reason behind the striking energy consumption in data centers is the rapid growth of computing and storage capacity in recent years. Two main contributors to high energy cost of data centers are computing cost and cooling cost. Computing cost refers to the energy consumption caused by computing infrastructures; cooling cost is the energy consumption of cooling systems that lower data center temperatures.

Various energy conservation techniques have been developed to reduce computing energy consumption by redistributing workload or turning off the power of disks or data nodes. For instance, MAID specifies a subset of disks as cache disks that are in charge of processing I/O requests; MAID aggressively spins down other disks to conserve energy [13]. The PDC (Popular Data Concentration) technique migrates frequently accessed data to a subset of disks [14]; the other disks storing non-frequently accessed data could be transitioned to the low-power mode. Both MAID and PDC substantially reduce total energy consumption of data nodes. Another handful of studies concentrate on energy-efficient resource management of data centers to minimize computing energy consumption. For example, Beloglazov and Buyya proposed an energy-efficient resource management system for virtualized Cloud data centers [15]. In this system, VMs are consolidated according to the utilization of resources, and virtual network topologies are built between VMs and node thermal statuses to save energy. Their technique reduces the operational costs of data centers and offers required Quality of Service (QoS). Furthermore, Beloglazov et al. investigated resource provisioning and allocation algorithms for energy-efficient Cloud computing [16]. The experimental results show that their framework has immense potential in lowering energy cost and improving energy efficiency under dynamic workload scenarios. Aksanli et al. designed an adaptive data center job scheduler, which utilizes prediction of solar and wind energy production [17]. Their scheduler improves the energy efficiency by a factor of three.

An increasing number of techniques have been developed to reduce cooling cost of data centers [18]. These effective strategies include managing airflow in data centers, locating cooling systems close to IT equipments, dynamically controlling thermal load of data centers, and maintaining high operating temperatures. Dynamic thermal control is the most important scheme among these strategies. In prior studies, thermal-aware workload management strategies have been proposed to reduce outlet temperatures of data nodes [19], to minimize heat recirculation [20], and to decrease node inlet temperatures [21]. For instance, XInt—a thermal-aware task scheduling algorithm—was proposed to minimize heat recirculation by balancing workload within a homogeneous data center [21]. The results show that cooling costs highly depend on peak inlet temperatures, which can be effectively reduced by the MPIT-TA task assignment policy. Simulation results reveals that MPIT-TA saves at least 20 % of cooling energy.

2.2 Thermal Modeling

Studying thermal impacts of computing resources is the first step toward characterizing thermal behaviors of data centers. The main components of a data node include CPU, disk, memory and network card, among which CPU and disk are key contributors to a data node's outlet temperatures.

For modeling CPU temperature, HotSpot was proposed to accurately and quickly predict CPU temperatures at the micro-architecture level [22]. HotSpot is based on an equivalent circuit of thermal resistances and capacitances that correspond to micro-architecture blocks and essential aspects of the thermal package. With HotSpot, in order to model the thermal behavior of different CPU types, one has to develop the micro-architecture and sophisticated models. Little work has been done to model CPU temperatures at the coarse-grained level.

A few studies have been conducted to characterize the thermal patterns of disks. Eibeck and Cohen proposed a thermal model to predict transient temperatures of IBM's fixed disk drive [23]. Tan et al. introduced a three-dimensional transient temperature model, which estimates disk temperatures under frequent seeking operations [24]. Gurumurthi et al. built a comprehensive thermal model that takes into account a disk's five components, including internal drive air, spindle motor, the base and cover of the disk, the voice-coil motor, and disk arms. Results showed that their model could predict disk temperatures very close to the real operating temperature of the disk [25]. A few studies also investigated the impact of seek time and inter-seek time on disks temperatures [26]; empirical findings indicated that either increasing inter-seek time or decreasing seek time can lower disk temperatures. In our recent study, we investigated disk temperatures as a function of CPU and I/O load; we proposed thermal models for hard drive disks as well as solid state disks [27, 28].

In addition, approaches were proposed for coordinating processors and memory to improve system performance and/or power efficiency during memory thermal emergency [29]. An adaptive core gating (DTM-ACG) and coordinated DVFS (DTM-CDVFS) schemes, as well as a thermal model, were designed to predict DRAM temperatures. Experimental results show that these two schemes exhibit 6.7 and 15.3 % of improvements in terms of performance.

2.3 Thermal Management

Improving energy efficiency increasingly becomes important and challenging for data centers. Strategies that reduce cooling cost of data centers make a great contribution to advance energy-efficient data centers. Growing attention has been paid to thermal management of computing and storage resources in data centers [30]. Thermal-aware resource management relies on data placement or load distribution policies to achieve balanced temperature distribution among data nodes.

Sharma et al. proposed a thermal-aware load-balancing framework to dynamically distribute loads across data nodes in a data center [31]. Their simulation results showed that equipment reliability can be substantially improved by placing an asymmetric workload and uniformly distributing temperatures in the data center.

A handful of studies proposed temperature-aware load balancing strategies [32, 33, 34]. For example, a simple yet effective way to ensure thermal control is setting a customized threshold to limit CPU temperatures. If CPU temperatures exceed the threshold, then CPU voltage [32] and frequency [33] are dynamically adjusted to conserve CPU energy consumption at the cost of increasing execution times. Another research proposed an on-line thermal prediction model for 3D chips [34]. Then the peak temperature on the chip was reduced through novel task scheduling algorithms. In this research, inter-iteration dependencies were considered, and simulations showed that these algorithms could reduce the peak temperature up to 8.1 °C. However, most existing thermal-aware load balancing strategies have not taken full consideration of disks as a critical factor affecting data nodes' outlet temperatures.

Predictive thermal management strategies have captured the attention of the research community. A performance-effective dynamic thermal management system called DTM for multimedia applications was designed to reduce energy consumption [35]. DTM proposed by Srinivasan and Adve performs significantly better than the existing reactive DTM algorithms. Ramos and Bianchini built a software structure - C-Oracle - for Internet services [36]. C-Oracle chooses the best reaction by predicting the temperature and performance impacts of multiple thermal management reactions. C-Oracle effectively deals with thermal emergencies without unnecessary performance degradation. Very recently, we proposed a framework called PEAM to reduce energy consumption induced by data transmissions [37]. At the heart of PEAM is a model that predicts energy cost of various data transfer policies. Experimental results show that given runtime information, PEAM is able to select the best data transmission policy to minimize energy consumption.

3 Thermal Modeling

Thermal models for data centers can be generally divided into two camps: models estimating thermal behaviors of CPU and disks; models characterizing outlet temperatures of data nodes. Models capturing thermal impacts of disks on data nodes are in their infancy. In this chapter, we conduct extensive experiments to illustrate the thermal behaviors of disks and CPU as well as their impacts on data nodes.

The experimental testbed is equipped with a server containing a Celeron(R) 2.2 GHz CPU, 1.0 GBytes RAM, and a 160 GBytes SATA disk. Temperature sensors and watchdog are applied to monitor the disk and inlet/outlet temperatures of the

Table 1 Testbed configurations

Hardware	Software
1 × Intel(R) Celeron(R) 450@2.2 GHz	Ubuntu 10.04
1 × 1.0 GBytes of RAM	Linux kernel 2.6.32
1 × WD 160 GBytes Sata disk	
(WD1600AAJS-75M0A0 [38])	

server. The configuration parameters of the testbed are summarized in Table 1. The ambient temperature is set to 23.2 °C.

3.1 CPU Thermal Model

We use the CPU's interior temperature sensor to monitor CPU temperatures. To study the thermal behavior of CPU, we first let the CPU remain in the idle state, in which the CPU temperature is around 39∼40 °C. Then, we run the whetstone benchmark—a float computation benchmark—to generate various experiment scenarios. In these scenarios, we slightly modify the whetstone source code in a way to deliver multiple CPU-utilization levels. More specifically, we set the number of iterations to 4000, 8000, 10000, 11900, 11950, and 12000, respectively. The CPU utilization in these cases can be found in Fig. 1; the CPU temperatures are shown in Fig. 2.

Figure 1 reveals that when we change the number of iterations in whetstone, the CPU utilization is relatively steady staying around specific values during the entire testing phase. Figure 2 shows that the CPU temperature trend can be modeled into three stages, namely, the heat up stage, the steady stage, and the cool down stage. In

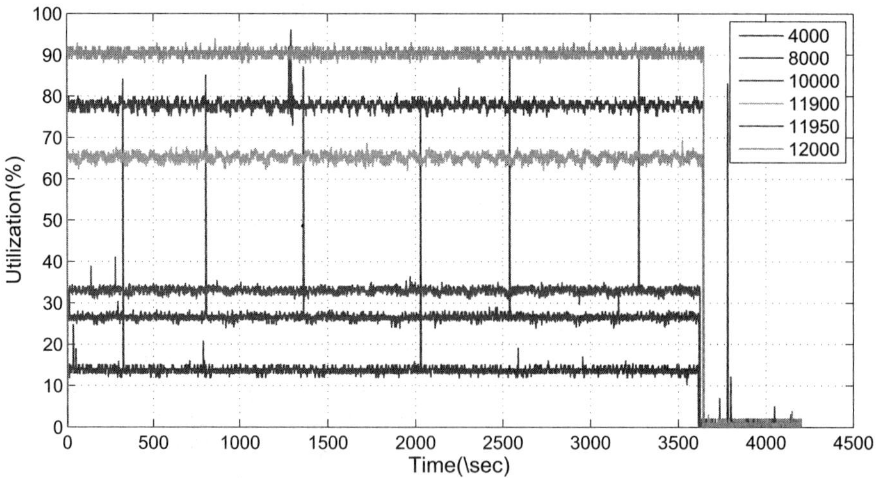

Fig. 1 CPU utilization under different scenarios

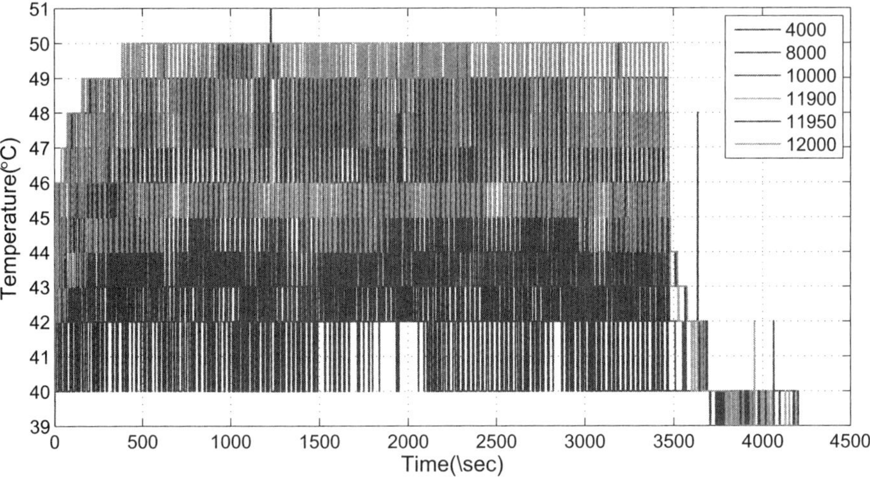

Fig. 2 CPU temperature under different scenarios

Table 2 CPU utilizations and temperatures in the steady stage under various number of loops

Scenarios	1	2	3	4	5	6
Loop number	4000	8000	10000	11900	11950	12000
Average utilization (%)	13.8	26.7	33.1	65.2	77.9	90.5
Average temperature (°C)	41.3	42.9	43.7	46.7	48	49.1
Max temperature (°C)	48	50	49	49	51	50
Min temperature (°C)	40	40	42	44	46	47

the heat up stage, the CPU temperature goes up very quickly until it reaches a peak value. In the steady stage, CPU temperature is held constant since the CPU remains busy. And in the cool down stage, CPU utilization is dropping; CPU temperature cools down to its initial temperature, which is equal to the idle state's CPU temperature.

Observing the above two figures, we conclude that increasing the number of iterations leads to high CPU utilization, which in turn increases CPU temperature. In all the experiments when CPU is active, CPU temperatures go up very quickly in the first 600 s (or 10 min), and then CPU remains in the steady state. And CPU temperature cooling speed is faster than its heat-up speed. We observe that the CPU heat up time is about 10 min; the CPU cool down time is less than 10 min.

We summarize the detailed CPU utilizations and temperatures in Table 2. The average CPU temperatures in the steady stage are 41.3, 42.9, 43.7, 46.7, 48.0, and 49.1 °C, respectively. The maximum CPU temperatures in the steady stage are anywhere between 48 and 51 °C, and the minimum CPU temperatures in the steady stage are increasing when the number of iterations increases.

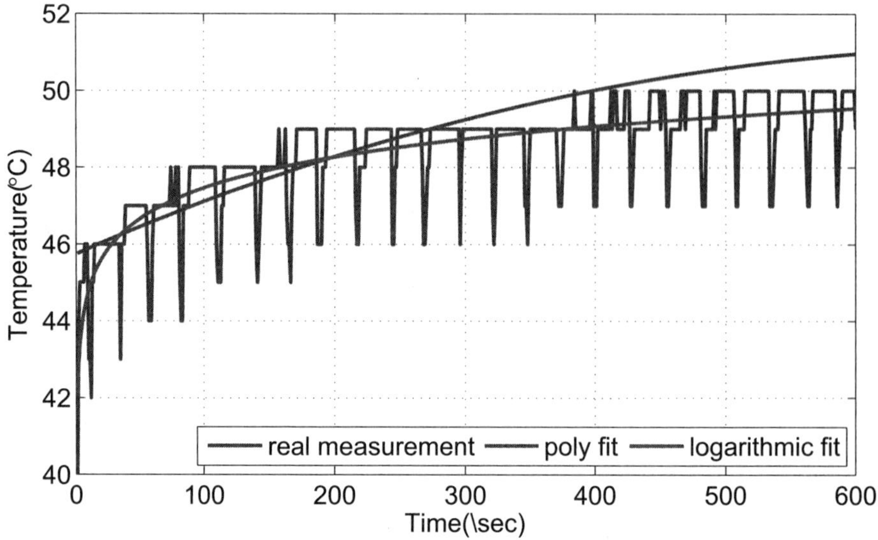

Fig. 3 CPU temperature model validation (12000LOOPS)

We use both the polynomial model and logarithmic model to capture the characteristics of CPU temperatures during the heat up stage under various CPU utilizations. Figure 3 shows a comparison of CPU temperature estimated by two types of models, where the number of iterations in the benchmark is set to 12000. The precision error of the polynomial model is 1.87 %, which is higher than that of the logarithmic model (1.31 %). The logarithmic model fit the CPU temperature curve better than the polynomial model when CPU is in the heat up stage. When it comes to the other experiments, we also collect evidence showing that the logarithmic model performs better than the polynomial model in most cases. Thus, we decide to choose the logarithmic model to estimate CPU temperatures in our study.

3.2 Disk Thermal Model

To study disk thermal behaviors, we conduct a second group of experiments using Postmark, which launches five I/O-intensive tasks. Before each task starts running, the disk is sitting idle in the steady state, where the initial disk temperature is 27.62 °C. The number of files accessed by Postmark is set to 100; file sizes are in a range between 1.E+6 and 1.E+8 Byte. We alter disk utilization by varying the write block size and buffer settings of Postmark. If the buffer of Postmark is enabled, then the buffered *stdio* function calls should be used instead of the lower level raw system calls [8]. All the other parameters of Postmark are set to their default values.

Table 3 Postmark configuration of experiments on disk

Scenarios	1	2	3	4	5
Buffer enabled	N	N	N	N	Y
Write block size(KB)	16	32	64	128	256

The disk utilization is periodically assessed by the *iostat* utility program. The disk temperature is monitored by a temperature sensor embedded on the disk. The experiment settings are summarized in Table 3.

The disk utilization and temperature in the five different experimental settings are shown in Figs. 4 and 5.

Figure 4 suggests that a large write block size leads to high disk utilization. As shown in Fig. 5, the disk temperature trend is also comprised of three stages (i.e., the heat up, steady, and cool down stages) when Postmark is running on the testbed under various settings.

Table 4 summarizes the average disk utilization in these five experiments. The results indicate that we are able to control disk utilization by choosing an appropriate write block size with Postmark.

We use the polynomial model and logarithmic model to fit the disk temperatures during the heat up stage under different disk utilizations. Figure 6 shows disk temperature in the heat up stage under the utilization of 80.57 % estimated by the two models. A comparison between the modeled and measured temperatures also can be found in Fig. 6. The precision errors of the polynomial model and the logarithmic model are 0.61 and 0.21 %, respectively. Here, we observe that the logarithmic model outperforms the polynomial model. The other four experiments also confirm that the

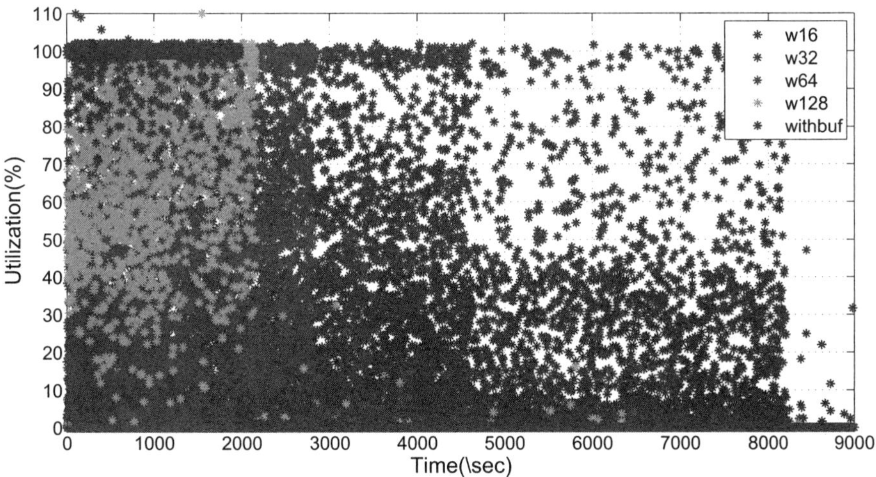

Fig. 4 Disk utilization with various write block size

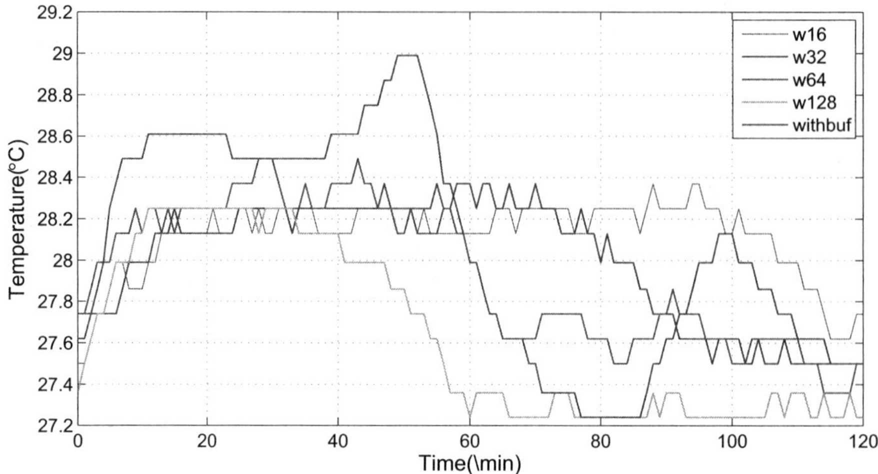

Fig. 5 Disk temperature with various write block size

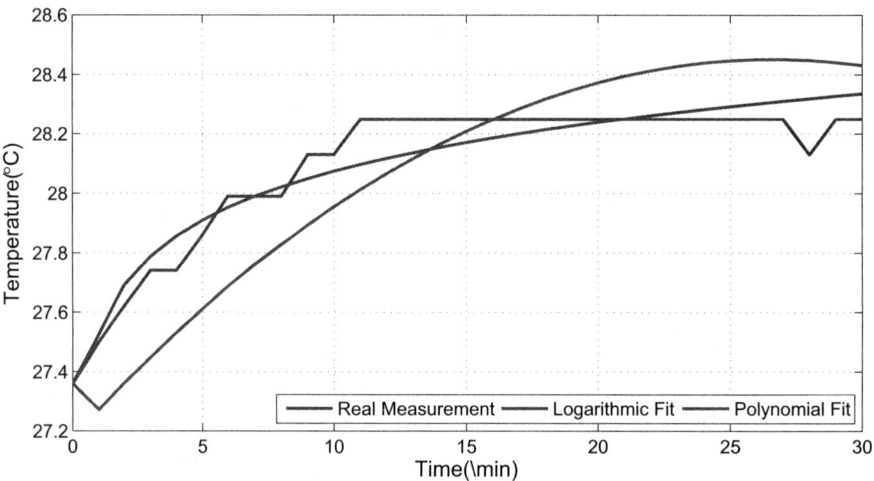

Fig. 6 Disk temperature model validation (Write block size: 128 Byte)

logarithmic model is superior to the polynomial model in terms of estimating disk temperatures.

3.3 *Thermal Model of Data Nodes*

In previous subsection, we have studied the thermal behaviors of CPUs and disks. Now, we are in a position to investigate the impacts of CPUs and disks on temperatures

Table 4 Impact of write block size on disk utilization

Scenarios	1	2	3	4	5
Average Util(%)	14.24	28.91	53.49	80.57	100

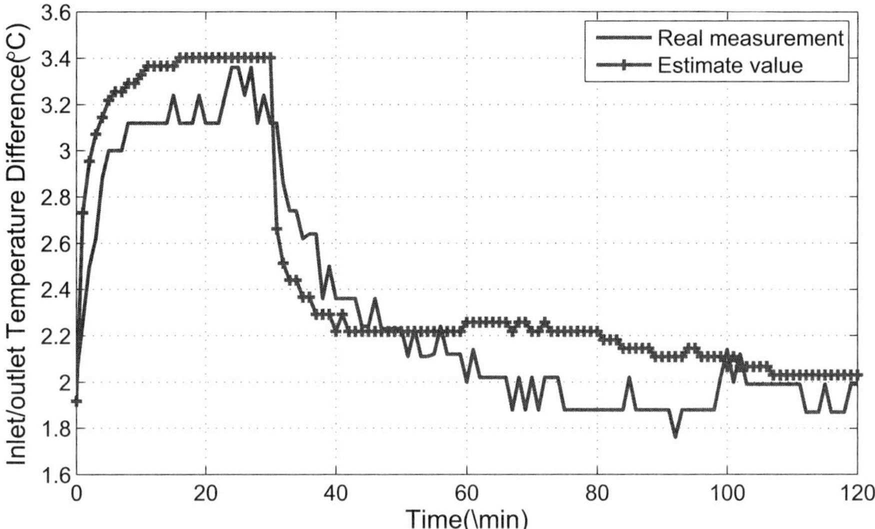

Fig. 7 Outlet temperature model validation [28]

of data nodes in a data center. A recent study [28] investigated the thermal behavior of two types of disks, the model of which was proposed as a linear one (i.e., $T_{outlet} = T_{inlet} + \alpha * T_{CPU} + \beta * T_{disk} + \gamma$) to predict outlet temperatures. A comparison of the modeled outlet temperatures and the real measured ones can be found in Fig. 7. In our study, we apply the outlet temperature model in our thermal management process.

3.4 Evaluation of Temperature Models

To verify the CPU, disk and outlet temperature models, we run the WordCount benchmark counting words of files randomly generated by Postmark. The total size of these files is around 10 GB. Figure 8 shows CPU and disk utilization of the testbed running WordCount. We observe that the CPU and disk utilization are relatively steady during the course of the benchmark's execution. The average CPU and disk utilizations are 92.48 and 18.6 %, respectively.

We built two models to predict CPU and disk temperatures. Figures 9 and 10 show the accuracies of the two models by comparing the modeled CPU and disk temperatures with the measured ones. The precision errors of the two models are as low as 1.52 and 0.48 %, respectively.

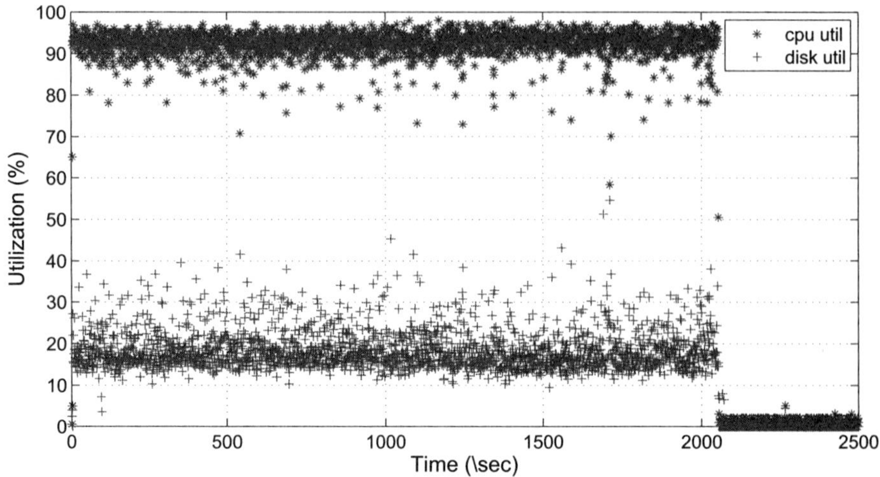

Fig. 8 CPU and disk utilizations for running WordCount

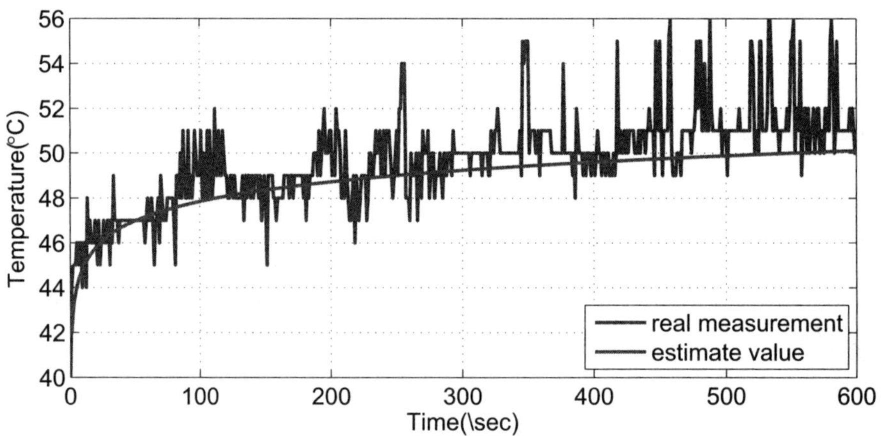

Fig. 9 CPU temperature model validation for WordCount

We also calculate the precision error of the outlet temperature model by obtaining measured temperatures and comparing them with the modeled ones. The experimental results show that the precision error of our outlet temperature is 3.77 %.

4 Thermal Management Strategies

Traditional thermal management strategies are focused on scheduling tasks in a way that temperatures of data nodes or components of data nodes are kept under a threshold. Most of these strategies distribute the load according to CPU temperatures, because processors are a top contributor to outlet temperatures of data nodes.

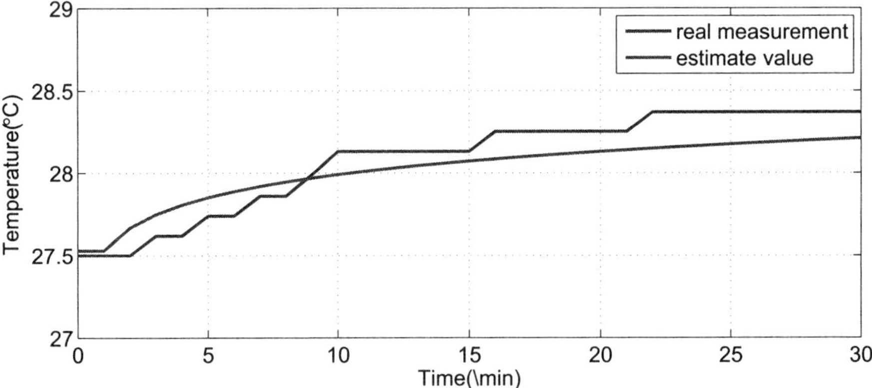

Fig. 10 Disk temperature model validation for WordCount

Our recent study shows that different combinations of CPU and disk workload lead to various thermal behaviors of data nodes [27]. In the previous section we observe that CPU and disk temperatures are related to their utilizations. With the CPU and disk temperature models in place [28], one can derive CPU and disk temperatures from resource utilization. Applying the outlet temperature model, a thermal management module is enabled to estimate data nodes' outlet temperatures. The thermal behaviors of disks have been investigated in prior studies; however, incorporating disk thermal models into task scheduler and data transmission mechanisms has not been addressed. In this section, we propose a task scheduling scheme and a predictive thermal manager. Both approaches incorporate CPU and disk activities into the thermal management.

4.1 *Task Scheduling*

We propose a thermal-aware task scheduling scheme to dispatch CPU-intensive or I/O-intensive tasks in a way to minimize negative thermal impact. Our preliminary results (see details in [27]) show that scheduling workloads could save energy consumption caused by both processors and disks. To demonstrate the strength of our scheme, we create workload scenarios that exhibit intensive CPU and disk load. Our design goal is to ensure that outlet temperatures do not exceed a predetermined threshold while minimizing the total energy consumption of the data storage systems.

Our thermal-aware scheduling policy follows three rules to dispatch tasks:

- to place as many data nodes as possible into the active mode,
- to make active data nodes run as busy as possible, and
- to ensure that each data node's outlet temperature does not exceed a threshold.

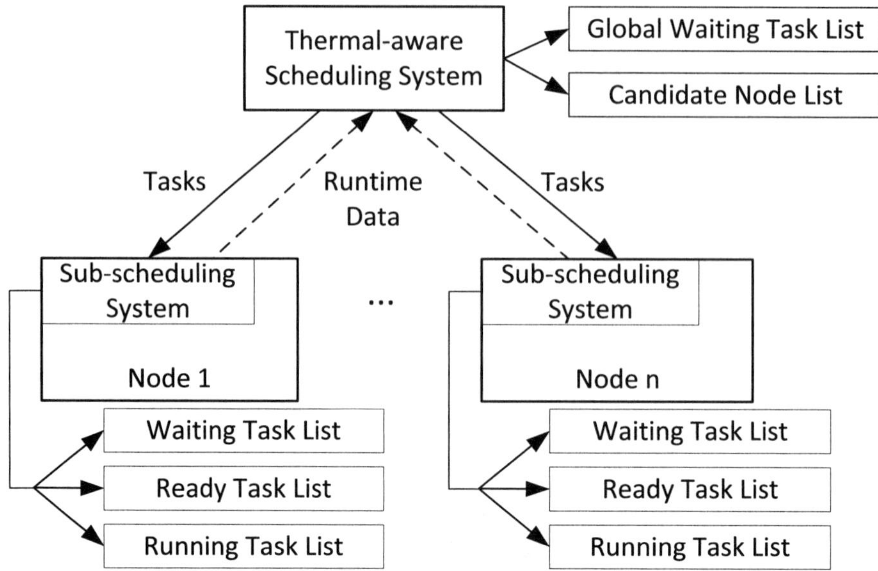

Fig. 11 The system framework of the thermal-aware task scheduler

Figure 11 plots the system framework of our thermal-aware task scheduler. In this framework, a storage system is comprised of *n* data nodes, a thermal-aware task scheduler that assigns tasks to the data nodes. At the heart of each data node, there is a sub-system, in which a monitor is responsible for detecting utilization and temperatures of multiple components (e.g., CPU and disks) in the data node. The task scheduler distributes CPU and I/O load so that data nodes make an effort to complete tasks in a short time period while keeping the outlet temperatures below a threshold. Under heavy workload conditions, our policy strives to keep as many nodes active as possible to finish tasks. As a result, the scheduling policy creates ample opportunities to transit nodes into the low-power mode to conserve energy after all the tasks are aggressively completed by the active nodes.

The task scheduler maintains the following two lists:

- a global waiting task list and
- a candidate node list

The global waiting task list holds pending tasks assigned to the storage system. The candidate node list maintains a group of data nodes that are either sitting idle or underutilized.

In the global task list, the tasks are arranged in the first-in-first-out order. To reduce the average response time of tasks, the scheduler may give higher priority to small tasks. Please note that other scheduling policies (e.g., small task first and earliest deadline first) can be seamlessly incorporated into our scheduling framework. The scheduling system monitors the behaviors of all data nodes, and assigns incoming tasks in the global task list to candidate data nodes. Before assigning a task to a specific data node, the monitor on each data node fetches runtime information, and temperature models are applied to estimate the thermal impacts of tasks on a list of candidate data nodes. Tasks are assigned to a data node if the tasks will not make the node a hot spot.

Three lists that are managed by the sub-scheduling system on each data node include:

- a waiting list,
- a ready list, and
- a running list.

The waiting list holds tasks that must be executed on this particular data node, the ready list contains tasks that are ready to run on the data node. The running list maintains tasks that are running on the node. The sub-scheduling system not only manages runtime information of all the running tasks, but also launches tasks stored in the ready list on each node. When the ready list becomes empty, tasks are moved from the waiting list into the ready list under the condition that new ready tasks placed in the ready list do not push the node's outlet temperature to a high level exceeding the specified threshold.

Upon the arrival of a new task, the scheduler first checks if the task has any preferred nodes (i.e., the task must be executed on a particular data nodes due to data availability or hardware constraints). If the task has no preferred node, the task will be pushed into the global waiting task list. Then, the system dispatches tasks in the global waiting task list to the list of candidate data nodes. After a task in the global waiting list has been dispatched to a data node, the task is moved from the global waiting list to the node's ready list.

If a task has preferred node, the system will check the monitoring information provided by the target data node. In addition, the system estimates the CPU and disk utilization induced by the new task. With the thermal models (see the previous section) in place, the scheduler predicts outlet temperature impact to be made by this new task. If the outlet temperature does not top the threshold, the task will be placed into the ready list of the data node; the task is ready to be executed. In case of multiple ready tasks, the round robin algorithm is used to offer all the ready tasks with shared CPU utilization. If the expected outlet temperature exceeds the threshold, the new task will be moved to the waiting list of the data node. Tasks in the waiting list will not be moved to the ready list until the sub-scheduling system confirms that the waiting tasks are not going to drive the outlet temperature to exceed the threshold.

When the waiting list of a data node is so long that tasks could not be finished within an expected time period, the scheduler will migrate some of these waiting tasks from the current node to other candidate data nodes with lighter load. Such a migration policy follows the following rules to determine candidate tasks to be migrated to other nodes:

- choose CPU-intensive tasks first, and then consider I/O-intensive tasks,
- choose tasks, input data of which could be accessed in the candidate data nodes, and
- choose tasks, input data of which is small.

The first rule aims to migrate CPU-intensive and I/O-intensive tasks to quickly alleviate the burden of the local node. The second and third rules help in reducing migration overhead by minimizing the amount of data to be migrated. After choosing candidate tasks and their migration destinations, the scheduling system checks the destination nodes' data availability. If the destination nodes have input data of migrated tasks, then the tasks will be moved directly to the waiting list of the destination nodes. If the data is not available on the destination nodes, the system will have to migrate the input data to the destination data nodes. While the data is being migrated, tasks will be moved to the destination nodes' waiting lists.

4.2 Predictive Thermal-Aware Data Transmission

Data transmission plays an important role in reducing operational cost in data centers. A data transmission procedure is composed of the following three phases:

- the pre-transmission phase,
- the transmission phase, and
- the after-transmission phase.

In the first phase, data are read from disks and cached on source data nodes. Then in the second phase, data are transferred from the source data nodes to destination data nodes through networks. In the last phase, data are written down to disks residing in the destination nodes.

Frequent data transmissions can have significant impacts on energy consumption of data centers. Data placement strategies and data replica management are proposed to reduce energy cost caused by data transmissions. Compressing data before transmission is effective in lowering energy consumption of data transmission. In one of our preliminary studies, we have investigated thermal behaviors of a data compression mechanism [37]. Our predictive thermal-aware data transmission system proposed in this section aims to reduce data centers' energy cost by minimizing adverse thermal impacts of data transmissions.

We implement the following three data transmission strategies in our predictive thermal-aware management system. We also study the impacts of various types of data on these data transmission policies.

- direct transmission,
- archived transmission, and
- compressed transmission.

In the direct transmission (or DT for short) policy, data are directly transferred over the network without being archived or compressed. During the transmission phase, the size of transferred files remain unchanged.

In the archived transmission (or AT for short) policy, data are archived into a single file in the pre-transmission phase before being transferred through the network. After

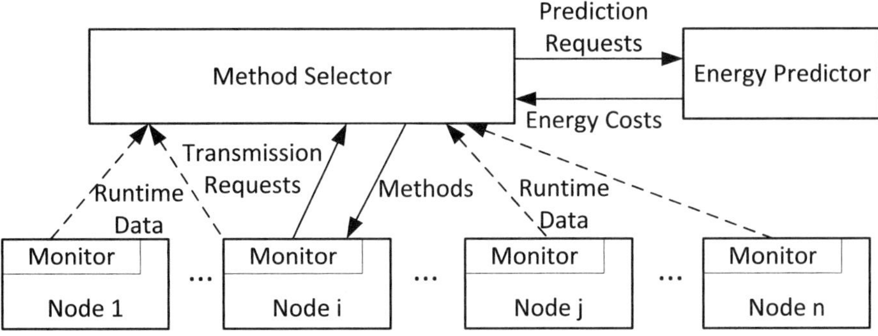

Fig. 12 The framework of the predictive thermal-aware management system (PTMS)

the archived file is received by a destination data node, the single file must be unarchived on a destination data node before being written to the node's disk in the after-transmission phase.

In the compressed transmission (or CT for short) policy, data are compressed into a single file in the pre-transmission phase. Then, the compressed file is transferred through the network. Finally, the file is decompressed by the receiving node before being stored to the node's disk during the after-transmission phase.

Compared with the direct data transmission policy, the compressed data transmission policy substantially reduces the amount of data being transferred. However, compressing data takes processing time and power, which in turn drives CPU utilization of the source node very high.

In data centers of large enterprises (e.g., Google, Amazon, and Facebook), the amount of data downloaded from the data centers is much larger than that of data uploaded to the data centers. Downloading processes are involved with transferring data from nodes in a data center to its clients. As such, in this study we are focusing on reducing energy cost of data transmission from the perspective of data centers.

Figure 12 illustrates the framework of predictive thermal-aware management system or PTMS for short. This system is centered around a storage system equipped with n data nodes. The monitor module gathers the runtime information pertinent to data transmissions, file metadata, and storage nodes (e.g., temperatures and utilizations). When a data transmission request is made, the module forwards the request to the data transmission method selector, which chooses the most thermal friendly and energy-efficient policy to transfer the requested data.

The method selector not only maintains the candidate data-transfer strategies, but also judiciously chooses the best strategy to reduce energy cost by minimizing negative thermal impacts. Figure 12 shows that upon the arrival of a data-transmission request, the method selector forwards the request along with all the candidate strategies to the energy predictor. According to energy cost estimated by the predictor, the method selector informs the data transmission module of the best strategy that will achieve the best energy efficiency during the data transfers.

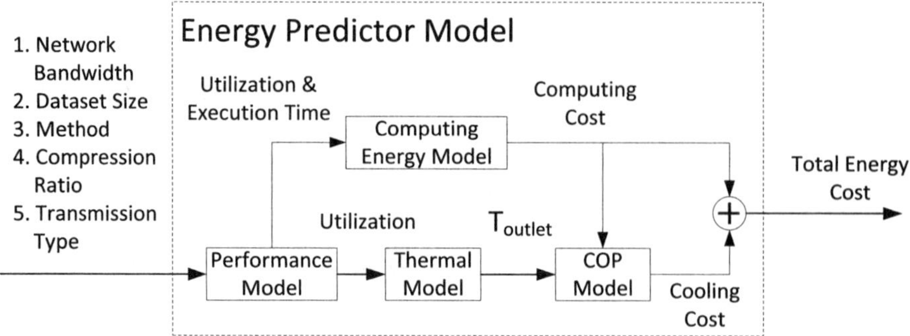

Fig. 13 The framework of the energy predictor module. (COP: Coefficient of Performance)

The energy predictor, as shown in Fig. 13, estimates the energy cost of data transmissions handled by a candidate strategy. In our predictive thermal-aware data transmission system, the predictor is focused on the energy consumption that includes both computing energy cost and cooling cost of data nodes in the storage system. Hence, before estimating energy cost, the data transmission type should be specified. Three types of data transmission from the perspective of data centers are upload, download, and data transmissions. In the energy predictor, we adopt the performance models and energy models employed in PEAM [37].

5 Results

5.1 *Task Scheduling*

To evaluate the performance of our task scheduling system, we conduct two groups of experiments, which resemble various real-world workload scenarios. Table 1 shows the parameters of a small-scale storage cluster of four data nodes. And throughout these experiments, we set the outlet temperature threshold for each data node to 27 °C.

For tasks without any preferred data nodes, it is flexible for our task scheduler to dispatch the tasks to any candidate nodes. While selecting the best candidate data node to assign tasks, the scheduler should address the following issue. The scheduler may assign tasks to the least loaded data nodes or data nodes with the highest utilization. For comparison purposes, we consider the following three scheduling policies:

- Distribute Evenly (DE): to evenly schedule tasks to all the data nodes in the first-in-first-out order, thereby well balancing the load among the nodes.
- Distribute based on Utilization (DU): to schedule tasks to as many as data nodes while keeping active nodes' utilization at a high level.
- Distribute to Minimum active Nodes (DMN): to schedule tasks in a way to minimize the number of active data nodes.

Table 5 Task configurations of CPU-intensive workloads

Tasks	Task 1	Task 2	Task 3	Task 4	Task 5
LOOPS(#)	4000	8000	10000	11820	11850
Avg Util(%)	13	25	32	44	52
Tasks	**Task 6**	**Task 7**	**Task 8**	**Task 9**	**Task 10**
LOOPS(#)	11900	11930	11980	12020	12050
Avg Util(%)	64	72	85	96	100

Table 6 Task scheduling schemes for CPU-intensive workload

Strategies	Node 1	Node 2	Node 3	Node 4
DE	Task 1, 5, 9	Task 2, 6, 10	Task 3, 7	Task 4, 8
DU	Task 1, 8, 9	Task 2, 7, 10	Task 3, 6	Task 4, 5
DMN	Task 1, 2, 3, 8	Task 4, 5, 9	Task 6, 10	Task 7

5.1.1 CPU-Intensive Workload

In the first group of experiments, we consider CPU-intensive workload. A total of ten CPU-intensive tasks are simultaneously running Whetstone on the cluster. These CPU-intensive tasks lead to various CPU utilizations. The configuration and average utilization for each task are summarized in Table 5.

Let us consider a baseline task scheduler that assigns all tasks to a single data node, thereby making use of the least number of active data nodes. We conduct experiments to assign all the ten tasks to one of the four available data nodes, and the tasks are sequentially executed on the node. The average time to complete the ten tasks scheduled by this baseline approach is 6131 s.

Table 6 lists the three task scheduling strategies under the CPU-intensive workload conditions. The DE strategy evenly assigns tasks to the four data nodes. For instance, on data node 1, tasks 1 and 5 are concurrently executed; task 9 is running after the completion of task 1 and 5. When the DU strategy is in charge of the scheduling, tasks 1 and task 8 are executed simultaneously on node 1, where the CPU utilization is as high as 98 %. After completing tasks 1 and 8, node 1 starts running task 9. With the DU strategy in place, each node keeps a high CPU utilization, while ensuing that its CPU is not overloaded. When it comes to the DMN strategy, new tasks are scheduled to minimize the number of active data nodes. Thus, tasks 1, 2, and 3 are all assigned to data node 1.

Figure 14 reveals the performance of the three scheduling strategies. Execution times are referred to as time spent in completing all submitted tasks; active times are defined as the accumulation of time intervals in which the four data nodes are staying in the active state. Experimental results show that the outlet temperatures of the data nodes do not exceed the specified threshold.

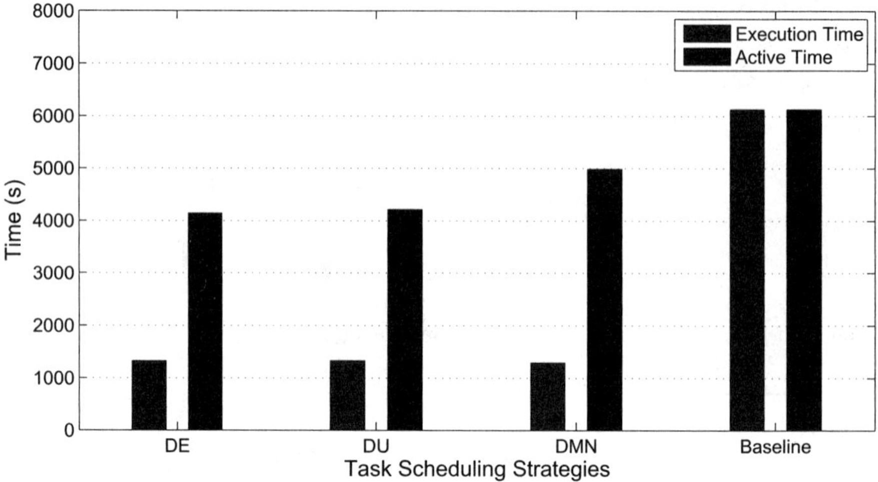

Fig. 14 Execution time and active time of data nodes

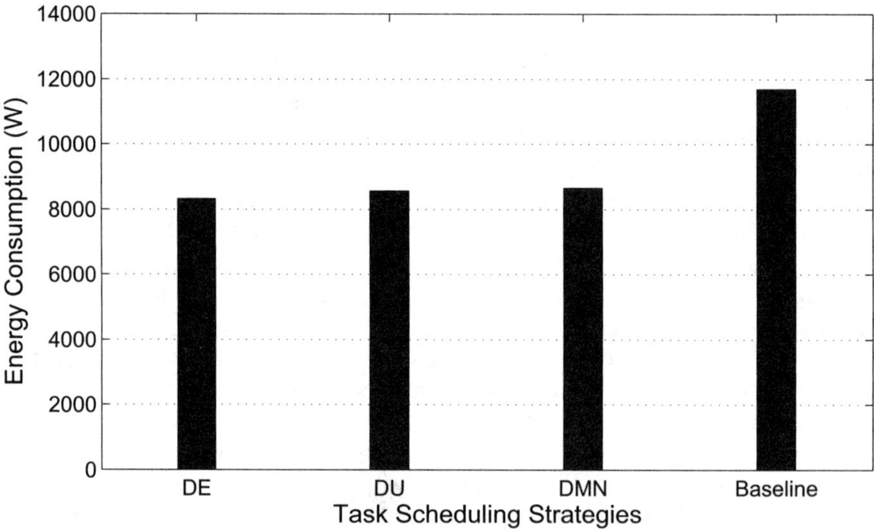

Fig. 15 Energy consumption of the four scheduling strategies

Figure 15 compared the baseline scheme with the three evaluated strategies in terms of energy consumption.

Among all the four scheduling strategies, the baseline one exhibits the longest execution time and consumes the most energy. By comparing the three evaluated strategies, the DMN strategy achieves the best performance, whereas DE delivers the highest energy efficiency. For example, DE saves the energy consumption of the other strategies by 3.8 %, and DE also conserves the energy consumption of the

Table 7 Configurations of I/O-intensive tasks

Tasks	Task 1,2	Task 3,4	Task 5,6	Task 7,8	Task 9,10
Write block size (Byte)	16	32	64	128	256
Avg Util(%)	14	29	54	81	100

Table 8 Task schedulers for I/O-intensive workload

Strategies	Node 1	Node 2	Node 3	Node 4
DE	Task 1, 5, 9	Task 2, 6, 10	Task 3, 7	Task 4, 8
DU	Task 1, 8, 9	Task 2, 7, 10	Task 3, 6	Task 4, 5
DMN	Task 1, 2, 3, 4	Task 5, 6	Task 7, 9	Task 8, 10

baseline scheme by 28.9 %. Thus, we could conclude that the DE strategy is the best scheduler for CPU-intensive load on storage clusters.

5.1.2 I/O-Intensive Workloads

In the second group of experiments, we assigned ten I/O-intensive tasks to the cluster. Each task generates 50 files and issues 200 transactions. We change the write block size to vary the disk utilization of each data node. The characteristics of these I/O-intensive tasks are shown in Table 7.

A baseline scheme assigns all the tasks to a single data node. We compare the aforementioned scheduling strategies with the baseline one. Table 8 shows the three task scheduling schemes.

Figure 16 shows the performance of the evaluated scheduling strategies. The results reveal that regardless of the tested schedulers, the outlet temperatures are kept below the pre-defined threshold. The energy consumption of the cluster managed by the three strategies compared with the baseline one can be found in Fig. 17.

Not surprisingly, the baseline strategy is outperformed by the three other schedulers in terms of execution time and energy consumption. The utilization-based scheduler is superior to the other three schemes in performance. The most energy efficient scheduler is the one (i.e., DE) that evenly distributes the load across all the four data nodes; this energy-efficient scheduler save the energy consumption of the baseline and the other schemes by 10.8 and 3.4%, respectively. Again, DE is the best scheduler for I/O-intensive workload.

In summary, under both CPU-intensive and I/O-intensive workload conditions, evenly distributing load across active data nodes is very energy efficient.

5.2 Predictive Thermal-Aware Management System

Massive amount of data are uploaded to and downloaded from data centers. For instance, 72 h of video are uploaded to Youtube every minute; 350 GB data are

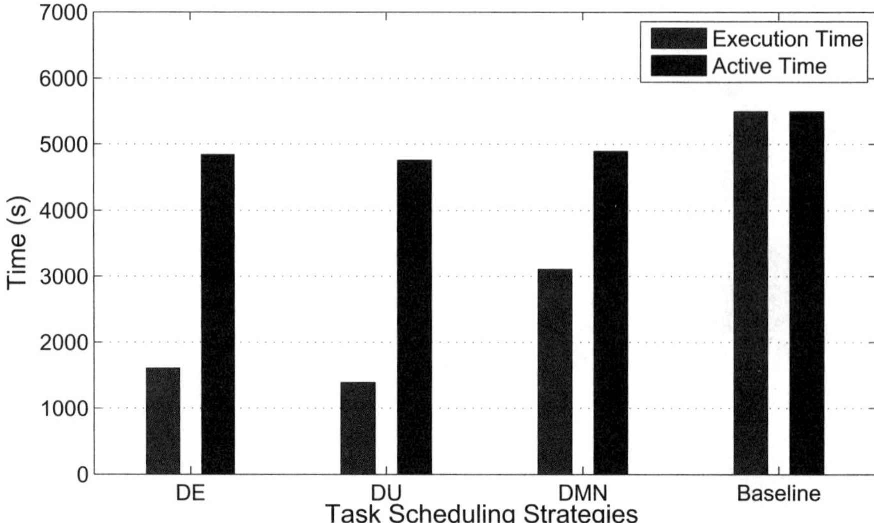

Fig. 16 Execution time and active time of data nodes under I/O-intensive workload

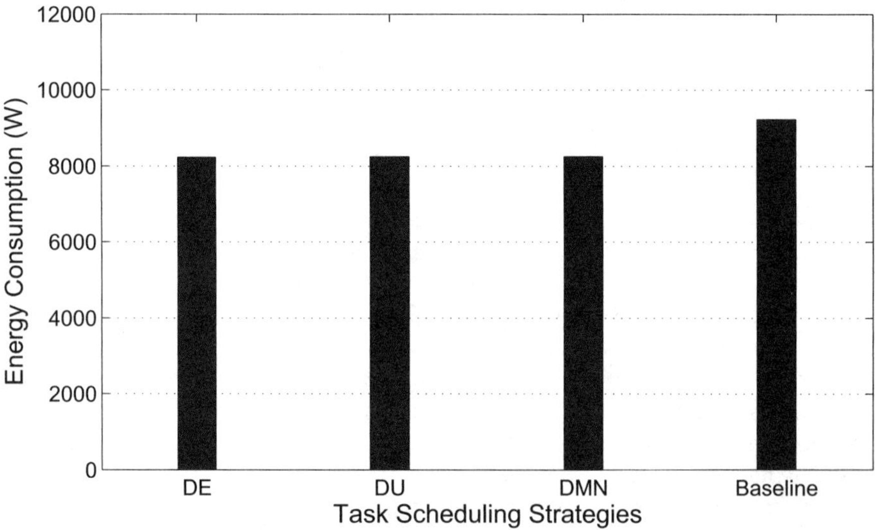

Fig. 17 Energy consumption of four different scheduling strategies under I/O-intensive workload

uploaded to Facebook every minute; 15,000 tracks are downloaded from iTunes every minute [39]. Uploading and downloading a large amount of data consume considerable energy and time; even worse, energy cost of data centers is rising dramatically with the increasing amount of data.

Table 9 Testbed configurations

	Node 1	**Node 2**
CPU	Intel(R) Celeron(R) 450@2.2GHz	
Network	1 GigaBit Ethernet network card	
Disk	WD-160GB Sata disk([38])	
Operating System	Ubuntu 10.04(lucid) Linux kernel 2.6.32-43	Ubuntu 10.04(lucid) Linux kernel 2.6.32-38

To evaluate the energy efficiency of our predictive thermal-aware management system designed for data centers, we conduct two sets of experiments. Table 1 summarizes the testbed used in the experiments:

In the first group of experiments, a pair of data nodes are transferring a dataset that contains hundreds of ASCII files generated by Postmark. The dataset's size is 1 GB; the file size of each is anywhere between 1 to 100 M. Among all the transferred files, small files are accessed more frequently than large files. It is important to study the energy consumption caused by transferring small files. For example, a report shows that there are 500 million files saved every 48 h on Dropbox as of May, 2012 [40]. A majority of Dropbox users use their free space to store small files. In most cases, uploaded files to the Dropbox servers are small in size.

We compare the performance of the four data transmission strategies (i.e., DT, AT, CT, and PTMS) transferring the two datasets. Figure 18 shows the energy cost of Node 1 that transfers the first dataset to Node 2. We observe that, compared to the other strategies, AT consumes less energy for both nodes 1 and 2 when the ASCII

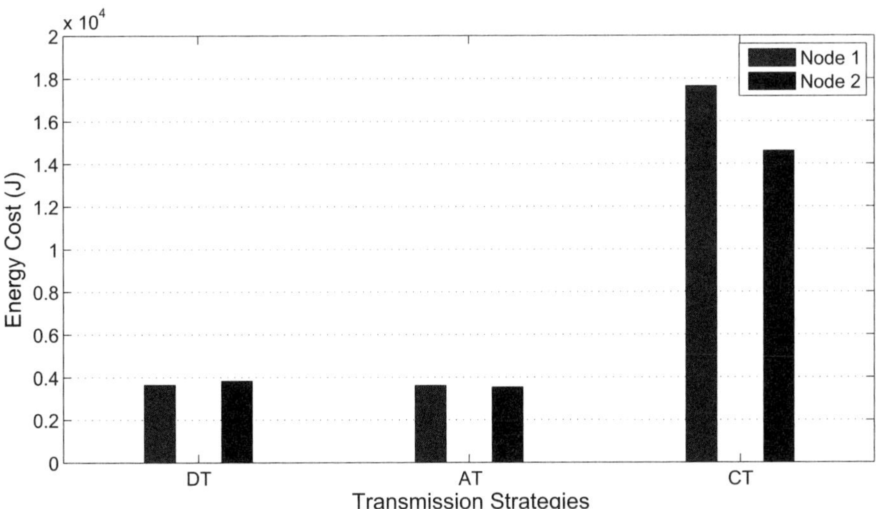

Fig. 18 Energy cost of moving the ASCII files

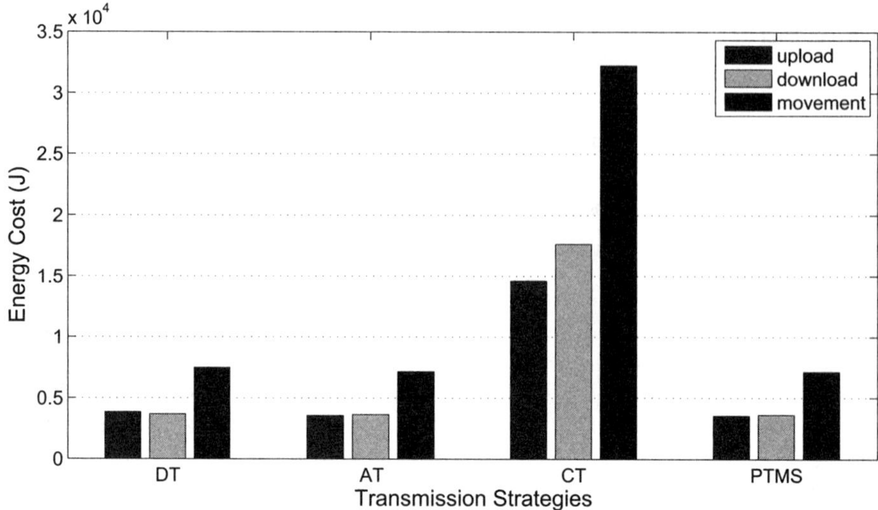

Fig. 19 Energy cost of the data transmission strategies transferring the ASCII files

files are transmitted. CT is the least energy-efficient scheme among all the evaluated strategies.

Now we are in a position to evaluate the energy efficiency of our PTMS. Figure 19 shows the energy cost of the four strategies under different transmission types. Not surprisingly, CT consumes more energy transferring this dataset than the other strategies. This is mainly because data compression or/and decompression requires extra CPU time and energy. Regardless of the transmission types, PTMS is the best one among all the tested strategies.

To resemble real-world cases where large files are transferred, in the second experiment group we choose to use a dataset of 60 GB Human Genome sequences. This dataset is available at NIH's (National Institutes of Health) NCBI website[1]. Each sequence file contains the DNA sequence of an entire chromosome. Most of the files in this dataset are larger than 3 GB.

Figure 20 shows the energy incurred by transferring the Human Genome dataset between nodes 1 and 2.

Figure 21 depicts the energy cost of transferring the Human Genome dataset with the four strategies under different transmission type.

We observe that regardless of data nodes, AT and PTMS outperform the other two strategies. The experimental results suggest that PTMS noticeably conserves energy for all the three data transmission types.

[1] ftp://ftp.ncbi.nih.gov/genomes/H_sapiens.

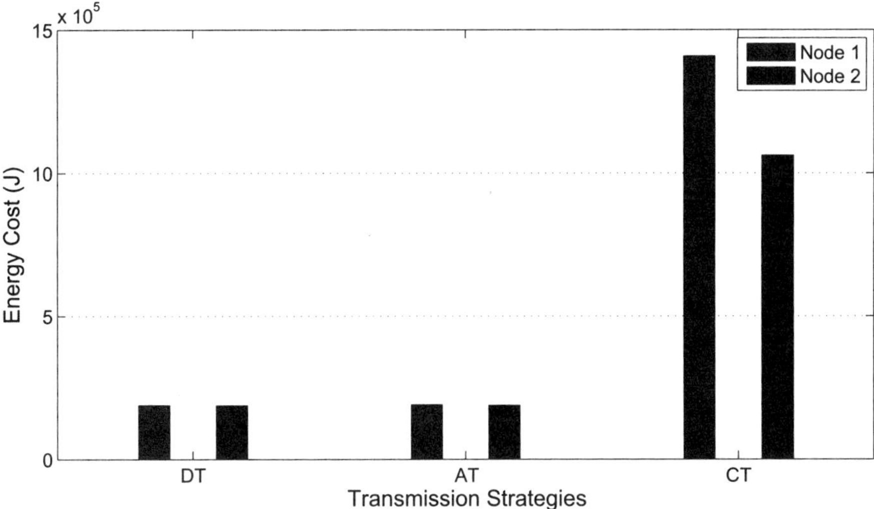

Fig. 20 Energy cost of transferring the Human Genome dataset between two nodes

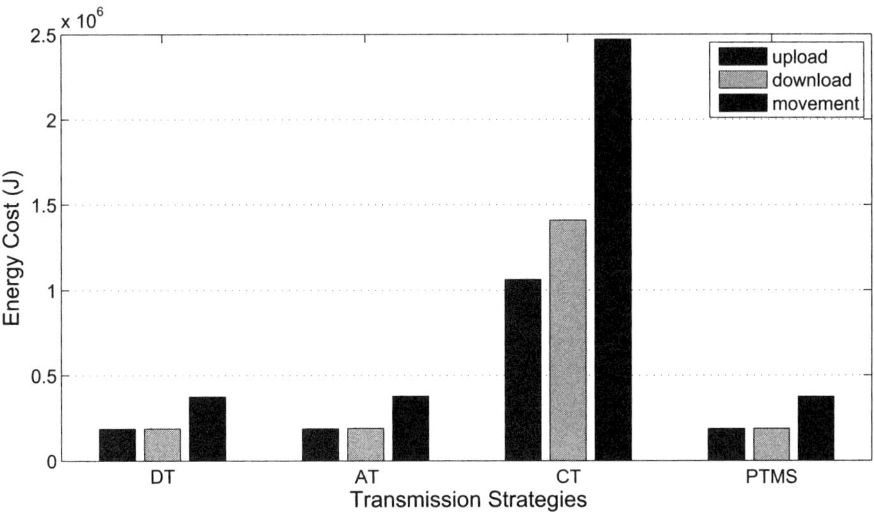

Fig. 21 Energy cost of transferring the Genome dataset under various data transmission type

6 Conclusion

Much attention has been paid to building energy-efficient data centers. Energy conservation techniques applied to data centers are classified into two categories: the first group is focused on reducing computing infrastructure energy consumption; the second group aims to reduce the cooling cost of data centers. For example, energy-aware task scheduling policies were proposed to redistribute workload in order to minimize the energy consumption of computing infrastructures. Thermal management strategies were proposed to reduce data centers' cooling cost. Recent studies show that cooling cost can be decreased by either reducing outlet temperatures of data storage nodes or minimizing heat recirculation in a data center. Thermal models play a significant role in thermal management. Unfortunately, most traditional thermal models do not holistically consider both CPUs and disks - an important contributor to outlet temperatures. In this study, we developed a thermal modeling approach that leads to new models applied to investigate thermal impacts of both CPUs and disks in data nodes. We demonstrated how to use our models to estimate the outlet temperatures of data nodes based on CPU and disk utilization.

We incorporated our thermal models into two thermal management strategies, which make data nodes thermal and energy friendly. The first strategy is integrated into a scheduler to dispatch and redistribute I/O tasks in a way to ensure that all the data nodes' outlet temperatures are below a threshold. The second one is a thermal-aware data transmission strategy, where data transfers are divided into three camps: uploads, downloads, and migrations within a data center. We implemented the thermal-aware data transmission strategy in a predictive thermal-aware management system or PTMS, which is conducive to estimating data nodes' energy consumption that guides the management of data transmissions. Among all the candidate data transmission policies, PTMS dynamically chooses the most appropriate one that meets the needs of a wide range of data-intensive applications coupled with various data transmission patterns.

Acknowledgments This research was supported by the U.S. National Science Foundation under Grants CCF-0845257 (CAREER), CNS-0917137 (CSR), CNS-0757778 (CSR), CCF-0742187 (CPA), CNS-0831502 (CyberTrust), CNS-0855251 (CRI), OCI-0753305 (CI-TEAM), DUE-0837341 (CCLI), and DUE-0830831 (SFS). Meikang Qiu's research was support by NSF CNS-1359557 and NSFC 61071061.

References

1. P. Thibodeau, "Data centers use 2 % of U.S. energy, below forecast," 2011. [Online]. Available: http://blogs.computerworld.com/18738/data_centers_use_2_of_u_s_energy_below_forecast.
2. IDC, "Annual it spending by Western European utilities to reach 12.7 billion by 2017, says IDC energy insights," 2013. [Online]. Available: http://www.idc-ei.com/getdoc.jsp?containerId=prUS24251013.
3. M. Baile, "The economics of virtualization: Moving toward an application-based cost mode," 2009. [Online]. Available: http://www.vmware.com/files/pdf/Virtualization-application-based-cost-model-WP-EN.pdf.
4. J. He, "Datacenter power management: Power consumption trend," 2008. [Online]. Available: http://communities.intel.com/community/datastack/blog/2008/02/20/datacenter-power-management-power-consumption-trend.

5. Statista, "Number of monthly active Facebook users worldwide from 3rd quarter 2008 to 2nd quarter 2013 (in millions)," 2013. [Online]. Available: http://www.statista.com/statistics/264810/number-of-monthly-active-facebook-users-worldwide/.
6. "What happens on Facebook in each day?" 2012. [Online]. Available: http://visual.ly/what-happens-facebook-each-day.
7. "Whetstone," http://www.netlib.org/benchmark/whetstones.
8. J. Katcher, "Postmark: A new file system benchmark," *System*, no. 3022, pp. 1–8, 1997. [Online]. Available: http://www.netapp.com/tech_library/3022.html.
9. P. Jones, "Industry census 2012: Emerging data center markets," 2012. [Online]. Available: http://www.datacenterdynamics.com/blogs/industry-census-2012-emerging-data-center-markets.
10. Y. Lee and A. Zomaya, "EnglishEnergy efficient utilization of resources in cloud computing systems," *EnglishThe Journal of Supercomputing*, vol. 60, no. 2, pp. 268–280, 2012. [Online]. Available: http://dx.doi.org/10.1007/s11227-010-0421-3.
11. Z. Liu, Y. Chen, C. Bash, A. Wierman, D. Gmach, Z. Wang, M. Marwah, and C. Hyser, "Renewable and cooling aware workload management for sustainable data centers," *SIGMETRICS Perform. Eval. Rev.*, vol. 40, no. 1, pp. 175–186, Jun. 2012. [Online]. Available: http://doi.acm.org/10.1145/2318857.2254779.
12. M. Al Assaf, X. Jiang, M. Abid, and X. Qin, "EnglishEco-storage: A hybrid storage system with energy-efficient informed prefetching," *EnglishJournal of Signal Processing Systems*, vol. 72, no. 3, pp. 165–180, 2013. [Online]. Available: http://dx.doi.org/10.1007/s11265-013-0784-9.
13. D. Colarelli and D. Grunwald, "Massive arrays of idle disks for storage archives," in *Proceedings of the 2002 ACM/IEEE conference on Supercomputing*, ser. Supercomputing '02. Los Alamitos, CA, USA: IEEE Computer Society Press, 2002, pp. 1–11. [Online]. Available: http://dl.acm.org/citation.cfm?id=762761.762819.
14. E. Pinheiro and R. Bianchini, "Energy conservation techniques for disk array-based servers," in *Proceedings of the 18th annual international conference on Supercomputing*, ser. ICS '04. New York, NY, USA: ACM, 2004, pp. 68–78. [Online]. Available: http://doi.acm.org/10.1145/1006209.1006220.
15. A. Beloglazov and R. Buyya, "Energy efficient resource management in virtualized cloud data centers," in *Proceedings of the 2010 10th IEEE/ACM International Conference on Cluster, Cloud and Grid Computing*, ser. CCGRID '10. Washington, DC, USA: IEEE Computer Society, 2010, pp. 826–831. [Online]. Available: http://dx.doi.org/10.1109/CCGRID.2010.46
16. A. Beloglazov, J. Abawajy, and R. Buyya, "Energy-aware resource allocation heuristics for efficient management of data centers for cloud computing," *Future Generation Computer Systems*, vol. 28, no. 5, pp. 755–768, 2012, <ce:title>Special Section: Energy efficiency in large-scale distributed systems</ce:title>. [Online]. Available: http://www.sciencedirect.com/science/article/pii/S0167739X11000689.
17. B. Aksanli, J. Venkatesh, L. Zhang, and T. Rosing, "Utilizing green energy prediction to schedule mixed batch and service jobs in data centers," *SIGOPS Oper. Syst. Rev.*, vol. 45, no. 3, pp. 53–57, Jan. 2012. [Online]. Available: http://doi.acm.org/10.1145/2094091.2094105.
18. "7 strategies to optimize data centre cooling," http://www.biztechmagazine.com/article/2011/01/keep-your-cool/.
19. J. Moore, J. Chasc, P. Ranganathan, and R. Sharma, "Making scheduling "cool": temperature-aware workload placement in data centers," in *Proceedings of the annual conference on USENIX Annual Technical Conference*, ser. ATEC '05. Berkeley, CA, USA: USENIX Association, 2005, pp. 5–5. [Online]. Available: http://dl.acm.org/citation.cfm?id=1247360.1247365.
20. Q. Tang, S. Gupta, and G. Varsamopoulos, "Thermal-aware task scheduling for data centers through minimizing heat recirculation," in *Cluster Computing, 2007 IEEE International Conference on*, sept. 2007, pp. 129–138.

21. Q. Tang, S. K. S. Gupta, and G. Varsamopoulos, "Energy-efficient thermal-aware task scheduling for homogeneous high-performance computing data centers: A cyber-physical approach," *IEEE Trans. Parallel Distrib. Syst.*, vol. 19, no. 11, pp. 1458–1472, Nov. 2008. [Online]. Available: http://dx.doi.org/10.1109/TPDS.2008.111.
22. K. Skadron, M. R. Stan, K. Sankaranarayanan, W. Huang, S. Velusamy, and D. Tarjan, "Temperature-aware microarchitecture: Modeling and implementation," *ACM Trans. Archit. Code Optim.*, vol. 1, no. 1, pp. 94–125, Mar. 2004. [Online]. Available: http://doi.acm.org/10.1145/980152.980157.
23. P. Eibeck and D. Cohen, "Modeling thermal characteristics of a fixed disk drive," *Components, Hybrids, and Manufacturing Technology, IEEE Transactions on*, vol. 11, no. 4, pp. 566–570, dec 1988.
24. C. Tan, J. Yang, J. Mou, and E. Ong, "Three dimensional finite element model for transient temperature prediction in hard disk drive," in *Magnetic Recording Conference, 2009. APMRC '09. Asia-Pacific*, jan. 2009, pp. 1–2.
25. S. Gurumurthi, A. Sivasubramaniam, and V. K. Natarajan, "Disk drive roadmap from the thermal perspective: A case for dynamic thermal management," *SIGARCH Comput. Archit. News*, vol. 33, no. 2, pp. 38–49, May 2005. [Online]. Available: http://doi.acm.org/10.1145/1080695.1069975.
26. Y. Kim, S. Gurumurthi, and A. Sivasubramaniam, "Understanding the performance-temperature interactions in disk i/o of server workloads," in *High-Performance Computer Architecture, 2006. The Twelfth International Symposium on*, feb. 2006, pp. 176–186.
27. X. Jiang, M. Alghamdi, J. Zhang, M. Assaf, X. Ruan, T. Muzaffar, and X. Qin, "Thermal modeling and analysis of storage systems," in *Performance Computing and Communications Conference (IPCCC), 2012 IEEE 31st International*, 2012, pp. 31–40.
28. X. Jiang, M. Al Assaf, J. Zhang, M. Alghamdi, X. Ruan, T. Muzaffar, and X. Qin, "EnglishThermal modeling of hybrid storage clusters," *EnglishJournal of Signal Processing Systems*, vol. 72, no. 3, pp. 181–196, 2013. [Online]. Available: http://dx.doi.org/10.1007/s11265-013-0787-6.
29. J. Lin, H. Zheng, Z. Zhu, and Z. Zhang, "Thermal modeling and management of dram systems," *IEEE Transactions on Computers*, vol. 99, no. PrePrints, 2012.
30. A. Shah, V. Carey, C. Bash, C. Patel, and R. Sharma, "EnglishExergy analysis of data center thermal management systems," in *EnglishEnergy Efficient Thermal Management of Data Centers*, Y. Joshi and P. Kumar, Eds. Springer US, 2012, pp. 383–446. [Online]. Available: http://dx.doi.org/10.1007/978-1-4419-7124-1_9.
31. R. Sharma, C. Bash, C. Patel, R. Friedrich, and J. Chase, "Balance of power: dynamic thermal management for internet data centers," *Internet Computing, IEEE*, vol. 9, no. 1, pp. 42–49, jan.-feb. 2005.
32. O. Sarood, A. Gupta, and L. Kale, "Temperature aware load balancing for parallel applications: Preliminary work," in *Parallel and Distributed Processing Workshops and Phd Forum (IPDPSW), 2011 IEEE International Symposium on*, may 2011, pp. 796–803.
33. O. Sarood and L. V. Kale, "A 'cool' load balancer for parallel applications," in *Proceedings of 2011 International Conference for High Performance Computing, Networking, Storage and Analysis*, ser. SC '11. New York, NY, USA: ACM, 2011, pp. 21:1–21:11. [Online]. Available: http://doi.acm.org/10.1145/2063384.2063412.
34. J. Li, M. Qiu, J.-W. Niu, L. T. Yang, Y. Zhu, and Z. Ming, "Thermal-aware task scheduling in 3d chip multiprocessor with real-time constrained workloads," *ACM Trans. Embed. Comput. Syst.*, vol. 12, no. 2, pp. 24:1–24:22, Feb. 2013. [Online]. Available: http://doi.acm.org/10.1145/2423636.2423642.
35. J. Srinivasan and S. V. Adve, "Predictive dynamic thermal management for multimedia applications," in *Proceedings of the 17th annual international conference on Supercomputing*, ser. ICS '03. New York, NY, USA: ACM, 2003, pp. 109–120. [Online]. Available: http://doi.acm.org/10.1145/782814.782831.
36. L. Ramos and R. Bianchini, "C-oracle: Predictive thermal management for data centers," in *High Performance Computer Architecture, 2008. HPCA 2008. IEEE 14th International Symposium on*, feb. 2008, pp. 111–122.

37. X.-F. Jiang, J. Zhang, M. I. Alghamdi, X. Qin, M.-H. Jiang, and J.-F. Zhang, “Peam: Predictive energy-aware management for storage systems,” in *Proceedings of 8th IEEE International Conference on Networking, Architecture, and Storage*.
38. “Wd1600aajs specification,” http://www.wdc.com/wdproducts/library/SpecSheet/ENG/2879-701277.pdf.
39. “What happens on line in 60 seconds?” http://www.mediabistro.com/alltwitter/online-60-seconds_b46813.
40. “Dropbox statistics,” http://techcrunch.com/2012/11/13/dropbox-100-million/.

Modeling and Simulation of Data Center Networks

Kashif Bilal, Samee U. Khan, Marc Manzano, Eusebi Calle, Sajjad A. Madani, Khizar Hayat, Dan Chen, Lizhe Wang and Rajiv Ranjan

1 Data Centers and Cloud Computing

Cloud computing is projected as the major paradigm shift in the Information and Communication Technology (ICT) sector [1]. In recent years, cloud market has experience enormous growth and adoption. The cloud adoption is expected to increase in coming years. As reported by Gartner [2], Software as a Service (SaaS) market is expected to rise to $ 32.3 billion in 2016 ($ 13.4 billion in 2011). Similarly, Platform as a Service (PaaS) and Infrastructure as a Service (IaaS) are projected to rise from $ 7.6 billion in 2011 to $ 35.5 billion in 2016. Cloud computing has been adopted in almost all of the major sectors, such as business, research, health, agriculture, e-commerce, and social life.

A data center is a repository to hold computation and storage resources interconnected to each other using network and communication infrastructure [3]. Data centers constitute the foundations and building blocks of cloud computing. Continuous evolution of cloud services, as well as their increased demand mandate growth in data center resources to deliver the expected services and required Quality of Service (QoS). Various cloud service providers already host hundreds of thousands of servers in their respective data centers. Google is estimated to host around

K. Bilal (✉) · S. U. Khan
North Dakota State University, Fargo, ND, USA
e-mail: kashifbilal@ciit.net.pk

M. Manzano · E. Calle
University of Girona, Girona, Spain

S. A. Madani · K. Hayat
COMSATS Institute of Information Technology, Islamabad, Pakistan

D. Chen · L. Wang
Chinese Academy of Sciences, Beijing, China

R. Ranjan
Australian National University, Canberra, Australia

S. U. Khan, A. Y. Zomaya (eds.), *Handbook on Data Centers*,
DOI 10.1007/978-1-4939-2092-1_31

0.9 million servers in their data centers. Similarly, Amazon is reported to have around 0.45 million servers to support Amazon Web Services (AWS). The number of servers in Microsoft data centers double every 14 months [4].

The projected number of resources required to accommodate future service demands within data centers are mounting. The "scale-out" or "scale-up" approaches alone cannot deliver a viable solution for escalating resource demands. Scale-out is the common approach adopted by data centers designers by adding inexpensive commodity hardware to the data center resources pool. Scale-up approach focuses on improving and adding more power and complexity to the enterprise-level equipment, which is expensive and power-hungry [6]. Increasing the computational and storage resources is currently not a major challenge in the data center scalability. However, how to interconnect these commodity resources together to deliver the required QoS is the major challenge. Besides scaling-out the data centers, energy consumption and resultant Operational expenses (OpEx) of data centers also pose serious challenges. Environmental aspects, enormous amount of Green House Gases (GHG) emissions by data centers, and increasing energy costs are worsening the problem. These aforementioned problems mandate the revisions in design and operation of data centers.

Data Center Networks (DCNs) play a pivotal role in asserting the performance bounds and Capital Expenditure (CaPex) of a data center. Legacy ThreeTier DCN architecture is unable to accommodate the growing demands and scalability within data centers [3]. Various novel DCN architectures have been proposed in the recent past to handle the growth trend and scalability demands within data centers [4]. Besides, electrical network technology, optical, wireless, and hybrid DCN architectures are also proposed [16, 17]. Moreover, intra-network traffic within the data center is growing. It has been estimated that around 70 % of the network traffic will flow within the data centers [7]. Various cloud and data center applications follow several communication patterns, such as one to one, one to many, many to many, and all-to-all traffic flows [8]. The traffic patterns within data centers are fairly different from the traffic patterns observed in other type of telecommunication networks. Therefore, the traffic optimization techniques proposed for such networks are inapplicable within data centers. Finally, it has been observed that the main DCN architectures have a low capacity to maintain an acceptable level of connectivity under different type of failures [5].

All of the aforementioned challenges require detailed analysis and quantification of various issues within a data center. In this particular case, simulation is an appropriate solution for detailed analysis and quantification of the aforementioned problems, since experiments comprising realistic DCNs scenarios are economically unviable. Simulation can help to quantify and compare the behavior of a network under a presented workload and traffic pattern. Unfortunately, network models and simulators to quantify the data center network and varying traffic patterns at a detailed level are scarce, currently. Moreover, current network simulators, such as ns-2, ns-3, or Omnet++ lack the data center architectural model and simulation capability. Therefore, we implemented the state-of-the-art DCN architectures in ns-3 simulator to carry out the DCN simulations and comparative analysis of major DCNs [3]. We implemented the three major DCN architectures namely, **(a)** legacy ThreeTier [10], **(b)** DCell [8], and **(c)** FatTree [19]. We implemented six traffic patterns to observe the behavior of the three DCN architectures under a specified workload. We carried out extensive simulations to perform a comparative analysis of the three considered DCN architectures.

2 DCN Architectures

Based on the packet routing model, the DCN architectures can be classified in two major categories, namely: **(a)** switch-centric and **(b)** server-centric networks. The switch-centric networks rely on network switches and routers to perform network packet forwarding and routing. The ThreeTier, FatTree, VL2, and JellyFish DCN architectures are the examples of the switch-centric networks [4]. The server-centric networks utilize computational servers to relay and route the network packets. The server-centric network may be pure server-based or hybrid (using an amalgam of network switches and computational server for traffic routing). The CamCube is a pure server based DCN architecture that relies solely on computational server for packet forwarding [9]. The DCell, BCube, and FiConn are examples of the hybrid server-centric DCN architectures [8].

The legacy ThreeTier architecture is the most commonly deployed network topology within data centers, currently. The ThreeTier architecture is comprised of a single layer of computational servers and a three layered hierarchy of network switches and routers (see Fig. 1) [10]. The computational servers are grouped in racks. Typically, around 40 servers within a rack are connected to a Top of the Rack (ToR) switch [11]. The ToRs connecting the servers within individual racks make the first layer of the network hierarchy called *access layer*. Multiple access layer switches are connected to the *aggregate layer* switches. The aggregate layer switches make the second layer of switches within the ThreeTier network hierarchy. A single access layer switch is connected to multiple aggregate layer switches. The high-end enterprise-level core switches make the topmost layer of the ThreeTier network hierarchy called *core layer*. A single core layer switch is connected to all of the aggregate layer switches within the data center. The intra-rack traffic flow is controlled by the access layer switches. The traffic flow among the racks with the ToRs connected to the same aggregate layer switch passes through the aggregate layer switches. The inter-rack traffic flow where the ToRs of the source and destination rack are connected to different aggregate layer switch passes through the core layer switches. Higher layers of the ThreeTier network architecture experiences higher oversubscription ratios. Oversubscription ratio is the worst-case available bandwidth among the end hosts and the total bisection bandwidth of the network topology [19].

The FatTree DCN architecture is a clos based arrangement of commodity network switches to deliver 1:1 oversubscription ratio [19]. The computational servers and commodity network switches are arranged in a hierarchical manner similar to the

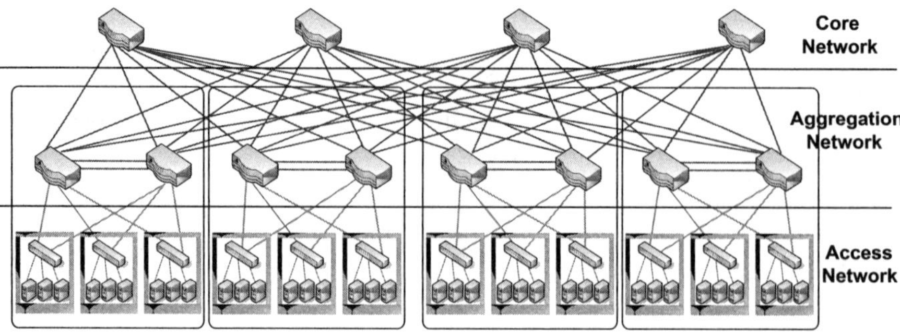

Fig. 1 ThreeTier DCN architecture

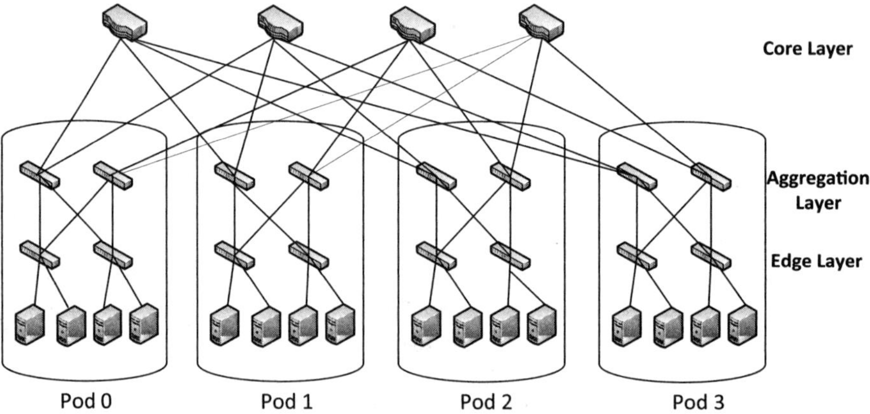

Fig. 2 FatTree DCN architecture

ThreeTier architecture. However, the number of the network devices and the interconnection topology is different from the ThreeTier architecture (please see Fig. 2). The number of pods (or modules) represented by 'k' decides the number of devices in each layer of the topology. There are total $(k/2)^2$ number of switches in the core layer of the FatTree architecture. The aggregate and access layers each contain $k/2$ number of switches in each pod. Each access switch is used to connect $k/2$ computational servers. Each pod in the FatTree contains k number of switches (arranged in two layers) and $(k/2)^2$ number of computational servers. The FatTree DCN architecture exhibit better scalability, throughput, and energy efficiency compared to the ThreeTier DCN. The FatTree architecture uses a custom addressing and routing scheme [AiL08].

The DCell is a hybrid server-centric DCN architecture [8]. DCell follows a recursively built topology where a server is directly connected to multiple servers in units called *dcells* (see Fig. 3). The $dcell_0$ constitutes the building block of the DCell architecture; where n servers are interconnected to each other using a commodity network switch (n is a small number usually less than eight). $n+1$ $dcell_0$ cells build

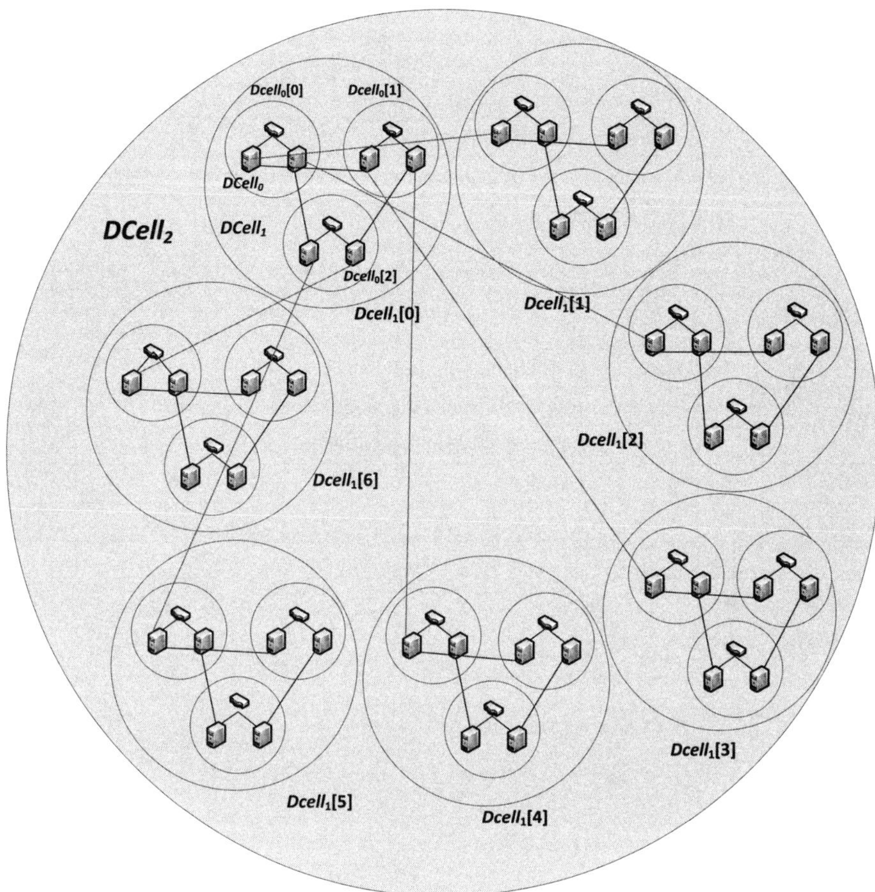

Fig. 3 DCell DCN Architecture

a *level-1* cell called $dcell_1$. A $dcell_0$ is connected to all other $dcell_0$'s within a $dcell_{1.}$ Similarly, multiple lower layer $dcell(_{L-1})$ cells constitute a higher level $dcell_{(L)}$ cell. DCell is an extremely scalable DCN architecture that may scale to millions of servers by having a *level-3* DCell with only six servers in each $dcell_0$.

3 DCN Graph Modeling

DCN architectures can be represented as multilayered hierarchical graphs [13]. The computational servers, storage devices, and network devices represent the vertices of the graph. The network links connecting the devices represent the edges of the graph. Table 1 presents the variables used in DCN models.

Table 1 Variables used in the DCN modeling

Variable	Represents
ν	Set of vertices (servers, switches, and routers) in the graph
ε	Network links connecting various devices
P_i	A *pod/module* in topology representing set of servers and middle layers switches
C	Core layer switches
δ	Servers
α	Access layer switch
γ	Aggregate layer switch
k	Total number of pods/modules in the topology
n	Total number of servers connected to a single access layer switch
s	Total number of servers connected to a switch in a $dcell_0$
m	Total number of the access layer switches in each pod
q	Total number of the aggregate layer switches in each pod
r	Total Number of the core layer switches in topology

3.1 ThreeTier DCN Model

The ThreeTier DCN can be represented as

$$DCN_{TT} = (\nu, \varepsilon\ ), \tag{1}$$

where ν represents the nodes in the ThreeTier graph (computational servers, network switches, and routers), and ε represent network links interconnecting the devices. The servers, access, and aggregate layer switches are arranged in k modules/pods (P_i^k) and a single layer of the core C_i^r switches (see Fig. 1 with four modules and a core layer)

$$\nu = P_i^k \cup C_i^r, \tag{2}$$

Each module or pod P_i is organized in three distinct layers of nodes, namely: **(a)** aggregate layer (l^g), **(b)** access layer (l^c), and **(c)** server layer (l^s). The nodes in each layer within a pod can be represented as

$$P_i = \{{l^s}_{m\alpha\ \times n\delta} \cup l^c_{m\alpha}\ \cup {l^s}_{q\gamma}\}, \tag{3}$$

where δ represents the servers, α represents the access layer switches, and the aggregate layer switches are represented by γ. $|P_i|$ represents the total number of nodes within a pod

$$|P_i| = \left(\sum_1^m n\ + m + q\right), \tag{4}$$

Total number of nodes in a Threetier architecture having n pods can be calculated as

$$|\nu| = \left\{ \sum_{i=1}^{k} |P_i| + |\mathrm{C}| \right\}, \tag{5}$$

There are three layers of edges (network links) interconnecting four layers of the ThreeTier architecture nodes

$$\varepsilon = \{\S, \acute{\alpha}, \text{₵}\}, \tag{6}$$

where § represent the edges that connect servers to access switches, $\acute{\alpha}$ are the edges used to connect aggregate and access layers switches, and ₵ represent the edges used to connect aggregate and core switches. Aggregate switches are also connected to each other within a pod, represented by $'\gamma$**.** The set of edges within the ThreeTier architecture can be represented by

$$\varepsilon = \{\S_{(\forall\delta,\alpha)}\ ,\ \acute{\alpha}_{(\forall\alpha,\ \forall\gamma)}, {}'\gamma_{(\forall\gamma,\ \forall\gamma)},\ \text{₵}_{(\forall\gamma\ \forall \mathrm{C})}\}. \tag{7}$$

Total number of edges within a ThreeTier DCN can be calculated as

$$|\varepsilon| = \sum_{1}^{k} \left(\sum_{1}^{m} n + \sum_{1}^{m} q + \frac{q\,(q-1)}{2} + \sum_{1}^{q} r \right). \tag{8}$$

3.2 *FatTree DCN Model*

As discussed in Sect. 2, the FatTree is also a multi-layered DCN architecture similar to the ThreeTier architecture. However, the number of devices and the interconnection pattern among the devices in various layers varies largely in both of the architectures. The FatTree architecture follows a Clos topology for network interconnection. The number of nodes in each layer within the FatTree topology is fixed and is based on the number of the pods 'k'

$$n = m = q = ({}^{k}/_{2}), \tag{9}$$

$$r = ({}^{k}/_{2})^2 \tag{10}$$

Similar to the ThreeTier architecture, the FatTree DCN can be modeled as

$$DCN_{FT} = (\nu, \varepsilon\ ), \tag{11}$$

where ν, P_i, $|P_i|$, and $|\nu|$ can be modeled by using Equation 2–5, respectively. However, the aggregate layers switches within a FatTree are not connected to each other. Moreover, the contrary to the ThreeTier architecture, each of the core layer switch is connected to a single aggregate layer switch from each pod

$$\varepsilon = \left\{ \S_{(\forall\delta,\alpha)} \cup \acute{\alpha}_{(\forall\alpha,\forall\gamma)} \cup \text{₵}_{(\forall,\gamma_i)} \right\}, \tag{12}$$

and the total number of edges can be calculated as

$$|\varepsilon| = \sum_{1}^{k}\left(\sum_{1}^{m} n + \sum_{1}^{m} q\right) + \sum_{1}^{R} k. \tag{13}$$

3.3 DCell DCN Model

DCell uses a recursively built topology, where a single server in a *dcell* is connected to servers in other *dcells* for server-based routing (see Fig 3). The graph model of the DCell DCN architecture can be represented as:

$$DCN_{DC} = (\nu, \varepsilon\ ), \tag{14}$$

$$\nu = \{\partial_i\ ,\ \partial_{i+1}\ ,\ \ldots\ ,\ \partial_L\}, \tag{15}$$

where $0\ \leq i\ \leq L$. ∂_0 represents $dcell_0$ and L denotes highest level.

$$\partial_0 = \left\{\delta \cup \alpha\right\}, \tag{16}$$

where δ represents the set of 's' servers with $dcell_0$ and α presents a single switch connecting the servers.

$$\partial_l = \{x_l.\partial_{l-1}\}, \tag{17}$$

where $1\ \geq l\ \leq L$, and x_l is the total number of ∂_{l-1} in ∂_l.

A $dcell_1$ can be represented by

$$\partial_1 = \{x_1.\partial_0\}, \tag{18}$$

$$x = s + 1 \tag{19}$$

similarly, for $l\ \geq 2$:

$$x_l = \left(\prod_{i=1}^{l-1} x_i \times s + 1\right). \tag{20}$$

A *3-level* DCell can accommodate around 3.6 million servers with s = 6. Total number of node in a *3-level* DCell can be calculated as:

$$\left|\nu_0^3\right| = \left(\sum_{1}^{x_3}\sum_{1}^{x_2}\sum_{1}^{x_1}(s+1)\right), \tag{21}$$

and the total number of edges in a *3-level* DCell can be calculated as:

$$\left|\varepsilon_0^3\right| = \sum_{1}^{x_3}\left(\sum_{1}^{x_2}\left(\left(\sum_{1}^{x_1} \mathrm{s}\right) + (x_1(x_1-1)/2)\right) + (x_2(x_2-1)/2)\right) + (x_3(x_3-1)/2) \tag{22}$$

The total number of vertices in the *l-level* DCell are:

$$|\nu| = \left(\prod_{i=1}^{n}\left(\sum_{1}^{x_i}(s+1)\right)\Big/(s+1)^{(l-1)}\right), \tag{23}$$

and the total number of edges can be calculated as:

$$|\varepsilon| = \left(\prod_{i=1}^{l}\left(\sum_{1}^{x_i}(s)\right)\Big/s^{(l-1)}\right) + 1/2\left[\sum_{j=1}^{l}\left(\left(\prod_{y=j}^{l}x_y\right)(x_j-1)\right)\right] \tag{24}$$

4 DCNs Implementation in ns-3

We implemented three major DCN architectures namely: **(a)** ThreeTier, **(b)** FatTree, and **(c)** DCell (see Sect. 2 for details). We used ns-3 discrete-event simulator to implement the three DCN architectures. We implemented **(a)** interconnection topology, **(b)** customized addressing scheme, and **(c)** customized routing logic for the three considered DCN architectures. Moreover, we implemented six traffic patterns to observe the behavior of the considered DCNs under various network conditions and traffic loads.

In the year 2003–2004, ns-2 was the most used network simulator for network research [14]. However, to address the outdated code design and scalability of ns-2, a new simulator called ns-3 was introduced. The ns-3 is a new simulator (not an evolution of ns-2) written from scratch. Some of the salient features of ns-3 simulator are as follows [3, 14], and [18]. The ns-3 simulator offers the modeling of realistic network scenarios. The ns-3 uses the implementation of real Internet Protocol (IP). Moreover, the ns-3 offers implementation of Berkeley Socket Distribution (BSD) sockets interface and installation of multiple network interfaces on a single node. Simulated packets in the ns-3 contain real network bytes. Furthermore, the ns-3 offers to capture the network traces, which can be analyzed by using various network tools, such as WireShark. We implemented our DCN models using the ns-3.13 release. Currently, ns-3.18 is the stable release of the ns-3 simulator [18].

One of the major drawbacks of the ns-3 simulator is that it does not provide Ethernet switch implementation. The BridgeNetDevice is the closely related bridge implementation that can be used to simulate an Ethernet switch. However, the BridgeNetDevice is used for simulating CSMA devices and does not work for the Point-To-Point devices. Therefore, we implemented a Point-To-Point based switch for simulations in ns-3 [18].

4.1 ThreeTier DCN Implementation Details

We offer a customizable implementation of the ThreeTier architecture. The simulation parameters can be configured to simulate the ThreeTier architecture with devices arranged in four layers having different oversubscription ratio. Users can define the number of pods/modules in the topology. Each pod contains **(a)** servers arranged in racks, **(b)** number of access layers switches, and **(c)** number of aggregate layer switches. Users can specify the required number of servers in each rack. All of the servers within a rack are connected by an access layer (ToR) switch. The bandwidth of the network link interconnecting the servers with the ToR switch can be configured. The default bandwidth for the links connecting servers to a ToR is 1Gbps. All of the ToRs, within a single pod/module are connected to all of the aggregate layer switches. The default bandwidth of the network links interconnecting the ToRs to the aggregate layer switch is 10Gbps, which is configurable. The number of devices and bandwidth of the links in each of the layers (server, access, and aggregate layer) of the ThreeTier architecture remains same in all of the pods.

The core layer of the ThreeTier architecture is the topmost layer that is used to connect various pods to each other. Users can specify the number of core switches in the topology. Each core layer switch is connected to all of the aggregate layer switches. The network switch used in the aggregate and core layer of the ThreeTier architecture are often high-end enterprise layer switches. One of the major features of the high-end switches is the support of Equal Cost Multi Path (ECMP) routing [15]. We have added the support for ECMP for the aggregate and core layer switches for realistic results.

We used ns-3 Ipv4GlobalRoutingHelper class for routing with the ECMP support in the ThreeTier architecture. It is worth mentioning that the performance and results of the ThreeTier architecture are heavily dependent on the oversubscription ratio at each layer and use of the ECMP. The throughput and network delay fluctuates substantially by varying the oversubscription ratio and use of ECMP routing. We configured each device with real IP address for simulation. The IP addressing scheme is also customizable and users can assign the network addresses of choice to devices. Figure 4a depicts the topology setup for the ThreeTier architecture with $k = 4$.

4.2 FatTree DCN Implementation Details

FatTree DCN is based on the Clos interconnection topology. The number of devices in each layer of the FatTree architecture is fixed based on the number of pods 'k'. Contrary to the ThreeTier architecture, the user only needs to configure the total number of pods for the FatTree simulation. The implementation of the FatTree architecture creates k pods and the required devices and interconnection within each pod. The value of k must be **(a)** greater than or equal to four and **(b)** an even number. The network bandwidth of the interconnecting links is configurable and the default value is 1Gbps. The entire network links uses same bandwidth value contrary to the ThreeTier architecture, where the bandwidth value of the network links connecting servers to ToR and the links connecting ToRs to aggregate layer switch is usually different.

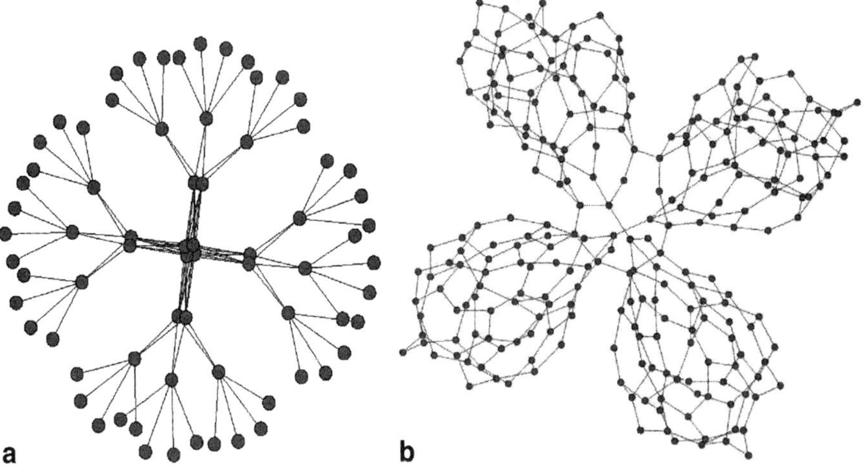

Fig. 4 DCN Topologies in ns-3 **a** ThreeTier Topology **b** DCell Topology

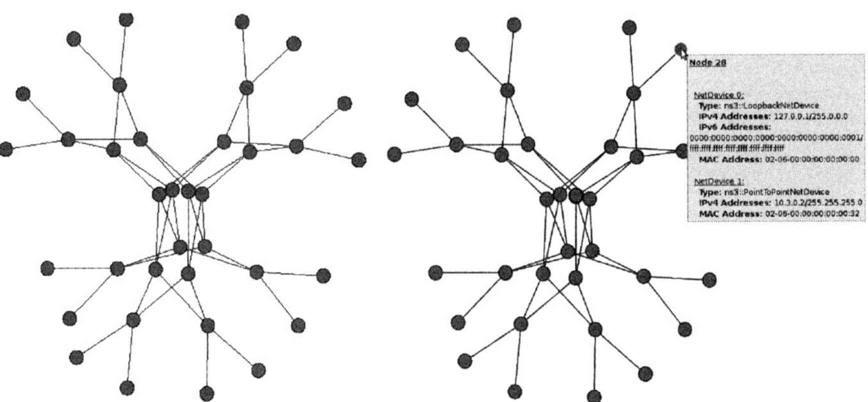

Fig. 5 Simulation of a $k = 4$ FatTree in ns-3

FatTree architecture uses a custom network addressing scheme. The network address of a server or a node is dependent on the location of the node. The network address of a server is based on the pod-number that contains the server, and the access switch number that connects the server. We have implemented the custom network addressing scheme of the FatTree architecture and each of the nods within the FatTree is assigned the addressing scheme as specified. Figure 5 and 6 depicts the assignment of the network addresses within each pod of a FatTree with $k = 4$.

The FatTree architecture uses a two-level routing scheme for packet forwarding. The packet forwarding decision is based on two-level prefix lookup scheme. The FatTree uses a primary prefix routing table and a secondary suffix table. Firstly, the longest prefix match is checked in the primary prefix table. If a match is found for the

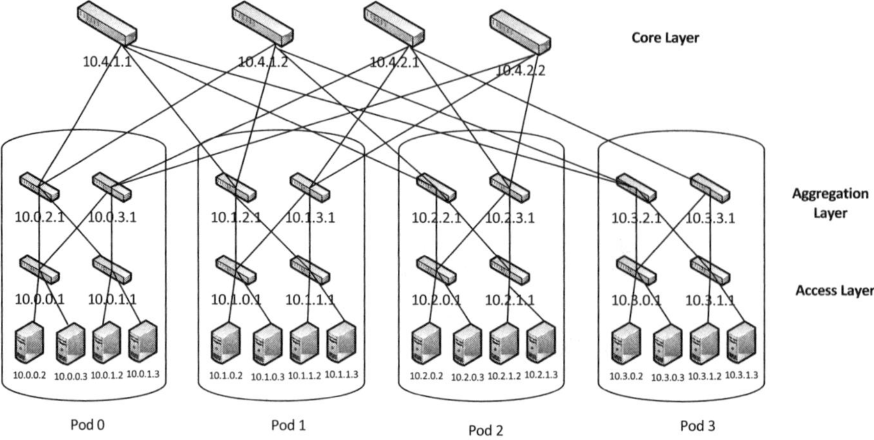

Fig. 6 Assignment of IP addresses in a $k=4$ FatTree

destination address of the packet, then the packet is forwarded to the port specified in routing table. If the longest prefix does not match, then the longest matching suffix in the secondary table is found, and the packet is forwarded to the port specified in the secondary routing table. The switches in the access and aggregate layers use same algorithm for routing table generation (please see Algorithm 1 in [12]). The core layer switches uses a different algorithm (please see Algorithm 2 in [12]) for core switch routing table generation. We implemented both of the algorithms for our FatTree DCN implementation.

4.3 DCell DCN Implementation Details

The DCell is a recursively built, highly scalable, and server-based hybrid DCN architecture. As detailed in Sect. 2 and 3, the building block of the DCell architecture are presented. We offer a customizable implementation of the DCell architecture. We have implemented *3-level* DCell topology that can accommodate millions of servers by having less than eight nodes in $dcell_0$. The user can configure the number of server in $dcell_0$. The number of servers in the DCell increases exponentially with the increase in number of servers in $dcell_0$. Table 2 presents the total number of servers in the DCell topology. As can be observed, by having only eight servers can lead to a DCell comprised of 27 million servers. Because of the exponential increase in the number of server and upper level *dcells*, we enabled the configuration of the number of the *dcells* at each level. The user can configure the number of servers, the number of $dcell_0$ in $dcell_1$, number of $dcell_1$ in each $dcell_2$, and number of $dcell_2$ in $dcell_3$ to control the number of servers in resulting DCell.

Each $dcell_0$ contains a Ethernet switch that is used for packet forwarding among the servers within the $dcell_0$. The traffic forwarding among the servers in different

Table 2 Number of Servers in DCell

Number of servers in $dcell_0$	Total number of servers in *3-level* DCell	Total number of nodes (including switch) in *3-level* DCell
2	1806	2709
3	24492	32656
4	176,820	221,025
5	865,830	1,038,996
6	3,263,442	3,80349
7	10,192,056	11,648,064
8	27,630,792	31,084,641

dcells is performed by servers, that are interconnected to each other. Each server node is equipped with multiple interface cards to directly connect switch and other servers. The default bandwidth of the network links is 1Gbps that is also configurable. We used a realistic IP address assignment to each server in the DCell architecture implementation.

The DCell does not specify any custom addressing scheme. However, the DCell routing scheme takes into account the placement of the server within the *dcells*. For instance, the NodeId (3, 1, 2) specifies the server number 2, in $dcell_0$ number 1, and $dcell_1$ number 3. We implemented programming routines that can find the IP address of a specified server number and vice versa. The DCell uses a custom routing protocol called DCell Fault-tolerant Routing (DFR) protocol for packet forwarding. The DFR is a recursive and source based routing protocol. When a node wants to initiate communication with some other node, the DFR is invoked to calculate the end-to-end path for the flow. The DFR first calculates the link connecting the *dcells* of the source and destination node. Then the DFR calculates the path from source to the link and from link to the destination. The combination of all paths is the *end-to-end* path. We implemented the DFR protocol. The output path from the DFR provides the NodeIds instead of the IP address. We use a custom programmed routine to convert the NodeIds based path to IP based path. We place the complete source to destination path in an extra header in each packet, as the DFR is a source based routing protocol. Each intermediate node parses the header and decides the forwarding port/link for the next hop.

Unfortunately, the algorithm listing for the DCell in the original paper was incomplete and erroneous (please see Fig. 3 in [8]). In Sect. 4.1.1 of the original paper [8], it is mentioned that if ($S_{k-m} < d_{k-m}$), then the link interconnecting the *sub-dcells* can be calculated as '([S_{k-m}, $d_{k-m} - 1$], [d_{k-m}, S_{k-m}])'. The 'else clause' for the aforementioned 'if statement' is not given in the original paper. Therefore, the implementation of the DFR was erroneous and incomplete. We figured the else clause for the aforementioned scenario and implemented the complete DFR algorithm for traffic routing. Moreover, the example for the path calculated using DFR also had a typographical mistake in the original paper.

References

1. IBM, IBM Data Center Networking Planning for Virtualization and Cloud Computing, 2011. Online: http://www.redbooks.ibm.com/redbooks/pdfs/sg247928.pdf
2. Gartner, Market Trends: Platform as a Service, Worldwide, 2012–2016, 2H12 Update, 2012.
3. K. Bilal, S.U. Khan, L. Zhang, H. Li, K. Hayat, S.A. Madani, N. Min-Allah, L. Wang, and D. Chen, "Quantitative Comparisons of the State of the Art Data Center Architectures," Concurrency and Computation: Practice and Experience, (DOI:10.1002/cpe.2963).
4. K. Bilal, S. U. R. Malik, O. Khalid, A. Hameed, E. Alvarez, V. Wijaysekara, R. Irfan, S. Shrestha, D. Dwivedy, M. Ali, U. S. Khan, A. Abbas, N. Jalil, and S. U. Khan, "A Taxonomy and Survey on Green Data Center Networks," Future Generation Computer Systems. (Forthcoming.)
5. M. Manzano, K. Bilal, E. Calle, S. U. Khan, "On the connectivity of Data Center Networks", IEEE Communication Letters. (Forthcoming)
6. K. Yoshiaki, and M. Nishihara. "Survey on Data Center Networking Technologies." IEICE transactions on communications, Vol. 96, No. 3, 2013.
7. P. Mahadevan, P. Sharma, S. Banerjee, and P. Ranganathan, "Energy aware network operations," IEEE INFOCOM Workshops 2009, pp. 1–6.
8. C. Guo, H. Wu, K. Tan, L. Shi, Y. Zhang, and S. Lu, "DCell: A Scalable and Fault-tolerant Network Structure for Data Centers." ACM SIGCOMM Computer Communication Review, Vol. 38, No. 4, 2008, pp. 75–86.
9. H. Abu-Libdeh, P. Costa, A. Rowstron, G. O'Shea, and A. Donnelly, "Symbiotic Routing in Future Data Centers," ACM SIGCOMM 2010 conference, New Delhi, India, 2010, pp. 51–62.
10. Cisco, Cisco Data Center Infrastructure 2.5 Design Guide, Cisco press, 2010.
11. A. Greenberg, J. Hamilton, N. Kandula, C. Kim, and S. Sengupta, "VL2: a scalable and flexible data center network," ACM SIGCOMM Communication Review, Vol. 39, No. 4, 2009, pp. 51–62.
12. M. Al-Fares, A. Loukissas, and A. Vahdat, "A scalable, commodity data center network architecture," ACM SIGCOMM 2008 conference on Data communication, Seattle, WA, 2008, pp. 63–74.
13. K. Bilal, M. Manzano, S. U. Khan, E. Calle, K. Li, and A. Y. Zomaya, "On the Characterization of the Structural Robustness of Data Center Networks," IEEE Transactions on Cloud Computing. (Forthcoming.)
14. G. Carneiro, ns-3, Network Simulator 3, 2010. Online: http://www.nsnam.org/tutorials/NS-3-LABMEETING-1.pdf
15. C. Hopps, Analysis of an Equal-Cost Multi-Path Algorithm. RFC 2992, Internet Engineering Task Force, 2000.
16. C. Kachris and L. Tomkos, "A Survey on Optical Interconnects for Data Centers," Communications Surveys & Tutorials, IEEE, Vol. 14, No. 4, 2012, pp. 1021–1036
17. X. Zhou, Z. Zhang, Y. Zhu, Y. Li, S. Kumar, and A. Vahdat, "Mirror Mirror On The Ceiling: Flexible Wireless Links For Data Centers," ACM SIGCOMM Computer Communication Review, Vol. 42, No. 4, pp. 443–454, 2012.
18. ns-3 Simulator, online: http://www.nsnam.org/.
19. Mohammad Al-Fares, Alexander Loukissas, and Amin Vahdat, "A Scalable, Commodity Data Center Network Architecture," ACM SIGCOMM, Seattle, Washington, USA, August 17–22, 2008, pp: 63–74.

Part VI
Security

C²Hunter: Detection and Mitigation of Covert Channels in Data Centers

Jingzheng Wu, Yanjun Wu, Bei Guan, Yuqi Lin, Samee U. Khan, Nasro Min-Allah and Yongji Wang

1 Introduction

Data centers provides both the applications, systems software and the hardware as services over the Internet, which is named cloud computing [1–3]. It is core infrastructure of cloud computing, supporting dynamic deployment and elastic resource management. With the powerful computing and storing capabilities, cloud computing has become increasingly popular [4, 5]. The fundamental mechanism of cloud

J. Wu (✉) · Y. Wu
Institute of Software, Chinese Academy of Sciences, Beijing, China
e-mail: jingzheng@nfs.iscas.ac.cn

Y. Wu
e-mail: yanjun@nfs.iscas.ac.cn

B. Guan · Y. Lin · Y. Wang
National EngineeringResearch Center for Fundamental Software, Beijing, China
e-mail: guanbei@nfs.iscas.ac.cn

Y. Lin
e-mail: yuqi@nfs.iscas.ac.cn

Y. Wang
e-mail: ywang@itechs.iscas.ac.cn

S. U. Khan
North Dakota State University, Fargo, ND, 58108-6050, USA
e-mail: samee.khan@ndsu.edu

N. Min-Allah
COMSATS Institute of Information Technology, Islamabad, Pakistan
e-mail: nasar@comsats.edu.pk

Y. Wang
State Key Laboratory of Computer Sciences, Beijing, China
e-mail: ywang@itechs.iscas.ac.cn

S. U. Khan, A. Y. Zomaya (eds.), *Handbook on Data Centers*,
DOI 10.1007/978-1-4939-2092-1_32

is virtualization which allows virtual machines (VM) instantiate stand-alone operating systems on demand based on a software layer called virtual machine monitor (VMM) or *hypervisor* [6]. Although the virtualization technology provides strong isolation for the cloud, security and privacy are always the open problems [7]. Some of the problems are essentially traditional web application and data-hosting ones, e.g., phishing, downtime, data loss, and password weakness. One of the new problems introduced by the shared environment to cloud computing is the covert channel attack [8]. By this way, information is leaked from the data centers and meanwhile the security provided by isolation is breaken down [9, 10].

To enhance the security of data centers, some protection mechanisms have been presented [11, 12]. sHype [13] is a Mandatory Access Control based (MAC) security extension to Xen hypervisor [6]. For example, sHype enables the Chinese Wall and the Type Enforcement policies to specify whether or not the resources can be accessed by the VMs. Lares presents a hybrid approach, giving security tools the ability to monitor actively while still benefiting from the increased security of an isolated virtual machine [14]. HyperSentry presents a novel framework to enable integrity measurement of a running hypervisor by introducing a software component [15]. Some other frameworks such as HyperSafe [11], Antfarm [16], HIMA [17], Vulcloud [18, 19] etc. all focus on providing integrity measurement of a hypervisor by introducing a software component.

However, these protection mechanisms may fail, because the data centers create numerous implicit high resolution clocks used to construct the covert channels [20]. Covert channel is a leakage mechanism used to transfer confidential information violating security policies specified by the information systems [21, 22]. It is the main threat to the multi-level secure systems, e.g., operating systems [23, 24], database systems [25], network [26, 27] and cloud computing [9]. TCSEC [21] and CC [28] secure criterions require covert channel analysis when building secure systems. The objectives of covert channel analysis are identification, estimation capacity, detection and handling [23, 29, 30, 31]. This paper concerns the detection and mitigation technology in data centers and presents a new framework termed as C^{2}Hunter (Covert Channel Hunter).

C^{2}Hunter in this paper detects the covert channels from operational track records and mitigates the threat. Cabuk et al. [32, 33] present an algorithm to detect TCP/IP network covert timing channel (IPCTC). They believe that the regularity of the packet intervals indicates the difference between the covert and normal channels. Therefore, they present two methods to measure the regularity. The first method examines whether the variance of the intervals remains constant using standard deviation. The second method computes the relative difference between each pair of the sorted intervals. Similar to Cabuk, Berk et al. [34] present a statistical algorithm to detect the covert channel whose packet intervals center around two different values. The algorithm compares the ratio of the mean packet count to the maximum packet count for normal traffic. The lower the ratio is, the higher the probability of having a covert channel hidden in packet intervals. Nagatou et al. [35] define a security automata and show the enforced properties. To detect several covert channels at run time, they use an extra structure to emulate the behavior of a system by running a subsequence from an interleaved state sequence of processes. However, these detection methods and some other ones (e.g., [26, 36]) are limited to the particular network or operating systems, unsuitable for detecting the covert channels in data centers.

The covert channels in data centers are induced by the massive shared computing resources, and they cannot be simply detected and mitigated by the above methods. In this paper, the threat model of covert channels is first analyzed, and the channels are classified into three categories. Only the channels between virtual machines new to data centers are considered, and three typical scenarios (e.g. CPU load based, Cache based, shared memory based channels) are demonstrated and their features are studied in-depth. C^2Hunter is presented to detect and mitigate the covert channels in data centers following some basic requirements: small modification to the protected cloud systems, flexible extension to detect and mitigate new channels and acceptable performance impact. An error-corrected four states automata is proposed to model the channel scenario, a two-phase synthesis algorithm is designed to detect the covert channels using Markov and Bayesian models, and a network pump like method is implemented to mitigate the threat of the timing channel. The prototype of C^2Hunter is implemented on Xen hypervisor and the performance to detect the three typical covert channels is investigated. The experiment results show that C^2Hunter is able to detect and mitigate these channels, and it is believed that C^2Hunter will detect and mitigate incoming covert channels in the future after small extension.

The distinguished contributions made in this paper are as follows:

- For the first time, the covert channels in data centers are classified into three categories, and only the channels new to data centers are concerned.
- The covert channel scenario in data centers is modeled into an error-corrected four states automata, which is the basis of the detection scheme.
- A flexible framework named C^2Hunter is designed, consisting of a controller in hypervisor and a synthesis analyzer in the virtual domain. The controller places hooks into the hypervisor, captures the hypercalls and adds some latencies to the potentially malicious operations. The performance to detect and mitigate the three typical covert channels is evaluated.
- A two-phase synthesis detection algorithm using Markov and Bayesian model is presented, and the Pessimistic Threshold is adopted to lower the false negative.

The rest of the paper is organized as follows. Section 2 introduces the research background, hot topics and key problems about covert channels. Section 3 discusses the threat model, typical scenarios and some assumptions. Section 4 describes the design of C^2Hunter by introducing the challenges and thc formal requirements. Sections 5 and 6 cover the details of the two-phase synthesis detection algorithm and mitigate algorithm. Section 7 shows the prototype implementation and evaluates the detection performance of C^2Hunter. Section 8 discusses the extendibility of C^2Hunter and Sect. 9 concludes this work finally.

2 Background

Lampson first introduced covert channel in 1973 [22]. The explicit definition is: *Given a non-discretionary (e.g. mandatory) security policy model M and its interpretation* $I(M)$ *in an operating system, any potential communication between two subjects* $I(S_h)$ *and* $I(S_i)$ *of* $I(M)$ *is covert if and only if any communication between the corresponding subjects* S_h *and* S_i *is illegal under* M[24]. Covert channel is the only mechanism to leak confidential messages in the secure operating systems [37].

Since Girling presented three types of covert channel in local network [38], the network covert channel becomes the hot issue in this field [26]. Two types of network covert channel exist [39]: embedding covert messages into the header fields [40, 41] and encoding information into the transmission time of packets [33]. They are called network covert storage channel and network covert timing channel respectively. Storage channel usually encodes messages into the unused or reserved bits of frames, such as IP Type of Service(TOS) field, Don't Fragment (DF), URGent (URG) or TCP Flags bits in the packet header [38, 40–42]. Timing channel encodes messages into the sending/receiving time or the packets interval time etc, which is much more difficult to detect or handle [33, 34, 43].

Timing channel in cloud computing is firstly studied by Ristenpart et al. [9]. They state that any physical machine resources multiplexed between the attacker and target may form a potentially leakage channel between the virtual machines. These resources include network access, CPU branch predictors, CPU instruction cache, DRAM memory bus, CPU pipelines, scheduling of CPU cores, disk access, etc. They implement some experimental timing channels in Amazon's EC2 based on cache access time and CPU load, pointing out that once the malicious VM and the target one are co-located the information leakage by the timing channel is possible. Yinqian et al. investigate the cache based timing channel from the viewpoint of co-residency detection, which can be seen as a quality measurement of cloud service [44].

Ristenpart etc. have referred to the load based covert channel, and Okamura and Oyama [45] analyze it in more detail. They quantitatively evaluate the channel performance and develop CCCV (Covert Channels using CPU loads between Virtual machines) which creates covert channel to communicate secretly. CCCV describes a scenario that the sender and receiver processes locate in different domains hosted by Xen hypervisor, and the vCPUs are mapped to the same physical CPU (core). Each process executes task *t* in its domains without interference. The receiver repeatedly executes task *t* and investigate the elapsed time. If the sender does not execute a task, the elapsed time keeps the same. If the sender issues a task and the physical CPU alternately schedule in every time slice, the elapsed time obtained by the receiver gets longer. The different time can be modeled as a timing channel, and the malicious users can transmit covert message according to the changes of CPU load. The capacity and accuracy are evaluated in detail under various conditions. The CPU load based channel achieves a better capacity than the cache based one, and only small errors when interfered by other processes.

Timing channel is more threatening than storage channel in data centers. During the past 40 years, researchers have paid much attention to the timing channel identification, scenario construction, elimination, mitigation and detection technologies. More and more timing channels have been found and handled, for example, the Event-Flag channel in operating system [23], the Data Conflict channel in secure database system [46, 47] and the IPCTC channel in network [32, 27]. When cloud computing and data center became mainstream, the timing channels have also been found in this new computing paradigm [48].

This paper focuses on the timing channels found in data center, whose infrastructure is based on Xen platform. Xen supports both full virtualization and paravirtualization and manages the resource of virtual machines using a reference hypervisors [6]. Xen hypervisor provides both isolation and inter-domain communication between the virtual machines, for example, the shared memory mechanism is a way for guests to communicate analogous to interprocess communication between user space processes. The interface of the shared memory mechanisms is implemented via the grant table operations including mapping or transferring pages. They both involve inserting the physical page(s) to or from the caller's address space. Mapping is used to create shared memory, whereas transferring is used to move data from one domain to another. The difference is whether the original reference is removed or not.

3 Threat Model, Scenarios and Assumptions

In this section, the covert channel threat model and scenarios to the data centers are investigated, especially to the lifetime of virtual machines. The section ends with some assumptions used to design the covert channel detection and mitigation framework.

3.1 Threat of Data Center

Date centers instantiate stand-alone operating systems on demand, and deploy software or offer service to the application layer. When the tasks have finished, the virtual machines are destroyed. Some threats are presented in the lifetime of virtual machines as shown in Fig. 1.

Run-Time attacks (A1,A2) Virtualization is the fundamental mechanism of the cloud, providing the abstract services of hardware resources to the operating systems in VMs. A1 denotes the information leakage attack to inter-VMs, e.g. the CPU load based and Cache based covert channel. Although the hypervisor allocates vCPU(virtual CPU) to each VM, the tasks will run in the physical CPU eventually inducing A1 attacks. A2 denotes the covert channels in VMs, e.g. the event-flag

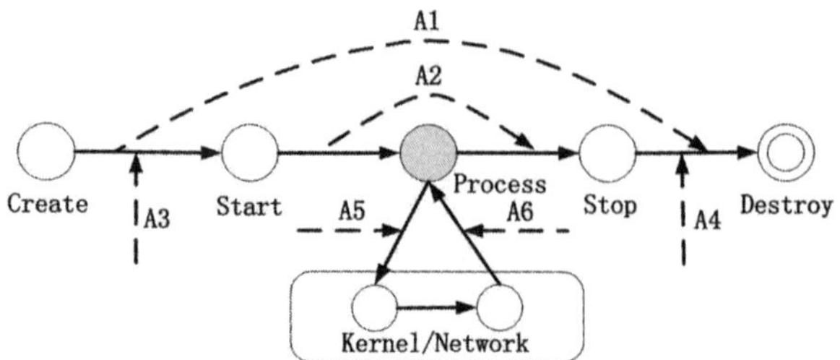

Fig. 1 Lifetime threats to cloud computing in data centers, including tamper, leakage and application attacks

channels in Linux [23, 30]. A1 and A2 leak confidential information in or between VMs, and cannot be eliminated by the deployed access control policies.

Start/Stop Attacks (A3,A4) Cloud platform allows users to start and deploy malicious hosts easily, which are used as DDoS and Botnet attacks. In A3 attack, hackers tamper the image of virtual machines and trojan the system before startup. In A4 attacks, hackers tamper the stored data to leak or steal the confidential data. A3 and A4 are induced by virtual machines, which lead to inside attacks becoming outside attacks.

Application Attacks (A5,A6) A5 attacks target and modifies the context information of operating system calls. A6 attacks alter the function responses by intercepting and modifying the return values. In network, A5 and A6 may lead Man-in-the-Middle attacks. However, A5 and A6 are traditional attacks, which have been studied in-depth [29].

3.2 Threat Categories of Covert Channels

The above attacks cover the lifetime threats of cloud, and the bottom-up attack targets are hypervisor, VMs, and applications. In this paper, only the covert channel attacks (A1 and A2) are concerned. They can be classified into three categories as shown in Fig. 2.

Intra-VM covert channels (CC1), i.e. processes level covert channels. Malicious processes P_i and P_j with different secure levels locate in the same operating system in DomU(Domain Unit). P_i with the higher secure level leaks confidential information to the lower secure level P_j using the covert channel. However, the threat of the covert channel is limited to the stand-alone operating system. Channels of CC1 have been studied for years, and some mature analysis methods can be referred to [23, 29].

Cross-platform covert channels (CC2), i.e. network level covert channels. Malicious processes P_k and P_x locate in different domains and different hardware platforms. P_k and P_x communicate with each other through the network, therefore the confidential information can be transmitted by the network storage and timing channels. Channels of CC2 are based on the network, which have been studied since 1987 [26, 38].

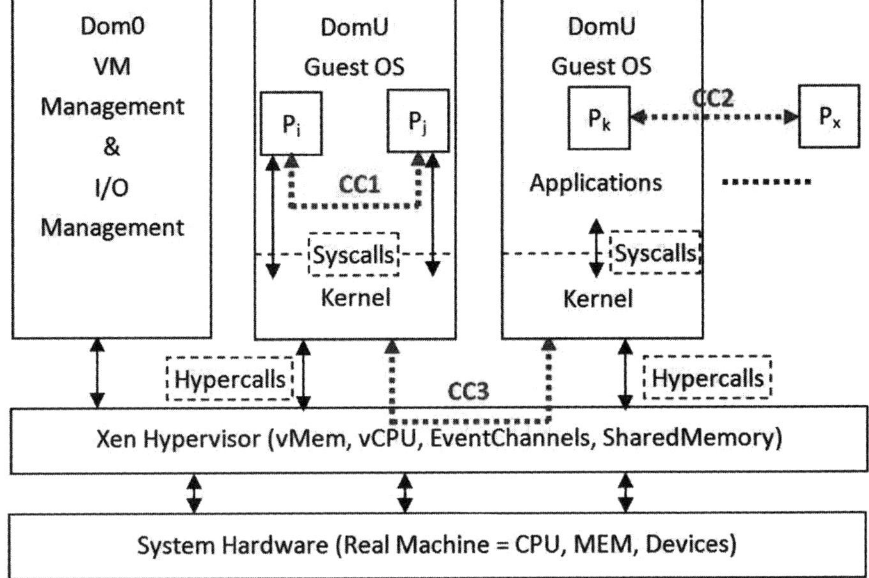

Fig. 2 Three categories of covert channels in Xen virtual machines

Inter-VM covert channels (CC3), i.e. operating system level covert channels. Malicious processes locate in different domains but the same hardware platform. CC3 is introduced by the shared resource managed by the hypervisor [9, 48, 49]. Confidential information may be leaked by CC3 among competitive companies that locate in the same hardware, will bring huge economic losses.

3.3 *Threat Scenarios of Covert Channels*

While there are a number of avenues to extract confidential information from cloud computing, the covert channel attack is more advanced. Only CC3 channels are new covert channels to cloud, and the other two types are operating systems and network covert channels. In this paper, three typical covert channel scenarios belonging to CC3 are described.

CPU load based channels [9, 45]. It has been considered that any physical resources which are multiplexed between the attacker and target host may form a potential covert channel in virtual machines. One of the most common resources is CPU load, which can be approximated by the amount of time taken for certain computations. The confidential information is pre-encoded into a binary sequence. The sender and receiver transfer information by changing and observing the CPU load according to a certain communication protocol, e.g. long wait time to complete a task means bit 1 is transmitted, otherwise bit 0 is transmitted.

Cache based channels [9, 44]. The Cached based covert channel takes the different cache access latencies as the different bits. The sender uses the idles as transmitting bit 0 and the frantic accesses to memory block as transmitting bit 1. The receiver accesses a memory block of her own and observes the access latencies. High latency denotes the sender is evicting the receiver's data from the caches and means bit 1 is transmitted, otherwise bit 0 is transmitted.

Shared memory based channels [48]. The shared memory based channel takes different memory access intervals as the different bits. The sender sends covert messages by controlling the data sending time, and the receiver obtains the message by observing the data arrival time. The confidential information is encoded into the different intervals. For example, longer and shorter intervals denote bit 1 and bit 0 respectively. This type of covert channel is named shared memory Covert Timing Channel, short for SMCTC.

3.4 Assumptions

It is usually considered that the data centers are based on cloud computing and the fundamental technology is virtualization. C^2Hunter framework runs on Xen platform, which manages the hardware, hypervisor, management domain and some guest domains [6]. It assumes that only three types of covert channels are concerned, which are CPU load based, Cache based and shared memory based channel. It is also assumed that the covert channels exist in the malicious domains along with some other innocent domains in the same cloud platform.

4 Overview of C^2Hunter

In this section, the design of C^2Hunter is discussed, starting off by describing the challenges. To guarantee the detection and mitigation of the three types of channels, the requirements are formally stated. Then, C^2Hunter is investigated in detail, requiring the covert channel to be modeled into an error-corrected four states automata.

4.1 Challenges

The first issue faced is that the traditional detection and mitigation technologies cannot be used in data centers. Although covert channel has been studied for almost 40 years, the researches mainly focus on operating systems, database systems, and networks [29]. Covert channel in cloud platform is a new topic, whose features have not been fully understood. In this paper, this new type of covert channel is modeled into an error-corrected four states automata, and the features are investigated.

The next issue is whether the framework can be flexibly extended to detect and mitigate the incoming covert channels. C^{2}Hunter addressed this problem by adopting a plug-in mechanism. To detect a new covert channel, the first step is to extract the features of the channel, e.g. finding the shared resource and the values or states which can be used by the channels. Then these features of the channel are normalized into detection parameters. Thirdly, some latencies are added into the hypercalls to reduce the threat. Finally, a XML(Extensible Markup Language) config file containing these parameters is automatically generated and loaded by C^{2}Hunter as a plug-in.

The last issue is how to solve the by-pass operations. Some operations that have super privileges cannot be captured and interrupted in the traditional operating systems, which may induces covert channels and hence leads to the failure of detection. However, hypervisor or VMM in virtualization platform has a higher privilege than the operating systems, where C^{2}Hunter resides in and captures all the operation tracks.

4.2 *Formal Requirements*

C^{2}Hunter is a virtualization based framework designed to detect and mitigate the CC3 type of covert channels by monitoring the operations in VMs. To counter the above challenges, the design of C^{2}Hunter is based on the following four high level formal requirements:

- Complete Detection. Detection is to determine the real covert channel in operation records. All of the operations may be potential covert channels, so they should be totally recorded and detected.
- Flexibility of Extension. A well-functioning detector and mitigator should have the flexibility to support detecting and mitigating the new covert channels and adding new detection algorithms.
- Acceptable Performance Impact. An additional performance impact introduced by C^{2}Hunter should be within acceptable limits. This requirement ensures that the overall performance of the virtualization platform should not be hurt. It also requires that C^{2}Hunter should detect and mitigate the covert channels in real time. Once any covert channel is detected, the alarm will be issued.
- Anonymous Detection and Mitigation. Operations of the virtual machines should be protected in privacy. C^{2}Hunter only investigates the features of shared resources, e.g. vCPUs, cache and shared memory. The precise operations will not be inferred.

4.3 *C^{2}Hunter Framework Summary*

Figure 3 shows the architecture of C^{2}Hunter. C^{2}Hunter consists of two-part components. One is a core active controller, locating in hypervisor and capturing all the

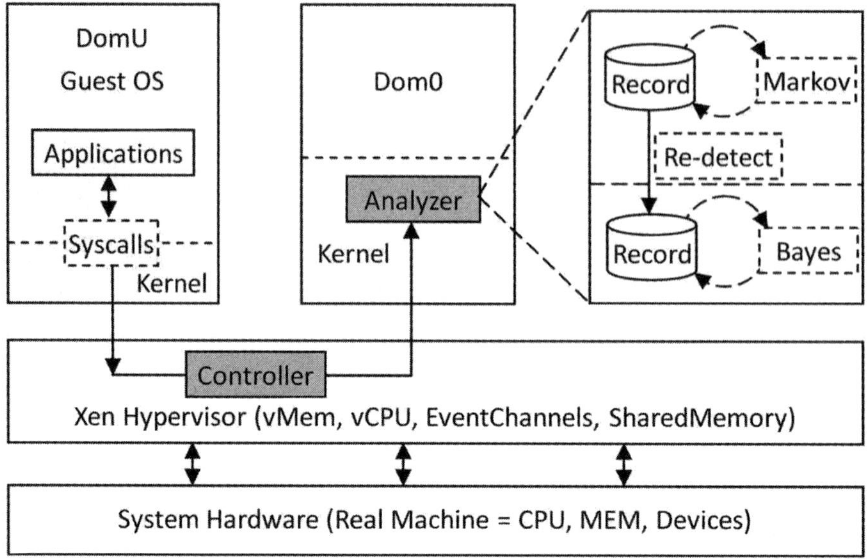

Fig. 3 High-level view of C^{2}Hunter Framework, which consists of two part components, including a core interrupter module located in hypervisor and an analyzer located in Dom0

operations triggered by the guest operating systems. The other one is a back-end analyzer locating in Dom0, used to analyze the captured operation records and detect the covert channels from them.

The controller places a hook inside hypervisor, monitors all the hypercalls trigged by VMs and adds latencies. A hypercall is conceptually similar to a system call. The hypercall interface allows domains to trigger a synchronous software trap into hypervisor to execute a privileged operation, and the communication from hypervisor to a domain is provided through an asynchronous event mechanism [50]. Hypervisor can intercept any instructions which change the states of the machine in a way that impacts other processes. Therefore, it is an ideal monitor place where any hypercalls cannot bypass. In CC3 channels, the malicious processes locating in different VMs cooperate with the shared resources to communicate indirectly. All the operations and each state of the shared resource will be recorded by the capture application. The records will be sent to the detector locating in Dom0.

The controller includes a detector and a interrupter. The detector is the module to detect in C^{2}Hunter, which is designed in a plug-in form. The detector locates in Dom0, which has elevated privileges and manages the other domains. The detector is made up of two main blocks presently: Markov detection module and Bayesian detection module. A two-phase synthesis algorithm is implemented in both modules sequentially according to the covert channel model, which can be extended through plug-in in the future. The records first flow into the Markov detector, all the covert

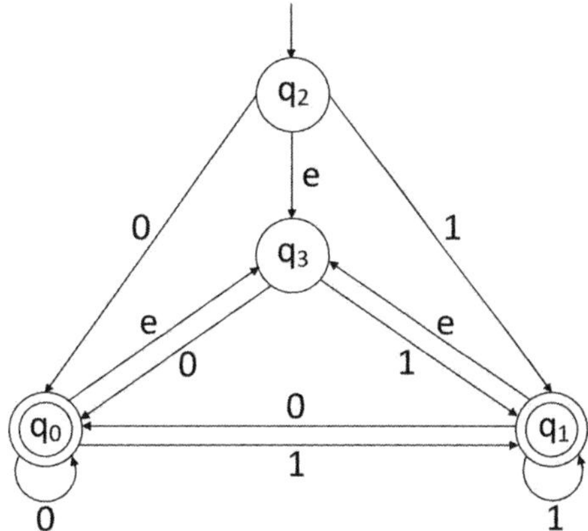

Fig. 4 Error-corrected four states automata model

channels will be detected with some false positive because of the Pessimistic Threshold. Then, the results flow into the Bayesian detector to refine. When the records have been detected by the detector, the outputs are the covert channels.

The interrupter is the module to mitigate the threat of covert channels in C²Hunter. It sends the logs of hypercalls to analyzer and adds latencies from the analyzer. The interrupter can intercept any instructions that change the state of the machine in a way that impacts other processes. The record stream is processed and saved concurrently by the controller. Then the records are normalized, for example, the controller only records the operation times in shared memory based channels. However, the real shared resource is the time intervals, so the intervals are calculated from the records. Finally, the intervals are inserted into the hypercalls.

4.4 *Covert Channel Modeling*

A covert channel consists of a shared resource, a sender process and a receiver process. The sender transmits confidential information by changing the properties of the shared resource, and the receiver receives the message by viewing the changes. In CPU load based, Cache based, and shared memory based channels, the properties are the CPU load, cache access time, and the memory write intervals. The changes of these resource properties are in certain regular patterns. An error-corrected four states automata is designed to model the patterns [51, 52].

Definition 1 (Error-corrected four states automata) Error-corrected four states automata is a five-tuple

$$(Q, \Sigma, \delta, q, F),$$

which is shown in Fig. 4, and the details of the tuple are described as follows.

- Q is a states set including the shared resources properties.
- Σ is a actions set abstracted from events involved in a system.
- $\delta : Q \times \Sigma \rightarrow Q$ is a transition functions set denoting that an action is triggered from one state to another.
- $q \in Q$ is the initial state.
- $F \subseteq Q$ is the final states set.

Taking shared memory based channel for example, the property of the channel is the memory write intervals [48]. The longer interval denotes bit 1 and shorter one denotes bit 0. SMCTC has only four states in the transmission process as shown in Fig. 4, and all the different states belong to set Q. Set δ includes all the transition functions.

At the beginning of the transmission cycle, the initial state is q_2. If bit 0 is sent, the state changes from q_2 to q_0, expressed as $q_2 \rightarrow q_0$. If bit 1 is sent, the state changes from q_2 to q_1, expressed as $q_2 \rightarrow q_1$. When an error occurs, the state changes from q_2 to q_3, expressed as $q_2 \rightarrow q_3$. When an error occurs, the state changes to q_0 or q_1 according to the error-corrected algorithm.

In SMCTC, the memory write intervals are not exactly consistent. For example, a sequence of intervals captured is expressed as $T = \{t_0, t_1, \ldots t_n\}$. We take ΔT_l, with $t_a < \Delta T_l < t_b$ as long interval and ΔT_s, with $t_c < \Delta T_s < t_d$ as short interval, $\Delta T_l > \Delta T_s$. Here t_a, t_b, t_c, t_d mean the ranges of the intervals. If an interval t_i is out of the ΔT_l and ΔT_s, an error occurs. The reason for this is that some interferences are running, and the transition are $q_2 \rightarrow q_3$, $q_0 \rightarrow q_3$ and $q_1 \rightarrow q_3$.

To deal with the errors, two simple error-corrected algorithms are presented in this paper.

- Value closer based error-corrected algorithm. This algorithm is based on the approximation. Take t_i as an example, if $|t_i - \Delta T_l| < |t_i - \Delta T_s|$, it is considered that a bit 1 is transmitted, and vice versa.
- Probability based error-corrected algorithm. This algorithm is based on the predetermined probability. If an error occurs, it is believed that a bit 1 is transmitted with the probability t, and a bit 0 with probability $(1 - t)$.

The first algorithm is easy to implement, and the second one needs to predetermine the probability. The Markov detector and the Bayesian detector adopt the second algorithm in this paper. The repeated errors may mean the channel environment has changed and the values of ΔT_l and ΔT_s should change correspondingly. The dynamic adjustment is complex, which is not considered in this paper.

Therefore, the automata model of SMCTC is instantiated as follows, shown in Fig. 4.

- $Q = \{q_0, q_1, q_2, q_3\}$ contains the four states of the shared resource property.
- $\Sigma = \{0, 1, e\}$ contains all the terminators normalized from the intervals captured.
- $\delta : Q \times \Sigma \rightarrow Q$ is a set of transition functions expressed as a state transition matrix as follows.

	0	1	e
q_0	q_0	q_1	q_3
q_1	q_0	q_1	q_3
q_2	q_0	q_1	q_3
q_3	x	y	ϕ

Triggers 0, 1, and *e* are in the first line of the matrix, and the current states are in the first row. The elements in the matrix are the next states transiting from the current states. For example, q_0 in the second line and the second row denotes the state changed from q_0 when bit 0 is transmitted, expressed as $q_0 \to q_0$.
In the last line of the transition matrix, the states transiting from q_3 depend on the error-corrected algorithms expressed by x and y. When an error occurs, the algorithm determines whether a bit 0 or 1 is transmitted and the corresponding state is q_0 or q_1. The state will not stay at p_3, denoted by φ in the matrix.

- q_2 is the initial state.
- $F = \{q_0, q_1\}$ contains all the final states.

Scenarios of CPU load based and Cache based channels can also be modeled into this error-corrected four states automata. The channel is assumed to take the binary encoding mechanism. If a multiple encoding mechanism is adopted, the corresponding automata just has more states. It will be discussed in Sect. 8.

5 Two-Phase Synthesis Detection Algorithm

In this section, a two-phase synthesis algorithm to detect the covert channels using Markov and Bayesian models is presented [53].

The error-corrected four states automata describes the channels. The real shared resources are operation intervals, which cannot be stored by the channel meaning memoryless. The next state in the automata just relates to the current state, which is triggered by the stochastic bit. According to these properties, a two-phase synthesis detection algorithm is presented to detect the covert channels.

The algorithm is synthesized by Markov detection algorithm and Bayesian detection algorithm. If the change pattern of the shared resource properties is closer to the Markov model, it is believed that the sequence of the operations is transferring confidential information through some covert channel. The normal operation sequences are modeled into Bayesian model. If an operation sequence deviates from the Bayesian model, it is believed that a covert channel occurs. Markov and Bayesian model in the two-phase synthesis algorithm are complementary. Markov detector detects the covert channels, and Bayesian detector distinguishes the covert channels from the normal operation sequences.

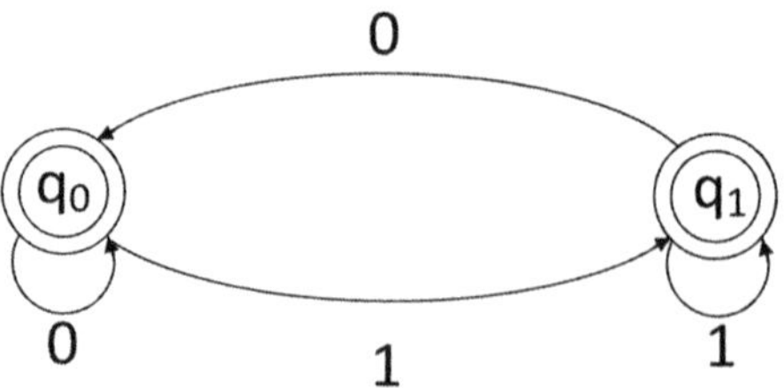

Fig. 5 Two-states Markov model of covert channel

5.1 Markov Detection Algorithm

Markov property refers to the memoryless property of a stochastic process [54, 55]. Covert channel model is such a stochastic process that the next state depends only upon the current state and has nothing to do with the previous status.

The error-corrected four states automata has only two final states q_0 and q_1. State q_2 in Fig. 4 is the start state, and no other states can transform to it. The automata will not stay at q_3, because of the error-corrected algorithms. To simplify the model, q_2 and q_3 are pruned, and a two-states Markov model is presented in Fig. 5. The trigger actions in the automata are stochastic, and the next state depends only upon the current state.

The distribution probabilities of the triggers 0, 1, and *e* are set as *p*, *q*, and *r*, and $p + q + r = 1$. Therefore, the probability of transition $q_2 \rightarrow q_0$ is *p*, the probability of $q_2 \rightarrow q_1$ is *q*, and the probability of $q_2 \rightarrow q_3$ is *r*. According to the second error-corrected algorithm, the state q_3 transits to q_0 and q_1 with probability *t* and 1-*t*. The Markov model is instantiated as follows.

- State Space. The Markov chain is $H(t), t = 1{,}2, \ldots$, and the state space is $\Phi = \{q_0, q_1\}$ denoting the confidential bits transmitting. The number of the states in this model is *N*, and $N = 2$.
- State transition probability distribution is shown as

$$A = [a_{ij}]_{N \times N} = \begin{bmatrix} p + rt & q + r(1-t) \\ p + rt & q + r(1-t) \end{bmatrix}.$$

  In the matrix, a_{00} denotes the probability of transition from state q_0 to q_0, where the error state is hidden. The reason for $a_{00} = p + rt$ is that the distribution probability of bit 1 is *p*, the error is *r* and transmits to bit 1 with probability of *t*. All the other elements are calculated in the same way.
- Observable symbols and distribution probability. The observation symbols correspond to the records modeled. The individual symbols are denoted as

$$V = \{v_0, v_1, \ldots, v_M\}.$$

  The observation symbol probability distribution in state *i* is

$$B = \{b_i(k)\},$$

where

$$b_i(k) = p[(v_k \ at \ t)|q_t = S_i],$$
$$1 \le i \le N, 1 \le k \le M.$$

$b_i(k)$ means the appearance probability of action v_k at time t under the state S_i. M is the number of distinct observation symbols per state. Therefore, values of B in the covert channel is

$$b_0(0) = p + rt, \quad b_0(1) = 0,$$
$$b_1(0) = 0, \qquad b_1(1) = q + r(1-t).$$

- Initial state distribution. This distribution is $\pi = \{\pi_0, \pi_1\}$, where $\pi = \{p+rt, q+r(1-t)\}$ in this model. Values of π is calculated with the distribution probability of the bits of 0,1 and e when initializing.
- Observation sequence is

$$O(t), t = 1,2,\ldots,T, O(t) \in V,$$

Where T is the total number of the observed signals.

The covert channel is described by the Markov model as follows,

$$\lambda = (A, B, \pi),$$

and the probability of the observation sequence $O = \{O_1, O_2, \ldots\}$ is calculated. Given the model λ, this probability is defined as $P(O|\lambda)$. $P(O|\lambda)$ denotes the observation sequence built from the model of λ, if the value is bigger enough, it is believed that a covert channel occurs.

To calculate $P(O|\lambda)$, the forward variable

$$a_t(i) = P\{O_1 O_2 \cdots O_t, H(t) = S_i|\lambda\}$$

must be considered first, and it is the probability of the partial observation sequence $(O_1 O_2 \ldots O_t)$ until time t and the state is S_i. $a_t(i)$ is calculated inductively as follows [54]:

- Initialization:

$$a_1(i) = \pi_i b_i(O_1), 1 \le i \le N.$$

- Induction:

$$a_{t+1}(i) = \left[\sum_{j=1}^{N} a_t(i) a_{ij}\right] b_j(O_{t+1}),$$
$$1 \le t \le T-1, \ 1 \le j \le N.$$

- Termination:

$$P(O|\lambda) = \sum_{i=1}^{N} a_T(i).$$

Based on the Markov model $\lambda = (A, B, \pi)$, $P(O|\lambda)$ is calculated as follows.

$$\begin{aligned}
P(O|\lambda) &= \sum_{i=1}^{N} a_T(i) = a_T(0) + a_T(1) \\
&= \left[\sum_{i=1}^{N} a_t(0)a_{i0}\right] b_0(O_{t+1}) + \left[\sum_{i=1}^{N} a_t(1)a_{i1}\right] b_1(O_{t+1}) \\
&= [a_t(0)a_{00} + a_t(1)a_{10}]b_0(O_{t+1}) \\
&\quad + [a_t(0)a_{01} + a_t(1)a_{11}]b_1(O_{t+1}) \\
&= (p + rt)[a_t(0) + a_t(1)]b_0(O_{t+1}) \\
&\quad + (q + r(1 - rt))[a_t(0) + a_t(1)]b_1(O_{t+1}).
\end{aligned}$$

Therefore, the following equation is obtained directly,

$$\begin{aligned}
a_{t+1}(0) + a_{t+1}(1) &= [a_t(0) + a_t(1)] \\
&\cdot [(p + rt)b_0(O_{t+1}) + (q + r(1 - t))b_1(O_{t+1})]
\end{aligned}$$

Finally, $P(O|\lambda)$ is obtained

$$P(O|\lambda) = \prod_{i=1}^{T} [(p + rt)b_0(O_i) + (q + r(1 - t))b_1(O_i)],$$

where p, q, r are the signal probabilities and $p + q + r = 1$.

The inputs of the Markov detector are the captured records, and the outputs are whether the records are covert channels or not. $P(O|\lambda)$ is the decision factor and calculated from the records. A Pessimistic Threshold is introduced to set the resulting boundary.

Definition 2 (Pessimistic Threshold) Pessimistic Threshold is calculated in the worst case. With this value, there are no false negative results, but some false positive ones. P_{thr} is used to denote the Pessimistic Threshold.

The Markov detector takes the situation that there is no covert channel as the worst case, so the Pessimistic Threshold is calculated as follows.

- The first step of calculating the Pessimistic Threshold is building a test bed, where all the operations are normal without covert channels.
- Then, the value is calculated by Markov detector under this situation, which is relative bigger.

- Thirdly, the operations in virtual machines is executed repeatedly and all the values are calculated and recorded.
- Finally, the smallest values is adopted and set as the Pessimistic Threshold.

When the Pessimistic Threshold is calculated, the decision policy of Markov detector is

$$P(O|\lambda) < P_{thr}.$$

If $P(O|\lambda)$ of a record is smaller than P_{thr}, it is considered as a covert channel. Because the P_{thr} is calculated in the worst case, some normal records may be mistaken as potential covert channels. Therefore, a Bayesian detector is needed to refine the results and lower the false positive.

5.2 *Bayesian Detection Algorithm*

Bayesian reasoning provides a probabilistic approach to inference. It is based on the assumption that the quantities of interest are governed by probability distributions and that the optimal decisions are made by reasoning about these probabilities together with observed data [56–58]. Naive Bayesian classifier is a highly practical Bayesian learning method whose performance is comparable to a neural network and decision tree learning in some domains.

In this paper, a naive Bayesian detector is designed and classifies the captured records into two classes, the covert channels and the normal operations. Each task processed by the Bayesian detector is described by a conjunction of attribute values, e.g. $x = < a_1, a_2, \ldots, a_n >$ (n is the number of the properties). The input of the detector is the records detected by Markov detector, including some false negative records. The output of the detector is a target function $f(x)$ whose value domain is $V = \{yes, no\}$. Values of *yes* and *no* indicate whether x is a covert channel or not. To the detector, the training samples are $X = \{x_1, x_2, \ldots, x_n\}$. Both the covert channels and normal samples are used to train the detector respectively. Then, the most probable target value v_{MAP} calculated as

$$v_{MAP} = \underset{v_j \in V}{\operatorname{argmax}} P(v_j | a_1, a_2, \ldots, a_n), \tag{1}$$

where $P(v_j|a_1, a_2, \ldots, a_n)$ denotes the probability to classify a certain sequence $x = < a_1, a_2, \ldots, a_n >$ into class v_j.

Then, v_{MAP} is rewritten into the following expression as

$$\begin{aligned} v_{MAP} &= \underset{v_j \in V}{\operatorname{argmax}} \frac{P(a_1, a_2, \ldots, a_n | v_j) P(v_j)}{P(a_1, a_2, \ldots, a_n)} \\ &= \underset{v_j \in V}{\operatorname{argmax}} P(a_1, a_2, \ldots, a_n | v_j) P(v_j), \end{aligned}$$

where $P(a_1, a_2, \ldots, a_n)$ is removed because it is a constant independent of v_j.

Bayesian detector is based on the simple assumption that the target attribute values are conditionally independent. Therefore, the probability of the conjunction $a_1, a_2, \ldots, a_n$ is calculated as follows.

$$P(a_1, a_2, \ldots, a_n | v_j) P(v_j) = \prod_{i=1}^{n} P(a_i | v_j).$$

Finally, the Bayesian detector is expressed as

$$v_{BD} = \underset{v_j \in V}{\operatorname{argmax}} P(v_j) \prod_{i=1}^{n} P(a_i | v_j),$$

where v_{BD} is the target value of the Bayesian detector.

All the captured records are detected by the two-phase synthesis algorithm. Both of the algorithms are easy to be implemented and applied online. The records are detected by Markov and Bayesian detector sequentially, and the final outputs are the detection results.

6 Mitigation Algorithm

The mitigator consists of two part components, including a core interrupter module located in hypervisor and an analyzer located in Dom0. The detailed design is shown in Fig. 3.

The analyzer estimates the capacity and the accuracy of the timing channel. Channel capacity is required by both TCSEC (Trusted Computer System Evaluation Criteria (Orange Book)) [21] and CC (Common Criteria for Information Technology Security Evaluation) [28, 46] criterions, which denotes the amount of information transferred by the timing channel per unit time (bits per second). Formal and non-formal methods are presented to calculate the capacity by Millen and Tsai [29]. The accurate capacity can be obtained by calculation, experiment and mathematical analysis. Generally, the capacity is calculated as follows,

$$C = \frac{N(t)}{t} \text{bits/s},$$

where $N(t)$ is the amount of the information transmitted in total time t.

Mitigation is to lower the capacity by adding latencies which is calculated by the analyzer. For example, a tuple of latencies $t_{add} = \{t_{add\_1}, t_{add\_2}, \ldots, t_{add\_N}\}$ is added into the channel. Therefore, the final capacity C_f is calculated as follows,

$$C_f = \frac{N(t)}{t + \sum_{i=1}^{N} t_{add_i}} \text{bits/s}.$$

The capacity is obviously lowered, because C_f is smaller than C expressed as $C_f < C$. TCSEC requires that the capacity of a timing channel should be under a certain threshold, from which the latencies tuple t_{add} can be calculated.

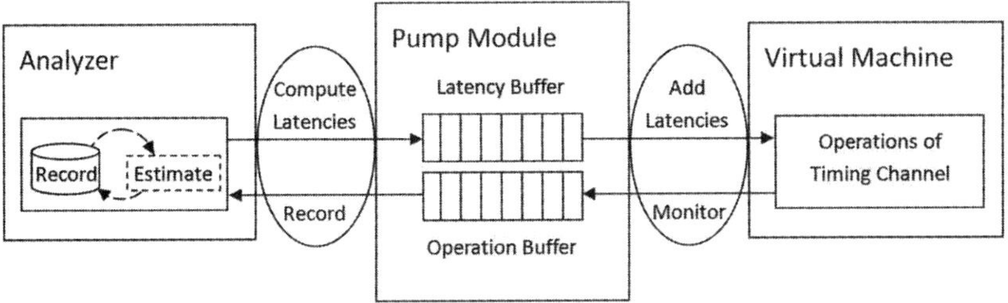

Fig. 6 An illustration of the interrupter module. The interrupter module receives the control latencies from the analyzer and adds them into the hypercalls. After execution, the records of hypercalls are sent to the analyzer and the latencies are calculated again. This cycle is repeated until the information has been sent out

Another metric to estimate the threat of the timing channel is transmission accuracy. If the confidential information can not be decoded correctly, the obtained information is valueless. The accuracy is measured by the percentage of correctly received bits. Cabuk et al. [32] use the edit distance [59] to measure the accuracy, which is the minimum distance between two strings. When the latencies have been added, the accuracy would decrease.

The analyzer calculates the latencies to be added into the timing channel, and sends a control message back to the pump module as shown in Fig. 6.

The interrupter is designed according to the Pump, which acts as a router that connects low secure level applications to high applications [60]. In particular, the Pump was designed to minimize the timing channel threat from the necessary message acknowledgements, without penalizing system performance and reliability. The basic Pump places a non-volatile buffer between Low and High secure level hosts, and adds latencies by sending acknowledgements (ACKs) to Low host at probabilistic times. The Network Pump services many senders and receivers from different applications simultaneously to lower the threat of the timing channels [61].

The interrupter module receives the tuple of latencies and places them into the latency buffer. The hypercalls are intercepted and delayed with the latencies, which are logged at the same time. The records of hypercalls are sent to the pump module and placed into the operation buffer. The analyzer receives the records and estimates the threat to determine whether the threat has been at an acceptable level. After estimation, the analyzer calculates the latencies and sends them to pump module. This cycle is repeated until all the information has been sent out.

7 Implementation and Evaluation

This section describes the implementation and evaluation of C²Hunter framework on Xen hypervisor. Firstly, the three types of covert channels concerned in this paper is introduced. Secondly, the controller in hypervisor and the analyzer with the two-phase synthesis detection algorithm in Dom0 are described. Then, the configurations

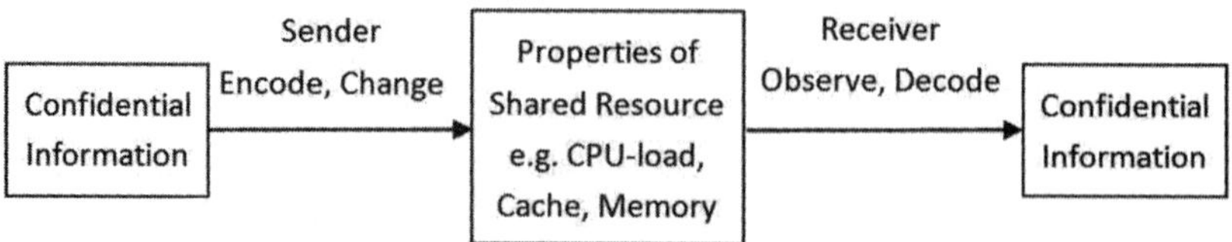

Fig. 7 Transmission cycle of a covert channel scenario

of the covert channels are demonstrated, and the operational records are captured. At last, the performance of the controller, the Markov detector and the Bayesian detector in C^{2}Hunter is evaluated.

7.1 Covert Channels Scenarios

A transmission cycle of a covert channel is shown in Fig. 7 [62]. The confidential information is transmitted from the sender to the receiver. The sender encodes the information into binary bits at first. Then, the properties of the shared resources are changed by the sender according to the bits. Finally, the receiver observes the changes and decodes the confidential information from these changes. The sender and receiver predetermine the parameters (e.g. decoding mechanism) and repeat the process until all the confidential information has been transmitted.

CPU load based, Cache based, and shared memory based channels are implemented according to the transmission cycle. The shared resources of the channels are CPU load, cache access time, and the memory writing intervals.

CPU load based covert channels use CPU load rate to denote the confidential information [9, 45]. A web server is running in the receiver VM. The sender encodes the confidential information into binary bits and issues numerous HTTP requests via *JMeter 2.4* (a utility for load testing HTTP servers). The receiver monitors the CPU utilization of the web server, and decodes the confidential information from the changes.

The shared resource in cache based covert channel is the cache access time [9, 44]. The sender shares a common cache with the receiver, and occupies the cache according to the encoded binary bits. The receiver observes the cache access time and decodes the confidential information from the time. A Prime-Probe protocol is used in the transmission cycle [63, 64]. A basic construction of the Prime-Probe protocol is described as follows.

- PRIME: The receiver fills an entire cache set *S* by reading memory region *M* from memory space.
- IDLE: The receiver waits for a pre-specified Prime-Probe interval while the cache is being utilized by the sender.
- PROBE: The receiver times the reading of the same memory region *M* to learn the sender's cache activity on cache set *S*.

The shared resource in shared memory based channel is the memory writing interval [48]. A memory sharing module is loaded into the guest OS. The sender encodes the confidential information and writes memory according to the different bits. The receiver observes the writing intervals and decodes the confidential information from them. The intervals may be affected by the VM loads, which will be investigated in the experiments. Shared memory based timing channel has much higher threat than the other types of timing channel, e.g. CPU load based and cache access time based timing channels. The threat metrics of the timing channel include channel capacity and transmission accuracy, both of which should be estimated in the real channel scenario [27]. In a cloud computing testbed, the capacity of the shared memory based timing channel is 174.98 bits/s and the error rate is 2 % [48]. XenPump presented in this paper is a new mitigation method to lower the capacity and/or increase the error rate of the timing channel, which is described in detail in the following section.

To setup a shared memory based timing channel, a shared ring structure is firstly initialized by the receiver, and then the grant table is sent to the sender. The sender maps the ring structure into his own space and sends the bits according to the computed intervals. When the sender has sent out all the bits, the receiver observes the write intervals and reverses the confidential information.

7.2 Captor and Detector

C^{2}Hunter framework is implemented in a desktop computer with an Intel® Core™ 2 Quad Q9400 2.66 Hz CPU, 320 GB disk and 4 GB main memory. Xen hypervisor is 4.0.0, and each VM is allocated 512 MB virtual memory running Fedora 8 Linux with kernel 2.6.34.1. The three types of covert channel programs used to leak the confidential information are implemented in C-Language.

The captor places the hook function *do_capture_init* in the hypervisor. *do_capture_init* first initializes a buffer *capture_buf*, and then calls the function *do_capture_op* to record all the operational events. When a hypercall is triggered, *do_capture_op* records the event information, including *vcpu_id*, *dom_id*, *grant_ref_t*, memory writing time, etc. The information recorded is filled into *capture_buf*, and send to the detector in Dom0 at last. This framework is similar to Lares [14], which mainly focuses on memory integrity protections. Comparing with Lares, the captor monitors the operations related to the shared resources.

The detector is implemented as a Linux kernel module (LKM), including the Markov and Bayesian detectors. Dom0 is the privileged domain. Therefore, the LKM is automatically loaded into Dom0 when C^{2}Hunter starts. LKM is module and running in the kernel space, which has a small impact on the hypervisor performance. The Markov and Bayesian detectors are implemented in C-Language. C^{2}Hunter is easy to be extended to support other detection algorithms, for example, the decision tree and neural network algorithms. The inputs of the detector are the records from the captor, and the outputs are the detection results. If the captured record set is too large to be processed in a single VM online, the detector can extend to multi-VMs and process the records in parallel. In this paper, the parallel process is out of consideration.

7.3 Interrupter in Hypervisor

Interrupter places the hook function *do_xenpump_init* in the hypervisor. *do_latency_init* first initializes a buffer *latency_buf*, and then the function *do_latency_op* is called to fill *latency_buf* with the computed latencies. After filling, *do_add_latency_init* is called to add the latencies to the hypercalls invoked by the timing channel.

On the other hand, *do_record_init* initializes a buffer *record_buf*, which is created to save the operation events. When a hypercall is invoked by timing channel, *do_record_op* records the event information, including *vcpu_id*, *dom_id*, *grant_ref_t*, memory writing time, etc. This information is filled into *record_buf*, and send to the analyzer in Dom0 at last. The analyzer estimates the threat of the timing channel by computing the capacity and accuracy, and calls *do_latency_op* to start a new cycle.

The analyzer is implemented as a Linux kernel module (LKM), recording and estimating the records. Dom0 is the privileged domain. Therefore, the LKM is automatically loaded into Dom0 when Interrupter starts. LKM is running in the kernel space, which has a small impact on the hypervisor performance.

7.4 Experimental Settings

The confidential information transmitted from the sender to the receiver is 50 letters text. The sender and receiver are located in the different VMs running Fedora 8 Linux, and use the same encode scheme. The text is encoded into 400 bits length binary string. The transmission cycle is shown in Fig. 7, where the bits denote the different operations.

1) CPU Load Based Covert Channel

In CPU load based covert channel, two VMs start up as the sender and the receiver. A web server *Apache 2.2.6* is running in the receiving VM, where a single 48K-byte text-only HTML page is made publicly accessible. On the other side, *JMeter* is installed in the sending VM, which is a pure Java desktop application designed to load test functional behavior and measure performance. Then, the sender controls the CPU loads of the receiving VM, and the transmission cycle is described as follows.

- First, when a bit 1 is transmitting, the sender in sending VM starts 100 threads simulating 100 single users and each thread takes 10 load samples. This burst HTTP requests take about $t = 10$ seconds, and the CPU load is obviously higher than 50%.

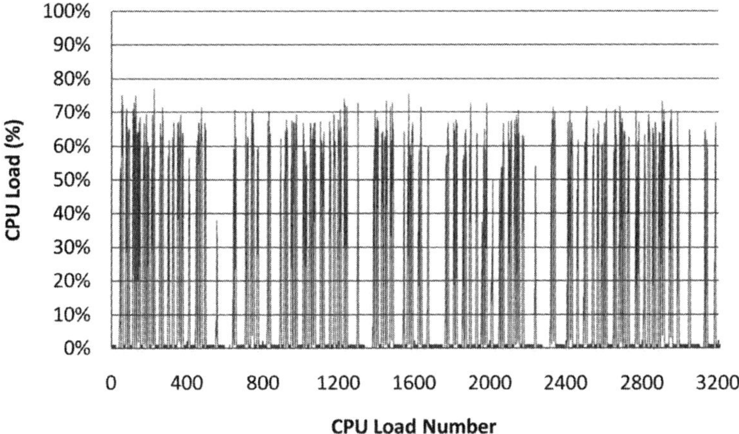

Fig. 8 CPU load of covert channel without noises

- Then, when a bit 0 is transmitting, the sender in sending VM pause for $t = 10$ seconds. The CPU load is obviously lower than 50% during this period.
- The sender starts HTTP requests and pauses for t seconds according to the transferred binary string, and each request is made as fast as possible. This transmission cycle is repeated until all the bits have been sent out.

An example of the CPU load based covert channel without noise is shown in Fig. 8. The load samples are obviously different when HTTP requests are performed or not. The burst HTTP requests consume almost 70% of the CPU load and almost 0% on the contrary. The transmission continues for 3200 s, where each $t = 10$ s denoting a bit. The receiver can easily decode the confidential information from the changes of CPU loads.

2) Cache Based Covert Channel

In cache based covert channel, sender and receiver are located in different VMs but the same hypervisor. Two L2 caches are in the test bed, each serving two CPU cores. Each L2 cache is 12-way set-associative with $m = 3072$ cache sets and a line size of $l = 64$ B, yielding a cache size of $c = 3$ MB. Each of the sending and receiving VM is allocated a single vCPU, and the Prime-Probe protocol is implemented as follows.

- Firstly, the receiver reads a 3072 bytes file to fill the entire cache and investigates the access time.
- At the same time, the sender starts tasks to occupy and release the cache in each $t = 1$ s period according to the transferred binary bits.
- If the cache is occupied, the receiver will take about 1 s to read the file. Otherwise, if the cache is free to use, the receiver will take about 0.5 s to read the file.
- This transmission cycle is repeated until all the bits have been sent out.

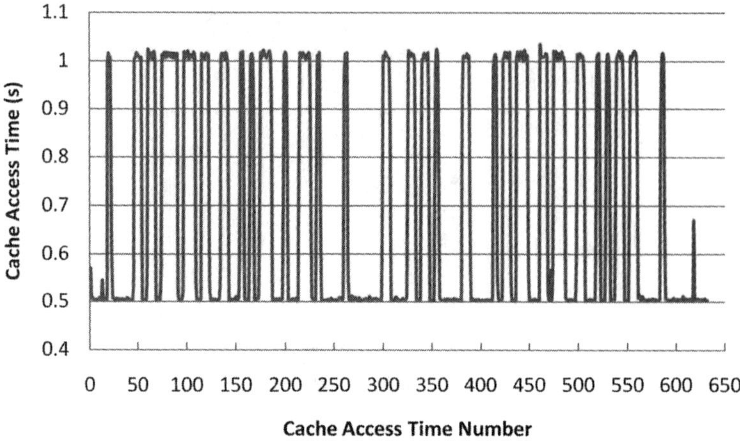

Fig. 9 Cache access time of covert channel without noises

An example of the Cache based covert channel without noise is shown in Fig. 9. The cache access time is obviously different from which the receiver decodes the confidential information.

3) Shared Memory Based Covert Channel

In shared memory based covert channel, the receiver and the sender locate in the different VMs but sharing the same memory. The sender and the receiver have predetermined the size of the *buffer_list* and cooperate in the producer and consumer model. The sender writes data into the memory buffer ring in certain intervals according to the binary bits. The explicit communication protocol is described as follows.

- The receiver setups a shared ring structure with a grant table and passes the grant reference to the sender.
- The sender computes the intervals and then sends data according to the intervals in each transmission cycle.
- The receiver obtains the data and computes the intervals on each interrupt.
- This transmission cycle is repeated until all the bits have been sent out.

An example of SMCTC without noise is shown in Fig. 10. The memory writing intervals are from almost 2000–10,000 μs. When a bit 0 (or 1) is transmitted, the interval is short (or long). When all the intervals have been obtained by the receiver, the confidential information can be decoded.

7.5 Detection Analysis

C^{2}Hunter is tested by the three types of covert channels. Taking shared memory based covert channel for example, a test sample includes 100 SMCTC and 100 normal operation sequences. Each sample is repeated 100 times to get the average detection

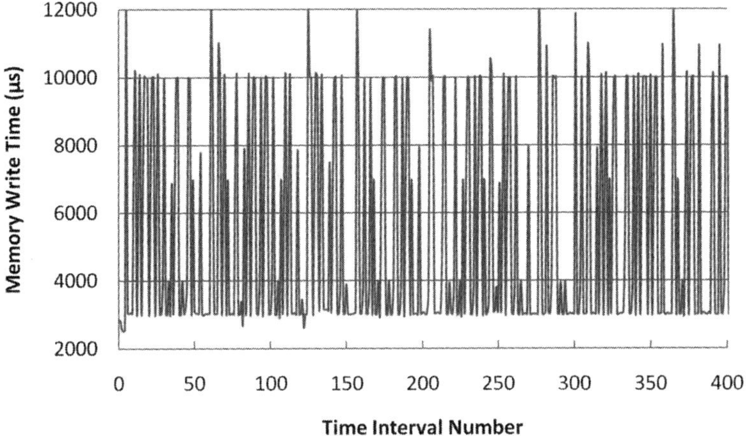

Fig. 10 Time intervals of SMCTC without noises

performance. When the experiment is executing, some operations are running in the third VM to bring some noise, which causes the receiver to decode the message with some errors.

In SMCTC, the sender process P_i runs in the user space, and then calls the shared ring using hypercall in Xen. P_i writes data into the shared ring and the receiver process P_j reads data from it. The call branches are expressed as follows, where $\odot$ means the action of operating, $\rightarrow$ means writing action, and $\leftarrow$ means reading action.

$$P_i \odot \{Share\ Ring\} = \{user\_write \rightarrow send\_request \rightarrow RING\_PUSH\_REQUESTS\_AND\_CHECK\_NOTIFY \rightarrow Shared\ Ring\},$$
$$P_j \odot \{Share\ Ring\} = \{user\_read \leftarrow get\_request \leftarrow RING\_GET\_REQUEST \leftarrow Shared\ Ring\}.$$

The above call branches are captured as the operational records and flow into the Markov detector. The results of the Markov detector are shown in Fig. 11. The value domain of $P(O|\lambda)$ is

$$0.078955 \leq P(O|\lambda) \leq 0.228742.$$

As described in Sect. 5, the covert channels have small value of $P(O|\lambda)$. To lower false negative rate, a Pessimistic Threshold is set as

$$P_{thr} = 0.228742.$$

Therefore, all the covert channels satisfying the following condition

$$P(O|\lambda) < P_{thr}$$

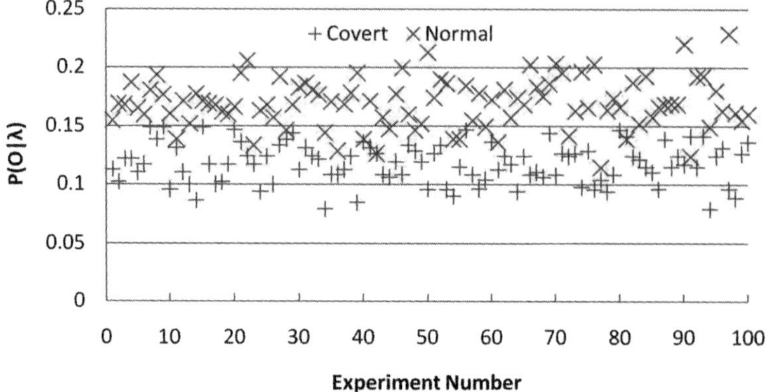

Fig. 11 Markov detector results in detecting SMCTC

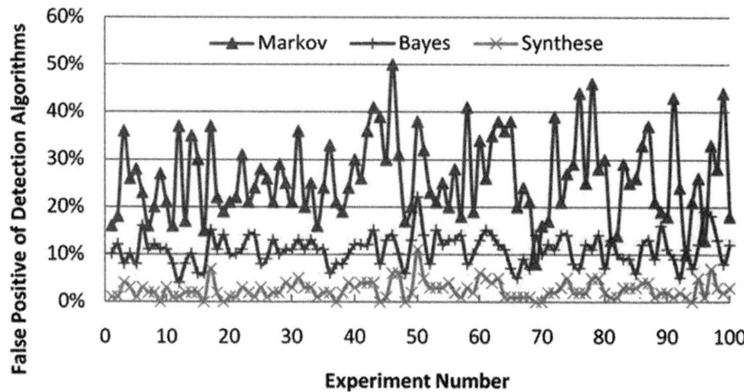

Fig. 12 False positives of the Markov detector in detecting SMCTC

are potential covert channels. Because of P_{thr} being pessimistic, some normal operations sequences are included in this result by mistaken.

The sample has been repeatedly 100 times as shown in Fig. 12. The false positive of the detectors in detecting SMCTC are shown in this figure. The false positive of the Markov detector is 16% meaning 16 normal sequences have been detected by mistaken. Comparing to the Markov detector, the false positive of Bayesian detector is 10% in detecting the same sample. It is believed that the performance of Bayesian detector is better than the Markov detector when used independently.

The detector detects the sample use of Markov detector and Bayesian detector sequentially. Therefore, the false positive falls to 1%. The reason for the reducing is that the Bayesian detector refines the resulting output by the Markov detector. Only the channels that have been taken as covert by both the detectors are output as covert channels.

Table 1 Detection results of the three channels

Channels	Capacity (bps)	Error rate (%)	Markov (%)	Bayesian (%)	False positive (%)
CPU load	0.098	4.76	29.92 %	8.14	4.46
Cache Based	0.297	6.67	24.51	15.61	1.89
shared memory	189.15	2.43	26.28	16.42	2.53

The experiments of CPU load based and cache based covert channels are detected in the same way. Table 1 shows the average detection results of all the three types of covert channels.

In Table 1, capacity means the threat performance of a covert channel. The bigger capacity means the covert channel is more threat to the cloud computing. It can be seen that the shared memory based covert channel is the most threatening to cloud computing. Error means the decoding error. In the experiments, it is calculated by edit distance [59], which is the minimum distance between two strings. The values of Markov and Bayesian are the false positive of each detector. In these experiments, the Bayesian detector is always better than the Markov detector. When the two detectors detect the channels sequentially, the final false positive is acceptable.

These experiments show that the two-phase synthesis algorithm is efficient to detect covert channels. C²Hunter framework detects all the three types of channels in cloud computing with no false negative and low false positive.

7.6 Mitigation Analysis

An example of shared memory based timing channel is shown in Fig. 13. A confidential information is encoded into a 200 bits string, which is repeatedly sent 100 times in each experiment.

In Fig. 13, the *No-Latency* denotes the original timing channel without mitigation. The memory write intervals are from almost 2 to 30 ms, most of which are around 10 ms. In this experiment, bit 0 is sent directly, but bit 1 is sent after sleeping 5 ms. The intervals are relatively small and the capacity is relatively high, which is about 140 bits/s as shown in Fig. 14. When all the intervals have been obtained by the receiver, the confidential information can be decoded from the intervals. In the experiment, the receiver decodes the confidential information with small error rate, where the error rate is about 5 % as shown in Fig. 15. The small transmission error rate means most of the confidential information can be transmitted correctly in the timing channel.

The mitigation performance of interrupter is tested by two experiments, which are named as Latency-I and Latency-II in this paper. In Latency-I experiment, a tuple of latencies $t_{add\_I} = \{t_{add\_I\_1}, t_{add\_I\_2}, \cdots, t_{add\_I\_N}\}$ is added into the timing channel by interrupter, where the values of t_{add_I} is produced from a uniform distribution

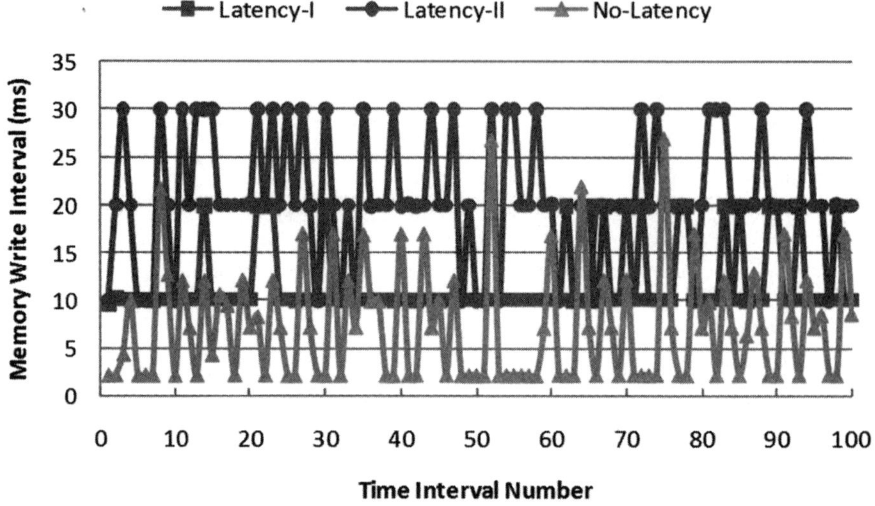

Fig. 13 Memory write intervals of the shared memory

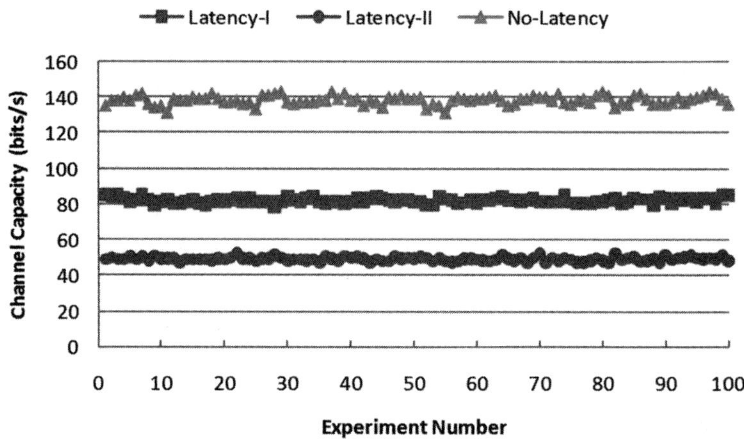

Fig. 14 Channel capacity of the shared memory based timing channel

$U(0, 0.01)$. Each of the interval is added a latency $t, 0 < t < 0.01$, which is shown in Fig. 13 denoted by *Latency-I*. The intervals of *Latency-I* is almost 10–20 ms. Therefore, the capacity is lowered to about 80 bits/s and shown in Fig. 14. Interrupter adds the latencies to the timing channel, which lowered the capacity and increased the error rate. The error rate is about 20%, meaning some of the confidential information cannot be decoded correctly.

Latency-II experiment produces the latencies $t_{add\_II} = \{t_{add\_II\_1}, t_{add\_II\_2}, \ldots, t_{add\_I\_N}\}$ by the uniform distribution $U(0, 0.025)$. When the latencies have been added to the timing channel, the memory write intervals is almost

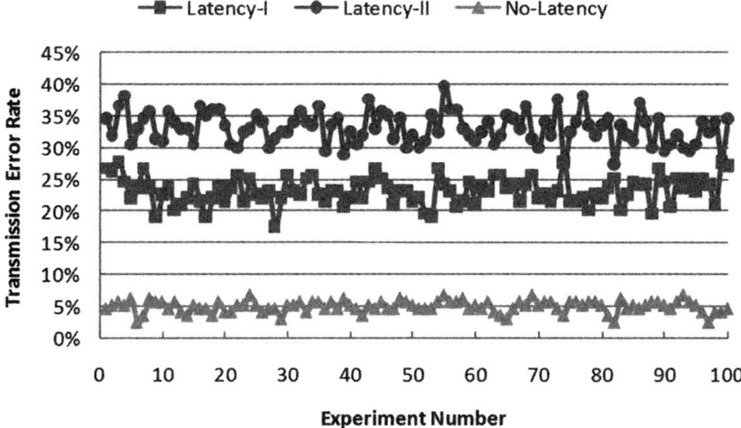

Fig. 15 Transmission error rate of the shared memory based timing channel

10–30 ms as shown in Fig. 13, lowering the capacity to about 50 bits/s as shown in Fig. 14. The error rate is increased to about 35%, meaning the confidential information decoded by the receiver is valueless.

In the experiments, interrupter adds latencies to the timing channel inducing the capacity decrease and the error rate increase. If the longer latencies are added into the timing channel, the values of the capacity will be smaller and the error rate will be bigger. With a bigger error rate, the decoded confidential information is valueless. Because of the latencies, transmitting the same confidential information needs longer time, causing the performance decrease. For example, the time used by *Latency-I* and *Latency-II* is longer than *No-Latency* by 63.87 % and 190.37 % respectively.

The experiments show that interrupter can mitigate the timing channel by adding the latencies which are produced by taking consideration of both the capacity and transmission accuracy.

8 Discussion

The covert channel is modeled into an error-corrected four states automata, adopting the probability based error-corrected algorithms referred in Sect. 4.4 to process the errors in the experiments. When an error occurs, it takes the error as bit 1 (or 0) depending on the probability of the property states in Markov detector and Bayesian detector. Although this error-corrected algorithm simplifies the detection process, the detection accuracy may be affected. A more sophisticated algorithm will be developed in the future.

It has been shown that all the three types of covert channels can be detected by C²Hunter framework. XML file is used as a plug-in to config the detector to detect

```
<covert channel>
  <channel name = "CPU_load_based_covert_channel">
    <dom>
      <sender id = "..."/>
      <receiver id = "..."/>
    </dom>
    <Markov>
      <Pessimistic Threshold="..."/>
    </Markov>
      ......
  </channel>
   ......
</covert channel>
```

Fig. 16 XML configuration file in C^{2}Hunter

covert channels. A typical XML file includes the channel name, the property domain values, and the Pessimistic Threshold. An example is shown as follows in Fig. 16.

This information is collected and appended to the XML file, so it is convinced that C^{2}Hunter can detect the incoming channels as long as they can be modeled into this model. More channels will be built to test whether C^{2}Hunter can detect or not in the future work.

To protect the privacy of the cloud customers in data centers, cloud providers should not collect the users' data. In C^{2}Hunter, only the change regularities of the shared resources are concentrated, e.g. CPU load, cache access time, and the memory writing intervals. There is no need to know which operation triggers these changes. The ignorance of the channel details protects the users' privacy and lower the performance influence to hypervisor.

Only the channels using the binary encoding scheme are investigated and C^{2}Hunter is easily extended to support multi-encoding channel [27]. A multi-encoding channel will be modeled into an error-corrected multi-states automata in the same way. Each codeword is modeled into a state in the automata and can be transited from all the other states. The corresponding Markov detector can be modified easily. There is no difference between multi-encoding and binary channels, and C^{2}Hunter detect this type of channels in the same way as the binary ones.

The interrupter mitigates the threat of the timing channel anonymously by adding latencies to the Hypercalls without collecting users' privacy data. In mitigation, only the Hypercalls are monitored and operated meaning it is no need to know which user operation triggers these changes. The ignorance of the channel details protects the users' privacy and lower the performance influence to hypervisor.

C^{2}Hunter is a plug-in framework, whenever a new efficient detection algorithm is discovered, it can replace the existing detector or be added as another detector. However, C^{2}Hunter is built to detect covert channels only and other intrusion methods cannot be detected. The mitigation is easy to control by the interrupter. The capacity and the error rate vary according to the variation of the latencies which are calculated by the analyzer. Take the Latency-I and Latency-II experiment as examples, if the threat of the timing channel is required to small, the latencies should be relatively bigger. The interrupter is built to mitigate the timing channels, which may be integrated into the monitoring tools in the future. C^{2}Hunter may be integrated into other intrusion detection tools in the future.

9 Related Work

Most of the prior works on cloud computing security have focused on the integrity protection. Several frameworks have been proposed to protect the integrity of the guest kernels using hypervisor. For example, HIMA [17], Lares [14], HyperSentry [15], HyperSafe [11], and Antfarm [16] are all designed to provide the integrity protection to VMs. These approaches are related to the virtualization security and integrity, but only C^{2}Hunter focuses on the covert channels.

Millen presents a covert channel detection approach, but it is limited to the covert storage channels [51]. This paper aims to detect timing channels in cloud computing, which is more difficult. Cabuk and Berk present the detection algorithms to detect network covert timing channels [32–34]. The covert channels they detected are classified into CC2 type of channels which has been studied in-depth for many years [26, 27, 39, 47, 65]. In this paper, only CC3 type of channels that are recently brought by cloud computing are considered.

Nagatou et al. [35] present a run-time detection approach and focus on enforced properties. C^{2}Hunter can also perform online detection. HomeAlone [44] is proposed to determine whether a VM is physically co-resident to another VM and the same functions are proposed in [9, 45]. Although these approaches are related to covert channels, they are not intended for detecting them.

Some introspections used as intrusion detection systems have been proposed upon virtual machines. For example, HyperSpector [66] is designed to monitor the actions without any additional hardware by using virtualization to isolate each IDS from the servers. Some other IDSes are presented in [67–69]. Covert channel can be seen as a type of intrusion and C^{2}Hunter is a special detector only detecting the covert channels. C^{2}Hunter is easy to extend and has better detection performance, which is a complement to these introspections.

The practical method for the handling of known timing channel is the mitigation, which requires the reduction of the capacity or increase of the error rate of the timing channel to a predefined acceptable level. It is obvious that adding latency and introducing noise into the channel can achieve this goal. In this paper, XenPump is proposed to mitigate the timing channel in cloud computing by adding latency.

Kang et al. proposed Pump for the first time in [60], which pushes messages to the high system and provides a controlled stream of acknowledgements to the low system. The Pump provides quantifiable security, acceptable reliability and minimal performance penalties to the communication from a low level to a high level system, which may cause a timing channel by the acknowledgements. They extended the Pump to a certain multi-level security network architecture in order to mitigate the threat of network timing channels [61, 70]. They even designed a custom hardware architecture of Network Pump^{TM} with separate microprocessors, memory, input/output (I/O) circuitry, etc [71]. Network Pump^{TM} connects the Low net and the High net with only a shared stable memory buffer in common to mitigate the threat of the timing channel.

Pump is proposed to limit the threat of the timing channel in network. In this paper, XenPump is designed to mitigate the threat of the timing channels in cloud computing. XenPump is located in the virtual machine management layer and has high privilege to monitor and operate the Hypercalls used by the timing channels. The main difference between Network Pump^{TM} and XenPump is that XenPump is designed for cloud computing.

Another mitigation method is proposed by Hu [72] using fuzzy time. Fuzzy time has proven to be highly effective against the timing channels in the VAX Security Kernel. Not only does fuzzy time close the high speed channels, it does so at a much lower than anticipated performance cost. Fuzzy time isolates a process from all precise timing information. This method is suitable for the timing channels using the same time clock, e.g. Bus-Contention timing channel. Compared with XenPump, fuzzy time mitigates the threat by reducing the accuracy and precision of the system clocks, which are not the main concerns in cloud computing.

Recent work introduced predictive mitigation, a new way to mitigate leakage through timing channels in interactive systems [73, 74]. This method bounds the amount of information leaked through the timing channel as a function of elapsed time, predicts timing from past behavior, and then enforces the predictions. Comparing with this method, XenPump is relatively easy to implement. Once the Hypercalls have been identified, the latencies can be added into the Hypercalls and the threat will decrease.

10 Conclusion

With the growing popularity of cloud computing, more confidential applications have been deployed in the data centers. Covert channel is a serious threat to the data of cloud customers. In this paper, the fundamentals to the covert channels in the cloud are argued and the channels are classified into three categories. Only the new channels brought by cloud computing are concerned and modeled into an Error-Corrected Four States Automata. Some formal requirements are presented to build a covert channel detector and a plug-in detection framework named C^{2}Hunter is designed.

C^{2}Hunter satisfies the four formal requirements proposed in Sect. 4.2 by using two key techniques. The first technique installs hooks inside hypervisor to monitor all the hypercalls, capture the operation tracks and add latencies into the hypercalls. The second key technique is the two-phase synthesis detection algorithm implemented as Markov detector and Bayesian detector. To evaluate C^{2}Hunter, a prototype is implemented on Xen hypervisor. The CPU load based, Cache based, and shared memory based covert channels have been implemented and detected by C^{2}Hunter.

The threat of shared memory based timing channel is mitigated by interrupting both the capacity and transmission accuracy. The results show that C^{2}Hunter can detect all the three types of the channels using the Pessimistic Threshold, mitigate the threat and has a small false positive rate. In addition, this research demonstrates that C^{2}Hunter is highly efficient and feasible to extend to detect and mitigate the incoming new channels.

Acknowledgements This work is supported by the National Science and Technology Major Project No.2012ZX01039-004, No.2010ZX01036-001-002, the National Natural Science Foundation of China No.61303057, No.61170072 and the Major Program of the National Natural Science Foundation of China No.91124014. Samee U. Khan's work was partly supported by the Young International Scientist Fellowship of the Chinese Academy of Sciences, (Grant No. 2011Y2GA01).

References

1. M. Armbrust, A. Fox, R. Griffith, A. D. Joseph, R. H. Katz, A. Konwinski, G. Lee, D. A. Patterson, A. Rabkin, I. Stoica, and M. Zaharia, "A view of cloud computing," *Commun. ACM*, vol. 53, no. 4, pp. 50–58, 2010.
2. A. Greenberg, J. Hamilton, D. A. Maltz, and P. Patel, "The cost of a cloud: research problems in data center networks," *SIGCOMM Comput. Commun. Rev.*, vol. 39, no. 1, pp. 68–73, Dec. 2008.
3. G. L. Valentini, W. Lassonde, S. U. Khan, N. Min-Allah, S. A. Madani, J. Li, L. Zhang, L. Wang, N. Ghani, J. Kolodziej, H. Li, A. Y. Zomaya, C.-Z. Xu, P. Balaji, A. Vishnu, F. Pinel, J. E. Pecero, D. Kliazovich, and P. Bouvry, "An overview of energy efficiency techniques in cluster computing systems," Cluster Computing, vol. 16, no. 1, pp. 3–15, 2013.
4. L. M. Vaquero, L. Rodero-Merino, J. Caceres, and M. Lindner, "A break in the clouds: towards a cloud definition," *SIGCOMM Comput. Commun. Rev.*, vol. 39, pp. 50–55, December 2008.
5. R. Buyya, C. S. Yeo, S. Venugopal, J. Broberg, and I. Brandic, "Cloud computing and emerging it platforms: Vision, hype, and reality for delivering computing as the 5th utility," *Future Gener. Comput. Syst.*, vol. 25, pp. 599–616, June 2009.
6. P. Barham, B. Dragovic, K. Fraser, S. Hand, T. L. Harris, A. Ho, R. Neugebauer, I. Pratt, and A. Warfield, "Xen and the art of virtualization," in *SOSP*, 2003, pp. 164–177.
7. H. Takabi, J. B. D. Joshi, and G.-J. Ahn, "Security and privacy challenges in cloud computing environments," *IEEE Security & Privacy*, vol. 8, no. 6, pp. 24–31, 2010.
8. Y. Chen, V. Paxson, and R. H. Katz, "What' s new about cloud computing security?" EECS Department, University of California, Berkeley, Tech. Rep. UCB/EECS-2010-5, Jan 2010.
9. T. Ristenpart, E. Tromer, H. Shacham, and S. Savage, "Hey, you, get off of my cloud: exploring information leakage in third-party compute clouds," in *ACM Conference on Computer and Communications Security*, 2009, pp. 199–212.
10. J. Wu, L. Ding, and Y. Wang, "Research on key problems of covert channel in cloud computing," *Journal of Communications*, vol. 32, no. 9A, pp. 184–203, 2011.
11. Z. Wang and X. Jiang, "Hypersafe: A lightweight approach to provide lifetime hypervisor control-flow integrity," in *IEEE Symposium on Security and Privacy*, 2010, pp. 380–395.
12. B. D. Payne, R. Sailer, R. Cáceres, R. Perez, and W. Lee, "A layered approach to simplified access control in virtualized systems," *Operating Systems Review*, vol. 41, no. 4, pp. 12–19, 2007.
13. R. Sailer, T. Jaeger, E. Valdez, R. Cáceres, R. Perez, S. Berger, J. L. Griffin, and L. van Doorn, "Building a mac-based security architecture for the XenXen open-source hypervisor," in *ACSAC*, 2005, pp. 276–285.

14. B. D. Payne, M. Carbone, M. I. Sharif, and W. Lee, "Lares: An architecture for secure active monitoring using virtualization," in *IEEE Symposium on Security and Privacy*, 2008, pp. 233–247.
15. A. M. Azab, P. Ning, Z. Wang, X. Jiang, X. Zhang, and N. C. Skalsky, "Hypersentry: enabling stealthy in-context measurement of hypervisor integrity," in *ACM Conference on Computer and Communications Security*, 2010, pp. 38–49.
16. S. T. Jones, A. C. Arpaci-Dusseau, and R. H. Arpaci-Dusseau, "Antfarm: Tracking processes in a virtual machine environment," in *USENIX Annual Technical Conference, General Track*, 2006, pp. 1–14.
17. A. M. Azab, P. Ning, E. C. Sezer, and X. Zhang, "Hima: A hypervisor-based integrity measurement agent," in *ACSAC*, 2009, pp. 461–470.
18. J. Wu, Y. Wu, Z. Wu, M. Yang, and Y. Wang, "Vulcloud: Scalable and hybrid vulnerability detection in cloud computing," in *Software Security and Reliability-Companion (SERE-C), 2013 IEEE 7th International Conference on*, 2013, pp. 225–226.
19. J. Wu, Y. Wu, M. Yang, Z. Wu, and Y. Wang, "Vulnerability detection of android system in fuzzing cloud," in *Proceedings of the 2013 IEEE Sixth International Conference on Cloud Computing*, ser. CLOUD '13. Washington, DC, USA: IEEE Computer Society, 2013, pp. 954–955.
20. A. Aviram, S. Hu, B. Ford, and R. Gummadi, "Determinating timing channels in compute clouds," in *CCSW '10: Proceedings of the 2010 ACM workshop on Cloud computing security workshop*. New York, NY, USA: ACM, 2010, pp. 103–108.
21. NCSC, "Trusted computer system evaluation criteria (orange book)," 1985.
22. B. W. Lampson, "A note on the confinement problem," *Commun. ACM*, vol. 16, no. 10, pp. 613–615, 1973.
23. J. Wu, L. Ding, Y. Wang, and W. Han, "A practical covert channel identification approach in source code based on directed information flow graph," in *IEEE SSIRI*, Jeju Island, Korea, 2011, pp. 98–107.
24. C.-R. Tsai, V. D. Gligor, and C. S. Chandersekaran, "A formal method for the identification of covert storage channels in source code," in *IEEE Symposium on Security and Privacy*, 1987, pp. 74–87.
25. T. F. Keefe, W.-T. Tsai, and J. Srivastava, "Database concurrency control in multilevel secure database management systems," *IEEE Trans. Knowl. Data Eng.*, vol. 5, no. 6, pp. 1039–1055, 1993.
26. S. Zander, G. J. Armitage, and P. Branch, "A survey of covert channels and countermeasures in computer network protocols," *IEEE Communications Surveys and Tutorials*, vol. 9, no. 1–4, pp. 44–57, 2007.
27. J. Wu, Y. Wang, L. Ding, and X. Liao, "Improving performance of network covert timing channel through huffman coding," *Mathematical and Computer Modelling*, vol. 55, no. 1–2, pp. 69–79, 2012.
28. ISO/IEC, "Common criteria for information technology security evaluation," 2005.
29. Y. Wang, J. Wu, H. Zeng, L. Ding, and X. Liao, "Covert channel research," *Journal of Software*, vol. 21, no. 9, pp. 2262–2288, 2010.
30. J. Wu, Y. Wang, L. Ding, and Y. Zhang, "Constructing scenario of event-flag covert channel in secure operating system," in *ICIMT*, Hongkong, 2010, pp. 371–375.
31. C.-R. Tsai and V. D. Gligor, "A bandwidth computation model for covert storage channels and its applications," in *IEEE conference on Security and privacy*, Oakland, California, 1988, pp. 108–121.
32. S. Cabuk, C. E. Brodley, and C. Shields, "IP covert timing channels: design and detection," in *ACM Conference on Computer and Communications Security*, 2004, pp. 178–187.
33. ——, "IP covert channel detection," *ACM Trans. Inf. Syst. Secur.*, vol. 12, no. 4, pp. 1–29, 2009.
34. V. Berk, A. Giani, G. Cybenko, and N. Hanover, "Detection of covert channel encoding in network packet delays," *Rapport technique TR536, de lUniversité de Dartmouth. Novembre*, 2005.

35. N. Nagatou and T. Watanabe, "Run-time detection of covert channels," in *ARES*, 2006, pp. 577–584.
36. L. Hélouët and A. Roumy, "Covert channel detection using information theory," in *SecCo*, 2010, pp. 34–51.
37. J. K. Millen, "20 years of covert channel modeling and analysis," in *IEEE Symposium on Security and Privacy*, 1999, pp. 113–114.
38. C. G. Girling, "Covert channels in LAN's," *IEEE Trans. Software Eng.*, vol. 13, no. 2, pp. 292–296, 1987.
39. L. Yao, X. Zi, L. Pan, and J. Li, "A study of on/off timing channel based on packet delay distribution," *Computers & Security*, vol. 28, no. 8, pp. 785–794, 2009.
40. T. G. Handel and M. T. S. II, "Hiding data in the osi network model," in *Information Hiding*, 1996, pp. 23–38.
41. K. Ahsan and D. Kundur, "Practical data hiding in TCP/IP," in *Proc. Workshop on Multimedia Security at ACM Multimedia*. Citeseer, 2002.
42. C. Rowland, "Covert channels in the TCP/IP protocol suite," *First Monday*, vol. 2, no. 5–5, 1997.
43. S. Gianvecchio and H. Wang, "Detecting covert timing channels: an entropy-based approach," in *CCS '07: Proceedings of the 14th ACM conference on Computer and communications security*. New York, NY, USA: ACM, 2007, pp. 307–316.
44. A. O. Yinqian Zhang, Ari Juels and M. K. Reiter, "Homealone: Co-residency detection in the cloud via side-channel analysis," in *IEEE Symposium on Security and Privacy*, Oakland, California, 2011, pp. 313–328.
45. K. Okamura and Y. Oyama, "Load-based covert channels between Xen virtual machines," in *SAC*, 2010, pp. 173–180.
46. H. Zeng, Y. Wang, L. Ruan, W. Zu, and J. Cai, "Covert channel mitigation method. for secure real-time database using capacity metric," *Journal on Communications*, vol. 29, no. 8, pp. 46–56, 2008.
47. Y. Wang, J. Wu, L. Ding, and H. Zeng, "Detecion approach for covert channel based concurrency conflict interval time," *Journal of Computer Research and Development*, vol. 48, no. 8, pp. 1542–1553, 2011.
48. J. Wu, L. Ding, Y. Wang, and W. Han, "Identification and evaluation of sharing memory covert timing channel in Xen virtual machines," in *IEEE CLOUD*, Washington DC, USA, 2011, pp. 283–291.
49. J. Wu, L. Ding, Y. Lin, N. Min-Allah, and Y. Wang, "Xenpump: A new method to mitigate timing channel in cloud computing," in *IEEE CLOUD*, Hawaii, USA, 2012, pp. 678–685.
50. D. Chisnall, *The definitive guide to the xen hypervisor*. Prentice Hall Press, 2007.
51. J. K. Millen, "Finite-state noiseless covert channels," in *CSFW*, 1989, pp. 81–86.
52. R. Lanotte, A. Maggiolo-Schettini, and A. Troina, "Time and probability-based information flow analysis," *Software Engineering, IEEE Transactions on*, vol. 36, no. 5, pp. 719–734, 2010.
53. J. Wu, L. Ding, Y. Wu, N. Min-Allah, S. U. Khan, and Y. Wang, "C^2detector: A covert channel detection framework in cloud computing," *Security and Communication Networks*, 2013.
54. L. R. Rabiner, "A tutorial on hidden markov models and selected applications in speech recognition," *Proceedings of the IEEE*, vol. 77, no. 2, pp. 257–286, feb 1989.
55. J. Hu, X. Yu, D. Qiu, and H.-H. Chen, "A simple and efficient hidden Markov model scheme for host-based anomaly intrusion detection," *IEEE Network*, vol. 23, no. 1, pp. 42–47, 2009.
56. T. M. Mitchell, *Machine learning*. McGraw-Hill, 1997.
57. A. W. Moore and D. Zuev, "Internet traffic classification using Bayesian analysis techniques," in *SIGMETRICS*, 2005, pp. 50–60.
58. T. Auld, A. W. Moore, and S. F. Gull, "Bayesian neural networks for internet traffic classification," *IEEE Transactions on Neural Networks*, vol. 18, no. 1, pp. 223–239, 2007.
59. E. S. Ristad and P. N. Yianilos, "Learning string-edit distance," *IEEE Trans. Pattern Anal. Mach. Intell.*, vol. 20, no. 5, pp. 522–532, 1998.

60. M. H. Kang and I. S. Moskowitz, "A pump for rapid, reliable, secure communication," in *ACM Conference on Computer and Communications Security*, 1993, pp. 119–129.
61. M. H. Kang, I. S. Moskowitz, and D. C. Lee, "A network pump," *IEEE Trans. Software Eng.*, vol. 22, no. 5, pp. 329–338, 1996.
62. J. Son and J. Alves-Foss, "A formal framework for real-time information flow analysis," *Comput. Secur.*, vol. 28, no. 6, pp. 421–432, 2009.
63. D. A. Osvik, A. Shamir, and E. Tromer, "Cache attacks and countermeasures: The case of aes," in *CT-RSA*, 2006, pp. 1–20.
64. E. Tromer, D. A. Osvik, and A. Shamir, "Efficient cache attacks on aes, and countermeasures," *J. Cryptology*, vol. 23, no. 1, pp. 37–71, 2010.
65. S. Chen, R. Wang, X. Wang, and K. Zhang, "Side-channel leaks in web applications: A reality today, a challenge tomorrow," in *IEEE Symposium on Security and Privacy*, 2010, pp. 191–206.
66. K. Kourai and S. Chiba, "Hyperspector: virtual distributed monitoring environments for secure intrusion detection," in *VEE*, 2005, pp. 197–207.
67. T. Garfinkel and M. Rosenblum, "A virtual machine introspection based architecture for intrusion detection," in *NDSS*, 2003.
68. X. Jiang and X. Wang, ""out-of-the-box" monitoring of vm-based high-interaction honeypots," in *RAID*, 2007, pp. 198–218.
69. J. Li, B. Li, T. Wo, C. Hu, J. Huai, L. Liu, and K. Lam, "Cyberguarder: A virtualization security assurance architecture for green cloud computing," *Future Generation Computer Systems*, vol. 28, no. 2, pp. 379–390, 2012.
70. M. Kang, I. Moskowitz, and D. Lee, "A network version of the pump," in *Security and Privacy, 1995. Proceedings., 1995 IEEE Symposium on*, 1995, pp. 144–154.
71. M. Kang, I. Moskowitz, and S. Chincheck, "The pump: a decade of covert fun," in *Computer Security Applications Conference, 21st Annual*, 2005, pp. 352–360.
72. W.-M. Hu, "Reducing timing channels with fuzzy time," in *IEEE Symposium on Security and Privacy*, 1991, pp. 8–20.
73. D. Zhang, A. Askarov, and A. C. Myers, "Predictive mitigation of timing channels in interactive systems," in *Proceedings of the 18th ACM conference on Computer and communications security*, ser. CCS '11. New York, NY, USA: ACM, 2011, pp. 563–574.
74. A. Askarov, D. Zhang, and A. C. Myers, "Predictive black-box mitigation of timing channels," in *ACM Conference on Computer and Communications Security*, 2010, pp. 297–307.

Selective and Private Access to Outsourced Data Centers

Sabrina De Capitani di Vimercati, Sara Foresti, Giovanni Livraga and Pierangela Samarati

1 Introduction

The increasing amount of information being generated, collected, shared, and disseminated nowadays is making the in-house management of data centers by private and public companies more and more difficult and economically expensive. The wide availability of cloud providers offering high-quality services for data storage and management is then a driving motivation for companies that more often move their data centers to the cloud. Although this trend has clear economic advantages, it also introduces novel security issues. In fact, when moving a data center to the cloud, the data are no more under the direct control of their owner who needs to rely on an external system for providing the same guarantees as in their in-house management (e.g., data availability, protection against external attacks, selective access to the data, fault tolerance management [32–34, 39]). However, being external third parties, cloud providers are often assumed to be *honest-but-curious*, and hence trusted to correctly manage the data they store but not trusted to access their content. This situation raises several concerns, especially with respect to the proper protection of the confidentiality of the data. An effective solution consists in encrypting the data before outsourcing them so that non-authorized parties (including the cloud provider), not knowing the encryption key, cannot access the data content in plaintext [9, 31]. Data encryption before outsourcing presents however some disadvantages.

S. De Capitani di Vimercati (✉) · S. Foresti · G. Livraga · P. Samarati
Dipartimento di Informatica, Università degli Studi di Milano,
Via Bramante 65, 26013 Crema, Italy
e-mail: sabrina.decapitani@unimi.it

S. Foresti
e-mail: sara.foresti@unimi.it

G. Livraga
e-mail: giovanni.livraga@unimi.it

P. Samarati
e-mail: pierangela.samarati@unimi.it

S. U. Khan, A. Y. Zomaya (eds.), *Handbook on Data Centers*,
DOI 10.1007/978-1-4939-2092-1_33

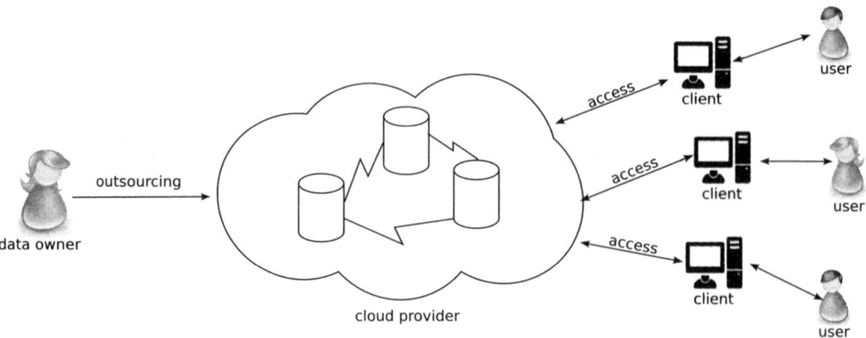

Fig. 1 Reference scenario

First, while effectively hiding plaintext data to the eyes of the provider, encrypting all data with a single key would require either all users to have complete visibility of the resources in the data collection, or the data owner to mediate access requests to the data to enforce selective access. Second, encryption complicates query evaluation since the cloud provider cannot directly evaluate users queries over encrypted data. Third, in cases where also the queries posed by users need to be protected, encryption might not provide sufficient protection guarantees.

To overcome such issues, different techniques have been proposed that aim at supporting selective and private access to outsourced data. These techniques are based on the use of *selective encryption*, meaning that different pieces of data are encrypted with different keys according to who can access them. *Indexes* are instead used by cloud providers to select the data to be returned in response to a query, possibly even without revealing the target of the query itself. While, singularly taken, these techniques represent effective solutions, the combined adoption of selective encryption and indexes may cause violations of confidentiality that still need to be carefully addressed. In this chapter, we present an overview of the techniques proposed for enabling data to self-enforce the access control policy defined by their owner, and for supporting query evaluation on encrypted data. Fig. 1 illustrates the reference scenario where a data owner outsources her data to a cloud provider and users access such data.

The remainder of this chapter is organized as follows. Sect. 2 shows how encrypted data can enforce access control restrictions, without requiring the intervention of the data owner or the collaboration of the storing server. Sect. 3 presents an overview of the techniques proposed for supporting query evaluation over encrypted data. Sect. 4 describes novel solutions for accessing outsourced data collections without revealing the target of the query to the storing server. Sect. 5 illustrates the privacy issues arising when combining solutions for access control enforcement with indexing techniques and introduces preliminary solutions to this problem. Finally, Sect. 6 presents our closing remarks.

2 Access Control Enforcement

The information stored in data centers can be of any type: relational databases, XML documents, multimedia files, and so on. For simplicity, but without loss of generality, in this chapter we assume the data stored in the cloud to be organized in a relational database, with the note that all approaches illustrated in the following can be easily adapted to operate on any logical data modeling. We then consider a relation r defined over schema $R(a_1,\ldots,a_n)$, where attribute a_i is defined over domain D_i, $i = 1,\ldots,n$. At the storing server, relation r is represented through an encrypted relation r^k, defined over schema R^k(tid, enc), with tid a numerical primary key added to the encrypted relation and enc the encrypted tuple. Each tuple t in r is represented as an encrypted tuple t^k in R^k, where t^k[tid] is randomly chosen by the data owner and t^k[enc]=$E_k(t)$, with E a symmetric encryption function with key k.

Different techniques have been proposed to enforce access control with the intervention of neither the storing server, for confidentiality reasons, nor the data owner, for efficiency reasons (e.g., [10, 12, 29]). These solutions are based on the idea that data *self-enforce* selective access restrictions through encryption, as illustrated in the following of this section.

2.1 Selective Encryption

A promising solution for enforcing access control to outsourced data is based on selective encryption, which adopts different encryption keys for different tuples, and selectively distributes keys to authorized users. Each user can decrypt and therefore access a subset of tuples, depending on the keys she knows. The authorization policy regulating which user can read which tuple is defined by the data owner before outsourcing relation r (e.g., [10, 12]). The authorization policy can be represented as a binary *access matrix* M with a row for each user u, and a column for each tuple t, where: M[u,t] = 1 iff u can access t; M[u,t] = 0 otherwise. To illustrate, consider relation Patients in Fig. 2. Figure 3 illustrates an example of access matrix regulating access to the tuples in relation Patients by users A, B, C, D, and E. The jth column of the matrix represents the access control list $acl(t_j)$ of tuple t_j, for each $j = 1,\ldots,|r|$. As an example, with reference to the matrix in Fig. 3, $acl(t_1) = ABC$. The encryption policy, which defines and regulates the set of keys used to encrypt tuples and the distribution of keys to the users, must be *equivalent* to the authorization policy, meaning that each user should be able to decrypt all and only the tuples she is authorized to access.

Solutions translating an authorization policy into an equivalent encryption policy (e.g., [12]) have two main design desiderata: *(i)* guarantee that each user has to manage only one key; and *(ii)* encrypt each tuple with only one key (i.e., no tuple is replicated). To fulfill these two requirements, selective encryption approaches rely on *key derivation techniques* that permit to compute an encryption key k_j starting from the knowledge of another key k_i and (possibly) a piece of publicly available

PATIENTS

	SSN	Name	ZIP	MarStatus	Illness
t_1	123456789	Ann	22010	single	gastritis
t_2	234567891	Barbara	24027	divorced	neuralgia
t_3	345678912	Carl	22010	married	gastritis
t_4	456789123	Daniel	20100	married	gastritis
t_5	567891234	Emma	21048	single	neuralgia
t_6	678912345	Fred	23013	married	hypertension
t_7	789123456	Gary	22010	widow	gastritis
t_8	891234567	Harry	24027	widow	hypertension

Fig. 2 An example of relation

	t_1	t_2	t_3	t_4	t_5	t_6	t_7	t_8
A	1	1	0	1	1	1	1	0
B	1	1	1	1	1	0	0	0
C	1	1	1	0	1	1	0	0
D	0	0	0	1	1	1	0	1
E	0	0	0	1	1	1	0	0

Fig. 3 An example of access matrix

information. To determine which key can be derived from which other key, key derivation techniques require the preliminary definition of a *key derivation hierarchy*. A key derivation hierarchy can be graphically represented as a directed graph with a vertex v_i for each key k_i in the system and an edge (v_i,v_j) from key k_i to key k_j iff k_j can be directly derived from k_i. Note that key derivation can be applied in chain, meaning that key k_j can be computed starting from key k_i if there is a path (of arbitrary length) from v_i to v_j in the key derivation hierarchy.

A key derivation hierarchy can have different shapes, as described in the following.

- *Chain of vertices* (e.g., [40]): the key k_j associated with vertex v_j is computed by applying a one-way function to the key k_i of its predecessor in the chain. No public information is needed.
- *Tree hierarchy* (e.g., [41]): the key k_j associated with vertex v_j is computed by applying a one-way function to the key k_i of its direct ancestor, and a public label l_j associated with k_j. Public labels are necessary to guarantee that different children of the same node in the tree have different keys.
- *DAG hierarchy* (e.g., [2, 3, 7, 19]): keys in the hierarchy can have more than one direct ancestor, and each edge in the hierarchy is associated with a publicly available *token* [3]. Given two keys k_i and k_j, and the public label l_j of k_j, token $t_{i,j}$ permits to compute k_j from k_i and l_j. Token $t_{i,j}$ is computed as $t_{i,j} = k_j \oplus f(k_i,l_j)$, where $\oplus$ is the bitwise XOR operator, and f is a deterministic cryptographic function. By means of $t_{i,j}$, all users knowing (or able to derive) key k_i can also derive key k_j.

Each of the proposed key derivation hierarchies has advantages and disadvantages. However, the token-based key derivation best fits the outsourcing scenario by minimizing the need of re-encryption and/or key re-distribution in case of updates to the authorization policy [12] (for more details, see Sect. 2.2).

The set containment relationship $\subseteq$ over the set U of users can nicely be used to define a DAG key derivation hierarchy suited for access control enforcement and able to satisfy the desiderata of limiting the key management overhead [12]. Such a hierarchy has a vertex for each of the elements of the power-set of the set U of users, and a path from v_i to v_j iff the set of users represented by v_i is a subset of that represented by v_j. The correct enforcement of the authorization policy defined by the data owner is guaranteed iff: *(i)* each user u_i is communicated the key associated with the vertex representing her; and *(ii)* each tuple t_j is encrypted with the key of the vertex representing $acl(t_j)$. With this strategy, each tuple can be decrypted and accessed by all and only the users in its access control list, meaning that the encryption policy is equivalent to the authorization policy defined by the data owner. Furthermore, each user has to manage one key only, and each tuple is encrypted with one key only. For instance, Fig. 4a illustrates the key derivation hierarchy induced by the set U= $\{A, B, C, D\}$ of users and the subset containment relationship over it (in the figure, vertices are labeled with the set of users they represent). Fig. 4b and Fig. 4c illustrate the keys assigned to users in the system and the keys used to encrypt the tuples in relation PATIENTS in Fig. 2, respectively. The encryption policy in the figure enforces the access control policy in Fig. 3 restricted to the set U= $\{A, B, C, D\}$ of users as each user can derive, from her own key, the keys of the vertices to which she belongs and hence decrypt the tuples she is authorized to read. For instance, user C can derive the keys used to encrypt tuples t_1, t_2, t_3, t_5, and t_6, and then access their content.

Even though this approach correctly enforces an authorization policy and enjoys ease of implementation, it defines more keys and more tokens than necessary. Since tokens are stored in a publicly available catalog at the server side, when a user u wants to access a tuple t she needs to interact with the server to visit the path in the key derivation hierarchy from the vertex representing u to the vertex representing $acl(t)$. Therefore, keeping the number of tokens low increases the efficiency of the derivation process, and then also of the response time to users. The problem of minimizing the number of tokens, while guaranteeing equivalence between the authorization and the encryption policies, is NP-hard (it can be reduced to the set cover problem) [12]. It is however interesting to note that: *(i)* the vertices needed for correctly enforcing an authorization policy are only those representing singleton sets of users (corresponding to users' keys) and the access control lists of the tuples (corresponding to keys used to encrypt tuples) in r; *(ii)* when two or more vertices have more than two common direct ancestors, the insertion of a vertex representing the set of users corresponding to these ancestors reduces the total number of tokens. Elaborating on these two intuitions to reduce the number of tokens, the following heuristic approach efficiently provides good results [12].

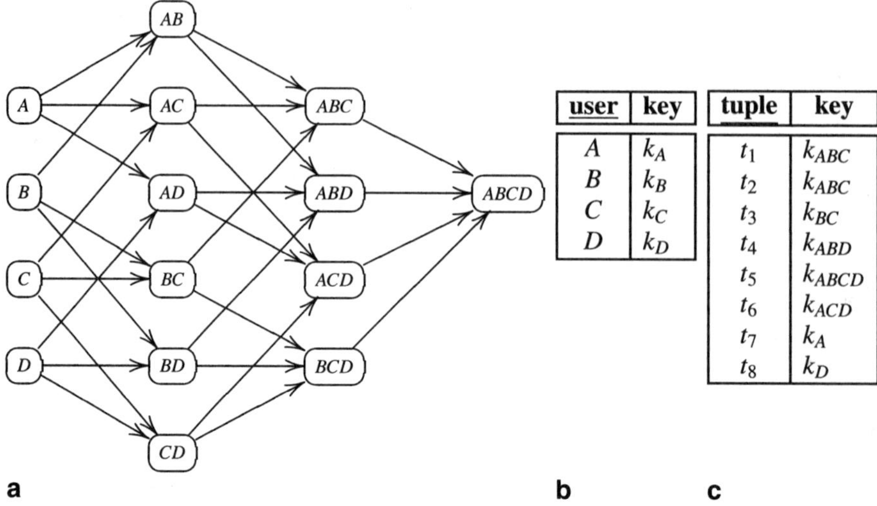

user	key
A	k_A
B	k_B
C	k_C
D	k_D

tuple	key
t_1	k_{ABC}
t_2	k_{ABC}
t_3	k_{BC}
t_4	k_{ABD}
t_5	k_{ABCD}
t_6	k_{ACD}
t_7	k_A
t_8	k_D

Fig. 4 An example of encryption policy equivalent to the access control policy in Fig. 3, considering the subset $\{A, B, C, D\}$ of users

1. *Initialization*. The algorithm first identifies the vertices necessary to implement the authorization policy, that is, the vertices representing: *(i)* singleton sets of users, whose keys are communicated to users and that allow them to derive the keys of the tuples they are entitled to access; and *(ii)* the access control lists of the tuples, whose keys are used for encryption. These vertices represent the set of *material* vertices of the system.
2. *Covering*. For each material vertex v corresponding to a non-singleton set of users, the algorithm finds a set of material vertices that form a *non-redundant set covering* for v, which become direct ancestors of v. A set V of vertices is a set covering for v if for each u in v, there is at least a vertex v_i in V such that u appears in v_i. It is non-redundant if the removal of any vertex from V produces a set that does not cover v.
3. *Factorization*. For each set $\{v_1, \ldots v_m\}$ of vertices that have $n > 2$ common ancestors $v'_1, \ldots, v'_n$, the algorithm inserts an intermediate vertex v representing all the users in $v'_1, \ldots, v'_n$ and connects each v'_i, $i = 1, \ldots, n$, with v, and v with each v_j, $j = 1, \ldots, m$. In this way, the encryption policy includes $n + m$, instead of $n \cdot m$ tokens in the catalog.

Figure 5 illustrates, step by step, the definition of the key derivation hierarchy through the algorithm in [12], for the authorization policy in Fig. 3. The initialization phase generates the set of (material) vertices in Fig. 5a. The covering phase generates the preliminary key derivation hierarchy in Fig. 5b, where each vertex is connected to a set of parents including all and only the users in the vertex itself. The factorization phase generates the key derivation hierarchy in Fig. 5c, which has an additional non-material vertex (i.e., *ADE*, denoted with a dotted line in the figure) representing the users that belong to both *ABDE* and *ACDE*. This factorization saves one token.

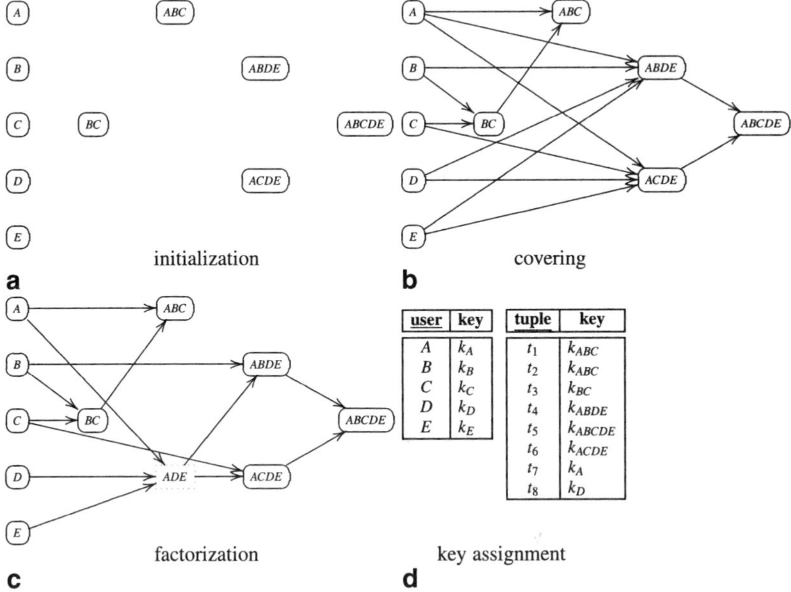

Fig. 5 Definition of an encryption policy equivalent to the access control policy in Fig. 3

Figure 5d illustrates the keys assigned to users in the system and the keys used to encrypt the tuples in relation PATIENTS in Fig. 2.

2.2 *Updates to the Access Control Policy*

In case of changes to the authorization policy, the encryption policy must be updated accordingly, to guarantee their equivalence. Since the key used to encrypt each tuple t in r depends on the set of users who can access it, it might be necessary to re-encrypt the tuples involved in the policy update with a different key that only the users in their new access control lists know or can derive. A trivial approach to enforce a grant/revoke operation on tuple t requires the data owner to: *(i)* download t^k from the server; *(ii)* decrypt it; *(iii)* update the key derivation hierarchy if it does not include a vertex representing the new set of users in $acl(t)$; *(iv)* encrypt t with the key k' associated with the vertex representing $acl(t)$; *v)* upload the new encrypted version of t on the server; and *(vi)* possibly update the public catalog containing the tokens. For instance, consider the encryption policy in Figs. 5c–d and assume that user D is granted access to tuple t_1. The data owner should download t_1^k; decrypt it using key k_{ABC}; insert a vertex representing $acl(t_1) = ABCD$ in the key derivation hierarchy; encrypt t_1 with k_{ABCD}; and upload the encrypted tuple on the server, together with the tokens necessary to users A, B, C, and D to derive k_{ABCD}. This approach, while effective and correctly enforcing authorization updates, leaves to the data owner the burden of managing the update. Also, re-encryption operations are computationally expensive. To limit the data owner overhead, in [12] the authors propose to use two layers of encryption (each characterized by its own encryption policy) to partially delegate to the server the management of grant and revoke operations.

- The *Base Encryption Layer* (BEL) is applied by the data owner before outsourcing the dataset. A BEL key derivation hierarchy is built according to the authorization policy existing at initialization time. In case of policy updates, BEL is only updated by possibly inserting tokens in the public catalog (i.e., edges in the BEL key derivation hierarchy). Note that each vertex v in the BEL key derivation hierarchy has two keys: a derivation key k (used for key derivation only), and an access key k^a (used to encrypt tuples, but that cannot be exploited for key derivation purposes).
- The *Surface Encryption Layer* (SEL) is applied by the server over the tuples that have already been encrypted by the data owner at BEL. It dynamically enforces the authorization policy updates by possibly re-encrypting tuples and changing the SEL key derivation hierarchy to correctly reflect the updates. Differently from BEL, vertices in the SEL key derivation hierarchy are associated with a single key k^s.

Intuitively, with the over-encryption approach, a user can access a tuple t only if she knows the keys used to encrypt t at BEL and SEL. At initialization time, the encryption policies at BEL and SEL coincide, but they immediately change and become different at each policy update. Grant and revoke operations are enforced as follows.

- *Grant*. When user u is granted access to tuple t, she needs to know the key used to encrypt t at both BEL and SEL. Hence, the data owner adds a token in the BEL key derivation hierarchy from the vertex representing u to the vertex whose key is used to encrypt t (i.e., the vertex representing $acl(t)$ at initialization time). The owner then asks the server to update the key derivation hierarchy at SEL and to possibly re-encrypt tuples. Tuple t in fact needs to be encrypted at SEL with the key of the vertex representing $acl(t) \cup \{u\}$ (which is possibly inserted into the hierarchy). Besides t, also other tuples may need to be re-encrypted at SEL to guarantee the correct enforcement of the policy update. In fact, tuples that are encrypted with the same key as t at BEL and that user u is not allowed to read must be encrypted at SEL with a key that u does not know (and cannot derive). The data owner must then make sure that each tuple t_i sharing the BEL encryption key with t are encrypted at SEL with the key of the vertex representing $acl(t_i)$. For instance, consider the access matrix in Fig. 3 and the encryption policies at BEL and SEL enforcing it in Fig. 6, and assume that user D is granted access to tuple t_1. Figure 7 illustrates the encryption policies at BEL and SEL after the enforcement of the grant operation. To enforce this change in the access control policy, the data owner must first add a token that permits user D to derive the access key of vertex ABC (k^a_{ABC}) used to encrypt t_1 at BEL (dotted edge in the figure). Also, she will ask the server to update the SEL key derivation hierarchy

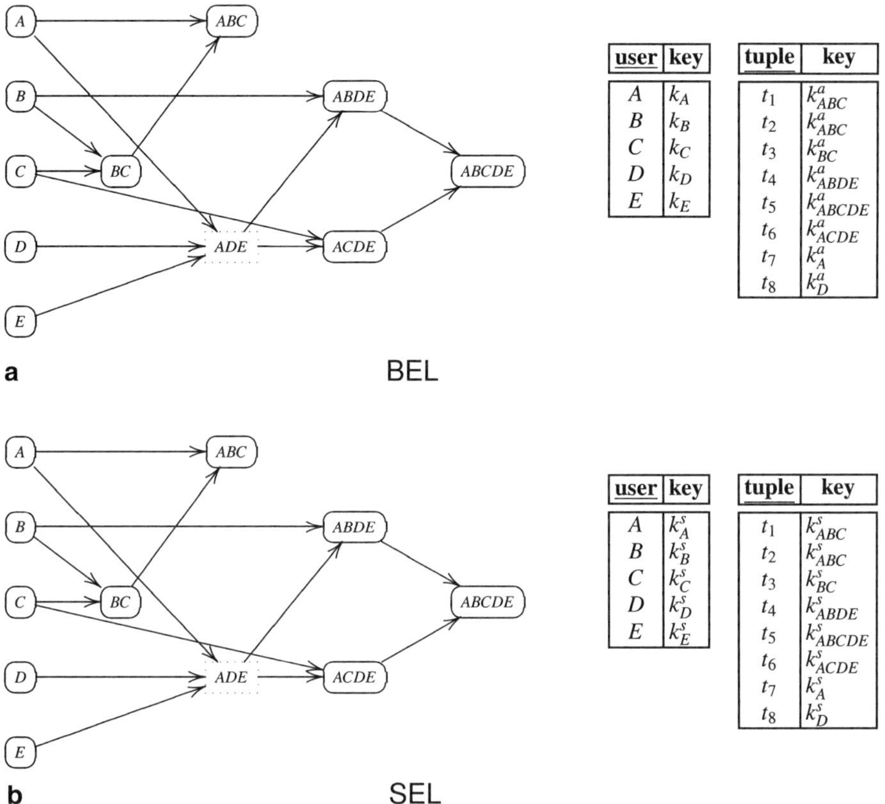

user	key
A	k_A
B	k_B
C	k_C
D	k_D
E	k_E

tuple	key
t_1	k^a_{ABC}
t_2	k^a_{ABC}
t_3	k^a_{BC}
t_4	k^a_{ABDE}
t_5	k^a_{ABCDE}
t_6	k^a_{ACDE}
t_7	k^a_{A}
t_8	k^a_{D}

user	key
A	k^s_A
B	k^s_B
C	k^s_C
D	k^s_D
E	k^s_E

tuple	key
t_1	k^s_{ABC}
t_2	k^s_{ABC}
t_3	k^s_{BC}
t_4	k^s_{ABDE}
t_5	k^s_{ABCDE}
t_6	k^s_{ACDE}
t_7	k^s_{A}
t_8	k^s_{D}

Fig. 6 Encryption policies at BEL and SEL, equivalent to the access control policy in Fig. 3

to add a vertex representing *ABCD*. Tuple t_1 is then over-encrypted at SEL with the key of this new vertex.

- *Revoke.* When user u loses the privilege of accessing tuple t, the data owner simply asks the server to re-encrypt (at SEL) the tuple with the key associated with the set $acl(t) \setminus \{u\}$ of users. If the vertex representing this set of users is not represented in the SEL key derivation hierarchy, the server first updates the hierarchy inserting the new vertex, and then re-encrypts the tuple. For instance, consider the encryption policies at BEL and SEL in Fig. 7 and assume that the data owner revokes B the privilege to access t_4. The data owner requires the server to change SEL (BEL is not affected by revoke operations) to guarantee that tuple t_4 is encrypted with a key that user B cannot derive. To this aim, t_4 is re-encrypted with key k^s_{ADE}. Figure 8 illustrates the encryption policies at BEL and SEL after the enforcement of the revoke operation. Note that vertex *ABDE* is removed from the hierarchy since it is neither necessary for policy enforcement nor useful for reducing the number of tokens.

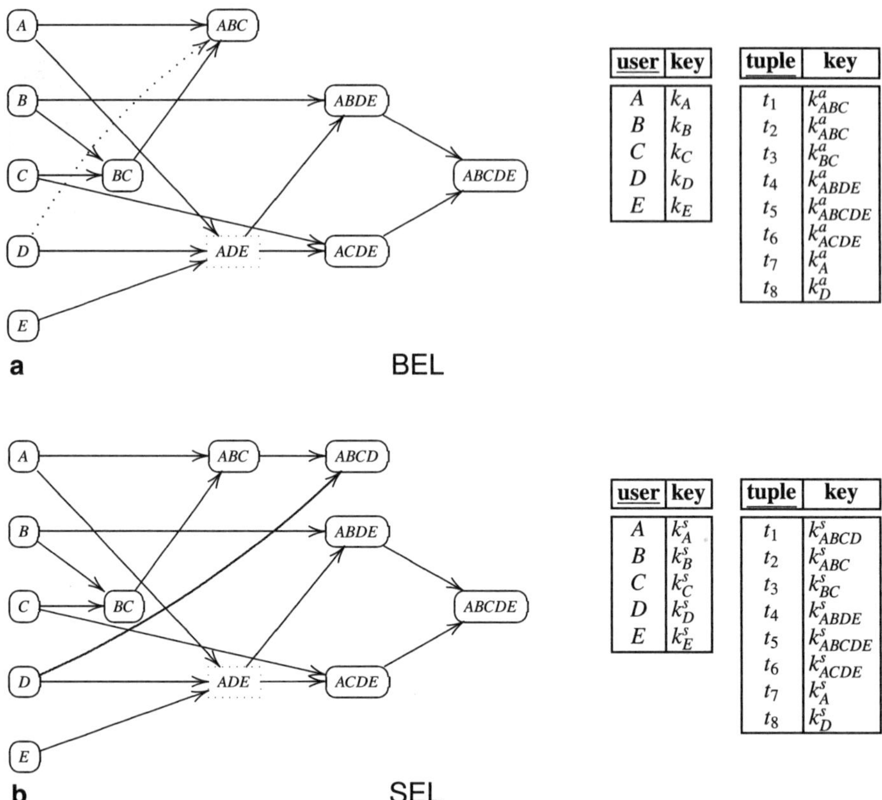

a BEL

user	key
A	k_A
B	k_B
C	k_C
D	k_D
E	k_E

tuple	key
t_1	k^a_{ABC}
t_2	k^a_{ABC}
t_3	k^a_{BC}
t_4	k^a_{ABDE}
t_5	k^a_{ABCDE}
t_6	k^a_{ACDE}
t_7	k^a_A
t_8	k^a_D

b SEL

user	key
A	k^s_A
B	k^s_B
C	k^s_C
D	k^s_D
E	k^s_E

tuple	key
t_1	k^s_{ABCD}
t_2	k^s_{ABC}
t_3	k^s_{BC}
t_4	k^s_{ABDE}
t_5	k^s_{ABCDE}
t_6	k^s_{ACDE}
t_7	k^s_A
t_8	k^s_D

Fig. 7 Encryption policies at BEL and SEL in Fig. 6 after granting D access to t_1

Since the management of (re-)encryption operations at SEL is delegated to the server, there is the risk of collusions with users. In fact, by combining their knowledge, a user and the server can possibly decrypt tuples that neither the server nor the user can access. For instance, with reference to the encryption policy in Fig. 8, the server and user D can access to tuple t_2 by combining their knowledge. In fact, this tuple is encrypted with access key k^a_{ABC} at BEL, known to user D as it is used to encrypt t_1, and with key k^s_{ABC} at SEL, known to the server. Collusion represents a risk to the correct enforcement of the authorization policy, but this risk is limited. In fact, collusion between a user u and the server permits them to decrypt a tuple t that they are not authorized to access only if u is granted the privilege to read a tuple t' (different from t) that is encrypted with the same key as t at BEL. Indeed, u knows the key with which t is encrypted at BEL (as it is necessary to access t') while the server knows the key with which it is encrypted at SEL (as it manages all the encryption keys at SEL). Collusion risk can then be mitigated at the price of using a higher number of keys at BEL, that is, by using the same encryption key at BEL only for tuples whose *acl*s are likely to evolve in the same way [12].

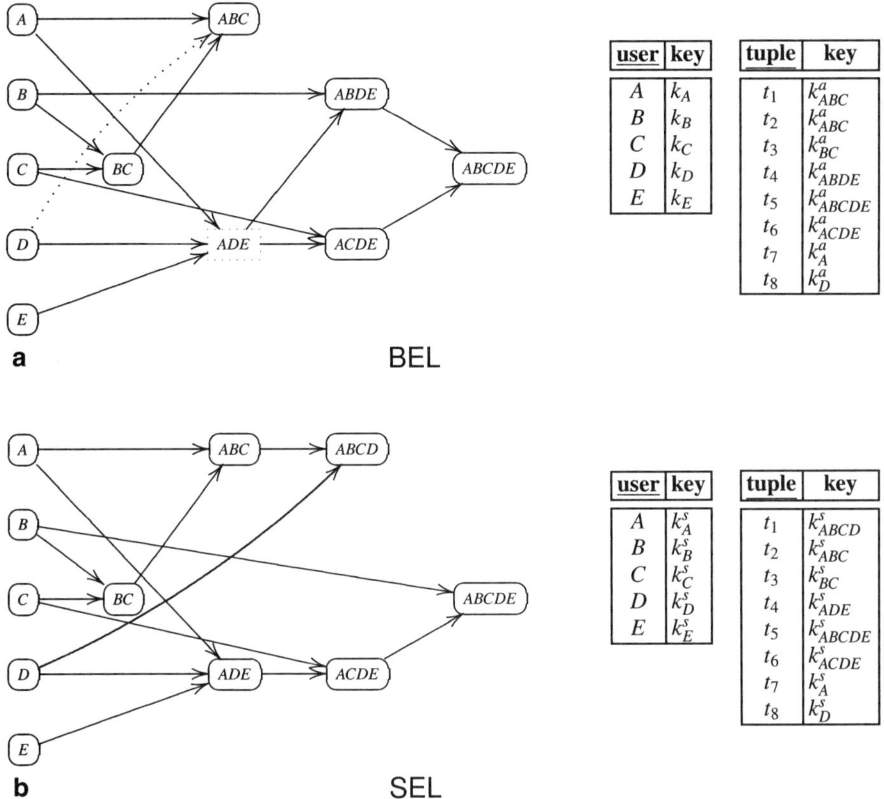

a BEL

user	key
A	k_A
B	k_B
C	k_C
D	k_D
E	k_E

tuple	key
t_1	k^a_{ABC}
t_2	k^a_{ABC}
t_3	k^a_{BC}
t_4	k^a_{ABDE}
t_5	k^a_{ABCDE}
t_6	k^a_{ACDE}
t_7	k^a_A
t_8	k^a_D

b SEL

user	key
A	k^s_A
B	k^s_B
C	k^s_C
D	k^s_D
E	k^s_E

tuple	key
t_1	k^s_{ABCD}
t_2	k^s_{ABC}
t_3	k^s_{BC}
t_4	k^s_{ADE}
t_5	k^s_{ABCDE}
t_6	k^s_{ACDE}
t_7	k^s_A
t_8	k^s_D

Fig. 8 Encryption policies at BEL and SEL in Fig. 7 after revoking B access to t_4

2.3 Write Privileges

The solution described in the previous section, while effectively enforcing read privileges and updates to them, assumes the outsourced relation to be read-only (i.e., only the owner can modify tuples). To allow the data owner to selectively authorize other users to update the outsourced data, this approach has been complemented with a specific technique to manage write privileges. The approach in [11] associates each tuple with a *write tag* (i.e., a random value independent from the tuple content) defined by the data owner. Access to write tags is regulated through selective encryption: the write tag of tuple t is encrypted with a key known only to the users authorized to write t (i.e., the users specified within its write access list, denoted $acl_w t$) and by the server. In this way, only the server and authorized writers have access to the plaintext write tag of each tuple. The server will then accept a write request on a tuple when the requesting user proves knowledge of the corresponding write tag.

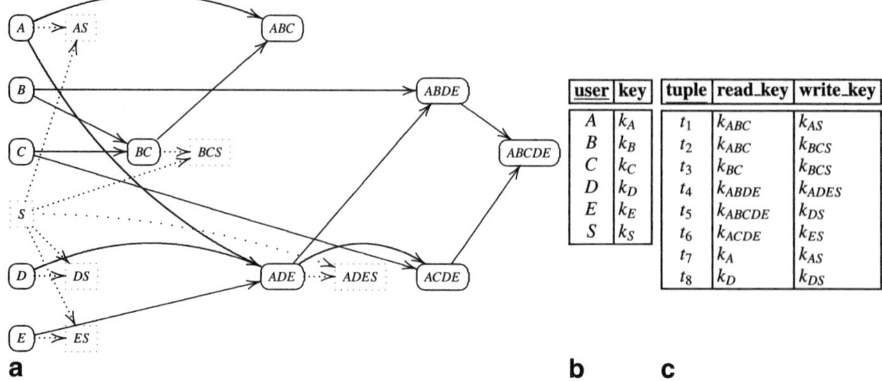

user	key
A	k_A
B	k_B
C	k_C
D	k_D
E	k_E
S	k_S

tuple	read_key	write_key
t_1	k_{ABC}	k_{AS}
t_2	k_{ABC}	k_{BCS}
t_3	k_{BC}	k_{BCS}
t_4	k_{ABDE}	k_{ADES}
t_5	k_{ABCDE}	k_{DS}
t_6	k_{ACDE}	k_{ES}
t_7	k_A	k_{AS}
t_8	k_D	k_{DS}

Fig. 9 Encryption policy in Figs. 5c–d extended to the enforcement of write privileges

Since the key used for encrypting the write tag of a tuple has to be shared between the server and the tuple writers, it is necessary to extend the key derivation hierarchy with the storing server. However, the server cannot access the outsourced tuples in plaintext, and hence it cannot be treated as an additional authorized user (i.e., with the ability of deriving keys in the hierarchy). The keys used to encrypt write tags are then defined in such a way that: *(i)* authorized users can compute them applying a secure hash function to a key they already know (or can derive via a sequence of tokens); and *(ii)* the server can directly derive them from a key k_S assigned to it, through a token specifically added to the key derivation hierarchy. Note that keys used for encrypting write tags cannot be used to derive other keys in the hierarchy. For instance, consider the encryption policy in Figs. 5(c–d) and assume that $acl_w(t_1) = acl_w(t_7) = A$, $acl_w(t_2) = acl_w(t_3) = BC$, $acl_w(t_4) = ADE$, $acl_w(t_5) = acl_w(t_8) = D$, and $acl_w(t_6) = E$. Figure 9a illustrates the key derivation hierarchy, extended with the key k_S assigned to the server S and the keys necessary to encrypt write tags (the additional vertices and edges are dotted in the figure). Figures 9b–c summarize the keys assigned to users and to the server, and the keys used to encrypt the tuples in relation PATIENTS and their write tags, respectively.

The over-encryption approach (Sect. 2.2), while effective for enforcing updates to a read authorization policy, cannot unfortunately be adopted to enforce grant and revoke of write authorizations. A possible way to enforce dynamic write privileges [11] operates as follows.

- *Grant.* When user u is granted the privilege to modify tuple t, the write tag of t is encrypted with a key known to the server and the users in $acl_w(t) \cup \{u\}$. If the key derivation hierarchy does not include it, such a key is created and properly added to the hierarchy. For instance, with reference to the encryption policy in Figure 9, assume that user B is granted the write privilege over t_4. The write tag of the tuple needs to be encrypted with key, k_{ABDES}, which is inserted into the key derivation hierarchy, while key k_{ADES} can be removed.

- *Revoke.* When user u is revoked the write privilege over tuple t, a fresh write tag must be defined for t, having a value independent from the former tag (e.g., it can be chosen adopting a secure random function). This is necessary to ensure that u, who is not oblivious, cannot exploit her knowledge of the former write tag of tuple t to perform unauthorized write operations. After the tag has been generated, it is encrypted with a key known to the server and to the users in $acl_w(t) \setminus \{u\}$. For instance, with reference to the encryption policy in Fig. 9, assume that user C is revoked the write privilege over t_3. The write tag of the tuple needs to be changed and encrypted with key k_{BS}, which should be inserted into the key derivation hierarchy.

Note that, since the server is authorized to know the write tag of each and every tuple to correctly enforce write privileges, the data owner can delegate to the storing server both the generation and encryption (with the correct key) of the write tag of the tuples [11].

2.4 Attribute-Based Encryption

An alternative solution to selective encryption for the enforcement of access restrictions is represented by *Attribute-Based Encryption* (ABE [29]). ABE is a particular type of public-key encryption that regulates access to tuples on the basis of policies defined on descriptive attributes, associated with tuples and/or users. ABE can be implemented as either *Ciphertext-Policy ABE* (CP-ABE [47]), or *Key-Policy ABE* (KP-ABE [29]), depending on how attributes and authorization policies are associated with tuples and users. Both the strategies have been recently widely investigated, and several solutions have been proposed, as briefly illustrated in the following.

CP-ABE CP-ABE associates with each user u a set of descriptive attributes, and a private key that is generated on the basis of these attributes. Each tuple t in a relation r is associated with an *access structure* modeling the access control policy regulating accesses to t. Graphically, an access structure is a tree whose leaves represent attributes and whose internal nodes represent logic gates (e.g., conjunctions and disjunctions). Figure 10 illustrates an example of access structure associated with tuple t_2 in relation PATIENTS in Fig. 2. This access structure corresponds to the Boolean formula (job = 'doctor' $\vee$ job = 'nurse') $\wedge$ $ward$ = 'neurology', meaning that only doctors or nurses working in the neurology ward can access the medical data of *Barbara* (i.e., tuple t_2). CP-ABE key generation technique guarantees that the key k of user u can decrypt tuple t only if the set of attributes used when generating k satisfies the access policy represented by the access structure considered when encrypting t. Although CP-ABE effectively and efficiently enforces access control policies, one of its main drawbacks is related to the management of attribute revocation. Intuitively, when a user loses one of her attributes, she should not be able to access tuples that require the revoked attribute for the access —which however is hard to enforce while guaranteeing efficiency. A solution to this problem is presented in

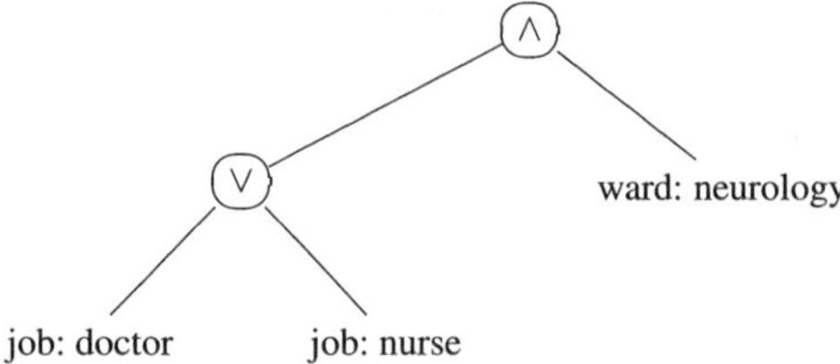

Fig. 10 An example of access structure associated with tuple t_2 of relation PATIENTS in Fig. 2

[51], where the authors illustrate an encryption scheme able to manage attribute revocation, ensuring the satisfaction of both backward security (i.e., a user cannot decrypt the tuples requiring the attribute revoked to the user) and forward security (i.e., a new user can access all the tuples outsourced before her join, provided her attributes satisfy the access control policy). In [44], the authors instead define a hierarchical attribute-based solution that relies on an extended version of CP-ABE in which attributes associated with users are organized in a recursive set structure, and propose a flexible and scalable approach to support revocations.

KP-ABE KP-ABE associates an access structure with each user and a set of descriptive attributes with each tuple. The key associated with each user is then generated on the basis of her access structure, while the key used to encrypt each tuple depends on its attributes. Thanks to the properties of KP-ABE key generation techniques, each user u can decrypt only tuples t such that the attributes of tuple t satisfy the access structure associated with user u. Since ABE is based on public-key encryption, to reduce the overhead caused by asymmetric encryption, the tuple content can be encrypted with a symmetric key, which is in turn protected through KP-ABE [53]. Only authorized users can remove the KP-ABE encryption layer to retrieve the symmetric key use to protect the content of the tuples. This solution also efficiently supports policy updates and couples ABE with proxy re-encryption to delegate to the storing server most of the re-encryption operations necessary to enforce policy updates.

The support of write privileges is provided by the adoption of Attribute-Based Signature (ABS) techniques. The proposal in [21] combines CP-ABE and ABS techniques to enforce read and write access privileges, respectively. This approach, although effective, has the disadvantage of requiring the presence of a trusted party for correct policy enforcement. A similar approach, based on the combined use of ABE and ABS for supporting both read and write privileges, is illustrated in [38]. This solution has the advantage over the approach in [21] of being suited also to distributed scenarios.

3 Efficient Access to Encrypted Data

Since data stored in the cloud are encrypted for confidentiality reasons, the storing server cannot directly evaluate users' queries since it is not trusted to access the data content. This makes access to outsourced data time consuming and computationally

PATIENTS

	SSN	Name	ZIP	MarStatus	Illness
t_1	123456789	Ann	22010	single	gastritis
t_2	234567891	Barbara	24027	widow	neuralgia
t_3	345678912	Carl	22010	married	gastritis
t_4	456789123	Daniel	20100	married	gastritis
t_5	567891234	Emma	21048	single	neuralgia
t_6	678912345	Fred	23013	married	hypertension
t_7	789123456	Gary	22010	widow	gastritis
t_8	891234567	Harry	24027	divorced	hypertension

a

PATIENTS^k

tid	enc	I_z	I_m	I_i
1	aD4%l	α	ζ	η
2	7Eoi)	β	κ	ξ
3	Gx?b3	α	θ	μ
4	dn4$z	γ	θ	η
5	2Cyl=	δ	ζ	ξ
6	Joi2s	ε	θ	ρ
7	(ceWm	α	κ	μ
8	w2!qk	β	κ	ρ

b

Fig. 11 An example of plaintext relation (**a**) and the corresponding encrypted and indexed relation (**b**)

expensive (the client would need to download the data and locally evaluate her query). To limit such an overhead, either *keyword search* or *index*-based approaches can be adopted, which enable query evaluation at the server side without the need to decrypt data [39]. Keyword search techniques (e.g., [6, 8, 25, 42, 45]) permit to search for documents including a keyword of interest in an encrypted data collection. Indexes are metadata that depend on the plaintext values of the attributes in the original relation, and are stored in the encrypted relation as additional attributes. Given a relation r, defined over schema $R(a_1, \ldots, a_n)$, the corresponding encrypted and indexed relation r^k has schema $R^k(\underline{\texttt{tid}}, \texttt{enc}, I_{i_1}, \ldots, I_{i_j})$, where $I_{i_l}, l = 1, \ldots, j$, is the index defined over attribute a_{i_l} in R. Note that not all the attributes in R need to have a corresponding index in R^k, but only those that are expected to be involved in queries. For instance, Fig. 11b represents an example of an encrypted version of relation PATIENTS in Fig. 2 (also reported in Fig. 11a for the reader' s convenience), where attributes `ZIP`, `MarStatus`, and `Illness` are associated with indexes I_z, I_m, and I_i, respectively. In this and in the following examples, for readability, we will denote index values with Greek letters.

To provide efficient access to the outsourced data collection, different indexing techniques have been proposed, aimed at supporting the server-side evaluation of a variety of conditions and clauses in SQL queries. The most important indexing approaches can be classified in three main categories, depending on how the index function ι maps the original attribute values to the corresponding index values, as illustrated in the following.

- *Direct Index*. Each plaintext value is represented by a different index value and vice versa. An example of direct index (e.g., [9]) is adopted by encryption-based indexes, which map plaintext value *val* to index value $E_k(val)$, where E is a symmetric encryption function with key k. Index I_z in Fig. 11b is an example of a direct index defined over attribute `ZIP` in Fig. 11a.
- *Bucket-based Index*. Each plaintext value is represented by one index value, but different plaintext values are mapped to the same index value, generating collisions. There are different approaches for defining which plaintext values are

represented by the same index value. The two most common techniques are partition-based and hash-based indexes. Partition-based indexes (e.g., [31]) partition the domain D of attribute a into subsets of contiguous values and associate a label with each of them. The index value representing a value in a partition is the label of the partition. Hash-based indexes (e.g., [5]) instead rely on a secure hash function h generating collisions. Given plaintext value *val*, its corresponding index value is computed as $h(val)$. Index I_m in Fig. 11b is an example of hash-based index over attribute `MarStatus` in Fig. 11a, where values divorced and widow generate a collision and are both represented by index value κ.

- *Flattened Index*. Each plaintext value is represented by different index values, each characterized by the same number of occurrences (flattening). Each index value, however, represents one plaintext value only. A flattened index can be obtained by properly combining encryption with a flattening post-processing that guarantees that the frequency of index values be the same (e.g., [46]). Index I_i in Fig. 11b is an example of a flattened index over attribute `Illness` in Fig. 11a, where plaintext value gastritis is represented by index values η and μ.

Intuitively, the fact that the outsourced relation is encrypted and enriched with indexes must be transparent to the final users. The basic indexing techniques illustrated above nicely support the server-side evaluation of simple SQL queries including equality conditions in the WHERE clause. Consider a query q submitted by a user of the form "SELECT *Att* FROM *R* WHERE *Cond*", where $Att \subseteq R$ and *Cond* is a set of equality conditions of the form $a = val$, with $a \in R$ and *val* a constant value in the domain D of a. To determine the query that should be submitted to the storing server, each condition $a = val$ in *Cond* is first translated into an equivalent condition of the form: $I = \iota(val)$, if I is a direct or a bucket-based index; and I IN $\iota(val)$, if I is a flattened index and hence $\iota(val)$ may return a set of values. The query q_s submitted to the server is then "SELECT `enc` FROM R^k WHERE $Cond^k$", where $Cond^k$ is obtained as illustrated above. The result returned by the server must then be decrypted by the client, to retrieve the plaintext content of the tuples. The client may also need to perform a projection over the attributes in *Att*, if they represent a proper subset of R, and to filter spurious tuples, that is, tuples that satisfy $Cond^k$ but that do not belong to the query result (i.e., they do not satisfy *Cond*). Note that the presence of spurious tuples may depend on collisions possibly caused by bucket-based indexes, where multiple plaintext values are mapped to the same index value. The client then evaluates a query q_c of the form "SELECT *Att* FROM $D(Res^k)$ WHERE *Cond*", where Res^k is the relation returned by the server as the result of the evaluation of query q_s. The result of query q_c is returned to the requesting user. Consider, as an example, a query q = SELECT `SSN`, `Name` FROM PATIENTS WHERE `MarStatus` = "widow" AND `Illness` = "gastritis" operating on relation PATIENTS in Fig. 11a. The query q_s to be sent to the server is SELECT `enc` FROM PATIENTS^k WHERE $I_m = \kappa$ AND I_i IN $\{\eta,\mu\}$, which returns tuple t_7. The client will then decrypt the result returned by the server and evaluate query SELECT `SSN`, `Name` FROM $D(Res^k)$ WHERE `MarStatus` = "widow" AND `Illness` = "gastritis" to check whether tuple t_7 satisfies both the conditions and to project the attributes of interest for the requesting user.

Besides the techniques illustrated and classified above, many other approaches have been proposed for efficiently delegating to the server the evaluation of complex conditions and/or SQL clauses. As an example, order preserving encryption has been proposed as an effective solution for supporting range conditions, as well as grouping and ordering clauses (e.g., [1, 46]). Aggregate functions can instead be computed if the index over the attribute of interest has been defined through homomorphic encryption techniques, which support the evaluation of arithmetic operators on encrypted data (e.g., [24, 30]). Different techniques, which do not fit into the classification above, have also been proposed to the aim of enjoying the advantages of traditional database indexing techniques also in the outsourcing scenario (e.g., in [9] the authors propose to use encrypted $B+$-trees for the efficient evaluation of range queries).

4 Protecting Access Privacy

Besides protecting the confidentiality of the outsourced data collection, it is also paramount to protect the privacy of the accesses to the data themselves. In fact, queries can be exploited for inference, making both users' and data privacy at risk. As an example, consider a scenario where the outsourced data contain medical information. Revealing that a user submitted a query looking for the symptoms of lung cancer implicitly reveals that (with high probability) either her or a person close to her suffers from such a disease. Also, users accesses may be exploited to infer the private content of the outsourced data collection. Indeed, by monitoring patterns of frequently accessed tuples, an observer can draw inferences on their specific values thanks to additional knowledge she may have on how frequently each piece of data in a given domain is accessed. In this case, it is necessary to protect both *access confidentiality* (i.e., each query singularly taken) and *pattern confidentiality* (i.e., the fact that two queries aim at the same target value). A first attempt to protect access confidentiality is represented by keyword search approaches (e.g., [6, 8, 25, 42, 45]), which do not reveal to the server any information about the outsourced data and the target keyword. A similar approach consists in defining a set of tokens that can be adopted by users to evaluate queries on outsourced data without disclosing the conditions in their queries to the storage server [18, 36]. Protection of accesses to a $B+$-tree index structure can instead be obtained by grouping the nodes in the tree into buckets [37]. The use of homomorphic encryption techniques then permits to access the node of interest in each bucket, while preventing the server from precisely identifying the node target of each access. These approaches represent a first step towards the definition of privacy-preserving indexing approaches, but they fall short in protecting the confidentiality of repeated accesses and, more in general, of patterns thereof. In the remainder of this section, we will illustrate some of the most important approaches recently proposed to address both access and pattern confidentiality in a scenario where data need to remain confidential (i.e., outsourced data are encrypted).

4.1 Oblivious RAM

One of the first approaches [49] proposed to protect access and pattern confidentiality in a scenario where also the confidentiality of the data must be protected is based on the *Oblivious RAM* (ORAM) data structure [26]. The outsourced database is organized as a set of n encrypted blocks, which are stored in a pyramid-shaped data structure. Each level l of the ORAM structure stores 4^l blocks and is characterized by a Bloom filter and a hash table that permit to quickly determine whether an index value is stored in the level and, if this is the case, to identify the block where it is stored. Access and pattern confidentiality are provided by guaranteeing that: *(i)* the search process does not reveal the level in the structure where the target block is stored, and *(ii)* a block in the hash table is never accessed more than once with the same search key.

The search algorithm visits the ORAM structure level by level, starting from the top of the pyramid. For each level l, the search algorithm uses the Bloom filter to determine whether the target of the search is stored in the level. If this is the case, the item of interest is extracted from the level (by accessing the block identified by the hash table), decrypted, re-encrypted with a different nonce, and inserted into a cache. Otherwise, the algorithm extracts a random element from the level and inserts it into the cache. We note that, even when the target element is retrieved, the search process visits all the lower levels in the ORAM structure extracting at each level a random (fake) element. This guarantees that, by observing accesses to the structure, the server is not able to identify the level where the target of the search was stored. The search process terminates when the last level in the ORAM structure is visited.

When the cache is full, it is merged with the first level of the ORAM structure and the items in the resulting new level are re-shuffled, to destroy any correspondence between old and new items in the level. As a consequence, the Bloom filter associated with the level is re-defined, to correctly refer to the new level content. The same process applies to each level in the structure: when level l is full, it is merged with level $l+1$, the blocks are re-shuffled, and the Bloom filter is redefined. The cost of accessing the ORAM clearly depends on the possible need to reorganize a level in the indexing structure while visiting it, and on the specific level that needs to be redefined. The amortized cost per query, which takes into consideration the impact of periodic reorganizations of the structure, is $O(\log n \log\log n)$, under the assumptions of $O(\sqrt{n})$ temporary client storage and of $O(n)$ server storage overhead. However, the cost of reorganizing the bottom level of the pyramid is $O(n)$, where n is the number of index values in the dataset. Response time of any access request submitted during the reordering of lower levels of the database is therefore high and not acceptable in many real-world scenarios.

To mitigate the cost of query evaluation when low levels in the ORAM structure need to be reorganized, the proposal in [20] puts forward the idea of limiting the shuffling operation to the blocks that store accessed tuples. This approach is based on the presence of a secure coprocessor on the server that locally manages a cache of size k, which is empty at initialization time. Each tuple in the dataset is associated with a label, initially set to value 'white'. Once a tuple is accessed, its label becomes 'black'. For each access to the dataset, the search algorithm fetches a black tuple and a white tuple. If the target tuple is already in cache, the algorithm retrieves two randomly chosen fake tuples (a black one and a white one), otherwise it accesses the target tuple and a randomly chosen fake tuple. When the cache is full, the secure coprocessor shuffles black tuples (performing a partial shuffling) only and re-encrypts them before re-writing the blocks on the server. Partial shuffling provides access and pattern confidentiality, since white tuples have not been accessed and hence it is not necessary to move their content to hide the traces that an access could have left. The amortized cost per query of this solution is $O(\sqrt{n \log n / k})$, which is lower than the proposal in [49]. It however relies on a secure coprocessor for guaranteeing access and pattern confidentiality.

Alternative techniques that can be adopted to limit the response time of the ORAM structure are based on the idea of minimizing the number of interactions between the client and the server [27, 48]. Indeed, the communication costs have a high impact on response times and reducing the number of interactions provides benefits to users. Other approaches instead rely on enhancing the support of concurrent accesses [28, 50]. These solutions basically define copies of the levels of the ORAM structure. Searches operate on a read-only copy of the level of interest, while the master copy of the same level is dynamically updated and used for reordering purposes only. In this way, exclusive locks blocking access to a level of the structure during its reorganization process do not delay users' accesses.

Path-ORAM has recently been proposed as an alternative approach to provide access and pattern confidentiality without paying the high price of re-shuffling, which characterizes traditional ORAM structures [43]. Path-ORAM is a tree-shaped data structure whose nodes are buckets storing a fixed number of blocks (which can either be dummy or contain actual data). Each block is mapped to a randomly chosen leaf in the tree and it is stored either at the client side (in a local cache, which is called stash) or in one of the buckets along the path to the leaf to which it is associated. Read operations download from the server and store in the stash all the buckets in the path from the root to the leaf to which the block of interest is mapped. The mapping of the target block is randomly updated (i.e., the block is mapped to a new, randomly chosen, leaf). The path read from the server is then written back, inserting into the buckets the blocks in the local stash (provided the bucket is along the path to the leaf to which the block is mapped).

4.2 Dynamically Allocated Data Structures

Dynamic data allocation solutions aim at destroying the otherwise static relationship between disk blocks and the information they store. These approaches are based on the definition of a dynamically allocated index structure (e.g., a B+-tree, a hash table, a flat index) that guarantees private and efficient access to the data.

If the data are organized in a tree-shaped index structure, access confidentiality is provided by guaranteeing that the storing server does not know (nor can infer) which is the node in the tree target of the access, as it would otherwise reveal the value target of the search. The first step to protect the confidentiality of the dataset content consists in encrypting the nodes in the tree before outsourcing, and in storing each encrypted node in a different disk block. However, repeated accesses to the same physical block inevitably represent repeated accesses to the same node content and hence queries aiming to the same value (or to values within a small interval). If the storing server knows the relative frequency of accesses to the plaintext values, it can reconstruct the correspondence between node contents and encrypted blocks, by simply matching access frequencies. A preliminary approach aimed at protecting access confidentiality through a privacy-preserving tree relies on the combined adoption of the following three protection techniques [35]:

- *access redundancy*: each access request visits, besides the node target of the access, m additional blocks (at least one of which should be empty) for each level in the tree to hide the target of the access in a set of $m+1$ equally-probable candidate nodes;
- *node swapping*: the node target of the access is swapped with one of the empty blocks downloaded from the server for the same level, meaning that the target node is stored in an empty block and viceversa;
- *node re-encryption*: all the nodes downloaded from the server are re-encrypted, to hide the swap.

Although effective for protecting content and access confidentiality, this proposal falls short in providing pattern confidentiality, since frequently accessed blocks can easily be identified by the server and then exploited for inference purposes.

An alternative approach, which does not operate on a tree-shaped index structure, is based on the adoption of a lightweight scheme that provides access and pattern confidentiality by combining the following three protection techniques [52]:

- *dummy data items*: each access request visits, besides the block target of the access, two additional blocks;
- *swapping*: the target of the access is swapped with one of the dummy data items downloaded from the server;
- *repeated patterns*: dummy data items are selected in such a way that, out of the three blocks downloaded from the server, two (and only two) are among the ones accessed during the previous search.

The goal of the combined adoption of these three protection techniques is to make each access to the outsourced data collection indistinguishable from the server's point of view. In fact, each access has two blocks in common with the previous one, while the third one is fresh. Swapping protects repeated accesses and is combined with re-encryption of the content of all the accessed blocks, to prevent the server from reconstructing which swap has been performed (thus possibly recognizing repeated accesses).

4.3 *Shuffle Index*

A recent technique addressing the need of providing efficient query execution, while protecting access and pattern confidentiality, is based on the definition of a *shuffle index* [14].

Data Structure A shuffle index is a privacy-preserving indexing technique, used for organizing data in storage and for efficiently executing users' queries. It can be seen at three different abstraction levels (i.e., abstract, logical, and physical), as illustrated in the following. At the abstract level, the shuffle index is an *unchained* B*+-tree* with fan-out F, built over a candidate key K of relation r. Each internal node of the tree represents the root of a sub-tree with $q \geq \lceil F/2 \rceil$ children (except for the root node, where $1 \leq q \leq F$), and stores $q-1$ ordered values $val_1 \leq \ldots \leq val_{q-1}$ of attribute K. The leaves store the tuples of the outsourced relation, together with their key value, but (in contrast to traditional $B+$-tree structures) are not connected in a chain, so not to allow the server storing the data to discover the relative order among the values in the leaves. Figure 12a illustrates an example of unchained $B+$-tree.

At the logical level, each node n of the unchained $B+$-tree is represented by a pair $\langle id, n \rangle$ where id is the *logical identifier* associated with the node and n is its content. Pointers to children of internal nodes of the abstract data structure are represented, at the logical level, through the identifier of child nodes. Figure 12b illustrates an example of logical representation of the unchained $B+$-tree in Fig. 12a. Note that the order of logical identifiers does not necessarily reflect the value-order relationship between the node contents. For readability, in the figure nodes are ordered according to their logical identifier (reported on the top of each node), whose first digit corresponds to the level of the node in the tree.

At the physical level, each node $\langle id, n \rangle$ is concatenated with a random salt, to destroy plaintext distinguishability, and then encrypted in CBC mode, using a symmetric encryption algorithm. The logical identifier of the node easily translates into the physical address where the block representing the encrypted node is stored at the server side (for simplicity, we assume that the physical address of a block coincides with the logical identifier of the corresponding node). Figure 12c illustrates the physical representation of the logical index in Fig. 12b. Note that the physical representation of the shuffle index coincides with the view of the storage server over the outsourced data collection. In fact, although the server does not have knowledge of the encryption key, it can establish the level in the tree associated with each block by observing a long enough history of accesses to the $B+$-tree structure, because accesses visit the tree level by level.

Protection Techniques To protect content, access, and pattern confidentiality, encryption is complemented with three protection techniques: *cover searches*, *cached searches*, and *shuffling*. These protection techniques apply to every access to the shuffle index, which proceeds by visiting the $B+$-tree level by level from the root to the leaves.

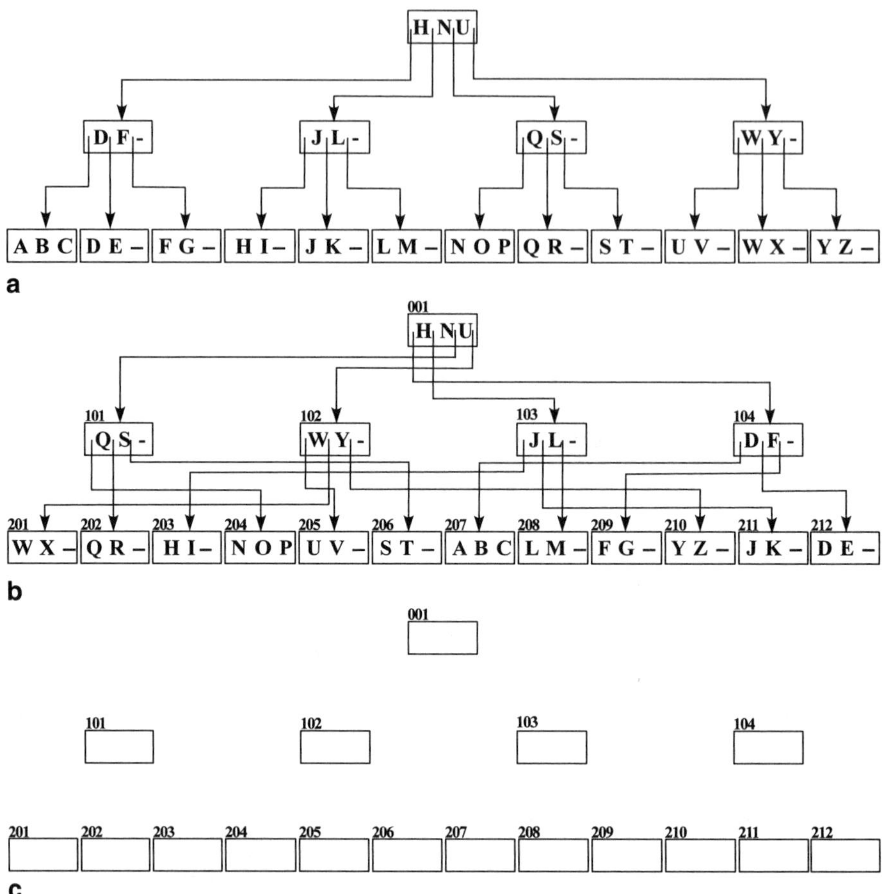

Fig. 12 An example of abstract (**a**), logical (**b**), and physical (**c**) representation of a shuffle index

- *Cover searches.* Cover searches aim at hiding the target of an access within a set of other potential targets, in such a way that the server cannot recognize the value of interest for the user. Cover searches are *fake* searches, which are not recognizable as such by the storage server, that are executed in conjunction with the search for the target value (i.e., the value of interest for the requester). For each level of the shuffle index (but the root level) the client downloads *num_cover* + 1 blocks: one for the node along the path to the target, and *num_cover* for the nodes along the paths to the covers. Hence, at the server's eyes, each of the *num_cover* + 1 leaf blocks accessed during a visit of the shuffle index has the same probability of storing the target. To provide sufficient protection to the target of the access, cover searches must guarantee: *(i)* indistinguishability with respect to target searches, meaning that the server should not be able to determine whether an accessed block is a cover or the target; and *(ii)* block diversity, meaning that paths to covers and to the target must be disjoint (except for the root node).

- *Cached searches.* Cached searches aim at protecting repeated accesses to a node content, by making them indistinguishable from non-repeated accesses to the eyes of the server. The cache is a layered structure with a layer for each level in the shuffle index. It is maintained at the client side and stores the nodes along the paths to the targets of the *num_cache* most *recent accesses* to the shuffle index. Being stored at a trusted party, the cache is maintained in plaintext. Each layer of the cache is managed according to the Least Recently Used (LRU) policy, which guarantees the property that the parent of each cached node (and hence also the path connecting it to the root of the tree) is also in cache. Whenever the target of an access is in cache, it is replaced by an additional cover for the access, to guarantee that *num_cover* + 1 blocks are downloaded for each level of the tree (but the root level). This makes repeated accesses look like accesses to nodes that have not been previously accessed. The adoption of a local cache prevents short-time intersection attacks, which could be exploited by the server to identify repeated accesses when subsequent searches download non-disjoint sets of blocks. In fact, accesses within a time frame of *num_cache* accesses do not have nodes in common.
- *Shuffling.* Shuffling aims at breaking the relationship between node content and block where it is stored, to avoid that accesses to the same physical block correspond to accesses to the same node content. By changing the node-block allocation, the server cannot draw conclusions on the content of the accessed nodes by observing accessed blocks. In fact, repeated accesses to the same block do not necessarily correspond to repeated accesses to the same node content. Shuffling consists in *moving* the content of accessed (either as target or as covers) and cached nodes to different blocks (i.e., shuffling assigns a different block address to each accessed node, choosing among the downloaded blocks). To prevent the server from reconstructing node shuffling, every time a node content is moved to a different block, it is re-encrypted using a different random salt. Its parent is also updated to guarantee that the parent-child relationship between them is preserved.

Search Process Each search operation for a value then combines these three protection techniques, as described in the following.

Given the value *target_value*, target of the access to the outsourced relation, the search algorithm (operating at the client side) first randomly chooses a set of *num_cover* + 1 cover values in the actual domain of the key attribute K on which the shuffle index has been defined. Since these values should act as cover searches for *target_value*, this choice must guarantee both indistinguishability and block diversity, as described above. Note that the algorithm chooses one additional cover (i.e., *num_cover* + 1 instead of *num_cover*) as it is needed if the target of the access is in the local cache. The search algorithm then visits the shuffle index level by level, starting from the root. For each level l of the shuffle index, the search algorithm first checks whether the node in the path to *target_value* is in the local cache and, if a cache miss occurs, it discards one of the cover searches initially chosen proceeding the search with *num_cover* covers. It then determines the address of the blocks storing the nodes along the paths to *target_value* and to the *num_cover* cover searches (*num_cover* + 1 covers if a cache hit occurred). These blocks are then downloaded from the server, decrypted to retrieve the content of the nodes they store, and randomly shuffled together with the nodes at level l in the local cache. To preserve the correctness of the shuffle index data structure, the parents of shuffled nodes are updated, guaranteeing that pointers refer to the blocks where the children of each node are stored. The search algorithm also updates the local cache structure, according to the LRU policy. If the node along the path to *target_value* is in cache, the algorithm simply refreshes its timestamp; otherwise, the node along the path to the target is inserted as the most recently accessed node and the least recently accessed node is removed from the cache. Before moving to the next level, the nodes shuffled during the previous iteration (i.e., accessed and cached nodes at level $l-1$) are encrypted with a fresh random salt and sent to the server for storage. Upon receiving the encrypted blocks, the server replaces the old block stored at each physical address with the new one received from the client. The process terminates when the visit of the shuffle index reaches the leaf level. The leaf along the path to *target_value* is then returned to the requesting user, since it contains the tuple with value *target_value* for attribute K, if such a tuple exists in r. For instance, consider a search for value 'W' on shuffle index in Fig. 12 that adopts one cover. Also, assume that the cache has size 2 and that it stores: the root node at level 0; nodes 103[*J*,*L*,–] and 102[*W*,*Y*,–] at level 1; and leaves 211[*J*,*K*,–] and 210[*Y*,*Z*,–] at level 2. The client first chooses two covers for the target 'W', say 'E' and 'Q', and visits the root node (block 001), which is stored in the local cache. It then identifies the block at level 1 along the paths to the target (i.e., 102) and to the two covers (i.e., 104 and 101, respectively). Since block 102 is in cache, the client downloads from the server blocks 104 and 101, decrypts their content, and shuffles the accessed and cached blocks (i.e., 101, 102, 103, and 104) as illustrated in Fig. 13b. It then updates the pointers to children in the root node, encrypts its content and sends it back to the server for storage. Moving to the next level, the client first identifies the leaf blocks along the path to the target (i.e., 201) and to the two covers (i.e., 212 and 202, respectively). Since block 201 is not in cache, one of the two covers is discarded, say 202, and the client downloads from the server and decrypts blocks 201 and 212. The client then shuffles blocks 201, 210, 211, and 212, updates the pointer to them in their parents, encrypts nodes 101, 102, 103, and 104, and sends the resulting blocks to the server for storage. Then, it updates the cache at level 2 inserting leaf node 212[*W*,*X*,–] and removing leaf node 211[*Y*,*Z*,–]. Finally, the client encrypts the shuffled leaves and sends the resulting blocks to the server. Figure 13c illustrates the logical shuffle index resulting after the access.

The search algorithm operates in logarithmic time in the size of the outsourced database (i.e., its computational complexity is $O((1+$ *num_cover* $+$ *num_cache*$)$ $\log_F(n))$, with n the number of tuples in r), since for each search the algorithm visits *num_cover* + 1 different paths of the shuffle index.

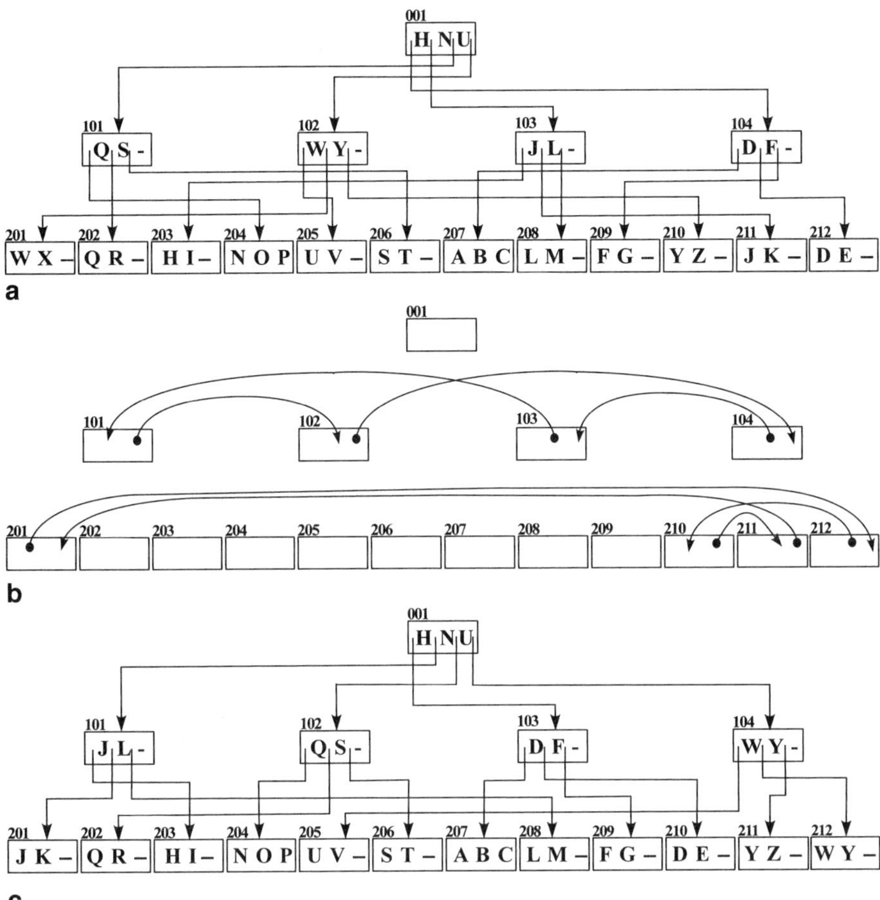

Fig. 13 An example of evolution of the logical shuffle index in Fig. 12b as a consequence of a search for value 'W' with 'E' and 'Q' as covers

Extensions of the Shuffle Index The original shuffle index proposal has been extended in several directions to support: *(i)* concurrent accesses to the data; *(ii)* searches over attributes different from K; and *(iii)* data storage at different servers. Concurrency is provided by the adoption of *delta versions* [17], which are copies of portions of the shuffle index that are dynamically created/updated by subsequent accesses. Each access to the shuffle index is assigned to a different delta version with exclusive write lock. Accessed blocks are downloaded from the delta version (if the delta version includes the node of interest) or from the shuffle index (otherwise), while shuffled blocks are written on the delta version. Periodically, delta versions are reconciled and applied to the shuffle index, to preserve the effects of the different shuffling operations performed by different users.

To efficiently support private accesses to r based on attributes different from K, in [17] the authors propose to complement the primary shuffle index with different *secondary shuffle indexes*, built on candidate keys that are expected to be often involved in query evaluation. A secondary index defined on attribute a is a shuffle index that stores, in association with value val for a, the values that attribute K has in tuple t (i.e., $t[K]$), such that $t[a]=val$. A search for the tuples in r with value val for attribute a proceeds then in two steps: *(1)* search for value val in the secondary index, retrieving the value val_K of attribute K in the tuples of interest; and then *(2)* search for value val_K in the primary index, retrieving the tuple of interest.

The distribution of the shuffle index over different servers, which are not aware one of each other, increases the protection offered to the confidentiality of users' accesses. According to the proposal in [16], in a distributed scenario cover searches, cached searches, and shuffling protection techniques can be complemented with *shadowing*. Shadowing guarantees that the observations by each server of accessed blocks make it believe to be the only server storing the whole data collection. In fact, each server observes the same number of blocks read (written, respectively) at each level of the tree.

5 Combining Access Control and Indexing Techniques

Access control enforcement and query evaluation over encrypted outsourced data has been widely studied, as testified by the different approaches illustrated in the previous sections of this chapter. However, the problem of combining them is still an open issue. The joint adoption of selective encryption (Sect. 2) and indexing techniques (Sect. 3) may permit authorized users to infer information they are not entitled to access. In fact, authorized users can infer the values that attributes have in tuples they should not be able to read, by exploiting their visibility over the index values for the tuples they are entitled to access. For instance, with reference to the encrypted relation in Fig. 11b and the access control policy regulating it in Fig. 3, user B can infer that $t_7[\texttt{ZIP}]=22010$ even if $B \notin acl(t_7)$, because $t_7^k[I_z] = t_1^k[I_z]$ and B knows that $t_1[\texttt{ZIP}]=22010$ since B belongs to $acl(t_1)$.

The problem of jointly adopting selective encryption and indexing techniques has recently been investigated, leading to the identification of different privacy risks that vary depending on the technique adopted for index definition (see Sect. 3) [13]. Before illustrating these risks, we summarize the knowledge of an authorized user u (i.e., a user who can access a subset of the tuples in r). Each authorized user knows: *(i)* index function ι used to define index I over attribute a (necessary for query evaluation); *(ii)* the plaintext tuples that the user can access; *(iii)* all the encrypted tuples in R^k (they are publicly available). For instance, consider the encrypted relation in Fig. 11b and the access control policy regulating it in Fig. 3. User A knows the index functions used by the data owner to define I_z, I_m, and I_i; all the plaintext tuples but t_3 and t_8; and the encrypted relation in Fig. 11b. The knowledge of user A

t	acl(t)
t_1	*ABC*
t_2	*ABC*
t_3	*BC*
t_4	*ABDE*
t_5	*ABCDE*
t_6	*ACDE*
t_7	*A*
t_8	*D*

a

PATIENTS

	SSN	Name	ZIP	MarStatus	Illness
t_1	123456789	Ann	22010	single	gastritis
t_2	234567891	Barbara	24027	widow	neuralgia
t_3					
t_4	456789123	Daniel	20100	married	gastritis
t_5	567891234	Emma	21048	single	neuralgia
t_6	678912345	Fred	23013	married	hypertension
t_7	789123456	Gary	22010	widow	gastritis
t_8					

b

PATIENTS^k

tid	enc	I_z	I_m	I_i
1	aD4%l	α	ζ	η
2	7Eoi)	β	κ	ξ
3	Gx?b3	α	θ	μ
4	dn4$z	γ	θ	η
5	2Cyl=	δ	ζ	ξ
6	Joi2s	ε	θ	ρ
7	(ceWm	α	κ	μ
8	w2!qk	β	κ	ρ

c

Fig. 14 Access control lists (**a**), knowledge of user *A* over relation PATIENTS (**b**), and over relation PATIENTS^k (**c**)

is summarized in Fig. 14, where gray cells denote plaintext values that user *A* is not authorized to read.

The inferences that an authorized user can draw on index *I* representing attribute *a* can be summarized as follows.

- *Direct index*. Since each plaintext value is associated with one index value and viceversa, if $t_i^k[I] = t_j^k[I]$ then also $t_i[a] = t_j[a]$ and viceversa. Hence, each user *u* can infer the plaintext value of attribute *a* for all those tuples in *r* that have the same value as a tuple that *u* is authorized to access. Consider, as an example, direct index I_z in relation PATIENTS^k in Fig. 14c. User *A* knows that t_1[ZIP] = t_3[ZIP] = t_7[ZIP] = 22010 even if she cannot read t_3, since all these tuples have the same value for index I_z.
- *Bucket-based index*. Since different plaintext values are mapped to the same index value, the information leakage illustrated for direct indexes is mitigated by the presence of collisions. Hence, if $t_i^k[I] = t_j^k[I]$ there is a certain (greater than zero) probability that also $t_i[a] = t_j[a]$, but there is no guarantee that this equality condition holds. Consider, as an example, index I_m in relation PATIENTS^k in Fig. 14c. Since the value for I_m is the same for t_2, t_7, and t_8, user *A* can infer that probably t_2[MarStatus] = t_7[MarStatus] = t_8[MarStatus] = widow. We note however that plaintext values 'widow' and 'divorced' are represented by the same index value κ.
- *Flattened index*. Although less straightforward, the inference risk caused by flattened indexes is the same as illustrated for direct indexes. In fact, each index value represents one plaintext value only and then if $t_i^k[I] = t_j^k[I]$, also $t_i[a] = t_j[a]$. The viceversa is instead not true, that is, not all the occurrences of a value *val* are represented by the same index value. However, each authorized user knows the index function ι adopted by the data owner and can then compute $\iota(val)$, retrieving all the index values representing *val*. For instance, consider flattened index I_i in relation PATIENTS^k in Fig. 14c. Although user *A* is not authorized to read tuple t_3, she can infer that t_3[Illness] = gastritis as t_3 and t_7 have the same value for

PATIENTS^k

tid	enc	I_z
1	aD4%l	α'_A, α_B, α_C
2	7Eoi)	β_A, β_B, β_C
3	Gx?b3	α'_B, α'_C
4	dn4$z	γ_A, γ_B, γ_D, γ_E
5	2Cyl=	δ_A, δ_B, δ_C, δ_D, δ_E
6	Joi2s	ε_A, ε_C, ε_D, ε_E
7	(ceWm	α_A
8	w2!qk	β_D

Fig. 15 An example of encrypted and indexed version of relation PATIENTS with index I_z over ZIP defined according to a salted user-dependent function

index I_i. Also, since she can compute ι(gastritis)= $\{\eta, \mu\}$, she can infer that also t_1 and t_4 have this value for attribute Illness.

From the observations above, it is easy to see that attribute values are exposed when the same index value appears in association with tuples characterized by different access control lists. Consider two tuples t_i and t_j in r such that $acl(t_i) \neq acl(t_j)$, and $t_i^k[I] = t_j^k[I]$. All the users in $acl(t_i)$ $(aclt_j$, respectively) can draw inferences on the value of $t_j[a]$ ($t_i[a]$, respectively). For instance, tuples t_1, t_3, and t_7 have the same value for attribute ZIP, but different *acl*s. This permits A to infer that t_3[ZIP]$=$22010 even if she should not be able to read such a tuple. A first solution to limit such a leakage of information is based on the idea that the index value representing value $t[a] = l$ should not only depend on l but also on $acl(t)$. In [13] the authors present a solution that operates on direct indexes, which represent the worst-case scenario. This approach associates a different index function ι_u with each user u (depending on a piece of secret information shared between u and the data owner). Function ι_u is salted (i.e., a randomly chosen salt is applied) to avoid that tuples with the same plaintext value v but different *acl* are associated with the same index value $\iota_u(v)$ for user u, which could easily be exploited for inferences. For instance, consider direct index I_z in relation PATIENTS^k in Fig. 14c. Figure 15 illustrates relation PATIENTS^k, where index I_z has been defined using a different (salted) index function for each user in the system. For readability, in the figure we use a subscript to indicate the user to which the index value refers (e.g., α_A is a value computed by ι_A) and symbol $'$ denotes the salted version of index values (e.g., α'_A is the salted version of α_A).

While interesting, the proposal illustrated in [13] and mentioned above considers one specific indexing technique only. Even if it can be easily extended to operate with bucket-based and flattened indexing functions, it cannot be combined with the privacy-preserving indexing approaches described in Sect. 4. Furthermore, user-based indexing techniques are suitable for static scenarios, as dynamic observations of repeated accesses to the data can reveal to an observer which index values represent the same plaintext value. In fact, index values representing the same plaintext value are often accessed together by authorized users. For instance, with reference to relation PATIENTS^k in Fig. 15b, every time user A needs to access all the tuples with ZIP =22010, she will query the encrypted relation with the condition $I_z = \alpha_A$ OR $I_z = \alpha'_A$. The server can then easily conclude that α_A and α'_A represent the same plaintext value.

6 Conclusions

Public and private organizations are more and more resorting to cloud systems for outsourcing their own data centers. While bringing intuitive benefits in terms of economies of scale, moving to the cloud raises new privacy risks, since data are no more under the direct control of their owner. The research and development communities have dedicated many efforts in the design and development of novel techniques for protecting outsourced data and accesses to them. In this chapter, we surveyed recent approaches that, while protecting confidentiality of the data to the eyes of the storing server through encryption, enforce access control restrictions and efficiently evaluate queries over encrypted data, possibly without even revealing to the server the target of accesses. We also described the main issues arising when these techniques are adopted in combination, illustrating a preliminary approach for their solution.

Acknowledgements This chapter is based on joint work with Sushil Jajodia, Gerado Pelosi, and Stefano Paraboschi. This work was supported in part by the Italian Ministry of Research within PRIN project "GenData 2020" (2010RTFWBH), and by Google, under the Google Research Award program.

References

1. Agrawal, R., Kierman, J., Srikant, R., Xu, Y.: Order preserving encryption for numeric data. In: Proc. of SIGMOD 2004. Paris, France(June 2004)
2. Akl, S., Taylor, P.: Cryptographic solution to a problem of access control in a hierarchy. ACM TOCS 1(3), 239–248 (August 1983)
3. Atallah, M., Blanton, M., Fazio, N., Frikken, K.: Dynamic and efficient key management for access hierarchies. ACM TISSEC 12(3), 18:1–18:43 (January 2009)
4. Bertino, E., Jajodia, S., Samarati, P.: Database security: Research and practice. Information Systems 20(7), 537–556 (November 1995)
5. Ceselli, A., Damiani, E., De Capitani di Vimercati, S., Jajodia, S., Paraboschi, S., Samarati, P.: Modeling and assessing inference exposure in encrypted databases. ACM TISSEC 8(1), 119–152 (February 2005)
6. Chang, Y., Mitzenmacher, M.:Privacy preserving keyword searches on remote encrypted data. In: Proc. of ACNS 2005. New York, NY, USA(June 2005)
7. Crampton, J., Martin, K., Wild, P.: On key assignment for hierarchical access control. In: Proc. of CSFW 2006. Venice, Italy (July 2006)

8. Curtmola, R., Garay, J., Kamara, S., Ostrovsky, R.: Searchable symmetric encryption: Improved definitions and efficient constructions. In: Proc. of ACM CCS 2006. Alexandria, VA, USA (October - November 2006)
9. Damiani, E., De Capitani di Vimercati, S., Jajodia, S., Paraboschi, S., Samarati, P.: Balancing confidentiality and efficiency in untrusted relational DBMSs. In: Proc. of ACM CCS 2003. Washington, DC, USA (October 2003)
10. De Capitani di Vimercati, S., Foresti, S., Jajodia, S., Livraga, G.: Enforcing subscription-based authorization policies in cloud scenarios. In: Proc. of DBSec 2012. Paris, France (July 2012)
11. De Capitani di Vimercati, S., Foresti, S., Jajodia, S., Livraga, G., Paraboschi, S., Samarati, P.: Enforcing dynamic write privileges in data outsourcing. Computers & Security 39, 47–63 (November 2013)
12. De Capitani di Vimercati, S., Foresti, S., Jajodia, S., Paraboschi, S., Samarati, P.: Encryption policies for regulating access to outsourced data. ACM TODS 35(2), 12:1–12:46 (April 2010)
13. De Capitani di Vimercati, S., Foresti, S., Jajodia, S., Paraboschi, S., Samarati, P.: Private data indexes for selective access to outsourced data. In: Proc. of WPES 2011. Chicago, IL, USA (October 2011)
14. De Capitani di Vimercati, S., Foresti, S., Paraboschi, S., Pelosi, G., Samarati, P.: Efficient and private access to outsourced data. In: Proc. of ICDCS 2011. Minneapolis, MN, USA (June 2011)
15. De Capitani di Vimercati, S., Foresti, S., Paraboschi, S., Pelosi, G., Samarati, P.: Supporting concurrency in private data outsourcing. In: Proc. of ESORICS 2011. Leuven, Belgium (September 2011)
16. De Capitani di Vimercati, S., Foresti, S., Paraboschi, S., Pelosi, G., Samarati, P.: Distributed shuffling for preserving access confidentiality. In: Proc. of ESORICS 2013. Egham, UK. (September 2013)
17. De Capitani di Vimercati, S., Foresti, S., Paraboschi, S., Pelosi, G., Samarati, P.: Supporting concurrency and multiple indexes in private access to outsourced data. JCS 21(3), 425–461 (2013)
18. De Cristofaro, E., Lu, Y., Tsudik, G.: Efficient techniques for privacy-preserving sharing of sensitive information. In: Proc. of TRUST 2011. Pittsburgh, PA, USA (June 2011)
19. De Santis, A., Ferrara, A., Masucci, B.: Cryptographic key assignment schemes for any access control policy. IPL 92(4), 199–205 (November 2004)
20. Ding, X., Yang, Y., Deng, R.: Database access pattern protection without full-shuffles. IEEE TIFS 6(1), 189–201 (March 2011)
21. Fangming, Z., Takashi, N., Kouichi, S.: Realizing fine-grained and flexible access control to outsourced data with attribute-based cryptosystems. In: Proc. of ISPEC 2011. Guangzhou, China (May-June 2011)
22. Gamassi, M., Lazzaroni, M., Misino, M., Piuri, V., Sana, D., Scotti, F.: Quality assessment of biometric systems: a comprehensive perspective based on accuracy and performance measurement. IEEE TIM 54(4), 1489–1496 (August 2005)
23. Gamassi, M., Piuri, V., Sana, D., Scotti, F.: Robust fingerprint detection for access control. In: Proc. of RoboCare Workshop. Rome, Italy (May 2005)
24. Gentry, C.: Fully homomorphic encryption using ideal lattices. In: Proc. of STOC 2009. Bethesda, MA, USA (May 2009)
25. Goh, E.J.: Secure indexes. Tech. Rep. 2003/216, Cryptology ePrint Archive (2003), http://eprint.iacr.org/
26. Goldreich, O., Ostrovsky, R.: Software protection and simulation on Oblivious RAMs. JACM 43(3), 431–473 (May 1996)
27. Goodrich, M., Mitzenmacher, M., Ohrimenko, O., Tamassia, R.: Practical oblivious storage. In: Proc. of CODASPY 2012. San Antonio, TX, USA (February 2012)
28. Goodrich, M., Mitzenmacher, M., Ohrimenko, O., Tamassia, R.: Privacy-preserving group data access via stateless Oblivious RAM simulation. In: Proc. of SODA 2012. Kyoto, Japan (January 2012)

29. Goyal, V., Pandey, O., Sahai, A., Waters, B.: Attribute-based encryption for fine-grained access control of encrypted data. In: Proc. of ACM CCS 2006. Alexandria, VA, USA (October-November 2006)
30. Hacigümüs, H., Iyer, B., Mehrotra, S.: Efficient execution of aggregation queries over encrypted relational databases. In: Proc. of DASFAA 2004. Jeju Island, Korea (March 2004)
31. Hacigümüs, H., Iyer, B., Mehrotra, S., Li, C.: Executing SQL over encrypted data in the database-service-provider model. In: Proc. of SIGMOD 2002. Madison, WI, USA (June 2002)
32. Jhawar, R., Piuri, V.: Fault tolerance management in IaaS clouds. In: Proc. of ESTEL 2012. Rome, Italy (October 2012)
33. Jhawar, R., Piuri, V.: Fault tolerance and resilience in cloud computing environments. Computer and Information Security Handbook 2nd Edition Vacca J. (ed.), Morgan Kaufmann (2013)
34. Jhawar, R., Piuri, V., Samarati, P.: Supporting security requirements for resource management in cloud computing. In: Proc. of CSE 2012. Paphos, Cyprus (December 2012)
35. Lin, P., Candan, K.: Hiding traversal of tree structured data from untrusted data stores. In: Proc. of WOSIS 2004. Porto, Portugal (April 2004)
36. Lu, Y., Tsudik, G.: Privacy-preserving cloud database querying. JISIS 1(4), 5–25 (November 2011)
37. Pang, H., Zhang, J., Mouratidis, K.: Enhancing access privacy of range retrievals over $B+$-trees. IEEE TKDE 25(7), 1533–1547 (July 2013)
38. Ruj, S., Stojmenovic, M., Nayak, A.: Privacy preserving access control with authentication for securing data in clouds. In: Proc. of CCGrid 2012. Ottawa, Canada (May 2012)
39. Samarati, P., De Capitani di Vimercati, S.: Data protection in outsourcing scenarios: Issues and directions. In: Proc. of ASIACCS 2010. Beijing, China (April 2010)
40. Sandhu, R.: On some cryptographic solutions for access control in a tree hierarchy. In: Proc. of the 1987 Fall Joint Computer Conference on Exploring Technology: Today and Tomorrow. Dallas, TX, USA (October 1987)
41. Sandhu, R.: Cryptographic implementation of a tree hierarchy for access control. IPL 27(2), 95–98 (February 1988)
42. Song, D., Wagner, D., Perrig, A.: Practical techniques for searches on encrypted data. In: Proc. of IEEE S&P 2000. Berkeley, CA, USA (May 2000)
43. Stefanov, E., van Dijk, M., Shi, E., Fletcher, C., Ren, L., Yu, X., Devadas, S.: ObliviStore: High performance oblivious cloud storage. In: Proc. of ACM CCS 2013. Berlin, Germany (November 2013)
44. Wan, Z., Liu, J., Deng, R.H.: HASBE: A hierarchical attribute-based solution for flexible and scalable access control in cloud computing. IEEE TIFS 7(2), 743–754 (April 2012)
45. Wang, C., Cao, N., Ren, K., Lou, W.: Enabling secure and efficient ranked keyword search over outsourced cloud data. IEEE TPDS 23(8), 1467–1479 (August 2012)
46. Wang, H., Lakshmanan, L.: Efficient secure query evaluation over encrypted XML databases. In: Proc. of VLDB 2006. Seoul, Korea (September 2006)
47. Waters, B.: Ciphertext-policy attribute-based encryption: An expressive, efficient, and provably secure realization. In: Proc. of PKC 2011. Taormina, Italy (March 2011)
48. Williams, P., Sion, R.: Single round access privacy on outsourced storage. In: Proc. of ACM CCS 2012. Raleigh, NC, USA (October 2012)
49. Williams, P., Sion, R., Carbunar, B.: Building castles out of mud: Practical access pattern privacy and correctness on untrusted storage. In: Proc. of ACM CCS 2008. Alexandria, VA, USA (October 2008)
50. Williams, P., Sion, R., Tomescu, A.: PrivateFS: A parallel oblivious file system. In: Proc. of ACM CCS 2012. Raleigh, NC, USA (October 2012)
51. Yang, K., Jia, X., Ren, K.: Attribute-based fine-grained access control with efficient revocation in cloud storage systems. In: Proc. of ASIACCS 2013. Hangzhou, China (May 2013)
52. Yang, K., Zhang, J., Zhang, W., Qiao, D.: A light-weight solution to preservation of access pattern privacy in un-trusted clouds. In: Proc. of ESORICS 2011. Leuven, Belgium (September 2011)
53. Yu, S., Wang, C., Ren, K., Lou, W.: Achieving secure, scalable, and fine-grained data access control in cloud computing. In: Proc. of INFOCOM 2010. San Diego, CA, USA (March 2010)

Privacy in Data Centers: A Survey of Attacks and Countermeasures

Luis Javier García Villalba, Alejandra Guadalupe Silva Trujillo and Javier Portela

1 Introduction

A Data Center collects, stores, and transmits huge dimensions of sensitive information of many types. Data Center security has become one of the highest network priorities as data thieves and crime cells look to infiltrate perimeter defenses through increasingly complex attack vectors with alarming success and devastating effects.

Today, organizations are placing a tremendous amount of collected data into massive repositories from various sources, such as: transactional data from enterprise applications and databases, social media data, mobile device data, documents, and machine-generated data. Much of the data contained in these data stores is of a highly sensitive nature and would trigger regulatory consequences as well as significant reputation and financial damage. This may include social security numbers, banking information, passport numbers, credit reports, health details, political opinions and anything that can be used to facilitate identity theft.

Our daily activities are developed in a digital society where the interactions between individuals and other entities are through technology. Now, we can organize an event and send the invitation using a social network like Facebook, sharing photos with friends using Instagram, listening to music through Spotify, asking for an address using Google Maps; all of these activities are just some of the ways in which many people are already working on the Internet every day. Personal information in real world is protected from strangers but it is different in the online world, where people disclose it [1]. All available information about a person gets cross-referenced,

L. J. García Villalba (✉) · A. G. Silva Trujillo · J. Portela
Group of Analysis, Security and Systems (GASS), Department of Software Engineering and Artificial Intelligence (DISIA), Faculty of Information Technology and Computer Science, Office 431, Universidad Complutense de Madrid (UCM), Madrid, Spain
Calle Profesor José García Santesmases 9, Ciudad Universitaria, 28040 Madrid, Spain
e-mail: javiergv@fdi.ucm.es

S. U. Khan, A. Y. Zomaya (eds.), *Handbook on Data Centers*,
DOI 10.1007/978-1-4939-2092-1_34

and the resulting dossier ends up being used for many purposes, lawful and otherwise. This practice has expanded over the years; the companies that compile and sell these dossiers are known as data brokers.

The communication systems behaviour has changed and it has been forced to improve its management in order to protect users privacy and satisfy the new requirements. Data centers provide a unique choice, rather than collecting data on network devices with limited capabilities for measurement, it offers measurements at the servers, even commodity versions of which have multiple cores besides other facilities. The essence of a data center is not based on concentration of data but rather the capacity to provide particular data or combinations of data upon request.

Governments and industry take advantage of sophisticated data storage tools and are using it to profile their users for financial, marketing, or just statistical purposes; organizations are able to acquire and maintain massive infrastructure at bargain prices and this derives to multiple benefits.

Individuals have the right to control their private information and only provide it to certain third parties. In the last decade users privacy concerns have grown [2–4] and since then several technologies have been developed to enhance privacy. Privacy enhancing technologies (PETs) are designed to offer mechanisms to protect personal information, and can be used with high level policy definition, human processes and training in the use of computer and communication systems [5–7]. PETs have been proposed to defend users privacy in user, network and server areas. Private and public organizations, as well as individuals should include the protection of privacy besides the typical aspects like integrity, confidentiality and availability of data. Privacy protection must avoid the disclosure of identities in a communication system. Motivations of these issues include censorship resistance, spies or law enforcement, whistleblowers, dissidents and journalists living under repressive regimes.

There are some technologies used to accelerate the transition to encryption as a service including hardware-based encryption key storage, centralized data protection schemes for applications, databases, storage and virtualized environments, as well as role-based access controls. Despite significant investment in security technology, organizations have a great hole in security effectiveness. This is due to the fact that conventional defenses rely on IP addresses and digital signatures. Signatures used in antivirus and intrusion prevention systems are effective at detecting known attacks at the time attacks are launched. They are not effective, however at detecting new attacks and are incapable of detecting hackers who are still in the reconnaissance phase, probing for weakness to attack. IP reputation databases, meanwhile, rely on the notion that attackers can be identified by their IP addresses, and so share this information across systems. Unfortunately, this is as ineffective method as it uses a postal address to identify someone. Network attacks are a serious threat to an organization. Next generation technologies are encouraged to improve the encryption solutions available at data center level. However, it has been proved that traffic and network topology analysis do not provide enough users privacy protection, even when anonymization mechanisms are applied. Using auxiliary information, adversaries can diminish anonymity properties.

In this chapter we focus on how the analysis of traffic data can compromise anonymity, showing the methods and techniques of how large amounts of traffic that has been routed through an anonymous communication system can establish communication relationships. In terms of information retrieved and considering these as leakages, designers in data centers will take them to build better capabilities to prevent attacks. Cloud computing and data centers have revolutionized the industrial world but have data protection implications which should be seriously looked into by all stakeholders to avoid putting people's privacy at risk. The solution to the previously mentioned privacy problems could be the adoption of appropriate privacy enhancing technologies.

2 Privacy

The definition of privacy according to [8] is "the right of the individual to decide what information about himself should be communicated to others and under what circumstances".

Economists, sociologists, historians, lawyers, computer scientists, and others have adopted their own privacy definitions, just as the value, scope, priority and proper course of study of privacy. Details about the background, law and history of privacy are showed in [9]. According to experts, privacy and intimacy are difficult concepts to define. However, we may consider personal health conditions, identity, sexual orientation, personal communications, financial or religious choices, along with many other characteristics. References from literature on how privacy solutions are applied from economic, social and technical areas are in [4, 10, 11].

Respect for privacy as a right includes undesirable interference, the abusive indiscretions and invasion of privacy, by any means, documents, images or recording. The legal foundations date back to 1948. In that year, the Universal Declaration of Human Rights was released, in which it was established that no person "shall be subjected to arbitrary interference with his privacy, family, home or correspondence, nor to attacks upon his honor and reputation". However, despite the legal and political developments that have taken place since then, it has not been possible to solve a fundamental problem to curb abuses every day. The lack of clarity and precision in the right to freedom of expression and information limits is an open issue; cases that threaten these rights are increasing.

The development of digital media, the increasing use of social networks, the easier access to modern technological devices, is perturbing thousands of people in their public and private lives. Examples abound, the most recent was the deputy mayor of a Flemish town, who was caught and recorded on a video while having sex with a man in the Town Hall offices. The recording was made and released for an unknown group of young boys. Another scandal was the president of the Guatemalan Institute of Social Security, who was shot in his office committing "lewd acts". Unlike the previous one, in this case there was a crime and the action given was justified publicly. All of this stuff is available on the Internet and traditional media, the videos that were leaked to the Vice Minister of Culture and Youth of Costa Rica, and the PSOE councilor in the Yébenes, Spain. Nobody seems to care about the effects which it continues to have on their lives. Indifference seems to be the constant. Participation of national and international human rights, government, media, and even the civil society organizations, seems to be far from this problem. However, the situation should be of concern. The scandal at the expense of the intrusion and dissemination of the private and intimate lives of people is unacceptable. It is a vicious circle that has its origin in the violation of a right, but when it is the social networks and hence most of the national and international media, on the pretext of being "news".

3 Privacy Enhancing Technologies

The European Commission define Privacy enhancing technologies [12] as "The use of PETS can help to design information and communication systems and services in a way that minimizes the collection and use of personal data and facilitates compliance with data protection rules. The use of PETs should result in making breaches of certain data protection rules more difficult and / or helping to detect them".

There is no widely accepted definition of the term PETs nor does there a distinguished classification exist. Literature about categorized PETs according to their main functions, privacy management and privacy protection tools [13–15].

In general PETs are observed as technologies that focus on:

- Reducing the risk of breaking privacy principles and legal compliance.
- Minimizing the amount of data held about individuals.
- Allowing individuals to maintain control of their information at all times.

Several researchers are centered on protection of privacy and personal data through sophisticated cryptology techniques. PET's applications such as individual digital safes or virtual identity managers have been proposed for trusted computing platforms.

PETs have traditionally been restricted to provide "pseudonymisation" [16]. In contrast to fully anonymized data, pseudonymisation allows future or additional data to be linked to the current data. These kind of tools are software that allow individuals to deny their true identity from those operating electronic systems or providing services through them, and only disclose it when absolutely necessary. Examples include: anonymous web browsers, email services and digital cash.

In order to give a better explanation about PETs applied in a data center, consider the Solove's Taxonomy [17] used to categorize the variety of activities to infringe privacy. We refer to [16] for further definitions of privacy properties in anonymous communication scenarios.

- *Information Collection*: Surveillance, Interrogation.
- *Information Processing*: Aggregation, Identification, Insecurity, Secondary Use, Exclusion.

- *Information Dissemination*: Breach of Confidentiality, Disclosure, Exposure, Increased Accessibility, Blackmail, Appropriation, Distortion.
- *Invasion*: Intrusion, Decisional Interference.

Collecting information can be a damaging activity, not all the information is sensitive but certain kinds definitely are. All this information is manipulated, used, combined and stored. These activities are labeled as Information Processing. When the information is released, this group of activities is called Information dissemination. Finally, the last group of activities is Invasion that includes direct violations of individuals. Data brokers are companies that collect information, including personal information about consumers, from an extensive range of sources for the purpose of reselling such information to their customers, which include private and public sector entities. Data brokers activities can fit in all of the categories above.

In other sub-disciplines of computer science, privacy has also been the focus of research, concerned mainly with how the privacy solutions are to be applied in specific contexts. In simple terms, they are concerned with defining the process of when and how to apply privacy solutions. Before choosing a technology for privacy protection, several questions have to be answered because there is no certainty that one type of technology solves one specific problem. One of the questions to consider is who defines what privacy is? (The technology designer, the organization's guidelines, or the users) [18].

4 Anonymous Communications

Anonymous communications aim to hide communications links. Since anonymity is the state of absent identity, anonymous communication can only be achieved by removing all the identifying characteristics from the anonymized network. Let's consider a system as a collection of actors, such as clients, servers, or peers, in a communication network. These actors exchange messages via public communication channels. Pitfzmann and Hansen [16] defined anonymity as "the state of being not identifiable within a set of subjects, the anonymity set".

One of the main characteristics of the anonymity set is its variation over time. The probability that an attacker can effectively disclose the message's sender is exactly $1/n$, with n as the number of members in the anonymity set. The research on this area has been focused on developing, analyzing and attacking anonymous communication networks. The Internet infrastructure was initially supposed to be an anonymous channel, but now we know that anyone can be spying in the network to reveal our data. Attackers have different profiles such as their action area, users volume capacity, heterogeneity, distribution and location. An outside attacker may identify traffic patterns to deduce who has communication with whom, when, and its frequency.

There are three different perspectives on anonymous communication: (i) Sender anonymity: Sender can contact receiver without revealing its identity; (ii) Receiver anonymity: Sender can contact receiver without knowing who the receiver is; (iii)

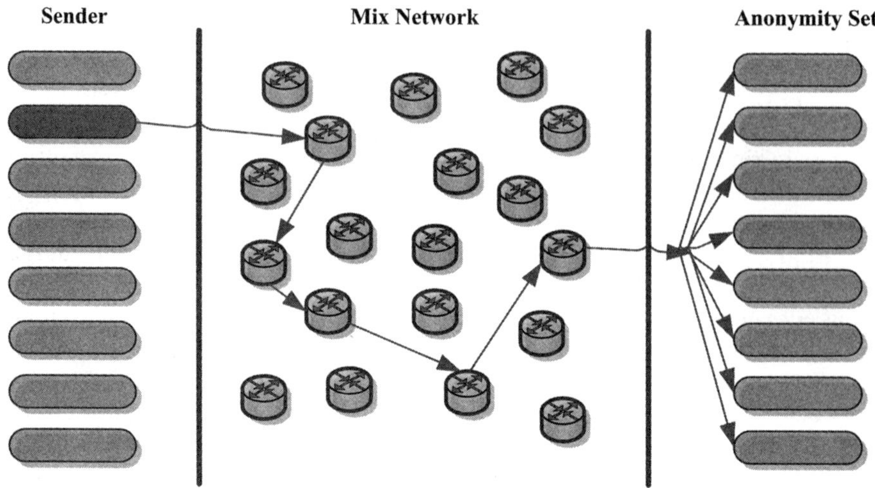

Fig. 1 Anonymous communications network

Unlinkability: Hide your relationships from third parties. According to [16] unlinkability between two items of interest occurs when an attacker of the system cannot distinguish if the two items of interest (in a system) are related or not.

Over the past years, anonymous communications has been classified by two categories: high latency systems and low latency systems. The first ones aim to provide a strong level of anonymity but are just applicable for limited activity systems that do not demand quick responses, such as email systems. On the other hand, low latency systems offer a better performance and are used in real-time systems. Examples include web applications, secure shell and instant messenger. Both systems are built on a reflection of Chaum's proposal [19]. Unlinkability is provided in a similar way in both cases using a sequence of nodes between a sender and its receiver, and using encryption to hide the message content. An intermediate node knows only its predecessor and its successor.

The mix networks systems are the basic building blocks of all modern high latency anonymous communication systems [19]; On the other hand, several designs have been developed to provide anonymity in recent years with for low latency systems, such as Crowds [20], Hordes [21], Babel [22], AN.ON [23], Onion routing [24], Freedom [25], I2P [26] and Tor [27]. Nowadays, the most widely used anonymous communication network is Tor; allowing anonymous navigation on the web. A comparison of the performance of high latency and low latency anonymous communication systems is showed in [28].

5 Mix Networks

In 1981, Chaum [19] introduced the concept of Mix networks whose purpose is to hide the correspondences between the items in its input and those in its output. A mix network collects a number of packets from distinct users called anonymity set, and then it changes the incoming packets appearance through cryptographic operations. This makes it impossible to link inputs and outputs taking into account timing information. Anonymity properties are strongest as well as the anonymity set is bigger, and these are based on uniform distribution of the actions execution of the set subjects. A mix is a go-between relay agent that hides a message's appearance, including its bit pattern and length. For example, say Alice generates a message to Bob with a constant length, a sender protocol executes several cryptographic operations through Bob and Mix public keys. After that, a mix hides the message's appearance by decoding it with the Mix private key.

The initial process for Alice to be able to send a message to Bob using a Mix system is to prepare the message. The first phase is to choose the path of the message transmission; this path must have a specific order for iteratively sending before the message gets its final destination. It is recommended to use more than one mix in every path to improve the security of the system. The next phase is to use the public keys of the chosen mixes for encrypting the message, in the inverse order that they were chosen. In other words, the public key of the last mix initially encrypts the message, then the next one before the last one and finally the public key of the first mix will be used. Every time that the message is encrypted, a layer is built and the next node address is included. This way when the first mix gets a message prepared, this will be decrypted with his correspondent private key and will get the next node address.

External attacks are executed outside the network, while internal attacks are from compromised nodes, which are actually part of the network. Mix networks are a powerful tool to mitigate outside attacks by making the sender and receiver path untraceable. The participant nodes in a mix network relay and delay messages in order to hide the route of the individual messages through the mix. However, they can be corrupted nodes that perform inside attacks. This kind of problem is addressed [20] by hiding the sender or the receiver from the relay nodes.

6 Traffic Analysis

Traffic analysis belongs to a family of techniques used to deduce information from patterns in a communication system. It has been demonstrated that encryption by itself does not provide proper anonymity; different works utilize traffic analysis techniques to uniquely identified encrypted entities. Even if communication content is encrypted, routing information has to be clearly sent because routers must determine the next network point to which a packet should be forwarded. For example, various traffic analysis techniques have been used to disclose identities in an anonymity communication network [29].

However, there is very little information about network-level traffic characteristics of recent data centers. A data center refers to any large, dedicated cluster of computers that is owned and operated by a single organization. Data center of various sizes are being built and employed for a diverse set of purposes today. On the one hand, large universities and private enterprises are increasingly consolidating their IT services within on-site data centers containing a few hundred to a few thousand servers. Furthermore, large online service providers, such as Microsoft, Google and Amazon, are rapidly building data centers to accomplish their requirements.

Very few studies of data center traffic have been published since the challenge of instrumentation and the confidentiality of the data create significant obstacles for researchers. According to literature, there are a few that contain traffic data from corporate data centers [30]. An overview of enterprise and Internet traffic based on traces captured at Lawrence Berkeley National Laboratory appears in [31]. How using the data collected from end hosts to assess the number of unsuccessful connection attempts in an enterprise network has been applied is found in [32]. A survey showing data center components and management challenges, including: power, servers, networking and software is presented in [33]. Finally, [34] examines congestion in a data center network, but only [35] focused on the design and implementation of protocols to provide reliable communication on data centers, but recognizes that more work need to be done in order to protect privacy.

7 Mix Systems Attacks

The attacks against mix systems are intersection attacks [36]. They take into account a message sequence through the same path in a network, it means performing traffic analysis. The set of most likely receivers is calculated for each message in the sequence and the intersection of the sets will make it possible to know who the receiver of the stream is. Intersection attacks are designed based on correlating the times when senders and receivers are active. By observing the recipients that received packets during the rounds when Alice is sending, the attacker can create a set of Alice's most frequent recipients, this way diminishing her anonymity.

Next, we present the family of statistical disclosure attack, which is based in executing traffic analysis techniques.

8 The Disclosure Attack

The beginning of this family is the disclosure attack [37, 38]. The attack was modeled by considering a bipartite graph $G = (A \bigcup B, E)$. The set of edges E represents the relationship between senders and recipients A and B. Mixes assume that all networks links are observable. So, the attacker can determine anonymity sets by observing messages to and from an anonymity network; the problem arises for how long the observation is necessary. The attack is global, in the sense that it retrieves information about the number of messages sent by Alice and received by other users, and passive, in the sense that the attacker cannot alter the network (sending false messages or delaying existent ones). Authors assume a particular user, Alice, sends messages to limited m recipients. A disclosure attack has a learning phase and an excluding phase. The attacker should find m disjoint recipients set by observing Alice's incoming and outgoing messages. In this attack, authors make several strategies in order to estimate the average number of observations for achieve the disclosure attack. They assume that: i) Alice participates in all batches; ii) only one of Alice's peer partners is in the recipient sets of all batches. In conclusion, this kind of attack is very expensive because it takes an exponential time taking into account the number of messages to be analyzed trying to identify mutually disjoint sets of recipients. This is the main bottleneck for the attacker, and it derives from an NP-complete problem. Test and simulations showed it only works well in very small networks.

9 The Statistical Disclosure Attack (SDA)

The SDA proposed by Danezis [39] is based on the previous attack. It requires less computational effort by the attacker and gets the same results. The method tries to reveal the most likely set of Alice's friends using statistical operations and approximations. It means that the attacks applies statistical properties on the observations and recognize potential recipients, but it does not solve the NP-complete problem presented in previous attack. Consider $\overrightarrow{v}$ as the vector with N elements corresponding to each potential recipient of the messages in the system. Assume Alice has m recipients as the attack above, so $\frac{1}{m}$ might receive messages by her and it's always $|\overrightarrow{v}| = 1$. The author also defines $\overrightarrow{u}$ as the uniform distribution over all potential recipients N. In each round the probability distribution is calculated, so recipients are ordered according to its probability. The information provided to the attacker is a series of vectors representing the anonymity sets The highest probability elements will be the most likely recipients of Alice. Variance on the signal and the noise introduced by other senders is used in order to calculate how many observations are necessary. Alice must demonstrate consistent behaviour patterns in the long term to obtain good results, but this attack can be generalized and applied against other anonymous communication network systems. A simulation over pool mixes are in [40]. Distinct to the predecessor attack, SDA only show likely recipients and does not identify Alice's recipients with certainty.

10 Extending and Resisting Statistical Disclosure

One of the main characteristics in Intersection Attacks relies on a fairly consistent sending pattern or a specific behaviour for users in an anonymity network. Mathewson and Dingledine in [41] make an extension of the original SDA. One of the more significant differences is that they consider that a real social network has a scale-free network behaviour, and also such behaviour changes slowly over time. They do not simulate these kinds of attacks.

In order to model the sender behaviour, authors assume Alice sends n messages with a a probability $Pm(n)$; and the probability of Alice sending to each recipient is represented in a vector $\overrightarrow{v}$. First the attacker gets a vector $\overrightarrow{u}$ whose elements are: $\frac{1}{b}$ the the recipients that have received a message in the batch, and 0 for recipients that have not. For each round i in which Alice sent a message, the attacker observes the number of messages m_i sent by Alice and calculates the arithmetic mean.

Simulations on pool mixes are presented, taking into account that each mix retains the messages in its pool with the same probability every round. The results show that increasing variability in the message makes the attack slower by increasing the number of output messages. Finally they examine the degree to which a non-global adversary can execute a SDA. Assuming all senders choose with the same probability all mixes as entry and exit points and attacker is a partial observer of the mixes. The results suggest that the attacker can succeed on a long-term intersection attack even when it partially observes the network. When most of the network is observed the attack can be made, and if more of the network is hidden then the attacker will have fewer possibilities to succeed.

11 Two Sided Statistical Disclosure Attack (TS-SDA)

[42] Danezis et al. provide an abstract model of an anonymity system considering that users send messages to his contacts, and takes into account some messages sent by a particular user are replies. This attack assumes a more realistic scenario regarding the user behaviour on an email system; its aim is to estimate the distribution of contacts of Alice, and to deduce the receivers of all the messages sent by her.

The model considers N as the number of users in the system that send and receive messages. Each user n has a probability distribution D_n of sending a message to other users. For example, the target user Alice has a distribution D_A of sending messages to a subset of her k contacts. At first the target of the attack, Alice, is the only user that will be model as replying to messages with a probability r. The reply delay is the time between a message being received and sent again. The probability of a reply r and the reply delay rate are assumed to be known for the attacker, just as N and the probability that Alice initiates messages. Based on this information the attacker estimates: (i) the expected number of replies for a unit of time; (ii) The expected volume of discussion initiations for each unit of time; (iii) The expected volume of replies of a particular message.

Finally authors show a comparative performance of the Statistical Disclosure Attack (SDA) and the Two Sided Disclosure Attack (TS-SDA). It shows that TS-SDA obtains better results than SDA. The main advantage of the TS-SDA is its ability to uncover the recipient of replies. on reveal discussion initiations. Inconvenient details for application on real data is the assumption that all users have the same number of friends to which they send messages with uniform probability.

12 Perfect Matching Disclosure Attack (PMDA)

The PMDA [8] is based on graph theory, it considers all users in a round at once, instead of one particular user iteratively. No assumption on the users behaviour is required to reveal relationships between them. Comparing with previous attacks where Alice sends exactly one message per round, this model permits users to send or receive more than one message in each round. Bipartite graphs are employed to model a threshold mix, and through this, they show how weighted bipartite graphs can be used to disclosure users communication. A bipartite graph $G = (S \bigcup R, E)$ considers nodes divided in two distinct sets S (senders) and R (receivers) so that every edge E links one member in S and one member in R. It is required that every node is incident to exactly one edge. In order to build a threshold mix, it is thought that t messages sent during one round of the mix form the set S, and each node $s \in S$ is labeled with the sender's identity $\sin(s)$. Equally, the t messages received during one round form the set R where each node r is labeled with the receiver's identity $rec(r)$. A perfect matching M on G links all t sent and received messages. Additionally P' is $t \times t$ matrix containing weights w_s, r, representing probabilities for all possible edges in G.

The procedure for one round is: (i) sent messages are nodded in S, and marked with their senders identities; (ii) received messages are nodes in R, and marked with their receivers identities; (iii) derive the $t \times t$ matrix: first estimating user profiles when SDA and then de-anonymize mixing round with $P'(s,r) = \widetilde{P}_{\sin(S),SDA}(rec(r)), s \in S_i, r$; iv) replace each element of the matrix $P'(s,r)$ with $log_{10}(P'(s,r))$; v) having each edge associated with a log-probability, a maximum weighted bipartite matching on the graph $G = (S \bigcup R, E)$ outputs the most likely sender-receiver combination. This work shows that it is not enough to take the perspective of just one user of the system.

Results of experimentation show that this attack does not consider the possibility that users send messages with different frequencies. An extension proposal considers a Normalized SDA. Another related work concerning perfect matchings is perfect matching preclusion [43, 44] where Hamiltonian cycles on the hypercube are used.

13 Vida: How to Use Bayesian Inference to De-anonymize Persistent Communications

A generalization of the disclosure attack model of an anonymity system applying Bayesian techniques is introduced by Danezis et al. [45]. Authors build a model to represent long term attacks against anonymity systems, which are represented as N_{user} users that send N_{msg} messages to each other. Assume each user has a sending profile, sampled when a message is to be sent to determine the most likely receiver. The main contributions are two models: (1) Vida Black-box model represents long term attacks against any anonymity systems; (2) Vida Red-Blue allows an adversary to performance inference on selected target through traffic analysis.

Vida Black Box model describes how messages are generated and sent in the anonymity system. In order to perform inference on the unknown entities they use Bayesian methods. The anonymity system is represented by a bipartite graph linking input messages i_x with its correspondent output messages o_y without taking into account their identities. The edges are labelled with its weight that is the probability of the input message being sent out. Senders are associated with multinomial profiles, which are used to choose their correspondent receivers. Through Dirichlet distribution these profiles are sampled. Applying the proposed algorithm will derive a set of samples that will be used for attackers to estimate the marginal distributions linking senders with their respective receivers.

Vida Red-Blue model tries to respond to the needs of a real-world adversary, considering that he is interested in particular target senders and receivers. The adversary chooses Bob as a target receiver, it will be called "Red" and all other receivers will be tagged as "Blue". The bipartite graph is divided into two sub-graphs: one containing all edges ending on the Red target and one containing all edges ending on a Blue receiver. Techniques Bayesian are used to select the candidate sender of each Red message: the sender with the highest a-posterior probability is chosen as the best candidate.

The evaluation includes a very specific scenario which considers: (i) messages sent by up to 1000 senders to up to 1000 receivers; (ii) each sender is assigned 5 contacts randomly; (iii) everyone sends messages with the same probability; (iv) messages are anonymized using a threshold mix with a batch of 100 messages.

14 SDA with Two Heads (SDA-2H)

One of the most used strategies to attempt against SDA is sending cover traffic which consists of fake or dummy messages mixed with real ones that can hide Alice's true sending behaviour. SDA-2H [46] is an extension of SDA [39] and takes its predecessor as a baseline to improve it as it considers background traffic volumes in order to estimate the amount of dummy traffic that Alice sends. Dummy traffic serves as a useful tool to increase anonymity and they are classified based on their origin: (i) user cover, generated by the user Alice; (ii) background cover, generated by senders other than Alice in the system; (iii) receiver-bound cover, generated by the mix. This work is centered on background cover which is created when users generated false messages along with their real ones. The objective for the attacker is to estimate how much of Alice's traffic is false based on the observations between the volume of incoming and outgoing traffic. Authors make several simulations and find that for a specific number of total recipients, the increase in the background messages makes it harder for the attacker to succeed having total recipients and Alice's recipients unchanged. They also find that when Alice's recipients stay and the number of total recipients increases, the attacker would need few rounds of observations to find Alice's recipients. A comparative between SDA and SDA-2H shows that SDA-2H may not be better than SDA in all cases, but SDA-2H takes into account the effect of background cover to achieve a successful attack.

15 Conclusions

In spite of widespread interest in datacenter networks, little has been published that reveals the nature of their traffic, or the problems that arise in practice. This chapter first shows how traffic analysis can be used to disclosure information, even considering patterns such as which servers talk to each other, when and for what purpose; or characteristics as duration streams or statistics. Although modern technologies have enhanced the way we conduct everyday business—these same technologies create new risks as they are deployed into the modern IT environment. The digital environment is changing and the focus must be on attackers, more work should be done to provide a useful guide for datacenter network designers. The real problem: Not only have attacks against the entire data center infrastructure increased, they've also become much more sophisticated. The influx of advanced attacks has become a serious issue for any data center provider looking to host modern technologies. As privacy research advances, we observe that some of our assumptions about the capabilities of privacy solutions also change. Risk reduction to acceptable levels should be taken into account to develop measures against internal and external threats.

Acknowledgment Part of the computations of this work were performed in EOLO, the HPC of Climate Change of the International Campus of Excellence of Moncloa, funded by MECD and MICINN.

References

1. Krishnamurthy, B.: Privacy and Online Social Networks: Can Colorless Green Ideas Sleep Furiously? IEEE Security Privacy **11**(3) (May 2013) 14–20
2. Dey, R., Jelveh, Z., Ross, K.: Facebook Users Have Become Much More Private: A Large-Scale Study. In: IEEE International Conference on Pervasive Computing and Communications Workshops. (19–23 March 2012) 346–352

3. Christofides, E., Desmarais, A.M.S.: Information Disclosure and Control on Facebook: Are They Two Sides of the Same Coin or Two Different Processes? CyberPsychology & Behavior **12**(3) (June 2013) 341–345
4. Gross, R., Acquisti, A.: Information Revelation and Privacy in Online Social Networks. In: 2005 ACM Workshop on Privacy in the Electronic Society, ACM (2005) 71–80
5. Goldberg, I., Wagner, D., Brewer, E.: Privacy-enhancing technologies for the Internet. In: IEEE Compcon'97. (February 23–26 1997) 103–109
6. Goldberg, I.: Privacy-Enhancing Technologies for the Internet, II: Five Years Later. In: Second International Workshop on Privacy Enhancing Technologies. (April 14–15 2003) 1–12
7. Goldberg, I.: Privacy Enhancing Technologies for the Internet III: Ten Years Later. In: Digital Privacy: Theory, Technologies and Practices, Auerbach Publications (December 2007) 3–18
8. Westin, A.F.: Privacy and Freedom. The Bodley Head Ltd (1997)
9. R. Gellman, P.D.: Online Privacy: A Reference Handbook. ABC-CLIO (2011)
10. Berendt, B., Günther, O., Spiekermann, S.: Privacy in e-Commerce: Stated Preferences vs. Actual Behavior. Communications of the ACM **48**(4) (April 2005) 101–106
11. Narayanan, A., Shmatikov, V.: De-Anonymizing Social Networks. In: IEEE Symposium on Security and Privacy, Washington, DC, USA, IEEE Computer Society (2009) 173–187
12. Commission, E.: Privacy Enhancing Technologies (PETs): The Existing Legal Framework (May 2007)
13. Fritsch, L.: State of the Art of Privacy-Enhancing Technology (PET). Technical report, Norsk Regnesentral, Norwegian Computing Center (2007)
14. Group, M.: Privacy Enhancing Technologies". Technical report, Ministry of Science, Technology and Innovation (March 2005)
15. Adams, C.: A Classification for Privacy Techniques. University of Ottawa Law & Technology Journal **3**(1) (July 2006) 35–52
16. Pfitzmann, A., Hansen, M.: Anonymity, Unlinkability, Undetectability, Unobservability, Pseudonymity, and Identity Management: A Consolidated Proposal for Terminology. http://dud.inf.tu-dresden.de/Anon_Terminology.shtml (February 2008) v0.31.
17. Solove, D.J.: A Classification for Privacy Techniques. University of Pennsylvania Law Review **154**(3) (January 2006) 477–560
18. Diaz, C., Gürses, S.: Understanding the Landscape of Privacy Technologies. In: The 13th International Conference on Information Security (Information Security Summit). (2012) 1–6
19. Chaum, D.L.: Untraceable Electronic Mail, Return Addresses, and Digital Pseudonyms. Communications of ACM **24**(2) (February 1981) 84–90
20. Reiter, M.K., Rubin, A.D.: Crowds: Anonymity for Web Transactions. ACM Transactions on Information and System Security **1**(1) (November 1998) 66–92
21. Levine, B.N., Shields, C.: Hordes: A Multicast Based Protocol for Anonymity. Journal of Computer Security **10**(3) (September 2002) 213–240
22. Gulcu, C., Tsudik, G.: Mixing Email with Babel. In: Symposium on Network and Distributed System Security, Washington, DC, USA, IEEE Computer Society (1996) 1–15
23. Berthold, O., Federrath, H., Kopsell, S.: Web MIXes: A System for Anonymous and Unobservable Internet Access. In: International Workshop On Designing Privacy Enhancing Technologies: Design Issues In Anonymity And Unobservability, Springer-Verlag New York, Inc. (2001) 115–129
24. Goldschlag, D.M., Reed, M.G., Syverson, P.F.: Hiding Routing Information. In: First International Workshop on Information Hiding, London, UK, UK, Springer-Verlag (May 30 - June 1 1996) 137–150
25. Back, A., Goldberg, I., Shostack, A.: Freedom Systems 2.1 Security Issues and Analysis (May 2001)
26. Back, A., Goldberg, I., Shostack, A.: I2P (2003)
27. Dingledine, R., Mathewson, N., Syverson, P.: Tor: The Second-generation Onion Router. In: 13th Conference on USENIX Security Symposium - Volume 13, Berkeley, CA, USA, USENIX Association (2004) 21–21

28. Loesing, K.: Privacy-Enhancing Technologies for Private Services. PhD thesis, University of Bamberg (2009)
29. Edman, M., Yener, B.: On Anonymity in an Electronic Society: A Survey of Anonymous Communication Systems. ACM Computing Surveys **42**(1) (December 2009) 1–35
30. Benson, T., Anand, A., Akella, A., Zhang, M.: Understanding Data Center Traffic Characteristics. ACM SIGCOMM Computer Communication Review **40**(1) (January 2010) 92–99
31. Pang, R., Allman, M., Bennett, M., Lee, J., Paxson, V., Tierney, B.: A First Look at Modern Enterprise Traffic. In: 5th ACM SIGCOMM Conference on Internet Measurement, Berkeley, CA, USA, USENIX Association (October 19-21 2005) 2–2
32. Guha, S., Chandrashekar, J., Taft, N., Papagiannaki, K.: How Healthy Are Today's Enterprise Networks? In: 8th ACM SIGCOMM Conference on Internet Measurement, New York, NY, USA, ACM (October 20-22 2008) 145–150
33. Kandula, S., Sengupta, S., Greenberg, A., Patel, P., Chaiken, R.: The Nature of Data Center Traffic: Measurements & Analysis. In: 9th ACM SIGCOMM Conference on Internet Measurement Conference, New York, NY, USA, ACM (November 4-6 2009) 202–208
34. Greenberg, A., Maltz, D.A.: What Goes Into a Data Center? (2009)
35. Balakrishnan, M.: Reliable Communication for Datacenters. PhD thesis, Cornell University (September 2008)
36. Raymond, J.F.: Traffic Analysis: Protocols, Attacks, Design Issues and Open Problems. In: International Workshop On Design Issues In Anonymity And Unobservability, Springer-Verlag New York, Inc. (July 25-26 2000) 10–29
37. Kedogan, D., Agrawal, D., Penz, S.: Limits of Anonymity in Open Environments. In: 5th International Workshop on Information Hiding, London, UK, UK, Springer-Verlag (October 7–9 2002) 53–69
38. Agrawal, D., Kesdogan, D.: Measuring Anonymity: The Disclosure Attack. IEEE Security Privacy **1**(6) (2003) 27–34
39. Danezis, G.: Statistical Disclosure Attacks: Traffic Confirmation in Open Environments. In: IFIP Advances in Information and Communication Technology, Kluwer (2003) 421–426
40. Danezis, G., Serjantov, A.: Statistical Disclosure or Intersection Attacks on Anonymity Systems. In: 6th Information Hiding Workshop. (May 23–25 2004) 293–308
41. Mathewson, N., Dingledine, R.: Practical Traffic Analysis: Extending and Resisting Statistical Disclosure. In: 4th International Conference on Privacy Enhancing Technologies. (May 23-25 2004) 17–34
42. Danezis, G., Diaz, C., Troncoso, C.: Two-sided Statistical Disclosure Attack. In: 7th International Conference on Privacy Enhancing Technologies, Berlin, Heidelberg, Springer-Verlag (June 20–22 2007) 30–44
43. Brigham, R., Harary, F., Violin, E., Yellen, J.: Perfect-Matching Preclusion. Congressus Numerantium **174** (2005) 185–192
44. Park, J.H., Son, S.H.: Conditional Matching Preclusion for Hypercube-like Interconnection Networks. Theoretical Computer Science **410**(27–29) (June 2009) 2632–2640
45. Danezis, G., Troncoso, C.: Vida: How to use Bayesian Inference to De-anonymize Persistent Communications. In: 9th International Symposium of Privacy Enhancing Technologies, Springer Berlin Heidelberg (August 5-7 2009) 56–72
46. Al-Ameen, M., Gatz, C., Wright, M.: SDA-2H: Understanding the Value of Background Cover Against Statistical Disclosure. In: 14th International Conference on Computer and Information Technology. (December 22-24 2011) 196–201

Part VII
Data Services

Quality-of-Service in Data Center Stream Processing for Smart City Applications

Paolo Bellavista, Antonio Corradi and Andrea Reale

1 Introduction

The wide diffusion of cheap, small, and portable sensors integrated in an unprecedented large variety of devices—from smartphones to household appliances, from cars to fixed monitoring stations—, and the availability of almost ubiquitous Internet connectivity through Wi-Fi hotspots or cellular networks, makes it possible to collect and use valuable real-time information about many fundamental aspects of the environment we live in. If properly understood and used, this information has the potential to bring important improvements to cross-concerning areas that have strong and direct impact on the quality of people's life, such as healthcare, urban mobility, public decision making, and energy management. This continuous collection and exploitation of real-time data from people and objects of the real world is at the foundations of the *Smart City* vision [26], where people, places, environment, and administrations become closer and get connected through novel ICT services and networks. In the last years, several projects from academia, industries and governments have started to work toward the actual implementation of this vision in big urban areas. Examples of these initiatives are the many European funded projects such as European Digital Cities [21], Smart Cities Stakeholder Platform [42], SafeCity [41], or EUROCITIES [23], or industry-led activities, such as the IBM Smarter Cities project [29], or the Intel Collaborative Research Institute for Sustainable Connected Cities [30].

In order to implement novel and useful smart services for the city, it is not only sufficient to collect the raw content of these *Big Data Streams*, but is also crucial to distill interesting and usable knowledge from them. However, the unprecedented

P. Bellavista (✉) · A. Corradi · A. Reale
Department of Computer Science and Engineering (DISI), Università di Bologna, Italy
e-mail: paolo.bellavista@unibo.it

A. Corradi
e-mail: antonio.corradi@unibo.it

A. Reale
e-mail: andrea.reale@unibo.it

S. U. Khan, A. Y. Zomaya (eds.), *Handbook on Data Centers*,
DOI 10.1007/978-1-4939-2092-1_35

heterogeneity in data *representation* and *semantics*, and in the *goals* and *quality requirements* of analysis tasks is a hard technical challenge to face while developing applications that deal with huge and continuous streams of information. Distributed Stream Processing Systems (DSPSs) represent very relevant technological support frameworks for the industrial and cost-effective implementation of Smart City applications: for instance, by efficiently leveraging the distributed resources available in data centers with limited impact on the complexity of the application logic, they answer the requirements of *performance* and *scalability* that continuous data streams analysis impose.

In this chapter we analyze the state-of-the-art of DSPSs, with a strong focus on the characteristics that make them more or less suitable to serve the novel processing needs of Smart City scenarios. In particular, we concentrate on the ability to offer differentiated *Quality of Service* (QoS). A growing number of Smart City applications, in fact, including those in the security, healthcare, or financial areas, require configurable and predictable behavior. For this reason, a key factor for the success of new and original stream processing supports will be their ability to efficiently meet those needs, while still being able to scale to fast growing workloads.

The chapter is organized as follows: Section 2 introduces the class of big data analysis platforms known as DSPSs. It does so by providing a simple framework for their description and comparison. Section 3 presents three state-of-the-art and widely used DSPSs and compares their specific characteristics by using the framework presented in Sect. 2. Section 4 focuses on the problem of integrating QoS-aware behavior in DSPSs by emphasizing the reasons why they would be especially useful in Smart City scenarios. Our QoS-aware DSPS, called Quasit, is presented in Sect. 5; Quasit has been specifically designed to allow a rich customization of quality-related stream processing parameters, and is able to enforce them at runtime in a scalable and cost-effective way. Finally, in Sect. 6, we look at a special kind of *weak* QoS specifications, which can be flexibly and adaptively enforced by DSPSs: to this purpose, we present LAAR, a technique for adaptive DSPS operator replication, which allows to trade "perfect" fault-tolerance guarantees off for reduced execution cost, while being able to handle variable load conditions and to offer guaranteed lower bounds on the achievable system reliability.

2 Distributed Stream Processing Systems

A stream processing application is a collection of software components whose goal is to process, analyze, or transform streams of information to produce continuous results in the form of output streams. A *Stream Processing System* (SPS) is a middleware that provides support for both the development and the execution of stream processing applications, and is labeled as *distributed* (DSPS) if it deploys them on a set of distributed computing resources, such as, very relevant nowadays, the ones in a data center.

We propose an original representation model for DSPSs that helps analyzing them according to a simple three layer scheme. The layers are complementary, each

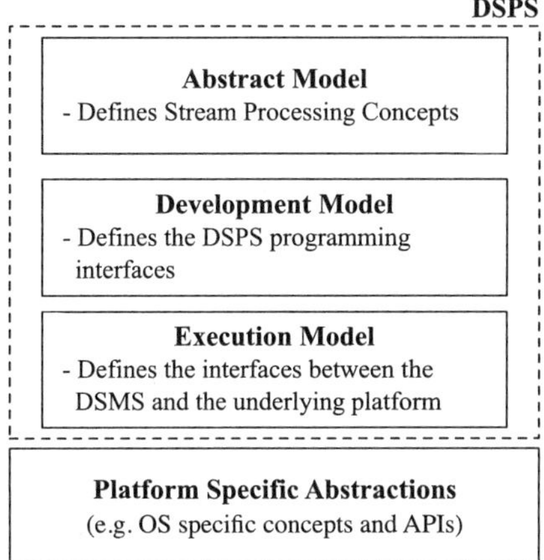

Fig. 1 A three-layer model of Distributed Stream Processing Systems

describing a different aspect of the stream processing system, and are called *abstract model*, *development model*, and *execution model*, respectively (Fig 1).

- *The abstract model* defines the high-level stream processing concepts adopted by the system. For instance, it gives precise definitions of data streams and relevant system events; it determines the characteristics of data processing flows, and the type, role, and granularity of processing components.
- *The development model* defines the set of interfaces that are exposed to developers to build the stream processing components defined in the abstract model. A development model, for example, could map abstract concepts on syntactic constructs of special-purpose stream processing languages, or on ad-hoc Application Programming Interfaces (APIs) and libraries developed for existing general-purpose languages.
- *The execution model* determines how abstract model components are mapped on runtime entities executed by the distributed servers on which the DSPS is deployed. For example, an entire application could be mapped, at execution time, on a single process of the host operating system, or it could be split into several interacting processes.

While the three models may, in theory, largely vary from system to system, in practice, it is easy to identify many recurring aspects among the most common solutions. In the remainder of this section, we discuss the three models and overview how they are commonly realized in existing state-of-the-art solutions.

2.1 Abstract Model

The abstract model of a DSPS defines the high level concepts on which the system is based, including the system-dependent concepts of stream, stream processing application, and the processing workflow that the system adopts. While development and

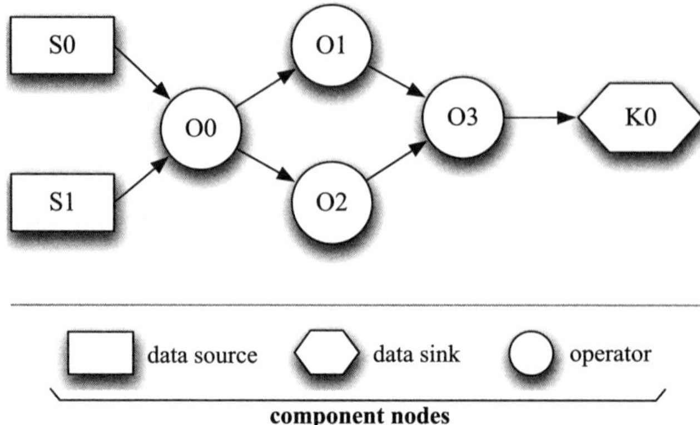

Fig. 2 A generic processing graph in Distributed Stream Processing Systems

execution models usually can be significantly different from one system to another, abstract models tend to be very similar and based on the common abstraction of *processing graph* [2, 4, 6, 10, 16, 24, 35, 44].

A processing graph (Fig. 2) is a Directed and Acyclic Graph (DAG) whose nodes represent data processing and transformation steps, and whose edges represent streams flowing between components. A *stream* is an unbounded sequence of discrete elements, often called *samples* or *tuples*. The *type* of a sample defines its structure, and every stream contains samples all of the same type. Depending on the system, a sample type could be a primitive type—such as an integer or floating point number—or it could be a composite type, similar to a structure in the C language, or, in some cases, to objects of an object oriented type system. Every processing graph is always fed by one or more input streams, and produces one or more output streams as a result. The origin and destination of input and output streams can be highly heterogeneous, such as for example, a file, a network socket, a PUB/SUB endpoint, or a relational database. Since input streams are unbounded, a characterizing feature of stream processing applications is that, once started, they continue to execute forever, unless explicitly stopped.

A graph node can be of three different kinds, i.e., *data source*, *data sink*, and *operator*. A *data source* node identifies a data stream that is conceptually out of the application: its role is to abstract from the actual nature of the stream producer. It can represent either an external stream source or the output of another application running on the same system. A *data sink* node, conversely, represents the destination of an application output; data sinks can be used either to redirect output streams to other systems for additional processing steps or storage, or to connect the output of an application with the input of another one. An *operator* node is associated with one or more input data streams and *generates* one or more output streams. Operators are the core of stream processing applications: they define the set of operations that can be performed on streams. Operators can implement, for example, relational manipulations of single or moving windows of samples, such as projections or joins;

they can perform aggregation or filtering actions, or realize more complex, arbitrary, algorithms. Operators, data sources, and data sinks are collectively called *graph components* or, more simply, components.

Samples are received and produced by stream components on their *input and output ports*, each having its own type, which corresponds to the type of the stream it receives or produces. Every graph component performs its processing operations on data samples according to an asynchronous processing model; conceptually all the components operate in parallel and perform their processing actions as soon as data samples are available at their input ports.

2.2 Development Model

A development model maps the concepts defined in the abstract model on the programming-level constructs offered to developers to write their stream processing applications. These constructs should allow to:

1. Define new applications, by describing which source, operator, and sink components should be instantiated and how they should be connected into a processing graph.
2. Customize component instances, in order to adapt their behavior to specific application needs (e.g., to bind graph source nodes to actual external sources).
3. Develop new components with custom functionalities.

Any DSPS development model should at least define the tools to achieve the first two goals of the list; in fact, in many cases, it is not necessary to create new or custom components, especially in the common scenario where the DSPS comes bundled with collections of ready-to-use components (also referred to as *toolkits*) that can satisfy most common application requirements.

In the available literature, two families of application development models are common. The first includes models whose mappings are based on *special-purpose languages*; the second relates to the exploitation of *general purpose languages* for that.

Special-purpose stream processing languages are usually tightly bound to the system they have been designed for. They normally allow a very concise definition of applications and components, by having stream processing concepts mapped one-to-one to language-level concepts. For example, the Stanford STREAM system [6] defines the so called *Continuous Query Language* (CQL), which permits to develop stream processing applications by writing continuous queries in a syntax that strongly resembles SQL queries. These queries are processed by the underlying system and decomposed in a processing graph of pre-defined operators. If ad-hoc languages permit faster and easier application development, they generally lack the flexibility of general-purpose languages and, more importantly, they force developers to learn new languages and new development processes.

Development models based on *general-purpose* languages, instead, usually have a less steep learning curve, as system-specific stream processing concepts are expressed using familiar constructs offered by common general-purpose programming languages, such as C++, Java, or Python. For example, in Apache S4 [35], operators are defined by writing corresponding Java classes, all subclasses of a common abstract superclass. The developer has to "fill-in" the methods that implement the operator logic, which the system automatically invokes when corresponding events of interest occur. Using general-purpose languages has several benefits, including the possibility to seamlessly reuse existing libraries and software modules in new stream processing applications. However, this usually comes at the expense of conciseness and prototyping speed because those APIs can be verbose and sometimes cumbersome.

2.3 Execution Model

An execution model maps the elements defined in the abstract model and described through the development model onto runtime objects that the hosting platform is able to run directly. An execution model defines:

1. The characteristics of platform specific execution units, and the high-level policies adopted for scheduling local resources, such as CPU and memory.
2. The distribution of the execution units on the cluster of available servers.
3. The mapping of graph edges on communication channels, such as shared memory, pipes, or network sockets.

The first important aspect of an execution model is the mapping of operators, sources, and sinks on host platform concepts, such as processes or threads. With a *process-per-operator* allocation, each operator is individually instantiated as a separate process, with one or possibly more concurrent threads of execution (e.g., one per input port). This grants the maximum isolation, since problems with one component cannot affect other concurrently running ones. With this choice, the local scheduling of resources is demanded to the standard facilities of the host operating system CPU and memory schedulers. The first implementations of the Stream Processing Core (SPC) [4] (at the basis of the more recent IBM InfoSphere Streams system— see the following) used a similar approach, by isolating single components into their own containers, corresponding to standard UNIX processes.

A *process-per-server* allocation creates just one process per server. Within this process, components are hosted as separate software modules, for example as instances of a given class in case of a class-based object oriented implementation. While, on the one side, this arrangement does not grant the same execution isolation that a process-per-operator allocation does, on the other side, it permits a tighter control on resource scheduling policies. For example, every in-process component could be given a dedicated execution thread or, more interestingly, a pool of threads could be used to execute groups of components according to internal policies or QoS requirements (e.g., for a priority-proportional scheduling of resources). Moreover, the communication of components running within the same process is usually faster and cheaper, as shared memory based channels can be used. The process-per-server allocation is used, for example, by Apache S4 [35] and Quasit [10], which start a Java virtual machine on each cluster server, and deploy sources, operators, and sinks as objects running within the local VM.

Somewhere in the middle between the two previous solutions, the *cluster-of-operators* approach *fuses* subsets of tightly coupled components (e.g., operators with strong reciprocal communication dependencies) into one process. Again, within every single process, very flexible resource scheduling approaches and faster communication channels can be used. Different operator clusters, however, are still mapped onto different processes, granting a better isolation to each group. IBM InfoSphere Streams [24] uses a similar hybrid approach thanks to a technique, called *operator fusion* [33], that groups multiple operators into single execution units automatically.

Knowing what the execution units are, the application processing graph can be rewritten in the corresponding *runtime graph* where nodes represent individual runtime objects (e.g., processes) and edges represent inter-process communication channels. A further role of the execution model is the definition of an *assignment strategy* for runtime objects. An assignment strategy decides the distribution of runtime objects on the available cluster servers: a good solution should take into account the resource requirements of every object (e.g., CPU and memory), the resources availability of each server, and the expected/declared application communication patterns, and it should find an assignment that satisfies the application resource and quality requirements while minimizing its execution cost. The assignment can be static-only, or can have dynamic phases as well. During the static phase an initial assignment is decided based on a-priori knowledge of the application and input streams characteristics. Due to changing load conditions, for example caused by load spikes in some input streams, the initial assignment could be no longer adequate to satisfy the application QoS requirements; in these cases, a dynamic phase can be performed at runtime to adaptively deal with load variations. For example, [39] and [38] propose two-phase algorithms to perform both static and dynamic assignment phases, while [40] tries to find an initial static assignment that maximizes the system robustness to possible variations.

Finally, an execution model should decide how communication channels are implemented at runtime. For in-process communication, *function calls* or *shared memory-based message passing* are the most commonly chosen alternatives. While the first binds the execution thread of the caller to that of the callee, the second allows an independent execution of the two components. For what concerns inter-process communication, the choice depends on whether the channel endpoints reside on the same host or on remote hosts. In the first case, solutions such as system-level shared memory or pipes can be adopted to implement faster and cheaper communication solutions; in case of remote communication, the choice of the protocol depends very much on the desired communication cost and QoS level. For example, if cheap and unreliable communication is enough, UDP-based channels are a possible solution.

3 Platforms for Distributed Stream Processing

In the last decade, the problem of effectively processing continuous information flows has been faced in several projects. In the early 2000s, systems like TelagraphCQ [8, 17], Aurora [1, 16], Borealis [2], and Stream [6, 7] have started recognizing the ineffectiveness of using traditional database management systems (DBMSs) for the real-time analysis of continuous data, and have proposed their own alternative solutions. More recently, as a result of the industrial success of scalable and parallel batch data processing systems like MapReduce [20], solutions such as Map-Reduce-Merge [46] or MapReduce Online [18] have tried to reuse its successful scalable processing model in stream processing scenarios, by enhancing it with continuous and dynamic data analysis capabilities. In this section, we have selected three prominent state-of-the-art DSPSs, and we discuss their design and architectural features under the light of the three-layers modeling framework introduced in the previous section. The three systems described in the following are IBM InfoSphere Streams, Apache S4, and Storm. The particular choice of these systems over the different alternatives available in the literature is motivated by the fact that, to our knowledge, the selected DSPSs are the most widely adopted in real-world large scale production systems, including large data center deployments from important industry players such as IBM, Yahoo! and Twitter: for this reason, we believe that their analysis can provide important insights about the common requirements of real stream processing workloads, including those of Smart City scenarios. It is not the goal of this work to provide an extensive survey of existing stream processing solutions: for a comprehensive work, the interested reader is referred to [19].

In the following three subsections, we briefly overview each selected DSPS by analyzing its abstract, development, and execution model; for each of them, we also emphasize peculiar QoS-related features, when supported.

3.1 IBM InfoSphere Streams

IBM InfoSphere Streams [24] is a DSPS evolved from the SPC research project [4]. In Streams, application processing graphs are defined in an ad-hoc special-purpose Stream Processing Language (SPL) that is used to describe *operators* and their stream connections. The language is very flexible, as it permits to define new data types, or to customize the behavior of existing operators by changing, for example, their number of input/output ports or their output logic. Besides defining simple operators, SPL also enables to combine them in composite ones, which encapsulate more complex behavior.

In addition, the system exposes two sets of general-purpose APIs that can be used to build user-defined custom operators. The first is a mixed C++ and Perl API that, by using a two-steps code generation process, gives them maximum customization flexibility and execution efficiency [25]. The second is a simpler Java API, based on runtime reflection techniques rather than code generation. Due to the cost of reflection, however, this API is in general less efficient than its C++/Perl counterpart. Operators built through either API can be exported to reusable toolkits and used directly within SPL source files.

At compile time, an optional operator fusion process can be manually or automatically started in order to cluster groups of correlated operators. Each group is then transformed into its corresponding runtime object, called Processing Element (PE), whose execution is mapped onto an operating system process. Hence, InfoSphere Streams follows an operator-per-process or cluster-of-operators approach, depending on whether the fusion step is performed or not. Depending on configuration parameters, operators inside the same process are executed by dedicated threads—in this case they communicate to other in-process operators through message passing—or by shared threads—in this case they communicate via function calls. At the time of writing, the only explicit QoS parameter supported by Streams is a loose form of fault-tolerance, based on checkpointing [32]: periodically or driven by events, runtime objects can save their current state on secondary memory; whenever a crash occurs, that state is restored, but all the processing operations performed between the last checkpoint and the failure are lost.

3.2 Apache S4

Apache S4 [5] is a DSPS initially developed and maintained by Yahoo! [35] and currently part of the Apache Incubator project umbrella.

In S4 processing graphs, there is no distinction between sources, sinks, and operators, but all the components are uniformly modeled and called PEs[1]. PEs can import streams coming from other applications running on the same platform, process them, and possibly export output streams either to external destinations or to other applications concurrently running on the platform. External streams of data (i.e., coming from sources external to the platform itself) can be transformed into internal streams by developing and running special S4 applications, called adaptors.

To develop PEs or adaptors, S4 offers its own Java API. Developer create new PE types by writing classes that inherit from the `ProcessingElement` superclass, whose methods are automatically invoked by the framework whenever new samples to process are available or in a time-driven fashion with customizable rate.

The S4 execution model follows a process-per-server approach: S4 instantiates a JVM container on each cluster server and PE instances are executed within these containers. The VM execution threads are not directly associated with PE instances, but with streams: within a VM container, the platform instantiates one thread for each stream feeding a hosted PE, and this thread executes all the methods of the PE instances served by that stream. Very peculiarly, every S4 stream can be optionally

---

[1] Note that, while in IBM InfoSphere Streams (Sect. 3.1) the concept of PE belongs to the execution model, in S4, it represents an abstract model concept.

keyed. In a *keyed* stream, every data sample has a unique key: the S4 runtime support dynamically creates a new PE instance for each different key in a stream, so that every instance processes all the stream samples for one key in a sort of functional *map* operation. This allows an easy parallelization of PEs and, consequently, an easy scale-up mechanism. S4 permits to choose inter-VM (and hence remote) communication transports among UDP- or TCP-based ones (by default UDP is employed). Similarly to IBM InfoSphere Streams, the only QoS policy supported by S4 is a weak form of fault-tolerance based on periodic or event-driven checkpointing of PE state.

3.3 Storm

Storm [44] is a DSPS developed by BackType and recently released under the Eclipse Public License by Twitter after its acquisition of BackType. As IBM InfoSphere Streams and Apache S4, Storm is based on the processing graph abstract model presented in Sect. 2.1. In the Storm model, data sources are called *spouts* and operators *bolts*; there is no explicit concept of sink, but destinations can be realized through *bolts* themselves since they can perform arbitrary actions on received samples (including saving them on files or forwarding them to external systems).

According to the Storm development model, the main method to define new spouts and bolts is through a Java API. As in Apache S4, in Storm, custom bolts and spouts are defined by writing classes that extend specific base classes, which, in turn, provide basic functionalities to newly built components. Graph instances are defined by creating instances of component classes and by defining their connection edges via specific API calls.

The Storm execution model is rather articulated. There are three parameters that influence how a particular graph is instantiated on the hosting platform, i.e, (i) the number of worker processes, (ii) the number of per-component tasks, and (iii) the number of per-component threads. The first parameter determines the total number of processes instantiated in the Storm cluster; the second, associated with every spout or bolt, determines the number of instances per component (also called *tasks* in Storm terminology) instantiated across all the cluster; the third determines the total number of threads dedicated to serve a component's set of tasks. At runtime, every worker is instantiated in a different Java VM, which can host one or more tasks (and execute one or more threads) from the same application. When multiple tasks for a single component are running, the routing of stream samples to different component instances is based on a further parameter, configurable at development time, which determines a grouping policy: for example, samples can be randomly shuffled among tasks for load balancing purposes, or can be routed using a modulo hashing of some sample fields.

Very peculiarly, Storm puts a strong focus on fault-tolerance by optionally providing at-least-once processing semantics: this means that it guarantees that every sample produced by any of the graph spouts is processed at least once. To do so, for each *root sample* (i.e., a sample generated by a spout), Storm keeps track of all the samples whose creation has been *caused by* its processing, and buffers it until all the tracked samples are acknowledged by their final destinations. Given the highly customizable nature of stream processing functionalities, Storm is not able to automatically keep track of *caused by* relationships between samples, but it requires explicit developer intervention for that: at code-level, in fact, Storm developers have to explicitly mark every new sample as caused by another sample if willing to avail of Storm fault-tolerance facilities.

4 QoS-Aware Stream Processing

In every kind of IT infrastructure serving mission-critical application scenarios, such as healthcare, finance, or transportation, it is very important that services behave in conformance to a well-defined *Service Level Agreement* (SLA) that determines the required QoS level. An SLA normally puts constraints on the functional behavior of the service (e.g., it should produce *all and correct* results in normal conditions) but also, and more importantly, constraints on how the service is expected to behave according to a set of performance indicators (non-functional requirements). The range of possible performance indicators is, in general, very large and application-dependent: two common and simple examples are *latency*—measuring the maximum time interval between a service request and the corresponding response—or *availability*—measuring the fraction of time the service generates correct results, even in spite of possible failures. Other indicators can refer to lower-abstraction details of the service, by measuring, for example, platform-specific parameters such as *memory* or *CPU* usage. Every constraint in an SLA that binds a specific performance indicator to some value is said to represent a *QoS policy* for the service.

In general, the implementation of *QoS-aware* services, i.e., services that are guaranteed to deterministically operate according to a set of associated QoS policies, is a very difficult task, and maps to the ability of the runtime platform to allocate (both statically and dynamically) the proper amount of computational resources where they are needed to satisfy the specified quality requirements. The technical challenge is even harder in the case of stream processing applications. In fact, differently from simple request-response or batch-oriented processing scenarios, where characteristics of computational tasks are known a-priori and thus easier to reason about, in stream processing, the properties of input streams (e.g., their data rate) change continuously and their behavior is not completely known in advance and difficult to predict. The consequent high variability in the load that applications have to sustain during long provisioning times, makes it very challenging to implement effective and adaptive resource scheduling techniques.

Nonetheless there is a growing number of real-world applications that has to deal with the analysis of large data streams and that requires, at the same time, predictable performance guarantees. This is often the case in Smart City scenarios, where a common goal is to use the results of stream analysis to trigger real-time feedback actions on real-world aspects of the city itself and of the urban life. These actions can be responses to emergency conditions, such as the activation of alarms in smart telecare systems [45], or the computation of emergency rescue plans in a smart traffic management system [22], which must be performed in a timely and reliable fashion. To better emphasize the importance of properly handling application-specific QoS requirements, let us briefly expand on this second scenario.

Consider a Traffic Management System (TMS) deployed in a Smart City. In this system every car periodically reports its position and speed to ad-hoc collection points using vehicle-to-vehicle and vehicle-to-infrastructure communications [34]. In their turn, each collection point relays these data to the data center-hosted stream processing application that processes these data in order to realize the TMS services. The TMS generally has the following three high-level functions:

- *Traffic flow control.* By analyzing short term and long term variations in car speeds along different roads, the system adapts the traffic lights timings to current road network conditions.
- *Management of road emergencies.* In case of car accidents, the vehicles involved and other cars passing immediately route messages about the event to on-road collection points, which, in their turn, relay them to the data center application. By analyzing these messages, the TMS detects the emergency condition, notifies the appropriate emergency service (e.g., ambulances), and tries to adapt the traffic flow to the new conditions, for example, by suggesting alternative navigation paths to other drivers (see next point).
- *Real-time navigation.* Cars traveling in the city can query the TMS for advanced navigation services. The system will answer with an always up-to-date route that takes into account road load conditions and possible emergency situations.

The three tasks of the TMS service, although based on the same input data streams, have very different quality requirements. For example, the traffic light timers must be promptly and quickly adapted to new road load conditions, meaning that the related processing actions should be performed with bounded *latency*. Similarly, processing of emergency notifications should be performed within deterministic time limits, in order to allow immediate rescue actions to take place. For the same reason, the management of all the emergency situations must take *priority* over other computations; this is especially useful during periods of high computational load (e.g., during traffic peaks) when the available DSPS resources might not be enough to satisfy all the processing flows. Accident notification messages should be transferred and processed *reliably* becuase the consequences of information loss can be very severe. On the other hand, the analysis of vehicles' position and speed to determine road load conditions can be performed *best-effort*: the related processing tasks can be executed with lower priority; in addition, data loss can be largely tolerated in this case given the implicit spatial and temporal information redundancy present in the corresponding data streams.

This simple but, we believe, very representative example shows how important can be for DSPSs to provide a rich and native support for *QoS-aware* stream processing. By using this type of support, developers of Smart City applications could focus their attention on application-level modeling and implementation problems, while delegating the realization of complex QoS enforcing mechanisms to the underlying DSPS. However, to the best of our knowledge, the most widely used and state-of-the-art DSPSs have only limited QoS-based configuration capabilities, which in most cases include only the selection of various reliability mechanisms (see Sect. 3).

On the contrary, we claim that QoS should be introduced as a first class concept in DSPSs at all the three abstract, development, and execution layers of our model. *QoS in the abstract model* should permit to specify, with different levels of granularity, the QoS policies required for graphs, single components, or groups of components directly in the application models. At this layer, different DSPSs should define their own quality-related vocabulary and determine which are the performance aspects controllable through their QoS policies, to which specific components they can apply, and how they interact with each other. *QoS in the development model* should define the programming constructs (either as extensions of ad-hoc stream processing languages or as specific APIs for general-purpose languages) that can be used to annotate application code with the quality requirements expressed at the model level. Finally, *QoS in the execution model* should support the execution of applications specified according to the other two layers. In this layer, each different DSPS should map different QoS policies to different mechanisms for runtime admission, monitoring, enforcement, and management, and should develop proper resource scheduling algorithms to satisfy the required QoS specifications.

In the following section, as a practical example of this kind of approach, we introduce Quasit, an original DSPS designed and implemented by following the above QoS-related guidelines.

5 Quasit

Quasit [10, 11] is a distributed stream processing system whose main design goal is to support QoS-aware stream analysis. To do so, it incorporates the concept of QoS at all the abstract, development, and execution model layers. Quasit is designed to run on large data centers made of commodity hardware, exploiting all the available processing power, and automatically handling various types of failures.

The Quasit abstract model is based on the processing graph concepts presented in Sect. 2.1. Originally, every element in Quasit processing graphs, called *streaming information graphs* (SIGs) in Quasit terminology, can be augmented with a QoS specification (a collections of QoS policies applied to that element); collectively, QoS specifications are used to dynamically adapt to variable load conditions and to the quality requirements of different parts of the stream processing flow. The Quasit development model is based on a simple Scala API, which lets developer (i) write, compose, and reuse custom operators, sources, and sinks, (ii) arrange components in SIGs to deploy on the infrastructure, and (iii) define the required QoS configurations for components, channels, or graphs as a whole. The API is designed to support a functional-like programming style that clearly separates operator behavior and state, thus making it easier for the runtime to support advanced QoS provisioning strategies. The Quasit runtime model maps the SIG components to runtime objects that run on all the available data center resources, and implements the set of QoS mechanisms that make it possible to execute application SIGs while enforcing their QoS requirements.

In the following subsections, we will concentrate on the three model-levels and overview the main ideas behind Quasit QoS-aware stream processing.

5.1 *Quasit Abstract Model*

The basic modeling unit in Quasit is the *Streaming Information Graph* (SIG), a weakly connected acyclic and directed graph representing the transformations that, applied to one or more input streams, produce an output data stream. Similarly to the model described in Sect. 2.1, three kinds of nodes can be used in a SIG graph i.e., *operators*, *data sources* and *data sinks*.

SIG nodes and edges can be labeled with QoS specifications, which define non-functional configuration parameters or constraints. Depending on the type of node or edge, a QoS specification can consist of several *QoS policies*, each policy influencing a different quality aspect. For example, through QoS specifications on SIG edges, it is possible to control the characteristics of the protocol used to exchange data among nodes they connect, or, through QoS Specifications on operator nodes, it is possible to configure their reliability guarantees.

The processing core of the Quasit abstract model is the *simple operator* component, whose structure is shown in Fig. 3.

A simple operator can be *stateless* or *stateful*. When stateful, the operator processing behavior is defined by the combination of the value of its *state* and its *processing function*; when stateless, by the processing function alone. The processing function is executed asynchronously whenever a sample from any of the operator input ports is available and its result may depend on the current value of the operator state. The role of the processing function is to describe how input streams are combined to produce an operator's output stream, and, if necessary, to update the operator internal processing state.

Quasit also allows to combine operators into more complex ones, by defining *composite operators*: existing operators (either simple or composite) can be arranged in a special SIG type, called *operator definition* SIG (OD-SIG), whose source and sink nodes are *virtual*, i.e., they do not correspond to real streaming data producers/consumers. Quasit defines a mapping between this kind of SIG and the associated composite operators: each data source in the OD-SIG defines a typed input port of the composite operator, and the data sink in the graph determines the type of the operator output port. Without digging into formal details, the behavior of a composite operator is defined by the internal structure of its defining OD-SIG: when a sample arrives to an input port of the composite operator it is processed as if it was processed by the OD-SIG operators graph. This composition mechanism provides

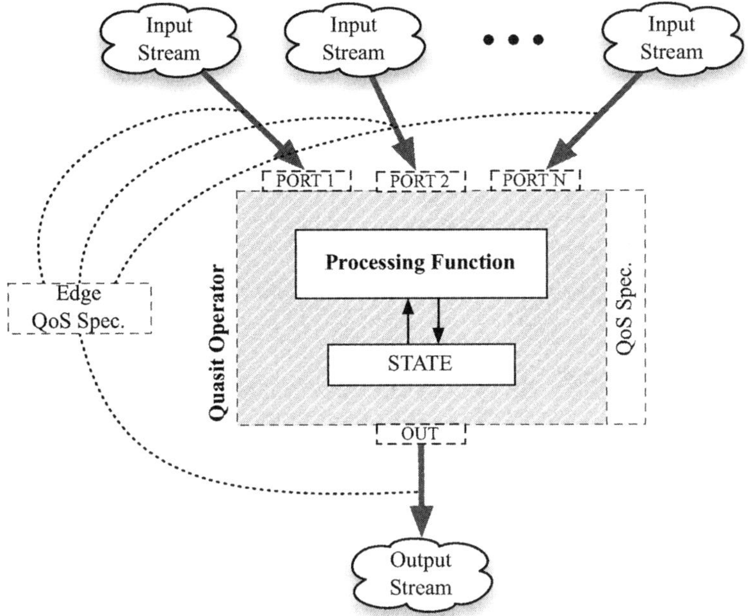

Fig. 3 Structure of a Quasit simple operator

an easy way to create new complex processing functionalities in terms of simpler ones, and promotes sharing and reusing existing and well-tested components.

5.2 *Quasit Development Model*

Quasit offers a very simple Scala API to let developers write their stream processing applications, create new data sources, data sinks and operators (either simple or composite), arrange components in SIGs, and associate QoS specifications to SIG elements.

In order to define new components, developers write *descriptor* classes that contain all the information that the framework needs to instantiate component instances at runtime. Depending on the type of component (i.e., source, sink, simple or composite operator), the descriptor class must extend an appropriate superclass which acts as a sort of "template" for the new descriptor. For example, operator descriptor classes must inherit either from `StatefulOperatorDescriptor[O,S]` or `StatelessOperatorDescriptor[O]`, depending on whether the operator needs to maintain some state between subsequent processing operations or not. Any component descriptor class must implement the appropriate life-cycle methods that are asynchronously invoked when relevant events occur. For example, an operator

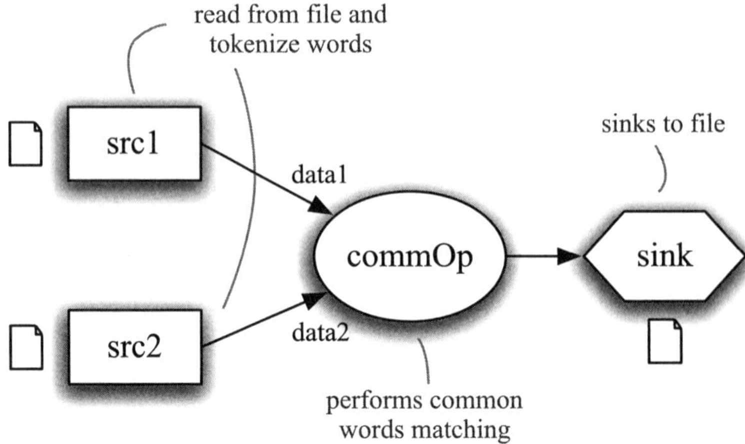

Fig. 4 A Quasit SIG implementing a simple, `comm`-like application

descriptor class must implement the *processingFunction* method, which realizes the main operator processing logic. This method, in fact, is called whenever samples are available at any operator input port. If the operator is stateless, the optional return value of the processing function is a list of samples to produce in the operator output stream; otherwise, it is a pair of objects, the first being a list of output samples, the second (optional) the new state the operator should transition to. A `SIGDescriptor` describes how components are arranged in the processing graph. It lists instances of component descriptors and the edges that connect them. While instantiating component descriptors, edges, or SIG descriptors, it is possible to associate to each of these elements specific `QosSpecification` objects, which will be used by the Quasit runtime to enforce the required quality levels.

Let us see, with a brief concrete example, how it is possible to create component descriptors and arrange them in SIGs. To keep the discussion self-contained we will consider a very simple scenario, where two sources continuously read lines each from a different file, tokenize them into words and send them to an operator (henceforth referred to as the comm operator) that determines and outputs words found in both the input files; this application can be thought as a sort of distributed and scalable implementation of the UNIX `comm` utility. A representation of the corresponding SIG is shown in Fig. 4.

Listing 1 shows how the descriptor class for the comm operator is defined. `CommOpDescritpor` inherits from `StatefulOperatorDescriptor [WordMsg,Map[String,Short]]`. The two type parameters passed to the base class (i.e., `WordMsg` and `Map[String,Short]` respectively represent the output type of comm operators and the type of their processing state. In fact, `WordMsg` is a simple container for words, while the state of a comm operator is a map that associates every word ever met with three possible values: one, if the word has been found on the first source only, two, if the word has been found on the second source only, and three if it has been found on both. Every instance of `CommOpDescriptor` has to specify two parameters, i.e., a symbolic name and an instance of `OperatorQosSpec` that will determine the set of QoS policies associated to the operator instance. Lines 3–7 determine the parameters passed to the `StatefulOperatorDescriptor` constructor; in particular, it is interesting to pay attention to the definition of the two operator input ports and their types (line 5) and to the definition of the operator initial state, i.e., an empty map (line 7). The comm operator processing function is defined from line 10 to line 15: by leveraging the expressiveness of Scala partial functions, it is possible to express the actions to perform in case a sample is received from the first data source ("`data1`" port) or the second one ("`data2`" port) very concisely. The private function `processWord` (line 17, determines the actual behavior of the operator, and the return values of its processing function. Note that, for example, once a word is found in both sources, a sample is produced on the output and the operator state updated accordingly (line 33).

```
1 class CommOpDescriptor( name: String, qos: OperatorQosSpec)
2  extends StatefulOperatorDescriptor[WordMsg,Map[String,Short]](
3   name, qos,
4   // Define the operator ports
5  Map( "data1" -> classOf[WordMsg], "data2" -> classOf[WordMsg]),
6    // The initial state of the operator is an empty map
7    Map[String,Short]()) {

8
9
10   override def processingFunction = {
11    case (msg: WordMsg, "data1", state: Map[_,_]) =>
12     processWord(msg.word, 1, state)
13    case (msg: WordMsg, "data2", state: Map[_,_]) =>
14     processWord(msg.word, 2, state)
15 }
16
17   private def processWord(word: String, srcMask: Short,
18    state: Map[String, Short]):
19     (Option[WordMsg], Option[Map[String,Short]]) = {
20
21    val wordState = state.get(word)
22   wordState match {
23    case None =>
24      // There is no entry in the map
25      val newState = state + (word -> srcMask)
26      // Produce no output but a new state
27      (None, Some(newState))
28     case Some(mask) =>
29 if ((mask & srcMask) != 0) { // already seen from this source
```

```
        (None, None)
      } else {
  val newState = state + (word -> 3.toShort) // seen from both
        (Some( new WordMsg(word)), Some(newState))
      }
    }
  }
}
```

Listing 1: Definition of a simple Quasit operator.

Listing 2, instead, shows how to instantiate operator descriptors and how to represent a SIG through a SIG descriptor. First, in lines 3 and 4, the two source descriptors are instantiated, pointing to the input files. After that, in lines 7–11, the comm operator descriptor instance is instantiated as well; notice that it is explicitly given the name "commOp" and that it is assigned a *queuing* QoS policy, which determines the type of queues used to buffer the operator input samples. After instantiating a file sink (line 14), the actual `SigDescriptor` instance is created in lines 18–29. The SIG descriptor instance is given a unique name, and the graph components are listed one by one through references to their descriptors. Graph edges connecting components are described through a sequence of triples (line 23), each defining an edge through its source node, its target node, and the QoS specification associated to the corresponding stream channel (default in this case). Finally, the SIG descriptor is associated with a SIG-wise QoS specification (lines 28–29): in this particular case, a *fault-tolerance* related policy, called *internal completeness*, is requested. As we will see in more detail in Sect. 6, this policy trades off perfect resiliency to failures and the ability to handle load spikes, while steel guaranteeing that a given fraction of samples are correctly processed (in the example, 70 % of the samples).

```
1 def  main(args: Array[String]): Unit = {
2  // Instantiate the data source descriptors
3  val src1 = new FileSourceDescriptor("src1", "/pathto/f1")
4  val src2 = new FileSourceDescritpr("src2", "/pathto/f2")
5
6   // Instantiate a CommOp descriptors
7   val comOpA = new CommOpDescriptor("commOp",
8      OperatorQosSpec().withPolicy(
9         QueingPolicy(QueueingPolicy.Unbounded,
10             QueueingPolicyKind.Fifo)),
11   )
12
13  // Instantiate the sink
14  val sink = new FileNativeSink[WordMsg]("sink",
     DataSinkQosSpec(),
15     "/pathto/comm.txt")
16
17   // Define the graph
18  val sig = SigDescriptor(
```

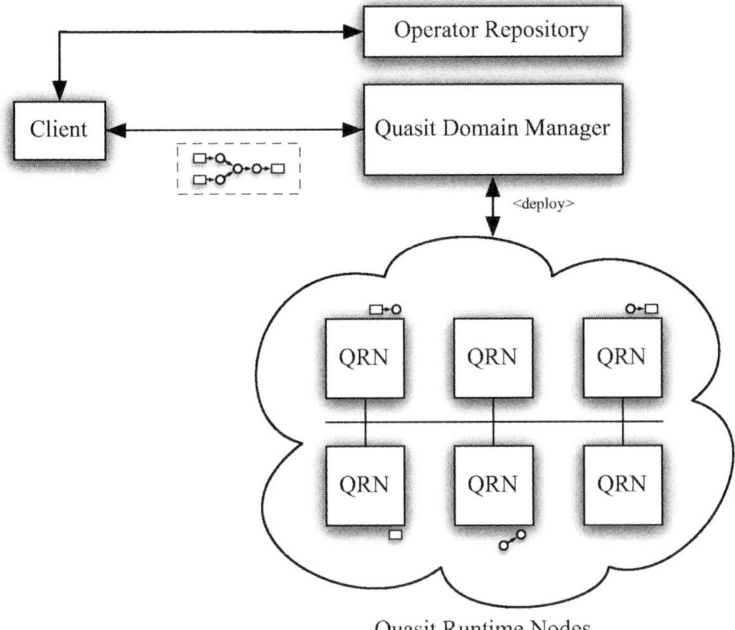

Fig. 5 Distributed architecture of a Quasit domain

```
     name = "CommSig",
     dataSources = Seq(src1, src2),
    dataSink = sink,
    operators = Seq(comOpA),
     edges = Seq(
       ("src1" -> "commOp", "data1", ChannelQosSpec()),
        ("src2" -> "commOp", "data2", ChannelQosSpec()),
       ("commOp" -> "sink", ChannelQosSpec()))
    )
  ).withSigQosSpec(SigQosSpec()
     .withPolicy(FTPolicy(FTPolicy.IC, 0.7)))


 ...
}
```

Listing 2: Creation of a SIG descriptor instance.

5.3 *Quasit Execution Model*

The Quasit abstract model, combined with the corresponding development model, offers a flexible and intuitive way to express stream analysis needs through the composition of small processing stages, and allows to customize these stages by means of several QoS policies. The Quasit execution model supports the execution of Quasit components at runtime by leveraging the computing power of a cluster of commodity computers within large-scale data centers.

A running Quasit deployment is called *domain*. A domain handles the distributed and QoS-aware execution of one or more user-defined SIGs. Similarly to other scalable data processing architectures (e.g., [20, 24, 31, 44]), the distributed architecture of a Quasit domain follows the *master-workers* pattern, with a central component with management and monitoring responsibilities and several distributed nodes performing the actual data processing operations. Figure 5 shows the three core distributed components running in one Quasit domain:

- Several *Quasit Runtime Nodes* (QRN), the *workers*;
- One *Quasit Domain Manager* (QDM), the *master* node;
- One optional *Quasit Operator Repository* (QOR).

The main QoS-aware execution services of our Quasit framework are provided by the co-operation of the QRN and QDM components. A typical Quasit deployment includes one QDM node and a cluster of QRN nodes, usually interconnected by a high-speed local area network (LAN). A QRN is in charge of providing the execution environment for Quasit *simple* operators and implement *threading*, *networking*, and local *QoS management* services. The QDM has management and control responsibilities over a Quasit domain. It does not take any direct role in stream processing tasks: for this reason, its centralized architecture does not represent a relevant bottleneck to the overall system scalability. Finally, the QOR is a repository of simple and composite operator types, and users can use its services to publish their operator definitions and to search for previously published ones.

In the current Quasit prototype implementation, every cluster server hosts one QRN, which is executed within a Java Virtual machine process (i.e. *process-per-server* model). Within this JVM, operator instances are modeled as distributed actors [3] managed by the Akka Actors framework [27]. All the actors running within the same JVM are managed by a pool of threads of configurable size, which executes operators processing functions at need, i.e., when there are samples to process at their input ports. This threading schema gives tremendous flexibility because it permits easily implementation of custom scheduling strategies, such as, for example, priority-based ones.

Operators deployed on different QRNs are connected via channels realized by leveraging the OMG DDS standard for high-performance PUB/SUB data exchange [36, 37]. Concretely, the PUB/SUB communication module maps the output port of every stream source (either operator or data source) to a unique destination topic and, symmetrically, every input port (of either an operator or a data sink) to a topic subscription. This solutions provides, at the same time, (i) strong decoupling between data producers and consumers, (ii) reduced space and time overhead thanks to the very efficient serialization mechanisms of the DDS middleware, and— most importantly— (iii) possibility to exploit the very fine-grained control of low-level QoS-related parameters that the DDS standard permits to associate with its topics, readers, and writers.

QoS policies defined at model-level on Quasit SIGs are enforced at runtime thanks to a two level QoS-management architecture, realized through the interaction of one *domain QoS manager*, running within the QDM, and several *node QoS managers*, one for each QRN. The domain QoS manager performs global admission control and QoS-based system configuration, while node QoS managers leverage the computational resources of the QRNs on which they execute to implement and enforce the requested QoS policies on locally running operators and I/O ports.

6 Load-Adaptive Active Replication (LAAR)

We have seen that the nature of stream applications poses several different and hard challenges to platform providers, including the ability to offer, at the same time, performance *elasticity* in spite of load variations, and *resiliency* to failures, while keeping *costs* limited. Handling load fluctuations due to sudden and possibly temporary variations in the data rates of input streams is a very complex task: in general, it maps to the ability to plan and allocate, statically and—more importantly—dynamically, the available computing resources to different parts of the hosted applications.

As stream processing applications usually run for (indefinitely) long time intervals, failures become very likely to occur. Many proposals in the literature have investigated possible *fault-tolerance* techniques—including active replication [14, 43], checkpointing [15, 32], replay logs [9, 28], or hybrid solutions [47] —each providing different trade-offs between *best-case* runtime cost and *recovery* cost. Whichever the adopted technique, maintaining some form of replication at some level (software/hardware components, state, or messages) imposes non-negligible overhead in terms of computing and communication resources.

In this section we present a possible solution to deal with temporary load variations in stream processing applications. This original approach trades off reliability guarantees and execution cost in actively replicated stream processing applications by temporarily claiming computational resources back from the fault-tolerance layer and by using them to handle possible load spikes. Our technique, called LAAR (Load-Adaptive Active Replication) [12, 13], dynamically deactivates and activates redundant replicas of processing components in order to claim/release resources and accommodate temporary load variations. At the same time, LAAR provides a-priori guarantees about the achievable levels of fault-tolerance, expressed in term of an *internal completeness* metric that captures the maximum amount of information that can be lost in case of failures.

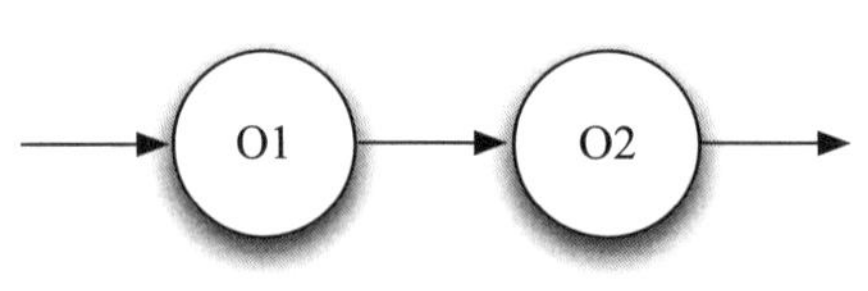

	O1	O2
Selectivity	1	1
CPU Cost	0.1 s/smpl	0.1 s/smpl

	Low Rate	High Rate
Source	4 smpls/s	8 smpls/s
PE 1	4 smpls/s	8 smpls/s
Prob.	0.8	0.2

Fig. 6 A simple processing scenario. On the *left*, the application graph. On the *right*, concise characteristics of the application and of its data source

LAAR builds on top and significantly extends existing static replication techniques that have been previously proposed for DSPSs [28]: for every component in the application processing graph, k replicas are deployed at runtime. At any moment in time, one of the k replicas of each component has the role of *primary*, the others are called *secondary*. Primary and secondary replicas all receive samples from the primaries of their predecessors, and all process them advancing through the same sequence of internal states. However, only the primary outputs samples to the replicas of its successors. LAAR continuously monitors the input rate of application sources. It automatically activates and deactivates replicas in order to satisfy two goals:

1. The application deployment is never *overloaded*;
2. The *internal completeness* constraint expressed in the SLA is satisfied.

For the sake of simplicity, an application deployment is said to be overloaded when, for any host, the total CPU cycles per second that would be needed to execute the components assigned to it is greater than the available CPU cycles per second. Note that, in an overloaded system, samples accumulate at operator or sink input queues (increasing latency) and are eventually dropped when the corresponding buffers fill.

Let us illustrate the basic intuition upon which our approach is based in a minimal application scenario. Consider the application in Fig. 6: it consists of two operators connected in a very simple pipeline; the first operator (O1) processes data from a single data source (not reported in the figure for the sake of simplicity) and forwards its output to O2, which, in turn, sends the results of its computations to an external data sink (also not depicted in the figure). The selectivity of both operators is 1, meaning that for every received input sample they produce one output sample; moreover, it takes 100 ms for both operators to process an incoming sample, considering the CPU architecture of the hosts where the application is going to be deployed. The single data source can produce samples at two different rates, "Low" and "High": the "Low" rate is 4 samples per second and is active on average for 80 % of the time (0.8 probability), while the "High" rate is 8 samples per second and is active in the remaining time intervals (0.2 probability).

The application is replicated and deployed on two servers, each hosting a copy of each operator, as shown in Fig. 7a. It is straightforward to see that, when the input configuration is "Low", 80 % of the CPU time available at both hosts will be

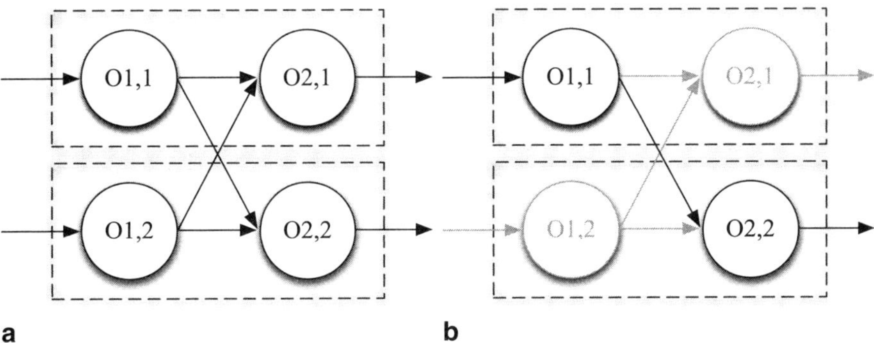

Fig. 7 **a** Replicated deployment of the application of Fig. 6 on two hosts. **b** Dynamic deactivation of replicas by LAAR during a "High" input configuration

occupied for processing samples. More importantly, when the input configuration is "High", the application would need 160 % of the total CPU time available, which—of course—is available only by adding extra resources to the deployment environment (with an increased cost).

The basic idea behind LAAR is to monitor the data sources and, according to the current data rates, dynamically deactivate replicas in order to release the resources necessary to face load variations. Figure 7b, for example, shows how LAAR could deactivate replicas of O1 and O2 during a load peak, so that the total CPU available will become enough to handle the new load.

Figure 8a and b show this behavior in a real stream processing deployment. We implemented and executed the replicated pipeline stream processing application shown in Fig. 7a on a deployment environment consisting of two servers equipped with a single core CPU. Figure 8a shows the execution of the application using static active replication: when the input passes to the "High" configuration (around 50 s from the beginning of the experiment), the CPU of the two hosts saturates, and the output is not able to keep up with the input rate; on the contrary, Figure 8b shows how, by temporarily deactivating replicas during the "High" input configuration, it is possible to save enough resources to allow the output stream to follow the input.

Obviously, if a failure of one of the active operators occurred during a "High" period, part of the input would not be processed as expected. However, the unique and strong aspect of LAAR is its ability to quantify *a-priori* these effects on the overall application reliability. As anticipated, LAAR defines the concept of *internal completeness*, a metric that tries to capture the amount of samples that are guaranteed to be processed in a *pessimistic* failure scenario, i.e., a scenario where all the active operator replicas fail. Without digging into formal details, the Internal Completeness metric (IC) is defined as follows:

$$IC(s) = \frac{\text{no. of samples processed in a pessimistic failure scenario P}}{\text{no. of samples processed with no failures}} \qquad (1)$$

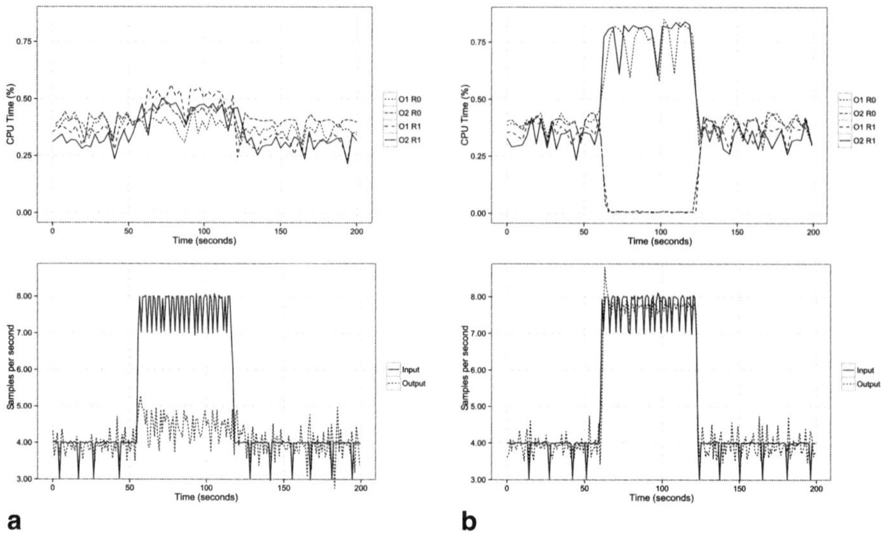

Fig. 8 **a** CPU time used by the two couples of replicated operators—top—and corresponding input and output rate—bottom. **b** CPU time and input/output data rate when O1 replica 1 and O2 replica 0 are deactivated by LAAR

where s represents a particular *replica activation strategy*, which associates the activation/deactivation status of application operators to each possible input rates configuration. For instance, in the example scenario presented above, during a period of T seconds and in absence of any failure, the application would process a total of $T(0.8 \cdot 8 + 0.2 \cdot 16)$ samples (considering both operators). On the contrary, considering a very pessimistic failure scenario P, where the active replica of each operator (respectively $O1, 1$ and $O2, 2$) is crashed all the time, the total number of samples processed would be $T(0.8 \cdot 8 + 0)$, for a total IC value of $\frac{6.4}{9.6} = 0.\overline{6}$. This means that, even in case of failures, at least 60 % of the total processing operations would be correctly performed.

However, finding a replica activation strategy that, at the same time, is able to keep the system in a non-overloaded condition despite load variations and to satisfy a user-defined IC requirement while keeping costs limited, is a very hard problem, especially when the processing graphs are much more complex than the one presented before. In order to solve this problem, LAAR performs a static optimization phase where the problem is modeled as a Mixed Integer Programming (MIP) instance. Although a precise formal description of the problem model is out of the scope of this chapter, in the following we sketch its formulation, in order to help the readers understand the main ideas behind the approach.

$$\underset{s}{\text{minimize}} \quad cost\,(s) \tag{2}$$

subject to:

$$IC(s) \geq \textit{SLA Constr.} \tag{3}$$

$$load(h, s, c) < Thres. \qquad \forall \text{ server } h \text{ and input conf. } c \tag{4}$$

$$nreplicas(o, s, c) \leq 1 \qquad \forall \text{ operator } o \text{ and input conf. } c \tag{5}$$

In the equations above, the *cost* function in the minimization term represents the cost, in terms of resources (e.g., CPU, memory, and bandwidth), for a service provider to run the application during a billing period of length T using replica activation strategy s. Equation 3 constraints IC to satisfy the value required in the application SLA, while Eq. 4 states that each host in the deployment should never be overloaded. *Thres.* is a constant expressing the number of CPU cycles per second available at the deployment servers. The last constraint, expressed in Eq. 5, requires that there is at least one active replica of every operator in every input configuration, to ensure that the measured IC value is always one in absence of failures.

To have a rough idea of the complexity of the above problem, consider that the solution space has a size that is exponential in the number of operators, number of replicas per operator, and number of possible input configurations. In addition, the computation of IC values, resource usage, and server load levels require exponential time with respect to the number of operators, since they depend on the number of samples processed by different operators in different configurations, which in turn recursively depend on the number of samples processed by their predecessors. To deal with this complexity, LAAR solves the problem using an original constraint programming based algorithm, called *FT-Search*, that is able to find optimal or sub-optimal solutions to problem instances of reasonable size (i.e., graphs with tens of operators) in limited time, largely compatible with practical industrial data center constraints and application-specific requirements.

After having found solutions to the above optimization problem before application deployment time, LAAR performs its dynamic replica activation operations at runtime by inserting a special operator in the application graph, which continuously monitors the input rates and, according to the measured values, sends ad-hoc activation/deactivation commands to operator replicas.

If compared to active replication techniques, LAAR is able to handle load spikes by completely avoiding increased latency or sample drops due to full operator buffers. Moreover, by using weaker fault tolerance specifications through the IC metric, it can also reduce the cost of running stream processing applications proportionally to the required fault-tolerance guarantees. A large corpus of experiments, performed on a LAAR implementation built on top of IBM InfoSphere Streams and executed on a 60 cores IBM BladeCenter Cluster deployment, confirms the above claims. In particular, we have executed 100 different artificially generated stream processing applications using four different fault tolerance techniques. A *No Replication* (NR) approach runs the streaming applications without instantiating any operator replica. A *Static Replication* (SR) approach creates two replicas for each operator and keeps them

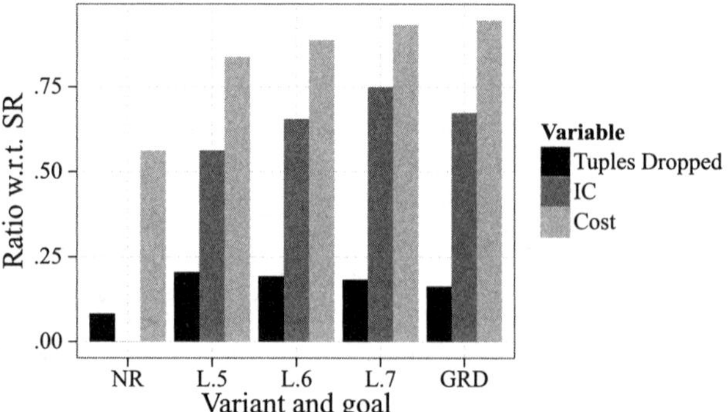

Fig. 9 Comparison of the different replication strategies; average values normalized w.r.t. SR

active all the time, independently on the input configuration. The *LAAR* replication approach uses the previously described techniques to run the streaming applications with three different IC requirements, 0.5, 0.6, and 0.7 (labelled L.5, L.6 and L.7, respectively). Finally, a *greedy* (GRD) approach uses techniques similar to those adopted by LAAR, but instead of deactivating replicas according to the results of a static optimization phase, it uses a simple runtime heuristic (i.e., it deactivates the most resource greedy component first).

Figure 9 shows a concise summary of the results collected in these experiments, by showing the average numbers of samples dropped, the average IC value achieved, and the average cost of different replication strategies as a fraction of the same values measured using the SR approach. It is immediate to see that, by using LAAR, it is possible to control the desired IC guarantees to directly influence the associated deployment cost, which is considered highly valuable and relevant in many business application scenarios.

7 Conclusions

The interest around the Smart City paradigm has been growing at an increasing pace in the last years and it is very likely that, thanks to the technical advances in computing devices and wireless, mobile, and wearable sensing, it will continue to grow in the next years. The efficient and effective exploitation of the unprecedented amounts of real-world data generated and injected every day inside IT infrastructures is a crucial step toward a real improvement in people's quality of life through smart computing technologies. In this context, DSPSs are a key technology for their ability to analyze information "on-the-fly" and produce continuous feedback that can be exploited to adapt real-world processes to their dynamically varying conditions. The tremendous heterogeneity of data and applications, together with their often unpredictable dynamic requirements, pose additional challenges for these systems. Developers of stream processing applications should be allowed to express, in a flexible way, the QoS requirements of their application scenarios, and DSPSs should understand these requirements and automatically adapt their internal mechanisms to meet them in a way that is as much transparent as possible to the streaming applications and their logic. However, only a few modern DSPSs expose QoS-based customization features, and, in most cases, their are not first-class elements in all the three abstract, development, and runtime models, oppositely from the role we believe they should have. At the best of our knowledge, Quasit represents the most prominent exception, by allowing to express and enforce a large variety of QoS policies at each of the three levels.

We claim that future research on DSPSs should focus on QoS-related open issues with much stronger attention. In particular, it should: (i) improve the existing stream processing models to give application developers the opportunity to integrate rich QoS requirements in their applications; (ii) study efficient mechanisms to implement QoS policies on large scale deployments of DSPSs inside data centers. About this last point, a particularly promising research direction is the development of a novel class of *weak or probabilistic* QoS requirements that, in contrast with more traditional *strong* requirements, give runtime platforms additional degrees of freedom in their enforcement and more possibilities to adapt to highly variable system workloads. We believe (and the first preliminary results already collected are confirming our claim) that the *internal completeness* reliability metric adopted in the LAAR replication technique well represents this new class of QoS requirements for stream processing applications.

Acknowledgements We would like to thank the IBM Research Dublin Lab, and in particular Spyros Kotoulas, for his valuable work and feedback on LAAR (overviewed in Sect. 6), designed and implemented within a joint research collaboration.

References

1. Abadi, D.J., Carney, D., Çetintemel, U., Cherniack, M., Convey, C., Lee, S., Stonebraker, M., Tatbul, N., Zdonik, S.: Aurora: a new model and architecture for data stream management. The VLDB Journal, 12, 2, pp. 120–139 (2003).
2. Abadi, D.J., Ahmad, Y., Balazinska, M., Çetintemel, U., Cherniack, M., Hwang, J.-H., Lindner, W., Maskey, A.S., Rasin, A., Ryvkina, E., Tatbul, N., Xing, Y., Zdonik, S.: The Design of the Borealis Stream Processing Engine.Proceedings of the 2nd Biennial Conference on Innovative Data Systems Research (CIDR). IEEE, Asilomar, CA (2005).
3. Agha, G. A.: Actors: a model of concurrent computation in distributed systems, Ph.D. dissertation, Artificial Intelligence Laboratory, Cambridge MA, USA (1985).
4. Amini, L., Andrade, H., Bhagwan, R., Frank Eskesen and Richard King and Yoonho Park and Chitra Venkatramani: SPC: A distributed, scalable platform for data mining. Proceedings of the Workshop on Data Mining Standards, Services and Platforms (DM-SS 2006). pp. 27–37. ACM, Philadelphia, PA (2006).
5. Apache S4 Project Web Site. Available, http://incubator.apache.org/s4. Last visited in September 2013.
6. Arasu, A., Babcock, B., Babu, S., Cieslewicz, J., Ito, K., Motwani, R., Srivastava, U., and Widom, J.: STREAM : The Stanford Data Stream Management System, Technical report, Stanford InfoLab (2004).

7. Arasu, A., Babu, S., Widom, J.: The CQL continuous query language: semantic foundations and query execution. The VLDB Journal. 15, 2, pp. 121–142 (2005).
8. Avnur, R., Hellerstein, and J.M.: Eddies: continuosly adaptive query processing. Proceedings of the ACM SIGMOD international conference on Management of data (SIDMOD 2000). pp. 261–272. ACM, Dallas, TX, USA (2000).
9. Balazinska, M., Balakrishnan, H., Madden, S.R., Stonebraker, M.: Fault-tolerance in the borealis distributed stream processing system. ACM Trans. Database Syst. 33, 1, Article 3, 44 pages (2008).
10. Bellavista, P., Corradi, A., Reale, A.: The QUASIT Model and Framework for Scalable Data Stream Processing with Quality of Service. Proceedings of the 5th International Conference on Mobile Wireless Middleware, Operating Systems, and Applications (MOBILWARE 2012). Springer Berlin-Heidelberg, Berlin, Germany (2012).
11. Bellavista. P., Corradi, A., Reale, A.: Design and Implementation of a Scalable and QoS-aware Stream Processing Framework: the Quasit Prototype. Proceedings of the IEEE International Conference on Cyber, Physical and Social Computing (CPSCOM 2012). IEEE, Besançon, France (2012).
12. Bellavista. P., Corradi, A., Kotoulas, S., Reale, A.: Dynamic datacenter resource provisioning for high-performacne distributed stream processing with adaptive fault-tolerance. Proceedings of the 14th ACM/IFIP/USENIX International Middleware Conference—Demo & Poster Track, ACM, Beijing, China (2013).
13. Bellavista. P., Corradi, A., Kotoulas, S., Reale, A.: Adaptive fault-tolerance for dynamic resouce provisioning in distributed stream processing systems. Proceedings of the 17th International Conference on Extending Database Technology (EDBT 2014), ACM, Athens, Greece (2014). To appear.
14. Brito, A., Fetzerm C., Felber, P.: Multithreading-enabled active replication for event stream processing operators. In: 28th Symposium on Reliable Distributed Systems, pp. 22–31, IEEE, Niagara Falls, NY, USA (2009).
15. Cai, Z., Kumar, V., Cooper, B.F., Eisenhauer, G., Schwan, K., Strom, R.E.: Utility-driven proactive management of availability in enterprise-scale information flows. In: ACM/IFIP/USENIX 7h International Middleware Conference, Springer, Melbourne, Australia (2006).
16. Carney, D., Çetintemel, U., Cherniack, M., Convey, C., Lee, S., Seidman, G., Stonebraker, M., Tatbul, N., Zdonik, S.: Monitoring streams: a new class of data management applications. Proceedings of the 28th international conference on Very Large Data Bases (VLDB 2002). The VLDB Endowment, Hong Kong, PRC (2002).
17. Chandrasekaran, S., Shah, M.A., Cooper, O., Deshpande, A., Franklin, M.J., Hellerstein, J.M., Hong, W., Krishnamurthy, S., Madden, S.R., Reiss, F.: TelegraphCQ. Proceedings of the ACM SIGMOD international conference on on Management of data (SIGMOD 2003). pp. 668. ACM, San Diego, CA, USA (2003).
18. Condie, T., Conway, N., Alvaro, P., Hellerstein, J.M., Elmeleegy, K., Sears, R.: MapReduce Online. Proceedings of the 7th USENIX conference on Networked systems design and implementation (NSDI 2010). USENIX Association, San Jose, CA, USA (2010).
19. Cugola, G., Margara, A.: Processing flows of information: From Data Stream to Complex Event Processing. ACM Comput. Surv.. 44, 3, pp. 1–62 (2012).
20. Dean, J., Ghemawat, S.: MapReduce : Simplified Data Processing on Large Clusters. Commun. ACM, vol. 51, no. 1, pp. 107–113 (2008).
21. Digital Cities Project Web Site. Available, http://www.digital-cities.eu. Last visited in September 2013.
22. Djahel, S., Salehie, M., Tal, I., Jamshidi, P.: Adaptive Traffic Management for Secure and Efficient Emergency Services in Smart Cities. Proceedings of the IEEE International Conference on Pervasive Computing and Communicatino (PerCom 2013)—WiP Session. pp. 340–343, IEEE, San Diego, CA, USA (2013).
23. EUROCITIES Web Site. Available, http://www.eurocities.eu/. Last visited in September 2013.

24. Gedik, B., Andrade, H.: A model-based framework for building extensible, high performance stream processing middleware and programming language for IBM InfoSphere Streams. Softw. Pract. Exper. 42, 11, 1363–1391 (2012).
25. Gedik, B., Andrade, H., Wu, K.-L.: A code generation approach to optimizing high-performance distributed data stream processing. Proceeding of the 18th ACM conference on Information and knowledge management (CIKM 2009). p. 847, ACM, Hong Kong, PRC (2009).
26. Giffinger, R., Fertner, C., Kramar, H., Kalasek, R., Pichler-Milanovic, P., Meijers, M.: Smart cities – Ranking of European medium-sized cities. Vienna UT, Centre of Regional Science (2007) Available, http://www.smart-cities.eu/download/smart_cities_final_report.pdf. Last visited in September 2013.
27. Haller, P. and Odersky, M.: Scala Actors: Unifying thread-based and event-based programming. Theoretical Computer Science, vol. 410, no. 2–3, pp. 202–220 (2009).
28. Hwang, J.-H., Balazinska, M., Rasin, A., Çetintemel, U., Stonebraker, M., Zdonik, S.: High-availability algorithms for distributed stream processing. In: 21st International Conference on Data Engineering, pp. 779–790, IEEE, Tokyo, Japan (2005).
29. IBM Smarter Cities Project Web Site. Available, http://www.ibm.com/smarterplanet/us/en/smarter_cities/. Last visited in September 2013.
30. Intel Collaborative Research Institute for Sustainable Connected Cities. Available, http://www.intel-university-collaboration.net/?page_id=1420. Last visited in September 2013.
31. Isard, M., Budiu, M., Yu, Y., Birrell, A., and Fetterly, D.: Dryad: distributed data-parallel programs from sequential building blocks. In: 2nd ACM SIGOPS/EuroSys European Conference on Computer Systems, vol. 41, no. 3, p. 59–72, ACM New York, NY, USA (2007).
32. Jacques-Silva, G., Gedik, B., Andreade, H., Wu, K.-L.: Language level checkpointing support for stream processing applications. In: 2009 International Conference on Dependable Systems & Networks, pp. 145–154, IEEE, Estoril, Portugal (2009)
33. Khandekar, R., Hildrum, K., Parekh, S., Rajan, D., Wolf, J.: COLA : Optimizing Stream Processing Applications Via Graph Partitioning. Proceedings of the ACM/IFIP/USENIX 10th International Middeware Conference. pp. 308–327, Springer Berlin Heidelberg, Urbana Champagin, IL, USA (2009).
34. Martinez, F., Toh, C.-K., Cano, J., Calafate, C., Manzoni, P.: Emergency Services in Future Intelligent Transportation Systems Based on Vehicular Communication Networks. IEEE Intelligent Transportation Systems Magazine 2,2, pp. 6–20 (2010).
35. Neumeyer, L., Robbins, B., Nair, A., Kesari, A.: S4: Distributed Stream Computing Platform. In: 2010 IEEE International Conference on Data Mining Workshops (ICDMW '10), pp. 170–177, IEEE Los Alamitos, USA (2010).
36. OMG: Data Distribution Service for Real-time Systems – Version 1.2, Specification. Object Management Group (2007).
37. Pardo-Castellote, G.: OMG data-distribution service: Architectural overview. Proceedings of the 23rd International Conference on Distributed Computing Systems Workshops, pp. 200–206. IEEE, Providence, RI, USA (2003).
38. Wolf, J., Bansal, N., Hildrum, K., Parekh, S., Rajan, D.: SODA : An Optimizing Scheduler for Large-Scale Stream-Based Distributed Computer Systems. Proceedings of the ACM/IFIP/USENIX 9th International Middleware Conference. pp. 306–325. Springer Berlin Heidelberg, Leuven, Belgium (2008).
39. Xing, Y., Zdonik, S., Hwang, J.-H.: Dynamic Load Distribution in the Borealis Stream Processor. Proceedings of the 21st International Conference on Data Engineering (ICDE 2005). pp. 791–802. IEEE, Tokyo, Japan (2005).
40. Xing, Y., Hwang, J.-H., Zdonik, S.: Providing Resiliency to Load Variations in Distributed Stream Processing. Proceedings of the 32nd international conference on Very large data bases (VLDB 2006). pp. 775–786. The VLDB Endowment, Seoul, Korea (2006).
41. Safe City Project Web Site. Available, http://www.safecity-project.eu/. Last visited in September 2013.

42. Smart Cities Stakeholder Platform. Available, http://www.eu-smartcities.eu/. Last visited in September 2013.
43. Shah, M., Hellerstein, J., Brewer, E.: Highly available, fault-tolerant parallel dataflows. In: ACM International Conference on Management of Data, pp. 827–838, ACM, Paris, France (2004).
44. The Storm Project Web Site. Available, http://storm-project.net/. Last visited in September 2013.
45. Tang, P., Venables, T.: "Smart" homes and telecare for independent living. J. Telemed. Telecare. 6, 1, pp. 8–14 (2000).
46. Yang, H.-c., Dasdan, A., Hsiao, R., Parker, D.: Map-reduce-merge: simplified relational data processing on large clusters. In: Proceedings of the ACM SIGMOD international conference on Management of data (SIGMOD 2007). pp. 1029–1040, Beijing, PRC (2007).
47. Zhang, Z., Gu, Y., Ye, F., Yang, H., Kim, M., Lei, H., Liu, Z.: A hybrid approach to high availability in stream processing systems. In: 30th IEEE International Conference on Distributed Computing Systems, pp. 138–148, Genoa, Italy (2010).

Opportunistic Databank: A context Aware on-the-fly Data Center for Mobile Networks

Osman Khalid, Samee U. Khan, Sajjad A. Madani, Khizar Hayat, Lizhe Wang, Dan Chen and Rajiv Ranjan

1 Introduction

In recent years, significant advancement in the wireless communication technologies, such as Bluetooth, 802.11/WiFi, and ZigBee, has been seen in mobile ad hoc networks (MANETs). Such technologies enable mobile devices to form *on-the-fly data centers* where nodes opportunistically participate in data storage and sharing applications [3, 8, 11]. In such a setup, the basic assumption is that there must exist an

O. Khalid (✉) · Samee U. Khan
Department of Computer Sciences,
COMSATS Institute of Information Technology, Abbottabad, Pakistan, University Road, COMSATS, Abbottabad
e-mail: osman@ciit.net.pk

Samee U. Khan
e-mail: samee.khan@ndsu.edu

Sajjad A. Madani
COMSATS Institute of Information Technology, Islamabad, Pakistan
e-mail: madani@ciit.net.pk

K. Hayat
Computer Sciences Section, University of Nizwa, Birkat Al Mawz, Oman
e-mail: khizar.hayat@unizwa.edu.om

L. Wang
Center for Earth Observation and Digital Earth, Chinese Academy of Sciences, Beijing, China
e-mail: lzwang@ceode.ac.cn

D. Chen
School of Computer Science, China University of Geosciences, Wuhan, China
e-mail: dan.chen@ieee.org

R. Ranjan
Computer and Information Technology Building, Australian National University, Canberra, Australia
e-mail: raj.ranjan@csiro.au

S. U. Khan, A. Y. Zomaya (eds.), *Handbook on Data Centers*,
DOI 10.1007/978-1-4939-2092-1_36

end-to-end communication path between a source and a destination node [8, 11, 19]. Every mobile host acts as a router and communicates with other mobile hosts. Even if source and destination mobile hosts are not in each other's communication range, data is still forwarded to the destination mobile host by relaying transmission through other mobile hosts that exist between the source and the destination nodes. The scenarios when there are frequent disruptions and delays in message transfer due to network partitioning, higher degree of variation in network topology, and sparsity of nodes, such network environments are known as Delay Tolerant Networks (DTNs) [2, 12]. The DTNs lack end-to-end communication paths between source and destination nodes. Numerous DTN scenarios that correspond to opportunistic data storage/sharing applications include [3, 34]: **(a)** disaster/emergency response systems, **(b)** battlefield networks, **(c)** sensor networks, **(d)** road traffic information dissemination systems, **(e)** content dissemination systems, and **(f)** cellular traffic data offloading [27]. In the aforementioned scenarios, the cellular 3G infrastructure may usually be unavailable, or if available, provide too limited bandwidth to transmit data traffic. Instead, mobile users rely on their opportunistic contacts for storing, sharing, and accessing data. For instance, nodes may participate in forming an on-the-fly network to store information extracted from sensors, which may be the information regarding live weather data, road traffic condition, or about any upcoming disaster.

Wireless radio range variations, limited energy resources, sparsity of mobile nodes, continuous mobility, and noise, to name a few, are the reasons due to which DTNs suffer from frequent disconnections. This phenomenon is undesirable when mobile hosts are accessing data from each other. As it is not possible to control randomly occurring network disconnections, an alternative solution to this problem is to replicate multiple copies of data onto various mobile hosts so that when disconnections/disruptions occur, mobile hosts can still access data [13, 15]. Replication process distributes additional copies of primary data items into the network in order to increase accessibility and decrease communication costs. In the past few years, data replication has been studied extensively for both the MANETs and DTNs environments [2, 8, 11, 12, 13, 23, 28, 30, 33]. Figures 1a and 1b illustrate the replication scenario in MANETs. If the central link between nodes C and E fails, then the set of mobile hosts E, F, and D will not be able to access the data item M_1. Similarly, the data item M_2 will be inaccessible by the nodes A, B, and C. To cope with the problem of data inaccessibility due to network division, one possible solution is to create replicas of original messages M_1 and M_2, and place these replicas at the opposite sides of the ad hoc network. In this way, every mobile host can access both data items even after the network division, as indicated in Fig. 1b. Due to the existence of end-to-end communication paths, the aforementioned mechanism of proactively placing replicas on nodes before the link failures is possible only in the case of MANETs. However, replica placement exhibits more complexity when the network is sparse with no end-to-end communication paths among nodes, as in the case of DTNs.

Figure 2 illustrates an example scenario of replica allocation in DTNs. For the sake of simplicity, we consider allocation of a single message M_1 from node A to node D. At a reference time T_1, node A is not in the communication range of node D. On making contact with nodes B and C, suppose the node A is not sure which of the two nodes B and C will make contact with node D in future. Therefore, node A

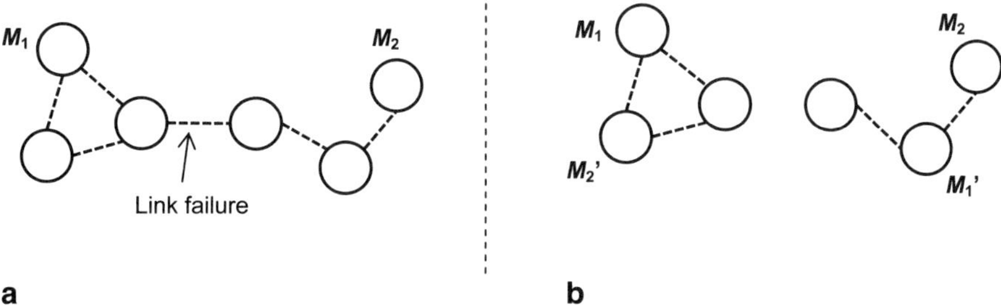

Fig. 1 Replication example in MANETs: **a** network division and data access, and **b** effective data replication for continued data access

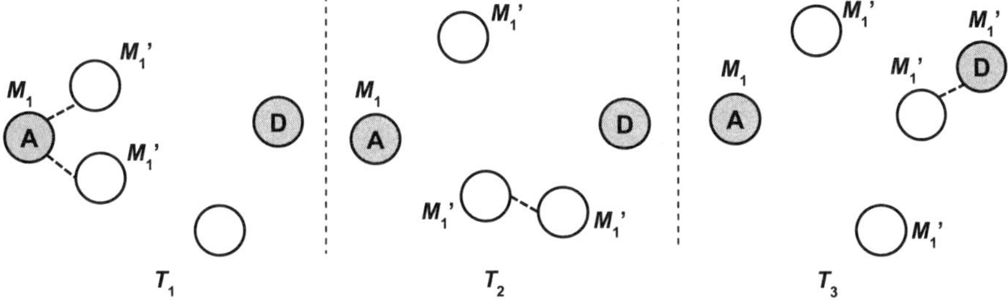

Fig. 2 Replication example in DTNs. Data is replicated from node to node during opportunistic contacts at time slots T_1, T_2, and T_3, without any global knowledge of network topology

places a replica of M_1 on both the mobile nodes B and C. During time T_2, only the node C makes contact with another node, and replicates M_1 on node F. Finally, node F transfers replica on node D on making contact at time slot T_3.

It is evident from the given example that replica placement in DTNs is dependent on the occurrence of opportunistic contacts among mobile nodes. Therefore, the main difference in data replication between MANETs and DTNs is the absence of any centralized mechanism and/or global knowledge of network for DTNs. Moreover, in any of MANET/DTN network, the decision of where to place replica must trade off the cost of accessing data that is reduced by additional copies with the cost of storing and updating the replicas [8, 13]. These costs have severe implications in ad hoc network environments because mobile hosts have limited resources (energy, storage, and processing power). Therefore, efficient and effective replication schemes strongly depend on how many replicas to be placed in the system, and more importantly where [13].

In this chapter, our primary focus is to address the data replication in DTNs. Despite different natures of DTNs and MANETs, the data replication strategies in both types of networks share some commonalities. Therefore, we also perform a selective study of the data replication schemes in MANETs. The rest of the chapter is organized as follows. We give a brief overview of some of the well known strategies for replica placement in MANETs in Sect. 2. In Sect. 3, we address the data replication problem in DTNs, and present a cost efficient solution for DTN replica placement. Sect. 4 concludes the chapter.

2 Data Replication in Manets—A Brief Overview

In MANETs, it is very important issue to prevent deterioration of data accessibility at the point of network division. In general, mobile hosts would experience reduced access latencies provided that data is replicated within their close proximity. However, such replication is practically useful in cases when data updates are not the main focus. If updates of the contents are also considered, then the locations of the replicas have to be: **(a)** in close proximity to the mobile hosts, and **(b)** in close proximity to the primary copy (assuming a "master" replication environment [13, 25]).

Some initial work on ad hoc network data replication was performed by Hara [5]. The author discussed various mechanisms of replica placement during a *relocation period*. The relocation period was a time interval during which each mobile node broadcasts its host identifier as well as information about access frequencies to its data items [5]. Based on the broadcasted information, the author [5] proposed three schemes for replica placement: **(a)** Static Access Frequency (SAF), **(b)** Dynamic Access Frequency and Neighborhood (DAFN), and **(c)** Dynamic Connectivity based Grouping (DCG). In the SAF method, only the access frequency to each data item is taken into account. During relocation period, access frequencies to all data items in the network are computed, and each mobile host allocates data items in descending order of access frequencies. The drawback of such approach is that same data item might be allocated on two neighbor nodes, which may also reduce data diversity and accessibility in the network. Alternatively, the DAFN method not only takes into account the data item access frequency, but also considers the replica status on neighbor nodes. The first phase of DAFN is similar to SAF. In the second phase, the system sequentially starts with the node with smallest identifier, and replaces data item of a neighbor node, which is not the primary owner of the data. The data item is replaced with the one having next higher global access frequency in the network. If both the nodes are not primary owner of the data item, then the host whose access frequency to the data item is lower than the other host, changes the replica with another next highly accessed replica. Compared to the SAF method, the DAFN approach enhances data diversity in the network. However, the DAFN method does not completely eliminate replica duplication among neighboring hosts. This is because, the DAFN method only executes the replica elimination process by scanning the network once based on the breadth first search. In contrast to SAF and DAFN, the DCG method creates groups of nodes in the network in the form of bi-connected graph components. The grouping is performed based on nodes' degree of network connectivity. The access frequencies to data items are calculated at group level by summing up access frequencies of individual nodes in the group. Beginning with the smallest host identifier within a group, replicas are allocated to nodes in the descending order of access frequencies. During replica allocation, it is ensured that not two neighbors are allocated same replica, similar to what implemented in SAF and DAFN. Because DCG method involves heavy computations and control signaling, this method takes the largest time among the three methods to relocate replicas. Large processing time leads to even more problems if network topology changes during replica relocation period. Moreover, the above methods do not incorporate mechanisms for replica updates. The author [5] broadened his work in [6] and [7] by incorporating various network connectivity related issues. Although the above-mentioned works are plausible in the sense that they advance the study of replica allocation in MANETs, none involves reasoning via a concrete mathematical model.

Khan et al. in their celebrated work have rigorously addressed replica allocation problem in distributed systems [13–26, 31, 32]. Specifically, in [13, 25], Khan *et al* pioneered in applying game theory to ad hoc network replica allocation problem (ADRP). In [13], the authors proposed a novel scheme that seeks to strategically balance energy, bandwidth, and storage space through a cooperative game-theory approach for replication in a mobile environment. In the presented work [13], the authors: **(a)** derived a mathematical problem formulation for ADRP, **(b)** proposed an optimization technique that allocates replicas so as to minimize the network traffic under storage constraints with "read from the nearest" and "push based update through the primary mobile server" policies, and **(c)** used a strict consistency model as opposed to an opportunistic consistency model. The authors addressed selfish behavior of mobile servers in the proposed solution. In ad hoc networks, resources may belong to different self-interested servers. These servers may manipulate the resource (replica) allocation mechanism for their own benefit by misrepresenting their preferences, which may result in severe performance degradation. The proposed technique involved players (mobile hosts) that compete through bids in a non-cooperative environment to replicate data objects that are beneficial to themselves and the system as a whole. It is always possible that the players in order to satisfy local queries, replicate data objects that are not beneficial to the system as a whole in terms of saving communication cost (although it may be productive from the players' point of view). To counter such negative notions, a referring body was introduced (termed as the mechanism). The aim of the mechanism was to direct the competition in such a fashion that a global optimal was achieved even though the agents are competing against one another. Moreover, the basic objective of the proposed work was to make the system robust against incorrect dissemination of information by the players. To cater for the possibility of collusive behavior of the players, the scheme used the *Vickrey* payment protocol [4] that leaves the players with no option other than to bid in such a fashion that is beneficial to the system as a whole. The goal of a player is to maximize its profit, which is payment minus cost. The goal of the mechanism is to minimize the total data item transfer cost in the network due to the read and update accesses. In Mosaic-Net [13], the authors used side payments to encourage players to tell the truth. The authors [13] proved that the ADRP problem in general is NP-complete and also identified some useful properties of the proposed scheme and the necessary conditions of optimality.

Hirsch et al. [8] proposed a game-theory based model for ADRP, where all nodes were assumed to be cooperative. The authors applied ideas from *Volunteer's dilemma* [1, 29] in area of game theory. Under the Volunteer's dilemma approach, a node volunteers to store replicas that will incur some cost to the node in terms of its resources, but in return will benefit the resource conservation and lifespan of the whole network. The proposed approach performed volunteer nodes' selection for replica assignment in such a manner that a global utility function defining the network cost is optimized. In the proposed algorithm, named as Cooperative Altruistic Data Replication (CADR), the net global benefit (NGB) is calculated for each node on the return path of requested replica, where NGB depends on two parameters: **(a)** global savings (GS) and **(b)** global cost (GC). The GS is the global network savings when the node makes a local replica of data item to minimize traffic through read requests. The GC is the cost incurred when data item is updated, or displaced from primary node to other node when primary node is low on resources. The CADR algorithm proceeds as illustrated in the following. On return path of a requested replica k, each node i calculates NGB as $NGB_i^k = GS_i^k - GC_i^k$, and stores NGB_i^k into a matrix appended in the header of response replica. When replica is received by the requesting node r the node computes NGB_r^k. Then, if $NGB_r^k > NGB_i^k, \forall i$, then the node r stores a copy of data item in its buffer. Otherwise, replica is placed on a node i on the request/response path, such that $NGB_r^k > NGB_i^k, \forall j$.

It is noteworthy from the above described techniques addressing ADRP that most of the approaches utilize a common assumption of availability of global network knowledge. Such global network knowledge constitutes the following information: **(a)** number of replicas of original data items, **(b)** the identifiers of nodes having original and copies of data items, **(c)** frequency of access of each replica, and **(d)** frequency of contacts among various nodes. However, it is formally proved by Khan et al. [13] that despite the availability of global information, the varying dynamics of network topology in MANETs make replica allocation problem NP hard. The things get further complicated in DTNs, due to lack of global network state information, as well as scarcity of end-to-end communication paths. In the following section we address the replica allocation challenges in DTNs and propose our solution to the problem.

3 Data Replication in DTNs

DTNs are resource-constrained networks in terms of transfer bandwidth, energy, and storage. Formally, the data replication problem in DTNs can be stated as: "*How to perform data replication during an opportunistic contact, such that, it contributes to overall improvement of network performance parameters, such as communication cost, delivery ratio, and delay*?" It is quite challenging to find a precise answer to this question as data replication in DTNs is affected by numerous overlapping factors, such as: **(a)** data item size, **(b)** data item life-time, **(c)** buffer size, **(d)** bandwidth, **(e)** transmission range, **(f)** interference, **(g)** node speed, **(h)** node energy, **(i)** mobility pattern, **(j)** node's sleep intervals, **(k)** contacts frequency, **(l)** contact duration, **(m)** inter contact times, and **(n)** network size. Due to inherent uncertainty about network conditions, the multiple copies of data item needs to be replicated at different locations to ensure data accessibility. The difficulty in coordinating multiple replica nodes makes it hard to optimize the tradeoff between data accessibility, transmission delay, and network overhead. Generally, the following are the challenges faced by data replication approaches in DTNs, as discussed in various literatures [3, 11, 34]:

1. In a highly disruptive network, it is challenging to propagate link state information due to the link delays/interruptions during transit. Moreover, the link state information may be inconsistent and stale, as after the disruption is over, the next hop reached may not be the same as the one before the disruption occurred.
2. Unlike MANETs, it is difficult to propagate topology information in DTNs due to frequent disconnections, and latencies in message propagations. The topology information may become outdated, as the information varies with a higher degree topology variation in DTNs.
3. In DTN scenarios, where nodes have low duty cycling to conserve battery powers, the topology information may not be able to reach the nodes that are in sleep state. As long as nodes stay in sleep state or power off mode, they are unaware of the recent changes in the network, such as creation and removal of spatiotemporal routes among sources and destinations.
4. One of the major challenges faced by DTN data replication is the update of replicas. A replication scheme may allocate greater life times to replicas so as to increase the availability and propagation time of replicas. However, the primary data item might be updated in the meantime. Because the communication is based on opportunistic contacts, many nodes might be left with outdated information.
5. Due to sparsity of nodes, it is also difficult to derive global utility of placing a replica over a node. Therefore, in some cases there might be too many replicas in the network, and in other cases replicas are not in sufficient numbers to satisfy nodes. To achieve optimality in replica placement, every node might need to store a lot of network state information. This will cause overload on buffer, as well as on processor for performing heavy computations during nodes' contacts.

To address the aforementioned challenges, in the next subsection we present our replica placement scheme for DTNs. The proposed scheme intends to perform replica placement in a way that not only restricts the number of replicas in the network, but also improves the average delivery probability of replicas to the requesting nodes. We compare our scheme with the selected replica placement approaches in DTNs, and test the proposed model on real-world and as well as synthetic trace datasets.

3.1 System Model

We consider a DTN of a set of N mobile nodes. As indicated in Fig. 3, we divide the network nodes (represented by filled circles) into three types: **(a)** producers, **(b)**

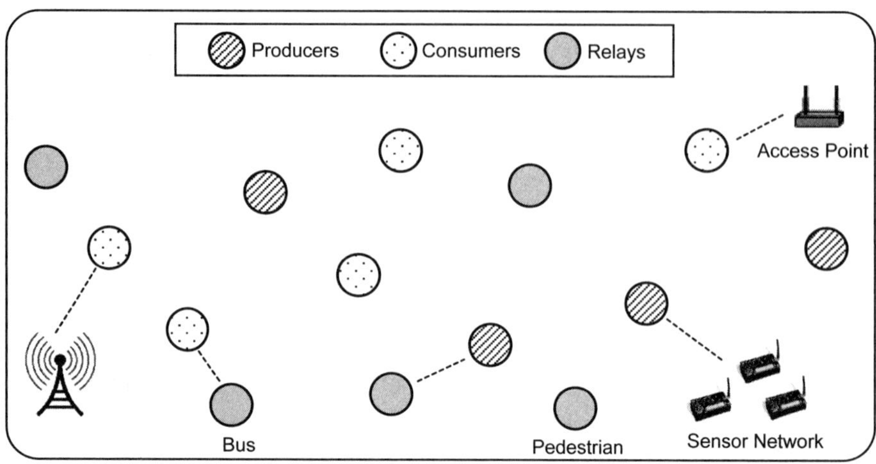

Fig. 3 An example of heterogeneous DTN network

Table 1 Notations and their meanings

Notation	Meaning
m^k	Message k
T_L^k	Life time of kth
T_w^k	Time the kth message spent waiting in buffers
T_t^k	Message transmission time
X^k	Random variable indicating the additional time that kth message might wait before delivery
τ	Time when a node makes contact with another node
$Z_i^j(\tau)$	Mean inter-contact time between node i and j computed at time τ

consumers, and **(c)** relays. Producers hold the original data item, which may be a measurement from a sensor network, or any piece of information generated by a node, such as emergency information or weather information, etc. Consumers are the information requesters, which are the nodes that act as sink for the information item. A relay node holds replicas of data items on behalf of other nodes. The nodes are able to change their roles in our particular network scenario. Moreover, when two nodes make contact, both nodes make use of in-band control signaling to exchange their locally maintained network state information. In the following, we elaborate few assumptions we make for the proposed model. The most frequent notations used in the chapter are shown in Table 1.

Each mobile node has a unique network identifier. The producer nodes generate the data items, with each data item having a unique identity. A data item can have many replicas in the network. For simplicity, we assume the same meaning of 'message' and 'replica'. A producer can directly serve the consumer node on making an opportunistic contact, or the replica can be relayed through relay nodes towards consumers. When two nodes make contact, the sender node may retain the replica, or delete it from its local buffer, after transferring it to the neighbor. Such decision is based on probability measures that we illustrated in our model. Few relay nodes, such as buses, may follow scheduled mobility patterns, while others, such as pedestrians may follow scheduled, as well as random mobility models. The nodes share only a portion of their buffer capacity for the opportunistic data storage. Moreover, during the opportunistic contacts, the mobility of nodes restricts the amount of data transferred due to limited contact duration. Therefore, we formally state the replication problem in DTNs as "*Given a limited duration opportunistic contact between two nodes, what replicas must be selected for exchange between the nodes, so that they contribute in the global optimization of network overhead and message delivery percentage*?"

As discussed earlier, in DTNs, nodes cannot maintain global network state. Therefore, we assume that nodes exchange network information during the opportunistic contacts. We denote the contact durations and inter-contact times between any two nodes i and j as C_i^j *and* I_i^j, respectively. Each node is maintaining a 2-tuple time-series information given as $< C_i^j[t],\ I_i^j[t] >$, where $t = 1,2,3,\ldots\omega$. The parameter ω is the index of last entry in the time-series data.

3.2 *Hybrid Scheme for Message Replication (HSM) for DTNs*

In this section, we present our scheme for message replication in DTNs. Suppose a node i has a message m^k that is requested by node d. At a time instant τ, the node i makes contact with a relay node j. At this occasion, the node i has to decide whether or not to replicate m^k on node j (in a hope that node j, might carry forward replica to node d). The node i will replicate m^k on node j if and only if for the replica m^k, node j's utility is greater than node i's utility value. The aforementioned utility value depends on: **(a)** the probability that a node will deliver message to destination before the life time expiry of message, and **(b)** the probability that the node will stay in contact with a message's destination for a duration greater than time required to transfer the message. If node j exhibits greater values of **(a)** and **(b)** as compared to the node i, the replica will be transferred to node j, and subsequently node i will delete the replica from its buffer. Otherwise, after transferring replica to node j, the node i will retain its local copy of replica. The motivation behind this strategy is to remove the excessive replicas from the network to conserve storage by placing replicas on nodes that appear to be more central in the network.

Let T_w^k be the time the message m^k has spent waiting in buffers since its creation, and T_L^k be the life time of message m^k. We denote X^k to be a random variable representing the additional time that m^k might wait before reaching destination. Then, we define message's utility as the probability that the message will be delivered to the destination d before the life time expiry [2], given as $U^k = P[T_W^k + X^k < T_L^k]$.

This can also be represented as:

$$U^k = P[X^k < T_L^k - T_W^k]. \tag{1}$$

In the above equation, we need to find the probability that additional wait time of replica is less than its remaining life time. As the message is transferred only during an opportunistic contact, the probability in Eq. (1) is same as the probability that the node i will make a contact with node d, before the expiry of message. We call such probability as utility value of node i for the current message:

$$U_{i,d}^k = P[Z_i^d(\tau) < T_L^k - T_W^k]. \tag{2}$$

In the above equation, $Z_i^d(\tau)$ is mean inter-contact time between nodes i and d at time τ. The network nodes are cumulating their inter-contact time information in the form of bounded time series data. Moreover, a few nodes (such as buses) are following partially scheduled mobility patterns. Therefore, we can apply exponential smoothing to forecast the value of inter-contact time between the node i and node d, as given below:

$$Z_i^d(\tau) = (1-\alpha)^{\tau-1}.s[\tau] + \sum_{k=0}^{\tau-1} \alpha.(1-\alpha)^{\tau-k-1}.I_i^d[k] \tag{3}$$

In the above equation, the parameter $0 \leq \alpha \leq 1$ is time-series smoothing constant, $I_i^d[k]$ is inter-contact time of node i with d at time instant k, $s[\tau]$ is the base value of recursion, and $Z_i^d(\tau)$ is the forecasted inter-contact time node i with d. As the mobile devices are limited in memory and processing, we cannot store unlimited time-series data of the past meetings. Therefore, we set a limit on the maximum number of entries stored per node in the form of a sliding time window $1 \leq t \leq \omega$, where the entry at $\tau = \omega$ represents the latest meeting. The more recent entries within the range $[1, \omega]$ must be given higher weightage than the others to ensure information freshness. Therefore, we assign progressively decreasing weights to the older entries, such that, as the entry becomes older it contributes less to the overall forecasting. The base case value of recursion $s[\tau]$ computed at time instant τ is given as:

$$s[\tau] = \frac{1}{n}.\sum_{j=0}^{n-1} I_i^d[\tau - j] \tag{4}$$

The above equation is the simple moving average of latest n entries of the inter-contact times I_t^d between i and d.

Let T_t^k be the time required to transfer a message m^k when two nodes make contact. Assuming that neighbour node has sufficient buffer space, the message will be successfully transferred to neighbour if and only if the contact duration of sender and receiving neighbour is greater than the required message transfer time T_t^k. Therefore, we compute the utility $V_{i,d}^k = P[T_t^k < C_i^d(\tau)]$, indicating the probability that the message will be transferred between node i and d in contact duration C_i^d.

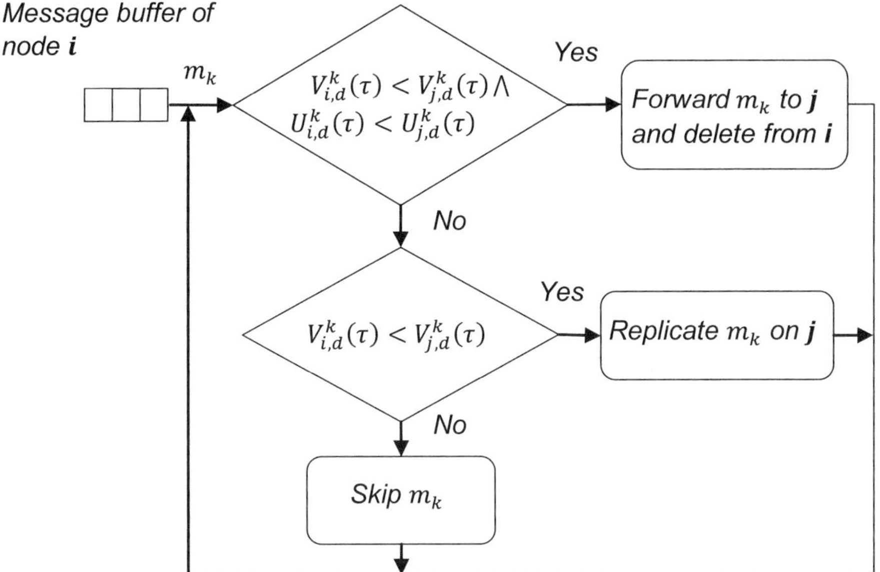

Fig. 4 Flow chart of proposed replication scheme

To compute the aforementioned probability, we need to find the estimated value of contact durations between nodes i and d. By replacing C_i^d with I_i^d in Eq. (3 and 4), we get the forecasted value of contact duration $C_i^d(\tau)$.

Figure 4 illustrates the flow of our replica placement scheme *HSM*. We call the scheme as hybrid, as we are also considering the occasional presence of MANET like environments in our network, when for example, the pedestrians stay closer to each other for longer durations, or the nodes are communicating with road side base stations. As reflected from Fig. 4, the procedure attempts to remove the redundant copies of messages from nodes' buffers, and attempts to allocate replicas on more appropriate nodes in terms of contact durations and inter-contact times with the destinations.

3.3 Empirical Setups and Results

In this section we present the performance analysis of the presented replication scheme *HRM*. Simulations are performed with the Opportunistic Network Environment (ONE) simulator [10] by using synthetic mobility model as well as real connection traces of participants of the INFOCOM 2006 conference [9]. The synthetic mobility model consists of several independent groups of mobile nodes including pedestrians, buses, cars, and access points. Some mobile nodes, such as pedestrians, follow random mobility patterns, whereas buses follow scheduled mobility patterns. The car nodes follow paths representing roads on map. The parameters considered for simulations are: **(a)** nodes' range 50–100, **(b)** world size 4250×3900 m, **(c)** time per simulation run 12 h, **(d)** transmission range 20 m, **(e)** message size 500KB–1 MB, **(f)** message time to live (TTL) 500 min, and **(g)** buffer size 10–100 MB. The world size is taken large enough to ensure that nodes are far enough to represent a DTN environment. The *HRM* scheme gives best performance for values of $a = 0.6, \omega = 50,\ and\ n = 10$ determined empirically under numerous simulation runs. In the following subsection we discuss the performance metrics considered for evaluation.

3.3.1 Performance Metrics

To evaluate the performance of the presented scheme, we consider the following three performance metrics: **(a)** message delivery ratio, **(b)** latency, and **(c)** overhead.

- **Message delivery** ratio is the percentage of messages delivered successfully. The maximization of message delivery ratio is the major goal of any DTN replication scheme. Message delivery ratio is calculated as

$$Message\ Delivery\ Ratio = \frac{1}{M}\sum_{k=1}^{M} R_k. \tag{5}$$

In above equation, $R_k = 1$ if and only if message is delivered, otherwise $R_k = 0$.

- **Message latency** is the total time spent between message creation and delivery to the destination. The average latencies of messages contribute to the overall latency measure of replication scheme. A scheme must minimize latency but without compromising message delivery ratio. The latency (in seconds) is given by

$$Latency\ average = \frac{1}{N}\sum_{k=1}^{N} Receive\ Time_k - Creation\ Time_k. \tag{6}$$

- **Overhead** is the approximate measure of the consumption of bandwidth, energy, and storage by a replication scheme due to message transmissions. We calculate overhead as relative estimate of number of message transmissions:

$$Overhead = \frac{Total\ msgs\ relayed - Total\ msgs\ delivered}{Total\ msgs\ delivered}. \tag{7}$$

The overhead ratio indicates extra transmissions for each delivered message.

3.3.2 Related DTN Replication Schemes

To perform comparisons, we selected the following related replication schemes for DTNs: **(a)** *PRoPHET* [28], **(b)** *Epidemic* [33], **(c)** *Random* [2], and **(d)** *Wave* [30]. These schemes utilize various strategies and heuristics to replicate messages in the network. As the data routing in DTNs is based on data replication during opportunistic contacts, the aforementioned schemes can also be called as the routing schemes for DTNs. We briefly describe the selected schemes in the following.

- ***PRoPHET***
  The PRoPHET scheme [28] calculates the delivery predictability for every node in the system. The delivery predictability value depends on the number of recent interactions of a node with other nodes in the network. Nodes perform the transitive updates of delivery predictabilities by sharing network-state tables on making contacts with each other. A node replicates a message to the neighbor, if and only if the delivery predictability of the neighbor is greater than the sender node. This way the PRoPHET scheme attempts to reduce the overhead. However, when the network size is large, such as a city-wide DTN network, it may take significant time in building up of delivery predictabilities. Therefore, in such cases PRoPHET may experience increase in overhead due to higher number of replications. The overhead also results in the increased buffer overflows, which may lead to the reduced message delivery ratio for PRoPHET.
- ***Epidemic***
  The Epidemic scheme [33] is an uncontrolled replication-based scheme that functions like epidemic. A node having a message copy is said to be infected. When this node makes contact with another node, the infection is transferred to the other node such that at the end of communication both nodes are having same infection (similar copies of a message). The Epidemic scheme spreads greater number of message copies in the network to improve message delivery probability. However, increasing message copies may cause greater overhead, higher utilization of buffers, and increased network congestion. Therefore, the Epidemic scheme is ideal for scenarios that have higher bandwidth and greater buffer storage available. The scenarios where nodes have limited buffer capacity, the Epidemic scheme may result in packet drops due to buffer shortages.
- ***Random***
  The Random scheme [2] allows a node to forward a message towards a randomly selected neighbor. After the message is successfully forwarded, the sender node deletes local copy of message. This makes the Random a single-copy multi-hop forwarding scheme. A message maintains a list of hops it has passed through to avoid visiting the same hops again. The Random scheme does not exhibit optimal performance in terms of message delivery ratio. This is because the randomly selected neighbor may not appear to be a best candidate to forward message towards destination.

- ***Wave***
  The Wave strategy [30] utilizes tracking lists to track messages that were recently relayed by a node. The idea is to prevent a node from receiving the same message replica again in short time duration. When a node receives a message, the message entry (e.g., message ID and receiving time) is maintained in the tracking list. During message exchange, the sender node transfers the message to the neighbor, but does not remove the message entry from the tracking list. This prevents the node from receiving the same message replica within a short time span. Such reductions in message replications minimize the overall overhead. However, decrease in the message replicas also decreases the message delivery probability of the Wave scheme.

3.3.3 Simulation Results

Simulation results with synthetic mobility model are indicated in Figs. 5a, b, c, whereas Figs. 5d, e, f present the simulation results with real-world connectivity traces. As reflected in Figs. 5a, b, c, the *HSM* scheme outperforms the rest of the replication schemes in terms of delivery ratio and overhead. This is because *HSM* accurately forecasts future contacts by performing online analysis of limited sized time-series data of previous contacts with varying qualities. On the contrary, *PRoPHET* performs future contact estimation on the basis of number of contacts without considering the time varying pattern of contact duration and inter-contact times.

The *Epidemic* scheme maximizes the flooding to improve message delivery. However, higher flooding causes increased overhead and higher message drop rate in resource constrained network scenarios. Alternatively, *HSM* performs selective message replication resulting in decreased overhead (Fig. 5c) and higher delivery ratio (Fig. 5a). The *Random* scheme forwards single message copy to any randomly selected neighbor, whereas the *Wave* scheme performs replica flooding in a controlled manner. Despite that both the aforementioned schemes are resource conservative, they exhibit low performance than *HSM* as reflected in Fig. 5. This is because these schemes do not utilize the past meeting patterns to perform a node's utility estimations.

The simulation results with real-world connectivity traces indicated that *HSM* performed better for delivery ratio and overhead. As reflected in Figs. 5d, e, f *HSM* precisely utilized the meeting patterns of conference participants to perform future contact forecasts. The latency metric of *Epidemic*, *Random*, and *Wave* is better than *HSM* (Fig. 5e). However, this is at the expense of their low delivery ratio. Moreover, *HSM* exhibits minimum overhead as compared to *PRoPHET* and *Epidemic*, despite being multiple copy replication scheme.

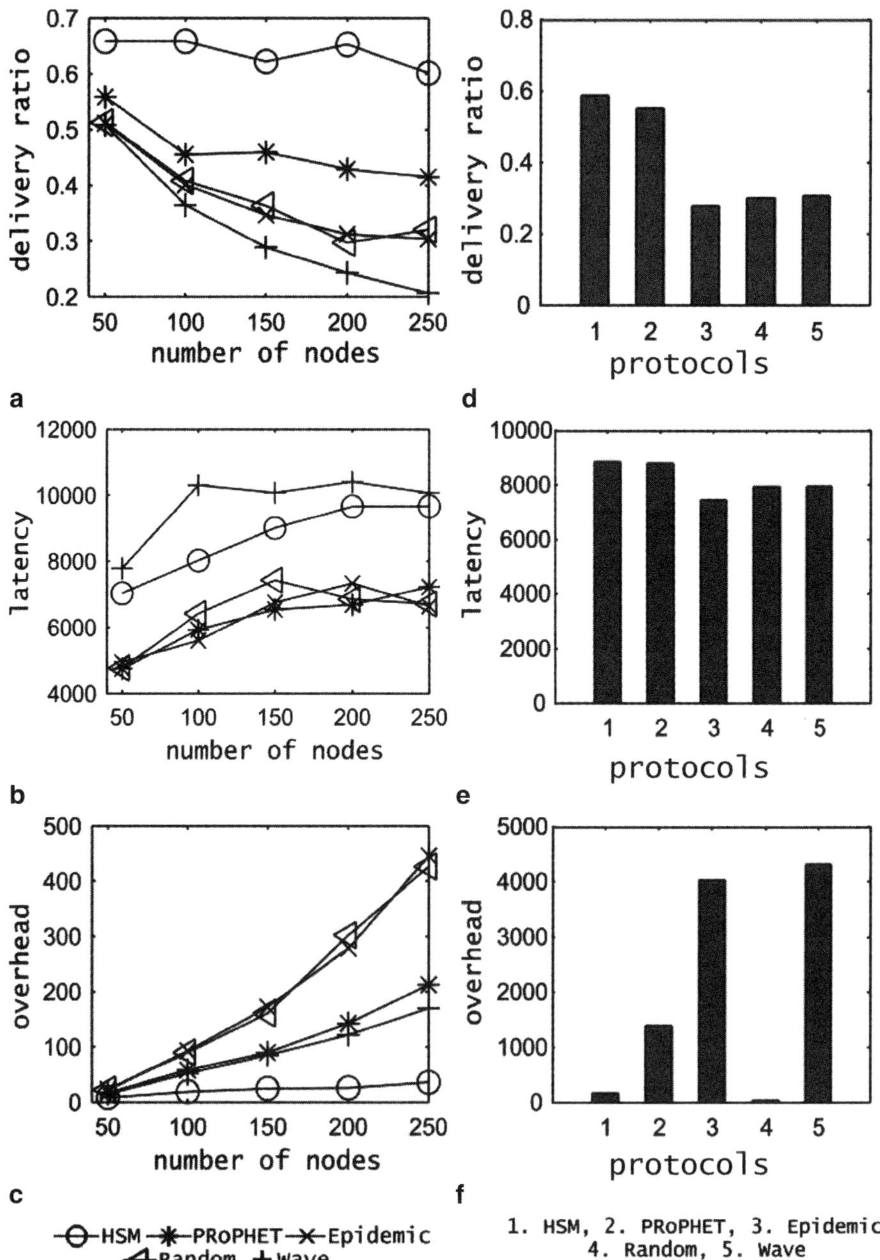

Fig. 5 Performance comparison results with synthetic mobility (Figures 1a, b, c) and real mobility trace (Figure 1d, e, f). The schemes compared are: *1* HSM, *2* PRoPHET, *3* Epidemic, *4* Random, and *5* Wave

4 Conclusions

In this chapter, we examined various challenges faced by the network when nodes are willing to participate in opportunistic data sharing and storage applications, to construct on-the-fly data centers. We addressed replica placement as one of the major challenges in ad hoc based data storage. In these networks, the decision of where to replicate data must trade off the cost of accessing data. The data access cost can be reduced by replications of data items, but with additional cost of storing and updating the replicas. These costs have severe implications in ad hoc networks because mobile hosts have limited resources (energy, storage, and processing power). Therefore, efficient and effective replication schemes strongly depend on how many replicas to be placed in the system, and more importantly where. We performed a comparative study of some of the well-known data replication schemes for MANETs, and discussed various pros and cons of the studied schemes. We observed that data replication is quite challenging in DTN like environments due to the non-existence of end-to-end communication paths. Unlike MANETs, the lack of end-to-end connectivity in DTNs prevents the global network information propagation. We formulated the data replication problem in DTNs and proposed a utility based replication scheme. The aforementioned utility value was based on two things: **(a)** probability that the node will be able to deliver message before life time expiry, and **(b)** probability that node will stay in contact with message's destination long enough to compensate message transfer time. Our results from synthetic mobility as well as real-world traces indicated that the proposed scheme produced the minimum network cost, and maximum delivery ratio. As a future work, we intend to explore numerous opportunistic message storage and sharing applications for bus based DTNs and vehicular ad hoc networks (VANETs).

References

1. M. Archetti, "Cooperation as a volunteer's dilemma and the strategy of conflict in public." *Journal of Evolutionary Biology*, vol. 22, no. 11, page. 2192, 2009.
2. A. Balasubramanian, B. Levine, and A. Venkataramani, "Replication Routing in DTNs: A Resource Allocation Approach," *IEEE/ACM Transactions on Networking*, vol. 18, no. 2, pp. 596–609, 2010.
3. M. Conti and M. Kumar, "Opportunities in Opportunistic Computing," *IEEE Computer*, vol. 43, no. 1, pp. 42–50, Jan. 2010.
4. D. Grosu and A. T. Chronopoulos, "Algorithmic mechanism design for load balancing in distributed systems," *IEEE Transactions on Systems, Man, and Cybernetics, Part B*, vol. 34, no. 1, pp. 77–84, January 2004.
5. T. Hara, "Effective replica allocation in ad hoc networks for improving data accessibility," In *Proceedings of IEEE INFOCOM*, 2001, pp. 1568–1576.
6. T. Hara, "Replica allocation in ad hoc networks with data update," *Mobile Network Applications*, vol. 8, pp. 343–354.
7. T. Hara and S. K. Madria, "Dynamic data replication using aperiodic updates in mobile ad hoc networks," In *Proceedings of 9th international conference on database systems for advance applications*, 2004, pp. 869–881.

8. D. Hirsch and S. Madria, "Data Replication in Cooperative Mobile Ad-Hoc Networks," *Mobile Networks and Applications*, vol. 18, no. 2, pp. 237–252, 2013.
9. Infocom06 connectivity traces on CRAWDAD website. [Online] http://crawdad.cs.dartmouth.edu/meta.php?name=cambridge/haggle#N100C4, 2013.
10. A. Keränen, J. Ott, and T. Kärkkäinen, "The ONE simulator for DTN protocol evaluation," in *Proc. of 2nd International Conference on Simulation Tools and Techniques*, 2009, doi.10.4108.
11. M. J. Khabbaz, C. M. Assi, and W. F. Fawaz, "Disruption-tolerant networking: A comprehensive survey on recent development and persisting challenges," *IEEE Communications Surveys Tutorials*, vol. 14, no. 2, pp. 607–640, Second Quarter, 2012.
12. O. Khalid, S. U. Khan, J. Kolodziej, L. Zhang, J. Li, K. Hayat, S. A. Madani, L. Wang, and D. Chen, "A checkpoint based message forwarding approach for opportunistic communication," *European Conference of Modeling and Simulation*, 2012.
13. S. U. Khan, "Mosaic-Net: A Game Theoretical Method for Selection and Allocation of Replicas in Ad Hoc Networks," *Journal of Supercomputing*, vol. 55, no. 3, pp. 321–366, 2011.
14. S. U. Khan and I. Ahmad, "Replicating Data Objects in Large Distributed Database Systems: An Axiomatic Game Theoretical Mechanism Design Approach," *Distributed and Parallel Databases*, vol. 28, nos. 2–3, pp. 187–218, 2010.
15. S. U. Khan and I. Ahmad, "A Pure Nash Equilibrium based Game Theoretical Method for Data Replication across Multiple Servers," *IEEE Transactions on Knowledge and Data Engineering*, vol. 21, no. 4, pp. 537–553, 2009.
16. S. U. Khan and C. Ardil, "A Frugal Bidding Procedure for Replicating WWW Content," *International Journal of Information Technology*, vol. 5, no. 1, pp. 67–80, 2009.
17. S. U. Khan, A. A. Maciejewski, and H. J. Siegel, "Robust CDN Replica Placement Techniques," in *23rd IEEE International Parallel and Distributed Processing Symposium (IPDPS)*, Rome, Italy, May 2009.
18. S. U. Khan, "A Multi-Objective Programming Approach for Resource Allocation in Data Centers," in *International Conference on Parallel and Distributed Processing Techniques and Applications (PDPTA)*, Las Vegas, NV, USA, July 2009, pp. 152–158.
19. S. U. Khan, "On a Game Theoretical Methodology for Data Replication in Ad Hoc Networks," in *International Conference on Parallel and Distributed Processing Techniques and Applications (PDPTA)*, Las Vegas, NV, USA, July 2009, pp. 232–238.
20. S. U. Khan, "A Frugal Auction Technique for Data Replication in Large Distributed Computing Systems," in *International Conference on Parallel and Distributed Processing Techniques and Applications (PDPTA)*, Las Vegas, NV, USA, July 2009, pp. 17–23.
21. S. U. Khan, "A Game Theoretical Resource Allocation Technique for Large Distributed Computing Systems," in *International Conference on Parallel and Distributed Processing Techniques and Applications (PDPTA)*, Las Vegas, NV, USA, July 2009, pp. 48–54.
22. S. U. Khan and C. Ardil, "A Fast Replica Placement Methodology for Large-scale Distributed Computing Systems," in *International Conference on Parallel and Distributed Computing Systems (ICPDCS)*, Oslo, Norway, July 2009, pp. 121–127.
23. S. U. Khan and C. Ardil, "A Competitive Replica Placement Methodology for Ad Hoc Networks," in *International Conference on Parallel and Distributed Computing Systems (ICPDCS)*, Oslo, Norway, July 2009, pp. 128–133.
24. S. U. Khan and I. Ahmad, "Comparison and Analysis of Ten Static Heuristics-based Internet Data Replication Techniques," *Journal of Parallel and Distributed Computing*, vol. 68, no. 2, pp. 113–136, 2008.
25. S. U. Khan, A. A. Maciejewski, H. J. Siegel, and I. Ahmad, "A Game Theoretical Data Replication Technique for Mobile Ad Hoc Networks," in *22nd IEEE International Parallel and Distributed Processing Symposium (IPDPS)*, Miami, FL, USA, April 2008.
26. H. S. Kia and S. U. Khan, "Server Replication in Multicast Networks," in *10th IEEE International Conference on Frontiers of Information Technology (FIT)*, Islamabad, Pakistan, December 2012, pp. 337–341.
27. Y. Li, M. Qian, D. Jin, P. Hui, Z. Wang, and S. Chen, "Multiple Mobile Data Offoading Through Disruption Tolerant Networks," *IEEE Transactions on Mobile Computing*, vol. PP, no. 99, page. 1.

28. A. Lindgren, A. Doria, A. Davies, and S. Grasic, "Probabilistic Routing Protocol for Intermittently Connected Networks," http://tools.ietf.org/html/rfc6693, 2012, accessed on March 2014.
29. S. Lee, D. Levin, V. Gopalakrishnan, B. Bhattacharjee, "Backbone construction in selfish wireless networks." In *ACM SIGMETRICS*, New York, 2007, pp. 121–132.
30. J. Ott, A. Keränen, and E. Hyytiä, "BeachNet: Propagation-based Information Sharing in Mostly Static Networks," in *Proc. of ACM ExtremeCom*, 2011.
31. J. Taheri, A. Y. Zomaya, P. Bouvry, and S. U. Khan, "Hopfield Neural Network for Simultaneous Job Scheduling and Data Replication in Grids," *Future Generation Computer Systems*, vol. 29, no. 8, pp. 1885–1900, 2013.
32. J. Taheri, A. Y. Zomaya, and S. U. Khan, "Grid Simulation Tools for Job Scheduling and Datafile Replication," in *Scalable Computing and Communications: Theory and Practice*, S. U. Khan, L. Wang, and A. Y. Zomaya, Eds., John Wiley & Sons, Hoboken, NJ, USA, 2013, ISBN: 978–1–1181–6265–1, Chapter 35.
33. A. Vahdat and D. Becker, "Epidemic routing for partially connected ad hoc networks," *Technical Report CS-200006, Duke University*, 2000.
34. A. G. Voyiatzis, "A Survey of Delay- and Disruption-Tolerant Networking Applications," *Journal of Internet Engineering*, vol. 5, no. 1, pp. 331–344, 2012.

Data Management: State-of-the-Practice at Open-Science Data Centers

Ritu Arora

1 Introduction

The amount of data involved in computational research is continuously growing. There are activities like scientific simulations, analyses of high-resolution data from latest instrumentation, and creation of digital archives that result in the generation of data and data products worth terabytes on a daily basis. In addition to storing, sharing, and accessing data for achieving their own scientific goals, scientists are being required by various funding agencies to capture data provenance. Hence, the long-term preservation of scientific data is becoming more critical than before. In order to support data-intensive computing and storage needs of their users, the data centers supporting open-science research are continuously expanding their storage and networking infrastructure.

At open-science data centers, because resources are shared amongst multiple projects and users, the disk-space owned by any user-account is limited. Therefore, maintaining large datasets (in the range of TBs and above) for in-place processing and analyses is usually not an option. Instead, the data for input to and output from a computational job needs to be staged to a filesystem associated with a computational resource. Let us consider the example of the Stampede system at the Texas Advanced Computing Center (TACC) [1]. The user-owned storage on Stampede is available in three directories, identified by the $HOME, $WORK and $SCRATCH environment variables. These directories are separate filesystems. The $HOME, which is backed up, has a quota limit of 5 GB and 150 K files. Such a quota is often not adequate when dealing with large volumes of input and output data with respect to a computational job. Therefore, the $WORK filesystem provides limit of 400 GB and 3 million files. However, it is not backed up. The $SCRATCH filesystem has no such quota limits but like $WORK, it is not backed up either.

R. Arora (✉)
Texas Advanced Computing Center, University of Texas at Austin, Ste 1.408, Bldg. 196, J.J. Pickle Research Campus, 10100 Burnet Road, 78758 Austin, TX, USA
e-mail: rauta@tacc.utexas.edu

S. U. Khan, A. Y. Zomaya (eds.), *Handbook on Data Centers*,
DOI 10.1007/978-1-4939-2092-1_37

For the storage of large volumes of data that could be of short-term, mid-term or long-term retention value, large capacity online disks and/or tape-based archival systems are used. In case of applications writing large amounts of data to a filesystem on a computational resource, staging data out of a computational resource to a storage resource and vice versa can be challenging. Data has to be moved out at a fast rate from the filesystem to secondary storage to accommodate new incoming data. However, the data movement is constricted due to network bandwidth and latency issues [2]. Retrieving very large datasets from secondary or tertiary storage systems can be time-consuming as well especially if these storage systems are tape-based archival systems. This is because before the data can be staged to a computational-resource for processing, it has to be de-migrated from tapes to the disk subsystem.

An HPC platform's I/O subsystems are typically slow as compared to its other parts (*viz*., processors) and one can easily saturate the bandwidth during large data transfers. As an example, the latency associated with the different levels of memory hierarchy on Stampede's dual-socket compute nodes equipped with Intel Xeon E5-2680 processors, is specified below:

- L1 data cache latency is typically 4–6 cycles, depending on data type and addressing mode.
- L2 latency is about 12 cycles.
- L3 latency is 26–31 cycles (depending on how far away the L3 slice containing the data is from the core requesting it).
- Local memory latency is about 77 cycles.
- Remote (other socket) memory latency is about 137 cycles.

The clock frequency of Intel Xeon E5-2680 processors is 2.7 GHz (or 2.7×10^9 Hz). Given that 1 Hz is equivalent to 1 cycle per second, 1 cycle roughly translates to approximately 0.3 ns on Stampede. The hard-disk access time is roughly around 1–2 μs, and hence we can see that the I/O gap between the memory speed and average disk access stands at roughly 10^{-3}. This example shows that it is important to develop an understanding of the characteristics of the I/O infrastructure in order to adapt HPC applications for doing efficient I/O. Efficient I/O will result in performance gains without overloading the filesystem.

There are four major storage levels in an I/O infrastructure: (1) internal—the processor registers and cache, (2) main—the system RAM and controller cards, (3) online mass storage—the secondary storage on disks, and (4) off-line bulk storage—the tertiary and off-line storage on tapes. The coordination of data movement from application layer to the I/O layer is needed for the optimal usage of the I/O bandwidth without burdening the shared HPC resources. By optimal usage, we mean how to achieve minimal completion time when we need to transfer across the hierarchies. The online mass storage and off-line bulk storage infrastructures are discussed in Sect. 2 of this chapter. A discussion on data movement in the context of the two storage levels is described in Sect. 3 of this chapter. Efficient I/O is also presented in Sect. 3.

Because HPC resources at open-science data centers are limited and certain datasets have mid-term or long-term retention value, it is important to formulate data archiving and preservation strategies. As a function of data archiving, the data is moved between different disk systems and tape libraries to ensure that valuable data is retained and accessed at a low-cost. As a function of data preservation, it is also important that the storage media for archived data is inspected periodically such that the data can be migrated to a new storage media before the older media is past its prime. Further discussion on data archiving and data preservation are presented in Sect. 4 and 5 respectively.

2 Data Storage Infrastructure

In this section we discuss the data storage media and the general high performance storage hierarchies that are created by using these media in open-science data centers. We also briefly discuss the usage of databases for handling structured and semi-structured datasets.

2.1 Data Storage Media

Hard Disk Drives (HDDs) provide random access storage using rapidly spinning disks that are coated with magnetic material. Once written, data is retained on an HDD even when it is powered off. HDDs are popular secondary storage devices mainly due to its price per unit of storage. From this point forward, when we mention "disk" in this chapter, it would mean HDD.

Flash or Solid-State Drives (SSDs) are silicon-based storage media that store data electronically instead of magnetically. Because flash or solid-state storage does not contain any mechanical parts, the data transfer to and from this storage media takes place at a high-speed. The seek-time is nearly uniform in the case of SSDs and hence the data can be read from any location on the drive at the same speed. The latency involved in access and retrieval of data stored on SSDs is very low as compared to the latency involved in the case of data access and retrieval from a spinning disk. Due to the low-latency associated with them, SSDs are ideal for applications that need fast random I/O to large files. A 64 GB SSD can support 712 I/O operations per watt (of power) whereas a 2 TB spinning HDD can support only 35 I/O operations per watt. Hence, SSDs are power-efficient in comparison to HDDs [3]. It should be noted here though that the read and write speeds may vary for an SSD.

A tape drive reads and writes data sequentially to magnetic tapes that are packaged in cartridges. Unlike a disk drive that provides random access storage, a tape drive provides sequential access storage. A tape drive physically winds tape in a reel to read any data and hence has a long latency for random access. However, for sequential access, once the tape is positioned, the tape drives can access data as fast as disk for contiguous blocks of data. Tape-based data storage is mostly used for offline storage of data archives and does not have issues like firmware corruption that are seen in the case of SSDs. It is a reliable, power-efficient, and a cost-effective option for mass storage systems.

2.2 General Architecture of a Data Storage System

In general, a data storage system for managing large volumes of data comprises of several levels of hardware and software components that are connected with a high-speed interconnect like Infiniband or Ethernet.

The secondary-level in a data storage hierarchy is often an array of disks (a few hundred or up to tens of thousands for parallel filesystems) to load and store the most recently used files. SSDs are also becoming a popular choice of storage at this level. A filesystem makes the multiple disks or SSDs appear as one to the end-user. More detail on this is presented in Sect. 3.

If a computational job is accessing or generating a large amount of data, the disk-space at the secondary-level can get filled quickly during single or multiple runs of the job. Therefore, the files need to be transferred to the tertiary-level which is usually meant for long-term data storage or is a storage silo. The tertiary-level often comprises of a few disks and a large number of tapes. The storage silos have thousands of slots containing magnetic tapes. They also have four to eight robotic hands, known as "handbots", for searching the appropriate tapes and inserting them in the readers (or tape drives) that are assigned to them by the data management software.

The data or storage management software like High Performance Storage System (HPSS) [4] or Sun's Storage Archive Manager Filesystem (SAM-FS) [5] is used to control the hardware components—storage disks, tapes and the handbots. Depending upon the file-access pattern, the files on disks in the tertiary-level are automatically transferred onto a high-capacity and low-cost media like magnetic tapes with the help of the data management software.

2.3 Supporting Databases for Structured and Semi-Structured Datasets

Some of the data-intensive and compute-intensive jobs running at data centers require or produce structured or semi-structured datasets and hence need a database for storing and accessing results. The databases can be as simple as a collection of flat files or could be an advanced relational database management system like DB2 [6].

If the number of users increases beyond a couple of thousand, replicas of the database are created on multiple nodes for load-balancing purposes. For small datasets (below few hundred terabytes), the performance of all the following relational database management systems is equivalent: DB2, Postgres and MySQL.

However, for hundreds of TBs of data, practitioners recommend DB2 over Postgres or MySQL because it scales well.

Besides having distributed databases, there are other techniques that are used to write scalable applications involving databases. As an example, consider mpiBLAST [7] which is a parallel implementation of an algorithm used by computational biologists to find similarity between biological sequences. This scalable software uses distributed computational resources and techniques like database fragmentation, query segmentation, intelligent scheduling, and parallel I/O through MPI [8], to speed up the process of finding similar biological sequences. The database fragmentation technique is used to partition a large database of FASTA files (a bioinformatics format for specifying sequences) into multiple small fragments. The fragments are then distributed across many compute nodes and each fragment can be searched in parallel. The query segmentation technique is used to segment a query into multiple independent searches such that, multiple searches for sequences can be performed simultaneously. From this example, it should be noted that data partitioning and data layout are crucial factors on which the performance of application software like mpiBLAST depends.

2.4 Examples of Notable Storage Systems at Open-Science Data Centers

The Gordon Supercomputer at the San Diego Supercomputing Center (SDSC) has over 300 Terabytes (TB) of SSDs and achieves over 36 million IOPS (I/O operations per second) [9]. With latency lower than spinning disks, Gordon's I/O nodes are well-suited for applications having random data access patterns or I/O involving frequent small reads.

The Ranch system at TACC is a tape-based archival system with 60 PB Sun StorageTek mass storage solution [10]. Ranch's user-access disk cache is built on DELL MD3600i disk array containing approximately 1150 TB of spinning disks. This disk array is controlled by a SAM-FS Metadata server, which has 16 CPUs and 72 GB of RAM. Two Sun StorageTek SL8500 automated tape libraries are combined to serve as the offline archival storage. Each SL8500 library contains 10,000 tape slots and 64 tape drive slots. Each SL8500 library also contains four handbots to manage tapes and move them to or from the tape drives. In the current Ranch configuration, one library contains 10,000 tapes, each having 1 TB uncompressed data capacity and the other has 10,000 tapes of 5 TB each, equaling 60 PB of total combined capacity. When archiving or retrieving multiple terabytes of datasets on Ranch, up to four tape drives are used automatically in parallel, thereby, leading to almost four times the speed of using only a single tape drive. With a balanced distribution of data sets, one could achieve close to 1 GB/s transfer rate.

Yet another example of a tape-based archival system is the HPSS-based system that is part of the Blue Waters project at the National Center for Supercomputing Applications (NCSA). HPSS is built using multiple automated tape libraries, data movers, 40 Gigabit Ethernet connection, multiple high-performance tape drives, and about a 100,000 tape cartridges. NCSA developed and deployed an HPSS capability for Redundant Arrays of Independent Tapes (RAIT) [11], that is, the RAID idea carried over to tapes instead of disk drives. In principle it achieves the data protection as well as the speed-boost. The HPSS hierarchical filesystem software can be used for the management of data life-cycle and accessing large volumes of data at a high speed.

Open-science data centers also support research data repositories for the storage and sharing of research data. An example of one such system is Corral [12] at TACC. Corral encompasses a collection of storage systems and services for data management and provides about 5 PB of online storage to researchers. There are over 4000 hard drives in Corral and its filesystems are accessible from computational resources at TACC. Corral is appropriate for data with long-term retention value. Corral also supports the iRODS [13] data management system, database management systems like Postgres and MySQL, and web-hosting services. It offers both replicated and non-replicated storage. With replicated storage, the data is synchronously stored over two storage installations, resulting in two copies of all data and metadata, while non-replicated means that only one storage site is used, and only one copy of the data and metadata are stored within the system. Users with general data management and storage needs can opt for replicated storage. The users who wish to replicate their data across multiple systems, including the Ranch tape archive, are encouraged to utilize the iRODS data management tool.

3 Data Movement

HPC applications often involve I/O (data movement) for activities like reading initial conditions or datasets for processing, writing numerical data from simulations, and saving application-level checkpoints. Though often an afterthought for most HPC programmers, the time spent in doing I/O should also be optimized in addition to optimizing the time spent in doing computations and communication for getting the best performance. As a starting point, the programmers might want to consider balancing I/O (both serial and parallel). They can also consider using parallel I/O such that multiple processes can participate in reading data from or writing data to a common file in parallel (see Fig. 1). Such strategies can improve performance when dealing with large datasets by optimizing the usage of the I/O bandwidth.

3.1 Parallel File-System Associated with Computational Resources—Secondary Storage

In order to manage the storage hardware and allow concurrent data access for parallel I/O, a parallel filesystem is needed. The parallel filesystem makes hundreds of

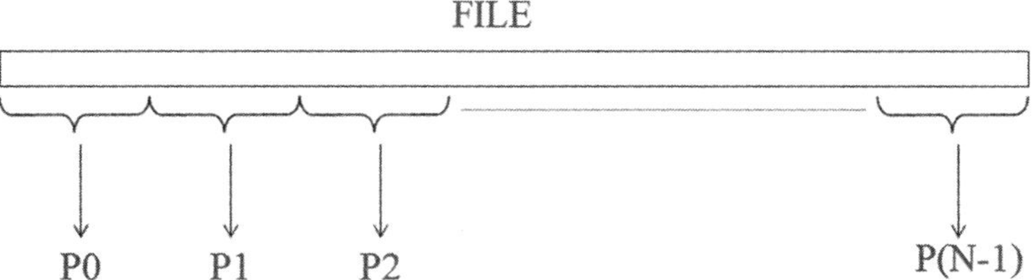

Fig. 1 N number of processes participating in reading or writing a portion of a common file

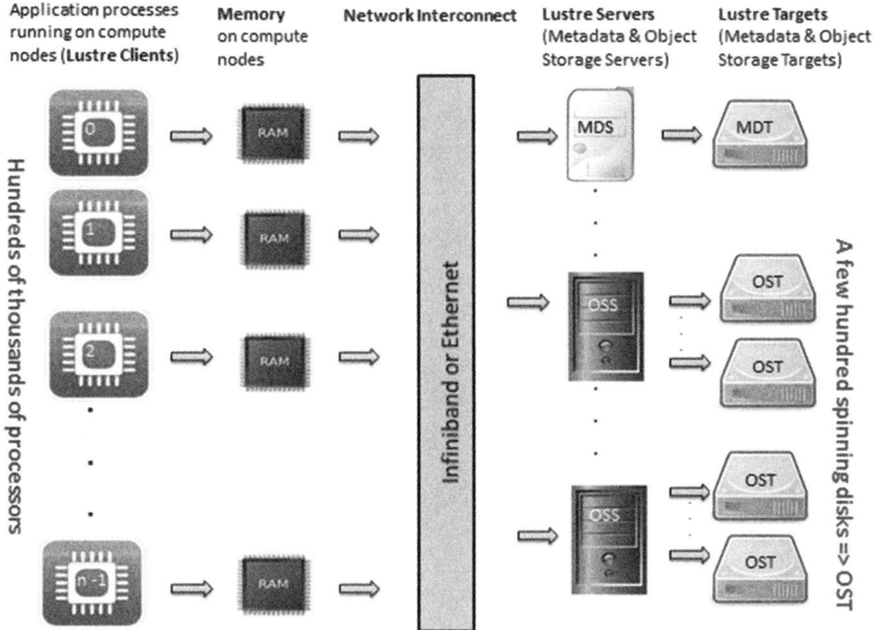

Fig. 2 Lustre file system

spinning disks act like a single disk that can then be accessed by thousands of processors that could be present in an HPC platform. One example of such a filesystem is Lustre [14, 15]. Lustre maintains logical file space and provides efficient access to data. The different components that are part of the Lustre-based secondary storage are explained below and are pictorially shown in Fig. 2. The path of data movement from the memory of application process to disk is shown with the help of arrows in Fig. 2.

The set of I/O servers called Object Storage Servers (OSSs) and disks called Object Storage Targets (OSTs) together make the Lustre File System (LFS). The OSTs are block storage devices (HDDs and/or SSDs) that store the data associated with the files of the users. An OST may be thought of as a single virtual disk even though it often consists of several physical disks, possibly in a RAID configuration.

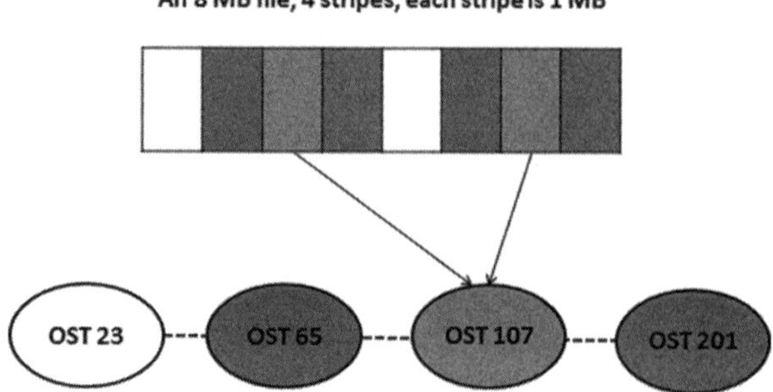

Fig. 3 Lustre supporting striping of files across several I/O servers

The data associated with user files is stored in one or more objects, with each object stored on a separate OST. The number of objects per file is user configurable and can be tuned to optimize performance for a given workload. The OSSs manage a small set of OSTs by controlling I/O access and handling network requests to them. The OSSs contain metadata about the files stored on their OSTs. They typically serve between 2 and 8 OSTs, where each OST is generally up to 24 or 32 TB in size.

Each user file has metadata associated with it like filename, permissions on the file and directories. The metadata associated with user files is stored on a MetaData Target (MDT) and is controlled by a MetaData Server (MDS). The MDS used to be a single service node that would assign and track all of the storage locations associated with each file in order to direct file I/O requests to the correct OSSs and then eventually to the correct set of OSTs. However, with the recently included feature of Distributed Namespace (DNE), the Lustre namespace can be divided across multiple metadata servers. This enables the size of the namespace and metadata throughput to be scaled with the number of servers. Once a file is opened, no further involvement of the MDS is needed for I/O to or from the file. This enables the scalability of bandwidth in Lustre. Lustre permits Remote Direct Memory Access (RDMA) for Infiniband (OFED) and hence enables multiple, bridging RDMA networks to use Lustre routing, which in turn leads to increased performance.

The application processes running on compute nodes communicate with the Lustre servers and hence the Lustre targets via the available network. On Stampede, there are three instances of LFS, each with different usage policies: $HOME, $WORK, and $SCRATCH [1]. Lustre supports the "striping" of files across several I/O servers (similar to RAID 0) where each stripe is a fixed size block (see Fig. 3). This means that chunks of a file will exist on multiple OSTs. Application programmers could consider balancing serial or parallel I/O load by setting the appropriate stripe-count and stripe-size that is commensurate to their data load. In order to match the I/O bandwidth with the switch bandwidth, the stripe-count can be set such that the switch bandwidth is saturated. It should be noted that the system administrators set a default stripe-count and stripe-size that applies to all newly created files on a system. For example on Stampede, the default is 2 stripes/1 MB in $SCRATCH and 1 stripe/1 MB in $WORK. However, the users can reset the default stripe-count or stripe-size

```
% lfs getstripe ./testfile
./testfile
lmm_stripe_count:   2
lmm_stripe_size:    1048576
lmm_stripe_offset:  50
        obdidx           objid          objid            group
            50         8916056       0x880c58                0
            38         8952827       0x889bfb                0

% lfs setstripe -c 4 -s 4M testfile2
% lfs getstripe ./testfile2
./testfile2
lmm_stripe_count:   4
lmm_stripe_size:    4194304
lmm_stripe_offset:  21
        obdidx           objid          objid            group
            21         8891547       0x87ac9b                0
            13         8946053       0x888185                0
            57         8906813       0x87e83d                0
            44         8945736       0x888048                0
```

Fig. 4 Example of `getstripe` and `setstripe` on Lustre

using the "`lfs setstripe`" command. See Fig. 4 for an example of using this command.

3.2 *Optimizing Data Movement in Context of Secondary Storage System*

Parallel I/O can provide a single file for storage and transfer purposes. However, it can be hard to coordinate and optimize I/O if working directly at the level of Lustre API. Therefore, a number of intermediate layers (sitting between low-level hardware layer and the top-level application layer) have been implemented, over which, parallel applications involving parallel I/O can be developed. Examples of such intermediate layers are MPI-I/O and parallel HDF5 [16].

When using MPI-I/O, the MPI should be built with Lustre support so that the Lustre stripe-count, stripe-size, and # of writers can be set using MPI-I/O Hints. The MPI-I/O Hints are extra information supplied to the MPI implementation through the following function calls for improving the I/O performance: `MPI_File_open`, `MPI_File_set_info`, and `MPI_File_set_view`. The MVAPICH2 and the OpenMPI implementations of MPI support Lustre. Parallel HDF5 is a layer of abstraction over MPI-I/O and hence all MPI-I/O techniques apply to parallel HDF5.

While doing parallel I/O, it is best to have a small subset of the total number of application processes (tasks) participate in the I/O. For example, if there are 1024 nodes that are participating in a computation and 16 processes are launched on each node, then having all the processes participate in I/O (1024×16 processes $= 16384$ I/O clients) can result in an oversubscribed system. In such a situation having a subset of processes, let us say, one task per node can result in a better I/O balance (1024 I/O clients).

Contiguous access of data in memory and on disk provides best performance. A performance improvement is seen if instead of frequently writing small amounts of data to a file, the output from an application can be buffered till a reasonable size of data is collected for writing to an output file. As mentioned previously, the number of processes that are involved in accessing a file at a given point should be restricted by gathering the data to be written to a subset of processes. In order to avoid burdening the MDS, the frequent opening and closing of files from an application should be avoided. This is because, each file open or close operation has an overhead cost that adds up, especially if multiple tasks are opening and closing files at the same time.

If the size of the file to be read is small (i.e. < 1 GB), it is advisable to have the parallel application involve only one process to read the file and then broadcast the data that it read to all the other processes. While writing small files and directories (i.e. < 1 GB), it is advisable to write to a single OST.

In our experience, directories that contain large number of files (in the order of hundreds or thousands) will be slow to respond to I/O operations due to overhead involved in the indexing of the files. Therefore, a directory should be broken into subdirectories with fewer files. The number of files in a directory should be limited and one should avoid creating deeply nested directories. The impact of deeply nested directories and multiple small files in a directory can lead to slow system response time even with basic commands like "`ls -ltr`". The users are advised to avoid "`ls -l`" too frequently in order to avoid unnecessary overheads in communicating with multiple OSTs on which a single file might be stored. Instead of using "`ls -l`", the users might want to consider using "`ls -U`" or "`lfs find`" where possible.

3.3 Optimizing Data Movement in Context of Tertiary Storage System

On a tape-based archival system like Ranch, if the average size of the files to be archived is less than 1 GB, it is better to combine them into a tar file before storing them on an archival system. Staging tens of thousands of files could take multiple days, though it could be done in hours if optimized to reduce the mounting of tape cartridges. Attempts to transfer unstaged data from tape drives to a computational resource could be slow for most users, leading to atomic calls per file (retrieval of one file per mount) even when the tape has all of them in a single media. To make things worse, if users add threads (like in scp, rsync, and gridftp) in an attempt to parallelize the atomic process, they could make the transfer even slower by creating more threads in wait state. It is not uncommon to witness one user with a single GridFTP session [17] achieving 10 TB transfer overnight while another user having 10 GridFTP processes moving order of magnitudes less. This deplorable practice stems partly from the asymmetric nature of data movement to and from Ranch. When the users put data on Ranch, multiple threads help because Ranch is more like a disk based filesystem during the data ingestion stage. On retrieval, however, Ranch is far from a filesystem, and it is nothing but an array of tape drives. Unless staged first, none of the typical transfer methods can achieve the expected performance, and having multiple threads aggravates the situation by blocking other users.

If the average size of the data to be retrieved is more than 10 GB, the tape mount overhead is minimal no matter what transfer method is used. Users in this group can ignore the necessity of staging without significant penalty. The most interesting case is when the average data size is 1–10 GB. Some users may be able to avoid the necessity of pre-stage like in the 10 GB case, but can also achieve significant improvement with aligned staging.

Also of vital importance for users planning to transfer more than million files totaling tens of Terabytes is the physical limit of disk cache. If one tries to stage the whole set, some staged files will be released back to the tape while the staging is still going, creating a never-ending movement of blocks. Depending on the other users' activities, release period can be as short as within a day while staging a million files can take up to 3 days. For such massive migration it is better to coordinate with the system administrators, rather than trying it on their own.

4 Data Archiving

Data explosion is being considered as the biggest hardware infrastructure challenge for data centers and calls for investments in data archiving. Often, data archiving is confused with data backup. While the process of backing up the data results in creating an additional copy of the data that can be stored at a different location, data archiving implies that the data, which is currently in the primary storage, is moved to a secondary or tertiary storage system. Thus, data archiving helps in freeing the costly space on primary storage media. It should be noted that data archiving is used for the long-term storage of the primary and the only copy of the data. In contrast to the backed up data which is mainly created for disaster recovery and which may never be accessed, the archived data is accessed as the need arises.

An important requirement for data archives is that they should be searchable. The users should themselves be able to find the needed files and access them in the original format. There are off-the-shelf archiving solutions that come with file management software and can help in automating the process of data archival in a cost-effective manner. For example, Ranch utilizes Oracle's Storage Archive Manager Filesystem (SAM-FS) for migrating files to and from a tape archival system.

In this era of Big Data, tape-based archival is a reliable, cost-effective and energy-efficient way for managing large datasets for archival purposes. A tape drive has a shelf-life of roughly 30 years if stored under optimal environmental conditions. Unlike disks, tape drives do not need persistent power and cooling. Once the data is written on tapes, the tapes do not need any power.

5 Data Preservation

Data preservation entails the management of digital archives so that the archived information and the associated metadata can be accessed and used at any point of time in future. The preservation of digital legacy requires careful planning that begins at the time of designing the data archival system. Because the resources and technologies keep changing at a fast pace, planning is required to address issues like maintenance of the current storage media, transitioning to the new storage media, and the backward compatibility of new systems to support data archives created on old systems [18].

It should be noted that different datasets need different software environments in which they can be rendered. One challenge for open-science data centers is to secure sustained funding for the management of repositories and collections such that even after the life-cycle of a computational or storage resource is over, software support can still be provided for accessing and using the data in the repositories. The software used for rendering the archived data might become obsolete for the hardware infrastructures in future. Therefore, advance planning is needed for developing new software to emulate the old infrastructure on latest hardware. Because of the need to handle a wide variety of datasets and their associated formats, data preservation becomes a complex endeavor for open-science data centers.

In order to standardize the functions related to digital preservation practice and provide a set of recommendations to data preservationists, an Open Archival Information System (OAIS) reference model was developed [19]. All the technical aspects of data life-cycle—ingestion, archival, data management, administration, access, and preservation planning—are addressed by OAIS. The model recommends that five types of metadata is stored for all data that is meant to be preserved—reference information, provenance, context, authenticity indicators, and representation (formatting, file structure, and semantics).

The data centers avoid the loss of digital information by various strategies like replication, refreshing the storage media, planned conversions of data from old file-formats to new ones while the old format is still recognized by the latest infrastructure, and metadata extraction. The concept of persistent archives as developed at SDSC states that both the data and the information required to assemble the units of data into a data collection must be archived [20].

6 Conclusion

Data Centers are continuously expanding their storage capacity to meet the needs of data-intensive computing. In this chapter, we have provided an overview of the data storage infrastructure at open-science data centers with the focus on handling large datasets in an efficient and cost-effective manner.

The general data characteristics or patterns of data-usage vary immensely across a given set of users at an open-science data center. At times, the size of data files is too large to manage while at other times there are large numbers of files to manage. Some datasets might be stored for long-term usage while other datasets or subsets might be required for short-term and rapid processing. While some datasets are natively suitable for database processing, others (semi-structured and raw datasets) are not. It is also observed that while some users might want to draw different inferences from large datasets using different subsets, there are other users who want to quickly find the correlation across terabytes of datasets collected at regular intervals to find the underlying cause of unusual problems. Given such a diverse range of data analyses and processing needs of the users at an open-science data center, the recommendations for improving I/O performance are usually provided on a case-to-case basis. However, some of the generally recommended best practices and strategies for efficient I/O were discussed in this chapter.

Acknowledgement We would like to thank Chris Hempel, Jim Foster, Junseong Heo, John Cazes, John McCalpin, Robert McClay, and Sukrit Sondhi for their help in development of this chapter. The authors acknowledge the Texas Advanced Computing Center (TACC) at The University of Texas at Austin for providing HPC resources that have also contributed to the development of this book chapter.

References

1. Stampede User-Guide, accessed on 2nd October 2013: http://www.tacc.utexas.edu/user-services/user-guides/stampede-user-guide.
2. Henry M. Monti, Ali Raza Butt, Sudharshan S. Vazhkudai. 2010. CATCH: A Cloud-Based Adaptive Data Transfer Service for HPC. In the proceedings of 25th IEEE International Symposium on Parallel and Distributed Processing, 2010, Anchorage, Alaska, U S A.
3. Jiahua He, Arun Jagatheesan, Sandeep Gupta, Jeffrey Bennett, and Allan Snavely. 2010. DASH: a Recipe for a Flash-based Data Intensive Supercomputer. In Proceedings of the 2010 ACM/IEEE International Conference for High Performance Computing, Networking, Storage and Analysis (SC '10). IEEE Computer Society, Washington, DC, USA, 1–11. DOI = 10.1109/SC.2010.16 http://dx.doi.org/10.1109/SC.2010.16
4. High Performance Storage System, accessed on 2nd October 2013: http://www.hpss-collaboration.org/
5. Sun™ SAM-FS and Sun™ SAM-QFS, Storage and Archive Management Guide, accessed on 2nd October 2013: http://docs.oracle.com/cd/E19598-01/816-2544-10/816-2544-10.pdf
6. IBM DB2 Software, accessed on 2nd October 2013: http://www-01.ibm.com/software/in/data/db2/.
7. Aaron E Darling, Lucas Carey, and Wu-Chun Feng. 2003. The Design, Implementation, and Evaluation of mpiBLAST. In Proceedings of 4th International Conference on Linux Clusters:

The HPC Revolution 2003 in conjunction with the ClusterWorld Conference & Expo, San Jose, CA, U S A.
8. William Gropp Ewing Lusk Anthony Skjellum, "Using MPI: Portable Parallel Programming with the Message-Passing Interface", MIT Press, 1999, pp. 1–371.
9. Gordon User-Guide, accessed on 2nd October 2013: http://www.sdsc.edu/us/resources/gordon/
10. Ranch User-Guide, accessed on 2nd October 2013: http://www.tacc.utexas.edu/user-services/user-guides/ranch-user-guide
11. HPSS RAIT Architecture, accessed on 15th January 2014: http://www.hpss-collaboration.org/documents/HPSS_RAIT_Architecture.pdf
12. Corral User-Guide, accessed on 2nd October 2013: http://www.tacc.utexas.edu/user-services/user-guides/corral-user-guide
13. iRODS—Integrated Rule-Oriented Data System, accessed on 2nd October 2013: https://www.irods.org/
14. Lustre File System, accessed on 2nd October 2013: http://wiki.lustre.org/index.php/Main_Page (We may want to replace this obsolete page with the currently maintained one in http://build.whamcloud.com/job/lustre-manual/lastSuccessfulBuild/artifact/lustre_manual.xhtml.)
15. Lustre Filesystem (Scratch Directory), accessed on 2nd October 2013: http://www.nics.tennessee.edu/computing-resources/file-systems/io-lustre-tips.
16. Parallel HDF5, accessed on 2nd October 2013: http://www.hdfgroup.org/HDF5/PHDF5/
17. GridFTP, accessed on 15th January 2014: http://toolkit.globus.org/toolkit/docs/latest-stable/gridftp/.
18. Greg Janee Justin Mathena James Frew. 2008. A data model and architecture for long-term preservation. JCDL 2008, pp. 134–144.
19. ISO Archiving Standards, accessed on 2nd October 2013: http://nssdc.gsfc.nasa.gov/nost/isoas/
20. Report on Collection Based Persistent Archives, accessed on 2nd October 2013, http://www.sdsc.edu/NARA/Publications/col-rep.html

Data Summarization Techniques for Big Data—A Survey

Z. R. Hesabi, Z. Tari, A. Goscinski, A. Fahad, I. Khalil and C. Queiroz

1 Introduction

In current digital era according to (as far) massive progress and development of internet and online world technologies such as big and powerful data servers we face huge volume of information and data day by day from many different resources and services which was not available to human kind just a few decades ago. This data comes from available different online resources and services that are established to serve customers. Services and resources like Sensor Networks, Cloud Storages, Social Networks and etc., produce big volume of data and also need to manage and reuse that data or some analytical aspects of the data. Although this massive volume of data can be really useful for people and corporates it could be problematic as well. Therefore big volume of data or big data has its own deficiencies as well. They need big storage/s and this volume makes operations such as analytical operations, process operations, retrieval operations real difficult and hugely time consuming. One resolution to overcome these difficult problems is to have big data *summarized* so they would need less storage and extremely shorter time to get processed and retrieved. The summarized data will be then in "compact format" and still informative version

Z. R. Hesabi (✉) · Z. Tari · A. Fahad · I. Khalil
School of Computer Science and IT, RMIT University, Melbourne, Australia
e-mail: zhinoos.razavi@rmit.edu.au

Z. Tari
e-mail: zahir.tari@rmit.edu.au

I. Khalil
e-mail: ibrahim.khalil@rmit.edu.au

A. Goscinski
School of Information Technology, Deakin University, Melbourne, Australia
e-mail: andrzej.goscinski@deakin.edu.au

C. Queiroz
IBM Research Laboratory, Melbourne, Australia
e-mail: caxqueiroz@gmail.com

S. U. Khan, A. Y. Zomaya (eds.), *Handbook on Data Centers,*
DOI 10.1007/978-1-4939-2092-1_38

of the entire data. *Data summarization* techniques aim then to produce a "good" quality of summaries. Therefore, they would hugely benefit everyone from ordinary users to researches and corporate world, as it can provide an efficient tool to deal with large data such as news (for new summarization).

The aim of this chapter is to provide an overall view of different data summarization techniques found in the literature including clustering, sampling, compression, histograms, wavelets and micro-cluster with respect to their applications in the variety of fields such as data mining. However the focus here will be the selection of those techniques that are suitable for *big data*. Some aspects then need a careful attention when dealing with big data, as therefore help selecting those techniques that are suitable for big data. *Volume* of the data is the first and obvious important characteristic to deal with when summarizing big data comparing to conventional data summarization, as this requires substantial changes in the architecture of storage systems. The other important characteristic of big data is *Velocity*. This requirement leads to highly demand for on line processing of data where processing speed is required to deal with data flows. *Variety* is the third characteristic, where different data types such as text, image, and video are produced from various sources such as sensors, mobile phones, etc. These three "V" (Volume, Velocity, and Variety) are the core character tics of big data [1] which must be taken into account when selecting data summarization techniques.

Summarization can be performed in various ways. We selected here the following summarization techniques that are applicable for big data:

- *Clustering* is an unsupervised summarization technique that aims to gather similar objects into a group called *clusters*. The similarity among objects can be determined through different metrics, such as distance. Clustering algorithms can be categorized into different models such as hierarchal, partitioning, density-based, and grid-based and so on.
- *Sampling* is another summarization technique that provides a concise and still informative representation of the entire data set. A *sample* is a representative of a larger group (population) which preserves the same characteristics of the population and study is conducted on the sample instead of the population. There are two main categories of sampling techniques, namely probability-based and nonprobability-based, provide an efficient way to summarize big data.
- *Compression* is a well-known technique that represents data in a compact way to save time and space. Lossless and lossy are two different compression methods which are broadly used in different areas such as video and image coding. We will explain a compression method based on a minimum description length principle and its applications for data summarization in this chapter.
- *Wavelets* can be considered as a summarization technique that is mostly used in image and query processing applications. Different wavelet transformations such as Haar, and dimensional wavelets are used to transform data from one domain to another one. We will show in this chapter how wavelet transformations can be applied to summarize data.

- *Histogram* is a method used to represent a large volume of data in a compact manner so that can be considered as a data reduction or summarization technique. In fact, data distribution can be shown in a synopsis structure through histograms. Since, we will discuss some various type of histograms, however we will not elaborate on it further.
- *Micro-clustering* is a method to construct a synopsis model of data stream that considers evolving behavior of data streams. In this chapter, we will explain more details about micro-clustering as another summarization technique for data streams.

The roadmap of this chapter is as follows. Next section discusses some applications of big data summarization. Details about some of the well-known data summarization techniques available in the literature will be discussed. In this order, first each technique will be explained and then a number of research directions that are conducted on each technique will be briefly reviewed. However, readers referee to each subject for more consideration because of the lack of space in this paper. Section 5 concludes our study.

2 Applications of Data Summarization

Summarization is considered as a descriptive task in data mining to provide a synopsis representation of data. It makes many tasks such as preprocessing, analysis, and management of data easier and faster and therefore overcoming the space and time limitations. It reduces the size of data that may leads to obtain approximate results in comparison with the exact results achieved from the original data that most of them satisfy the user requirements considering saving time, cost and space. A major feature of summarization is that reduced data is still informative and the approximate (or sometimes the same interference) can be obtained from reduced data over the original data. Therefore, with growing data in a phenomenal rate, big data summarization has attracted many attentions to obtain compact data considering its accuracy and performance.

In recent years, we are witness of producing the ever increasing amount of data from different sources such as wireless sensors networks, Internet, mobiles devices, RFID readers and so on. Many social networks and companies like Facebook, Twitter, Yahoo, and Google are the main sources of generating big data. Mining, processing, analyzing, and presenting these huge amount of data require of exponential time and space. In this context, data summarization plays has played a key role in recent years to reduce time, space, cost, accuracy and data content based on user's applications.

Data summarization has been applied in a wide variety of fields with the number of different applications, and below are just some of them.

- **Medical informatics**: Biology, genetics, clinic are examples of medical science that apply some summarization techniques to meet different objectives. As an application of summarization in medical science, we can mention to the clinical pathway analysis. Analyzing aggregated data from clinical path leads to focus on process mining techniques towards discovering clinical pathway models from data. Since collected medical data may be very large, the number of minded patterns can be huge. Therefore, it is of interest to present a synopsis data describing the whole structure of clinical path, meanwhile exposing crucial medical information in less time over the total time of complete pathway.

- **Astronomy and earth science**: Since very large data is produced through geology, geography and astronomy sciences, summarization is brought up remarkably in these areas to analyze, categorize and process extracted information from aforementioned scientific fields.
- **Social networks**: In recent years, social networks such as Facebook and twitter find a popular place among users through which vast amount of data are generated daily. Analyzing and processing such large volume of data concludes spending lots of time and space. Therefore summarization is concerned to tackle this issue which leads to save time and space. Text summarization, exploratory search and topic summarization for Twitter, sampling from diffusion networks, summarization of communication patterns in large-scale social networks can be regarded as examples of summarization application for big data.
- **Business and marketing**: Ecommerce is getting more and more widespread leading to generate the large number of customer reviews. Mining vast amounts of reviews will be ended to use summarization techniques. Purchasing pattern recognition and stock trend analysis also take benefit of summarization.
- **World Wide Web**: The importance of data summarization will be revealed in today's World Wide Web. For example, Web page summarization plays an important role in web analysis. These summaries may satisfy the required information of users by means of fast and accurate browsing and search. Network traffic analysis and report generation are of applicability of summarization concept.
- **Sensor Networks**: Sensor networks produce large amount of data which require to be analyzed to extract interesting information. Query processing is a common task in sensor networks which need to process a large volume of data to transmit the answer to the user. However, sometime approximate answers with reasonable accuracy satisfy the requirement of user so that summarization is of interest in data stream mining.

As shown above, data summarization has a broad applicability in different fields. Therefore, it is of interest of many researchers to design and develop various summarization techniques to present a compact and yet informative representation of large data. In the remaining parts of this paper, we will provide an overview of the most ever used summarization techniques and their applications in the literature stressing on large, very large or big data. The presented methods are including clustering, sampling, compression, wavelets, histograms, and micro-clustering.

3 Clustering Algorithms

3.1 Background

One of the mostly ever used techniques with the purpose of data summarization is *clustering*. This is an unsupervised process of collecting similar data objects in a group, where data objects within a same cluster are more similar to each other than to the objects in other clusters. The goal of clustering is to group similar objects together to simplify further processing such as data mining, summarization, and analysis.

To cluster data points, the following questions need to be addressed: What are the similarity metrics that can be used to cluster data points in the same group? How the similarities are measured? What type of data can be used in different clusters? How the discovered clusters are evaluated? Is it possible to cluster all data points of the entire data sets? Are the clusters having the same shape? How many clusters are required to present data objects? And the last but not least question is how large a data set could be to cluster its data points?

Answering these questions is a requirement of classifying clustering algorithms into different categories. Amongst those that could be related to big data, one can find *Hierarchal*, *Partitional*, *Density-based* and *Grid-based* algorithms. We also like to note that in clustering data points, some issues as requirements of clustering algorithms should be considered. Some examples are including scalability, dealing with different type of attributes, finding arbitrary shape clusters, ability to cope with outliers and noise, considering high dimensional data sets, interpretability and usability.

To better understand of clustering algorithms, first we will identify different types of data and then explain a preliminary principle of clustering algorithm, which is a proximity measure, a common phrase to represent similarity $s(i, j)$ and dissimilarity $d(i, j)$ measure between two data points, two clusters or a data point and a cluster.

Data Type In order to universal interpretation of data, data has been categorized into two scale of measurement: qualitative and quantitative. The former includes nominal scales and ordinal scales. The latter includes interval and ratio scales. In addition to the mentioned measurement scales, there are other words to describe types of data including categorical, numerical, binary, continuous, and discrete.

Similarity/Dissimilarity Measures Based on type of data, various similarity/dissimilarity measures can be defined. Therefore, some similarity measures are reviewed quickly in the following. One of the most used similarities metric is distance. There are many distance measures which are mostly used in clustering algorithms such as Minkowski, Manhattan or City block, Euclidean, Mahalanobis, etc. which are described below.

Minkowski Distance For two numerical points of $x_i = (x_{i1}, x_{i2}, \ldots, x_{i1})$ and $x_j = (x_{j1}, x_{j2}, \ldots, x_{j1})$, Minkowski distance or L_p norm is calculated as following:

$$D_{ij} = \left[\sum_{l=1}^{d} |x_{il} - x_{jl}|^{1/n}\right]^n$$

An example of application of Minkowski distance can be found in [1].

Euclidean Distance Euclidean distance or L_2 norm is a mostly common used measure distance in clustering algorithms such as [2] to find similar numerical objects and tend to find hyper spherical clusters. It is a particular instance of Minkowski at n = 2. It measures distance between two points of $x_i = (x_{i1}, x_{i2}, \ldots, x_{i1})$ and $x_j = (x_{j1}, x_{j2}, \ldots, x_{j1})$ as follows:

$$D_{ij} = \left[\sum_{l=1}^{d} |x_{il} - x_{jl}|^{1/2}\right]^2$$

K-means, CURE and BIRCH algorithms are some examples of clustering algorithms that measure similarities between data points based on closeness via Euclidean distance measure.

Manhattan/City Block Distance Considering Minkowski at n = 1 for two numerical points of $x_i = (x_{i1}, x_{i2}, \ldots, x_{i1})$ and $x_j = (x_{j1}, x_{j2}, \ldots, x_{j1})$ gives Manhattan or City block distance or L_1 norm like in [3] as follows:

$$D_{ij} = \left[\sum_{l=1}^{d} |x_{il} - x_{jl}|\right].$$

Manhattan or City block distance normally causes finding hyper rectangular clusters.

Mahalanobis Distance Mahalanobis distance considers the correlation between variables or the variance-covariance matrix. Hyper ellipsoidal clusters can be discovered through applying Mahalanobis distance which is formulated as following:

$$D_{ij} = (x_i - x_j)^T S^{-1} (x_i - x_j)$$

where S is the within cluster covariance matrix. Examples of Mahalanobis distance can be found in [4, 5].

There are also different similarity and dissimilarity measures for categorical data. The simple matching distance proposed in [6] is one of the well-known one to measure dissimilarity between categorical data. Let be x and y as two categorical data points. The simple matching distance between x and y is computed as follows:

$$\delta(x, y) = \begin{cases} 0 \; if \; x = y \\ 1 \; if \; x \neq y \end{cases}$$

For two categorical data points of x and y with l attributes, dissimilarity metric based on the simple matching distance is calculated:

$$d_{sim}(x, y) = \sum_{j=1}^{l} \delta(x_j, y_j)$$

Note that there are far more similarity/dissimilarity measures in the literature. However, due to the lack of space, we have only explained the most common ones and listed some of them in the following.

Some other dissimilarity measures for numerical data are *Mean character difference* [7], *index of association* [8], *Canberra metric* and *Coefficient of divergence* [9], and *Czekanowski coefficient* [10]. Some matching coefficients measures for nominal data are Russell and Rao [11], simple matching [12], Jaccord [13], Rogers-Tanimoto [14], Kulczynski [15]. There are also some similarity measures for binary data such as Jaccord, Dice, Pearson, Sokal-Sneatha/b/c/d, Yule, Ochiai which some of them are summarized in [16]. The other metric that consider the similarity between groups of objects is linkage criterion or connectivity between them.

After presenting some of the concepts related to data clustering, we will consider some of the most prominent clustering algorithms found the literature, relevant to the topic of the handbook, namely big data stored and managed in data centers. T rest of this section focuses on clustering of very large data sets; therefore we will only describe those clustering algorithms that can be applied on very large data sets with the aim of summarization.

3.2 *Hierarchical Clustering*

Hierarchical clustering, also called *Connectivity-based clustering*, is one of the classical approaches for data summarisation. It creates clusters in the form of a tree in which each cluster is represented as a node. The main idea is the structured-based on proximity measure, where the nearby objects are clustered into a group. Therefore, distance between objects plays a pivotal role in the clustering of the data objects.

Tree-based hierarchal clustering algorithms can be formed into two types: Bottom up (Divisive) and Top down (Agglomerative). In former approach, the clustering process starts from root in such as way the entire data set is considered as a large cluster in root; later it iteratively splits data into partitions, where it terminates at leaves level. There are two good examples of divisive clustering algorithms, namely MONA and DIANA. These are detailed in [17] and some applications of such algorithms are given in [18]. In the agglomerative clustering algorithm, each data point is considered as a single cluster at the leaf level, and then every two closest clusters based on proximity measures will be merged up together to achieve a single cluster at the root tree.

Although a broad range of agglomerative hierarchal clustering algorithms exist in the literature such as *single linkage clustering* [19], *complete linkage clustering* [20], *group average clustering* [21], and *centroid method* [21], just some of them can be applied for very large data sets which is the focus of this chapter. The prominent reason that the above clustering algorithms are not selected to cluster large data sets is their quadratic computational complexity which is a function of number of data points.

As there are several deficiencies with hierarchical clustering algorithms, new algorithms were proposed to cover their shortcomings, such as the high degree of sensitivity to noise and outliers, incapability of correcting previous misclassification and unclear termination criterion. It should be noted that highlighting defects of hierarchal clustering algorithms does not contravene their important benefits, such as handling any forms of similarity or distance, covering different type of attribute and not requiring knowing the number of clusters in advance. Considering all pros and cons of hierarchal clustering algorithms, this has led to describe some other hierarchal clustering algorithms that could be applied on large data sets, such as BIRCH, CURE and ROCK.

- **BIRCH** [22] employs a tree structure, called CF Tree (Clustering Feature Tree). This tree is a height-balanced one and consists of leaf and intermediate nodes where each of them has certain entries. The number of entries is constrained by two branching factors, noted as B and L. The B factor is a maximum number of entries for each intermediate node, and the factor L represents the maximum number of entries for each leaf node. Each entry of intermediate node is in the form of $[CF_i, Child_i]$, in which CF_i is a summary information consisting a 3-tuple <N, LS, SS>, where N is the number of data points in a cluster, LS is the linear sum and SS is the square sum of the N data points in a cluster and $Child_i$ is a pointer to its ith child node. Entries of the leaf nodes are also in the form of $[CF_i]$. The number of leaf entries is controlled by a threshold T which is set to 0 by default. The height of tree is also defined by T. The larger T leads to the smaller tree.
  This algorithm is a local one since it does not scan all the data points once and it starts with sub-clustering of leaf entries via closeness metric. Five alternative metrics were used to measure closeness of clusters, and these include the followings: centroid Euclidian distance, centroid Manhattan distance, average inter-cluster distance, average intra-cluster distance and variance increase distance. The algorithm clusters dense area as a single cluster and remove sparse area as an outlier. It starts by building a CF tree dynamically and incrementally based on available memory and adjustable threshold of T. Each entry is inserted to CF tree based on the closest child node metric in leaves. If the number of entries of a leaf node does not exceed L, then a new entry is inserted to this leaf node otherwise the leaf node is divided and this division will be continued in ascend trend in the CF tree till a node, whether leaf or intermediate node, is found that has capacity to add more entries. If this trend presumes up to the root, the root of CF tree will be split and therefore the height of tree will be increased by one.
  Since CF tree is built based on agglomerative algorithm which is a bottom up approach as previously explained therefore, a new cluster in an upper level of CF-tree is constructed through merging two sub-clusters in lower level of CF Tree. In order to do that, Additivity theorem is used to merge two clusters. Based on Additivity theorem, CF vectors of two clusters are computed as below if two clusters are merged

$$\mathrm{CF_m} = \mathrm{CF_1} + \mathrm{CF_2} = (\mathrm{N_1} + \mathrm{N_2}, \mathrm{LS_1} + \mathrm{LS_2}, \mathrm{SS_1} + \mathrm{SS_2})$$

where CF_m is a clustering feature of newly merged cluster. The insertion operation in CF Tree is similar to B + -tree and also an Additivity Theorem is used to build a CF Tree.

BIRCH's main goal is to minimizing running time, memory and data scans. It also makes clustering decisions without scanning the whole data, group dense area as a single cluster and handle sparse area as outliers by removing them. Hence, it can handle outliers. Despite BIRCH's advantageousness, it has some deficiencies. One of the major problems of BIRCH is that it cannot perform well in face of non-spherical shape clusters since boundary of a cluster is controlled by the notion of radius or diameter. As it has been mentioned earlier, BIRCH clusters data points by using clustering features of the original data instead of using the whole data and consequently it causes to reduce storage space and frequent I/O operations, and also its computational complexity of O(N) makes it to be used to cluster very large data sets by making the time and memory constraint explicit.

Several extensions of BIRCH were proposed, which we briefly described here. A clustering algorithm is proposed [23] where its pre-clustering phase is similar to BIRCH. In this way that, the whole data set is scanned to find the dense areas and then clustering of dense regions are performed by applying a hierarchal clustering algorithm and making CF tree. In contrast with traditional clustering algorithms (that can deal with one of those attribute type), this algorithm can handle both continuous and categorical attributes. Therefore, clustering feature CF has been changed to $CF_j = (N_j, S_{Aj}, S^2_{Aj}, N_{Bj})$, where N_j shows the number of data spots in cluster C_j, S_{Aj} is the sum of consecutive features, S^2_{Aj} is the square sum of consecutive features of N_j data spots, and N_{Bj} is a d-dimensional vector representing the value of categorical attributes and distance of pair of clusters is measured by log-likelihood function.

BIRCH is generalized in [24] into a wider framework, called BIRCH*, in distance spaces. This framework is based on two algorithms named as BUBBLE and BUBBLE-FM. Parameters of CF vector in BIRCH* are a sum of the squared distance of a data point to other data points, centroid of the cluster which is determined based on minimum squared distance and the radius r of the cluster which are components to build the CF-Tree. BIRCH has been extended in many more studies such as [25–27] and [28].

- **CURE** (Clustering Using REpresentatives) [25] is another hierarchal clustering algorithm. Unlike BIRCH, CURE is robust against outliers and can deal with arbitrary-shape clusters. The handling of arbitrary shape clusters is due to the fact that each cluster is represented by a set of representative points instead of a single centroid or all-points. Therefore, it can find non-convex shape clusters. Furthermore, this set of representatives is shrunk towards centroid through an adjustable parameter $\propto = [0,1]$ to deal with outliers. Shrinkage causes outliers come closer to the centroid of the cluster to avoid wrong clustering.

  CURE is designed to apply for large data sets by using random sampling and partitioning. First, a sample of data set is chosen randomly, and then this sample is partitioned to K equal partitions. To reduce time complexity, these partitions are pre-clustered like pre-clustering phase of BIRCH and then an agglomerative hierarchal clustering is applied on each pre-cluster partition. At the end of the process, a label is assigned to each data points based on its distance from representatives. CURE applies two data structures in its algorithm, namely *kd-tree* and *heap-tree*. CURE stores its representatives in *kd-tree* and clusters are stored in the *heap-tree* date structure. The time complexity of CURE is O (N^2_{sample}) which depends on the number of sampling data and the number of partitions.

- **ROCK** (RObust Clustering using linKs) [26] is an agglomerative hierarchical clustering algorithm that groups categorical data points through the non-metric measures. The two metrics that measure either Euclidian distance in hierarchal clustering algorithms or a criterion function (such as square error) in partition clustering algorithm cannot properly measure similarity for categorical data points. Therefore, two similarities metrics were introduced in ROCK to enable accurate merging as well as clustering of data points. These metrics are: *sim* (p_i,p_j) to consider neighbors of a point and *link* (p_i,p_j) to define the number of common neighbors between two points p_i and p_j. The similarity measure is defined as follow:

$$\text{Sim }(T_1, T_2)=\frac{|T_1 \cap T_2|}{|T_1 \cup \ T_2|}$$

where | T_i| is the number of items in the transaction T_i and $Sim(T_1, T_2) \geq \theta$; $0 \leq \theta \leq 1$. Data points are considered as transactions in market basket. If no similarity is found between transactions, then $\theta = 0$; meaning means that any pair of transactions can be neighbors of each other. If θ= 1, therefore only identical transactions can be considered as neighbors of each other. Thus, it is important to properly define θ, which is a user-specified parameter based on desired closeness. The number of links between a pair of points also indicates the probability whether data points are presented in a same cluster or not. The larger link is, the more probable two points belonging to the same cluster. Using links allows ROCKS to be robust. It is important to note that ROCK clusters data points in a similar way to CURE does, however the difference is that ROCKS used links and different similarity measures instead of distance measure. It can also handle outliers as well. Finally, ROCK uses random sampling and labeling techniques, which makes it a good approach to deal with very large data sets.

Table 1 provides a summary of characteristics of these three well-known algorithms described earlier, namely BIRCH, CURE and ROCK. Our investigation regarding such clustering approaches for big data can be simply summarized as follows: BIRCH is suitable for large data sets where finding spherical shape clusters in a linear time is required. CURE can be used to search arbitrary-shape clusters of numeric data in large data set. CURE is robust against outliers. Furthermore, CURE can be fit in available memory since a random sample of large data set is chosen to perform clustering. ROCK will tackle with the presence of categorical data in large data sets.

Table 1 Characteristics of BIRCH, CURE and ROCK clustering algorithms

Algorithms for large data sets	Type of data	Cluster shape	Time complexity	Space complexity
BIRCH	Numerical	Spherical	O (N)	–
CURE	Numerical	Arbitrary	$O(N_{sample}^2 \log N_{sample})$	$O(N_{sample})$
ROCK	Categorical	–	$O(N_{sample}^2 \log N_{sample} + N_{sample}^2 + kN_{sample})$	$O(\min\{n^2, nm_m m_a)$

3.3 *Partitioning Clustering*

In Partitioning clustering algorithms, a data set is partitioned into *k* partitions with *n* objects within each partition using a predefined objective function. Minimizing square error function is as an objective function which is computed as follows:

$$E = \sum \sum ||p - m_i||2$$

where *p* is a data point in a cluster and m_i is the mean of the cluster. As the centroid/medoid-based algorithm considers all possible partitions, this is not practical for large data sets due to its high computational complexity. Hierarchal clustering algorithms cannot undo in their clustering phases, meaning that if two clusters are merged, it is not possible to obtain the two original clusters before merge operation by splitting the merged cluster. Therefore, a few heuristic methods are used to deal with this issue, such as *k-means* and *k-medoids*. In partitioning algorithms, it is possible to move an object from one cluster to another cluster to improve the clustering quality conversely hierarchal clustering algorithms. Howbeit, if a point is nearby to the center of another cluster, it maybe causes overlapping problem. Here we summarize some of the well-known partitioning algorithms and we briefly explain how they deal with very large data sets.

- **K-Means** [27, 28] is probably one of the most known partitioning algorithm. It divides data objects into *k* partitions in such a way that each object is assigned to the nearest cluster center. This operation is continued till visiting all data objects, and then the centroid is recalculated to achieve better clustering. The number of clusters (namely *k*), cluster initialization and distance metric are user-specified parameters, in which selection of *k* is the most challengeable task. Therefore, K-means is a heuristic algorithm and run several times to find better partitions with the smallest squared error since it aims to minimize the within-cluster sum of square.
  K-means is a greedy algorithm with time complexity of *O(TKN),* where *N* is the number of objects, *K* is the number of clusters and *T* is the number of iterations. *T* and *K* can be ignored since they are negligible in comparison with *N*. Therefore, K-means algorithm is scalable and suits for large data sets because of its linear complexity. However, the numbers of clusters are needed to be defined in advance; K-means has limitations when dealing with outliers and discovering non-convex

cluster's shape. K-means is also not suitable for categorical data and usually terminates at local optimum.

K-means utilizes the Euclidean distance so spherical shaped clusters are founded in this way. However, it is based on Mahalanobis distance to discover hyperellipsoidal clusters [29] with a higher computational cost.

Several extensions of K-means were later proposed to deal various aspects, such as cluster size, merge and split operations. ISODATA (Iterative Self-Organizing Data Analysis Technique) and FORGY are proposed in [30] and [31] respectively are some examples which were proposed in the field of pattern recognition. In [32] and then [33] some changes in K-means are made in terms of type of clustering (hard, in which each object belongs to just one cluster and in contrast in soft, each object can belongs to multiple clusters). This is called *Fuzzy c-means*. Another approach is proposed in [34] to make Fuzzy c-means and K-means faster through data reduction by replacing group examples with their centroids before clustering. Bisecting K-means [35] is another example, where it divides data recursively into two clusters at each phase. Another interesting extension of K-means is presented in [36], and it applies *kd-tree* to discover the closest cluster centers. In [37], *x-means* is proposed to defines k using Akaike Informatio Criterion (AIC) or Bayesian Information Criterion (BIC). Finally, Kernel K-means [38] and K-medoid [39] are the other extensions of K-means.

- **CLARA** (Clustering LARge Applications) [39] deals with the deficiencies of the clustering algorithms previously above using PAM (Partitioning Around Medoid) algorithm [39], which has a time complexity of O (k $(n - k)^2$), where k is the number of medoid objects and n is the number of non-medoid objects. Despite the attempt to fix the limitations of clustering algorithms, PAM is not an appropriate algorithm to be used for large data sets because of its time complexity.

  PAM is a medoid-based clustering algorithm. *Medoid* is a data point located roughly in the center of a cluster. PAM starts by finding k medoids randomly as representatives of each cluster and form k clusters. Then through the use of a brute force approach, it finds the best k medoids between all pairs of the entire data set to perfectly cluster k partitions. Obviously this is the reason for its high complexity. CLARA takes benefit of PAM algorithm by applying it on a random sample of the data set instead of the whole set. CLARA takes multiple samples from the data set and then applies PAM on each sample to find the best k medoids among the sampled data. After that, CLARA attempts to discover the most similar data points to each k medoids from the entire data set to form k clusters. However, there is not guarantee that CLARA can find the best k medoids during the sampling process and also does not achieve the best clustering.

  As it is mentioned, the problem with PAM is that it stores all pair-wise distances between objects which is space consuming and it is also not an option to apply for large data sets. However CLARA does not consider the whole dissimilarity matrix through sampling, which leads to achieve linear complexity in terms of time and space. So CLARA can be applied in large data sets.
- **CLARANS** (Clustering Large Applications based upon Randomized Search) [40] is an improved version of CLARA in terms of quality and scalability. This can

Table 2 Summaries of some of the characteristics of K-means, CLARA and CALARANS

Algorithms for large data sets	Type of data	Cluster shape	Time complexity	Space complexity
K-means	Numerical	Spherical	O (NKd)	O (N + k)
CLARA	Numerical	Arbitrary	$O(k(40+k)^2+k(N-k))^+$	–
CLARANS	Numerical	Arbitrary	Quadratic	–

be applied for large and high dimensional data sets since it uses a randomize search to cluster data points. CLARANS is also suitable to find polygon objects. The clustering process in CLARANS is similar to a search process in a graph. Each node in the graph is a representative of a set of *k* medoids. Two nodes are neighbors if their set of medoids differs by one. The algorithm starts with a random node and max-neighbors are checked in a random way to find better partition. If the neighbors provide a better partition, this process resumes with a new node; otherwise, the search stops by finding a local minimum. This iteration continues to find several local optimums, and the "best" local optimum is considered as clustering output.

CLARANS and CLARA are similar in terms of sampling. However, there is a difference between them when choosing samples from a data set. Although CLARANS does sampling for a set of neighbors of a node and does not consider all neighbors of a node, it does not restrict a search to a localized area. This means that CLARA draws a sample from the whole data set and then works on the selected sample, while CALARANS draws a sample of neighbors and dynamically changes this sample and so works on all data set not just on a particular sample of the entire data set. Since CLARANS considers local area at each step, it can detect outliers more precise than CLARA and it is more resistant to deal with increasing dimensionality.

Table 2 summarises some features of K-medoids, CLARA, and CLARANS in terms of complexity, data type, and cluster shape. These tree aforementioned partitional clustering algorithms can be applied for large numerical data sets. However, K-means is appropriate to find clusters of spherical shape, while CALAR and CLARANS can find any arbitrary shape clusters. For clustering large data sets, CLARANS demonstrates better quality and efficiency than CLARA in discovering clusters, however it fails to enables clustering very large data set because of its quadratic time complexity.

3.4 *Density-Based Clustering Algorithms*

In Density-based algorithms, clusters are created based on highly dens areas over the remainder areas and the sparse area are classified as noise or border area. In this way, they can deal with outliers and non-convex shape clusters. Some of the mostly used density-based algorithms for large and high dimensional data sets are

DBSCAN, DBCLASD, GDBSCAN, DENCLUE, and OPTICS which are briefly explained below.

DBSCAN (Density-Based Spatial Clustering of Applications with Noise) [41] defines clusters using the concept of *density reachability*. Simply, a point q is directly density-reachable from a point p if this is not farther away than a given distance. N_{eps} and the minimum number of points (MinPts) are critical factors for DBSCAN to generate a cluster. This algorithm starts with a random or arbitrary point, and if sufficient neighbors are surrounded within the range of *eps*-neighborhood of a selected node, a cluster is then formed. Otherwise, the point is considered as noise. However, it is possible that a rejected point (noise) to be reconsidered as a part of a cluster if this meets specific conditions. If a point is found to be a dense part of a cluster, its *eps*-neighborhood is also part of that cluster. Hence, all points that are found within the *eps*-neighborhood are added, as is their own-neighborhood when they are also dense. This process continues until the density-connected cluster is completely found. Then, a new non-visited point is retrieved and processed, leading to the discovery of a further cluster or noise.

DBCLASD (Distribution Based Clustering of Large Spatial Databases) [42] is an incremental density based clustering algorithm that uses a uniform distribution of data points in a cluster. Nearest neighbor distance is a key parameter through which clusters are formed. This algorithm builds clusters incrementally, meaning that it does not require loading the whole dataset into the memory and it processes each data point on time. It is also named online clustering. Arbitrary shape clusters are discovered in this algorithm and it does not need any input parameter because of this is called independent user-specified parameters algorithm. However, the problem is that it is an order dependent algorithm.

GDBSCAN (Generalized Density Based Spatial Clustering of Applications with Noise) [43] is a generalized version of DBSCAN. Two definitions were changed in this algorithm. First, it changes the definition of neighborhood by a symmetric and reflexive binary predicate. This means that any binary predicate which is symmetric and reflexive can define a neighborhood like intersect predicate to identify neighborhood in polygon. The second change, to obtain cardinality of a neighborhood, other measures such as non-spatial attribute are used instead of direct enumerating data objects of a neighborhood's object.

DENCLUE (DENsity-based CLUstEring) [44] is build based on the idea that every data point has an impact within its neighborhood, which is mathematically modeled through *influence* function. In addition, sum of influence functions are computed to obtain density attractors that are local maxima of the overall density function to define clusters. DENCLUE is robust against noise and outliers. It can handle arbitrary shape clusters in high dimensional data set. It is faster than DBSCAN since it uses grid cells and just keeps information of grid cells in a tree-structure access. In spite of all these advantageous, DENCLUE needs to choose the density parameter and noise threshold carefully since they have a remarkable impact on the clustering quality.

OPTICS (Ordering Points To Identify the Clustering Structure) [45] is an algorithm for finding density-based clusters in spatial data. The rationale behind OPTICS

Table 3 A comparative study of the various clustering algorithms

Algorithms for large data sets	Type of data	Cluster shape	Time complexity
DBSCAN	Numerical	Arbitrary	O(NlogN)
DBCLASD	–	Arbitrary	Roughly 3 times of DBSCAN
GDBSCAN	–	–	O(n* runtime of a neighborhood query)
DENCLUE	Numerical	Arbitrary	O(NlogN)
OPTICS	Numerical	Arbitrary	O(NlogN)

* stands for multiply

is similar to DBSCAN, however it addresses one of DBSCAN's major weaknesses: the detection of meaningful clusters in variable density data set. To do so, the points of the database are (linearly) ordered such that points are spatially closest become neighbors in the ordering. Additionally, a special distance is stored for each point that represents the density that needs to be accepted for a cluster in order to have both points belong to the same cluster. This is represented as a *dendogram*.

Table 3 summarizes some the features of the described density based clustering algorithms. As seen, they can be applied for large data sets including numerical data. All of them are capable of finding arbitrary shaped clusters.

3.5 Grid-Based Clustering Algorithms

Grid-based clustering algorithms generate a finite number of cells by quantizing data space and make a grid structure to perform clustering process on it. Since these methods are dependent on the number of cells in each dimension and not on the number of data objects, their processing time is very fast. STING, CLIQUE, GRIDCLUS, Wave Cluster, FC and OptiGrid are examples of well-known grid-based clustering methods applicable in large and high dimensional data sets.

The main idea behind Grid-based clustering methods is taken from [46, 47] and can be summarized as follows:

1. *"Creating a grid structure, i.e., partitioning the data space into a finite number of non-overlapping cells;*
2. *Calculating the cell density for each cell;*
3. *Sorting of the cells according to their densities;*
4. *Identifying cluster centers;*
5. *Traversal of neighbor cells".*

Also mentioned in [48], some of the main characteristics of the grid-based methods are as follows: (1) no distance computations; (2) clustering is performed on summarized data points; (3) shapes are limited to union of grid-cells; and (4) the complexity of the algorithm is usually O (# *populated-grid-cells*).

STING (Statistical Information Grid-based clustering) [49] decomposes the spatial data into rectangular cells and represented by a hierarchical tree. Likewise BIRCH, STING makes data summaries in this way that statistical information such as mean, maximum and minimum values, standard variation and distribution type are stored in each cell. STING is a query independent method since grid-cells store statistical information as summary information which is independent of the query. Incremental updating and parallelization are suitable for this grid structure. In addition, Time complexity of STING is O (K), where K is the number of grid cells at the lowest level. Despite all of advantageous of STING, its performance depends on the granularity of the bottom layer of grid structure. Moreover, created clusters are enclosed horizontally or vertically not diagonally which impacts on the quality of clustering.

WAVECLUSTER [50] is originated from signal processing and it transforms spatial data into a frequency domain to find a dense area in the frequency domain. In this way, different clusters with different resolutions and scales are obtained. The computational complexity of wavelet transformation is O (N), where N is number of objects in the data space. WaveCluster can handle outliers and works very well with high dimensional spatial data. It is able to find arbitrary shape clusters. Moreover, it is not required to know the number of clusters in advance.

GRIDCLUS [51] is used on the space surrounding the data values instead of the data by taking benefits of multidimensional data grid. A neighbor search algorithm is applied to cluster blocks organizing patterns. This algorithm consists of five main steps: (1) insertion of points into the grid structure, (2) calculation of density indices, (3) sorting the blocks with respect to their density indices, (4) identification of cluster centers, and (5) traversal of neighbor blocks.

FC [52] is a self-similar clustering algorithm in which self-similarity is measured by applying concept of fractal dimension through Hausdorff dimension. FC incrementally adds points into the cluster and after clustering, there is not any radical change in the cluster's fractal dimension. Since the space is partitioned into the cells of a grid, it is counted as a grid based clustering algorithm. FC scans the data once and it is a suitable clustering algorithm for large data sets and high dimensional one. It can handle noise and also discover clusters of arbitrary shape.

OptiGrid [53] uses a grid clustering algorithm that is applied for high dimensional data sets. It runs in the way that the whole data set is recursively partitioned into different subsets to find optimal grid partitioning. "Optimal" grid partitioning is achieved by finding good cutting plane for each cluster recursively through a set of contracting projections.

CLIQUE (Clustering in QUEst) is proposed in [54] in which subspaces of k-dimensional data set are defined to find their dense areas to present a cluster in k-dimensional data space. It identifies subspaces of a high dimensional data space to achieve better clustering than original space. To find dense regions in a subspace, each dimension is divided into equal intervals. A dense area is found when the number of data points in this area exceeds a defined threshold. Also a cluster in a subspace is a maximal set of connected dense units. Therefore, after identifying subspaces containing clusters, it finds dense areas and connected dense areas in all subspaces

Table 4 Comparisons of time complexity and applicability of clustering algorithms for high dimensional data

Clustering Algorithms for large data sets	Capability to apply in high dimensional data	Time complexity
Wave Cluster	No	O(N)
STING		O(number of cells at the bottom layer)
FC	Yes	O(N)
CLIQUE	Yes	Linear with the number of objects and quadratic with the number of dimensions
OptiGrid	Yes	Between O(Nd) and O(NlogN)

of interest, and then through MDL principle clustering is terminated. CLIQUE automatically identifies subspaces, it is not sensitive to the size of input and the number of dimensions and it can scale linearly. However, it is a simple method that causes to lose the accuracy of the clustering (Table 4).

To conclude on clustering algorithms, the reader may refer to some other interesting [55–57]. This section however had a focus on analyzing some specific algorithms that can be used for very large data sets.

4 Sampling

The definition of sampling from Merriam Webster dictionary is "*the act, process, or technique of selecting a representative part of a population for the purpose of determining parameters or characteristics of the whole population.*" Based on this definition, sampling can be considered as a summarization technique that could reduce time and space by just observing a part of the whole data set that is still informative instead of the entire data set.

With the advent of digital technology, many data storages bear a huge volume of data that need to be process and analyze to meet user's requirements. However, considering this huge amount of data demands a lot of time and cost. So for tackling these issues, sampling techniques have been applied widely in many research areas such as data mining, data management, query optimization, approximate query answering, statistics estimation and data stream processing which meet the purpose of summarization.

Therefore, before going through explaining different sampling techniques, it might be better to describe some preliminaries useful for the reader to understand sampling techniques.

- *What is a sample?* A sample is a representative of a larger group (population) which preserves the same characteristics of the population and study is conducted on the sample instead of the population.
- *What is population?* Population is the large group of data from which sample is taken to do study on it.

- *What is a frame?* Sampling is performed on a special set of population which is called frame.
- *What is the aim of sampling?* Generalization of an induction derived from a data collection (sample) to the population is the main goal of sampling.
- *What is sampling error?* Since statistical characteristics of a population are illustrated by a sample, it is expected to encounter sampling error which is the difference between sample and population or in the other words, from statistics and optimization study, statistical error is the difference between observed value (sample) and unobserved value (population).
- *What is sampling bias?* Sampling with unequal probability of being selected individual data points (sample) from a data set (population) is indicated as bias sampling which is a non-random sampling.

After explanation of these preliminaries, let us consider the different sampling methods. There are various sampling techniques to meet different aims of wide range of applications; however most of them use the following steps to take sample from a data set.

- Define population (N) to be sampled.
- Determine sample size (n).
- Control for bias and error.
- Select sample.

As a matter of fact, some factors should be concerned in selection of a sampling technique such as degree of accuracy, research objectives, resources, time frame, knowledge of population, research scope and statistical analysis requirements.

This section aims at reviewing some of the primary sampling techniques and their extensions looking back their summarization aspect. To avoid confusion, it should be mentioned that we will use the terms data set and population interchangeably throughout this section. In general, sampling techniques are categorized into two main groups of probability-based and non-probability-based sampling. Sampling algorithms that give an equal chance of being selected to all data points are considered as probability or unbiased sampling. While, biased sampling algorithms consider data points with different probability and sampling rate. Based on study of the literature on probability and non-probability sampling techniques, the most common methods of probability-based sampling are *simple random sampling*, *systematic sampling*, *stratified sampling*, and *clustering sampling*. Meanwhile the most common non-probability sampling is *accidental sampling*, *quota sampling*, *snowball* and *purposive sampling*. All these techniques are described below.

4.1 Probability Sampling

A) Simple random sampling Simple random sampling [58] is one of the basic sampling techniques in which probability of being chosen of each individual data point as a sample is as equal as other data points in the data set. Simply speaking, every data point has an equal chance to be selected as a sample. The data points are numbered from *1* to n and then a sample including some random number is chosen from them. Simple random sampling can be performed into two ways: with replacement and without replacement. In the former, every time a data is drawn from the data set, this is replaced to the data set and gets a chance to be re-selected with the same probability in the next round. In the latter means, every data point can be selected once and after a selection, it is removed from data set and it is not considered any more.

The advantage of random sampling is that it is really easy to perform with minimum insight from data set in advance. However, it needs to have a list of all population. There is a broad study of random sampling for various applications in the literature. We briefly review some of them in this section. In the context of large, very large or big data set, analyzing and processing such large data takes time and sometime it is not possible to store the whole data set such as data stream. Therefore, to accelerate performing these tasks, random sampling (which does not require pre-knowledge of data) can be helpful in this way to efficiently process and analysis data and be performed on a small part of the entire data (*sample*) which is still informative and accurate. Approximate answer can be obtained in a faster way instead of considering the massive volume of data. Therefore sampling can be considered as a good summarization technique for big data.

Random Sampling with Reservoir [59] solves the issue of selecting a sample size of m without replacement, randomly from a data set of size N (N *elements*), where N is not known in advance. It is an extension of random sampling that is one of the classical uniform schemes, and also is an infrastructure for many uniform sampling methods such as concise sampling, dynamic inverse sampling, chain sampling, and distinct sampling.

In the sampling algorithm, a *reservoir* maintains a fixed size uniform random sample of k that is drawn during a sequential pass through the data set. This means that the first n data points are added to a reservoir. Then, by arriving $n + 1^{th}$ data point, one of the existing data point in the reservoir is randomly chosen to be deleted and therefore makes space for new data points in the reservoir since the size of the reservoir is fixed and it is required to keep it constant.

The unbiased reservoir random sampling is performed with average CPU time of $O\ (n\ (1 + \log N/n))$.

In [60], the authors also proposed an online algorithm to choose a sequential random sample of n from a data set of size N with minimum memory requirement.

A sampling method is described [61] to summarize data traffic in vehicle to vehicle (V2V) space. Previous sampling methods (such as sliding Window, Reservior Sampling and Exponentially biased reservoir sampling) consider incoming traffic flows that are increasingly ordered based on data arrivals; therefore there was a limitation in this context. In fact, they investigated the case the data traffic is disorder because of transmission delays and multiple sources. They extended the early sampling method and made some changes to make them compatible with disorder data streams. They also proposed another sampling method, called Polynomially Biased Reservoir Sampling (PBRS), which is applicable for multi-dimensional sampling tasks. In this way, a huge volume of traffic data can be summarized to be considered as data stream and used this summary to predict upcoming traffic data and its conditions.

The reservoir sampling method is improved within DSS (distance-based sampling) algorithm [62] for transactional data stream to cover deficiency of low performance of reservoir in dealing with small sampling. They enhanced accuracy of reservoir sampling by using and comparing Euclidean distance function and re-ranking step in arriving new transaction to decide whether include to sample or not.

Acceptance/Rejective sampling is based on Bernoulli design, where every data point can be included in the sample. It is subjected to an independent Bernoulli trial that has an outcome that could be Success (1) or Failure (0). Every data point can be randomly included in the sample. Therefore, if the probability of success is shown by p, the probability of failure is $q = 1-p$. Given n independent Bernoulli trials, then the probability of m success is mathematically shown as follows, which is called Binomial distribution.

$$\mathrm{P(m)} = \binom{\mathrm{n}}{\mathrm{m}}\mathrm{p^m q^{(n\text{-}m)}}.$$

In Acceptance/Rejective sampling [63], a candidate is obtained where acceptance or rejection of candidate depends on meeting some user-specified conditions. If it meets condition, then it is accepted to be included in sample; otherwise, it is rejected and next candidate will be selected repeatedly.

Chain sampling and priority sampling are two extensions of reservoir sampling [64]. The problem is how to select a sample from a moving window of recent data. The chain sampling deals with expired data in this way that a constant size sample of k is taken from window size of W. Whenever a new data i arrives, its chance to be taken as sample is 1/min (n, W) and if so, then an index from domain of ($i+1, \ldots,$ $i+n$) is candidate to be swap with i, when i^{th} data point is expired and this process of finding a substitute for newly arrived item is continued like a chain. This approach is applied on sequence based windows with space complexity of O(kLog$n$).

They also considered the case that window size is not constant and it is time stamp based. A priority between 0 and 1 is allocated to newly arrived data points and the highest priority will be chosen to be included in the sample. The space complexity of priority algorithm is also not more than O(kLog$n$) without any prior knowledge of size n.

Biased Reservoir Sampling Reservoir sampling is an unbiased sampling algorithm, where data points have equal chance of being selected as a sample. This unbiased sampling approach may have some deficiency in coping with evolving data streams. Indeed, after a period of time, some parts of a sample may be less related and therefore get "useless" because of evolution of the data stream. Then, since recent history of evolving data streams are more considered than the rest of data stream, the probability of their appearance in the sample should be changed and they don't have the same probability over the other part of data streams for sampling. A biased reservoir sampling is proposed in [65] and employs a memory-less bias functions to use replacement algorithm in the occurrence of stream evolution.

Random Pairing As mentioned earlier, Reservoir sampling method cannot deal with expired data. This means that reservoir sampling can handle only updates and insertions, and not the deletions. To deal with this issue, a new sampling method is proposed in [66], called Random Pairing (RP), to cope with deletions in a data set with stable size. In this way that, they add new data points to the sample to keep the size of sample constant when a deletion occurs in the sample. They also considered growing data sets whose size increase over time and proposed a resizing algorithm to control growing the sample over time.

Concise and Counting Sampling These are uniform random sampling algorithms [67]. Concise sampling is similar to the reservoir sampling having this difference that the values which are appeared frequently in the sample are displayed as a couple of < value, count> to save more space. This approach inserts new data point to the sample with a probability of 1/T. If a newly arrived item has been visited beforehand in the sample, then the count increased. By overpassing predefined sample size bound, the new bound T' is defined such that $T' >$ T, and the deletion of each data point with p (T/ T') and insertion of subsequent data points to the sample with p (1/ T') is performed to achieve uniform sampling with lower overhead.

Counting sampling [67] is an alteration of concise sampling with different treatment to deal with exceeding the predefined threshold of sample size. It is more accurate than concise sampling. Later, Counting sampling was extended in [68] through applying a tracking counter as an estimator to count and discover the high frequency item set, sum and average and employing Bernoulli samples over evolving multisets.

Weighted Random Sampling (WRS) [69]. Unlike uniform random sampling, where each data point has equal chance of being selected, data points do not have the same probability. Therefore, data points are weighted and they are selected based on their assigned weights. For example, WRS can be applied over data stream that is considered as big data, to take a sample from the recent data streams since based on weighting different part of data streams, it is possible to choose a part of data streams which has high weight according to recent data streams. There are various extensions of WRS in the literature such as [70] and [71].

Congressional sampling method [72] is composed of biased and unbiased sampling techniques which are called senate and house respectively. Data are divided into groups and then a uniform sampling is performed on each group (house) and a biased sampling (senate) is applied on the entire data set and then two taken samples are combined. They applied their proposed approach on group-by approximate query to enhance accuracy of group by query. Generating fast approximate answers to complex queries in very large data warehouses can be achieved through pre-computed summaries, samples, instead of considering the whole data warehouses that are very large and takes too much time to find answers from them.

B) Systematic Sampling Let assume one wants to choose n samples from a data set with N data points. First, systematic sampling [58] computes an interval through this way that $K = N/n$, where K is the size of interval. Then, a random starting point is selected. Thereafter, the starting point and K^{th} data point from starting point are picked as the first and second sample. This process continues to pick every K^{th}data point based on the predefined interval till n data points are selected as samples. For example, for a data set with 20 data points, and 4 samples, the interval will be $20/4 = 5$. Therefore, a starting point will be selected randomly from the first interval [1, 5]. Suppose that third data point is chosen as starting point. Afterward, every 5th data point is picked as sample. So, 3, 8, 13, 18 are chosen as the sample set. It should be noticed that inappropriate selection of intervals may be caused that some patterns in data are stayed hidden.

An advantage of systematic sampling method is in the simplicity of sample selection as well as its accuracy in comparison with random sampling. However, the chance of being chosen for all data points is not equal and depends on the starting point and interval. Systematic sampling can prepare enough samples if there is not any pattern in data. It is a good option for web query analysis where fixed interval sampling is performed.

Some studies concentrated their efforts to improve systematic sampling. Linear systematic sampling in [73], circular systematic sampling in [74] are also considered in the literature. In [75], a modified balanced circular systematic sampling when $N \neq nk$ is suggested. Another modification of systematic sampling, called FCFS-SS (First Come First Serve following Systematic Sampling) [76], is proposed. This works based on conventional systematic sampling with having difference of taking more than a sample at each time.

C) Stratified Sampling Stratified random sampling [58] divides population into L non-overlapping subpopulations called strata. Then, a sample is taken from each stratum individually and if random sampling is employed in drawing sample from each stratum, the method will be called stratified random sampling. It is noted that the size of a sample is usually computed in different ways for each stratum. One way is the proportional to the size of the stratum, called *optimal allocation*, aiming to maximize precision with minimum cost.

$$n_s = n^*[(N_s * \sigma_s)/sqrt(c_s)\,]\,/\,[\Sigma(N_i * \sigma_i)/sqrt(c_i)]$$

where n_s is the sample size for stratum s, n is total sample size, N_s is the population size for stratum s, σ_s is the standard deviation of stratum s, and c_s is the direct cost to sample an individual element from stratum s.

Another way is *Neyman allocation*, where size of sample for each stratum is defined based on the stratum size and its standard deviation with the aim of maximizing precision with a given fixed sample size, which is defined as

$$n_s = n^*(N_s * \sigma_s)/\ [\Sigma(N_i * \sigma_i)]$$

where n_s is the sample size for stratum s, n is total sample size, N_s is the population size for stratum s, σ_s is the standard deviation of stratum s.

Although categorizing and identifying proper strata is not easy and analysing results is complicated in stratified sampling, it covers the population better than the simple random sampling. Stratifying a sample is easy and helpful to better analysis data for each group with different characteristics. Accuracy of stratified sampling can be regarded. Stratified sampling drawn many attentions to be observed and extended in many studies. Some of these studies are cited accordingly.

- In a heterogeneous data stream, stratified sampling could be a good option to take a sample from every sub-stream with different statistical properties. In this way, a data stream is clustered as strata and then a random sample is taken from each obtained homogeneous cluster or strata. In [77], an adaptive size reservoir sampling method is proposed to regulate the size of reservoir since it is mentioned earlier that the size of reservoir in conventional reservoir sampling is constant. Therefore, they proposed the method to maintain constant size of reservoir in some situations that the size of reservoir varies. Then, they extended their proposed solution to adaptive multi-reservoir sampling.
- In [78], an adaptive stratified reservoir sampling (ASRS) is offered in which two issues of optimal size determination of sub-samples of each sub-stream and uniformity maintenance of each sub-sample is addressed. They considered the first issue by applying power allocation from [79] and for the second issue, they employed an alteration of their previously proposed adaptive-size reservoir sampling technique [77].
- In [80], sampling algorithm is illustrated and called strata which is based on the stratified sampling with the aim of drawing a sample to minimize the workload error and applicable to approximately answering aggregate queries.
- A recursive stratified sampling method is presented in [81] to apply association rule and differential rule mining on the deep web. So for, they draw a testable sample from deep web to discover rules. After that, a learning phase is established to find out data distribution and their relationship. At the end, the recursive stratified sampling is executed on the deep web. Their approach outperforms simple random sampling in terms of accuracy and cost.

D) Clustering Sampling In cluster sampling method, the population is grouped into the mutually exclusive and collectively exhaustive clusters, and later some clusters not individual points are picked through random sampling. There are two kind of clustering sampling: *single-stage* and *multi-stage*. So if clustering sampling is a single-stage, all data points of all selected clusters are seen as a sample; otherwise, in case of multi-stage, a random sampling method is employed to select the data points from each chosen clusters in each stage. If clusters are similar, then sampling error will be reduced. In case of very different clusters, sampling error gets larger and cluster sampling is not a suitable method in this situation.

It is worth to note that in stratified sampling, an individual data point is drawn from each stratum as a sample but in cluster sampling, a cluster is selected and then treated as a sample. Stratified sampling aims to increase accuracy while cluster sampling aims to reduce cost with respect to boosting efficiency of sampling. The prominent advantage of cluster sampling method rather than the other methods is that it is cheap but with the expense of higher sampling error.

Other Improvements and Applications of Probability Sampling Techniques Many studies are conducted within scope of random sampling techniques with different applications that are quickly reviewed some of them in the following.

In [82], random sampling method improved by taking advantage of generating a decision tree from a data set. In fact, decision tree presents a knowledgeable structure of data set which may be very large. On the other hand, random sampling is a way to summarize the data set but it may be, cannot present the general picture of the whole data set very well via taken samples. Therefore, they took benefits of both methods of random sampling and decision tree to present a better picture of the entire data set in a concise way.

A method for online maintenance of arbitrary sample size is investigated in [83], in which sample is drawn through random sampling without replacement, by applying a suggested geometric file with the expense of O $(\omega \times log\,|B|\,/\,|B|)$ random disk head movements for newly sampled record.

In [84], a sampling method is proposed to take a representative sample from a relational database considering data correlation. They named their new sampling method as CoDs (Chains of Dependencies-based sampling) through which a link of dependencies between data is extracted by considering foreign key constraints. They employed histograms in order to simply depict these relationships in distributed data, then they were analyzed these discovered dependencies to do sampling.

In [85], a sampling framework for parallel data mining was suggested. They aimed to mine useful information from a large data base by finding frequent item sets and sequential patterns. To achieve their purpose, they employed a pattern-growth algorithm which is categorized as divide and conquer algorithm since it projects and segments data base based on discovered patterns. Then, they tried to balance distribution of work load of mining tasks across processors. In order for parallel data mining, they required to estimate time mining of different tasks to achieve load balance, therefore they addressed this issue by proposing a selective sampling. They tested their parallel mining algorithm on a selective sample to estimate required time for each task and also identifying large items.

Their proposed selective sampling takes a sample from frequent item set by casting off a fraction of the most and the less frequent items and not considering the last m ending items of each sequence and also infrequent ones.

There are some studies on sampling for approximate query answering applications such as [86]. There is more research considering maintenance of dynamic data streams such as [87]. Discovery of association rules through sampling has been investigated in many studies by [88–90] sampling approaches for database files are reviewed.

4.2 Non-Probabilistic Sampling

A few approaches relate to non-probabilistic sampling:

- *Accidental sampling* [91], also called *convenience* or *opportunity sampling*, takes those data points from data set as a sample that are more available or close to hand. Therefore, taken sample through this approach could not be a good representative of the entire data set.

- *Quota sampling* [92] is a non-random sampling in which data set or population is divided into mutually exclusive groups. Then through a judgment, samples are drawn from each group satisfying determined proportion. In fact, quota sampling is the no-probability case of stratified sampling. Quota sampling can be evoked in some cases including when factor of time is more momentous than accuracy, when budget is limited or when sampling frame is not accessible.
- In *purposive sampling* [58], samples are taken from a specified population. It means that the research focuses on sampling from a particular population because of that it is named purposive. [93] is an example of applying purposive sampling in social networks with the purpose of recruitment.
- In *snowball sampling*, a data point or group of data points which are sampled provide more data points to be sampled. Snowball sampling is useful in data mining of social network such as [94].

5 Compression

Data compression aims to represents data using fewer bits than the raw data through which resource's usage like storage space or transmission capacity is reduced. This is categorized into lossless and lossy techniques. In former, compression is achieved by removing statistical redundancy and original data is retrieved after decompression, while in lossy compression, compression is obtained by throwing away inessential data and recovering original data is impossible. Since many redundant data and similar patterns in data can be extracted in compression, this can be considered as one of the summarization techniques that can reduce the size of data and present a compact version that is still informative and accurate. In other words, compression reduces the data size to save more space, communication cost and fast data transfer. Meanwhile, summarization tries to give a compact version of the entire data to be analyzed. Therefore, this compact version can be obtained through compression techniques. Since compression is considered as a data reduction techniques in the context of data mining, many researches are focused on applying compression as a summarization technique on big data sets.

Although there is a broad range of studies around data compression such as video coding, image coding, audio coding, and text coding, we consider briefly compression of event sequences based on Minimum Description Length (MDL) considering summarization aspects.

First, we consider definition of MDL, and then we review some studies focusing on compression of event sequences based on MDL (Table 5).

Minimum Description Length (MDL) traces backs to Occam's Razor principle. Occam's razor declares that among competing hypotheses, hypothesis with the least number of assumptions should be picked from which it results that simplicity generally causes correctness. MDL [95] formalized Occam's Razor principle in the way that the best hypothesis for a given set of data is the one that can achieve the best

Table 5 Advantages and disadvantageous of each probability and non-probability sampling. (Source: Black, T. R. (1999). Doing quantitative research in the social sciences: An integrated approach to research design, measurement, and statistics. Thousand Oaks, CA: SAGE Publications, Inc. (p. 118))

Technique	Descriptions	Advantages	Disadvantages
Simple random	Random sample from whole population	Highly representative if all subjects participate; the ideal	Not possible without complete list of population members; potentially uneconomical to achieve; can be disruptive to isolate members from a group; time-scale may be too long, data/sample could change
Stratified random	Random sample from identifiable groups (strata), subgroups, etc	Can ensure that specific groups are represented, even proportionally, in the sample(s) (e.g., by gender), by selecting individuals from strata list	More complex, requires greater effort than simple random; strata must be carefully defined
Cluster	Random samples of successive clusters of subjects (e.g., by institution) until small groups are chosen as units	Possible to select randomly when no single list of population members exists, but local lists do; data collected on groups may avoid introduction of confounding by isolating members	Clusters in a level must be equivalent and some natural ones are not for essential characteristics (e.g., geographic: numbers equal, but unemployment rates differ)
Stage	Combination of cluster (randomly selecting clusters) and random or stratified random sampling of individuals	Can make up probability sample by random at stages and within groups; possible to select random sample when population lists are very localized	Complex, combines limitations of cluster and stratified random sampling
Purposive	Hand-pick subjects on the basis of specific characteristics	Ensures balance of group sizes when multiple groups are to be selected	Samples are not easily defensible as being representative of populations due to potential subjectivity of researcher
Quota	Select individuals as they come to fill a quota by characteristics proportional to populations	Ensures selection of adequate numbers of subjects with appropriate characteristics	Not possible to prove that the sample is representative of designated population

Table 5 (Continued)

Technique	Descriptions	Advantages	Disadvantages
Snowball	Subjects with desired traits or characteristics give names of further appropriate subjects	Possible to include members of groups where no lists or identifiable clusters even exist (e.g., drug abusers, criminals)	No way of knowing whether the sample is representative of the population
Volunteer, accidental, convenience	Either asking for volunteers, or the consequence of not all those selected finally participating, or a set of subjects who just happen to be available	Inexpensive way of ensuring sufficient numbers of a study	Can be highly unrepresentative

compressed data. Rissanen [96] stated that rationale behind MDL is finding regularities in the visited data that prosperity of detecting these regularities is evaluated through the length with which the data can be explained.

MDL principle is defined as "*a relatively recent method for inductive inference. The fundamental idea behind the MDL Principle is that any regularity in a given set of data can be used to compress the data, i.e. to describe it using fewer symbols than needed to describe the data literally.*" [97]

After definition of MDL, we intend to describe event sequence and then we will explore some studies that are designed a compression method based on MDL for the event sequences. The reason for discussing compression of event sequences based on MDL in this chapter is because event sequences are generated massively through monitoring of systems and users' activities such as network traffic or logging systems. In order to easily analyse of these massive event sequences, a summarized version of this large volume of data can be concerned.

Event sequences are produced by monitoring user/system activities such as logging systems, network traffic data and so on. In order to handle and have a general picture of the entire system behavior, some research attempts focus on finding some comprehensive and short summaries of the entire event sequence. However, studies on presenting a local picture of a system's behavior have been done. In general, there are two points of view: local structure and global structure of a system. It is also noted that in data analysis, event summaries have meet some properties such as brevity and accuracy, global statement of data, local pattern recognition, and parameter free.

A nice solution to find short and accurate summaries based on MDL principle is described in [98]. This meets aforementioned properties. The authors proposed a summarization method, which findings an optimal segmentation and local models that are derived from MDL principles, as an optimization problem. Each sequence is segmented into n intervals where events with similar frequencies are grouped together. Also, the probability of occurrence of different event types at each timestamp is independent of the probability of occurrence of other event types and their segment. This means that different event types can appear in a same time. The authors tried to segment an interval of event sequence into contiguous and non-overlay intervals. After segmentation, a local model is computed as the one that can best describe data with a fewer bits in each segment. Two dynamic programming solutions were applied to find the minimum total cost through greedy algorithm, and the proposed segmentation method reduces the compression ratio as well as achieves the minimum overall description length in polynomial time.

An extension of [98] is proposed in [99] by considering overlapping segments, segments separated by gaps and presenting an event summarizer tool. However this extended approach has some limitations. First of all, the segmentation approach does not consider the relation between different models, and this is not a good option for predicting future patterns. Moreover, it stores same copies of models in an occurrence of long event sequences having many duplicated models leading to low compression ratio.

An event summarization method using MDL and Hidden Markov Model (HMM) is given in [100], and this captures both the global and local view of a system. An HMM is learnt to portray the global relationships among the segments. An event sequence is divided into disjoint segments based on the frequency changes of the events and then modeled each segment in a way that overall description has being the minimum length. Two types of models were considered: *independent* (M_{ind})and *dependent* (M_{dep}). In the former, each segment is isolated from other segments conversely in the latter, segments are correlated. Two different costs were considered to compute the final cost of encoding an event sequence: (1) the cost of encoding segmentation and (2) the cost of encoding event occurrences. The quality of summarization is evaluated through an objective function. The problem of this method is that only the intra-correlation among event types of adjacent segments was considered and the temporal information between event types in a segment is not taken into account. For this reason, they took benefits of the concept of machine state with the knowledge that system behavior within each state is stable.

To address limitations of the methods suggested in [100, 101], *Natural Event Summarization* (NES) was designed where inter-arrival histograms are used to exploit pairwise temporal correlations among events. By applying disjoint histograms and using MDL to encode histograms, event sequences are summarized. The temporal patterns of the events, which can be periodic or correlation, are discovered through histograms and then by employing multi-resolution characteristics of wavelet transformation, the size of discovered histograms has been shortened and the summary is visualized by event relationship network (ERN). However, there are different possibilities of drawing histograms to present correlation or periodic patterns between events, and employing MDL pave the way to choose the most suitable histograms to explain event sequence. Inter-arrival histograms are applied to find correlations amongst event types. Inter-arrival histograms facilitate the way to discover periodic and correlation patterns to describe temporal dynamics of event sequences. Therefore, in this way they depict a histogram graph presenting relationship among event types. Then the histogram graph is encoded through finding shortest path from it.

Dijkstra algorithm is used to find shortest path in a polynomial time O ($|D|^2$). To speed up the summarization process, the histogram graph is pruned by employing a wavelet transformation relying on multi-resolution analysis (MRA).

Conversely to previous work, inter-arrival histograms are used as they can define various boundaries of the segmentations of different event types. Thus pattern discovery become more efficient and produces *better* summarization. After discovering patterns in a sequence, the summarization results are summarized through ERN.

Pattern mining has attracted many attentions in the field of data mining. Specially, finding a small set of patterns is a concern, as the characteristics of a large data base could be presented in a small set of patterns instead of the whole discovered patterns. This becomes an aspect of summarization-compression. Two kinds of long pattern mining approaches are considered in the literature: *maximal item sets* [102, 103] and *closed item sets* [104]. The former is considered as lossy compression, while the latter is considered as lossless compression.

In [105], an algorithm called Clo_episode is offered to pick closed episode effectively through pruning methods and minimal occurrence. Furthermore, some studies take benefits of the MDL principle to find short, beneficial and high quality set of patterns to summarize and compress data that are shortly reviewed as follows. A two-phase MDL-based code table mining approach is investigated in [106], named as KRIMP, in which a set of frequent items is discovered. Then a pattern is selected from this set to improve compression ratio. In [107], a new version of KRIMP considered classification, where the issue of having large frequent items set is verified at low threshold by using specific heuristic methods. A set of item sets, which compresses the database in a lossless and good manner, is mined. The goodness of compression is determined by employing the MDL principle. They achieved a set of frequent item sets with four orders of magnitude shorter than the entire frequent item sets. Experimental results are extended in [108] for evaluating other methods, and proposed STREAMKRIMP in [109] as an extension of KRIMP to discover changes in data streams.

An alternative pattern mining of [106] is proposed in [110] as one-pass approach, called SLIM. In contrast to KRIMP that finds the pattern set from a chosen set of candidates, SLIM finds the best pattern set from data. In [111], an event summarization algorithm is proposed in which serial episodes are discovered, and a set of patterns is mined instead of individual patterns. They took benefits of MDL to choose the best set of patterns that can describe a short and accurate summary of a database of events. Sequential data are encoded through set of patterns and employed two algorithms: SQS-Candidates search (to choose an appropriate set of patterns from a set of candidates) and SQS-Search (to find the appropriate set of patterns from database). They showed that event sequences can be summarized through finding a set of patterns which are short and non-redundant. Finally GoKRIMP [112] improves KRIMP by using two heuristic methods to compress sequential patterns. There are more work on mining compressing sequential patterns such as [113].

6 Wavelets

Generally speaking, these transformations are mathematical functions (that project a set X to another set Y) with the aim to make the "work" easier with the transformed set instead of the original one. There are different well-known transformations, such as Fourier transform and wavelet transform that can be applied on a set of large data. In the context of summarization, a wavelet transformation is mostly used to transform data and then make it possible to construct a compact representation of the data in a transformed domain. As a wavelet transformation can truncate the wavelet transformed data through saving strongest wavelet coefficient and setting the rest of coefficients to zero, then a compact version of data could be achieved. This section provides required background on transformation techniques and then it describes the wavelet transformations as a tool to build a summarized version of massive data sets to achieve fast approximate answers.

From signal processing, the purpose of transformation is access to information in signal that is not easily attainable from the original signal [114]. Therefore, many studies focused on the application of various transformations in different fields. One of the most famous transformations is *Fourier Transform* (FT) [115]. FT can plot frequency-amplitude curve, meaning that spectrum frequency of a signal can be observed in signal by FT. However, FT is not able to present time when the spectrum frequency becomes visible in the signal. Thereby, FT is a good choice in applications where "time" is not an important factor such as stationary signal. Stationary signal is the one that its frequency does not change over time conversely non-stationary signal in which frequency is varied over time. Hence, other transformations through which time resolution of frequency is observable were required.

Wavelet transformation is the one that time-frequency of signal can be represented simultaneously through it. We first explain briefly what wavelet transformation is and then review some studies that have applied wavelet transformations in their works considering the aspect of summarization for big data.

The definition of *wavelet* is a small wave and the *wavelet transform* is the process of converting a signal into a series of wavelets through which signals can be stored more efficiently than FT. As it is mentioned, FT only considers the frequency of signals, namely frequency content and its amount are shown through FT, whereas time-frequency of signals can be represented simultaneously by wavelet transforms. Therefore, a wavelet decomposition is considered as another compression technique that can be used to create a summary of large data sets. In fact, wavelet decomposition is a mathematical tool that is widely used in compression field especially for image compression.

The rationale behind wavelet is that a data vector V is transformed to a numerically different vector of wavelet coefficients. Then, the higher coefficients which have most compact energy will be retained and the other ones will be set to zero and cut up that leads to achieve compressed data. In other words, wavelet transform provide a time-frequency representation of signal by decomposing a signal into a set of basis functions (wavelets) which are orthonormal. Wavelets are produced from mother

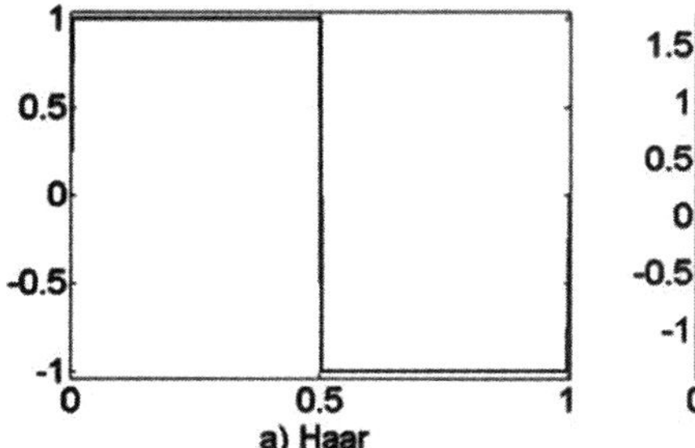

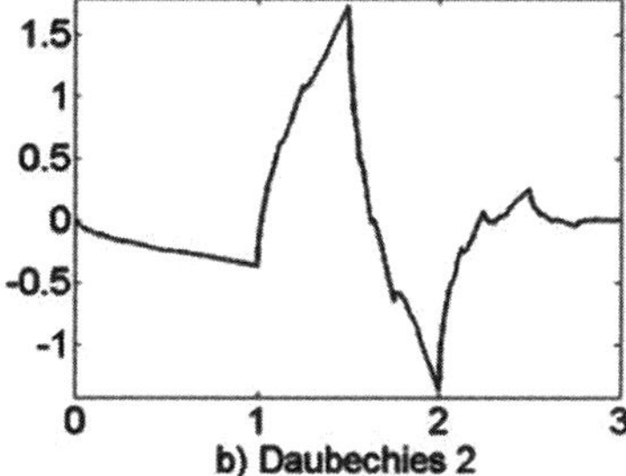

Fig. 1 Example of Haar and Daubechies Wavelet

wavelet by dilation and shifting [114]

$$\psi_{a,b}(t) = \frac{1}{\sqrt{a}}\Psi(\frac{t-b}{a})$$

where "a" denotes scaling parameter and "b" denotes shifting parameter.

DWT is a wavelet transform which is mostly used in data mining applications. The properties of wavelet help data mining to present data in an efficient manner. These properties could be considered such as hierarchal decomposition, multiresolution decomposition, vanishing moment, linear complexity, and decorrelated coefficients. Haar wavelet [116] (1D and 2D), Daubechies [117] and multi-resolution transform are most popular transforms derived from DWT as seen in Fig. 1. DWT works in a way that it decomposes a signal with length of L into high and low frequency parts by applying low and high pass filters and down and up sampling. Simply speaking, Haar wavelet also works in this manner that first signal is halved and average of each pair of samples is computed. Then the difference between the average and sample is calculated and first half is replaced with average and the second half is replaced with the difference as detail coefficients. This process continues till full decomposition is achieved.

For example suppose that we have a 1D signal as [9 7 3 5]. The Haar wavelet transform is calculated as following.

First half is $(9+7)/2=8$ and second half is $(3+5)/2=4$ so that we have [8 4]. Then, $(8+4)/2=6$ and $(8-4)/2=2$ so we get [6 2]. On the other hand, we have $(9-7)/2=1$ and $(3-5)/2=\ -1$. Therefore, the Haar wavelet decomposition of signal [9 7 3 5] is [6 2 1 -1], where 1 and -1 are detail coefficients in order to reconstruct original signal.

After briefly explanation of wavelet transform, we will provide an overview of the major studies that take advantage of wavelets in data mining applications with the aim of summarization and make data synopsis in the following. Because of space limitation lack, we will not explain each method however a general view is given to readers about the various techniques and their applications with purpose of data reduction and data summarization.

In large scale decision support systems (DSS), query processing plays an important role. Sometimes, answering to the queries does not need to be exact and approximate answer but fast answer satisfies the user requirements. Therefore many studies focused on proposing some data reduction mechanism to achieve compact sets of data to give approximate answers to the queries from these synopsis sets which results in achieving fast approximate answers. Some of the proposed methods rely on wavelet based methods to attain these compact sets.

In [118], probabilistic wavelet decomposition is proposed to find precise approximate answers to queries. Since, approximate answers provided by wavelet decomposition are widely different and there is not any guarantee that obtained answer is accurate. Therefore, in contrast to conventional wavelet transform, each coefficient is allocated a probability that shows its importance to preserve it for reconstruction. In [119], a Haar-wavelet based histogram creates a synopsis of data to obtain accurate selectivity estimations for query optimization. In [120], optimality of the heuristic method in [119] is also demonstrated.

A synopsis OLAP data cube is proposed in [121] and it applies multi-resolution wavelet decomposition. They retained a compact set of wavelet coefficients over a data cube for the approximate range sum queries considering space limitation. An extension of aforementioned work regarding approximate query answering through wavelet can be found in [122]. In [123], a general wavelet technique is presented to calculate a small space representation for data streams.

In [124], a new method to create wavelet synopses is proposed, called hierarchically compressed wavelet synopses (HCWS). To build optimal HCWS, a dynamic programming algorithm is presented to minimize the sum squared error considering space limitation and consequently increasing accuracy of the created synopses.

Haar wavelet decomposition can be used to minimize mean squared error and other metrics such as relative errors in data value reconstruction. However, the main purpose of Haar wavelet is minimizing mean square error. In [125, 126], the authors showed that these wavelet based synopsis approaches of different measures may cause reducing accuracy of approximate answering. Thus, they presented an idea of extended wavelet coefficient and proposed new algorithms for creating extended wavelet considering storage limitations and multi measure data (sum square and relative error norms). An extension of this work can also be found in [127].

The study in [128] presented some algorithms to create unrestricted wavelet synopses to achieve an "optimal" solution. A dynamic maintenance of wavelet-based histograms for data streams were considered in [129] because if underlying data distribution is changed then maintaining accuracy of histogram is not easy. Sampling and probabilistic counting are used this this approach.

In [130], the authors presented the first known streaming algorithms based on Group-Count Sketch (GCS) wavelet synopsis for both one and multi-dimensional data, satisfying polylogarithmic space usage, logarithmic update times and polylogarithmic query times for computing the top wavelet coefficients from the GCS. In [131], wavelet synopses are built for static and streaming massive data by using a greedy algorithm for maximum error metrics. U-HWT algorithm is suggested in

[132] to deal with uncertain data streams through applying Haar wavelet decomposition. The accuracy of the proposed algorithm was demonstrated via experiments and it was shown that a compact uncertain data stream can approximate the raw data stream. There is another study about compact representation of uncertain time series through hierarchical wavelet decomposition in [133].

Also in [136], the authors considered the issue of constructing data summaries through wavelet histogram in Map-Reduce. Haar wavelet-based synopses on probabilistic data is investigated in [135] through applying dynamic programming.

There are far more studies about constructing synopses through wavelet decomposition, and the reader can find some of the important ones in [134, 135].

7 Histograms

Histogram is a method used to represent a large volume of data in a compact manner so that can be considered as a data reduction or summarization technique. In fact, data distribution can be shown in a synopsis structure through histograms. Mathematically speaking, a histogram is a function x_i that represents how much data are within the disjoint ranges (bins/buckets) and represents the frequencies of data fall in these ranges. This function can be shown graphically. The function x_i satisfies the following condition

$$\mathrm{Y} = \sum\nolimits_{i=1}^{n} x_i$$

where Y is the total data and n is the total number of buckets or bins. Depending on the type of data attribute, histograms can be depicted. If an attribute is nominal, then a pole or vertical bar is displayed for each value of data. If the attribute is numerical, then data is divided into buckets in which buckets are disjoint subsets of data. In other words, data is divided into successive disjoint sub-ranges. For instance, a data attribute value within a range of 5–45 can be partitioned into 8 equal sub-ranges, as shown in Fig. 2.

Each sub-range is plotted with a bucket or bin in which width of bucket is the size of sub-range and the height of the bucket indicates the frequency of observed item within the sub-range.

There are different types of histograms that some of the popular ones are categorized as follows.

- **Equi-sum** [138], also known as *Equal-width histogram*, categorizes continuous ranges of attribute values into N equal intervals (buckets). The width of intervals is calculated based on the maximum (Max) and minimum (Min) values of the attribute as follows: $W = (\text{Max-Min})/N$. Equal-width histograms have been employed in many commercial systems. However, they are not suitable for Skewed data.

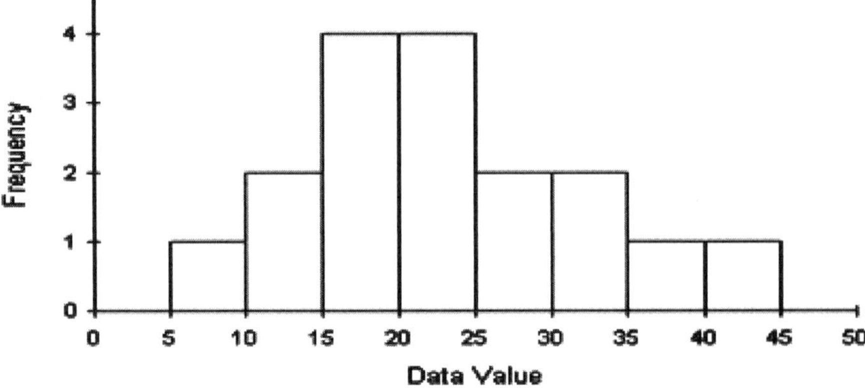

Fig. 2 An example of Histogram based on frequency and data value

- **Equal-depth (frequency) histogram**, also known as *Equi-height histogram*, is similar to Equal-width but with equal frequency in each bucket. In other words, the range is divided into *N* intervals with approximately constant frequency for each bucket which is a good option for range queries with low skew data distribution but not a proper option for commercial systems since bucket boundaries computation is an expensive task [138, 139].
- **V-optimal histogram** categorizes the continuous set of frequencies into a set of buckets to achieve minimum variance of the entire frequency approximation. Simply speaking, V-optimal considers all types of histogram for a given number of buckets to pick the one with the least variance [140].
- **V-Optimal-End-Biased** [140] groups the highest and lowest frequencies into individual buckets and the rest of frequencies are located in a single bucket. The advantageous of V-optimal over Equi-depth and Equi-width is that it can give a better approximation of original data with fewer errors. However, updating V-optimal histogram is not as easy as the two other histograms and sometimes it is required to change the whole histogram and build it again.
- In **MaxDiff histogram** [141], first data are sorted and then the margin of each bucket is computed considering adjacent values. The margins of buckets are determined where the difference between neighbor values is Maximum.
- **Spline histogram** [142] groups attribute values into contiguous buckets in which width of bucket can be varied and it is not fixed. Data distribution in buckets is not uniformed and data distribution in buckets is presented as a spline function instead of flat value.

Note that the aforementioned histogram methods are considered as one-dimensional summarization techniques. There are some multi-dimensional histograms, and a few of them keep a one-dimensional histograms for each dimension based on the attribute value independence assumption (AVI) [143] and in the other ones, the data is divided into d-dimensional buckets such as GENHIST[144]. Many studies are conducted to apply histograms with the aim of fast approximate query answering which an example can be found in [145].

8 Micro-Clustering

Mining data streams have attracted many attentions in recent years. Specific characteristics of data streams such as being infinite and real time leads to be processed as they arrived from different sources such as sensor networks or mobile devices. The clustering of data streams is studied here in a separate section of summarization methods because micro-clustering techniques deal with real time summarization of data. One of the early work done in this area is described in [146]. Many studies considered one-pass clustering over entire data stream not therefore on user-defined time slices. Also, since a data stream is infinite, it was impossible to store the whole data streams because of memory limitation so that it would be beneficial to store a compact representation of data streams. Therefore, a two-phase micro-clustering algorithm was investigated in [146] for infinite and evolving data streams. The algorithm has two phases relating to online and offline situations. The summary statistics of data are collected in online phase and then a clustering algorithm is performed on these summary data. The proposed algorithm, called CluStream, enables micro clusters to store summary statistics in a pyramidal time frame. The summary statistics are obtained as a temporal extension of cluster feature vector of BIRCH [22]. They added timestamps to the feature vector as ($\overrightarrow{CF2^X}$, $\overrightarrow{CF1^X}$, $CF2^t$, $CF1^t$, n),where $\overrightarrow{CF2^X}$ and $\overrightarrow{CF1^X}$ are the same as SS and LS in CF in BIRCH and $CF2^t$ and $CF1^t$ are the sum of squares of timestamps $T_{i1} \ldots T_{in}$ and the sum of timestamps $T_{i1} \ldots T_{in}$, respectively.

The pyramidal time frame is used to store micro clusters that are captured at specific instant time and named as snapshot in order to answer the queries of user over different time horizon. K-means is used to perform clustering in offline mode.

After CluStream, other micro-clustering algorithms over data streams were proposed. Some of these studies considered micro-clustering frameworks based on a density feature. A density-based micro clustering algorithm for data stream, called DenStream, is proposed [147]. Like CluStream, this has two online and offline components. It made some changes in concept of density that was used in DBSCAN by weighting areas of points in the neighborhood. The proposed algorithm can find arbitrary shaped clusters and outliers by using p-micro-cluster and core-micro-cluster and outlier micro cluster. They also applied a pruning strategy with the purpose of emerging new clusters.

Another instance of two phase components is investigated in [148] and proposed D-Stream. A grid is built for each input data point in online component. Then arbitrary shaped clusters are formed based on the grid density in offline component. In [149], SDStream as another online-offline framework based on CluStream is proposed. Since the framework focuses on the most recent data, so sliding windows model [150] is used. SDStream finds arbitrary shape clusters like DenStream. Therefore, they modified and used the core micro cluster and outlier micro cluster which are recorded as Exponential Histogram of Cluster Feature (EHCF) in main memory. Micro clusters are discovered and removed through the value of t in Temporal Cluster Feature (TCF). Clustering of discovered potential micro clusters through DBSCAN in online mode is performed in offline mode.

rDenStream is suggested in [151] considering the concept of outlier retrospect. It is a developed version of Denstream with three phases. rDenStream is a good option for applications with large amounts of outliers since it stores rejected outliers in outside temporary memory in order to give them an opportunity to include in clustering process with the aim of increasing accuracy of clustering. This phase is called retrospect as a third phase of this algorithm. The others two phases are the same as DenStream. It is obvious that by adding the third phase to process the historical buffer, the time complexity and memory usage will be increased in comparison with DenStream which are also demonstrated through experimental results. However, its performance is better than DenStream.

In [152], C-DenStream is studied as a density based clustering algorithm for data stream based on DenStream. They suggested their algorithm based on the concept of static semi-supervised through and domain information in order to achieve highly satisfactory results. More studies related to clustering data streams are conducted in the literature such as AclueStream [153], OPClueStream [154], and ClusTree[155]. Also two extensive surveys on clustering data stream can be found in [156] and [157]. More detailed micro clustering of data stream is available in [136].

9 Conclusion

In this paper, we described the concept of summarization. We also presented some the important applications of summarization techniques to illustrate the urgent need of big data summarization in future. We provided an overview of some of the well-known summarization techniques that could be useful for big data. Specifically, clustering, sampling, compression, wavelets, histograms, and micro-cluster are discussed in details. We hope that the reader has been provided with enough technical depth in this area that could give him/her a tool to make a decision to which technique to be used for his/her specific area/application.

References

1. A. Hathaway, J. Bezdek, and Y. Hu, "Generalized fuzzyc-means clustering strategies using Lnorm distances," IEEE Transaction on Fuzzy Systems, 8(5):576–582, October 2000.
2. J. MacQueen, "Some methods for classification and analysis of multivariate observations," in Proc. 5th Berkeley Sympium, 1:281–297, 1967.
3. G. Carpenter, S. Grossberg, and D. Rosen, "Fuzzy ART: Fast stable learning and categorization of analog patterns by an adaptive resonance system," Neural Network, 4:759–771, 1991.

4. G. Anagnostopoulos and M. Georgiopoulos, "Ellipsoid ART and ARTMAP for incremental unsupervised and supervised learning," Proceedings of IEEE International Joint Conference Neural Networks (IJCNN'01), Washington DC, pp. 1221–1226, 2001.
5. J. Mao and A. Jain, "A self-organizing network for hyperellipsoidal clustering (HEC)," IEEE Transactions Neural Networks, 7(1):16–29, January 1996.
6. C. Van Rijsbergen, "Information Retrieval," Butterworth-Heinemann, 1979.
7. J. Cezkanowski, "Zur differentialdiagnose der neandertalgruppe. KorrespondenzBlatt deutsch. Ges. Anthropol," Ethnol. Urgesch, 40:44–47, 1909.
8. R. Whittaker, "A study of summer foliage insect communities in the Great Smoky Mountains," Ecological Monographs, 22:1–44, 1952.
9. L. Legendre and P. Legendre, "Numerical ecology," New York: Elsevier Scientific, 1983.
10. R. Johnson and D. Wichern, "Applied multivariate statistical analysis," Englewood Cliffs, NJ: Prentice–Hall, 1998.
11. P.F. Russel and T. R. Rao, "On habitat and association of species of anopheline larvae in south-eastern Madras," Journal of Malaria India Institute (3):153–178, 1940.
12. R.R. Sokal and C. D. Michener, "A statistical method for evaluating systematic relationships," Bulletin of the Society of University of Kansas, 38:1409–1438, 1958.
13. P. Jaccard, "Étude comparative de la distribuition florale dans une portion des Alpes et de Jura," Bulletin de la Societé Voudoise des Sciences Naturelles, 37:547–579, 1901.
14. J.S. Rogers and T. T. Tanimoto, "A computer program for classifying plants," Science, 132:1115–1118, 1960.
15. S. Kulczynski, "Classe des Sciences Mathématiques et Naturelles," Bulletin International de l'Acadamie Polonaise des Sciences et des Lettres Série B (Sciences Naturelles) (Supplement II), pp. 57–203, 1927.
16. J. Tubbs, "A note on binary template matching," Pattern Recognition, 22(4):359–365, 1989.
17. L. Kaufman and P. Rousseeuw, "Finding Groups in Data: An Introduction to Cluster Analysis," Wiley, 1990.
18. B. Everitt, S. Landau, and M. Leese, "Cluster Analysis," London:Arnold, 2001.
19. P. Sneath, "The application of computers to taxonomy," J. Gen. Microbiology, 17:201–226, 1957.
20. T. Sorensen, "A method of establishing groups of equal amplitude in plant sociology based on similarity of species content and its application to analyzes of the vegetation on Danish commons," Biologiske Skrifter, 5:1–34, 1948.
21. A. Jain and R. Dubes, "Algorithms for clustering data," Englewood Cliffs, NJ: Prentice–Hall, 1988.
22. T. Zhang, R. Ramakrishnan, and M. Livny, "BIRCH: An efficient data clustering method for very large databases," Proceedings of ACM International Conference Management of Data (SIGMOD), pp. 103–114, 1996.
23. T. Chiu, D. Fang, J. Chen, Y. Wang and C. Jeris, "A robust and scalable clustering algorithm for mixed type attributes in large database environment," Proceedings of 7th ACM SIGKDD International Conference on Knowledge Discovery and Data Mining, pp. 263–268, 2001.
24. V. Ganti, R. Ramakrishnan, J. Gehrke, A. Powell, and J. French, "Clustering large datasets in arbitrary metric spaces," Proceedings of the 15th International Conference on Data Engineering (ICDE), pp. 502–511, 1999.
25. S. Guha, R. Rastogi, and K. Shim, "CURE: An efficient clustering algorithm for large databases," Proc. ACM SIGMOD International Conference Management of Data, pp. 73–84, 1998.
26. S. Guha, R. Rastogi, and K. Shim, "ROCK: A robust clustering algorithm for categorical attributes," Information Systems, 25(5):345–366, 2000.
27. E. Forgy, "Cluster analysis of multivariate data: efficiency vs. interpretability of classifications," Biometrics, 21:768–780, 1965.
28. J. MacQueen, "Some methods for classification and analysis of multivariate observations," Proceedings of 5th Berkeley Symposium, 1:281–297, 1976.

29. J. Mao and A.K. Jain, "A Self-organizing network for hyperellipsoidal clustering (HEC)," IEEE Transactions on Neural Networks, 7(1):16–29, 1996.
30. J. Dunn, "A fuzzy relative of the ISODATA process and its use in detecting compact well separated clusters," Journal of Cybernetic, 3(3):32–57, 1974.
31. E. Forgy, "Cluster analysis of multivariate data: Efficiency versus interpretability of classification," Biometrics, 21:768–780, 1965.
32. J. Dunn, "A fuzzy relative of the ISODATA process and its use in detecting compact well separated clusters," Journal of Cybernetics, 3(3):32–57, 1974.
33. J. Bezdek, "Pattern Recognition with fuzzy objective function algorithms," New York: Plenum, 1981.
34. S. Eschrich, J. Ke, J. Hall and D. Goldgof, "Fast accurate fuzzy clustering through data reduction," IEEE Transactions on Fuzzy Systems, 11 (2):262–270, 2003.
35. M. Steinbach, G. Karypis, and V. Kumar, "A comparison of document clustering techniques," KDD Workshop on Text Mining, 2000.
36. D. Pelleg and A. Moore, "Accelerating exact K-means algorithms with geometric reasoning," Proceedings of the 5th ACM SIGKDD International Conference on Knowledge Discovery and Data Mining, pp.277–281, 1999.
37. D. Pelleg and A. Moore, "X-means: extending K-means with efficient estimation of the number of clusters," Proceedings 17th International Conference on Machine Learning (ICML), Stanford University, 2000.
38. B. Schölkopf, C. Burges, and A. Smola, "Advances in kernel methods: support vector learning," The MIT Press, 1999.
39. L. Kaufman and P. Rousseeuw, "Finding groups in data: an introduction to cluster analysis," John Wiley and Sons, New York, NY, 1990.
40. R. Ng and J. Han, "Efficient and effective clustering methods for spatial data mining," Proceedings of the 20th International Conference on Very Large Databases (VLDB), pp.144–155, Santiago, Chile, 1994.
41. M. Ester, H-P. Kriegel, J. Sander, and X. Xu, "A density-based algorithm for discovering clusters in large spatial databases with noise," Proceedings of the 2nd ACM SIGKDD International Conference on Knowledge Discovery and Data Mining, pp. 226–231, Portland, Oregon, 1996.
42. X. Xu, M. Ester, H-P. Kriegel, and J. Sander, "A distribution-based clustering algorithm for mining in large spatial databases," Proceedings of the 14th International Conference on Data Engineering (ICDE), 324–331, Orlando, FL, 1998.
43. J. Sander, M. Ester, H-P. Kriegel, and X. Xu, "Density-based clustering in spatial databases: the algorithm GDBSCAN and its applications," Data Mining and Knowledge Discovery, 2(2):169–194, 1998.
44. A. Hinneburg and D. Keim, "An efficient approach to clustering large multimedia databases with noise," Proceedings of the 4th ACM SIGKDD International Conference on Knowledge Discovery and Data Mining, pp. 58–65, 1998.
45. M. Ankerst, M. Breunig, and H-P. Kriegel, K. Sander, "OPTICS: Ordering points to identify clustering structure," Proceedings of the ACM SIGMOD International Conference on Management of Data, pp. 49–60, 1999.
46. P. Grabusts and Borisov, "A Using grid-clustering methods in data classification," Proceedings of the IEEE International Conference on Parallel Computing in Electrical Engineering (PARELEC), 2002.
47. F. Murtagh and P. Contreras, "Methods of Hierarchical Clustering," CSIR, 2011.
48. S.A. Elavarasi, J. Akilandeswari, B. Sathiyabhama, "A survey on partition clustering algorithms," International Journal of Enterprise Computing and Business Systems, 2011.
49. W. Wang, J. Yang, and R. Muntz, "STING: a statistical information grid approach to spatial data mining,", Proceedings of the 23rd International Conference on Very Large Databases (VLDB), pp. 18–195, 1997.
50. G. Sheikholeslami, S. Chatterjee, and A. Zhang, "Wavecluster: a wavelet based clustering approach for spatial data in very large databases," The VLDB Journal, 8(3–4):289–304, 2000.

51. E. Schikuta, "Grid-clustering: An efficient hierarchical clustering method for very large data sets," Proceedings of the 13th IEEE International Conference on Pattern Recognition, pp. 101–105, 1996
52. D. Barbar and P. Chen, "Using the fractal dimension to cluster datasets," Proceedings of the 6th ACM SIGKDD International Conference on Knowledge Discovery and Data Mining, pp. 260–264, 2000.
53. A. Hinneburg and D. Keim, "Optimal grid-clustering: towards breaking the curse of dimensionality in high-dimensional clustering," Proceedings of the 25th International Conference on Very Large Data Bases (VLDB), pp. 506–517, 1999.
54. R. Agrawal, J. Gehrke, D. Gunopulos, and P. Raghavan, "Automatic subspace clustering of high dimensional data for data mining applications," Proc. ACM SIGMOD Int. Conf. Management of Data, pp. 94–105, 1998.
55. P. Berkhin, "Survey of clustering data mining techniques," Technical report, Accrue Software, San Jose, California, 2002.
56. P. Kaur and S. Aggrawal, "Comparative study of clustering techniques," International Journal on Advanced Research in Engineering and Technology, 1:69–75, 2013.
57. R. Xu and D. Wunsch, "Survey of clustering algorithms," IEEE Transactions on Neural Networks, 16(3):645–678, 2005.
58. W.G. Cochran, "Sampling techniques," 3rd Ed. John Wiley, 1977.
59. J.S. Vitter. "Random sampling with a reservoir," ACM Transactions on Mathematical Software, pp.37–57, 1985.
60. J.S. Vitter, "Faster methods for random sampling," Communication of the ACM (CACM), 27(7), July 1984.
61. J. Zhang, J. Xu, and S. Liao, "Sampling methods for summarizing unordered vehicle-to-vehicle data streams", Transportation Research Part C—Emerging Technologies, 23:56–67, 2012.
62. M. Dash. And W. Ng, "Efficient reservoir sampling for transactional data streams," Proceedings of IEEE International Conference on Data Mining (ICDM), pp. 662–666, 2006.
63. D. Ghosh, and A. Vogt, "A modification of Poisson sampling," Proceedings of the American Statistical Association, Survey Research Methods Section, pp.198–199, 1999.
64. B. Babcock, M. Datar, and R. Motwani, "Sampling from a moving window over streaming data," Proceedings of the 13th Annual ACM-SIAM Symposium on Discrete *Algorithms* (SODA). Society for Industrial and Applied Mathematics, Philadelphia, pp. 633–634, 2002.
65. C.C. Aggarwal. "On biased reservoir sampling in the presence of stream evolution," Proceedings of the 32nd International Conference on Very large Data Bases (VLDB), pp.607–618, 2006.
66. R. Gemulla, W. Lehner, and P.J. Haas, "A Dip in the reservoir maintaining sample synopses of evolving datasets," Proceedings of the 32nd International Conference on Very large Data Bases (VLDB), pp. 595–606, 2006.
67. P.B. Gibbons and Y. Matias, "New sampling-based summary statistics for improving approximate query answers," Proceedings of the ACM International Conference on Management of Data (SIGMOD), New York, NY USA, pp. 331–342, 1998.
68. R. Gemulla, W. Lehner, and P.J. Haas, "Maintaining Bernoulli samples over evolving multisets," In: Proc. ACM International Conference on Principles of Database Systems (PODS), pp. 93–102, 2007.
69. S. Chaudhuri, G. Das, M. Datar, R. Motwani, and V. Narasayya, " Overcoming limitations of sampling for aggregation queries," Proceedings of the IEEE International Conference on Data Engineering (ICDE), 2001.
70. C. Hua-Hui and L. Kang-Li, "Weighted random sampling based hierarchical amnesic synopses for data streams,"Proceedings of the 5th International Conference on Computer Science and Education (ICCSE), pp.1816–1820, 2010.
71. P.S. Efraimidis and P.G. Spirakis, "Weighted random sampling with a reservoir," Information Processing Letters, 97(5):181–185, 2006.

72. S. Acharya, P.B. Gibbons, and V. Poosala, "Congressional samples for approximate answering of group-by queries," ACMSIGMOD Record, 29(2):487–498, 2000.
73. H.J. Chang and K.C. Huang, "Remainder linear systematic sampling," Sankhya B 62, pp. 249–256, 2000.
74. N. Uthayakumaran, "Additional circular systematic sampling methods". Biometrical Journal, 40 (4):467–474, 1998.
75. C.-H. Leu and F.F. Kao, "Modified balanced circular systematic sampling," Statistics & Probability Letters, 76(4):373–383, 2006.
76. M.A. Bujang et al., "Modification of systematic sampling: a comparison with a conventional approach in systematic sampling," Proceedings of the International Conference on Statistics in Science, Business, and Engineering (ICSSBE), pp.1–4, 2012.
77. M. Al-Kateb, B.S. Lee, and X.S. Wang, "Adaptive-size reservoir sampling over data streams," Proceedings of the 19th IEEE International Conference on Scientific and Statistical Database Management, Banff, Canada, pp. 22–33, 2007.
78. M. Al-Kateb and B.S. Lee, "Adaptive stratified reservoir sampling over heterogeneous data streams," Information Systems, Available online, 2012.
79. M.D. Bankier, "Power allocations: determining sample sizes for subnational areas," The American Statistician, 42:174–177, 1988.
80. S. Chaudhuri, G. Das, and V. Narasayya, "Optimized stratified sampling for approximate query processing," ACM Transactions on Database Systems (TODS), 32(2), p.9-es, June 2007.
81. T. Liu and G. Agrawal, "Stratified k-means clustering over a deep web data source," Proceedings of the 18th ACM International Conference on Knowledge Discovery and Data Mining (KDD), pp.1113–1121, 2012.
82. H. Sug, "A structural sampling technique for better decision trees," Proceedings of the 1st Asian Conference on Intelligent Information and Database Systems (ACIIDS), pp.24–27, 2009.
83. A. Pol, C. Jermaine, and S. Arumugam, "Maintaining very large random samples using the geometric file," The VLDB Journal, 17:997–1018, 2008.
84. T.S. Buda, J. Murphy, and M. Kristiansen, "Towards realistic sampling: generating dependencies in a relational database". Proceedings of the 7th International Conference on Ubiquitous Information Management and Communication (ICUIMC), 2013.
85. S. Cong, J. Han, J. Hoeflinger, and D. Padua, "A sampling-based framework for parallel data mining," Proceedings of the 10th ACM SIGPLAN Symposium on Principles and Practice of Parallel Programming (PPoPP), pp. 255–265, 2005.
86. B. Babcock, S. Chaudhuri, and G. Das, "Dynamic sample selection for approximate query processing," Proceedings of the ACM International Conference on Management of Data (SIGMOD), pp. 539–550, 2003.
87. R. Gemulla, W. Lehner, and P. J. Haas, "Maintaining bounded-size sample synopses of evolving datasets," The VLDB Journal, 17:173–201, 2008.
88. R. Agrawal, H. Mannila, R. Srikant, H. Toivonen, and A. I. Verkamo, "Fast discovery of association rules," In *Advances in Knowledge Discovery and Data Mining*, 1996.
89. B. Chen, P. Haas, and P. Scheuermann, "A new two-phase sampling based algorithm for discovering association rules," Proceedings of the eighth ACM SIGKDD International Conference on Knowledge Discovery and Data Mining (KDD), 2002.
90. F. Olken, "Random sampling from databases," Ph. D. Dissertation, 1993.
91. I. Boxill, C. Chambers, and W. Eleanor, "Introduction to social research with applications to the Caribbean," University of the West Indies Press, Chapter 4, page 36, 1997.
92. C.A. Moser, "Quota sampling," Journal of the Royal Statistical Society, 115(3):411–423, 1952.
93. C. Sibona and S. Walczak, "Purposive sampling on Twitter: a case study," Proceedings of the 45th Hawaii International Conference System Science (HICSS), pp. 3510, 3519, 2012.
94. D.F. Nettleton, "Data mining of social networks represented as graphs," Computer Science Review, 7:1–34, 2013.

95. P.D. Grünwald, "Minimum description length tutorial," In: Advances in Minimum Description Length, P. Grünwald and I. Myung I (eds), MIT Press, Cambridge, 2005.
96. J. Rissanen, "Modeling by shortest data description," Automatica, 14(1):465–471, 1978.
97. P.D. Grunwald, "The Minimum description length principle and reasoning under uncertainty," cwi.nl, 1998.
98. J. Kiernan and E. Terzi,"Constructing comprehensive summaries of large event sequences," Proceedings of the 14th ACM SIGKDD International Conference on Knowledge Discovery and Data Mining (KDD), pp. 417–425, 2008.
99. J. Kiernan and E. Terzi, "Constructing comprehensive summaries of large event sequences," ACM Transactions on Knowledge and Data Discovery Data, 3(4), 2009.
100. P. Wang, H. Wang, M. Liu, and W. Wang, "An algorithmic approach to event summarization," Proceedings of the ACM International Conference on Management of data (SIGMOD), pp.183–194, 2010.
101. Y. Jiang, C.-S. Perng, and T. Li, "Natural event summarization," Proceedings of the 20th ACM International Conference on Information and Knowledge Management (CIKM), pp.765–774, 2011.
102. R. Agrawal, C. Aggarwal, and V.V.V. Prasad, "Depth first generation of long patterns," Proceedings of 7th International Conference on Knowledge Discovery and Data Mining, 2000.
103. D. Burdick, M. Calimlim, and J. Gehrke, "MAFIA: a maximal frequent itemset algorithm for transactional databases," Proceedings of the International Conference on Data Engineering (ICDE), April 2001.
104. J. Pei, J. Han, and R. Mao, "Closet: An efficient algorithm for mining frequent closed itemsets," Proceedings of the ACM SIGMOD Workshop on Data Mining and Knowledge Discovery, May 2000.
105. W. Zhou, H. Liu, and H. Cheng, "Mining closed episodes from event sequences efficiently," Proceedings of the 14th Pacific-Asia Conference on Advances in Knowledge Discovery and Data Mining (PAKDD), pp. 310–318, 2010.
106. S. A. Vreeken and M. van Leeuwen, "Item sets that compress," Proceedings of SIAM International Conference on Data Mining (SDM), pp.393–404, 2006.
107. M. van Leeuwen, J. Vreeken, A. Siebes, "Compression picks the item sets that matter," Proceedings of the European Conference on Machine Learning and Knowledge Discovery in Databases (ECML-PKDD), pp 585–592, 2006.
108. J. Vreeken, M. van Leeuwen, and A. Siebes, "Krimp: mining itemsets that compress," Data Mining and Knowledge Discovery, 23(1):169–214, 2011.
109. M. Leeuwen and A. Siebes, "StreamKrimp: detecting change in data streams," Proceedings of the European Conference on Machine Learning and Knowledge Discovery in Databases (ECML-PKDD), pp: 672–687, 2008.
110. K. Smets and J. Vreeken, "Slim: directly mining descriptive patterns," Proceedings of SIAM International Conference on Data Mining (SDM), pp. 236–247, 2012.
111. N. Tatti and J. Vreeken, "The long and the short of it: summarising event sequences with serial episodes," Proceedings of the 18th ACM SIGKDD international conference on Knowledge Discovery and Data Mining (KDD), pp: 462–470, 2012.
112. L.H. Thanh, M. Fabian, F. Dmitriy, and C. Toon, "Mining compressing sequential patterns," Statistical Analysis and Data Mining, 2013.
113. F. Moerchen, M. Thies, and A. Ultsch, "Efficient mining of all margin-closed itemsets with applications in temporal knowledge discovery and classification by compression," Knowledge Information Systems, 29:55–80, 2011.
114. R. Polikar, "The wavelet tutorial," http://engineering.rowan.edu/polikar/WAVELETS/WTtutorial.html.
115. G. Strang, "Wavelet transforms versus fourier transforms," Bulletin of American Mathematic Society, (new series 28):288–305, 1990.
116. A. Haar, "Zur Theorie der orthogonalen Funktionensysteme,"Mathematische Annalen, 69(3):331–371, 1910.

117. I. Daubechies, "Ten lectures on wavelets," SIAM publications, 1992.
118. M. Garofalakis and P. B. Gibbons, "Probabilistic wavelet synopses," ACM Transactions on Database Systems (TODS), 29:43–90, 2004.
119. Y. Matias, J.S. Vitter, and M. Wang, "Wavelet-based histograms for selectivity estimation," Proceedings of the ACM International Conference on Management of Data (SIGMOD), pp. 448–459, 1998.
120. Y. Matias and D. Urieli, "Inner-product based wavelet synopses for range-sum queries," Proceedings of the 14th Annual European Symposium on Algorithms (ESA), pp. 504–515, 2006.
121. J. S. Vitter and M. Wang, "Approximate computation of multidimensional aggregates of sparse data using wavelets", Proceedings of the ACM International Conference on Management of Data (SIGMOD), pp. 193–204, 1999.
122. K. Chakrabarti, M. Garofalakis, R. Rastogi, and K. Shim, "Approximate query processing using wavelets," The VLDB Journal, 10(2–3):199–223, 2001.
123. A.C. Gilbert, Y. Kotidis, S. Muthukrishnan, and M. Strauss, "Surfing wavelets on streams: One-pass summaries for approximate aggregate queries". The VLDB Journal, pp. 79–88, 2001.
124. D. Sacharidis, A. Deligiannakis, and T. Sellis, "Hierarchically compressed wavelet synopses," The VLDB Journal, 18:203–231, 2009.
125. A. Deligiannakis and N. Roussopoulos, "Extended wavelets for multiple measures," Proceedings of ACM International Conference on Management of Data (SIGMOD), pp. 229–240, 2003.
126. A. Deligiannakis, M. Garofalakis, and N. Roussopoulos, "Extended wavelets for multiple measures," ACM Transactions on Database Systems (TODS), 32(2), 2007.
127. S. Guha, C. Kim, and K. Shim, "Xwave: Approximate extended wavelets for streaming data," Proceedings of the International Conference on Very Large Data Bases (VLDB), pp. 288–299, 2004.
128. S. Guha and B. Harb, "Approximation algorithms for wavelet transform coding of data streams," Proceedings of the ACM-SIAM Symposium on Discrete Algorithms (SODA), 2006.
129. Y. Matias, J.S. Vitter, and M. Wang, "Dynamic maintenance of wavelet-based histograms," Proceedings of International Conference on Very Large Data Bases (VLDB), pp. 101–110, 2000.
130. G. Cormode, M. Garofalakis, and D. Sacharidis, "Fast approximate wavelet tracking on streams," Proceedings of the International Conference on Extending Database Technology (EDBT), 2006.
131. P. Karras and N. Mamoulis, "One-pass wavelet synopses for maximum-error metrics," Proceedings of the International Conference on Very Large Data Bases (VLDB), pp. 421–432, 2005.
132. K.-L. Liao, H.-H. Chen, J.-B. Qian, and Y.-H. Dong, "Wavelet decomposition algorithm for uncertain data streams,"Proceedings of the 6th International Conference on Computer Science & Education (ICCSE), pp.965–970, 2011.
133. Y. Zhao, C. Aggarwal, and P. Yu, "On wavelet decomposition of uncertain time series data sets," Proceedings of the 19th ACM International Conference on Information and Knowledge Management (CIKM), pp.129–138, 2010.
134. C.C. Aggarwal (ed.), "Data streams: models and algorithms", Springer, 2007.
135. M. Stern, E. Buchmann, and K. Böhm, "A wavelet transform for efficient consolidation of sensor relations with quality guarantees," Proceedings of the International Conference on Very Large Databases (VLDB), pp.157–168, 2009.
136. J. Jestes, K. Yi, and F. Li, "Building wavelet histograms on large data in MapReduce," Proceedings of the International Conference on Very Large Databases (VLDB), pp.109–120, 2011.
137. G. Cormode and M. Garofalakis, "Histograms and wavelets on probabilistic data, "Proceedings of the IEEE 25th International Conference on Data Engineering (ICDE), pp.293–304, 2009.

138. R. P. Kooi, "The optimization of queries in relational databases," PhD thesis, Case Western Reserver University, Sept. 1980.
139. M. Muralikrisbna and D.J. Dewitt, "Equi-depth histograms for estimating selectivity factors for multidimensional queries," Proceedings of ACM International Conference on Management of Data (SIGMOD), pp. 28–36, 1988.
140. Y. Ioannidis and V. Poosala. "Balancing histogram optimality and practicality for query result size estimation". Proceedings of ACM International Conference on Management of Data (SIGMOD), pp. 233–244, 1995.
141. V. Poosala, Y.E. Ioannidis, P.J. Haas, E.J. Shekita, "Improved histograms for selectivity estimation of range predicates," Proceedings of ACM International Conference on Management of Data (SIGMOD), pp. 294–305, 1996.
142. A.C. Konig and G. Weikum, "Combining histograms and parametric curve fitting for feedback-driven query result-size estimation," Proceedings of the International Conference on Very Large Data Bases (VLDB), Edinburgh, pp. 423–434, 1999.
143. V. Poosala and Y. Ioannidis, "Selectivity estimation without the attribute value independence assumption," Proceedings of the International Conference on Very Large Data Bases (VLDB), Athens, pp: 486–495, 1997.
144. D. Gunopulos, G. Kollios, V.J. Tsotras, and C. Domeniconi, "Approximating multi-dimensional aggregate range queries over real attributes," Proceedings of the ACM International Conference on Management of Data (SIGMOD), pp.463–474, 2000.
145. N. Bruno and S. Chaudhuri, "Exploiting statistics on query expressions for optimization," Proceedings of the ACM International Conference on Management of Data (SIGMOD), pp. 263–274, 2002.
146. C. C. Aggarwal, J. Han, J. Wang, and P. S. Yu, "A framework for clustering evolving data streams," Proceedings of the 29th International conference on Very Large Data Bases (VLDB), pp. 81–92, 2003.
147. F. Cao, M. Ester, W. Qian, and A. Zhou, "Density-based clustering over an evolving data stream with noise," Proceedings of SIAM Conference on Data Mining (SDM), pp. 328–339, 2006.
148. Y. Chen, "Density-based clustering for real-time stream data," Proceedings of the Knowledge Discovery and Data Mining (KDD), San Jose, California, USA, pp. 133–142, 2007.
149. J. Ren, R. Ma, and J. Ren, "Density-based data streams clustering over sliding windows," Proceedings of the 6th International Conference on Fuzzy systems and Knowledge Discovery (FSKD), Piscataway, NJ, USA, pp. 248–252, 2009.
150. W. Ng and M. Dash, "Discovery of frequent patterns in transactional data streams," Transactions on Large-Scale Data- and Knowledge-Centered Systems II,. Springer Berlin/Heidelberg, 6380:1–30, 2010.
151. L.-X. Liu, H. Huang, Y.-F. Gu, and F.-C. Chen, "rDenStream—a clustering algorithm over an evolving data stream,"Proceedings of CIECS International Conference on Information Engineering and Computer Science, pp.1–4, 2009.
152. C. Ruiz, E. Menasalvas, and M. Spiliopoulou, "C-DenStream: using domain knowledge on a data stream," Proceedings of the 12th International Conference on Discovery Science, pp. 287–301, 2009.
153. W.-H. Zhu, Y. Yin, Y.-H. Xie, "Arbitrary shape cluster algorithm for clustering data stream," Journal of Software, 17(3):379–387, 2006.
154. H. Wang, Y. Yu, Q. Wang, and Y. Wan, "A density-based clustering structure mining algorithm for data streams," Proceedings of the 1st ACM International Workshop on Big Data, Streams and Heterogeneous Source Mining: Algorithms, Systems, Programming Models and Applications (BigMine), pp. 69–76, 2012.
155. P. Kranen, I. Assent, C. Baldauf, and T. Sei, "The ClusTree: indexing micro-clusters for anytime stream mining," Knowledge Information Systems, 29(2):249–272, 2011.

156. A. Amini, T.Y. Wah, M.R. Saybani, and S.R.A.S. Yazdi, "A study of density-grid based clustering algorithms on data streams," Proceedings of 18th International Conference Fuzzy Systems and Knowledge Discovery (FSKD), 3:1652–1656, 2011.
157. A. Amini and T.Y. Wah," Density micro-clustering algorithms on data streams: a review," Proceeding of the International Multiconference of Engineers and Computer scientists (IMECS), 2011.

Part VIII
Monitoring

Central Management of Datacenters

Babar Zahoor, Bibrak Qamar and Raihan ur Rasool

1 Introduction

A centrally managed data center allows administration & management from a single location. In this chapter we discuss various functions related to administration and management of centrally managed data centers. Some of the major functions that we intend to discuss are summarized below. Monitoring of data traffic, thwarting attacks, monitoring activities of hardware and software that include resource utilization and alarming systems which helps in diagnosing and fixing any faults that arise during the operation of the data center.

Provisioning the servers and configuration of network devices not only include installation of operating systems, configuration of various services, patches management, software lifecycle management, but also includes the management of the inventory of hardware and software as well. Centrally logging various kinds of logs (activity logs, events logs, errors logs, and traffic logs, debug logs and alert logs) generated by applications and devices to help make system administration efficient.

Preventing from external threats over network traffic whether inbound or outbound is inspected by intrusion detection system.

B. Zahoor (✉)
ICT Department, Oxfam Novib, The Hague, The Netherlands
e-mail: babar.zahoor@oxfamnovib.nl

B. Qamar
Center for High Performance Scientific Computing, School of Electrical Engineering and Computer Science—NUST, Islamabad, Pakistan
e-mail: bibrak.qamar@seecs.edu.pk

R. ur Rasool
School of Electrical Engineering and Computer Science—NUST, Islamabad, Pakistan
e-mail: raihan.rasool@seecs.edu.pk

S. U. Khan, A. Y. Zomaya (eds.), *Handbook on Data Centers*,
DOI 10.1007/978-1-4939-2092-1_39

2 Organization of the Chapter

This chapter is organized into following sections:

1. **Management Layer Network**
2. **Provisioning of Servers**
3. **Platform Configuration Management System**
4. **Resource Utilization Monitoring**
5. **Alerting & Alarming System**
6. **Central Logging System**
7. **Intrusion Detection & Prevention System**
8. **Data Center Backup and Restore**
9. **Security Management System**

2.1 Management Layer Network

The Management layer network in datacenter is a dedicated & isolated network layer; from where datacenter administrator can easily manage & monitor the system. Management layer network is the network where management interfaces of all devices are connected. This layer is not accessible by users or outer world except the system admins the datacenter.

Since a typical datacenter can contain hundreds of servers that makes it impossible to manage every machine physically; we recommend creating separate dedicated network layer where every machine will be connected to be monitored & managed afterwards.

Connecting a keyboard/mouse with dedicated monitor to each machine is almost impossible, which is also wastage of resources and money. As technology has grown tremendously, few years back intelligent platform management interface (IPMI) protocol was introduced to manage server machines remotely. It allows system administrator to remotely install operating systems on machines to manage and monitor servers' hardware remotely.

IPMI allows datacenter admins to perform these tasks remotely by using IPMI based dedicated network port on servers. Almost every server has special dedicated network port which is designed to manage & monitor servers remotely known as IPMI i.e. Integrated lights out (iLO) is trademark of HP, Dell uses DRAC for their servers machines and few vendors use KVM over IP terminology based on IPMI protocol.

Storage devices are integral part of any datacenter. For smooth operations of datacenter proper monitoring and management of storage systems is required. In enterprise level storage devices; dedicated network interfaces (management module) are available to manage those devices. We recommend connecting management modules of SAN/NAS system into MLN where datacenter admins can easily monitor the performance and manage the devices on daily basis. Similarly to manage

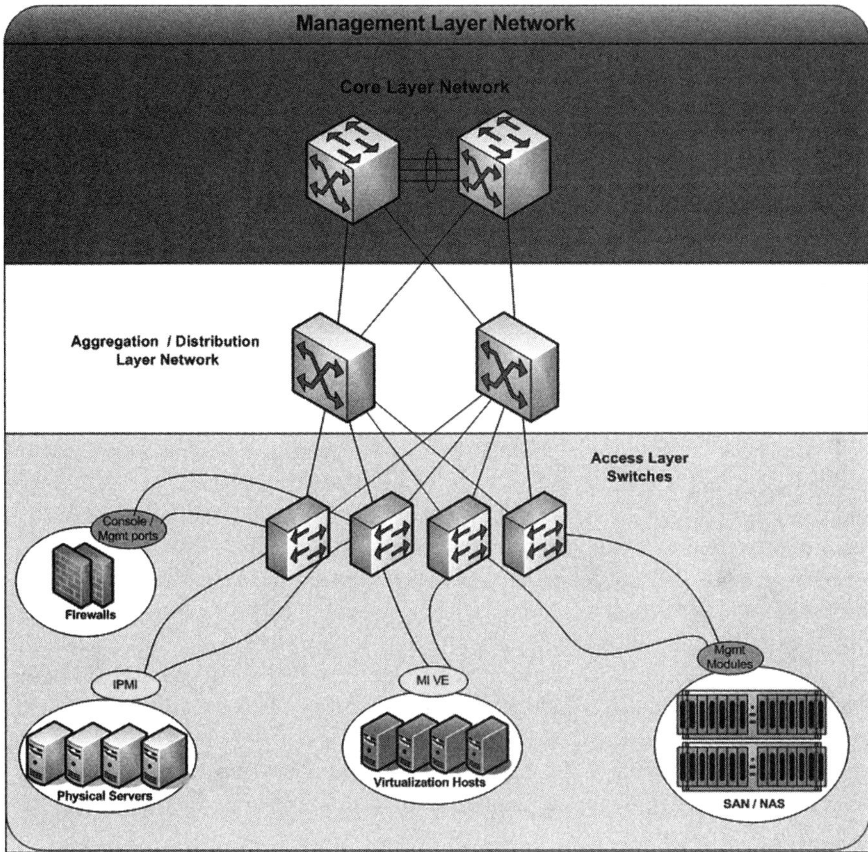

Fig. 1 The MLN- connected with the management interfaces of devices

routers; there is console ports or Mgmt interfaces available in the routers; those ports or interface shall be connected to MLN for maintenance & administration tasks when needed.

To secure LAN from WAN based threats in datacenter we recommend using firewalls and to manage firewalls; there are dedicated console ports or management interfaces which shall be connected with MLN for daily maintenance/administration and monitoring tasks.

In datacenters for connectivity of servers or server racks with each other we use multiple switches and to manage enterprise switches; dedicated management interfaces can be connected to MLN for maintenance/administration & monitoring of traffic flows. The following Fig. 1 shows MLN connected with the management interface of devices.

In this epoch of cloud computing organizations are deploying or already adopted a virtualization technologies layer in their datacenters for consolidation & optimization of hardware resources; save power, space with better performance and low costs. Management of virtualization layer is also critical task. Most of the virtualization layer engines provide dedicated management network interface; which allows system admins to perform administration, maintenance monitoring tasks by using the same interface. Management interfaces of virtualization engines shall be connected to management layer network for daily administration, maintenance and monitoring tasks [1].

2.2 *Provisioning of Servers*

Provisioning of servers is a set of actions to prepare a server with appropriate systems, data and software and to make it ready for network operations.

Typical tasks when provisioning a server are: selecting a server from a pool of available servers, load the appropriate software (operating systems, device drivers, middleware, and applications), appropriately customize and configure the system and the software to create or change a boot image for this server, and then change its parameters, such as IP addresses, IP Gateway to find associated network and storage resources.

The installation of provisioning servers reduces the workload of data center administrators and technicians as it automates the installation of operating systems, patch management, software up-gradations and many more functions in the data center environment [6].

There are many provisioning-servers software available in the market [2–4]. Selection of software can be done according to below mentioned suggested functionalities in the intended software:

- Inventory management system for hardware and software information.
- Provisioning and maintaining computer systems and virtual machines
- Installation and maintenance of software.
- Distributing custom software packages into manageable groups.
- Deploying and managing configuration files to computer systems.
- Distributing content across multiple physical or geographical sites in an efficient manner.

2.2.1 Reason to Use Provisioning Servers

A small IT environment having few servers can be easily managed by system admin in case of any disaster; the time for reconfiguration is minimal in such a scenario. But in a datacenter with multiple servers without an automated provisioning service the job of the system administrator becomes tedious and less flexible [5].

Normally in datacenters operating systems need security patches, important software updates, automated configuration of applications on servers and management of inventory. It is difficult to update the systems and manage these tasks in timely manner at large-scale environments. This is where automated provisioning servers come into play, where by the datacenter administrator can easily manage the entire infrastructure setup using automated processes configured according to needs. Hence automation of daily routine tasks using provisioning servers is vital for proper datacenter management.

2.3 *Platform Configuration Management System*

The Platform Configuration management is the process of standardizing resource configurations and enforcing their state across IT infrastructure in an automated manner. Configuration management is critical to the success of IT processes that include; provisioning of servers, change management, release management, patch management, compliance and security. Many IT organizations rely on manual tasks, custom scripts driven configurations, and customized OS images to accomplish the replication of activities. In large environments with multiple IT professional's team these methods are difficult to balance, track and continue which can create several issues. Including configuration flow, non-compliance, decrease productivity and responsiveness. This is where platform configuration management system becomes required for automated systems.

Configuration management is a best practice in any security plan. Configuration management can help us to Enforce Standard Operating Environment (SOE) standards by eliminating the configuration drifts.

It also supports the vulnerability management requirements by quickly identifying resources that need to be patched and then distributing patches. It satisfies auditing requirements by providing a complete audit trail. Configuration management system is being used in IT Automation process, physical or virtual machines and Cloud computing platforms and networked resources. It can be configured on storage devices as well as on network hardware & software, firewalls, smart phones and tablets, Phablet, out of band/IPMI management products, monitoring & alarm systems and routers and switches.

The required features of configuration management systems & provisioning server can also be available in a single software [2, 4]. Figure 2 shows the working of a typical provisioning & configuration management system.

2.4 *Resource Utilization Monitoring*

The Resource utilization & Monitoring is a software solution to monitor hardware health status and performance of devices in a datacenter. It aims to monitor resource utilizations of servers, storage devices (SAN/NAS) performance & hardware health status, firewalls inbound/outbound network traffic & hardware resource utilization status. It is used to monitor routing devices monitoring/WAN traffic status, network

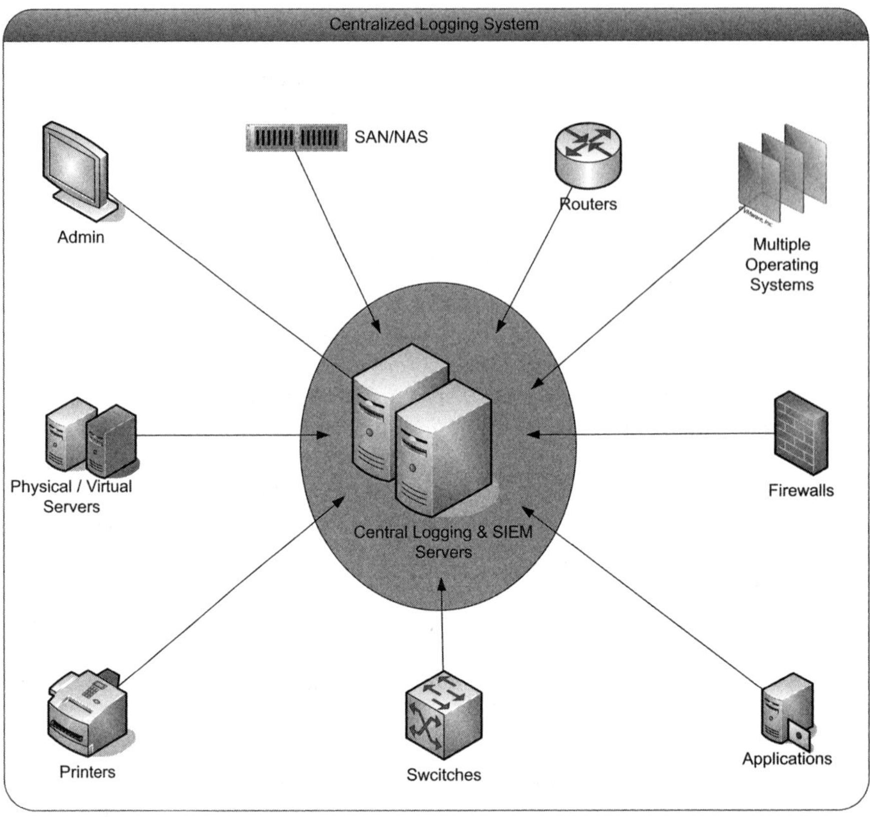

Fig. 2 Provisioning and configuration management system

switches activities & traffic flows, UPS systems for health, servers hard disk drives usage & their health status, fan speeds & temperature status of devices and high loads of CPUs of servers.

Normally we can connect Network Monitoring System with Management layer network where all Servers' IPMI interfaces are connected for management purposes and NMS (Network Monitoring Systems) can get servers' hardware information using SNMP string. On the same network layer storage devices' management modules can also be connected for efficient management. Using management modules we can monitor storage devices hardware resources & health status.

Other network devices like routers, firewalls, gateways are also connected to the management layer network for management purposes using management ports or interfaces, which can easily be utilized for resource utilization monitoring purposes; because these devices provide support for SNMP strings using same interfaces.

We can easily generate reports or view graphs to monitor resource utilization at hardware layer like CPU, RAM, and storage health.

The resource utilization monitoring systems can generate live graphs of operating systems stats; it can be configured to monitor different Operating systems i.e. various distributions of Linux, Windows, Mac OS, Solaris, AIX and HP-UX using Simple Network Management Protocol (SNMP), Secure Shell (SSH), Windows Management Instrumentation (WMI), Java Management Extensions (JMX) and Syslog protocol.

There are plenty of resource utilization monitoring systems available in the market, Resource utilization monitoring software can be configured to monitor different applications performance graphs and can also generate alerts to system admins during issues with running applications. Virtualization engines can easily be configured to send resource utilization monitoring updates to NMS & Alerting/alarming system via SNMP protocol using management network interfaces of virtualization engines (MI VE). There are plenty of Open Source network monitoring systems (NMS) available in the market. We can use Open Source network monitoring system NMS i.e. Nagios, Zenoss Core, Groundworks, Ansible, Gangila and many more for datacenter resource utilization and monitoring purposes.

2.5 *Alerting and Alarming System*

The alerting & alarming system is a software program which sends alerts/alarms using SMS, E-mail & voice messages to datacenter administrators and managers about the critical issues with network devices, servers, applications (web or desktop) database services, network services and many more.

The alerting & alarming system are able to send true alarms or alerts, restart the services in case of failures of those services, escalate the alerts, and report the system activities.

There is plenty of software available that can easily perform the resource utilization monitoring and alerting/alarming system tasks together [9, 10]. It will reduce hardware & software overhead by buying, using & configuring one system for resource utilization monitoring and alerting & alarms tasks. Following Fig. 3 shows how the resource utilization monitoring and alarming systems work.

2.6 *Central Logging System*

The Central Logging server is a system that allows machines, physical or virtual on the network to write logs information on its logging system. It also allows other devices such as; storage devices (SAN/NAS), network devices (routers, firewalls, and switches), printers, MFP devices to send logs to its logging center.

These logs are used for computer system management and security auditing [7] as well as generalized informational, analysis, and debugging purposes of systems & applications. Centralized logs show the information about the activities of different

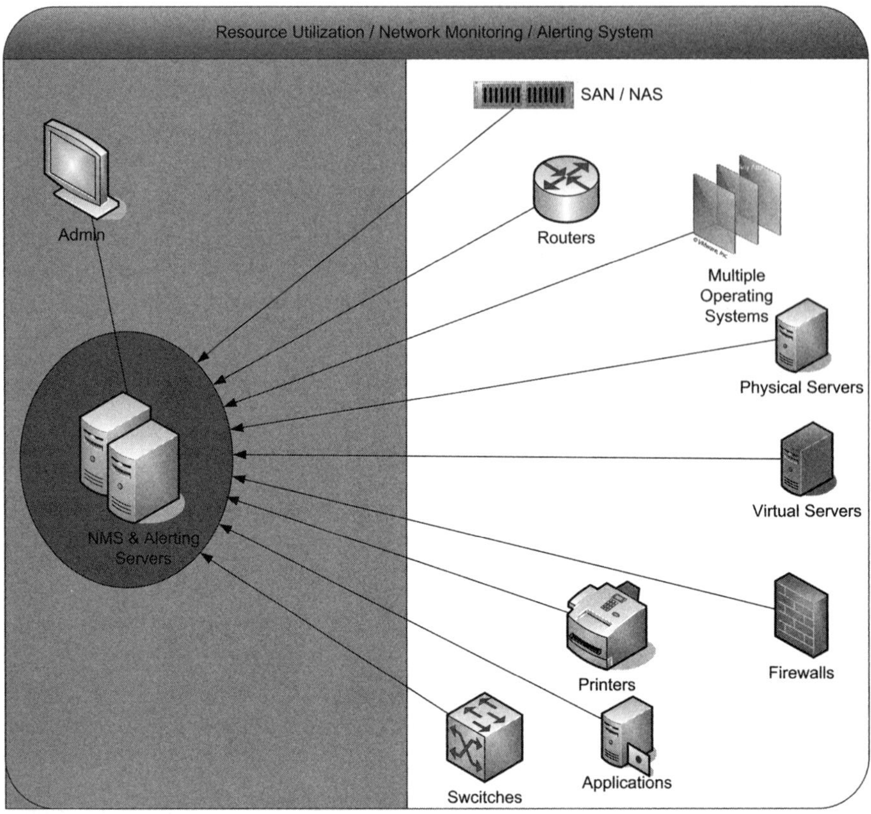

Fig. 3 Resource utilization, network monitoring and alarming system–the network monitoring, resource utilization and alerting/alarming servers get information from network devices

processes on servers, start/stop timing of services, errors in services/applications, and causes of errors.

System admins can easily perform audit on logs from outside the production environment or live systems.

It is always better to do the search in a centralized logging system. Without centralized logging, it becomes a logistical nightmare to research a single transaction that may have been processed on any one of an array of application servers–since datacenter administrators then have to log into each server and start searching through each. There are many available central log server software some of them include Syslog, Syslog-ng, RSyslog, Kibana, Logstash, Graylog2 and many more.

2.6.1 Security Information Event Management

A security event manager (SEM) or Security Information Event Management (SIEM) or Security Information Management (SIM) is a computerized tool for collecting logs & events information generated by servers (physical/virtual), applications, firewalls, routers, and switches in data center networks to generate events reports and inform datacenter administrators or infrastructure managers for further actions.

The key feature of a Security Information Event Management tool is the ability to analyze the collected logs to highlight events or behaviors of attention, for example system admin or admin/Super User logon, outside of normal working hours. Central Logging management system & SIEM are closely related; Central Logging systems focus on gathering logs, whereas SEM focuses on data analysis of logs.

Many applications working on a computer networks generate activities logs or events log. Protocols, such as Syslog and SNMP, can be used to transport these events, when they occur, to logging software on a centralized logging system explained in Central logging system. There are many proprietary and Open Source SIEMs or SEMs are available in the market that provides a support to many communication protocols to collection information from system. Figure 4 shows the working of a Centralized Logging System & SIEM.

2.7 Intrusion Detection and Prevention System

The Intrusion Detection & Prevention Systems are basically two different systems or functions, which work in digital communication.

i. Intrusion Detection System
ii. Intrusion Prevention System

An intrusion detection system (IDS) is a function performed by hardware device or software application that monitors network or system activities for malicious activities or organizational policy violations and produces alert to a datacenter administrator on management station.

Next level is automated set of actions to stop those malicious activities on a system; called Intrusion Prevention System.

Intrusion detection and prevention systems (IDPS) are primarily focused on identifying possible incidents, logging information about them, preventing malicious activities to harm the systems and reporting attempts. In addition, organizations use IDPS for identifying problems with security policies, documenting existing threats and deterring individuals from violating security policies. IDPS has become a crucial addition to the security of infrastructure in every organization.

IDPS typically record malicious activities information related to observed events notify data center administrators or Information security officers. Many IDPS can also respond to a detected threat by attempting to prevent it from succeed. They use several response techniques, which involve the IDPS stopping the attack itself,

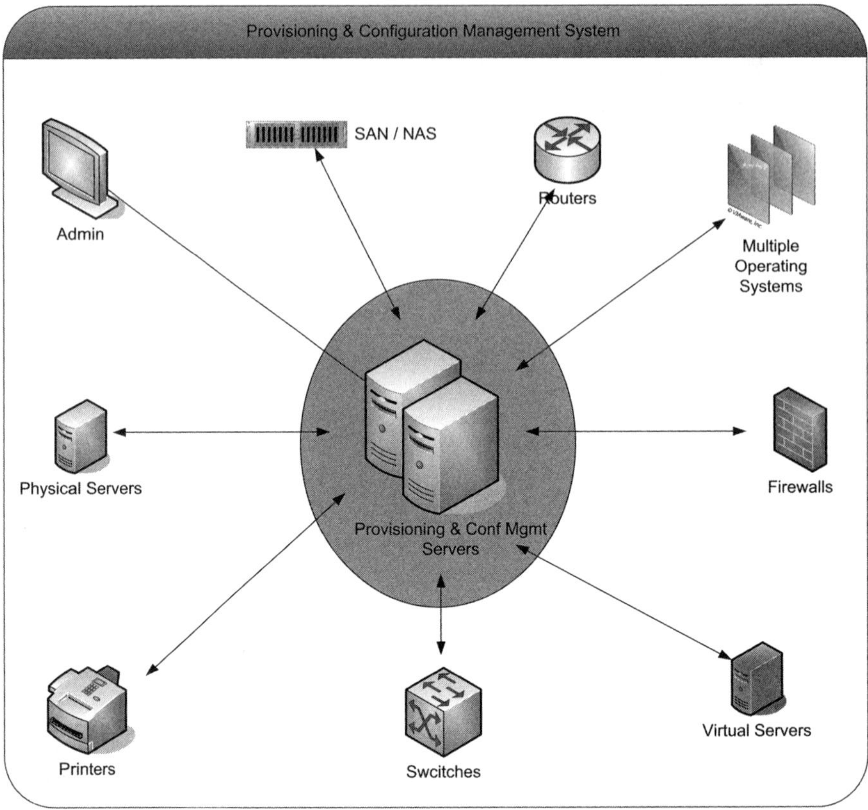

Fig. 4 Central logging system & SIEM- cluster of machines are configured to host logs from different devices with in a datacenter; routers, switches, firewalls, printers, physical & virtual servers, storage devices such as storage area networks (*SAN*) & Network attached storage (*SAN/NAS*), and applications and operating systems can also send logs to these servers. Admin can easily use these logs to audit the systems & applications & other purposes

changing the security setting (e.g. reconfiguring a firewall) or changing the attack's content.

2.7.1 Types of Intrusion Detection System (IDS)

There are two types of Intrusion Detection Systems that can be used to inspect the network traffic:

Network-Based Intrusion Detection System (NIDS)

A Network-Based Intrusion Detection System or NIDS is typically a standalone hardware appliance that is placed at the network perimeter along with the firewall. It monitors all the network traffic that enters or leaves the network. An NIDS contains hardware sensors located at various points within the network, which inspect the data packets from all devices reside inside the local area network [8].

Host-Based Intrusion Detection System (HIDS)

Host-Based Intrusion Detection System or HIDS is a software application that is installed on every host and device that resides in the internal network. HIDS analyzes the inbound and outbound network traffic only from the specific device on which it is installed, and alerts the datacenter administrator once the security violation or intrusion even occur. Host-Based Intrusion Detection System enables datacenter administrator to specify the well-known attacks which then make it easier for them to monitor the intrusion events if they occur. Moreover, HIDS also prevents the Trojans, backdoors, etc. from getting installed into the specific host, and monitors the key system files as well. If it is configured correctly on the host, it can also provide real-time detection of suspicious activities on that host.

2.7.2 How Intrusion Detection System Works?

To determine an attack, Intrusion Detection System follows one of the two detection methods discussed as below:

Anomaly-Based Intrusion Detection System

An Anomaly-Based Intrusion Detection System monitors the network traffic, and compares it against the security baseline establish by datacenter administrator. The security baseline defines the criteria such as used bandwidth, protocols, ports, and the types of devices that can be connected to each-other. If the network traffic is detected abnormal or different from the defined criteria in the baseline of system, Anomaly-Based Intrusion Detection System immediately sends alerts or alarms the administrators about the incident. Datacenter administrator can then take appropriate actions according to the type and sternness of the breach or intrusion. An Anomaly-based ID is also known as Behavior-based IDS.

Signature-Based Intrusion Detection System

A Signature-Based Intrusion Detection System checks the signatures or patterns of data packets, and compares them against the well-known network attack patterns that are store in its database system. As the pattern matches to any one of the patterns, the Signature-Based Intrusion Detection System generates a report and sends to data center administrator via email, short message on mobile devices or any other communication medium. A Signature-based ID is also referred to as Knowledge-based IDS.

2.8 Datacenter Backup and Restore

The Data Center Backup and Recovery is mandatory for business continuity process. In critical environments large enterprises spend thousands or even hundreds of thousands dollars to build a dense data warehouse to store data backups for future data recovery needs. Pieces of digital information which make financial records, online transactions, client's information, data mining, internal project reports, project process & data, and huge sensitive business information, all of these are decisive to running a booming business. If anything went wrong with data or it is stolen, it could severely put in danger the entire enterprise and damage its reputation. That is why many enterprises spend in a secure data warehouses to store their sensitive data and sign up for remote backup and security services to increase redundancies. If that data files become corrupted or get deleted, recovery or restore services will jump into action to restore the files to their original state.

The Remote data back-up services for organizational datacenter can add substantial value to its datacenter's efficient functionality. For planning datacenter backup/restore procedures, business continuity planning plays an important role because in Business continuity planning we can decide recovery point time objectives (RTO) & Recovery point objectives (RPO) which define for how long we can work without datacenters operations. Two technical indicators are used to measure disaster recovery:

- **Recovery point objective (RPO)**: maximum acceptable amount of data loss
- **Recovery time objective (RTO)**: acceptable longest duration of time within which services are interrupted or the shortest duration between the time when a disaster occurs and the time when services are restored

RPO measures data loss, while RTO measures service loss. RPO and RTO are not necessarily related. RTO and RPO vary according to services and enterprises, and are calculated based on service requirements after risk analysis and service influence analysis are performed.

If the RTO is shorter and the RPO is newer, the service loss will be less. The costs in developing and building the system, however, will become higher. Both factors are an important consideration.

2.8.1 The Components of Data Backup and Recovery

Simply put, data backup is designed to mitigate the loss and risk of sensitive business data while being able to recover or reconstruct the files in case of failure. Datacenters follow numerous protocols when backing up sensitive business data to ensure the maximum level of security.

Cold and Hot Backup

The "Cold" data backup means running a nonstop backup of files process with constant backup of the entire facility. This can be basis of quite a bit of trouble because it is always updating and take up a huge amount of storage. It is also a fatal process and, since it cannot differentiate old information from new, new information does not have priority update over old files that remain unchanged. However, "Hot" backup will backup files that are being updated or changed. When a data files gets updated or added into the system, another copy of the file is created and stored in case the file become degraded. This requires sophisticated software but takes up fewer resources than running a regular organism wide backup.

Enterprise Backup and Restore Software

Enterprise backup & restore software for backup services ranges from quick storage backup to secure data storage backup. There are many different types of software available in the market which can be selected according to organizational needs can facilitate the backup procedures. This software can back up your data and safeguard it against outside threat that seeks to corrupt or mine important data files. Some software requires scanning of the entire system before it begins to back up the files for security while others are optimized for speed.

Online and Offline Storage

Online Storage backup facilities store files online and there are more chances of being tampered with as compared to files stored offline. Online files are usually those that are accessed more frequently and edited by authorized users. Files that are offline contain old files that are not being used recently. With online and offline storage, users can benefit from having fast access to important files while still being able to store old files in a secure & proper location.

To get a proper concept of chapter; few protocols need consideration to be studied first. In the following section basic information about those protocols are available. Section 9 also explains some standards & protocols which are being used to connect NMS & Alerting/Alarming & central logging server with network devices, servers, applications to get information and generated reports and alarm notifications.

2.9 *Security Management Systems*

The main goal of security management systems is to manage access controls on datacenter network resources aligned with organizational policy & SOP guidelines, so that, the network cannot be damaged (purposely or accidentally). A security management system can also monitor users logging on to a network resource, refusing access for those who entered wrong access codes or pass-keys. A good security management implementation starts with sound security policies and procedures in place. It is important to create platform specific configuration standards for all network devices (e.g. routers and switches, firewalls) aligned with the industry best practices for security and performance. There are various methods of controlling access on network resources, some of them include, Access Control Lists (ACL) and UserIDs and passwords local to the device.

3 Conclusion

This chapter explained how central management of datacenters' resources can increase productivity and efficiency by allowing the administrators to manage and monitor all devices from a single location. Some open source and proprietary tools have also been listed for the central management of whole datacenter.

It is recommend to create network layer (MLN) in a datacenter, to manage devices using network connections, getting performance details of devices, load average stats of CPU usage, uptime/downtime details of hardware, alerts of server or services downtimes, alerts of critical issues in datacenter operations. There are so many tasks that be performed by the suggested management for example; provisioning of servers, patch management, release management and inventory management.

Appendix

i. **Simple network management protocol**

   Simple Network Management Protocol (SNMP) is an Internet-standard protocol for managing devices on IP networks. Devices that typically support SNMP including routers, switches, servers, workstations, printers, modem racks and more. It is used mostly in network management systems to monitor network-attached devices for conditions that warrant administrative attention.

ii. **Secure shell**

   The Secure Shell (SSH) is network communication protocols for secure data communication; it is used for remote command-line login. Computer runs SSH server program called server and allow connection from clients running SSH client from remote location via IP networks to execute commands on server command shell. It is used Unix/Linux based Operating systems & network devices.

iii. Windows management instrumentation

Windows Management Instrumentation (WMI) is a set of extensions to the Microsoft Windows Driver Model that provides an operating system interface through which instrumented components provide information and notification. WMI prescribes enterprise management standards and related technologies for Windows that work with existing management standards such as Desktop Management Interface (DMI) and SNMP.

iv. Java management extensions

The Java Management Extension (JMX) technology provides the tools for building distributed, Web-based, modular and dynamic solutions for managing and monitoring devices, applications, and service-driven networks. By design, this standard is suitable for adapting legacy systems, implementing new management and monitoring solutions, and plugging into those of the future.

v. Syslog

Syslog is a standard for computer message logging. It allows separation between applications or services or software those generates messages from the system and stores them for the software which can generate reports after analyzing them. Syslog can be used for computer system management and security auditing as, it is supported by a wide variety of devices for example servers (physical or virtual) other devices run on the network such as; storage devices (SAN/NAS), network devices (routers, firewalls, and switches), printers, MFP devices and receivers across multiple platforms.
There is multiple a levels of messages in syslog; which classify the criticality level of devices logs. Messages are also labeled with a facility codes from 0 to 7 & more.
0 Emergency: system is not usable
1 Alert: Immediately action required
2 Critical: critical conditions
3 Error: error conditions
4 Warning: warning conditions
5 Notice: normal but significant condition
6 Informational: informational messages
7 Debug: debug-level messages

vi. Intelligent platform management interface

The Intelligent Platform Management Interface (IPMI) is a standardized computer system interface available in almost all type of server machines and enterprise network switches used by system administrators for out of band management of computer systems and monitoring of servers. It is a way to manage a computer that may be powered off or otherwise unresponsive by using a network connection to

the hardware rather than to an operating system or login shell weather the computer is power off or on. By using this protocol a machine can be easily managed, it can also be powered-up using IP. Intelligent platform management interface IPMI protocol allows administrator to configure HDD RAID levels on servers, monitor components performance of servers i.e. fan, rpm, CPU temperature 0 °C. It can be used to provide SNMP strings to resource utilization monitoring & alerting/alarming servers (NMS) to plot server's hardware in graphs or to generate alerts for informing system admins about hardware problems.
NMS will be discussed in detail during Resource Utilization Monitoring Systems section. Using IPMI protocol can also do remote installation of Operating Systems.
As a message-based, hardware-level interface specification, IPMI operates independently of the Operating system (OS) to allow administrators to manage & administrate a machine remotely.

References

1. Managing VMware ESXi, Information guide VMware http://www.vmware.com/files/pdf/ESXi_management.pdf, Accessed on Nov 2013
2. Cobbler Deployment System http://www.cobblerd.org/, Accessed during Apr 2013
3. OpenQRM http://www.openqrm-enterprise.com/, Accessed during Feb 2013
4. Spacewalk http://spacewalk.redhat.com/, Accessed during Aug 2013
5. Cory Lueninghoener "Getting Started with Configuration Management". APRIL 2011 Getting Started with Configuration Management
6. Thomas Delaet Wouter Joosen and Bart Vanbrabant "A survey of system configuration tools", 2012
7. Christopher S. Duffy "Creating a Bastioned Centralized Audit Server with GroundWork Open Source Log Monitoring for Event Signatures", Jan 2013
8. Snort http://www.aboutdebian.com/snort.htm, Accessed during Jan 2014
9. Zenoss http://www.zenoss.com/, Accessed during March 2013
10. "Zenoss Enterprise Architecture Overview" http://docs.huihoo.com/zenoss/Zenoss-Enterprise-Architecture-Overview.pdf, Accessed during Oct 2013

Monitoring of Data Centers using Wireless Sensor Networks

Cláudia Jacy Barenco Abbas, Ana Lucila Sandoval Orozco and Luis Javier García Villalba

1 Introduction

As data center energy densities, measured in power per square foot, increase, energy savings for cooling can be carried out by applying WSN technology and using the gathered information to efficiently manage the data center.

Data centers' energy consumption has attracted global attention because of the fast growth of the information technology industry. Up to 60 % of the energy consumed in a data center is used for cooling in wasteful ways as a result of lack of environmental information and overcompensated cooling systems [1].

Data centers consume $1 < 2\ \%$ of today's world electricity production, increasing at a rate of 12 % per year due to high demand for these resources. As a consequence, the amount of heat generated by data centre equipment is growing rapidly. Currently, about half of data centers' energy is used for cooling.

Continuous monitoring of the spatial temperature distribution in a data center DC (Data Center) is important for reliably operating the computing equipment and minimizing the required cooling energy [2].

Typical data centers use equipment that cannot operate in high temperatures, resulting in extensive use of energy-consuming air cooling infrastructure. This infrastructure consists of cooling and computing components that typically use low-efficiency, single-speed fans and do not allow for dynamic shifting of cool air to where it is needed most.

L. J. García Villalba (✉) · A. L. Sandoval Orozco · C. J. Barenco Abbas
Group of Analysis, Security and Systems (GASS), Department of Software Engineering and Artificial Intelligence (DISIA), Faculty of Information Technology and Computer Science, Office 431, Universidad Complutense de Madrid (UCM),
Calle Profesor José García Santesmases 9 Ciudad Universitaria, 28040 Madrid, Spain
e-mail: javiergv@fdi.ucm.es

C. J. Barenco Abbas
Universidade de Brasília, Campus Universitário Darcy Ribeiro, Faculdade de Tecnologia, Departamento de Engenharia Elétrica, Barsília - D.F. - Brazil

S. U. Khan, A. Y. Zomaya (eds.), *Handbook on Data Centers*,
DOI 10.1007/978-1-4939-2092-1_40

A 25 % savings in energy use can easily save a data center $ 8–16 per square feet in total annual energy costs, while delivering a power usage effectiveness (PUE or total facility power/information technology [IT] equipment power) rating of 1.25, a dramatic improvement over the 1.7 or higher rating for most data centers (a rating closer to 1 is preferred) [3].

To enable data center operators to run their cooling system closer to the economically attractive upper limit, continuous temperature monitoring at thermally critical locations in the data center is required. In addition, thermal models based on measurement data can be used to analyze and optimize layout, air flow and workload distribution in the data center. Changing operating parameters with sophisticated control concepts based on real-time temperature information can optimize the cooling-efficiency even further.

It is argued that the combined computational and networking capability of a sensor network enables it to interact with the clusters in a much more sophisticated way and enhance essential functions in a data center [4].

The first step to limit the waste of energy in the operation of data centers is to standardize energy efficiency metrics and distinguish the most inefficient parts of the system. The green grid association [5] has reported metrics for measuring energy efficiency in data centers. Similar steps are also presented in [6].

Sensor network technology has recently been adopted for data center thermal monitoring because of its nonintrusive nature for the already complex data center facilities and robustness to instantaneous CPU or disk activities [7].

Data center operators tend to further decrease the CRAC's temperature settings when servers issue thermal alarms because they lack the information to accurately diagnose the problem [8]. Historical and real time data about the environmental conditions inside a data center are invaluable not only for diagnosing problems but for improving the data center's efficiency [9, 10].

Thermal and air dynamics in data centers can be complex. Fig. 1 allows us to gain an understanding of the underlying spatial variability through a thermal image captured by an infrared camera. This picture exposes the temperature variations that exist over the air intakes of multiple server racks. One can observe temperature differences larger than 10 °F across various heights of the same rack, as well as significant differences in the temperature distribution patterns across different racks [8].

There are seemingly several options for measuring the temperature and humidity distributions inside a data center. For one, thermal images such as the one shown in Fig. 1 visualize temperature variations over the camera's view frame. However, continuously capturing thermal images throughout the data center is prohibitively expensive. Alternatively, modern servers have several onboard sensors that monitor the thermal conditions near key server components, such as the CPUs, disks, and I/O controllers. These sensors are used to detect and prevent hardware failures due to overheating rather than sense the data center's ambient environment. Some recent servers also have temperature sensors at the air intake, and administrators can estimate room conditions from these sensors. However, for servers that do not have sensors at the air intake, it is difficult to accurately estimate the room temperature and humidity from another onboard sensors [8].

Fig. 1 A row of computer racks inside a data center (*left*) and the corresponding infrared image representing the spatial temperature distribution (*right*) [8]

The communication mechanism used to retrieve the collected measurements is the other crucial aspect in the system design. Options in this case are divided in two categories: in-band vs. out-of-band. In-band data collection routes measurements through the server's operating system (OS) to the data center's (wired) IP network. The advantage of this approach is that the network infrastructure is, in theory, available and the only additional hardware necessary are relatively inexpensive USB-based sensors. However, data center networks are in reality complex and fragile. They can be divided into several independent domains not connected by gateways. Traversing across network boundaries can lead to serious security violations. Finally, the in-band approach requires the host OS to be always on to perform continuous monitoring. Doing so however would prevent turning off unused servers to save energy [8].

Out-of-band solutions use separate devices to perform the measurements and a separate network to collect them. Self contained devices provide higher flexibility in terms of sensor placement, while a separate network does not interfere with data center operations. However, deploying a wired network connecting each sensing point is undesirable as it would add thousands of network endpoints and miles of cables to an already cramped data center [8].

For this reason, wireless networks are the only feasible option. Moreover, networks based on IEEE 802.15.4 radios are more attractive compared to Bluetooth or WiFi radios. The key advantage is that a 15.4 network has a simpler network stack compared to alternative solutions. This simplicity has multiple implications. First, sensing devices need only a low end MCU such as the MSP430 [11] thus reducing the total cost of ownership and implementation complexity. Second, the combination of low power 15.4 radios and low power MCUs leads to lower overall power consumption. The need for low power consumption will become apparent when we present the mechanism used to power multiple sensing devices from the same power source.

2 Survey Study

FEMP (U.S. Federal Energy Management Program) [12] presents a wireless sensor technology that provides real time data center conditions needed to optimize energy use and achieve substantial savings all with minimal impact on day to day operations. This technology includes branch circuit power monitors, temperature sensors, humidity sensors, and pressure sensors, along with an integrated software product to help analyze the collected data.

This wireless sensor technology provides a cost effective and facilitates a friendly way of helping data center operators visualize and implement system changes that reduce overall energy consumption.

In order to evaluate the real world effectiveness of wireless sensor technology, GSA's Green Proving Ground (GPG) program worked with the Energy Department's Lawrence Berkeley National Laboratory (LBNL) as a demonstration project location. Sensors using a wireless mesh network and data management software to capture and graphically display real time conditions for energy optimization were installed.

The study showed that providing real time, floor to ceiling information on humidity, air pressure, and temperature conditions is feasible. This data, when combined with power use, leak detection, and equipment status, could enable data center operators to significantly improve the energy efficiency of even well managed data centers.

The main benefits of this technology are:

- *The Bottom Line*: Efficiency measures implemented as a result of information provided by the wireless sensor network reduced the demonstration facility's cooling load by 48 %, and reduced the total data center power usage by 17 %.
- *Simplified Assessment Tools Limit Power Interruption*: The data center operator at the demonstration facility found that full deployment of the permanently installed wireless sensor network provides valuable real time information needed for the ongoing optimization of data center performance. However, permanent installation of the sensor network required multiple interruptions of facility power.

The LBNL evaluation team concluded that broad deployment represents a best practice that could help agencies meet mandated targets cost effectively.

The major advantages found for wireless sensor networks were: reduce operating expenses; reduce capital expenses; increase capacity and reduce failures.

Microsoft's Data Center Genome project [13] presents a data center monitoring system using WSN. ENVM is used to monitor and control air conditioners in a data center using WSNs. With this SENVM, a solution to make a small data center to be "Green" is feasible. SENVM is designed to directly measure temperature at servers and send a control signal to air-conditioners whether a server is too hot or too cold; therefore, SENVM can make sure that temperature at servers will be in an appropriate condition all time [14].

This technical bulletin [15] presents an overview of wireless sensor technology and a wireless network implementation of the installation project at LBNL.

First, is recommended three basic issues to be addressed when specifying and installing a WSN:

- *Overall reliability*: It is recommended to perform a field test of sensor locations to ensure reliable operations.
- *Sensor battery life*: Battery life must be considered during the configuration of the sensor device because it is primarily affected by the network latency.
- *Interoperability of gateway*: The interoperability of the protocol over a common, non-proprietary interface would allow future upgrades and installations that would include legacy installations.

Many practical benefits can be realized by using a WSN for:

- Cooling performance visualization through software.
- Humidification requirements.
- Floor tile tuning.
- Hotspot identification.
- Historical data trending.
- Preventative Maintenance prediction.
- Multiple computer room air conditioner/computer room air handler (CRAC/CRAH) unit operational control and coordination.
- Real-time Power Usage Effectiveness (PUE) calculation.

At LBNL, the project included installing a WSN with approximately 800 monitored points including air temperature, relative humidity, under-floor air pressure, and electrical current and power. The selected wireless sensor manufacturer installed the wireless sensors into a meshed network for increased reliability.

With the WSN installation, operators of this LBNL data center achieved energy efficiency by:

- Increasing data center set point temperature.
- Optimizing control coordination of CRAC units.
- Eliminating humidification systems, which can have unintended, simultaneous operations.
- Improving floor tile arrangements and server blanking.
- Installing hot-aisle or cold-aisle isolation systems.

The demonstrable results and benefits achieved at LBNL included:

- *Visualizing thermal performance*: observing thermal profiles above and below floor in real time; heat mapping.
- *Learning from sensors*: instant feedback when installing blanking panels.
- *Tuning floor-tile locations*: balancing under-floor airflow to eliminate hot spots.
- *Focusing on a single sensor*: watching impact of floor tile opening.
- *Verifying the impact of maintenance*: trending data during maintenance.
- *Determining the need for humidification*: monitoring relative humidity and power consumption.
- *Providing instant feedback*: real-time information on data center anomalies.

- *Real-time PUE calculation*: calculates Total power consumption including thermal BTU monitoring and compares IT power consumption at 15 min intervals.

Results obtained with this project originate the following recommendations that may be relevant to other WSN installations:

- *Maintain Airflow Devices*: Regular preventative maintenance, inspection, and tune-ups are highly effective in reducing energy waste in data centers.
- *Manage Energy with Metering*: The LBNL energy efficiency project clearly validated the old energy axiom that generally states that you cannot manage energy without monitoring energy. Essential metering and monitoring is provided by the wireless sensor system.
- *Supervise Performance with EMCS*: An energy monitoring and control system (EMCS), or other building monitoring system, should be used in conjunction with a wireless sensor system.
- *Optimize Rack Cooling Effectiveness*: Data center operators should consider the following items to maximize rack cooling effectiveness:
- Match under-floor airflow to IT equipment needs.
- Locate higher density racks near CRAC units and verify airflow effectiveness.
- Locate severs to minimize vertical and horizontal empty space in racks.
- Consolidate cable penetrations to reduce leaks.
- Load racks bottom first in under-floor distribution systems.
- Use blanking plates and panels.
- Eliminate floor openings in hot aisles.
- Establish hot and cold airstream isolation.

The paper presents [16] a suite of assessment tools (DC Pro Software Tool Suite) that is useful for assessing data center energy use and identifying potential energy efficiency measures. These tools were developed by DOE Industrial Technologies Program's (DOE ITP). The technology that was evaluated consists of a network of wireless sensors.

The wireless sensor technology installed and evaluated as part of this project consists of sensor nodes, gateways, routers, server platforms, and software applications. To measure and validate performance claims for this technology, the study team selected the USDA's National Information Technology Center (NITC) Data Center in St. Louis, Missouri. The NITC facility is a Tier 3 data center located at the GSA Goodfellow Federal Complex.

The demonstration consisted of deploying a self configuring, multipath network of wireless sensors that provide real time measurement of server inlet temperature and sub floor pressure differential. Analytics based on mapping of the sensor data helped identify improvements for more energy efficient cooling of the IT equipment.

A total of 588 environmental sensors were installed throughout the demo room: 16 temperature and 16 humidity sensors were installed in the CRAC units, and 420 temperature sensors were located at the top, middle, and bottom of computer racks measuring air intake and exhaust conditions at the IT equipment. Additional sub floor reference temperatures were monitored at selected racks. The balance of the system included routers and a gateway that was connected to a server.

After the network was fully commissioned, data was gathered and analyzed by a qualified assessor. Results were used to create an accurate understanding of the data center operation, and the measured data was input into the DC Pro Software Tool Suite. The output from the assessment tools provided recommendations on specific potential energy savings opportunities.

The evaluation team recommends that this type of technology be used in facilities that wish to achieve energy savings using non intrusive/non interruptive equipment. This technology supports a rapid assessment to identify energy efficiency measures, one of which would likely be to install a full wireless monitoring system. In summary, the study validates the effectiveness of a dense network of wireless sensors to provide a reliable, facility friendly, cost effective source of real time information that enables data center operators to achieve 10 % or greater improvements in overall data center efficiency. Dissemination of these findings should encourage the adoption of this technology throughout GSA and the data center industry.

This study confirmed that data center operators and analysts can accurately baseline their facility's energy performance using a mesh network of sensors to measure environmental parameters and electrical power. It also demonstrated that analysts can input this data into energy assessment software to quickly identify energy efficiency opportunities, even at a data center that is relatively efficient, well operated, and well designed. In addition, data obtained by this technology can be input into assessment tools that can identify additional best practice measures applied to: Air Management, Optimize Cooling, Humidity Control and Optimizing Sub Floor Pressure. Energy savings result from the implementation of this best practices.

Applying the findings from the evaluation of the demonstration facility to all tenant operated data centers in the GSA portfolio yields the following potential reductions: applying a 17 % overall reduction in overall data center energy use at a typical federal data center with a 69 watts-per-square-foot (W/sf) IT load and a PUE of 1.94 (average for all of GSA's data centers) will result in annual savings of \$ 21.50/sf (assuming \$ 11/kWh).14 The PUE for this example was reduced to 1.51.

The installation of the environmental sensors was non intrusive and non interrupting to data center operations; however, this was not the case for power meters. The shutdowns required to install power meters in electrical panels interrupted the data center operations and delayed the assessment. While installing temperature sensors in front and back of the servers, the evaluation team was careful not to disturb access to the servers by the IT staff.

The high cost and logistical constraints of deploying a wired sensor network provided a significant barrier to capturing such data. The most significant barrier posed by this technology was the multiple interruptions of facility power required to safely connect power monitoring equipment at the demonstration facility.

In [12] the development and impacts of a new wireless sensor technology is described for data centers called *SynapSense Wireless Green Data Center Solution* leased and installed in the Sacramento Municipal Utility District's (SMUD) data center with the intent to investigate the technology. SynapSense employs a wireless mesh network to monitor everything from specific equipment to environmental conditions. The SynapSense solution is used to baseline energy efficiency, identify and alert staff to environmental issues and manage operational improvement opportunities.

A major barrier to improved energy efficiency is the difficulty of collecting data on the energy consumption of individual components of data centers, as well as the lack of data collection for data centers overall. Better energy data collection would not only help to quantify the energy load of data center operations, thus highlighting the importance of energy-efficiency improvements and facilitating right-sizing of equipment to match the energy load, but it would also allow data center managers and facility managers to monitor and evaluate the energy savings and corresponding GHG reductions SynapSense humidity sensors on each rack could allow SMUD to widen the minimum and maximum acceptable RH % resulting in additional energy savings.

There are clear operational and financial benefits for SMUD to get on the path to a Green Data Center. Specifically, there are a number of opportunities to improve the energy efficiency and overall operation of the SMUD data center without significant capital outlay. SMUD can improve airflow loss from 50 to 21 % by better managing the data center airflow.

The site assessment findings suggest that while there is air conditioning (cooling capacity) to spare, the airflow in the SMUD data center must be improved in order to meet the data processing needs of the next few years if more high density servers are part of the growth plan.

The study in [17] describes the implementation of the workload monitoring, analysis, and emulation toolkit to enable the automated collection and analysis of workload traces from data centers, and use those traces as the basis for repeatable and verifiable experiments and workload emulation. This toolkit has three tools:

- *Splice*: Is a data collection tool, enables us to correlate observations across the IT/facilities boundaries and understand the location-dependent aspects of data center management, such as the temperature throughout the data center. In addition, it aggregates sensor and performance data in a relational database using a database schema that has been designed to treat information that rarely changes in much the same way as those that frequently change.
- *SeASR (Sensor Analysis and Synthetic Reproduction)*: Helps us understand how objects respond and change during experiments, and provides feedback to Splice, enabling more efficient data collection and retention. Examines how readings change, how often sensors update readings, and into what range of values attribute readings fall. SeASR uses one-dimensional Expectation Maximization (EM) clustering.
- *Sstress*: Enables finegrained and repeatable control over server resource utilization, allowing it to explore the IT/facilities relationships in one machine, or emulate workload playback across the data center. Sstress is an application for selectively utilizing parts of a single machine or networked servers. The desired

functionality is the ability to take a sequence of CPU, memory, disk, and network utilization figures for one or more servers and force another set of servers to recreate those conditions.

The proposed architecture includes a data communication and filtering engine and a database schema implemented on top of a relational database and is designed to support easy extensibility, scalability, and support for the notion of higher-level object views and events in the data center.

In order to evaluate the effectiveness of the instrumentation and analysis components, and the flexibility of the emulation toolkit. They examined the results of running Splice on two clusters: HP's Utility Data Center (UDC) and the Duke Computer Science "Devil Cluster". Results show that cooling costs in a moderately-sized data center are significantly lower when using temperature-aware workload placement. These savings can represent tens to hundreds of thousands of dollars per year.

Data Center Infrastructure Management (DCIM) [18] is a tool that monitors, measures, manages and/or controls data center use and energy consumption of all IT related equipment. The primary components of a DCIM solution are Input, Process and Output. Various sensors and other system feeds comprise the input. This raw data then sent through an analysis process to create actionable data. The processed data is then presented as output to the user, perhaps in the form of a dashboard or trend graph, and is also used as control data back into the input component.

DCIM proves the following benefits:

- Access to accurate, actionable data about the current state and future needs of the data center
- Standard procedures for equipments changes
- Single source of truth for asset management
- Better predictability for space, power and cooling capacity means increased time to plan
- Enhanced understanding of the present state of the power and cooling infrastructure and environment increases the overall availability of the data center
- Reduced operating cost from energy usage effectiveness and efficiency

The data collected from sensors, as well as knowledge of how environmental variables affect the conditions in the room, can be used to design control systems that can adjust the cooling resources, such as the fans and outlet temperatures, to maintain the room in its operating range [1]. The prototype wireless sensor network is used to monitor environmental data relevant to the cooling processes at a data center. For this purpose, Sensirion SHT15 temperature and humidity sensors are selected. The wireless sensor network is developed based on XBee 2.5 RF modules, which were engineered to operate within the IEEE 802.15.4/ZigBee protocol and support the needs of low cost, low power wireless sensor networks. This study demonstrates that wireless sensor networks can be an effective tool for environment monitoring in a data center. Such a network offers the advantage of easy deployment throughout the computer racks because there is no need for wiring for power and data transmission. This network also offers freedom in deployment, as the sensor modules can be placed in locations where wired sensors would be unfeasible for technical or safety reasons.

To enable data center operators to run their cooling system closer to the economically attractive upper limit, continuous temperature monitoring at thermally critical locations in the data center is required. In addition, thermal models based on measurement data can be used to analyze and optimize layout, air flow and workload distribution in the data center. Changing operating parameters with sophisticated control concepts based on real-time temperature information can optimize the cooling efficiency even further [19].

For reliably operating all data processing equipment in the data center, the temperatures at the air inlets of these devices have to be monitored continuously to ensure that the maximum admissible inlet air temperature specified by the device manufacturers are not exceeded.

There are several possible approaches to monitor temperatures in data centers. For example, most data processing devices are equipped with internal temperature sensors. To detect thermal problems in the data center, however, data from these internal sensors is generally not the first choice because it reflects the activity of the device rather than the environmental conditions of the data center. Furthermore, high installation and configuration effort is required to collect and aggregate this data, especially in environments with heterogeneous devices from different manufacturers. Several solutions do exist with external wired sensors, but in practice they are not widely adopted, mainly because of the difficulty of dealing with changes in the data center layout. A wireless sensor network, on the other hand, offers a low cost non intrusive way to gather temperature data at key locations in the data center. The sensors can be quickly deployed and easily repositioned if data processing equipment is relocated or replaced [19].

CFD (*Computational Fluid Dynamics*) simulations and experiments show that the cold air supply temperature can be significantly increased if the air flow is adapted dynamically based on measured inlet air temperatures of the data processing equipment. In data centers with significantly varying workload, this approach can yield cooling energy savings of up to 20 % [20].

Air flow control and thermal aware workload scheduling concepts rely on real-time temperature information collected in the data center. High reliability, frequent sampling, and low latency are key requirements for wireless sensor networks used in real time control applications [8]. A prototype was successfully deployed in the ZRL Data Center Wireless Sensor Network in production data centers. The data center houses 400 racks with heterogeneous data processing equipment and is cooled by 40 computer room air conditioners. The temperature changes in the cold aisles of the data center were tracked with 108 sensors during an upgrade of the cooling system.

Results showed that the reference temperature of the computer room air conditioning units in the data center was increased by 3°C, thus achieving a significant cooling energy reduction without risking device overheating [8].

Deployments in production data centers have shown that the DCWSN (*Data Center Wireless Sensor Network*) performs well in terms of configuration effort,

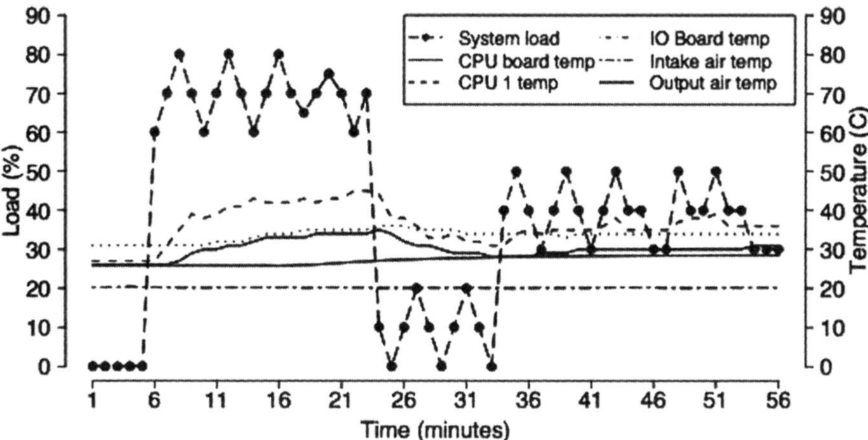

Fig. 2 Temperature measured at different locations in and around an HP DL360 server. Also shown is the server's CPU load. Internal sensors reflect the server's workload instead of ambient conditions [8]

reliability, and power efficiency. Moreover, it was demonstrated in a data center with 400 racks that the cooling efficiency of a data center can be significantly increased by improving the air flow and temperature distribution based on measurement data from the DCWSN [8].

Figure 2 plots the temperature measured at various points along with the CPU utilization for an HP DL360 server with two CPUs. Air intake and output temperatures are measured with external sensors near the server's front grill and its back cover. It is evident from this figure that internal sensors are quickly affected by changes in the server's workload, rather than reflecting ambient conditions [8].

The RACNet system [8] is among the first attempts to provide visibility into a data center's cooling behavior, a problem of increasing importance as cooling comprises a large percentage of a data center's energy consumption. At the same time this compelling application challenges wireless sensor network technology in terms of reliability and scalability. The WRAP (*Wireless Reliable Acquisition Protocol*) protocol tackles these challenges by combining three mechanisms: channel diversity, decoupling of tree maintenance from data gathering and congestion avoidance via a token passing mechanism.

Evaluation results from a medium size testbed and pilot deployments at a data center suggest that WRAP favorably compares to existing data collection protocols. Specifically, as the aggregate amount of traffic grows, WRAP achieves higher data yields than open loop protocols such as CTP and higher total throughput than rate control protocols such as RCRT [21]. Furthermore, results from a large scale production deployment show that WRAP offers stable performance with data yields consistently higher than 99 %.

3 Conclusion

Wireless sensor networks offer a low cost non intrusive solution to gather environmental information in data centers. The sensors can be quickly deployed and easily repositioned. Examples of applications include continuous temperature monitoring, data collection for thermal modeling and temperature sensing for real time control.

Continuous temperature monitoring is essential to prevent device overheating while operating the cooling system close to the upper temperature limit for increased energy efficiency. Collecting data to understand temperature and humidity distribution is a first step toward improving a data centers energy efficiency.

Deployments in production data centers have shown that the Wireless Sensor Network performs well in terms of configuration, reliability and power efficiency.

Acknowledgment Part of the computations of this work were performed in EOLO, the HPC of Climate Change of the International Campus of Excellence of Moncloa, funded by MECD and MICINN.

References

1. Rodriguez, M.G., Uriarte, L.E.O., Jia, Y., Yoshii, K., Ross, R., Beckman, P.H.: Wireless Sensor Network for Data-Center Environmental Monitoring. In: Fifth International Conference on Sensing Technology. (2011) 533–537
2. Weiss, B., Truong, H.L., Schott, W., Scherer, T., Lombriser, C., Chevillat, P.: Wireless Sensor Network for Continuously Monitoring Temperatures in Data Centers. IBM Research Report RZ3807, IBM Research (April 2011)
3. Department, U.E.: Federspiel Controls' Data Center Energy Efficient Cooling Control System. Technical report (2011)
4. Hong, K., Yang, S., Ma, Z., Gu, L.: A Synergy of the Wireless Sensor Network and the Data Center System. In: IEEE 10th International Conference on Mobile Ad-Hoc and Sensor Systems. (14-16 October 2013) 263–271
5. Green Grid: The Green Grid,http://www.thegreengrid.org/ (2013)
6. Green IT. Promotion Council: GIPC: Concept of New Metric for Data Center Energy Efficiency: Introduction to Datacenter Performance per Energy DPPE. Technical report, Green IT. Promotion Council (February 2010)
7. Wang, X., Wang, X., Xing, G., Chen, J., Lin, C.X., Chen, Y.: Intelligent Sensor Placement for Hot Server Detection in Data Centers. IEEE Transactions on Parallel and Distributed Systems **24**(8) (August 2013) 1577–1588
8. Liang, C.J.M., Liu, J., Luo, L., Terzis, A., Zhao, F.: RACNet: A High-fidelity Data Center Sensing Network. In: 7th ACM Conference on Embedded Networked Sensor Systems, New York, NY, USA (2009) 15–28
9. Chen, G., He, W., Liu, J., Nath, S., Rigas, L., Xiao, L., Zhao, F.: Energy-Aware Server Provisioning and Load Dispatching for Connection-intensive Internet Services. In: 5th USENIX Symposium on Networked Systems Design and Implementation, Berkeley, CA, USA, USENIX Association (2008) 337–350
10. Patel, C.D., Bash, C.E., Sharma, R., Beitelmal, M.: Smart Cooling of Data Centers. In: Pacific RIM/ASME International Electronics Packaging Technical Conference and Exhibition, Maui, Hawaii, USENIX Association (July 2003)

11. Texas Instruments: MSP430x1xx Family User's Guide (Rev. F), http://www.ti.com/litv/pdf/slau049f/ (2006)
12. Oberg, R., Sanchez, E., Nealon, P.: Wireless Sensor Technology for Data Centers. Conference paper, American Council for an Energy-Efficient Economy (2008)
13. Liu, J., Zhao, F., O'Reilly, J., Souarez, A., Manos, M., Liang, C.J.M., Terzis, A.: Expand Project Genome: Wireless Sensor Network for Data Center Cooling. The Architecture Journal (2008)
14. Choochaisri, S., Niennattrakul, V., Jenjaturong, S., Intanagonwiwat, C., Ratanamahatana, C.A.: SENVM: Server Environment Monitoring and Controlling System for a Small Data Center Using Wireless Sensor Network. CoRR If (2011)
15. U. S. Department of Energy: Wireless Sensors Improve Data Center Energy Efficiency Technology Case Study Bulletin. Technology case study bulletin, U. S. Department of Energy (September 2010)
16. Mahdavi, R., Tschudi, W.: Wireless Sensor Network for Improving the Energy Efficiency of Data Centers. Technical report, Lawrence Berkeley National Laboratory (March 2012)
17. Moore, J., Chase, J.: Data center workload monitoring, analysis, and emulation. In: Eighth Workshop on Computer Architecture Evaluation using Commercial Workloads. (2005)
18. Cole, D.: Data Center Knowledge Guide to Data Center Infrastructure Management (DCIM). Technical report, Data Center Knowledge (May 2012)
19. Scherer, T., Lombriser, C., Schott, W., Truong, H.L., Weiss, B.: Wireless Sensor Network for Continuous Temperature Monitoring in Air-cooled Data Centers: Applications and Measurement Results. In: 11th International Conference on Ad-hoc, Mobile, and Wireless Networks, Berlin, Heidelberg, Springer-Verlag (2012) 235–248
20. Biller, P., Chevillat, P., de Lorenzi, F., Scherer, T., Schott, W., Ullmann, R., Vömel, C.: Efficient cooling of Data Centers. In: 4th World Engineer's Convention. (September 2011)
21. Paek, J., Govindan, R.: RCRT: Rate-Controlled Reliable Transport for Wireless Sensor Networks. In: 5th International Conference on Embedded Networked Sensor Systems, New York, NY, USA (2007) 305–319

Network Intrusion Detection Systems in Data Centers

Jorge Maestre Vidal, Ana Lucila Sandoval Orozco and Luis Javier García Villalba

1 Introduction

Access to Data Centers must be protected by perimeter defense systems such as firewalls, access lists or intrusion detection systems. Despite the importance of each of them, the NIDS (Network-based Intrusion Detection Systems) are the most sophisticated and accurate measure to deal with external attacks. Therefore, it is essential to know the characteristics of this kind of system, and each of its variants. In this chapter the most relevant aspects of the NIDS are described in detail, in order to improve their integration into networks operating on Data Centers.

- **Denial of service attacks**: The denial of service attacks usually present centralized or distributed features, and aims to deny the service of the protected system and the defensive systems. To do this they focus on depleting their computing resources.
- **Enumeration attacks**: The enumeration attacks gather information about the Data Center in order to identify vulnerabilities in their security perimeters and exploit those that allow better access.
- **Unauthorized access**: Unauthorized accesses are usually produced from theft of user credentials, theft of sessions or by exploiting vulnerabilities that facilitate gain privileges.
- **Malware**: Malware are malicious applications that when executed, are able to compromise the system security.
- **Attacks on network structure**: The attacks against the network structure undertake the critical elements of its topology. Usually focus on the DNS (Domain Name Systems), by exploiting the spread of malware servers (such as botnets) and its denial of service.

L. J. García Villalba (✉) · A. L. Sandoval Orozco · J. Maestre Vidal
Group of Analysis, Security and Systems (GASS), Department of Software Engineering and Artificial Intelligence (DISIA), Faculty of Information Technology and Computer Science, Office 431,
Universidad Complutense de Madrid (UCM), Calle Profesor José García Santesmases 9 Ciudad Universitaria, 28040 Madrid, Spain
e-mail: javiergv@fdi.ucm.es

S. U. Khan, A. Y. Zomaya (eds.), *Handbook on Data Centers*,
DOI 10.1007/978-1-4939-2092-1_41

In order to combat these threats, Data Centers deploy a perimeter defense that incorporates different preventive systems. These include firewalls, ACLs (Access Lists) and IDS (Intrusion Detection Systems). Firewalls are mainly used to filter traffic directed to certain addresses or ports, and to define the security perimeters. They act based on a set of guidelines or rules, and do not have the ability to identify their nature. Moreover, the access lists are also deployed to filter traffic. They are usually based on using blacklists and whitelists of IP addresses which consider the source and destination of every packet. Finally, unlike the two previous, IDS provide prevention capacity and the ability to communicate in advance about malicious activities. This kind of system usually delegated the task of preventing the operator itself, although there is a certain type of IDS, well known as IPS (Intrusion Prevention System), which also has the capacity to decide on countermeasures to be taken. When evaluating the contribution of an ID, the following features are considered:

- **Accuracy**: The IDS accuracy is the ability to detect attacks and distinguish them from the legitimate usage of the system. The IDS classifies events produced in the monitored system, labeling them as positives or negatives. The evaluation of labels is divided into four sub-categories: true positives, true negatives, false positives and false negatives.
  Figure 1 shows the different labeling evaluations issued by the IDS. The alerts in green boxes (true positives and true negatives) are correct labels and the others correspond to error detections.
  The following describes each of them:
  - *True positives*: The true positives are properly labeled intrusion attempts.
  - *True negatives*: The true negatives are events corresponding to common and legitimate usage of the system, properly labeled.
  - *False positives*: False positives are legitimate events incorrectly classified as attacks. Besides posing a problem for the system quality of service, such errors can be exploited by attackers by inducing the IDS to issuing large amounts of alerts, causing the depletion of its resources.
  - *False negatives*: False negatives are malicious activities that the IDS have confused with legitimate activities.

  At present there are different functional standards for evaluating the accuracy of the IDS. For IDS deployed on networks, the MIT Lincoln Labs DARPA [1] and KDD Cup [2] datasets are the most commonly used. Locally it is usual generating own datasets [3, 4]
- **Performance**: The performance of an ID determines its ability to process information in real time. This parameter usually depends on the detection strategy, and on the processing capability of implementation environment (software or hardware). In [5] is explained in detail each of the computational costs involved in the detection process, as well as the mechanisms to achieve the balance between accuracy and performance.
- **Robustness**: The robustness of an ID is its fault tolerance and ability to confront attempts of nonuse. There are methodologies [6] to measure the impact of the most common mistakes issued by network-based IDS. Moreover, the appearance

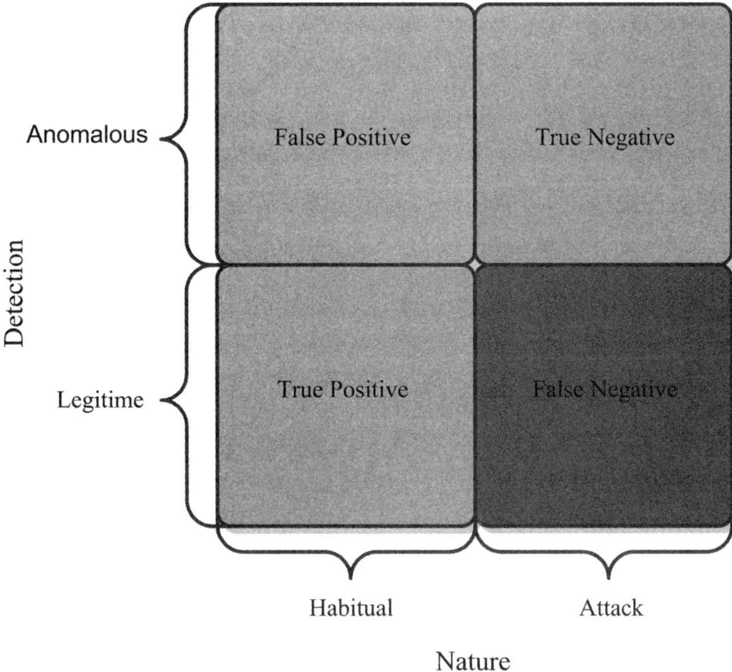

Fig. 1 Labeling evaluation

of specific evasion techniques [7] involves the need to design IDS with higher fault tolerant, and capable to withstand this type of threats.

- **Scalability**: The scalability of an ID is its ability for been adapted to different monitoring environments without losing quality nor be compromised. Certain characteristics of the monitor environment, such as the use of security protocols for data encryption [8], can degrade its capacity analysis. Therefore it is desirable that the IDS deployed on Data Centers are compatible with the characteristics of the monitored environment.

The intrusion detection systems have evolved to adapt to the need arising from changes in the monitor environment. This has led to a large variety of systems, specializing in strengthening different characteristics. Figure 2 summarizes some of the most common classification criteria.

- **Detection strategy**: Initially the IDS behavior was based on identifying patterns corresponding to known threats, which are commonly referred to as signature-based detection systems, such as Snort [9] and Bro [10]. Later, as a result of the rapid proliferation of intrusion techniques, as well as the daily emergence of thousands of signatures with unknown threats (zero-day attacks) [11, 12], defensive systems in addition to acknowledge signatures, must be able to detect new threats to be fully effective. Because of this, intrusion detection systems were

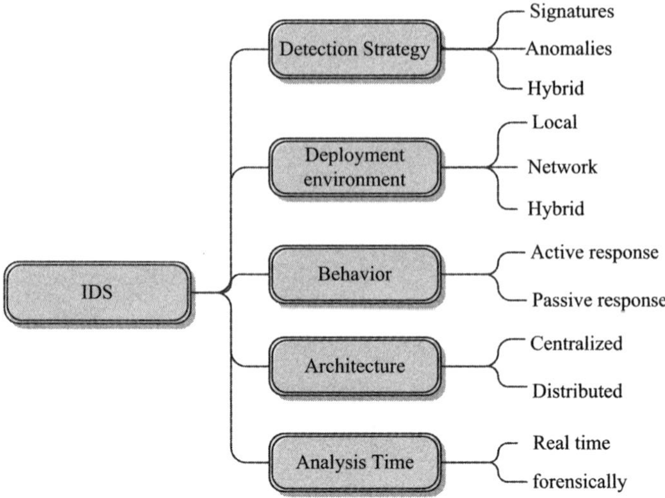

Fig. 2 IDS Classification

proposed based on detection of anomalies in the usual and legitimate usage of the systems. The anomaly-based detection is effective against unknown threats, but vulnerable to evasion techniques, given its tendency to generate high rates of false positives. Therefore alert correlation techniques have been proposed [13] [14] and hybrid detection strategies [15].

- **Deployment environment**: When the IDS operates on a networked environment it is known as NIDS (Network-based Intrusion Detection System) and if it operates on a local environment (host) it is known as HIDS (Host-based Intrusion Detection System). Additionally, there are hybrid schemes that exploit the advantages of each of them. This situation is common in IDS with distributed architecture, such as [16].
  In Fig. 3 an example of hybrid IDS is shown. The system considers both the information provided by local sensors, and the information provided by network sensors.
- **Behavior**: The IDS behavior is determined by the characteristics of the monitored information, and the allowed reaction time once a threat has been detected. The IDS with active behavior have the ability to decide and implement countermeasures once an intrusion is detected. However, the IDS with passive behavior merely inform the supervisor of the identified threats. This implies a slower reaction against intruders, but it guarantees the stability of the system in case of large amounts of false positives. Normally the IDS with active behavior are known as IPS (Intrusion Prevention System), and the IDS with passive behavior are simply called IDS.
- **Architecture**: The IDS architecture is determined according to the characteristics of the deployment environment. Currently there are two types of architectures:

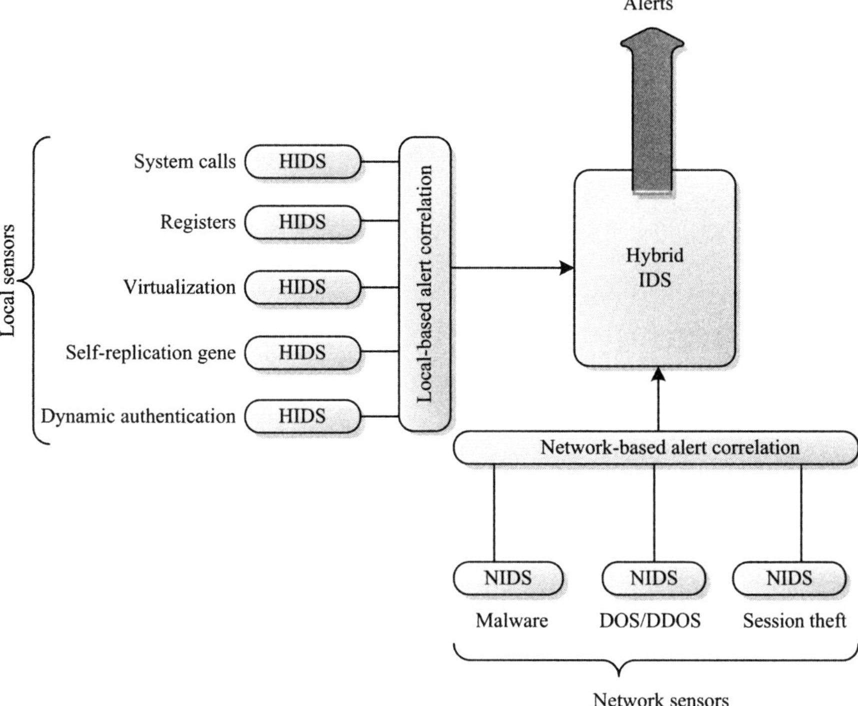

Fig. 3 Hybrid IDS example

centralized and distributed. The IDS with centralized architecture are characterized by being composed from a single node. Moreover, the IDS with distributed architecture are composed of various nodes, which are spread along the protected environment. This implies the need for a more complex design process and the establishment of communication between the different components of the IDS. In [17] a more extensive comparison of these types of architectures is proposed.

- **Analysis Time**: The frequency with which the IDS analyze the information determines the analysis time. The information processing can be performed in real time (online) and forensically (offline). The main advantage of data processing in real time is that it can detect attacks in progress and respond before they cause major damage. Occasionally, the need to achieve greater accuracy, or the need to protect the system from a wider variety of threats has led to IDS proposals based on the analysis of information in time intervals. The problem of these systems is that if the intervals are too long, preventive measures can be applied too late. Finally, when the only purpose of the IDS is detecting and extracting the features of a succeeded intrusion, a forensics approach (offline) is required. Such systems are slow, but very accurate. In [18] the benefits of applying a forensic study in the analysis of network flows are detailed.

Moreover, the IDS tend to generate large amounts of alerts, situation that hinders the prevention efforts. Therefore, current systems must incorporate alert correlation mechanisms capable of sorting, classifying, verifying, and normalize them in order to facilitate management. In addition to these tasks, it is common for these systems to extend the information provided by sensors, by detailing the nature of the detected threat. For these reasons the alert correlation systems are an indispensable complement to an IDS.

Below there will be an explanation of the main aspects to take into consideration when deploying NIDS on the security perimeter of Data Centers. Thus, this chapter contains seven sections. The first section is this introduction, which simply addresses the concept of intrusion detection system and its features. The second section is devoted entirely to the description of the NIDS, their origin and the description of a classical architecture that shows the organization of their various components. The third section describes the most frequent architectures when they are deployed on Data Centers. The fourth section discusses the subjects of study of the NIDS, and the benefits and disadvantages they lead. In the fifth section the different detection strategies are explained and the consequences arising from the choice of each of them. The sixth section describes the main features of the alert correlation systems and the alerts management. Finally, in the seventh section, the conclusions will be presented.

2 Origin and Standardization

In the same way that the IDS have adapted to current needs, the NIDS have evolved from the very latest trends in the field of information networks. In the late 90s, the U.S. research agency DARPA (Defense Advanced Research Projects Agency) created the group known as CIDF (Common Intrusion Detection Framework) which focused on the development of a framework for the intrusion detection [19]. In 2000 the group joins the IETF (Internet Engineering Task Force) under the acronym IDWG (Intrusion Detection Working Group). The architecture of the proposed framework considers a division of the NIDS functionality in different blocks, whether they are in the same machine or distributed on different computers. In 2006 the CIDF architecture has resulted in that the International Organization for Standardization known as ISO, standardize the IDS concept [20]. Figure 4 shows the basic scheme of the CIDF architecture contained in the ISO standard. It is noteworthy that the division between each component is functional and not physical. This means that the modules can be located on different machines with very different internal representations of events, so that the use of an independent format for communication is necessary between them. This need has led to creating the CISL (Common Intrusion Specification Language) which has the following features: expressiveness to define any type of intrusion, univocal characterization of the intrusions, flexibility to accommodating new kinds of intrusions, simplicity in constructing their representations and portability when are implemented on a wide variety of platforms [21]. Subsequently,

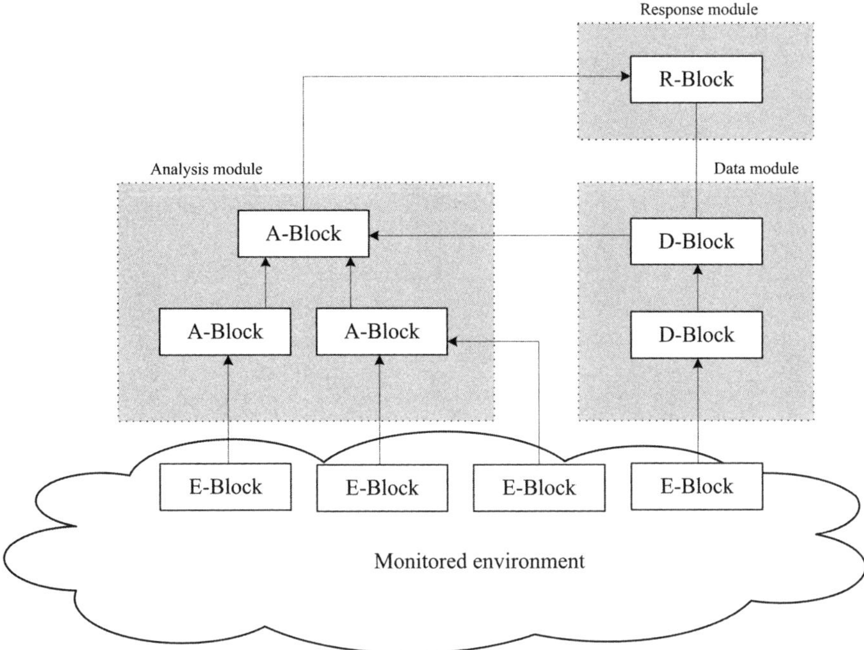

Fig. 4 Schematic of the architecture proposed by the CIDF

the need to standardize the flowing information between components, resulted in the standard IDMEF (Intrusion Detection Message Exchange Format) [22] for the IDS information exchanges.

Below the different CIDF architecture components are briefly described.

- **Monitored environment**: The election of the monitoring environment affects the design phase of the remaining IDS components. There is a great difference between monitored network events which involves parameters such as IP addresses, ports, protocols or traffic payloads, with monitored local events such as processes, memory addresses or registers.
- **Event blocks**: (E-blocks) Event blocks provide information about environmental events to the remaining IDS components. They are sensors that extract information from the environment, and express the results in form of the communication objects known as GIDO (General Intrusion Detection Object) specified in the CISL language.
- **Analysis blocks**: (A-blocks) Analysis blocks are responsible for analyzing the data collected by event blocks looking for potentially malicious activity. Just like the event blocks, they consider GIDO communication objects to exchange information. These modules are capable of synthesizing input events in order to lighten the speed of information processing. They will issue alerts if malicious events are detected.

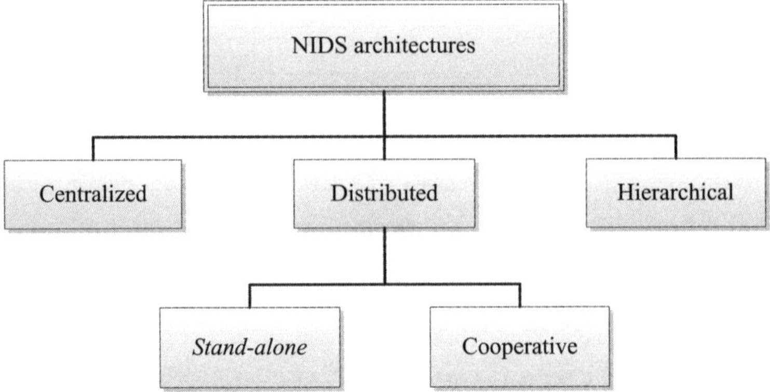

Fig. 5 NIDS architectures

- **Database blocks**: (D-blocks) Data blocks assist the event blocks and response blocks by storing the GIDO objects corresponding to pending events to be processed.
- **Response blocks**: (R-blocks) Response blocks processed GIDO objects corresponding to the events labeled as malicious, and decide the preventive measures. The response generated may be passive or active. If it is passive, prevention tasks are delegated to an operator. When the answer is active, prevention actions will be automatically decided.

Although the CIDF was unsuccessful as a framework for developing new intrusion detection systems, it laid the foundation for the ISO standardization. Most current NIDS consider its basis, and the block division which its authors proposed. In addition, over the years it has been considered as a classic example of the IDS component division and the relationships between them. The CIDF is a good introductory example, before explaining the most relevant NIDS features.

3 Architecture

At the present time the design of NIDS tends to consider centralized, distributed and hierarchical architectures (see Fig. 5). Centralized architectures grouped all the NIDS components on the same level, which provides a simple design process and an efficient performance. But because of the diversity and the specialization of the intrusion strategies, it is inaccurate and has low scalability, so that gradually has been replaced by distributed and hierarchical architectures, designed to protect different types of networks. The following explains in detail each one of them.

- **Centralized architecture**: Centralized architecture was the first to be used. It is the simplest implementation of a general purpose NIDS, and combines its different components in a same node. Its design simplicity is commonly penalized with a lower accuracy, but its efficiency is better, because it doesn't require the interaction processes between its various components. However, currently most of the NIDS are deployed in a distributed or hierarchical way, due to the existence of two major problems: scalability and the presence of a single point of failure.

- **Distributed architecture**: Distributed architecture spreads the NIDS components throughout the protected system. The NIDS deployed under this architecture are usually referred with the acronym DIDS (Distributed Intrusion Detection Systems), although it is noteworthy that the ISO [20] standard does not collect this terminology. On the basis of the relationship between their nodes, two operating modes are considered: stand-alone mode and cooperative mode.
  - *Stand-alone mode*: When distributed architecture operates in stand-alone mode, the NIDS components are spread over different nodes, acting independently and without sharing information. Each node usually has a specific purpose, and is responsible for detecting a particular threat. Although its accuracy is quite high, the lack of coordination between sensors can overwhelm the system, being prone to suffer high rates of false positives. Further whenever the NIDS operates in active response mode, it may apply preventive measures to mitigate the intrusions attempts, which can drain system resources or create behavior inconsistencies. Therefore, it is architecture for systems that only enable low resource consumption. In [23] an application example of this architecture is presented for intrusion detection in MANET (Mobile Ad Hoc Network) by monitoring the battery consumption of different mobile devices. The choice of the stand-alone mode for this purpose is the reliability of the measurement by the hardware parameters which trigger the events, and the need that the energy consumption impact on the protected devices is minimal.
  - *Cooperative mode*: When distributed architecture operates in cooperative mode, the set of nodes acts as a giant spider web. As in the stand-alone mode case, it purposes is specific, but this time its nodes share information. The various warnings issued are pooled and the response module takes into consideration the information provided by each of them. It is a precise strategy that sacrifices some system performance in the detection stage (due to the latency of communications between nodes), in order to optimize the prevention stages. In [24] (Fig. 6) a distributed IDS example is shown working on cooperative mode for detecting botnet's infection sequences. Such a system analyzes bidirectional communications between the target system and the infected system, and identifies sequences which involve intrusion attempts. For this purpose the reports issued are considered by three types of sensors, designed for detecting each one of the infection stages.
- **Hierarchical architecture**: Hierarchical architecture is an extension of the distributed architectures which operates on cooperative mode. It is inspired by infrastructure networks with multiple layers, where the subnets are grouped into clusters. In them, the Clusterheand nodes are those that have the greatest impact on the proper functioning of the network, because they typically act as control points or gateway for the other cluster members. In the hierarchical architecture,

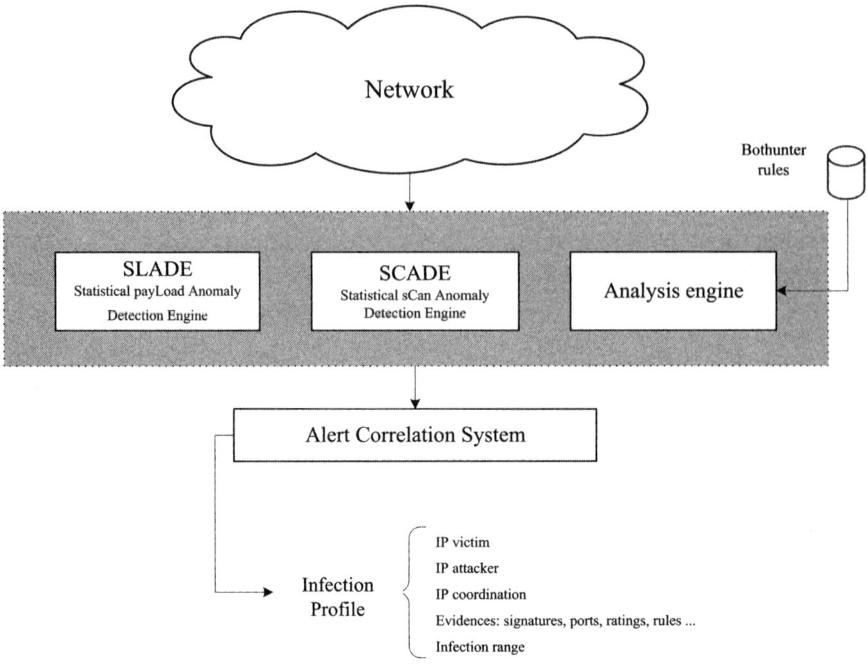

Fig. 6 Example of distributed NIDS on cooperative mode

the sensors are distributed so that some of them act as Clusterheand for the remaining nodes. In this way different levels of information processing are formed. Each level performs the pre-processing of the issued alerts, and the labeling that will facilitate its processing on the higher levels. Hierarchical architecture is the most accurate architecture, since different processing layers allow a deeper analysis of the events. A larger number of processing levels, the greater the accuracy and scalability, but also a greater impact on the protected system.

The IDS with GNU license known as Snort [9] (Fig. 7) is a good example of a hierarchical architecture system. Snort is composed of a collection of preprocessing modules which analyze the network traffic, for different purposes. A first layer of modules is responsible for preparing the information. Once processed, a set of modules is responsible for the detection. Such modules are specialized in protecting certain protocols, such as HTTP Inspect for HTTP, POP3 for POP or SMTP Preprocessor for the SMTP protocol. The information is delivered to a higher layer in which the Snort rule-base engine processes the results and emits the final alerts.

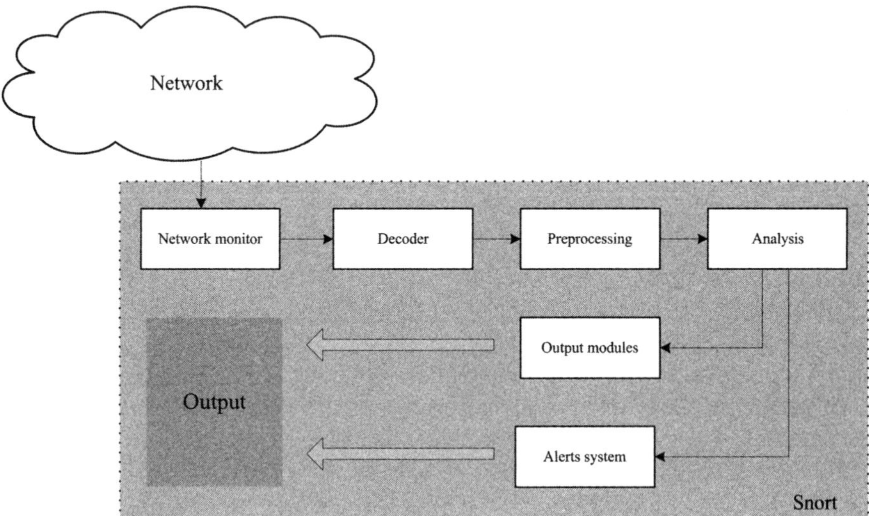

Fig. 7 Snort architecture

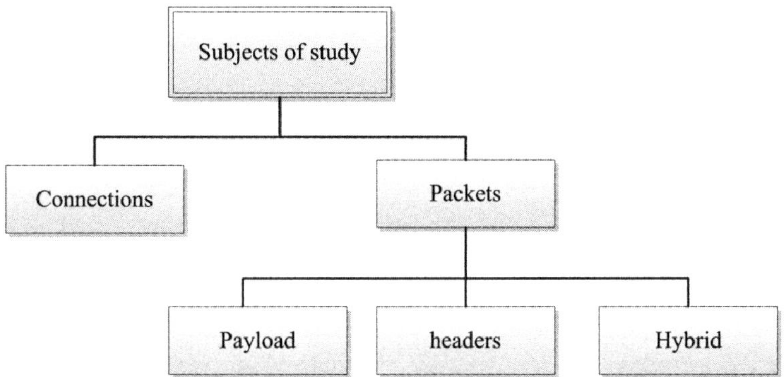

Fig. 8 Subjects of study

4 Subjects of Study

To carry out the study of traffic flowing through the network, the NIDS may focus on analyzing the content of each of the packages, or the complete information that occurs in an exchange of information between two systems. Figure 8 shows the most frequent subjects of analysis in such systems. Each of them has some advantages and some disadvantages that are important to know before its deployment on a defensive perimeter. The choice of an NIDS that takes as source of information, packets or connections, directly influences the accuracy and performance of the system. Below the features of each of them are explained.

- **Packet-oriented**: The packet-oriented NIDS extract information by processing the packets flowing through the network. The study aims to identify signs of malicious activity in their payload, their header, or both of them. Below their features are explained.
  - *Payload analysis*: The payload analysis is especially effective for detecting threats which pretend the exploitation of application level vulnerabilities, as long as the communication was established in a customary manner [25]. Its design generally focuses on the detection of malware [26] and on the detection of attempts to exploit vulnerabilities at the HTTP protocol [27]. Today it is also particularly effective against SQL injection attacks [28] cross-domain [29] and emerging threats, such as those concerning mobile devices [30].
  - *Header analysis*: The header analysis is especially effective for detecting threats which attempt to exploit vulnerabilities in the implementation of network protocols, and enumeration attacks. Its application provides a quickly and inexpensive identification of repetition-based threats such as the denial of service attacks [31], and is very useful when establishing mechanisms for alert correlation, allowing you to determine the source of the threat and the time of the issue.
  - *Hybrid analysis*: The hybrid analysis combines the advantages of the payload analysis with the header analysis. By this way it is possible to detect application-level threats, attacks against network protocols, and replay-based attacks. In contrast, the information processing is computationally more expensive, and requires a larger base of knowledge in the design phases. A classic example of hybrid NIDS is PAYL [25]. The PAYL system is based on the anomalies detection at the network traffic. For them, it builds a model of the normal and legitimate traffic from the protected network. Such model draws on the content of the payload header and features three parameters: port, length and direction of the traffic flow (input and output). Despite the fact that showed excellent results in terms of accuracy, its deployment was too computationally expensive, which subsequently led to different optimizations, highlighting POSEIDON [32], which proposed the use of the artificial neural networks SOM (Self-organizing Maps).
- **Connection-oriented**: The connection-oriented analysis extracts the connections status information. It considers parameters such as the number of bytes sent and received, the connection lifetime, protocol and intermediate nodes. The necessity of applying this type of analysis has come from the performance problems caused by the packet analysis, and the current needs of the IDS deployment at high speed networks of several Gigabits per second [33]. In addition, it must be considered as an arising problem when analyzing packets from encryption protocols [34]. Today, the connection-oriented analysis is carried out from the traces defined as IP Flows, published by the group IPFIX (IP Flow Information Export) and specified by the IETF [35, 36]. The Netflows traces [37] are also used, when previously used by the manufacturer Cisco. Each information flow is identified by the quintet (IP source, IP destination, port source, port destination, protocol), whose parameters are known as Flow Keys and contains a collection of packets.

It is noteworthy that the connection concept at traffic traces is different from that used at the TCP protocol. By this way, flows belonging to protocols that do not require connection may be issued, such as UDP. In [38] the use of flow traces is explained in more details, as well as the trends derived from it.

There is controversy between connection-oriented based NIDS and packet-oriented based NIDS. Some authors advocate the idea of developing hybrid proposals that combine both of them [38]. In particular, two-level systems were posed, where a first layer is responsible for processing the information Flows of information flowing through the network, and a second layer is responsible for analyzing the packets flowing to the critical regions of the protected system. Other studies, such as [25] conclude in the light of their experiments, that the performance results obtained from packet-oriented systems are slightly significant compared to the precision obtained by analyzing traffic traces. Subsequent studies support this idea, as [38] which clearly show the lack of precision at the connection-oriented systems, as well as the high computational cost of analyzing packets. The current trend towards the NIDS optimization techniques by exploiting parallelism, applying concurrency, using GPU processing [39] or implementing on reprogrammable hardware, coupled with the growth of the high speed networks, could end up tipping the balance toward any of the two trends.

5 Detection Strategies

Another important aspect of the NIDS is its strategy of detection. It is a key issue, when designing the NIDS, as it is when deploying at a Data Center security perimeters. Initially, intrusion detection focused on recognizing the most representative features of the attacks, known as signature based detection. However, the rapid proliferation of malware and the intrusion strategies makes it unfeasible to maintain a database containing information about all the existing threats. This prompted the design of new proposals which can detect unknown threats based on the recognition of anomalies in the usual and legitimate system usage. Because of the complexity of the techniques for modeling the legitimate usage, such systems tend to generate a high false positive rate, which have led to raise hybrid proposals that leverage the best of each of them. The following explains in detail each detection strategy (see Fig. 9), and the advantages and disadvantages involved in its implementation.

- **Signature based detection**: Signature based detection is to contrast the analyzed traffic characteristics with the signatures stored in a database of known threats. The IDS which implements this strategy are known as SBS (Signature-Based Systems), where the contrast signature is a pattern matching process which finish when ending successfully, by issuing alerts. The greatest advantages of this type of system are its excellent results in the false positive rate. However, SBS are usually not capable of detecting new threats or mutations of known attacks, which involves its constant updating by skilled operators. At the time of its deployment,

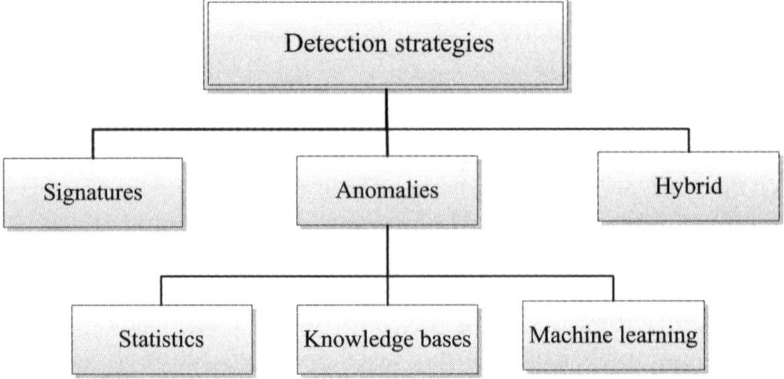

Fig. 9 Detection strategies

a SBS does not require too much effort and allows the operator to make a selection of signatures for identifying specific threats. Another advantage is the accurate classification of detected attacks, which will allow its treatment prioritization based on the impact factor. Actually, the SBS are particularly accurate in the detection of threats based on the exploitation of vulnerabilities in programming languages, as in [40–42].

- **Anomaly based detection**: The anomaly based detection arose from the need to deal with the unknown threats that are appearing continuously. This strategy consists on building a model of the usual and legitimate usage of the protected network (anomaly detection), building a model of the usual attacks against the protected network (misuse detection) or combining both of them (hybrid approaches). From these models it is possible to identify the network anomalous use, and match it with potential threats, dispensing the need to use complex databases that require constant updates. The IDS which implement this strategy are known as ABS (Anomaly-Based Systems). In [43] it is explained in detail the various anomalies on networks. The ABS deployment is not as intuitive as the SBS: development of modeling has to be supervised by experienced operators and precisely adapted to the characteristics of the network. The three most criticized aspects of using such systems are the tendency to generate high rates of false positives, the tendency to behave as a black box and its imprecise classification. Based on the modeling strategy, the ABS can be categorized as follows [44] (illustrated in Fig. 10).
  - *Statistical modeling*: The ABS modeled statistically capture legitimate network or misuse traffic, and develops a stochastic model of their behavior. The detection process is to compare the characteristics of the monitored traffic with the models, and determine whether it is potentially dangerous depending on the degree of similarity. The first approaches using statistical modeling, posed univariate models by applying independent Gaussian random variables. But were replaced by multivariate models that took into account various metrics,

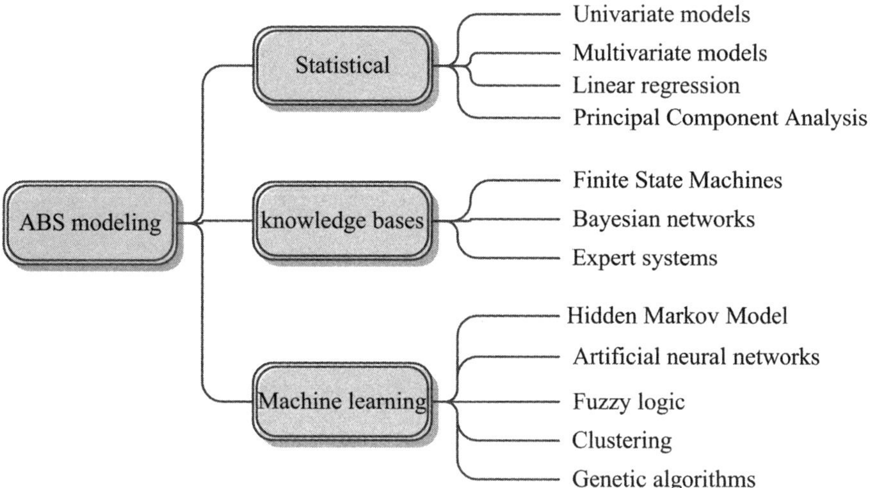

Fig. 10 Most used techniques for modeling ABS

achieving more accurate results [45]. Currently it is usual apply linear regression and PCA (Principal Component Analysis) for reducing the treatment complexity by ordering the traffic characteristics by relevance [44].

- *Modeling by using knowledge bases*: A knowledge base is a type of database adapted for the knowledge management and representation. The ABS which incorporates these mechanisms will require a training phase able to identify the most representative parameters of the collections of legitimate and malicious traffic. Once identified, a rule base is generated, and it is possible to determine the nature of the monitored traffic. The ABS modeling by expert systems is one of the most used today [25, 46]. However, there are also approaches based on the use of FSM (Finite State Machines) and the use of Bayesian networks [47]. Figure 11 shows an example of ABS based on expert systems. It is shown a clear distinction between the training and modeling phases, and also how by modeling are inferred rules that will be considerate by the analysis engine.
- *Modeling by machine learning techniques*: In contrast to statistical strategies, the machine learning techniques aim to establish models focusing on the feasibility of their application and in reducing the impact of computer use. The training and detection strategies of each of them are different, so its choice will depend on the characteristics of the monitored traffic and their deployment restrictions. Below the most used techniques for designing this class of ABS are summarized.
  - *Hidden Markov Model (HMM) and Markov Chains (MC)*: Markov chains are usually used on HIDS by analyzing sequences of system calls [48]. On the other hand, the hidden Markov model is most often applied on modeling the monitored network traffic.

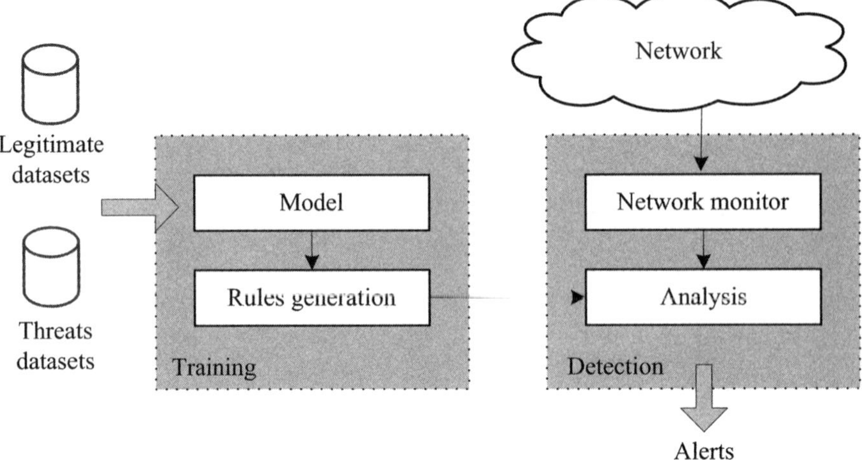

Fig. 11 Example of ABS based on expert systems

- *Artificial neural networks*: Artificial neural networks are another common modeling strategy based on the classification [49, 50]. Their choice is due to its flexibility and adaptability to the network changes, and its excellent performance. It allows the NIDS run in high-speed networks and a good fault tolerance.
- *Fuzzy logic*: The fuzzy logic application on the NIDS field is particularly intuitive, since as [51] pointed out the main features to analyze the network traffic can be interpreted as fuzzy variables.
- *Clustering*: Although several clustering techniques have been applied, the most used strategy on the NIDS field is the isolation of samples (outlier). When applying this technique, the monitored traffic that does not fit any of the established clusters is considered abnormal [52]. In [53] is shown an example of clustering by SVM (Support Vector Machines)
- *Genetic algorithms*: Genetic algorithms allow the identification of anomalies based on searches that estimate the optimal solution of problems. They are especially effective for deriving classification rules and for defining the characteristics applied in the model [54].

- **Detection based on hybrid proposals**: The hybrid-detection based NIDS combine techniques based on signatures recognition and anomalies recognition. This allows avail the benefits of implementing each of them, as well as an accurate detection of threats, achieving a low false positive rate. A classic example of hybrid detection is performed by the EMERALD [55] system, which also was one of the pioneering proposals at the IDS field. EMERALD is a distributed and hierarchical system based on combining host-level and network-level sensors. They are deployed over the network and operate through various blocks by applying different analysis techniques aimed at exploring the use of different services. To

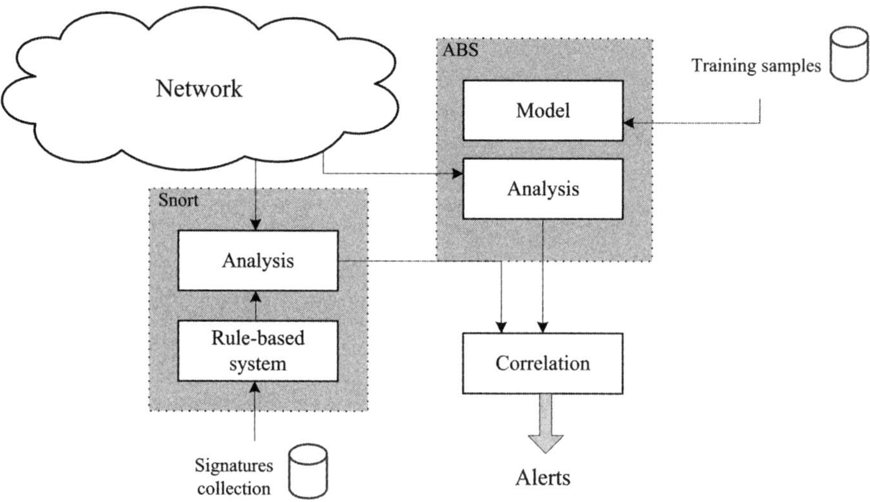

Fig. 12 Example of hybrid architecture

this is added the incorporation of an alert correlation system, and protocols for the blocks communications. In [56] the anomaly detection is performed based on a self-organizing map SOM, while detection of signatures is supported by the use of decision trees. In that proposal the intermediate alerts are combined into a rule-based decision module known as DSS (Decision Support System), responsible for issuing the final warnings that reach the operator. On the other hand, some studies integrate the anomaly detection module on freely distributed NIDS. Figure 12 shows a NIDS example that combines the information provided by an ABS with the information provided by signature-based IDS. In a similar way, in [15] the abnormality detection module is combined using the set of rules used by Snort and Bro. This takes advantage of having large communities support, and of have been specifically designed for new pre-processing modules installation. Despite the good results obtained in the different hybrid proposals, some researchers warn that results obtained by hybrid systems are not always better than those obtained by applying the strategies separately.

6 Alert Correlation

Another important aspect to consider when deploying NIDS is how they will manage the alerts issued. Under normal conditions the NIDS tends to report thousands of events in short periods of time. From the viewpoint of a human operator, the alerts analysis is unfeasible if there are not mechanisms that allow their classification. On the other hand, when the response process is automated, the huge amount of alerts can dramatically affect the quality of service. For these reasons, the use of alert correlation systems is necessary to complement the NIDS, and by this way bring an effective protection of Data Centers.

The alert correlation systems are designed for facilitating the management of the alerts issued from the various IDS sensors, allowing grouping and providing additional information about the events that generate them. Moreover, the tendency to deploy distributed systems and their hierarchization on different pre-processing modules specialized in detecting particular threats, emphasizes their importance when complementing NIDS. For these reasons, sometimes this type of system is incorporated into their design phases. The alert correlation systems are composed of different modules. In [57] the operation is explained, highlighting the presence different stages: normalization, verification, correlation and aggregation. The following briefly explains each.

- **Normalization**: The normalization phase aims at converting the format of the warnings issued by various NIDS sensors to a single format. Today the most common used format is the standard known as IDMEF (Intrusion Detection Message Exchange Format) [22]. This format unifies their representation and solves the problem of clock synchronization thanks to its compatibility with NTP (Network Time Protocol).
- **Verification**: The alert verification is the analysis of each of the alerts generated by the sensors and the establishment of the probability that the attack fulfill its purpose. Thus, the alerts associated with attacks with less probability to succeed are marked as low risk threats, and therefore its influence on the correlation process will be lesser. This leads that if, for example, the protected system runs under a GNU-Linux operating system, and has been detected sending a malware that affects only the Windows operating system; obviously the priority of treatment and the level of risk are marked as very low. However, if the protected system works under a Windows operating system, the alert must be marked as a higher priority, which involves special importance when deciding preventive measures.
- **Correlation**: The alert correlation phase aims to discover the similarities between the alerts content that have been issued by the different NIDS sensors. This will more precisely determine their nature and the risk. It will also be possible to indicate the priority of treatment to be considered when deciding preventive measures.
- **Aggregation**: The alert aggregation phase performs the fusion of the alerts issued with temporal proximity and similar characteristics. This avoids the multiple processing of the same event, and saturation of the operator.

Figure 13 shows an example of alert correlation system. In it, the warnings issued by various NIDS are normalized in a single format. Then they are verified, classified and grouped. The order of these phases could be different, and it depends on the needs of system design and on what extent should complement the NIDS.

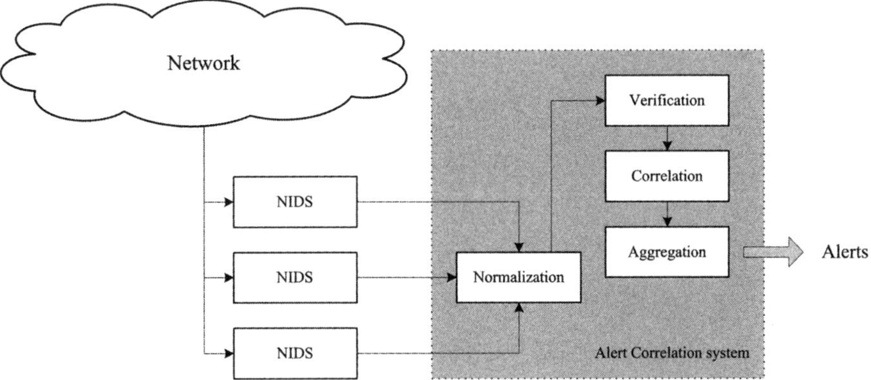

Fig. 13 Example of alert correlation system

7 Summary

In the same way that the IDS have adapted to current needs, the NIDS have evolved from the very latest trends in the information networks field. This has involved the adoption of specific strategies for operating in different environments, and the adaptation of the complementary measures to obtain better results. The aim of this chapter was to highlight the most important aspects to take into consideration about the NIDS when facing their deployment on the Data Center defense perimeters. After describing the general aspects of the IDS, the first described feature was their architecture. To do this, one of the classic examples of modular NIDS was briefly explained, known as CIDF (Common Intrusion Detection Framework). Its structure clearly shows the different steps to be carried out for processing information, which have served as a reference in most of the current architectures. Then, the most frequently applied architectures were explained: centralized, distributed, and hierarchical, the latter being an extension of the increasingly used distributed architecture. It is noteworthy that the choice of architecture will especially affect the accuracy, robustness and performance of defensive mechanisms. Another of the most important aspects of a NIDS are the characteristics of the processed information. As a result, their different subjects of analysis were explained. The choice of packet-oriented analysis or connectionless (connection-oriented) analysis will also be relevant for obtaining a better behavior, directly influencing the accuracy and the system performance. Likewise, the choice of a strategy based on signatures recognition, anomalies detection or both of them, it is also important in order to achieve good accuracy. Following this, another section has been devoted to the explanation of the detection strategies, and its benefit against certain kinds of threats. Finally, were briefly explained the characteristics of the most common alert correlation systems.

Acknowledgment Part of the computations of this work were performed in EOLO, the HPC of Climate Change of the International Campus of Excellence of Moncloa, funded by MECD and MICINN.

References

1. Lippmann, R.P., Cunningham, R.K.: Improving Intrusion Detection Performance Using Keyword Selection and Neural Networks. Computer Network **34**(4) (October 2000) 597–603
2. University of California, Irvine: KDD Cup 1999 Data. http://kdd.ics.uci.edu/databases/kddcup99/kddcup99.html (Accessed August 2013)
3. Yeung, D.Y., Ding, Y.: Host-Based Intrusion Detection using Dynamic and Static Behavioral Models. Pattern Recognition **36**(1) (January 2003) 229–243
4. Shiravi, A., Shiravi, H., Tavallaee, M., Ghorbani, A.A.: Toward Developing a Systematic Approach to Generate Benchmark Datasets for Intrusion Detection. Computers & Security **31**(3) (May 2012) 357–374
5. Lee, W., Miller, M., Stolfo, S.J., Fan, W., Zadok, E.: Toward Cost-Sensitive Modeling for Intrusion Detection and Response. Journal of Computer Security **10** (August 2002) 5–22
6. K. Killourhy, R.M.: Why Did My Detector Do That?! In: Proceedings of the 13th International Symposium on Recent Advances in Intrusion Detection. (September 15–17 2010) 256–276
7. Cheng, T.H., Lin, Y.D., Lai, Y.C., Lin, P.C.: Evasion Techniques: Sneaking through Your Intrusion Detection/Prevention Systems. IEEE Communications Surveys Tutorials **14**(4) (October 2012) 1011–1020
8. Kumar, M., Hanumanthappa, M., Suresh Kumar, T.V.: Encrypted Traffic and IPsec Challenges for Intrusion Detection System. In: Proceedings of the International Conference on Advances in Computing. (August 9–11 2012) 721–727
9. Sourcefire and CTO Martin Roesch: Snort: Open Source Network Intrusion Detection System. http://www.snort.org (Accessed August 2013)
10. Paxson, V.: Bro: A System for Detecting Network Intruders in Real-Time. Computer Networks **31** (December 1999) 2435–2463
11. Thonnard, O., Bilge, L., O'Gorman, G., Kiernan, S., Lee, M.: Industrial Espionage and Targeted Attacks: Understanding the Characteristics of an Escalating Threat. In: Proceedings of the 15th International Conference on Research in Attacks, Intrusions, and Defenses, Berlin, Heidelberg, Springer-Verlag (September 12–14 2012) 64–85
12. Wang, L., Jajodia, S., Singhal, A., Noel, S.: K-zero Day Safety: Measuring the Security Risk of Networks Against Unknown Attacks. In: Proceedings of the 15th European Conference on Research in Computer Security, Berlin, Heidelberg, Springer-Verlag (September 2010) 573–587
13. Salah, S., Maciá-Fernández, G., Díaz-Verdejo, J.E.: A Model-Based Survey of Alert Correlation Techniques. Computer Networks **57**(5) (April 2013) 1289–1317
14. Elshoush, H.T., Osman, I.M.: Alert Correlation in Collaborative Intelligent Intrusion Detection Systems–A Survey. Applied Soft Computing **11**(7) (October 2011) 4349–4365
15. Hwang, K., Cai, M., Chen, Y., Qin, M.: Hybrid Intrusion Detection with Weighted Signature Generation over Anomalous Internet Episodes. IEEE Transactions on Dependable and Secure Computing **4**(1) (February 2007) 41–55
16. Dreger, H., Kreibich, C., Paxson, V., Sommer, R.: Enhancing the Accuracy of Network-based Intrusion Detection with Host-based Context. In: Proceedings of the Second International Conference on Detection of Intrusions and Malware, and Vulnerability Assessment, Berlin, Heidelberg, Springer-Verlag (July 7–8 2005) 206–221
17. Nehinbe, J.: Log Analyzer for Network Forensics and Incident Reporting. In: Proceedings of the International Conference on Intelligent Systems, Modelling and Simulation. (January 27–29 2010) 356–361

18. Spafford, E.H., Zamboni, D.: Intrusion Detection Using Autonomous Agents. Computer Networks **34**(4) (October 2000) 547–570
19. Porras, P., Schnackenberg, D., Staniford-Chen, S., Stillman, M., Wu, F.: The common Intrusion Detection Framework Architecture. CIDF Working Group. http://gost.isi.edu/cidf/drafts/architecture.txt (Accessed August 2013)
20. Standard, I.: Information technology - Security Techniques - Selection, Deployment and Operations of Intrusion Detection Systems. Technical Report ISO/IEC 18043:2006, ISO/IEC (June 2006)
21. Feiertag, R., Kahn, C., Porras, P., Schnackenberg, D., Staniford-Chen, S.: A Common Intrusion Specication Language (CISL). http://gost.isi.edu/cidf/drafts/language.txt (Accessed August 2013)
22. H. Debar, D. Curry, B.F.: The Intrusion Detection Message Exchange Format (IDMEF). Requests for Comments RFC 4765, Internet Engineering Task Force (March 2007)
23. Jacoby, G.A., Davis, N.J.: Mobile Host-Based intrusion Detection and Attack Identification. IEEE Wireless Communications **14**(4) (August 2007) 53–60
24. Gu, G., Porras, P., Yegneswaran, V., Fong, M., Lee, W.: BotHunter: Detecting Malware Infection Through IDS-driven Dialog Correlation. In: Proceedings of the 16th USENIX Security Symposium, Berkeley, CA, USA, USENIX Association (August 6–10 2007) 167–182
25. Wang, K., Stolfo, S.J.: Anomalous Payload-based Network Intrusion Detection. In: Proceedings of the 7th International Symposium on Recent Advances in Intrusion Detection. (September 15–17 2004) 203–222
26. Wang, K., Cretu, G., Stolfo, S.J.: Anomalous Payload-based Worm Detection and Signature Generation. In: Proceedings of the 8th International Conference on Recent Advances in Intrusion Detection, Berlin, Heidelberg (September 20–22 2006) 227–246
27. Ingham, K.L., Inoue, H.: Comparing Anomaly Detection Techniques for HTTP. In: Proceedings of the 10th International Conference on Recent Advances in Intrusion Detection, Berlin, Heidelberg, Springer-Verlag (September 5–7 2007) 42–62
28. Chandrashekhar, R., Mardithaya, M., Thilagam, S., Saha, D.: SQL Injection Attack Mechanisms and Prevention Techniques. In: Proceedings of the International Conference on Advanced Computing, Networking and Security, Berlin, Heidelberg, Springer-Verlag (2012) 524–533
29. Lekies, S., Nikiforakis, N., Tighzert, W., Piessens, F., Johns, M.: DEMACRO: Defense against Malicious Cross-Domain Requests. In: Proceedings of the 15th International Symposium on Recent Advances in Intrusion Detection, Berlin, Heidelberg, Springer-Verlag (September 12–14 2012) 254–273
30. Zhou, Y., Jiang, X.: Dissecting Android Malware: Characterization and Evolution. In: Proceedings of the IEEE Symposium on Security and Privacy. (May 20–23 2012) 95–109
31. Park, K., Lee, H.: On the Effectiveness of Probabilistic Packet Marking for IP Traceback under Denial of Service Attack. In: Proceedings of the Twentieth Annual Joint Conference of the IEEE Computer and Communications Societies. Volume 1. (April 22–26 2001) 338–347
32. Bolzoni, D., Etalle, S., Hartel, P.: POSEIDON: A 2-Tier Anomaly-Based Network Intrusion Detection System. In: Proceedings of the Fourth IEEE International Workshop on Information Assurance. (April 13–14 2006) 144–156
33. Lin, P.C., Lee, J.H.: Re-Examining the Performance Bottleneck in a NIDS with Detailed Profiling. Journal of Network and Computer Applications **36**(2) (March 2013) 768–780
34. Puzis, R., Klippel, M.D., Elovici, Y., Dolev, S.: Optimization of NIDS Placement for Protection of Intercommunicating Critical Infrastructures. In: Proceedings of the 1st European Conference on Intelligence and Security Informatics, Berlin, Heidelberg, Springer-Verlag (2008) 191–203
35. Quittek, J., Zseby, T., Claise, B., Zander, S.: Requirements for IP Flow Information Export (IPFIX). Requests for Comments RFC 3917, Internet Engineering Task Force (October 2004)
36. Claise, B.: Specification of the IP Flow Information Export (IPFIX) Protocol for the Exchange of IP Traffic Flow Information. Requests for Comments RFC 5101, Internet Engineering Task Force (July 2008)
37. Claise, B.: Cisco Systems NetFlow Services Export Version 9. Requests for Comments RFC 3954, Internet Engineering Task Force (October 2004)

38. Brauckhoff, D., Tellenbach, B., Wagner, A., May, M., Lakhina, A.: Impact of Packet Sampling on Anomaly Detection Metrics. In: Proceedings of the 6th ACM SIGCOMM Conference on Internet Measurement, New York, NY, USA (October 25–7 2006) 159–164
39. Vasiliadis, G., Antonatos, S., Polychronakis, M., P, E., Ioannidis, S.: Gnort: High Performance Network Intrusion Detection using Graphics Processors. In: Proceedings of the 11th International Symposium on Recent Advances in Intrusion Detection. (September 15–17 2008) 116–134
40. Egele, M., Wurzinger, P., Kruegel, C., Kirda, E.: Defending Browsers Against Drive-by Downloads: Mitigating Heap-Spraying Code Injection Attacks. In: Proceedings of the 6th International Conference on Detection of Intrusions and Malware, and Vulnerability Assessment, Berlin, Heidelberg, Springer-Verlag (July 9–10 2009) 88–106
41. Heiderich, M., Frosch, T., Holz, T.: IceShield: Detection and Mitigation of Malicious Websites with a Frozen DOM. In: Proceedings of the 14th International Conference on Recent Advances in Intrusion Detection, Berlin, Heidelberg, Springer-Verlag (September 20–21 2011) 281–300
42. Pietraszek, T., Berghe, C.V.: Defending Against Injection Attacks Through Context-sensitive String Evaluation. In: Proceedings of the 8th International Conference on Recent Advances in Intrusion Detection, Berlin, Heidelberg, Springer-Verlag (September 7–9 2005) 124–145
43. Chandola, V., Banerjee, A., Kumar, V.: Anomaly Detection: A Survey. ACM Computing Surveys **41**(3) (July 2009) 1–58
44. Shyu, M.L., Chen, S.C., Sarinnapakorn, K., Chang, L. In: Principal Component-based Anomaly Detection Scheme. Volume 9. Springer Berlin Heidelberg (2006) 311–329
45. Guo, Z., Chung, S.L., Gu, M., Sun, J.G.: Efficient Presentation of Multivariate Audit Data for Intrusion Detection of Web-Based Internet Services. In: Proceedings of the 1st International Conference on Applied Cryptography and Network Security. (October 16–19 2003) 63–75
46. Wang, K., Parekh, J.J., Stolfo, S.J.: Anagram: A Content Anomaly Detector Resistant to Mimicry Attack. In: Proceedings of the 9th International Conference on Recent Advances in Intrusion Detection, Berlin, Heidelberg, Springer-Verlag (September 20–22 2006) 226–248
47. Howard, G.M., Bagchi, S., Lebanon, G.: Determining Placement of Intrusion Detectors for a Distributed Application through Bayesian Network Modeling. In: Proceedings of the 11th International Symposium on Recent Advances in Intrusion Detection, Berlin, Heidelberg, Springer-Verlag (September 15–17 2008) 271–290
48. Xu, X., Sun, Y., Huang, Z.: Defending DDoS Attacks Using Hidden Markov Models and Cooperative Reinforcement Learning. In: Proceedings of the 2007 Pacific Asia Conference on Intelligence and Security Informatics, Berlin, Heidelberg, Springer-Verlag (April 11–12 2007) 196–207
49. Ramadas, M., Ostermann, S., Tjaden, B.: Detecting Anomalous Network Traffic with Self-organizing Maps. In: Proceedings of the 6th International Symposium on Recent Advances in Intrusion Detection, Berlin, Heidelberg, Springer-Verlag (September 8–10 2003) 36–54
50. Golovko, V., Bezobrazov, S., Kachurka, P., Vaitsekhovich, L.: Neural Network and Artificial Immune Systems for Malware and Network Intrusion Detection. In Koronacki, J., Raś, Z., Wierzchoń, S., Kacprzyk, J., eds.: Advances in Machine Learning II. Volume 263 of Studies in Computational Intelligence. Springer Berlin Heidelberg (2010) 485–513
51. Bridges, S.M., Vaughn, R.B.: Fuzzy Data Mining And Genetic Algorithms Applied To Intrusion Detection. In: Proceedings of the 23rd National Information Systems Security Conference. (October 16–19 2000) 13–31
52. Bridges, S.M., Vaughn, R.B., Professor, A., Professor, A.: Data Mining for Intrusion Detection: From Outliers to True Intrusions. In: Proceedings of the 13th Pacific-Asia Conference on Advances in Knowledge Discovery and Data Mining. (April 27–30 2009) 891–898
53. Nassar, M., State, R., Festor, O.: Monitoring SIP Traffic Using Support Vector Machines. In: Proceedings of the 11th International Symposium on Recent Advances in Intrusion Detection, Berlin, Heidelberg, Springer-Verlag (September 15–17 2008) 311–330
54. Kim, J., Bentley, P.J., Aickelin, U., Greensmith, J., Tedesco, G., Twycross, J.: Immune System Approaches to Intrusion Detection – a Review. Natural Computing **6**(4) (December 2007) 413–466

55. Porras, P.A., Neumann, P.G.: EMERALD: Event Monitoring Enabling Responses to Anomalous Live Disturbances. In: Proceedings of the 20th National Information Systems Security Conference. (October 1997) 353–365
56. Zhang, J., Zulkernine, M.: A Hybrid Network Intrusion Detection Technique using Random Forests. In: Proceedings of the First International Conference on Availability, Reliability and Security. (April 2006) 262–269
57. Zang, T., Yun, X., Zhang, Y.: A Survey of Alert Fusion Techniques for Security Incident. In: Proceedings of the Ninth International Conference on Web-Age Information Management. (July 20–22 2008) 475–481

Software Monitoring in Data Centers

Chengdong Wu and Jun Guo

In recent years, thousands of commodity servers have been deployed in Internet data centers to run large scale Internet applications or cloud computing services. How to continuously monitor the availability, performance and security of data centers in real-time operational environments becomes a daunting task. In this chapter, a comprehensive solution for software monitoring is discussed in Internet data centers.

1 Introduction

Internet users require more today, not only the wealth of information, but also the high efficiency and stability of the information service. The reliability of the network application service has become one of the most attention performances. The server and service performance monitoring system emerges from the traditional network monitor system, and gradually becomes one of the main means of providing reliable network services. Monitor and control the web service integrated has become an important issue.

Production clusters often include thousands of nodes. Large distributed systems, such as Hadoop, can fail in complicated and subtle ways. As a result, Hadoop is extensively instrumented. A two-thousand nodes cluster configured for normal operation generates nearly half a terabyte of monitoring data in one day, which are mostly the application-level log files. This data is invaluable for debugging, performance measurement, and operational monitoring. However, processing this data in real time at scale is a formidable challenge. A good monitoring system ought to scale out to very large deployments and handle crashes gracefully [1].

C. Wu (✉) · J. Guo
Northeastern University, Shenyang, People's Republic China
e-mail: wuchengdong@ise.neu.edu.cn

J. Guo
e-mail: guojun@ise.neu.edu.cn

S. U. Khan, A. Y. Zomaya (eds.), *Handbook on Data Centers*,
DOI 10.1007/978-1-4939-2092-1_42

1.1 Performance Degradation

The primary reason monitoring data center is performance degradation. There are many reasons for the performance degradation, such as long-term continuous use of the system, file fragmentation, space debris, large concurrent or software aging.

1. Long-term continuous use of the system
   A lot of the software are used long-term for the system. These software are not completely cleared from memory and often have data stranded in cache when closed. The longer the system is running, the more the software are used. The more data is stranded, the greater resource will be consumed. As the result, the system's operation speed and the performance are greatly reduced. It will also cause accessories accelerated aging and damaging because of work overload.
2. File fragmentation and space debris
   A lot of movies, music, pictures, documents, software, etc are saved in the computer. User often manages them by categories, removes some unwanted application, and downloads some new movies and so on. These works will clutter computer disk and large files tend to run out of memory. Therefore, the operating system on the disk generally produces temporary swap file and the disk space occupied by the file is used into the virtual memory. Virtual memory management program would be hard to read and write frequently, which will produce a large amount of debris and affect the performances of computer.
3. Big system
   Today the computer is powerful and users tend to install a lot of software. But the software installed too much will make the system mast. Like a big fat it is difficult to move quickly. Therefore it is necessary to clean and delete the useless software frequently [2].
4. Large concurrent
   The number of concurrent users is defining as the online user interactive with the servers. The main feature of these users is that they produced interaction with server. This interaction can be one-way or two-way transmission. Instantaneous high concurrency may result in system performance degradation, even paralysis. The number of concurrent users to monitor and forecast software can alleviate this situation to a certain extent.
5. Software aging
   Software aging refers to progressive performance degradation or a sudden hang/crash of a software system due to exhaustion of operating system resources, fragmentation and accumulation of errors [3].

The performance degradation will make a great impact on the services provided by data center, which may not guarantee SLA (Service-Level Agreement) and meet the requirements of the users. Therefore, monitoring the performance degradation is an indispensable task for the manager.

1.2 Function Failure

As the key point to evaluate cloud service quality, dependability has become a hot research topic of cloud computing, which refers to the probability of successful implementation of user submission service, reflects the ability of cloud service to complete the user submission service from the user point of view.

Unfortunately, the definition of what constitutes a failure of a node is ambiguous. For example, is it a failed disk-request or retried by the operating system? What is the operating system (OS) memory leak that makes socket connections painfully slow? This ambiguity is circumvented by defining failure as the shutdown or crash of a node, regardless of the cause [4]. Function failure is typically caused by the following three aspects.

1. Software failure

   The subtask sets in cloud service running in different cloud computing nodes. These nodes may include failure error software [4].

   At the macro granularity, there are dense blade-systems which are packed in a rack as a cluster. With a high load imposed on these dense systems both on CPUs and on disks, heat dissipation becomes a very important concern. Potentially leading to thermal instability can cause system/node breakdowns. System software and applications become more complex. Such complexity makes them more prone to bugs and other software failures (e.g., memory leaks, state corruption, etc.). These bugs/failures can cause the system breakdown [5].

   The computer virus should also be included. There is no doubt about the damage caused by virus. For example, when a computer is infected some kinds of worms, the worms copy themselves and occupy system resources. The running speed of the computer slows down. Finally system resource exhausted and collapse. Some of the system files have been damaged even if the computer virus was found and removed, which also make computer operation performance greatly decrease.
2. Hardware failure

   There are a lot of hardware resources in the data center. Therefore, cloud computing system will have the hardware failures [4].

   Denser integration of semiconductor circuits, though preferable for performance, makes them more susceptible to strikes by alpha particles and cosmic rays. At the same time, there is an increasing tendency to lower operating voltages in order to reduce power consumption. Such reduction in voltage levels can increase the likelihood of bit-flips, when circuits are bombarded by cosmic rays and other particles so that lead to transient errors. Memory structures are typically the target for protection against errors, but more recent studies have pointed out that the error rates in combinational circuits are likely to surpass those of memory cells in the next decade [5].

   Computer is composed by a variety of boards and integrated circuit chip. When the computer runs, it will generate a lot of heat. The high temperature not only affects computer performance but also accelerates the aging of the machine. Hence, the computer is equipped with a heat sink which is connected with a small fan to reject the heat generated in the chassis.

- Fan question
  Usually, the fan is the inexpensive non-precision devices. It is the one of the most easily damaged parts in computer accessories. After a period of time running, the power of the fan begins to drop and heat weakness. The chassis cannot effectively dissipate the heat generated, so that all parts are in an abnormal high temperature work environment. This will cause the performance declined significantly, and often produce operation errors. More seriously, the fan will be completely "strike". After this occurring, the computer freezes frequently, shut down and start automatically, even the chip is burned.
- Dust problem
  It is well known that static electricity will attract dust when electrical appliances are used. Small computer chassis, sophisticated computer accessories work together will consume more power, generate more static magnetic field. It is particularly easy to gather much dust. The accumulation of dust will not only be difficult to distribute the heat, but also affect the cooling fan rotation.

3. Communication link failure
   When the sub-tasks access to remote data, logical or physical link may be destroyed, especially in long-distance and large-scale data access. Therefore, the communication link may be failed [6].

Networking (whether it is the Internet, or a local/system/storage area network) has made it convenient to deploy systems that are inherently parallel in nature (whether it is functional different systems performing different operations, or data parallelism different systems performing same operations but on different pieces of data). This can not only be performance-efficient, but also make it easier to write and deploy distributed programs/systems.

However, the growing reliance on each other makes nodes within a parallel/distributed system more susceptible to another failure/error [5].

Taking the PSN (Sony Play station Network) as an example, down time has appeared several times which will cause millions of users unable to successfully log in the game during this period. Service problems do not occur frequently and often last for a short time. But users are incapable of action, except no play PSN [7]. Function failure will make the serious result. Therefore, real-time monitoring system is very necessary in order to take immediate measures.

1.3 Energy Conservation

The goal of data center network is to interconnect the massive number of data center servers, and provide efficient and fault-tolerant routing service to upper-layer applications [8]. Now the energy consumed by power-hungry devices becomes a headache problem for many data center owners. According to figures, the total energy consumption of network devices in data centers of the US in 2006 was 3 billion KW/h. It has been shown that network devices consume $20\,\% \sim 30\,\%$ energy in the whole data center, and the ratio will grow with the rapid development of power-efficient hardware and energy-aware scheduling algorithm on the server side. Ideally, any idle switch would consume no power, and energy consumption would grow with increasing network load. Unfortunately, today's network devices are not energy proportional. The fixed overheads such as fans, switching fabric, and line-cards, have the waste energy at low network load. The energy consumption of network devices at low network load still accounts for more than 90 % of that at busy-hour load. As the result, the large number of idle network devices in high-density networks waste significant amount of energy [9].

One of the major causes of energy inefficiency in data centers is the idle power wasted when servers run at low utilization. Even at a very low load, such as 10 % CPU utilization, the power consumed is over 50 % of the peak power. Similarly, if the disk, network, or any such resource is the performance bottleneck, the idle power wastage in other resources goes up. In the cloud computing approach, multiple data center applications are hosted on a common set of servers. This allows for consolidation of application workloads on a smaller number of servers that may be kept better utilized, as different workloads may have different resource utilization footprints and may further differ in their temporal variations. Consolidation allows amortizing the idle power costs more efficiently [9].

Energy-saving method for data center:

1. turning off idle equipment;
2. storage consolidation, tiering and virtualization;
3. using efficient IT equipment to supply power;
4. using UPS (uninterruptible power supply device);
5. choosing the best method of refrigeration;
6. estimation the utilization and the efficiency of the data center.

The construction of green energy-efficient data center is the social development needs, as well as the needs of enterprise development. It is not only for the environment, but also for the cost reducing of IT operation and maintenance. From a technical point, our goal is to improve the PUE (power consumption of the whole data center/power consumption of IT equipment) value of data center. Google uses customized evaporative cooling to significantly reduce the energy consumption of data center [10].

2 Monitoring Content

In cloud computing environment, monitoring is an essential link. Monitoring could help the user to know the information of the hardware and software. Some hardware and software, such as basic software, middleware, database, application software, PM(physical machine), VM(virtual machine) and user behavior, is introduced in this section. After knowing this information, the following work should be done, such as user behavior analysis, hot-spot evaluation, performance prediction, advanced warning and performance bottlenecks analysis.

Table 1 Basic software major monitoring indicators

Monitoring object	Monitoring describe	Monitoring indicators
CPU	The usage rate of CPU	CPU usage rate (%)
Disk	The usage rate of Disk	Disk usage rate (%)
		The free space (MB)
Memory	NT Service	Memory usage rate (%)
		The free space (MB)
		Error pages/second
		The total space (MB)
Ping	The status of the server	Package success rate (%)
		The round-trip time of data (ms)
		Status value
Windows service	The status of the service	Monitoring results
		Disk space usage
Windows event log	NT event log	The total number of rows
		The number after Filter
		The total number of rows
Windows process	The information of the host	CPU usage rate (%)
		Total number of threads
		Using memory space(KB)
Port	The connect with the TCP port	The round-trip time of data (ms)

2.1 Basic Software

In general, basic software includes operating system, database, office software, and middleware. But operating system is the basic software here.

The state and behavior of software system always keep up with its requirements specification when running. It is an important way to continue to provide the high quality service [11]. Monitoring technology of software system has been paid more and more attention from the researchers, because it is an important method to make sure the software quality [12] (Table 1).

2.2 Middleware

Middleware is the computer software that connects software components and applications. The software consists of a set of enabling services that allow multiple processes running on one or more machines to interact across a network. This technology evolves to provide for interoperability in support of the move to coherent distributed architectures, which are used often to support and simplify complex, distributed applications. It includes web servers, transaction monitors, and messaging-and-queuing software [13].

According to different middleware, different information should be monitored. The following seven kinds of middleware will be introduced. They are Data Access Middleware (DAM), Remote Process Call Middleware (RPCM), Message Oriented Middleware (MOM), Object Oriented Middleware (OOM), Task Process Middleware (TPM), Grid Middleware (GM) and Terminal Emulation Middleware (TEM) [14] (Table 2).

2.3 Database

With the recent development of cloud computing, the importance of cloud databases has been widely acknowledged. Here, the features, influence and related products of cloud databases are discussed firstly. Then, research issues of cloud databases are presented in detail, which includes data model, architecture, consistency, programming model, data security, performance optimization, benchmark, and so on. Finally, the future development trends in this area is discussed [15].

With the rising of cloud computing technology, well-known large companies that did not engage in product development database before have released their products, such as the Simple DB of Amazon and the BigTable of Google. But traditional database still has their useful values [16]. And the indicators of the traditional database and the cloud database should be monitored are the same. The monitoring indicators of Microsoft SQL Server database is introduced here (Table 3).

2.4 Application Software

Service-Oriented Architecture (SOA) shifts the focus of software development from product-oriented program to dynamic service-based composition. It has emerged as a major computing approach of distributed software architecture [17].

Runtime monitoring is an effective technique for quality assurance. It tracks software execution, observes the system behavior, verifies the service dependability properties, and responds to policy violations [18].

Monitoring collects the information for behavior diagnosis, defect detection, and status recovery [19]. Various approaches are available to monitor the service invocation, process and interactions, using different frameworks, requirements and constraints, and specifications [20, 21] (Table 4).

Table 2 Middleware major monitoring indicators

Monitoring object	Monitoring indicators
Memory	The use space(KB)
	The use space of progress(KB)
	The shared space(KB)
	Physical memory(KB)
	The status of memory(KB)
Disk	The free space(KB)
	The size of disk(KB)
	The type of disk
	The disk of local(KB)
	The disk of remote(KB)
The status of task	The number of task
	The number of event/second
	The peak number of user
File database space	The use space(KB)
	The free space(KB)
	Surplus rate (%)
Execute queue	The total number of execution threads
	Execute queue name
	The free number of execution threads
	Pending request oldest time
	The number of requests untreated in queue
	The number of requests handled in queue
The status of JVM	The total space of memory(KB)
	The free space(KB)
	The used space(KB)
	Usage rate (%)(KB)
The pool of status of the connection	The number of connect created
	The number of connect destroyed
	The number of connect assigned
	The number of connect returned
	The size of the pool
	The waiting time
	The average waiting time
	The number of error
	Used rate
	Maximum used rate

Table 3 Microsoft SQL server database major monitoring indicators

Monitoring object	Monitoring indicators
Database performance	Content of database
	Status of database
	Volume of business
	Unallocated space
	Reserved space
	Data space
	Index space
	Unused space
	Transaction log size
SQL buffer manager	Buffer cache hit rate
	Lazy writes/second
	The number of database page/second
SQL memory manager	Amount of memory used
	Dynamicmemory used to connect
	The total memory used to select and Optimize
	The total of dynamicmemory
SQL user manager	The number of user connections
	The number of login/second
	The number of logout action/second
SQL server cache manager	Hit the target rate
	The number of page used
	The number of user in cache
	The number of the object used
SQL server static manager	The number of transact-SQL task/second
	The number of trying parameterization/second
	The number of SQL compile/second
	The number of SQL compile again/second

Table 4 Application software major monitoring indicators

Monitoring object	Monitoring indicators
Downtime	The downtime of software
Software response time	Software response time
Software throughput	Software throughput
Error rate	Error rate
CPU used	CPU used rate of the software
Memory used	Memory used rate of the software
Disk used	Disk used rate of the software
I/O	the I/O of device
Network port	The speed of network port in/out
Threads	The threads of host
User registered	The information of user registered
User login	The information of user login
The status of connection	The status of connection

Table 5 PM major monitoring indicators

Monitoring object	Monitoring indicators
CPU	Total, used and used rate
Memory	Total, used and free space
Disk	Total, used, free space and I/O speed
Network	Total, used, free space and I/O speed
Swap	Total, used and used rate
VM	The number of total and active
Time	Local time

2.5 PM (Physical Machine) and VM (Virtual Machine)

PM may be a real machine or one node in a large server and has a set of complete hardware and software devices. But its primary task is to provide the resources for some tool virtualization. There lists some commonly used tools, for example, VMware [22], Connectix [23], Xen [24] and KVM [25]. On the PM, one of them should be set up to manage resource for VM (Table 5).

VM is set up on the virtualization resources and has its own software devices sharing the same hardware devices. All VMs share resources with others. But VM cannot compete with others for resources, because their resources are independent and isolated. The others' resources cannot be seen in their system. Therefore, each VM could be monitored to record the information that directly relates to the relationship between VM and resources (Table 6).

The total resource of all VM is the part of the resources of the PM, because the resource cannot be enlarged. For example, the memory space is four GB, it could

Table 6 VM major monitoring indicators

Monitoring object	Monitoring indicators
CPU	Total number, used rate
Memory	Max, total, used and used rate
Disk	Total, used, free space and I/O speed
Network	Total, used, free space and I/O speed
Swap	Total, used and free space
Time	Local time
Service	The number and type of service

not be five GB after virtualization. The information for CPU of PM and VM are not same. For the PM, the number of real and virtual CPU can be known, but for the VM, only the virtual CPU. Sometimes the number of real CPU and virtual CPU are different. In general, the number of virtual CPU is more than that of real CPU. The fact is that some VM will use the same real CPU.

2.6 User Behavior Analysis

Millions of users interact with search engines daily. They query problems, follow some of the links in the results, click on ads, spend time on pages, reformulate their queries and perform other actions. These interactions can serve as a valuable source of information for tuning and improving web search result ranking, and can complement more costly explicit judgment [26].

User behavior analysis, in the case of site visits to the basic data obtained, refers to statistics and analysis of the related data. The laws of users visit the web site may be discovered, These laws will combine with network marketing strategy to find the possible problems in network marketing activities, and to make further correction, or to provide evidence for the network marketing strategy [27].

Through analysis of the data obtained by monitoring user behavior, it can let enterprise understand in more detail and more clearly to the user behavior to find the site promotion channels problems existing in the enterprise marketing situation, and to make enterprise marketing more accurate, effective and to improve the business transformation [28] (Table 7).

User behavior analysis includes data record and finishing, keywords analysis, data analysis, website usability analysis, communication and performance. For example, in the keywords analysis, because of the great amount of data, keywords should be extracted for analysis.

2.7 Hot-Spot Evaluation

There are hot-spot problems in various fields. For example, in computer network, because of a switch having a big load, it leads to communication congestion. This

Table 7 User behavior analysis monitoring indicators

Monitoring object	Monitoring indicators
User	The space user come from
	IP address
	Page
	Register or not
Page	Dwell time
	Jump rate
	Return visitor
	New visitor
	The number of return
	Interval of return
Relation	Page and page
	User and user
	Page and user
Way	Way of user login

switch will be the hot spot of the system bottleneck. Different definition will appear according to different needs. For example, a VM which has little ability to deal with current work becomes a hot spot. And a module which occupies mostly resources and influences the performance of others in VM becomes a hot spot.

The method of Hot spot problem finding can be change for different solutions [29, 30]. The first method is virtualized resources allocation, another is migration. The effectivity of two methods are proved in a lot of papers.

Virtualized data centers enable sharing of resources among hosted applications. However, it is difficult to satisfy service-level objectives (SLOs) of applications on shared infrastructure, as application workloads and resource consumption patterns change over time. Two modules to achieve the function are designed [31]. AppControler and NodeControler are the major contribution. AppControler analyzes the information of VM and computes the resources needed next time. NodeControler receives the information from AppControler and decides a project to solve the hot spot problem.

Another method is the migration. Migration technology could make VMs migrates from one PM to another online quickly and keep running during the period of migration [32–34]. The resources that can be migrated are so many, but not everyone is feasible, for example, a disk. In general, a disk is bigger than 20 GB. Thus, it cannot migrate disk from one PM to another. Nowadays, most methods will use memory for migration. How to transmit memory is a key problem. Someone proposes an iteration method. The first step is to transmit the total memory and not to shut down the VM. The second step is to transmit the memory which is altered during the first step. When the memory is altered in the recently transmittal, it is smaller than the

Table 8 Hot spot evaluation monitoring indicators

Monitoring object	Monitoring indicators
CPU	Total
	Used
	Used rate
Memory	Total
	Used
	PM free
Swap	Total
	Used
Network	Bandwidth
	Speed
Disk	I/O rate
	Speed

set value. But this method needs the FTP(File Transfer Protocol) technology which helps VM's disk storage on a public PM. On the other hand, this technology keeps VMs use large network resource (Table 8).

2.8 *Performance Prediction and Advanced Warning*

The performance analysis can be carried out analytically or through experiments. In general, an effective parallel program development cycle may iterate many times before achieving the desired performance. In parallel programming, the goal of the design process is not to optimize a single metrics, for example the speed. A good design has to take into considering the execution time of a specific function, memory requirements, implementation cost, and the others [35]. During performance evaluation of parallel programs, different metrics are used [36].

The information about application and system behavior is used to predict the application runtime in the environment conditions where the application was not previously run. The changes of the hardware environment can include only the difference in the number of processors, changes in the number and speed of the processors, and variations in the general architecture.

In the application domain the prediction is needed when the application is run with new input data, execution parameters or on different hardware. The hardware and applications prediction can be based either on the precise analytical calculations or approximations. The precise calculations can be performed only if the full analysis of the application or full hardware specification is present. If either of the conditions is not met, the approximation techniques are needed. The most popular approximation techniques are based on the utilization of the historical data available. If the historical

Table 9 Performance bottlenecks analysis monitoring indicators

Monitoring object	Monitoring indicators
The number of user	Average
	The max
	The min
Response time	Average
	Longest
	Shortest
Server resource	Memory
	CPU
	Disk I/O
	Network I/O
Database	The number of concurrent users
	Speed

data is present, the prediction is based on the information extrapolation to the situation in the new environment. In the more general case, the prediction is based on the utilization of either statistical or heuristic techniques (e.g. genetic algorithms or simulated annealing) to find the similarities between analyzed situation and the data previously collected [35].

Advanced warning is the final target of performance prediction. If a good result of performance prediction could be obtained, the purpose of advanced warning will be achieved. There are many of performance prediction tools, such as Dimemas, Network Weather Service and Grid performance PREdiction System.

2.9 The Performance Bottlenecks Analysis

In the previous section, the performance prediction has been mentioned. When performance bottlenecks occurred, the causes to this situation should be analyzed.

Bottleneck generally refers to the key limiting factor in the overall. Bottlenecks have different meanings in different areas. Bottleneck in production refers to the overall level limit workflow, including workflow completion time, the workflow quality of a single or a few factors. Usually, the slowest part of the process is the bottleneck (Table 9).

The performance bottlenecks analysis contributes to do better in solving the problem. For example, when our system faces a performance declining, there will be some solutions for us to choose instead of helpless. But not every method will be better to solve the problem. Every method has its own performance values. After performance bottlenecks analysis, the key factors of performance declining would be known. An appropriate method to solve the problem should be adapted and make the least available resource to achieve the most performance.

3 Monitoring Timing

Some monitor technologies are used to monitor timing, such as the resource-oriented monitoring and the business-oriented monitoring. For the resources-oriented monitoring, the resources are monitored in 7*24 h, and the business-oriented monitoring, the whole life cycle of business is monitored to identify and obtain the system's performance. In this way, the impact of system crashed will be reduced

3.1 Resource-Oriented Monitoring

Why do we need to monitor the resources? As you know, in recent years, the cloud computing has been developed quickly and become the topics at the forefront of today's computer research. The monitoring system of the cloud data center has also gradually got attentions. Due to real-time, security, scalability, and the volume of traffic overload, traditional monitoring systems are not suitable to the present cloud monitoring. If some resources cannot meet the requirements, the system may be crashed. Hence, runtime monitoring is necessary for runtime system status tracking and anomaly detecting [37, 38].

What resources should be monitored? Generally, when you identify server performance degradation, the usual suspects are CPU, memory, and the disk. Therefore, these system resources should be early monitored on Windows and Unix-based servers [39].

A tool was required to monitoring resources-source monitor. The resource monitoring tools are immature compared to traditional distributed computing and grid computing [40]. What is the resource monitor? Resource monitor is a system application in Microsoft Windows operating systems. It is used to view information about the use of hardware (CPU, memory, disk, and network) and software (file handles and modules) resources in real time [41]. Resource monitor is available in Windows Vista and onwards only (In Windows Vista, it is known as reliability and performance monitor).Resource monitor can be launched by executingresmon.exe (perfmon.exe in Windows Vista). It was also on Windows 95, 95 OSR, 95 OSR2, 98, 98SE, ME, and NT [38].

Let us take CPU as an example.

The following is CPU monitor configuration screen, as shown in Fig. 1. A threshold can be set to test the monitor to gauge the current performance statistics in a click.

The CPU utilization is monitored in 7*24 h, and the result of the CPU utilization graph is gotten. The monitoring results are shown in Fig. 2 and 3 respectively. When

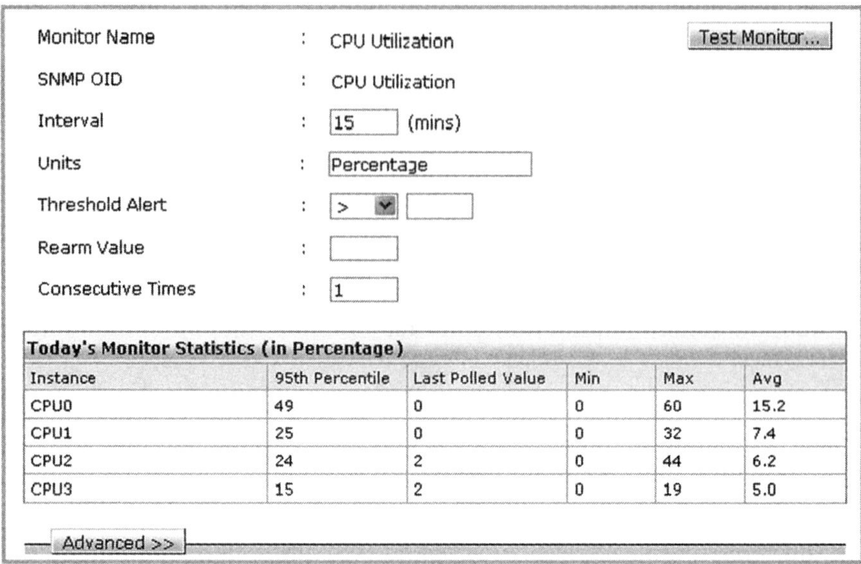

Today's Monitor Statistics (in Percentage)

Instance	95th Percentile	Last Polled Value	Min	Max	Avg
CPU0	49	0	0	60	15.2
CPU1	25	0	0	32	7.4
CPU2	24	2	0	44	6.2
CPU3	15	2	0	19	5.0

Fig. 1 CPU monitor configuration screen

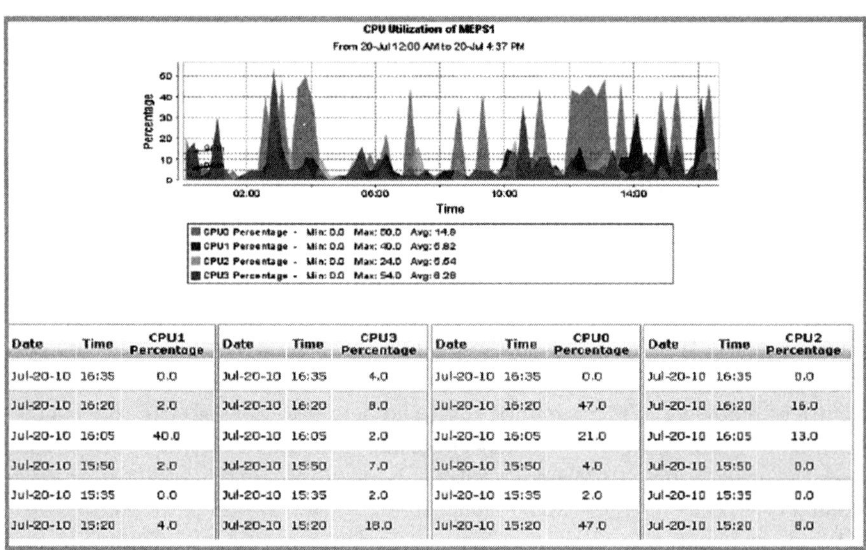

Date	Time	CPU1 Percentage
Jul-20-10	16:35	0.0
Jul-20-10	16:20	2.0
Jul-20-10	16:05	40.0
Jul-20-10	15:50	2.0
Jul-20-10	15:35	0.0
Jul-20-10	15:20	4.0

Date	Time	CPU3 Percentage
Jul-20-10	16:35	4.0
Jul-20-10	16:20	8.0
Jul-20-10	16:05	2.0
Jul-20-10	15:50	7.0
Jul-20-10	15:35	2.0
Jul-20-10	15:20	18.0

Date	Time	CPU0 Percentage
Jul-20-10	16:35	0.0
Jul-20-10	16:20	47.0
Jul-20-10	16:05	21.0
Jul-20-10	15:50	4.0
Jul-20-10	15:35	2.0
Jul-20-10	15:20	47.0

Date	Time	CPU2 Percentage
Jul-20-10	16:35	0.0
Jul-20-10	16:20	16.0
Jul-20-10	16:05	13.0
Jul-20-10	15:50	0.0
Jul-20-10	15:35	0.0
Jul-20-10	15:20	8.0

Fig. 2 CPU utilization graph for a quad-core CPU

the utilization is higher than a certain threshold, the system will raise an alarm to the Alert File, and the Alert File will find some solutions to solve the alerts.

The following is the result of monitor, including the response times, the packet loss, the CPU utilization, the memory utilization, the disk utilization, and the process count, as shown in Fig. 4.

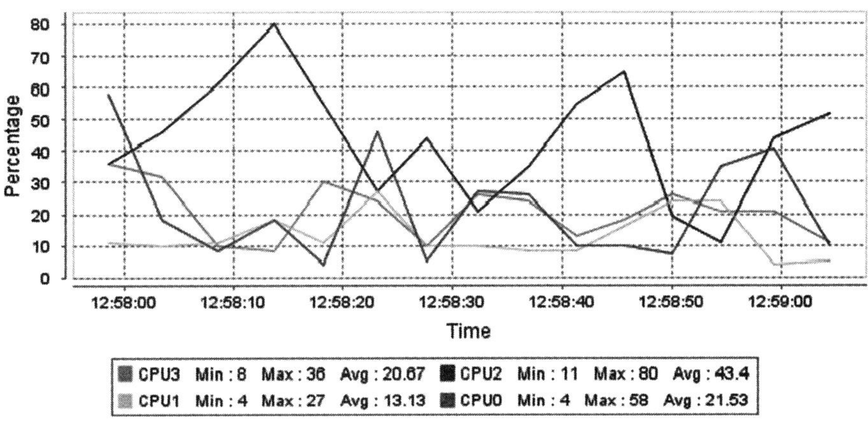

Fig. 3 A real time CPU utilization monitoring graph for a quad core server processor

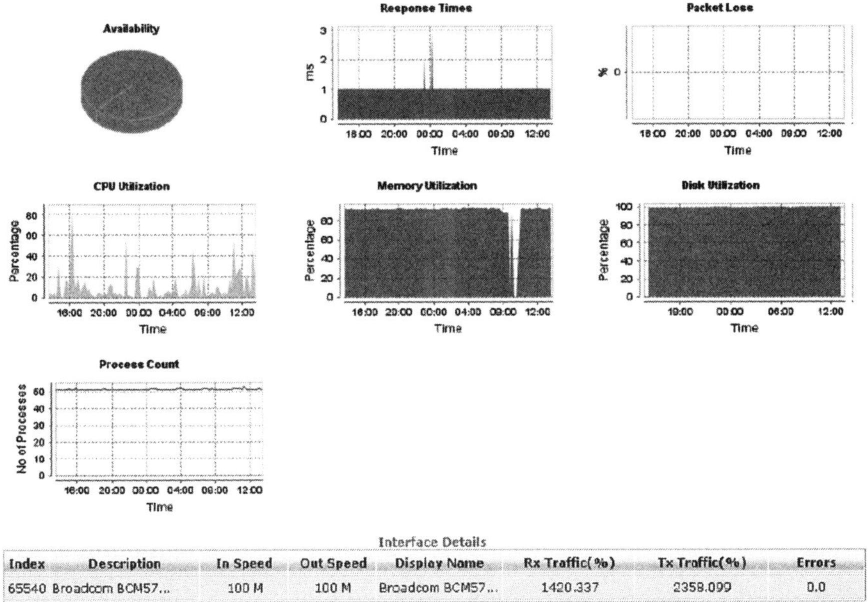

Interface Details

Index	Description	In Speed	Out Speed	Display Name	Rx Traffic(%)	Tx Traffic(%)	Errors
65540	Broadcom BCM57...	100 M	100 M	Broadcom BCM57...	1420.337	2358.099	0.0

Fig. 4 The result of monitoring

The system uses monitor technology to monitor the process to identify and obtain the system's performance. It reduced the impact of system crashed. The actual results of monitoring have shown that the monitoring system not only can complete the work of traditional monitoring system, but also can provide the information of many resource performances [42].

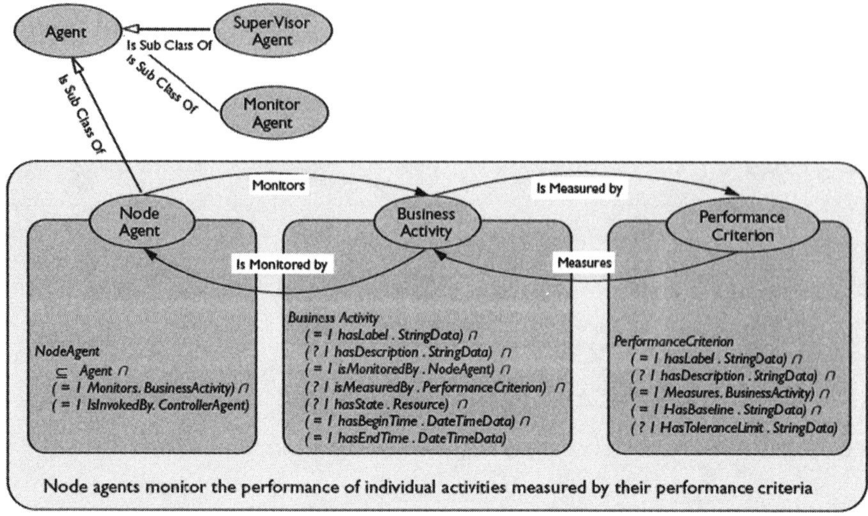

Fig. 5 DL model of business activity performance criteria monitored by Node agents

3.2 Business-Oriented Monitoring

Resource-oriented monitoring is a crucial issue for guaranty service delivery. However, most resource-oriented monitoring approaches are specific and focus on IT level. The challenge of how to monitor the whole life cycle of business simply and flexibly needs to be overcome. It attracts most researchers to focus on the business-oriented monitoring. A model-driven approach for service monitoring from business perspective was proposed [43]. A business-oriented service monitoring meta-model is put forward to define various monitoring models on demand. The model can flexibly specify the monitored information in both business level and IT level and the monitoring process.

In this section, we introduce a semantic architecture which is proposed by Thomas M, Redmond R and Yoon V to do the resource-oriented monitoring [44].

Performance criteria are described using the OWL (Web Ontology Language). Agents monitor the whole life cycle of business and activities. Tools like Racer and Protégé verify the conformance to model requirements.

Figure 5 shows the DL model of business activity performance criteria monitored by Node agents to monitor the whole life cycle of business.

The architecture and roles of each tier are described as following:

1. Supervisory agent (SA)
   SA is the main interactive agency management decision-making. Inputs to SA are process performance goals.
2. Monitoring agent (MA)
   MAs are instantiated and dispatched by SA for each specific goal.

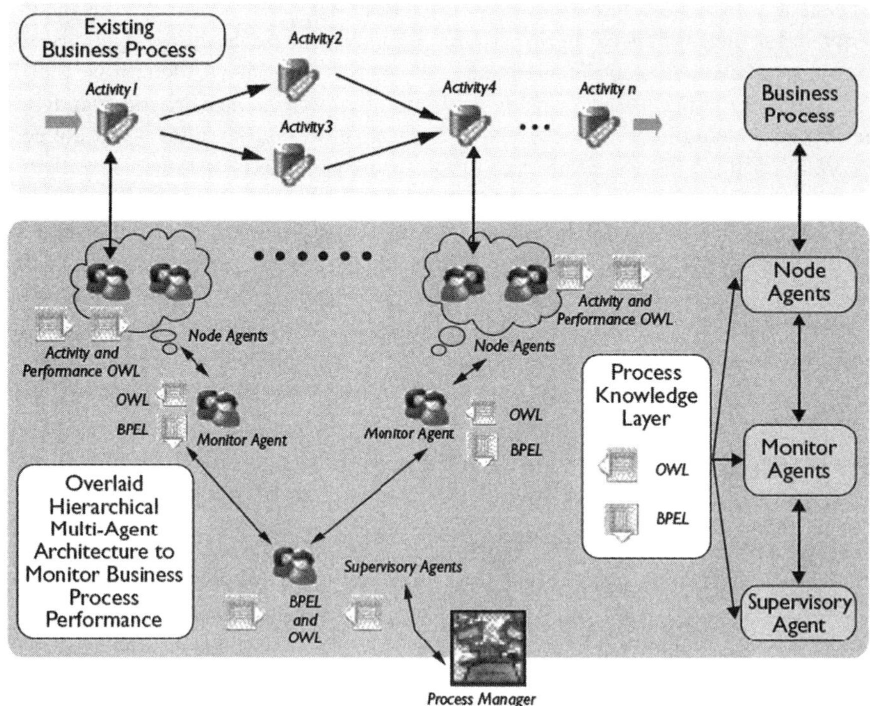

Fig. 6 A hierarchical multi-agent architecture to monitor business process performance

3. Node agent (NA)
   NA is data gatherers and transformers. A NA is instantiated by MAs to monitor and assess the performance for each business activity.

A SA operates management level and invokes MAs to support process performance. A SA invokes a MA and provides the MA information about the business process. MAs support decision models utilizing process performance inputs from NAs. The NAs transform process performance observations into information to support the decision models of its parent MA. The NA transformation function includes the capability to transform continuous and discrete activity observations, including the state of the process with respect to established performance criteria, based on established tolerance limits.

The process manager architecture for interacting and monitoring the whole life of business is shown in Fig. 6.

The SA invokes the specific-goal monitor agents that use process knowledge to invoke NAs for each business activity. If new goals are defined for which no MA exits, the measurement standards are defined as inputs through the SA API interface. Therefore, the SA modifies the SA's OWL to create definitions for new MAs to address the new process performance measurement criteria and creates a new goal-based service reference to deploy a MA to accommodate this need.

NAs are invoked to monitor every business activity, with knowledge of the activity and its performance criteria contained. Business logic described by the BPEL (Business Process Execution Language) is used by the MA to determine global process performance measures using the monitored performance criteria of individual business activities reported by NAs. Each goal-specific monitor agent collects information from individual NAs using performance criteria specified in activity-performance ontology, and reports the process performance to the SA. The SA can provide aggregated and goal-specific process performance information to the process manager to support decision making using business process performance measures [43].

The semantic architecture to monitor the whole life cycle of business is used. It is shown its simply and flexibly.

4 Participators

Software products also have life cycles. They go through inoculation, birth, growing up, mature, and decline stage. In software engineering, the whole software life cycle is commonly divided into several phases. Each phase has its specific tasks, and makes the large scale, complex structure and complex software development management easy to control. In general, feasibility analysis and software life cycle including development plan, analyzing, design, coding, test and maintenance activities. These activities can be allocated to different phases to finish in proper way [45]. The points is an idea in software engineering principles according to time schedule way of thinking. In order to improve the quality of software, each stage should be defined, worked, reviewed and formed document for communication or for future reference.

Therefore, the software also needs to be monitored to avoid affecting the performance of the services after its aging. In the process of software monitoring, four main objects are involved.

4.1 Resource Managers

Leonard Kleinrock [46], one of the chief scientists of the original Advanced Research Projects Agency Network (ARPANET) which seeded the Internet, said, "As of now, computer networks are still in their infancy, but as they grow up and become sophisticated, we will probably see the spread of 'computer utilities' which, like present electric and telephone utilities, will service individual homes and offices across the country". This vision of computing utilities based on a service provisioning model anticipated the massive transformation of the entire computing industry in the twenty-first century, whereby computing services will be readily available on demand, like other utility services available in today's society. Similarly, users (consumers) need to pay providers only when they access the computing services.

In addition, consumers no longer need to invest heavily or encounter difficulties in building and maintaining complex IT infrastructure [47].

In such a model, users access services based on their requirements without regard to where the services are hosted. This model has been referred to as utility computing, or recently as cloud computing [48]. The latter term denotes the infrastructure as a cloud from which businesses and users can access applications as services from anywhere in the world on demand. Hence, cloud computing can be classified as a new paradigm for the dynamic provisioning of computing services supported by state-of-the-art data centers that usually employ Virtual Machine (VM) technologies for consolidation and environment isolation purposes [49].

Due to the expansion of scale and the diversity of the function in the data centers, we should consider the huge cost of maintenance management as well as reliability in the data center. Traditional data center is focus on the stability of the application, data security and reliability in operation, while the problems such as resource utilization, energy efficiency are not too concerned about. Based on monitoring software in the data center, it not only can improve the reliability of the data center, but also can improve resource utilization of the data centers and achieve the energy conservation goal, through the resource manager to dynamic adjustment of resources.

Today's enterprise data centers are designed with a silo oriented architecture in mind: each application has its own dedicated servers, storage and network infrastructure, and a software stack tailored for the application controls these resources as a whole. Due to the stringent requirements placed on the enterprise applications and the time-varying demands that they experienced, each application silo is vastly over-provisioned to meet the application service goals. As a result, data centers are often under-utilized, while some nodes may become heavily-loaded sometimes, resulting in SLA violations due to that poor application performance [50].

Cloud computing is offering utility-oriented IT services to users worldwide. Based on a pay-as-you-go model, it enables hosting of pervasive applications from consumer, scientific, and business domains. However, data centers hosting cloud applications consume huge amounts of energy, contributing to high operational costs and carbon footprints to the environment. Therefore, we need Green Cloud computing solutions that can not only save energy for the environment but also reduce operational costs [48].

Therefore, if the development of dynamic resource provisioning and allocation algorithms is focused, based on the performance of the software, the data center will represent not only higher resource availability, either be grossly under-utilized or experience poor application-level QoS due to insufficient resources under peak loads, but also more energy efficient.

4.2 Service Operators

In a cloud computing environment, the service operators are effectively subscribers, who now only pay for the software needed from the providers on an operational expense basis. Corporate users of cloud computing have an active role to play in ensuring that cloud computing ends up delivering on its promise of revolutionizing corporate computing, by liaising with industry groups as well as national and international regulators. Effective use of cloud computing will reduce the stress on the IT departments as they spend less time maintaining software and more in developing innovative applications for the organization [51].

Cloud computing is a style of computing paradigm in which typically real-time scalable resources such as files, data, programs, hardware, and third party services can be accessible from a Web browser via the Internet to users (or called customers alternatively). These customers pay only for the used computer resources and services by means of customized service level agreement (SLA), as well as not to be aware that how a service provider uses a underlying computer technological infrastructure to support them. The SLA is a contract negotiated and agreed between a customer and a service provider. That is, the service provider is required to execute service requests from a customer within negotiated quality of service (QoS) requirements for a given price. Thus, accurately predicting customer service performance based on system statistics and a customer's perceived quality allows a service provider to not only assure quality of services, but also avoid over provisioning to meet a service level agreement [52].

4.3 Data Owner

All data in your organization should have an owner. The owner is responsible to determining how much risk to accept, and makes decisions that who will be permitted to access the information and how they will use it.

In small businesses, the business owner might possess all types of information. As organizations growing, ownership is typically distributed. For example, financial data ownership might fall to the vice president of the accounting department. Intellectual property might belong to the head of engineering. In any case, the data owner should understand the sensitivity of the information and undertake his or her responsibilities for protecting it.

Data owners don not perform these tasks alone. They work closely with information services and security to determine risk levels, current controls, and next steps. Security and information services delivery teams then takes steps to ensure handling, distribution, regular usage of electronic information and appropriate controls that have been implemented and managed in the storage [53].

4.4 Software Developers

Software development is the process by which a company, team, or individual devises and implements an overall plan to create a new software program. This process can also be applied to an established program to create a new version of this software, though this is usually an abridged version of the process unless the new version is largely different from the previous one. Numerous steps are involved in this process, beginning with understanding what is needed from software, developing a plan for creating it, writing the code, and bug testing prior to launch. Software development can be a process that involves anything from a single programmer to dozens or hundreds of individuals. The process of the software development usually begins with research or a general understanding of what type of software is needed in the market place. This may be an entirely new program that addresses an unfulfilled need or a new piece of software in an existing market. As software development begins, this research establishes the purpose of the software being developed and the overall goals of the development [54].

Software maintenance is the longest stage in software life cycle, after developing the software and putting it into use. Due to various of reasons, the software cannot continue to meet the needs of users. Software maintenance in software engineering is the modification of a software product after delivery to correct faults, to improve performance or other attributes [55]. A common perception of maintenance is that it merely involves the fixing defects. However, one study indicated that the majority, over 80 %, of the maintenance effort is used for non-corrective actions. This perception is perpetuated by users submitting problem reports that in reality are functionality enhancements to the system. More recent studies put into the bug-fixing proportion closer to 21 % [56].

The key software maintenance issues are both managerial and technical. Key management issues are as following: alignment with customer priorities, staffing, which organization does maintenance, estimating costs. And the key technical issues are as follows: limited understanding, impact analysis, testing, and maintainability measurement. Software maintenance is a very broad activity that includes error corrections, enhancements of capabilities, deletion of obsolete capabilities, and optimization. Because change is inevitable, mechanism must be developed for evaluation, controlling and making modifications.

Some works done to change the software is considered to be maintenance work. The purpose is to preserve the value of software over the time. The value can be enhanced by expanding the customer base, meeting additional requirements, becoming easier to use, more efficient and employing newer technology. Maintenance may span for 20 years, whereas development may be 1–2 years.

It's need to repeatedly test software for the software developers to find performance bottlenecks and function of software flaws, in order to improve the performance of the software.

5 Monitoring Site

It is divided into on-site monitoring and off-site monitoring in the system monitoring. On-site monitoring can guarantee the safety of monitoring. However, off-site monitoring can reduce the user cost. Currently used monitoring system is the combination of on-site monitoring and off-site monitoring.

5.1 On-Site Monitor

Formal on-site study monitoring involves overseeing the progress of systems via regular, ongoing site-level quality checks that are conducted by an appropriately qualified individual. Investigators should specify who, with what frequency, how and to whom monitoring staff [57].

On-site monitoring is in a relatively independent and closed environment monitoring, so that it can provide more security monitoring. On-site monitoring can provide centralized monitoring programmer, which is beneficial to the large-scale monitoring system. Firstly, the obvious advantage of online monitoring is the monitoring scope. Secondly, on-site monitoring makes monitoring more reliable. Finally, it can be easy to manage and save administrative costs due to its centralization.

However, there are also disadvantages for on-site monitoring. First, on-site monitoring is a large testing environment that will lead to expensive costs, which will increase the extra burden of users. For example, the user costs for using the on-site monitoring are more than the off-site monitoring. In the selection, most users will choose off-line monitoring. Second, on-site monitoring results in non-professional self- management. In the process, users in the use of on-line monitoring cannot interact well with monitoring system. It is not conducive to personal management and maintenance.

On-site study monitors should physically check the conduct and documentation of study activities to ensure followings [57]:

1. All necessary approvals are in place prior to commencement of recruitment, screening or enrollment activities.
2. All necessary approvals remain in place throughout the duration of research activities.
3. Recruitment, screening, enrollment and the informed consent process are being conducted per conditions of CHR [58] (Contact Hole Roughness) approval.
4. Documentation is on file demonstrating that all participants meet all inclusion and no exclusion criteria.
5. Documentation is on the file demonstrating protocol adherence (and documentation when protocol adherence is not met).
6. No changes are made to the CHR approved study protocol or consent form unless immediate changes are required to protect the safety of study participants.
7. Adverse events and pertinent safety information is being captured, assessed, reported and followed as required by CHR, sponsor and/or regulatory agencies.
8. Any other unanticipated problems (including violations, incidents and/or research-related concerns or complaints) are reported as required by CHR.

5.2 Off-Site Monitor

Instead of using production lines and instrumentation, Off-site monitor, which differs from on-site monitor, takes artificial random testing on status of the production and equipment, by using multiple test instruments.

Off-site monitoring's advantages are obvious. First, off-site monitoring enables users to reduce the cost, and is within the range that users can afford. Second, off-site monitoring is suitable for small and medium-sized monitoring system and for ordinary users, while on-site monitoring is suitable for the large-scale monitoring system and for large companies.

It should be pointed out that, off-line monitoring cannot guarantee security. User's unpredictable behavior often results in the decreases of the system security. Therefore, as soon as off-site monitoring is chosen, the security should be also taken into account.

Off-site monitoring methods have been developing because of an ambiguity about whether their purpose is long-run or short-run forecasting. The monitoring can be also useful to support measuring and evaluation of Quality of Services (QoS) metrics that are required by Service Level Agreements (SLA) [59]. A SLA defines the agreed level of performance for a particular service between a service provider and a service customer [60, 61].

It is necessary to clarify the primary purpose of off-site monitoring. The answer probably lies in the examination system. If one believes that on-site exams provide the most useful and accurate forecasts of long-term bank health, then the purpose of off-site monitoring would be seen as complementary to the exam system and very short-run in nature. If exams are thought to be useful primarily in assessing the current condition of a bank, then it might be appropriate to use off-site methods for long-term forecasts. The issue of the willingness to act on this information needs to be broached [62].

The monitoring of real-time distributed computing systems involves the collection and interpretation of information, such as event time stamps, synchronization sequences, race conditions, register status, transaction identifications, and interrupt activities. These characteristics make the monitoring of the global states and the execution flows of a real-time distributed computing system almost impossible. Snodgrass views monitoring as an information-processing activity and asserted that the relational model is an appropriate formalism for structuring the information generated by a distributed computing system [63].

6 Monitoring Methods

There are many monitoring methods, such as visualization monitoring, hot-spot evaluation, performance prediction, analyzing user's habits and tools. These methods, as well as some performance test tools, are introduced in this section.

6.1 Visualization Monitoring

De Chaves S. A. et al designed an abstract general-purpose monitoring architecture [64], as shown in Fig. 7. The three-layer architecture addresses the monitoring's

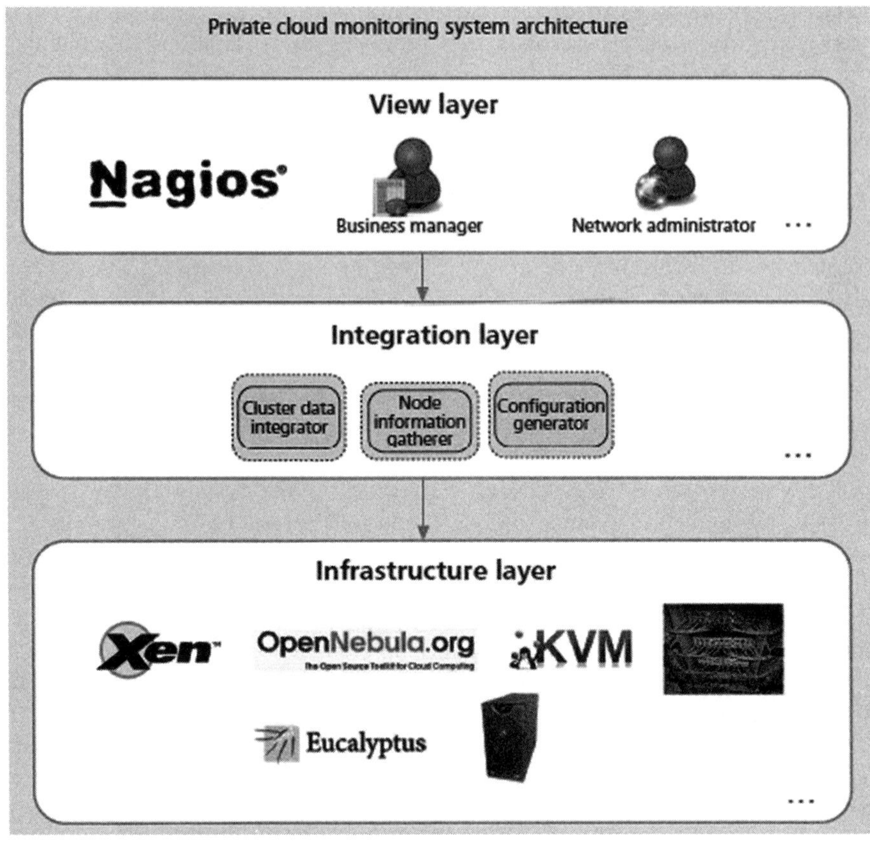

Fig. 7 Private cloud monitoring system architecture

needs in a private cloud, which is extensible, modular, and simple. The middle integration layer provides a clear separation by abstracting the infrastructure details and the monitoring information required by cloud users.

The infrastructure layer contains basic facilities, services, installations and available sofware, which are mainly heterogenerous resources. For example, a request cloud is managed by Xen or KVM, which depends on the deployment of the infrastructure layer. Therefore, the integration layer is applied to abstract infrastructure details. The view layer is regarded as the monitoring interface, which can analyze the fulfillment of organizational policies and service level agreements (SLAs). Based on the needs of the different enterprises, this layer may accomplish different views.

An extensible modular monitoring framework called PCMONS was developed, which acts principally on the integration layer by retrieving, gathering, and preparing relevant information for the visualization layer [64]. The framework tries to incorporate many tools and management practices applied in cloud management, in order to integrate into organizations' existing management infrastructure and operations. PC-MONS exploits the installed software and hardware base, besides the experience and skills of IT administrators. Developing the PCMONS considered the four high-level components of IaaS tools, such as cloud manager, node controller, storage controller and cluster controller. The control centralization adapts a client/server model to be monitored. Considered specific phase of a VM life cycle, the system is divided into the following modules [64].

1. Node Information Gatherer
   This module gathers different local information on a cloud node according to specific demands. The module gathers local VMs information for the Cluster Data Integrator in the current version.
2. Cluster Data Integrator
   Because most of the cloud tools group nodes to the cluster, a specific agent gathers and prepares the data for the next layer, avoiding transferring unnecessary data to the Monitoring Data Integrator.
3. Monitoring Data Integrator
   The module collects cloud data in the database, stores and provides them to the Configuration Generator.
4. VM Monitor
   This module makes the VMs transfer useful data to the monitoring system by injecting scripts into the VMs. The useful data may be memory and CPU usage.
5. Others
   Configuration Generator, Monitoring Tool Server, User Interface and Database, are also important modules. The Configuration Generator retrieves information from database, such as the necessary configuration files. Monitoring Tool Server receives monitoring data and takes actions, for example, storing the information in the database, although it has imperfections. Nagios has a sufficient interface, no need of developing specific ones. Database stores information for the Configuration Generator and the Monitoring Data Integrator.

6.2 Hot-Spot Evaluation

Based on the hierarchy structure of hot-spot evaluation indexes in Fig. 8, the subjective weight and objective weight of each index should be calculated to get the comprehensive evaluating index of hots-pot degree. Due to the ambiguity of index evaluation, triangular fuzzy number in fuzzy AHP (Analytic Hierarchy Process) [65] is deployed to determine the subjective weight of each index influencing hot degree, then the multi-objective decision-making method is used to determine the target type of each index. Dimensionless processing is to get the optimal size of index and its matrix, and maximizing deviation method is to determine the objective weight of index. Finally combining subjective weights with objective weights, the order and hierarchy of monitored spots (virtual machines) "hot degree" can be obtained by linear

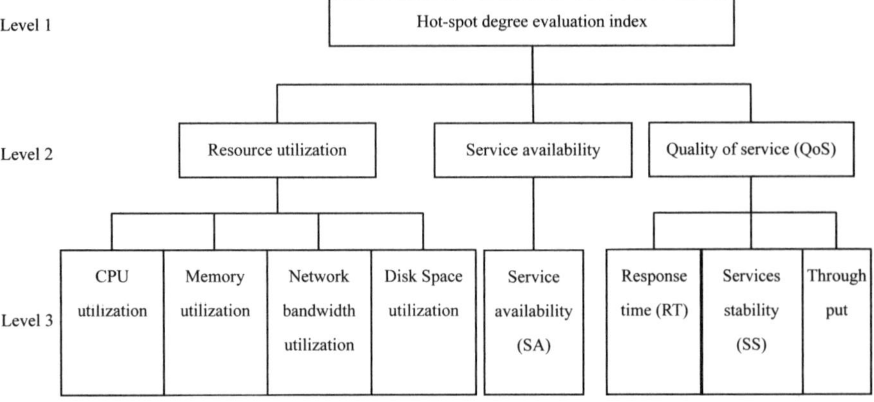

Fig. 8 The hierarchy structure of hot-spot evaluation indexes

weighted sum method of multi-objective decision. As shown in Fig. 9, the specific steps are elaborated. Here, the hot-spot evaluation question is handled through four sections. They are the optimal relative and classification of index target type, the comprehensive weight determining, the hot-spot degree comprehensive evaluation value and hot-spot sorting and hot-spot or cold spot judging.

1. Optimal relative and classification of index target type

Optimal relative is the degree of relative to "optimal", which is similar to the concept of membership degree in fuzzy mathematics, and is determined according to the target type and characteristics. There are some common types of attributes, such as *efficiency type*, *cost type*, *fixed type*, *range type*, *deviation type* and *deviation interval type* [66], and the target types of single hot indexes can be divided into *fixed type*, *range type*, *cost type* and *efficiency type*. μ_{ij} is the optimal relative.

Fixed type—regards stabilizing at a fixed value as the best type of indexes.

$$\mu_{ij} = \begin{cases} 1, & f_{ij} = f_i^* \\ 1 - \left| f_{ij} - f_i^* \right| / \sigma_i, & f_{ij} \neq f_i^* \end{cases} \tag{1}$$

f_{ij} is the measured value of i index in virtual machine j. f_i^* is the optimal value of f_i, which is the i index given in advance. $\sigma_i = \max\left\{ f_{ij} - f_i^* \right\}$ is the absolute differences maximum of f_{ij} and f_i^* among n observed spots.

Range type—regards property values falling in a fixed interval as the best type of indexes.

$$\mu_{ij} = \begin{cases} 1 - (f_i^L - f_{ij})/\eta_i, & f_{ij} < f_i^L \\ 1, & f_{ij} \in [f_i^L, f_i^R] \\ 1 - (f_{ij} - f_i^R)/\eta_i, & f_{ij} > f_i^R \end{cases} \tag{2}$$

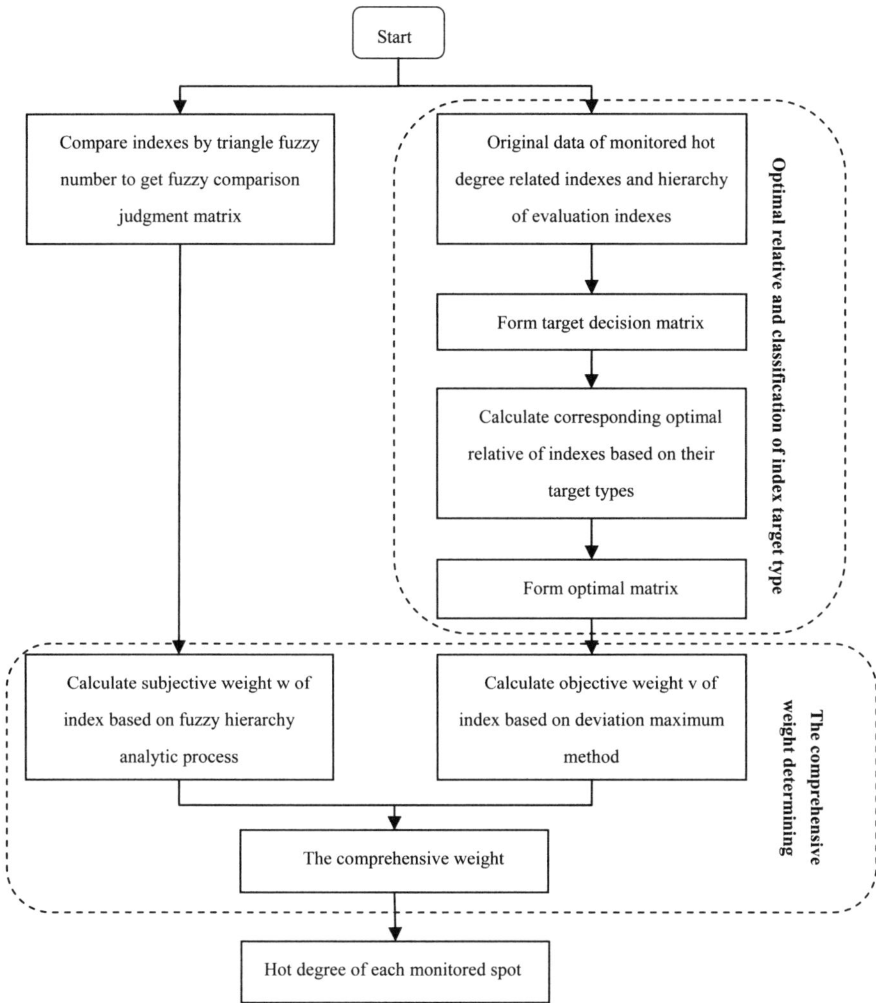

Fig. 9 Comprehensive evaluation strategy for hot-spot

f_i^L and f_i^R are the best lower bound and upper bound of i given index, respectively. $\eta_i = \max\{f_i^L - f_{i\min}, f_{i\max} - f_i^R\}$ is the absolute maximum of f_{ij} deviating the optimal range. $f_{i\min}$ and $f_{i\max}$ are the maximum and minimum of measured values in index f_i monitored spots, respectively.

Cost type—regards the smallest attribute value as the best index.

$$\mu_{ij} = 1 - f_{ij}/(f_{i\max} + f_{i\min}) \tag{3}$$

Efficiency type—regards the biggest attribute value as the best index, just contrary to *cost type*.

$$\mu_{ij} = f_{ij}/(f_{i\max} + f_{i\min}) \tag{4}$$

Table 10 Objective types of hot-spot degree indices

Level 1 type	Level 2 type	Level 3 type	Target type
Hot degree	Resource utilization	CPU utilization	Range type
		Memory utilization	Range type
	Service performance	Response time	Cost type
		Throughput	Efficiency type
		Service stability	Efficiency type
	Availability	Service availability	Efficiency type

Therefore, some representative index types of hot degree evaluation can be summarized as shown in Table 10.

To evaluate m hot degree of n monitored spots, firstly the target decision matrix F is formed according to the measured value of monitoring spots.

$$F = \begin{bmatrix} f_{11} & f_{12} & \cdots & f_{1n} \\ f_{21} & f_{22} & \cdots & f_{2n} \\ \vdots & \vdots & \vdots & \vdots \\ f_{m1} & f_{m2} & \cdots & f_{mn} \end{bmatrix} \tag{5}$$

Using formula (Eq. 1) $\sim$ (Eq. 5) and combined with the target type of hot index, the target decision matrix converts to optimal relative matrix μ:

$$\mu = \begin{bmatrix} \mu_{11} & \mu_{12} & \cdots & \mu_{1n} \\ \mu_{21} & \mu_{22} & \cdots & \mu_{2n} \\ \vdots & \vdots & \vdots & \vdots \\ \mu_{m1} & \mu_{m2} & \cdots & \mu_{mn} \end{bmatrix} \tag{6}$$

2. The comprehensive weight determining

The judgment matrix represents the relative importance of an upper layer element and this layer element (or the upper factors). Importance scale is a fairly good digital measurement in compassion to index importance. There are commonly used importance scale at present: 1–9, 9/9–9/1, 10/10–18/2, 90/9–98/9, 20/2–28/2, e0/4–e8/4 and e0/5–e8/5.

Saaty had compared these 27 kinds of indices in experiment and concluded that 1–9 scale is not possessed of optimal performance but also not worth than more complex scale [65]. Currently, 1–9 type is widely used in AHP due to its capacity of consistency and uniformity of distribution. The 1–9 scale type importance scale is adopted and its specific meaning is shown in Fig. 10.

Based on the index hierarchy shown in Fig. 8, triangular fuzzy numbers are introduced to form the fuzzy judgment matrix according to the uncertainty of subjective

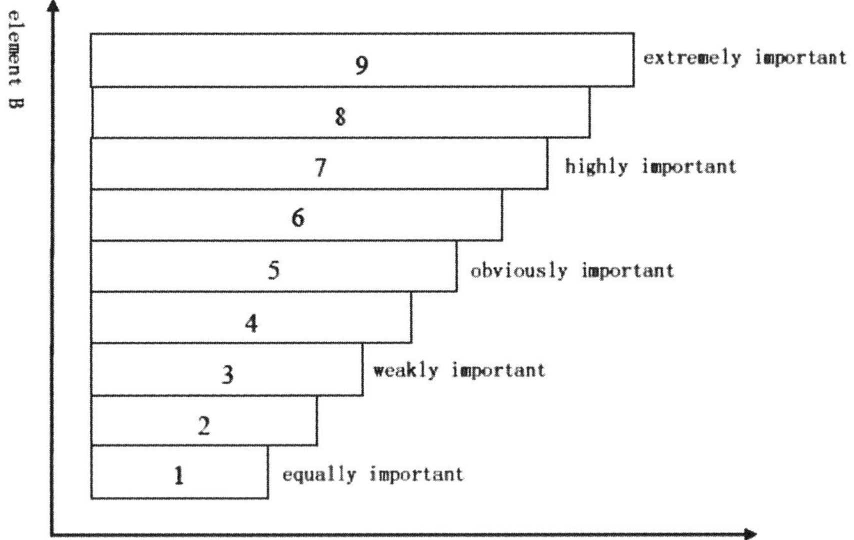

Note: 2, 4, 6, 8 are the middle values between adjacent levels.

1, 1/2, ···, 1/9 are the reciprocal numbers of a_{ij}, that is the important rate of B relative to A.

Fig. 10 Meaning of the importance of scale a_{ij} for 1–9 scale type (A relative to B)

judgment experts built comparative judgment matrix [66]. These scales order and obtain weighted value could be determined by using the theory of fuzzy number comparison size.

Assume that there are n hot-spots comprehensive evaluation index of this layer related to upper layer, indices sets $A = \{a_1, a_2, \cdots, a_n\}$. Triangular fuzzy number, $b_{ij} = [l_{ij}, m_{ij}, u_{ij}]$, is an importance fuzzy judgment of index i relative to index j experts making. l_{ij} and u_{ij} represent the fuzzy extent of judgment, the grater $u_{ij} - l_{ij}$ means the higher comparative fuzzy degree. Finally, the fuzzy comparison judgment matrix B via comparison is achieved as following.

$$B = (b_{ij})_{n \times n} = \begin{bmatrix} [l_{11}, m_{11}, u_{11}] & \cdots & [l_{1n}, m_{1n}, u_{1n}] \\ [l_{21}, m_{21}, u_{21}] & \cdots & [l_{2n}, m_{2n}, u_{2n}] \\ \vdots & \vdots & \vdots \\ [l_{n1}, m_{n1}, u_{n1}] & \cdots & [l_{nn}, m_{nn}, u_{nn}] \end{bmatrix} \tag{7}$$

Similarly, the application of above methods can construct the fuzzy comparison judgment matrix of each layer relative to it's upper among index systems.

In the fuzzy comparison judgment matrix, the fuzzy relative weight of index i compared with other index in this layer is:

$$Q_i = \left[\frac{\sum_{j=1}^{n} l_{ij}}{\sum_{i=1}^{n} \sum_{j=1}^{n} u_{ij}}, \frac{\sum_{j=1}^{n} m_{ij}}{\sum_{i=1}^{n} \sum_{j=1}^{n} m_{ij}}, \frac{\sum_{j=1}^{n} u_{ij}}{\sum_{i=1}^{n} \sum_{j=1}^{n} l_{ij}} \right] \tag{8}$$

Each triangle fuzzy number in the fuzzy relative weight vector is required to clarity before sorting the current layer index. The corresponding subjective weight of $Q_i = (l_i, m_i, u_i)$ can be determined as follows [67–70]:

$$w_i = \frac{l_i + 2m_i + u_i}{4} \tag{9}$$

The index i will slight impact on the hot-spot degree evaluation and should be given less weight corresponding if there are small differences in the hot-spot degree i of the monitoring, otherwise given a greater weight.

Based on optimal relative matrix μ, multi-objective decision-making method is employed to get objective weight v_i of index i:

$$v_i = \frac{\sum_{j=1}^{n} \sum_{k=1}^{n} \left|\mu_{ij} - \mu_{ik}\right|}{\sum_{i=1}^{m} \sum_{j=1}^{n} \sum_{k=1}^{n} \left|\mu_{ij} - \mu_{ik}\right|}, i = 1, 2, \cdots, m \tag{10}$$

Here, j and k represent different monitoring sites, respectively. $\left|\mu_{ij} - \mu_{ik}\right|$ refers to the absolute value of the membership degree. The weight reflects the differentiation and decisive of index i in hot-spot degree evaluation sorting process.

The comprehensive weight of index i is obtained by the subjective weight and objective weight.

$$S_i = \frac{w_i v_i}{\sum_{i=1}^{n} w_i v_i}, i = 1, 2, \cdots, n \tag{11}$$

3. The hot-spot degree comprehensive evaluation value

The linear combination of the individual index evaluation value reflects the quality of comprehensive index, but there are one more indexes not qualified in the index evaluation process. An agreement is made that the comprehensive evaluation value is not qualified when any of the indicators of evaluation value obviously unqualified. Standard of qualified or not should be based on the actual situation. Service performance indexes, such as response time, can get the relevant limit threshold in the SLA agreement. Resource utilization threshold has uncertain setting in some extent.

Different virtual machines and performance service reflect the different resource utilization and performance. Hence it can judge degree of unqualified according to the way of beyond percentage. For example, setting the CPU resource utilization threshold is 75 %, if the resource utilization rate reached 90 %, which is beyond the threshold of 20 %, comprehensive evaluation value is judged to be unqualified because the CPU resource utilization index is not qualified.

The linear weighted method among multi-objective decision method is applied to hot degree comprehensive evaluation value of monitoring point j, as shown in the following.

$$y_i = \begin{cases} \sum_{i=1}^{n} S_i \mu_{ij}, & \textit{all indexes are qualified} \\ 0, & \textit{one or more indexes are not qualified} \end{cases} \tag{12}$$

4. Hot-spot sorting and hot-spot or cold spot judging

Based on the hot-spot comprehensive evaluation value, hot-spot degrees of all monitoring point can be linear ordering. Hierarchy is also obtained in the sorting process. In particular, firstly, original index grade line standard of hot-spot degree evaluation index system is gotten by experts judgment method. Secondly, each of level value is regarded as a "monitoring" of the original data, then hot-spot degree is evaluated according to the previous steps as the same other ordinary monitoring points. Finally, a corresponding comprehensive evaluation value is achieved based on every level value. Hot-spot degree range of each monitoring point can be obtained after uniformly ranking these hot-spot comprehensive evaluation values and all levels of comprehensive evaluation values. Then hot-spot level of each monitoring point will be correspondingly obtained based on removing principle. The cold/ hot judgment of the monitoring will come out in line with the level of ownership rules in advance.

6.3 Performance Prediction

1. Prediction process
   The DaaS (Data as a Software) service performance prediction process is shown in Fig. 11. The concrete steps can be summarized as follows.
   - Data Gathering
     Some indicators are gathered and recorded, such as DaaS service transactions, service occupied physical resources and response time of each key business monitoring spot. These indicators can be gathered by deploying monitoring plugins in each virtual machine. Then, three crucial performance indexes are defined.
     I. DaaS Service Transactions Index (STI)
        The number of transactions that require the DaaS service to process at each moment. STI reflects service loads at each moment directly. According to a monitoring plugin, it can be recorded as a time series: $x_1, x_2, \ldots, x_{n-1}, x_n$. In many cases, most of the STI time series are nonlinear sequence and their trends can be predicted using nonlinear time series forecasting methods.

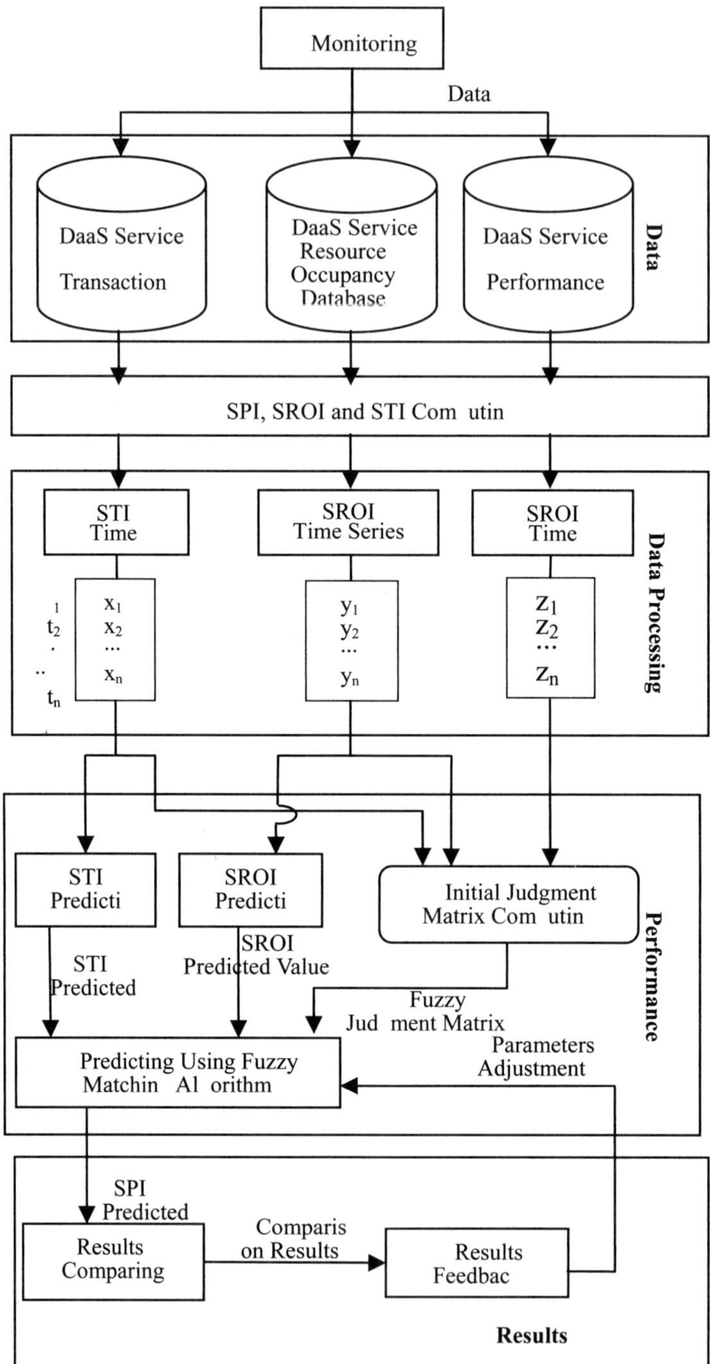

Fig. 11 Prediction process of DaaS performance

II. DaaS Service Resource Occupancy Index (SROI)
The comprehensive calculation value of DaaS service occupied resources at each moment. In cloud environment, DaaS resource occupancy means these physical resources that allocated to the DaaS service. The physical resources include CPU, memory, storage, flash storage, network bandwidth, etc.

III. DaaS Service Performance Index (SPI)
DaaS services response time at each moment. Response time values can directly reflect DaaS service performance. The value of SPI is the comprehensive calculation result of the response time of some key businesses monitoring points.

- Index Calculation and Prediction
After defining three key indexes and gathered the information, different methods or algorithms can be used to calculate three indexes. In particular, chaos time series prediction algorithm is adopted to predict STI's trend, radar char comprehensive method is used to calculate SROI, and weighted average method is applied to calculate SPI.
- Performance Prediction
A fuzzy judgment matrix cloud be formed by STI and SROI. A fuzzy matching matrix may be gained according to the definition of fuzzy closeness degree, and calculate the best matching value of STI and SROI of the predicted moment using lattice similarity matching algorithm. SPI corresponding to the best matching value is regarded as the prediction result DaaS service performance.

2. Computing methods of each index
   - STI Time Series Prediction Algorithm
   DaaS service system is a nonlinear system, which contains a wealth of kinetic information, and STI time series are mostly nonlinear time series on cloud platform. Therefore, it proposes to verify the STI time series' chaotic characteristics using chaos theory, and presents a prediction model of STI time series based on the largest Lyapunov exponent.
   - Reconstruct STI time series' phase space
   According to Takens embedding theorem [70], let $x_1, x_2, \ldots, x_{n-1}, x_n$, to be the STI time series, construct m dimensional phase space with N phase points, then find a proper embedding dimension. If the delay coordinate dimension $m \geq 2d + 1$, d is power system dimension, then every point in space phase can be defined as:

$$Y(t_i) = [x(t_i), x(t_i + \tau), \cdots x(t_i + (m - 1)\tau)] \tag{13}$$

   Where, $i = 1, 2, \ldots, N$, $N = n - (m - 1)\,\tau$.
   - Calculate STI chaos characteristics
   STI time series' chaotic property can be verified in a qualitative analysis way or a quantitative calculation way. Quantitative calculation method is to judge by calculating some chaotic characteristic quantities, such as embedded delay,

embedding dimension, Lyapunov index, and so on. Embedded delay refers to these phase space parameters that determine the differences among variables in STI time series, and have significant influence on the information of original DaaS system contained in the reconstructed phase space. Embedding dimension refers to the minimal space dimensions which could completely contain all characteristics of attractors in DaaS service system. Embedding dimension and embedded delay are dependent on each other in most cases. The correlation function for choosing time delays algorithm was used to simultaneously solve embedding dimension and embedded delay [71].

The Lyapunov index is an important parameter to judge whether a system is a chaotic systems. The basic characteristic of a chaotic system is sensitive to the initial conditions, i.e. two resource variable curves, with very close initial values, separate very fast with time in exponential way. Lyapunov index is the chaotic invariant to describe the phenomenon, expressing the convergence average rate and divergence rate in phase space between adjacent tracks.

A small amount of data algorithm is used to compute the largest Lyapunov exponent [72, 73]. If the largest Lyapunov exponent is greater than 0, the system is a chaotic system.

- Prediction algorithm based on largest Lyapunov index

  For STI time series x_1, $x_2, \ldots, x_{n-1}$, x_n if it needs to predict x_{n+k}, then selects a point X_p in a phase space, and X_p is the prediction center: $X_p = (x_n - (m - 1)\tau, x_{n+1} - (m - 1)\tau, \ldots, x)$. Let X_p be the nearest neighbor points X_l, $X_l \in \{X_1, X_2, \ldots, X_{p-1}\}$, the distance between X_p and X_l is d, then $\mathrm{d} = \min_j |X_p - X_l| = |X_p - X_l|$. After determining the nearest neighbor points X_l, further evolution X_{p+1} and X_{l+1} can be done according to X_p and X_l, then:

$$\left|X_p - X_{p+1}\right| = |X_l - X_{l+1}|\, e^{\lambda} \tag{14}$$

  In formula (Eq. 14), λ_1 is the largest Lyapunov index of the time series x_1, $x_2, \ldots, x_{n-1}$, x_n, where the last component of X_{p+1}, and x_{n+1} is unknown, so x_{n+1} is predictable. If it needs to predict x_{n+k}, k step prediction can be analogized based on one step prediction. Formula (Eq. 14) is the prediction model of largest Lyapunov index λ_1, and the prediction time is determined by$1/\lambda_1$.

- SROI computing method

  DaaS service occupied resources occupancy refers to resources allocated to DaaS service by the cloud platform, including CPU, memory, storage, flash storage, network bandwidth, etc. However, measuring DaaS service resources occupancy is a difficult problem because DaaS services occupied resources are distributed on different virtual machines or physical machines, and the measurement unit of various resources is different. An irregular polygon (radar chart) is presented to describe DaaS resources occupancy. A radar chart is a graphical method of displaying multivariate data in form of a two-dimensional chart of three or more quantitative variables represented on exes starting from

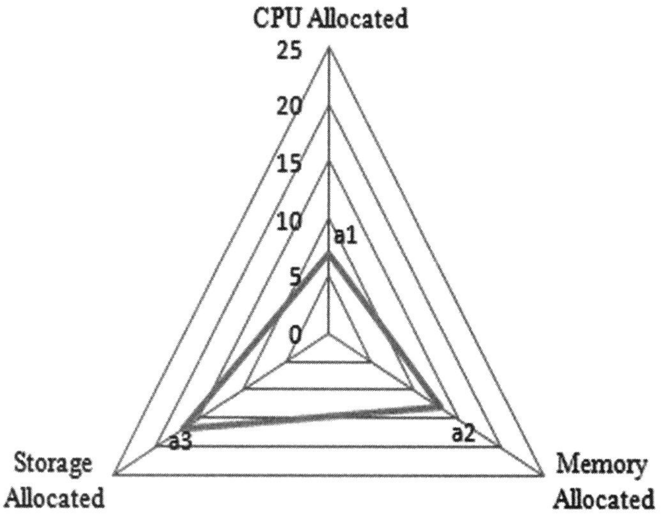

Fig. 12 DaaS service resources occupancy

the same point [74, 75]. According to the drawn radar chart, the radar characteristic value can be calculated, and work out SROI. Three key physical resources (CPU, storage, memory) are selected to form radar char, an example (at time t) is shown in Fig. 12.
In Fig. 12, the radar chart characteristics can be worked out: the radar chart area SRA (t) can be gotten as follows:

$$SRA(t) = \frac{1}{2}\sin\left(\frac{2\pi}{3}\right)[a1 \times a2 + a2 \times a3 + a1 \times a3] \tag{15}$$

- SPI computing method
  For a DaaS service, there are many kinds of performance evaluation indexes, as well as a lot of methods to evaluate these indexes. The n service performance monitoring points are selected to record DaaS service performance. The response time $t_1, t_2, \cdots, t_n$ are corresponding to monitoring points $a_1, a_2, \cdots, a_n$. Therefore, at any moment i, the value of SPI equals to P_i:

$$P_i = \frac{1}{n}\sum_{i=1}^{n} t_i \tag{16}$$

3. DaaS service performance prediction method based on the fuzzy nearness
   The DaaS service performance is influenced by STI and SROI. Meanwhile, DaaS service resources occupancy and DaaS transactions are dynamic change real-time. So these two factors should be taken into consideration during DaaS service performance. A fuzzy matching algorithm based on fuzzy nearness by using STI and SROI historical data is proposed. The nearness degree of STI (X_n) and SROI

(Y_n) of the prediction time n in the historical data matrix are calculated to work out the results X_i and Y_i, which are the closest to X_n and Y_n in historical data. Then the Z_i of SPI is used, corresponding to X_i and Y_i, as the predicted performance value Z_n in predicted time n.

– Calculating membership degree
  Membership function is a function that describes some elements belonging to some characteristics. The value of membership function is called a membership degree, and its values ranges is from 0 to 1. When the value is closer to 1, it means its membership degree is higher. An eigenvalue matrix using SRI (Xi) and SROI (Yi) is formed:

$$\begin{pmatrix} X_1\ X_2 & \cdots & X_i & \cdots & X_n \\ Y_1\ Y_2 & \cdots & Y_i & \cdots & Y_n \end{pmatrix}$$

  Membership degree calculating formulas are as follows:

$$F(X_i) = \frac{X_i - \min(X_n)}{\max(X_n) - \min(X_n)} \tag{17}$$

$$F(Y_i) = \frac{Y_i - \min(Y_n)}{\max(Y_n) - \min(Y_n)} \tag{18}$$

– Calculating close degree
  According to formula (Eq. 17) and formula (Eq. 18), the following fuzzy matrix can be obtained:

$$\begin{pmatrix} F(X_1)F(X_2)\cdots F(X_i)\cdots F(X_n) \\ F(Y_1)F(Y_2)\cdots F(Y_i)\cdots F(Y_n) \end{pmatrix}$$

For the predicting time n, the nearness degree of STI (X_n) and SROI (Y_n) in the historical data matrix could be calculated one by one with lattice close degree formula:

$$\begin{aligned} N(A,B) = {} & [(F(X_i) \wedge F(X_n)) \vee (F(Y_i) \wedge F(Y_n))] \wedge \\ & \times \{[(1 - F(X_i)) \wedge (1 - F(X_n))] \vee [(1 - F(Y_i)) \wedge (1 - F(X_n))]\} \end{aligned} \tag{19}$$

In formula (Eq. 19), A represents the matrix in any moment i, B represents the matrix in predicting moment n.

$$A = \begin{pmatrix} F(X_i) \\ F(Y_i) \end{pmatrix} \tag{20}$$

$$B = \begin{pmatrix} F(X_n) \\ F(Y_n) \end{pmatrix} \tag{21}$$

6.4 Analyzing User's Habits

User's habits in the user access sequence prediction can be analyzed. PrefixSpan algorithm is applied to handle the question of user access sequential mining [75]. KMP pattern matching algorithm is used to predict the user access sequence.

1. User access sequential pattern mining based on PrefixSpan algorithm
   - Definition 1
     In the user access pattern library, the sequence collection $U = \{u_1, u_2, \ldots, u_i, \ldots, u_n\}$, which $u_i = \{p_i, t_i\}$, where p_i refers to user access patterns, t_i refers to the mean residence time sequences of a page.
   - Definition 2
     User access patterns $P_i = \{x_1, x_2, \ldots, x_k, \ldots, x_n\}$, $1 \leq k \leq n$, where x_k refers to the page label.
     It is not random for users to visit the web site's pages, because there is a certain correlation among pages. The user's access also has some models. These models can be mined by PrefixSpan algorithm. Based on this algorithm, steps of the user access pattern P_i are as follows:
     Finding the access patterns sequence with the length of one in user access sequence collection, dividing the search space *Sub*, recursive mining the subset of sequential patterns.
   - Definition 3
     The mean residence time sequences of a page $T_i = \{t_1, t_2, \ldots, t_{n-1}\}$, $1 \leq k \leq n-1$, which t_k refers to the mean residence time from page x_k to x_{k+1}.

When users access a page, the residence time is different because the degree of interest. The user's operating proficiency and reading speed are different. Some data may have significant differences with other data. Therefore, the noise data need to be removed. After the noise removal, the mean residence time of a page is as following:

$$t_k = \frac{Sum(x_k, x_{k+1})}{Count(x_k, x_{k+1})} \tag{22}$$

Sum (X_k, X_{k+1}) refers to the total time from page X_k to page X_{k+1} in all users access record after the noise removal, Count (X_k, X_{k+1}) refers to the number of times of containing a sequence (X_k, X_{k+1}) in users' record.

2. The prediction of user access sequence based on KMP pattern matching algorithm

KMP algorithm is a classical pattern matching algorithm. Its core idea is the process which uses part of known match information to match the later information. The KMP pattern matching algorithm is used to predict user access intention. The algorithm includes the follows steps:

Before carrying out pattern matching, the pattern and text need to be changed into the reverse form, which means compare the pattern with text forward one by one from the last character.

In matching process, if a complete subsequence of the patterns in text is not found, it needs to find a match sequence as long as possible. Regard the longest match sequence found from each text as the matching of the pattern subsequence under the text, marked as $P|_T$ (i).

The returning results are $P|_T$ (i) and the predicted follow-up visits sequence. For example:

A record in users access pattern library is T：1 3 4 6 4 8 1 2 The current users access sequence is P：5 4 8

In the current access sequence, the last visited page class 8 has the greatest impact to the follow-up prediction result, and then followed by 4 and 5. Therefore, before applying the KMP algorithm, the current access sequence is set and the library record is reversed. The results are as follows:

A record in users access pattern library is T：2 1 8 4 6 4 3 1 The current users access sequence is P：8 4 5

T does not exist 5, so string P cannot be found in T, which is considered that the matching is unsuccessful in general KMP algorithm. However, the improved KMP algorithm finds the longest matching string as much as possible if there is not a complete match. The record 5 has impact in the user intention in P, so it needs to remove 5, and continue to search for 8 and 4 in T. When finding the string '8 4', the match succeeds. The general KMP algorithm returns the position 2, where string '8 4' is located in T.

The improved KMP algorithm returns the follow-up sequence and matching length: '2 1 2', which '2 1' is the reverse order sequence of the predicted follow-up visits, and the last '2' means the length of matching is 2. The string '4 8' in P matches the string '4 8' in T, and then it shows that this user probably accesses as this pattern. Therefore, it can be predicted that the user's follow-up visit sequence is '1 2' according to the follow-up sequence is '1 2' in T.

6.5 *Tools*

Recently, there are many performance testing products, such as HP Load Runner, IBM RPT (Rational Performance Tester), Apache JMeter, Segue Silk Performer, Red View Web Load, Compuware QA Load and so on. Here HP Load Runner, IBM RPT and Apache JMeter will be introduced as the following.

1. HP Load Runner
   HP Load Runner is the industry-standard performance testing product for predicting system behavior and performance [76]. Load Runner emulates thousands of concurrent user requests to put the application through the rigors of real-life user requests. An application can be stressed from end-to-end and the response times of key business processes can be measured. Simultaneously, Load Runner collects system and component-level performance information through a comprehensive array of system monitors and diagnostics modules. These metrics are combined into a sophisticated analysis module that allows teams to drill down to isolate bottlenecks within the architecture. Load Runner is widely, customized, and certified to work with ERP/CRM applications from PeopleSoft, Oracle, SAP, and Siebel.
   - Three smaller applications [77]
   The Virtual User Generator enables you to determine what actions you would like the Vusers, or virtual users, to perform under stress within the application. Some scripts generate a series of actions, such as logging on, navigating through the application, and exiting the program, which can be taken to run through a schedule pre-setup by controller. The controller should know how many Vusers to activate, when to activate them, and how to group the Vusers and keep track of them. The Results and Analysis program returns the results of the load test in various forms.
   - Load Runner Testing Process [77]
   The process includes planning the test, creating the Vuser scripts, creating the scenario, running the scenario, and analyzing test results.
   - HP Load Runner in the Cloud [77]
   In May 2010, HP announced that an on-demand version of the application performance testing software would be available via Amazon Elastic Compute Cloud. HP Load Runner in the Cloud is offered as beta software in the U.S and is available with pay-as-you-go pricing, which is for businesses of very big size.
2. IBM RPT
   IBM Rational Performance Tester [78] is a test creation, execution, and analysis tool that validate application scalability and reliability under multiple user loads. Rational Performance Tester enables teams to pinpoint system bottlenecks before application deployment. Teams need little even no programming knowledge to understand and modify the tests. Using an intuitive graphical test scheduler and data pooling capability, teams can accurately organize their tests to simulate the different types of users and their activities. During test execution, while emulating the desired number of concurrent users, Rational Performance Tester generates reports that highlight poorly performing Web pages, URLs, and transactions. Teams can expose performance problems for problem identification and repair before the system goes live.
3. Apache JMeter
   The Apache JMeter™ desktop application [79] is open source software, a 100 % pure Java application designed to load test functional behavior and measure performance. It was currently expanded to other test functions, besides Web

Applications. Apache JMeter may be used to test performance both on static and dynamic resources (files, Servlets, Perl scripts, Java Objects, Data Bases and Queries, FTP Servers and more). It can be used to simulate a heavy load on a server, network or object to test its strength or to analyze overall performance under different load types. It can be used to make a graphical analysis of performance or to test the server/script/object behavior under heavy concurrent load.

References

1. Boulon J, Konwinski A, Qi R, et al. Chukwa: a large-scale monitoring system//Proceedings of IEEE International Conference on Control Applications, 2008, 8.
2. http://blog.sina.com.cn/s/blog_5433a9f80100byg0.html
3. http://en.wikipedia.org/wiki/Software_aging
4. Cheng S, Pan Y, Analysis of cloud service reliability model based on node failure recovery. Journal of Software Guide, 2012, 11 (5): 90–92.
5. Sahoo R K, Squillante M S, Sivasubramaniam A, et al., Failure data analysis of a large-scale heterogeneous server environment//IEEE International Conference on Dependable Systems and Networks, 2004: 772–781.
6. Heath T, Martin R P, Nguyen T D, Improving cluster availability using workstation validation. ACM SIGMETRICS Performance Evaluation Review, 2002, 30(1): 217–227.
7. http://www.d1net.com/cloud/xaas/88623.html
8. Shang Y, Li D, Xu M, Energy-aware routing in data center network//Proceedings of the first ACM SIGCOMM Workshop on Green Networking, 2010: 1–8.
9. Srikantaiah S, Kansal A, Zhao F., Energy aware consolidation for cloud computing//Proceedings of the Conference on Power Aware Computing and Systems. USENIX Association, 2008, 10.
10. HOOPER A. Green computing. Communications of the ACM, 2008, 51(10): 1–13.
11. Avizienis A, Laprie J C, Randell B, et al., Basic concepts and taxonomy of dependable and secure computing. IEEE Transactions on Dependable and Secure Computing, 2004, 1(1): 11–33.
12. Dawson D, Desmarais R, Kienle H M, et al., Monitoring in adaptive systems using reflection//Proceedings of the International Workshop on Software Engineering for Adaptive and Self-managing Systems, 2008: 81–88.
13. http://baike.baidu.com/view/23710.html .
14. Zhang Y, Li, Zhang J, et al., Reviewed on middleware technology. Computer Engineering and Applications, 2002, 15(1): 80–82.
15. Lin Z, Lai Y, Lin C, et al., The research on cloud database. Journal of Software, 2012, 23(5):1148–1166.
16. Chang F, Dean J, Ghemawat S, et al., Bigtable: A distributed storage system for structured data. ACM Transactions on Computer Systems (TOCS), 2008, 26(2): 4–11.
17. Web Services Architecture [S],W3C Working Draft, 2002. http://www.w3.org/TR/ws-arch.
18. Bai X, Liu Y, Wang L, et al., Model-based monitoring and policy enforcement of services. Simulation Modelling Practice and Theory, 2009, 17(8): 1399–1412.
19. Delgado N, Gates A Q, Roach S. A taxonomy and catalog of runtime software-fault monitoring tools. IEEE Transactions on Software Engineering, 2004, 30(12): 859–872.
20. Da Cruz S M S, Campos M L M, Pires P F, et al., Monitoring e-business Web services usage through a log based architecture//Proceeding of IEEE International Conference on Web Services, 2004: 61–69.
21. Sahai A, Graupner S, Machiraju V, et al. Specifying and monitoring guarantees in commercial grids through SLA//Proceedings of 3rd IEEE International Symposium on Cluster Computing and the Grid, 2003: 292–299.

22. Devine S W, Bugnion E, Rosenblum M. Virtualization system including a virtual machine monitor for a computer with a segmented architecture: U.S. Patent 6,397, 242, 2002-5-28.
23. http://www.connectix.com/ products/vs.html.
24. Barham P, Dragovic B, Fraser K, et al., Xen and the art of virtualization. ACM SIGOPS Operating Systems Review, 2003, 37(5): 164–177.
25. Kivity A, Kamay Y, Laor D, et al., KVM: the Linux virtual machine monitor//Proceedings of the Linux Symposium, 2007(1): 225–230.
26. Agichtein E, Brill E, Dumais S., Improving web search ranking by incorporating user behavior information//Proceedings of the 29th Annual International ACM SIGIR Conference on Research and Development in Information Retrieval, 2006: 19–26.
27. Shin J, Narayanan S S, Gerber L, et al. Analysis of user behavior under error conditions in spoken dialogs//INTERSPEECH. 2002.
28. Granka L A, Joachims T, Gay G., Eye-tracking analysis of user behavior in WWW search//Proceedings of the 27th Annual International ACM SIGIR Conference on Research and Development in Information Retrieval. 2004: 478–479.
29. Bellare M, Miner S K. A forward-secure digital signature scheme//Advances in Cryptology—CRYPTO'99. Springer Berlin Heidelberg, 1999: 431–448.
30. Günther C G. An identity-based key-exchange protocol//Advances in Cryptology—Eurocryp '89. Springer Berlin Heidelberg, 1990: 29–37.
31. Padala P, Hou K Y, Shin K G, et al. Automated control of multiple virtualized resources//Proceedings of the 4th ACM European Conference on Computer Systems, 2009: 13–26.
32. Lowe S. Mastering VMware vSphere 4. Indianapolis: Wiley Publishing, 2009.
33. Matthews J N, Dow E M, Deshane T, et al., Running Xen: a hands-on guide to the art of virtualization. Prentice Hall PTR, 2008.
34. Williams D E. Virtualization with Xen (tm): Including XenEnterprise, XenServer, and XenExpress: Including XenEnterprise, XenServer, and XenExpress.Syngress, 2007.
35. SzymańskaKwiecień A, Kwiatkowski J, Pawlik M, et al., Performance prediction methods//Proceedings of the International Multiconference. 2006, 1896: 7094.
36. Kwiatkowski J. Evaluation of parallel programs by measurement of its granularity//Parallel Processing and Applied Mathematics. Springer Berlin Heidelberg, 2006: 145–153.
37. Jin'an Hu. The data centers in the cloud computing resources monitoring system research and design. Journal of University of Electronic Science and Technology. 2012.5.
38. ZHAO Fang, LI Lan-ying. Study of web application monitoring system based on business process. Journal of Beijing Forestry University. 2013.6.
39. Wikipedia, the resource monitor[OL]. http://en.wikipedia.org/wiki/Resource_Monitor .
40. Han F, Peng J, Zhang W, et al. Virtual resource monitoring in cloud computing. Journal of Shanghai University (English Edition), 2011, 15: 381–385.
41. Fastest VPN for Asia. How to use the resource monitor in vista. http://www.vistax64.com/tutorials/111020-resource-monitor.html
42. Manage Engine-IT Enterprise Management. CPU, Memory and Disk Monitoring [OL]. http://www.manageengine.com/network-monitoring/cpu-memory-disk.html
43. Zhuohao Wang, Zhuofeng Zhao. A Model-Driven Approach for Business-Oriented Monitoring of Service Operation//Proceeding of IEEE International Conference on Service Sciences (ICSS). 2010, 5:13–14.
44. Thomas M, Redmond R, Yoon V, et al. A semantic approach to monitor business process. Communications of the ACM, 2005, 48(12): 55–59.
45. Ince D C, Andrews D. The Software Life Cycle. Butterworth-Heinemann, 1990.
46. Kleinrock L. A vision for the Internet. ST Journal of Research, 2005, 2(1): 4–5.
47. Buyya R, Beloglazov A, Abawajy J., Energy-efficient man-agement of data center resources for cloud computing:a vision, architectural elements, and open challenges//Proceeding of the 2010 International Conference on Parallel and Distributed Processing Techniques and Applications,2010:1–12.
48. Weiss A. Computing in the clouds. Networker, 2007, 11(4):16–25.

49. Barham P, Dragovic B, Fraser K, et al. Xen and the art of virtualization. ACM SIGOPS Operating Systems Review, 2003, 37(5): 164–177.
50. Padala P, Shin K G, Zhu X, et al. Adaptive control of virtualized resources in utility computing environments. ACM SIGOPS Operating Systems Review, 2007, 41(3): 289–302.
51. Marston S, Li Z, Bandyopadhyay S, et al. Cloud Computing—The business perspective. Decision Support Systems, 2011, 51(1): 176–189.
52. Xiong K, Perros H. Service performance and analysis in cloud computing//Proceeding of IEEE 2009 World Conference on Services-I, 2009: 693–700.
53. http://www.brighthub.com/computing/smb-security/articles/11337.aspx.
54. http://www.wisegeek.com/what-is-software-development.htm.
55. http://www.iso.org/iso/iso_catalogue/catalogue_tc/catalogue_detail.htm?csnumber=39064
56. Eick S G, Graves T L, Karr A F, et al. Does code decay? assessing the evidence from change management data. IEEE Transactions on Software Engineering, 2001, 27(1): 1–12.
57. FormalOn-SiteMonitoringApril7, 2006. http://www.research.ucsf.edu/chr/Forms/ chrOnsite-Monit.asp
58. Contact Hole Roughness (CHR). http://www.research.ucsf.edu/chr/Forms/chrOnsiteMonit.asp
59. Vaculín R, Sycara K. Semantic web services monitoring: An OWL-S based approach//Proceedings of the 41st Annualon IEEE Hawaii International Conference on System Sciences, 2008: 313–313.
60. Dan A, Davis D, Kearney R, et al. Web services on demand: WSLA-driven automated management. IBM Systems Journal, 2004, 43(1): 136–158.
61. Lundy L, Pradeep R. On the migration from enterprise management to integrated service level management. IEEE on Network, 2002, 16(1): 8–14.
62. Avery, Robert B., Off-Site Surveillance Systems. History of the Eighties: Lessons for the Futur,1997,2:25–29.
63. Snodgrass R. A relational approach to monitoring complex systems. ACM Transactions on Computer Systems (TOCS), 1988, 6(2): 157–195.
64. De Chaves S A Uriarte R B Westphall C B. Toward an Architecture for Monitoring Private Clouds. IEEE on Communication Magazine, 2011, 49(12): 130–137.
65. Saaty T L. What is the analytic hierarchy process? [M]. Springer Berlin Heidelberg, 1988.
66. Sun Z, Xu Z, Da Q. A Model Based on Alternative Similarity Scale for Uncertain Multi-Attribute Decision-Making. Journal of Management Science, 2001, 9(6): 58–62.
67. Wu D, Cheng H, Xi X, et al. Annual Peak Power Load Forecasting Based on Fuzzy AHP//Proceedings of the Chinese Society of Universities for Electric Power System and Automation, 2007, 1: 009.
68. Keufmann A, Gupta M M. Introduction to Fuzzy Arithmetic:Theory and Application. NY: Van Nostrand Reinhold, 1991.
69. Liu H, Kong F. A new MADM algorithm based on fuzzy subjective and objective integrated weights. International Journal of Information System and Sciences, 2005, 1(3–4): 420–427.
70. Kukavica I, Robinson J C. Distinguishing smooth functions by a finite number of point values, and a version of the Takens Embedding Theorem. Physica D: Nonlinear Phenomena, 2004, 196(1): 45–66.
71. Kember G, Fowler A C. A Correlation Function for Choosing Time Delays in Phase Portrait Reconstructions. Physics Letters A, 1993, 179(2): 72–80.
72. Rosenstein M T, Collins J J, De Luca C J. A Practical Method for Calculating Largest Lyapunov Exponents From Small Data Sets. Physica D: Nonlinear Phenomena, 1993, 65(1): 117–134.
73. Chambers J M. Graphical Methods for Data Analysis.Spring Science, 1983.
74. Gao J, Pattabhiraman P, Bai X, et al. SaaS Performance and Scalability Evaluation in Clouds//Proceeding of IEEE the 6th International Symposium on Service Oriented System Engineering (SOSE), 2011: 61–71.
75. Guo J, Huang H, Wang B, et al. Research on the prediction of web application system aging trend oriented to user access intention.Instrumentation, Measurement, Circuits and Systems. Springer Berlin Heidelberg, 2012: 983–991.

76. HP LoadRunner—Free download and Software Reviews—CNET Download.com. http://download.cnet.com/HP-LoadRunner/3000-2383_4-10306263.html#ixzz2a39AvT61 .
77. Load Runner Load Testing Tools Resources. http://www.load-testing-tools.com/loadrunner.html.
78. Brown A, Johnston S K, Larsen G, et al. SOA Development Using the IBM Rational Software Development Platform: APractical Guide.Rational Software, 2005.
79. Apache JMeter, ApacheCon North America Portland, Oregon 26th–28th February 2013, http://jmeter.apache.org

Part IX
Resource Management

Usage Patterns in Multi-tenant Data Centers: a Large-Case Field Study

Robert Birke, Lydia Chen and Evgenia Smirni

1 Introduction

Data centers are nowadays ubiquitous and have become a commonplace computing platform for corporations as well as individuals, providing a diverse array of services. Data centers may be universal and prevalent, but so are their administrative challenges that include how to best use them, as well as how to optimize their power and cooling costs. The sheer diversity of customer demands (e.g., one may expect very different needs and performance expectations between individual users of cloud-based data centers versus corporate customers) make data center administration challenging and without clear solutions. Studying the workload that typical data centers experience can provide many useful insights for the better usage of data centers, for the design of autonomic management policies for various resources, even for more efficient power and/or cooling management policies.

Most of existing data center studies can be roughly classified as those that focus on power and thermal management [4–7], capacity planning and resource provisioning [11–13], and traffic engineering [1, 2, 8–10]. The above works either aim at a specific architectural component (e.g., network) or evaluate resource provisioning via simulation or via small scale prototypes at a laboratory setting. To the best of our knowledge, there is very little information on *how exactly* corporate data centers are used by clients, how their workload demands change across time, and how customer demands on different resources fluctuate across time. In this paper we fill this gap by providing the first very detailed resource allocation study that considers two specific

R. Birke (✉) · L. Chen
IBM Research Zurich Lab, 8803 Rüschlikon, Switzerland
e-mail: bir@zurich.ibm.com

L. Chen
e-mail: yic@zurich.ibm.com,

E. Smirni
College of William and Mary, Williamsburg, VA 23187, USA
e-mail: esmirni@cs.wm.edu

S. U. Khan, A. Y. Zomaya (eds.), *Handbook on Data Centers*,
DOI 10.1007/978-1-4939-2092-1_43

data centers and focuses on the usage patterns of several corporate customers across different time scales.

Birke et al, [3] present a workload characterization study of corporate data centers consisting of several thousand servers that are geographically dispersed across the entire globe. The collected data represent the evolution of cloud workload in a time span of 24 months and give a view on how data center workloads evolve across different enterprises, countries, even continents. The focus of [3] is on the time evolution and seasonal characteristics of resource demands from a capacity planning viewpoint. The holistic view that is adopted in [3] gives an excellent perspective for an economics analysis but does not shed any light into the interdependency of workloads that are collocated within the same datacenters, how specific enterprises utilize specific data center resources, or how the demands on the various resources are correlated.

In this study, we select two specific data centers, which host multiple enterprises from the data set collected by Birke et al. [3]. One is a “small” data center that consists of 393 servers and hosts service applications from three enterprises and the other one is a “large” datacenter that consists of 3681 servers and hosts more than 10 enterprises. Here, we focus on how specific enterprises use specific data centers. We adopt a per-enterprise view and concentrate on how six different enterprises, three from each data center, utilize each center’s resources. While we give an overview of resource utilization for CPU, disk, memory, and file system, we especially focus on the per-server CPU utilization by each enterprise. We illustrate the importance of the granularity of the observation by showing the per day utilizations of the servers used by each enterprise across a time period of 1 month and then we focus on a random weekday and a random weekend day to further untangle utilization patterns but at a much finer time granularity.

The presented study aims at characterizing the workload diversity by looking at how specific enterprises utilize data center resources across time. We especially focus on CPU utilizations because we find that CPUs are consistently underutilized. This characterization can be incorporated in the design of autonomic solutions for workload consolidation and load balancing. Beyond that, this characterization could be further used in combination with temperature/cooling information for better power management in data centers, but we believe that it can be primarily used in the design of autonomic policies for data center resource allocation.

Despite the fact that this study is based on a very rich data set that reveals how corporate data centers use data center resources, it does not provide any information on the type of applications that different enterprises use or on the response times of these applications. Nevertheless, from the utilization values across the different servers and across time, one could speculate about the effects of over provisioning as well as the efficacy of designing better autonomic policies for workload consolidation and resource management in today’s data centers.

2 Multi-tenant Datacenters

We collect resource utilization statistics from several thousands of servers from two in-production data centers. These systems are used by different industries, including banking, pharmaceutical, IT, consulting, and retail industries. The collected samples contain a rich set of representative server statistics reflecting the current practices of resource management in corporate data centers and contain mainly resource utilizations.

The average utilization values over base periods are collected via prevailing utilities such as `vmstat` and `df`. All data is stored in a database and aggregated into different time scales ranging from 1 min to 1 month. For each timescale only a fixed number of the most recent records is kept. Consequently, recent data is available at a higher time resolution than older data in order for the database to maintain an upper limit of the space footprint per server. In particular, we consider fine grained 15-min data on Wednesday, May 23, 2012 and Sunday, May 27, 2012. In most cases, we focus on the comparison between Wednesday and Sunday, i.e., the workload difference between a working day and a weekend day.

We focus on the basic physical resources per server: CPU, network, memory, disk, and file system statistics that are collected in units of resource utilization which is just a percentage value. Since a server can have multiple disks, the disk utilization is defined by the sum of all used space divided by the sum of all disk sizes. The file system includes both local and remote data storage, which can be on the media of disks and memory. Similar to the disk, when there are multiple file systems, the utilization is computed from the sum of all used space divided the sum of all file system sizes. The CPU utilization is defined by the percentage of time the CPU is active over an observation period. Utilization values of memory, disk, and file system are defined by the volume usage, i.e., used space divided by the total available space.

2.1 Evolution of Resource Demands

Before focusing on how the workload evolves across time, we present some aggregate measures in Table 1. The table presents the mean and standard deviation of utilization values of the servers used by the six selected enterprises across an observation period of 3 years, and gives a per enterprise view, as well as a per data center view. Note that across all resource utilization values, standard deviations are significant, especially in light of the fact that utilizations values are bounded between 0 and 100.

We also present how the workload diversity evolves across the observation time. We depict the time series of monthly utilization across 3 years for all resources in Fig. 1. The utilization evolution for most of resources and across all enterprises are rather stable albeit with a slightly increasing trend, which indicates that the growth of demand of a particular resource is greater than the growth of its supply. A decreasing trend indicates the opposite, i.e., the growth of supply is greater than the demand. Dips in the time series correspond to hardware/software upgrades. In general, Fig. 1

Table 1 Overview of resource utilization by different enterprises

Small DC	*CPU*		*Memory*		*Disk*		*File Sys.*	
	Mean	Std	Mean	Std	Mean	Std	Mean	Std
All	23.53	23.17	78.17	9.93	76.03	22.34	51.55	20.62
Ent 1	38.22	23.15	84.18	9.51	82.37	19.63	47.39	15.60
Ent 2	21.64	21.24	84.33	6.61	69.08	23.15	48.54	20.23
Ent 3	20.73	22.86	73.70	11.16	77.86	21.93	53.25	21.22
Large DC								
All	20.82	23.84	80.33	10.52	68.89	26.08	40.15	19.82
Ent 4	10.56	14.02	72.80	5.67	70.05	28.00	33.88	18.64
Ent 5	16.10	20.75	85.37	9.47	58.09	26.08	38.14	20.96
Ent 6	37.81	27.89	83.12	13.63	60.34	25.74	43.32	18.95

consistently shows that the CPU is the *least* utilized resource across all datacenters and across all enterprises. In the following, we further investigate how CPUs are utilized by the various data centers.

2.2 CPU Load Balancing

In Fig. 2 we illustrate the time series of daily average CPU utilization for the six enterprises across the entire month of May 2012. On this figure we also plot the standard deviation of the average server utilization. Notice that the number of servers that are assigned to each enterprise range from 130 to 1380, therefore the significant standard deviation can be viewed as a strong indication of very unbalanced utilization values across servers, which is a consistent trend across all enterprises. The figure also illustrates another interesting behavior: there are clear daily patterns, i.e., weekends can be clearly seen across most enterprises (see the distinct dips in the plots for enterprises 1, 3, 5, and 6). Weekend utilizations can be as low as ten percentage points comparing to working day utilizations. Enterprises 2 and 4 have a more even utilization pattern across time, with only slight ripples in their averages across time.

Motivated by these observations, we now turn to the time series of CPU utilizations calculated every 15 min. We focus on one representative week day, Wednesday May 23, and one representative weekend day, Sunday May 27. Figure 3 illustrates the average utilizations together with the standard deviation. On each graph, we plot data for Wednesday and Sunday. The figure clearly illustrates that there are severe utilization imbalances across servers within 15-min time periods. In addition, it shows that some enterprises utilize the CPUs during weekdays very differently from the weekends (e.g., look in Fig. 3a and 3e where there is a significant change in values, especially during mid-day) while some others have a quite similar usage on

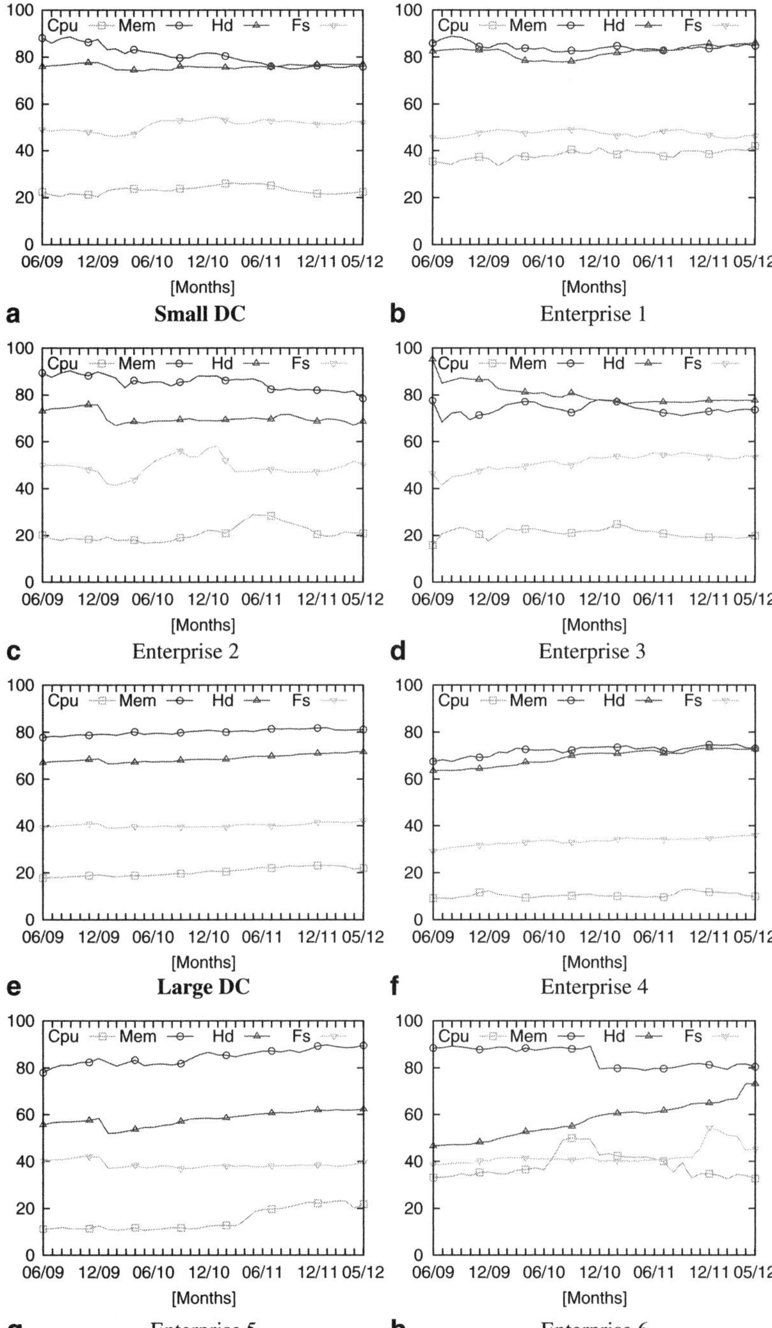

Fig. 1 Time series of resource utilizations of monthly averages over 3 years. Note that memory, disk, and file system values correspond to space utilization

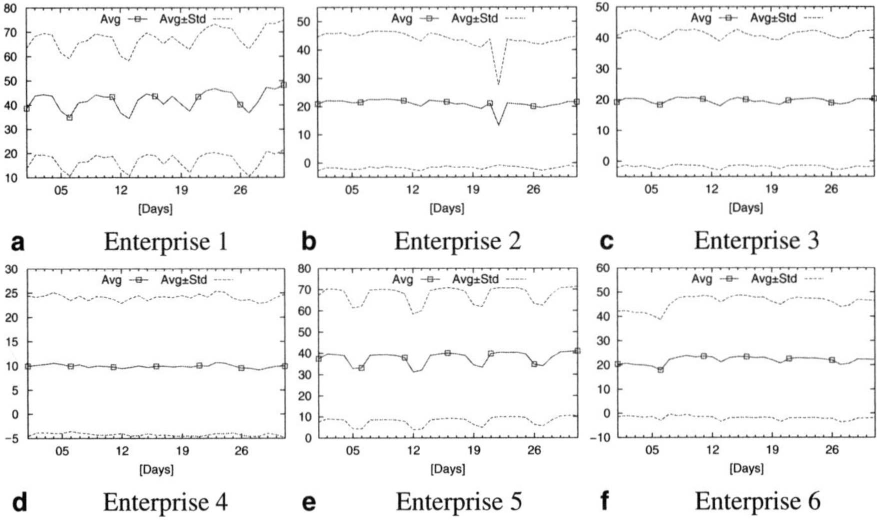

Fig. 2 Time series of daily CPU utilization over during the entire month of May 2012

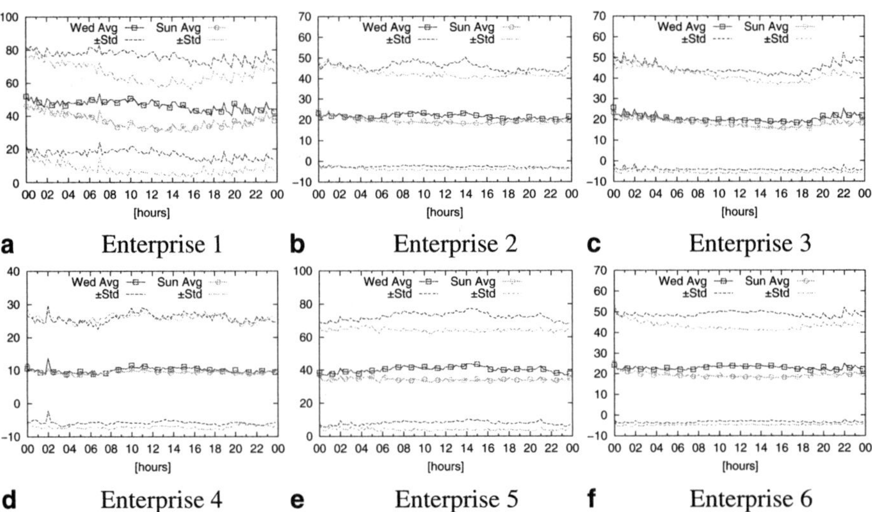

Fig. 3 Time series of 15-min CPU utilization during a representative workday (Wednesday, May 23 2012) and during a representative weekend day (Sunday, May 27 2012)

both Wednesday and Sunday. Across several enterprises, we also observe a slightly spiky pattern at the beginning of each hour. This pattern is consistent across most graphs in Fig. 3 and is especially prominent during off-peak hours. Close observation on the data and on what CPU does during these periods shows that this is due to the scheduling of batch jobs on the hour, commonly done via a cron utility. In addition,

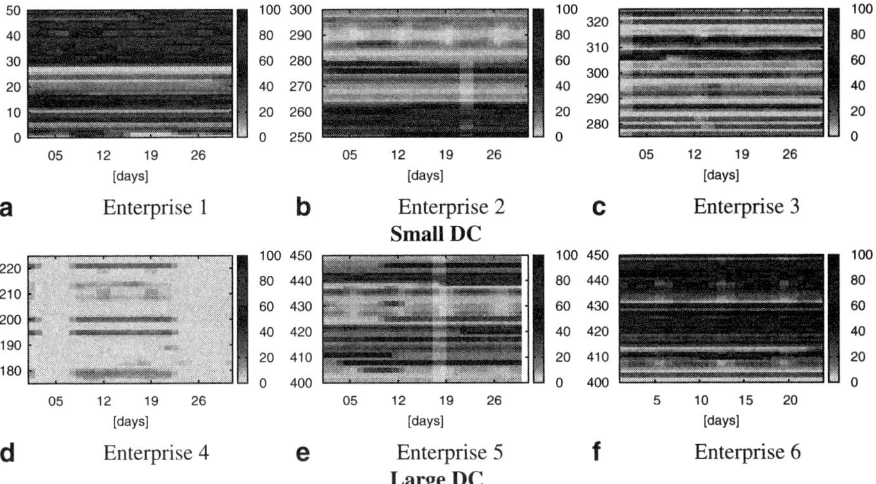

Fig. 4 CPU daily utilization of different industries for May 2012

the figure illustrates the tendency to schedule batch jobs during off-peak hours. In general, we observe clear load unbalances across enterprises and at the same time similar practices in resource usage across time and across all enterprises.

2.3 *The Impact of Time Scales*

To illustrate fluctuations of resource demands across time and also across servers, we also present the daily utilizations for 31 representative days, namely May 2012 for the 6 enterprises, see Fig. 4. In the interest of a clearer presentation, we present heatmaps that illustrate the CPU utilization levels for only a subset of the servers used by each enterprise, namely only for 50 servers which are mapped on the y-axis of each graph.[1] The x-axis corresponds to the specific days within the selected 31-day period, (i.e., the entire May). The y-axis corresponds to the server identification number. We note that despite the fact that we only show daily utilizations for 50 servers, the patterns shown are representative. Dark colors in the figure correspond to high utilization values, while light colors to low utilizations.

Figure 4 clearly shows that across all enterprises, there are well defined cyclic patterns that correspond to week- and weekend-days, see for example the illustration of three vertical lighter regions that signify weekend (lighter) loads across all enterprises. In addition, a striking similarity across all enterprises is that there is a clear load unbalancing, as shown by the presence of multiple colors/shades as well as the

[1] Each enterprise uses different number of servers, with the majority of them using more than 130 servers, which unfortunately results in an illegible heat map.

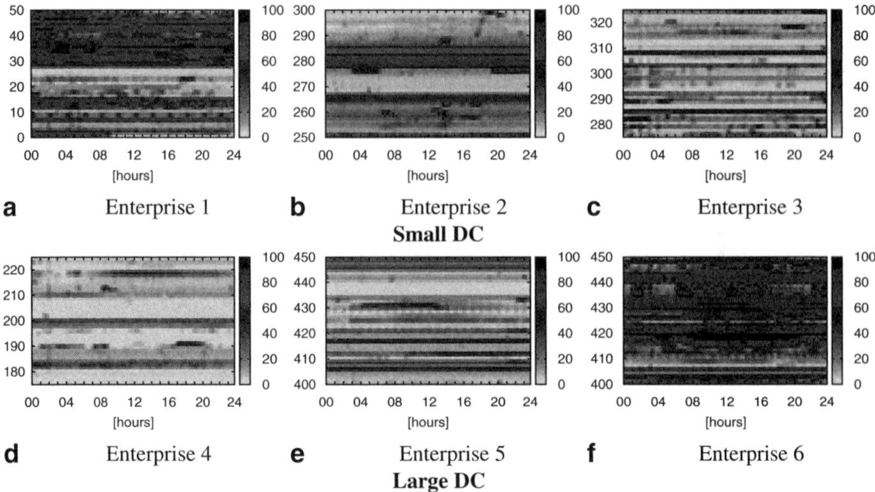

Fig. 5 CPU utilizations for 15-min time periods for a typical working day (Wednesday, May 23, 2012)

fact that some servers are mostly always highly loaded, even during low weekend cycles. The graph illustrates opportunities for power savings, e.g., see Fig. 4e as well as load balancing to improve system usage. Overall, these graphs can be used to demonstrate the current state-of-the-practice and can be used to evaluate the need of new load balancing or consolidation techniques.

Figure 5 selects a single day, Wednesday, May 23 2012, and zooms into the utilization of the same set of selected CPUs as in Fig. 4 but looking at utilizations at 15-min time periods. This new figure allows us to see more clearly fluctuations in utilization on the selected CPUs during a 24-h period. Observe that in most of enterprises, there is a surge in utilizations on the first 15-min period of every hour. In addition, it appears that load is not steady across time but could fluctuate significantly.

Figure 6 selects the exact same servers as in Fig. 5 but plots 15-min utilizations for a typical weekend day. For some enterprises there is a dramatic drop in utilizations (see for example enterprise 6) comparing to the typical working day, while for some others (e.g., enterprises 2, 4, and 5) the reduction in utilization is only moderate. In general, comparing Figs. 4, 5, and 6, we observe that there are dramatic load imbalances both across servers and across time, pointing to excellent opportunities for intelligent workload consolidation and improved load balancing. This observation, in addition to the fact that CPUs tend to be lowly utilized, should drive the development of more effective workload management in data centers.

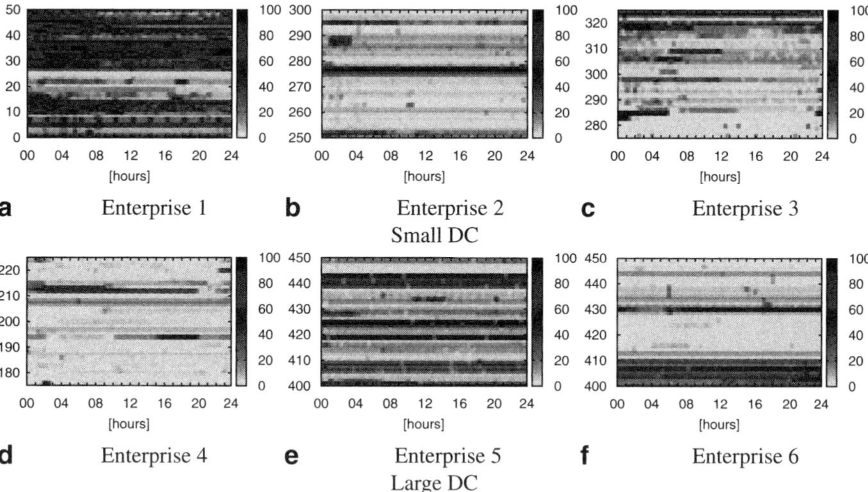

Fig. 6 CPU utilizations for 15-min time periods for a typical weekend day (Sunday, May 27, 2012)

3 Summary

We characterize workloads collected from two data centers, each of which host multiple enterprises, for the span of 3 years. We report on the utilization of four resources, i.e., CPU, memory, disk, and file system, but focus mainly on the CPU utilizations across different time scales. We show the workload diversity, long-term evolution, CPU load balancing, and discuss the needs for developing autonomous resource management at different time scales. This study can be used as a baseline against which autonomic policies for data center management can be evaluated.

Acknowledgments This paper with the title "Usage Patterns in Multi-Tenant Data Centers: a Temporal Perspective", appeared in the Proceedings of International conference on Autonomic Computing i(ICAC) 2012, San Jose, CA, September 2012, pp. 161–166.

We thank Nishi Gupta for granting us access to the data and share our insights with the scientific community. Part of this work has been done while Evgenia Smirni was on sabbatical leave at IBM Research, Zurich Lab. Evgenia Smirni is partially supported by NSF grants CCF-0937925 and CCF-1218758.

References

1. M. Al-Fares, A. Loukissas, and A. Vahdat. A scalable, commodity data center network architecture. In *SIGCOMM*, pages 63–74, 2008.
2. T. Benson, A. Akella, and D. A. Maltz. Network traffic characteristics of data centers in the wild. In *IMC*, pages 267–280, 2010.

3. R. Birke, L. Y. Chen, and E. Smirni. Data centers in the cloud: A large scale performance study. In *IEEE CLOUD*, pages 336–343, 2012.
4. G. Chen, W. He, J. Liu, S. Nath, L. Rigas, L. Xiao, and F. Zhao. Energy-aware server provisioning and load dispatching for connection-intensive internet services. In *NSDI*, pages 337–350, 2008.
5. Y. Chen, A. Das, W. Qin, A. Sivasubramaniam, Q. Wang, and N. Gautam. Managing server energy and operational costs in hosting centers. In *SIGMETRICS*, pages 303–314, 2005.
6. N. El-Sayed, I. A. Stefanovici, G. Amvrosiadis, A. A. Hwang, and B. Schroeder. Temperature management in data centers: why some (might) like it hot. In *SIGMETRICS*, pages 163–174, 2012.
7. S. Govindan, J. Choi, B. Urgaonkar, A. Sivasubramaniam, and A. Baldini. Statistical profiling-based techniques for effective power provisioning in data centers. In *EuroSys*, pages 317–330, 2009.
8. C. Guo, H. Wu, K. Tan, L. Shi, Y. Zhang, and S. Lu. Dcell: a scalable and fault-tolerant network structure for data centers. In *SIGCOMM*, pages 75–86, 2008.
9. D. Halperin, S. Kandula, J. Padhye, P. Bahl, and D. Wetherall. Augmenting data center networks with multi-gigabit wireless links. In *SIGCOMM*, pages 38–49, 2011.
10. B. Heller, S. Seetharaman, P. Mahadevan, Y. Yiakoumis, P. Sharma, S. Banerjee, and N. McKeown. Elastictree: Saving energy in data center networks. In *NSDI*, pages 249–264, 2010.
11. M. Kutare, G. Eisenhauer, C. Wang, K. Schwan, V. Talwar, and M. W. Matthew. Monalytics: online monitoring and analytics for managing large scale data centers. In *ICAC*, pages 141–150, 2010.
12. R. Singh, U. Sharma, E. Cecchet, and P. Shenoy. Autonomic mix-aware provisioning for non-stationary data center workloads. In *ICAC*, pages 21–30, 2010.
13. X. Zhu, D. Young, B. J. Watson, Z. Wang, J. Rolia, S. Singhal, B. McKee, C. Hyser, D. Gmach, R. Gardner, T. Christian, and L. Cherkasova. 1000 islands: an integrated approach to resource management for virtualized data centers. *Cluster Computing*, 12(1):45–57, 2009.

On Scheduling in Distributed Transactional Memory: Techniques and Tradeoffs

Junwhan Kim, Roberto Palmieri and Binoy Ravindran

1 Introduction

Data centers have been increasingly employed in distributed services to support a vast of amount of consumer requests. The requests range from web services to gaming for computation intensive applications. In order to process these requests, the data centers exploit in-memory data for high performance and ensure transactional properties for concurrent requests such as atomicity, consistency, and isolation. Traditionally lock-based synchronization has been used for the consistency of data, but is inherently error-prone. For example, coarse-grained locking, in which a large data structure is protected using a single lock is simple and easy to use, but permits little concurrency. In contrast, with fine-grained locking, in which each component of a data structure (e.g., a hash table bucket) is protected by a lock, programmers must acquire only necessary and sufficient locks to obtain maximum concurrency without compromising safety, and must avoid deadlocks when acquiring multiple locks. Both these situations are highly prone to programmer errors. The most serious problem with locks is that it is not easily *composable*—i.e., combining existing pieces of software to produce different functionality is not easy. This is because, lock-based concurrency control is highly dependent on the order in which locks are acquired and released. Thus, it would be necessary to expose the internal implementation of existing methods, while combining them, in order to prevent possible deadlocks. This breaks encapsulation, and makes it difficult to reuse software.

Transactional memory (TM) is an alternative synchronization model for shared in-memory data objects that promises to alleviate the difficulties of lock-based synchronization (i.e., scalability, programmability, and composability issues). As TM

J. Kim (✉) · R. Palmieri · B. Ravindran
The Department of Electrical and Computer Engineering,
Virginia Tech, Blacksburg, VA 24061, USA
e-mail: junwhan.kim@udc.edu

S. U. Khan, A. Y. Zomaya (eds.), *Handbook on Data Centers*,
DOI 10.1007/978-1-4939-2092-1_44

code is composed of read/write operations on shared objects, it is organized as *memory transactions*, which optimistically execute while logging any changes made to accessed objects. Two transactions *conflict* if they access the same object and one access is a write. When that happens, a contention manager (CM) resolves the conflict by aborting one and allowing the other to commit, yielding (the illusion of) atomicity. Aborted transactions are re-started, often immediately, after rolling-back the changes. Sometimes, a *transactional scheduler* is also used, which determines an ordering of concurrent transactions so that conflicts are either avoided altogether or minimized.

We first consider the the single object copy DTM model (i.e., SV-STM). A distributed transaction typically has a longer execution time than a multiprocessor transaction, due to communication delays that are incurred in requesting and acquiring objects, which increases the likelihood for conflicts and thus degraded performance [4]. We present a novel transactional scheduler called Bi-interval [14] that optimizes the execution order of transactional operations to minimize conflicts. Bi-interval focuses on read-only and read-dominated workloads (i.e., those with only early-write operations), which are common transactional workloads [11]. Read transactions do not modify the object; thus transactions do not need exclusive object access. Bi-interval categorizes concurrent requests for a shared object into read and write intervals to maximize the parallelism of read transactions. This reduces conflicts between read transactions, reducing transactional execution times. Further, it allows an object to be simultaneously sent to nodes of read transactions, thereby reducing the total object traveling time.

With a single object copy, node/link failures cannot be tolerated. If a node fails, the objects held by the failed node will be simply lost and all following transactions requesting such objects would never commit. Additionally, read concurrency cannot be effectively exploited. Thus, an array of DTM works—all of which are cluster DTM—consider object replication. These works provide fault-tolerance properties by inheriting fault-tolerance protocols from database replication schemes, which rely on broadcast primitives (e.g., atomic broadcast, uniform reliable broadcast) [3, 5–7, 17]. Broadcasting transactional read/write sets or memory differences in metric-space networks is inherently non-scalable, as messages transmitted grow quadratically with the number of nodes [21]. Thus, directly applying cluster DTM replication solutions to data-flow DTM may not yield similar performance.

We therefore consider a cluster-based object replication model for data-flow DTM. In this model, nodes are grouped into clusters based on node-to-node distances: nodes which are closer to each other are grouped into the same cluster; nodes which are farther apart are grouped into different clusters. Objects are replicated such that each cluster contains at least one replica of each object, and the memory of multiple nodes is used to reduce the possibility of object loss, thereby avoiding expensive brute-force replication of all objects on all nodes. Cluster-based transactional scheduler (CTS) [16] focuses on how to schedule memory transactions in the cluster-based partial replication model for high performance. Each cluster has an object owner for scheduling transactions. In each object owner, CTS enqueues live transactions and identifies some of the transactions that must be aborted to avoid future conflicts, resulting in the concurrency of the other transactions.

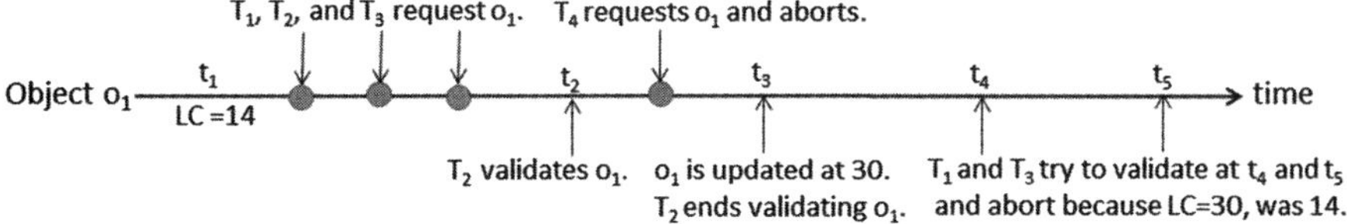

Fig. 1 An example of TFA

2 Preliminaries and System Model

2.1 Distributed Transactions

A set of *distributed transactions* $T = \{T_1, T_2, \ldots\}$ is assumed. The transactions share a set of objects $O = \{o_1, o_2, \ldots\}$, which are assumed to be distributed in the network. A transaction contains a sequence of requests, each of which is a read or write operation request to an individual object. An execution of a transaction is a sequence of timed operations. An execution ends by either a commit (success) or an abort (failure). A transaction is in one of three possible states: *live*, *aborted*, or *committed*. Each transaction has a unique identifier (id), and is invoked by a node in the system.

We use the *Transactional Forwarding Algorithm* (TFA) [21] to provide *early validation* of remote objects, guarantee a consistent view of shared objects between distributed transactions, and ensure atomicity for object operations in the presence of asynchronous clocks. As an extension of the Transactional Locking 2 (TL2) algorithm [9], TFA replaces the central clock of TL2 with independent clocks for each node and provides a means to reliably establish the "happens-before" relationship between significant events. TFA is responsible for caching local copies of remote objects. Without loss of generality, objects export only read and write methods (or operations).

For completeness, we illustrate TFA with an example. In Fig. 1, a transaction updates object o_1 at t_1 (i.e., local clock (LC) is 14) and four transactions (i.e., T_1, T_2, T_3, and T_4) request o_1 from the object holder. Assume that T_2 validates o_1 at t_2 and updates o_1 with LC = 30 at t_3. Any read or write transaction (e.g., T_4), which has requested o_1 between t_2 and t_3 aborts. When write transactions T_1 and T_3 validate at times t_4 and t_5, respectively, T_1 and T_3 that have acquired o_1 with LC = 14 before t_2 will abort, because LC is updated to 30.

TFA ensures atomicity, consistency, and isolation of transactions and has been extensively evaluated with competitor DTM implementations, resulting in enhanced performance as much as 4× [20]. We implemented Bi-interval and CTS in the HyFlow DTM framework [21] and measured the transactional throughput—i.e., the number of committed transactions per second under increasing number of requesting nodes, for the different schemes. Thus, Bi-interval and CTS on TFA are evaluated with only TFA to show the effectiveness of transactional schedulers.

2.2 Definitions

For the purpose of analysis, we consider a symmetric network of N nodes scattered in a metric space. The metric $d(n_i, n_j)$ is the distance between nodes n_i and n_j, which determines the communication cost of sending a message from n_i to n_j. We consider three different *models*: no replication (NR), partial replication (PR), and full replication (FR) in data-flow DTM to show the effectiveness of Bi-interval and CTS in NR and PR, respectively.

Definition 1 Given a scheduler A and N transactions in DTM, $makespan_A^N(Model)$ is the time that A needs to complete N transactions on $Model$.

Definition 2 The competitive ratio (CR) of a scheduler A for N transactions in $Model$ is $\frac{makespan_A^N(Modle)}{makespan_{OPT}^N(Model)}$, where OPT is the optimal scheduler.

Definition 3 The relative competitive ratio (RCR) of schedulers A and B for N transactions on $Model$ in DTM is $\frac{makespan_A^N(Model)}{makespan_B^N(Model)}$.

Also, the RCR of model 1 and 2 for N transactions on scheduler A in DTM is $\frac{makespan_A^N(Model1)}{makespan_A^N(Model2)}$. Given schedulers A and B for N transactions, if RCR (i.e., $\frac{makespan_A^N(Model)}{makespan_B^N(Model)}$) < 1, A outperforms B. Thus, RCR of A and B indicates a relative improvement between schedulers A and B if $makespan_A^N(Model) < makespan_B^N(Model)$.

The execution time of a transaction is defined as the interval from its beginning to the commit. In distributed systems, the execution time consists of both communication delays to request and acquire a shared object and the time duration to conduct an operation on a processor, so the local execution time of T_i is defined as γ_i, $\sum_{i=1}^{N} \gamma_i = \Gamma_N$ for N transactions.

If only a transaction T_i invoking in n_i exists and T_i requests an object from n_j on NR, it will commit without any contention. Thus, $makespan_A^1(NR)$ is $2 \times d(n_i, n_j) + \gamma_i$ under any scheduler A.

2.3 Transactional Scheduler

As mentioned before, a complementary approach for dealing with transactional conflicts is transactional scheduling. Broadly, a transactional scheduler determines the ordering of concurrent transactions so that conflicts are either avoided altogether or minimized. Two kinds of transactional schedulers have been studied in the past: reactive [1, 10] and proactive [4, 23]. When a conflict occurs between two transactions, the contention manager determines which transaction wins or loses, and then the loosing transaction aborts. Since aborted transactions might abort again in the future, *reactive schedulers* enqueue aborted transactions, serializing their future

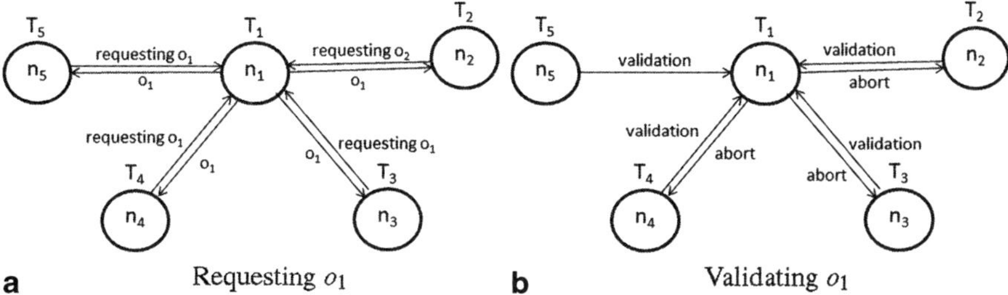

Fig. 2 A scenario consisting of four transactions

execution [1, 10]. *Proactive schedulers* take a different strategy. Since it is desirable for aborted transactions to be not aborted again when re-issued, proactive schedulers abort the loosing transaction with a backoff time, which determines how long the transaction is stalled before it is re-started [4, 23]. Both reactive and proactive transactional schedulers have been studied for multiprocessor TM. However, they have not been studied for DTM, which is the focus of this chapter.

3 Bi-interval

3.1 Motivation

Unlike multiprocessor transactions, data flow-based DTM incurs communication delays in requesting and acquiring objects. Figure 2 illustrates a scenario on data flow DTM consisting of five nodes and an object o_1. Figure 2a shows that nodes n_2, n_3, n_4, and n_5 invoke T_2, T_3, T_4, T_5, respectively and request o_1 from n_1. In Fig. 2b, T_5 validates o_1 first. T_2, T_3, and T_4 abort when they validate.

Contention managers deal with only conflicts, determining which transaction wins or not. Past transactional schedulers (e.g., proactive and reactive schedulers) serialize aborted transactions but do not consider moving objects in data flow DTM. In DTM, the aborted transactions request an object again, increasing communication delays. Motivated by this observation, the transactions requesting o_1 are enqueued and the transactions immediately abort when one of these validate o_1. As soon as o_1 is updated, o_1 is sent to the aborted transactions. The aborted transaction will receive the updated o_1 without any request, reducing communication delays. Meanwhile, we focus on which order of the aborted transactions lead to improved performance. Read transaction defined as read-dominated workloads will simultaneously receive o_1 to maximize the parallelism of read transactions. Write transactions including write operations will receive o_1 according to the shortest delay to minimize object moving time.

3.2 Scheduler Design

Bi-interval is similar to the BIMODAL scheduler [1] in that it categorizes requests into read and write intervals. If a transaction aborts due to a conflict, it is moved to a scheduling queue and assigned a backoff time. Bi-interval assigns two different backoff times defined as read and write intervals to read and write transactions, respectively. Unless the aborted transaction receives the requested object within an assigned backoff time, it will request the object again.

Bi-interval maintains a scheduling queue for read and write transactions for each object. If an enqueued transaction is a read transaction, it is moved to the head of the scheduling queue. If it is a write transaction, it is inserted into the scheduling queue according to the shortest path visiting each node invoking enqueued transactions. When a write transaction commits, the new version of an object is released. If read and write transactions have been aborted and enqueued for the version, the version will be simultaneously sent to all the read transactions and then visit the write transactions in the order of the scheduling queue. The basic idea of Bi-interval is to send a newly updated object to the enqueued-aborted transactions as soon as validating the object completes.

There are two purposes for enqueuing aborted transactions. First, in order to restart an aborted transaction, the cache-coherence (CC) protocol will be invoked to find the location of an object, incurring communication delays. An object owner holds a queue indicating the aborted transactions and sends the object to the node invoking the aborted transactions. The aborted transactions may receive the object without the help of the CC protocol, reducing communication delays. Second, Bi-interval schedules the enqueued aborted transactions to minimize execution times and communication delays. For reduced execution time, the object will be simultaneously sent to the enqueued read transactions. In order to minimize communication delays, the object will be sent to each node invoking the enqueued write transactions in order of the shortest path, so the total traveling time for the object in the network decreases.

Bi-interval determines read and write intervals indicating when aborted read and write transactions restart, respectively. This intends that an object will visit each node invoking aborted read and write transactions within read and write intervals, respectively. As a backoff time, a read interval is assigned to aborted read transactions and a write interval is assigned to aborted write transactions. A read interval is defined as the local execution time γ_i of transaction T_i. All enqueued-aborted read transactions will wait for γ_i and receive the object that T_i has updated. A write interval is defined as the sum of the local execution times of enqueued write transactions and a read interval. The aborted write transaction may be serialized according to the order of the scheduling queue. If any of these transactions do not receive the object, they will restart after a write interval.

Figure 3 shows a scenario consisting of four transactions based on Bi-interval. Node n_1 holds o_1 and write transactions T_2, T_3, T_4, and T_5 request object o_1 from n_1. n_1 has a scheduling queue holding requested transactions T_2, T_3, T_4, and T_5. If T_5 validates o_1 first as being illustrated by Fig. 3b, T_2, T_3, and T_4 abort. If n_2 is closest

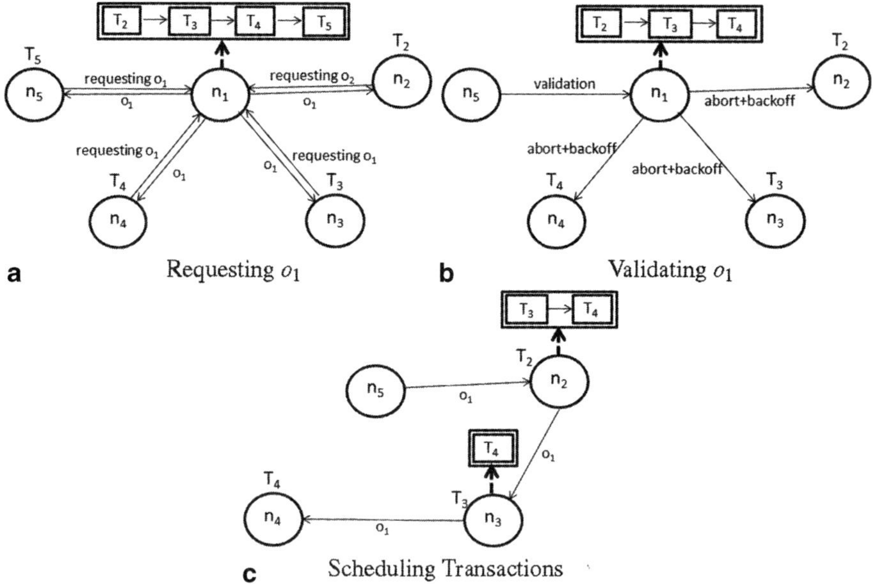

Fig. 3 A scenario consisting of four transaction on bi-interval

to n_5, o_1 updated by T_5 is sent to n_2, and two backoff times are sent to T_3 and T_4, respectively. Figure 3c shows one write interval.

While T_5 validates o_1, let us assume that other read transactions request o_1. The read transactions will be enqueued and simultaneously receive o_1 after T_5 completes its validation. Thus, once the scheduling queue holds read and write transactions, a read interval will start first. The write transactions will be serialized according to the shortest object traveling time.

3.3 Analysis

The object moving cost is defined as $\eta_A(u, V_{T_{N-1}})$, which is the total communication delay for visiting each node from node u holding an object to $N-1$ nodes in $V_{T_{N-1}}$, under scheduler A. $V_{T_{N-1}}$ represents a set of $N-1$ nodes invoking transactions.

Theorem 1 *Bi-interval's execution makespan competitive ratio is* $1+\frac{I_r}{N-k+1}$ *for N transactions including k read transactions, where* I_r *is the number of read intervals.*

Proof The optimal off-line algorithm concurrently executes all read transactions. So, Bi-interval's optimal execution for N transactions including k read transactions is $\sum_{m=1}^{N-k+1} \gamma_m$.

$$CR_{Biinterval} \leq \frac{\gamma_\omega \cdot I_r + \sum_{m=1}^{N-k+1} \gamma_m}{\sum_{m=1}^{N-k+1} \gamma_m} \approx \frac{I_r + N - k + 1}{N - k + 1},$$

where γ_ω is γ of a read transaction. The theorem follows.

Theorem 2 *Bi-interval's traveling makespan competitive ratio for k reads of N transactions is* $\log(N + I_r - k - 1)$.

Proof Bi-interval follows the nearest neighbor path to visit each node in the scheduling list. We define the *stretch* of a transactional scheduler as the maximum ratio of the moving time to the communication delay—i.e., $Stretch_\eta(u, V_{T_{N-1}}) = \max \frac{\eta_{Biinterval}(u, V_{T_{N-1}})}{d(u, V_{T_{N-1}})} \leq \frac{1}{2}\log(N-1) + \frac{1}{2}$ from [19]. Hence, $CR_{Biinterval} \leq \log(N + I_r - k - 1)$. The theorem follows.

Theorem 3 *The total worst-case competitive ratio* $CR^{Worst}_{Biinterval}$ *of Bi-interval for N transactions is* $O(\log(N))$.

Proof In the worst-case, $I_r = k$. This means that there are no consecutive read intervals. Thus, $\text{makespan}^{\text{N}}_{\text{OPT}}$ and $\text{makespan}^{\text{N}}_{\text{Biinterval}}$ satisfy the following, respectively:

$$makespan^N_{OPT} = \Gamma_{N-k+1} + \min d(u, V_{T_{N-k+1}}) \tag{1}$$

$$makespan^N_{Biinterval} = \Gamma_{N-1} + \log(N-1)\max d(u, V_{T_{N-1}}) \tag{2}$$

Hence, $CR^{Worst}_{Biinterval} \leq \log(N-1)$. The theorem follows.

We now focus on the case $I_r < k$.

Theorem 4 *When* $I_r < k$, *Bi-interval improves the traveling makespan (i.e.,* $makespan^N_{Biinterval}(NR)$*) as much as* $O(|\log(1 - (\frac{k-I_r}{N-1})|)$ *for k reads of N transactions.*

Proof

$$\begin{aligned} &\max \frac{\eta(u, V_{T_{N+I_r-k-1}})}{d(u, V_{T_{N-1}})} \\ &= \max\left(\frac{\eta(u, V_{T_{N-1}})}{d(u, V_{T_{N-1}})} + \frac{\varepsilon}{d(u, V_{T_{N-1}})}\right) \\ &\leq \frac{1}{2}\log(N - k + I_r - 1) + \frac{1}{2} \end{aligned} \tag{3}$$

When $I_r < k$, a read interval has at least two read transactions. We are interested in the difference between $\eta(u, V_{T_{N-1}})$ and $\eta(u, V_{T_{N+I_r-k-1}})$. Thus, we define ε as the difference between two η values.

$$\max \frac{\varepsilon}{d(u, V_{T_{N-1}})} \leq \frac{1}{2}\log\left(\frac{N - k + I_r - 1}{N-1}\right) \tag{4}$$

In (4), due to $I_r < k$, $\frac{N-k+I_r-1}{N-1} < 1$. Bi-interval is invoked after conflicts occur, so $N \neq k$. Hence, ε is a negative value, improving the traveling makespan. The theorem follows.

The average-case analysis (or, probabilistic analysis) is largely a way to avoid some of the pessimistic predictions of complexity theory. Bi-interval improves the competitive ratio when $I_r < k$. This improvement depends on the size of I_r—i.e., how many read transactions are consecutively arranged. We are interested in the size of I_r when there are k read transactions. We analyze the expected size of I_r using probabilistic analysis. We assume that k read transactions are not consecutively arranged (i.e., $k \geq 2$) when N requests are arranged according to the nearest neighbor algorithm. We define a probability of actions taken for a given distance and execution time.

Theorem 5 *The expected number of read intervals $E(I_r)$ of Bi-interval is* $\log(k)$.

Proof The distribution used in the proof of Theorem 5 is an independent uniform distribution. p denotes the probability for k read transactions to be consecutively arranged.

$$\begin{aligned} E(I_r) &= \int_{p=0}^{1} \sum_{I_r=1}^{k} \binom{k}{I_r} \cdot p^k (1-p)^{k-I_r} dp \\ &= \sum_{I_r=1}^{k} \left(\frac{k!}{I_r! \cdot (k-I_r)!} \int_{p=0}^{1} p^k (1-p)^{k-I_r} dp \right) \\ &\approx \sum_{I_r=1}^{k} \frac{k!}{I_r!} \cdot \frac{k!}{(2k - I_r + 1)!} \approx \log(k) \end{aligned} \tag{5}$$

We derive Eq. 5 using the beta integral. The theorem follows.

Theorem 6 *Bi-interval's total average-case competitive ratio ($CR^{Average}_{Biinterval}$) is $\Theta(\log(N-k))$ for k reads of N transactions.*

Proof We define $CR^m_{Biinterval}$ as the competitive ratio of node m. $CR^{Average}_{Biinterval}$ is defined as the sum of $CR^m_{Biinterval}$ of $N + E(I_r) - k + 1$ nodes.

$$\begin{aligned} CR^{Average}_{Biinterval} &\leq \sum_{m=1}^{N+E(I_r)-k+1} CR^m_{Biinterval} \\ &\leq \log(N + E(I_r) - k + 1) \approx \log(N-k) \end{aligned}$$

Since $E(I_r)$ is smaller than k, $CR^{Average}_{Biinterval} = \Theta(\log(N-k))$. The theorem follows.

3.4 Evaluation

We developed a set of four distributed applications as benchmarks (Fig. 4). These include distributed versions of the Vacation benchmark of the Stanford STAMP

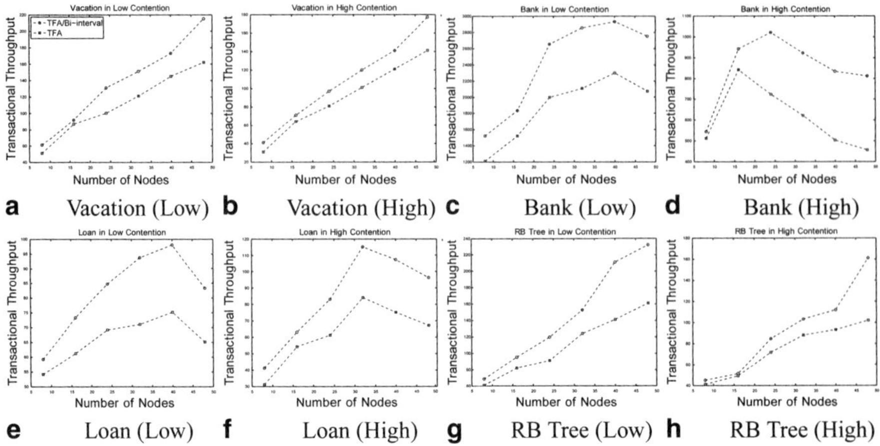

Fig. 4 Throughput under four benchmarks in low and high contention

(multiprocessor STM) benchmark suite [18], two monetary applications (Bank and Loan). and Red/Black Tree (RB-Tree) [11] as microbenchmarks. We created 10 objects, distributed them equally over the 48-nodes, and executed hundred transactions at each node. We used low and high contention levels, which are defined as 90 % read transactions and 10 objects, and 10 % read transactions and 5 objects, respectively.

A transaction's execution time consists of inter-node communication delay, serialization time, and execution time. Communication delay between nodes is limited to a number between 1 and 10 ms to create a static network. Serialization delay is the elapsed time to ensure correctness of concurrent transactions. This delay also includes waiting time in a scheduling queue and Bi-interval's computational time.

In low contention, Bi-interval produces high concurrency due to the large number of read-only transactions. In high contention, Bi-interval reduces object moving time. In both cases, Bi-interval improve throughput, but concurrency of read-only transactions improves more throughput than reduced object moving time. Our experimental evaluation shows that Bi-interval enhances throughput over TFA as much as $1.77 \sim 1.65\times$ speedup under low and high contention, respectively.

4 Cluster-Based Transactional Scheduler

4.1 Motivation

Directory-based CC protocols (e.g., Arrow and Ballistic) [8, 12] in the single-copy model often keep track of the single writable copy. In practice, not all transactional requests are routed efficiently; possible locality is often overlooked, resulting in high communication delays. A distributed transaction consumes more execution time, which include the communication delays that are incurred in requesting and

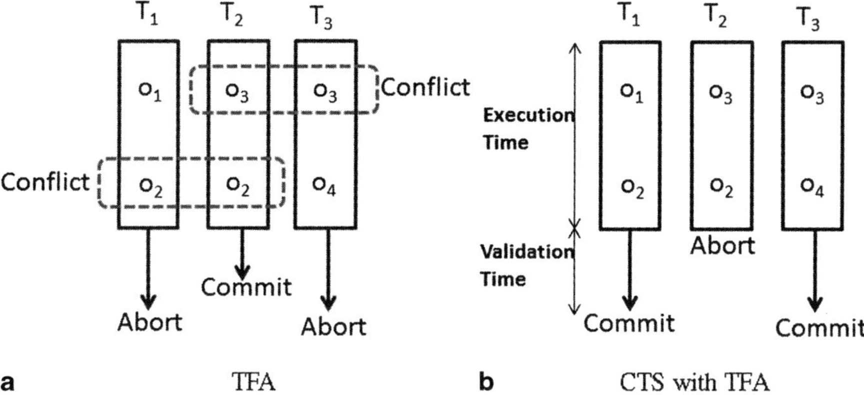

Fig. 5 Executing T_1, T_2, and T_3 concurrently

retrieving objects than a transaction on multiprocessors [15]. Thus, the probability for conflicts and aborts is higher.

Even though a transaction in a full replication model does not request and retrieve objects, maintaining replicas of all objects at each node is costly. Increasing locality (and availability) by brute-force replication while ensuring one-copy serializability [2] can lead to communication overhead. Motivated by this, we consider a k-cluster-based replication model for cc DTM. In this model, multiple copies of each object are distributed to k selected nodes to maximize locality and availability and to minimize communication overhead.

Moreover, a transaction may execute multiple operations with multiple objects, increasing the possibility of conflicts. Figure 5 shows a scenario two conflicts occurring with three concurrent transactions, T_1, T_2, and T_3 using two objects. Under TFA, a conflict over o_2 between T_1 and T_2 occurs and another conflict over o_3 between T_2 and T_3 occurs. If T_2 commits first, T_1 and T_3 will abort because T_2 will update o_3 and o_2 even though T_1 and T_3 do not contend. If T_2 aborts as shown in Fig. 5b, T_1 and T_3 will commit. Motivated by this, CTS aborts T_2 in advance and allows T_1 and T_2 to commit concurrently. A contention manager resolves a conflict between two transactions, but CTS avoids two conflicts among three transactions and guarantees the concurrency of two transactions of them.

4.2 Scheduler Design

In the case of an off-line scheduling algorithm (all concurrent transactions are known), a simple approach to minimize conflicts is to check the *conflict graph* of transactions and determine a *maximum independent set* of the graph, which is *NP-complete*. However, as an on-line scheduling algorithm, CTS checks for conflicts between a transaction and other ongoing transactions accessing an object whenever the transaction requests the object.

Let node n_x belong to cluster z. When transaction T_x at node n_x needs object o_y for an operation, it sends a request to the object owner of cluster z. When another transaction may have requested o_y but no transaction has validated o_y, there are two possible cases. The first case is when the operation is read. In this case, o_y is sent to n_x without enqueuing, because the read transaction does not modify o_y. In the second case, when the operation is write, CTS determines whether o_y is sent to the requester (i.e., n_x) or not by considering previously enqueued transactions and objects. Once CTS allows T_x to access o_y, CTS moves x and y representing T_x and o_y respectively to two scheduling queues. The object owners for each cluster maintain the following two queues, $\mathbb{O}$ and $\mathbb{T}$. Let $\mathbb{O}$ denote the set of enqueued objects and $\mathbb{T}$ denote the set of transactions enqueued by the object owners. If the object owner of cluster z enqueues x and y, it updates its scheduling queues to the other object owners'.

If $x \in \mathbb{T}$ and $y \notin \mathbb{O}$, x and y are enqueued and o_y is sent to n_x. This case indicates that T_x has requested another object from the object owner and o_y has not been requested yet. However, if $x \notin \mathbb{T}$ and $y \in \mathbb{O}$, CTS has to check for whether $\mathbb{T} \mid \beta$ includes more than two transactions or not, where $\beta = \mathbb{O} \mid \alpha$ and $\alpha = \mathbb{T} \mid y$. $\mathbb{O} \mid \alpha$ indicates objects requested by T_α and $\mathbb{T} \mid y$ represents transactions requesting o_y. This case shows when o_y is being used by other transactions and the transactions share an object with another transaction. CTS does not consider a conflict between two transactions because a contention manager aborts one of them when they validate. Thus, the transactions involved in $\mathbb{T} \mid y \cap \mathbb{T} \mid \beta$ abort, x and y are enqueued, and o_y is sent to n_x. The aborted transactions are dequeued.

If $x \in \mathbb{T}$ and $y \in \mathbb{O}$, CTS has to check for whether $\mathbb{T} \mid \gamma$ is distinct from $\mathbb{T} \mid y$ or not, where $\gamma = \mathbb{O} \mid x$. This case means that T_x has requested an object requested by another transaction and also o_y has been requested by another transaction. If two different transactions are using different objects that T_x has requested and is requesting, respectively, CTS aborts T_x to protect two transactions from aborting. Thus, if $\mathbb{T} \mid \gamma$ is distinct from $\mathbb{T} \mid y$, x and y also are enqueued and o_y is sent to n_x. Otherwise, o_y will not be sent to n_x, aborting T_x. In this case, the object owner knows that T_x aborts. Thus, the objects that T_x has requested will be sent to n_x after the objects are updated.

Figure 6 illustrates an example of CTS after applying the three-clustering algorithm on a six-node network. The black circles represent object owners. The scheduling queue includes live transactions T_1 and T_2, and each transaction indicates its objects in use. If T_3 requests o_3, CTS checks for conflicts between T_3 and the enqueued transactions (i.e., T_1 and T_2). CTS aborts T_2 because of two conflicts among T_1, T_2 and T_3. T_2 restarts after T_1 and T_3 commit. The committed transactions are dequeued, and T_2 is enqueued.

We consider two effects of CTS on clusters. First, when a transaction requests an object, CTS checks for conflicts between the transaction and the previous requesting transactions and aborts some transactions in advance to prevent other transactions from aborting. This results in a reduced number of aborts. Second, in TFA, if a transaction aborts, the transaction will restart and request an object again, incurring

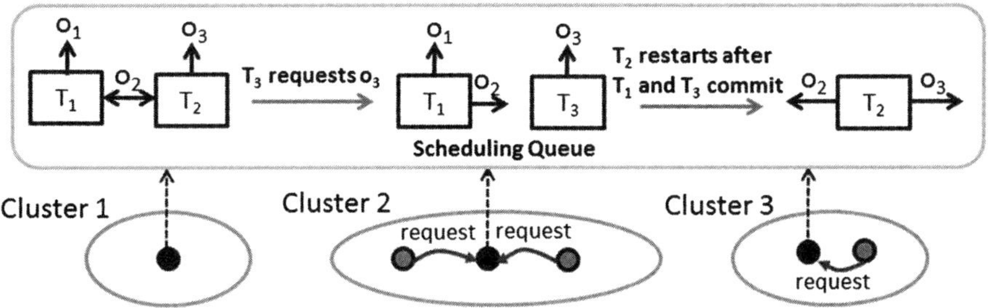

Fig. 6 An example of CTS

communication delays. However, in CTS, object owners hold aborted transactions. When validation of an object completes, the object is sent to the nodes invoking the aborted transactions. Thus, CTS lets the aborted transactions use newly updated objects without requesting the object again, reducing communication delays.

4.3 Analysis

In the worst case, N transactions are simultaneously invoked to update an object. Whenever a conflict occurs between two transactions, let scheduler B abort one of these and enqueue the aborted transaction (to avoid repeated aborts) in a distributed queue. The aborted transaction is dequeued and restarts after a backoff time. Let the number of aborts of T_i be denoted as λ_i. We have the following lemmas.

Lemma 1 *Given scheduler B and N transactions,* $\sum_{i=1}^{N} \lambda_i \leq N - 1$.

Proof Given a set of transactions $T = \{T_1, T_2, \cdots T_N\}$, let T_i abort. When T_i is enqueued, there are ζ_i transactions in the queue. T_i can only commit after ζ_i transactions commit if ζ_i transactions have been scheduled. Hence, if a transaction is enqueued, it does not abort. Thus, one of N transactions does not abort. The lemma follows.

Lemma 2 *Given scheduler B and N transactions,* $makespan_B^N(NR) \leq 2(N-1)\sum_{i=1}^{N-1} d(n_i, n_j) + \Gamma_N$.

Proof Lemma 1 gives the total number of aborts on N transactions under scheduler B. If a transaction T_i requests an object, the communication delay will be $2 \times d(n_i, n_j)$ for both requesting and object retrieving times. Once T_i aborts, this delay is incurred again. To complete N transactions using scheduler B, the total communication delay will be $2(N-1)\sum_{i=1}^{N-1} d(n_i, n_j)$. The theorem follows.

Lemma 3 *Given scheduler B, N transactions, k replications,* $makespan_B^N(PR) \leq (N-k)\sum_{i=1}^{N-k} d(n_i, n_j) + (N-k+1)\sum_{i=1}^{N-1}\sum_{j=1}^{k-1} d(n_i, n_j) + \Gamma_N$.

Proof In PR, k transactions do not need to remotely request an object, because k nodes hold replicated objects. Thus, $\sum_{i=1}^{N-k} d(n_i, n_j)$ is the requesting time of N transactions and $\sum_{i=1}^{N-1} \sum_{j=1}^{k-1} d(n_i, n_j)$ is the validation time based on atomic multicasting for only k nodes of each cluster. The theorem follows.

Lemma 4 *Given scheduler B and N transactions,* $makespan_B^N(FR) \leq \sum_{i=1}^{N-1} \sum_{j=1}^{N-1} d(n_i, n_j) + \Gamma_N$.

Proof Transactions request objects from their own nodes, so their requesting times do not occur in FR, even when the transactions abort. The basic idea of transactional schedulers is to minimize conflicts through enqueueing transactions when the transactions request objects. Thus, the transactional schedulers (i.e, B and CTS) do not affect $makespan_{x\in\{B,CTS\}}^N(FR)$. Thus, when a transaction commits, FR takes $\sum_{i=1}^{N-1} \sum_{j=1}^{N-1} d(n_i, n_j)$ for only atomic broadcasting to support one-copy serializability.

Theorem 7 *Given scheduler B and N transactions,* $makespan_B^N(FR) \leq makespan_B^N(PR) \leq makespan_B^N(NR)$.

Proof Given k PR, $\lim_{k\to 1} makespan_B^N(PR) \leq 2(N-1)\sum_{i=1}^{N-1} d(n_i, n_j) + \Gamma_N$, and $\lim_{k\to N} makespan_B^N(PR) \leq \sum_{i=1}^{N-1} \sum_{j=1}^{N-1} d(n_i, n_j) + \Gamma_N$. The theorem follows.

Theorem 8 *Given N transactions and M objects, the RCR of schedulers CTS on PR and scheduler B on FR is less than 1, where* $N > 3$.

Proof Let $\sum_{i=1}^{N-1} d(n_i, n_j)$ denote δ_{N-1}. To show that the RCR of CTS on PR and B on FR is less than 1, $makespan_{CTS}^N(PR) < makespan_B^N(FR)$. CTS detects potential conflicts and aborts a transaction incurring the conflicts. The aborted transaction does not request objects again. Thus we derive $makespan_{CTS}^N(PR) \leq 2M\delta_{N-k} + M\sum_{i=1}^{N-1} \delta_{k-1} + M\Gamma_N$. $2\delta_{N-k} + (N-1)\delta_{k-1} \leq (N-1)\delta_{N-1}$, so that $2\delta_{N-k} \leq (N-1)\delta_{N-k}$. Only when $N \geq 3$, PR is feasible. Hence, $makespan_{CTS}^N(PR) < makespan_B^N(FR)$, where $N > 3$. The theorem follows.

Theorem 8 shows that CTS in PR performs better than FR. Even though PR incurs requesting and object retrieving times for transactions, CTS minimizes these times, resulting in less overall time than the broadcasting time of FR.

4.4 Evaluation

To select k nodes for distributing replicas of each object, we group nodes into clusters, such that nodes in a cluster are closer to each other, while those between clusters are far apart. Recall that the distance between a pair of nodes in a metric-space network determines the communication cost of sending a message between them. We use a k clustering algorithm based on METIS [13], to generate k clusters with small intra-cluster distances i.e., k nodes may hold the same objects. Our partial replication relies on the usage of a *total order multicast* (TOM) primitive to ensure

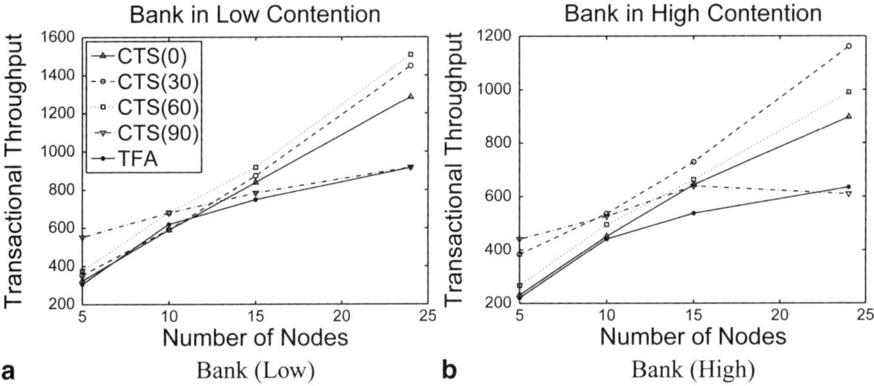

a Bank (Low) **b** Bank (High)

Fig. 7 Throughput of bank benchmark with no node failures

agreement on correctness in a genuine multicast protocol [22]. The object owners for each cluster update objects through a TOM-based protocol.

Our experiments were conducted on 24-node testbed. Each node is an AMD Opteron processor clocked at 1.9 GHz. We use Ubuntu Linux 10.04 server OS and a network with a private gigabit ethernet. Each experiment is the average of 10 repetitions. The number of objects for a transaction is selected randomly from 2 to 20. We considered CTS(30) and CTS(60), meaning CTS over 30 and 60 % object owners of the total nodes, respectively. For instance, CTS(30) under 10 nodes means CTS over three-clustering algorithm.

Figure 7 intends to show two effects of scheduling by CTS and the improvement of object availability by increasing object locality. To show the effectiveness of CTS, TFA is compared with CTS(0)—the combination of CTS and TFA with no replication. CTS(0) improves throughput over TFA as much as 1.5× under high contention because the number of conflicts decreases. CTS(0) outperforms CTS(90) in throughput, but it is non-fault-tolerant. The throughput produced by CTS(90) is degraded due to the large number of broadcasting messages needed to update all replicas. Due to high object availability on CTS(90), the requesting times of aborted transactions are less reduced. Meanwhile, due to low object availability on CTS(0), the requesting times are more reduced but object retrieving times increase. Thus, CTS(30) and CTS(60) achieve decreased object requesting and retrieving times, resulting in a better throughput than CTS(0) and CTS(90).

5 Summary and Conclusion

Bi-interval shows that the idea of grouping concurrent requests into read and write intervals to exploit concurrency of read transactions—originally developed in BI-MODAL for multiprocessor TM—can also be successfully exploited for DTM. Doing so poses a fundamental trade-off, however, one between object moving times and concurrency of read transactions. Bi-interval's design shows how this trade-off can be effectively exploited towards optimizing throughput.

CTS uses multiple clusters to support partial replication for fault-tolerance. The clusters are built such that inter-node communication within each cluster is small. To reduce object requesting times, CTS partitions object replicas into each cluster (one per cluster), and enqueues and assigns backoff times for aborted transactions. CTS's design shows that such an approach yields significant throughput improvement.

Data centers have to manage a vast amount of concurrent requests to access data objects. Given our results, distributed transactional schedulers are a viable strategy to ensure object consistency and enhance performance in the data centers.

References

1. Hagit Attiya and Alessia Milani. Transactional scheduling for read-dominated workloads. In *Proceedings of the 13th International Conference on Principles of Distributed Systems*, OPODIS '09, pages 3–17, Berlin, Heidelberg, 2009. Springer-Verlag.
2. Philip A. Bernstein and Nathan Goodman. Multiversion concurrency control—theory and algorithms. *ACM Trans. Database Syst.*, 8:465–483, December 1983.
3. A. Bieniusa and T. Fuhrmann. Consistency in hindsight: A fully decentralized stm algorithm. In *Parallel Distributed Processing (IPDPS), 2010 IEEE International Symposium on*, pages 1–12, 2010.
4. Geoffrey Blake, Ronald G. Dreslinski, and Trevor Mudge. Proactive transaction scheduling for contention management. In *Proceedings of the 42nd Annual IEEE/ACM International Symposium on Microarchitecture*, MICRO 42, pages 156–167, New York, NY, USA, 2009. ACM.
5. Robert L. Bocchino, Vikram S. Adve, and Bradford L. Chamberlain. Software transactional memory for large scale clusters. In *Proceedings of the 13th ACM SIGPLAN Symposium on Principles and practice of parallel programming*, PPoPP '08, pages 247–258, New York, NY, USA, 2008. ACM.
6. N. Carvalho, P. Romano, and L. Rodrigues. A generic framework for replicated software transactional memories. In *Network Computing and Applications (NCA), 2011 10th IEEE International Symposium on*, pages 271–274, aug. 2011.
7. Maria Couceiro, Paolo Romano, Nuno Carvalho, and Luís Rodrigues. D2STM: Dependable distributed software transactional memory. In *Proceedings of the 2009 15th IEEE Pacific Rim International Symposium on Dependable Computing*, PRDC '09, pages 307–313, Washington, DC, USA, 2009. IEEE Computer Society.
8. Michael J. Demmer and Maurice Herlihy. The arrow distributed directory protocol. In *Proceedings of the 12th International Symposium on Distributed Computing*, DISC '98, pages 119–133, London, UK, UK, 1998. Springer-Verlag.
9. David Dice, Ori Shalev, and Nir Shavit. Transactional locking II. In *DISC*, 2006.
10. Shlomi Dolev, Danny Hendler, and Adi Suissa. CAR-STM: scheduling-based collision avoidance and resolution for software transactional memory. In *Proceedings of the twenty-seventh ACM symposium on Principles of distributed computing*, PODC '08, pages 125–134, New York, NY, USA, 2008. ACM.
11. Rachid Guerraoui, Michal Kapalka, and Jan Vitek. STMBench7: a benchmark for software transactional memory. *SIGOPS Oper. Syst. Rev.*, 41(3):315–324, 2007.
12. Maurice Herlihy and Ye Sun. Distributed transactional memory for metric-space networks. In *Proceedings of the 19th international conference on Distributed Computing*, DISC'05, pages 324–338, Berlin, Heidelberg, 2005. Springer-Verlag.

13. George Karypis and Vipin Kumar. A fast and high quality multilevel scheme for partitioning irregular graphs. *SIAM J. Sci. Comput.*, 20:359–392, 1998.
14. Junwhan Kim and Binoy Ravindran. On transactional scheduling in distributed transactional memory ystems. In *Proceedings of the 12th international conference on Stabilization, safety, and security of distributed systems*, SSS'10, pages 347–361, Berlin, Heidelberg, 2010. Springer-Verlag.
15. Junwhan Kim and Binoy Ravindran. Scheduling closed-nested transactions in distributed transactional memory. In *Parallel Distributed Processing Symposium (IPDPS), 2012 IEEE 26th International*, pages 179–188, 2012.
16. Junwhan Kim and Binoy Ravindran. Scheduling transactions in replicated distribute software transactional memory. In *Proceedings of the 2013 13th IEEE/ACM International Symposium on Cluster, Cloud and Grid Computing (CCgrid 2013)*, CCGRID '13, Delft , The Netherlands, 2013. IEEE Computer Society.
17. Christos Kotselidis, Mohammad Ansari, Kim Jarvis, Mikel Luján, Chris Kirkham, and Ian Watson. DiSTM: A software transactional memory framework for clusters. In *Proceedings of the 2008 37th International Conference on Parallel Processing*, ICPP '08, pages 51–58, Washington, DC, USA, 2008. IEEE Computer Society.
18. Chi Cao Minh, Jaewoong Chung, C. Kozyrakis, and K. Olukotun. STAMP: Stanford transactional applications for multi-processing. In *Workload Characterization, 2008. IISWC 2008. IEEE International Symposium on*, pages 35–46, 2008.
19. Daniel J. Rosenkrantz, Richard Edwin Stearns, and Philip M. Lewis II. An analysis of several heuristics for the traveling salesman problem. *SIAM J. Comput.*, 6(3):563–581, 1977.
20. M.M. Saad and B. Ravindran. Transactional forwarding: Supporting highly-concurrent stm in asynchronous distributed systems. In *Computer Architecture and High Performance Computing (SBAC-PAD), 2012 IEEE 24th International Symposium on*, pages 219–226, 2012.
21. Mohamed M. Saad and Binoy Ravindran. Supporting STM in distributed systems: Mechanisms and a Java framework. In *Sixth ACM SIGPLAN workshop on Transactional Computing*, 2011.
22. N. Schiper, P. Sutra, and F. Pedone. P-store: Genuine partial replication in wide area networks. In *Reliable Distributed Systems, 2010 29th IEEE Symposium on*, pages 214–224, 2010.
23. Richard M. Yoo and Hsien-Hsin S. Lee. Adaptive transaction scheduling for transactional memory systems. In *Proceedings of the twentieth annual symposium on Parallelism in algorithms and architectures*, SPAA '08, pages 169–178, New York, NY, USA, 2008. ACM.

Dependability-Oriented Resource Management Schemes for Cloud Computing Data Centers

Ravi Jhawar and Vincenzo Piuri

1 Introduction

A major factor in the growth of the Information and Communications Technology has been the widespread use of data centers for deploying and executing web services, business processes, and scientific and e-commerce applications [12]. While some data centers are designed to operate a specific business (e.g., Google's search engine), others are used as the backbone infrastructure to deliver computing resources as services to hundreds of users (e.g., Amazon's EC2 service). Relying on data centers for running applications, particularly when resources are delivered to the users as a service, offer significant benefits [13, 33]. For example, applications can benefit significantly from the economy of scale, and users are relieved from buying expensive hardware and software licenses and from maintaining the computing infrastructure.

To ensure resource availability, Quality-of-Service (QoS) and dependability to hundreds of applications in the modern data centers under fluctuating workloads, server failures, and network congestion, dedicated servers are allocated to applications and server capacity is often over-provisioned. However, the use of dedicated hardware not only leads to poor energy usage, but also made it difficult to react to system changes. Furthermore, the growing number of under-utilized servers increases the data center's operating costs such as system management, energy consumption of servers, and network and cooling infrastructure costs.

In the last decade, the virtualization technology has emerged as a very effective approach to address these issues by de-coupling physical servers from the resources needed by applications. In particular, virtualization provides an efficient way to insulate and partition server's resources so that only a portion of them can be utilized

R. Jhawar (✉) · V. Piuri
Dipartimento di Informatica, Università degli Studi di Milano,
Via Bramante 65, 26013 Crema, Italy
e-mail: ravi.jhawar@unimi.it

V. Piuri
e-mail: vincenzo.piuri@unimi.it

S. U. Khan, A. Y. Zomaya (eds.), *Handbook on Data Centers*,
DOI 10.1007/978-1-4939-2092-1_45

by an application. It also provides a greater flexibility and control over resource management, allowing for dynamic adjustment of CPU and memory usage, and live migration of virtual machines among physical servers (e.g., [20, 21]). Nowadays, Cloud computing is the most popular approach to create virtualized environments for application execution in distributed data centers.

Deploying virtualized services in data centers, including those based on Cloud computing, create new resource management problems, such as optimal placement of virtual machines. Existing solutions focus mainly on placing and adaptive managing virtual machines in order to (i) reduce energy consumption costs and maximize profits for the data center owner, and (ii) improve the performance and QoS of applications. Only a few approaches are available in the literature to improve dependability—fault tolerance and security—of applications (e.g., [2–5, 7, 9–11, 14, 15]). This chapter aims to investigate resource management schemes that improve dependability (both fault tolerance and security) and performance of applications in virtualized data centers. In particular, we mainly focus on initial virtual machine placement and runtime adaption schemes developed for the Infrastructure-as-a-Service (IaaS) paradigm in Cloud computing data centers.

The remainder of this chapter is organized as follows. Section 2 briefly outlines the failure characteristics of data centers' components. Section 3 introduces the analysis of resource management in data centers by formalizing the virtual machine placement problem and by defining constraints for representing data center owner and users perspectives. In Sect. 4 we discuss initial virtual machine placement approaches that satisfy dependability and performance constraints at the activation of the applications. Finally, in Sect. 5 we discuss runtime adaption schemes that balance performance and availability of applications during system operation with possible occurrence of dependability threats. Section 6 provides some concluding remarks.

2 System Model and Failure Behavior of Data Center Components

To appreciate the resource management schemes which improve dependability and performance of applications deployed in Cloud computing data centers, in this Section we first review the typical architecture of modern data centers and, then, we analyze the behavior of the various data center components, the main causes of failures, and the impact of failures on applications. This is very useful to analyze the efficacy and efficiency of resource management schemes.

2.1 Overview of the Data Center Architecture

A data center interconnects a large number of physical hosts H to form a resource pool that is partitioned into a set of clusters [19]. A cluster C is formed by grouping the hosts that have identical resource characteristics or administrative parameters (e.g., hosts that belong to the same network latency class or geographical location). Each host or server contains multiple processors, storage disks, memory modules and network interfaces. The resource characteristics of each physical host $h \in H$ can be represented using a D dimensional vector T_{max}, where each dimension represents the amount of host's residual capacity corresponding to a distinct resource type (e.g., CPU, memory, storage, network bandwidth). For simplicity, the resource capacity of hosts can be denoted using normalized values between 0 and 1. For example, host h characterized as $\vec{h}$ = (CPU,Mem) = (0.6,0.5) implies that 60 % of CPU, 50 % of memory on h is available for use. A *hypervisor* is deployed on each host to virtualize its resources: virtual machines of required size are created on appropriate hosts to allocate resources for users' applications.

Hosts are connected using several network switches and routers. The most common network topology of data centers is as follows [19]. Physical hosts, servers or racks of servers are first connected to a Top of Rack switch (ToR), which is in turn connected to two (primary and backup) aggregation switches (AggS). The subsystem formed by the group of servers under an aggregate switch can be viewed as a cluster. An AggS then connects tens of ToRs to redundant access routers (AccR). Thus, each AccR handles traffic from thousands of servers and routes it to Core routers that connect different data centers to the Internet. All links in the data centers commonly use Ethernet as the link layer protocol and redundancy is applied to all network components at each layer in the network topology (except for ToRs). In addition, redundant pairs of load balancers (LBs) are connected to each AggS and mapping between static IP address presented to the users and dynamic IP addresses of internal servers that process user's requests is performed.

Servers and network components are subject to failures during their life, thus affecting the correct operation of applications. In this Section, we review their typical failure behavior in order to understand the redundancy that is needed to overcome critical situations and ensure successful completion of applications.

2.2 *Failure Behavior of Servers*

Let's consider first the servers' life and their possible failures. Vishwanath and Nagappan [37] have studied server failures and hardware repair behavior using a large collection of servers (nearly 100,000 servers) and corresponding data on part replacement. For example, they mine data relevant to server configuration, when a hard disk has been marked for replacement and when it has been actually replaced, to generate statistical reports on failure behavior. The data repository for their study includes server collection spanning multiple data centers distributed across different countries. Some interesting outcomes of this study are as follows:

- The annual failure rate (AFR) of servers is approximately 8 %, and the average number of repairs per each failure-prone server is 2. The remaining 92 % of servers do not see any repair events (20 repair/replacement events have been identified in 9 machines over a 14 months period).

- For 8 % AFR, repair costs that amount to 2.5 million dollars are approximately spent for 100,000 servers.
- Hard disks are the most failure-prone hardware components and the most significant reason behind server failures. In particular, about 78 % of total faults/replacements have been detected on hard disks, 5 % on RAID controllers, and 3 % due to memory failures. 13 % of replacements have been due to a collection of components (not particularly dominated by a single component failure).
- About 5 % of servers experience a disk failure in less than 1 year from the date when it is commissioned (young servers), 12 % when they are 1 year old, and 25 % when they are 2 years old.
- Comparing the number of repairs per machine (RPM) against the number of disks per server in a group of servers (clusters) indicates that (i) there is a relationship in the failure characteristics of servers that have already experienced a failure, and (ii) the number of RPM has a correspondence to the total number of disks on that machine.

This analysis can be used to develop robust fault tolerance mechanisms (e.g., by improving the reliability of hard disks to substantially reduce the number of failures) and resource management schemes (e.g., to improve the availability of a given application, its tasks must not be allocated on the server whose hard disks have already experienced a failure).

2.3 *Failure Behavior of Network Components*

Let's consider now the possible network failures. Gill et al. [16] have performed a large scale study in data centers. A link failure happens when the connection between two devices on a specific interface is down. A device failure happens when the device is not routing/forwarding packets correctly (e.g., due to power outage or hardware crash). Some interesting outcomes from this study are as follows:

- The overall data center network reliability is about 99.99 % for 80 % of the links and 60 % of devices.
- Among all the network devices, ToRs are most reliable (with a failure rate of less than 5 %) and load balancers are least reliable (with failure probability of 1 in 5). The root causes for failures in LBs are dominantly software bugs and configuration errors; moreover, LBs demonstrate frequent but short failures. This observation clearly indicates that low-cost commodity switches such as ToRs and AggS provide sufficient reliability.
- The links forwarding traffic from LBs have highest failure rates; links higher in the topology (e.g., connecting AccRs) and links connecting redundant devices have second highest failure rates.

- Network redundancy reduces the median impact of failures (in terms of number of lost bytes) by only 40 %, as opposed to the common belief that redundancy almost completely masks failures from applications.
- The estimated median number of packets lost during a failure is 59K and median number of bytes is 25MB (average size of lost packets is 423Bytes). Based on prior measurement studies (that observe packet sizes to be bimodal with modes around 200Bytes and 1400Bytes), it is estimated that most lost packets belong to the lower-level functions (e.g., ping messages or ACKs).

2.4 *Analysis of the Impact of Failures on Applications*

A detailed analysis of the impact of various component failures on applications as well as the definition of the impact boundary of a given failure (e.g., fault region) are useful to improve the dependability of applications [18, 25, 28, 36]. In [27] such analysis has been integrated within the resource management schemes to provide fault tolerance support to applications in the Cloud computing paradigm. For example, if a switch failure disconnects a rack of servers, then replicas of a given applications can be allocated in different clusters in order to increase the application's failure independence.

Jhawar and Piuri [23] use the notion of fault trees and provide a hierarchical model to analyze the impact of component failures. As an example, let's consider the impact of power failures (other failures can similarly be analyzed in a straightforward manner). Assume that a data center receives power via an uninterrupted power network, and a redundant distribution unit (DU) is deployed for each cluster within the data center. A DU hence provides power to all the servers within a cluster. In this context, a failure in the DU is independent of other DUs and the central power supply. The fault tree for power failures include the conditions ($Power1 \wedge Power2$) for redundant power units of a server, ($DU1 \wedge DU2$) for a cluster, and ($Power1 \wedge Power2$) $\vee$ ($DU1 \wedge DU2$) $\vee$ ($Central\ Power\ Supply$) for the power failure of the whole system. An application failure happens when this latter condition evaluates *true*. We note that the boundary (server, cluster, data center) of the impact of each failure can also be identified using this approach.

The fault tree model can also be extended to understand the failure behavior of applications by integrating it with Markov chains, where the probability values permit to quantitatively analyze the fault tolerance of the given application in the envisioned data center. Building on the notion of fault trees and Markov chains, Jhawar and Piuri [23] identify three main Deployment Levels DL for applications. A deployment level is the smallest subsystem within the infrastructure where the replicas of a users' application are deployed to ensure dependability. Deployment levels correspond to fault regions in the data center, and describe how replicas of applications can be allocated, depending on its fault tolerance, performance, availability, and reliability goals. The three deployment levels are as follows.

- *Multiple machines within a cluster*: Replicas of an application can be placed on hosts that are connected by a ToR switch, that is, in a LAN. This deployment provides benefits in terms of low latency and high bandwidth but offers least failure independence. Replicas cannot communicate and execute the dependability protocol upon a single switch failure, or a failure in the power distribution unit results in an outage of the entire application.
- *Multiple clusters within a data center*: Replicas of an application can be placed on hosts that belong to different clusters in the same data center, that is, connected via a ToR switch and AggS. This deployment still provides moderate benefits in terms of latency and bandwidth, and offers higher failure independence. The replicas are not bound to an outage with a single power distribution or switch failure.
- *Multiple data centers*: Replicas of an application can be placed on hosts that belong to different data centers, that is, connected via a ToR switch, AggS and AccR. This deployment has a drawback with respect to high latency and low bandwidth, but offers a very high level of failure independence. A single power failure has least effect on the availability of the application.

A partially-ordered hierarchy can be defined between the deployment levels. For example, a data center is a larger subsystem or deployment level when compared to a cluster. A transitive closure indicating that "contains-in" relationship also exists on the hierarchy of deployment levels. For example, a host can be part of a cluster that in turn exists in the data center. Intuitively, availability (failure independence) of an application increases with the increase of the deployment level; that is, the availability of an individual host is smaller than the availability of a cluster, which is still smaller than the availability of a data center. On the other hand, the network latency increases with the increase of the deployment level; that is, hosts in the same rack have lower network latency than hosts across different clusters. Hence, if $Lat(DL)$ denotes the maximum latency between two hosts in the deployment level DL, then the virtual machine placement algorithm can decide a suitable DL based on users desired performance goals (e.g., in terms of the expected response time). In the next Sections, we discuss how to integrate this analysis within resource management schemes to improve fault tolerance, security, and performance of applications.

3 Resource Management in Data Center Environments

The problem of resource management in data center environments has gained a wide attention from the research community and the industry. In this Section, we model the resource management problem with specific reference to improving dependability and performance of users applications deployed in the data centers using virtual machines.

Resource management for Cloud-based data centers can be modeled as a placement problem in which virtual machines are allocated on the data center's hosts [20, 21, 24], having been the applications mapped on the appropriate virtual machine templates available in the data center environment. Existing solutions, particularly for services that provision on-demand resources to users, primarily focus on making virtual machine *placement decisions* at two distinct levels: (i) *initial* virtual machine *allocation* and (ii) *runtime adaption* of current virtual machine allocation [30].

Based on user's requirements and failure characteristics of the envisioned data center, a set of dependability and performance constraints are specified (e.g., constraint specifying that replicas of the user's application be allocated on two different physical hosts to avoid single points of failure). The initial resource allocation process identifies the physical hosts on which the requested virtual machines can be allocated such that all the placement constraints are satisfied. Once the required virtual machines are created and delivered to the user, the runtime adaption process monitors the system and resizes virtual machines or migrates them to other physical hosts in order to meet the predefined goals (e.g., energy conservation), while satisfying the placement constraints. While the objective of most resource management algorithms in this context has been to maximize the service provider's goals (e.g., through resource consolidation, load balancing, satisfaction of SLAs), we will provide a broader perspective which encompasses both the provider's and the users' views, balancing all needs in a comprehensive way.

In this Section, we formulate and categorize various placement constraints that can be used to specify dependability—fault tolerance and security—and performance related conditions during resource management. Within this framework, in Sect. 4 we will study resource management schemes for placing applications in the Cloud computing environment at the initial deployment. Then in Sect. 5 we will discuss dynamic adaptation of the applications placement to deal with changing working status of the architecture components, balancing dependability and performance of users' applications.

Let us start by defining a mapping function $p : V \to H$ that maps each virtual machine $v \in V$ on a physical host $h \in H$ in the data center. Notation $p(v) = h$ denotes that virtual machine v has been allocated on physical host h.

Placement constraints can be distinguished in three main categories.

- *Global constraints* that must be satisfied across all the hosts and virtual machines in the data center. These constraints essentially specify the conditions necessary to maintain a consistent system state.
- *Infrastructure-oriented constraints* that are specified by the data center owners and service providers who build their services on top of the data center network (e.g., a Cloud-based IaaS provider). These constraints aim to ensure security and quality of service.
- *Application-oriented constraints* that are specified by the users who use data center's resources to execute their applications (e.g., application owners who use the IaaS service). These constraints allow users to specify conditions relevant to the specific configuration of their dependability mechanisms, and consequently, to improve availability, reliability, and response time of their applications.

These constraints have been introduced and formalized in [23]. Other solutions also formulate the resource allocation problem by including one or more constraints discussed in this Section (e.g., [1, 6, 32]). Some additional constraints have also been considered in the past, depending on the specific formulation of the allocation problem. For example, Shi et al. [32] include FULL and SEC constraints: FULL specify that either all or none of the virtual machines in a given request must allocated, while SEC requires that a given physical host can be assigned only the virtual machines from the same request and no other request. Similarly, Zheng et al. [ZLX2013] include constraints relevant to the maintenance schedule of the physical hosts. In particular, they include (i) a constraint that confines each host to finish its maintenance activity before a specified deadline and (ii) a constraint specifying that any host does not execute any maintenance activity after its deadline.

For simplicity, in the rest of this chapter, we define constraints using virtual machine and host identifiers. However, these constraints can also be specified in a straight-forward manner for groups of virtual machines or hosts, clustered by means of some of their properties (e.g., all the hosts that belong to a given cluster or a deployment level).

3.1 Global Constraints

Global constraints include the classical resource capacity constraints. Similarly to the representation of the resource characteristics $\vec{h} = (h[1], \ldots, h[d], \ldots, h[D])$ of a host $h \in H$, the amount of each resource type d in a virtual machine $v \in V$ can be represented by $\vec{v} = (v[1], \ldots, v[d], \ldots, v[D])$. The resource capacity constraint states that the amount of resources consumed by all the virtual machines mapped on a single physical host cannot exceed the total capacity of that host in any dimension d. In general, this constraint is formulated as follows: for all the virtual machines $v \in V$ and physical hosts $h \in H$ in the system, mapping function $p : V \rightarrow H$ must satisfy

$$\forall\, h \in H,\ d \in [1..D] \qquad \sum_{v \in V \mid p(v)=h} v[d] \leq h[d]$$

This constraint is taken into account by most solutions existing in the literature, and is necessary to ensure that the data center operates correctly. Some variants of this constraint have also been considered in the literature, depending on the problem context. For example, solutions for server consolidation require that the amount of resources consumed by a virtual machine when placed in isolation on a host or with other co-hosted virtual machines may be different. When multiple virtual machines are placed on a host, several memory pages can be shared between them thus reducing the overall memory requirements; conversely, the hypervisor or host operating system may consume additional CPU cycles or I/O bandwidth for resource scheduling. Similarly, virtual machines may interfere with each other and consume higher amounts of shared resources (e.g., the L2 cache during context switching).

To avoid performance degradation and inconsistent system state due to the above factors, the data center owner can define an upper bound on the resource capacity of each host that can be used by users' virtual machines. The resource capacity constraint then ensures that virtual machines are allocated on a host only if its capacity in any dimension does not exceed the upper bound or *threshold* value, that is,

$$\forall\, h \in H,\ d \in [1..D] \qquad \sum_{v \in V \mid p(v)=h} v[d] \leq (h[d] * threshold[d])$$

The $threshold[d]$ can be specified using percentage or normalized values between 0 and 1.

3.2 Infrastructure-Oriented Constraints

The data center owner may need to impose restrictions on the mapping function to improve the security, operational performance and reliability of the data center. While a number of conditions can be introduced depending on the specific system architecture, here we discuss two representative infrastructure-oriented constraints: *forbid* and *count*.

Forbid To improve security, the data center owner may need to dedicate a set of hosts only to execute system-level services (e.g., the access control engine or reference monitor) and, as a consequence, it may need to specify that the mapping function does not allocate users' virtual machines on those hosts. In this context, the forbid constraint prevents a virtual machine v from being allocated on a physical host h. When the data center owner defines a set $Forbid = \{(v_i, h_j) | \, v_i \in V,\ h_j \in H\}$ specifying the virtual machines $v_i \in V$ that must be forbidden from being allocated on hosts $h_j \in H$, the allocation algorithm guides the mapping function $p : V \rightarrow H$ to satisfy the following condition:

$$\forall\, v \in V,\ h \in H \qquad (v, h) \in\ Forbid \Rightarrow p(v) \neq h$$

Count The performance of a host degrades as the number of virtual machines co-hosted on it increases. For example, the performance of a storage disk decreases if the number of I/O intensive applications in the virtual machines increases; similarly, the network traffic from a host, virtual machine management costs, and CPU utilization costs gradually increase. To avoid such conditions, the count constraint allows the data center owner to limit the number of virtual machines that can be allocated on a given host. When the data center owner defines $count_h$ as the maximum number of virtual machines allowed on host h, the mapping function $p : V \rightarrow H$ ensures that the following condition is satisfied:

$$\forall\, v \in V,\ h \in H \qquad |\{v \in V\}\colon p(v) = h\}| \ \leq\ count_h$$

3.3 Application-Oriented Constraints

Based on the dependability policy, users may need to impose a set of restrictions on the placement of their virtual machines in the data center. For example, suppose that a user applies a replication technique to increase the reliability and availability of her application. She may then need to impose a set of conditions on the system parameters and relative placement of her virtual machines to correctly implement the dependability policy while satisfying her performance goals. We discuss three representative application-oriented constraints that can be used to realize the aforementioned conditions: *restrict*, *distribute* and *latency*.

Restrict To ensure survival and possible continuous operation of an application even in the case of failures, it can be replicated in the data center so as to have at least one replica able to proceed in the application activities. A user may wish to allocate each replica of her application in a specific region (e.g., cluster, availability zone) of the data center. To achieve this, the user may leverage the restrict constraint which limits a virtual machine $v \in V$ on being allocated only on a specified group of physical hosts $H' \in H$. We note that the analysis of the failure behavior of the data center components may be beneficial in this context since each replica can be placed in a different failure zone or deployment level (see Sect. 2.4), thus increasing the failure independence of the application replicas. When a user defines the set $Restr = \{(v_i, H'_j) \,|v_i \in V \wedge H'_j \subseteq H\}$, the mapping function $p : V \to H$ ensures the following condition:

$$\forall\, v_i \in V, H'_j \in 2^H \qquad (v_i,\ H'_j) \in Restr \Rightarrow p(v_i) \in H'_j$$

The restrict constraint is also beneficial in other scenarios such as enforcement of privacy policies and improvement of applications performance. For example, a user may leverage the restrict constraint to satisfy mandatory government enforced obligations (e.g., EU Data Protection 95/46/EC Directive) requiring virtual machines to be always located within a given community area (e.g., within EU countries). Similarly, to improve the application's performance, a user may require the mapping function to place her virtual machines on the hosts whose geographical location is closest to her customers.

Distribute A replication-based fault tolerance scheme inherently requires that each replica be placed on different physical hosts in order to avoid single points of failure. If a user replicates her application on two virtual machines, and if both the virtual machines are allocated on the same host, then a failure in the host results in an unavailability of the user's application. To avoid such situations, the distribute constraint allows a user to specify that two virtual machines v_i and v_j must never be allocated on the same host at the same time. Given the set $Distr = \{(v_i,\ v_j)|v_i,\ v_j \in V' \subseteq V\}$ of pairs of virtual machines that cannot be deployed on the same host, the mapping function $p : V \to H$ ensures the following:

$$\forall\, v_i,\ v_j \in V' \subseteq V, h \in H \qquad (v_i,\ v_j) \in Distr \Rightarrow p(v_i) \neq p(v_j)$$

Latency To ensure that the response time of an application does not exceed a maximum value, a user may want to specify the latency allowed between each application replica. For instance, in a checkpoint-based reliability mechanism, the state of the backup virtual machine instance must be frequently updated with that of the primary instance to maintain the system in a consistent state. This task involves high amounts of message exchanges, and hence an upper bound in the network delay is essential; otherwise, the wait-time of the primary instance during which the state transfer to the backup takes place may increase significantly and the overall availability of the application may be reduced.

For simplicity's sake, let's assume that the network latency between two virtual machines is equal to the network latency between the physical hosts on which they are deployed. The latency constraint forces the mapping function to allocate two virtual machines $v_i,\ v_j \in V$ such that the network latency $latency(p(v_i),\ p(v_j))$ between them is less than a specified value T_{max}. Given the set $MaxLatency = \{(v_i,\ v_j,\ T_{max}) | v_i, v_j \in V\}$ that specifies the acceptable network latency T_{max} between two virtual machine instances v_i and v_j, the mapping function $p : V \rightarrow H$ ensures the following condition:

$$\forall\, v_i,\ v_j \in V,\ (v_i,\ v_j,\ T_{max}) \in MaxLatency \quad latency(p(v_i),\ p(v_j)) \leq T_{max}$$

4 Initial Allocation of Virtual Machines in Data Center Environments

In this Section, we discuss resource allocation schemes that, for each user request, identify the physical hosts on which the requested virtual machines can be allocated while satisfying the placement constraints. The solutions discussed here perform initial allocation considering the fault tolerance, security, and performance constraints, and aiming at maximizing the data center efficiency. We first analyze the solution by Jhawar et al. [27] in detail (Sect. 4.1) since it satisfies all the constraints described in the previous Section, and then provide an overview of other state-of-art schemes (Sect. 4.2).

4.1 A Comprehensive Scheme for Virtual Machines Allocation

To address the needs of optimum data center management in Cloud-based environments, while satisfying both data center owner's constraints and users' goals, the approach presented in [27] performs the initial placement of virtual machines by designing the mapping function $p : V \rightarrow H$ to meet the following objectives:

- reduce the energy consumption and operational costs by consolidating virtual machines on physical hosts such that the number of *free* hosts is maximized;

- reduce the load variance of physical hosts across all the clusters in the Cloud in order to improve the performance and resilience of the data center.

Besides, the mapping function is also structured to satisfy the resource capacity, forbid, count, restrict, distribute, and latency constraints, thus satisfying application's fault tolerance, security and performance goals.

The allocation algorithm works as follows. Each time a request to allocate new virtual machines arrives, the data center is analyzed to identify the clusters and the physical hosts that can be used for resource allocation. For each shortlisted physical host, the set of virtual machines that can be allocated on that host are identified. In particular, *to reduce the load variance between different clusters*, the new virtual machines are created in the cluster that has highest amount of available resources.

A priority queue CL is built based on the resource availability in each cluster, and then, the cluster C with maximum resource availability is extracted from CL. *To reduce the energy consumption costs,* each host within the selected cluster is analyzed to allocate as many virtual machines on that host as possible. This heuristics maps virtual machines on fewer physical hosts, thus leaving the remaining hosts to operate in the *cold-standby* mode.

A priority queue $VMreq$ is created based on the vector dot-product [35] value of virtual machines with respect to the current host and entries from $VMreq$ are extracted in the decreasing order of the dot-product values and analyzed for performing the final allocation.

Therefore, the algorithm can be viewed as a two-step process: (i) select the least-used cluster and (ii) allocate virtual machines on its hosts using the dot-product method, and satisfies both the aforementioned objectives.

In this allocation algorithm appropriate checks are introduced to guide the mapping of virtual machines on physical hosts as follows. Since the *forbid* and *restrict* constraints define conditions on the association between virtual machines and physical hosts, corresponding checks are applied mainly while building the $VMreq$ priority queue (i.e., when analyzing the suitability of allocating a given virtual machine on the current physical host). Hence, a temporary set V' that contains all the virtual machines that must be allocated is created, and a virtual machine v from set V' is discarded if: (i) an entry $(v, h) \in Forbid$ exists or (ii) a set of hosts is specified in the $Restr$ set for the virtual machine v but the considered host h does not belong to that set. This check allows the allocation algorithm to enforce the forbid and restrict constraints.

The *capacity* and *count* constraints deal with the resource usage of the individual physical hosts. To enforce the capacity constraint, the residual resource capacity of each physical host is maintained, based on the threshold values specified by the data center owner. A virtual machine is allocated on a physical host if the resource requirements of the virtual machine are less than the residual capacity of that host in all the dimensions. Each time a virtual machine is allocated on a host, its residual capacity value is updated.

To ensure the count constraint, the vector representation of hosts and virtual machines is extended with respect to Sect. 2.1 by adding a new dimension on each physical host that denotes the number of virtual machines that can be allocated on that host $h[D+1] = count_h$. Similarly, the $[D + 1]^{th}$ dimension of each virtual machine is initialized to 1, and the count control is enforced with the capacity constraint.

The *distribute* constraint is enforced by verifying whether a given host h already contains a virtual machine v_j for which the user has specified a condition $(v, v_j) \in Distr$. The algorithm simply does not consider the virtual machine v when working on the allocation for the host h if $p(v_j) = h$ is true.

Finally, the allocation algorithm enforces the *latency* constraint based on the notions of *forward allocation* and *reserve list*. If a virtual machine v_i satisfies all other constraints when considered for allocation on a host h, but is related to other virtual machines $v_j \in V_i \subset V$ by latency constraints, then the virtual machine v_i cannot be actually allocated on the host h until an allocation for all $v_j \in V_i$ is found. Therefore, the allocation algorithm tentatively allocates the virtual machines by saving pair (v_i, h) in the $Reserve\_list$ and calls the function $Forward\_Allocate$ to find an allocation for the other virtual machines $v_j \in V_i$. The $Forward\_Allocate$ function first determines the virtual machines $v_j \in V_i$ that are related to v_i by the latency constraints and stores them in a priority queue. Each virtual machine from the priority queue is then extracted in the increasing order of T_{max}, and the set of hosts that can be reached from the current host h within the specified network threshold time (and not conflicting w.r.t. the restrict and forbid constraints) is selected. The capacity, count and distribute constraints are then verified for each shortlisted host, and the corresponding allocation is saved in the $Reserve\_list$. The $Forward\_Allocate$ function is recursively called until an allocation for all the virtual machines is determined. If an allocation is not obtained, the entries from the $Reserve\_list$ are removed, and the algorithm resumes from another host.

4.2 *Other Schemes for Virtual Machines Allocation*

In the literature there are other approaches for the initial allocation of virtual machines in data centers. Existing solutions either use Constraints Programming (CP) solvers (e.g., [6]) or design heuristics (e.g., [32]) to obtain the placement solutions. In general, while the goal is to maximize the goals of the data center owner, each solution takes a different approach in modeling the context of the system and, consequently, defines different objective functions and placement constraints. In this sub-section, we discuss four representative solutions to understand different dimensions in which the overall problem has been studied.

Bin et al. [6] combine the Hardware Predicted Failure Analysis alerts (HwPFA) and live migration techniques to provide a high availability solution. On predicting hardware failure alerts, a trigger to the cluster management system is provided so as to move the virtual machines from the failing host to other working hosts. Depending on the allowed response time, either a complete live relocation of the virtual machine is performed so that continuous operation of the applications is ensured, or a cold relocation is performed by starting a new virtual machine on a working host with a small interruption.

The goal of their solution is to provide k-resiliency to users applications while reducing the resource consumption costs. We note that k-resiliency allows a given application or virtual machine to tolerate up to k host failures. In general, to ensure k-resiliency, a feasible solution should dedicate at least k hosts for the given virtual machine (in addition to the virtual machine itself). The proposed approach introduces the notion of *shadow virtual machines* that denotes the location or host where a virtual machine can be evacuated (i.e., a shadow serves as a placeholder) and aim to construct shadow placement constraints so to reduce the overall resource requirements to a value less than $(k + 1)$. To achieve this, they transform the placement problem with k-resiliency constraints into a constraint satisfaction problem including the notion of *shadow virtual machines*, and solve it using a constraint programming engine. All the shadows of a given virtual machine and the virtual machine itself are *anti-colocated* (equivalent to the Distribute constraint discussed in Sect. 3.3). In addition, they employ a scheme of numbering shadows and failures in a way that identifies the possible overlaps of actual virtual machine evacuations. In particular, each failing host is assigned a unique index (1 through k) and each shadow of a virtual machine is assigned a unique index. Upon failure of a host indexed i, the virtual machines on that host are evacuated to the location of their i-th shadow. The placement constraints are defined to specifically numbered shadows and virtual machines that may overlap following host failures, thus reducing the number of backup hosts required. For example, virtual machines that are placed on different hosts cannot be evacuated together to shadows with the same index (as each host would be assigned a different failure index); therefore, their shadows with same index can overlap.

Machida et al. [31] consider consolidated server systems and present a method to redundant configuration of virtual machines, in anticipation of host server failures for online applications. They estimate the requisite minimum number of virtual machines according to the performance requirements of the given application, and compute the virtual machine placement solution so that the configuration can survive k host server failures. The overall problem is defined as a combinatorial optimization problem and a greedy algorithm for determining the placement solution is provided with the aim of minimizing the number of required hosting servers. Their method performs better than the conventional $N + M$ redundant configuration in terms of the number of hosting servers required.

Shi et al. [32] formulate the problem of virtual machine placement as an Integer Linear Programming (ILP) problem and provide a twofold solution. First, they use solvers to obtain optimal results. Second, since the scalability of this approach is limited, they also provide a modified version of the first fit decreasing heuristic to generate sub-optimal results. In particular, they classify the requests for virtual machine placements into different categories and satisfy the following constraints using the first fit decreasing heuristic, in the form of a multidimensional vector packing problem: (i) the *full deployment* constraint that ensures either all the virtual machines requested by the user are allocated or none; (ii) the *anti-colocation* constraint requiring all the virtual machines to be placed on different physical hosts; and (iii) the *security* constraint requiring a physical host only be assigned virtual machines from the same user request and not be assigned any virtual machines from other requests.

The three aforementioned resource allocation techniques consider only fault tolerance and security constraints. The solution by Jayasinghe et al. [26] that also take into account various performance attributes while performing initial allocation of virtual machines. In particular, they propose a structural constraints-aware virtual machine placement approach to improve the performance and availability of applications deployed in the data centers. They integrate the structural information of users applications within the algorithm for initial placement of virtual machines by means of three constraints: (i) *demand constraint*, that defines the lower bound of resource allocations that each virtual machine requires from the service to meet its SLA; (ii) *availability constraint*, that improves the overall availability of given applications using a combination of anti-collocation/collocation constraints; and (iii) *communication constraint*, that represents the communication requirement between two virtual machines. The objective of the proposed algorithm is to minimize the communication cost while satisfying both the demand and availability constraints. Their solution uses the divide-and-conquer technique which involves the following steps: (1) the group of virtual machines requested by the user is divided into a set of smaller virtual machine groups and the upper bound of the virtual machine group size is determined by the average capacity of a server rack; (2) a suitable server rack is identified for each virtual machine group such that the mapping minimizes the total communication cost and guarantees the satisfaction of availability constraints; (3) a physical host satisfying the demand constraint is identified for each virtual machine.

5 Runtime Adaption of Virtual Machine Allocation in Data Center Environments

Data centers are highly dynamic in terms of task activation, bandwidth availability, component failures and recovery. This implies that static deployment strategies for virtual machines that perform only initial allocation (such as the $p : V \rightarrow H$ function) may not provide satisfactory results at runtime and application's dependability and performance requirements may not be satisfied all along their lives. A naïve approach is to re-compute the allocation from *scratch* each time system changes affect an application. However, since this method may not scale well during runtime [24], a number of solutions have been proposed in the literature to adapt the current allocation of applications using fewer actions. In this Section, we first discuss the solution that balances application's performance and availability goals at runtime (Sect. 5.1) and then present other state-of-art adaptive resource management schemes (Sect. 5.2).

5.1 *Runtime Adaption to Balance Availability and Performance*

A heuristics-based approach that minimizes the performance and availability degradation of applications due to various system changes has been presented in [24]. To dynamically manage the adaptation of the allocation of applications' virtual machines on the hosts of a data center a dedicated monitoring process (called *online controller*) is introduced in the data center itself. The online controller uses the monitoring information (e.g., application workload, server's failure behavior, processor and bandwidth usage) to create a comprehensive view of the working status of the whole system. When events may violate the applications' availability or performance goals, the online controller re-deploys the applications so as to maintain the dependability and performance characteristics of the data center. For example, when the availability of an application (i.e., the probability that the application is working correctly and is not affected by hardware failures or security threats) is below the desired value, the current allocation is adapted by deploying new application task replicas as a response to server failures and/or by migrating individual tasks on (other working hosts) across different deployment levels in the system. The initial allocation algorithm (see Sect. 4) is executed only if the output generated by the online controller is unfeasible for the users' applications.

The online controller is based on the notion of deployment levels discussed in Sect. 2.4 and on the following remarks. First, the availability of the application increases and the performance (response time) decreases as the number of replicas increase. Second, the availability of the application increases as the deployment level in which its replicas are placed increase. On the contrary, response time decreases as the deployment level increases. Using this analysis, the online controller changes the current allocation of a given application and redeploys it by applying the following allocation actions.

- *Launch* (v, h): The online controller may have to create new application replicas when a system failure happens. To achieve this, it uses action *Launch* (v, h) to create a new virtual machine v on a physical host h, in which deploys a new replica of the failed application.
- *Migrate* (v, h_i, h_j): The online controller may have to locate a subset of application's replicas on different physical hosts as a response to performance or availability degradation. For example, to respond to network congestion in cluster C_1, the online controller may want to move replicas initially allocated in C_1 to another cluster C_2. Action *Migrate* (v, h_i, h_j) specifies that the virtual machine v deployed on host h_i must be moved to host h_j.
- *Delete* (v, h): Due to performance overhead, the online controller may need to reduce the number of replicas of the application, even though dependability might be reduced. Action *Delete* (v, h) removes virtual machine v, hosting application's replica, from host h.

On the bases of the current allocation, system status, application tasks and the sets specifying allocation constraints as input, the online controller generates the sequence of allocation actions that brings the system to a new allocation state in which the constraints are again satisfied. The online controller is invoked when a failure or performance degradation is identified for an application. It is worth noting that it is therefore useful for maintaining the availability and performance goals for long-running applications which are more exposed to possible component failures and overloads, while short-running applications can practically be managed by the initial deployment alone.

The algorithm for online adaptation of the applications' allocation consists of two main conditions, one concerning availability violation due to system failures and the other concerning performance degradation. If the real availability of an application is less than the minimum desired value, the online controller first identifies the task replica failures and tentatively launches new replicas at the same deployment level. This *Launch* action is performed only until the current replication level is same as that of the original level and as long as performance goals are not violated. If the addition of application's replicas does not satisfy the user's requirements, the online controller tries to move task replicas to a higher deployment level using the *Migrate* action. We note that the availability increases with increasing deployment levels. This action allows the online controller to generate the new allocation solution without increasing the resource consumption costs. If performance is degraded by moving tasks to higher deployment levels, additional replicas must be created to improve the availability. To create new replicas, the online controller starts from higher deployment levels and moves gradually to lower levels, creating the replicas at the level where availability and performance goals are fulfilled. These actions are realized using *Migrate* and *Launch* actions.

When real performance is less than desired minimum value, virtual machines are deleted instead of launching new replicas, and migration takes place to lower deployment levels instead of moving higher in the hierarchy. We note that the response time of the application improves as the number of replicas and the deployment level decreases.

5.2 Other Schemes for Runtime Virtual Machines Allocation Adaption

In the literature other approaches for dynamic adaptation of virtual machines' allocation in data centers have been studied, even though often they focus only on some of the constraints discussed above. The placement of application replicas to achieve dependability becomes especially challenging when they consist of communicating components (e.g., multi-tier web applications). Recent works on performance optimization of such applications (e.g., [8, 22]) address the performance impact of resource allocation, but does not combine performance modeling with availability requirements and dynamic regeneration of failed components.

The trade-off between availability and performance is considered in the literature on dependability since increasing availability (by using more redundancy) typically increases response time. In fact, the well-known Brewer's theorem states that consistency, availability, and partition tolerance are the three commonly desired properties by a distributed system, but it is impossible to achieve all three [17]. Examples of work that explicitly address this issue include [22, 24, 29, 34]. Among these solutions, [34] considers the problem of when to invoke a (human) repair process to optimize various metrics of cost and availability defined on the system. The optimal policies that specify when the repair should be invoked (as a function of system state) are computed off-line via Markov decision process models of the system. Similarly, Jung et al. [22] study how virtualization can be used to provide enhanced solutions to the classic problem of ensuring high availability while maintaining performance of multi-tier web services. Software components are restored whenever failures occur and component placement is managed using information about application control flow and performance predictions.

Addis et al. [1] devise a resource allocation policy for virtualized Cloud computing environments aiming at identifying performance and energy trade-off, with a priori availability guarantees for the users. They model the problem as a mixed integer non-linear programming problem and propose a heuristic solution based on non-linear programming and local-search techniques. In particular, the availability requirements are introduced as constraints, and the objective is to determine a resource allocation that allows improving the total profit (the difference between the revenues from SLA contracts and the total costs). Their solution defines the following four actions to take into account availability constraints: (i) increase/decrease the working frequency for each server according to the application loads (a frequency change compatible with the new application loads does not affect availability); (ii) switch a server to low power sleep state and allocate all applications of that server on the other available servers (some applications from the overloaded servers can be duplicated and allocated on another active server to satisfy availability constraints); (iii) reallocation of class-tier applications on a server with sufficient availability assurance to satisfy the performance constraints; (iv) servers exchange is used to move all applications from a server which should be switched to low the power sleep state to a different active server if the availability of the new server guarantees the availability requests of all the moved applications.

Zheng et al. [ZLX2013] consider the lack of maintenance to be the root cause for downtime events in a Cloud computing data center. To address this issue, they first present a heuristics for resource provisioning under a given maintenance schedule and, then, build on the heuristics to solve the joint resource provisioning and maintenance scheduling problem. Yang et al. [38] propose an algorithm that uses Markov chain models to schedule tasks so that they get the best value of utility. A job being executed on the Cloud possesses the following factors: deadline, data, and reward factor for completing it on time. The reward is assigned to each task depending on the time it takes to complete the job with respect to its deadline. The earlier it completes, the higher is the reward. If a task fails during execution, then the time required for the task to recover is also added to the total time it takes to complete. The proposed algorithm includes reliability while calculating the value of the reward for each task. The impact of the amount of time which is spent in recovery from a failure and the useless waiting time in the task queue are added in this model. The proposed rules are: execute a job as soon as possible when the resources are available, and pausing a job in the queue till a resource is available or assigning a new task to free resources. This approach has been tested against well-known task scheduling algorithms: if there are not system failures results have quality similar to other conventional scheduling techniques, while in the case of system failures the proposed algorithm produces better efficiency and stability.

6 Conclusions

This chapter has discussed the state-of-art resource management schemes that are designed to improve dependability and performance of applications deployed in data centers based on Cloud computing environments. First, we briefly discussed the failure characteristics of data center components and presented an approach to evaluate the impact of system failures on users' applications. We have formalized the applications placement problem and formulated various placement constraints, involving fault tolerance, security and performance conditions of both users and data center owners. Then, we discussed virtual machine placement algorithms that satisfy the placement constraints and reduce the management costs for the data center owner. In particular, the approaches to virtual machines placement discussed in this chapter apply to two different operating phases: the initial allocation of applications at their activation, and the runtime adaption of their placement to deal with dependability threats and performance changes. These strategies ensure dependable data center based on Cloud computing environment for users' applications.

References

1. B. Addis, D. Ardagna, B. Panicucci, and L. Zhang, "Autonomic management of cloud service centers with availability guarantees," in Proc. of 3rd International Conference on Cloud Computing, Miami, FL, USA, July 2010, pp. 220–227
2. M. Albanese, S. Jajodia, R. Jhawar, and V. Piuri, "Reliable Mission Deployment in Vulnerable Distributed Systems," in Proc. of the 43rd Annual IEEE/IFIP International Conference on Dependable Systems and Networks Workshops, Budapest, Hungary, June 24–27, 2013, pp. 1–8
3. M. Albanese, S. Jajodia, R. Jhawar, and V. Piuri, "Securing Mission-Centric Operations in the Cloud," in *Secure Cloud Computing*, S. Jajodia, K. Kant, P. Samarati, V. Swarup, C. Wang (eds.), Springer, 2013
4. C. A. Ardagna, R. Jhawar, and V. Piuri, "Dependability Certification of Services: A Model-Based Approach," in Computing, Springer, Oct, 2013, pp. 1–28
5. G. Bertoni, L. Breveglieri, I. Koren, P. Maistri, V. Piuri, "On the propagation of faults and their detection in a hardware implementation of the Advanced Encryption Standard", in Proc. of

the 2002 IEEE International Conference on Application-Specific Systems, Architectures and Processors, San Jose, CA, USA, July 2002, pp. 303–312
6. E. Bin, O. Biran, O. Boni, E. Hadad, E. Kolodner, Y. Moatti, and D. Lorenz, "Guaranteeing high availability goals for virtual machine placement," in Proc. of 31st International Conference on Distributed Computing Systems, Minneapolis, USA, June 2011, pp. 700–709
7. C. Blundo, S. Cimato, S. De Capitani di Vimercati, A. De Santis, S. Foresti, S. Paraboschi, P. Samarati, "Efficient Key Management for Enforcing Access Control in Outsourced Scenarios," in Proc. of the 24th IFIP TC-11 International Information Security Conference (SEC 2009), Cyprus, Greece, May 2009
8. I. Cunha, J. Almeida, V. Almeida, and M. Santos, "Self-adaptive capacity management for multi-tier virtualized environments," in Proc. of 10th IFIP/IEEE International Symposium on Integrated Network Management, Munich, Germany, May 2007, pp. 129–138
9. E. Damiani, S. De Capitani di Vimercati, S. Foresti, S. Jajodia, P. Samarati, "Key Management for Multiuser Encrypted Databases," in Proc. of the International Workshop on Storage Security and Survivability, Fairfax, Virginia, USA, Nov 2005
10. S. De Capitani di Vimercati, S. Foresti, S. Jajodia, S. Paraboschi, P. Samarati, "Controlled Information Sharing in Collaborative Distributed Query Processing," in Proc. of the 28th International Conference on Distributed Computing Systems (ICDCS 2008), Beijing, China, June 2008
11. S. De Capitani di Vimercati, S. Foresti, S. Ja jodia, and G. Livraga, "Enforcing Subscription-based Authorization Policies in Cloud Scenarios," in Proc. of the 26th Annual IFIP WG 11.3 Working Conference on Data and Applications Security and Privacy, Paris, France, July 11–13, 2012
12. S. De Capitani di Vimercati, S. Foresti, S. Jajodia, G. Livraga, S. Paraboschi, and P. Samarati, "Enforcing Dynamic Write Privileges in Data Outsourcing," in Computers & Security, 2013
13. S. De Capitani di Vimercati, S. Foresti, P. Samarati, "Managing and Accessing Data in the Cloud: Privacy Risks and Approaches," in Proc. of the 7th International Conference on Risks and Security of Internet and Systems (CRiSIS 2012), Cork, Ireland, Oct 2012
14. S. De Capitani di Vimercati, G. Livraga, V. Piuri, F. Scotti, "Privacy and Security in Environmental Monitoring Systems," in Proc. of the 1st IEEE-AESS Conference in Europe about Space and Satellite Communications (ESTEL 2012), Rome, Italy, Oct 2012
15. S. De Capitani di Vimercati, A. Genovese, G. Livraga, V. Piuri, F. Scotti, "Privacy and Security in Environmental Monitoring Systems: Issues and Solutions", in Computer and Information Security Handbook, 2nd Edition, J. Vacca (ed.), Morgan Kaufmann, Boston, 2013, pp. 835–853
16. P. Gill, N. Jain, and N. Nagappan, "Understanding Network Failures in Data Centers: Measurement, Analysis and Implications", ACM Computer Communication Review, vol. 41, no. 4, 2011, pp. 350–361
17. S. Gilbert and N. Lynch, "Brewer's conjecture and the feasibility of Consistent, Available, Partition-tolerant web services," SIGACT News, vol. 33, no. 2, Jun 2002, pp. 51–59
18. R. Guerraoui and M. Yabandeh, "Independent faults in the cloud," in Proc. of 4th International Workshop on Large Scale Distributed Systems and Middleware, Zurich, Switzerland: ACM, 2010, pp. 12–17
19. U. Helzle and L. A. Barroso, "The Datacenter as a Computer: An Introduction to the Design of Warehouse-Scale Machines", 1st ed. Morgan and Claypool Publishers, 2009
20. F. Hermenier, S. Demassey, and X. Lorca, "Bin repacking scheduling in Virtualized datacenters", in Proc. of Constraints Programming, Perugia, Italy, 2011
21. F. Hermenier, J. Lawall, J.M. Menaud, and G. Muller, "Dynamic Consolidation of Highly Available Web Applications", INRIA, Tech. Rep. RR-7545, 2011
22. G. Jung, K. Joshi, M. Hiltunen, R. Schlichting, and C. Pu, "Performance and availability aware regeneration for cloud based multitier applications," in Proc. of 2010 IEEE/IFIP International Conference on Dependable Systems and Networks, Chicago, IL, USA, July 2010, pp. 497–506
23. R. Jhawar and V. Piuri, "Fault tolerance management in IaaS Clouds," in Proc. of the 1st IEEE-AESS Conference in Europe about Space and Satellite Telecommunications, Rome, Italy, Oct 2012, pp. 1–6

24. R. Jhawar and V. Piuri, "Adaptive resource management for balancing availability and performance in Cloud computing," in Proc. of the 10th International Conference on Security and Cryptography, Reykjavik, Iceland, Jul 2013, pp. 254–264
25. R. Jhawar and V. Piuri, "Fault Tolerance and Resilience in Cloud Computing Environments," in *Computer and Information Security Handbook*, 2nd ed., J. Vacca (ed.), Morgan Kaufmann, 2013, pp. 125–141
26. D. Jayasinghe, C. Pu, T. Eilam, M. Steinder, I. Whally, and E. Snible, "Improving performance and availability of services hosted on IaaS Clouds with structural constraint-aware virtual machine placement," in Proc. of IEEE International Conference on Services Computing, Washington, DC, USA, Jul 2011, pp. 72–79
27. R. Jhawar, V. Piuri, and P. Samarati, "Supporting security requirements for resource management in cloud computing," in Proc. of the 15th IEEE International Conference on Computational Science and Engineering, Paphos, Cyprus, Dec 2012, pp. 170–177
28. R. Jhawar, V. Piuri, and M. Santambrogio, "Fault Tolerance Management in Cloud Computing: A System-Level Perspective," in IEEE Systems Journal 7 (2), June, 2013, pp. 288–297
29. S. Kim, F. Machida, and K. Trivedi, "Availability modeling and analysis of virtualized system," in Proc. of 15th IEEE Pacific Rim International Symposium on Dependable Computing, Shanghai, China, Nov 2009, pp. 365–371
30. K. Mills, J. Filliben, and C. Dabrowski, "Comparing VM-Placement Algorithms for on-demand Clouds", in Proc. of CLOUD'11, Washington, DC, USA, 2011, pp. 91–98
31. F. Machida, M. Kawato, and Y. Maeno, "Redundant Virtual Machine Placement for Fault Tolerant Consolidated Server Clusters", in Proc. of IEEE/IFIP NOMS'10, Osaka, Japan, 2010, pp. 32–39
32. L. Shi, B. Butler, D. Botvich, and B. Jennings, "Provisioning of requests for virtual machine sets with placement constraints in IaaS clouds," in Proc. of IFIP/IEEE International Symposium on Integrated Network Management, Ghent, Belgium, May 2013, pp. 499–505
33. P. Samarati and S. De Capitani di Vimercati, "Data Protection in Outsourcing Scenarios: Issues and Directions", in Proc. of the 5th ACM Symposium on Information, Computer and Communications Security, Beijing, China, 2010, pp. 1–14
34. K. Shin, C. M. Krishna, and Y. Hang Lee, "Optimal dynamic control of resources in a distributed system," IEEE Transactions on Software Engineering, vol. 15, no. 10, 1989, pp. 1188–1198
35. A. Singh, M. Korupolu, and D. Mohapatra, "Server-storage virtualization: Integration and load balancing in data centers," in Proc. of SC'08, Austin, TX, USA, Nov 2008, pp. 53:1–53:12
36. W.E. Smith, K.S. Trivedi, L.A. Tomek, and J. Ackaret, "Availability Analysis of Blade Server Systems", IBM Systems Journal, vol. 47, no. 4, 2008, pp. 621–640
37. K. Vishwanath and N. Nagappan, "Characterizing Cloud Computing Hardware Reliability", in Proc. of SoCC'10, Indianapolis, IN, USA, 2010, pp. 193–204
38. B. Yang, X. Xu, F. Tan, and D.-H. Park, "An utility-based job scheduling algorithm for cloud computing considering reliability factor," in Proc. of International Conference on Cloud and Service Computing, Hong Kong, Dec 2011, pp. 95–102

Resource Scheduling in Data-Centric Systems

Zujie Ren, Xiaohong Zhang and Weisong Shi

1 Introduction

Effective resource scheduling is a fundamental issue for achieving high performance in various computer systems. The goal of resource scheduling is to arrange the best location of each resource and determine the most appropriate sequence of job execution, while satisfying certain constraints or optimizations. Although the topic of resource scheduling has been widely investigated for several decades, it is still a research hotspot as new paradigms continue to emerge, such as grid computing [1, 2], cloud computing [3, 4], big data analytics [5, 6], and so on.

With the explosive growth of data volumes, more and more organizations are building large-scale data-centric systems (DCS). These systems are hosted by one or more data centers, where they serve as IT infrastructures for data processing, scientific computing, and a variety of other applications involving "big data". Data-centric systems offer new solutions for existing applications and promote warehouse-scale data businesses such as cloud computing, cloud storage services, and so on.

Unfortunately, there is no widely accepted standard definition for data-centric systems. However, in general, if a computing system involves large volumes of data which are hosted by data centers, it can be labeled as "data-centric systems". Examples include large-scale web search engine, data management systems, data mining systems. Particularly, we focus on three kinds of data-centric systems in this chapter:

Z. Ren (✉)
School of Computer Science and Technology,
Hangzhou Dianzi University, Hangzhou, China
e-mail: renzju@gmail.com

X. Zhang
Shenzhen Institutes of Advanced Technology,
Chinese Academy of Science, Shenzhen, China
e-mail: xh.zhang@siat.ac.cn

W. Shi
Department of Computer Science, Wayne State University, Detroit, USA
e-mail: weisong@wayne.edu

S. U. Khan, A. Y. Zomaya (eds.), *Handbook on Data Centers*,
DOI 10.1007/978-1-4939-2092-1_46

- *Cloud computing platforms.* A cloud computing platform is depicted as a large pool of computing and storage resources, which provides various services (IaaS, PaaS and SaaS) and elastic resources [3] to public users via the Internet. Recent years have witnessed a rapid growth in the number of cloud computing platforms, such as Amazon EC2 [7], IBM Blue Cloud [8], Google AppEngine [9], RackSpace [10] and Microsoft Azure [11].
- *Data-Intensive Super Computing (DISC) systems.* DISC systems are new forms of high-performance computing (HPC) systems that concentrate on high-volume data, rather than computation. DISC is responsible for the acquisition, updating, sharing, and archiving of the data. In addition, DISC supports data-intensive computation over high-volume data [12–14].
- *MapReduce-style systems.* MapReduce-style processing systems are designed to deal with big data volume in parallel on large-scale clusters. A traditional and popular example is Hadoop [15], an open-source implementation of the MapReduce framework [16]. Hadoop can easily scale out to thousands of nodes and work with petabyte data.

In the context of DCS, effective resource scheduling is notoriously difficult due to the complexity and diversity of DCS. More specifically, the challenges for scheduling optimization include the following: (1) the software/hardware stack in data-centric systems is composed of many layers [17, 18]. The entities and objectives of scheduling may be completely different across these software/hardware layers. (2) the workload running the data-centric systems is significantly miscellaneous. The workload is usually comprised of long-running applications, Web services, MapReduce jobs, HPC jobs, and so on. Therefore, compared with the traditional distributed systems like distributed file systems and DBMS, data-centric systems pose many more challenges for improving resource efficiency by scheduling due to the system complexity and workload diversity.

To address these challenges, various resource scheduling methods in the context of DCS have been proposed in recent years. For example, motivated by the market behaviors in the field of economics, some literature has focused on regulating the supply and demand of resources in cloud environments, using such as commodity-based [19, 20] or auction-based strategies [21, 22]. These resource scheduling polices are designed for reducing cost for resource consumers and maximizing profits for resource providers. Other literature focuses on optimizing the system throughput by allocating resources based on various heuristics. For example, the scheduler may concentrate on system utilization [23], job completion time [24, 25], load balance [26], energy consumption [27–29], data locality [30, 31], or real-time satisfaction [32, 33].

While the topic of resource scheduling in data-centric systems is broad enough to provide enough content for a book, those existing techniques are scattered and poorly organized. A systematic survey on the existing research advances is necessary and helpful to further improvement and performance optimization. In this chapter, we classify the resource scheduling approaches into three categories according to the scheduling model: resource provision, job scheduling and data scheduling. We give a systematic review of the most significant techniques for these categories, and present some open problems to be addressed. We believe this systematic and comprehensive analysis can help researchers and engineers to better understand those scheduling techniques and inspire new developments within this field.

The chapter is organized as follows. Section 2 presents the definitions of a list of terminologies used in the chapter. A taxonomy of existing works on resource scheduling is presented in Sect. 3. In Sect. 4, we will look at four case studies, each of which is derived from practical or productional systems. In Sect. 5, we outline interesting future trends and challenges of resource scheduling.

2 Terminology

Due to the diversity of data-centric systems, the terminology used in this field is often inconsistent. To clarify the description in this chapter, we define the following necessary terminology.

Resource. Resource is a collection of components that can be scheduled to perform an operation. Some traditional examples of resources are CPU cores for computing, memory spaces for storage, network links for transferring, electrical power, and so on.

Task. A task is an atomic action from the scheduler's point of view. A Task is defined by a collection of input data and corresponding operations.

Job. A job is a group of tasks that will be executed on a set of resources. The definition of jobs is recursive, which means that jobs are composed of sub-jobs and/or tasks, and sub-jobs can be decomposed further into atomic tasks.

Service. A service is a program to enable access to one or more resources, where the access is provided by a predefined interface. For instance, cloud computing, which is provisioned as services, are broadly divided into three categories: software-as-a-service (SaaS), platform-as-a-service (PaaS), and infrastructure as-a-service (IaaS).

Data-Centric Systems. Although there is no de-facto standard definition for the term "data-centric systems", they are very common in various forms. In most cases, they are characterized (partially or fully) by the following features:

- managing of large volumes of data, in range of petabyte-level and beyond
- hosted by one or more data centers
- involving complex software/hardware stacks
- serving for multiple users and execute diverse workloads

Generally, many computing systems can be labeled as "data-centric systems", such as web search engines, data management systems, and data mining systems. To summarize, this chapter concentrates on three kinds of traditional data-center systems, including cloud computing platforms, data-intensive super computing systems, and MapReduce-style systems.

3 Classification and State-of-the-Art

In this section, we present a broad view of resource scheduling issues in data-centric systems. As data-centric systems involve multiple software layers, scheduling operations take place on multiple layers. For example, assume that a set of MapReduce jobs are submitted to a data processing application, which is hosted on a cloud platform like Amazon EC2, the scheduling operations will be conducted multiple times. Firstly, when the application for processing MapReduce jobs, such as Hadoop, is loaded on the cloud, the application needs to be provisioned with a certain amount of resources, which is often referred to as *resource provision* (aka. *resource allocation*). Secondly, when the set of job requests are submitted to the application, the scheduler in the application needs to map the set of jobs to multiple servers in a certain manner, which is also known as *job scheduling*. Thirdly, to improve the resource utilization or job execution efficiency, the scheduler within storage systems needs to schedule the data transfer, replication, distribution, either during the job execution or in advance, which is often referred to as *data scheduling*.

3.1 Hierarchy of Resource Scheduling in DCS

In fact, similar as in the context of a data processing system, the resource scheduling issue in DCS also can be generally divided into the problems of resource provision, job scheduling and data scheduling. Although data-centric systems come with variable implementations, we still can abstract a common hierarchical architecture of various data-centric systems from the perspective of scheduling, which is depicted in Fig. 1.

- **Resource provision.** Resource provision is to allocate resources to satisfy multiple applications efficiently. In one aspect, on the top layer of data-centric systems, various applications, VM instances, Web services etc., run on the data-centric systems. They demand a certain type and amount of resources when they are loaded. In the other aspect, the data-centric systems is a unified resource platform, which holds massive computation and storage resources in the data centers. The resources are allocated to users based on a certain policy to satisfy the requirements of resource providers and users. Therefore, the scheduling issue in this layer is often also referred to as *resource allocation* [34–37].
- **Job scheduling.** Within a data-centric system, various jobs, such as HPC and MapReduce-style jobs, will be submitted in parallel by many applications (or users). Simple scheduling algorithms such as FIFO, are hard to satisfy performance requirements in most cases. To improve the job's execution performance, a scheduling algorithm is needed to assign jobs or tasks to appropriate nodes in a certain order. Therefore, the scheduling issue in this layer is also known as *job scheduling*.

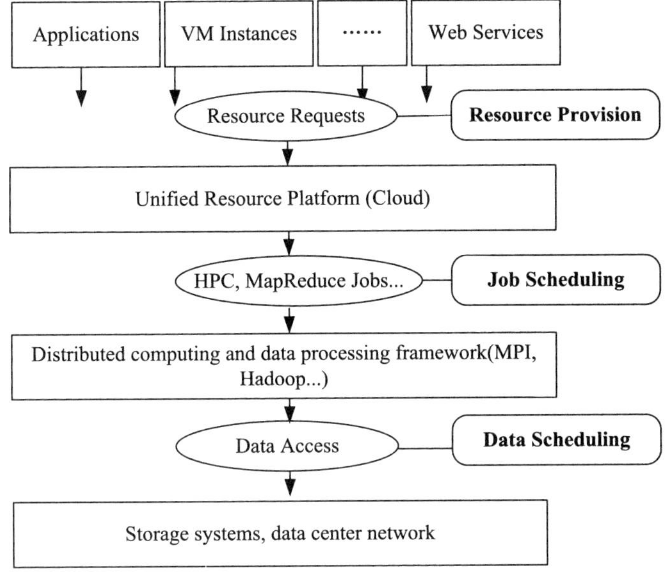

Fig. 1 Resource scheduling hierarchy in DCS

- **Data scheduling.** On the storage layer of data-centric systems, there are large volumes of data stored and managed by distributed nodes. The goal of data scheduling is two-fold: in one aspect, data-centric systems employ various data placement and migration techniques to increase the storage resource utilization, data reliability and availability; in the other aspect, data-centric systems apply online scheduling techniques for data prefetching, data transfer etc., for accelerating the job (request) execution, aiming to reduce to data access/transfer latency Fig. 2.

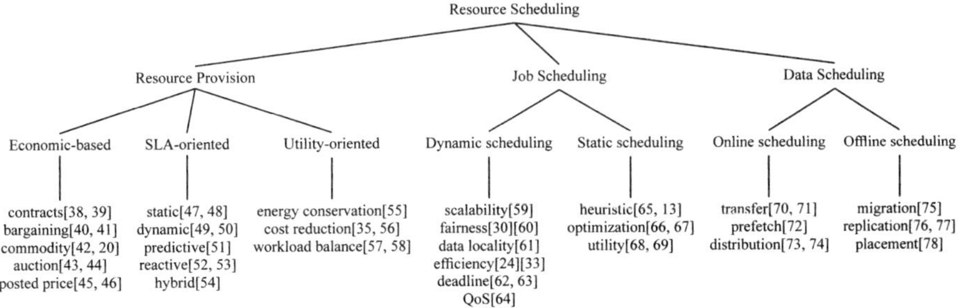

Fig. 2 Resource scheduling taxonomy in DCS

3.2 *Resource Provision*

During the past few years, cloud computing has become a main trend in delivering IT services, where the computing and storage capabilities are shared among multiplex many users. In a cloud computing platform, the resources are available on-demand, charged on a pay-as-you-go basis. In one aspect, cloud providers hold enormous computing resources in their data centers, while in the other aspect, cloud users lease the resources from cloud providers to run their applications. Usually, the resource requirement imposed by cloud users are heterogeneous [18] and time-varying [79, 80], which makes the scheduling much more complicated.

According to the provision model, we classify the resource provision techniques into three groups: economic-based, SLA-oriented, and utility-oriented. The first and second groups focus on resource provision issues between providers and consumers using economical models or SLA contracts, while the third group concentrates on high-efficiency resource management from the perspective of the data center owner.

3.2.1 Economic-Based Resource Provision

To maximize benefits on cloud platforms, many researchers proposed various economic models to effectively solve the issues of scheduling problems in the grid or cloud environments, such as commodity market [81], posted price [45, 82], tendering/contract [38], bargaining [83, 84], auction [43, 21], and so on. Economics-based methods are very suitable for handling the provision issues in a cloud environment, as they have been effectively utilized in the field of economics to regulate the supply and demand of limited resources.

The concept of a *commodity market* model is similar to commodity trade in real markets in our daily life. Resource providers specify their service prices and charge users according to the amount of resources they use. The users can freely choose a proper service, but the price is unable to change. The prices can be generated based on the resource supply and demand. Generally, the resources are priced in such a way that supply and demand equilibrium is reached.

The *posted price* model is similar to the commodity market model. The only difference is that the posted price model advertises special offers in order to attract consumers. The posted-price offers will have usage conditions, but they might be attractive for some users because the posted prices are generally cheaper compared to regular prices.

Although some economic-based resource allocation are non-price-based [85], most of the economic-based schedulers emphasize the schemes for establishing an appropriate price based upon their users' demands. They in turn determine a proper price that keeps supply and demand in equilibrium. Several market principles are considered in the process of figuring out the price scheme, including equilibrium price [82], Pareto efficiency [86], individual rationality [87], stability [88], and communication efficiency [55].

3.2.2 SLA-Oriented Resource Provision

Although *economic-based methods* achieve impressive performances for allocating resources, there exist some limitations in some cases. The limitations lie in the difficulty for users to determine an quantized resource demand. When a user sends a request for resources to a provider, the provider looks for resources to satisfy the request and assigns the resources to the requesting users, usually as a form of virtual machines with different capabilities. However, for the users of the systems it would be difficult, even unable, to make a decision about the number and types of resources needed, especially when the request is time-varying. The ultimate concern of the user is to meet application-level requirements, instead of determining resource allocation needs.

To address such problems, many researchers proposed dynamic provisioning of resource using virtualization. The amount of provisioned resource can be adjusted with the workload fluctuates over time. Meng et al. [47] proposed a joint-VM provisioning approach in which multiple VMs are consolidated and provisioned together, based on an estimate of their aggregate capacity needs. This approach exploits statistic multiplexing among the workload patterns of multiple VMs to improve the overall resource utilization.

Garg et al. [49] proposed a dynamic resource provision strategy that considers SLAs of different types, particularly transactional and non-interactive applications. Both types of applications have different types of SLA requirements and specifications. For transactional workload, the placement decisions are made dynamically to respond to the workload variation. For non-interactive workload, the resource manager predicts the future resource availability and schedules the jobs by stealing CPU cycles.

Cloud providers such as Amazon EC2, usually offer differentiable QoS guarantees for users, which are essential for ensuring the service quality users received. The QoS guarantees are defined in the form of SLA (Service Level Agreement). Under such circumstances, cloud providers are delegated to make the decisions about the number and types of resources allocated. *SLA-oriented methods* are proposed to allocate resources to each user with the fulfillment of SLA [89]. Besides satisfying the SLA, these methods also concern other system performance metrics, such as improving the resource utilization [35], energy conservation [90], and cost reduction [56].

3.2.3 Utility-Oriented Resource Provision

Besides these two kinds of resource provision, there are some provision techniques that neglect actual levels of services required by different users and assume all requests are of equal importance. These provision techniques focus on the system utilization, rather than the profit and SLA contracts, so they are labeled as utility-oriented resource provision.

Paragon [35] is a heterogeneity and interference-aware data center scheduler, which supports the classification of an unknown application with respect to heterogeneity and interference. Paragons classification engine utilizes existing data from previously scheduled applications and offline training and requires only a minimal signal about a new workload. It uses singular value decomposition to perform collaborative filtering and identify similarities between incoming and previously scheduled workloads.

Researchers [90] improve the service scheduling by historical workload traces characterization. The long-term workload patterns are derived by workload discretization. The resources are allocated predictively by the predicted base load at hour-level scale and reactively allocated to handle any excess workload at minute-level scale. The combination of predictive and reactive provisioning contributes to meeting SLA requirements, conserve energy, and reduce allocation cost.

Beloglazov et al. [55] proposed resource provisioning and allocation algorithms for energy-efficient management in cloud computing environments. The proposed energy-aware allocation heuristics provision data center resources to client applications in a way that improves energy efficiency of the data center, while delivering the negotiated Quality of Service (QoS).

Birke et al. [91] characterized the evolution and the elasticity of workload demands in several thousands of servers at geographically distributed data centers, to improve the effectiveness of capacity planning and resource provision in data centers.

Xiong et al. [56] proposed a SLA (Service Level Agreement)-based approach for allocating resources to satisfy the quality of service (QoS) constraints while minimizing the total cost of computational power. These QoS metrics include percentile response time, cluster utilization, packet loss rate and cluster availability.

Economic-based methods are very suitable for scheduling resources in cloud environments, for regulating the supply and demand of resources at market equilibrium. With the advent of economic-based methods, SLA-oriented methods are promoted to differentiate QoS guarantees for users. SLA-oriented methods are suitable for the users that are only concerned with application-level requirements, rather than the amounts and types of involved resources. Utility-oriented methods aim to improve the system utilization, regarding all resource requests as having equal importance. Therefore, utility-based methods are applicable in cluster computing systems that do not have to consider customer-driven service managements.

3.3 Job Scheduling

Once the resources are provisioned to applications (or VM instances), each application needs to schedule the allocated resources to perform various computation jobs. In this context, the scheduling problem concerns matching the jobs to the available resources for maximization of system throughput, execution efficiency, and so on. The optimal matching is an optimization problem with NP-complete complexity.

Due to the high diversity of jobs and situations, there is no general job scheduling algorithm that can fit for all jobs. The most widely-used methods are heuristic methods, such as genetic algorithms, tabu search and simulated annealing. These methods have been successfully applied as approximately optimal algorithms to solve the job scheduling problem.

In this chapter, we classify these job scheduling methods into *static scheduling* and *dynamic scheduling*. Static scheduling techniques are suitable for the environments where the details of all jobs and resources are known prior to the scheduling being performed. On the contrary, dynamic job scheduling is performed on the fly each time a job arrives. Dynamic scheduling techniques are applied in the environments where job information and resource states cannot be available in advance.

3.3.1 Static Job Scheduling

Static scheduling techniques are commonly used in HPC and computing grid environments. In order to minimize the turnaround time, many approximation algorithms have been proposed, such as genetic algorithms [92], simulated annealing algorithms [93], and ant colony algorithms [94]. Some of these approximation methods are inspired by nature's phenomena. They do not guarantee an absolute optimal solution, but they are guaranteed to find an approximate optimal solutions in a timely manner. The quality of these solutions can be tuned by a series of parameters.

Genetic Algorithm is an evolutionary technique for solving job scheduling problem where the solution space is large. Using a genetic algorithm, the scheduling problem is represented as a genome, while a scheduling genome can be defined by the sequence of tasks. Each task and its corresponding start time represents a gene, which is a unit of genome.

The Simulated Annealing (SA) is a well-known greedy method where the search process is simulated by the thermal procedure of obtaining low-energy crystalline states of a solid. To avoid falling local optimum, SA results in a worse solution in some cases, however in most cases it results in a better solution. Analogous to the thermal procedure of metal smelting, the probability is based on the temperature that decreases for each iteration. This means, as the search progresses, a worse solution is increasingly difficult to be generated.

So far, static scheduling has been widely applied in the field of grid computing. Braun et al. [95] evaluated and compared the efficiency of 11 heuristics, including GA, SA, Tabu, Minimum Execution Time (MET), Minimum Completion Time (MCT), and so on. This study gives valuable guidelines for choosing a technique which outperforms another under a specific circumstance. More details on static scheduling can also be found in [96].

3.3.2 Dynamic Job Scheduling

Dynamic scheduling are applicable to the situation when the jobs arrive one after another, rather than being fixed. During the jobs execution, available resources can be scheduled on the fly to handle the new coming jobs. The goals of various dynamic job scheduling methods differ greatly. Besides system throughput, many job scheduling methods are designed to emphasize other metrics in certain environments, including fairness, load balance, QoS guarantee, energy consumption, and so on.

Schedulers in Hadoop are a representation of the implementation of dynamic job scheduling. The original default scheduler in Hadoop uses FIFO policy to schedule jobs. Later significant research efforts have been devoted to developing more effective and efficient schedulers. Now, the default scheduler in Hadoop is replaced by FAIR scheduler [60]. Moreover, a variety of alternative job schedulers, i.e. Delay Scheduler, Dynamic Proportional Scheduler, Capacity Scheduler etc., have been proposed.

Zaharia et al. [60] proposed FAIR Scheduler, with a rational of allocating every job a fair share of the slots over time. In fair scheduler, jobs are assigned to pools, which are assigned a guaranteed minimum quota of logic units of resources, aka. *slots*. Slots are first allocated into pools and then allocated to individual jobs within each pool. Each pool is given a minimum share and the sum of minimum quota of all pools does not exceed the system capacity. Idle slots are shared among jobs and assigned to the job with the highest slot deficit. Due to its simplicity and high performance, FAIR scheduler has gained a high popularity in Hadoop community. However, some recent work [97] has shown that the FAIR scheduler is not very well-suited for scheduling diverse workloads with considerably small jobs.

Similar as FAIR scheduler, Capacity Scheduler was also developed to ensure a fair allocation of computing resources among large number of users. The jobs from these users are submitted to different queues. Each queue is configured with a fraction of resource capacity, and free resources can be shared among the queues. Within each queue, the share of resources allocated to a user is limited, this is to guarantee that no user occupies or controls the resources exclusively. In addition, jobs can be configured with priorities. Jobs with high priorities can be allocated resources preferentially.

Delay scheduling method proposed by Zaharia et al. [31] preferentially schedule jobs to nodes where these jobs have good data locality. The method would schedule the job of which the input data is available on a node with free slots, rather than the job with the highest priority. Delay scheduling performs well in typical Hadoop workloads because there are multiple locations where a task can run to access each data block.

YARN [59], known as the next generation of Hadoop compute platform, separates resource management functions from the programming model. This separation makes various alternative programming models besides MapReduce applicable on YARN, such as Dryad [98], Spark [99], and so on.

In YARN, the functionalities of the *JobTracker* node in traditional Hadoop is split and performed by two components: a global *ResourceManager* and per-application *ApplicationMasters*. The *ResourceManager* allocates resources among all the applications in the system. The *ResourceManager* cooperates with per-node slaves, and form the data-computation framework. The *ApplicationMaster* is responsible for negotiating resources from the *ResourceManager* and working with the computing slaves to execute and monitor the tasks.

Chang et al. [24] proposed a theoretical framework for optimal scheduling in MapReduce. The authors formulate a linear program which minimizes the job completion times to solve the problem. Given the hardness at solving the linear program, approximate algorithms are designed to achieve feasible schedules within a small constant factor of the optimal value of the objective function.

Sandholm et al. [100] developed a dynamic priority (DP) scheduler, which allows users to bid for task slots or quality of service levels dynamically. For a given user, the budget of slots is proportional to the spending rate at which a user has previously bid for a slot and inversely proportional to the aggregate spending rate of all existing users. When a group of slots have been allocated to one user, that same spending rate is deducted from the users budget. Using this mechanism, the scheduler allows users to optimize and customize their slots allocation according to job requirements and system overhead.

Sparrow [33] provides low response times for parallel sub-second jobs that are executed on a large-scale cluster. The authors focus on short task workload scheduling for low-latency and high throughput. The schedulers run on a set of machines that operate autonomously and without shared state. Such a decentralized design offers attractive properties of high scalability and availability.

Energy-aware methods aim to optimize energy consumption by job dispatching. The method proposed by Wang et al. [29] and the one proposed by Kliazovich et al. [26] belong to this group of methods. Wang et al. [101] presents a thermal aware scheduling algorithm for data centers to reduce the temperatures inside of the data center. An analytical model, which describes data center resources with heat transfer properties and workloads with thermal features are used to guide the scheduler to find suitable resources for workload execution.

Nguyen et al. proposed a reputation-based resource selection scheme to reduce the energy waste caused by failures. They introduced a reputation model, called Opera, combined with a vector representation of the reputation and the just-in-time feature that represents the real-time system status. Opera enables the scheduler in Hadoop to select appropriate nodes which helped to reduce not only the number of re-executed tasks, but also improve the energy efficiency of the whole system.

Job scheduling focused on matching multiple jobs to multiple nodes using various heuristics. Scheduling techniques for MapReduce jobs usually use dynamic heuristics, such as fairness, data locality, and execution efficiency. While for HPC jobs, the scheduling techniques use static heuristics, such as OLB (Opportunistic Load Balancing), MET, MCT, GA, SA, and so on.

3.4 Data Scheduling

In the early stage of distributed computing systems, such as data grids, the data scheduling was coupled with job scheduling. In this mechanism, the cost for data access and movement are taken into considerations when deciding job scheduling. However, due to the increased growth of data size, data scheduling was gradually decoupled from job scheduling [102], and became an important issue in large-scale distributed systems.

There have been several recent studies investigating new approaches for data management and data transfer in distributed systems. These approaches can be classified into two categories: *online* data scheduling and *offline* data scheduling. The former focuses on scheduling data for serving the job (request) execution. The main goal is to reduce to data access latency and improve the job (request) execution efficiency. The latter handles the data scheduling for improving the storage resource utilization or improving the data reliability. These data scheduling approaches are offline because they are not directly performed for the online job execution.

3.4.1 Online Data Scheduling

Balman et al. [70] developed data scheduling methodologies and the key attributes for reliability, adaptability and performance optimization of distributed data placement tasks. An adaptive scheduling of data placement tasks is proposed for improving end-to-end performance. The adaptive scheduling approach includes dynamically tuning data transfer parameters over wide area networks for efficient utilization of available network capacity and optimized end-to-end data transfer performance.

To optimize the performance of data transfer, Chowdhury et al. [71] proposed a global data transfer management architecture and a set of network resource scheduling algorithms. Guo et al. [103] decrease the network traffic via inter-flow data aggregation with an efficient incast tree.

Al-Fares et al. [104] proposed a dynamic flow scheduling system, called Hedera, for multi-stage switch topologies found in data centers. Hedera collects flow information from constituent switches, computes non-conflicting paths for flows, and instructs switches to re-route traffic accordingly. The design goal of Hedera is to maximize aggregate network utilization-bisection bandwidth and to do so with minimal scheduler overhead or impact on active flows.

Seo et al. [72] proposed prefetching and pre-shuffling optimization to improve the MapReduce performance. The prefetching scheme involves the intra-block prefetching and the inter-block prefetching. The prefetching scheme exploits data locality, while the pre-shuffling scheme significantly reduces the network overhead required to shuffle key-value pairs.

3.4.2 Offline Data Scheduling

To improve data locality, Abad et al. [76] observed the correlation between benefits of data locality and data access patterns. They propose a distributed adaptive data replication algorithm, called DARE, that aids the scheduler to achieve better data locality. DARE addresses two problems, how many replicas for each file and where to place them. DARE makes use of probabilistic sampling and a competitive aging algorithm independently at each node. It takes advantage of existing remote data accesses in the system and incurs no extra network usage.

To save the energy consumption caused by communication fabric, DENS [26] combines energy efficiency and network awareness to achieve the balance between job performance, QoS requirement, traffic demands and energy consumed by the data center. DENS is designed to avoid hotspots with a data center while minimizing the number of computing servers required for job execution. DENS is particulary relevant in data centers running data-intensive jobs which produce heavy data transfer.

Ranganathan et al. [102] developed a data scheduling framework to satisfy various and general metrics and constraints, including resource utilization response times. The data movement operations may be either tightly bound to a job, or performed by a decoupled, asynchronous process on the basis of historical data access patterns.

In the context of traditional data storage systems, such as data grids, various offline data scheduling have been proposed and implemented. Offline data scheduling focuses on data storage, transfer, copy and replication management, aiming to improve the utilization ratio of storage resources and data access QoS, instead of directly serving the process of task execution. Online data scheduling focuses on job execution acceleration, and explores the strategies of data prefetch, parallel transfer and distribution for task execution procedure on a massive data processing framework. Compared with offline data scheduling, online data scheduling overcomes the limitation of lack-responsivity to job execution, and limits data I/O latency during the job execution.

4 Case Studies

Section 3 reviewed recourse scheduling techniques from three aspects, resource provision, job scheduling and data scheduling. In this section, we present how these techniques work in practical production systems. Particularly, we have chosen Amazon EC2, Dawning Nebulae, Taobao Yunti, and Microsoft SCOPE as the cases for study.

4.1 Amazon EC2

Amazon EC2 is one of the most popular IaaS cloud platforms which allow users to rent computing and storage resources to run applications, typically in forms of virtual machines. EC2 enables users to create virtual machines, each of which is called an instance. EC2 defines several type of instances,and configures each type with different computing power, memory and storage capacity.

EC2 applies *commodity market* and *posted pricing* models for provisioning the resource to users. More specifically, EC2 creates separate resource pools and has separate capacities for each type of VM. The market price for each VM type can fluctuate periodically to reflect the balance between demand and supply. Using the *commodity market* model, EC2 announces its service price according to the resource capacity and configuration. Customers can choose an appropriate service that meet their objective. The pricing policy can be derived from the resource supply and demand. In general, services are priced in such a way that achieves a supply and demand equilibrium. Using the *posted price* model, EC2 announces the special offers as a supplement of regular prices. The scheduling compares whether special offers can meet the requirement of users, and match the supply and demand if they are matched. If not, the scheduling apply commodity strategy as usual.

In addition, EC2 offers three purchasing models to facilitate the cost optimization for users. The models provide different guarantees regarding when instances can be launched and terminated.

1. On-Demand instances, which allow users to pay an hourly fee with no guarantee that launching will be possible at any given time.
2. Reserved instances, which allow users to pay a low, one-time fee and in turn receive a significant discount on the hourly usage charge for that instance.Paying a yearly fee buys clients the ability to launch one reserved instance whenever they wish.
3. Spot instances, which enable users to bid for unused Amazon EC2 capacity. The Spot Price changes periodically based on supply and demand, and customers whose bids meet or exceed it gain access to the available Spot Instances.

4.2 Dawning Nebulae

Supercomputers are regarded as the important infrastructure to carry out high performance computing. They are expected to run not only computation-intensive applications but also data-intensive applications, which challenges the job scheduling softwares on these supercomputers. To satisfy the requirements of different users, the scheduling softwares must exploit various policies, and assign different kinds of jobs flexibly. Here, we use Dawning Nebulae as a case of the job scheduling techniques applied to supercomputers.

Dawning Nebulae is a supercomputer developed by Chinese Academy of Sciences. It includes more than 9200 multi-core CPUs, and more than 4600 NVIDIA GPUs. It achieves a performance of more than 1270 trillion operations per second or 1.27 petaflops [105]. It ranked second in the TOP 500 list of the world's most powerful supercomputers released in June 2010 [106]. Dawning Nebulae has been set up in NSCC-Shenzhen[http://www.nsccsz.gov.cn]. It provides about 200 user groups and research entities with application services such as weather forecast, ocean data simulation, gene research, universe evolution, and so on.

Dawning Nebulae includes huge computing resource and storage resource, and has to depend on a special and powerful software platform to manage these resource. Platform LSF (Load Sharing Facility) [107] is such a platform. It contains multiple distributed resource management softwares, and it can connect computers into a cluster, monitor loads of systems, schedule and balance workload and so on. Here, we only focus on the scheduling software of Platform LSF, and take it as the scheduler of Dawning Nebulae.

The scheduler provides several scheduling policies like first-come-first-service (FCFS), preemption, fair share, and so on. It supports multiple policies co-existing in the same cluster. For convenience of description, we introduce these policies one by one.The first policy is FCFS. According to this policy, the scheduler attempts to assign jobs in the order submitted. However, the shorter jobs with higher priorities will be pending for a long time if a long job with low priority was submitted earlier.

The second policy is the preemption policy. Preemption is not enabled until all the job slots in a cluster are occupied. After receiving the job with high priority, the scheduler suspends one job with low priority to free the slots occupied by the job. And then, it assigns the job with high priority to these slots. It resumes the suspended job if free job slots are available.

The third policy is the fair share policy. According to this policy, the scheduler divides cluster resources into shares, and assign shares to users. The policy can avoid the cluster resources monopolized by one user. The forth policy is exclusive policy. With this policy, the scheduler allows a job exclusive use of specified server hosts, and does not preempt the exclusive jobs. The last policy is the backfill policy. Under the policy, the scheduler allows small jobs to use the slots reserved for other jobs. However, it will kill those small jobs if they cannot be finished within their run limit.

4.3 Taobao Yunti

With the rapid growth of data volume in many enterprises, effective and efficient analytics on large-scale data becomes a challenging issue. Large-scale distributed computing systems, such Hadoop, have been applied by more and more organizations. Here, we take a Hadoop production cluster in Taobao [108] as another example to illustrate job scheduling techniques.

Taobao is the biggest online e-commerce enterprise in Asia, ranked 10th in the world as reported by Alexa. The Yunti cluster is an internal data platform in Taobao for processing petabyte-level business data mostly derived from the e-commerce web site of "www.taobao.com". The total volume of data stored in the Yunti has exceeded 25 PB, and the data volume grows with the speed of 30 TB per day.[1] The goal of the Yunti cluster is to provide multi-user businesses with large-scale data analysis service for some online applications. Yunti is built on Hadoop 0.19, with some slight modifications.

In the early stage, the Yunti cluster directly employed FAIR [60] to allocate the slots because FAIR achieves high performance and supports multi-user clusters. However, after several months of system running, it is observed that FAIR is not optimal for scheduling small jobs within a miscellaneous workload. The goal of FAIR is to assure the fairness among all jobs. FAIR always reassigns idle slots to the pool with the highest slot deficits. However, small jobs usually apply fewer slots, thus the slot deficits of small jobs are often smaller than the ones of normal jobs. Therefore, small jobs are more likely to suffer from long waits than the other jobs.

[1] These statistics were released on the year of 2012.

The users of Yunti submitting small jobs, including application developers, data analysts and project managers from different departments in Taobao, will complain about the long-waits.

As new workloads which feature short and interactive jobs are emerging, small jobs are becoming pervasive. Many small jobs are initiated by interactive and online analysis, which requires instant and interactive response. Ren et al. [97] proposed and implemented a job scheduler called Fair4S, to optimize the completion time of small jobs. Fair4S introduces pool weights and extends job priorities to guarantee the rapid response for small jobs. It is verified that Fair4S accelerates the average waiting times by a factor of 7 compared with FAIR scheduler for small jobs.

4.4 Microsoft SCOPE

SCOPE [109] is a distributed computation platform in Microsoft for processing large-scale data analysis jobs and serving a variety of online services. Tens of thousands of jobs are executed on SCOPE everyday. Scope integrates parallel databases with MapReduce systems, achieving both good performance and scalability.

SCOPE relies on a distributed data platform, named COSMOS, for storing large volumes of data sets. COSMOS is designed to run on tens of thousands of servers and has similar goals to other distributed storage systems, like Google File System [110] and Hadoop Distributed File System [111]. COSMOS is an append-only file system optimized for large sequential I/O. All writes are append-only, and concurrent writers are serialized by the system. Data are distributed and replicated for fault tolerance and compressed to save storage and increase I/O throughput.

In SCOPE, the executions of jobs are scheduled by a centralized job manager. The job manager constructs the job graph (directed acyclic graph) and schedules the tasks across the available servers in the cluster. The job manager simplifies job management by classifying distinct types of vertices into separate stages. Like JobTracker in Hadoop, the job manager maintains the job graph and monitors the status of each vertex (task) in the graph.

As SCOPE is deployed on globally distributed data centers, an automated mechanism to place application data across these datacenters is quite necessary. SCOPE employs a data placement algorithm, called Volley [112], to minimize the bandwidth cost and data access latency. Volley analyzes the logs using an iterative optimization algorithm based on data access patterns and client locations, and outputs migration recommendations back to the cloud service.

Volley periodically analyzes COSMOS to determine whether the migration should be executed. To perform the analysis, Volley relies on the SCOPE to accelerate the analysis efficiency. The analysis procedure is composed of three phases. In Phase 1, a reasonable initial placement of data items based on client IP addresses is computed. In Phase 2, the placement of data items by moving them freely over the surface of the earth is improved iteratively, which consumes the dominant computational time. Phase 3 iteratively collapses data with the satisfaction of capacity constraints of data centers.

5 Future Trends and Challenges

The topic of resource scheduling has been investigated in a great deal of literature, however, this is still an emerging field and there are many open problems in the area of data-centric systems. In this section, we enumerate a few such challenges that may help to inspire new developments in the field.

Increasing System Heterogeneity. With the progress of IT technologies, new software and hardware products emerge increasingly. In order to improve system performance and satisfy users' requirements, data centers have to adopt timely new products such as SSD and SDN [113], and hence they always include different types of equipment, even multiple generation equipments of the same type. Data centers are heterogeneous inevitably, and their heterogeneity grows with the adoption of new equipments. The ever-growing heterogeneity challenges resource provision especially when considering the different requirements from users.

It's very common that some tasks are designed to run on some machines for special purposes, i.e. the machines with special accelerators for an expected performance goal. Users define the constraints or preference of the machines to run their tasks by task specifications, which provide detailed requirements of users, meanwhile this makes resource provision more difficult and complicated. In addition, such resource affinity and constraints also complicate task migration.

Scalable Decentralized Scheduling. In a system with a centralized architecture, scheduling decision are made by a master node. The node maintains all information about tasks and keeps track of all available resources in the system. A centralized scheduler can be deployed easily, while its performance is limited by the master node. However, in a decentralized system, a master node and multiple slave nodes cooperate to schedule tasks. Hence, the scheduler in such a system can assign tasks with higher performance and scalability.

Decentralized schedulers have begun to attract more and more attentions as the scales of data centers grow. In decentralized schedulers, the nodes involved in co-scheduling are assumed to be autonomous, and responsible for their own scheduling decisions. However, if these nodes make these decisions independently, they can only optimize their performance rather than the performance of the whole system. New techniques and models need to be designed to schedule jobs, and hence optimize the performance of the whole system.

Enhancing Information Sharing. In data-centric systems of which the resources belong to multiple providers, users request resources to run their applications, while providers respond to these requests, and allocate resource for the users. If providers and users can share detailed information about resources and applications, schedulers can make efficient decision, and optimize system performance. However, providers and users only reveal limited information about resources and applications due to security concerns as well as other reasons. Some works were carried out to capture characters of workloads by analyzing historical trace, which makes it feasible to optimize job schedulers according to workloads. For periodic jobs, if we can derive their characters, we can optimize the scheduling of these kind of jobs by pre-scheduling. Unfortunately, there exist few examples of such work.

Schedulability Analysis. When processing real-time jobs like interactive queries, periodic jobs and so on, a data-centric system must satisfy the time constraints of them. However, it is challenging to satisfy the time constraints because the system has to respond to the requirements from multiple users with relative QoS, especially when the job scales increase dramatically. And hence, an efficient and smart scheduler is needed to handle these kind of real-time jobs. Unfortunately, not all data-centric systems are suitable for real-time jobs. So it is very important to analyze whether a system can process real-time jobs with the specified time constraints before submitting real-time jobs to the system. There exist some research works which carry out scheduability analysis, however they only apply to multiprocessors [114] and virtualized platforms [115]. Besides, the models in these works are simple and only suitable for computing resources. Therefore, these works cannot be exploited to do scheduability analysis in data-centric systems, and new scheduability analysis techniques should be investigated as soon as possible with the consideration of computing resources, storage resources, network bandwidth, job scale, data distribution, resource competition, dynamic load, and so on.

Predictive Resource Allocation. Resource demand prediction [116] plays an essential role in dynamic resource allocation and job scheduling. For example, if a user has a job that needs to be finished within a certain deadline, an adequate amount of computing resources must be allocated. To determine whether or not a certain amount of resources are "adequate", the user needs to predict the completion time of the job with the resources. However, due to the heterogeneity and dynamism of the workload, the prediction of future resource demands would be hardly accurate. Reiss et al. [18] analyzed Google trace data [117] to reveal several insights which are helpful for improving the resource scheduling in a cloud infrastructure. The most notable characteristics of workload are heterogeneity and dynamism, which make the resource demand prediction very difficult.

6 Conclusions

In this chapter, we gave a survey of the scheduling techniques used in the three kinds of data-centric systems, including cloud computing platforms, data-intensive super computing systems, and MapReduce-style systems. According to the scheduling model, we categorized these techniques into three groups, including resource provision, job scheduling and data scheduling. We reviewed the new techniques systematically and outlined the open problems in each level. Further more, four practical systems selected from the industrial field are discussed to further understand the scheduling techniques and their applications. Finally, we concluded with some open problems in resource scheduling, aiming to inspire new developments within this field.

Acknowledgement We thank Raymond Darnell Lemon for his valuable comments on the early version of this chapter. This research is supported by NSF of Zhejiang (LQ12F02002), NSF of China (No. 61202094), Science and Technology Planning Project of Zhejiang Province (No.2010C13022). Xiaohong Zhang is supported by Ph.D. foundation of Henan Polytechnic University (No. B2012-099). Weisong Shi is in part supported by the Introduction of Innovative R&D team program of Guangdong Province (NO. 201001D0104726115), Hangzhou Dianzi University, and the NSF Career Award CCF-0643521.

References

1. Schwiegelshohn, U., Badia, R.M., Bubak, M., Danelutto, M., Dustdar, S., Gagliardi, F., Geiger, A., Hluchy, L., Kranzlmüller, D., Laure, E., et al.: Perspectives on grid computing. Future Generation Computer Systems **26**(8) (2010) 1104–1115
2. Xhafa, F., Abraham, A.: Computational models and heuristic methods for grid scheduling problems. Future generation computer systems **26**(4) (2010) 608–621
3. Armbrust, M., Fox, A., Griffith, R., Joseph, A.D., Katz, R., Konwinski, A., Lee, G., Patterson, D., Rabkin, A., Stoica, I., et al.: A view of cloud computing. Communications of the ACM **53**(4) (2010) 50–58
4. Foster, I., Zhao, Y., Raicu, I., Lu, S.: Cloud computing and grid computing 360-degree compared. In: Grid Computing Environments Workshop, 2008. GCE'08, Ieee (2008) 1–10
5. Dittrich, J., Quiané-Ruiz, J.A.: Efficient big data processing in hadoop mapreduce. Proceedings of the VLDB Endowment **5**(12) (2012) 2014–2015
6. Madden, S.: From databases to big data. Internet Computing, IEEE **16**(3) (2012) 4–6
7. Amazon Elastic Compute Cloud: http://aws.amazon.com/ec2/
8. Irwin, D., Chase, J., Grit, L., Yumerefendi, A., Becker, D., Yocum, K.G.: Sharing networked resources with brokered leases. resource **6** (2006) 6
9. Ciurana, E.: Developing with Google App Engine. Apress (2009)
10. Rackspace: http://www.rackspace.com
11. Windows Azure: http://www.windowsazure.com/
12. Bryant, R.E.: Data-intensive supercomputing: The case for disc. (2007)
13. Garg, S.K., Yeo, C.S., Anandasivam, A., Buyya, R.: Environment-conscious scheduling of hpc applications on distributed cloud-oriented data centers. Journal of Parallel and Distributed Computing **71**(6) (2011) 732–749
14. Gorton, I., Gracio, D.K.: Data-intensive computing: A challenge for the 21st century. Data-Intensive Computing: Architectures, Algorithms, and Applications (2012) 3
15. White, T.: Hadoop - The Definitive Guide. O'Reilly (2009)
16. Dean, J., Ghemawat, S.: Mapreduce: Simplified data processing on large clusters. In: OSDI. (2004) 137–150
17. Chen, Y.: Workload-driven design and evaluation of large- scale data-centric systems (May, 09 2012)
18. Reiss, C., Tumanov, A., Ganger, G.R., Katz, R.H., Kozuch, M.A.: Heterogeneity and dynamicity of clouds at scale: Google trace analysis. In: SoCC. (2012) 7
19. Macías, M., Guitart, J.: A genetic model for pricing in cloud computing markets. In: SAC, ACM (2011) 113–118
20. Niyato, D., Vasilakos, A.V., Zhu, K.: Resource and revenue sharing with coalition formation of cloud providers: Game theoretic approach. In: CCGRID, IEEE (2011) 215–224
21. Lin, W.Y., Lin, G.Y., Wei, H.Y.: Dynamic auction mechanism for cloud resource allocation. In: CCGRID, IEEE (2010) 591–592
22. Lucas-Simarro, J.L., Moreno-Vozmediano, R., Montero, R.S., Llorente, I.M.: Dynamic placement of virtual machines for cost optimization in multi-cloud environments. In: HPCS, IEEE (2011) 1–7

23. Wolf, J., Balmin, A., Rajan, D., Hildrum, K., Khandekar, R., Parekh, S., Wu, K.L., Vernica, R.: On the optimization of schedules for mapreduce workloads in the presence of shared scans. The VLDB Journal **21**(5) (2012) 589–609
24. Chang, H., Kodialam, M.S., Kompella, R.R., Lakshman, T.V., Lee, M., Mukherjee, S.: Scheduling in mapreduce-like systems for fast completion time. In: INFOCOM, IEEE (2011) 3074–3082
25. Wolf, J.L., Rajan, D., Hildrum, K., Khandekar, R., Kumar, V., Parekh, S., Wu, K.L., Balmin, A.: Flex: A slot allocation scheduling optimizer for mapreduce workloads. In: Middleware. (2010) 1–20
26. Kliazovich, D., Bouvry, P., Khan, S.U.: DENS: data center energy-efficient network-aware scheduling. Cluster Computing **16**(1) (2013) 65–75
27. Chen, Y., Alspaugh, S., Borthakur, D., Katz, R.H.: Energy efficiency for large-scale mapreduce workloads with significant interactive analysis. In: EuroSys, ACM (2012) 43–56
28. Wang, L., Khan, S.U.: Review of performance metrics for green data centers: a taxonomy study. The Journal of Supercomputing **63**(3) (2013) 639–656
29. Wang, L., Khan, S.U., Chen, D., Kolodziej, J., Ranjan, R., Xu, C.Z., Zomaya, A.Y.: Energy-aware parallel task scheduling in a cluster. Future Generation Comp. Syst **29**(7) (2013) 1661–1670
30. Isard, M., Prabhakaran, V., Currey, J., Wieder, U., Talwar, K., Goldberg, A.: Quincy: fair scheduling for distributed computing clusters. In: SOSP, ACM (2009) 261–276
31. Zaharia, M., Borthakur, D., Sarma, J.S., Elmeleegy, K., Shenker, S., Stoica, I.: Delay scheduling: a simple technique for achieving locality and fairness in cluster scheduling. In: EuroSys. (2010) 265–278
32. Borthakur, D., Gray, J., Sarma, J.S., Muthukkaruppan, K., Spiegelberg, N., Kuang, H., Ranganathan, K., Molkov, D., Menon, A., Rash, S., Schmidt, R., Aiyer, A.S.: Apache hadoop goes realtime at facebook. In: SIGMOD Conference. (2011) 1071–1080
33. Ousterhout, K., Wendell, P., Zaharia, M., Stoica, I.: Sparrow: Scalable scheduling for sub-second parallel jobs. Technical Report UCB/EECS-2013-29, EECS Department, University of California, Berkeley (April 2013)
34. Buyya, R., Yeo, C.S., Venugopal, S., Broberg, J., Brandic, I.: Cloud computing and emerging IT platforms: Vision, hype, and reality for delivering computing as the 5th utility. Future Generation Comp. Syst **25**(6) (2009) 599–616
35. Delimitrou, C., Kozyrakis, C.: Paragon: QoS-aware scheduling for heterogeneous datacenters. In: ASPLOS. (2013) 77–88
36. Vasic, N., Novakovic, D.M., Miucin, S., Kostic, D., Bianchini, R.: Dejavu: Accelerating resource allocation in virtualized environments architectural support for programming languages and operating systems, (17th ASPLOS'12). In: Proceedings of the 17th International Conference on, ACM Press (2012) 423–436
37. Zhu, X., Young, D., Watson, B.J., Wang, Z., Rolia, J., Singhal, S., McKee, B., Hyser, C., Gmach, D., Gardner, R., Christian, T., Cherkasova, L.: 1000 islands: an integrated approach to resource management for virtualized data centers. Cluster Computing **12**(1) (2009) 45–57
38. Kale, L.V., Kumar, S., Potnuru, M., DeSouza, J., Bandhakavi, S.: Faucets: Efficient resource allocation on the computational grid. In: Proceedings of the 2004 International Conference on Parallel Processing (33th ICPP'04), Montreal, Quebec, Canada, IEEE Computer Society (August 2004) 396–405
39. Rodero-Merino, L., Caron, E., Muresan, A., Desprez, F.: Using clouds to scale grid resources: An economic model. Future Generation Computer Systems **28**(4) (2012) 633 – 646
40. Kang, Z., Wang, H.: A novel approach to allocate cloud resource with different performance traits. In: Proceedings of the 2013 IEEE International Conference on Services Computing. SCC '13, Washington, DC, USA, IEEE Computer Society (2013) 128–135
41. Sim, K.M.: Towards complex negotiation for cloud economy. In: Advances in Grid and Pervasive Computing. Springer (2010) 395–406
42. Garg, S.K., Vecchiola, C., Buyya, R.: Mandi: a market exchange for trading utility and cloud computing services. The Journal of Supercomputing **64**(3) (2013) 1153–1174

43. Izakian, H., Abraham, A., Ladani, B.T.: An auction method for resource allocation in computational grids. Future Generation Comp. Syst **26**(2) (2010) 228–235
44. Zaman, S., Grosu, D.: Combinatorial auction-based allocation of virtual machine instances in clouds. In: CloudCom, IEEE (2010) 127–134
45. Samimi, P., Patel, A.: Review of pricing models for grid & cloud computing. In: Computers & Informatics (ISCI), 2011 IEEE Symposium on, IEEE (2011) 634–639
46. Wang, Q., Ren, K., Meng, X.: When cloud meets ebay: Towards effective pricing for cloud computing. In Greenberg, A.G., Sohraby, K., eds.: INFOCOM, IEEE (2012) 936–944
47. Meng, X., Isci, C., Kephart, J.O., Zhang, L., Bouillet, E., Pendarakis, D.E.: Efficient resource provisioning in compute clouds via VM multiplexing. In Parashar, M., Figueiredo, R.J.O., Kiciman, E., eds.: ICAC, ACM (2010) 11–20
48. Zhang, W., Qian, H., Wills, C.E., Rabinovich, M.: Agile resource management in a virtualized data center. In Adamson, A., Bondi, A.B., Juiz, C., Squillante, M.S., eds.: WOSP/SIPEW, ACM (2010) 129–140
49. Garg, S.K., Gopalaiyengar, S.K., Buyya, R.: SLA-based resource provisioning for heterogeneous workloads in a virtualized cloud datacenter. In Xiang, Y., Cuzzocrea, A., Hobbs, M., Zhou, W., eds.: ICA3PP (1). Volume 7016 of Lecture Notes in Computer Science., Springer (2011) 371–384
50. Urgaonkar, B., Shenoy, P., Chandra, A., Goyal, P.: Dynamic provisioning of multi-tier internet applications. In: Autonomic Computing, 2005. ICAC 2005. Proceedings. Second International Conference on, IEEE (2005) 217–228
51. Gong, Z., Gu, X., Wilkes, J.: Press: Predictive elastic resource scaling for cloud systems. In: Network and Service Management (CNSM), 2010 International Conference on, IEEE (2010) 9–16
52. Padala, P., Hou, K.Y., Shin, K.G., Zhu, X., Uysal, M., Wang, Z., Singhal, S., Merchant, A.: Automated control of multiple virtualized resources. In: Proceedings of the 4th ACM European conference on Computer systems, ACM (2009) 13–26
53. Xu, J., Zhao, M., Fortes, J., Carpenter, R., Yousif, M.: Autonomic resource management in virtualized data centers using fuzzy logic-based approaches. Cluster Computing **11**(3) (2008) 213–227
54. Gmach, D., Krompass, S., Scholz, A., Wimmer, M., Kemper, A.: Adaptive quality of service management for enterprise services. ACM Transactions on the Web (TWEB) **2**(1) (2008) 8
55. Beloglazov, A., Abawajy, J., Buyya, R.: Energy-aware resource allocation heuristics for efficient management of data centers for cloud computing. Future Generation Computer Systems **28**(5) (2012) 755–768
56. Xiong, K., Perros, H.G.: SLA-based resource allocation in cluster computing systems. In: IPDPS, IEEE (2008) 1–12
57. Gu, J., Hu, J., Zhao, T., Sun, G.: A new resource scheduling strategy based on genetic algorithm in cloud computing environment. Journal of Computers **7**(1) (2012) 42–52
58. Hu, J., Gu, J., Sun, G., Zhao, T.: A scheduling strategy on load balancing of virtual machine resources in cloud computing environment. In: Parallel Architectures, Algorithms and Programming (PAAP), 2010 Third International Symposium on, IEEE (2010) 89–96
59. Vavilapalli, V.K., Murthy, A.C., Douglas, C., Agarwal, S., Konar, M., Evans, R., Graves, T., Lowe, J., Shah, H., Seth, S., Saha, B., Curino, C., O'Malley, O., Radia, S., Reed, B., Baldeschwieler, E.: Apache hadoop YARN: Yet another resource negotiator. In: SoCC. (2013)
60. Zaharia, M., Borthakur, D., Sarma, J.S., Shenker, S., Stoica, I.: Job scheduling for multi-user mapreduce clusters. Technical Report No. UCB/EECS-2009-55, Univ. of Calif., Berkeley, CA (April 2009)
61. Zhang, X., Zhong, Z., Feng, S., Tu, B., Fan, J.: Improving data locality of mapreduce by scheduling in homogeneous computing environments. In: Parallel and Distributed Processing with Applications (ISPA), 2011 IEEE 9th International Symposium on, IEEE (2011) 120–126
62. Kc, K., Anyanwu, K.: Scheduling hadoop jobs to meet deadlines. In: Cloud Computing Technology and Science (CloudCom), 2010 IEEE Second International Conference on, IEEE (2010) 388–392

63. Tang, Z., Zhou, J., Li, K., Li, R.: MTSD: A task scheduling algorithm for mapreduce base on deadline constraints. In: IPDPS Workshops, IEEE Computer Society (2012) 2012–2018
64. Schwiegelshohn, U., Tchernykh, A.: Online scheduling for cloud computing and different service levels. In: Proc. 9th High-Performance Grid & Cloud Computing – 9th HPGC'12, Proc. IEEE International Parallel and Distributed Processing Symposium Workshops & PhD Forum (26th IPDPS'12), IEEE Computer Society (2012) 1067–1074
65. Venugopal, S., Buyya, R.: An scp-based heuristic approach for scheduling distributed data-intensive applications on global grids. Journal of Parallel and Distributed Computing **68**(4) (2008) 471–487
66. Chang, R.S., Chang, J.S., Lin, P.S.: An ant algorithm for balanced job scheduling in grids. Future Generation Computer Systems **25**(1) (2009) 20–27
67. Kolodziej, J., Khan, S.U., Xhafa, F.: Genetic algorithms for energy-aware scheduling in computational grids. In: P2P, Parallel, Grid, Cloud and Internet Computing (3PGCIC), 2011 International Conference on, IEEE (2011) 17–24
68. Lee, Y.H., Leu, S., Chang, R.S.: Improving job scheduling algorithms in a grid environment. Future generation computer systems **27**(8) (2011) 991–998
69. Samuel, T.K., Baer, T., Brook, R.G., Ezell, M., Kovatch, P.: Scheduling diverse high performance computing systems with the goal of maximizing utilization. In: High Performance Computing (HiPC), 2011 18th International Conference on, IEEE (2011) 1–6
70. Balman, M.: Failure-awareness and dynamic adaptation in data scheduling (November 14 2008)
71. Chowdhury, M., Zaharia, M., Ma, J., Jordan, M.I., Stoica, I.: Managing data transfers in computer clusters with orchestra. In: SIGCOMM, ACM (2011) 98–109
72. Seo, S., Jang, I., Woo, K., Kim, I., Kim, J.S., Maeng, S.: Hpmr: Prefetching and pre-shuffling in shared mapreduce computation environment. In: Cluster Computing and Workshops, 2009. CLUSTER'09. IEEE International Conference on, IEEE (2009) 1–8
73. Çatalyürek, Ü.V., Kaya, K., Uçar, B.: Integrated data placement and task assignment for scientific workflows in clouds. In: Proceedings of the fourth international workshop on Data-intensive distributed computing, ACM (2011) 45–54
74. Xie, J., Yin, S., Ruan, X., Ding, Z., Tian, Y., Majors, J., Manzanares, A., Qin, X.: Improving mapreduce performance through data placement in heterogeneous hadoop clusters. In: Parallel & Distributed Processing, Workshops and Phd Forum (IPDPSW), 2010 IEEE International Symposium on, IEEE (2010) 1–9
75. Zeng, W., Zhao, Y., Ou, K., Song, W.: Research on cloud storage architecture and key technologies. In: Proceedings of the 2nd International Conference on Interaction Sciences: Information Technology, Culture and Human, ACM (2009) 1044–1048
76. Abad, C.L., Lu, Y., Campbell, R.H.: DARE: Adaptive data replication for efficient cluster scheduling. In: Proc. '11 IEEE International Conference on Cluster Computing (13th CLUSTER'11), Austin, TX, USA, IEEE Computer Society (September 2011) 159–168
77. Castillo, C., Tantawi, A.N., Arroyo, D., Steinder, M.: Cost-aware replication for dataflows. In: NOMS, IEEE (2012) 171–178
78. Chervenak, A.L., Deelman, E., Livny, M., Su, M.H., Schuler, R., Bharathi, S., Mehta, G., Vahi, K.: Data placement for scientific applications in distributed environments. In: GRID, IEEE Computer Society (2007) 267–274
79. Chen, Y., Ganapathi, A.S., Griffith, R., Katz, R.H.: Analysis and lessons from a publicly available google cluster trace. Technical Report UCB/EECS-2010-95, EECS Department, University of California, Berkeley (Jun 2010)
80. Chen, Y., Ganapathi, A.S., Griffith, R., Katz, R.H.: Towards understanding cloud performance tradeoffs using statistical workload analysis and replay. University of California at Berkeley, Technical Report No. UCB/EECS-2010-81 (2010)
81. Stuer, G., Vanmechelen, K., Broeckhove, J.: A commodity market algorithm for pricing substitutable grid resources. Future Generation Comp. Syst **23**(5) (2007) 688–701
82. Teng, F., Magoulès, F.: Resource pricing and equilibrium allocation policy in cloud computing. In: CIT, IEEE Computer Society (2010) 195–202

83. Eymann, T., Reinicke, M., Villanueva, O.A., Vidal, P.A., Freitag, F., Moldes, L.N.: Decentralized resource allocation in application layer networks. In: CCGrid, IEEE (May 12 2003) 645–650
84. Padala, P., Harrison, C., Pelfort, N., Jansen, E., Frank, M.P., Chokkareddy, C.: OCEAN: The open computation exchange and arbitration network, A market approach to meta computing. In: Proc. 2nd International Symposium on Parallel and Distributed Computing (2nd ISPDC'03), Ljubljana, Slovenia, IEEE Computer Society (October 2003) 185–192
85. Peterson, L., Anderson, T., Culler, D., Roscoe, T.: PlanetLab: A Blueprint for Introducing Disruptive Technology into the Internet. In: First ACM Workshop on Hot Topics in Networks, Association for Computing Machinery (October 2002) Available from http://www.planet-lab.org/pdn/pdn02-001.pdf.
86. Ghodsi, A., Zaharia, M., Hindman, B., Konwinski, A., Shenker, S., Stoica, I.: Dominant resource fairness: Fair allocation of multiple resource types. Technical report, University of California, Berkeley (2011)
87. Mihailescu, M., Teo, Y.M.: Dynamic resource pricing on federated clouds. In: CCGRID, IEEE (2010) 513–517
88. Dutreilh, X., Rivierre, N., Moreau, A., Malenfant, J., Truck, I.: From data center resource allocation to control theory and back. In: Proc. IEEE International Conference on Cloud Computing (3rd IEEE CLOUD'10). (2010) 410–417
89. Buyya, R., Garg, S.K., Calheiros, R.N.: SLA-oriented resource provisioning for cloud computing: Challenges, architecture, and solutions. In: Cloud and Service Computing (CSC). (January 21 2012)
90. Gandhi, A., Chen, Y., Gmach, D., Arlitt, M.F., Marwah, M.: Minimizing data center SLA violations and power consumption via hybrid resource provisioning. In: IGCC, IEEE Computer Society (2011) 1–8
91. Birke, R., Chen, L.Y., Smirni, E.: Data centers in the cloud: A large scale performance study. In: Proc. 2012 IEEE Fifth International Conference on Cloud Computing (5th IEEE CLOUD'12). (June 2012) 336–343
92. Gao, Y., Rong, H., Huang, J.Z.: Adaptive grid job scheduling with genetic algorithms. Future Generation Computer Systems **21**(1) (2005) 151–161
93. Fidanova, S.: Simulated annealing for grid scheduling problem. In: Modern Computing, 2006. JVA'06. IEEE John Vincent Atanasoff 2006 International Symposium on, IEEE (2006) 41–45
94. neng Chen, W., 0003, J.Z.: An ant colony optimization approach to a grid workflow scheduling problem with various qoS requirements. IEEE Transactions on Systems, Man, and Cybernetics, Part C **39**(1) (2009) 29–43
95. Braun, T.D., Siegel, H.J., Beck, N., Bölöni, L., Maheswaran, M., Reuther, A.I., Robertson, J.P., Theys, M.D., Yao, B., Hensgen, D.A., Freund, R.F.: A comparison of eleven static heuristics for mapping a class of independent tasks onto heterogeneous distributed computing systems. J. Parallel Distrib. Comput **61**(6) (2001) 810–837
96. Dong, F., Akl, S.G.: Scheduling algorithms for grid computing: State of the art and open problems. School of Computing, Queens University, Kingston, Ontario (2006)
97. Ren, Z., Wan, J., Shi, W., Xu, X., Zhou, M.: Workload analysis, implications and optimization on a production hadoop cluster: A case study on taobao. IEEE Transactions on Services Computing (2013)
98. Isard, M., Budiu, M., Yu, Y., Birrell, A., Fetterly, D.: Dryad: distributed data-parallel programs from sequential building blocks. In: EuroSys, ACM (2007) 59–72
99. Zaharia, M., Chowdhury, M., Franklin, M.J., Shenker, S., Stoica, I.: Spark: cluster computing with working sets. In: Proceedings of the 2nd USENIX conference on Hot topics in cloud computing. (2010) 10–10
100. Sandholm, T., Lai, K.: Dynamic proportional share scheduling in Hadoop. In Frachtenberg, E., Schwiegelshohn, U., eds.: Job Scheduling Strategies for Parallel Processing. Springer Verlag (2010) 110–131
101. Wang, L., von Laszewski, G., Dayal, J., He, X., Younge, A.J., Furlani, T.R.: Towards thermal aware workload scheduling in a data center. In: ISPAN, IEEE Computer Society (2009) 116–122

102. Ranganathan, K., Foster, I.T.: Decoupling computation and data scheduling in distributed data-intensive applications. In: HPDC, IEEE Computer Society (2002) 352–358
103. Guo, D., Li, M., Jin, H., Shi, X., Lu, L.: Managing and aggregating data transfers in data centers (2013)
104. Al-Fares, M., Radhakrishnan, S., Raghavan, B., Huang, N., Vahdat, A.: Hedera: Dynamic flow scheduling for data center networks. In: NSDI, USENIX Association (2010) 281–296
105. Sun, N.H., Xing, J., Huo, Z.G., Tan, G.M., Xiong, J., Li, B., Ma, C.: Dawning nebulae: a petaflops supercomputer with a heterogeneous structure. Journal of Computer Science and Technology **26**(3) (2011) 352–362
106. : Top500 list
107. Lumb, I., Smith, C.: Scheduling attributes and platform lsf. In: Grid resource management. Springer (2004) 171–182
108. Taobao: http://www.taobao.com
109. Chaiken, R., Jenkins, B., Larson, P.Å., Ramsey, B., Shakib, D., Weaver, S., Zhou, J.: Scope: easy and efficient parallel processing of massive data sets. Proceedings of the VLDB Endowment **1**(2) (2008) 1265–1276
110. Ghemawat, S., Gobioff, H., Leung, S.T.: The google file system. In: ACM SIGOPS Operating Systems Review. Volume 37., ACM (2003) 29–43
111. Shvachko, K., Kuang, H., Radia, S., Chansler, R.: The hadoop distributed file system. In: Mass Storage Systems and Technologies (MSST), 2010 IEEE 26th Symposium on, IEEE (2010) 1–10
112. Agarwal, S., Dunagan, J., Jain, N., Saroiu, S., Wolman, A., Bhogan, H.: Volley: Automated data placement for geo-distributed cloud services. In: NSDI. (2010) 17–32
113. McKeown, N.: Software-defined networking. INFOCOM keynote talk, Apr (2009)
114. Liu, D., Lee, Y.H.: Pfair scheduling of periodic tasks with allocation constraints on multiple processors. In: IPDPS. (2004)
115. Lee, J., Easwaran, A., Shin, I.: LLF schedulability analysis on multiprocessor platforms. In: IEEE Real-Time Systems Symposium. (2010) 25–36
116. Islam, S., Keung, J., Lee, K., Liu, A.: Empirical prediction models for adaptive resource provisioning in the cloud. Future Generation Computer Systems **28**(1) (2012) 155–162
117. Wilkes, J., Reiss, C.: Details of the clusterdata-2011-1 trace (2011)

Index

S. U. Khan, A. Y. Zomaya (eds.), Handbook on Data Centers,
DOI 10.1007/978-1-4939-2092-1